Lecture Notes in Computer Science 4522

Commenced Publication in 1973
Founding and Former Series Editors:
Gerhard Goos, Juris Hartmanis, and Jan van Leeuwen

Bjarne Kjær Ersbøll
Kim Steenstrup Pedersen (Eds.)

Image Analysis

15th Scandinavian Conference, SCIA 2007
Aalborg, Denmark, June 10-14, 2007
Proceedings

Springer

Volume Editors

Bjarne Kjær Ersbøll
Technical University of Denmark
Department of Informatics and Mathematical Modelling
Richard Petersen Plads, DTU-Building 321, 2800 Lyngby, Denmark
E-mail: be@imm.dtu.dk

Kim Steenstrup Pedersen
University of Copenhagen, Department of Computer Science
Universitetsparken 1, 2100 Copenhagen, Denmark
E-mail: kimstp@diku.dk

Library of Congress Control Number: 2007928118

CR Subject Classification (1998): I.4, I.5, I.3

LNCS Sublibrary: SL 6 – Image Processing, Computer Vision, Pattern Recognition, and Graphics

ISSN	0302-9743
ISBN-10	3-540-73039-7 Springer Berlin Heidelberg New York
ISBN-13	978-3-540-73039-2 Springer Berlin Heidelberg New York

Springer is a part of Springer Science+Business Media

springer.com

Printed in Germany

Typesetting: Camera-ready by author, data conversion by Scientific Publishing Services, Chennai, India
Printed on acid-free paper SPIN: 12075904 06/3180 5 4 3 2 1 0

Preface

The present volume contains the proceedings of the Scandinavian Conference on Image Analysis, SCIA 2007, held at Hotel Hvide Hus, Aalborg, Denmark, June 10–14, 2007.

Initiated in 1979 by Torleiv Orhaug in Sweden, SCIA 2007 represented the 15th in the biennial series of conferences. It is arranged in turn by the Scandinavian countries of Sweden, Finland, Denmark, and Norway, making it a regional conference. However, judging by the nationalities of contributing authors and participants, it is also an international conference.

Worldwide, there is no lack of conferences on image analysis. Indeed, hardly a day passes without an announcement of yet another conference. Therefore the pattern recognition societies of the Scandinavian countries take particular pride in being able to continue the SCIA tradition. SCIA has indeed matured over the many years it has been in existence, but in our opinion SCIA has maintained flexibility and has been able to adopt and incorporate necessary changes and adjustments over that time span. An important key to the success of SCIA is the constant and continuing high quality of the scientific content. Furthermore, the relaxed and friendly atmosphere of the conference itself is well known within the community. The objective to keep in mind must be to continue along those lines.

The number of submissions for this year's event was an impressive 228. Of these, 99 can be found in the present proceedings, leading to an acceptance rate of 43%. In order to optimize the outcome for the participants, the conference was organized as a single track event. Thirty-three papers were presented in the oral sessions and 66 were presented in the poster sessions. Each paper was reviewed by at least two peers. Acceptance was based solely on these reviews. The papers can roughly be grouped into the following topics: computer vision; 2D and 3D reconstruction; classification and segmentation; medical and biological applications; appearance and shape modeling; face detection, tracking and recognition; motion analysis; feature extraction and object recognition. Two tutorials preceded the conference.

A conference is the result of careful planning and lots of work from numerous volunteers. It is important that we acknowledge these important contributions. We thank the invited speakers for enlightening us in their areas of research and the contributing scientists for their presentations. Furthermore, we thank the reviewers for the pleasant interaction during the review process and for the excellent work in helping to maintain the high quality of SCIA.

It is our sincere hope that the participants had an enjoyable and fruitful experience, both scientifically and socially, in Aalborg.

June 2007

Bjarne Kjær Ersbøll
Kim Steenstrup Pedersen

Organization

SCIA 2007 was organized jointly by Aalborg University, the Technical University of Denmark (DTU), the Department of Computer Science at University of Copenhagen, and the IT University of Copenhagen.

Executive Committee

General Chair	Peter Johansen, University of Copenhagen, Denmark
Program Chairs	Bjarne Kjær Ersbøll, Technical University of Denmark
	Kim Steenstrup Pedersen, University of Copenhagen, Denmark
	Søren Ingvor Olsen, University of Copenhagen, Denmark
Local Chair	Claus Brøndgaard Madsen, Aalborg University, Denmark

Program Committee

Bjarne Kjær Ersbøll (Chairman)	Technical University of Denmark
Ingela Nyström	Uppsala University, Sweden
Ivar Austvoll	University of Stavanger, Norway
Janne Heikkilä	University of Oulu, Finland

Invited Speakers

Paul Debevec	University of Southern California, USA
Henrik Wann Jensen	University of California at San Diego, USA and Technical University of Denmark
Jitendra Malik	University of California at Berkeley, USA
Henrik I. Christensen	Georgia Institute of Technology, USA and Royal Institute of Technology, Sweden
Aapo Hyvärinen	University of Helsinki, Finland

Tutorials

Henrik Wann Jensen	University of California at San Diego, USA and Technical University of Denmark
Jitendra Malik	University of California at Berkeley, USA

Reviewers

Lukas Ahrenberg
Fritz Albregtsen
Hans Jørgen Andersen
Mats Andersson
Ulf Assarsson
Ivar Austvoll
Adrien Bartoli
Ewert Bengtsson
Magnus Borga
Gunilla Borgefors
Marleen de Bruijne
Anders Brun
Jakob Andreas Bærentzen
Stefan Carlsson
Jens Michael Carstensen
Niels Jørgen Christensen
Anne Cuzol
Anders Bjorholm Dahl
Ola Dahl
Sune Darkner
Jan-Olof Eklundh
Søren G. Erbou
Kenny Erleben
Bjarne Kjær Ersbøll
Søren Forchhammer
Robert Forchheimer
Vincent Gay-Bellile
Magnus Gedda
Mads Fogtmann Hansen
Lars Kai Hansen
Michael Sass Hansen
Markku Hauta-Kasari
Janne Heikkilä
Knud Henriksen
Anders Heyden
Jouni Hiltunen
Bjarke Jakobsen
Peter Johansen

Fredrik Kahl
Juho Kannala
Sune Høgild Keller
Hans Knutsson
Anders Wang Kristensen
Volker Krüger
Norbert Krüger
Iivari Kunttu
Heikki Kälviäinen
Bent Larsen
Rasmus Larsen
Reiner Lenz
Leena Lepistö
Joakim Lindblad
Marco Loog
Claus Brøndgaard Madsen
J. Birgitta Martinkauppi
Brian Mayoh
Thomas B. Moeslund
Allan Aasbjerg Nielsen
Henning Nielsen
Mads Nielsen
Michael Nielsen
Ingela Nyström
Erkki Oja
Matti Okkonen
Hildur Ólafsdóttir
Søren Ingvor Olsen
Niels Chr. Overgaard
Jussi Parkkinen
Kim Steenstrup Pedersen
Mathieu Perriollat
Johanna Pettersson
Daniel Pizarro
Esa Rahtu
Joakim Rydell
Pekka Sangi
Sigurjon A. Sigurdsson
Karl Sjöstrand
Arnold Skimminge
Örjan Smedby
Jon Sporring
Rasmus Stenholt
Björn Svensson
Stina Svensson
Jean-Philippe Tardif
Martin Vester-Christensen
Erik Vidholm
Ole Winther
Henrik Aanæs
Kalle Åström

Sponsoring Organizations

The International Association for Pattern Recognition

IAPR

Danish Pattern Recognition Society (DSAGM).

Table of Contents

Computer Vision

2D and 3D Reconstruction

Motion Analysis

Feature Extraction and Object Recognition

Classification and Segmentation

Poster Session 1

Poster Session 2

Accurate Interpolation in Appearance-Based Pose Estimation

Erik Jonsson and Michael Felsberg*

Computer Vision Laboratory
Dept. of Electrical Engineering, Linköping University
erijo@isy.liu.se, mfe@isy.liu.se

Abstract. One problem in appearance-based pose estimation is the need for many training examples, i.e. images of the object in a large number of known poses. Some invariance can be obtained by considering translations, rotations and scale changes in the image plane, but the remaining degrees of freedom are often handled simply by sampling the pose space densely enough. This work presents a method for accurate interpolation between training views using local linear models. As a view representation local soft orientation histograms are used. The derivative of this representation with respect to the image plane transformations is computed, and a Gauss-Newton optimization is used to optimize all pose parameters simultaneously, resulting in an accurate estimate.

1 Introduction

Object recognition and pose estimation can be done in several ways. In the bag-of-features approach, local coordinate frames are constructed around points of interest [5], [9], and features from each local frame vote for a certain object and pose hypothesis. In the model-based approach [2], [11], a geometrical model is fitted to the observed image. This approach is often very accurate, but requires a good initial guess and a manually constructed 3D model. Global appearance-based methods extract features from the appearance of the entire object and match these to training views in memory. Ever since [10], [7], the most common approach seems to be using PCA.

In this paper, we use an appearance-based method using full object views, but avoid PCA due to the global nature of this representation. The main goal is to maximize the accuracy of the pose estimate by interpolating between a limited number of training views. The interpolation method is based on representing the views with *channel-coded orientation* [3], [4], and optimizing all pose parameters (including position, rotation and scale in the image plane) simultaneously using a Gauss-Newton method. The method requires an initial guess, which in a real system could be obtained using your favorite fast but inaccurate bag-of-features approach.

* This work has been supported by EC Grant IST-2003-004176 COSPAL. This paper does not represent the opinion of the European Community, and the European Community is not responsible for any use which may be made of its contents.

B.K. Ersbøll and K.S. Pedersen (Eds.): SCIA 2007, LNCS 4522, pp. 1–10, 2007.

The motivation for using full object views is two-fold. The first reason is that once we have formed an initial object hypothesis, it makes sense to use as much image information as possible in order to get an accurate estimate. The second reason is that using full views, we can focus on the interpolation and view representation, and ignore other aspects like how to choose interest points and construct local frames in a bag-of-features approach. This makes it easier to compare different view representations. Similar interpolation techniques as proposed here should however be possible to integrate also in a bag-of-features framework.

In contrast to model-based methods, our approach requires no knowledge of 3D geometry in the system, and is in no way specific to 3D pose estimation. The training set could consist of any parameterized image set, e.g. a robotic arm in different configurations etc.

2 Algorithm

2.1 Pose Estimation

The appearance of an object is determined by the object state $\mathbf{p} = [\theta, \phi, s, \alpha, x, y]$. The parameters s, α, x, y represent the scale, rotation and position of the object in the image plane and will be referred to as the *image parameters* $\mathbf{p}_{\text{img}}$. The two auxiliary angles θ and ϕ cover all pose variations not explained by rotation in the image plane and will be referred to as the *pose angles* $\mathbf{p}_{\text{pose}}$.

During training, we learn the appearance of the object given (θ, ϕ) using canonical image parameters. The result of the learning can be seen as a function $\mathbf{f}$ that maps the pose angles to a predicted feature vector:

$$\hat{\mathbf{c}} = \mathbf{f}(\theta, \phi) \ . \tag{1}$$

During operation of the system, we maintain a current hypothesis of the object state, and cut out an image patch around the current (x, y) with rotation α and size s. This can be formalized by a function

$$\mathbf{c} = \mathbf{g}(s, \alpha, x, y) \tag{2}$$

producing an observed feature vector from the image given certain image parameters. The pose estimation problem is now to find an object state $\mathbf{p}_*$ which minimizes the difference between the observed and predicted feature vectors:

$$\mathbf{p}_* = \arg\min_{\mathbf{p}} \|\mathbf{r}(\mathbf{p})\|^2 \tag{3}$$

where

$$\mathbf{r}(\mathbf{p}) = \mathbf{f}(\theta, \phi) - \mathbf{g}(s, \alpha, x, y) \ . \tag{4}$$

This can be solved using your favorite optimization method. We use a Gauss-Newton method, with a simple backtracking line search [8]. The update step direction $\mathbf{s}$ is given by

$$\mathbf{J}\mathbf{s} = -\mathbf{r} \ , \tag{5}$$

where $\mathbf{J}$ is the Jacobian of $\mathbf{r}$:

$$\mathbf{J} = [\mathbf{f}', \; -\mathbf{g}'] = \left[\mathbf{f}'_\theta, \mathbf{f}'_\phi, -\mathbf{g}'_s, -\mathbf{g}'_\alpha, -\mathbf{g}'_x, -\mathbf{g}'_y\right] \tag{6}$$

The derivative of $\mathbf{g}$ with respect to transformations in the image plane depends on the choice of view representation and will be discussed in Sect. 3. The derivative of $\mathbf{f}$ can be approximated by a local linear approximation of the training manifold, discussed in Sect. 2.3.

In each step of the iterations, we measure $\mathbf{g}(\mathbf{p}_{\text{img}})$ directly in the query image, i.e. we cut out a new patch using the current $\mathbf{p}_{\text{img}}$ and extract a new feature vector from the image. A faster but less accurate option would be to keep the original feature vector and Jacobian, and use them as a linear approximation of $\mathbf{g}$ throughout the entire solution procedure.

2.2 Geometrical Interpretation of Gauss-Newton

To fully understand the method, it is useful to have a geometrical image in mind. The output of the functions $\mathbf{f}$ and $\mathbf{g}$ define two manifolds in feature space. The first manifold contains all expected appearances of the object, learned from training examples, and the second one contains all observable feature vectors at different positions in the query image. The objective is to find one point on each manifold such that the distance between the two points is minimal.

What the Gauss-Newton method does is to approximate each manifold with its tangent plane and find the points of minimal distance on these hyperplanes. Let $\mathbf{f}(\mathbf{p}_{\text{pose}} + \mathbf{s}_{\text{pose}}) \approx \mathbf{f}(\mathbf{p}_{\text{pose}}) + \mathbf{f}'(\mathbf{p}_{\text{pose}})\mathbf{s}_{\text{pose}}$ and $\mathbf{g}(\mathbf{p}_{\text{img}} + \mathbf{s}_{\text{img}}) \approx \mathbf{g}(\mathbf{p}_{\text{img}}) + \mathbf{g}'(\mathbf{p}_{\text{img}})\mathbf{s}_{\text{img}}$. The minimum-distance points are given by the over-determined equation system

$$\mathbf{f}(\mathbf{p}_{\text{pose}}) + \mathbf{f}'(\mathbf{p}_{\text{pose}})\mathbf{s}_{\text{pose}} = \mathbf{g}(\mathbf{p}_{\text{img}}) + \mathbf{g}'(\mathbf{p}_{\text{img}})\mathbf{s}_{\text{img}} \;, \tag{7}$$

which is solved by (5) with $\mathbf{s} = [\mathbf{s}_{\text{pose}} \; \mathbf{s}_{\text{img}}]$. If the linear approximation is good enough, we can expect good results even after a single iteration.

2.3 Local Linear Approximation

In this section, we describe how to approximate the value and derivative of $\mathbf{f}$ by a variety of *locally weighted regression* [1], [6]. To simplify the notation and avoid double subscripts, let $\mathbf{p} = [\theta, \phi]^{\mathrm{T}}$, and let $\mathbf{p}_0$ be the current guess of $\mathbf{p}$, i.e. we consider only the auxiliary pose angles. The system is given a set of training views with pose angles $\mathbf{p}_i$, from which features $\mathbf{c}_i$ are extracted. The learning consists simply of storing all these training samples $\{\mathbf{p}_i, \mathbf{c}_i\}$. In operation mode we need the value and derivative of $\mathbf{f}$ at the current hypothesis $\mathbf{p}_0$, which is computed by fitting a linear model to the training samples closest to $\mathbf{p}_0$. The basic strategy is to weight all training samples according to their distance to $\mathbf{p}_0$:

$$w_i = K(\|\mathbf{p}_i - \mathbf{p}_0\|) \;. \tag{8}$$

Here K is a smooth Gaussian-looking weighting kernel with compact local support; in our case a second-order B-spline. We then solve the weighted least-squares problem

$$\min_{\mathbf{A},\mathbf{b}} \sum_i w_i \|(\mathbf{A}(\mathbf{p}_i - \mathbf{p}_0) + \mathbf{b} - \mathbf{c}_i\|^2 \ . \tag{9}$$

This produces an interpolation using neighboring points only. From the Taylor expansion of $\mathbf{f}$, we can identify $\mathbf{b}$ and $\mathbf{A}$ as the approximated function value and derivative respectively:

$$\mathbf{f}(\mathbf{p}) \approx \mathbf{f}(\mathbf{p}_0) + \mathbf{f}'(\mathbf{p}_0)(\mathbf{p} - \mathbf{p}_0) \approx \mathbf{b} + \mathbf{A}(\mathbf{p} - \mathbf{p}_0) \ . \tag{10}$$

If the training views are irregularly distributed in (θ, ϕ)-space, the number of samples included within the support of K is arbitrary and may even be zero. In contrast, if the weighting kernel is large, the linear approximation may be poor. Instead of using a fixed kernel, we could always select the k nearest neighbors, but without any sample weighting this would produce a discontinuity when set of neighbors changes. Since we expect $\mathbf{f}$ to be a smooth function, and since we are going to use the approximation in an iterative optimization, it is important that our approximation is also smooth.

Our method does something in between, by using a weighting kernel that is scaled according to the nearest training samples. We let $K(r)$ be scaled such that $K(r) = 0$ iff $r > 1$ and sort the training samples by their distance to $\mathbf{p}_0$, such that $\|\mathbf{p}_i - \mathbf{p}_0| \leq \|\mathbf{p}_{i+1} - \mathbf{p}_0\|$ for all i. We now weight our samples according to

$$w_i = K\left(\beta\|\mathbf{p}_i - \mathbf{p}_0\|/\|\mathbf{p}_k - \mathbf{p}_0\|\right) \ . \tag{11}$$

If $\beta = 1$, this gives zero weight to the k'th sample and non-zero weight to all samples strictly closer to $\mathbf{p}_0$. However, if there are several samples with the same distance to $\mathbf{p}_0$ as $\mathbf{p}_k$, all these samples will get zero weight as well, which may produce too few active samples for a reliable model fitting. This is solved by choosing β slightly larger than 1, giving at least k active samples. Each w_i is now a continuous function of $\mathbf{p}_0$, and $\mathbf{A}, \mathbf{b}$ depend continuously on the w_i's. This ensures that our approximation responds continuously to changes in $\mathbf{p}_0$.

3 View Representation

Given the pose estimation procedure in Sect. 2.1, we wish to find a good description of each view in terms of a feature vector $\mathbf{c}$. To make the representation depend continuously on the input image, we avoid too complicated things like region detection etc. What we want is a simple non-linear transformation of the view. In order to be invariant to lighting, we use local orientation instead of image intensity. For robustness against occlusion and background clutter, we avoid global view representations like PCA or DCT. One simple option could be to simply downsample the gradient magnitude. Note however that in order for the interpolation between training views to be successful, the same edge of an object must be visible within the receptive field of one pixel in the downsampled image for several views (see Fig. 1). Since our training views are rather coarsely spaced, we would need very heavy smoothing or downsampling, which would destroy much image information. Instead, we use *channel coded orientation* or *soft orientation histograms* as view representation.

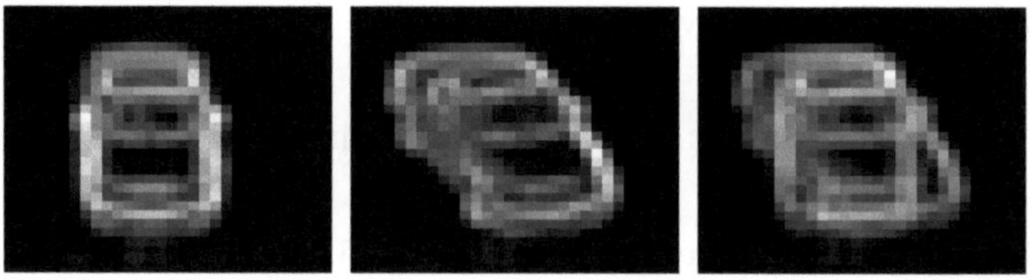

Fig. 1. Top: Gradient magnitude of two adjacent training views and an interpolated intermediate view. Since the spatial resolution of the feature representation is too large, linear interpolation does not produce the expected features of intermediate views.

3.1 Channel Coding

Given a scalar-valued feature image $z(x, y)$ and a weight function $w(x, y)$, the *channel coded feature map* is a 3D array $\mathbf{c}$ with elements

$$c_{ijk} = \int_{\mathbb{R}^2} B_{ijk}(x, y, z(x, y))w(x, y) \, \mathrm{d}x \, \mathrm{d}y \ , \tag{12}$$

where

$$B_{ijk}(x, y, z) = B_1(x - \tilde{x}_i)B_2(y - \tilde{y}_j)B_3(z - \tilde{z}_k) \tag{13}$$

are smooth, localized but overlapping basis functions, causing each pixel to smoothly contribute to several channels. Each channel c_{ijk} measures the presence of the feature value $\tilde{z}_k$ around image position $(\tilde{x}_i, \tilde{y}_j)$. The points $(\tilde{x}_i, \tilde{y}_j, \tilde{z}_k)$ are called the *channel centers.* We can think of it as first representing the feature image as a set of points (x, y, z) in a 3D *spatio-featural* space, and then downsampling this space in all three dimensions simultaneously.

In our case, the feature z is local orientation taken modulo π, such that we do not distinguish between positive and negative edges. The gradient magnitude is used as weight $w(x, y)$. If the kernels are chosen as rectangular and non-overlapping, we get a simple 3D histogram. If we create a binary weight $w_{\mathrm{B}}(x, y)$ by thresholding $w(x, y)$ at 10% of the maximum value and use first order (linear) B-spline [12] as basis function, we something similar to the SIFT descriptor [5]. By increasing the overlap and smoothness of the basis functions, we expect to get a smoother behavior. In the evaluations, second order B-spline kernels will be used.

The expected advantage comes from the fact that we can use a coarse spatial resolution but still maintain much useful information. For example, we can represent the presence of multiple orientations in a region without averaging them together. The low spatial resolution and the smoothness of the basis functions makes it more likely that the view representation transforms smoothly between the training poses, which makes it suitable for view interpolation.

3.2 Derivatives of the Channel Coding

In updating the image parameters iteratively according to Sect. 2.1, the derivatives of the view representation with respect to the image parameters are required. For simplicity of notation, we ignore the weight function $w(x, y)$ for a moment, and rewrite (12) in vector notation:

$$c_{ijk} = \int_{\mathbb{R}^2} B_{ijk}(\mathbf{x}, z(\mathbf{x}))\mathrm{d}\mathbf{x} \ . \tag{14}$$

The weights will be considered again in Sect. 3.3. We consider now a certain channel coefficient c_{ijk}, and to further simplify the notation, the integral limits and indices ijk will be dropped. We are interested in what happens when the feature image is rotated, scaled and translated:

$$c = \int B(\mathbf{x}, z(\mathbf{A}\mathbf{x} + \mathbf{b})) \ \mathrm{d}\mathbf{x} \tag{15}$$

where

$$\mathbf{A} = e^s \mathbf{R} = e^s \begin{pmatrix} \cos\alpha & -\sin\alpha \\ \sin\alpha & \cos\alpha \end{pmatrix}, \ \mathbf{b} = \begin{pmatrix} b_x \\ b_y \end{pmatrix} \tag{16}$$

Substituting $\mathbf{u} = \mathbf{A}\mathbf{x} + \mathbf{b}$ where $\mathbf{u} = [u, v]^{\mathrm{T}}$ gives $\mathbf{x} = \mathbf{A}^{-1}(\mathbf{u} - \mathbf{b})$ and

$$c = |\mathbf{A}^{-1}| \int B(\mathbf{A}^{-1}(\mathbf{u} - \mathbf{b}), z(\mathbf{u})) \ \mathrm{d}\mathbf{u} \ . \tag{17}$$

We now want to differentiate (17) with respect to α, s, b_x, b_y, and start with α. For compactness, we leave out the arguments to B and its derivatives. These arguments are always as in (17). First note that $|\mathbf{A}^{-1}|$ is constant with respect to α. Since everything is smooth and well-defined, we can replace the order of the integration and differentiation.

$$\frac{\mathrm{d}c}{\mathrm{d}\alpha} = |\mathbf{A}^{-1}| \int \frac{\mathrm{d}}{\mathrm{d}\alpha} [B(\ldots)] \ \mathrm{d}\mathbf{u} \ = \ |\mathbf{A}^{-1}| \int B'_{\mathbf{x}}(\ldots) \frac{\mathrm{d}\mathbf{A}^{-1}}{\mathrm{d}\alpha} \mathbf{u} \ \mathrm{d}\mathbf{u} \tag{18}$$

where

$$\frac{\mathrm{d}\mathbf{A}^{-1}}{\mathrm{d}\alpha} = e^{-s} \begin{bmatrix} -\sin\alpha & \cos\alpha \\ -\cos\alpha & -\sin\alpha \end{bmatrix} \tag{19}$$

$$B'_{\mathbf{x}} = [B'_x, \ B'_y] \tag{20}$$

The differentiation with respect to $\mathbf{b}$ proceeds similarly. We get

$$\frac{\mathrm{d}c}{\mathrm{d}\mathbf{b}} = |\mathbf{A}^{-1}| \int \frac{\mathrm{d}}{\mathrm{d}\mathbf{b}} [B(\ldots)] \ \mathrm{d}\mathbf{u} \ = \ -|\mathbf{A}^{-1}| \int B'_{\mathbf{x}}(\ldots) \mathbf{A}^{-1} \ \mathrm{d}\mathbf{u} \tag{21}$$

In differentiating with respect to s, $|\mathbf{A}^{-1}|$ is no longer constant. The product rule gives us

$$\frac{\mathrm{d}c}{\mathrm{d}s} = \frac{\mathrm{d}|\mathbf{A}^{-1}|}{\mathrm{d}s} \int B(\ldots)\,\mathrm{d}\mathbf{u} \; + \; |\mathbf{A}^{-1}| \int \frac{\mathrm{d}}{\mathrm{d}s} [B(\ldots)] \;\mathrm{d}\mathbf{u} = \tag{22}$$

$$= -2|\mathbf{A}^{-1}| \int B(\ldots)\,\mathrm{d}\mathbf{u} \; + |\mathbf{A}^{-1}| \int B'_{\mathbf{x}}(\ldots) \frac{\mathrm{d}\mathbf{A}^{-1}}{\mathrm{d}s} \mathbf{u} \;\mathrm{d}\mathbf{u} = \tag{23}$$

$$= -\,|\mathbf{A}^{-1}| \int 2B(\ldots) + B'_{\mathbf{x}}(\ldots)\mathbf{A}^{-1}\mathbf{u} \;\mathrm{d}\mathbf{u} \tag{24}$$

Setting $s = 0, \alpha = 0, \mathbf{b} = 0$ gives $\mathbf{A}^{-1} = \mathbf{I}$, and the derivatives in (18), (21) and (24) become

$$\frac{\mathrm{d}c}{\mathrm{d}b_x} = -\int B'_x(\mathbf{u}, z(\mathbf{u}))\;\mathrm{d}\mathbf{u} \tag{25}$$

$$\frac{\mathrm{d}c}{\mathrm{d}b_y} = -\int B'_y(\mathbf{u}, z(\mathbf{u}))\;\mathrm{d}\mathbf{u} \tag{26}$$

$$\frac{\mathrm{d}c}{\mathrm{d}s} = -\int 2B(\mathbf{u}, z(\mathbf{u})) + uB'_x(\mathbf{u}, z(\mathbf{u})) + vB'_y(\mathbf{u}, z(\mathbf{u}))\;\mathrm{d}\mathbf{u} \tag{27}$$

$$\frac{\mathrm{d}c}{\mathrm{d}\alpha} = \int vB'_x(\mathbf{u}, z(\mathbf{u})) - uB'_y(\mathbf{u}, z(\mathbf{u}))\;\mathrm{d}\mathbf{u} \tag{28}$$

By computing these derivatives for each channel and stacking them into vectors, we get $\mathbf{g}'_x, \mathbf{g}'_y, \mathbf{g}'_s, \mathbf{g}'_\alpha$ required in (6). Note that if the basis functions are B-splines, all terms in the above integrals are just piecewise polynomial functions with the same support, so the amount of computation required to evaluate each of the derivatives is in the same order of magnitude as computing the channel-coded feature map itself.

3.3 Weighted Data

In the previous section, the weights from (12) were not considered. By introducing these weights again, the results are similar. Since the weights are defined for each pixel in the feature image, they transform with the features, i.e. (15) becomes in the weighted case

$$c = \int B(\mathbf{x}, z(\mathbf{A}\mathbf{x} + \mathbf{b}))w(\mathbf{A}\mathbf{x} + \mathbf{b})\;\mathrm{d}\mathbf{x}\;. \tag{29}$$

After the variable substitution, we have

$$c = |\mathbf{A}^{-1}| \int B(\mathbf{A}^{-1}(\mathbf{u} - \mathbf{b}), z(\mathbf{u}))w(\mathbf{u})\,\mathrm{d}\mathbf{u} \tag{30}$$

In this expression, the weighting function is independent of the transformation parameters $\alpha, s, \mathbf{b}$ and is left unaffected by the differentiation. The complete expressions for the derivatives in the weighted case are just (25)-(28) completed with the multiplicative weight $w(\mathbf{u})$ inside the integrals.

3.4 Normalization

The method has shown to work better if the channel vectors are normalized using $\tilde{\mathbf{c}} = \mathbf{c}/\|\mathbf{c}\|$, where $\|\cdot\|$ is the L_2 norm. In this case, we should change the derivatives from previous section accordingly. From the quotient rule, we have

$$\frac{\mathrm{d}\tilde{\mathbf{c}}}{\mathrm{d}\alpha} = \|\mathbf{c}\|^{-2}\left(\frac{\mathrm{d}\mathbf{c}}{\mathrm{d}\alpha}\|\mathbf{c}\| - \mathbf{c}\frac{\mathrm{d}\|\mathbf{c}\|}{\mathrm{d}\alpha}\right) = \|\mathbf{c}\|^{-1}\left(\frac{\mathrm{d}\mathbf{c}}{\mathrm{d}\alpha} - \tilde{\mathbf{c}}\tilde{\mathbf{c}}^{\mathrm{T}}\frac{\mathrm{d}\mathbf{c}}{\mathrm{d}\alpha}\right) \tag{31}$$

where we have used that $\frac{\mathrm{d}\|\mathbf{c}\|}{\mathrm{d}\alpha} = \tilde{\mathbf{c}}^{\mathrm{T}}\frac{\mathrm{d}\mathbf{c}}{\mathrm{d}\alpha}$. The derivatives with the respect to the other variables are derived analogously.

4 Experiments

The method is evaluated on a set of objects scanned with a turn-table, producing training images like in Fig. 2. Views are available for every 5° in both the θ and ϕ domain. The method was trained on views 20° apart, and evaluated on all intermediate views. This gives 50 training views and 629 evaluation views.

In the first experiment, we assume that the image parameters are known, and optimize the pose angles $[\theta, \phi]$. The iterations were initialized at the closest training view, which is similar to what can be expected from an inaccurate object detector. The best parameter settings were found by an exhaustive search. The results using different varieties around the best option are shown in Table 1. As we see, the performance is rather insensitive to parameter variations in this order of magnitude.

In the second experiment, all 6 pose parameters were optimized simultaneously. One problem here is that the set of angles $[\alpha, \theta, \phi]$ is ambiguous such that two distinct set of angles can represent the same pose. Because of this, we combine the pose angles and image rotation into a rotation quaternion and measure the error in the quaternion domain. An RMS quaternion error of 0.015 corresponds to around 2° error in each angle. The method was initialized using the

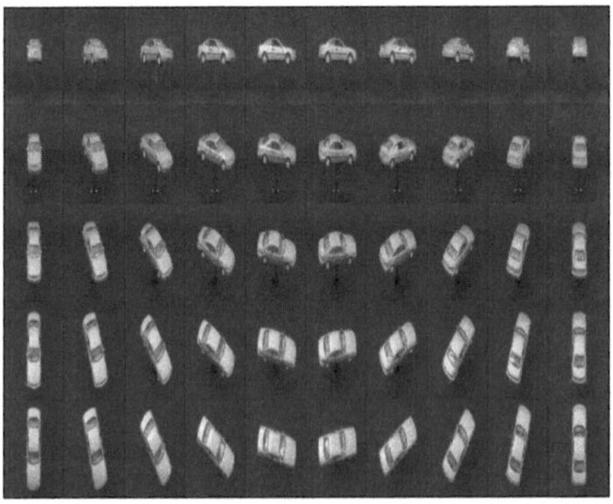

Fig. 2. Training views. Y-axis: θ, X-axis: ϕ.

Table 1. RMS error in degrees for pose angles only, around the manually selected option $8 \times 8 \times 6$ channels, 4 neighbors. Left: Varying spatial resolution. Middle: Varying number of orientation channels. Right: Varying number of neighbors.

$n_x \times n_y$	error
6x6	1.2
8x8	1.2
10x10	1.3
12x12	1.4
14x14	1.4

n_f	error
4	1.2
6	1.2
8	1.3
10	1.3

neighbors	error
3	1.3
4	1.2
5	1.2
6	1.2
7	1.3
8	1.6
9	1.9

Table 2. RMS error for all parameters (x,y,s error is in pixels) around the manually selected option $8 \times 8 \times 6$ channels, 4 neighbors. Left: Varying spatial resolution. Middle: Varying number of orientation channels. Right: Varying number of neighbors.

$n_x \times n_y$	x,y	s	q
6x6	3.5	6.8	0.016
8x8	3.5	6.2	0.016
10x10	4.2	6.7	0.017
12x12	4.1	7.0	0.017
14x14	4.5	7.6	0.018

n_f	x,y	s	q
4	3.55	6.6	0.015
6	3.5	6.8	0.016
8	3.5	7.0	0.015
10	3.6	7.5	0.016

neighbors	x,y	s	q
3	3.5	6.1	0.016
4	3.5	6.2	0.016
5	3.4	6.0	0.016
6	3.1	6.0	0.015
7	3.6	6.3	0.017
8	3.6	6.6	0.020

true image parameters and the closest pose parameters from the training set. The results for different options are shown in Table 2. Here s is the radius in pixels of a box containing the object, and the errors in x, y, s are measured in pixels. The size of the car in the images is around 300 pixels.

The current implementation runs at a few frames per second on an AMD Athlon 3800+.

5 Discussion

This paper has described an accurate interpolation method for view-based pose estimation using local linear models and Gauss-Newton optimization. Some varieties in the view representation have been compared in terms of fitness to this framework. However, the evaluation is in no way complete. There are several more parameters to play around with, e.g. the amount of overlap between basis functions, different soft tresholdings of the orientation image etc. It would also be interesting to perform a more detailed study on the performance of this representation compared to other approaches like PCA, wavelet representations etc.

The most critical fact is however that the evaluation dataset is too simple, without occlusion and difficult backgrounds. To verify that the method works in the real world, it has been run on hand-camera video sequences, but without any quantitative error measures. Some video clips are available online [1]. A more rigorous evaluation should be performed on real image sequences with ground truth. In the near future, we plan to create datasets for this purpose and make them publicly available.

Recall that the full view setting was chosen mainly to be able to focus on the interpolation and feature representation. The goal of our future research is to transfer these techniques to methods based on local patches. This creates new problems, e.g. how to handle the fact that patches may be selected from different positions on the object in different views. Solving these problems will be challenging but hopefully rewarding in terms of greater performance.

References

1. Atkeson, C.G., Moore, A.W., Schaal, S.: Locally weighted learning. Artificial Intelligence Review 11, 11–73 (1997)
2. Comport, A., Marchand, E., Chaumette, F.: A real-time tracker for markerless augmented reality. In: Proc. The Second IEEE and ACM International Symposium on Mixed and Augmented Reality, pp. 36–45 (2003)
3. Felsberg, M., Forssén, P.-E., Scharr, H.: Channel smoothing: Efficient robust smoothing of low-level signal features. IEEE Transactions on Pattern Analysis and Machine Intelligence 28(2), 209–222 (2006)
4. Granlund, G.H.: An associative perception-action structure using a localized space variant information representation. In: Sommer, G., Zeevi, Y.Y. (eds.) AFPAC 2000. LNCS, vol. 1888, Springer, Heidelberg (2000)
5. Lowe, D.G.: Object recognition from local scale-invariant features. In: IEEE Int. Conf. on Computer Vision (September 1999)
6. Moore, A.W., Schneider, J., Deng, K.: Efficient locally weighted polynomial regression predictions. In: Proc. 14th International Conference on Machine Learning, pp. 236–244. Morgan Kaufmann, San Francisco (1997)
7. Murase, H., Nayar, S.K.: Visual learning and recognition of 3-d objects from appearance. International Journal of Computer Vision 14(1), 5–24 (1995)
8. Nocedal, J., Wright, S.J.: Numerical Optimization. Springer, Heidelberg (1999)
9. Obdrzalek, S., Matas, J.: Object recognition using local affine frames on distinguished regions. In: British Machine Vision Conf. (2002)
10. Pentland, A., Moghaddam, B., Starner, T.: View-based and modular eigenspaces for face recognition. In: CVPR (1994)
11. Cipolla, R., Drummond, T.: Real-time visual tracking of complex structures. IEEE Transactions on Pattern Analysis and Machine Intelligence,vol. 24(7) (July 2002)
12. Unser, M.: Splines: A perfect fit for signal and image processing. IEEE Signal Processing Magazine 16(6), 22–38 (1999)

[1] `http://www.cvl.isy.liu.se/Research/Object/ChannelCodedFeatureMaps/`

Automatic Segmentation of Overlapping Fish Using Shape Priors

Sigmund Clausen[1], Katharina Greiner[2], Odd Andersen[1], Knut-Andreas Lie[1], Helene Schulerud[1], and Tom Kavli[1]

[1] SINTEF ICT, Oslo, Norway
[2] University of Applied Sciences, Wiesbaden, Germany

Abstract. We present results from a study where we segment fish in images captured within fish cages. The ultimate goal is to use this information to extract the weight distribution of the fish within the cages. Statistical shape knowledge is added to a Mumford-Shah functional defining the image energy. The fish shape is represented explicitly by a polygonal curve, and the energy minimization is done by gradient descent. The images represent many challenges with a highly cluttered background, inhomogeneous lighting and several overlapping objects. We obtain good segmentation results for silhouette-like images containing relatively few fish. In this case, the fish appear dark on a light background and the image energy is well behaved. In cases with more difficult lighting conditions the contours evolve slowly and often get trapped in local minima

Keywords: Segmentation, Overlapping objects, Mumford-Shah, Shape priors.

1 Introduction

In the fish-farming industry it is of major importance to know the total biomass and weight distribution of the fish within the fish cages. Such knowledge is important in order to monitor the growth of the fish, to optimize the subsequent logistics and the process of providing a correct level of feeding. One can think of several measurement techniques that could be used for this particular problem; however, few have been tested in detail. A rough estimate of the total biomass could possibly be obtained by using acoustic techniques like echo sounding and sonars. In order to estimate the weight distribution, more detailed information and higher resolution is needed. One possible approach towards solving this problem consists in using a stereo-based underwater video surveillance system [1]. In this case, stereo images of the fish swimming around are captured at a certain frame-rate and a stereo mapping is conducted. Based upon the stereo mapping and the size of corresponding fish within a stereo pair, the weight of the fish can be estimated using simple empirical calibration models. Such a system has the possibility of monitoring the fish continuously and build up adequate statistics over time.

For this approach to work effectively the same fish within one stereo pair has to be properly segmented. This is a difficult task due to a highly cluttered image background, inhomogeneous lighting and several overlapping objects. Figure 1 shows

B.K. Ersbøll and K.S. Pedersen (Eds.): SCIA 2007, LNCS 4522, pp. 11–20, 2007.

an example of a medium complex image within a fish cage. The image contains approximately 20 fish and our goal is to segment as many of these as possible. Preferably we would like to segment fish where the complete fish is imaged within the image frame. We will use the image in Fig. 1 as an example image throughout this paper.

Fig. 1. An example of a medium complex image of fish swimming around in a fish cage. Several overlapping fish are observed. In addition the net of the cage is clearly visible, creating a non-trivial background. The lighting is inhomogeneous with background intensities ranging from 90 in the lowest part and up to 255 (saturation) in the upper part. We consider 8-bit monochrome images. The average intensity within a fish will often lie somewhere in between these values, causing a non-trivial region-based segmentation.

Due to the high complexity of the images, we have chosen to use a variational segmentation approach, where information about the shape is added to the functional defining the image energy. Variational image segmentation has become increasingly popular during the last years and there exist several approaches towards solving this problem. Some approaches make use of edges [2], whereas others define homogeneity requirements on the statistics of the regions being segmented [3]. In many cases the images contain a lot of noise, the contrast may be low and the background is often highly cluttered. All these effects cause a non trivial behavior of the image energy. To resolve some of these issues various efforts have been made to include prior information about the shape of the objects of interest. In this paper we follow Cremers et al. [4], who have proposed to include information about shape in a modified Mumford-Shah functional. The modified version is a hybrid model combining the external energy of the Mumford-Shah functional with the internal energy of the snakes into one single functional. Following [4], we use an explicit representation of the contour, but instead of using a spline, we represent the contour by a closed polygonal curve. The statistical shape model of a fish is based upon a set of training shapes, all extracted from real images of fish in a semi-automatic manner.

In Section 2 we briefly describe Cremers' approach. The energy functional including the shape energy is defined together with a description of the minimization process using gradient descent. In addition, we describe a simple initialization procedure that has been proven to work well for the fish segmentation problem. In Section 3 we present several segmentation results for images of varying degree of complexity. Finally, in Section 4 we conclude and discuss possible future improvements of the procedure.

2 Variational Image Segmentation Using Shape Priors

In this section we describe the energy functional following Cremers et al. [4]. The image information and shape information are combined into one variational framework. For a given contour *C*, the energy is defined by

$$E(u,C) = E_i(u,C) + \alpha \cdot E_c(C) . \tag{1}$$

Here E_i measures how well the contour *C* and the associated segmentation *u* approximate the image. The term E_c favors statistically familiar contours by a learning process, where the parameter α defines the relative strength of the shape prior. We refer to [4] and [5] for detailed information about the theory underlying the model and aspects regarding the implementation of the model.

2.1 Image Energy from Polygonal-Based Mumford-Shah

Cremers' modified Mumford-Shah functional for the image energy E_i is given by the following expression [4]:

$$E_i(u,C) = \frac{1}{2}\int_{\Omega}(f-u)^2 dx + \frac{\lambda^2}{2}\int_{\Omega-C}|\nabla u|^2 dx + \nu\int_0^1 C_s^2 ds, \tag{2}$$

where *f* is the input image and *u* is an approximation of *f* within the image plane Ω. The image *u* may be discontinuous across the segmenting contour *C*. The parameter λ defines the spatial scale on which smoothing is performed. In all the results presented here we have used λ=7. For this value, the diffusion process obtained when minimizing (2) produces a medium smooth image *u* maintaining some of the local information in the original image. In this way, gradient information is allowed to propagate towards the fish contour. The last term is due to Cremers et al. [4], who replaced the original contour length norm in the Mumford-Shah functional with the squared L_2-norm to restrict the contour points from clustering in one place.

2.2 Shape Energy from Gaussian Shape Statistics

The explicit representation of the contour allows for a statistical treatment of the different shapes. This is obtained by extracting several *training* shapes from high contrast images containing non-overlapping fish. The images of the training shapes are binarized and the fish contour is extracted and represented by a polygonal curve. All the training contours are aligned by similarity transformation and cyclic permutation of the contour points. The contour is represented by a polygonal curve *z* with *N* points:

$$z = (x_1, y_1, \ldots, x_N, y_N)^T. \tag{3}$$

For a given set of training shapes, the distribution of all contours given by (3) is assumed to have a Gaussian shape probability, with μ denoting the mean contour point vector and Σ the regularized covariance matrix [4]:

$$\mathrm{P}(z) = \exp\left(-\frac{1}{2}(z-\mu)^T \Sigma^{-1}(z-\mu) \right). \tag{4}$$

The Gaussian shape probability corresponds to the following energy measure:

$$E_c(z) = -\log(\mathrm{P}(z)) + const = -\frac{1}{2}(z-\mu)^T \Sigma^{-1}(z-\mu). \tag{5}$$

Figure 2 shows 20 training fish used to collect information about the fish shape statistics. All these contours are aligned with respect to similarity transformations and a cyclic permutation of the contour points. Based upon the aligned contours, the mean shape μ and the regularized covariance matrix Σ are extracted.

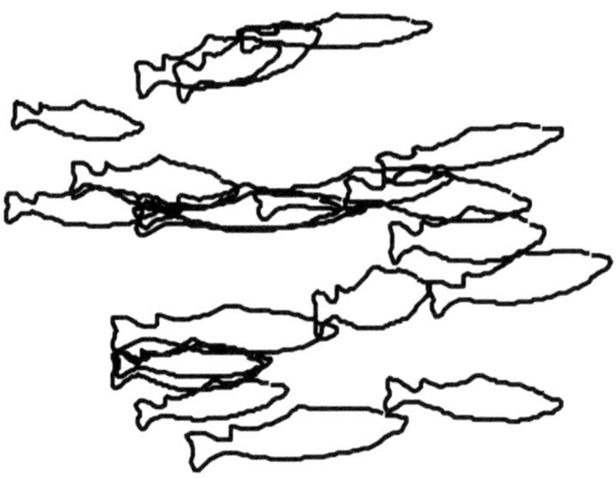

Fig. 2. 20 fish training shapes extracted from images of real fish. The Gaussian shape statistics is built from these 20 training shapes and incorporated in the Mumford-Shah energy functional.

2.3 Energy Minimization by Gradient Descent

We briefly present the resulting evolution equations and refer to Cremers et al. [4,5] for details. Note that the total energy (1) is minimized with respect to both the segmenting contour C and the segmented image u. The Euler-Lagrange equations corresponding to the minimization are then solved by gradient descent.

It can be shown that the x-component of the evolution equations for the segmenting contour is given by the following expression

$$\frac{dx_i(t)}{dt} = \left[e^+(s_i,t) - e^-(s_i,t)\right] \cdot n_x(s_i,t) + \nu(x_i - 2x_i + x_{i+1}) - \alpha\left[\Sigma^{-1}(z-\mu)\right]_{2i-1}. \tag{6}$$

where n_x is the x-component of the outer normal vector and the index $2i$-1 in the last term is associated with the x-component of contour point number i. $e^{+/-}$ is the energy density in the regions s_i adjoining the contour C at contour point number i

$$e^{+/-} = (f-u)^2 + \lambda^2 \left|\nabla u\right|^2. \tag{7}$$

The same equations apply for the y-component of the segmenting contour C. Now, the first term in (6) forces the contour towards the object boundaries. The second term forces equidistant spacing of the contour points, whereas the last term causes a relaxation towards the most probable shape. As pointed out by Cremers et al. [4], the most probable shape is weighted by the inverse of the covariance matrix causing less familiar shape deformations to decay faster.

Now, let us turn to the evolution equation for the segmenting image u. Minimizing (1) with respect to the segmenting image u is equivalent to the steady state of the following diffusions process [4]:

$$\frac{\partial u}{\partial t} = \nabla \cdot (\omega_c \nabla u) + \frac{1}{\lambda^2}(f-u). \tag{8}$$

Here the contour defines a boundary giving rise to an inhomogeneous diffusion process. The indicator function ω_c equals zero at the boundary C and one otherwise. Cremers [5] suggested both an explicit finite-difference scheme for (8) and a multigrid method for solving the steady-state equation directly. In this work we use the explicit approximation of (8).

2.4 Automatic Initialization of the Contours

The initialization of the contours is of major importance towards a successful segmentation. In this work we have developed a simple algorithm to provide an initial estimate of the segmenting contour. First we locate the potential tails of the fish by thresholding a low-pass filtered difference image between to consecutive video frames. These consecutive frames are separated 20 ms in time and the motion of the fish usually creates a significant difference between the two frames, especially for the tail region of the fish. This difference is caused by the fish swimming in a direction perpendicular to the edge of the tail. Figure 3 shows the potential tail regions marked with black diamonds in our example image from Fig. 1.

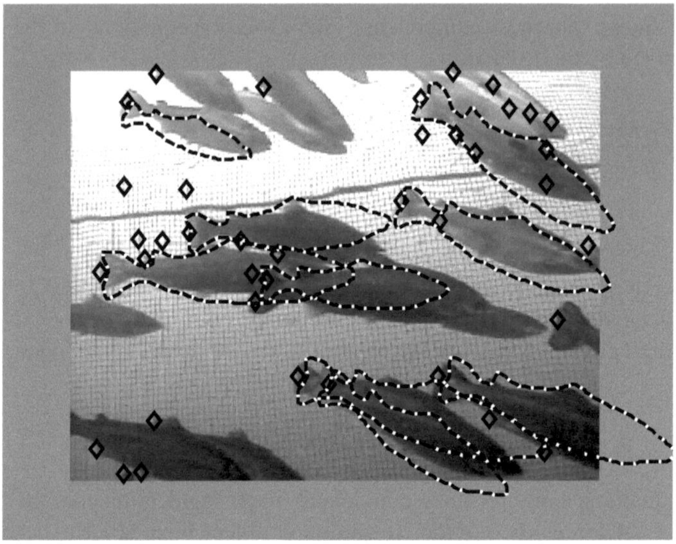

Fig. 3. Potential fish tail candidates marked by black diamonds in our example image from Fig.1. The final initial contours in black and white lines are given by minimizing the projection of the local image orientation along the normal vector of the mean-shape fish μ..

As can be seen from the figure all the true fish tails are correctly located. However, in addition there are quite a few erroneously detected tails related to other parts of the fish or movements of the camera and/or background. All these wrongly classified tails are ruled out by the next step in the initialization process. This step provides a coarse estimate of the size and the rotation of the fish by minimizing the projection of the local image orientation along the normal vectors of the mean-shape fish μ. The projections are obtained by calculating the average scalar product of the fish normal vectors and the local image orientation vectors for several possible fish sizes and angles of rotation. During this process the contour of the mean shape fish is locked to the tail position.

3 Segmentation Results

In this section we present several segmentation results. We start by considering our example image in Fig. 1, where we focus on two different cases.

Figure 4 shows the resulting segmentation of one fish located in the upper left corner of our example image for 4 different initializations of the fish contour. The figure illustrates the sensitivity to the initialization and the problem of the evolving contour being trapped in local minima.

Figure 5 shows a similar behavior for the large silhouette-like fish in the middle of the example image in Fig. 1.

In [4], Cremers et al. discuss the modified Mumford-Shah model and its cartoon limit, which is obtained as $\lambda \to \infty$ in (2). In this case, the diffusion process for u in (8) is replaced by a simple averaging process across the two regions separated by the contour. Thus, the cartoon model is more affected by global gray-level information. Actually, the cartoon model is equally affected by the gray-level information in any part of the image. The cartoon model works poorly for the fish-segmentation problem due to the highly varying gray-level information within both the object and the background. In many cases, the average image intensity inside the fish is similar to the average background intensity causing the segmentation to fail totally. By using the full modified Mumford-Shah model, more local image information is taken into account and in many cases we obtain good segmentation results. This is especially the case for silhouette-like images, where the fish appear dark on a light background. However, the previous two examples show that the segmentation might still fail if the initial contour is too far from the object of interest. Parts of the contour are easily locked to edges belonging to other fish, thereby causing the segmentation to get trapped in local minima. A good initialization is therefore a prerequisite for a correct segmentation.

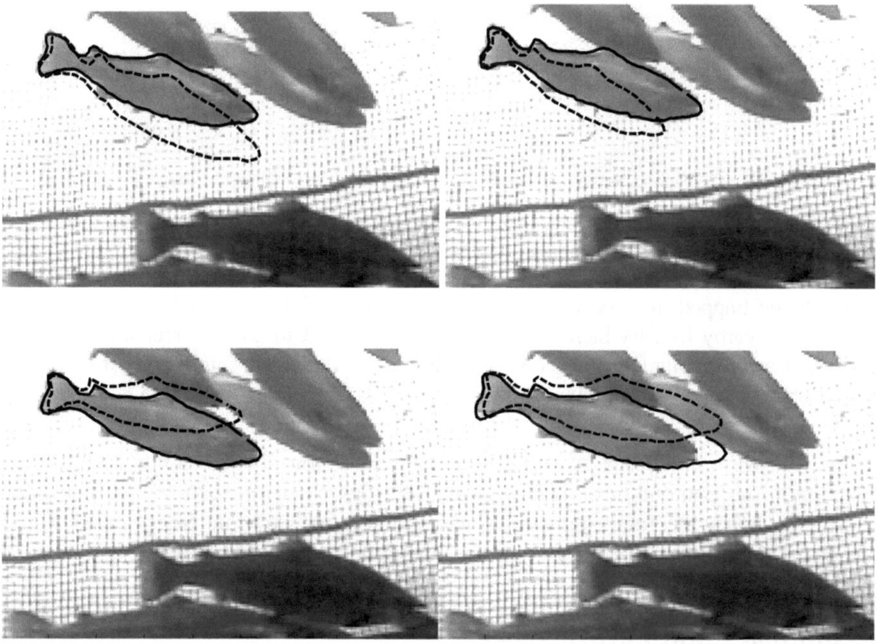

Fig. 4. Segmentation results for one fish in the upper-left corner of the example image in Fig. 1. The images show the final segmentation results (full black lines) for four different contour initializations (dashed black lines). The lower-right plot shows an example of the segmentation process being trapped in a local minimum, where a part of the fish contour is stuck to the edge of a nearby fish.

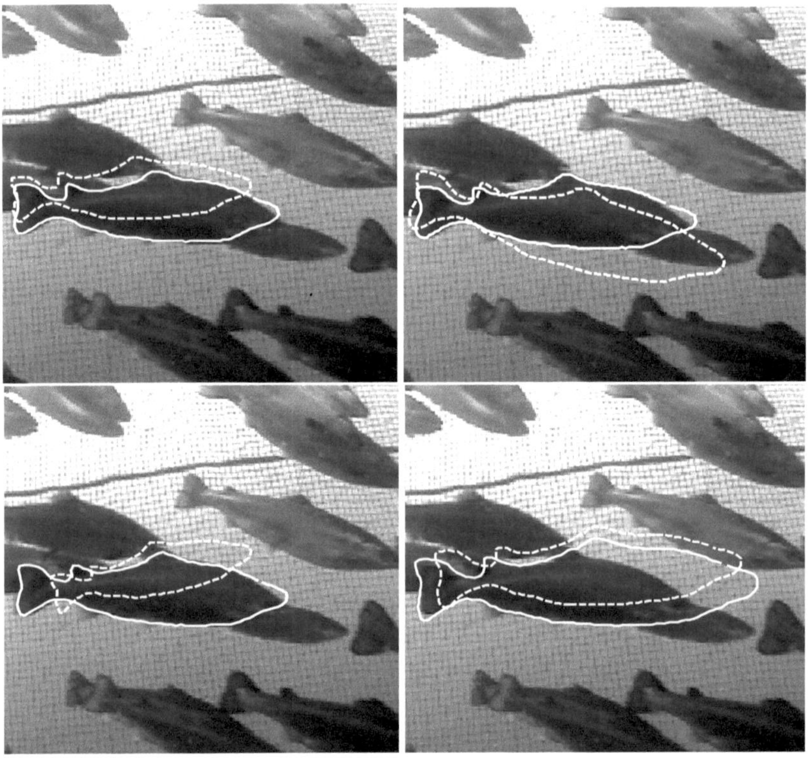

Fig. 5. Segmentation results for one fish in the central region of the example image in Fig. 1. The images show the final segmentation results (full white lines) for four different contour initializations (dashed white lines). The lower-right plot shows an example of the segmentation process being trapped in a local minimum, where a part of the fish contour is stuck to the contour of a nearby fish. By increasing the scale parameter λ to 20, a correct segmentation is obtained also for this initialization. For this value of λ the gradient information from the true fish edge is propagated all the way to the fish contour.

The following examples in Fig. 6 show segmentation results on complete images based upon the contour initialization procedure described in Section 2.4. The upper left figure shows the final segmentation result for our example image from Fig. 1. Recall that the initial fish contours for this example image are shown in Fig. 3.

4 Summary and Future Work

We have presented results from a study where we segment fish in images captured within fish cages. Our ultimate goal has been to use this information as part of a stereo-based video surveillance system to extract the weight distribution of the fish within the cages. Statistical shape knowledge was added to the Mumford-Shah functional following Cremers et al. [4]. The fish shape was represented explicitly by a polygonal curve and the energy minimization was done by gradient descent. We

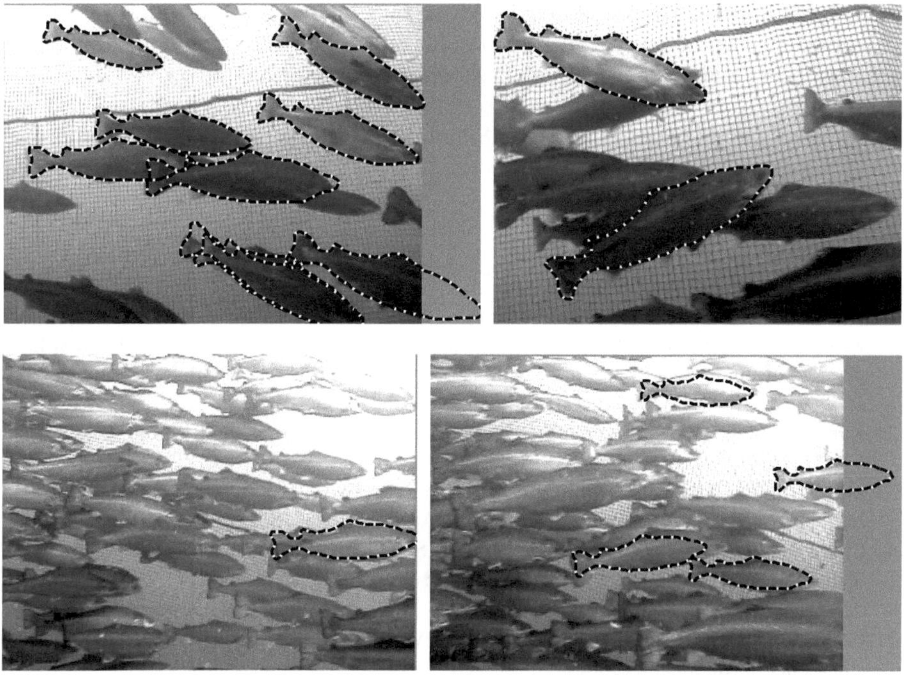

Fig. 6. The figure shows the automatic segmentation results for four different images with a varying degree of complexity. The upper-left image shows the segmentation result for the complete example image from Fig. 1. All the fish are correctly segmented except one of the fish in the bottom part of the image. Here two fish are overlapping almost completely and the segmented fish hiding behind is probably larger than ground truth. In the upper-right image both fish are correctly segmented. The upper fish has a more complex intensity distribution than the silhouette-like fish we have looked at so far. The two lowermost images show examples of automatic segmentation results in rather complex images, where the background is practically similar to the objects of interest. In the lower-left image only one fish has passed the initialization criterion. The segmented contour misses the tail somewhat. In the lower-right image the lowermost fish is correctly segmented. The contour of the uppermost fish misses the upper part of the fish, whereas the contour of the fish in the central region of the image is a bit too small. The rightmost fish seems to be correctly segmented; however, it is hard to tell since half the fish is outside the image frame.

obtained good segmentation results for silhouette-like images containing relatively few fish. In this case, the fish appears dark on a light background and the image energy is well behaved. In cases with more difficult lighting conditions, the contours evolve slowly and often get trapped in local minima. Here the interior of the fish often contains many edges and the image intensity varies in a non-trivial manner. In this case, the Mumford-Shah hypothesis of a piecewise homogeneous gray level value is violated.

To improve the existing segmentation, we intend to incorporate local gray-level models for the interior of the fish. In addition, we have the possibility to include motion cues and stereo information to make the segmentation more robust. Moreover, adding balloon forces to the functional (1) could in some cases resolve the problem of the fish contour being locked to nearby edges. Finally, by using more global methods, like graph-cuts and dynamic programming for minimizing the energy, the problem of contours being trapped in local minima should be reduced even further.

Acknowledgments. We thank AKVASmart for providing images of fish within fish cages.

References

1. The AKVA group has developed Vicass – a stereo-based imaging system for size and weight estimation currently used in more than 100 fish farms and distributed and sold in more than eight countries world wide. For more information about an existing stereo system, see: www.akvasmart.com
2. Kass, M., Witkin, A., Terzopoulos, D.: Snakes: Active contour models. Int. J. of Comp. Vis. 1(4), 321–331 (1988)
3. Mumford, D., Shah, J.: Optimal Approximations by piecewise smooth functions and associated variational problems. Comm. Pure Appl. Math. 42, 577–685 (1989)
4. Cremers, D., Tischhäuser, F., Weickert, J., Schnörr, C.: Diffusion Snakes: Introducing Statistical Shape Knowledge into the Mumford-Shah functional. Int. J. of Comp. Vis. 50(3), 295–313 (2002)
5. Cremers, D.: Statistical Shape Knowledge in Variational Image Segmentation PhD thesis. University of Mannheim (2002)

Automatic Feature Point Correspondences and Shape Analysis with Missing Data and Outliers Using MDL

Kalle Åström, Johan Karlsson, Olof Enquist, Anders Ericsson, and Fredrik Kahl

Centre for Mathematical Sciences, Lund University, Sweden

Abstract. Automatic construction of Shape Models from examples has recently been the focus of intense research. These methods have proved to be useful for shape segmentation, tracking, recognition and shape understanding. In this paper we discuss automatic landmark selection and correspondence determination from a discrete set of landmarks, typically obtained by feature extraction. The set of landmarks may include both outliers and missing data. Our framework has a solid theoretical basis using principles of Minimal Description Length (MDL). In order to exploit these ideas, new non-heuristic methods for (i) principal component analysis and (ii) Procrustes mean are derived - as a consequence of the modelling principle. The resulting MDL criterion is optimised over both discrete and continuous decision variables. The algorithms have been implemented and tested on the problem of automatic shape extraction from feature points in image sequences.

1 Introduction

Inspired by the successful methods for finding feature correspondences along curves and on surfaces for deriving shape variation models, [4], this paper develops these methods further for the problem of shape modelling of unordered point sets with outliers and possibly missing data. The main idea is that an information criterion, such as Minimum Description Length (MDL), is well suited for determining which points should be considered as outliers and which points should be considered to be in correspondence.

One motivation for this study is that, although it is straightforward to detect interesting features in images, e.g., using corner detectors such as [8], it is not at all straightforward to solve the correspondence problem. Many methods for finding correspondences are based on either continuity assumptions, e.g., [18] or that a model is a priori known, e.g., [16].

For unordered points sets, as opposed to what is usually assumed for curves, it is often the case that points are frequently missing and that there are outliers. Traditional methods for shape analysis have problem with missing data. As a by-product of the development here, we derive novel methods for Procrustes analysis and principal component analysis for missing data, based on principles

B.K. Ersbøll and K.S. Pedersen (Eds.): SCIA 2007, LNCS 4522, pp. 21–30, 2007.

for model selection. Although generalisations to missing data have been done before both for Procrustes analysis, [9] and for principal component analysis, [11], in the present formulation they are shown to be a logical consequence of the modelling principle.

The underlying modelling principle we will use is MDL, that is, choose the representation or model with shortest description, cf. [15]. It has previously been successfully applied in computer vision to model selection problems, e.g., [12] and as mentioned curve and surface modelling [4].

MDL has also been used on sets of images to unify groupwise registration and model building, [20], although these mostly work on appearance based models. This is related to our approach, but here we include decisions about what to include in the modelling and what to consider as outliers explicitly in calculating the description length. The algorithm therefore decides what in the images to build the model from and what to consider as irrelevant background data.

There are several algorithms for matching one point cloud to another by determining both a deformation and correspondences, see for example [2], and many handle outliers in some way. However, these algorithms do not work with a whole sequence of point sets at once. Although it is conceivable that such methods could be used to construct shape variation models, the model built from the corresponding points might not be optimal at all. In this paper the model itself is an integral part of the algorithm for determining the correspondences.

Another line of research that is related to our work is the area of non-rigid factorisation methods, e.g., [3]. In addition to reconstructing a model for the point set, they also try to estimate the camera motion. However, they assume that feature correspondences are given or obtained by some heuristics. Similarly, outliers and missing data are handled with ad-hoc techniques. Our framework is well-suited to be applied to these problems as well.

2 MDL for Feature Point Selection and Correspondence

The main problem that we formulate and solve in this paper is the following. Assume that a number of examples $S = \{S_1, \ldots, S_n\}$ are given, where each example S_i is a set of unordered feature points

$$S_i = \{z_{i,1}, \ldots, z_{i,k_i}\}, \qquad (1)$$

and each point $z_{i,j}$ lies in $\mathbf{R}^p$ for some dimension p, typically $p = 2$ or $p = 3$. An example of such sets is the output of a feature detector, e.g., [10].

(i) How should one determine which points in each set is considered to be outliers?
(ii) How should one determine the correspondences of feature points across the examples?
(iii) What is a suitable shape variation model for the inliers?

Such a procedure would be useful for unsupervised learning of shape variation in image sequences. The learnt models could then be useful for a number of applications, including *tracking*, *recognition*, *object pose* etc.

The idea here is to transfer similar ideas from shape variation estimation for curves and surfaces, [19,4], where the correspondence problem is a crucial problem, to the concept investigated here.

As opposed to the theory for curves and surfaces, however, we do not here have the problem of how to weight different parts of the curves relative to the other. On the other hand we will here allow outliers and missing data. In this paper we will use the minimum description length paradigm, [1], to select a suitable model.

We pose the problem as that of selecting: (i) outliers, (ii) correspondences and (iii) model complexity, so that the description length is minimised. As we shall see this becomes a mixed combinatorial and continuous optimisation problem, where for each of a discrete set of possible outliers/correspondences, there is a continuous optimisation problem which has to be solved. These continuous optimisation problems involve both the problem of missing data Procrustes and missing data principal component analysis. In contrast to other ad-hoc methods dealing with outliers and missing data, the way we define missing data Procrustes and missing data principal component analysis is a natural consequence of the way we model the whole problem.

3 Unordered Point Set Shape Analysis

The formulation of the problem is as follows. Assume a set S of n unordered point sets, $S = \{S_1, \ldots, S_n\}$ is given. For simplicity we order the points in each set S_i of k_i points arbitrarily, cf. (1).

The object is now to find a reordering of such points. Assuming that the model contains N points, such a reordering can be represented either as a matrix O of size $n \times N$ whose entries are either 0 - representing that a model points is not visible in an image or the identity number between 1 and k_i telling which image point correspond to the model points, i.e. $O_{i,j} = 0$ if model point j is not visible in image i or $O_{i,j} = k$ if model point j in image i is $z_{i,k}$. Also introduce the set I of indices (i,j) such that model point j is visible in image j, i.e. $I = \{(i,j) \mid O_{i,j} \neq 0\}$. Outliers are then not represented in O.

Given an ordering O the data can be reordered, possibly with missing data into a structure T of N points in n images, i.e.

$$T_{i,j} = \begin{cases} z_{i,O_{i,j}} & \text{if } (i,j) \in I \\ \text{undefined} & \text{if } (i,j) \notin I. \end{cases}$$

For such a ordered point set T with missing data one can do a Procrustes mean shape analysis with respect to a transformation group G. In loose terms the aim is to find a mean shape m and a number of transformations $\{g_1, \ldots, g_n\}$ with $g_i \in G$ such that $g_i(m) \approx T_i$.

The usual method is then to perform a Principal Component Analysis on the residuals between $g_i^{-1}(T_i)$ and m, from which a number of shape variational modes, denoted v_l, can be determined. New shapes can then be synthesised as $g(m + \sum_{l=1}^{d} \lambda_l v_l)$, where λ_l are scalar coordinates.

Here we need to assess a number of different choices: the number of model points N, the ordering O, the mean shape m, the transformation group G, the transformations g_i, the number of variational modes d, the shape variation modes v_l and the coordinates λ_l. The approach we make here is that a common framework such as the minimum description length framework could be used to determine all of these choices [13]. This would put the whole chain of difficult modelling choices on an equal footing and would make it possible to use simple heuristics for making fast and reasonable choices, while at the same time have a common criterion for evaluating different alternatives.

The whole process can thus be seen as an optimisation problem

$$\min_{\mathcal{M}} \mathrm{dl}(\mathrm{S}, \mathcal{M}),$$

over the unknowns $\mathcal{M} = (O, m, \{g_i\}, d, \{v_l\}, \{\lambda_{li}\})$ given data S.

4 Calculating the Description Length

A number of sets are given. Each set of points comes typically from images, where a number of interesting points have been detected. In order to determine a model that explains points that can be seen in many of the images, the goal is to minimise the description length that is needed to transmit all the interesting points of all views, in hope that a model will be able to make a cheaper description than simply sending the data bit by bit. Here we will derive the description length for the data and the model. For the outliers one must simply send the information bit by bit. For the points that are modelled, the idea is that it is cheaper to send the model with parameters and residuals etc. to explain the data. For the modelled points one must send: the model, the model parameters, information if a certain point is missing, the transformation and the residuals.

Preliminaries on information theory. To transmit a continuum value α it is necessary to quantify the value. The continuum value α quantified to a resolution of Δ is here denoted $\hat{\alpha}$, $\alpha_{min} \leq \hat{\alpha} \leq \alpha_{max}, \hat{\alpha} = m\Delta, m \in \mathbf{Z}$.

The ideal coding codeword length for a value $\hat{\alpha}$, encoded using a statistical model $\mathcal{P}(\hat{\alpha})$ is given by the Shannon codeword length [17]. Using Shannon's codeword length the description length of a given value, $\hat{\alpha}$, encoded using a probabilistic model, is $-log(\mathcal{P}(\hat{\alpha}))$, where $\mathcal{P}$ is the probability-density function.

Coding data with uniform distribution. Assume α is uniformly distributed and quantified to $\alpha_{min} \leq \hat{\alpha} \leq \alpha_{max}$, $\hat{\alpha} = m\Delta$, $m \in \mathbf{Z}$. Then α can take $\frac{\alpha_{max}-\alpha_{min}}{\Delta}$ different values. Since uniform distribution is assumed, the probability for a certain value of $\hat{\alpha}$ is $\mathcal{P}(\hat{\alpha}) = \frac{\Delta}{\alpha_{max}-\alpha_{min}}$. This gives Shannon's codeword length for $\hat{\alpha}$, $-log(\mathcal{P}(\hat{\alpha})) = -log(\frac{\Delta}{\alpha_{max}-\alpha_{min}})$. If the parameters α_{min}, α_{max} and Δ are unknown to the receiver, these need to be coded as well.

Coding data with assumed Gaussian distribution. Since the mean of our data μ is zero, the 1-parameter Gaussian function can be used. The frequency function of the 1-parameter Gaussian function is $f(x;\sigma) = \frac{1}{\sigma\sqrt{2\pi}}\exp(-\frac{x^2}{2\sigma^2})$. The derivation for sending a number of normally distributed 1 dimensional data sets was done in Davies [5]. The derivation gives the following expression: $\mathcal{L}_{guassian} = \tilde{F}(n_s,R,\Delta)+\sum_{i=1}^{n_g}(n_s-2)\log(\sigma_i)+\frac{n_s}{2}+\sum_{j=n_g+1}^{n_g+n_{min}}(n_s-2)\log(\sigma_{min})+\frac{n_s}{2}(\frac{\sigma_j}{\sigma_{min}})^2$, where σ_{min} is a cutoff constant, n_g is the number of directions where $\sigma > \sigma_{min}$ holds and n_{min} is the number of directions where $\sigma \leq \sigma_{min}$ holds. $\tilde{F}(n_s,R,\Delta)$ is a function that only depends on the number of shapes n_s, the range of the data R, and the resolution Δ. It is assumed constant for a given training set, i.e. it does not depend on decisions about outliers or correspondences.

$$\mathcal{L}_{guassian} = F(n_s,R,\Delta) + \sum_{i=1}^{n_g}(n_s-2)\log(\sigma_i) + \frac{n_s}{2} + \sum_{j=n_g+1}^{n_g+n_{min}}(n_s-2)\log(\sigma_{min}) + \frac{n_s}{2}(\frac{\sigma_j}{\sigma_{min}})^2, \quad (2)$$

The total description length of the interesting points in all images. The description length for a point $\hat{x}$ equally distributed over the image is

$$dl_{rect} = -log(\mathcal{P}(\hat{x})) = -2log(\frac{dx}{X}) \ .$$

Here X is the range, typically 100 pixels in our examples, and dx is the resolution, which has been set to 0.5 pixels. The factor 2 comes from that an image point is two-dimensional.

The outliers are assumed to be uniformly distributed over all the image, so with n_o number of outliers $dl_{outliers} = n_o dl_{rect}$. For each model point we need to know if the point is missing in an image or not. This means one bit for each n_p model points in all n_v images. Conversion to nats gives $dl_{index} = log(2)n_v n_p$, where n_p is the number of landmarks in the model and n_v is the number of images. For each image the transformation g of the model has to be encoded. The transformations are assumed equally distributed within the size of the image. For translations this gives the following expression $dl_{trans} = (n_v-1)n_{dof}dl_{rect}$, where n_v is the number of images and n_{dof} is the degrees of freedom in the transformation group, e.g. with 10 images and using 2D translations, 18 translation parameters has to be encoded. The coordinates of the mean shape and the coordinates of the shape modes are also assumed equally distributed within the size of the image, thus the cost is

$$dl_{meanshape} = n_p dl_{rect}$$

$$dl_{shapemodes} = n_p n_m dl_{rect} \ ,$$

where n_m is the number of shape modes used by the model. The residuals and the λ-parameters are assumed Gaussian. The cost for these are

$$dl_\lambda + dl_{res} = \mathcal{L}_{guassian} \ .$$

So the full cost for sending the data is

$$DL_{tot} = dl_\lambda + dl_{res} + dl_{meanshape} + dl_{shapemodes} + dl_{trans} + dl_{index} + dl_{outliers}$$

Given a shape model that describes part of the data for a situation we can now calculate the description length for this data and model. For each suggested model one needs to calculate the description length of sending all the outliers and all the data modelled with that particular model. The number of shape modes can vary between zero to $n_s - 1$ and all these models must be evaluated. Note here that since the shape modes calculated when using missing data PCA depends on the number of modes used, the model needs to be calculated over and over as the number of shape modes increase. In the optimisation procedure the tested model with least description length is then compared to previous solutions.

5 Optimising DL

The whole optimisation process over all unknowns can be divided into two parts: (1.) Optimisation over the discrete ordering matrix O and (2.) optimisation over the remaining parameters $\tilde{\mathcal{M}} = (m, \{g_i\}, d, \{v_l\}, \{\lambda_{li}\})$.

Assume that a reordering O is given, then it is straightforward to reorder the inlier points into the data structure T as described above. Each ordering also determines the number of inliers n_{inlier} and the number of outliers $n_{outlier}$. The description length for the outliers is then independent of $\tilde{\mathcal{M}}$. Assume now also that a transformation group G and the number of shape variational modes d are given. The description length now depends more or less on the remaining residuals of the inliers. Minimising the description length is then a question of minimising

$$\min_{m,\{g_i\},\{v_l\},\{\lambda_{li}\}} \sum_{(i,j)\in I} \left| T_{i,j} - g_i \left(m_j + \sum_{l=1}^{d} v_{j,l}\lambda_{l,i} \right) \right| .$$

We solve this minimisation problem explicitly. To get an initial estimate we solve first for missing data Procrustes by

$$\min_{m,\{g_i\}} \sum_{(i,j)\in I} |T_{i,j} - g_i(m_j)|^2$$

and then use the residuals missing data residuals $r_{i,j} = g_i^{(-1)}(T_{i,j}) - m_j$ to obtain initial estimates on v and λ. These initial estimates are used as a starting point in a Gauss-Newton optimisation scheme to find the nearest local minima.

It is straightforward to search through the number of nodes d and the different transformation groups G and as described above, it is then possible to find optimal Procrustes, missing data PCA and model order that gives minimal description length, thereby determining the solution with the best description length for this particular choice of point reordering O. Thus, the minimal description length can be seen as a function of the ordering O.

Optimising description length with respect to O is a combinatorial optimisation problem. We suggest the following algorithm that (1) finds a reasonable initial guess by heuristics and (2) searches for a local minima in a combinatorial optimisation sense by adding/removing inliers and adding/removing model points.

We approach this optimisation by a local search methods with the following four types of perturbations: (i) Change of a point from outlier to inlier, (ii) Change of a point from inlier to outlier, (iii) Deletion of a model point, (iv) Addition of a model point.

The final algorithm for determining minimal description length is then

1. Make an initial guess on point ordering O based on heuristics or randomness.
2. Calculate optimal description length for that ordering.
3. See if any of the perturbations above lowers the description length.
4. – If it does, make those changes and continue with step 3.
   – If not, then we are at a local minima, stop.

5.1 Initial Guess

One way of picking a reasonable initial guess is the following. The initial guess is made by using the points in one image as the model. For each new image matching is done as follows. Given n points in the model and m points in the new image. Form an $n+1 \times m+1$ cost matrix C whose first $n \times m$ elements C_{ij} is the Euclidean distance between model point i and feature point j after the model points are translated or transformed according to the best fit in the previous frame. This step of the algorithms, thus assumes that there is a smooth motion of the feature point. The last row $C_{n+1,j}$ is set to a constant representing the cost of not associating a feature point j with a model point. Similarly the last column $C_{i,m+1}$ is set to a constant representing the cost of not associating a model point j to any of the feature points. The matching is then done by solving the transport problem with supply of $s = [1, \ldots, 1, m]$ and demand $t = [1, \ldots, 1, n]$. Here we used standard algorithms for solving the transport problem, cf. [14].

6 Experimental Validation

6.1 Feature Point Selection, Model Extraction and Shape Recognition

In this experiment we have taken a digital film recording of a persons face as it moves in the scene. A sequence of 944 frames was captured and a standard interest point detector [10], was run on all of the frames. In each frame a face detector was run and those interest points that were within the rectangular frame of a face detector [6], was kept.

The first 100 frames were used for model estimation. This gave roughly 880 feature points (between 5 and 13 points in each frame). Three such frames are shown in Figure 1 together with the extracted feature point shown as small rectangular points.

The initial guess ordering resulted in 584 of the 880 feature points being associated with any of the 9 model points. The description length for this ordering was 10 826 bits.

After local optimisation the description length lowered to 9 575 bits for a model with 12 model points. Here 740 of the 880 points were associated to a model point. In Figure 2 is shown three frames out of the 100 overlayed with feature points and best fit of the 12 model points obtained after minimising description length. Notice that certain points in Figure 1 are classified as outliers and are not shown in Figure 2.

Fig. 1. Three out of 100 frames used for testing. Detected feature points are shown as white squares.

Fig. 2. Three out of 100 frames used for testing. For measured points (in white) the fitted model points are shown in black. For missing data the fitted model points are shown in gray.

Recognition using the shape model

The model can also be used to find the object in a new image without any prior knowledge about its position. This is accomplished by the RANSAC algorithm [7], where the consensus was based on the description length of the matching. For simple transformation groups, such as translation, it is enough to randomly match one point in the model to a feature point in the image.

Current limitations. Although the theory presented in this paper is quite general in the sense that any transformation group G can be used, our current implementation handles only the cases of 'no transformation' and 'pure translation'. Another limitation is the way the combinatorial optimisation scheme is implemented. It happens that the algorithms gets stuck in local minima, so that some points that are inliers are associated to the wrong object point or are considered as outliers. More perturbation types could be allowed, for example moving an image point from one model point to another. Yet another limitation of the scheme is that it is relatively slow. Each evaluation of a selection of inliers and outliers involves several steps, including a singular value decomposition with missing data for the PCA.

7 Conclusions

In this paper we have studied the problem of automatic feature point correspondence determination and shape analysis with missing data and outliers using MDL.

The modelling problem is posed as a combined combinatorial/continuous optimisation problem. The continuous part involves missing data Procrustes and missing data PCA. The combinatorial part is solved by an initial guess based on heuristics followed by local search. Although not the main focus of this paper, missing data Procrustes and missing data principal component analysis are defined and algorithms for their determination are developed. The definitions are natural consequences of the modelling principles followed.

The result is an algorithm that given a number of unordered point sets determines (i) the number of model points, (ii) the mean shape and shape variational modes of the model, (iii) the outliers in the data sets, (iv) the transformations g, and (v) the inliers with correspondences. We envision this algorithm being used on a set of images after extraction of feature points. The fact that the model can be learnt automatically makes it possible to acquire models on the fly, without manual interactions. Such models can then be used for tracking, pose determination, recognition.

In this paper we have focused on the point positions. This makes the method relatively stable to lighting variations. However, it would be interesting to extend the ideas to that of feature points including local descriptors, that capture the local variations in intensity in patches around the feature points. This would probably make the system better at recognising and tracking under most circumstances.

Acknowledgements

This work has been supported by the Swedish Knowledge Foundation through the Industrial PhD programme in Medical Bioinformatics at Karolinska Institutet, Strategy and Development Office, the European Commission's Sixth Framework Programme under grant no. 011838 as part of the Integrated Project SMErobot, by the Swedish Foundation for Strategic Research (SSF) through the programme Vision in Cognitive Systems (VISCOS) and by the Swedish Road Administration (Vägverket) and by the Swedish Governmental Agency for Innovation Systems (Vinnova).

References

1. Barron, A., Rissanen, J., Yu, B.: The minimum description length principle in coding and modeling. IEEE trans. on information theory 44(6), 2743–2760 (1998)
2. Belongie, S., Malik, J., Puzicha, J.: Shape matching and object recognition using shape contexts. IEEE Trans. Pattern Analysis and Machine Intelligence 24(24), 509–522 (2002)
3. Bregler, C., Hertzmann, A., Biermann, H.: Recovering non-rigid 3d shape from image streams. Proceedings IEEE Conference on Computer Vision and Pattern Recognition. CVPR 2000 (Cat. No.PR00662) 2, 690–696 (2000)
4. Davies, R.H., Twining, C.J., Cootes, T.F., Waterton, J.C., Taylor, C.J.: A minimum description length approach to statistical shape modeling. IEEE Trans. medical imaging 21(5), 525–537 (2002)
5. Davies, R.H., Cootes, T.F., Taylor, C.J.: A minimum description length approach to statistical shape modeling. In: Information Processing in Medical Imaging (2001)
6. Eriksson, A.P.: K. Åström. Robustness and specificity in object detection. In: Proc. International Conference on Pattern Recognition, Cambridge, UK (2004)
7. Fischler, M.A., Bolles, R.C.: Random sample consensus: a paradigm for model fitting with applications to image analysis and automated cartography. Communications of the ACM 24(6), 381–395 (1981)
8. Förstner, W.: A feature based correspondence algorithm for image matching. In: ISP Comm. III, Rovaniemi 1986, International Archives of Photogrammetry, 26-3/3 (1986)
9. Gower, J.C.: Generalized procrustes analysis. Psychometrica 40, 33–50 (1975)
10. Harris, C., Stephens, M.: A combined corner and edge detector. In: Proc. of the 4th Alvey Vision Conference, pp. 147–151 (1988)
11. Jacobs, D.: Linear fitting with missing data: Applications to structure-from-motion and to characterizing intensity images. In: Proc. Conf. Computer Vision and Pattern Recognition, pp. 206–212 (1997)
12. Kanatani, K.: Geometric information criterion for model selection. Int. Journal of Computer Vision 26(3), 171–189 (1998)
13. Karlsson, J., Ericsson, A.: Aligning shapes by minimising the description length. In: Scandinavian Conf. on Image Analysis, Juuensu, Finland (2005)
14. Luenberger, D.G.: Linear and Nonlinear Programming. Addison-Wesley, London (1984)
15. Rissanen, J.: Modeling by shortest data description. Automatica 14, 465–471 (1978)
16. Rohr, K.: Recognizing corners by fitting parametric models. Int. Journal of Computer Vision 9(3), 213–230 (1992)
17. Shannon, C.E.: Communication in the presence of noise. Proc. IRE,vol. 37 (1949)
18. Shi, J., Tomasi, C.: Good features to track. In: Proc. Conf. Computer Vision and Pattern Recognition, CVPR'94 (1994)
19. Thodberg, H.H.: Minimum description length shape and appearance models. In: Taylor, C.J., Noble, J.A. (eds.) IPMI 2003. LNCS, vol. 2732, Springer, Heidelberg (2003)
20. Twining, C.J., Cootes, T.F., Marsland, S., Petrovic, V.S., Schestowitz, R.S., Taylor, C.J.: Information-theoretic unification of groupwise non-rigid registration and model building. Proceedings of Medical Image Understanding and Analysis 2, 226–230 (2006)

Variational Segmentation of Image Sequences Using Deformable Shape Priors

Ketut Fundana, Niels Chr. Overgaard, and Anders Heyden

Applied Mathematics Group
School of Technology and Society
Malmö University, Sweden
{ketut.fundana,nco,heyden}@ts.mah.se

Abstract. The segmentation of objects in image sequences is an important and difficult problem in computer vision with applications to e.g. video surveillance. In this paper we propose a new method for variational segmentation of image sequences containing nonrigid, moving objects. The method is based on the classical Chan-Vese model augmented with a novel frame-to-frame interaction term, which allow us to update the segmentation result from one image frame to the next using the previous segmentation result as a shape prior. The interaction term is constructed to be pose-invariant and to allow moderate deformations in shape. It is expected to handle the appearance of occlusions which otherwise can make segmentation fail. The performance of the model is illustrated with experiments on real image sequences.

Keyword: Variational formulation, segmentation, tracking, region-based, level sets, interaction terms, deformable shape priors.

1 Introduction

In this paper we address the problem of segmentation in image sequences using region-based active contours and level set methods. Segmentation is an important and difficult process in computer vision, with the purpose of dividing a given image into one or several meaningful regions or objects. This process is more difficult when the objects to be segmented are moving and nonrigid. The shape of nonrigid, moving objects may vary a lot along image sequences due to, for instance, deformations or occlusions, which puts additional constraints on the segmentation process.

There have been a number of methods proposed and applied to this problem. Active contours are powerful methods for image segmentation; either boundary-based such as geodesic active contours [1], or region-based such as Chan-Vese models [2], which are formulated as variational problems. Those variational formulations perform quite well and have often been applied based on level sets. Active contour based segmentation methods often fail due to noise, clutter and occlusion. In order to make the segmentation process robust against these effects, shape priors have been proposed to be incorporated into the segmentation

B.K. Ersbøll and K.S. Pedersen (Eds.): SCIA 2007, LNCS 4522, pp. 31–40, 2007.

process. In recent years, many researchers have successfully introduced shape priors into segmentation methods such as in [3,4,5,6,7,8,9].

We are interested in segmenting nonrigid moving objects in image sequences. When the objects are nonrigid, an appropriate segmentation method that can deal with shape deformations should be used. The application of active contour methods for segmentation in image sequences gives promising results as in [10,11,12]. These methods use variants of the classical Chan-Vese model as the basis for segmentation. In [10], for instance, it is proposed to simply use the result from one image as an initializer in the segmentation of the next.

The main purpose of this paper is to propose and analyze a novel variational segmentation method for image sequences, that can both deal with shape deformations and at the same time is robust to noise, clutter and occlusions. The proposed method is based on minimizing an energy functional containing the standard Chan-Vese functional as one part and a term that penalizes the deviation from the previous shape as a second part. The second part of the functional is based on a transformed distance map to the previous contour, where different transformation groups, such as Euclidean, similarity or affine, can be used depending on the particular application.

This paper is organized as follows: in Sect. 2 we discuss region-based segmentation, the level set method, and gradient descent procedures. In Sect. 3 we describe the segmentation model proposed. Experimental results of the model are presented in Sect. 4. We end the paper with some conclusions and future work plans.

2 Theoretical Background

2.1 Region-Based Segmentation

We begin with a brief review of the classical Chan-Vese segmentation model [2]. In this model a gray scale image is considered to be a real valued function $I : D \to \mathbf{R}$ defined on the *image domain* $D \subset \mathbf{R}^2$, usually a rectangle. A point $\mathbf{x} \in D$ is often referred to as a pixel, and the function value $I = I(\mathbf{x})$ as the *pixel value*, or the *gray scale value*. The Chan-Vese model is an active contour model. The idea is to find a contour Γ, by which we mean a finite union of disjoint, simple, closed curves in D, such that the image I is optimally approximated by a single gray scale value μ_{int} on $\text{int}(\Gamma)$, the *inside* of Γ, and by another gray scale value μ_{ext} on $\text{ext}(\Gamma)$, the *outside* of Γ. The optimal contour Γ^* and the corresponding pair of optimal gray scale values $\boldsymbol{\mu}^* = (\mu^*_{\text{int}}, \mu^*_{\text{ext}})$ are defined as the solution of the variational problem,

$$E_{CV}(\boldsymbol{\mu}^*, \Gamma^*) = \min_{\boldsymbol{\mu}, \Gamma} E_{CV}(\boldsymbol{\mu}, \Gamma), \tag{1}$$

where E_{CV} is the well-known Chan-Vese functional,

$$E_{CV}(\boldsymbol{\mu}, \Gamma) = \alpha \int_\Gamma d\sigma + \beta \Big\{ \frac{1}{2} \int_{\text{int}(\Gamma)} (I(\mathbf{x}) - \mu_{\text{int}})^2 \, d\mathbf{x} + \frac{1}{2} \int_{\text{ext}(\Gamma)} (I(\mathbf{x}) - \mu_{\text{ext}})^2 \, d\mathbf{x} \Big\}. \tag{2}$$

Here $d\sigma$ denotes the arc length element, and $\alpha, \beta > 0$ are weight parameters. The first term in E_{CV} is the total length of the contour: It serves to regularize the optimal contour. The second term is the fidelity term, which penalizes deviations of the piecewise constant image model from the actual image I.

For any fixed contour Γ, not necessarily the optimal one, it turns out that the best choice of the gray scale values $\boldsymbol{\mu} = (\mu_{\text{int}}, \mu_{\text{ext}})$ corresponds to the mean value of the pixel values inside and the outside Γ, respectively:

$$\mu_{\text{int}} = \mu_{\text{int}}(\Gamma) = \frac{1}{|\,\text{int}(\Gamma)|} \int_{\text{int}(\Gamma)} I(\mathbf{x})\, d\mathbf{x}, \tag{3}$$

$$\mu_{\text{ext}} = \mu_{\text{ext}}(\Gamma) = \frac{1}{|\,\text{ext}(\Gamma)|} \int_{\text{ext}(\Gamma)} I(\mathbf{x})\, d\mathbf{x}. \tag{4}$$

Here the symbol $|A|$ denotes the area of the subset $A \subset \mathbf{R}^2$. Now, if we introduce the so-called "reduced" Chan-Vese functional

$$E_{CV}^R(\Gamma) := E_{CV}(\boldsymbol{\mu}(\Gamma), \Gamma), \tag{5}$$

then the optimal contour Γ^* can be found by solving the simpler minimization problem

$$E_{CV}^R(\Gamma^*) = \min_{\Gamma} E_{CV}^R(\Gamma). \tag{6}$$

Once Γ^* is found we have $\boldsymbol{\mu}^* = \boldsymbol{\mu}(\Gamma^*)$, of course. The minimization problem in (6) is solved using a gradient descent procedure, which will be recalled in the next section, after the material on the level set representation and the kinematics of moving surfaces have been presented.

2.2 The Level Set Method and Gradient Descent Evolutions

A simple closed curve Γ can be represented as the zero level set of a function $\phi : \mathbf{R}^2 \to \mathbf{R}$ as

$$\Gamma = \{\mathbf{x} \in \mathbf{R}^2 \; ; \; \phi(\mathbf{x}) = 0\} \; . \tag{7}$$

The sets $\text{int}(\Gamma) = \{\mathbf{x} \; ; \; \phi(\mathbf{x}) < 0\}$ and $\text{ext}(\Gamma) = \{\mathbf{x} \; ; \; \phi(\mathbf{x}) \geq 0\}$ are then the inside and the outside of Γ, respectively. Geometric quantities such as the outward unit normal $\mathbf{n}$ and the curvature κ can be expressed in terms of ϕ as

$$\mathbf{n} = \frac{\nabla\phi}{|\nabla\phi|} \quad \text{and} \quad \kappa = \nabla \cdot \frac{\nabla\phi}{|\nabla\phi|} \; . \tag{8}$$

The function ϕ is usually called the *level set function* for Γ, cf. e.g. [13].

A curve evolution, that is, a time dependent curve $t \mapsto \Gamma(t)$ can be represented by a time dependent level set function $\phi : \mathbf{R}^2 \times \mathbf{R} \to \mathbf{R}$ as $\Gamma(t) = \{\mathbf{x} \in \mathbf{R}^2 \; ; \; \phi(\mathbf{x}, t) = 0\}$. Let us consider the kinematics of curve evolutions. It does not make sense to "track" points as there is no way of knowing the tangential motion of points on $\Gamma(t)$. The important notion is that of *normal velocity*. The normal velocity of a curve evolution $t \mapsto \Gamma(t)$ is the scalar function defined by

$$v(\Gamma)(\mathbf{x}) = \frac{d}{dt}\Gamma(t)(\mathbf{x}) := -\frac{\partial\phi(\mathbf{x}, t)/\partial t}{|\nabla\phi(\mathbf{x}, t)|} \qquad (\mathbf{x} \in \Gamma(t)) \; . \tag{9}$$

The normal velocity is independent of the curve representation, in particular of the choice of level set function ϕ for Γ, and is therefore a geometric property of the evolution, cf. [14]. The set of possible normal velocities $v = v(\Gamma)$ of moving contours $t \mapsto \Gamma(t)$ passing through the contour Γ at time $t = 0$ is an infinite dimensional vector space. This vector space can be endowed with a natural scalar product and a corresponding norm, cf. [14],

$$\langle v, w \rangle_\Gamma = \int_\Gamma v(\mathbf{x}) w(\mathbf{x})\, d\sigma \quad \text{and} \quad \|v\|_\Gamma^2 = \langle v, v \rangle_\Gamma \ , \tag{10}$$

where v, w are normal velocities and $d\sigma$ is the arc length element. In the following we therefore denote the vector space of normal velocities at Γ by $L^2(\Gamma)$.

The scalar product (10) is important in the construction of gradient descent flows for energy functionals $E(\Gamma)$ defined on curves. Suppose $v \in L^2(\Gamma)$ is a fixed normal velocity, and let $t \mapsto \Gamma(t)$ be any moving contour which satisfies $\Gamma(0) = \Gamma$, and $(d/dt)\Gamma(0) = v$. Then the Gâteaux variation $dE(\Gamma)v$ of the functional $E = E(\Gamma)$ at the contour Γ is defined as the derivative,

$$dE(\Gamma)v := \frac{d}{dt} E(\Gamma(t))\Big|_{t=0} . \tag{11}$$

Suppose there exists a function $\nabla E(\Gamma) \in L^2(\Gamma)$ such that E's Gâteaux variation $dE(\Gamma)v$ at Γ can be expressed in terms of the scalar product (10) in the following manner,

$$dE(\Gamma)v = \langle \nabla E(\Gamma), v \rangle_\Gamma \quad \text{for all } v \in L^2(\Gamma) \ . \tag{12}$$

Then the vector $\nabla E(\Gamma)$ it is called the L^2-*gradient* of E at Γ. It is unique if it exists. The gradient descent flow for the problem of minimizing $E(\Gamma)$ can now be defined as the initial value problem:

$$\frac{d}{dt}\Gamma(t) = -\nabla E(\Gamma(t)), \qquad \Gamma(0) = \Gamma_0, \tag{13}$$

where Γ_0 is an initial contour specified by the user.

As an example, relevant for the application in this paper, notice that the L^2-gradient of the reduced functional E_{CV}^R defined in (5) is given by:

$$\nabla E_{CV}^R(\Gamma; \mathbf{x}) = \alpha\kappa + \beta\big[\frac{1}{2}(I(\mathbf{x}) - \mu_{\text{int}}(\Gamma))^2 - \frac{1}{2}(I(\mathbf{x}) - \mu_{\text{ext}}(\Gamma))^2\big], \quad (\mathbf{x} \in \Gamma), \tag{14}$$

where $\kappa = \kappa(\mathbf{x})$ is the curvature at $\mathbf{x} \in \Gamma$. If we combine the definition of gradient descent evolutions in (13) with the formula (9) for the normal velocity, then we get the gradient descent procedure in the level set framework:

$$\frac{\partial}{\partial t}\phi(\mathbf{x}, t) = \Big(\alpha\kappa + \beta\big[\frac{1}{2}(I(\mathbf{x}) - \mu_{\text{int}}(\Gamma))^2 - \frac{1}{2}(I(\mathbf{x}) - \mu_{\text{ext}}(\Gamma))^2\big]\Big)|\nabla\phi(\mathbf{x}, t)|, \tag{15}$$

with $\phi(\mathbf{x}, 0) = \phi_0(\mathbf{x})$, where ϕ_0 is the level set function for the initial contour Γ_0. It is understood that the gray scale values $\mu_{\text{int}}(\Gamma)$ and $\mu_{\text{ext}}(\Gamma)$ are given by (3) and (4), respectively.

3 Segmentation of Image Sequences

3.1 A Variational Updating-Model

In this section we are going to present the basic principles behind our variational model for updating segmentation results from one frame to the next in an image sequence.

Let $I_j : D \to \mathbf{R}$, $j = 1, \ldots, N$, be a succession of frames from a given image sequence. Also, for some integer k, $1 \le k \le N$, suppose that all the frames $I_1, I_2, \ldots, I_{k-1}$ have already been segmented, such that the corresponding contours $\Gamma_1, \Gamma_2, \ldots, \Gamma_{k-1}$ are available. In order to take advantage of the prior knowledge obtained from earlier frames in the segmentation of I_k, we propose the following method: If $k = 1$, i.e. if no previous frames have actually been segmented, then we just use the classical Chan-Vese model, as presented in Sect. 2. If $k > 1$, then the segmentation of I_k is given by the contour Γ_k which minimizes an *augmented* Chan-Vese functional of the form,

$$E_{CV}^{A}(\Gamma_{k-1}, \Gamma) := E_{CV}^{R}(\Gamma) + \gamma E_I(\Gamma_{k-1}, \Gamma), \tag{16}$$

where E_{CV}^{R} is the reduced Chan-Vese functional defined in (5), $E_I = E_I(\Gamma_{k-1}, \Gamma)$ is an *interaction term*, which penalizes deviations of the current active contour Γ from the previous one, Γ_{k-1}, and $\gamma > 0$ is a coupling constant which determines the strength of the interaction. The precise definition of E_I is described in the next section.

3.2 The Interaction Term

The interaction $E_I(\Gamma_0, \Gamma)$ between a fixed contour Γ_0 and an active contour Γ, used in (16), may be chosen in several different ways. Two common choices are the so-called pseudo-distances, cf. [6], and the area of the symmetric difference of the sets $\mathrm{int}(\Gamma)$ and $\mathrm{int}(\Gamma_0)$, cf. [3]. We have found that none of the mentioned contour interactions satisfy our needs, and we have therefore chosen to introduce a completely new pose-invariant interaction term.

To describe this interaction term, let $\phi_0 : D \to \mathbf{R}$ denote the *signed distance function* associated with the contour Γ_0, that is, the function:

$$\phi_0(\mathbf{x}) = \begin{cases} \mathrm{dist}(\mathbf{x}, \Gamma_0) & \text{for } \mathbf{x} \in \mathrm{ext}(\Gamma_0), \\ -\mathrm{dist}(\mathbf{x}, \Gamma_0) & \text{for } \mathbf{x} \in \mathrm{int}(\Gamma_0). \end{cases} \tag{17}$$

Then the interaction $E_I = E_I(\Gamma_0; \Gamma)$ is defined by the formula,

$$E_I(\Gamma_0, \Gamma) = \min_T \int_{\mathrm{int}(\Gamma)} \phi_0(T^{-1}\mathbf{x})\, d\mathbf{x}, \tag{18}$$

where the minimum is taken over the *group of Euclidean transformations* $T : \mathbf{R}^2 \to \mathbf{R}^2$ which preserves the orientation of the plane, that is, transformations T which are compositions of translations and rotations (but not reflections).

Minimizing over groups of transformations is a standard devise to obtain pose-invariant interactions, see [3] and [6].

For any given contour Γ, let $T = T(\Gamma)$ denote the transformation which minimizes the expression on the right hand side of (18). Since this is an optimization problem $T(\Gamma)$ can be found using gradient descent. For simplicity of presentation, suppose we only consider the group of translations $T_{\mathbf{a}} : \mathbf{x} \mapsto \mathbf{x}+\mathbf{a}$, $\mathbf{a} \in \mathbf{R}^2$, and want to determine the optimal translation vector $\mathbf{a} = \mathbf{a}(\Gamma)$. Then we have to solve the optimization problem

$$\min_{\mathbf{a}} \int_{\mathrm{int}(\Gamma)} \phi_0(\mathbf{x} - \mathbf{a})\, d\mathbf{x} \ .$$

The optimal translation $\mathbf{a}(\Gamma)$ can then be obtained as the limit, as time t tends to infinity, of the solution to initial value problem

$$\dot{\mathbf{a}}(t) = \int_{\mathrm{int}(\Gamma)} \nabla\phi_0(\mathbf{x} - \mathbf{a}(t))\, d\mathbf{x} \ , \qquad \mathbf{a}(0) = 0 \ . \tag{19}$$

Similar gradient descent schemes can be devised for rotations and scalings (in the case of similarity transforms), cf. [3], but will not be written out explicitly here.

3.3 The Gradient Descent Equations

The augmented Chan-Vese functional (16) is minimized using standard gradient descent as described in Sect. 2. That is, we solve the initial value problem

$$\frac{d}{dt}\Gamma(t) = -\nabla E_{CV}^{A}(\Gamma_{k-1}, \Gamma(t)) := -\nabla E_{CV}^{R}(\Gamma_{k-1}, \Gamma(t)) - \gamma\nabla E_I(\Gamma_{k-1}; \Gamma(t)), \tag{20}$$

with the initial contour $\Gamma(0) = \Gamma_{k-1}$, and pass to the limit $t \to \infty$. Here ∇E_{CV}^{R} is the L^2-gradient of the reduced Chan-Vese functional, see Eq. (14), and ∇E_I is the L^2-gradient of the interaction E_I, which is given by the formula,

$$\nabla E_I(\Gamma_{k-1}, \Gamma; \mathbf{x}) = \phi_{k-1}(T(\Gamma)\mathbf{x}), \qquad (\text{for } \mathbf{x} \in \Gamma), \tag{21}$$

as is easily verified. Here ϕ_{k-1} is the signed distance function for Γ_{k-1}.

4 Numerical Implementation and Experiments

In this section we present the results obtained from experiments using three different image sequences. In the first image of the sequence we use the Chan-Vese model to segment a selected object with approximately uniform intensity. Then the proposed method is applied to segment the image sequences sequentially frame-by-frame, where the segmentation in one frame is used as the initial contour in the next one. The minimization of the functional, giving the optimal contour, is obtained from the gradient descent procedure (20) which has been implemented in the level set framework outlined in Sect. 2. See also [13].

As illustrated in Fig. 1, the original Chan-Vese model is capable of segmenting a selected object in an image sequence without any problems. Further such results can be found in [10].

Fig. 1. Segmentation of a person in human walking sequence using the classical Chan-Vese model

Another experiment is given in Fig. 3, where a walking person is being segmented. Here the proposed method prevents the segmentation of the wrong objects, as is clearly shown.

However, as pointed out in the above reference, the classical Chan-Vese method will have problems segmenting an object if occlusions appear in the image which cover the whole or parts of the selected object. In Fig 2, we show the segmentation results for a car (the white van) in a traffic sequence, where occlusions occur. The classical Chan-Vese method fails to segment the selected object when it reaches the occlusion (first column). Using the proposed method, including the frame-to-frame interaction term, we obtain much better results (second column).

In both experiments the coupling constant γ is varied to see the influence of the interaction term on the segmentation results. The contour is only slightly affected by the prior if γ is small. On the other hand, if γ is too large, the contour will be close to a similarity transformed version of the prior.

5 Conclusions and Future Works

We have presented a new method for variational segmentation of image sequences containing nonrigid, moving objects. The proposed method is formulated as variational problem, with one part of the functional corresponding to the Chan-Vese model and another part corresponding to the pose-invariant interaction with a shape prior based on the previous contour. The optimal transformation as well as the shape deformation are determined by minimization of an energy functional using a gradient descent scheme. Preliminary results are shown on several real image sequences and its performance looks promising.

Fig. 2. Segmentation of a car which passes occlusions in the traffic sequence. Left Column: without interaction term, and Right Column: $\gamma = 80$.

Fig. 3. Segmentation of a person covered by an occlusion in the human walking sequence. Left Column: without interaction term, Middle Column: $\gamma = 20$, and Right Column: $\gamma = 70$.

Acknowledgements

This research is funded by EU Marie Curie RTN FP6 project VISIONTRAIN (MRTN-CT-2004-005439). The human walking sequences were downloaded from EU funded CAVIAR project (IST 2001 37540) website and the traffic sequence from KOGS/IAKS Universität Karlsruhe.

References

1. Caselles, V., Kimmel, R., Sapiro, G.: Geodesic active contours. International Journal of Computer Vision 22(1), 61–79 (1997)
2. Chan, T., Vese, L.: Active contour without edges. IEEE Transactions on Image Processing 10(2), 266–277 (2001)
3. Chan, T., Zhu, W.: Level set based prior segmentation. Technical Report UCLA CAM 03-66, Department of Mathematics, UCLA (2003)
4. Cremers, D.: Statistical Shape Knowledge in Variational Image Segmentation. Phd thesis, Department of Mathematics and Computer Science, University of Mannheim (July 2002)
5. Cremers, D., Sochen, N., Schnörr, C.: Towards recognition-based variational segmentation using shape priors and dynamic labeling. In: Griffin, L.D, Lillholm, M. (eds.) Scale Space Methods in Computer Vision. LNCS, vol. 2695, pp. 388–400. Springer, Heidelberg (2003)
6. Cremers, D., Soatto, S.: A pseudo-distance for shape priors in level set segmentation. In: Faugeras, O., Paragios, N., (eds.): 2nd IEEE Workshop on Variational, Geometric and Level Set Methods in Computer Vision (2003)
7. Cremers, D., Funka-Lea, G.: Dynamical statistical shape priors for level set based sequence segmentation. In: Paragios, N., Faugeras, O., Chan, T., Schnörr, C. (eds.) VLSM 2005. LNCS, vol. 3752, pp. 210–221. Springer, Heidelberg (2005)
8. Rousson, M., Paragios, N.: Shape priors for level set representations. In: Heyden, A., Sparr, G., Nielsen, M., Johansen, P. (eds.) ECCV 2002. LNCS, vol. 2351, pp. 78–92. Springer, Heidelberg (2002)
9. Leventon, M., Grimson, W., Faugeras, O.: Statistical shape influence in geodesic active contours. In: CVPR (2000)
10. Moelich, M., Chan, T.: Tracking objects with the chan-vese algorithm. Technical Report UCLA CAM 03-14, Department of Mathematics, UCLA (March 2003)
11. Paragios, N., Deriche, R.: Geodesic active contours and level set methods for the detection and tracking of moving objects. IEEE Trans. PAMI 22(3), 266–280 (2000)
12. Paragios, N., Deriche, R.: Geodesic active regions and level set methods for motion estimation and tracking. Computer Vision and Image Understanding 97, 259–282 (2005)
13. Osher, S., Fedkiw, R.: Level Set Methods and Dynamic Implicit Surfaces. Springer, Heidelberg (2003)
14. Solem, J.E., Overgaard, N.C.: A geometric formulation of gradient descent for variational problems with moving surfaces. In: Kimmel, R., Sochen, N.A., Weickert, J. (eds.) Scale-Space 2005. LNCS, vol. 3459, pp. 419–430. Springer, Heidelberg (2005)

Real-Time Face Detection Using Illumination Invariant Features

Klaus Kollreider, Hartwig Fronthaler, and Josef Bigun

Halmstad University, Box 823, SE-30118, Halmstad, Sweden
{klaus.kollreider,hartwig.fronthaler,josef.bigun}@ide.hh.se
http://www.hh.se

Abstract. A robust object/face detection technique processing every frame in real-time (video-rate) is presented. A methodological novelty are the suggested quantized angle features ("quangles"), being designed for illumination invariance without the need for pre-processing, e.g. histogram equalization. This is achieved by using both the gradient direction and the double angle direction (the structure tensor angle), and by ignoring the magnitude of the gradient. Boosting techniques are applied in a quantized feature space. Separable filtering and the use of lookup tables favor the detection speed. Furthermore, the gradient may then be reused for other tasks as well. A side effect is that the training of effective cascaded classifiers is feasible in very short time, less than 1 hour for data sets of order 10^4. We present favorable results on face detection, for several public databases (e.g. 93% Detection Rate at 1×10^{-6} False Positive Rate on the CMU-MIT frontal face test set).

Keywords: Object detection, Face Detection, Biometrics, Direction Field, Orientation Tensor, Quantized Angles, Quangles, AdaBoost.

1 Introduction

When attempting to *detect* faces (or *locate* a single face) in a visual representation, image-based and landmark-based methods may be primarily distinguished between [1,2]. This paper focuses on the detection of frontal faces in 2D images and is assigned to the former category. Features here represent measurements made by means of some basis functions in a multidimensional space which should be contrasted to the term "facial features" sometimes used in the published studies to name subparts of a face, e.g. the eyes, mouth, etc., which we refer to as "landmarks". Challenges in face detection are generally comprised of varying illumination, expression changes, (partial) occlusion, pose extremities [1] and requirements on real-time computations.

The main characteristics of still image-based methods is that they process faces in a holistic manner. Faces are learned by training on roughly aligned portraits as well as non-face-like images, and no parts of the face are intentionally favored to be used for face detection. The specific statistical pattern recognition method employed characterizes published studies. A popular approach uses

B.K. Ersbøll and K.S. Pedersen (Eds.): SCIA 2007, LNCS 4522, pp. 41–50, 2007.

the so-called Eigenfaces [3], or the PCA (Principal Component Analysis) coordinates, to quantify the "faceness" of an image (region). More recent face detection systems employed neural networks [4,5], or Support Vector Machines (SVM) as [6] in [7] to classify image regions as face or non-face. Also, a naive Bayes scheme was implemented in [8], and recently in [9], whereas an AdaBoost procedure [10] was concurrently adapted in [11,12]. In principle, these techniques are not specific to detect faces in an image, but can be trained in an analogous manner to detect other objects, e.g. cars. The AdaBoost based face detection in [11,12] has been suggested as being real-time, and has been followed up with other studies extending it to multi-poses, and reducing classifier complexity, e.g. [13,14]. However, the employed features play a decisive role besides the used classifiers. Of all the published methods, only a few use the gray values directly for classification, but rather features. However, almost all approaches use a preprocessing of the gray values (e.g. histogram equalization or normalization) to minimize the effect of adverse light conditions, at the expense of computational processing. The methods suggested by [11,13,14] use Haar-like rectangle features, translating into a high detection speed whereas [12,9] employed edge features with arguably lower execution speed. The recent method of [15] proposed binary coded local histograms (LBF) as features. A novelty in this study is the use of gradient angles only driven by the observation, that the gradient angle as opposed to the magnitude is, simply put, naturally robust to illumination changes. Gray value preprocessing becomes redundant, and we extend the illumination resilience by two contributions: First, the use of hierarchical and adaptive quantization levels improves the detection performance. Second, we do not only exploit the gradient angle, but also the structure tensor direction [16], encoding local orientation. Because we use quantized angle features, we term the latter "quangles" for expediency. Furthermore, these quangles are boosted in layers of a decision cascade as in [11], enabling also small classifiers. We achieve scale invariance through signal theoretically correct downsampling in a pyramidal scheme. The usefulness of our technique is shown in the context of face detection. A methodological advantage of the suggested scheme is the readily availability of some filtered signals for differential algorithms, for example, optical flow estimation, exploited for immediate person "liveness" [17] assessment. In comparison, the rectangle features suggested in [11], despite their value in pure object detection in still images, have limited reusability when it comes to other tasks. For a survey of landmark-based methods, which focus on a few salient parts, landmarks, e.g. the single eyes, mouth, nose of the face, we refer to [18,1]. We present experimental results on several public databases, namely the MIT-CMU [5] and the YALE [19] face test sets.

2 Object Detection

2.1 The Quantized Angle Features (Quangles)

In this section we present the features for object detection, which we call "quangles", representing quantized angle features. The gradient of an image is given

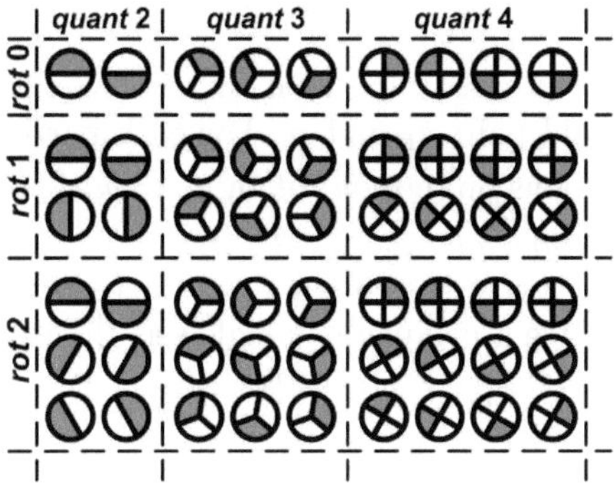

Fig. 1. Example of a set of quangle masks (angle displayed in polar form). The gray shaded areas correspond to the partitions yielding value 1 in equation (2).

in equation 1,

$$\nabla f = \begin{pmatrix} f_x \\ f_y \end{pmatrix} \tag{1}$$

where f_x and f_y denote the derivatives in x and y direction respectively. Furthermore, $|\nabla f|$ indicates the magnitude of the gradient and $\angle \nabla f$ refers to its angle. For the sake of object detection, we disregard the magnitude or intensity since it is highly affected by undesired external influences like illumination variations. The key instrument of our quangle features are the *quangle masks*, which are denoted as follows:

$$Q(\tau_1, \tau_2, \phi) = \begin{cases} 1, & \text{if } \tau_1 < \phi < \tau_2, \\ 0, & \text{otherwise} \end{cases} \tag{2}$$

The thresholds τ_1 and τ_2 constitute the boundaries of a partition in $[0, 2\pi]$. The quangle mask yields 1 if an angle ϕ, is located within such a partition and 0 otherwise. In order to produce a set of quangle masks, we divide the full angle range $[0, 2\pi]$ into an increasing number of quantizations (partitions), which are additionally rotated. An example is depicted in figure 1. A set of quangle masks $\{Q\}_{maxQuant,numRot}$ is fully determined by the maximum number of quantizations *maxQuant* and rotations *numRot*. The parameter *maxQuant* has to be interpreted cumulatively, meaning that all quangle masks with less quantization steps are included in the set as well. The second parameter, *numRot*, indicates the number of rotations included in addition to each basic quangle mask. For example the final row in figure 1 corresponds to $\{Q\}_{4,2}$, which consists of 27 different quangle masks. In order to create such a set of quangle masks the thresholds τ_1 and τ_2 of each partition need to be determined. This can be done in a three step procedure:

1. First we define a sequence of threshold pairs α_1 and α_2 delimiting the desired number of partitions *nQuant* in the interval $[0, 2\pi]$, disregarding the

rotational component, by $\alpha_1 = \frac{2\pi}{nQuant} \cdot quant$ and $\alpha_2 = \frac{2\pi}{nQuant} \cdot (quant+1)$, where $quant \in \{0, ..., nQuant-1\}$.

2. In the second step we create the final threshold sequence for $nQuant$ containing the threshold pairs τ_1 and τ_2. For each partition $quant$ we include the number of rotations up to $numRot$, by $\tau_k = mod\left(\alpha_k - \frac{\pi}{nQuant} \cdot rot, 2\pi\right)$, where $rot \in \{0, ..., numRot\}$ and $k \in \{1, 2\}$.
3. Performing the first two steps corresponds to creating a single cell in figure 1. In order to produce a complete quangle set, the two steps above need to be repeated for $nQuant = \{2, ..., maxQuant\}$.

To detect objects in a single scale we use a sliding window approach, where an image is scanned by a so-called search or detection window. In order to look for candidates, the quangle masks need to be assigned to positions (i, j) within the detection window x. This defines at the same time our *quangle features.* We furthermore distinguish between two different types: Equation (3a) describes a quangle feature using the original gradient angle, whereas in equation (3b) double angle representation is employed.

$$q_1(x, i, j, \tau_1, \tau_2) = Q(\tau_1, \tau_2, \angle\nabla x(i, j)) \tag{3a}$$

$$q_2(x, i, j, \tau_1, \tau_2) = Q(\tau_1, \tau_2, mod(2 \cdot \angle\nabla x(i, j), 2\pi)) \tag{3b}$$

Both quangle feature types in the equations above take the detection window x, the position (i, j) within x and a particular quangle mask out of $\{Q\}$. Using both, q_1 and q_2, the number of possible features is determined by the size of the detection window and the number of employed quangle masks. We include both, single and double angle representation in our set of quangle features since they are meaningful at different sites within the search window. The original gradient is more informative within the object, e.g. between landmarks of a face, because it distinguishes between dark-light and light-dark transitions. The double angle representation maps ϕ to 2ϕ, and has been shown to represent the structure tensor eigenvector directions [16]. Thereby ∇f and $-\nabla f$ are equivalent and represent orientations of linear structures, e.g. lines. The double angle representation is more resistant to illumination changes, especially helpful at object boundaries (background changes). Accordingly, both single angle and double angle features are complementary and meaningful features to represent objects.

2.2 Classifier Building

A good classification (yielding low error rate) cannot be obtained with a single quangle feature, but obviously, it is neither meaningful nor practical to evaluate all of them within the detection window. In order to find the most suitable quangles we employ AdaBoost [10]. In the process, a number of good features (termed weak classifiers) are combined, yielding a so-called strong classifier. Following the discrete AdaBoost algorithm, we select the weak classifier

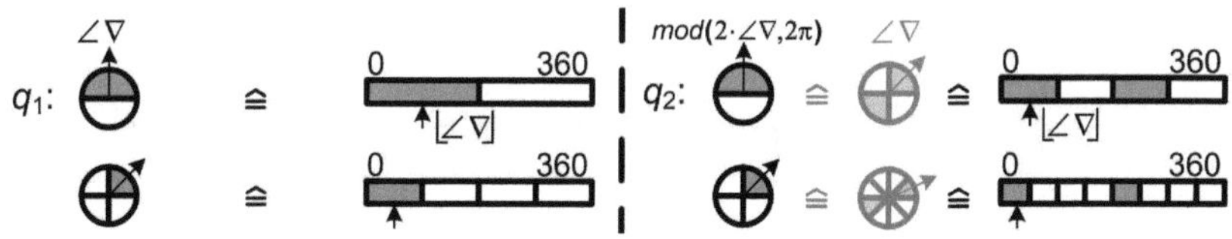

Fig. 2. Example lookup tables (stripes) representing quangle masks in case of single angle (left hand side) and double angle (right hand side) representation

$h_t(x) = q_{k_t}(x, i_t, j_t, \tau_{1_t}, \tau_{2_t})$ which minimizes the error $e_t = min_{i,j,\tau_1,\tau_2,k} \sum_l w_l \cdot |q_k(x, i, j, \tau_1, \tau_2) - y_l|$ over the training set (indexed by l) in round t of the feature selection (y denotes the true class "1" or "0" and w weights the training examples). Eventually, we obtain a strong classifier which is composed of T weak classifiers, and each of the latter has a say in the final decision depending on the individual error $\alpha_t = \log \frac{1-e_t}{e_t}$. While generally improving the detection/false positive rates, adding more and more weak classifiers unfortunately directly affects the classification time. An alternative to a single strong classifier is the so-called cascaded classifier scheme, a series of less complex strong classifiers, which is computationally efficient [11]. A single negative decision at any level of such a cascade leads to an immediate disregard of the concerned candidate image region. When training a strong classifier and adding it to the cascade, we apply a bootstrapping strategy. Previously rejected negative class examples are replaced by new ones, which the current cascade would (wrongly) classify as positive examples.

Another important factor for the training of such a cascade is time. Viola&Jones, for example, reported that the training time of their final classifier was in the order of weeks on a single machine. This was due to the large amount of rectangle features, necessary there, in combination with finding an eligible threshold for each of them. Employing our features, the training of a comparable cascade takes about an hour on an ordinary desktop computer, because less features suffice (quangles build upon derivative features) and the expensive calibration is skipped.

2.3 Implementation

Cascaded classifiers favor processing time in that only a few strong classifications accrue per image site. Furthermore, we can reduce the number of operations needed to calculate and classify a single feature. In this study, we employ so-called lookup tables to speed up this process. Recalling the quangle features of type q_1 and particularly q_2 in equations (3a) and (3b), lookup tables provide an effective solution for both of them. Figure 2 depicts two exemplary lookup tables for both q_1 and q_2. Each quangle mask is represented by a binary lookup table, generated off-line. Gray shaded areas correspond to 1 and are defined by the respective quangle mask. The original gradient angle is used as table index, therefore we floor it to integer values in $[0, 360[$. However, the quangle features of

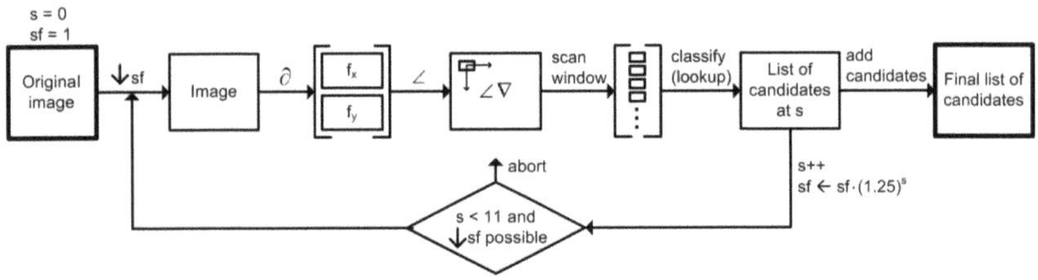

Fig. 3. The proposed object detection process

type q_2, need some further attention. As visualized, $mod\,(2 \cdot \angle\nabla, 2\pi)$ corresponds to $\angle\nabla$, by means of "helper quangles" (displayed alleviated), only existing in the form of lookup tables. To construct one such, we half the thresholds of the original quangle mask. The resulting partition together with a 180° shifted version define the lookup table for q_2. As a consequence, 1 array access is needed per weak classification for any quangle.

The whole detection process is illustrated in figure 3. The image to be analyzed serves as a starting point at scale $s = 0$ and scale factor $sf = 1$. We approximate the gradient (see equation (1)) of the whole image using separable Gaussians and their derivatives and extract the angle information. After this, we scan the image with the detection window to be classified using a trained cascade and the lookup tables introduced above. Having the candidates of the first scale, we successively reduce the image size by a factor of 1.25 and start over with the gradient calculation and window scanning, repeating like this for 10 times. The candidates from each scale are integrated. In order to eliminate multiple detection, neighboring candidates in position and scale are grouped.

2.4 Face Detection

In this section we apply the object detection system introduced in sections 2.1 and 2.2 to face detection. The size of the search window for face detection is 22×24. Our system operates in real-time at a resolution of 640×480 using 11 scales on a standard desktop computer. We have been collecting approximately 2000 faces of varying quality from online newspapers for training purposes. All face images were aligned and artificially rotated in the interval $[-10°, ..., +10°]$ for pose resilience. Some background is included in a typical positive (face) example. On the other hand, the negative examples are chosen randomly from a large amount of images, which do not contain any faces.

In order to strengthen the argument in section 2.1, where we suggest the use of both single and double angle features, we train a strong classifier employing both feature types. This will also help us to pre-confine the feature space, in an attempt to prevent overtraining and to support feature selection a priori. Empirical tests on a subset of positive examples (900) and 9000 negative examples revealed that $\{Q\}_{8,5}$ is an eligible set of quangle masks for face detection. The least number of quangle features to separate this 9900 examples errorfree

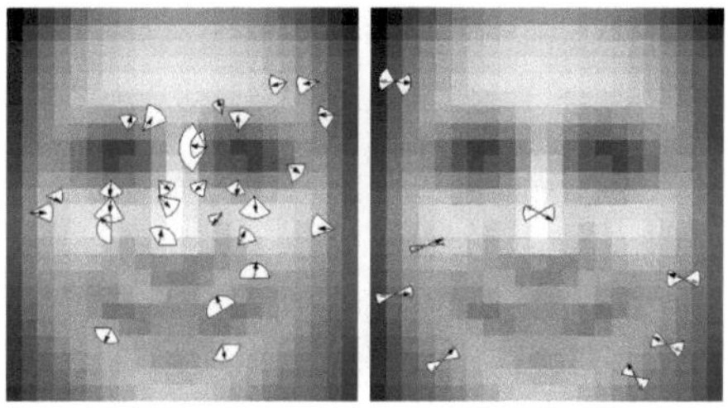

Fig. 4. A strong classifier employing 36 both, single and double angle features, which are displayed (projected lookup-tables) side by side

served as a criterion, besides economic parameters for the set $\{Q\}$. By doing so, we also advanced to reduce the complexity of strong classifiers. A number of 36 quangle features (28 of type q_1 and 8 of type q_2) were selected in the case of using $\{Q\}_{8,5}$. Figure 4 visualizes the selected features in separate detection windows. In both cases the black arrows indicate $\angle\nabla$ of the underlying average training face. The white partitions show the range, the respective angle is supposed to be in. The hourglass-shaped partitions in the second image indicate double angle features, where the gradient could also have pointed in the opposite direction (gray arrows). The radii are modulated by α_t, the weight of the corresponding weak classifiers. It can be observed that single angle features frequently occur in the inner facial regions, whereas features of the second type are situated in the bounding regions. In a further step, we trained two strong classifiers using the same training setup, yet employing either features of type q_1 or q_2. Error-free separation of the training data involved 49 single angle or 83 double angle features, thus clearly favoring the combined setup. Other studies have suggested schemes for reducing classifier complexity [13, 14], which we did not investigate yet, because our features resulted in small classifiers. In a related study, [9], using a naive Bayes classifier, the single gradient angle was quantized into 7 partitions without a further study of flexible and lower quantization levels. In [12], no quantization but integer conversion was done and only the doubled gradient angle was used. Furthermore, the weak classifiers were different there, involving significantly more operations. We show an example of face (and mouth) detection by our method in figure 5.

3 Experiments

For the experiments, the face detector was configured as follows: The size chosen for the detection window was 22×24 and the employed quangle masks were in $\{Q\}_{8,5}$. A cascaded classifier, comprising 22 levels, was trained on 2000 faces and 4000 non-faces (refilled). The total number of weak classifiers in the cascade was 700. Such a classifier complexity is very small compared to a couple of thousands as reported in [11,13]. In operation, the first two levels of the cascade, comprising only 3 and 5 quangle features, respectively, are already able to reject 75% of all

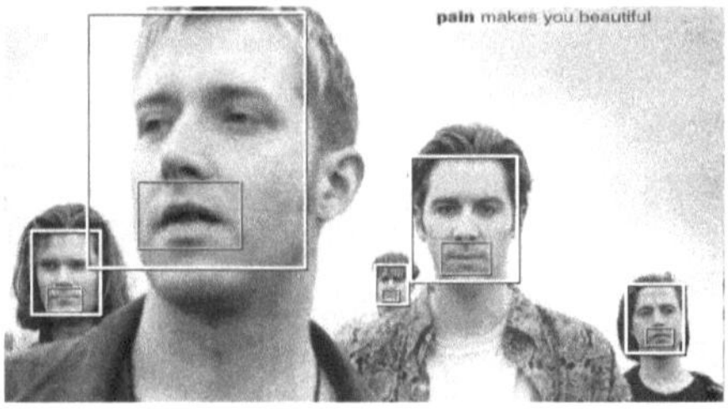

Fig. 5. A (cropped) example image from the CMU-MIT face test set, with faces and mouths detected by the proposed method

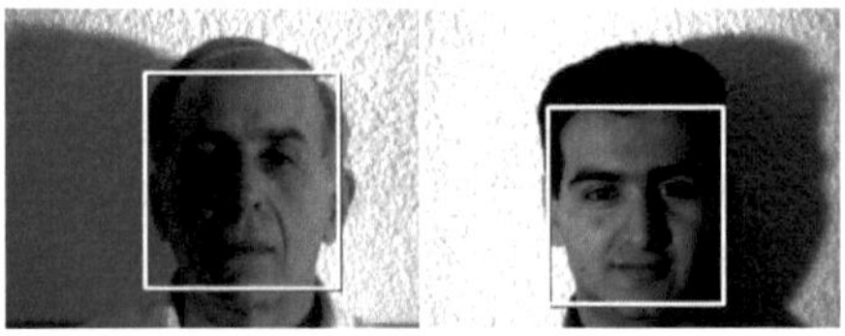

Fig. 6. Two images from the YALE face test set, illustrating "severe" illumination changes, managed by the proposed method though

non-faces. Furthermore, we used $s = 0$ (original resolution) as the starting scale and $sf = 1.1$ as the factor for downsizing.

The performance of the face detection system detailed above is benchmarked on two publicly available databases, namely the YALE [19] and the CMU-MIT [5] face test sets. Extreme illumination and expression changes are the main challenge of the former test set, which consists of 165 frontal face images of 15 subjects. Table 1 shows the detection rates and the number of false positives of our method together with the ones for the face detection algorithm proposed in [9], on the YALE face test set. The results on the YALE test set confirm

Table 1. Detection rates and the number of false positives on the YALE face test set

Method	Detection Rate	False Positives
Nguyen [9]	86,6%	0
Proposed method	100%	0

that our face detection method is resistant to substantial illumination changes without performing any (histogram related) preprocessing. Note, that the latter is actually done in all methods we compare our results to. In figure 6, two "YALE faces" are shown, with indicated detections by the proposed method. Note the severity of the illumination conditions.

The CMU-MIT frontal face test set is among the most commonly used data sets for performance assessment of face detection systems. It is composed of 130 images, containing 507 frontal faces in total. The quality of the images, as well as the scale of faces (compare figure 5) vary substantially here. In addition to the detection rate, this set also permits to give representative numbers for the false positives, because of many high resolution images. In table 2, the results of our technique on the CMU-MIT frontal test set are related to those of two prominent face detectors [11,5], by adjusting the false positive rate to a common level. Also, the detection rate achieved by our method for 1 false detection per million evaluated windows is given, constituting our best result.

Table 2. Detection and false positive rates on the CMU-MIT frontal face test set

Method	Detection Rate	False Positive Rate
Rowley [5]	89,2%	$1,27 \times 10^{-6}$
Viola&Jones [11]	92,9%	$1,27 \times 10^{-6}$
Proposed method	94,2%	$1,25 \times 10^{-6}$
Proposed method	93%	1×10^{-6}

4 Conclusion

In this study, we presented a novel real-time method for face detection. However, the technique is possible to be used as a general image-object detector, as current experiments indicate. The introduced quantized angle ("quangle") features were studied experimentally and we presented evidence for their richness of information measured by their discriminative properties and their resilience to the impacts of severe illumination changes. They need no preprocessing, e.g. histogram equalization, histogram normalization, adding to their computational advantage. This is achieved by considering both the gradient direction and orientation, yet ignoring the magnitude. A quantization scheme is presented to reduce the feature space prior to boosting, i.e. it enables fast evaluation (1 array access). Scale invariance was implemented through an image pyramid. The training excels in rapidness, which enables the use of our object detector for changing environments and application needs. The practicability of the proposed methods and ideas was corroborated by satisfying experimental results for face detection (e.g. 93% Detection Rate at 1×10^{-6} False Positive Rate on the CMU-MIT frontal face test set).

References

1. Yang, M.H., Kriegman, D., Ahuja, N.: Detecting Faces in Images: A Survey. IEEE-PAMI 24(1), 34–58 (2002)
2. Hamouz, M., Kittler, J., Kamarainen, J., Paalanen, P., Kalviainen, H., Matas, J.: Feature-Based Affine-Invariant Localization of Faces. PAMI 27(9), 1490–1495 (2005)
3. Turk, M., Pentland, A.: Eigenfaces for recognition. J. Cognitive Neuroscience (Winter) 3(1), 71–86 (1991)
4. Sung, K.K., Poggio, T.: Example Based Learning for View-Based Human Face Detection. Technical report, Cambridge, MA, USA (1994)

5. Rowley, H., Baluja, S., Kanade, T.: Human Face Detection in Visual Scenes. In: Advances in Neural Information Processing Systems 8. pp. 875 – 881 (1996)
6. Cortes, C., Vapnik, V.: Support-Vector Networks. Machine Learning 20, 273–297 (1995)
7. Osuna, E.: Freund, R., Girosi, F.: Training Support Vector Machines: an Application to Face Detection. In: IEEE Conference on Computer Vision and Pattern Recognition (1997)
8. Schneiderman, H., Kanade, T.: Probabilistic Modeling of Local Appearance and Spatial Relationships for Object Recognition. cvpr 00,45 (1998)
9. Nguyen, D., Halupka, D., Aarabi, P., Sheikholeslami, A.: Real-Time Face Detection and Lip Feature Extraction Using Field-Programmable Gate Arrays. SMCB 36(4), 902–912 (2006)
10. Freund, Y., Shapire, R.E.: A decision-theoretic generalization of online learning and an application to boosting. Journal of Computer and System Sciences 5(1), 119–139 (1997)
11. Viola, P., Jones, M.: Rapid object detection using a boosted cascade of simple features. In: CVPR, vol. 1, pp. 511–518 (2001)
12. Froeba, B., Kueblbeck, C.: Real-Time Face Detection Using Edge-Orientation Matching. In: Bigun, J., Smeraldi, F. (eds.) AVBPA 2001. LNCS, vol. 2091, pp. 78–83. Springer, Heidelberg (2001)
13. Li, S.Z., Zhu, L., Zhang, Z., Blake, A., Zhang, H., Shum, H.: Statistical Learning of Multi-view Face Detection. In: Heyden, A., Sparr, G., Nielsen, M., Johansen, P. (eds.) ECCV 2002. LNCS, vol. 2350, pp. 67–81. Springer, Heidelberg (2002)
14. Sochman, J., Matas, J.: WaldBoost Learning for Time Constrained Sequential Detection. In: CVPR '05: Proceedings of the 2005 IEEE Computer Society Conference on Computer Vision and Pattern Recognition (CVPR'05), vol. 2, pp. 150–156. IEEE Computer Society Press, Washington, DC, USA (2005)
15. Hadid, A., Pietikäinen, M., Ahonen, T.: A Discriminative Feature Space for Detecting and Recognizing Faces. In: CVPR (2). pp. 797–804 (2004)
16. Bigun, J.: Vision with Direction. Springer, Heidelberg (2005)
17. Kollreider, K., Fronthaler, H., Bigun, J.: Evaluating Liveness by Face Images and the Structure Tensor. In: Fourth IEEE Workshop on Automatic Identification Advanced Technologies AutoID 2005, Buffalo, New York, pp. 75–80 (2005)
18. Smeraldi, F., Bigun, J.: Retinal vision applied to facial features detection and face authentication. Pattern Recognition Letters 23, 463–475 (2002)
19. Belhumeur, P.N., Hespanha, J.P., Kriegman, D.J.: Eigenfaces vs.Fisherfaces: Recognition using class specific linear projection. PAMI 19(7), 711–720 (1997)
20. Li, J., Wang, Y., Tan, T., Jain, A.K.: Live face detection based on the analysis of Fourier spectra. In: Biometric Technology for Human Identification, SPIE vol. 5404, pp. 296–303 (2004)

Face Detection Using Multiple Cues

Thomas B. Moeslund, Jess S. Petersen, and Lasse D. Skalski

Laboratory of Computer Vision and Media Technology
Aalborg University, Denmark

Abstract. Many potential applications exist where a fast and robust detection of human faces is required. Different cues can be used for this purpose. Since each cue has its own pros and cons we, in this paper, suggest to combine several complimentary cues in order to gain more robustness in face detection. Concretely, we apply skin-color, shape, and texture to build a robust detector. We define the face detection problem in a state-space spanned by position, scale, and rotation. The state-space is searched using a Particle Filter where 80% of the particles are predicted from the past frame, 10% are chosen randomly and 10% are from a texture-based detector. The likelihood of each selected particle is evaluated using the skin-color and shape cues. We evaluate the different cues separately as well as in combination. An improvement in both detection rates and false positives is obtained when combining them.

1 Introduction

The "Looking at people" research field covers applications where cameras observe humans. This ranges from surveillance, HCI, to motion capture and analysis of athletes' performance. All applications require the humans to be segmented in the image and a tremendous amount of research has been conducted in this field due to the potential applications [11]. For some applications only one or a few body parts are required. For example the human face for applications such as identity recognition, facial expression recognition, head pose estimation or simply to detect the presence of a person. The core technology for such applications is first of all to detect human faces in an image or video sequence. Different methods for doing this have been suggested and they can roughly be grouped according to the type of data they operate on.

Many operate by finding skin pixels in the image and group then into head-shaped objects [6,10,11]. Such methods are sensitive to other skin-color objects present in the scene. A different approach is to look for head-shaped objects in either a silhouette or edge version of the input image [7,9,10,11,18]. But again, background clutter can disrupt this data type. Yet another approach is to find features inside the face, e.g., eyes, nose, mouth, hair-line, cheeks, etc., and build a classifier based on their structural relationships [10,14,17]. Such approaches are indeed affected by background clutter and therefore work best as a verification tool for possible head candidates [10]. Furthermore they tend to operate best on frontal images and can be quite heavy computational-wise. A related approach

B.K. Ersbøll and K.S. Pedersen (Eds.): SCIA 2007, LNCS 4522, pp. 51–60, 2007.

is to apply the appearance of the entire face, for example using the texture of the face [10,16,17].

No matter which cues one uses background clutter and other noise sources will challenge the detector. A combination of multiple cues can therefore be applied to make a detector more robust. Often this is done by using either a skin-color detector followed by a verification using facial features [6,10] or using an appearance-based detector followed by a verification using facial features [10,14].

In this work we do a parallel fusion of multiple cues in other to benefit directly the complimentary characteristics of the different cues. This is similar to approaches followed in other domains. For example in [13] where persons are detected using color and texture, and in [15] where hands are detected using color and motion, and in [1] where arms are detected using color, motion and shape. Concretely we combine color, shape and texture to build a robust and fast face detector. Furthermore, since we are interested in detecting faces in video, we apply the temporal context to improve the detections. In section 2, 3, and 4, we describe the shape, color and texture cues, respectively. In section 5 we describe how they are integrated and how the temporal context is applied. In section 6 results are presented and in section 7 a conclusion is given.

2 Shape-Based Detection

Shape-based detection is based on the fact that the contour of the head is a rather distinct feature in a standard image and the face can hence be found by finding this shape.

Shape-based detection (and recognition) is based on as least two key elements: 1) a definition and representation of the shape and 2) a measure for the similarity between the shape model projected into the current image and the edges found in the current image.

2.1 Shape Model

A correct model of human heads and their variations can be trained and modeled using for example a Point Distribution Model. However, for the purpose of detecting the human face a rough model will suffice. A rough and very simple model is an ellipse, which in many cases is a rather accurate match, see figure 1. The elliptic shape is not necessarily unique, i.e., other objects in a scene might have this shape. It has therefore been suggested to enhance the uniqueness by including the shoulder-profile [18], see figure 1. While the contour of a head seldom deform this is not the case for the shoulder and the head-shoulder profile is therefore often modeled using some kind of dynamic contour represented by a Spline. Due to the extra parameters such methods require extra processing and can be sensitive to arm/shoulder movements. We therefore use an elliptic model.

We note that the neck can be hard to identify resulting in a poor match for the lower part of the ellipse. We therefore only apply the elliptic arc seen in figure 1.

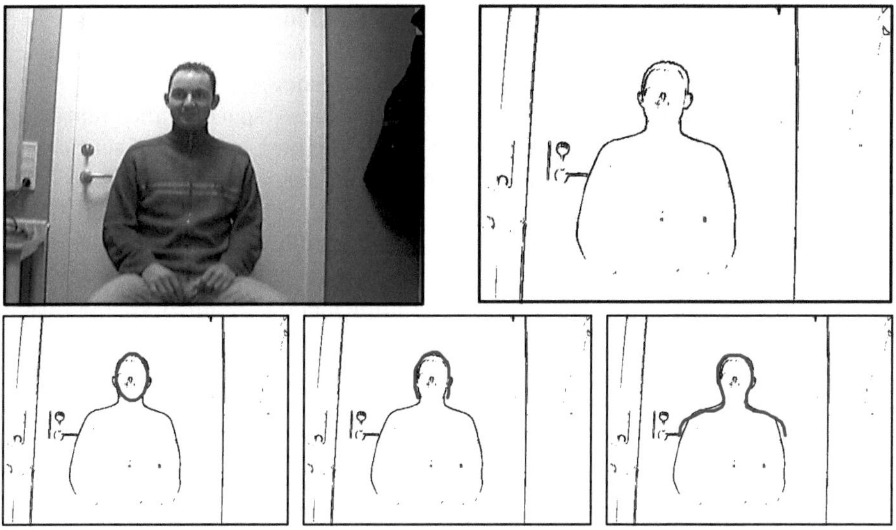

Fig. 1. An illustration of matching different shape types with the human head in the image

2.2 Shape Matching

Matching is here based on edges extracted from the images. When matching a shape to an image, it is desired that a smooth search space is present. This ensures that not only a perfect match results in a high similarity measure but also solutions in the proximity. This is vital since a perfect match is virtually impossible due to noise and an imperfect shape model. We apply Chamfer matching [2] to generate a smooth search space from an edge image.

Chamfer matching can be used to find occurrences of a shape in an image based on edges extracted from the image. The matching is based on a distance map, which is created from the edge image using a distance transformation. This distance map is an image, in which each pixel contains the distance to the nearest edge. The matching is done by projecting the shape into the distance map, and sum the values of the overlapping pixels in the distance map. The sum is normalized by the number of pixels resulting in the average distance, $\overline{d}$, for a particular shape x. This average distance is converted into a likelihood measure as

$$P_{Shape}(x) = \begin{cases} 0, & \overline{d} \geq 10; \\ 1 - \frac{\alpha \cdot \overline{d}}{10}, & \text{Otherwise.} \end{cases} \quad (1)$$

where α is a constant learned during training.

3 Color-Based Detection

Detecting a face based on color relays on the notion that skin-color is a rather distinct feature in a standard image. Skin color is in fact a strong cue for finding faces, but obviously flawed by the fact that other skin regions, e.g., hands and arms, are often present. As for the shape cue, this cue also requires the choice of an appropriate representation and matching scheme.

We have assessed different color representations (spaces) and matching schemes and found the one suggested by [8] to be the best in terms of sensitivity. Besides, it operates in the RGB color space, which means that there is no computational overhead in transforming the colors from the input image (R,G,B). Furthermore, the method is chosen, because it is able to detect skin color in indoor scenes with changing illumination. The matching scheme is shown below. The likelihood measure for the color cue is implemented as an AND operation

Table 1. Skin classification rules from [8]

Lighting conditions	Uniform daylight	Flashlight or lateral daylight
Skin color classification rule	R > 95, G > 40, B > 20 Max{R,G,B}-Min{R,G,B} > 15 |R-G| > 15, R > G, R > B	R > 220, G > 210, B > 170 |R-G| ≤ 15, B < R, B < G

between the thresholded and filtered skin-color image and an ellipse representing a candidate face, see figure 2. When representing a face using an ellipse, not all face pixels will be classified as skin color, due to eyes, hair, and mouth. The likelihood measure is therefore calculated by counting the number of skin pixels within the ellipse and dividing by β of the total number of pixels within the ellipse.

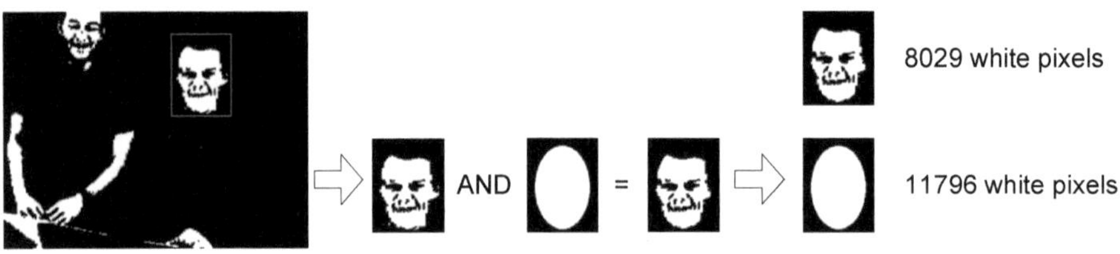

$$\text{Likelihood} = \frac{8029}{11796 \cdot 0.9} = 0.7563$$

Fig. 2. An example of how the skin color likelihood is calculated for $\beta = 0.9$

4 Texture-Based Detection

A texture-based detector looks for templates having face-like appearance. In its most simple form template matching is applied. However, in recent years a different strategy has been very successful, namely to use a number of simple *and generic* templates as opposed to merely one specific template. The basic idea was first proposed in [16], where two key ideas are presented: 1) create a face detector based on a combination of weak classifiers and combine them to a boosted classifier using machine learning, and 2) create a cascade of boosted classifiers resulting in the final face detector.

A weak classifier is constructed of a single Haar-like feature, a boosted classifier is a weighted combination of weak classifiers, and a cascaded classifier is a sequence of boosted classifiers as illustrated in figure 3.

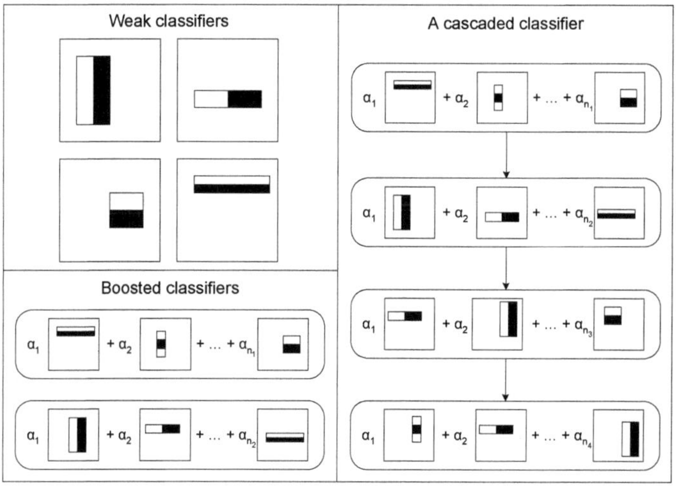

Fig. 3. Examples of weak classifiers, boosted classifiers and a cascaded classifier

To detect faces in images a subwindow is moved across the image in different scales. Each subwindow is then processed by the cascaded classifier. Each boosted classifier in the cascade is denoted a layer, for each of these the subwindow is evaluated by the corresponding boosted classifier. If the subwindow is classified as a face, it is passed to the next layer. A subwindow must be classified as a face by all layers of the cascade to be detected. A cascaded classifier is trained to consists of increasingly more complex boosted classifiers. Each boosted classifier has a very high detection rate and a moderate false acceptance rate, e.g. 99% detection rate and 50% false acceptance rate. This enables the first few layers of the cascaded classifier to reject a majority of the non-face subwindows in the input image. In this way more computation time is spent on more difficult samples.

The detector can process an image relatively fast due to the simple nature of the features and the clever invention of an integral image [16]. However, training the detector takes a very long time due to the massive amount of possible features, positions and scales and a large training set of normally several 1000s positive and negatives samples.

We apply a subwindow size of $24x24$ pixels resulting in around 160.000 possible features. We use the AdaBoost algorithm [16] to do the training using around

5000 face images and 7000 non-face images (which took approximately one month of constant processing!) resulting in a detector consisting of 11 boosted classifiers with a total of 587 weak classifiers.

The output of the detector is a number of subwindows likely to contain a face. Overlapping subwindows are merged into only one output.

5 Combining the Cues

The different detectors can each analyze a particular position, scale, and rotation in the image. In order to represent all possible solutions we define a state-space spanned by the different degrees-of-freedom. These are the two translations in the x- and y-direction, rotation in the image plane and scale. The first two have a resolution of one pixel and are limited to the image plane. The rotation parameter has a resolution of 15° degrees and is limited to ±30° [12]. The scale parameter is linked to the two primary axes of the ellipse in the following manner.

The size of an ellipse is defined by the major and minor axes. In order to limit the number of parameters we use the 2006 anthropometric data from NASA [12] to find the ratio between these axes for average humans: Male: 25,7cm(Height) / 16,5cm(Width) = 1,558. Female: 24,3cm(Height) / 16,8cm(Width) = 1,446. Based on this we define a general average ratio of 1.5 and use that to scale the ellipse. The resolution of the scale-factor is 5 and it is limited by height $\in [48; 144]$ pixels, corresponding to $1m - 3m$ from the camera.

The state-space is spanned by four axes and each point in the space corresponds to one particular position, rotation and scale of the head. A detector can now operate by trying each possible state and see how well it matches the current image using the detectors described above. If a state has a high match then a face is located. Due to the size and resolution of the state-space such a brute force approach is not realistic due to the heavy processing.

For most applications where face detection is required the movement of people between frames will typically be limited, hence, temporal knowledge can be used to reduce the state-space. A well-documented framework for this is a Particle Filter [11].

When a particle filter is used to reduce the state-space, the space is limited to the states, denoted particles, with a high likelihood in the previous frame. This allows for approximating the entire state-space using several magnitudes of fewer possible solutions. A low number of particles may however cause problems if new persons enter the scene or if a person is not detected in all frames. We therefore only sample 80% of the particles from the state-space in the previous frame and as suggested in [4] we randomly sample 10% of the particles to cover random events. The last 10% are sampled from the output of the texture-based detector. This means that the likelihood function used to evaluate each particle will only be based on color and shape. We have tried different combinations and found this to be the most suitable solution since the texture-based detector is the best stand-alone detector and tends to produce many false positives making it suitable to detect new objects and lost objects. Furthermore, the texture-based

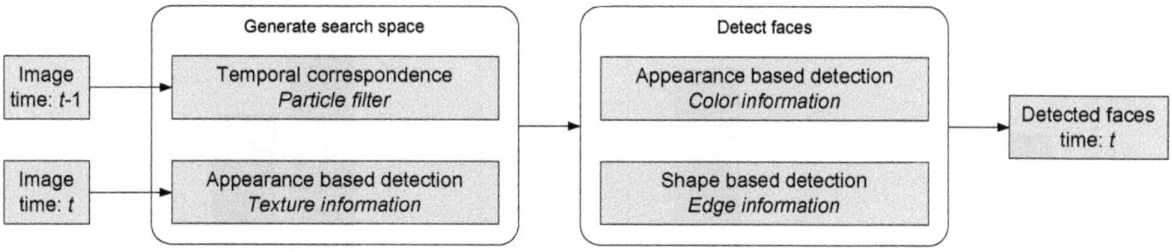

Fig. 4. An illustration of the cue integration

detector only estimates three of the four state-space parameters. In figure 4 the cue integration is illustrated.

To summarize, the particle filter has two main steps for each frame. First a search space is generated, i.e., likely particles are defined. 80% are predicted from the previous frame using Random Walk. 10% are chosen randomly, and the last 10% are taking from the output of the texture-based detector and diffused to create slight variations. This results in N particles likely to contain a correct state of a face in the current image. Each of these N states are now evaluated regarding color and shape using both the color-based detector and the shape-based detector. The output will be a state-space where each entry contains a likelihood of this particular state being present in the current image. By finding the maximum peaks the faces are detected. Since the particle filter is defined in a Bayesian framework the peaks are equal to the MAP (maximum a posteriori).

6 Results

We define a correct detection as a situation where minimum 50% of the face is inside the ellipse defined by a state and minimum 50% of the pixels inside the ellipse are skin-pixels from the face. A face is defined as the visible region of the head with hair, but without the neck. We use 250 manually annotated frames from a complicated scene containing background clutter, non-human skin color objects and motion in the background. The algorithm is evaluated using 1000 particles corresponding to 0.2% of the total number of possible solutions in the state-space. The framerate is 5.1Hz on a Pentium 1300MHz Centrino.

Five different evaluations are performed:

A: Texture-based detection
B: Particle filter using color detection, and 10% randomly sampled particles in each image
C: Particle filter using shape detection, and 10% randomly sampled particles in each image
D: Particle filter using color and shape detection, and 10% randomly sampled particles in each image
E: Particle filter using color and shape detection, and 10% randomly sampled particles in each image, and 10% samples distributed using the texture-based detection

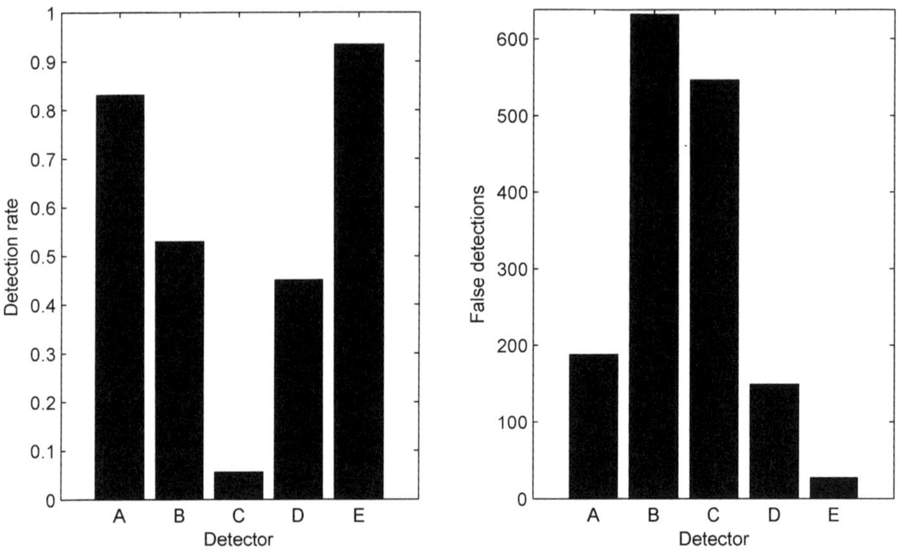

Fig. 5. The results of evaluating the five detectors

The particle filter is initialized by random sampling all particles. However, in evaluation **E** the particle filter is initialized by randomly sampling 50% of the particles and distributing the remaining 50% based on the detections from the texture-based detection. In figure 5 the detection rates and the number of false detections for each of the detectors are shown.

The results show that detector **E** has the highest detection rate and has significantly fewer false detections than the other detectors, i.e., using multiple cues outperforms any of the individual detectors. Detector **A** has significantly higher detection rate than **B**, **C**, and **D**, which underline the current trend in computer vision of using machine learning based on massive training data. The detection rate of detector **B** is slightly higher than the detection rate of detector **D** which is unexpected. However, the number of false detections using detector **D** is remarkably lower than using detector **B** meaning that detector **C** is too sensitive as a detector (primarily due to background clutter), but can contribute to eliminating false detections as it compliment detector **B**. Comparing detector **D** and **E** shows the significance of introducing particles selected by the texture-based detector **A**. Detector **A** does introduce more false detections, but using the color and shape cues allow us to prune non-supported states resulting in relative few false detections. Figure 6 shows example images containing for detector **E**.

7 Discussion

The tests clearly show the benefit of combining several complimentary cues. Most of the errors of detector **E** are very close to the actual face, see figure 6, meaning that the detection rate can be increased and the false detections lowered

Fig. 6. Example images containing both true and false detections using detector **E**

if the definition of detection in section 6 is relaxed. Another possible improvement is to increase the number of particles, but experiments show that significantly more particles are required resulting in a somewhat slower system. Other alternatives are either some kind of postprocessing, e.g., a Mean Shift tracker [3] or to make each particle converge to a local maximum see e.g., [5]. But again, such improvements will reduce the speed of the system.

References

1. Azoz, Y., Devi, L., Yeasin, M., Sharma, R.: Tracking the Human Arm Using Constraint Fusion and Multiple-Cue Localization. Machine Vision and Applications 13(5-6), 286–302 (2003)
2. Borgefors, G.: Hierarchical Chamfer Matching: A Parametric Edge Matching Algorithm. IEEE Transaction on Pattern Analysis and Machine Intelligence 10(6), 849–865 (1988)
3. Cheng, Y.: Mean Shift, Mode Seeking, and Clustering. IEEE Transactions on Pattern Analysis and Machine Intelligence 17(8), 790–799 (1995)
4. Davis, L., Philomin, V., Duraiswami, R.: Tracking Humans from a Moving Platform. In: International Conference on Pattern Recognition, Barcelona, Spain (September 3-8, 2000)
5. Deutscher, J., Reid, I.: Articulated Body Motion Capture by Stochastic Search. International Journal of Computer Vision 61(2), 185–205 (2005)
6. Hsu, R.L., Abdel-Mottaleb, M., Jain, A.K.: Face Detection in Color Images. IEEE Transactions on Pattern Analysis and Machine Intelligence 24(5), 696–706 (2002)
7. Hu, W., Hu, M., Zhou, X., Tan, T., Lou, J., Maybank, S.: Principal Axis-Based Correspondence between Multiple Cameras for People Tracking. IEEE Transactions on Pattern Analysis and Machine Intelligence 28(4), 663–671 (2006)
8. Kovac, J., Peer, P., Solina, F.: 2D Versus 3D Colour Space Face Detection. In: EURASIP Conference on Video / Image Processing and Multimedia Communications EC-VIP-MC'03, Zagreb, Croatia (July 2-5, 2003)

9. Lanitis, A., Taylor, C.J., Cootes, T.F.: Automatic Interpretation and Coding of Face Images Using Flexible Models. IEEE Transactions on Pattern Analysis and Machine Intelligence 19(7), 743–756 (1997)
10. Kriegman, D.J., Yang, M.H., Ahuja, N.: Detecting Faces in Images: A Survey. IEEE Transactions on Pattern Analysis and Machine Intelligence 24(1), 34–58 (2002)
11. Moeslund, T.B., Hilton, A., Kruger, V.: A Survey of Advances in Vision-Based Human Motion Capture and Analysis. Computer Vision and Image Understanding 104(2-3), 90–126 (2006)
12. NASA - National Aeronautics and Space Administration. Anthropometry and Biomechanics,`http://msis.jsc.nasa.gov/sections/section03.htm` (2006)
13. Okuma, K., Taleghani, A., de Freitas, N., Little, J., Lowe, D.: A Boosted Farticle Filter: Multitarget Detection and Tracking. In: European Conference on Computer Vision, Prague, Czech Republic (May 11-14, 2004)
14. Santana, M.C., Suárez, O.D., Artal, C.G., González, J.I.: Cue Combination for Robust Real-Time Multiple Face Detection at Different Resolutions. In: International Conference on Computer Aided Systems Theory, Las Palmas de Gran Canaria, Spain (February 7-11, 2005)
15. Stenger, B.: Template-Based Hand Pose Recognition Using Multiple Cues. In: Asian Conference on Computer Vision, Hyderabad, India (January 13-16, 2006)
16. Viola, P., Jones, M.: Rapid Object Detection using a Boosted Cascade of Simple Features. In: Computer Vision and Pattern Recognition, Kauai Marriott, Hawaii (December 9-14, 2001)
17. Wu, J., Pedersen, J.M., Putthividhya, P., Norgaard, D., Trivedi, M.M.: A Two-level Pose Estimation Framework Using Majority Voting of Gabor Wavelets and Bunch Graph Analysis. In: ICPR workshop on Visual Observation of Deictic Gestures, Cambridge, UK (August 22, 2004)
18. Zhao, T., Nevatia, R.: Stochastic Human Segmentation from a Static Camera. In: MOTION '02: Proceedings of the Workshop on Motion and Video Computing, Orlando, Florida, USA (December 5-6, 2002)

Individual Discriminative Face Recognition Models Based on Subsets of Features

Line H. Clemmensen[1], David D. Gomez[2], and Bjarne K. Ersbøll[1]

[1] Informatics and Mathematical Modelling, Technical University of Denmark, DK-2800 Lyngby, Denmark
lhc@imm.dtu.dk and be@imm.dtu.dk

[2] Computational Imaging Lab, Pompeu Fabre University, Barcelona, Spain
david.delgado@upf.edu

Abstract. The accuracy of data classification methods depends considerably on the data representation and on the selected features. In this work, the elastic net model selection is used to identify meaningful and important features in face recognition. Modelling the characteristics which distinguish one person from another using only subsets of features will both decrease the computational cost and increase the generalization capacity of the face recognition algorithm. Moreover, identifying which are the features that better discriminate between persons will also provide a deeper understanding of the face recognition problem. The elastic net model is able to select a subset of features with low computational effort compared to other state-of-the-art feature selection methods. Furthermore, the fact that the number of features usually is larger than the number of images in the data base makes feature selection techniques such as forward selection or lasso regression become inadequate. In the experimental section, the performance of the elastic net model is compared with geometrical and color based algorithms widely used in face recognition such as Procrustes nearest neighbor, Eigenfaces, or Fisherfaces. Results show that the elastic net is capable of selecting a set of discriminative features and hereby obtain higher classification rates.

1 Introduction

Historical facts (New York, Madrid, London) have put a great emphasis on the development of reliable and ethically acceptable security systems for person identification and verification. Traditional approaches such as identity cards, PIN codes, and passwords are vulnerable to falsifications and hacking, and such security breaks thus also appear frequently in the media.

Another traditional approach is biometrics. Biometrics base the recognition of individuals on the intrinsic aspects of a human being. Examples are fingerprint and iris recognition [1][2]. However, traditional biometric methods are intrusive, i.e. one has to interact with the individual who is to be identified or authenticated. In some cases, however, iris recognition is implemented as a standard security check in airports (e.g. New York JFK). Recognition of people from facial

B.K. Ersbøll and K.S. Pedersen (Eds.): SCIA 2007, LNCS 4522, pp. 61–71, 2007.

images on the other hand is non-intrusive. For this reason, face recognition has received increased interest from the scientific community in the recent years.

Face recognition consists of problems with a large number of features (of geometrical or color related information) in relation to the number of face images in the training sets. In order to reduce the dimensionality of the feature space we propose to use *least angle regression - elastic net* (LARS-EN) model selection to select discriminative features that increase the accuracy rates in facial identification. LARS-EN was introduced by Zou et. al in 2005 [3]. It regularizes the *ordinary least squares* (OLS) solution with both the Ridge regression and Lasso constraints. The method selects variables into the model where each iteration corresponds to loosening the regularization with the Lasso constraint. The ridge constraint ensures that the solution does not saturate if there are more variables in the model than the number of observations.

The rest of the paper is organized as follows: In section two, a review of the standard face recognition techniques is presented. Section three describes the LARS-EN algorithm. In section four, we describe and state the results for several experiments which we conducted to test the discriminative capacity of the obtained features. Finally, section 5 gives a conclusion of the conducted experiments and discusses some future aspects of the research.

2 Face Recognition Review

The first techniques developed for face recognition aimed at identifying people from facial images based on geometrical information. Relative distances between key points such as mouth or eye corners were used to characterize faces [4][5]. At this first stage of facial recognition, many of the developed techniques focused on automatic detection of individual facial features. The research was notably strengthened with the incursion of the theory of statistical shape analysis. Within this approach, faces were described by landmarks or points of correspondence on an object that matches between and within populations. In a 2D-image, a landmark $\mathbf{l}$ is a two dimensional vector $\mathbf{l} = (x, y)$ that, to obtain a more simple and tractable mathematical description, is expressed in complex notation by $\mathbf{l} = x + iy$, where $i = \sqrt{-1}$. In this framework, a face in an image is represented by a configuration or a set of n landmarks $[\mathbf{l}_1, \mathbf{l}_2, ..., \mathbf{l}_n]$ placed on meaningful points. Geometrical face recognition based on landmarks is conducted by evaluating the similarity of the configuration of a test face with respect to the configurations in a facial database. In order to achieve this, different measures of similarity have been proposed, see e.g. [6]. Among all the proposed metrics, the Procrustes distance has been the most frequently used. Given two configurations w and z, the Procrustes distance between them is defined by

$$D_P(w, z) = \inf_{\beta,\theta,a,b} \| \frac{z}{\|z\|} - \frac{w}{\|w\|}\beta e^{i\theta} - a - ib\| \quad , \tag{1}$$

where $\|\cdot\|$ represents the l_2 norm, and the parameters β, θ, a, and b, which denotes a scaling, a rotation, and a translation of configuration w, are chosen to minimize the distance between w and z. Several extensions of this measure have been proposed. For instance, Shi et. al [7] has recently proposed a refined Procrustes distance based on principal component analysis. The configurations (the landmark representations of the faces) are first centered at the origin and transformed to have unit size. Then a complex principal component analysis is conducted to reduce the dimensionality. The similarity measure is defined in this lower m-dimensional space by

$$D_{RP}(w,z) = \sum_{k=1}^{m} \| \frac{\hat{z}_k}{\sqrt{\lambda_k^{(z)}}} - \frac{\hat{w}_k}{\sqrt{\lambda_k^{(w)}}} \| \quad , \tag{2}$$

where $\hat{z}_k$ is the kth eigenvector of configuration y, $\hat{w}_k$ is the kth eigenvector of configuration w, and $\lambda_k^{(z)}$ and $\lambda_k^{(w)}$ the corresponding eigenvalues.

The publication of Eigenfaces by Turk and Pentland [8] showed that it was possible to obtain better classification rates by using the color intensities. Since then, geometrical face recognition was gradually declining until the extent that, nowadays, it principally remains to support color face recognition. The appearance of Eigenfaces provided an excellent way of summarizing the color information of the face. The facial images in a training database were first registered to obtain a correspondence of the pixels between the images. Then, a principal component analysis was conducted to reduce the high data dimensionality, to eliminate noise, and to obtain a more compact representation of the face images. When a new test image was desired classified, the same data reduction was applied to obtain a comparable compact test image representation. The similarity of the compact test image representation was measured with each of the compact training image representations based on the Euclidean distance. The test image was associated with the training image with the smallest Euclidean distance. Based on Eigenfaces, Fisherfaces obtained higher classification rates by applying a Fisher Linear discriminant on the obtained principal components. As a result of the publication of Fisherfaces a considerable percentage of the current research in the field is devoted to find more discriminative projections [9][10].

In this paper, an approach to increase the discrimination among individuals is proposed. However, instead of looking for more discriminative projections as the previous methods, it aims at finding more discriminative features. This is in line with the face detector of Viola and Jones [11] that selects Haar features which are important for the face detection task. Basing the identification on only a subset of the features will make the system work faster for future identifications. The approach is described in next section.

3 Elastic Net Model Selection

We consider the linear model:

$$\boldsymbol{y} = \mathbf{X}\boldsymbol{\beta} + \boldsymbol{\epsilon} \quad , \tag{3}$$

where each $\epsilon_i \sim N(0, \sigma^2)$. We assume $\boldsymbol{y}$ centered (i.e. $\sum_{i=1}^n y_i = 0$) and the columns of $\mathbf{X}$ normalized to zero mean and unit length.

The LARS-EN method is used to make multiple individual discriminative models by the use of dependent variables with ones and zeros discriminating one individual from the remaining people in the data set. In the case of one image per individual the kth individual model is:

$$\text{center}\left(\begin{bmatrix}\mathbf{0}_{k-1}\\ 1\\ \mathbf{0}_{n-k}\end{bmatrix}\right) = \text{normalize}\left(\begin{bmatrix} x_{11} & \dots & x_{1p}\\ \vdots & \ddots & \vdots\\ x_{n1} & \dots & x_{np}\end{bmatrix}\right)\begin{bmatrix}\beta_1\\ \vdots\\ \beta_p\end{bmatrix} + \begin{bmatrix}\epsilon_1\\ \vdots\\ \epsilon_n\end{bmatrix} \quad , \tag{4}$$

where n is the number of individuals (there are $n-1$ individuals distinct from individual k), and p is the number of features. $\mathbf{0}_{k-1}$ denotes a vector of $k-1$ zeros. The geometrical features used in this work were the x and the y coordinates of the landmarks. The color based features were the gray scale intensities of the facial images after warping.

3.1 The Elastic Net

Least angle regression - elastic net (LARS-EN) model selection was proposed by Zou et. al [3] to handle $p \gg n$ problems. The method regularizes the *ordinary least squares* (OLS) solution using two constraints, the 1-norm and the 2-norm of the coefficients. These constraints are the ones used in the *least absolute shrinkage and selection operator* (Lasso) [12] and Ridge regression [13], respectively. The naive elastic net estimator is defined as

$$\hat{\boldsymbol{\beta}} = \text{argmin}_{\boldsymbol{\beta}}\{\|\boldsymbol{y} - \mathbf{X}\boldsymbol{\beta}\|_2^2 + \lambda_1\|\boldsymbol{\beta}\|_1 + \lambda_2\|\boldsymbol{\beta}\|_2^2\} \quad , \tag{5}$$

where $\|\boldsymbol{\beta}\|_1 = \sum_{i=1}^p |\beta_i|$, $|\cdot|$ denoting the absolute value, and $\|\boldsymbol{\beta}\|_2^2 = \sum_{i=1}^p \beta_i^2$. Choosing $\lambda_1 = 0$ yields Ridge solutions, and likewise choosing $\lambda_2 = 0$ yields Lasso solutions. For the Lasso method it is likely that one or more of the coefficients is zero at the solution, while for the Ridge regression it is not very likely that one of the coefficients is zero. Hence, we obtain a sparsity in the solution by using the Lasso constraint. The Ridge constraint ensures that we can enter more than n variables into the solution before it saturates.

We can transform the naive elastic net problem into an equivalent Lasso problem on the augmented data (c.f. [3, Lemma 1])

$$\mathbf{X}^* = (1+\lambda_2)^{-1/2}\begin{bmatrix}\mathbf{X}\\ \sqrt{\lambda_2}\mathbf{I}_p\end{bmatrix} \quad , \quad \boldsymbol{y}^* = \begin{bmatrix}\boldsymbol{y}\\ \mathbf{0}_p\end{bmatrix} \quad . \tag{6}$$

The normal equations, yielding the OLS solution, to this augmented problem are

$$\left(\frac{1}{\sqrt{1+\lambda_2}}\right)^2\begin{bmatrix}\mathbf{X}\\ \sqrt{\lambda_2}\mathbf{I}_p\end{bmatrix}^T\begin{bmatrix}\mathbf{X}\\ \sqrt{\lambda_2}\mathbf{I}_p\end{bmatrix}\hat{\boldsymbol{\beta}}^* = \frac{1}{\sqrt{1+\lambda_2}}\begin{bmatrix}\mathbf{X}\\ \sqrt{\lambda_2}\mathbf{I}_p\end{bmatrix}^T\begin{bmatrix}\boldsymbol{y}\\ \mathbf{0}_p\end{bmatrix} \Leftrightarrow$$

$$\frac{1}{\sqrt{1+\lambda_2}}\left(\mathbf{X}^T\mathbf{X} + \lambda_2\mathbf{I}_p^T\mathbf{I}_p\right)\hat{\boldsymbol{\beta}}^* = \mathbf{X}^T\boldsymbol{y} \quad . \tag{7}$$

We see that $\frac{1}{\sqrt{1+\lambda_2}}\hat{\boldsymbol{\beta}}^*$ is the Ridge regression estimate with parameter λ_2. Hence, performing Lasso on this augmented problem yields an elastic net solution. The *least angle regression* (LARS) model selection method proposed by [14] can be used with advantage to compute the Lasso solution on the augmented problem. The LARS algorithm obtains the Lasso solution with a computational speed comparable to computing the OLS solution of the full set of covariates.

The algorithm uses the LARS implementation with the Lasso modification as described in the following section. Hence, we have the parameter λ_2 to adjust, but also the number of iterations for the LARS algorithm can be used. The larger λ_2, the more weight is put on the Ridge constraint. The Lasso constraint is weighted by the number of iterations. Few iterations corresponds to a high value of λ_1, and vice versa. The number of iterations can also be used to ensure a low number of active variables like the forward selection procedure.

3.2 Least Angle Regression

The least angle regression selection (LARS) algorithm method proposed by Efron et. al [14] finds the predictor most correlated with the response, takes a step in this direction until the correlation is equal to another predictor, then it takes the equiangular direction between the predictors of equal correlation (*the least angle direction*) and so forth.

By ensuring that the sign of any non-zero coordinate β_j has the same sign as the current correlation $\hat{c}_j = \boldsymbol{x}_j^T(\boldsymbol{y} - \hat{\boldsymbol{\mu}})$, the LARS method yields all Lasso solutions[1]. This result is obtained by differentiating the Lagrange version of the Lasso problem. For further details see [14].

3.3 Distance Measure

By introducing a distance measure we obtain a measure of how close a new image is to the different individuals in the database. We used the absolute difference between the predicted value $\hat{y}_k$ for model k and the true value y_k for an image belonging to individual k as a measure of the distance between the new image and individual k.

4 Results and Comparison

In order to test the performance of LARS-EN with respect to the previously commented geometrical and color face recognition technique, two identification experiments were conducted. The difference of the experiments is in the used features. In the first experiment, only the landmarks were used. The second experiment considered only the color. In order to conduct the experiments, the XM2VTS database was used [15]. Eight images for each of the first 50 persons were selected. For all experiments a 4-2-2 strategy was chosen: 4 images of each

[1] $\boldsymbol{y}$ is centered and normalized to unit length, $\mathbf{X}$ is normalized so each variable has unit length, and $\hat{\boldsymbol{\mu}} = \mathbf{X}\hat{\boldsymbol{\beta}}$.

person to train the model, 2 images of each person to adjust the parameters in the model, and 2 images of each person to verify the model.

To evaluate the performance of the algorithms we used rank plots of the cumulative match scores as proposed in [16]. The horizontal axis of the rank plots is the rank itself (referring to the sorted distance measure) and the vertical axis is the cumulated probability of identification. Hence, we obtain an answer to the question: "Is the correct match in the top n matches?".

4.1 Geometrical Face Recognition

In order to conduct this first experiment, a set of 64 landmarks were placed along the face, eyes, nose and mouth of each of the 400 selected images. Figure 1 displays the landmarks used in the experiment.

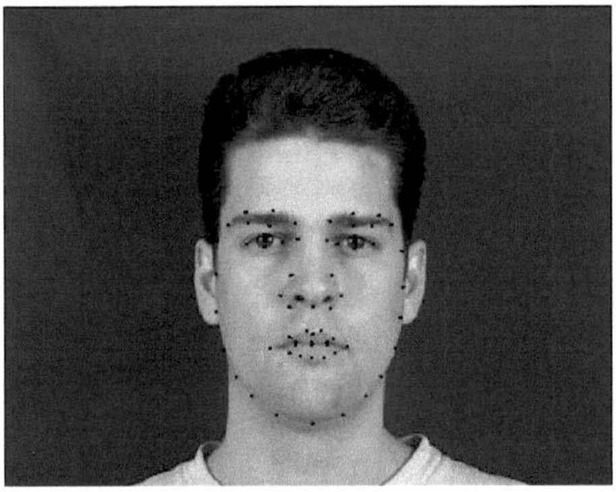

Fig. 1. Illustration of the landmarks used in the experiment

Table 1 summarizes the classification rates obtained using only the landmarks. The LARS-EN and the Fisher methods are comparable in validation error. However, the test error of the Fisher method is increased by 7%, which might indicate an overfitting of the training and validation images.

The LARS-EN models included on average 55 of the 128 shape features (x and y coordinates of the landmarks). It should be noted that the mean square error of both the training and the test set in LARS-EN were of the same size, i.e. no severe

Table 1. Summary of the classification rates for the models based solely on the landmarks

Method/Classification rate	Training	Validation	Test
Procrustes	1.00	-	0.67
Refined Procrustes	1.00	0.76	0.52
PCA	1.00	0.73	0.63
PCA+Fisher	1.00	0.88	0.81
LARS-EN	1.00	0.91	0.91

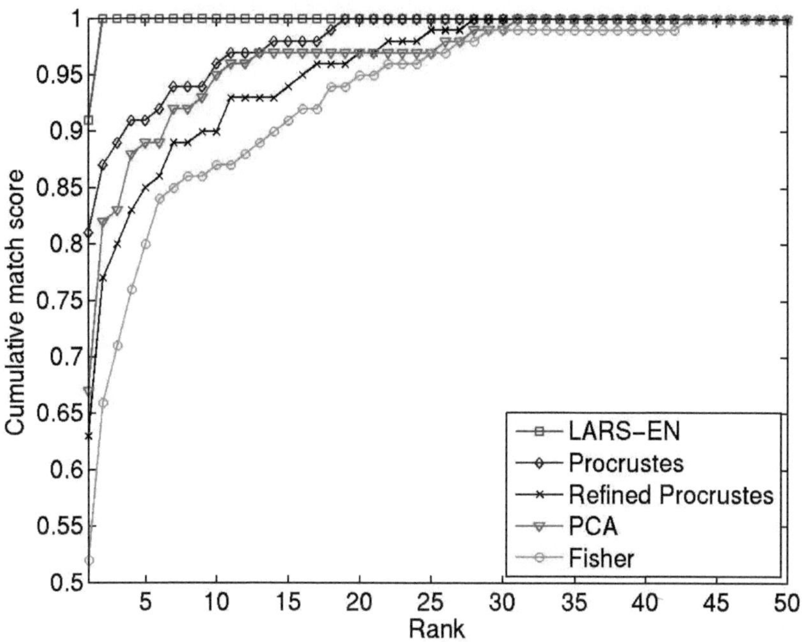

Fig. 2. Identification performance of the models based solely on landmarks

overfitting was observed. Figure 2 illustrates a rank plot of the performances of the landmark models. We see a very good performance for LARS-EN as all persons were identified correctly using the top two matches.

Figure 3 illustrates which landmarks are selected for four of the individual models. Observe how the selected landmarks depend on the facial characteristics of each person.

4.2 Color Face Recognition

In order to obtain a one to one correspondence of pixels between the images the faces were aligned with warping. The same 4-2-2 validation strategy as before was applied and the Eigenfaces, Fisherfaces, and LARS-EN methods were compared. Table 2 summarizes the results.

Table 2. Summary of the classification rates for the models based solely on the color information

Method/Classification rate	Training	Validation	Test
Eigenfaces	1	0.87	0.85
Fisherfaces	1	0.96	0.94
LARS-EN	1	0.97	0.92

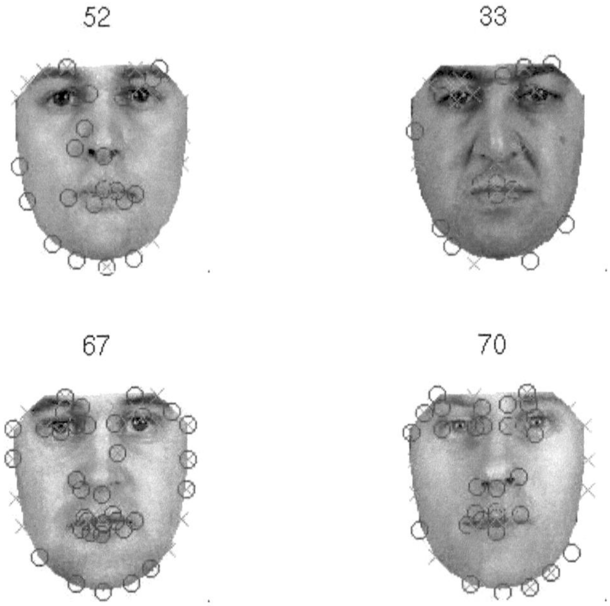

Fig. 3. Illustration of four persons and the selected landmarks in the individual LARS-EN models. x-coordinates are marked with crosses, and y-coordinates are marked with circles. From left to right the person are: No. 1, no. 13, no. 36, and no. 44.

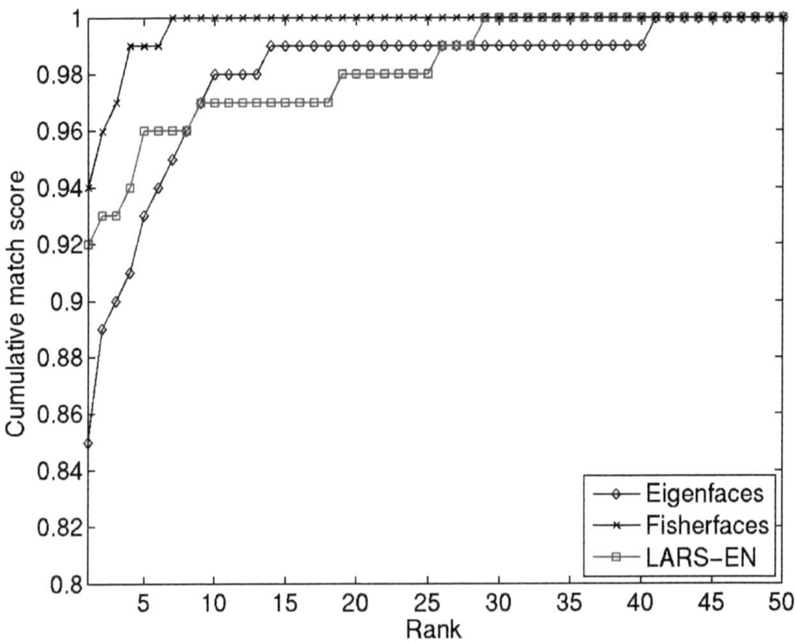

Fig. 4. Identification performance of the models based solely on color information

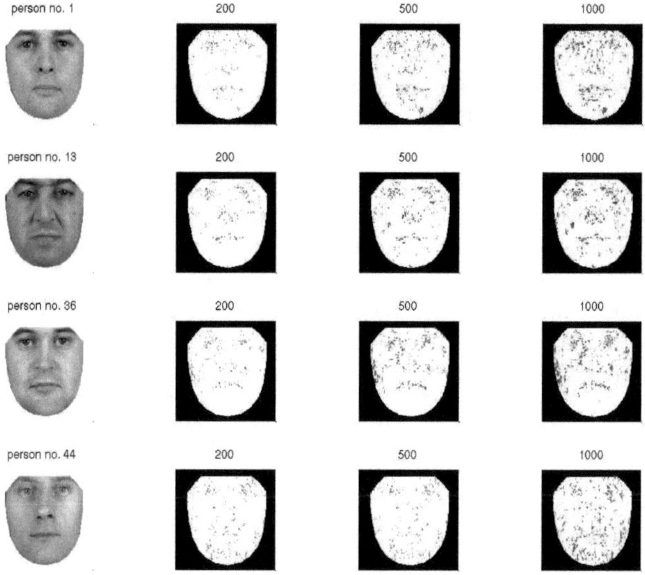

Fig. 5. Illustration of four persons with the first 200, 500, and 1000 selected pixels marked

Based on color information we observed only slightly higher classification rates than those for LARS-EN based on geometrical information. Furthermore, we observed some over fitting of the test and validation set for the LARS-EN method, which was also reflected in the mean square error of the regression analysis. The LARS-EN models included around 2000 features (pixels). Figure 4 illustrates the performance of the color based methods. The performance of Fisherfaces was slightly better than for the other two methods which were comparable in performance. Notably, the performance of Fisherfaces was not better than that for LARS-EN based on geometrical information.

Similar to what was done for the geometric features we now examine which features were selected in experiment two. Figure 5 shows the selected color pixels on four different persons. The selected pixels are to a high degree situated around the eyebrows, the eyes, the nose, and the mouth, but also on e.g. the cheeks and the chin. Furthermore, the features are individual from person to person. Observe e.g. the different selection of pixel features on and around the noses of the individuals.

5 Discussion and Conclusion

The LARS-EN method performed better than the reference methods (Procrustes, refined Procrustes, PCA, and PCA+Fisher) when based solely on information from landmarks. LARS-EN identified all persons in the top two matches.

Based on color information the LARS-EN models obtained slightly better classification rates than the geometrically based models. However, the rank identification performance was poorer. Here, Fisherfaces performed better.

Additionally, we identified important features via the feature selection. For the landmarks, only 55 features were needed on average for the individual models. The color models were based on around 2000 features which were situated around the eyes, the nose, the mouth, and the eyebrows, but also on the cheeks and the chin. Furthermore, the selected features differ from individual to individual.

Consequently, our results show that a limited number of geometrical features can suffice for face recognition, and emphasize that geometrical information should not be disregarded. There are several other possibilities of feature extraction from geometrical information of faces, such as ratios and angles between landmarks, which would be interesting to explore. The LARS-EN algorithm is a good tool for exploring new feature spaces and finding the more interesting ones.

In future work, it is furthermore of interest to examine the methods for a larger database.

Acknowledgements

The authors would like to thank Karl Sjöstrand who has implemented the LARS and LARS-EN methods in Matlab. The implementations are available at his homepage[2].

References

1. Daugman, J.: How iris recognition works. In: Proceedings of 2002 International Conf. on Image Processing,vol. 1 (2002)
2. Daugman, J.: High confidence visual recognition of persons by a test of statistical independence. IEEE Transactions on Pattern Analysis and Machine Intelligence 15(11), 1148–1161 (1993)
3. Zou, H., Hastie, T.: Regularization and variable selection via the elastic net. J. R. Statist. Soc. B. 67(2), 301–320 (2005)
4. Goldstein, A.J., Harmon, L.D.B.A.: Lesk and identification of human faces. Proc. IEEE 59(5), 748–760 (1971)
5. Craw, I., Kato, a.T., C, N., Akamatsu, S.: How should we represent faces for automatic recognition. IEEE Trans. Pattern Anal. Mach. Intell. 21(8), 725–736 (1999)
6. Dryden, I., Mardia, K.: Statistical Shape Analysis. Wiley series in probability and statistics (1998)
7. Shi, J., Samal, A., Marx, D.: How effective are landmarks and their geometry for face recognition? Computer Vision and Image Understanding 102, 117–133 (2006)
8. Turk, M., Pentland, A.: Face recognition using eigenfaces. IEEE Conf. Computer Vision and Pattern Recognition (1991)

---

2 www.imm.dtu.dk/~kas

9. Belhumeur, P., Hespanha, J., Kriegman, D.: Eigenfaces vs. fisherfaces: Recognition using class specific linear projection. IEEE Trans. Pattern Analysis and Machine Intelligence 19(7), 711–720 (1997)
10. Cevikalp, H., Neamtu, M., Wilkes, M., Barkana, A.: Discriminative common vectors for face recognition. IEEE Transactions on Pattern Analysis and Machine Intelligence 27(1), 4–13 (2005)
11. Viola, P., Jones, M.: Robust real-time object detection. In: Proc. of IEEE Workshop on Statistical and Computational Theories of Vision (2001)
12. Tibshirani, R.: Regression shrinkage and selection via the lasso. J. R. Statist. Soc. B. 58(1), 267–288 (1996)
13. Hoerl, A.E., Kennard, R.W.: Ridge regression: Biased estimation for nonorthogonal problems. Technometrics 12, 55–67 (1970)
14. Efron, B., Hastie, T., Johnstore, I., Tibshirani, R.: Least angle regression. Ann. Statist. 32, 407–499 (2004)
15. Messer, K., Kittler, J.M.J., Luettin, J., Maitre, G.: Xm2vtsbd: The extended m2vts database. In: Proceedings 2nd Conference on Audio and Video-base Biometric Personal Verification (AVBPA99) (1999)
16. Philip, P.J., Moon, H., Rizvi, S.A., Rauss, P.J.: The feret evaluation methodology for face-recognition algorithms. IEEE Trans. on Pattern Analysis and Machine Intelligence 22(10), 1090–1104 (2000)

Occluded Facial Expression Tracking

Hugo Mercier[1], Julien Peyras[2], and Patrice Dalle[1]

[1] Institut de Recherche en Informatique de Toulouse
118, route de Narbonne, F-31062 Toulouse Cedex 9
[2] Dipartimento di Scienze dell'Informazione
via Comelico 39/41, I-20135 Milano
{mercier,peyras,dalle}@irit.fr

Abstract. The work presented here takes place in the field of computer aided analysis of facial expressions displayed in sign language videos. We use Active Appearance Models to model a face and its variations of shape and texture caused by expressions. The *inverse compositional* algorithm is used to accurately fit an AAM to the face seen on each video frame. In the context of sign language communication, the signer's face is frequently occluded, mainly by hands. A facial expression tracker has then to be robust to occlusions. We propose to rely on a robust variant of the AAM fitting algorithm to explicitly model the noise introduced by occlusions. Our main contribution is the automatic detection of hand occlusions. The idea is to model the behavior of the fitting algorithm on unoccluded faces, by means of residual image statistics, and to detect occlusions as being what is not explained by this model. We use residual parameters with respect to the fitting iteration *i.e.*, the AAM distance to the solution, which greatly improves occlusion detection compared to the use of fixed parameters. We also propose a robust tracking strategy used when occlusions are too important on a video frame, to ensure a good initialization for the next frame.

Keywords: Active Appearance Model; occlusion; facial expression; tracking; inverse compositional.

1 Introduction

We use a formalism called Active Appearance Models (AAM – [1,2]) to model a face and its variations caused by expressions, in term of deformations of a set of vertex points of a shape model. These points can be tracked with a good accuracy along a video when the face is not occluded and when it has been learned beforehand.

We focus here on the analysis of sign language videos. In sign language, facial expressions play an important role and numerous signs are displayed near the signer's face. Furthermore, the signer's skull frequently performs out-of-plane rotations.This implies, from the interlocutor's point of view (here replaced by the video acquiring system) that face might often be viewed only partially.

Past works mainly focused on robust variants of AAM fitting algorithms ([3], [4]) able to consider outlier data. We follow here the approach developed in [5],

B.K. Ersbøll and K.S. Pedersen (Eds.): SCIA 2007, LNCS 4522, pp. 72–81, 2007.

where parametric models of residual image are used in order to automatically detect the *localization* of occlusions. The main idea is here to learn various parameters computed from various fitting contexts and to select one in particular at each iteration, which greatly improves occlusion detection compared to the use of only one fixed parameter in earlier work,

In section 2 are presented Active Appearance Models and the way they are used to extract facial deformations of a face with an accurate optimization algorithm that can take occlusions into account by means of a pixel confidence map. In section 3 we show, through experiments, how to optimally compute the pixel confidence map to detect occlusions. Section 4 describes a robust tracking strategy that we use to track facial deformations along a video sequence.

2 Active Appearance Models

An Active Appearance Model (AAM) describes an object of a predefined class as a shape and a texture. Each object, for a given class, can be represented by its shape, namely a set of 2D coordinates of a fixed number of interest points, and a texture, namely the set of pixels lying in the convex hull of the shape.

The shape can be described by:

$$\mathbf{s} = \mathbf{s}_0 + \sum_{i=1}^{n} \mathbf{p}_i^s \mathbf{s}_i \tag{1}$$

where $\mathbf{s}_0$ is the mean shape, $\mathbf{s}_i$ are deformation vectors and $\mathbf{p}_i^s$ are weighting coefficients of these deformations. It can be written in matrix notation by $\mathbf{s} = \mathbf{s}_0 + \mathbf{S}\mathbf{p}^s$.

The texture is described by:

$$\mathbf{t} = \mathbf{t}_0 + \sum_{i=1}^{m} \mathbf{p}_i^t \mathbf{t}_i \tag{2}$$

or, in matrix notation $\mathbf{t} = \mathbf{t}_0 + \mathbf{T}\mathbf{p}^t$

The model is built upon a training set of faces, where a shape *i.e.*, 2D coordinates of a fixed set of interest points, is associated to each image. All the shapes are extracted from the training set and global geometric deformations are differentiated from facial deformations by a Procrustes analysis. It results a mean shape $\mathbf{s}_0$ and shapes that differ from the mean only by internal deformations.

Pixels that lie inside the shape of each face is then extracted and piecewise-affine-warped to the mean shape $\mathbf{s}_0$ to build the (shape-free) texture associated to a face.

Principal Component Analysis is applied both to aligned shapes and aligned textures and the eigen-vectors form the matrices $\mathbf{S}$ and $\mathbf{T}$. In our case, we retain enough eigen-vectors to explain 95% of the shape and texture variance (corresponding to 12 shape deformation vectors and 15 texture variation vectors).

A face close to the training set can then be represented by a vector of shape parameters $\mathbf{p}^s$ and a vector of texture parameters $\mathbf{p}^t$.

2.1 Weighted Inverse Compositional Algorithm

The goal of the AAM fitting algorithm is to find $\mathbf{p}^s$ and $\mathbf{p}^t$ that best describes the face seen on an input image. The shape and texture parameters are optimized by means of a residual image that represents differences between a current face estimation and the face seen on the input image I:

$$E(\mathbf{x}) = \mathbf{t}_0(\mathbf{x}) + \sum_{i=1}^{m} \mathbf{p}_i^t \mathbf{t}_i(\mathbf{x}) - I(W(\mathbf{x};\mathbf{p}^s)), \forall \mathbf{x} \in \mathbf{s}_0 \tag{3}$$

$I(W(\mathbf{x},\mathbf{p}^s))$ is the projection of the input image onto the mean shape $\mathbf{s}_0$, obtained by a piecewise affine warp. Instead of the Euclidean norm classically used in optimization, we can use a weighted distance:

$$\sum_{\mathbf{x}} Q(\mathbf{x}) E(\mathbf{x})^2$$

where $Q(\mathbf{x})$ weights the influence of the pixel $\mathbf{x}$.

We use the optimization scheme presented in [2], called the *inverse compositional* algorithm, which is efficient and accurate. Its main advantage is the fact that the jacobian matrix can be analytically derived, rather than learned by numerical differentiation (like in [1]).

Among all the variants proposed by the authors, we choose the *simultaneous inverse compositional* algorithm with a weighted distance. The *simultaneous* is a variant that can optimize both shape and texture parameters in an accurate manner. This is not the most efficient variant of the *inverse compositional* algorithms that can deal with texture variations (see for instance the *project-out* algorithm in [2]), but the most accurate.

Iterative update is given by (computation details can be found in [3] and [6]):

$$[\Delta\mathbf{p}^s, \Delta\mathbf{p}^t] = -H_Q^{-1} \sum_{\mathbf{x}} Q(\mathbf{x}) \left[G_s(\mathbf{x}), G_t(\mathbf{x})\right] E(\mathbf{x}) \tag{4}$$

with

$$\begin{aligned}
G_s(\mathbf{x}) &= \left[\left(\nabla\mathbf{t}_0(\mathbf{x}) + \textstyle\sum_{i=1}^{m} \mathbf{p}_i^t \nabla\mathbf{t}_i(\mathbf{x})\right) \frac{\partial W}{\partial \mathbf{p}_1^s}, \ldots, \left(\nabla\mathbf{t}_0(\mathbf{x}) + \textstyle\sum_{i=1}^{m} \mathbf{p}_i^t \nabla\mathbf{t}_i(\mathbf{x})\right) \frac{\partial W}{\partial \mathbf{p}_n^s}\right] \\
G_t(\mathbf{x}) &= \left[\mathbf{t}_1(\mathbf{x}), \ldots, \mathbf{t}_m(\mathbf{x})\right] \\
H_Q &= \textstyle\sum_{\mathbf{x}} Q(\mathbf{x}) \left[G_s(\mathbf{x}), G_t(\mathbf{x})\right]^T \left[G_s(\mathbf{x}), G_t(\mathbf{x})\right]
\end{aligned}$$

Shape parameters are then updated by inversion and composition:

$$W(\mathbf{x};\mathbf{p}^s) \leftarrow W(\mathbf{x};\mathbf{p}^s) \circ W(\mathbf{x};\Delta\mathbf{p}^s)^{-1}$$

And texture parameters are updated in an additive way by $\mathbf{p}^t \leftarrow \mathbf{p}^t + \Delta\mathbf{p}^t$. Essential steps of the algorithm are summarized on Fig. 1.

This algorithm performs accurately. For our experiments, we use what is called a person-specific AAM, meaning that the training set is composed by expressions of only one person. A more generic AAM would be less accurate and hard to control.

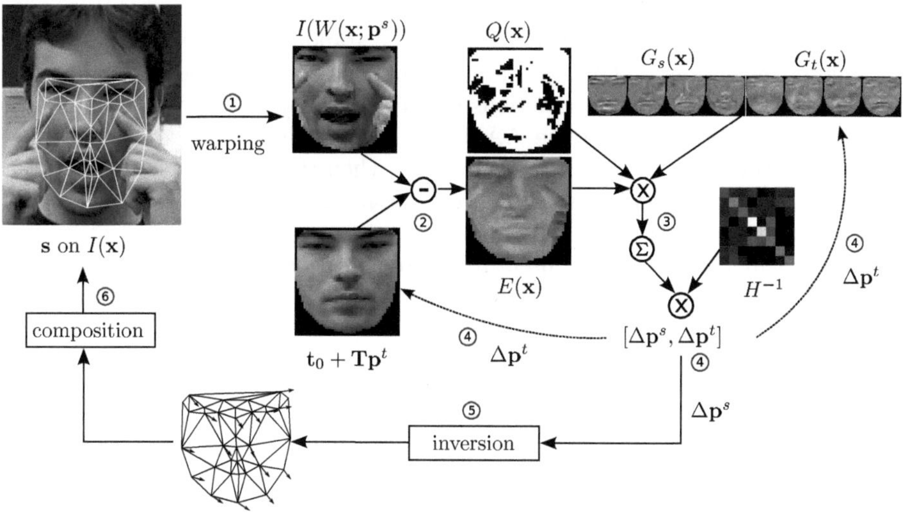

Fig. 1. Essential steps of the weighted simultaneous inverse compositional algorithm. Numbers give chronology of the steps for one iteration.

3 Occlusion Detection

The confidence map $Q(\mathbf{x})$ used in the weighted variant of the AAM fitting algorithm has to be as close as possible to the real occlusion map.

Our problem is to compute the best confidence map without knowledge on the localization of real occlusions. We propose here to model the behavior of the residual image in the unoccluded case and to detect occlusions as being what is not explained by the model, following the approach presented in [5].

3.1 Parametric Models of Residuals

We rely on parametric models of the residual image. We propose to test different confidence map computation functions:

$$Q_1(\mathbf{x}) = \begin{cases} 1 \text{ if } min(\mathbf{x}) \leq E(\mathbf{x}) \leq max(\mathbf{x}) \\ 0 \text{ else} \end{cases}$$

$$Q_2(\mathbf{x}) = \frac{1}{\sigma(\mathbf{x})\sqrt{2\pi}} e^{\left(-\frac{E(\mathbf{x})^2}{2\sigma(\mathbf{x})^2}\right)}$$

$$Q_3(\mathbf{x}) = \begin{cases} 1 \text{ if } |E(\mathbf{x})| \leq 3\sigma(\mathbf{x}) \\ 0 \text{ else} \end{cases}$$

$$Q_4(\mathbf{x}) = \begin{cases} 1 \text{ if } |E(\mathbf{x})| \leq 4\sigma(\mathbf{x}) \\ 0 \text{ else} \end{cases}$$

$$Q_5(\mathbf{x}) = e^{\left(-\frac{E(\mathbf{x})^2}{2\sigma(\mathbf{x})^2}\right)}$$

where $min(\mathbf{x})$ is the minimum value of the pixel $\mathbf{x}$ over all the residual images, $max(\mathbf{x})$ is the maximum value and $\sigma(\mathbf{x})$ is the standard deviation. One of each parameter (min, max and σ) are computed for each pixel $\mathbf{x}$ of the residual image.

The parameters of the Q_i functions could be learned from a random amount of residuals generated when the AAM fitting algorithm is run on unoccluded images. However, a residual image generated when the shape model is far from the solution is totally different from a residual image generated when the model is close to the solution.

That is why the parameters used in the computation of the Q_i functions have to depend on the distance to the solution: they have to be high (resulting in a permissive toward errors Q_i function) when the model is far from the solution and low when it gets closer (resulting in a strict function).

3.2 Partitioned Training Sets

To explicit the link between the model parameters and the distance to the solution, we conducted the following experiment.

A set of residual images are generated: the (non-weighted) AAM fitting algorithm is launched from perturbed ground truth shapes 15 iterations until convergence. To initialize the AAM, each vertex coordinate is perturbed by a Gaussian noise with 10 different variances (between 5 and 30), and the $\mathbf{p}^s$ parameters are obtained by projecting the perturbed shape model onto the shape basis $\mathbf{S}$. It is launched 4 times on 25 images that belong to the AAM training set. The distance to the solution, computed by the average Euclidean distance of the shape model vertices to the optimal ground truth shape vertices, and the residual image are stored at each iteration.

Instead of computing the model parameters ($min(\mathbf{x})$, $max(\mathbf{x})$ and $\sigma(\mathbf{x})$) on all the residual images, we form 15 partitions by regrouping residual images according to their corresponding distance to the solution. Each partition P_i contains 210 residual images and can be characterized by its minimum d_i^- and maximum distance d_i^+ to the solution. The model parameters are then learned, for each pixel $\mathbf{x}$, on all the residuals of each partition.

On figure 2 are represented standard deviations $\sigma(\mathbf{x})$ learned on each partition. For visualization purpose, only the average standard deviation σ, computed over all the pixels $\mathbf{x}$ is plotted.

3.3 Model Parameter Approximation

When the fitting algorithm is run on test face images, that might be occluded, the model distance to the solution is difficult to estimate. In the unoccluded case, a rough estimate of the distance to the solution can be extracted from the residual image. Such an information is not reliable anymore in the occluded case, because the residual image reflects either errors caused by a misplacement of the model or errors caused by occlusions.

However, we assume that we can rely on the iteration number of the fitting algorithm to select the appropriate partition, especially if the model distance to

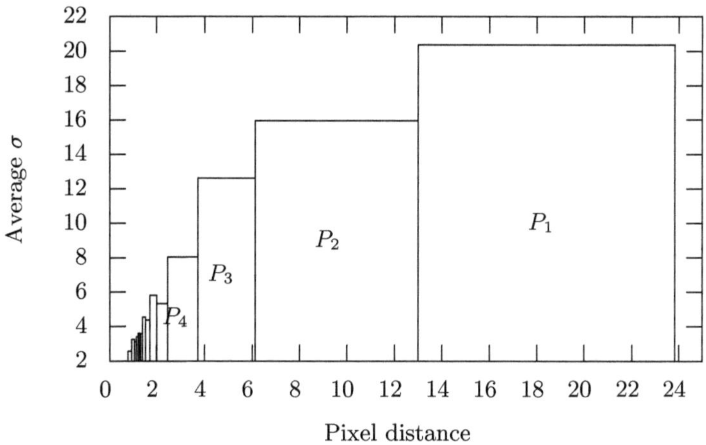

Fig. 2. Average standard deviation learned for each partition

the solution in the occluded case is lower than the maximum distance used to collect residuals in the first partition.

To validate this assumption, we proceed to the test that follows. Using variances computed for each of the 15 partitions, we test the weighted fitting algorithm launched for 20 iterations from Gaussian perturbed optimal positions (with a variance of 20) on occluded (known) images (25% of the input image is covered with 8×8 blocks of pixels of random intensity). Note that the amount of shape perturbations is less important than the amount used in the partition construction. Among all the $Q_i(\mathbf{x})$ functions, we use only $Q_3(\mathbf{x})$ to compute the confidence map at each iteration, for we are only interested in how to select its parameter, not how it performs. Different ways of getting the variance at each iteration are tested:

- S_{real}: selection from P_i where the real distance to the solution d_{model} is bounded by the distance range of P_i: $[d_i^-, d_i^+]$; for comparison purpose;
- S_{it}: selection from P_i where i is the current iteration (and $i = 15$ for iterations 15 to 20);
- S_f: selection from P_1;
- S_m: selection from P_7;
- S_l: selection from P_{15}.

The results on Fig. 3 show clearly that the best choice for the residual model parameter computation is S_{real}. It is not usable in practice (the ground truth shape is not *a priori* known), but we can rely on the S_{it} approximation. As a comparison, results are given for the unoccluded case and for fixed variances (S_f, S_m and S_l).

In [5], variances are fixed and computed on residual images obtained from the converged AAM, which corresponds here to the S_l selection strategy. When observing the mean distance obtained after 20 iterations, the proposed S_{it} variance selection strategy results in a distance divided by about 2 compared to S_l.

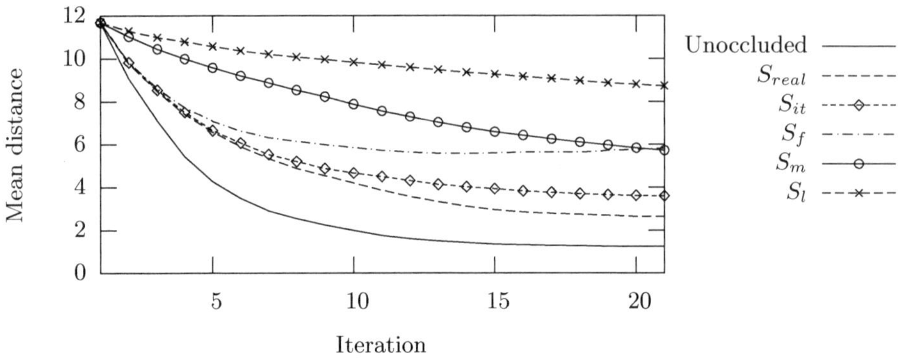

Fig. 3. Average behavior of the fitting algorithm for the reference unoccluded case, and for the occluded case with different computations of the variance

3.4 Choice of the Parametric Model

With the previous result we can then test what is the best way to compute the confidence map at each iteration.

For that purpose, we proceed to the following experiment: the weighted AAM fitting algorithm is launched on images of known faces, covered with a varying amount of occlusions, from a Gaussian perturbed shape (we use a variance of 20 for each vertex coordinate). We test each of the Q_i confidence map computation functions with a parameter choosed using S_{it}.

The convergence frequency is determined by computing the number of fittings that result in a shape with a distance to the ground truth lower than 2 pixels.

Results are summarized on Fig. 4. The Q_4 function clearly shows the best results. All the other functions perform worse, except for the function Q_1 that seems to be a good detector in the case of low occlusion rate and a very bad one in the case of high occlusion rate. Q_1 relies on computation of minimum and maximum value, which are very robust measures compared to variance, that is why the behavior of Q_1 is not always reliable.

4 Robust Tracking Strategy

The goal of the tracking algorithm is to take occlusion into consideration as much as possible. However, on some video frame, occlusions are too important to expect good fitting results, because too many pixels are considered unreliable. In such a case, the fitting algorithm is prone to divergence and the resulting shape configuration could be a bad initialization if used directly in the following frame.

That is why we propose to rely on a measure of divergence and on a rigid AAM to initialize the model.

The goal is to avoid bad configurations of the shape model in order not to perturb the fitting process on subsequent frames. We detect such bad configurations by detecting shapes that are not statistically well explained. For that

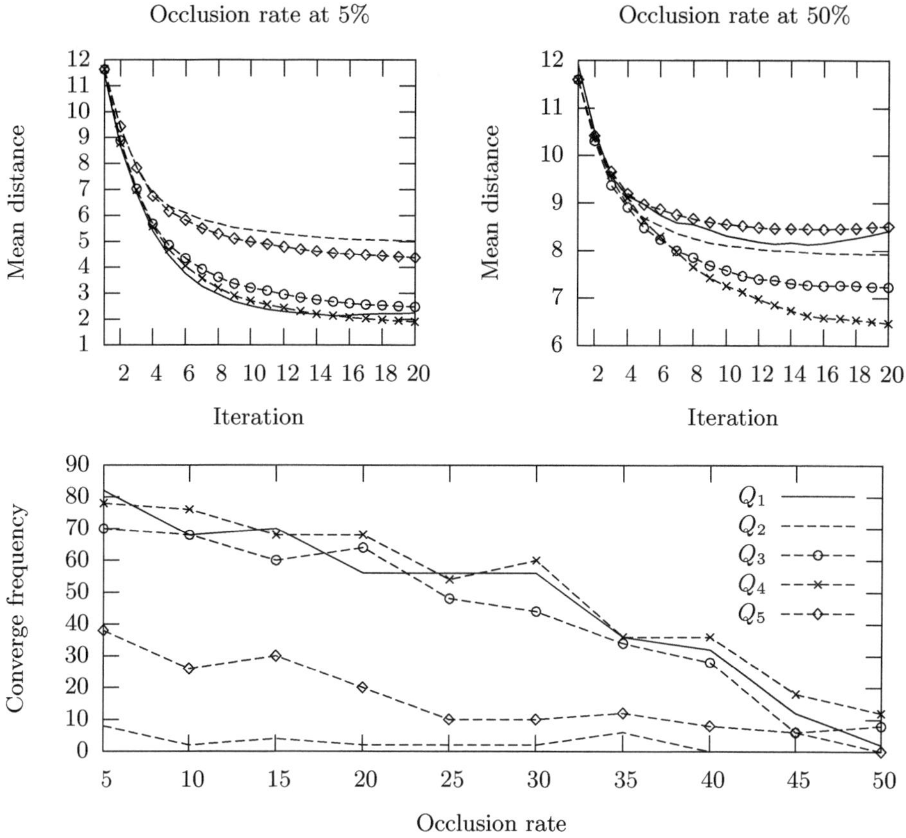

Fig. 4. Characterization of the confidence map computations. Average distance to the solution across iterations for 5% and 50% of occlusions (top curves) and convergence frequency (bottom curve).

purpose, we compare the shape parameters $\mathbf{p}^s$ to their standard deviations σ_i, previously learned from the shape training set. The divergence is decided, if:

$$\frac{1}{n}\sum_{i=1}^{n}\frac{|\mathbf{p}_i^s|}{\sigma_i} > \rho_1 \text{ or } \max_{i=1,\dots,n}\left\{\frac{|\mathbf{p}_i^s|}{\sigma_i}\right\} > \rho_2$$

The thresholds ρ_1 and ρ_2 are determined empirically and can be high (here we choose $\rho_1 = 2.5$ and $\rho_2 = 7.0$). The divergence is only tested after ten iterations, for the model deformations that occur during the first iterations can lead to convergence.

On a frame, if convergence is detected, the final shape configuration is stored and serves as an initialization for the next frame.

If divergence is detected, we rely for the following frame on a very robust tracker: an AAM build by retaining only the geometric deformation vectors. It is represented by the mean shape that can only vary in scale, in-plane rotation

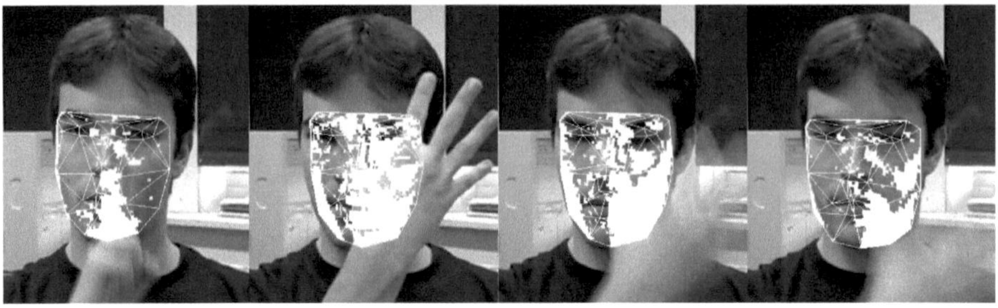

(a) (b) (c) (d)

Fig. 5. Video tracking results. (a) Example of a good occlusion detection. (b) Example of a divergence. Divergence on a frame (c) and convergence on the next frame (d).

and position but not in facial deformations. Such a model gives a rough estimate of the face configuration that can be used as an initialization for the non-rigid AAM. It avoids the non-rigid shape model to being attracted by local minima. The rigid AAM fitting algorithm uses also a confidence map to take occlusions in consideration. However, the confidence maps computed for the non-rigid AAM are too strict for the rigid AAM, we thus use a coarse occlusion detector (for example, the confidence map computed over the second partition for the non-rigid AAM).

The rigid AAM fitting is launched for 5 iterations from the last configuration that converged. The non-rigid AAM fitting algorithm is then launched from the resulting position.

We test this tracking algorithm on a video sequence of about 500 frames where signs frequently occlude the signer's face.

We show some typical results on selected frames (see figure 5). Blocks of white pixels represent areas of occlusions detected by our method. Compared to a naive tracker, the AAM always converges to an accurate configuration on unoccluded frames that occur after an occluded one.

5 Conclusion

We have presented a way to track facial deformations that occur on a video, taking into account hand occlusions by means of an Active Appearance Model of a face, a robust optimization scheme that down-weights pixel contributions in the presence of occlusions, an optimal way to compute the pixel confidence map and a robust tracking strategy based on a measure of divergence and a rigid AAM.

The pixel confidence map is computed based on a model of residual images. We use one model per iteration of the fitting algorithm, rather than one fixed model. This is clearly a better choice that improves occlusion detection, compared to earlier work.

Concerning the tracking test, experiments on convergence frequency of the algorithm with respect to the occlusion rate have still to be conducted.

The video sequence used to test the tracking algorithm contains only weak out-of-plane rotations. This is why the rigid 2D AAM can give a good initialization configuration for the non-rigid AAM fitting algorithm. On realistic sign language videos however, out-of-plane rotations may be important and we would have to rely on a rigid AAM that can take 3D pose into consideration.

We use the most accurate and most time-consuming robust variant of the *inverse compositional* algorithm. We have to investigate if approximations presented in [3], [6] or [5] could be applied to obtain an accurate and efficient facial deformation tracker.

References

1. Cootes, T.F., Edwards, G.J., Taylor, C.J.: Active appearance models. IEEE Transactions on Pattern Analysis and Machine Intelligence 23, 681–685 (2001)
2. Matthews, I., Baker, S.: Active appearance models revisited. ternational Journal of Computer Vision 60(2), 135–164 (2004)
3. Baker, S., Gross, R., Matthews, I., Ishikawa, T.: Lucas-Kanade 20 years on: A unifying framework: Part 2. Technical Report CMU-RI-TR-03-01, Robotics Institute, Carnegie Mellon University, Pittsburgh, PA (February 2003)
4. Gross, R., Matthews, I., Baker, S.: Active appearance models with occlusion. Image and Vision Computing 24(6), 593–604 (2006)
5. Theobald, B.J., Matthews, I., Baker, S.: Evaluating error functions for robust active appearance models. In: Proceedings of the International Conference on Automatic Face and Gesture Recognition,pp. 149 – 154 (April 2006)
6. Baker, S., Gross, R., Matthews, I.: Lucas-Kanade 20 years on: A unifying framework: Part 3. Technical Report CMU-RI-TR-03-35, Robotics Institute, Carnegie Mellon University, Pittsburgh, PA (November 2003)

Model Based Cardiac Motion Tracking Using Velocity Encoded Magnetic Resonance Imaging*

Erik Bergvall[1,2], Erik Hedström[2], Håkan Arheden[2], and Gunnar Sparr[1]

[1] Centre for Mathematical Sciences, Lund Institute of Technology, Lund, Sweden
[2] Department of Clinical Physiology, Lund University Hospital, Lund, Sweden

Abstract. This paper deals with model based regularization of velocity encoded cardiac magnetic resonance images (MRI). We extend upon an existing spatiotemporal model of cardiac kinematics by considering data certainty and regularity of the model in order to improve its performance. The method was evaluated using a computer simulated phantom and using in vivo gridtag MRI as gold standard. We show, both quantitatively and qualitatively, that our modified model performs better than the original one.

1 Introduction

Cardiovascular disease is the main cause of death in the western world with left ventricular infarction as the predominant contributor to this phenomenon. The use of non-invasive and non-ionizing imaging techniques, such as echocardiography and MRI have aided in the diagnosis through the ability to directly visualize cardiac structure and function.

Quantitative assessment of regional myocardial function is a challenging but important task as subjective assessment of regional wall motion may suffer from poor inter-observer agreement [1]. Several approaches have been developed for the quantitation of regional myocardial function using MRI. Saturation grid-tagging allows direct evaluation of myocardial deformation of the heart [2],[3], [4],[5] but is limited by relatively low spatial resolution and tag fading late in diastole [6],[7]. Further, specialized software for identification of tag lines is needed [8].

Another approach utilizes the velocity information present in phase contrast sequences providing velocity fields with high spatial and temporal resolution [9]. This velocity data may be used to directly calculate myocardial strain rate [10],[11],[12],[13] or can be integrated with respect to time, giving the motion of the myocardium. From this basal kinematic descriptor, a number of interesting mechanical properties, such as Lagrangian strain, may be derived, which may provide understanding of cardiac mechanics and diagnosis of disease.

The measured velocity field is subject to noise, imaging artifacts and degradation by sampling in time and space. The 'forward–backward' integration [14],[15]

* This work was supported by the Swedish Heart-Lung foundation and the Region of Scania.

B.K. Ersbøll and K.S. Pedersen (Eds.): SCIA 2007, LNCS 4522, pp. 82–91, 2007.

ensures a periodic motion but does not model any spatial coherence of the myocardium. Other approaches have used a deformable mesh guided by a Kalman filter, or combined with Fourier analysis to obtain both periodic motion and spatial smoothness [16], [17], [18],[19].

The purpose of this work is to develop a model based regularization strategy for velocity data measured using phase contrast MRI in order to be able to measure cardiac deformation and strain in a quantitative way. Zhu proposed the use of a cyclic spatiotemporal finite element model [19]. The elements in the model were constructed by piecewise linear functions and harmonics with varying frequency to ensure a periodic motion. The parameters in the model were determined by an iterative scheme consisting of updating the mesh configuration and projection of the sampled velocity onto the elements.

We extended upon this spatiotemporal model in a number of ways. Zhu et al used a moving mesh to define the spatial elements. These elements needed to be redefined and the stiffness matrix was reconstructed in every iteration. We instead described the deformation in Lagrangian coordinates which results in a fixed mesh and also a fixed stiffness matrix. Secondly, the use of piecewise continuous elements may be inappropriate as the velocity field in a solid in motion must be not only be continuous but also differentiable [20]. We therefore investigated other elements. Further, the projection of measured velocities onto a model does not necessarily imply smoothness if no constraints are imposed on the parameters. The measured velocity field may be locally corrupted by e.g. partial volume averaging. We therefore estimated the certainty of a given measurement and included this information when determining the model parameters.

2 Methods

2.1 Image Acquisition

Velocity data was acquired in both in-plane directions in long-axis slices in human subjects for a total of 20 acquisitions. A 1.5 T Gyroscan Intera Scanner (Philips Medical Systems) was used for the acquisitions. Acquisition time varied between 30-90 s. Spatial and temporal resolution was 1.5 mm × 1.5 mm × 10 mm, with 32 time frames covering the whole cardiac cycle. Typical imaging parameters were repetition time (TR) = 24 ms, echo time (TE) = 5.3 ms, v_{enc} = 0.20 m/s, flip angle = 20°, matrix size = 256 × 192 pixels and field of view (FOV) = 400 mm × 300 mm. Saturation bands (30 mm thickness, 30 mm gap to image plane) superior and inferior to the imaging slice were applied to reduce signal from blood [21]. Retrospective gating was used when reconstructing the images. The boundary of the left ventricle was manually delineated in the first frame of each image sequence.

For validation purposes, saturation tagged images were acquired as a single breathhold sequence in the same imaging plane. Typical imaging parameters were: TR = 3.8 ms, TE = 1.8 ms, flip angle = 15°, saturation tag gap = 7 mm, matrix size = 256 × 192 pixels, FOV 400 mm × 300 mm, resulting in a temporal resolution of 64 ms.

2.2 Spatiotemporal Model

Let $\mathbf{x}$ be the coordinate vector in $\mathbf{R}^2$, and time $t = [0, T]$. We assume that we have acquired an image sequence $I(\mathbf{x}, t) : \mathbf{R}^2 \times (0, T) \to [0, 1]$ and a velocity field $\mathbf{v}(\mathbf{x}, t) : \mathbf{R}^2 \times (0, T) \to \mathbf{R}^2$. The measured image sequence and the velocity field will only be given at discrete points in spacetime but can be treated as functions defined on all of $\mathbf{R}^2$ by interpolation. We let Ω be the set of points, henceforth called particles, that occupies the left ventricle at time $t = 0$ and let $\mathbf{x}$ denote the Lagrangian coordinate vector. The set Ω is given by manual delineation of the left ventricle in the images.

We are interested in the motion $\phi(\mathbf{x}, t) : \Omega \to \mathbf{R}^2$ for all particles. It is given by the particle trace equation

$$\frac{d\phi(\mathbf{x}, t)}{dt} = \mathbf{v}(\phi(\mathbf{x}, t), t). \tag{1}$$

The right hand side $\mathbf{v}(\mathbf{x}, t)$ is given by measurements by velocity encoded MRI, and is subject to noise and artifacts. Therefore we do not attempt to solve (1) for individual particles. Following an approach similar to the one of Zhu *et al* [19], we construct a spatiotemporal model of the deformation of the left ventricle. We construct a vector $\mathbf{G}(\mathbf{x})$ of length N with spatial basis elements $g_i(\mathbf{x})$ to be defined shortly, and a vector $\mathbf{H}(t)$ of length K with temporal basis elements of the form $h_k(t) = \exp 2\pi j k / T$. The spatiotemporal model is constructed as

$$\phi(\mathbf{x}, t) = \mathbf{x} + (\mathbf{G}(\mathbf{x}) \otimes \mathbf{H}(t))\, \mathbf{c}, \tag{2}$$

where $\otimes$ denotes the Kronecker direct product and $\mathbf{c}$ is a coefficient matrix of size $NK \times 2$. By construction, $\phi(\mathbf{x}, t)$ will be periodic in t which is useful when describing cardiac motion. The Lagrangian velocity is given by

$$\frac{d\phi(\mathbf{x}, t)}{dt} = (\mathbf{G}(\mathbf{x}) \otimes (\mathbf{H}(t)\omega))\, \mathbf{c},$$

with ω a diagonal matrix with elements $2\pi j k / T$. The goal is to adapt this model to data by choosing the coefficient matrix $\mathbf{c}$ in an appropriate way.

If (1) holds, then

$$\mathcal{E} = \int_0^T \int_\Omega \left| \frac{d\phi(\mathbf{x}, t)}{dt} - \mathbf{v}(\phi(\mathbf{x}, t), t) \right|^2 d\mathbf{x} dt = 0,$$

so the particle trace equation can be reformulated as minimization of $\mathcal{E}$. If we consider $\mathbf{v}(\phi(\mathbf{x}, t), t)$ as a fixed function, not depending on $\mathbf{c}$, we can differentiate $\mathcal{E}$ with respect to $\mathbf{c}$ and equate to zero to obtain a linear system

$$\left(\int_0^T \int_\Omega (\mathbf{G} \otimes (\mathbf{H}\omega))^\top (G \otimes (\mathbf{H}\omega)) d\mathbf{x} dt \right) \mathbf{c} = \int_0^T \int_\Omega (\mathbf{G} \otimes (\mathbf{H}\omega))^\top \mathbf{v}(\phi(\mathbf{x}, t)) d\mathbf{x} dt. \tag{3}$$

For ease of notation we let the left hand side stiffness matrix be denoted by $\mathbf{K}$ and the right hand side by $\mathbf{b}$ so that (3) reads

$$\mathbf{Kc} = \mathbf{b}. \tag{4}$$

Now, minimization of $\mathcal{E}$ can be accomplished by iterative sampling of $\mathbf{v}(\mathbf{x}, t)$ at $\mathbf{x} = \phi(\mathbf{x}, t)$, and updating the model parameters by solving (4).

2.3 Choice of Spatial Elements

There is a large selection of basis elements to choose from. A common choice of elements are the piecewise linear interpolation functions, as used by Zhu *et al* [19]. This choice may be inappropriate as the resulting function is not differentiable which violates the property that a deformation of a body is required to be in C^2 with respect to both space and time [20].

Our use of the elements is quite different from the use in e.g. the finite element method (FEM) where the purpose is to approximate a function as well as possible. In FEM applications it is common to refine a solution by adding and/or reshaping elements. In this application, a refinement would potentially lead to a less regular solution as the increased number of degrees of freedom will make it easier for the model to adapt to noise and artifacts. A similar reasoning applies to elements with a small support.

Thin plate splines (TPS) are a class of widely used non-rigid mappings, and are often used in computer vision tasks such as image registration or warping. The TPS is given by [22],

$$g(\mathbf{x}) = |\mathbf{x}|^2 \ln \mathbf{x},$$

which is a C^2 function outside the origin. We construct the vector $\mathbf{G}$ in (2) as $\mathbf{G}(\mathbf{x})_i = g(\mathbf{x} - \mathbf{x}_i)$, where the points $\mathbf{x}_i$, $i = 1, \ldots, N$ are placed on the boundary of Ω. By this we obtain a model that generates a C^2 mapping that can be locally controlled by the coefficient matrix $\mathbf{c}$.

2.4 Extension of the Spatiotemporal Model

The right hand side in (4) is constructed using measurements. It would therefore be a good idea to estimate a certainty of $\mathbf{v}(\phi(\mathbf{x}, t))$. We formally construct it as $w(\mathbf{x}, t) : \mathbf{R}^2 \to [0, 1]$, and redefine $\mathcal{E}$ as a weighted functional

$$\mathcal{E} = \int_0^T \int_\Omega \left(\frac{d\phi(\mathbf{x}, t)}{dt} - \mathbf{v}(\phi(\mathbf{x}, t), t) \right)^2 w(\phi(\mathbf{x}, t), t) d\mathbf{x} dt = 0.$$

The explicit construction of $w(\mathbf{x}, t)$ will be discussed below. This will also lead to a linear system, which will be similar to (3). We construct the weight function by considering the local variation of the vector field around a given point $\phi(\mathbf{x}_k, t)$ and let

$$w(\phi(\mathbf{x}_k, t) = \exp\left(- \int_{B_r \cap \Omega} \left(\mathbf{v}(\phi(\mathbf{x}, t)) - \mathbf{v}(\phi(\mathbf{x}_k, t))\right)^2 d\mathbf{x}/\sigma^2 \right),$$

where $B_r(\mathbf{x}_k)$ is a neighborhood of $\mathbf{x}_k$ and σ is a tunable parameter. Here we note that the variation is defined for the measured velocities sampled along the estimated particle traces. As these velocities are supposed to be sampled within the myocardium, we do not expect them to have a large variation within a given neighborhood.

Even if we use a motion model, it does not necessarily generate what we, in some sense, mean by a regular deformation. If the coefficient matrix can be chosen arbitrarily it would be easy to generate deformations that are unreasonable from both a physiological and a physical point of view. To avoid such unwanted behavior we propose an additional regularity term in the energy functional. We construct it as

$$\mathcal{E}_{\text{regularity}} = \int_0^T \int_\Omega |\Delta\phi(\mathbf{x}, y)|^2 d\mathbf{x}dt,$$

where Δ is the spatial Laplace operator. This regularity term only affects the spatial part of the deformation. The temporal part can be regularized by e.g. using only few harmonics in the model. Further, the regularizing term does not penalize affine transformations and as a consequence does not penalize rigid transforms. This is an important property, as otherwise the regularity term would depend on our choice of coordinate system. The addition of this term will transform (4) to

$$\mathbf{Kc} + \lambda\mathbf{Lc} = \mathbf{b},$$

where $\mathbf{L}_{ij} = \int_\Omega \Delta g_i(\mathbf{x})\Delta g_j(\mathbf{x})d\mathbf{x}$ and $\lambda > 0$ is a tunable parameter that determines the influence of the regularity term.

2.5 Experimental Validation

In a first experiment we constructed a computer generated phantom based on a kinematic model of the left ventricle described by Arts *et al* [23]. This model was used to generate Lagrangian motion as a gold standard and also Eulerian velocities which was used as input data. The pixel dimensions and temporal resolution of the model–generated data was chosen to be similar to measured velocity data. The Eulerian velocities were corrupted by Gaussian noise with standard deviation ranging from 0 to 5 pixels/frame which should be compared to the peak velocities of about 2 pixels/frame. The velocities outside the deforming body were generated as zero mean Gaussian noise with a standard deviation of 5 pixels/frame.

In a second experiment we used the acquired gridtag images as gold standard. Gridtag sequences were analyzed manually and tagline intersections were tracked manually through all time frames and for all image series acquired.

We investigated several versions of the spatiotemporal model. As a reference we used the case with piecewise linear elements with no data certainty estimate or regularization parameter. This will be close to the method proposed by Zhu *et al* [19]. In the other experiments we used TPS elements and combinations of weighted fit and regularity term. The parameters used were $\sigma = 0.1$ pixels/frame,

$\lambda = 100$ and the size of the neighborhood B_r was 3 pixels. We typically use $N = 25$ spatial basis elements and 10 iterations in the algorithm.

The root mean square (RMS) was used to measure the error between the particle traces estimated by our proposed method and the Lagrangian motion given by the kinematic model in the first experiment and the manually constructed particle traces in the second experiment.

3 Results

Figure 1 shows the result of motion tracking using simulated data. The linear element model and the TPS model without adjustments perform in a similar way and the best model is the TPS model with weighted fit and a regularity term.A qualitative comparison of the case with piecewise linear elements and the case with TPS elements with regularity term are shown in Figure 2. The TPS based model generates a smoother deformation than the other model. It is possible to see drastic changes in the deformation in the linear elements case which are due to the tessellation of the domain. In absolute terms the methods perform in a

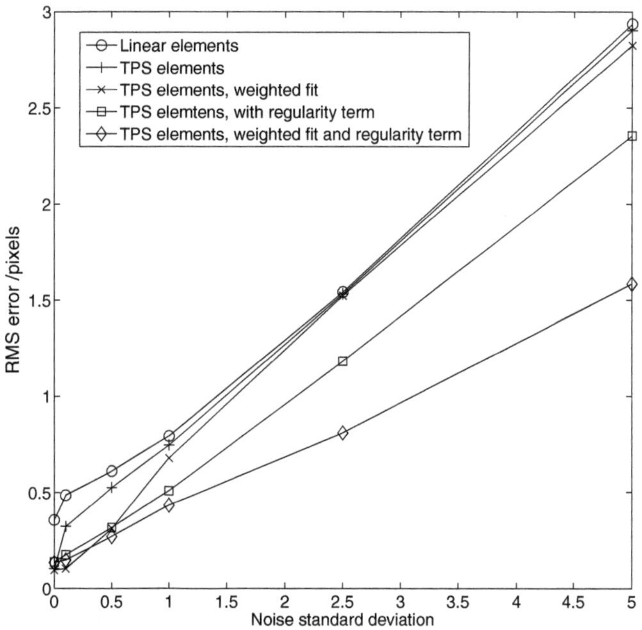

Fig. 1. The figure shows the motion tracking error for the simulated phantom for different levels of Gaussian noise. It can be seen that linear element model and the TPS model without adjustments are the worst performers. For higher levels of noise the weighting term flattens out and treats all noisy measurements in the same way which explains the appearance of the curve. Thus, it becomes essential that a regularity term is added when the noise level is high.

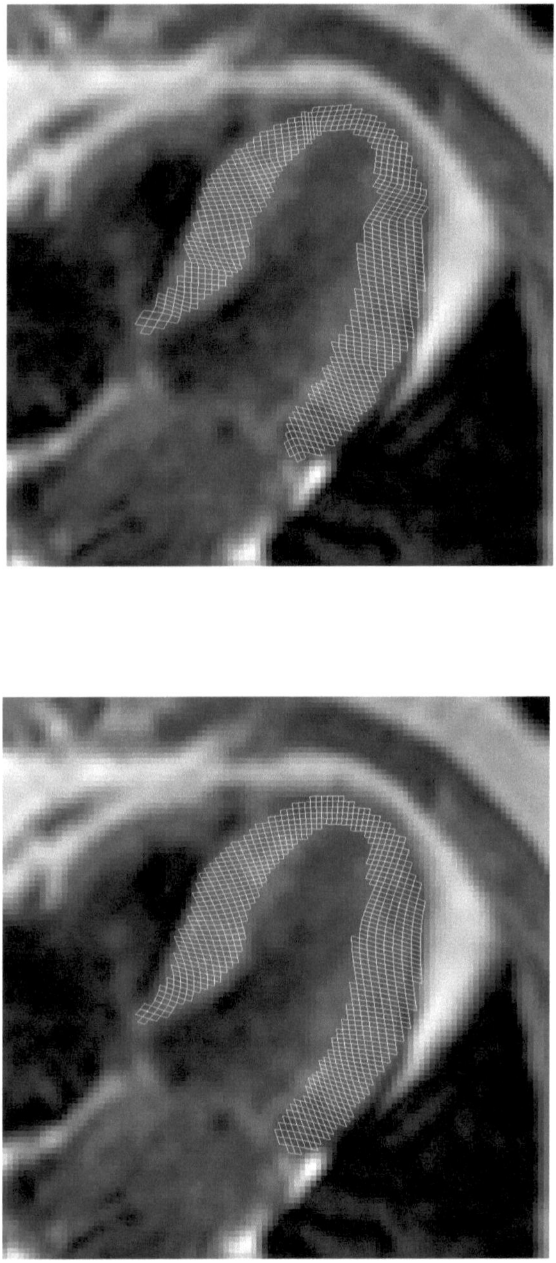

Fig. 2. A qualitative comparison of a spatiotemporal model with piecewise linear elements (top) and a model with TPS elements and additional regularity term (bottom). The deformation is shown at end systole (peak of contraction). The TPS based model generated a smoother deformation than the other model. It is possible to see drastic changes in the deformation in the top figure which are due to the tessellation of the domain.

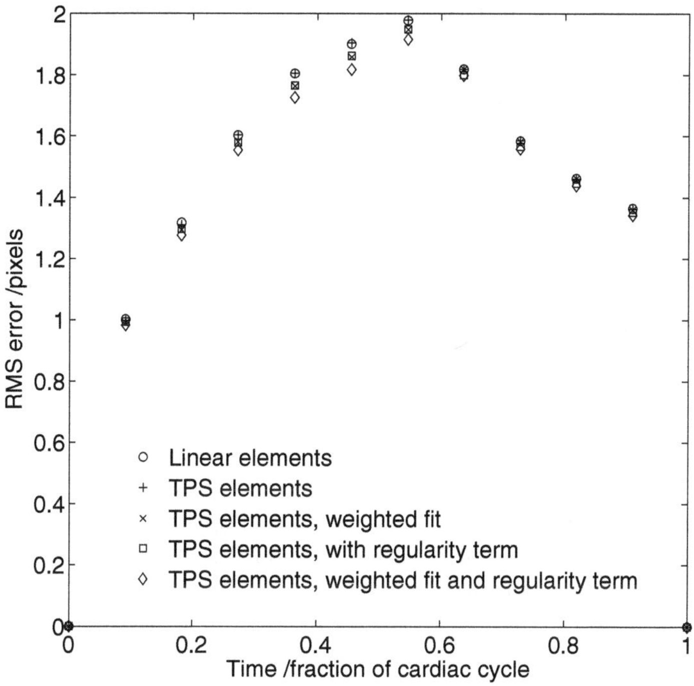

Fig. 3. A quantitative comparison of the different spatiotemporal models. The figure gives the mean RMS error for the 20 cases compared to grid tag MRI. The absolute difference between the different variants is small which is due to the sparsity of the displacement obtained from the grid tag images. The error is higher during the second half of the cardiac cycle which is due to grid tag fading. Note that the model using linear elements has a higher error than all the other methods. The model with the smallest error is the one with both a data certainty term and a regularity term.

similar way, the difference between particle traces generated by the two models will often be on the order of a pixel or less. This difference is, however, enough to generate quite different deformations, if evaluated in a qualitative way.

Figure 3 shows the RMS error for the different models. The best performer was the model with both data certainty and regularity terms. Again, the absolute differences between the models were quite small.

4 Discussion

In this paper we have extended a spatiotemporal model of cardiac deformation proposed by Zhu *et al* [19] in several ways. First, we made the observation that piecewise linear elements are inappropriate as they are not differentiable. Therefore we used the TPS as a spatial element. Second, we estimated a data certainty term so that the model does not adapt to noisy measurements. Third, we also added a regularity term so that smooth deformations are encouraged.

We also defined the model in Lagrangian coordinates which is beneficial from a computational point of view.

Based on the qualitative and quantitative results presented we conclude that all variants of our proposed model performs better than, or equally well, as the model of Zhu *et al.* The adjustments made to the model have proven to be beneficial as showed using a computer generated motion phantom where the RMS error can be up to 50% lower for our proposed model. There are however, only small quantitative differences between the model variants when evaluated in vivo. Part of this can be explained by our validation procedure where we used gridtag MRI as a gold standard. Grid tag MRI will only provide a sparse displacement field which means that the regularity of the calculated deformation will not be reflected in the error. The grid tag images were manually analyzed which is an additional source of error, in particular during diastole where tag fading complicates the analysis. This effect can be seen in Figure 3 where the error is higher during the second half of the cardiac cycle. Figure 2 showed that the absolute difference between the deformations generated by the model will be small, which is reflected in the error shown in Figure 3. The use of in vitro measurements would be helpful in order to determine the best model variant and is therefore the focus of future work.

References

1. Picano, E., Lattanzi, F., Orlandini, A., Marini, C., L'Abbate, L.: Stress echocardiography and the human factor: the importance of being expert. Journal of the American College of Cardiology 83, 1262–1265 (1991)
2. McVeigh, E., Ozturk, C.: Imaging myocardial strain. IEEE Signal Processing Magazine 18(6), 44–56 (2001)
3. O'Dell, W.G., Moore, C.C., Hunter, W.C., Zerhouni, E.A., McVeigh, E.R.: Three-dimensional myocardial deformations: Calculation with displacement field fitting to tagged MR images. Radiology 195(3), 829–835 (1995)
4. Axel, L., Dougherty, L.: MR imaging of motion with spatial modulation of magnetization. Radiology 171, 841–845 (1989)
5. Zerhouni, E.A., Parish, D.M., Rogers, W.J., Yang, A., Shapiro, E.P.: Human-heart-tagging with mr imaging - a method for noninvasive assessment of myocardial motion. Radiology 169(1), 59–63 (1988)
6. McVeigh, E.: MRI of myocardial function: Motion tracking techniques. Magnetic Resonance Imaging 14, 137–150 (1996)
7. Masood, S., Yang, G., Pennell, D.J., Firmin, D.N.: Investigating intrinsic myocardial mechanics: The role of MR tagging, velocity phase mapping and diffusion imaging. Journal of Magnetic Resonance Imaging 12(6), 873–883 (2000)
8. Osman, N.F., McVeigh, E.R., Prince, J.L.: Imaging heart motion using harmonic phase MRI. IEEE Transactions on Medical Imaging 19(3), 186–202 (2000)
9. Pelc, N.J., Herfkens, R.J., Shimakawa, A., Enzmann, D.R.: Phase contrast cine magnetic resonance imaging. Magnetic Resonance Quarterly 7(4), 229–254 (1991)
10. van Wedeen, J.: Magnetic resonance imaging of myocardial kinematics. techniques to detect, localize and quantify strain rates of the active human myocardium. Magnetic Resonance in Medicine 27(1), 52–67 (1992)

11. Robson, M.D., Constable, R.T.: Three-dimensional strain-rate imaging. Magnetic Resonance in Medicine 36(4), 537–546 (1996)
12. Arai, A.E., Gaither III, C.C., Epstein, F.H., Balaban, R.S., Wolff, S.D.: Myocardial velocity gradient imaging by phase contrast MRI with application to regional function in myocardial ischemia. Magnetic Resonance in Medicine 42(1), 98–109 (1999)
13. Selskog, P., Heiberg, E., Ebbers, T., Wigström, L., Karlsson, M.: Kinematics of the heart: Strain-rate imaging from time-resolved three-dimensional phase contrast MRI. IEEE Transactions on Medical Imaging 21(9), 1105–1109 (2002)
14. Constable, R.T., Rath, K.M., Sinusas, A.J., Gore, J.C.: Development and evaluation of tracking algorithms for cardiac wall motion analysis using phase velocity MR imaging. Magnetic Resonance in Medicine 32(1), 33–42 (1994)
15. Pelc, N.J., Drangova, M., Pelc, L.R., Zhu, Y., Noll, D.C., Bowman, B.S., Herfkens, R.J.: Tracking of cyclic motion using phase contrast cine mri velocity data. Journal of Magnetic Resonance Imaging 5(3), 339–345 (1995)
16. Meyer, F.G., Constable, T., Sinusas, A.J., Duncan, J.S.: Tracking myocardial deformation using phase contrast MR velocity fields: A stochastic approach. IEEE Transactions on Medical Imaging 15(4), 453–465 (1996)
17. Zhu, Y., Drangova, M., Pelc, N.J.: Fourier tracking of myocardial motion using cine-PC data. Magnetic Resonance in Medicine 35(4), 471–480 (1996)
18. Zhu, Y., Drangova, M., Pelc, N.J.: Estimation of deformation gradient and strain from cine-PC velocity data. IEEE Transactions on Medical Imaging 16(6), 840–851 (1997)
19. Zhu, Y., Pelc, N.J.: A spatiotemporal model of cyclic kinematics and its application to analyzing nonrigid motion with MR velocity images. IEEE Transactions on Medical Imaging 18(7), 557–569 (1999)
20. Chadwick, P.: Continuum Mechanics. Dover Publications, Mineola (1999)
21. Drangova, M., Zhu, Y., Pelc, N.J.: Effects of artifacts due to flowing blood reproducibility of phase-contrast measurements of myocardial motion. Journal of Magnetic Resonance Imaging 7(4), 664–668 (1997)
22. Bookstein, F.L.: Principal warps: thin-plate splines and the decomposition of deformations. IEEE Transactions on Pattern Analysis and Machine Intelligence 11(6), 567–585 (1989)
23. Arts, T., Hunter, C., Douglas, A., Muijtens, M.M., Reneman, R.: Description of the deformation of the left ventricle by a kinematic model. Journal of Biomechanics 25(10), 1119–1127 (1992)

Fractal Analysis of Mammograms

Fredrik Georgsson*, Stefan Jansson, and Christina Olsén

Department of Computing Science,
Umeå University, SE-901 87 Umeå, Sweden
fredrikg@cs.umu.se

Abstract. In this paper it is shown that there is a difference in local fractal dimension between imaged glandular tissue, supporting tissue and muscle tissue based on an assessment from a mammogram. By estimating the density difference at four different local dimensions (2.06, 2.33, 2.48, 2.70) from 142 mammograms we can define a measure and by using this measure we are able to distinguish between the tissue types. A ROC-analysis gives us an area under the curve-value of 0.9998 for separating glandular tissue from muscular tissue and 0.9405 for separating glandular tissue from supporting tissue. To some extent we can say that the measured difference in fractal properties is due to different fractal properties of the unprojected tissue. For example, to a large extent muscle tissue seems to have different fractal properties than glandular or supportive tissue. However, a large variance in the local dimension densities makes it difficult to make proper use of the proposed measure for segmentation purposes.

1 Introduction

It has been put forward that knowledge regarding properties of tissue can be assessed by estimating the fractal properties of an image, or more specifically an x-ray image, of the tissue [1,2]. The results have been hard to reproduce by other researchers [3,4,5] and the approach has been rightfully questioned.

We have, however, shown [6] that by estimating the box dimension (or any equivalent measure of fractal dimension) for a projected set, bounds for the dimension of the unprojected set can be imposed. From a theoretical point of view this tells us that information regarding the three dimensional object, e.g. the tissue composition of the breast, might be found by performing fractal analysis of an image, e.g. on a mammogram. This is of course encouraging, but it still remains to be shown if we are able to exploit this theoretical advance in practice.

The aims of this paper are to evaluate whether it is possible to use fractal analysis of x-ray images (mammograms) to classify them into tissue types and to draw conclusions regarding the fractal properties of the depicted object (the breast).

* Part of this research was funded by the Swedish Research Council under grant number 621-2002-3795.

B.K. Ersbøll and K.S. Pedersen (Eds.): SCIA 2007, LNCS 4522, pp. 92–101, 2007.

In [3] an attempt to differentiate tumor from healthy tissue using different fractal properties was carried out. The conclusion was that fractal properties might be useful for segmentation within an image, but it is not possible to compare fractal properties between images. In [1] the authors match the fractal dimension to Wolfe-grades [7] thus enabling a predictability on the probability of breast cancer. In [2] fractal geometry was used to describe the pathology of tumors but the approach was questioned in [8]. It has been argued that fractal properties alone are not sufficient for effective texture segmentation, thus it is quite common to use fractal features as part of feature vectors in texture classification [9,10]. However, we aim at investigating the correspondence between the property of the projected image and the properties of the unprojected organ and pure texture approaches are of little interest for us. Our work differ from all of the above since we make use of knowledge of the imaging system when estimating fractal properties from an image in order to gain knowledge of the structures that generated the image.

The paper is organized as follows; In section 1.1 some mathematical background to fractal geometry is given. In section 1.2 a brief description of mammography and mammographic imaging is given. In section 2 the method is accounted for and in section 3 the data set used is described. The results are presented in section 4, the results are discussed in section 5 and conclusions are drawn in section 6.

1.1 Fractal Geometry and Some New Projective Properties of Fractals

Fractal geometry has been put forward as being suitable for analyzing irregular sets and one of the most used measure of fractality is the fractal dimension. In the literature it is not always clear what is meant by fractal dimension but the most frequent interpretation seems to be that the fractal dimension is identical to the Box-dimension as defined in equation (1)[1] We let $N_\varepsilon(F)$ be the smallest number of sets of a diameter smaller than ε that is needed to cover the set F. The Box-dimension is then defined as

$$dim_B(F) = \lim_{\varepsilon \to 0} \frac{\log N_\varepsilon(F)}{-\log \varepsilon}. \tag{1}$$

Since we aim at applying an estimate of the Box-dimension on a digital (discrete) image we have to modify equation (1) slightly. The reason for this is, of course, that we cannot let $\varepsilon \to 0$ since $\varepsilon \geq \Delta$ where Δ is the pixel size. The solution is to rewrite equation (1) according to

$$\log N_\varepsilon(F) + \log \varepsilon \cdot dim_B(F) + h(\varepsilon) = 0 \tag{2}$$

where $h(\varepsilon)$ is assumed to be approximately constant.

[1] From a mathematical point of view the Box-dimension is a simplification of the dimensionality concept, since the dimension should be based on a measure to be interesting. Examples of such dimensions are the Hausdorff and Packing dimensions.

Equation (2) shows that there is a linear relationship between $\log\varepsilon$ and $\log N_\varepsilon(F)$ where $dim_B(F)$ is the constant of proportionality. Thus, the box dimension $dim_B(F)$ is found as the slope of a straight line fitted to the points $\{(\log\varepsilon_i, \log N_{\varepsilon_i}(F))\}_{i=1}^N$, where $\varepsilon_1 < \varepsilon_2 < \cdots < \varepsilon_N \leq w_1$. The fitting of the line to the points is done in a least square sense. The upper boundary w_1 is defined in section 2.

As an extension of work in [6] we have proved that

$$dim_B(F) \leq dim_B(\mathcal{F}) \leq \frac{12 \cdot dim_B(F)}{12 - dim_B(F)} \tag{3}$$

for almost all directions of projection. In equation (3), $F \subset \mathbb{R}^2 \times \mathbb{R}$ is the projection[2] of a set $\mathcal{F} \subset \mathbb{R}^3 \times \mathbb{R}$. In fact, we can show that it is not possible to find tighter bounds than the ones presented in equation (3). By using equation (3) we find bounds for the dimension of the unprojected set (for instance the breast) by estimating the Box-dimension of the projected set (for instance the mammogram). By using the upper and lower bounds from equation (3) it is possible to establish the required difference in box-dimensions for us to be able to say anything regarding the difference in the original set. This is illustrated in Fig. 1. Furthermore, by using equation (3) and an estimated dimension $dim_B(F)$ we can find an interval of dimensions for which we cannot exclude the possibility that the fractal properties of the projected volumes do overlap. This interval is given by

$$I_F = \left[\max\left(0, \frac{12 \cdot dim_B(F)}{12 + dim_B(F)}\right), \min\left(3, \frac{12 \cdot dim_B(F)}{12 - dim_B(F)}\right)\right]. \tag{4}$$

By using I_F we can make assertions like: If $dim_B(F_2) \notin I_{F_1}$ then $dim_B(\mathcal{F}_2) \neq dim_B(\mathcal{F}_1)$. The reverse is not true, i.e. if $dim_B(F_2) \in I_{F_1}$ we cannot conclude that $dim_B(\mathcal{F}_1) = dim_B(\mathcal{F}_2)$.

There are many versions of the Box-dimension that differ in what type of covering sets are used and how they are allowed to be arranged etc. In the continuous case they all have identical mathematical properties [11] while this is not necessarily the case in the discrete case, a fact that many users of fractal geometry do not seem to be aware of.

We denote a method for estimating a fractal dimension as an estimator and in [12,13] we have evaluated the performance of a substantial number of estimators of intensity images. Based on these results we have picked the method most suitable for determining the local dimension. This was determined to be a method known as the probability method [14] with some modifications. The modifications consists mainly of performing the calculations in local neighbourhoods of a maximal size $w_1 \times w_1$. The estimated local dimensions are then averaged over a neighborhood of size $w_2 \times w_2$. A third parameter, *scale*, is the number of discrete intensity levels used in the calculations. That is, the original m levels of intensity are linearly mapped to *scale* levels.

[2] We have shown this for parallel, central and realistic x-ray projection.

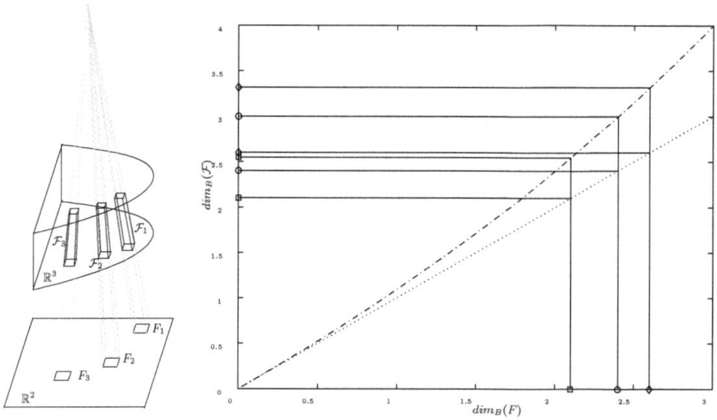

(a) Projective setting

(b) Relationship between dimensions

Fig. 1. (a) A schematic illustration of the projective setting used in mammography. A volume $\mathcal{F}_i$ is projected down to an area F_i. (b) The relationship between an estimated fractal dimension $dim_B(F)$ from a projection of a set and the upper boundary (dash-dotted line) and the lower boundary (dotted line) of the original, unprojected set $dim_B(\mathcal{F})$ as given by equation (3).

1.2 Mammography

As mentioned earlier, we have used mammography as the domain for our study. A mammogram is an x-ray image of breast tissue and since the difference in contrast between tissue types (glandular tissue, fat tissue and supporting tissue) is very low the technical requirements imposed on a mammographic system are very hard.

The reasons for using mammograms for a study of this kind are two-fold. Firstly, a thorough fractal analysis requires rather high resolution and contrast and we have shown in [15] that of the more popular medical modalities, mammography is the most suitable one for fractal analysis. Since part of our aim is to evaluate the usefulness for fractal geometry as a method for segmentation it is important to know that the data we apply it to are as good as possible. Secondly, the proper method to use for the segmentation of mammograms into tissue types is an open research question and, furthermore, the relative proportion of dense tissue can be mapped to a probability of risk for developing tumours [7].

2 Method

Given an image $F : D \to \{0, \ldots, m-1\}$ we use an estimator to calculate an image of the same size but the intensity at each spatial point corresponds to the local fractal dimension. We denote this image $F_{loc} : D \to [0, \ldots, 3]$. In our data, see section 3, we have binary markings $M^k \subset D$ corresponding to an anatomical feature M given by expert k. Let $F_{loc}[M]$ denote the local dimensions at the

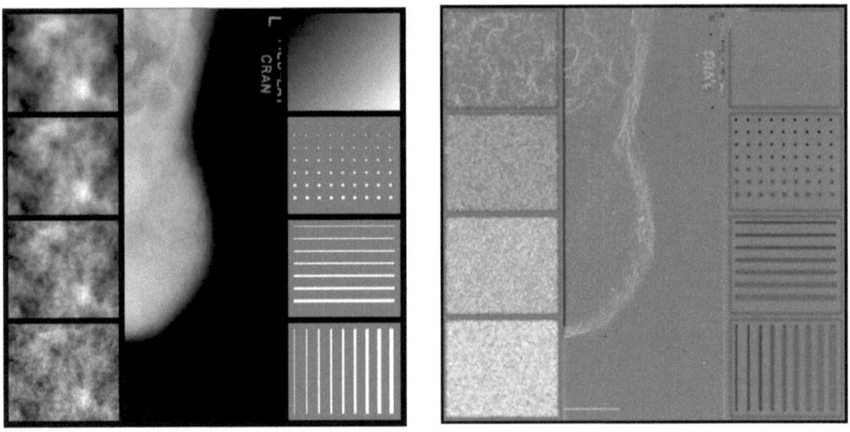

(a) Mammogram with calibration surfaces (b) Local dimensions

Fig. 2. Examples of a mammogram and its local dimensions. The calibration is only carried out once and the calibration surfaces can be dismissed after completing the calibration.

spatial coordinates given in M and let $H_{loc}[M]$ denote the histogram over these dimensions. The hypothesis is that the estimated distributions $H_{loc}[M^k]$ and $H_{loc}[N^k]$ differ if $M \neq N$. It is also of interest to see if there is a difference between the domain experts, that is if $H_{loc}[M^k]$ and $H_{loc}[M^l]$ if $k \neq l$.

Optimal values for the three parameters (w_1, w_2 and *scale*) that control the performance of the estimator were found by a simple optimization procedure. We inserted patches into the mammogram with known fractal properties, Fig. 2(a). The four patches to the left of the mammogram were generated by approximating a fractal Brownian motion [16] and have from the top dimensions of 2.1, 2.3, 2.5 and 2.7 respectively. We also added patches with a plane (dimension 2), points (dimension 0) and lines (dimension 1) to the right of the mammogram. The points and lines were not used in the optimization of parameters.

By using the calibration areas we could determine what sets of parameters that yield a linear mapping between estimated dimension and true dimension. The initial parameter space contains three discrete parameters and it is possible to impose upper and lower boundaries for each parameter and thus we have a finite number of parameter combinations to try out. A coarse discretization of the parameter space allowed us to determine where in the space the optimal parameter combinations were situated. Within this area an exhaustive search was carried out yielding the chosen parameter values. The optimization of parameters was only performed on the calibration patches in Fig. 2. The actual mammogram did not take part in this process.

Next we calculated an F_{loc} version for every mammogram in our database. For each F_{loc} we extracted $H_{loc}[M^k]$, $\forall M, k$ and used these for further analyses.

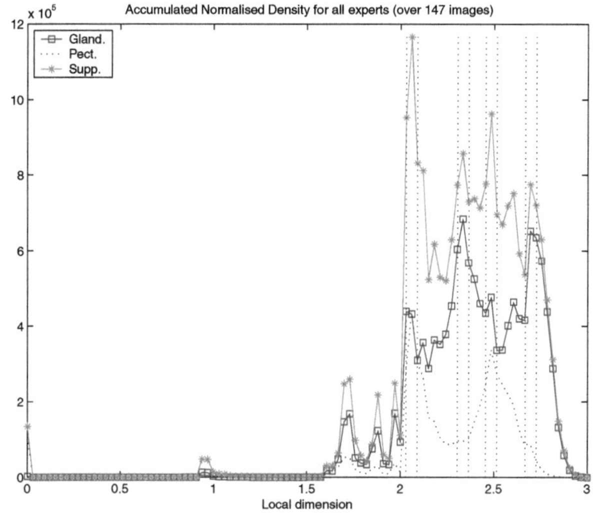

Fig. 3. A distribution over local fractal dimensions ($H_{loc}[Gland]$, $H_{loc}[Pect]$, $H_{loc}[Supp]$). The peaks $dim_{loc} = 2.060 \pm 0.03$, $dim_{loc} = 2.335 \pm 0.03$, $dim_{loc} = 2.485 \pm 0.03$ and $dim_{loc} = 2.700 \pm 0.03$ have been marked.

3 Data

The mammograms used in this study are randomly selected from the MIAS-database (8 bit, 1024×1024 images). Three domain experts were asked to mark the pectoralis muscle, the mamilla, the breast border and the glandular disc on the images. In total we have used 142 different mammograms and we have $3 \cdot 142$ different markings of regions. In this paper "Gland" denotes the glandular tissue, "Pect" denotes the pectoral muscle and "Supp" denotes the supporting tissue (including fat) that surrounds the glandular tissue. Each image in the MIAS-database is adjusted to be of square size, and since a mammogram is rectangular we always have empty black areas to the left and right of the actual mammogram. In these areas we inserted the calibration patches described in section 2. Thus we did not alter any mammographic information when adding the patches. The mammograms where all adjusted so that the pectoral muscle is to the left in the image, but no other preprocessing took place.

4 Results

For $D = \{1, \ldots, 1024\} \times \{1, \ldots, 1024\}$ images with intensity levels $L = \{0, \ldots, 255\}$ the optimal set of parameters was found to be $w_1 = 47$, $w_2 = 5$ and $scale = 51$. Using these parameters an F_{loc} was calculated for each image in our database, and $H_{loc}[M^k]$ were extracted from each F_{loc}.

The accumulated distributions $H_{loc}[M]$ are shown in Fig. 3 (based on 142 images) and here we clearly see that there is a difference in the relative shape

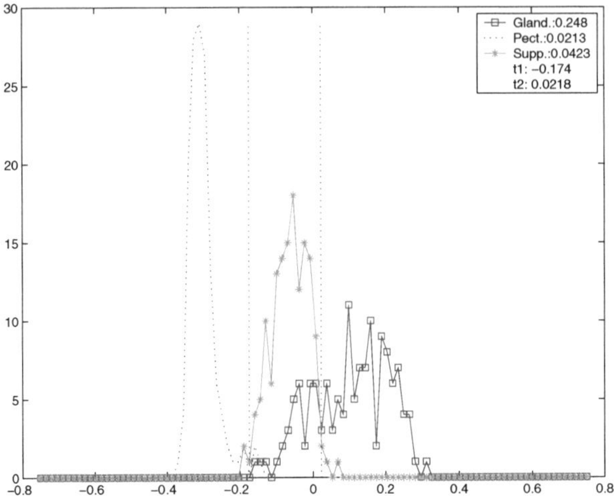

Fig. 4. Distributions over ρ-values for glandular tissue, muscle tissue and supportive tissue as calculated from 142 images. Optimal segmentation is obtained by thresholding at $\rho_1 = -0.174$ and $\rho_2 = 0.0218$ yielding an lowest combined probability of misclassification $P(err|Gland) = 0.248$, $P(err|Pect) = 0.0213$ and $P(err|Supp) = 0.0423$.

of the distributions, specially between the pectoral muscle vs. glandular and supportive tissue. The pectoral muscle has clear peaks at $dim_{loc} = 2.060$ and $dim_{loc} = 2.485$ whilst glandular and supportive tissue have peaks at $dim_{loc} = 2.335$ and $dim_{loc} = 2.700$. The peaks were estimated to have a width of $2\delta = 0.06$. We define a measure

$$\varphi_\ell(x) = \frac{\sum_{|\ell - x| < \delta} H_{loc}[\cdot](x)}{\sum_x H_{loc}[\cdot](x)}$$

and calculate

$$\boldsymbol{\rho} = \begin{bmatrix} \varphi_{2.060}(x) & \varphi_{2.335}(x) & \varphi_{2.485}(x) & \varphi_{2.700}(x) \end{bmatrix}^T . \tag{5}$$

The vector $\boldsymbol{\rho}$ describes the relative density in the histogram at each of the identified peaks. Since we have observed that glandular tissue has peaks at the second ($dim_{loc} = 2.485$) and fourth ($dim_{loc} = 2.700$) values we project the vector $\boldsymbol{\rho}$ onto the base vector $\mathbf{b} = [-1 \quad 1 \quad -1 \quad 1]^T$ with the inner product

$$\rho = \mathbf{b}^T \boldsymbol{\rho}. \tag{6}$$

It is easily shown that the value of ρ will be in the -1 to 1 interval, where a value of 1 corresponds to a histogram with all observed data at $dim_{loc} = 2.485$ and/or $dim_{loc} = 2.700$. A value of -1 corresponds to a histogram with all observed data at $dim_{loc} = 2.060$ and/or $dim_{loc} = 2.485$.

By calculating ρ for all 142 images we are able to estimate a distribution $p(\rho)$ for different tissue types. In Fig. 4 we can see $p(\rho)$ given the three tissue

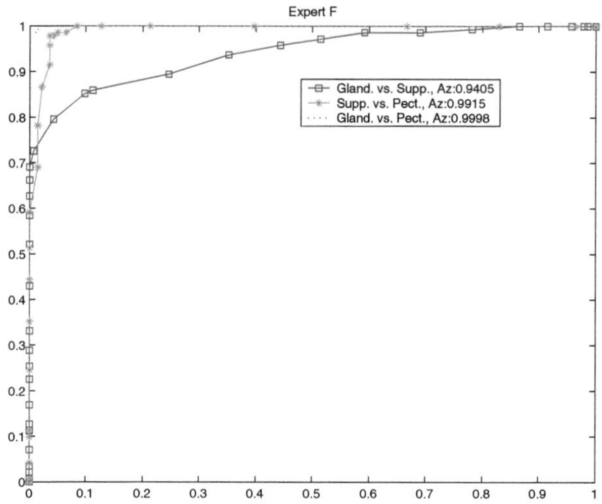

Fig. 5. Receiver operating characteristic (ROC) curve for using the measure to differentiate tissue types when using ground truth from a specific domain expert (F)

types. The optimal, in the sense least probability for misclassification, thresholds are $\rho_1 = -0.174$ and $\rho_2 = 0.0218$. The thresholds were found by a simple linear search of ρ (exhaustive). The discriminating power obtainable by using equations (5) and (6) is illustrated in Fig. 5 where we can see receiver operating characteristic for the tissue types. The performance differs slightly depending on what expert we used for the ground truth markings, see Table 1. The best A_z values (area under the ROC-curve) obtained for separating tissue types were 0.9998 for glandular vs. pectoralis, 0.9915 for supporting tissue vs. pectoralis and 0.9405 for glandular tissue vs. supporting tissue.

5 Discussion

Based on the ROC-curves from the $p(\rho)$ data (Fig. 5) we conclude that by using the local fractal dimension it is possible to separate tissue types. The ROC analyses are based on accumulated data from all 142 images but to be useful for segmentation, we have to use an estimate $\widehat{H_{loc}[\cdot]}$ from a neighbourhood in F_{loc}. It turns out that the variance is quite substantial which makes it rather difficult use the local density distribution. Thus equations (5) and (6) are not necessarily useful for segmentation purposed.

We have observed that there is a significant peak at about dimension 2, see Fig. 3. To some extent this can be explained by over and under exposure of the images. Similarly it seems that fractal analysis is very good at detecting film scratches and other artifacts. Both these observations are interesting, but since mammography as a modality slowly is going digital, where problems with

Table 1. Intra variance amongst domain experts regarding the area under the ROC-curve

Expert	Area under ROC-curve		
	Gland. vs. Pect	Gland vs. Supp	Supp vs. Pect
A	0.9995	0.8463	0.9994
C	0.9983	0.8874	0.9882
F	0.9998	0.9405	0.9915

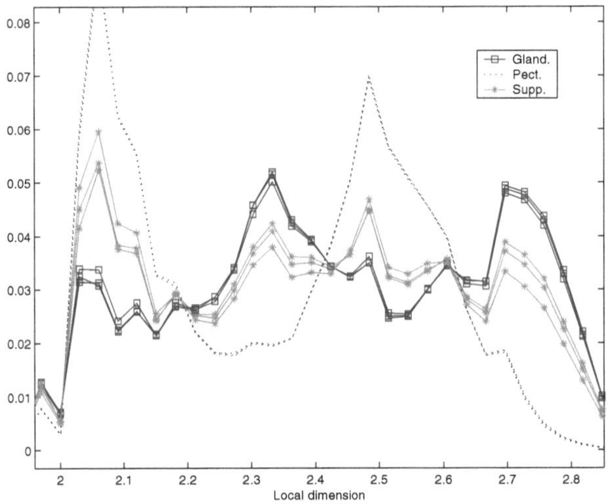

Fig. 6. Enlarged part of the normalised histograms for $H_{loc}[Gland^k]$, $H_{loc}[Pect^k]$ and $H_{loc}[Supp^k]$, $k \in \{A, C, F\}$. The observable intra variance is very low for the pectoral muscle and the highest intra variance is found in the supportive tissue.

exposure and film scratches are significantly smaller, it is not deemed interesting to investigate this further.

By making use of equation (3) we can say that the structures yielding local dimensions of $dim_{loc} = 2.060$ and $dim_{loc} = 2.485$ must have different dimensions in three dimensions. We cannot say the same for $dim_{loc} = 2.335$ and $dim_{loc} = 2.700$. Thus we conclude that muscle tissue has slightly different fractal properties than glandular tissue and supportive tissue. The difference could to some extent be explained by the fact that the muscle quite often is under exposed in the images.

From Fig. 4 it appears that supporting tissue and glandular tissue have more in common than muscle tissue and any other two tissue types. This might be explained by the fact that glandular tissue and supportive tissue to a larger extent are projected on top of each other. That is, within each projective cone, see Fig. 1 (a) there are both tissue types.

In Fig. 6 we can observe the intra variance amongst the domain experts. Although the observed difference is relatively low it will give rise to a significant difference in the area under the ROC curve values, see Table 1.

6 Conclusion

We conclude that there is a difference in local fractal dimension estimated by the box dimension from a mammogram. To part, we can say that this difference is due to difference in the dimension of the three dimensional unprojected tissue. However, the high variance of the local density function of the local dimension makes it difficult to make use of this difference in tissue segmentation.

References

1. Caldwell, C.B., Stapleton, S.J., Holdsworth, D.W., Jong, R.A., Weiser, W.J., Cooke, G., Yaffe, M.J.: Characterization of mammographic parenchymal pattern by fractal dimensions. Physics in Medicine and Biology 35, 235–247 (1990)
2. Baish, J.W., Jain, R.: Fractals and cancer. Cancer Research 60, 3683–3688 (2000)
3. Veenland, J., Grashuis, J.L., van der Meer, F., Beckers, A.L.D., Gelsema, E.S.: Estimation of fractal dimension in radiographs. Medical Physics 23, 585–594 (1996)
4. Huang, Q., Lorch, J., Dubes, R.: Can the fractal dimension of images be measured. Pattern Recognition 27, 339–349 (1994)
5. Jelink, H., Fernandez, E.: Neorons and fractals: how reliable and useful are calculations of fractal dimensions? Journal of Neoroscience Methods 81, 9–18 (1998)
6. Nilsson, A., Georgsson, F.: Projective properties of fractal sets. Chaos, Solitons and Fractals, Accepted for publication (2006)
7. Wolfe, J.N.: Risk for breast cancer development determined by mammograhic parenchymal pattern. Cancer 37, 2486–2492 (1976)
8. Chung, H.W., Chung, H.J.: Letter to the editor; correspondence re: J. W. Baish and R. K. Jain, Fractals and Cancer. Cancer Res. 61, 8347–8348 (2001)
9. Lee, W.L., Chen, Y.C., Chen, Y.C., Hsieh, K.S.: Unsupervised segmentation of ultrasonic liver images by multiresolution fractal feature vectors. Information Sciences 175, 177–199 (2005)
10. Chen, C., Daponte, J., Fox, M.: Fractal feature analysis and classification in medical imaging. IEEE Transactions on Medical Imaging 8, pp. 133–142 (1989) nr: 30
11. Falconer, K.: Fractal Geometry. Wiley, Chichester (1990)
12. Jansson, S., Georgsson, F.: Evaluation of methods for estimating the fractal dimension of intensity images. In: Georgsson, F., Börlin, N., (eds.): Proceedings of SSBA. Swedish Society for Automated Image Analysis, Number UMINF-06.11 in ISSN 0348-0542, pp. 69–72 (2006)
13. Jansson, S.: Evaluation of methods for estimating fractal properties of intensity images. Umnad: 699/06, Dept. of Comp. Sc. Umeå University (2006)
14. Keller, J., Chen, S., Crownover, R.: Texture description and segmentation through fractal geometry. Computer Vision, Graphics, and image processing 45, 150–166 (1989)
15. Georgsson, F.: Fractal dimensions and the geometry of x-ray imaging. In: Bengtsson, E., (ed.): Proccedings of SSBA 2004,pp. 150–153 (2004)
16. Peitgen, H.O., Saupe, D. (eds.): The Science of Fractal Images. Springer, Heidelberg (1988)

Reconstructing Teeth with Bite Information

Katrine Hommelhoff Jensen and Jon Sporring

Department of Computer Science, University of Copenhagen, Denmark
{katrine,sporring}@diku.dk

Abstract. We propose a method for restoring the surface of a tooth crown so that the pose and anatomical features of the tooth will work well for chewing. The system of teeth has been modeled with a 3D statistical multi-object shape model build from 3D scans of dental cast models. The restoration is carried out using the shape model statistics in a Bayesian framework to calculate the most probable tooth crown shape(s), given the fragments of one or more neighboring and opposing tooth crowns. The modeling of and reconstruction with the multi-object shape model has been realized by extending the model with a concept of elasticity that generalizes better to new teeth. The elasticity has been calculated from the surface curvature relations within and between each tooth sample, simulating a prior knowledge of the shape variation.

1 Tooth Reconstruction

In the dental industry, the design and construction of restorations to be inserted in a patient's mouth is carried out by dental technicians, that are highly trained experts in tooth anatomy and the function of the bite. The task can be to model the missing part of a broken tooth crown, model the crown of a whole missing tooth or even several missing teeth. The restorations are traditionally constructed directly from the materials by hand, but the use of software to model the construction elements of a restoration has been growing rapidly the last couple of years [1,2,3]. Other than saving money on the temporary building materials, the software solution saves time, as some of the traditional production steps can be skipped and proper customized 3D modeling tools and automatic routines can speed up the construction work. The existing dental software systems combines the software with a scanning device to produce a 3D surface model of the patients remaining teeth, on which the restoration is to be designed. The modeled restoration can then be exported as a 3D surface model and milled or printed directly in the final material.

One of the most challenging steps towards an automation of dental restoration modeling software is the anatomical deformations of the tooth crowns to be reconstructed. An anatomical correct deformation of a tooth crown surface cannot be calculated exclusively from the size and location of the surrounding surfaces of the scanned data. Some prior knowledge must be added to the system, which describes the shapes and legal deformations of the teeth. It is our goal to develop a system that can learn and describe the complex shape system of the bite, and

B.K. Ersbøll and K.S. Pedersen (Eds.): SCIA 2007, LNCS 4522, pp. 102–111, 2007.

with this knowledge reconstruct the surfaces of missing tooth crown parts, whole teeth, or several teeth from information extracted from scanned data. The work reported in this article is based on [4], and the plaster casts have been scanned by the 3Shape laser scanner, some shown in Figure 1.

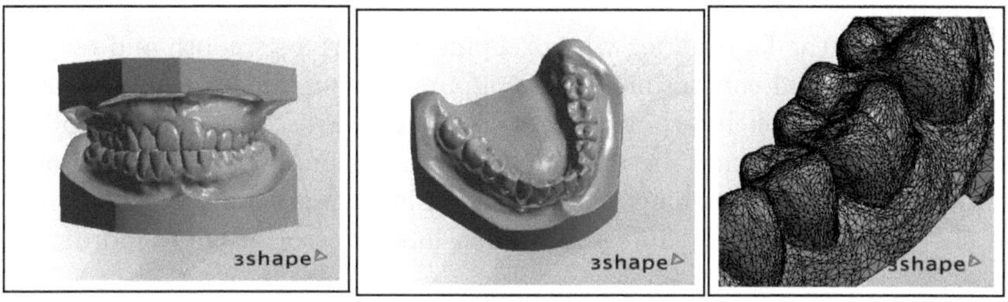

(a) (b) (c)

Fig. 1. Cast models scanned by the 3Shape laser scanner, aligned (a) and the lower jaw (b), and a triangulated surface mesh of a scanned cast model (c). Notice on (c) that the mesh has been decimated to a much lower solution than the usual quality.

2 Shape Modeling

We operate with the notion of shape as that which is left, when translation, rotation, and scaling is removed. To reconstruct the shape of an incomplete object, we need a model description of the object shape and variability. The classical approach is based on representing and modeling shape as a set of *landmarks* (see [5,6,7,8] and references herein). On each training shape, a finite number of landmarks are located on surface features that corresponds between the shapes. This representation is directly applicable in the *Active Shape Model* (ASM). The biggest disadvantage is the time-consuming manual labor needed. Alternative approaches, that carry a smooth surface implicitly in the shape description, are the *Level-set function* representation [9] and the *Medial representation* (M-rep) [10]. However, both representations are problematic, since they cannot robustly handle non-closed shape surfaces such as scans of plaster casts of teeth.

Shape Warping has also been studied in the literature: In [11], each training shape landmark is warped to a template shape, where a template shape mesh is projected onto the shape before warping landmarks and mesh vertices back. Based on this idea, [12] introduced a 3D morphable model that use 3D meshes as training data rather than images. Each mesh vertex achieves the same status as a landmark, and the correspondence between the training meshes and the template mesh is estimated from a sparse set of correspondence points, manually marked by the user on each training mesh. In [3] the reconstructions were adjusted for neighboring teeth using local mathematical morphology. The quality of the warping depends on both the complexity of the surfaces and the unknown variation of the samples. Further, when only extracting data from the occlusal, frontal, and back sides of the teeth, a significant amount of surface is left unknown. We are thus lead to the idea of keeping the surface representation separated from the data. Further, the reconstruction situations depends on the amount of a tooth that must be reconstructed. If the areas of reconstruction are somehow marked in the process, a better approach is to let the surface conform to the reconstructed data, while respecting the borders between the reconstructed parts and the original data. The task is to create a mesh that can be guided through the landmarks, while keeping the surface smooth and respecting the local and global constraints. This coupling we implement using Variational Implicit Surfaces [13] interpolated between landmarks, and thus we obtain a flexible method, that may be extended with additional information about the anatomical features, and allow us to focus on landmarks only.

A shape will be described as a $3n$ dimensioned vector of point coordinates

$$\boldsymbol{x} = [\boldsymbol{p}_1^T, \boldsymbol{p}_2^T, \ldots, \boldsymbol{p}_n^T]^T = [x_1, y_1, z_1, x_2, y_2, z_2, \ldots, x_n, y_n, z_n]^T$$

with each point representing the position of a landmark and n is the number of landmarks in the shape. Following [8], we use Procrustes analysis to align each shape $\boldsymbol{x}$ to the *mean shape* $\bar{\boldsymbol{x}}$, where $\bar{\boldsymbol{x}} = \frac{1}{N}\sum_{i=1}^{N} \boldsymbol{x}_i$, and N is the number of training samples. To align shapes, we first normalize for translation and size, and then normalize orientation by minimizing the *Procrustes distance*,

$$d^2_{\text{Procrustes}} = ||\boldsymbol{x} - \bar{\boldsymbol{x}}||^2 \tag{1}$$

We will misuse notation and use $\boldsymbol{x}$ for the normalized shape coordinates in the following. After having aligned the shapes we write,

$$\mathbf{X} = [(\boldsymbol{x}_1 - \bar{\boldsymbol{x}}), (\boldsymbol{x}_2 - \bar{\boldsymbol{x}}), \ldots, (\boldsymbol{x}_N - \bar{\boldsymbol{x}})] \tag{2}$$

Then we compute the covariance matrix as

$$\mathbf{C} = \frac{1}{N}\mathbf{X}\mathbf{X}^T \tag{3}$$

and its eigenvalues λ_i and eigenvectors ϕ_i,

$$\mathbf{C}\boldsymbol{\Phi} = \boldsymbol{\Lambda}\boldsymbol{\Phi} \tag{4}$$

where $\boldsymbol{\Lambda}$ is the diagonal matrix of eigenvalues and the columns of the matrix $\boldsymbol{\Phi}$ contains the corresponding eigenvectors or *modes*. Thus any shape $\boldsymbol{x}$ from the training set can be reproduced by a linear computation of the mean and the principal components as

$$\boldsymbol{x} = \bar{\boldsymbol{x}} + \boldsymbol{\Phi}\boldsymbol{b} \tag{5}$$

where $\boldsymbol{b}$ is the vector of shape model parameters. The strength of the model is that the eigenvectors corresponding to the largest eigenvalues model the training set with an error equal to the sum of the neglected eigenvalues.

Our training set for each tooth consist of 12 samples. The landmarks were set by a non-expert with expert assistance and are primarily anatomical, with the exception of some pseudo-landmarks. 11 principal components were calculated for each shape model. The relative small training set size could potentially introduce problems regarding the generality, if the dependence on the model statistics is not relaxed in the reconstruction procedure. We will attempt to add artificial eigenmodes to the models, to improve flexibility without hazarding the object shape or overruling the existing eigenmodes. The rationale is that landmarks on the same side of a surface are expected to be correlated proportional with their distance. Thus, if one landmark were to be moved, then we expect that the neighboring landmarks will be effected. We will refer to this as *elasticity*, which will be described in the following.

3 Model Elasticity

The modes of variations were calculated from the covariance matrix of the combined data samples. Consider the general $3n \times 3n$ covariance matrix with covariances c_{ij}

$$\mathbf{C} = \begin{bmatrix} c_{1\,1} & c_{1\,2} & \cdots & c_{1\,3n} \\ c_{2\,1} & c_{2\,2} & \cdots & c_{2\,3n} \\ \vdots & \vdots & \ddots & \vdots \\ c_{3n\,1} & c_{3n\,2} & \cdots & c_{3n\,3n} \end{bmatrix} \tag{6}$$

If the covariance matrix of the data sample matrix was replaced by the identity covariance matrix $\mathbf{C}_{Id} = \mathbf{I}_{3n}$, then all landmark coordinates would be independent of each other. This means that combinations of the resulting eigenmodes could move the landmarks of a modeled shape in any direction. In that sense, $\mathbf{C}_{Id}$ defines an under-constrained, lower limit to the shape models. In [14] it has been shown how to add smoothness constrained deformations to a shape model by increasing the correlation between neighboring points in 2D shapes. The idea is, that when adding a small value to neighboring points in $\mathbf{C}_{Id}$, a covariation between the points is artificially created. Visually, moving a point in the shape will have an elastic effect on the neighbors. The effect of moving a point in a 2D shape with $\mathbf{C}_{Id}$ and $\mathbf{C}_{Id}$ augmented with with a positive value, e.g. 0.5, in the covariances between neighboring points, is illustrated in Figure 2.

The actual smoothness used in [14] was, however, not controlled in the relation to the model statistics, they were implemented to *substitute* model statistics. We need to control the amount of smoothness so that its function is a deformation *supplement*. Furthermore, we need to re-think the concept of neighbors in 3D, so that the elastic deformation added makes sense and respect the object shape.

In order to relax the tooth shape models we add a small value to all the neighbor-landmark covariances in the covariance matrix. Defining 'neighborhood' is a little more difficult in 3D, though. Neighborhood should be defined more in terms of distance than a number of closest neighbors, and should furthermore be measured over the surface and not necessarily as the shortest distance between two landmarks. As the solid objects teeth are, the smaller artifacts on

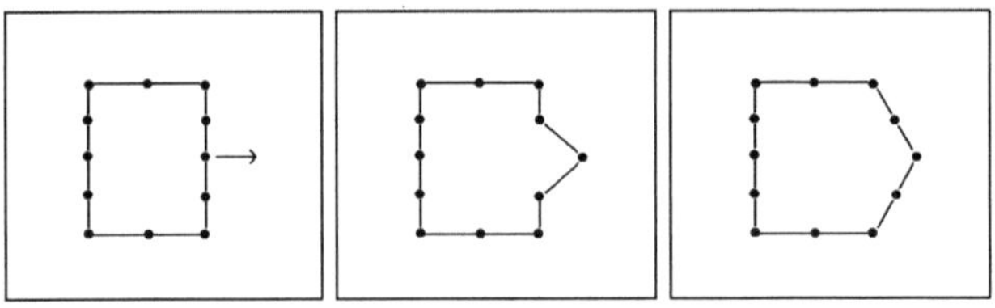

Fig. 2. The original 2D shape (left), the effect on the neighbors when moving a point with $\mathbf{C}_{Id}$ as covariance matrix (middle) and with $\mathbf{C}_{Id}$ augmented with a small value in the covariances of neighboring points (right)

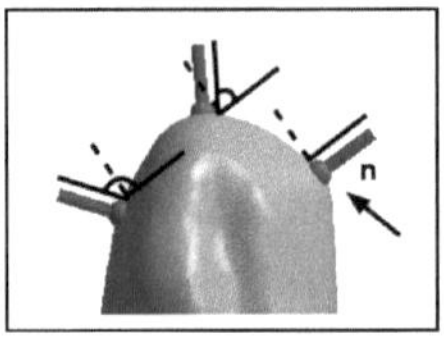

Fig. 3. Three points on a tooth (green) together with their surface normal (red), and angle difference to the rightmost normal. The neighbors to the rightmost landmark is a weighted sum of the surface-geodesic distance and the angle differences.

one side of a given tooth type doesn't have any effect on the smaller surface variations on the other side, disregarding the distance from landmarks on one side to landmarks on the other. In fact, when teeth have more unusual artifact's on the surface and is thus a difficult subject for an over-constrained shape model, it is usually due to abnormal chewing or smaller damages, both of which only have a local effect. The neighborhood of a landmark $\boldsymbol{p}$ could be measured as the landmarks within some fixed distance from $\boldsymbol{p}$. This, however, leaves us with the problem of deciding such a distance. Furthermore, it could introduce some problems regarding the scale of the individual tooth samples – we must choose a method that determines a well defined neighborhood of all shape samples in a model. Let d_{ab} be the mean surface-geodesic distance between landmark $\boldsymbol{p}_a$ and $\boldsymbol{p}_b$ over the training data. Then, a general way of calculating how much they should affect each other is to calculate this as a weight $0 < w_{ab} < 1$, where it is our experience that the following function is useful,

$$w_{ab} = \left(3^{1-\frac{\beta\, d_{ab}}{d_{\max}}} - 1\right) \frac{1 + \boldsymbol{n}_a \cdot \boldsymbol{n}_b}{2} \tag{7}$$

and where $\boldsymbol{n}_a$ are the normal at point $\boldsymbol{p}_a$ as shown in Figure 3, $d_{\max}$ is the maximum surface-geodesic distance between points on a shape over the training set, and β is a locality parameter typically 2 or 3. For a given pair of hypothetical landmarks $\boldsymbol{p}_a$ and $\boldsymbol{p}_b$, where $a \neq b$, we modify the 9 corresponding entries of the covariance matrix, $\mathbf{C}_{\text{elastic}} = \{c_{ij}^{\text{elastic}}\}$, as follows

$$c_{ij}^{\text{elastic}} = c_{ij} + w_{ab}\alpha \tag{8}$$

Table 1. Leave-one-out experiments without and with elasticity of $v = 0.2$ and locality $\beta = 2$

Shape model	Mean residual error	Mean residual error with elasticity
Upper 1st molar	0.010820445605	0.010042073205
Upper 2nd premolar	0.017800014466	0.016395261511
Upper 1st premolar	0.015877468511	0.014705151320
Upper canine	0.024195164442	0.022415077314
Lower 1st molar	0.011680148542	0.010770596564

where $\alpha = \frac{vN}{v_{\text{reg}}n}$ being a parameter to control the amount of regularization, and v_{reg} is found experimentally such that $v = 0$ implies no elasticity added and $v = 1$ implies maximum elasticity. The elasticity influences the least significant eigenmodes the most, and should be kept sufficiently small in order not to destroy the statistical properties of the training data.

The parameters will experimentally be found by performing the leave-one-out (LOU) experiments on the corresponding shape models based on PCA of $\mathbf{C}_{elastic}$. The goal is thus to find a set of parameters that decreases the residual error for all models in the experiment. The following results were achieved with $v_{\text{reg}} = 100$. Table 1 demonstrates the generally better results. With higher values of v than $v = 0.2$, the most significant eigenvectors slowly started changing direction, until the corresponding eigenmode changed significance at around $v = 0.9$, and therefore we accept $v = 0.2$ as the maximum. The amount of non-zero eigenmodes created from $\mathbf{C}_{elastic}$ are typically as many as kn, but a big amount of the least significant eigenmodes can be removed while still keeping the model more general than in the pure statistical model.

4 Reconstruction with Elasticity

We now wish to solve the problem of missing data for crown construction. Let $\boldsymbol{y}$ be an incomplete shape vector with $l < n$ points, and $\boldsymbol{x}$ be the corresponding full shape. Then, we wish to find a linear transformation $\mathbf{L} : \mathbb{R}^n \mapsto \mathbb{R}^l$ such that

$$\boldsymbol{y} = \mathbf{L}\boldsymbol{x} \tag{9}$$

This system is an overdetermined system of equations. Since we cannot expect to find a linear combination of the training samples that solves (9) exactly. Instead the values of $\boldsymbol{x}$ can be found by minimizing an energy functional,

$$E(\boldsymbol{x}) = ||\mathbf{L}\boldsymbol{x} - \boldsymbol{y}||^2 \tag{10}$$

which may be solved using the linear least squares method. Assume now that $\boldsymbol{y}$ has been subtracted with the (dimension reduced) mean $\bar{\boldsymbol{x}}$, so that a model approximation can be calculated as $\boldsymbol{x} = \boldsymbol{\Phi}\boldsymbol{b}$. Inserting this into (10) we get

$$E(\boldsymbol{b}) = ||\mathbf{L}(\boldsymbol{\Phi}\boldsymbol{b}) - \boldsymbol{y}||^2 \tag{11}$$

Table 2. Reconstruction of molar training sample for the normal model and the elastic model that includes regularized elasticity

Removed landmarks out of 44 total	Normal model $E_{\mathbf{r}_{rec}}$	Elastic model $E_{\mathbf{r}_{rec}}$
1	0.00159	0.00075
5	0.02201	0.00676
12	0.04915	0.01711
36	0.65755	0.65539
36 distr	0.19626	0.11631

However, a fundamental problem with least square fitting of a model to data is that of *overfitting*. Basically, there is a lot of uncertainties in the data, and the model is only an approximation to the real physical system. Hence, we formulate the problem in the Bayes setting: Given an incomplete shape vector $\boldsymbol{y}$, the reconstruction problem consist of finding the optimal model coefficients $\boldsymbol{b}$ for $\boldsymbol{y}$. In terms of probability:

$$P(\boldsymbol{b}|\boldsymbol{y}) = \frac{P(\boldsymbol{y}|\boldsymbol{b})P(\boldsymbol{b})}{P(\boldsymbol{y})} \propto P(\boldsymbol{y}|\boldsymbol{b})P(\boldsymbol{b}) \tag{12}$$

This states that the optimal coefficients $\boldsymbol{b}$ will be the ones with maximum probability, conditioned to $\boldsymbol{y}$. Both the prior probability $P(\boldsymbol{b})$ and the likelihood $P(\boldsymbol{y}|\boldsymbol{b})$ can be derived from the shape model definition. We use a normally distributed on $\boldsymbol{b}$ with a zero mean and covariance matrix equal to the identity, $\boldsymbol{b} \sim N(0, \mathbf{I})$, the probability density can then be written as:

$$P(\boldsymbol{b}) = (2\pi)^{-m/2} \exp\left(-\frac{||\boldsymbol{b}||^2}{2}\right) \tag{13}$$

and the probability density of the likelihood

$$P(\boldsymbol{y}|\boldsymbol{b}) = (2\pi\sigma^2)^{-l/2} \exp\left(-\frac{||(\mathbf{L}(\boldsymbol{\Phi}\boldsymbol{b}) - \boldsymbol{y}||^2}{2\sigma^2}\right) \tag{14}$$

The point of maximum posteriori is found to be [15]

$$\boldsymbol{x} = \bar{\boldsymbol{x}} + \boldsymbol{\Phi}\mathbf{V}\,\mathrm{diag}\left(\frac{w_i}{w_i^2 + \sigma^2}\right)\mathbf{U}^T(\boldsymbol{y} - \mathbf{L}\bar{\boldsymbol{x}}) \tag{15}$$

The reconstruction error for various missing landmarks is shown in Table 2. The reduced elastic model results in a significantly better reconstruction, compared to the normal covariance reconstruction.

The reconstruction with the elastic models showed that some kind of regularization is necessary, due to the increased number of eigenmodes and the effect of the neighborhood relations on the reconstruction. As the complexity of the reconstruction problem increases, the resulting shape quickly becomes very distorted. Our experiments indicate that only a small number of artificial

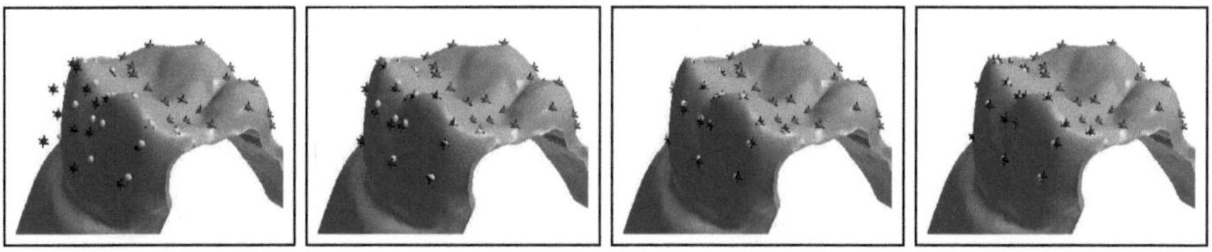

Fig. 4. Reconstruction molar training sample with elastic model, and different amount of elastic eigenmodes removed. Three sets of landmarks are shown: origional (green), the reconstructed (blue), and the ground truth (yellow). From left: 0 %, 50 %, 60 % and 80 % eigenmodes removed.

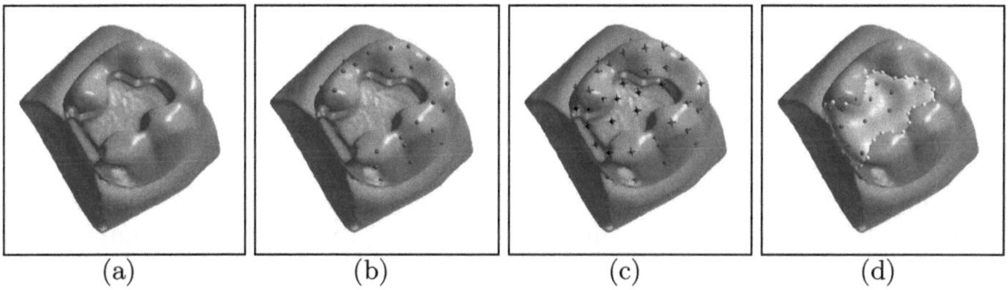

Fig. 5. Reconstruction molar from elastic shape model, with elasticity $v = 0.1$ and locality parameter $\beta = 3$. (a) Original tooth with surface part to be reconstructed. (b) Landmarks to reconstruct from. (c) Reconstructed landmarks (blue). (d) Reconstruction with surface mesh.

eigenmodes should be used for optimal reconstruction. This is show in Figure 4. With this approach we are now able to reconstruct hitherto unseen teeth such as shown in Figure 5.

5 Bite Constrained Reconstruction

One of the limitations of a PDM-based shape model is the undefined limit of shape variation in each eigenmode. We assume limits of three standard deviations of the mean, assuming a Gaussian data distribution. In PCA, when fitting the data to an affine subspace, we cannot guarantee the displacement vectors in some eigenmode not to overlap. The approximated limits on the modes of variation makes it a common problem that the PDM produces shapes with illegal border-overlaps. This is a problem of particular importance when modeling multiple tooth shapes with one model, since the surfaces of neighboring and antagonist teeth in a natural bite are not only close, they also share one or more contact points, and thus inter-model border overlapping is very likely.

In the reconstruction, we wish to maximize the posterior probability of the model parameters $\boldsymbol{b}$ given the incomplete shape vector $\boldsymbol{y}$, by minimizing $||(\mathbf{L\Phi})\boldsymbol{b} - \boldsymbol{y}||^2$. Let $\boldsymbol{p}_y$ be the landmark in $\boldsymbol{y}$ that was just given a new position away from the collision, and let $\boldsymbol{p}_m$ be the corresponding landmark in the

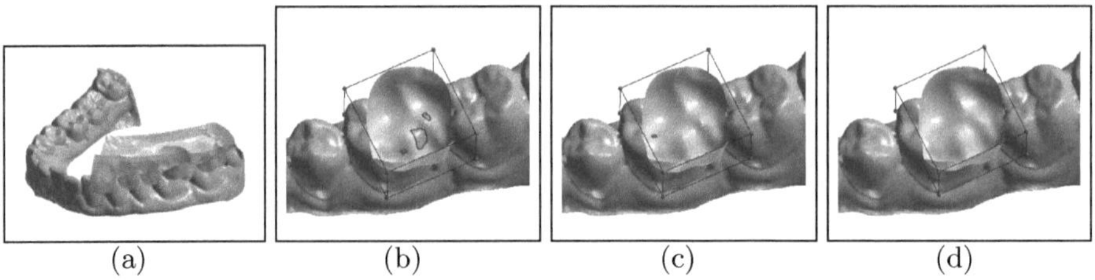

Fig. 6. Reconstruction of upper molar with collision with antagonist. Iterations in the collision response algorithm. (a) Overview. (b)-(d) iterative reduction in collisions.

model to approximate $\boldsymbol{p}_y$. Let Let $\boldsymbol{n}$ be a normal vector in the direction of the penetration. Then, by minimizing the distance between the landmarks as

$$||\boldsymbol{n}^T \boldsymbol{p}_m - \boldsymbol{n}^T \boldsymbol{p}_y||^2 \tag{16}$$

the distance between the landmarks is only measured in the direction of $\boldsymbol{n}$. This expression increases the possibility of estimating a $\boldsymbol{p}_m$ in a non-collision position, by relaxing the movement of $\boldsymbol{p}_m$ in the plane through $\boldsymbol{p}_y$ with the normal $\boldsymbol{n}$, thus motivating the most probably $\boldsymbol{p}_m$ close to this plane. We will also refer to this as a *plane constraint*. To respond to collisions, we apply an iterative algorithm, where we in each step, seek the landmark of deepest penetration. This landmark is then pushed out along the surface normal and apply the plane constraint, i.e. require, that this landmark no longer can move in the normal direction. Steps from this algorithm is illustrated in Figure 6.

6 Conclusion

We have presented a system for reconstructing teeth based on an extension of the Principal Component Analysis. Our extensions include both an elasticity term for the covariance matrix and collision avoidance for antagonist teeth. The conclusion is, that the reconstruction generalize well in terms of missing data, collisions are minimized for improved biting, and preliminary clinical evaluation indicate that the resulting models visualized by variational implicit surfaces are more natural looking that standard reconstructions.

References

1. Modgil, S., Hutton, T., Hammond, P., Davenport, J.: Combining biometric and symbolic models for customised, automated prosthesis design. Artificial Intelligence in Medicine 25, 227–245 (2002)
2. Gürke, S.: Restoration of teeth by geometrically deformable models, `http://citeseer.comp.nus.edu.sg/gurke97restoration.html` (1997)
3. Hayashi, T., Tsuchida, J., Kato, K.: Semi-automatic design of tooth crown using a 3-D dental CAD system, Vocs-1B. In: Proceedings of the 22nd Annual EMBS International Conferenc, Chicago IL, USA,pp. 565–566 (July 2000)

4. Jensen, K.H.: A constrained 3D statistical shape model for automatic reconstruction of teeth in a human bite. Master's thesis, University of Copenhagen (October 2006) ftp://ftp.diku.dk/diku/image/publications/jensen.061023.pdf
5. Dryden, I., Mardia, K.: Statistical Shape Analysis. John Wiley & Sons, New York (1998)
6. Cootes, T., Taylor, C., Cooper, D.H., Graham, J.: Active shape models - their training and application. Computer Vision and Image Understanding 61(1) (1995)
7. Cootes, T., Taylor, C.: Statistical models of appearance for computer vision. Technical report, University of Manchester (March 2004)
8. Bookstein, F.L.: Shape and the information in medical images: A decade of morphometric synthesis. Computer Vision and Image Understanding 66(2), 97–118 (1997)
9. Sethian, J.A.: Level Set Methods and Fast Marching Methods. Cambridge University Press, Cambridge (1999)
10. Pizer, S., Thall, A., Chen, D.: M-reps: A new object representation for graphics. Technical report, University of North Carolina (1999)
11. Hutton, T.J., Buxton, B.F., Hammond, P.: Dense surface point distribution models of the human face. IEEE Workshop on Mathematical Methods in Biomedical Image Analysis (MMBIA),153 (2001)
12. Blanz, V., Vetter, T.: A morphable model for the synthesis of 3d faces. In: Proc. of SIGGRAPH '99, Los Angeles,pp. 187–194 (August 1999)
13. Turk, G., O'Brien, J.F.: Variational implicit surfaces. Technical report, Georgia Institute of Technology (May 1999) Tech Report GIT-GVU-99-15
14. Wang, Y., Staib, L.H.: Boundary finding with correspondence using statistical shape models. In: Proc. IEEE Conf. Computer Vision and Pattern Recognition, pp.338 – 345 (1998)
15. Blanz, V., Vetter, T.: Reconstructing the complete 3d shape of faces from partial information. Technical report, University of Freiburg (2001) Computer Graphics Technical Report No. 1

Sparse Statistical Deformation Model for the Analysis of Craniofacial Malformations in the Crouzon Mouse

Hildur Ólafsdóttir[1,2], Michael Sass Hansen[1], Karl Sjöstrand[1], Tron A. Darvann[2], Nuno V. Hermann[2,3], Estanislao Oubel[4], Bjarne K. Ersbøll[1], Rasmus Larsen[1], Alejandro F. Frangi[4], Per Larsen[2], Chad A. Perlyn[5], Gillian M. Morriss-Kay[6], and Sven Kreiborg[2,3,7]

[1] Informatics and Mathematical Modelling, Technical University of Denmark, Lyngby, Denmark
[2] 3D-Laboratory, School of Dentistry, University of Copenhagen; Copenhagen University Hospital; Informatics and Mathematical Modelling, Technical University of Denmark, Copenhagen, Denmark
[3] Department of Pediatric Dentistry and Clinical Genetics, School of Dentistry, Faculty of Health Sciences, University of Copenhagen, Copenhagen, Denmark
[4] Computational Imaging Lab, Department of Technology - D.326, Pompeu Fabra University, Barcelona, Spain
[5] Division of Plastic Surgery, Washington University School of Medicine, St. Louis, MO, USA
[6] Department of Physiology, Anatomy and Genetics, Oxford University, Oxford, UK
[7] Department of Clinical Genetics, The Juliane Marie Centre, Copenhagen University Hospital, Copenhagen, Denmark

Abstract. Crouzon syndrome is characterised by the premature fusion of cranial sutures. Recently the first genetic Crouzon mouse model was generated. In this study, Micro CT skull scannings of wild-type mice and Crouzon mice were investigated. Using nonrigid registration, a wild-type craniofacial mouse atlas was built. The atlas was registered to all mice providing parameters controlling the deformations for each subject. Our previous PCA-based statistical deformation model on these parameters revealed only one discriminating mode of variation. Aiming at distributing the discriminating variation over more modes we built a different model using Independent Component Analysis (ICA). Here, we focus on a third method, sparse PCA (SPCA), which aims at approximating the properties of a standard PCA while introducing sparse modes of variation. The results show that SPCA outperforms both ICA and PCA with respect to the Fisher discriminant, although many similarities are found with respect to ICA.

1 Introduction

Crouzon syndrome was first described nearly a century ago when calvarial deformities, facial anomalies, and abnormal protrusion of the eyeball were reported

B.K. Ersbøll and K.S. Pedersen (Eds.): SCIA 2007, LNCS 4522, pp. 112–121, 2007.

in a mother and her son [1]. Later, the condition was characterised as a constellation of premature fusion of the cranial sutures (craniosynostosis), orbital deformity, maxillary hypoplasia, beaked nose, crowding of teeth, and high arched or cleft palate. Identification of heterozygous mutations in the gene encoding *fibroblast growth factor receptor type 2* (*FGFR2*) have been found responsible for Crouzon syndrome [2]. Recently a mouse model was created to study one of these mutations ($FGFR2^{Cys342Tyr}$)[3]. Incorporating advanced small animal imaging techniques such as Micro CT, allows for detailed examination of the craniofacial growth disturbances. Studying the craniofacial shape differences in detail contributes to the understanding of the syndrome, surgery planning and diagnosis in humans. A recent study, performing linear measurements on Micro CT scans, proved the mouse model applicable to reflect the craniofacial deviations occurring in humans with Crouzon syndrome [4]. Previously, we have extended this study to assess the local deformations between the groups by constructing a deformable shape and intensity-based atlas of wild-type (normal) mouse skulls. Deforming this atlas to all mice, the craniofacial shape differences can be analyzed [5].

To analyse and interpret these deformations in a meaningful way, it is desirable to reduce the large number of dimensions and at the same time localise the growth deviations with respect to the atlas. This leads us to statistical deformation models (SDMs). These are closely related to statistical shape models but the fact that the whole correspondence field is modelled makes them more powerful. A standard PCA has been a popular approach to build SDMs (e.g. [6,7,8]) but recently different techniques have been applied, e.g. wavelet-based PCA [9].

With respect to the mouse study, PCA was previously performed [10]. This analysis revealed only one discriminating mode of variation, mainly reflecting global differences between the groups. This kind of variation can be hard to interpret and in a recent study, we showed that applying Independent Component Analysis (ICA) to the deformation fields resulted in several discriminating modes, revealing the local differences between the groups. Sparse Principal Components Analysis (SPCA) [11] has proven successful when applied in shape modelling [12]. In this paper we introduce the use of SPCA to build a Sparse Statistical Deformation Model and provide a comparison to a standard PCA and ICA with focus on the discriminative ability. We believe this is the first time SPCA is applied to statistically model deformation fields.

2 Data Material

Production of the $Fgfr2^{C342Y/+}$ and $Fgfr2^{C342Y/C342Y}$ mutant mouse (Crouzon mouse) has been previously described [3]. All procedures were carried out in agreement with the United Kingdom Animals (Scientific Procedures) Act, guidelines of the Home Office, and regulations of the University of Oxford.

For three-dimensional (3D) CT scanning, 10 wild-type and 10 $Fgfr2^{C342Y/+}$ specimens at six weeks of age (42 days) were sacrificed using Schedule I methods and fixed in 95% ethanol. They were sealed in conical tubes and shipped to the

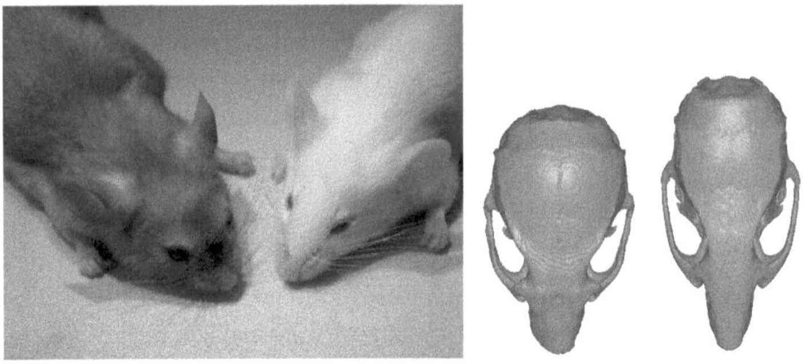

(a) (b) (c)

Fig. 1. (a) Photo of a Crouzon mouse (left) and a wild-type mouse (right). Skulls Extracted from CT images of (b) a Crouzon mouse, (c) wild-type mouse.

Micro CT imaging facility at the University of Utah. Images of the skull were obtained at approximately $46\mu m \times 46\mu m \times 46\mu m$ resolution using a General Electric Medical Systems EVS-RS9 Micro CT scanner. Fig. 1 shows an example of the living mice and the imaging data appearance.

3 Methods

The steps taken to automatically assess the local shape deviations between groups, statistically, from the Micro CT images are the following.

1. Build a craniofacial wild-type mouse atlas from the Micro CT's using non-rigid image registration
2. Match atlas to all 20 cases (wild-type and Crouzon mice) using nonrigid image registration
3. Use the resulting deformation parameters as input to a SPCA

3.1 Atlas Building and Registration

The first two steps of the procedure were presented in [5]. The nonrigid registration algorithm based on B-splines [13,14] was applied. This algorithm uses a transformation model which is a combination of a global and a local transformation model, $\mathbf{T}(\mathbf{x}) = \mathbf{T}_{\text{global}}(\mathbf{x}) + \mathbf{T}_{\text{local}}(\mathbf{x})$. The global transformation model consists in our case of a rigid transformation matrix (with 6 degrees of freedom). The local transformation model describing the nonrigid part of the model is written by the tensor product of the 1D cubic B-splines,

$$\mathbf{T}_{local}(x,y,z) = \sum_{l=0}^{3}\sum_{m=0}^{3}\sum_{n=0}^{3} B_l(u)B_m(v)B_n(w)\mathbf{c}_{i+l,j+m,k+n} \qquad (1)$$

where $\mathbf{c}$ are the parameters of the B-splines ordered in a $p_x \times p_y \times p_y$ lattice. u, v and w are the (x,y,z) image coordinates translated into the lattice coordinates.

3.2 A Sparse Statistical Deformation Model

The third step of the procedure listed above is the main focus of this paper. The control points (parameters) of the B-splines in Equation 1 provide a compact representation of the correspondence fields. As shown in [6] it is sufficient to perform a statistical analysis on these control points to obtain a compact description of the deformations. Using a common reference frame, e.g. an atlas, as the origin of the registrations, the control points for a subject reflect its local deviation from this reference frame. Concatenating the 3D control points for subject i into a row vector $\mathbf{C}_i = [c_1, ..., c_p]$, where $p = 3p_x p_y p_z$, gives the ith row of the $n \times p$ data matrix to analyse (n is the number of observations).

SPCA approximates the properties of a standard PCA while introducing sparsity in the modes of variation. Zou et al. [11] take advantage of formulating PCA as a regression problem leading to the *SPCA criterion*

$$(\hat{\mathbf{A}}, \hat{\mathbf{B}}) = argmin_{\mathbf{A},\mathbf{B}} \sum_{i=1}^{n} ||\mathbf{x}_i - \mathbf{A}\boldsymbol{B}^T\mathbf{x}_i||^2 + \lambda \sum_{j=1}^{k} ||\mathbf{b}_j||^2 + \sum_{j=1}^{k} \delta_j ||\mathbf{b}_j||_1 \quad s.t.\ \mathbf{A}^T\mathbf{A} = \mathbf{I} \tag{2}$$

Here $\mathbf{x}_i$ denotes the ith column of $\mathbf{X}^T$. This formulation assumes k modes to be retained in the model. The columns of $\mathbf{B}$ represent the principal axes (loading vectors $\mathbf{b}_j$, $j = 1, ..., k$) and $\mathbf{B}$ projects observation i onto those axes. The matrix $\mathbf{A}$ takes the observation back to the original space. Hence, the first term measures the reconstruction error of the model. The second term, the L2 penalty is included to ensure a unique solution, also in cases where $p > n$, and the third term, L1 penalty, introduces sparsity. These two latter terms are adopted from Elastic Net regression [15]. The constraint weight, λ, must be chosen beforehand, and has the same value for all PCs, while δ may be set to different values for each PC, providing good flexibility.

The problem in Equation 2 is usually solved iteratively by fixing $\mathbf{A}$ in each iteration, solving for $\mathbf{B}$ using the LARS-EN algorithm [15] and recalculating $\mathbf{A}$. However, when we have $p \gg n$ as in our case, Zou et al. have shown that by letting $\lambda \to \infty$, $\mathbf{B}$ can be determined by soft thresholding[1]

$$\mathbf{b}_j = (|\mathbf{a}_j^T\mathbf{X}^T\mathbf{X}| - \frac{\delta_j}{2})_+ \cdot \text{sign}(\mathbf{a}_j^T\mathbf{X}^T\mathbf{X}), \qquad j = 1, 2, ..., k \tag{3}$$

where k is the number of modes and $\mathbf{a}_j$ is the jth column of $\mathbf{A}$. This approach was taken here enforcing the same fixed level of sparsity in each loading vector by dynamically changing (δ_j) in each iteration. To maximise the total adjusted variance [11] explained by the SPCA, the modes were ordered allowing for perturbations as suggested in [12].

Since the aim of our sparse deformation model is to discriminate between the two groups of mice the final ordering of modes was defined with respect to the Fisher discriminant. That is, the observations were projected onto the principal directions,

[1] $(z)_+$ denotes that if $z < 0$, z is set to 0 and if $z >= 0$, z is kept unchanged. The term is denoted hinge-loss.

the Fisher discriminant between the groups calculated for each mode and the principal directions ordered with respect to decreasing Fisher discriminant score. In general, for class 1 and 2, the Fisher discriminant is defined as

$$F = \frac{(\mu_1 - \mu_2)^2}{\sigma_1^2 + \sigma_2^2}, \tag{4}$$

where μ_i is the mean of class i and σ_i^2 is the variance of class i.

4 Experimental Results

The accuracy of the image registration algorithm (registering the atlas to each of the 20 cases) is essential for the deformation model to be valid. In [5], the

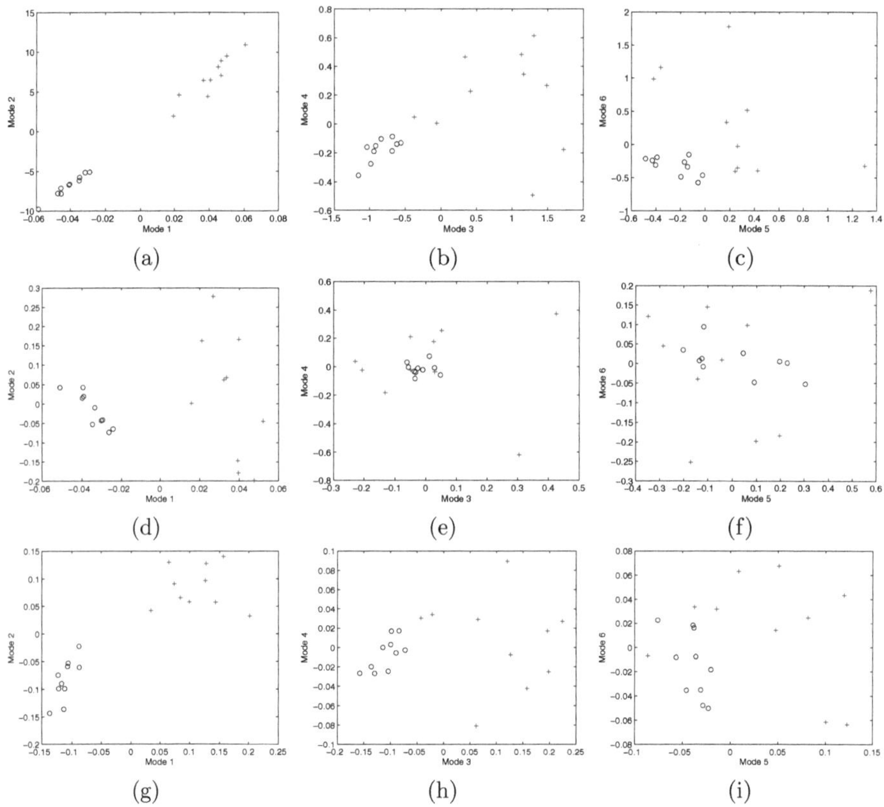

Fig. 2. Projection of observations into the space of the first six components (ordered by Fisher discriminant) using (a-c) SPCA, (d-f) PCA and (g-i) ICA. Crosses denote Crouzon cases while circles denote wild-type cases. (a,d,g) Mode 2 vs. mode 1; (b,e,h) Mode 4 vs. mode 3; (c,f,i) Mode 6 vs. mode 5.

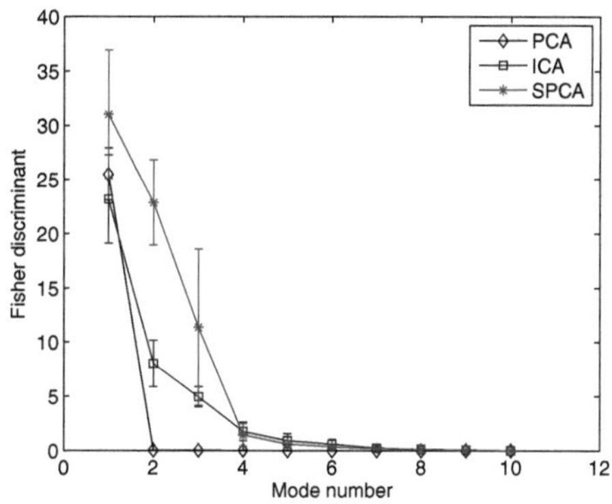

Fig. 3. The Fisher discriminant plotted vs. deformation mode number for PCA, ICA and SPCA. The values are obtained in a leave-one-out experiment providing the error bars (one standard deviation).

manual annotations from two observers were used to assess the registration accuracy. Using the optimal transformations from the image registrations, landmarks were obtained automatically. The landmark positions were statistically compared to those annotated by the human observers. This showed that the automatic method provided as good accuracy as the human observers and, moreover, it was more precise, judged from the significantly lower standard deviation.

The SPCA was applied to the matrix of control points ($p = 21675$). A threshold of 2000 points was used to obtain equal sparsity in each mode of variation. Fig. 2 (a-c) shows the observations projected onto the first six sparse principal directions (ordered by Fisher discriminant score). To evaluate the ability of the sparse SDM to assess the local group differences, it was compared to a standard PCA and our previous approach [16] using ICA [17]. Fig. 2(d-i) shows scatter plots of the first six modes for ICA and PCA, sorted with respect to the Fisher discriminant.

The score plots already give an idea about the discrimination ability of the different approaches. To give a more quantitative measure, the Fisher discriminant was assessed in a leave-one-out fashion for all three approaches. This is plotted with error bars for each of the approaches in Fig. 3.

With emphasis on the group differences, each mode of the sparse model was visualised by selecting the extremes from each group in model space (Fig. 2) and project back into the space of control points. This set of control points generated from the model was then applied to the atlas to obtain the deformed volumes of the two extremes. Subsequently the surfaces were extracted for visualisation. Fig. 4 shows mode 1,3,4 and 6. Mode 2 was excluded from this visualisation due to an overlap in variation with mode 1.

Deforming the atlas along the discriminating modes of the ICA model reveals many similarities between ICA and SPCA. To give an example Fig. 5 shows IC 5 which is closely related to SPC 4.

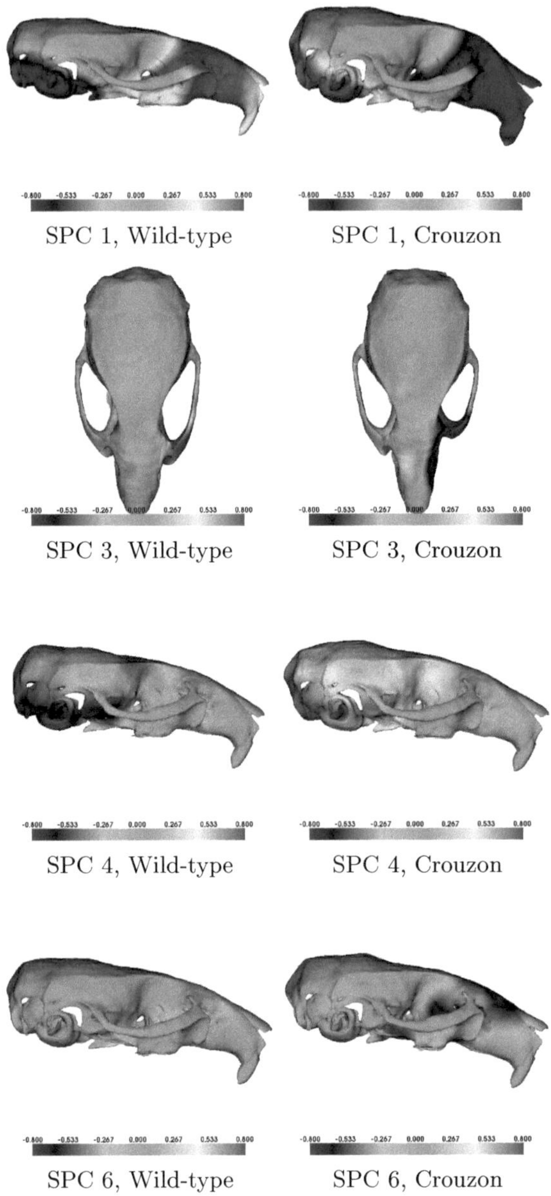

Fig. 4. Sparse Principal Deformation modes 1,3,4 and 6, visualised on surfaces after deforming atlas to the extremes of each mode. The colors are intended to enhance the regions where changes have occurred in the deformed surfaces. The colors denote displacement with respect to atlas (in mm), with positive values (red) pointing outwards.

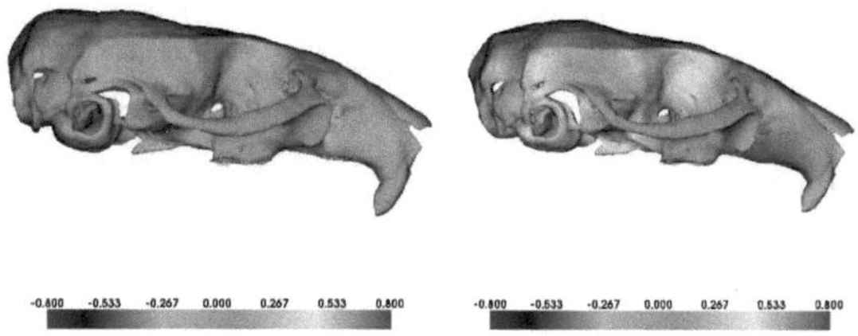

IC 5, Wild-type IC 5, Crouzon

Fig. 5. Independent Deformation mode 5 visualised on surfaces after deforming atlas to the extremes of the mode. The colors are intended to enhance the regions where changes have occurred in the deformed surfaces. The colors denote displacement with respect to atlas (in mm), with positive values (red) pointing outwards.

5 Discussion and Conclusions

The score plots in Figure 2 indicate that both SPCA and ICA are capable of discriminating between the two groups in up to six deformation modes. The standard PCA only discriminates between the groups in the first mode. Figure 3 confirms these speculations. It is evident that PCA is only capable of discriminating between the groups in one mode of variation. SPCA performs slightly better than the ICA, but the ICA seems to be more robust judged from the error bars. Considering the low number of points in the sparse model, this is understandable.

Visualising the sparse deformation modes in Figure 4 indicates that compared to wild-type mice, the skulls of Crouzon mice are higher and longer (SPC 1), are asymmetric with respect to zygoma and nose (SPC 3), have different shape of the middle ear and back of the head (SPC 4), and have an angulated cranial base (SPC 6). These observations correspond up to some degree with what has previously been seen in humans using manual measurements (see e.g. [18]). The asymmetric behaviour seen in SPC 3 can be explained by the full or partial fusion of cranial sutures at different sides and different times. The different shape of the middle ear and the increased angulation of the cranial base has not been reported in humans to our knowledge and may therefore be an important contribution to the understanding of the growth disturbances. The angulation was found in mice both using ICA [16] and PCA (with global transformation model extended to 9 DOFs) [10]. The difference in shape of the middle ear and back of the head was also captured by the ICA approach as seen in Figure 5. In fact SPC 4 and IC 5 are extremely similar, but SPCA seems to create slightly stronger evidence for the group difference. In general, the ICA modes introduce more noise than sparse PCA, since many elements are close to 0, while in SPCA, the sparsity property avoids this. Another advantage of SPCA is that it is solely based on second order statistics making it less committed than ICA, which uses higher order statistics.

In conclusion, with respect to discriminative ability, SPCA and ICA give similar results when applied to model deformations. Both of the approaches outperform a standard PCA. However, due to the simplicity and flexibility of SPCA, it should be the preferred method for this type of analysis.

Acknowledgements

For all image registrations, the Image Registration Toolkit was used under Licence from Ixico Ltd.

References

1. Crouzon, O.: Une nouvelle famille atteinte de dysostose cranio-faciale héréditère. Bull Mem. Soc. Méd Hôp Paris 39, 231–233 (1912)
2. Reardon, W., Winter, R.M., Rutland, P., Pulleyn, L.J., Jones, B.M., Malcolm, S.: Mutations in the fibroblast growth factor receptor 2 gene cause Crouzon syndrome. Nat. Genet. 8, 98–103 (1994)
3. Eswarakumar, V.P., Horowitz, M.C., Locklin, R., Morriss-Kay, G.M., Lonai, P.: A gain-of-function mutation of fgfr2c demonstrates the roles of this receptor variant in osteogenesis. Proc Natl Acad Sci, U.S.A. 101, 12555–12560 (2004)
4. Perlyn, C.A., DeLeon, V.B., Babbs, C., Govier, D., Burell, L., Darvann, T., Kreiborg, S., Morriss-Kay, G.: The craniofacial phenotype of the Crouzon mouse: Analysis of a model for syndromic craniosynostosis using 3D Micro CT. Cleft Palate Craniofacial Journal 43(6), 740–747 (2006)
5. Ólafsdóttir, H., Darvann, T.A., Hermann, N.V., Oubel, E., Ersbøll, B.K., Frangi, A.F., Larsen, P., Perlyn, C.A., Morriss-Kay, G.M., Kreiborg, S.: Computational mouse atlases and their application to automatic assessment of craniofacial dysmorphology caused by Crouzon syndrome. Journal of Anatomy (submitted) (2007)
6. Rueckert, D., Frangi, A.F., Schnabel, J.A.: Automatic construction of 3D statistical deformation models of the brain using nonrigid registration. IEEE Trans. on Medical Imaging 22(8), 1014–1025 (2003)
7. Mohamed, A., Zacharaki, E., Shen, D., Davatzikos, C.: Deformable registration of brain tumor images via a statistical model of tumor-induced deformation. Medical Image Analysis 10(5), 752–763 (2006)
8. Loeckx, D., Maes, F., Vandermeulen, D., Suetens, P.: Temporal subtraction of thorax CR images using a statistical deformation model. IEEE Transactions 22(11), 1490–1504 (2003)
9. Xue, Z., Shen, D., Davatzikos, C.: Statistical representation of high-dimensional deformation fields with application to statistically constrained 3D warping. Medical Image Analysis 10(5), 740–751 (2006)
10. Ólafsdóttir, H., Darvann, T.A., Ersbøll, B.K., Oubel, E., Hermann, N.V., Frangi, A.F., Larsen, P., Perlyn, C.A., Morriss-Kay, G.M., Kreiborg, S.: A craniofacial statistical deformation model of wild-type mice and Crouzon mice. In: International Symposium on Medical Imaging 2007, San Diego, CA, USA, The International Society for Optical Engineering (SPIE) (2007)
11. Zou, H., Hastie, T., Tibshirani, R.: Sparse principal component analysis. Technical report, Statistics Department, Stanford University (2004)
12. Sjöstrand, K., Stegmann, M., Larsen, R.: Sparse principal component analysis in medical shape modeling. In: International Symposium on Medical Imaging 2006, San Diego, CA, USA, vol. 6144. The International Society for Optical Engineering (SPIE) (2006)

13. Rueckert, D., Sonoda, L.I., Hayes, C., Hill, D.L.G., Leach, M.O., Hawkes, D.J.: Nonrigid registration using free-form deformations: application to breast MR images. IEEE Trans. on Medical Imaging 18(8), 712–721 (1999)
14. Schnabel, J.A., Rueckert, D., Quist, M., Blackall, J.M., Castellano-Smith, A.D., Hartkens, T., Penney, G.P., Hall, W.A., Liu, H., Truwit, C.L., Gerritsen, F.A., Hill, D.L.G., Hawkes, D.J.: A generic framework for non-rigid registration based on non-uniform multi-level free-form deformations. In: Niessen, W.J., Viergever, M.A. (eds.) MICCAI 2001. LNCS, vol. 2208, pp. 573–581. Springer, Heidelberg (2001)
15. Zou, H., Hastie, T.: Regularization and variable selection via the elastic net. Journal of the Royal Statistical Society: Series B (Statistical Methodology) 67(2), 301–320 (2005)
16. Hansen, M.S., Olafsdóttir, H., Darvann, T.A., Hermann, N.V., Oubel, E., Larsen, R., Ersbøll, B.K., Frangi, A.F., Larsen, P., Perlyn, C.A., Morris-Kay, G.M., Kreiborg, S.: Estimation of independent non-linear deformation modes for analysis of craniofacial malformations in crouzon mice. In: Miles Wernick, J.A.F. (ed.) 2007 IEEE International Symposium on Biomedical Imaging, IEEE, Los Alamitos (2007)
17. Hyvärinen, A.: Survey on independent component analysis. Neural Computing Surveys 2, 94–128 (1999)
18. Kreiborg, S.: Crouzon Syndrome - A Clinical and Roentgencephalometric Study. Doctorate thesis, Institute of Orthodontics, The Royal Dental College, Copenhagen (1981)

Monocular Point Based Pose Estimation of Artificial Markers by Using Evolutionary Computing

Teuvo Heimonen and Janne Heikkilä

University of Oulu, Information Processing Laboratory, Linnanmaa, Po Box 4500, 90014 University of Oulu, Finland
teuvo.heimonen@ee.oulu.fi
http://www.ee.oulu.fi/mvg/mvg.php

Abstract. Evolutionary computation techniques are being increasingly applied to a variety of practical and scientific problems. In this paper we present a evolutionary approach for pose estimation of a known object from one image. The method is intended to be used in pose estimation from only a few model point - image point correspondences, that is, in cases in which traditional approaches often fail.

1 Introduction

Pose estimation based on model point - image observation correspondences is an essential problem in many computer and robot vision applications. In our special interest are applications of computer aided surgery, where pose of, for example, surgical instrument with respect to the patient need to be determined accurately and robustly. In these applications it has been noted sensible to attach reflective, spherical fiducials rigidly to the object and thus obtain reliable (point) observations from different imaging directions.

Several different approaches to the point correspondence based pose estimation problem has been reported both in photogrammetry and computer vision literature (see. e.g. [1,2]). The proposed methods can be roughly divided to two groups: analytical (see. e.g. [3,4,5]) and iterative (see e.g. [6,7,5,8]). General deficiency with the analytic methods is the high sensitivity to the noise [5]. Because this the pose estimation results may be far from true values. Iterative approaches, on the other hand, need typically a good initial pose estimate. Even if the initial estimate is near the true solution, which is not always the case, these methods may find only a local optimum of the highly multi-modal solution space of the pose estimation problem [9], and thus not produce reasonable solution.

Evolutionary algorithms (EA), have also been applied to the pose estimation from one image [10,11,12,13]. They are reported to offer the advantages like autonomy, robustness against diverging and local optimums, and better noise and outlier immunity [9,11]. Common to the previous EA-approaches is that they have used six genes (parameters) of a chromosome (solution candidate) to

B.K. Ersbøll and K.S. Pedersen (Eds.): SCIA 2007, LNCS 4522, pp. 122–131, 2007.

define pose: three for position and three for orientation. In the previous EA-based pose estimation papers the usage of minimum or near minimum number of point correspondences has not been discussed.

In order to robustly estimate the pose of an artificial marker comprising only a few, usually coplanar, point fiducial, we propose here a two-phase, geometrically constrained approach in which

1. Mathematical framework of the problem is formulated so that minimal number of genes are needed, and solution space constrains are inbuilt.
2. Initial search space of the genes is not strictly limited. The restriction of the search space is performed by the algorithm.

The correspondence of the model points and image observations is solved automatically in the first phase of the approach using index genes like in [13]. Similar to [11] we use real number presentation for the pose genes and apply Gaussian mutator and Blend crossover as the genetic operators. We also utilize kick-out genetic operator suggested in [12]. During our procedure no other minimization technique than the evolution is used.

The rest of the paper is organized as follows: In section 2 mathematical framework of our method is presented. In section 3 outline of our genetic algorithm approach is described and genetic operators are presented. In section 4 test and results of these test are presented and paper is concluded in section 5.

2 Mathematical Framework of the Method

The pose estimation problem can be formulated as follows (see also Fig. 1 a)): Determine the rigid transformation between camera coordinate frame (**C**) and model coordinate frame (**M**) when some number of image observations $p_i^C = [u_i, v_i, f]^T$ and corresponding model points $P_i^M = [X_i, Y_i, Z_i]^T$ are available. The coordinates of a model point in the camera frame $P_i^C = [x_i, y_i, z_i]^T$ can be obtained from the rigid transformation

$$P_i^C = RP_i^M - T, \tag{1}$$

where R is a 3x3 rotation matrix and T 3x1 translation vector. The collinear relation of a model point P_i^C and its image p_i^C can be presented by

$$k_i p_i^C = P_i^C, \tag{2}$$

where k_i is scalar coefficient. By combining (1) and (2) we can compute the actual image coordinates in **C** from the model coordinates in **M** with

$$\begin{bmatrix} u_i \\ v_i \end{bmatrix} = \frac{f}{r^3 P_i^M - t_z} \begin{bmatrix} r^1 P_i^M - t_x \\ r^2 P_i^M - t_y \end{bmatrix}, \tag{3}$$

where f is the focal length known from the camera calibration (throughout of this paper we assume that intrinsic camera calibration parameters are known), r^i is the row i of R, and $[t_x, t_y, t_z]$ are the components of T.

According to (2) the coordinates of a model point P_i^C in camera frame can be, in a noise-free case, obtained from the image observations by multiplying the image observations by some factor k_i. The distance between two model points P_i and P_j should naturally be the same both in the model and in the camera frames:

$$D_{ij} + \eta = d_{ij} = \left\| k_i p_i^C - k_j p_j^C \right\| = \sqrt{k_i^2 (p_i^C)^T p_i^C - 2k_i k_j (p_i^C)^T p_j^C + k_j^2 (p_j^C)^T p_j^C}, \quad (4)$$

where $i = 1, .., n-1, j = 2, .., n, i \neq j$, D_{ij} is the known distance between the model points P_i^M and P_j^M, η is a noise term, and d_{ij} the distance between the computed points P_i^C and P_j^C. The distances D_{ij} are assumed to be exactly known. From n points $\frac{(n-1)n}{2}$ equations like (4) are obtained. The principal task of the pose estimation in this formulation is to solve coefficients k. Our approach is to fix one of the coefficients k and then solve the others from the equations (4). Fixing one of the coefficients k, let say k_i, gives us P_i^C (eq. 2) and two solutions for k_j (eq. 4). Related to the distance (d) between P_i^C and the ray from the projection center through the image point p_j^C to the infinity three cases for the solutions for k_j are possible (see Fig. 1b - 1d). If

1. $d < D_{ij}$, two different solutions for k_j are obtained. Other points available, say $P_{m \neq i,j}^C$, are considered in order to select the more feasible one of these two solutions.
2. $d = D_{ij}$, two equal solutions for k_j are obtained.
3. $d > D_{ij}$, two complex solutions for k_j are obtained. In this case we use $k_j = \frac{(u_i u_j + v_i v_j + f^2) k_i}{u_j^2 + v_j^2 + f^2}$ that yields minimum d_{ij} on condition $P_j^C = k_j \left[u_j, v_j, f \right]^T$.

Typically the 3D pose is defined with six parameters, three for position or translation $[t_x, t_y, t_z]$ and three for orientation or rotation $[\omega, \varphi, \kappa]$. In our approach these six pose parameters are extracted from the rotation matrix R and translation vector T of the rigid transformation (1) after the coefficients k_i and thus the model points P_i^C are obtained. We solve the rigid transformation

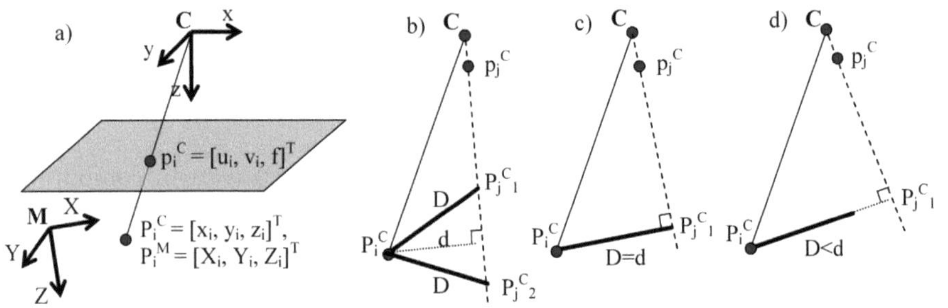

Fig. 1. a): Pose estimation framework. Because of the uncertainty in the image observations p_i^C and p_j^C b) two, c) one, or d) no real solution for the P_j^C is obtained depending on the distance D and the direction of the ray from **C** through p_j^C to infinity.

between the model coordinates in the model frame P_i^M and in the camera frame P_i^C by using method presented in [14].

3 Evolutionary Algorithm Approach

In evolutionary algorithms, a population of candidate solutions of the problem (chromosomes) is submitted repeatedly to the genetic operators (selection, reproduction, crossover and mutation) in order to create new generation of chromosomes with improved fitness with respect to the problem. We propose here a two-phase approach. In the first phase of our approach the correspondence of the points and a coarse estimate for k_1 are solved. In the second phase correspondence is assumed to be known and the final estimate for k_1 is searched for by letting also the image observations vary according to some predefined (noise) distribution. The chromosome encoding and the genetic operators, which are to be presented next, are partly different in the two phases.

3.1 Chromosome Encoding

In the first phase only the coefficient k_1 and the order of the image observations are revealed to the genetic operators. Thus we use chromosomes that comprise the coefficient (real number) and only an index list of the image observations. During the procedure each index (integers $1, 2, .., n$) may and should occur exactly once in the index list. So we have $n + 1$ genes in a chromosome (Fig. 2a). In the second phase the genes are the coefficient k and the image observations $[u_i, v_i]$. In this phase all of the genes can be varied. Thus the a chromosome includes $2n + 1$ real numbered genes (see Fig. 2b).

a)

k_1	$index_1$	$index_2$	...	$index_n$

b)

k_1	u_1	u_2	...	u_n	v_1	v_2	...	v_n

Fig. 2. Chromosome encoding a) in phase 1 and b) in phase 2

3.2 Initialization

In the first phase, the value of the first gene of each initial chromosome are randomly generated according to uniform distribution between such a values that the object can be anywhere between certain distance (along optical axis) from the camera center. The integer indexes of the image observations $[u_i, v_i]$ are initialized into random order. In the second phase, we limit the search base so that the gene values obey isotropic Gaussian distribution around the best solution from the first phase. Different standard deviation of the distribution may be used for different genes.

3.3 Genetic Operators

Cross-over. Cross-over occurs with a constant probability in both phases. The first gene in the first phase and all genes in the second phase of the off-springs resulting from the cross-over operation are linear combination of the corresponding genes of the parents (Blend cross-over). The weight of the linear combination (λ) is randomly generated real number between 0 and 1. Mathematically,

$$gene_i^{offspring} = (1-\lambda)gene_i^{father} + \lambda gene_i^{mother}. \tag{5}$$

For other genes in the first phase (indexes) partially matched cross-over (PMX) is applied in order to guarantee that all indexes are found exactly once in each offspring.

Mutation. Mutation occurs with a constant probability in both phases. The first gene in the first phase of the off-springs resulting from the mutation operation are random values from uniform distribution used also in the initialization phase. There is no mutation operation for other genes of the first phase (indexes). In the second phase the value of the first gene of the off-springs resulting from the mutation operation obeys Gaussian distribution around the best solution found in the first phase. Other genes are from the Gaussian distribution around the initial image observations. The standard deviations of these distributions are predetermined constants.

Selection. Parents for the reproduction are selected by using fitness-proportional tournament selection. In this method two chromosomes picked randomly from the population compete and the fitter is selected. In every generation a certain percent of the fittest choromosomes in current population are selected to form a basis for the population of the next generation (elite selection). Also off-springs are evaluated and the fittest of them are selected to fill in the population for the next generation. In this step of the algorithm the population is also pruned so that any certain chromosome occurs no more than once in the population. If the size of the population is about to decrease because of this pruning, we add new chromosomes into it. These new chromosomes are such that their genes are a little and randomly variated (in maximum $\pm 1\%$ of the current search space dimension) from some randomly chosen chromosome already in the population.

Kick-out. Kick-out occurs if the best solution has not changed in certain number of generations. In the kick-out operation population is re-initialized.

3.4 Fitness

In the first phase the fitness of a chromosome is the inverse of the sum of squared differences of the distances between the model points (D_{ij}) and back-projected points (the right-hand side of eq. (4)), that is

$$fitness^{-1} = \sum \left(D_{ij} - \left\|k_i p_i^C - k_j p_j^C\right\|\right)^2. \tag{6}$$

In order to evaluate the fitness of a chromosome, other coefficients k_j needed to back-project all the image observations to model coordinates are first solved (see section 2).

In the second phase the fitness of a chromosome is the inverse of the sum of distances between the image observations p_i^C and the model points projected to the image plane with (3). In order to incorporate the possible uncertainty knowledge of the p_i^C:s, we use the Mahalanobis distance as a distance measure. So the fitness measure in the second phase is

$$fitness^{-1} = \sum \left((p_i^C - p_i'^C)^T \Sigma^{-1} (p_i^C - p_i'^C) \right) , \qquad (7)$$

where $p_i'^C$ are the projected image coordinates. Different uncertainty estimates (variances) can be used for every p_i^C and also for x- and y-components of any p_i^C using the covariance matrix Σ obtained e.g. from the image feature extraction procedures. In our experiments we used identity matrix as Σ.

3.5 Stop Criteria

We use two different criteria in order to stop the evolution. The first one is the number of generations evaluated: both minimum and maximum number of generations are limited. The second criterion is the fitness of the best solution candidate: if this fitness is better than a specific limit the evolution is ended. Different values for these parameters is used in the first and the second phase of the algorithm.

4 Experiments

The method presented in this paper was evaluated with both synthetic and real image data. In both cases the basic test procedure was the following: The marker was positioned in some position and orientation in the field-of-view of the camera. The image coordinates were either computed according to a pinhole camera model (synthetic data) or extracted from the image of the real camera (real data).

The noisy image coordinates were inputted to the "Extcal" pose estimation method implemented in the calibration toolbox [15] and to the EA-method presented in this paper. The Extcal-method uses DLT for the pose initialization and Levenberg-Marquardt minimization for the refinement, and it can be considered as a traditional bundle-adjustment approach for pose estimation. However, this method does not include correspondence determination and thus the pose estimation with this method was performed using correct point correspondences.

The parameters used in the EA-method are presented in table 1. Different empirically determined fitness limits for the stop criteria were used for different amount of points, for example limits of 0.8 and 0.00001 mm were used for the four point cases for the phase 1 and phase 2 stop criteria, respectively. The search space was limited with $k_1 = [80, 300]$ (absolute values) in the first phase and the standard deviations of 4 and 2*noise standard deviation for k_1 and u_i and v_i, respectively.

Table 1. EA parameters, * = same value for both phases was used

parameter	value
Population size*	150
Minimum number of generations*	20
Maximum number of generations: Phase 1	100
Maximum number of generations: Phase 2	300
Kick-out frequency: Phase 1	5
Kick-out frequency: Phase 2	20
Cross-over probability: Phase 1	0.5
Cross-over probability: Phase 2	0.9
Mutation probability: Phase 1	0.5
Mutation probability: Phase 2	0.5
Elite selection probability: Phase 1	0.3
Elite selection probability: Phase 2	0.5
Selection method*	Tournament
Population initialization distribution: Phase 1	Uniform
Population initialization distribution: Phase 2	Gaussian
Mutation distribution: Phase 1	Uniform
Mutation distribution: Phase 2	Gaussian

4.1 Synthetic Data

In the synthetic data tests a virtual marker with three, four, or five points was positioned randomly inside the field-of-view of the camera and a volume with limits t_x = [-500, 500] mm, t_y = [-500, 500] mm, and t_z = [2000, 8000] mm. The rotations of the marker were limited to ω = [-70, 70] deg, φ = [-70, 70] deg, and κ = [-180, 180] deg. The size of the virtual image sensor was [4.8, 3.6] mm.

Table 2. Average pose parameter errors with different methods. The number after the method name indicates the number of points used in the pose estimation. The symbol 4b stands for the case where 4 four points were used and the Extcal-method failed (The Extcal-estimate was doomed to be a failure when the sum of the distances between image observations and the image coordinates computed with the estimated pose parameters was more than 20 times bigger than the standard deviation of the noise). Tabulated values are mean errors of all measurements (60 repeats x 9 noise levels = 540 measurements) of test in question.

Test	t_x [mm]	t_y [mm]	t_z [mm]	ω [deg]	ϕ [deg]	κ [deg]
Extcal-3	0.917	1.078	28.511	1.941	1.778	1.196
EA-3	0.898	0.997	26.781	2.068	1.738	1.224
Extcal-4	0.277	0.305	7.650	0.264	0.177	0.246
EA-4	0.260	0.275	6.950	0.213	0.151	0.137
Extcal-4b	1.853	2.417	80.900	4.352	3.335	3.866
EA-4b	0.343	0.389	7.338	0.399	0.214	0.341
Extcal-5	0.263	0.300	7.496	0.217	0.188	0.231
EA-5	0.288	0.311	8.027	0.203	0.167	0.139

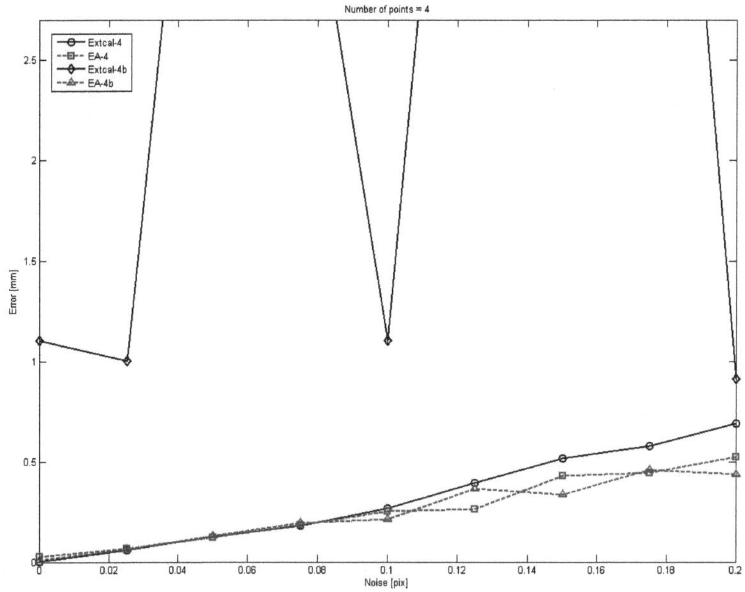

Fig. 3. Average errors in x and y directions with different methods. It should be noted that in the Extcal-4b and EA-4b cases the input data was chosen such that the Extcal-method failed. The EA-method succeed to remain robust also with these kinds of input data.

The model coordinates of the points in a marker were [0 0 0;0 50 0;113 0 0], [0 0 0;0 108 0;100 130 0;113 0 0], [0 0 0;0 108 0;100 130 0; 100 50 0;113 0 0] mm for three-, four-, and five-point marker, respectively. Model coordinates were projected to a image plane according to a pin-hole camera model with 25 mm focal length. Gaussian noise with standard deviations 0, 0.025, 0.05, ..., 0.2 pixels was added to the obtained image coordinates. For each noise level 50 random poses were evaluated.

From the results of this test it was observed that the methods succeeded almost equally in general, but in some special cases in which Extcal-method failed (the failure is caused by the faulty initial pose estimate obtained with the DLT), EA-method still performed satisfactorily. Compiled statistics of the results are presented in Table 2 and example plot is given in Fig. 3.

4.2 Real Images

A marker with four points (a passive planar rigid body from Northern Digital Inc.) was attached to a machine tool and moved to 19 poses. The poses were such that the marker was either translated in the x- direction (limits [-300, 300] mm) or rotated about y-axis ([-55, 40] deg). In every pose target was halted and 15 measurements were executed. One measurement comprised acquiring coordinates by using NDI Polaris tracking system and an image by using Sony

Table 3. Average errors in the position of the rotation axis and the rotation about it between different poses

Test	t_x [mm]	t_y [mm]	t_z [mm]	ω [deg]	ϕ [deg]	κ [deg]
Extcal-4	2.015	2.236	2.763	0.412	1.498	0.277
EA-4	2.036	2.089	2.776	0.461	1.490	0.187

XCD-X710 camera with Rainbow TV zoom lens. The model coordinates of the points in a marker were [0 0 0;0 108 0;100 130 0;113 0 0] mm.

From the images the centroids of the spherical markers were extracted by simply thresholding the image and by computing the center of masses of the spheres. The repeatability of the image formation and feature extraction was computed by comparing the image coordinates extracted from the images of same pose. The standard deviation of the extracted image coordinates was 0.030 pix.

The pose of the marker was estimated from every image using again both Extcal- and EA-methods. The input parameters of the EA-method were same as in synthetic data tests (see Tables 1), except that 0.030 pix for the noise standard deviation was used. The motion between of different poses was determined and compared to the known values and the motion information obtained with the Polaris tracker.

As in the synthetic data tests also here almost similar performance was observed (Table 3). The significantly larger errors especially in t_x, t_y, and ϕ in these tests than in the synthetic data tests was observed to be caused by the inaccuracy of the camera calibration and thus deficient correction of the geometric distortion.

5 Discussion

We proposed here a pose estimation method based on evolutionary computing. The method proved to be robust and reliable. It does not need accurate initial guess and finds point correspondences automatically.

We presented some examples of the test results with three, four, and five coplanar points. It should be noted that the proposed method can be directly used with 3D model points as well. The performance of the method is similar than we have demonstrated with the coplanar points.

As the evolution based methods in general, also the method proposed here is slow compared to analytical or more traditional, iterative pose estimation methods (computation of a new generation of ten chromosomes in our EA-method is about equal to the computation of one iteration in the Extcal-method). However, we believe that evolution based pose estimation is a feasible option for pose estimation in any application without a strict demand for real-time performance.

References

1. Faugeras, O.: Three-Dimensional Computer Vision: A Geometric Viewpoint. MIT Press, Cambridge, MA (1993)
2. Horaud, R., Dornaika, F., Lamiroy, B., Christy, S.: Object pose: The link between weak perspective, paraperspective and full perspective. International Journal of Computer Vision 22(2), 173–189 (1997)
3. Fischler, M., Bolles, R.: Random sample consensus: A paradigm for model fitting with applications to image analysis and automated cartography. Communications of the ACM 24(6), 381–395 (1981)
4. Horaud, R., Conio, B., Leboulleux, O., Lacolle, B.: An analytic solution for the perspective 4-point problem. CVGIP 47, 33–44 (1989)
5. Haralick, R., Joo, H., Lee, C., Zhuang, X., Vaidya, V., Kim, M.: Pose estimation from corresponding point data. IEEE Transactions on Systems, Man. and Cybernetics 19(6), 1426–1446 (1989)
6. Oberkampf, D., DeMenthon, D.F., Davis, L.S.: Iterative pose estimation using coplanar feature points. Journal of Computer Vision and Image Understanding 63(3), 495–511 (1996)
7. Lowe, D.: Fitting parameterized three-dimensional models to images. IEEE Transactions on Pattern Analysis and Machine Intelligence 13(5), 441–450 (1991)
8. Heikkilä, J.: Geometric camera calibration using circular control points. IEEE Transactions on Pattern Analysis and Machine Intelligence 22(10), 1066–1077 (2000)
9. Ji, Q., Zhang, Y.: Camera calibration with genetic algorithms. SMC-A. 31(2), 120–130 (2001)
10. Toyama, F., Shoji, K., Miyamichi, J.: Model-based pose estimation using genetic algorithm. In: ICPR '98: Proceedings of the 14th International Conference on Pattern Recognition, vol. 1, p. 198. IEEE Computer Society, Washington DC, USA (1998)
11. Hati, S., Sengupta, S.: Robust camera parameter estimation using genetic algorithm. Pattern Recogn. Lett. 22(3-4), 289–298 (2001)
12. C.Rossi, M.Abderrahim, J.C.Díaz: Evopose: A model-based pose estimation algorithm with correspondences determination. In: ICMA 2005: Proceedings of the IEEE International Conference on Mechatronics and Automation 2005, Niagara Falls, Canada, pp. 1551–1556 (2005)
13. Yu, Y., Wong, K., Chang, M.: Pose estimation for augmented reality applications using genetic algorithm. SMC-B. 35(6), 1295–1301 (2005)
14. Arun, K., Huang, T., Bolstein, S.: Least-Squares Fitting of Two 3-D Point Sets. IEEE Transactions on Pattern Analysis and Machine Intelligence 9, 698–700 (1987)
15. Heikkilä, J.: Camera calibration toolbox for matlab, `http://www.ee.oulu.fi/~jth/calibr/` (2000)

Camera-to-Camera Mapping for Hybrid Pan-Tilt-Zoom Sensors Calibration

Julie Badri[1,2], Christophe Tilmant[1], Jean-Marc Lavest[1], Quonc-Cong Pham[2], and Patrick Sayd[2]

[1] LASMEA, Blaise Pascal University,
24 avenue des Landais, Aubiere, F-63411 France
[2] CEA, LIST,
Boîte Courrier 65, Gif-sur-Yvette, F-91191 France
{julie.badri,christophe.tilmant,jean-marc.lavest}@univ-bpclermont.fr,
{quoc-cuong.pham,patrick.sayd}@cea.fr

Abstract. Video surveillance becomes more and more extended in industry and often involves automatic calibration system to remain efficient. In this paper, a video-surveillance system that uses stationary-dynamic cameras devices is presented. The static camera is used to monitor a global scene. When it detects a moving object, the Pan-Tilt-Zoom (PTZ) camera is controlled to be centered on this object. We describe a method of camera-to-camera calibration, integrating zoom calibration in order to command the angles and the zoom of the PTZ camera. This method enables to take into account the intrinsic camera parameters, the 3D scene geometry and the fact that the mechanism of inexpensive camera does not fit the classical geometrical model. Finally, some experiment results attest the accuracy of the proposed solution.

Keywords: Camera-to-camera calibration, zoom calibration, visual servoing, Pan-Tilt-Zoom camera, video surveillance.

1 Introduction

Surveillance companies want simultaneously to monitor a wide area with a limited camera network and to record identifiable imagery of all the people passing through that area. To solve this problem, it has been proposed to combine static cameras having a large field-of-view with Pan-Tilt-Zoom cameras. Indeed, it is possible to control the angle of rotation of the PTZ camera (pan and tilt angles) and the zoom. In practice, the system proceeds as follows. A scene event as a moving subject is detected and located using the static camera. The PTZ camera must be controlled with the information of the static camera in order to adjust its pan, tilt and zoom such as the object of interest remains in the field of view. Then, the high resolution image can be recorded in order to apply face or gesture recognition algorithm, for example.

The main problem to solve in this system is how to control the PTZ camera parameters from the information of the object position extracted in the static

B.K. Ersbøll and K.S. Pedersen (Eds.): SCIA 2007, LNCS 4522, pp. 132–141, 2007.

camera. These last years, two approaches emerged. Either, each camera is calibrated in order to obtain the intrinsic and extrinsic camera parameters before to find a general relation between 2D coordinates in the static camera and the pan and tilt angles, like Horaud *et al.* [5] and Jain *et al.* [6] or cameras are not calibrated like Zhou *et al.* [12] and Senior *et al.* [9]. They learned a look-up-table (LUT) linking several positions in the static camera with the corresponding pan-tilt angles. Then, for another point, they estimate the corresponding pan-tilt angles by interpolating using the closest learned values.

In order to position the presented paper, we develop the existing works. In the first case, wthe problem camera calibration, and in particular dynamic camera calibration has been extensively addressed. Either cameras calibration is based on the fact that it is a stereo-vision system like Horaud *et al.* [5] or each camera is calibrated separately like Jain *et al.* [6]. In this case, most existing methods for calibrating a pan-tilt camera suppose simplistic geometry model of motion in which axes of rotation are orthogonal and aligned with the optical center ([1], [2], [10], [4], [11]). If this assumptions can be suitable for expensive mechanisms, they are not to model the true motion of inexpensive pan-tilt mechanisms. In reality a single rotation in pan rotation induces a curved displacement in the image instead of straight lines.

Recently, Jain *et al.* [6] proposed a new calibration method with more degrees of freedom. As with other methods the position and orientation of the camera's axes can be calibrated, but it can be also calibrated the rotation angle. It is more efficient, more accurate and less computationally expensive than the previous works. Actually, Jain *et al.* [6] mean to be the only one to propose a method without simplistic hypothesis. The calibration step involves the presence of a person to deal with the calibration marks. So, this method can not be used in the goal of a turnkey solution for a no-expert public.

Now, methods based on the no-direct camera calibration are focused. Few people have explored this approach. The first were Zhou *et al.* [12] who used collocated cameras whose viewpoints are supposed to be identical. The procedure consisted of collecting a series of pixel location in the stationary camera where a surveillance subject could later appear. For each pixel, the dynamic camera was manually moved to center the image on the subject. The pan and tilt angles were recorded in a LUT indexed by the pixel coordinates in the stationary camera. Intermediate pixels in the stationary camera were obtained by a linear interpolation. At run time, when a subject is located in the stationary camera, the centering maneuver of dynamic camera used the recorded LUT. The advantage of this approach is that calibration marks are not used. This method is based on the 3D information of the scene but the LUT is learned manually.

More recently, Senior *et al.* [9] proposed a procedure more automatic than Zhou *et al.* [12]. To steer the dynamic camera, they need to know a sequence transformations to enable to link a position with the pan-tilt angles. This transformations are adapted to pedestrian tracking. An homography links the foot position of the pedestrian in the static camera with the foot position in the dynamic camera. A transformation links the foot position in the dynamic

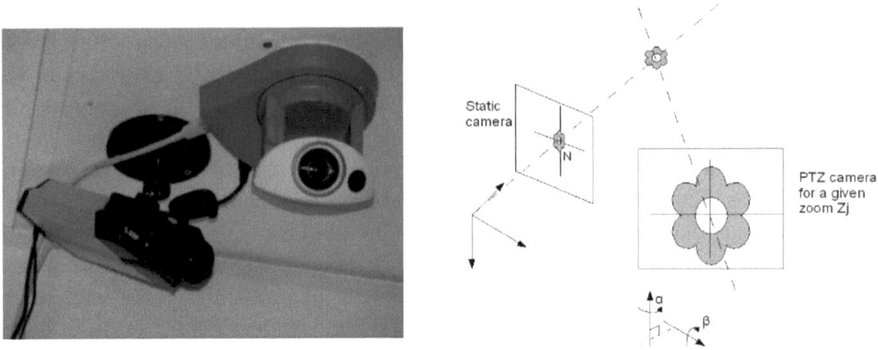

Fig. 1. Our system of collocated cameras : the static camera is on the left and the dynamic camera is on the right

camera with the position in the dynamic camera. Finally, a third transformation encoded in a LUT as Zhou *et al.* [12] links the head position in the dynamic camera with pan-tilt angles. These transformations are learned automatically from unlabelled training data. The default of this method is the step of the establishment of the training data. If this method is used for a turnkey solution for a no-expert public and unfortunately the scene changes, it is impossible that a no-expert public could constitute a good and complete training data in order to update the system.

A solution in the continuity of works of Zhou *et al.* [12] and Senior *et al.* [9] is proposed. Indeed, [6] need the depth information of the object in the scene. So they need to use stereo triangulation. But, like in figure 1, this system is composed of two almost collocated cameras.

Moreover, in the goal of an automatic and autonomous system, the solution of Jain *et al.* [6] and Senior *et al.* [9] are not used. In fact, they need an expert person knowing to use a calibration marks in the case of Jain *et al.* [6] or knowing to extract the good datas to make the training datas in the case of Senior *et al.* [9].

In this paper, an automatic and autonomous solution is presented for a no-calibrated pair of static-dynamic cameras. The solution adapts automatically to its environment. In fact, if the pair of cameras are in a changing environment, this solution can be restarted regularly.

2 Automatic Supervised Multi-sensor Calibration Method

The system used is composed of a static camera with a wide field of view and a Pan-Tilt-Zoom camera with a field-of-view 2.5 times smaller than that of the static camera at the minimal zoom. In the following, the images of the static and the PTZ camera are respectively noted I_s and $I_d(\alpha, \beta, Z)$. The parameters (α, β, Z) represent the pan, tilt and zoom parameters of the PTZ camera.

The method presented in this section enables to learn the relation ζ, for all zoom Z_j, between the pixel coordinates (x_s, y_s) of I_s and the pan-tilt parameters depending of the zoom Z_j :

$$(\alpha_{Z_j}, \beta_{Z_j}) = \zeta(x_s, y_s, Z_j) . \tag{1}$$

To learn the relation ζ, two steps are needed. The first step is the computation of an camera-to-camera mapping (LUT). The LUT gives the relation between n_s pre-defined pixels of I_s and the pan-tilt parameters such as the pixel is mapped to the center C_d of $I_d(\alpha, \beta, Z)$ for different samples of the zoom $Z_{j=0,10,...,m}$. In the following, the n_s pre-defined pixels of I_s are called nodes and noted $\mathbf{N} = \{N_s^0, N_s^1, ..., N_s^{n_s-1}\}$. The second step is the extension of the LUT for all the pixel of I_s and all values of the zoom Z.

2.1 Camera-to-Camera Calibration : 3D Scene Constraints Integration in LUT Computation

Computation of the LUT. The computation of the LUT integrates two loops: (1) computation of the LUT for a constant zoom Z_0 for all the nodes of $\mathbf{N}$ and (2) computation of the LUT for each zoom $Z_{j=0,1,...,m}$ for all the nodes of $\mathbf{N}$.

To begin the computation of the LUT at the zoom Z_0, we need to be in the neighbourhood V_0 of N_s^0. In order to move automatically the PTZ camera in V_0, pan-tilt parameters are randomly selected until the field-of-view of the PTZ is in a good neighbourhood of N_s^0.

The main steps of this procedure are :

1. Initialization on N_s^0;
2. For **each node** N_s^i in the static camera :
   (a) Selection of images I_s and $I_d(\alpha, \beta, Z_0)$ to be compared
   (b) Extraction and robust matching of interest points between I_s' and $I_d(\alpha, \beta, Z_0)$
   (c) Computation of an homography H between interest points of I_s' and $I_d(\alpha, \beta, Z_0)$
   (d) Computation of the $N_{i,s}$ coordinates in $I_d(\alpha, \beta, Z_0)$: $N_d^i = H \times N_s^i$
   (e) Command of the dynamic camera in order to N_d^i catch up with C_d
   (f) Process $N_{i,s}$ until the condition $|N_{i,d} - C_d| < \epsilon$ is reached. Otherwise we stop the loop after k loops
3. Go to the step (2) to process the node N_s^{i+1};

At the step (2a), a small image I_s' is extracted from the complete image I_s around the node N_s^i to process in order to optimize the matching result. In fact, the field-of-view of the PTZ camera is smaller than the one of the static camera. So, the size of I_s' is defined such as the field-of-view of I_s' is nearly the same that the field-of-view of the PTZ camera.

For the step (2b), the scale-invariant feature transform (SIFT) method proposed by Lowe [7] for extracting and matching distinctive features from images of static and PTZ cameras is used. The features are invariant to image scale, rotation, and partially invariant to changing viewpoints, and change in illumination.

At the step (2c), we assume that locally the area in the static and PTZ cameras can be approximated by a plane. Locally, the distortion in I_s can be considered insignificant. So, the homography H is searched such as it is the homography that best matches a set of points extracts of the lists of points obtained previously. The set of correspondences contains a lot of outliers. So, the homography H is robustly estimated with a RANSAC procedure. A homography is randomly computed from only four points, and test, how many other points satisfy it many times. The optimal solution is the homography which has the highest number of good points.

When the coordinates of N_s^i in I_d are known, the parameters (α, β) of the PTZ camera must be estimated in order to insure the convergence of N_d^i to the center C_d. We use a proportional controller based on the error between the coordinates of N_d^i and the coordinates of C_d, such as it minimizes the criterion of the step (2e). We assume that the pan-tilt axes and the coordinates axes are collocated. So we can write :

$$\begin{pmatrix} \alpha \\ \beta \end{pmatrix} = \begin{pmatrix} K_{x \to \alpha} & 0 \\ 0 & K_{y \to \beta} \end{pmatrix} \begin{pmatrix} \Delta x \\ \Delta y \end{pmatrix} \tag{2}$$

The error $(\Delta x, \Delta y)$ corresponds to the coordinates of $N_{i,d} - C_d$. As we are in a static training problem, a proportional controller is sufficient.

This procedure is repeated as long as $|N_d^i - C_d| < \epsilon$ is not achieved. If the system diverges, we stop after k loops.

After convergence, to steer the PTZ camera in the neighbourhood of the next point N_s^{i+1}, the needed angles are estimated with the knowledge of the previous learned points. For a new point N_s^{i+1}, we search the closest point among the previous processed points for each direction. The pan-tilt parameters of the best result are used to move the PTZ camera in the neighbourhood of N_s^{i+1}.

For the computation of the LUT for the other zooms, the same procedure is used. For a given zoom Z_j, instead of comparing a small image of I_s with an image I_d for a node N_s^i, an image I_d at the zoom Z_{j-1} centerred on the node N_s^i is used as the reference image for the visual servoing of the PTZ camera on the node N_s^i at the zoom Z_j.

Construction of N. The choice of the n_s nodes N_s^i depends on the information contained in the 3D scene. For an image I_s, interest points are extracted with SIFT method. For each pixel of I_s, we compute the density function of the interest points with the Parzen window method. The size of the window depends on the relation between each field-of-view of cameras. Then, we search the pixel in I_s with the maximal probability : it is the first node N_s^0 of $\mathbf{N}$. The value zero is given to the probability of the pixels around N_s^0 in order to obtain a best repartition of nodes. This procedure is repeated until the last pixel of I_s with a no-zero probability. The figure 2 shows the nodes obtained with this procedure.

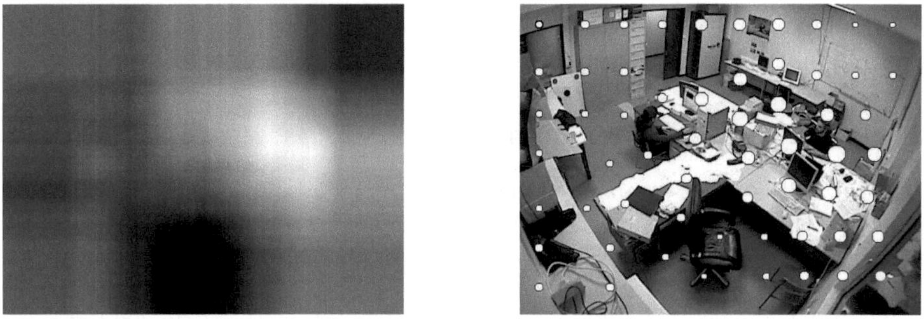

Fig. 2. The left figure represents the estimation of the density function of the interest points in I_s by using the Parzen window method. The right figure shows an example of the extracted grid to I_s, more the circle is important more the density function is important.

2.2 Expansion of LUT

After the previous section 2.1, we obtain a LUT for n_s pixels of the static image I_s for m different zooms. In order to complete the LUT for all the pixels of I_s and all values of the zoom, an approximation of this data is searched.

In such interpolating problems, Thin-Plate-Spline (TPS) interpolation, proposed by Bookstein *et al.* [3], is often preferred to polynomial interpolation because it gives similar results, even when using low degree polynomials and avoids Runge's phenomenon for higher degrees (oscillation between the interpolate points with a big variation). A TPS is a special function defined piecewise by polynomials.

The computation of the interpolation relation ζ resulting of the TPS method needs a training step. The correspondences between the coordinates of a pixel of I_s for a given zoom Z_j and the pan-tilt parameters for Z_j learned during the

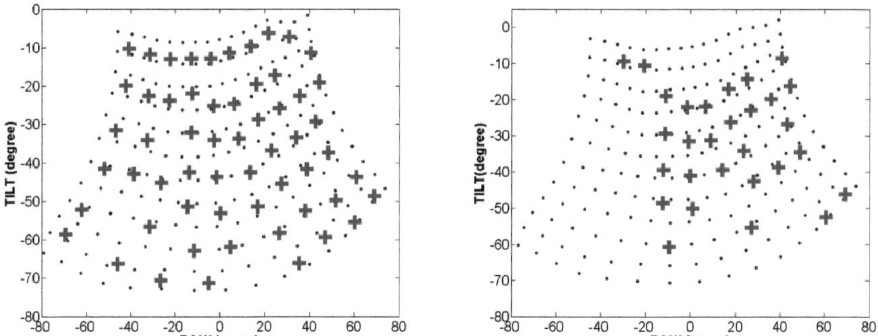

Fig. 3. Result of the TPS interpolation method : zoom Z_0 on the left figure and zoom Z_9 on the right figure. Plus correspond to the results of the computation of the LUT for the nodes of **N**. Points correspond to the use of ζ for unknown pixels of I_s.

LUT computation is used as training data for TPS method. So, for all triplet (x_s, y_s, Z_j), the pan-tilt parameters can be estimate with ζ, see figure 3.

3 Results and Experiments

Cameras of the AXIS company are used. The image resolution used is 640×480 pixels for the static camera and 704×576 pixels for the PTZ camera. The field of view of the static camera is around $90\,°$. In the case of minimal zoom, the field of view of the PTZ camera is of $42\,°$.

The PTZ camera has a 26x optical zoom. The difference of the field of view between the two cameras can be important. For example, the field of view of the PTZ camera is 2.5 times smaller than one of the static camera at the zoom Z_0, 5 times smaller at the zoom Z_4 and 12.5 times at the zoom Z_7.

The mechanical step of the PTZ camera is $0.11\,°$. Experimentally, we show that the mean displacement in the image I_d for the minimal mechanical step depends on the zoom, see table 1. At best, the accuracy of the solution is limited by this mechanical factor.

Table 1. Mean displacement in pixel in the image of the PTZ camera for different zooms

Zoom	0	1	2	3	4	5	6	7	8
mean in X (pixels)	1.64	2.03	2.42	2.59	3.75	4.76	6.93	9.47	15.71
mean in Y (pixels)	1.57	2.99	2.48	3.54	5.03	5.43	7.92	10.46	13.97

In order to estimate the accuracy of this supervised calibration method, it is necessary to know exactly the coordinates of a reference pixel P_s in I_s and to find precisely its coordinates in I_d. To solve this problem, a black ellipsis E which is visible in the two cases is used. To determine with accuracy the coordinates of the center of E, the binarization method of Otsu [8] is used. Pixels of a region of interest can be separated in two classes. Then, the coordinates of the center of gravity of black pixels are estimated with subpixelic precision.

3.1 Accuracy of the Visual Servoing Loop

For evaluating the accuracy of the visual servoing loop (see section 2.1) at the zoom Z_0, three positions of the ellipsis E are choosen : (1) E on a node issued of the Parzen window method with a high range, (2) a node with a middle range and (3) a node with a small range.

The result of the experimentation is given on the figure 4 with the absolute errors of pan and tilt. For the case 1, the standard deviation is small and of the same order of the mechanical step ($0.11\,°$). For the other two cases, the number of interest points around the nodes is less important so the error and the standard deviation increase. The estimation of the pan-tilt parameters is less accuracy. This result shows that it is important to choose an to classify the nodes in function of the density of interest points around the nodes.

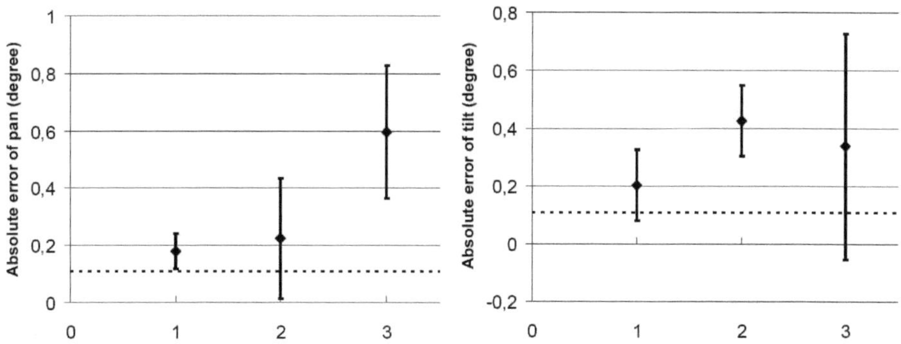

Fig. 4. Absolute errors of pan and tilt in degree resulting of the visual servoing loop for three positions n the scene. The dot line represents the mechanical step.

3.2 Accuracy of the Supervised Calibration Method

For evaluate the accuracy of the complete solution (see section 2), three cases are choosen : (1) E on a pixel which is on the same 3D plane that the closer learned nodes, (2) E on a pixel which is on a different 3D plane that each closer learned node and (3) E on an unknow object when ζ is learned, see figure 5.

The result of the experimentation is given on the figure 6 with the error normalized to mechanical step for the parameters pan-tilt. The case 1 (triangle) is the most ideal case because the 3D information is homogeneous. So, the error is small. The case 2 (diamond) is more complex because the 3D information of the scene presents big variations. So, the error is bigger than the case 1. In the case 3 (square), the unknown object modify the geometry of the scene. This variation was not learned during the learning step of ζ. So the error is more important than the previous cases. Moreover, we note on the figure 6 the error increases with the zoom. Along the learning of the LUT for several zoom, the

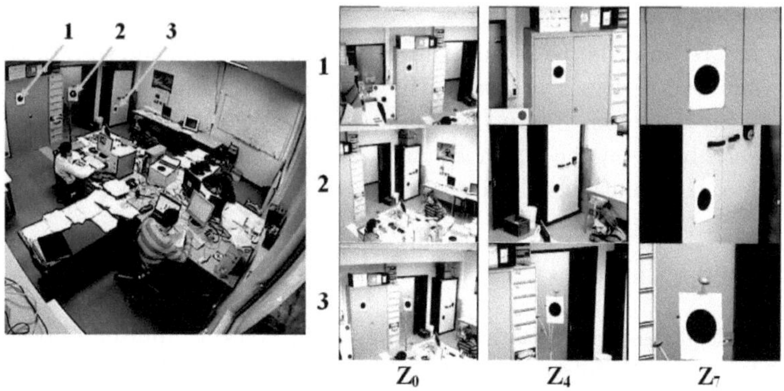

Fig. 5. Illustration of the method. The left figure represents the static camera with the three positions of the target noted. On the right figure, the result for the PTZ camera is shown for different levels of zoom.

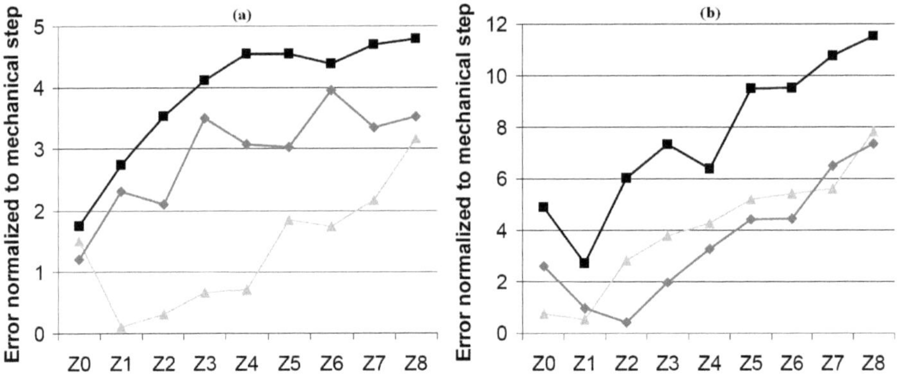

Fig. 6. Error normalized to mechanical step for pan (a) and tilt (b) parameters for several zoom for three cases in the 3D scene (see figure 5) : triangle for case 1, diamond for case 2 and square for case 3

comparison is made between two consecutive zooms. So, the error accumulates progressively. But, the figure 5 shows that even for high zoom, the result can be a good initialization to track a person.

4 Conclusion and Perspectives

In this paper, an automatic algorithm of camera-to-camera calibration integrating the zoom calibration was presented in order to steer a PTZ camera using information of the static camera. At the end, we obtain the relation ζ, for all zoom Z_j, between the pixel coordinates (x_s, y_s) of I_s and the pan-tilt parameters depending on the zoom Z_j : $(\alpha_{Z_j}, \beta_{Z_j}) = \zeta(x_s, y_s, Z_j)$.

All the parameters are automatically selected. The process includes a measure of grey level activity (SIFT). We want to apply the system with an automatic reconfiguration to develop a tracking survey system to focus on the human face along a sequence of displacement.

In the future, in order to reduce the error resulting from the step of the LUT learning for several levels of zoom, the envisaged solution is to compare the zoom Z_j with the zoom Z_0 and to change the zoom Z_0 for a high zoom when the difference between the zoom Z_j and Z_0 is too important. Then, after this amelioration, the solution will be tested in real condition of people tracking.

References

1. Barreto, J.P., Peixoto, P., Batista, J., Araujo, H.: Tracking Multiple Objects in 3D. IEEE Intelligent Robots and Systems. IEEE Computer Society Press, Los Alamitos (1999)
2. Basu, A., Ravi, K.: Active camera calibration using pan, tilt and roll. IEEE Transactions on Systems Man and Cybernetics. IEEE Computer Society Press, Los Alamitos (1997)

3. Bookstein, F.L.: Principal Warps: Thin-Plate Splines and the Decomposition of Deformations. IEEE Transactions on Pattern Analysis and Machine Intelligence. IEEE Computer Society Press, Los Alamitos (1989)
4. Davis, J., Chen, X.: Calibrating pan-tilt cameras in wide-area surveillance networks. ICCV '03: Proceedings of the Ninth IEEE International Conference on Computer Vision, p. 144. IEEE Computer Society, Washington (2003)
5. Horaud, R., Knossow, D., Michaelis, M.: Camera cooperation for achieving visual attention. Machine Vision Application, vol. 16, pp. 1–2. Springer, Heidelberg (2006)
6. Jain, A., Kopell, D., Kakligian, K., Wang, Y.-F.: Using Stationary-Dynamic Camera Assemblies for Wide-area Video Surveillance and Selective Attention. IEEE Computer Vision and Pattern Recognition, pp. 537–544. IEEE Computer Society, Los Alamitos (2006)
7. Lowe, D.G.: Object Recognition from Local Scale-Invariant Features. ICCV '99: Proceedings of the International Conference on Computer Vision, vol. 2, p. 1150. IEEE Computer Society, Washington (1999)
8. Otsu, N.: A threshold selection method from grey scale histogram. IEEE Transactions on Systems Man and Cybernetics, vol. 1, pp. 62–66. IEEE Computer Society Press, Los Alamitos (1979)
9. Senior, A.W., Hampapur, A., Lu, M.: Acquiring Multi-Scale Images by Pan-Tilt-Zoom Control and Automatic Multi-Camera Calibration. WACV-MOTION '05: Proceedings of the Seventh IEEE Workshops on Application of Computer Vision, vol. 1, pp. 433–438. IEEE Computer Society, Washington (2005)
10. Shih, S., Hung, Y., Lin, W.: Calibration of an active binocular head. IEEE Transactions on Systems Man and Cybernetics. IEEE Computer Society Press, Los Alamitos (1998)
11. Woo, D.C., Capson, D.W.: 3D visual tracking using a network of low-cost pan/tilt cameras. canadian Conference on Electrical and Computer Engineering Conference Proceedings. IEEE Computer Society Press, Los Alamitos (2000)
12. Zhou, X., Collins, R.T., Kanade, T., Metes, P.: A master-slave system to acquire biometric imagery of humans at distance. IWVS '03: First ACM SIGMM international workshop on Video surveillance, pp. 113–120. ACM Press, New York (2003)

Recursive Structure and Motion Estimation Based on Hybrid Matching Constraints*

Anders Heyden, Fredrik Nyberg, and Ola Dahl

Applied Mathematics Group
School of Technology and Society, Malmö University, Sweden
{heyden,fredrik.nyberg,ola.dahl}@ts.mah.se

Abstract. Motion estimation has traditionally been approached either from a pure discrete point of view, using multi-view tensors, or from a pure continuous point of view, using optical flow. This paper builds upon a novel framework of hybrid matching constraints for motion estimation, combining the advantages of both discrete and continuous methods. We will derive both bifocal and trifocal hybrid constraints and use them together with a structure estimate based on filtering techniques. A feedback from the structure estimate will be used to further refine the motion estimate. This gives a complete iterative structure and motion estimation scheme. Its performance will be demonstrated in simulated experiments.

1 Introduction

Structure from motion is one of the central problems in computer vision and has been extensively studied during the last decades. The objective is to compute the motion of the camera and the structure of the scene from a number of its two-dimensional images. The standard method is to first estimate the motion of the camera, based on matching tensors, obtained from corresponding points in an image sequence. Then, given the motion of the camera, the structure of the scene is obtained as a sparse set of 3D-points, which can be used as a starting point for surface estimation or texture mapping, cf. [1].

The most common method for estimation of the matching constraints is based on a discrete setting, where e.g. the fundamental (or essential) matrix is estimated between an initial view and another view obtained later in the sequence, c.f. [2]. In order to deal with long image sequences several matching constraints are then pasted together, giving a consistent set of matching constraints from which the motion of the camera can be estimated, cf. [3]. Another approach, closely related to optical flow, is to use a continuous setting and estimate the motion parameters from continuous time matching constraints based on image point positions and velocities, c.f. [4,5,6].

Attempts has been made to combine the discrete and the continuous methods. In [7], a number of differential matching constraints were derived and an algorithm for updating the fundamental matrix along an image sequence was

* This work was partially supported by the SRC project 621-2002-4831.

B.K. Ersbøll and K.S. Pedersen (Eds.): SCIA 2007, LNCS 4522, pp. 142–151, 2007.

indicated. However, no experimental evidence or details about the algorithm were given. These ideas have been taken up and extended to a form suitable for on-line structure and motion estimation, cf. [8], and preliminary results on an on-line structure and motion system have been reported in [9].

The main purpose of the present paper is to develop suitable methods for *on-line recursive structure and motion estimation* for long image sequences. By this, we mean methods that can update a current estimate of the position of the camera and the structure of the scene, when a new image in the sequence becomes available. Such methods have been presented in [10] and [11], where in both cases complex non-linear procedures are used to update the structure. We will propose a novel method where the motion estimation is separated from the structure estimation, enabling simpler and more stable update schemes.

In this work we derive and utilize two types of matching constraint, called *hybrid matching constraints* (HMC), for the estimation and update of motion parameters. The first one is an extension of the epipolar constraint to a *hybrid epipolar constraint*, where both corresponding points in two images as well as their motion in the second image are used. The second one is an extension of the trifocal constraint to a *hybrid trifocal constraint*, where both corresponding points in three images as well as their motion in the third image are used. These hybrid constraints will enable us to update the current motion estimate *linearly* based on at least *three corresponding points.* This will be shown theoretically, by proving the exact number of linearly independent constraints obtained from each corresponding point. The HMC are here fused with a continuous-discrete extended Kalman filter for the state estimation, in order to construct an algorithm for recursive estimation of both structure and motion. We also introduce a linear reprojection error constraint, where feedback of the structure estimates is used to recursively obtain corrections to the motion estimates. This constraint connects the structure and motion estimation processes in a consistent way, and is significantly improving the performance of the method.

2 Problem Description

2.1 Camera Model and Notation

We assume the standard pinhole camera model,

$$\lambda \mathbf{x} = P\mathbf{X} \ , \tag{1}$$

where $\mathbf{x}$ denotes homogeneous image coordinates, P the camera matrix, $\mathbf{X}$ homogeneous object coordinates and λ a scale factor. The camera matrix P is usually written as $P = K[\,R \mid -b\,]$, where K denotes the intrinsic parameters and (R, b) the extrinsic parameters (R being an orthogonal matrix). We will from now on assume that the camera is calibrated, i.e. K is known and that the image coordinates have been transformed such that P can be written as $P = [\,R \mid -b]$. When several images of the same point are available, (1) can be written as

$$\begin{cases} \lambda(t)\mathbf{x}(t) = P(t)\mathbf{X},\ t \in [0,T] \quad \text{or} \\ \lambda_i \mathbf{x}_i = P_i \mathbf{X},\ i = 1, \dots M \end{cases} \tag{2}$$

in the continuous time case and the discrete time case respectively. The camera matrix P is assumed to have the form

$$P(t) = [\, R(t) \mid -b(t) \,] \quad \text{or} \quad P_i = [\, R_i \mid -b_i \,] \ , \tag{3}$$

in the continuous case and in the discrete case respectively. We furthermore assume that the object coordinate system has been chosen such that $R(0) = R_1 = I$ and $b(0) = b_1 = 0$, implying that $P(0) = P_1 = [\, I \mid 0 \,]$.

2.2 Problem Formulation

A structure and motion estimation problem can now be formulated as the task of estimating both the structure $\mathbf{X}$ in (2) and the motion parameters $R(t)$ and $b(t)$ in (3) at the time t, given the set of perspective measurements $\mathfrak{M}_t = \{\mathbf{x}(t_i) \mid \forall i \ :\ t_i \leq t\}$. A *recursive* structure and motion problem can be formulated as given an estimate of the structure and motion parameters up to time t, update this estimate based on measurements obtained up to time $t + \Delta t$.

2.3 Discrete Matching Constraints

The *discrete matching constraints* are obtained by using the discrete version of (2), for several different i and eliminating the object coordinates $\mathbf{X}$ from the resulting equations. In the case of two views we obtain the well-known *epipolar constraint*

$$\mathbf{x}_1^T E \mathbf{x}_2 = 0, \quad \text{with} \quad E = R^T \hat{b} \ , \tag{4}$$

where we for simplicity have used the notation $R_2 = R$ and $b_2 = b$ and $\hat{b}$ denotes the skew-symmetric matrix corresponding to the vector b. The matrix E in (4) denotes the well-known *essential matrix*. This constraint can be used to estimate the motion parameters linearly from at least eight corresponding points. The three- and four-view constraints are obtained similarly, by starting with three (or four) images of the same point and eliminating the object coordinates $\mathbf{X}$, from the resulting system of equations, see [12,13,14]. In the three-view case the trifocal constraints and the trifocal tensor are obtained and in the four-view case, the quadrifocal constraints and the quadrifocal tensor are obtained. These can be used to estimate the camera motion linearly, from at least 7 and 6 corresponding points respectively.

2.4 Continuous Matching Constraints

The *continuous time matching constraints* are obtained from the camera matrix equation (2) in continuous form and its time derivative (where for simplicity the time dependency is expressed using an index):

$$\begin{aligned} \lambda_t \mathbf{x}_t &= [\, R_t \mid -b_t \,]\mathbf{X} = R_t \tilde{\mathbf{X}} - b_t, \\ \lambda_t' \mathbf{x}_t + \lambda_t \mathbf{x}_t' &= [\, R_t' \mid -b_t' \,]\mathbf{X} = \widehat{\omega}_t R_t \tilde{\mathbf{X}} - b_t' = \hat{\omega}_t(\lambda_t \mathbf{x}_t + b_t) - b_t' \ , \end{aligned} \tag{5}$$

where $\tilde{\mathbf{X}}$ denotes the first 3 components of the vector $\mathbf{X}$ and we have assumed that the $\mathbf{X}$ is normalized such that the fourth component is equal to 1. Furthermore, we have used the fact that the derivative of a rotation matrix can be written as $R_t' = \hat{w}_t R_t$, where w_t represents the momentary rotational velocity of the camera at time t. Similarly b_t' denotes the momentary translational velocity of the camera at time t. Define

$$\nu_t = b_t' - \widehat{w}_t b_t \tag{6}$$

(representing the momentary translational velocity in a local coordinate system) and multiply the last equation in (5) with $\nu_t \times \mathbf{x}_t$ giving

$$\mathbf{x}'^T \hat{\nu} \mathbf{x} - \mathbf{x}^T \widehat{w} \hat{\nu} \mathbf{x} = 0 \ , \tag{7}$$

which is the well known *continuous epipolar constraint.* This constraint can be used to estimate the motion parameters from at least eight corresponding points. Higher order continuous multi-view constraints are obtained by taking higher order derivatives of (2) and then eliminating the structure, cf. [4].

3 Hybrid Matching Constraints

In this section we will introduce a novel concept, called *the hybrid matching constraints*, that will be used to update the motion parameters in two different ways.

3.1 Epipolar Hybrid Matching Constraints

Write down the camera matrix equations for time 0, time t and time $t + \Delta t$:

$$\begin{cases} \lambda_0 \mathbf{x}_0 = [\, I \mid 0 \,]\mathbf{X} \\ \lambda_t \mathbf{x}_t = [\, R_t \mid -b_t \,]\mathbf{X} \\ \lambda_{t+\Delta t} \mathbf{x}_{t+\Delta t} = [\, R_{t+\Delta t} \mid -b_{t+\Delta t} \,]\mathbf{X} \ . \end{cases} \tag{8}$$

Using $R_t = e^{t\widehat{w}_t}$ as a first order approximation of R, valid between t and $t + \Delta t$, implying that

$$R_{t+\Delta t} = e^{\widehat{w}_t \Delta t} R_t \approx R_t + \widehat{w}_t R_t \Delta t \ ,$$

and

$$\begin{aligned} b_{t+\Delta t} &= b_t + d_t \Delta t, \text{ corresponding to } d_t = b_t' \ , \\ \lambda_{t+\Delta t} &= \lambda_t + \mu_t \Delta t, \text{ corresponding to } \mu_t = \lambda_t' \ , \\ \mathbf{x}_{t+\Delta t} &= \mathbf{x}_t + \mathbf{u}_t \Delta t, \text{ corresponding to } \mathbf{u}_t = \mathbf{x}_t' \ , \end{aligned} \tag{9}$$

and eliminating $\mathbf{X}$ using the first equation in (8) and expanding until the first order in Δt gives

$$\underbrace{\begin{bmatrix} R_t \mathbf{x}_0 & \mathbf{x}_t & 0 & b_t \\ \widehat{w}_t R_t \mathbf{x}_0 & \mathbf{u}_t & \mathbf{x}_t & d_t \end{bmatrix}}_{M_d} \begin{bmatrix} -\lambda_0 \\ \lambda_t \\ \mu_t \\ 1 \end{bmatrix} = \begin{bmatrix} 0 \\ 0 \end{bmatrix} \ , \tag{10}$$

implying that $\operatorname{rank} M_d < 4$, which will be called *the hybrid epipolar constraints.* Assuming normalization such that $\mathbf{x} = (x, y, 1)$, $\mathbf{u} = (u_x, u_y, 0)$, and expanding the minors of M_d give the following different constraints in the motion parameters:

1. Minors containing the first three rows give the epipolar constraint.
2. Minors containing two rows out of the first three give linear constraints in d_t and w_t, in total nine such linear constraints.
3. Minors containing the three last rows give non-linear constraints on the motion parameters.

For our purposes, only the second type of constraints are useful. In fact there are exactly two linearly independent constraints on the motion parameters from the nine constraints of the second type above. This implies that the essential matrix can be updated from at least three corresponding points, which is a huge improvement compared to the standard discrete approaches, where five corresponding points give highly non-linear constraints, and at least eight corresponding points are needed to obtain reasonable simple linear constraints.

In order to prove that there exist exactly two linearly independent constraints among the nine ones of the second type above, we have to prove that there are exactly seven independent linear dependencies between the constraints. Start by extending M_d with the second column, giving the following 6×5-matrix:

$$\begin{bmatrix} R_t\mathbf{x}_0 & \mathbf{x}_t & 0 & b_t & \mathbf{x}_t \\ \widehat{w}_t R_t \mathbf{x}_0 & \mathbf{u}_t & \mathbf{x}_t & d_t & \mathbf{u}_t \end{bmatrix} .$$

The 5×5 minors of this matrix are obviously identically zero. Expanding the three minors obtained by removing each one of the last three columns give

$$\begin{aligned} & x_t L_{11} - y_t L_{21} + L_{31} - u_x y_t E + u_y x_t E = 0 \\ & x_t L_{12} - y_t L_{22} + L_{32} - u_x E = 0 \\ & x_t L_{13} - y_t L_{23} + L_{33} + u_y E = 0 \ , \end{aligned} \tag{11}$$

where L_{ij} denotes the hybrid constraint obtained by removing row number i from the first three rows of M_d and row number j from the last three rows of M_d and E denotes the epipolar constraint between time 0 and t. Thus (11) contains exactly three linear dependencies among the epipolar hybrid constraints, assuming that the epipolar constraint between views 0 and t is fulfilled. Repeating the same procedure by instead adding the third column and expanding the minors obtained by removing each one of the first three columns give three further linear dependencies. Finally, adding both the second and the third column and expanding the determinant of the resulting 6×6 matrix, give the last linear dependency. Thus only two linearly independent hybrid constraints remain (assuming the epipolar constraint is fulfilled). For practical purposes, there might be more than two linearly independent constraints when the epipolar constraint is not exactly fulfilled. However, these are numerically unconditioned in the sense that they are close to spanning a two-dimensional space.

Observe that when d_t and w_t has been recovered, the new motion parameters and the new essential matrix can easily be obtained from

$$\begin{cases} R_{t+\Delta t} = e^{\widehat{w}_t \Delta t} R_t \\ b_{t+\Delta t} = b_t + d_t \Delta t \end{cases} \quad \Rightarrow \quad E_{t+\Delta t} = R^T_{t+\Delta t} \hat{b}_{t+\Delta t} \; . \tag{12}$$

The update in (12) guarantees that the new essential matrix fulfils the nonlinear constraints. Observe also that although the update is nonlinear in the motion parameters it can be performed efficiently using e.g. Rodriguez formula. Finally, observe that either ν_t or d_t may be used as a parameter for the translational velocity, since they are related according to (6).

3.2 Trifocal Hybrid Matching Constraints

Write down the camera matrix equations for time 0, s, t and $t + \Delta t$:

$$\begin{cases} \lambda_0 \mathbf{x}_0 = [\, I \mid 0 \,]\mathbf{X} \\ \lambda_s \mathbf{x}_s = [\, R_s \mid -b_s \,]\mathbf{X} \\ \lambda_t \mathbf{x}_t = [\, R_t \mid -b_t \,]\mathbf{X} \\ \lambda_{t+\Delta t} \mathbf{x}_{t+\Delta t} = [\, R_{t+\Delta t} \mid -b_{t+\Delta t} \,]\mathbf{X} \; . \end{cases} \tag{13}$$

Eliminating $\mathbf{X}$ using the first equation and expanding until the first order in Δt give

$$\underbrace{\begin{bmatrix} R_s \mathbf{x}_0 & \mathbf{x}_s & 0 & 0 & b_s \\ R_t \mathbf{x}_0 & 0 & \mathbf{x}_t & 0 & b_t \\ \widehat{w}_t R_t \mathbf{x}_0 & 0 & \mathbf{u}_t & \mathbf{x}_t & d_t \end{bmatrix}}_{N_d} \begin{bmatrix} -\lambda_0 \\ \lambda_s \\ \lambda_t \\ \mu_t \\ 1 \end{bmatrix} = \begin{bmatrix} 0 \\ 0 \\ 0 \end{bmatrix} \; , \tag{14}$$

implying that rank $N_d < 5$, which will be called *the hybrid trifocal constraints*. The minors of N_d gives the following different constraints in the motion parameters:

1. Minors containing only one row out of the first three give the previously derived differential-algebraic epipolar constraint.
2. Minors containing only one row out of the last three give the standard discrete trifocal constraints.
3. Minors containing two rows out of the first three, one row out of the middle three rows and two rows out of the last three rows give linear constraints in d_t and w_t, in total 27 such linear constraints.

For our purposes, only the last type of constraints are useful. It turns out that there only exist two linearly independent constraints on the motion parameters from the nine constraints of the second type above, which can be proven in the same way as above. This implies that the trifocal tensor can be updated from at least three corresponding points.

3.3 Motion Estimation Using HMC

Given a current estimate of the motion parameters and a new image with at least three corresponding points, the motion parameters can be updated using a linear system of equations of the type

$$M \begin{bmatrix} w_t \\ d_t \end{bmatrix} = m \ , \tag{15}$$

where $M = M(\mathbf{x}_0^k, \mathbf{x}_t^k, \mathbf{u}_t^k, R_t, b_t)$ for $k = 1, \ldots n$, $n \geq 3$ and $m = m(\mathbf{x}_0^k, \mathbf{x}_t^k, \mathbf{u}_t^k, R_t, b_t)$ in the epipolar case and similarly for the trifocal case, where also $\mathbf{x}_s^k$ appears as a parameter.

3.4 State Estimation Using the Continuous-Discrete EKF

Given the motion parameters it is possible to employ a number of algorithms for recursive structure recovery. Here we optionally select a continuous-discrete extended Kalman (EKF) filter for the state estimation process [15]. For further details, see [9] or [16].

3.5 Motion Estimation Refinement by Reprojection Constraints

Given motion estimates R_t and b_t obtained using the HMC through (15), the measurement $\mathbf{x}_t$, and the 3-D estimate $\mathbf{X}$ from the EKF, we seek correction vectors $\alpha, \beta \in \mathbb{R}^{3\times 1}$ of small magnitude, such that improved motion estimates R_t^+ and b_t^+ are given by the reprojection constraint

$$\lambda_t^+ \mathbf{x}_t = [\, R_t^+ \mid -b_t^+ \,] \mathbf{X} \ , \quad R_t^+ = e^{\widehat{\alpha}} R_t \ , \quad b_t^+ = b_t + \beta \ . \tag{16}$$

Expanding the first equation in (16) to a first order approximation gives

$$\lambda_t \mathbf{x}_t \approx R_t \tilde{\mathbf{X}} + \widehat{\alpha} R_t \tilde{\mathbf{X}} + b_t + \beta \Rightarrow \widehat{R_t \tilde{\mathbf{X}}} \alpha + \beta = \epsilon := R_t \tilde{\mathbf{X}} + b_t - \lambda_t \mathbf{x}_t \ , \tag{17}$$

where ϵ can be interpreted as the reprojection error. Observe that (17) is a linear constraint in the correction vectors α and β. Since (17) contains two linear constraints on these 6 parameters (λ_t is also a free parameter) in the correction vectors, a linear update on the motion parameters can be made from at least three corresponding points.

The inclusion of the reprojection constraint correction step significantly enhances the performance of the estimation procedure, leading to more accurate and robust estimates of both structure and motion.

3.6 Structure and Motion Algorithm

Using the results of the previous sections, the following algorithm can now be employed for recursive structure and motion recovery:

1. Preparations

- Assume that images are obtained sequentially at time instants t_i, i=$0, 1, 2\ldots$, equally spaced by Δt. Also assume some initial values for the state vector and the error covariance matrix in the EKF.
- Given the images at times $t_0 = 0$ and $t_1 = \Delta t$ with at least eight point correspondences, get initial parameter estimates w_0 and ν_0 using e.g. the continuous eight-point algorithm.
- Compute $R_{t_1} = e^{\widehat{w}_0 \Delta t}$ and $b_{t_1} = d_0 \Delta t$.

2. Estimation loop - for $i = 1, 2 \ldots$ do

- Using at least three point correspondences, set up the hybrid epipolar or trifocal matching constraints in (15).
- Solve the linear system (15) for the new parameter estimates w_{t_i} and d_{t_i}.
- Update the rotation matrix and the translation vector according to (12).
- Use w_{t_i} and ν_{t_i} in the EKF to get structure estimates over the time interval $[t_i, t_i + \Delta t]$.
- Refine the motion estimate according to (17).

Note that since we are estimating both structure *and* motion, the estimates are inherently subjected to a scale ambiguity. In the above algorithm the scale issue is resolved by assuming the translational velocity vector ν to be of unit length in the initialization procedure. This together with the assumption of normalized image coordinates fixes the scale for the subsequent parameter estimates through (15).

4 Experiments

Since the initial parameter values obtained by the initialization process generally can be assumed quite accurate, the truly interesting case will be when one or both of the parameter vectors w and ν are time varying. The hybrid-based method can then be evaluated by its ability to follow the time-variations in the parameters, as well as by its ability to correctly recover the 3-D structure.

For purpose of illustration we simulate images of an object consisting of eight points in a general configuration on a grid of stepsize 10^{-4}, and with the parameter vectors

$$w(t) = \frac{2}{3}\left(1,\ -1,\ 1\right)\ , \quad \nu(t) = \frac{1}{\sqrt{1.32}}\left(-1,\ -0.4,\ 0.4\right)^T + \frac{1}{2}\left(t,\ t,\ -0.5t\right)^T .$$

Perspective measurements were computed at time instants separated by $\Delta t = 0.01$. The estimates of the components of the rotational velocity w and the translational velocity ν together with the true values based on the hybrid epipolar constraints, are shown in Fig. 1(a) and Fig. 1(b) respectively. The resulting 3-D estimation error for one of the observed object points is shown in Fig. 1(c).

We also conducted a similar experiment on the same data based on the hybrid trifocal constraints. The same procedure as before, based on the initialization using the continuous epipolar constraint, and the recursive estimation using the epipolar hybrid constraints, was used until time $t = 0.4s$. After that, the trifocal hybrid constraints was used, with $s = t/2$, see Fig 2.

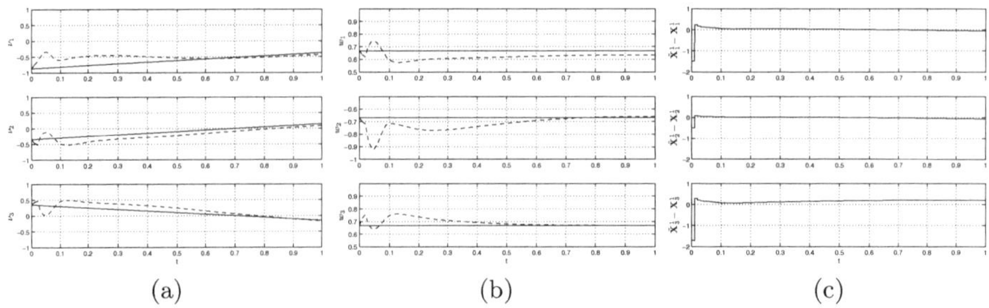

Fig. 1. Estimation results obtained using the hybrid epipolar constraints: (a) True (solid) and estimated (dashed) translational velocity ν, (b) True (solid) and estimated (dashed) rotational velocity w, (c) 3-D estimation errors for one of the observed object points

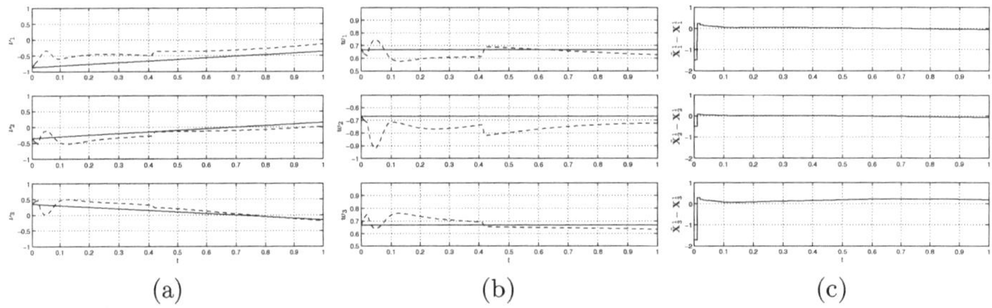

Fig. 2. Estimation results obtained using the hybrid trifocal constraints: (a) True (solid) and estimated (dashed) translational velocity ν, (b) True (solid) and estimated (dashed) rotational velocity w, (c) 3-D estimation errors for one of the observed object points

5 Conclusion

We have proposed an algorithm for recursive estimation of structure and motion from perspective measurements in a continuous-discrete setting, utilizing the novel concept of the hybrid epipolar and trifocal matching constraint for the estimation of the velocity parameters, combined with a state estimator, here optionally selected as the continuous-discrete EKF. The structure and motion estimation processes are connected by recursive feedback of the structure estimates, resulting in reprojection error constraints used to obtain refined motion estimates. Simulated experiments are included to illustrate the applicability of the concept. The main advantages of the presented method is that only three corresponding points are needed for the sequential update and correction of the velocity parameter estimates. Further, both these update schemes are linear.

Note that it is *not* necessary that the same three points are tracked throughout the whole image sequence. It is easy to change to any other triplet of point correspondences when needed.

References

1. Hartley, R., Zisserman, A.: Multiple View Geometry. Cambridge (2003)
2. Longuet-Higgins, H.C.: A computer algorithm for reconstructing a scene from two projections. Nature 293, 133–135 (1981)
3. Torr, P., Fitzgibbon, A.W., Zisserman, A.: Maintaining multiple motion model hypotheses over many views to recover matching and structure. In: Proc. International Conference on Computer Vision. pp. 485–491 (1998)
4. Åström, K., Heyden, A.: Continuous time matching constraints for image streams. International Journal of Computer Vision 28(1), 85–96 (1998)
5. Viévielle, R., Faugeras, O.: The first order expansion of motion equations in the uncalibrated case. Computer Vision and Image Understanding 64(1), 128–146 (1996)
6. Ma, Y., Košecká, J., Sastry, S.S.: Linear differential algorithm for motion recovery: A geometric approach. International Journal of Computer Vision 36(1), 71–89 (2000)
7. Triggs, B.: Differential matching constraints. In: Proc. of the International Conference on Computer Vision and Pattern Recognition. vol. 1, 370–376 (1999)
8. Heyden, A.: Differential-algebraic multiview constraints. In: Proc. International Conference on Computer Vision, vol. 1, pp. 159–162. IEEE Computer Society Press, Los Alamitos (2006)
9. Nyberg, F., Heyden, A.: Iterative estimation of structure and motion using hybrid matching constraints. In: 2nd Workshop on Dynamical Models for Computer Vision, DV2006 (at ECCV2006) (2006)
10. Soatto, S.: 3-D structure from visual motion: Modeling, representation and observability. Automatica 33(7), 1287–1312 (1997)
11. Azarbayejani, A., Pentland, A.P.: Recursive estimation of motion, structure, and focal length. IEEE Transactions on Pattern Analysis and Machine Intelligence 17(6), 562–575 (1995)
12. Triggs, B.: Matching constraints and the joint image. In: Proc. International Conference on Computer Vision. pp. 338–343 (1995)
13. Heyden, A.: Tensorial properties of multilinear constraints. Mathematical Methods in the Applied Sciences 23, 169–202 (2000)
14. Faugeras, O., Long, Q.T.: Geometry of Multiple Images. The MIT Press, Cambridge (2001)
15. Maybeck, P.S.: Stochastic Models, Estimation, and Control, vol. 2. Academic Press, San Diego (1982)
16. Huster, A.: Relative Position Sensing by Fusing Monocular Vision and Inertial Rate Sensors. PhD dissertation, Stanford University, Department of Electrical Engineering (2003)

Efficient Symmetry Detection Using Local Affine Frames

Hugo Cornelius[1], Michal Perd'och[2], Jiří Matas[2], and Gareth Loy[1]

[1] CVAP, Royal Institute of Technology, Stockholm, Sweden
[2] Center for Machine Perception, CTU in Prague, Czech Republic

Abstract. We present an efficient method for detecting planar bilateral symmetries under perspective projection. The method uses local affine frames (LAFs) constructed on maximally stable extremal regions or any other affine covariant regions detected in the image to dramatically improve the process of detecting symmetric objects under perspective distortion. In contrast to the previous work no Hough transform, is used. Instead, each symmetric pair of LAFs votes just once for a single axis of symmetry. The time complexity of the method is $n \log(n)$, where n is the number of LAFs, allowing a near real-time performance. The proposed method is robust to background clutter and partial occlusion and is capable of detecting an arbitrary number of symmetries in the image.

1 Introduction

Symmetry is a visual and physical phenomenon, occurring both naturally and in manufactured artefacts and architecture. In this paper we will concentrate on bilateral symmetry where features are reflected about a central axis.

Human perception of symmetry has been well-studied. Psycho-physical evidence points to symmetry detection being a pre-attentive process [1] and playing a role in both signalling the presence of objects and directing visual attention [1]. It is not only humans who can detect symmetry. Bees, for example, have been shown to naturally choose to visit more symmetrical flowers [2].

Symmetry seldom occurs by accident. If two symmetric image regions are detected it is likely that these regions are related in the real world, and there is a good chance that they belong to the same object. Thus by detecting symmetry it is possible to start grouping or segmenting the image without prior knowledge of image content. As many objects exhibit some degree of symmetry, this provides a context-independent mechanism for hypothesising the presence, location and extent of such objects in a scene. Thus computer vision systems can benefit from symmetry detection in a manner not dissimilar to the psycho-physical systems discussed above.

This paper builds on recently published results [3,4] that illustrated the effectiveness of symmetry detection using local features. In this work we use local affine frames (LAFs) [5] constructed on affine covariant distinguished regions [6,7] to dramatically improve the process of detecting symmetric objects under

B.K. Ersbøll and K.S. Pedersen (Eds.): SCIA 2007, LNCS 4522, pp. 152–161, 2007.

perspective distortion, resulting in a simpler, more efficient and more robust approach. The use of LAFs makes it possible to hypothesise the axis of symmetry from a single symmetric pair of features enabling a very computationally efficient algorithm. Each matching pair of reflected LAFs is hypothesising one bilateral symmetry axis. The symmetry axis hypotheses are grouped into symmetric constellations about common symmetry foci, identifying both the dominant bilateral symmetries present as well as a set of symmetric features associated with each foci. Our method simultaneously evaluates symmetry across all locations, scales and orientations under affine and perspective projection. An important observation is, that any affine covariant detection and matching process which provides richer than just point-to-point correspondences can be used in a similar way.

The remainder of this paper is organised as follows, Section 2 reviews previous work on symmetry detection and gives an introduction to distinguished region detectors and local affine frames, Section 3 describes the method, Section 4 presents experimental results and discusses the performance of the method, and Section 5 presents our conclusions.

2 Previous Work

Symmetry detection is a well-studied field in computer vision, and comprise a significant body of work spanning over 30 years. Comparatively, distinguished regions and local affine frames are relatively recent developments in the computer vision field. Here we present a brief review of symmetry detection focusing on methods that use local features, and also provide an introduction to distinguished regions and local affine frames which provide the essential tool for streamlining the solution to the symmetry detection problem.

2.1 Symmetry Detection

Symmetry detection has found use in numerous applications ranging from facial image analysis [8] and vehicle detection [9], to 3D reconstruction [10] and visual attention [11,12]. Existing symmetry detection techniques can be broadly classified into global and local feature-based methods. Global approaches treat the entire image as a signal from which symmetric properties are inferred, whereas local feature-based methods use local features, edges, contours or boundary points, to reduce the problem to one of grouping symmetric sets of points or lines. The local-feature based approaches offers numerous advantages in particular their ability to more efficiently detect local symmetries in images that are not globally symmetric.

Mukherjee *et al.* [13] used local features called distinguished points detected on curves that are preserved under perspective or affine transformation. Affine semi-local invariants are used to match pairs and form symmetry hypotheses. The method was succesfully used to find symmetries on contours of simple objects. Recent methods [14,4] have increased robustness and effectiveness by exploiting detection and matching of richer features – distinguished regions. However they use the Hough transform to accumulate symmetry hypotheses which slows down the detection process. Moreover, the methods do not utilise the full extent of the information provided by affine matching techniques. In this work we shall show that using this additional information significantly simplifies and streamlines the symmetry detection process.

Tuytelaars *et al.* [14] presented a method for the detection of regular repetitions of planar patterns under perspective skew using a geometric framework. The approach detected all planar homologies and could thus find reflections about a point, periodicities, and mirror symmetries. The method built clusters of matching points using a cascaded Hough transform, and typically took around 10 minutes to process an image. In contrast the new method proposed herein provides a simpler and more efficient approach. We utilise local feature information to establish robust symmetric matches that directly vote for single symmetry hypotheses.

Recently Cornelius and Loy [4] proposed a method for grouping symmetric constellations of features and detecting symmetry in perspective. This approach was developed from [3] which matched reflective feature pairs to detect symmetry in the image plane. The later method merged pairs of symmetrically matching features into feature quadruplets, so that each quadruplet defined a specific symmetry foci under perspective skew. The quadruplets were grouped into symmetric constellations of features about common symmetry foci, identifying the dominant symmetries present in the image plane. The disadvantage of this approach was the necessity to merge feature pairs into quadruplets. This was both time consuming and could easily result in spurious hypotheses.

Here we take the concept proposed by [4] and improve this by removing the need to form feature quadruplets, and thus solve symmetry detection under perspective in a cleaner and more efficient manner more concordant with the simple approach used to detect symmetry in the image plane [3]. That is, we show how to derive a unique symmetry hypothesis from a single pair of symmetric features, with the added challenge of an unknown perspective distortion.

2.2 Distinguished Regions and Local Affine Frames

The first step of the proposed algorithm is detection of distinguished regions followed by construction and description of local affine frames.

Distinguished regions are subsets of image pixels with some distinguishing property that allows their repeated detection over a range of image deformations, e.g. affine[1] transformations and photometric changes. Several affine covariant region detectors were presented and compared recently in [15]. Although any affine covariant region detector can be used, we have focused on only two kinds of regions in our work – Maximally Stable Extremal Regions (MSERs) [6] and Hessian-Affine regions [7]. MSERs (see Fig. 1) are invariant to photometric

[1] In the framework described in this work we assume a locally planar scene under perspective projection.

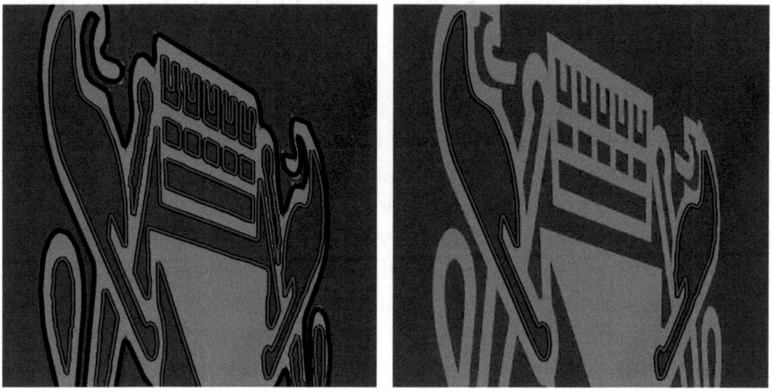

a) b)

Fig. 1. Distinguished Regions detection: a) detected maximally stable extremal regions, b) example of a reflected pair of MSERs

changes and covariant with all continuous transformations. The MSER detector generates simple polygons – contours that encompass interior pixels of stable regions. The Hessian-Affine detector first localises points in space and scale at local maxima of the Hessian and Laplacian-of-Gaussians. Affine covariant regions are then computed in an iterative affine adaptation process based on the second moment matrix of image gradients [16].

To resolve the affine transformations between distinguished regions in a single image a local affine frames approach [5] is exploited. Local affine frames (LAFs) are local coordinate systems constructed on distinguished regions covariantly with affine image transformations. LAFs provide an affine and photometric normalisation of local image patches into a canonical form. Geometrically and photometrically normalised patches can then be directly compared eliminating the need for invariants. The construction of the local affine frame resolves six free parameters of a two-dimensional affine normalisation. Two major groups of LAF constructions, based on shape and local image intensities were proposed in [5]. The shape based LAFs are formed by combining affine covariant measures, e.g. area, center of gravity and second moment matrix of the distinguished region, bi-tangent points, curvature extrema, point of inflections and other.

Intensity based LAF construction are computed from a center of gravity, covariance matrix and dominant gradient orientation (similar as in [17]) in the following way: First, a small image patch around the distinguished region is normalised using its centre of gravity and second moment matrix. Then a histogram of local gradient orientations is computed and dominant orientations are selected as local maxima of histogram bins. Each dominant orientation is used to fix the remaining parameter of an affine normalisation. In this work, the intensity based LAF construction is used with Hessian-Affine regions and shape based constructions with MSERs.

3 Detecting Bilateral Symmetry Using Local Affine Frames

As in [3] and [4], the method for detecting planar bilateral symmetry under perspective skew presented here is based on pairwise matching of local features within a single image. The features used here are local affine frames (LAFs). To allow matching of bilaterally symmetric LAFs, a mirrored version of each frame and its descriptor from the previous step is computed. The mirroring depends only on the type of the local affine frame construction. The local affine frame construction, assumes a right-handed coordinate system when ordering the points and forming the frame. When constructing the mirrored frame, a left-handed coordinate system is used e.g. if a frame is constructed from the centroid and two points on the concavity – entry and exit point – the points on the concavity are swapped as the ordering on the contour of distinguished region is changed from clockwise to anti-clockwise. Other LAF constructions are mirrored in a similar manner without any knowledge about the axis of symmetry. Examples of LAFs and their mirrored versions are shown in Fig. 2.

a) b)

Fig. 2. LAF constructions: a) Some of the local affine frames constructed on a maximally stable extremal region, b) Mirrored LAFs

In the next step LAFs are matched against their mirrored versions to find symmetric pairs. The naive approach would be to match every LAF to all mirrored ones and keep the k best matches for every frame. Such a matching scheme has time complexity $O(n^2)$, where n is the number of frames detected in the image. To avoid the $O(n^2)$ time complexity a decision tree is used as proposed in [18]. The decision tree is trained offline on a set of LAFs computed from training images[2]. In the matching phase frames are first entered into the tree in time proportional to $O(n \log(n))$ and then for each mirrored local affine frame a leaf

[2] A single random wide-baseline stereo pair of image was used to train the decision tree for all experiments.

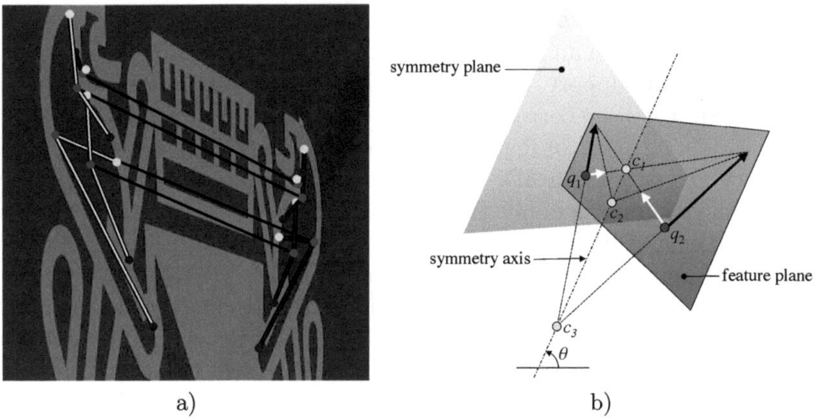

Fig. 3. a) Some matching LAF pairs. Each pair can be used to form a hypothesis. b) A reflective pair of LAFs (q_1, q_2) and the axis of symmetry around which the pair is symmetric. The intersections of the three lines defined by the three points in the LAFs all lie on the axis of symmetry (c_1, c_2, c_3).

with similar frames is located in the tree in $O(\log(n))$. Candidates from a leaf are sorted using correlation of normalised patches and the k best LAFs are returned. Throughout our experiments $k = 3$ was used. The total time complexity of this matching scheme is $O(n \log(n))$ which in practise means that it is possible to achieve real-time performance.

When pairs of reflected frames have been obtained, an axis of symmetry is calculated for all pairs where possible. A LAF is defined by three points and three lines can be obtained by connecting these points. For a symmetric pair of frames, the three pairs of corresponding lines will intersect on the axis of symmetry. This means that the axis of symmetry can be estimated from a single pair of matching LAFs (see Fig. 3) if intersections c_1, c_2, and c_3 exist. Please note that this holds not only for symmetric patterns that lie in a plane, but also if the two halves of the symmetric pattern lie in different planes whose intersection coincides with the axis of symmetry. For pair of frames reflected about an axis of symmetry a set of necessary conditions has to hold:

1. The intersection, c_i, between a pair of lines defined by two points in the LAFs has to lie on the same side of both LAFs (see Fig. 3).
2. The three intersections (c_1, c_2, and c_3) have to lie on a line (the axis of symmetry).
3. The axis of symmetry has to cross the line between the two LAFs.

Due to noise in the positions of the points in the LAFs, the second condition will in general not hold and small deviations from the line are allowed.

All matched pairs of frames for which these conditions hold, cast a vote for one axis of symmetry. Lines receiving many votes are identified as potential axes of symmetry and the pairs of frames that are consistent with the same axis of symmetry are grouped together. Even though just a small fraction of the feature pairs obtained in an image is correct, false symmetry axes will in general receive insignificant support. For a true symmetry axis the majority of the pairs consistent with it will be correct (inliers).

The lines joining pairs of features that are symmetric around the same axis of symmetry are parallel in 3D. This means that in an image, these lines will share a common vanishing point. For a planar bilaterally symmetric pattern, it is possible to estimate the vanishing point from the positions of the two features in a pair and the point where the axis of symmetry intersects the line between the features. The estimate will however be sensitive to small changes in the position of the axis of symmetry. Furthermore, if both halves of the symmetric pattern do not lie in the same plane but in two planes that intersect at the axis of symmetry, the intersections can not be used since they no longer correspond to the points that lie in the middle between the LAFs in 3D. It is however still possible to use the directions of the lines between the features and these will also be robust to small changes in the feature positions. We use the same approach as in [4] to estimate the vanishing point and reject the pairs that are not consistent with it. If ψ_i is the orientation of the line between the two features in the i:th pair of LAFs and θ is the direction perpendicular to the axis of symmetry, the following will hold:

$$\tan(\psi_i - \theta) = \frac{h_i - p}{L}$$

if h_i is the point where the line between the two features in the pair intersects the axis of symmetry, L is the distance from the vanishing point to the axis of symmetry and p is the point on the axis of symmetry closest to the vanishing point. Since $\tan(\psi_i - \theta)$ is a linear function of h_i, p and L (that determine the vanishing point) can be estimated from two (ψ_i, h_i)-pairs, i.e. from two symmetric pairs of LAFs. RANSAC [19] is used to find a good estimate of p and L and to reject the pairs not consistent with the vanishing point. Since the number of inliers will typically be much larger than the number of outliers, a good estimate of the vanishing point will be found quickly.

The final output from the method is one or more symmetry axes and vanishing points together with the LAF and region pairs that are consistent with the symmetries.

4 Experiments

We have tested our symmetry detection method on a range of images of symmetric objects and scenes. In Fig. 4 some results of detected symmetries are shown. The examples contain a wide variety of objects both man-made, such as the buildings and the carpet, and natural (the butterflies and the face). Many images contain a considerable amount of background clutter (see e.g. Fig. 4(h)), and some results with severe perspective skew are also shown (Fig. 4(a) and 4(b)). The butterflies are examples of symmetric patterns where the two halves of the patterns lie in different planes that intersect at the axis of symmetry. Especially in Fig. 4(i) the effect is quite clear. Symmetry detection on buildings is

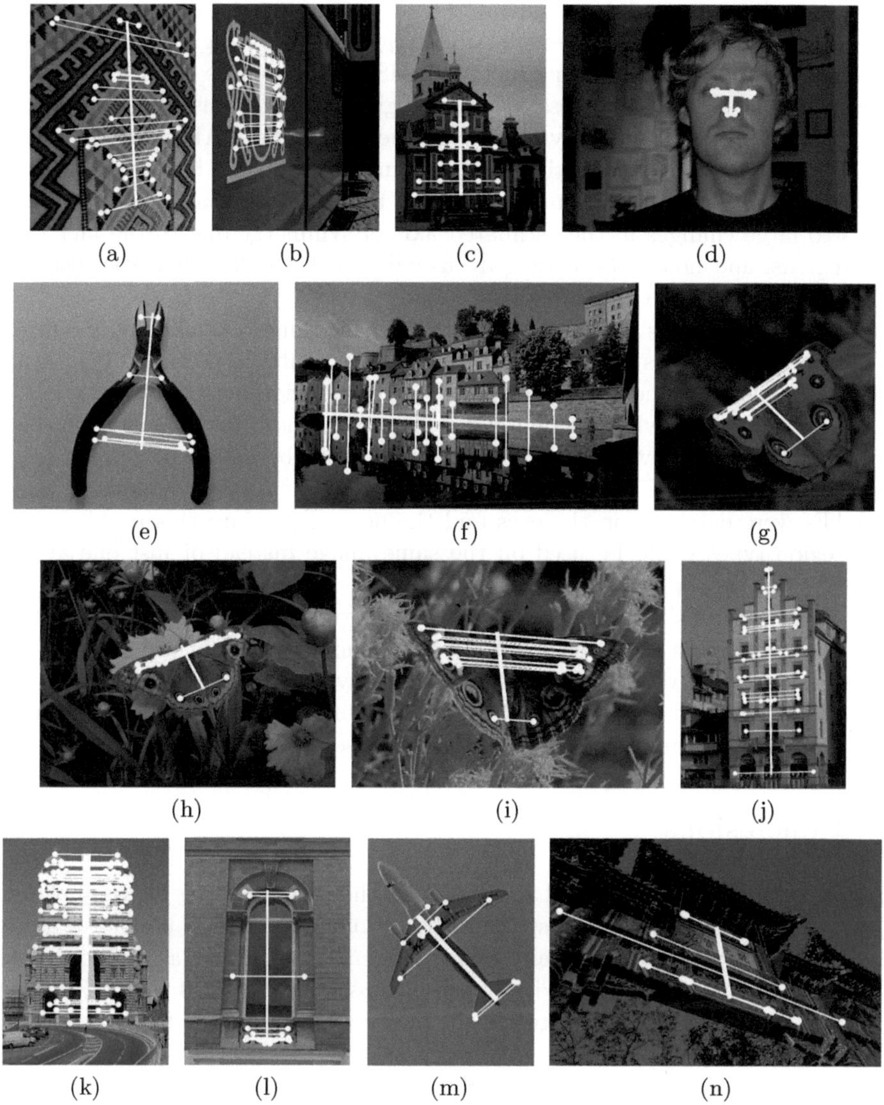

(a) (b) (c) (d) (e) (f) (g) (h) (i) (j) (k) (l) (m) (n)

Fig. 4. Symmetry detected in perspective images, showing the reflective matches and axes of symmetry detected

complicated by the fact that a part of a building will be symmetric with respect to many other parts on the building around different symmetry axes. In the results presented in Fig. 4 our algorithm managed to identify the strongest and most global symmetry axis. All axes of symmetry with support higher than a threshold depending on the support for the strongest axis of symmetry detected in the image are shown. If that threshold would be lowered more symmetry axes would appear on the buildings.

The results presented were generated using one of two different region detectors. Either MSERs [6] or Hessian-Affine regions [7]. During the experiments, we observed that the LAFs constructed on MSERs were more accurate than the ones constructed from the Hessian-Affine regions. This means that more accurate estimates of the axis of symmetry can be made from LAFs constructed on MSERs. Accuracy in the positions of the points in the LAFs is essential for the success of our algorithm. A small change in the position of a point sometimes leads to large changes in the estimated axis of symmetry due to the fact that the regions, and hence the LAFs, are usually much smaller than the distance between the regions in a pair.

Although the LAFs obtained from the Hessian-Affine regions were less accurate some good results can still be obtained. The results in the Figs. 4(a), 4(d), 4(f), 4(m) were obtained using the Hessian-Affine region detector.

A problem noted when testing the algorithm is that the symmetric surface often generates too few regions for the symmetry to be detected. Human faces are one example. In Fig. 4(d) it can be seen that only a few symmetric pairs of LAFs were detected on the eyes and the nose. To reduce this problem several region types could be used on the same image instead of just one at the time.

A natural extension to our current method would be to add a segmentation step at the end that would segment out the symmetric part of the image. Such a step would also verify the detected symmetry and could be used to reject incorrectly detected symmetries. A region growing algorithm like the one used in [14] could for example be used for this purpose.

5 Conclusions

In this paper we have presented an efficient method for detecting bilateral planar symmetry in images under perspective distortion. The method uses local affine frames constructed on maximally stable extremal regions or any other affine covariant regions. The complexity of the proposed algorithm is $n \log(n)$, where n is the number of LAFs detected in the image allowing in practice a near real-time performance.

Hypotheses are generated from only one corresponding reflected pair of local affine frames and verified using the rest of the corresponding pairs. Hence our method is very robust to background clutter and able to discover multiple symmetries in the image in near real-time.

Acknowledgements

Authors were supported by STINT Foundation under project Dur IG2003-2 062, by the Czech Ministry of Education projects 1M0567 and by Czech Science Foundation project 102/07/1317.

References

1. Dakin, S.C., Herbert, A.M.: The spatial region of integration for visual symmetry detection. Proc of the Royal Society London B. Bio. Sci. 265, 659–664 (1998)
2. Horridge, G.A.: The honeybee (apis mellifera) detect bilateral symmetry and discriminates its axis. J. Insect Physiol. 42, 755–764 (1996)
3. Loy, G., Eklundh, J.O.: Detecting symmetry and symmetric constellations of features. In: Leonardis, A., Bischof, H., Pinz, A. (eds.) ECCV 2006. LNCS, vol. 3952, Springer, Heidelberg (2006)
4. Cornelius, H., Loy, G.: Detecting bilateral symmetry in perspective. In: Proc of 5th Workshop on Perceptual Organisation in Computer Vision (POCV) (2006)
5. Obdrzalek, S., Matas, J.: Object recognition using local affine frames on distinguished regions. In: BMVC (2002)
6. Matas, J., Chum, O., Urban, M., Pajdla, T.: Robust wide baseline stereo from maximally stable extremal regions. In: BMVC. 3D and Video (2002)
7. Mikolajczyk, K., Schmid, C.: An affine invariant interest point detector. In: Heyden, A., Sparr, G., Nielsen, M., Johansen, P. (eds.) ECCV 2002. LNCS, vol. 2350, pp. 128–142. Springer, Heidelberg (2002)
8. Mitra, S., Liu, Y.: Local facial asymmetry for expression classification. In: CVPR (2004)
9. Zielke, T., Brauckmann, M., von Seelen, W.: Intensity and edge-based symmetry detection with an application to car-following. CVGIP 58(2), 177–190 (1993)
10. Hong, W., Yang, A.Y., Huang, K., Ma, Y.: On symmetry and multiple-view geometry: Structure, pose, and calibration from a single image. IJCV (2004)
11. Loy, G., Zelinsky, A.: Fast radial symmetry for detecting points of interest. PAMI 25(8), 959–973 (2003)
12. Sela, G., Levine, M.D.: Real-time attention for robotic vision. Real-Time Imaging 3, 173–194 (1997)
13. Mukherjee, D.P., Zisserman, A., Brady, J.M.: Shape from symmetry—detecting and exploiting symmetry in affine images. In: Philosophical Transactions of the Royal Society of London, Series A, vol. 351, pp. 77–106 (1995)
14. Tuytelaars, T., Turina, A., Gool, L.J.V.: Noncombinatorial detection of regular repetitions under perspective skew. PAMI 25(4), 418–432 (2003)
15. Mikolajczyk, K., Tuytelaars, T., Schmid, C., Zisserman, A., Matas, J., Schaffalitzky, F., Kadir, T., Gool, L.V.: A comparison of affine region detectors. IJCV (2006)
16. Lindeberg, T., Gårding, J.: Shape-adapted smoothing in estimation of 3-d cues from affine deformations of local 2-d brightness structure. ICV (1997)
17. Lowe, D.G.: Distinctive image features from scale-invariant keypoints. IJCV 60(2), 91–110 (2004)
18. Obdrzalek, S., Matas, J.: Sub-linear indexing for large scale object recognition. In: BMVC (2005)
19. Fischler, M., Bolles, R.: Random sample consensus: A paradigm for model fitting with applications to image analysis and automated cartography. Communications of the ACM 24(6), 381–395 (1981)

Triangulation of Points, Lines and Conics

Klas Josephson and Fredrik Kahl

Centre for Mathematical Sciences,
Lund University, Lund, Sweden
{klasj, fredrik}@maths.lth.se

Abstract. The problem of reconstructing 3D scene features from multiple views with known camera motion and given image correspondences is considered. This is a classical and one of the most basic geometric problems in computer vision and photogrammetry. Yet, previous methods fail to guarantee optimal reconstructions - they are either plagued by local minima or rely on a non-optimal cost-function. A common framework for the triangulation problem of points, lines and conics is presented. We define what is meant by an optimal triangulation based on statistical principles and then derive an algorithm for computing the globally optimal solution. The method for achieving the global minimum is based on convex and concave relaxations for both fractionals and monomials. The performance of the method is evaluated on real image data.

1 Introduction

Triangulation is the problem of reconstructing 3D scene features from their projections. Naturally, since it is such a basic problem in computer vision and photogrammetry, there is a huge literature on the topic, in particular, for point features, see [1,2]. The standard approach for estimating point features is:

(i) Use a linear least-squares algorithm to get an initial estimate.
(ii) Refine the estimate (so called *bundle adjustment*) by minimizing the sum of squares of reprojection errors in the images.

This methodology works fine in most cases. However, it is well-known that the cost-function is non-convex and one may occasionally get trapped in local minima [3]. The goal of this paper is to develop an algorithm which computes the globally optimal solution for a cost-function based on statistical principles [4].

In [3], the two-view triangulation problem for points was treated. The solution to the optimal problem was obtained by solving a sixth degree polynomial. This was generalized for three views in [5], but the resulting polynomial system turns out to be of very high degree and their solution method based on Gröbner bases becomes numerically unstable. In [6] convex linear matrix inequalities (LMI) relaxations are used to approximate the non-convex cost-function (again, in the point case), but no guarantee of actually obtaining the global minimum is provided. For line and conic features, the literature is limited to closed-form solutions with algebraic cost-functions and to local optimization methods, see [1] and the references therein.

B.K. Ersbøll and K.S. Pedersen (Eds.): SCIA 2007, LNCS 4522, pp. 162–172, 2007.

In this paper, we present a common framework for the triangulation problem for any number of views and for three different feature types, namely, points, lines and conics. An algorithm is presented which yields the global minimum of the statistically optimal cost-function. Our approach is most closely related to the work in [7], where fractional programming is used to solve a number of geometric reconstruction problems including triangulation for points. Our main contributions are the following. First, we show how a covariance-weighted cost-function - which is the statistically correct thing to consider - can be minimized using similar techniques as in [7] for the point case. For many point and corner detectors, e.g., [8,9], it is possible to obtain information of position uncertainty of the estimated features. Second, we present a unified framework for the triangulation problem of points, lines and conics and the corresponding optimal algorithms. Finally, from an algorithmic point of view, we introduce convex and concave relaxations of monomials in the optimization framework in order to handle Plücker constraints appearing in the line case.

2 Projective Geometry

In the triangulation of points, lines and conics, it is essential to have the formulation of the projection from the three dimensional space to the two dimensional image space in the same way as the standard projection formulation used in the point case. For that reason we begin with a short recapitulation of the projection of points with a standard pinhole camera. After that the methods to reformulate the projection of lines and quadrics into similar equations are considered. For more reading on projective geometry see [1].

2.1 Points

A perspective/pinhole camera is modeled by,

$$\lambda x = PX, \quad \lambda > 0, \tag{1}$$

where P denotes the camera matrix of size 3×4. Here X denotes the homogeneous coordinates for the point in the 3D space, $X = [U\ V\ W\ 1]^T$, and x denote the coordinates in the image plane, $x = [u\ v\ 1]$. The scalar λ can be interpreted as the depth, hence $\lambda > 0$ if the point appears in the image.

2.2 Lines

Lines in three dimensions have four degrees of freedom - a line is determined by the intersection of the line with two predefined planes. The two intersection points on the two planes encode two degrees of freedom. Even if lines only have four degrees of freedom, there is no universal way of representing every line in $\mathbb{P}^4$. One alternative way to represent a line is to use Plücker coordinates. With Plücker coordinates, the line is represented in an even higher dimensional space $\mathbb{P}^5$. The over parameterization is hold back by a quadratic constraint that has to be fulfilled for every line. In [1] definitions and properties for Plücker coordinates are described. The big benefit with Plücker coordinates is the Plücker camera that makes it possible to write the projection of lines as $\lambda l = P_{\mathcal{L}} \mathcal{L}$. A drawback on the other hand is that they have to fulfill the quadratic constraint

$$l_{12}l_{34} + l_{13}l_{42} + l_{14}l_{23} = 0, \tag{2}$$

otherwise projection of lines can be formulated in the same manner as point projections, but now it is a projection from projective space of dimension 5 to the image. Hence the line camera matrices are of dimension 3×6.

2.3 Conics

As for lines, we are interested in writing the projection of a quadric to an image conic in the form of the projection formula for points. To do that we use the projection formula of the duals to the quadric conics. These duals are the envelopes of the structures. For conics, the envelope consists of lines and for quadrics, the envelope consist of all planes tangent to the quadric locus. Provided the quadrics and conics are non-degenerate, one can show that the equations for the duals are, $U^T LU = 0$, where U are homogeneous plane coordinates and $L = C^{-1}$. Similar for conics, one gets, $u^T lu = 0$, where u are homogeneous line coordinates and $l = c^{-1}$. The projection for the envelope forms are,

$$\lambda l = PLP^T \quad \lambda \neq 0. \tag{3}$$

Now we want to reformulate (3) so it appears in a similar way as the point projection formula. This can be done in the form, $\lambda \tilde{l} = P_{\mathcal{C}} \tilde{L}$, where $\tilde{l}$ and $\tilde{L}$ are column vectors of length 6 and 10 obtained from stacking the elements in l and L. $P_{\mathcal{C}}$ is an 6×10 matrix. The entries in $P_{\mathcal{C}}$ are quadratic expressions in P.

As for the line case, it is not possible to make the interpretation that the scalar λ of the projected conic corresponds to the depth.

3 Triangulation

In triangulation the goal is to reconstruct a three dimensional scene from measurements in N images, $N \geq 2$. The camera matrices P_i, $i = 1 \ldots N$, are considered to be known for each image. In the point case, the camera matrix can be written $P = (p_1, p_2, p_3)^T$, where p_j is a 4–vector. Let $(u, v)^T$ denote the image coordinates and $X = (U, V, W, 1)^T$ the extended coordinates of the point in 3D. This gives the reprojection error

$$r = \left(\frac{u\, p_3^T X - p_1^T X}{p_3^T X}, \frac{v\, p_3^T X - p_2^T X}{p_3^T X} \right), \tag{4}$$

Further, $\sum_{i=1}^{N} \|r_i\|_2^2$ is the objective function to minimize if the smallest reprojection error is to be achieved in L_2-norm. To use the optimization algorithm proposed in this paper (see next section), it is necessary to write the error in each image as a rational function f/g where f is convex and g concave.

It is easy to see that the L_2-norm of the residual in (4) can be written as $\|r\|^2 = ((a^T X)^2 + (b^T X)^2)/(p_3^T X)^2$, where a, b are 4-vectors determined by the image coordinates and the elements of the camera matrix. By choosing $f = ((a^T X)^2 + (b^T X)^2)/(p_3^T X)$ (with the domain $p_3^T X > 0$) and $g = p_3^T X$, one can show that f is indeed convex and g concave. It is straight forward to form the same residual vectors in the line and conic cases - the only difference is that the dimension is different.

3.1 Incorporation of Covariance

The optimal cost-function is to weight the residual vector by its covariance [4] (at least to a first order approximation). Incorporating covariance weighted error transforms to,

$$\|Lr\| = \left\| L\left(\frac{x_1 p_n^T X - p_1^T X}{p_n^T X}, \dots\right) \right\|, \tag{5}$$

where L is the cholesky factorization of the inverse covariance matrix to the structure in each image. Notice that we have chosen to normalize by the last coordinate and in that case the covariance becomes a 2×2 symmetric matrix in the point and line cases and a 5×5 matrix in the conic case. The reason why the covariance matrix is one dimension lower than the image vector is that there is no uncertainty in the last element of the extended image coordinates.

3.2 Problem Formulation

In all of the above cases, the optimization problem to solve is the following:

$$\min \sum_{i=1}^{n} \|L_i r_i\|^2. \tag{6}$$

The only thing which differs (except for dimensions) in the different cases is that in the line case it is necessary to fulfill the quadratic Plücker constraint (2) for the coordinates of the three dimensional structure.

4 Branch and Bound Optimization

Branch and bound algorithms are used to find the global optimum for non-convex optimization problems. The algorithm gives a provable upper and lower bound of the optimum and it is possible to get arbitrary close to the optimum.

On a non-convex, scalar-valued objective function Φ at the domain Q_0 the branch and bound algorithm works by finding a lower bound to the function Φ on the domain Q_0. If the difference between the optimum for the bounding functions and the lowest value of the function Φ - calculated so far - is small enough, then the optimum is considered to be found. Otherwise the domain Q_0 is splitted into subdomains and the procedure is repeated in these domains.

If the lower bound on a subdomain has its optimum higher than a known value of the objective function in another subdomain it is possible to neglect the first subdomain since we know that the optimum in that region is greater than the lowest value obtained so far.

4.1 Bounding

The goal of the bounding function Φ_{lb} is that it should be (i) a close under-estimator to the objective function Φ and (ii) easy to compute the lowest value Φ_{lb} in given domain. Further, as the domain of the bounding functions is partitioned into smaller regions, the approximation gap to the objective function must converge (uniformly). A good choice of Φ_{lb} is the convex envelope [7].

Fractional Relaxation. Fractional programming is used to minimize/maximize a sum of $p \geq 1$ fractions subject to convex constraints. In this paper we are interested of minimizing

$$\min_{x} \sum_{i=1}^{p} \frac{f_i(x)}{g_i(x)} \qquad (7)$$
$$\text{subject to } x \in D,$$

where f_i and g_i are convex and concave, respectively, functions from $\mathbb{R}^n \to \mathbb{R}$, and the domain $D \subset \mathbb{R}^n$ is a convex set. On top of this it is assumed that f_i and g_i are positive and that one can compute a priori bounds on f_i and g_i. Even under these assumptions it can be shown that the problem is $\mathcal{NP}$-complete [10].

It is showed in [7] that if you have bounds on the domain D it is possible to rewrite (7) to a problem that is possible to find the convex envelope to for every single fraction by a *Second Order Cone Program* (SOCP) [11].

When Φ is a sum of ratios as in (7) a bound for the function can be calculated as the sum of the convex envelopes of the individual fractions. The summarized function will be a lower bound and it fulfills the requirements of a bounding function. This way of calculating Φ_{lb} by solving a SOCP problem can be done efficiently [12].

A more exhaustive description on fractional programming in multiple view geometry can be found in [7] where point triangulation (without covariance weighting) is treated.

Monomial Relaxation. In the line case the Plücker coordinates have to fulfill the Plücker constraint (2). This gives extra constraints in the problem to find lower bounds.

If we make the choice in the construction of the Plücker coordinates that the first point lies on the plane $z = 1$ and the second on the plane $z = 0$, remember that the Plücker coordinates are independent of the construction points, the two points $X = (x_1,\ x_2,\ 1,\ 1)^T$ and $Y = (y_1,\ y_2,\ 0,\ 1)^T$ gives the following Plücker coordinates for the line (2.2),

$$\mathcal{L} = (x_1 y_2 - x_2 y_1,\ -y_1,\ x_1 - y_1,\ -y_2,\ y_2 - x_2,\ 1)^T. \tag{8}$$

This parameterization involves that the last coordinate is one and that only the first one is nonlinear to the points of intersection. Hence it is only necessary to make a relaxation of the first coordinate (in addition to the fractional terms).

In [13] the convex and the concave envelopes are derived for a monomial $y_1 y_2$. The convex and the concave envelopes are given by,

$$\mathbf{convenv}(y_1 y_2) = \max \left\{ \begin{array}{l} y_1 y_2^U + y_1^U y_2 - y_1^U y_2^U \\ y_1 y_2^L + y_1^L y_2 - y_1^L y_2^L \end{array} \right\}, \tag{9}$$

$$\mathbf{concenv}(y_1 y_2) = \min \left\{ \begin{array}{l} y_1 y_2^L + y_1^U y_2 - y_1^U y_2^L \\ y_1 y_2^U + y_1^L y_2 - y_1^L y_2^U \end{array} \right\}. \tag{10}$$

Given bounds on x_1, x_2, y_1 and y_2 in the parameterization of a line, it is possible to propagate the bounds to the monomials $x_1 y_2$ and $x_2 y_1$.

4.2 Branching

It is necessarily to have a good strategy when branching. If a bad strategy is chosen the complexity can be exponentially but if a good choice is made it is possible to achieve a lower complexity - at least in practice.

A standard branching strategy for fractional programming [14] is to branch on the denominator s_i of each fractional term t_i/s_i. This limits the practical use of branch and bound optimization to at most about 10 dimensions [15] but in the case of triangulation the number of branching dimensions can be limited to a fixed number (at most the degree of freedom of the geometric primitive). Hence, in the point case is it enough to branch on three dimensions, in the line case four and in the cases of conics nine dimensions maximally.

In the line case, we choose not to branch on the denominators. Instead the coordinates of the points where the line intersect with the planes $z = 0$ and $z = 1$ are used for the parameterization (4.1). This gives four dimensions to split at, independent of the number of images. It is also important to choose a coordinate system such that the numerical values of these parameters are kept at a reasonable magnitude.

For strategies for branching and more on fractional programming see [15].

4.3 Initialization

It is necessary to have an initial domain Q_0 for the branch and bound algorithm. The method used for this is similar in the point and conic case but different in the line case due to the Plücker constraint.

Points and Conics. In order to get a bound on the denominators, we assume a bound on the maximal reprojection error. Ideally, with correctly weighted covariance, each such residual $L_i r_i$ should approximately be i.i.d. with unit variance. Thus the bounds are constructed from a user given maximal reprojection error. The bounds on the denominators $g_i(x)$ can then be calculated by the following optimization problem,

$$\begin{aligned} \text{for } i = 1, \ldots, p, \ \ \min/\max \quad & g_i(x) \\ \text{subject to} \quad & \frac{f_j(x)}{g_j(x)} \leq \gamma \quad j = 1, \ldots, p, \end{aligned} \tag{11}$$

where γ is the user given bound on the reprojection error. This is a quadratic convex programming problem. In the experiments, γ is set to 3 pixels.

Lines. In the case of lines, the Plücker constraint makes things a bit more problematic. Instead we choose a more geometric way of getting bounds on the coordinates of the two points defining the line.

For each image line l, two parallel lines are constructed with γ pixels apart (one on each side of l). Then, we make the hypothesis that the two points defining the optimal 3D line (with our choice of coordinate systems) are located on the planes $z = 0$ and $z = 1$, respectively. Now, finding bounds on x_1, x_2, y_1, y_2, see equation (8), becomes a simple linear programming problem. Again, it is important to choose the coordinate system such that the planes $z = 0$ and $z = 1$ are located appropriately. In addition, to avoid getting an unbounded feasible region, the maximum depth is limited to the order of the 3D point furthest away. In the experiments, we set γ to 5 pixels.

5 Experiments

The implementation is made in Matlab using a toolbox called SeDuMi [12] for the convex optimization steps.

The splitting of dimensions has been made by taking advantage of the information where the minimum of the bounding function is located.

While testing the various cases, we have found that the relaxation performed in the line case - a combination of fractions and monomials - the bounds on the denominators is a critical factor for the speed of convergence. To increase the convergence speed, a local gradient descent step is performed on the computed solution in order to quickly obtain a good solution which can be employed to discard uninteresting domains.

Two public sets of real data[1] were used for the experiments with points and lines. One of a model house with a circular motion and one of a corridor with a mostly forward moving motion. The model house has 10 frames and the corridor 11. In these two sequences there were no conics. A third real data sequence was used for conic triangulation. In Fig. 1 samples of the data sets are given.

[1] http://www.robots.ox.ac.uk/~vgg/data.html

Fig. 1. Image sets used for the experiments

Table 1. Reprojection errors for points and lines with three different methods on two data sets

	Points						Lines					
Data set	Optimal		Bundle		Linear		Optimal		Bundle		Linear	
	Mean	Std	Mean	Std	Mean	Std	Mean	Std	Mean	Std	Mean	Std
House	0.15	0.14	0.15	0.14	0.16	0.15	1.40	0.92	1.41	0.93	1.62	1.03
Corridor	0.13	0.11	0.13	0.11	0.13	0.11	3.42	4.29	3.30	4.34	4.02	5.45

Points and lines were reconstructed and then the reprojection errors between different methods were compared. The other methods compared are bundle adjustment and a linear method [1]. The covariance structure for the lines was computed by fitting a line to measured image points. In the reconstruction only the four first frames were used. In the house scene, there are 460 points and in the corridor 490. The optimum was considered to be found if the gap between the global optimum and the under-estimator was less than 10 %. The results are presented in Table 1.

In the house scene, the termination criterion was reached already in the first iteration for all points and for most of them the bounding functions was very close to the global minimum (less than the 10 % required). In the corridor scene, the average number of iterations were 3 and all minima were reached within at most 23 iterations.

In the line case, the under-estimators works not as well as in the point case. This is due to the extra complexity of the Plücker coordinates. Thus more iterations are needed. For the house scene with the circular motion the breakpoint is reached within at most 120 iterations for all the tested 12 lines. However, for the corridor sequence with a weaker camera geometry (at least for triangulation purposes) it is not even enough with 500 iterations for 6 of the tested 12 lines. Even if a lot of iterations are needed to certify the global minimum, the location of the optimum in most cases is reached within less than a handful of iterations.

It can be seen in Table 1 that both a linear method and bundle adjustment works fine for these problems. However, in some cases the bundle adjustment reprojection errors get higher than the errors for the optimal method. This shows that bundle adjustment (which is based on local gradient descent) sometimes gets stuck in a local minimum.

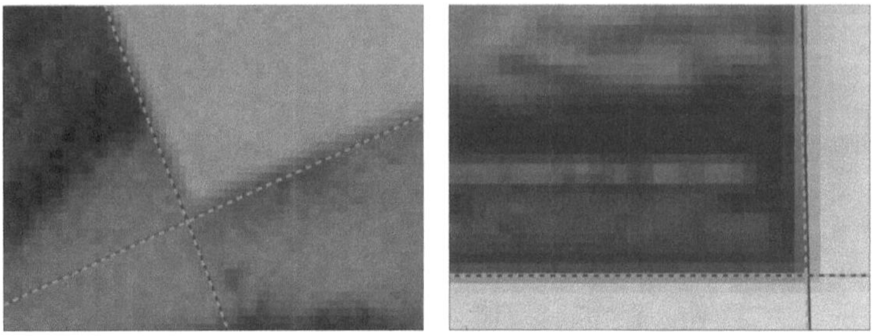

(a) (b)

Fig. 2. The result from reprojection of lines. The green dashed line is the original and the red solid line is the reprojected. Image (a) is from the house scene and (b) is from the corridor.

The result can also be seen in Fig. 2 where two lines from each data set are compared with reprojected line.

5.1 Conics

For conics, an example images can be found in Fig. 1. The covariance structure was estimated by fitting a conic curve to measured image points. The corresponding 3D quadrics were computed with the optimal and a linear method. The result of the reprojected conics from these two methods are imaged in Fig. 3.

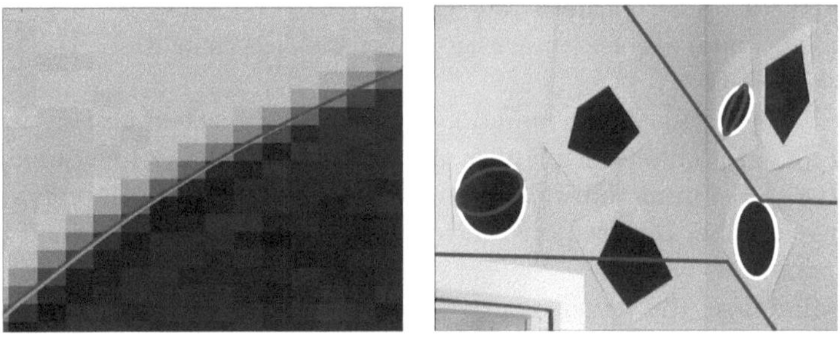

(a) (b)

Fig. 3. The result of the reprojected conics of the data set in Fig. 1. In image (a) a part of the reconstruction with optimal method is viewed. The light green is the reprojection and the dark red the original conic. In (b) the red lines are the reprojection after linear method and the white when the optimal method were used.

The number of iterations performed to reach the global minimum with a gap less than 5 % of the bounding function for the three conics were 3, 6 and 14. As can be seen from the images, the quadrics in the data set are planar and hence the condition number of the corresponding 4×4 matrix should be zero. For the three estimated quadrics with the optimal method, the condition numbers are

$1.2 \cdot 10^{-3}$, $3.7 \cdot 10^{-7}$ and $8.8 \cdot 10^{-6}$. This can be compared with the result for the linear estimate with condition numbers of $3.7 \cdot 10^{-4}$, $4.1 \cdot 10^{-5}$ and $1.1 \cdot 10^{-4}$.

Fig. 3 (a) shows the reprojected conic compared with the original for one of the conics. The fitting is very good and it is obvious from Fig. 3 (b) that the linear result is far from acceptable.

6 Discussion

A unified treatment of the triangulation problem has been described using covariance propagation. In addition to traditional local algorithms and algorithms based on algebraic objective functions, globally optimal algorithms have been presented for the triangulation of points, lines and conics. For most cases, local methods work fine (except for conics) and they are generally faster in performance. However, none of the competing methods have a guarantee of globality.

A future line of research is to include more constraints in the estimation process, for example, planar quadric constraints. This opens up the possibility to perform optimal auto-calibration using the image of the absolute conic.

Acknowledgments

The authors thanks Manmohan Chandraker and Sameer Agarwal at Department of Computer Science and Engineering, University of California, San Diego.

This work has been funded by the Swedish Research Council through grant no. 2004-4579 'Image-Based Localisation and Recognition of Scenes', grant no. 2005-3230 'Geometry of multi-camera systems' and the European Commission's Sixth Framework Programme under grant no. 011838 as part of the Integrated Project SMErobot.

References

1. Hartley, R.I., Zisserman, A.: Multiple View Geometry in Computer Vision, 2nd edn. Cambridge University Press, Cambridge (2004)
2. Slama, C.C.: Manual of Photogrammetry. American Society of Photogrammetry, Falls Church, VA (1980)
3. Hartley, R., Sturm, P.: Triangulation. Computer Vision and Image Understanding 68(2), 146–157 (1997)
4. Kanatani, K.: Statistical Optimization for Geometric Computation: Theory and Practice. In: Elsevier Science, Elsevier, North-Holland, Amsterdam (1996)
5. Stewénius, H., Schaffalitzky, F., Nistér, D.: How hard is three-view triangulation really? In: Int. Conf. Computer Vision, Beijing, China (2005)
6. Kahl, F., Henrion, D.: Globally optimal estimates for geometric reconstruction problems. In: Int. Conf. Computer Vision, Beijing, China (2005)
7. Agarwal, S., Chandraker, M., Kahl, F., Kriegman, D., Belongie, S.: Practical global optimization for multiview geometry. In: Leonardis, A., Bischof, H., Pinz, A. (eds.) ECCV 2006. LNCS, vol. 3951, Springer, Heidelberg (2006)

8. Harris, C., Stephens, M.: A combined corner and edge detector. In: Alvey Vision Conference. pp. 147–151 (1988)
9. Triggs, B.: Detecting keypoints with stable position, orientation, and scale under illumination changes. In: Pajdla, T., Matas, J(G.) (eds.) ECCV 2004. LNCS, vol. 3024, Springer, Heidelberg (2004)
10. Freund, R.W., Jarre, F.: Solving the sum-of-ratios problem by an interior-point method. Journal of Global Optimization 19, 83–102 (2001)
11. Tawarmalani, M., Sahinidis, N.V.: Semidefinite relaxations of fractional programs via novel convexification techniques. Journal of Global Optimization 20, 133–154 (2001)
12. Sturm, J.: Using SeDuMi 1.02, a Matlab toolbox for optimization over symmetric cones. Optimization Methods and Software 11(12), 625–653 (1999)
13. Ryoo, H.S., Sahinidis, N.V.: Analysis of bounds for multilinear functions. Journal of Global Optimization 19(4), 403–424 (2001)
14. Benson, H.P.: Using concave envelopes to globally solve the nonlinear sum of ratios problem. Journal of Global Optimization 22, 343–364 (2002)
15. Schaible, S., Shi, J.: Fractional programming: the sum-of-ratios case. Optimization Methods and Software 18, 219–229 (2003)

Robust Variational Reconstruction from Multiple Views

Natalia Slesareva[1], Thomas Bühler[1], Kai Uwe Hagenburg[1], Joachim Weickert[1], Andrés Bruhn[1], Zachi Karni[2], and Hans-Peter Seidel[2]

[1] Mathematical Image Analysis Group, Dept. of Mathematics and Computer Science, Saarland University, Building E1.1, 66041 Saarbrücken, Germany
{slesareva,buehler,hagenburg,weickert,bruhn}@mia.uni-saarland.de
[2] Max-Planck-Institut für Informatik, Stuhlsatzenhausweg 85, 66123 Saarbrücken, Germany
{karni,hpseidel}@mpi-sb.mpg.de

Abstract. Recovering a 3-D scene from multiple 2-D views is indispensable for many computer vision applications ranging from free viewpoint video to face recognition. Ideally the recovered depth map should be dense, piecewise smooth with fine level of details, and the recovery procedure shall be robust with respect to outliers and global illumination changes. We present a novel variational approach that satisfies these needs. Our model incorporates robust penalisation in the data term and anisotropic regularisation in the smoothness term. In order to render the data term robust with respect to global illumination changes, a gradient constancy assumption is applied to logarithmically transformed input data. Focussing on translational camera motion and considering small baseline distances between the different camera positions, we reconstruct a common disparity map that allows to track image points throughout the entire sequence. Experiments on synthetic image data demonstrate the favourable performance of our novel method.

Keywords: computer vision, variational methods, multi-view reconstruction, structure from motion, partial differential equations.

1 Introduction

Structure from motion is a challenging task in modern computer vision: Extraction of depth information from the images of a single moving camera is useful for such tasks as robot navigation, augmented reality [14], [13] or face recognition. In the latter case structure from motion allows to reconstruct a face from a set of images, obtained by a single moving camera. One of the typical scenarios in this context is a camera that moves horizontally with a constant speed and whose optical axis is fixed orthogonal to the path of motion. While such a setting simplifies the computation, it is still difficult to obtain dense reconstructions that are robust under noise and illumination changes and provide sharp object boundaries.

B.K. Ersbøll and K.S. Pedersen (Eds.): SCIA 2007, LNCS 4522, pp. 173–182, 2007.

All these demands can be satisfied by variational techniques, that have proved to be very useful in the context of optic flow estimation [2]. The reconstruction problem is formulated in an energy minimisation framework, under the assumption of global smoothness of the solution. Compared to other methods variational techniques offer a number of specific advantages: They allow transparent modeling without hidden assumptions or post-processing steps. Moreover, their continuous formulation enables rotationally invariant modeling in a natural way. The filling-in effect creates dense depth maps with sub-pixel precision by propagating information over the entire image domain. For these reasons we aim here at exploring the performance of variational methods in the context of 3-D reconstruction from multiple views.

Since the fundamental work of Faugeras and Keriven [5] many different methods for multi-view 3-D reconstruction have been proposed. In most cases a calibrated camera setup is assumed and locally constant intensity of objects in the scene is required. In the core of the minimisation procedure there lays either a gradient descent algorithm such as in [5], [15] or a sophisticated strategy of successive refinement of results as applied in [6]. The results are highly accurate, however the reported computational times take up to several hours. Comparing to other methods, that reconstruct 3-D objects from multiple views using variational framework, like for example in [9], [8] our method produces not a complete model, but only one disparity map. On the other hand the simplicity of our approach allows us to study more sophisticated models that help to improve the robustness of the method with respect to noise and varying illumination.

In this paper we focus on a prototypical scenario of a face recognition system that reconstructs the face surface from images taken by a camera that moves linearly with constant speed within an orthoparallel setting. This allows us to exploit a number of ideas that originate from the computation of optic flow fields. It is well-known that in the orthoparallel case the following relation holds:

$$Z = \frac{b \cdot f}{D} . \tag{1}$$

Here, Z denotes the depth of a point in the 3-D world, b is the baseline distance between successive camera positions, f specifies the focal length and D is the disparity, i.e. the distance between the projection of Z on two successive image planes. Formulating our problem in terms of disparity estimation, we obtain a scene reconstruction up to a scaling factor that depends on one intrinsic (focal length) and one extrinsic parameter (baseline).

Since the camera moves slowly with a constant speed, we obtain a series of consecutive disparity maps that are identical. Hence, it is sufficient to compute a single joint disparity map.

Our paper is organised as follows. The next section describes our variational model and its underlying assumptions in detail. Its PDE formulation is given by the Euler-Lagrange equation sketched in Section 3. Experiments in Section 4 illustrate the performance of our approach. The paper is concluded by a summary in Section 5.

2 Variational Framework

We assume a single camera that acquires images while moving slowly with a constant speed along the x-axis. Thus, approximately the same displacement field (disparity map) $\lambda(x,y)$ occurs between each pair of subsequent frames and can be recovered as minimiser of a single energy functional:

$$E(\lambda) = E_D(\lambda) + \alpha E_S(\lambda)\,, \tag{2}$$

where $E_D(\lambda)$ is a data term, $E_S(\lambda)$ is a smoothness term, and the regularisation parameter $\alpha > 0$ determines the desired amount of smoothness.

Let $f^i(\mathbf{x})$ denote the grey value of frame i at location $\mathbf{x} = (x,y)$. In order to render our method robust against noise, we first convolve with a Gaussian K_σ of standard deviation $\sigma > 0$. By applying a logarithmic transform to the result, the multiplicative effects of global illumination changes are transformed into additive perturbations. This leads to the images $g^i(\mathbf{x})$ for $i = 1,...,N$, which serve as input data for our variational approach.

For the data term $E_D(\lambda)$ we choose a gradient constancy assumption between corresponding structures within consecutive frames g^i and g^{i+1}:

$$\nabla g^{i+1}(x+\lambda, y) = \nabla g^i(x,y)\,. \tag{3}$$

It ignores any additive perturbations on $g^i(\mathbf{x})$ caused by global illumination changes between consecutive frames $f^i(\mathbf{x})$. Penalising deviations from this constancy assumption between all consecutive frame pairs in a statistically robust way [7] can be achieved by use of the data term

$$E_D(\lambda) = \int_\Omega \frac{1}{N} \sum_{i=1}^{N-1} \Psi\left(|\nabla g^{i+1}(x+\lambda,y) - \nabla g^i(x,y)|^2\right) d\mathbf{x}\,, \tag{4}$$

where $\Omega \subset \mathbb{R}^2$ denotes our rectangular image domain, and $\Psi(s^2) := \sqrt{s^2+\epsilon^2}$ is a L^1 penaliser with a small regularising constant $\epsilon > 0$ ensuring differentiability.

Since the baseline distance between consecutive frames is supposed to be small for our application, we can simplify our data term by the Taylor linearisations

$$\begin{aligned} \partial_x g^{i+1}(x+\lambda,y) &\approx \partial_x g^{i+1}(x,y) + \partial_{xx} g^{i+1}(x,y)\,\lambda\,, \\ \partial_y g^{i+1}(x+\lambda,y) &\approx \partial_y g^{i+1}(x,y) + \partial_{xy} g^{i+1}(x,y)\,\lambda\,. \end{aligned}$$

Introducing the matrices

$$J^i = \begin{pmatrix} (g_{xx}^{i+1})^2 + (g_{xy}^{i+1})^2 & (g_x^{i+1} - g_x^i)g_{xx}^{i+1} + (g_y^{i+1} - g_y^i)g_{xy}^{i+1} \\ (g_x^{i+1} - g_x^i)g_{xx}^{i+1} + (g_y^{i+1} - g_y^i)g_{xy}^{i+1} & (g_x^{i+1} - g_x^i)^2 + (g_y^{i+1} - g_y^i)^2 \end{pmatrix}$$

and the vector $\mathbf{w} := (\lambda(\mathbf{x}), 1)^\top$ allows to reformulate the data term in a compact way as a sum of robustified quadratic forms:

$$E_D(\lambda) = \int_\Omega \frac{1}{N} \sum_{i=0}^{N-1} \Psi\left(\mathbf{w}^\top J^i \mathbf{w}\right) d\mathbf{x}\,. \tag{5}$$

The role of the smoothness term $E_S(\lambda)$ in our energy functional is to penalise deviations from smoothness in the unknown disparity field $\lambda(\mathbf{x})$. Instead of a standard quadratic smoothness term (based on the L^2 norm), we use the anisotropic image-driven regulariser of Nagel and Enkelmann [12]:

$$E_S(\lambda) = \int_\Omega \nabla\lambda^\top D(\nabla g)\nabla\lambda \, d\mathbf{x} \,. \qquad (6)$$

Here, $D(\nabla g)$ is a normalised and regularised projection matrix orthogonal to ∇g. It is given by

$$D(\nabla g) = \frac{1}{|\nabla g|^2 + 2\nu^2} \begin{pmatrix} g_y^2 + \nu^2 & -g_x g_y \\ -g_x g_y & g_x^2 + \nu^2 \end{pmatrix}$$

with some small regularisation parameter ν.

Now we can write down the complete energy functional by combining the data term (5) and the smoothness term (6):

$$E(\lambda) = \int_\Omega \Big(\frac{1}{N} \sum_{i=0}^{N-1} \Psi\left(\mathbf{w}^\top J^i \mathbf{w}\right) + \alpha \nabla\lambda^\top D \nabla\lambda \Big) \, d\mathbf{x} \,. \qquad (7)$$

3 Euler-Lagrange Equation

From the calculus of variations [4] we know that a necessary condition for a function $\lambda(x, y)$ to be a minimiser of the energy functional (7) is given by the Euler-Lagrange equation

$$\sum_{i=0}^{N-1} \frac{1}{N} \Psi'\Big(\mathbf{w}^\top J^i \mathbf{w}\Big)(J_{11}^i \lambda + J_{12}^i) - \operatorname{div}(D(\nabla g)\nabla\lambda) = 0$$

with reflecting boundary conditions.

This nonlinear partial differential equation can be solved with the help of two nested fixed point iterations: The outer loop fixes nonlinearities with previously computed values of λ, while the inner loop solves the resulting linear problem with the well-known successive overrelaxation (SOR) method [16].

4 Experiments

We evaluate the performance of the algorithm with the help of two synthetic sequences created in OpenGL: The first one illustrates a female head, as shown in Figure 1, while the second one represents a more challenging task – reconstruction of a tree illustrated in Figure 2. The performance of the method was tested on original sequences and versions with varying illumination as well as variants with noise. Moreover, the results for the original sequences were compared to the publicly available two-frame graph cuts method of Kolmogorov and Zabih [10].

In our experiments both sequences contain up to 8 images with small displacements of up to one pixel between successive camera positions. The ground truth maps were obtained by rescaling and transforming the original OpenGL Z-buffers into disparity maps. Consequently, for comparison with a ground truth we compute the *average absolute disparity error (AADE)*

$$\mathbf{AADE} = \frac{1}{M}\sum_{i=1}^{M} |d_i^{\mathbf{truth}} - d_i^{\mathbf{estimate}}|\,,$$

where M denotes the number of pixels.

There are just two model parameters that require adjustment: A smoothness parameter α and a standard deviation of a Gaussian σ for the preprocessing step. Other numerical parameters were kept fixed and constant for all experiments. The computation in all cases was stopped when the normalised L^1 norm of the updates at a certain iteration k became sufficiently small:

$$\frac{\sum_i |\lambda_i^k - \lambda_i^{k-1}|}{\sum_i \lambda_i^k} < \eta$$

In our experiments values for η vary between 10^{-6} to 10^{-8}. The average time, required for the evaluation of our experiments on an Intel Pentium 4 CPU with 3.2 GHz is in the order of 10 to 40 minutes (for 8 images degraded with Gaussian noise). More sophisticated solvers such as multigrid methods, however, may allow even for runtimes of less than a second [3].

4.1 Face Sequence

In this experiment we were using 8 images of a head scene, created in 3DS Max 8.0 and imported to OpenGL. The performance of the algorithm was tested on the original sequence, its degraded version, contaminated with Gaussian noise of $\sigma = 25$ and a sequence made from the same scene but under conditions of varying illumination (see Figure 1). In all experiments we observed that the main details of the head have been reconstructed in a realistic way: One can recognise that the reconstructed object represents a human face with clearly shaped nose, lips and eye slots. All experiments in this subsection have been carried out with two slightly different error measures: Once, the overall AADE of the computed disparity map was used; the other time, the AADE measurement was restricted to the face region by using a mask shown in the Figure 1. Table 1 shows optimal parameters and error measures for the first setting, while Table 2 presents the results for the second setting. Further on we observe that the results are fairly robust under noise and varying illumination: All essential features of the face remain recognisable and in accordance with the ground truth map.

Additionally we have investigated the influence of the number of images on the reconstruction quality. The clear difference in error measurements confirms our expectations: A larger number of images produces more stable results, since the amount of correspondences and, therefore, the reliability of the result increases.

Table 1. Results for the *Head scene*. $AADE_f$ = Average absolute disparity error computed for the whole disparity map. Disparity values for these experiments vary in the interval (0.1, 1). The parameters α and σ have been optimised.

	2 Frames	4 Frames	6 Frames	8 Frames	Noise, 8 Fr.	Illum., 8 Fr.
α	0.5	0.12	0.06	0.05	4.9	0.6
σ	2.5	2.7	2.8	2.9	5.7	1.7
$AADE_f$	0.0357	0.0298	0.0284	0.0286	0.0819	0.0387

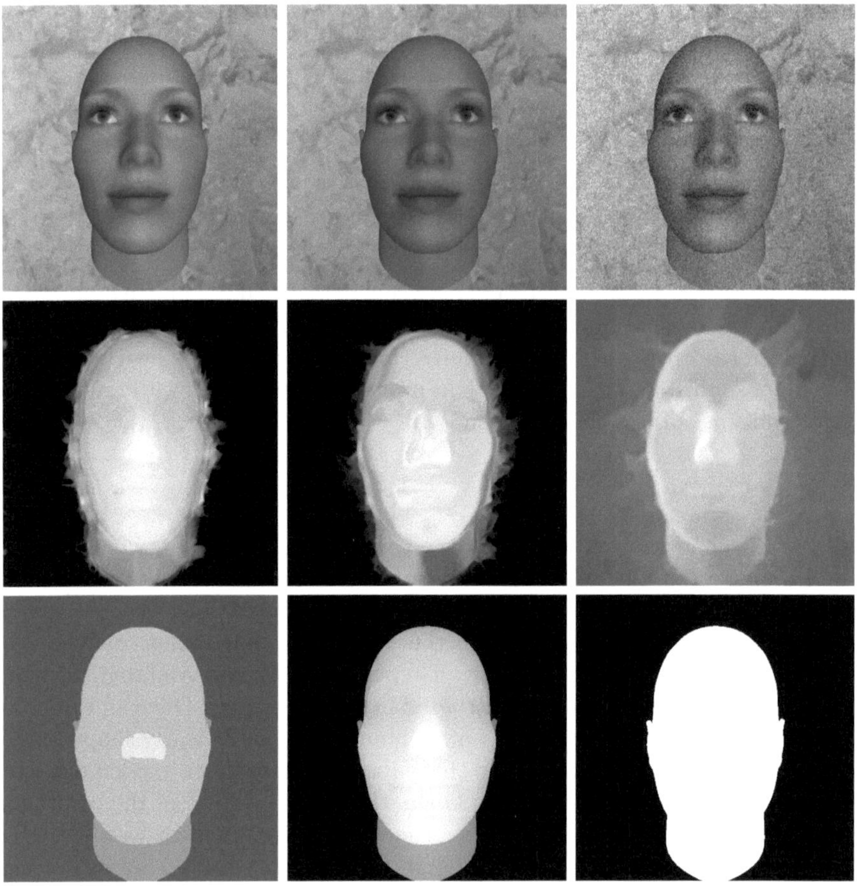

Fig. 1. *Head sequence*, top to bottom, left to right. *First row:* Original frame 1, frame 7 of a sequence with varying illumination, frame 1 of the sequence degraded with Gaussian noise of $\sigma = 25$. *Second row:* Typical results of reconstruction for 8 images of the original sequence, the sequence with illumination changes, and the noisy sequence. *Third row:* Graph cuts result (8 disparity levels), ground truth and mask.

Table 2. Results for the *Head scene*. $AADE_m$ = Average absolute disparity error computed for the face only. Disparity values for these experiments vary in the interval (0.1, 1). The parameters α and σ have been optimised.

	2 Frames	4 Frames	6 Frames	8 Frames	Noise, 8 Fr.	Illum., 8 Fr.
α	0.4	0.27	0.21	0.14	4.0	31
σ	4.1	3.9	4.0	4.1	10.1	2.2
$AADE_m$	0.0244	0.0204	0.0193	0.0192	0.0569	0.0371

Finally, let us compare our results to the one obtained by using the graph cuts method of Kolmogorov and Zabih [10]. Since this method relies on large displacements, we computed the disparity map between the first and the eighth image of the noise free sequence and divided the obtained result by 7 (number of images minus one). The corresponding disparity map which is presented in Figure 1 illustrates a very precise reconstruction of the silhouette of the head with clear distinction of the ears and the neck. However, the main features of the face were completely lost. Evidently, the algorithm is not able to reconstruct these features, because this would require to estimate the displacements at the corresponding locations with sub-pixel precision. But even for relatively large displacements it is well-known that reconstructions of graph cuts methods for such smoothly varying surfaces suffer from similar stair-casing effects [11], this time, however, due to the strong non-convexity of typical regularisers. Our observations are confirmed by the higher AADE for the graph cuts method for both the face region and the whole sequence which is given by $AADE_f = 0.0766$ and $AADE_m = 0.030$, respectively.

4.2 Tree Sequence

In this experiment we reconstruct an object of a very complex structure with fine level of details. Additional difficulty for the algorithm represents a homogeneous region, that corresponds to the sky above the landscape. As before, we make our task even more challenging by degrading the original sequence with Gaussian noise of $\sigma = 25$ and also varying the illumination in the scene.

For the original sequence we observe a very detailed reconstruction: Separate branches of the tree were estimated in accordance to the model, the overall silhouette of the tree was preserved quite well, even the difference in depth between neighbouring leaves appears to be very close to the ground truth map. The homogeneous region, corresponding to the sky was also estimated satisfactory: Since hardly any information is available in the sky region that allows for a direct estimation of the motion, our method propagates this information via the smoothness term. Again, the reconstruction process shows robustness with respect to noise and varying illumination: Both disparity maps show high similarity to the ground truth map with slightly higher values of AADE. In this experiment the difference between AADE values for the original sequence and those with noise and illumination change is not so large as in the previous experiment for the head sequence. This can be explained with the complexity of

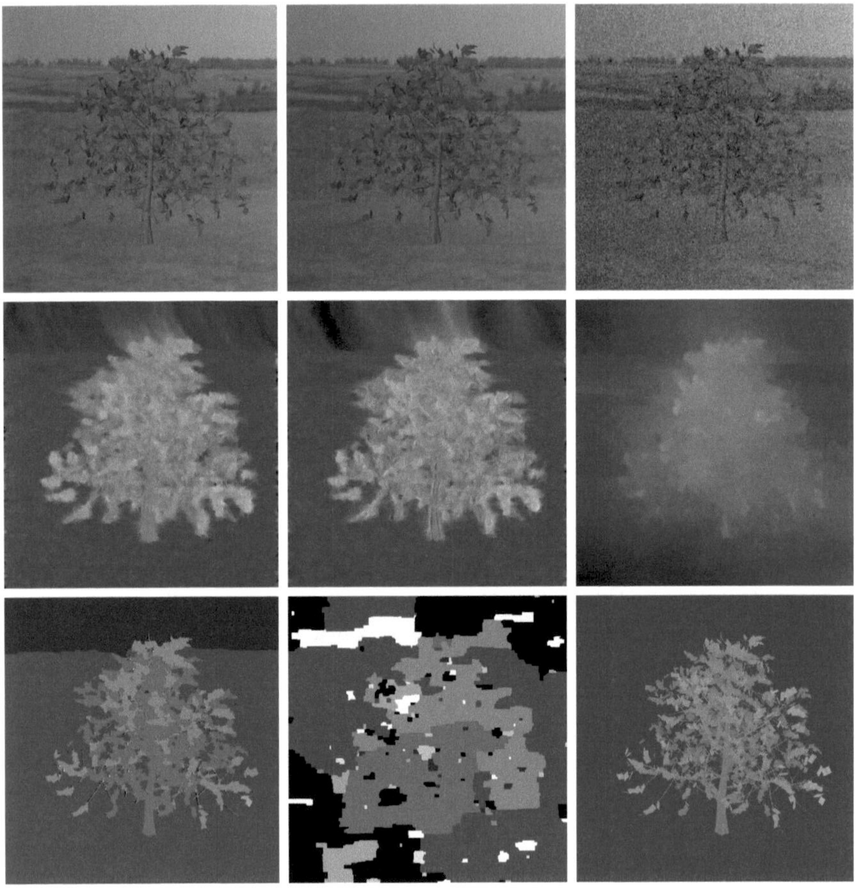

Fig. 2. *Tree sequence*, top to bottom, left to right. *First row:* Original Frame 1, frame 7 with varying illumination, frame 1 degraded with Gaussian noise of $\sigma = 25$. *Second row:* Typical results of reconstruction from 8 images of original sequence, with illumination changes and noise. *Third row:* Graph cuts results for noise free and noisy image sequence (8 disparity levels), ground truth.

the reconstructed object which leads to larger errors already in the undisturbed sequence.

The result of the graph cuts method for the noise free sequence between the first and eighth image (see Fig. 2) shows quite accurate reconstruction of the scene. Separate branches and overall shape of the tree were reconstructed very well and in accordance with the ground truth map. However, once again small variations of the disparity values cannot be estimated appropriately (the different disparity layers within the tree are very well visible). The corresponding AADE of $AADE = 0.0793$ for the graph cuts method is nevertheless close to ours. This is due to the accurate spatial reconstruction of the shape of the tree. For the noisy image sequence, however, the graph cuts method gives very poor results.

Table 3. Results for the *Tree sequence.* AADE = Average absolute disparity error. Disparity values for these experiments vary in the interval (0.1, 1).

	2 Frames	4 Frames	6 Frames	8 Frames	Noise, 8 Fr.	Illum., 8 Fr.
α	0.7	0.2	0.1	0.03	27.0	0.12
σ	1.8	2.0	2.3	2.66	2.7	2.18
AADE	0.0718	0.0644	0.0622	0.0616	0.0635	0.0650

Although we applied the same presmoothing strategy as for our stereo method, the disparity map contains many artifacts and the overall shape of the tree is very hard to recognise.

As before, we have experimented with smaller data sets of 2, 4 and 6 consequent images. Resulting error measures, presented in the Table 3 show consequent improvement as the number of images in the sequence grows.

5 Summary and Outlook

We have proposed a variational technique for a specific task of 3-D reconstruction for multiple views with small baseline distances. The method has been tailored towards applicability under more challenging conditions by incorporating various concepts that allow to handle data sets with varying illumination and noise. We have evaluated the performance of the approach with two sets of synthetic data with good results. The phenomena which are not taken into account so far are occlusions and specular reflections. This is a part of our ongoing work, whereby for the handling of specular reflections ideas from [9] and [1] are expected to be useful. An extension of our algorithm to arbitrary camera ego-motion is another topic of current research.

Acknowledgements

The authors thank Christian Morbach for providing his code for importing the 3ds files into OpenGL. Natalia Slesareva also gratefully acknowledges funding by the International Max-Planck Research School.

References

1. Birkbeck, N., Cobzas, D., Sturm, P., Jägersand, M.: Variational shape and reflectance estimation under changing light and viewpoints. In: Leonardis, A., Bischof, H., Pinz, A. (eds.) ECCV 2006. LNCS, vol. 3951, pp. 536–549. Springer, Heidelberg (2006)
2. Brox, T., Bruhn, A., Papenberg, N., Weickert, J.: High accuracy optic flow estimation based on a theory for warping. In: Pajdla, T., Matas, J(G.) (eds.) ECCV 2004. LNCS, vol. 3024, pp. 25–36. Springer, Heidelberg (2004)

3. Bruhn, A., Weickert, J., Kohlberger, T., Schnörr, C.: A multigrid platform for real-time motion computation with discontinuity-preserving variational methods. International Journal of Computer Vision 70(3), 257–277 (2006)
4. Elsgolc, L.E.: Calculus of Variations. Pergamon, Oxford (1961)
5. Faugeras, O., Keriven, R.: Variational principles, surface evolution, PDE's, level set methods, and the stereo problem. IEEE Transactions on Image Processing 7(3), 336–344 (1998)
6. Furukawa, Y., Ponce, J.: Carved visual hulls for image-based modeling. In: Leonardis, A., Bischof, H., Pinz, A. (eds.) ECCV 2006. LNCS, vol. 3951, pp. 564–577. Springer, Heidelberg (2006)
7. Huber, P.J.: Robust Statistics. Wiley, New York (1981)
8. Pons, R.K.J.-P., Faugeras, O.: Modelling dynamic scenes by registering multi-view image sequences. In: International Conference on Computer Vision and Pattern Recognition, vol. 2, pp. 822–827 (2005)
9. Jin, H., Yezzi, A.J., Soatto, S.: Variational multiframe stereo in the presence of specular reflections. In: Proceedings of the 1st International Symposium on 3D Data Processing Visualization and Transmission, June 2002, pp. 626–631 (2002)
10. Kolmogorov, V., Zabih, R.: Computing visual correspondences with occlusions using graph cut methods. In: Proc. Eighth International Conference on Computer Vision, Vancouver, Canada, July 2001, vol. 2, pp. 588–594. IEEE Computer Society Press, Los Alamitos (2001)
11. Li, G., Zucker, S.W.: Differential geometric consistency extends stereo to curved surfaces. In: Leonardis, A., Bischof, H., Pinz, A. (eds.) ECCV 2006. LNCS, vol. 3953, pp. 44–57. Springer, Heidelberg (2006)
12. Nagel, H.-H., Enkelmann, W.: An investigation of smoothness constraints for the estimation of displacement vector fields from image sequences. IEEE Transactions on Pattern Analysis and Machine Intelligence 8, 565–593 (1986)
13. Pressigout, M., Marchand, E.: Model-free augmented reality by virtual visual servoing. In: IAPR Int. Conf. on Pattern Recognition, ICPR'04, Cambridge, UK, August 2004, vol. 2, pp. 887–891 (2004)
14. Stricker, D.: Tracking with reference images: A real-time and markerless tracking solution for out-door augmented reality applications. In: Virtual Reality, Archaeology, and Cultural Heritage International Symposium (VAST01), Glyfada, Greece, November 2001 (2001)
15. Yezzi, A.J., Soatto, S.: Structure from motion for scenes without features. In: Proc. 2003 IEEE Computer Society Conference on Computer Vision and Pattern Recognition, June 2003, vol. 1, pp. 525–532. IEEE Computer Society Press, Los Alamitos (2003)
16. Young, D.M.: Iterative Solution of Large Linear Systems. Dover, New York (2003)

A Robust Approach for 3D Cars Reconstruction

Adrien Auclair[1], Laurent Cohen[2], and Nicole Vincent[1]

[1] CRIP5-SIP, University Paris-Descartes,
45 rue des Saint-Pères, 75006 Paris, France
{adrien.auclair,nicole.vincent}@math-info.univ-paris5.fr
[2] CEREMADE, University Paris-Dauphine,
Place du Maréchal De Lattre De Tassigny 75775 PARIS, France
cohen@ceremade.dauphine.fr

Abstract. Computing high quality 3D models from multi-view stereo reconstruction is an active topic as can be seen in a recent review [15]. Most approaches make the strong assumption that the surface is Lambertian. In the case of a car, this hypothesis is not satisfied. Cars contain transparent parts and metallic surfaces that are highly reflective. To face these difficulties, we propose an approach to robustly reconstruct a 3D object in translation. Our contribution is a one-dimensional tracker that uses the vanishing point computed in a first pass. We applied it to video sequences of cars recorded from a static camera. Then, we introduce a local frame for the car and use it for creating a 3D rough model. The final result is sufficient for some applications where it is needed to estimate the size of the vehicle. This model can also be used as an initialization for more precise algorithms.

Keywords: Structure From Motion, Feature Tracking, Surface Fitting, RANSAC.

1 Introduction

Automatic solutions for creating 3D digital representation of objects are needed in many domains. A practical approach is to use a video as input. Having a moving camera recording a static scene [14] can be useful to build a 3D model of outdoor scenes, or object like statues in archeology. In this article, we deal with the dual approach of having a static camera recording a moving rigid object. Some articles use this configuration to build a high quality model of an object that rotates on a turn-table in front of a static camera [5]. Many methods perform well in the case the object has a rich enough texture and does not contain much reflection or transparency. The recent review in [15] compares several algorithms that give sub-millimeter precision with these conditions.

Most of the structure-from-motion algorithms make the hypothesis that the object surface is Lambertian. This ensures that two images taken from two lightly different points of view are not very different. With this hypothesis, and assuming a rich texture on the surface object, a Lucas-Kanade based point tracker coupled with robust pose estimation method (e.g., RANSAC [6]) and non linear

B.K. Ersbøll and K.S. Pedersen (Eds.): SCIA 2007, LNCS 4522, pp. 183–192, 2007.

minimization (e.g., Bundle Adjustment step [18]) leads to good quality 3D point cloud. Several books detail precisely these methods ([9],[13]).

Motivated by some industrial application, we chose to reconstruct a 3D model of a car. Cars are challenging objects because the Lambertian assumption is strongly violated, parts of the objects are transparent and large parts have no texture. The lights for example are highly reflective surfaces behind a transparent plastic surface. The wheels are parts of the object that are moving relatively to the car. This leads to many difficulties for 3D reconstruction without priors. A feature tracker would not be able to make any difference between a moving part of the vehicle and an object of the environment reflected on the car surface. If there is much reflection, the number of outliers among the tracked points could exceed 50 percent. Moreover, many steps of tracking or reconstruction measures are based on photo-consistency over time. For example, the Lucas-Kanade feature tracker [17] assumes that a window around a point remains constant within a small interval of time. In some recent reconstruction algorithm, the depth of a point is correct if its corresponding 2D positions in movie frames have a high Normalized Cross Correlation score ([7],[5]). This photo-consistency measure cannot be used on cars because of highly reflective parts and transparency.

Our new approach is to use a two-pass features tracking to analyze the motion. The presented method can be used for building a 3D point cloud from video sequence of an object in translation in front of a static camera. The translation hypothesis makes the algorithm robust by using the vanishing point and achieves to track much more features. Then, we show that for cars, results still contain outliers. We propose methods to filter the 3D point cloud. Our second contribution is an approach to compute a local 3D frame for the car. Then, this local frame is used to fit the point cloud to a simple 3D car model. Figure 1 shows some images of an input sequence we used. The overall advantage of our method is to robustly build an approximate 3D model whereas other methods could fail because of the reflections and the transparencies.

Fig. 1. Frames 20,40,60 and 100 of an input movie

2 Structure from Motion for an Object in Translation

In the following section, we introduce a one dimensional feature tracker. Compared to a traditional Lucas-Kanade based feature tracker, this two-pass tracker is more robust and tracks many more points. The output is used in a classical structure from motion algorithm. At first, a two-views reconstruction is established and then, other views are added iteratively, in a manner close to [14].

Internal parameters of the camera are computed off-line by using a chessboard pattern.

2.1 Feature Tracking

The point tracker of Lucas-Kanade [17] is a well known algorithm. Its underlying assumption is that a window around a point stays constant between two frames. When applied to features on cars, this suffers from lighting changes, orientation changes, reflections. Moreover, good features to track must contain gradients in both directions, meaning they are corners. When using a Harris detector [8] on a car image, there are only few of these points. This is a consequence of the lack of texture on surfaces. Still, when using a pyramidal implementation of the Lucas Kanade, as described in [3], a proportion of the corners are correctly tracked.

To increase the number of tracked points, the translation hypothesis is used. In the case the object is translating in front of a static camera, and assuming that radial distortion has been removed, each feature track is ideally projected as a line on the focal plane of the static camera. All the feature lines are meeting at a vanishing point. Knowing this point would allow to apply a one dimensional tracking. In practice, the result of the bi-dimensional feature tracking does not lead to perfect lines meeting at a single point. Still, some parts of the tracks can easily be approximated by linear segments. A robust algorithm must then be used to approximate the vanishing point. Some high precision algorithms have been proposed [1]. In practice a RANSAC [6] framework is very efficient. Every pair of lines is a sufficient subsample to compute a potential vanishing point. The vanishing point is the one with the largest consensus.

In the traditional Lucas-Kanade tracker, the two dimensional displacement d of a feature is solution of the linear system

$$\left(\sum_{W_t} gg^\top\right).d = \sum_{W_t}(I_t - I_{t+1})g \ , \tag{1}$$

where W_t is the window of interest around a corner at time t, I_t the image intensity value at time t, I_{t+1} the image intensity at time $t+1$, g the image intensity gradient. This system is applied recursively in a Newton-Raphson manner to minimize the difference between windows at time t and $t+1$. But once the vanishing point is known, the bi-dimensional tracking is over-parametrized. With this knowledge, the equation (1) can be simplified (with a proof very close to the one from [17]) to a one dimensional equation :

$$\left(\sum_{W_t} (g.dir)^2\right).d = \sum_{W_t}((I_t - I_{t+1}).(g.dir)) \ , \tag{2}$$

where dir is the direction from the feature to the vanishing point and d is now a scalar representing the displacement along the vanishing direction.

In practice, we track features in two passes. At first, a bi-dimensional tracking is done to compute the vanishing point. In a second pass, the features are tracked by the one-dimensional tracker. As good features to track only need to have gradient along the vanishing direction, this leads to much more tracks, resulting in a better input for the structure from motion algorithm. The table 1 shows the number of inlier points that are valid for reconstruction between two frames. For the selected sequences and pairs of frames, the gain of our method is between 2.4 and 5.1. And on some sequences, the result of the bi-dimensional tracker is too poor to be used directly in a structure from motion algorithm.

Table 1. Number of tracked points declared inliers by the reconstruction algorithm

	2D tracker	1D tracker	gain
Seq 1	138	405	2.9
Seq 2	118	508	4.3
Seq 3	73	370	5.1
Seq 4	55	134	2.4
Seq 5	75	348	4.6
Seq 6	37	120	3.2

2.2 3D Point Cloud Reconstruction

The algorithm we implemented is close to [14], except that it is adapted to translation. At first, two frames are used for reconstruction. Then views are added iteratively to complete and refine the 3D point cloud.

Initial Reconstruction from Two Frames. A point x in a frame f_i and its correspondence x' in the frame f_j are linked by the fundamental equation :

$$x'^{\mathrm{T}} F x = 0 \, , \tag{3}$$

where F is called the fundamental matrix. It encodes the spatial relationship between the two cameras positions. In the particular case of translation, it can be seen [9] that the fundamental matrix has the particular form of a skew-symmetric 3x3 matrix. If the vanishing point, which is equivalent in this special case to the second epipole, is noted $e' = (x_e, y_e, z_e)$, the fundamental matrix $F$is:

$$F = [e']_x = \begin{bmatrix} 0 & -z_e & y_e \\ z_e & 0 & -x_e \\ -y_e & x_e & 0 \end{bmatrix}$$

Because computing the fundamental matrix can be unstable, this is very valuable to get it directly from the vanishing point. The problem is that all the points conform to the fundamental equation (3). And thus it is impossible to use the fundamental matrix as a filter between two views.

Using the internal calibration computed off-line, the fundamental matrix directly leads to external calibration ([9],[13]). Once the camera poses are known for two frames, using the two 2D positions of a point to find its 3D position is called triangulation. Again, because camera calibration is known, triangulation is achieved in Euclidean space and thus, a simple linear method leads to good quality 3D point cloud ([9], [10]).

Adding a View. From the cloud of 3D points and their corresponding 2D positions in a new frame, there are several methods to compute the camera pose for this frame. In case of pure translation, there are only three unknowns for the camera external calibration. That could be reduced to one parameter as the motion direction is known. But for not being too dependent of the vanishing point computation, we look for the full 3D translation. The projection equations lead to two equations for each couple 3D-2D. Thus we only need two 3D-2D correspondences to compute a pose. Because of the small number of data samples needed to estimate a model, a RANSAC approach [6] fits very well to this problem. Figure 2 shows a 3D point cloud with all the locations and orientations of cameras used for reconstruction.

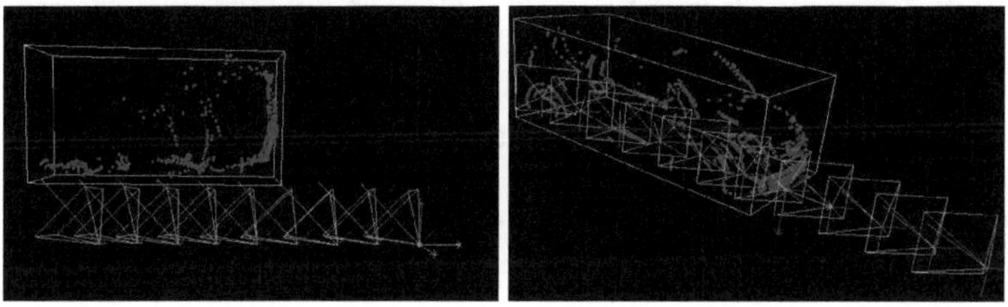

(a) (b)

Fig. 2. (a) Top view of the 3D point cloud and its bounding box. Cameras locations are on a line on the side of the car. (b) Another reconstruction example. For nicer display, point clouds shown here have been filtered using steps of section 3.

3 Cloud Filtering

Figure 3(a) shows the final result of the previous algorithm. There are still many outliers. This is because there is no filtering on points yet. A first cloud is computed from two views. Then other views are robustly added and the point positions are refined but none is rejected. In this section, we introduce two simples filter that achieve efficient filtering for our 3D clouds.

The first filter is to impose a threshold τ on the retro-projection error. For a 3D point X_i that has been reconstructed from a subset of views $\mathcal{F}$, the retro-projection error $err(X_i)$ is defined as :

$$err(X_i) = max_{c \in \mathcal{F}}(\|P_c(X_i) - x_c\|) ,$$

where P_c is the projection matrix for frame c and x_c the 2D position of the point in this frame. A 3D point X_i is declared outlier if $err(X_i) > \tau$.

Figure 3(b) shows the result with a threshold τ of one pixel. With our experiments, one pixel is the lowest acceptable value and setting a sub-pixel threshold remove too many points. This is mostly because there is no non-linear optimization in the presented algorithm. When using the bundle adjustment implementation of [12], the average value of the $err(X_i)$ defined above can get below 0.1

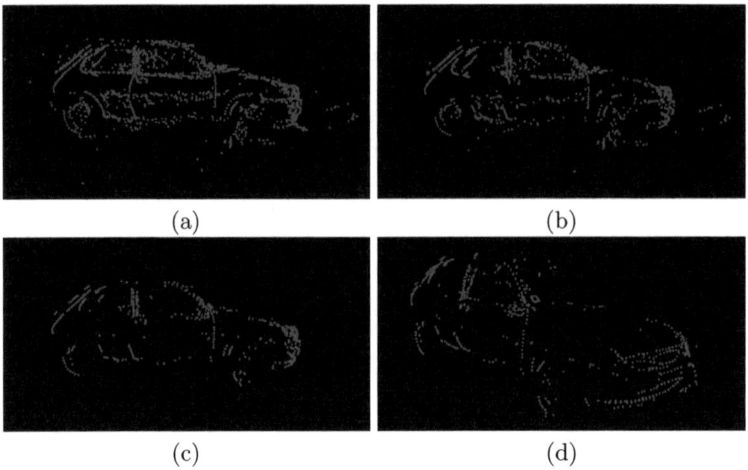

Fig. 3. (a) Initial cloud. (b) Cloud filtered by retro-projection error below one pixel (c) cloud filtered by retro-projection error below one pixel and feature tracked for at least three frames. (d) Perspective view of the final filtered cloud.

pixels for more than 200 points. But this optimization is time consuming and because our goal is to compute an approximate model, we skip it.

Filtering with the retro-projection error does not remove all the outliers. Figure 3(b) still contains large outliers. Analyzing these points leads to several potential sources of error. Some points are tracked only for the two initial reconstruction frames. Thus, their retro-projection error is zero. Other points belong to the shadow in front of the car. For a short distance, they have a displacement that is very close to the car's move. To filter these outliers, the chosen heuristic is to reject a point if its triangulation has been done from less than n frames. Figure 3(b) shows results for n=3. In practice, using n=4 or 5 ensures more robustness.

4 Building a 3D Car Model from the Point Cloud

The previous algorithm leads to a complex cloud of points. Some points on the car surface are correctly reconstructed. Some interior parts of the car were reconstructed as well. Several points on the wheels were forced to have a linear track and surprisingly, it happens that they pass the previous filters and are reconstructed at a wrong depth. Because of the complexity of the surface, some reflected parts cannot be filtered by algebraic criteria ([16]). Thus we need to robustly approximate the point cloud by a simple model. We propose an approach to establish first a local frame which has the same directions as the car. Working in this local frame, we approximate side part of the car by a second degree polynomial function and front part by a higher degree polynomial function.

4.1 Computing the Local Frame

Because the camera positions are the dual of the object positions, the vector between successive camera centers gives the inverse 3D direction of the translation of the car. We present here an approach to compute the missing horizontal direction. Our underlying hypothesis is that a front view of the car would contain many horizontal lines. At least, top and bottom lines of the license plate should be detected. Thus, using the 3D point cloud, we generate a textured synthetic front view of the car and then analyze it to find a major direction.

A front view virtual camera is positioned on an axis aligned with the motion direction, and far enough from the object to include the bounding box in its field of view. Its optical axis is pointing toward the point cloud (figure 4.a). To render a textured synthetic view, a 3D surface is needed. But approximating the 3D point cloud by a surface is a difficult problem. There exist general approaches based on Delaunay tetrahedrization ([2],[4]). For the cars, we consider that the 3D surface is a range image from focal plane of the first camera in the sequence. This allows to fit the points with a spline surface using the Multilevel B-Splines Approximation algorithm of [11]. Figure 4.b shows a line version of the 3D spline surface. Once this 3D surface is obtained, one can use any frame of the movie to texture it. Figure 4.c shows the textured 3D spline surface. To reduce texture projection error, it is a good choice to work with the camera whose optical axis is the most aligned with the vehicle motion. In general, this is the first camera of the movie. This figure shows that the result is good enough on the central part of the front of the car.

Once this 3D textured surface is obtained, we used the virtual front camera described above to render it for a front view. Edge detection is applied on the synthetic view and then lines are detected using an Hough transform on this edge image. By clustering the direction of the lines, the major direction (i.e., the horizontal one) is found. Figure 4.d shows a synthetic front view with the detected lines and the major direction.

4.2 Fitting a Polynomial Model on the Point Cloud

The 3D bounding box is now aligned on the frame of the car. Thus, its faces have a high-level meaning relatively to the car (e.g, front face, side face). Using this knowledge, our approach is to fit the side and front 2D profiles of the car by two polynomial functions.

First, points are projected in the front plane of the bounding box. The car side profile is then computed by fitting these 2D points by a second degree polynomial function. Fitting polynomial functions is made as a least square problem but because of the outliers in the cloud, it is required to use M-estimators (e.g., Tukey or Huber). These estimators are used in an iterative re-weighted least square algorithm. The obtained 2D profile is extruded in the car motion direction to obtain the side 3D surface of the car model. Then, we apply the same procedure for front profile. The 3D points are projected on the side face of the bounding box. The front profile is given by the fitting of these points with a higher degree

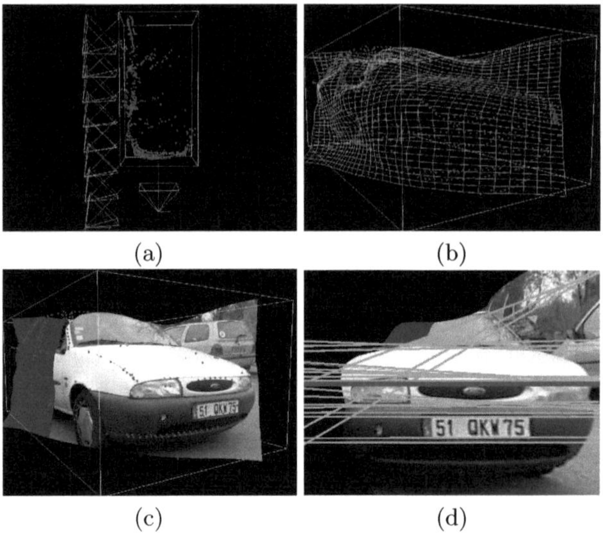

Fig. 4. (a) Virtual front view camera facing the point cloud. (b) The spline fitting on the point cloud. (c) Same view but the spline has been textured with one frame on the video sequence. Dark grey parts are spline surface nor reached by the projection of the texture. (d) The textured spline of (c) has been rendered with camera of (a). All detected lines are in green. Major direction is wider, in red.

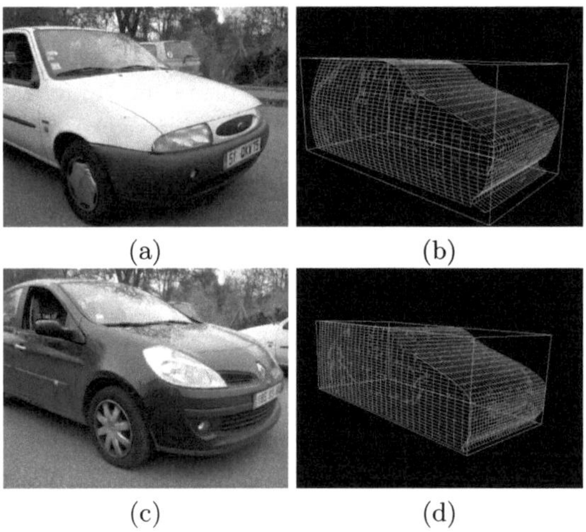

Fig. 5. A frame (a) of an input sequence and the reconstructed polynomial 3D model (b). Another frame (c) from another sequence with the corresponding 3D model (d).

polynomial function. This 2D profile is then extruded in the width direction to obtain the complete front car 3D surface. Final step is to merge the side and front surfaces to obtain the correct model (figure 5).

5 Conclusion

In this article, we introduced a complete algorithm to construct approximate 3D models of cars from video sequences. Our first contribution is to introduce a one-dimensional tracker that makes use of the vanishing point computed in a first pass. Then, we showed that some basic filters can clean-up the outliers that were introduced by the translation hypothesis. Our second contribution is to propose a method to work in a local frame for building the car model. This local frame is computed without any external informations as markers on the road. The obtained model is coarse but still useful for many applications. It can be used directly to classify vehicles according to their dimensions. Moreover, having established the local frame orientation is a strong and meaningful knowledge for further algorithms. For example, the cloud bounding box dimensions correspond to actual height, length and width of the car (at a certain scale). And one can generate synthetic views from a point of view relative to the car (i.e., front view, side view...) for higher-level interpretations. Our future work consists of exploring methods to use this coarse model in a deformable scheme to achieve high quality 3D cars reconstruction. It could be used as an initialization surface and also to generate a model-driven force to avoid collapsing through transparent surfaces of the car.

References

1. Almansa, A., Desolneux, A., Vamech, S.: Vanishing points detection without any a priori information. IEEE T.-PAMI 25(4), 502–507 (2003)
2. Amenta, N., Choi, S., Kolluri, R.K.: The power crust, unions of balls, and the medial axis transform. Computational Geometry 19(2-3), 127–153 (2001)
3. Bouguet, J.-Y.: Pyramidal implementation of the Lucas Kanade feature tracker (2000)
4. Dey, T., Goswami, S.: Tight cocone: A water-tight surface reconstructor (2003)
5. Esteban, C.H., Schmitt, F.: Silhouette and stereo fusion for 3d object modeling. Comput. Vis. Image Underst. 96(3), 367–392 (2004)
6. Fischler, M.A., Bolles, R.C.: Random sample consensus: A paradigm for model fitting with applications to image analysis and automated cartography. Communications of the ACM 24(6), 381–395 (1981)
7. Goesele, M., Curless, B., Seitz, S.M.: Multi-view stereo revisited. cvpr 2, 2402–2409 (2006)
8. Harris, C., Stephens, M.: A Combined Corner and Edge Detector. In: 4th ALVEY Vision Conference, pp. 147–151 (1988)
9. Hartley, R.I., Zisserman, A.: Multiple View Geometry in Computer Vision, 2nd edn. Cambridge University Press, Cambridge (2004)
10. Hartley, R.I., Sturm, P.: Triangulation. Computer Vision and Image Understanding: CVIU 68(2), 146–157 (1997)
11. Lee, S., Wolberg, G., Shin, S.Y.: Scattered data interpolation with multilevel B-splines. IEEE Transactions on Visualization and Computer Graphics 3(3), 228–244 (1997)

12. M.I.A. Lourakis and A.A. Argyros. The design and implementation of a generic sparse bundle adjustment software package based on the levenberg-marquardt algorithm. Technical Report 340, Institute of Computer Science - FORTH, Heraklion, Crete, Greece, Available from http://www.ics.forth.gr/~lourakis/sba (2004)
13. Ma, Y., Soatto, S., Kosecka, J., Sastry, S.S.: An Invitation to 3-D Vision. Springer Verlag, Berlin Heidelberg New York (2004)
14. Pollefeys, M., Van Gool, L., Vergauwen, M., Verbiest, F., Cornelis, K., Tops, J., Koch, R.: Visual modeling with a hand-held camera. Int. J. Comput. Vision 59(3), 207–232 (2004)
15. Seitz, S.M., Curless, B., Diebel, J., Scharstein, D., Szeliski, R.: A comparison and evaluation of multi-view stereo reconstruction algorithms. cvpr 1, 519–528 (2006)
16. Swaminathan, R., Kang, S.B., Szeliski, R., Criminisi, A., Nayar, S.K.: On the motion and appearance of specularities in image sequences. In: Heyden, A., Sparr, G., Nielsen, M., Johansen, P. (eds.) ECCV 2002. LNCS, vol. 2350, pp. 508–523. Springer, Heidelberg (2002)
17. Tomasi, C., Kanade, T.: Detection and tracking of point features. Technical Report CMU-CS-91-132, Carnegie Mellon University (1991)
18. Triggs, B., McLauchlan, P., Hartley, R., Fitzgibbon, A.: Bundle adjustment – A modern synthesis. In: Triggs, B., Zisserman, A., Szeliski, R. (eds.) Vision Algorithms: Theory and Practice. LNCS, vol. 1883, pp. 298–375. Springer, Heidelberg (2000)

Novel Stereoscopic View Generation by Image-Based Rendering Coordinated with Depth Information

Maiya Hori, Masayuki Kanbara, and Naokazu Yokoya

Graduate School of Information Science, Nara Institute of Science and Technology,
8916–5 Takayama, Ikoma, Nara 630–0192, Japan

Abstract. This paper describes a method of stereoscopic view generation by image-based rendering in wide outdoor environments. The stereoscopic view can be generated from an omnidirectional image sequence by a light field rendering approach which generates a novel view image from a set of images. The conventional methods of novel view generation have a problem such that the generated image is distorted because the image is composed of parts of several omnidirectional images captured at different points. To overcome this problem, we have to consider the distances between the novel viewpoint and observed real objects in the rendering process. In the proposed method, in order to reduce the image distortion, stereoscopic images are generated considering depth values estimated by dynamic programming (DP) matching using the images that are observed from different points and contain the same ray information in the real world. In experiments, stereoscopic images in wide outdoor environments are generated and displayed.

1 Introduction

A technology that enables users to virtually experience a remote site is called telepresence[1]. The telepresence system has to provide rich visual sensation so that user can feel like existing at the remote site. In general, methods to provide a user with rich visual sensation are divided into two approaches: A model-based rendering (MBR) approach[2,3,4] and an image-based rendering (IBR) approach[5,6,7,8]. In the MBR approach, since virtual scene images are generated from 3D model with the 3D shape and its reflectance properties of an object, it is difficult to automatically reconstruct large-scale virtual scene such as an outdoor environment for telepresence. On the other hand, the IBR approach can render a scene consisting of complicated shapes and reflectance properties because synthesized images are generated from captured images. Therefore, the IBR approach is often used for telepresence of an outdoor environment[9]. In the IBR approach, Ikeda et al.[10] have proposed an immersive telepresence system using high-resolution omni-directional videos. This system can show user a high-resolution virtual image, however the user's viewpoint is restricted on a camera path and user can see only monocular images.

B.K. Ersbøll and K.S. Pedersen (Eds.): SCIA 2007, LNCS 4522, pp. 193–202, 2007.

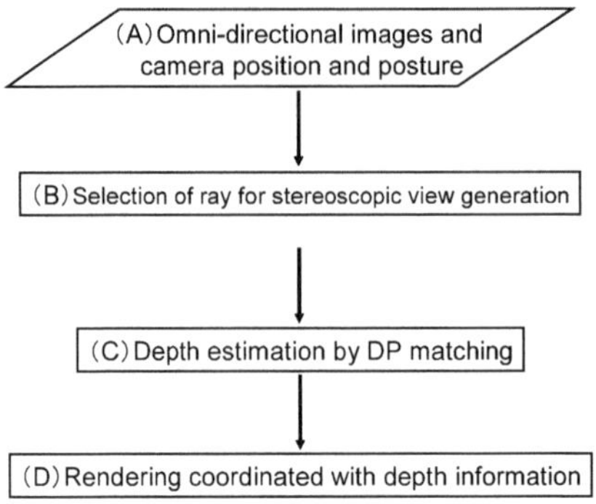

Fig. 1. Flow diagram of the proposed method

In this paper, we propose a method for generating stereoscopic images from omni-directional image sequences for telepresence in outdoor environments. In order to generate stereoscopic images, it is necessary to generate a novel view image from a set of images captured on the camera path. The conventional methods of stereoscopic view generation suffer from the distortion due to vertical parallax[11,12]. Our method employs the light field rendering[13] same as the conventional method, and tries to reduce the distortion of generated image by rendering coordinated with depth information. The depth value is estimated by dynamic programming (DP) matching[14,15,16] with sum of squared distances (SSD) between two images that are observed at different position and are captured a same ray in the world. We can generate stereoscopic images from omni-directional images that are captured along a free camera path in wide outdoor environments by using a vehicle equipped with a high accuracy position and posture sensor.

This paper is structured as follows; Section 2 explains a method for generating the stereoscopic view in detail. In Section 3, we demonstrate experimental results of the stereoscopic view generation in outdoor environments. Finally, Section 4 describes conclusion and future work.

2 Stereoscopic View Generation

This section describes a method of stereoscopic view generation from omni-directional images sequences in outdoor environments. Figure 1 shows a flow diagram of the proposed method. First, a pair of omni-directional image sequences and extrinsic camera parameters including position and posture of camera are acquired in outdoor environments(A). Next, positions of binocular viewpoint are determined in a process of stereoscopic view generation(B). Parts of images needed for stereoscopic image are collected from stored omni-directional images based on the relationship between the rays from the novel viewpoint and the captured omni-directional images. To reduce the distortion due to vertical parallax, the depth from novel viewpoint is estimated by DP matching(C). Finally,

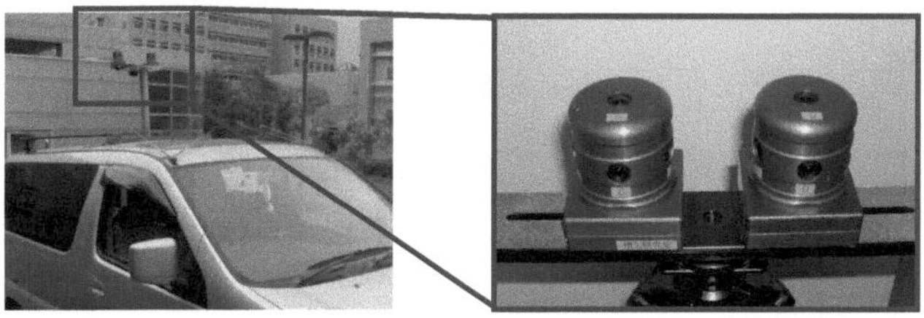

Fig. 2. Omni-directional multi-camera systems mounted on a vehicle

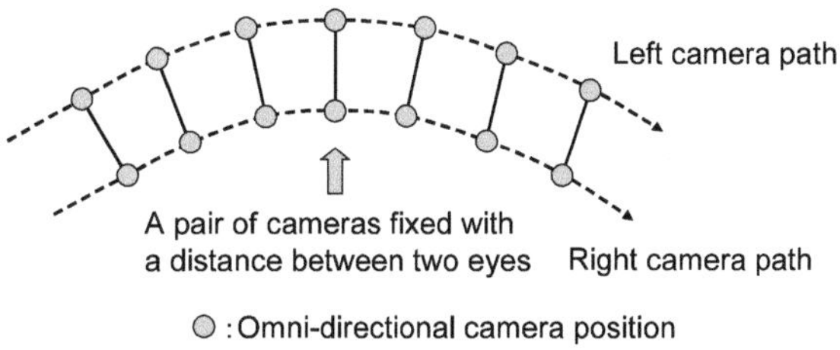

Fig. 3. Omni-directional camera position (top view)

stereoscopic images are rendered in the image plane coordinated with depth information(D).

2.1 Omni-Directional Images and Camera Positions

In this study, it is necessary to acquire a lot of rays required for stereoscopic view generation. We capture omni-directional image sequences in using a pair of omni-directional cameras fixed with a distance between two eyes as shown in Figure 2. Omni-directional cameras and their paths are illustrated in Figure 3. The images obtained from the fixed two omni-directional cameras enable us to generate the stereoscopic image when the view direction is parallel to the camera path. In addition, the configuration of cameras makes it possible to capture the images necessary for depth estimation mentioned in Section 2.3. The position and posture of each camera should be acquired at the same time with the high accuracy.

2.2 Selection of Ray for Stereoscopic View Generation

In this study, stereoscopic images at novel viewpoint are generated from pre-captured omni-directional images with a light field rendering approach. Omni-directional images exist at discrete points on the camera path as shown in Figure 4. In Figure 4, a perspective image can be generated from four omni-directional images captured from T_2 to T_5 and is vertically divided into four areas. As illustrated in Figure 5, we generate binocular stereoscopic images so that the center of left and right eyes is located on a circle whose diameter is the distance between two eyes. When the view direction is parallel to the camera

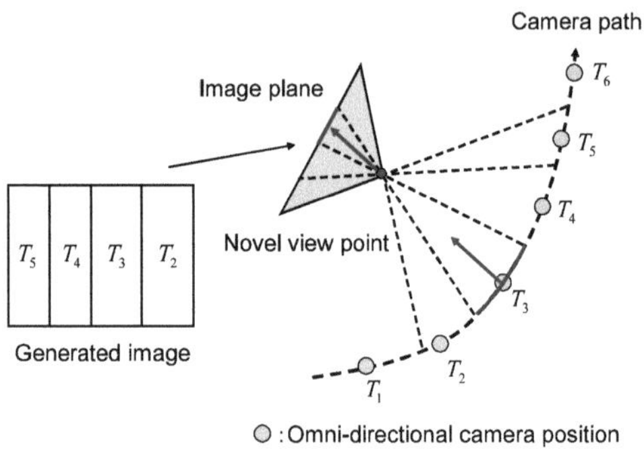

Fig. 4. Novel view image generated from pre-captured omni-directional images (top view)

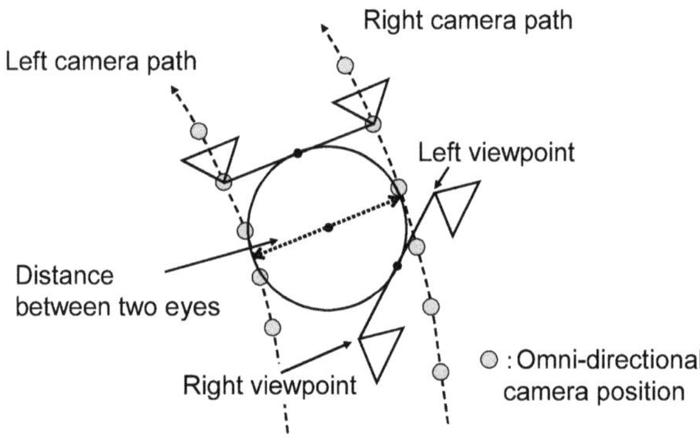

Fig. 5. Alignment of novel binocular viewpoints (top view)

path, each of stereo pair is made from a part of only one omni-directional image. When there is not ray to generate stereoscopic images because omni-directional images are captured at discrete positions, the omni-directional image of the nearest position is used. Some omni-directional images may contain the same ray to generate stereoscopic images. In this case, the omni-directional image captured at the near position from novel viewpoint is selected. In addition, mutual occlusions of camera bodies occur in the omni-directional images for generation of stereoscopic images because the omni-directional images are captured by using adjacent two omni-directional cameras. In this case, the omni-directional image captured by the other camera is used.

2.3 Depth Estimation by DP Matching

When a novel view is rendered considering without depth information, image distortion occurs in the boundary between subimages selected from different omni-directional images. The distortion appears in the generated image when the distances from an object to the omni-directional cameras differ from each other,

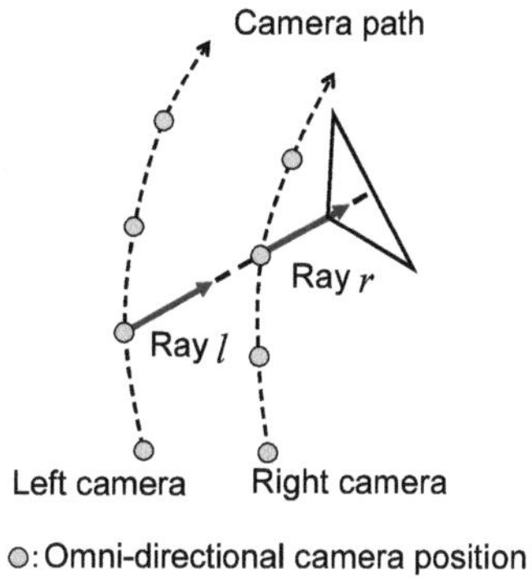

Fig. 6. Omni-directional images containing the same ray (top view)

because the cameras capture the object from different position. We reduce the distortion in generated image by rendering coordinated with depth information. The depth value is acquired by DP matching[14,15,16] between edges in two images that are captured at different points and containing the same ray as shown in Figure 6. Similarity measure of DP matching is defined as follows:

$$g(L_i, R_j) = min \begin{cases} g(L_{i-1}, R_j) + d(L_i, R_j) \\ g(L_{i-1}, R_{j-1}) + d(L_i, R_j) \\ g(L_i, R_{j-1}) + d(L_i, R_j) \end{cases} , \quad (1)$$

where $L_i(i = 1 \sim I)$ represents i-th edge in the image captured in the left camera and $R_j(j = 1 \sim J)$ does j-th edge in the image captured in the right camera. $d(L_i, R_j)$ denotes the distance between feature vectors of two edges. By calculating a path to minimize $d(L_i, R_j)$, both edges are matched. Here, the distance between feature vectors of two edges is defined by SSD. When the plural edges can be corresponded with one edge, the edge which has a smallest value of SSD in plural edges assumes the corresponding edge. In this study the window size of SSD is 25×25 pixels.

2.4 Rendering Coordinated with Depth Information

The depth values are obtained only on edges by the method above. Dense depth map is computed by linear interpolation using depth values on edges. A distortion of generated image due to a vertical parallax by rendering a novel view image with the depth value. Figure 7 illustrates rendering process of the conventional and the proposal methods. In conventional method as shown in Figure 7(a), since a real object is rendered without a depth value, the size of real object can not be correctly represented in the image. On the other hand, in the proposed method (Figure 7(b)), the real object can be correctly rendered because the real object is projected onto an image plane with a perspective projection whose center is a viewpoint of the novel image. Therefore, the proposed method can reduce the distortion of novel view image.

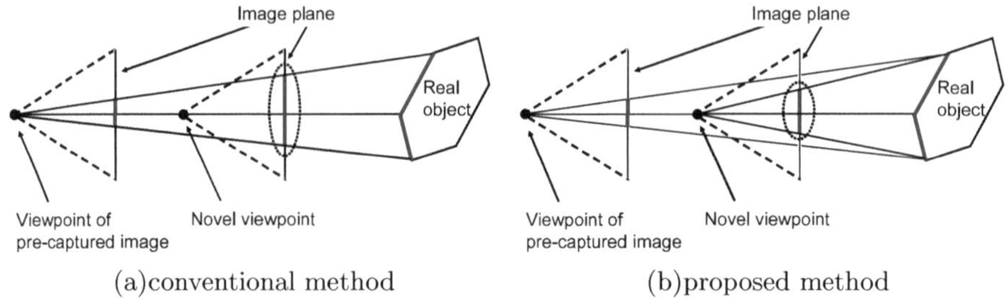

Fig. 7. Rendering process (side view)

3 Experiments

To verify the validity of the proposed method, we have actually generated novel stereoscopic views of outdoor environments. In experiments, omni-directional movies and extrinsic camera parameters including position and posture are acquired by vehicle-mounted two omni-directional multi-camera systems and a position and posture sensor. As an omni-directional multi-camera system, we use Ladybug2 (Point Grey Research). The camera unit consists of six cameras: Five radically configured on horizontal ring and one pointing vertically. The camera system can collect synchronized movies at 30fps covering more than 75%

(a)Left camera (b)Right camera

Fig. 8. Omni-directional images

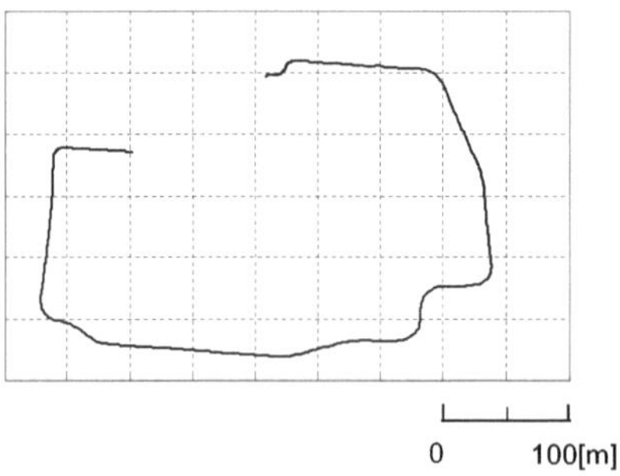

Fig. 9. Camera path

(a)Left camera (b)Right camera

Fig. 10. Images for depth estimation

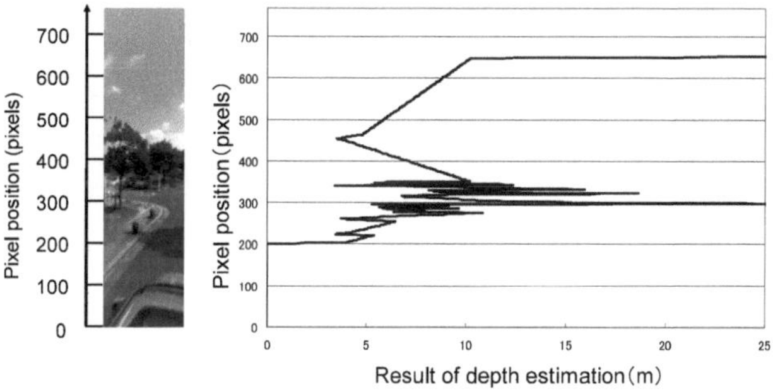

Fig. 11. Result of depth estimation

(a)Without depth information (b)With depth information

Fig. 12. Example of generated novel view images coordinated with and -out depth information

(i)View of going direction

(ii)View of 60 degrees turned from going direction

(iii)View of 120 degrees turned from going direction

(iv)View of 180 degrees turned from going direction

(a)View of left eye (b)View of right eye

Fig. 13. Generated stereoscopic images

of the full spherical view almost the same apparent point of view. The camera position and posture are measured by a hybrid sensor which consists of a real time kinematic global positioning system (RTK-GPS) LogPakII (Nikon-Trimble) and an inertial navigation system (INS) TISS–5–40 (Tokimec). This hybrid sensor can supplement the lowness of the measurement rate of GPS and the accumulation of measuring error of INS each other. The camera position and posture are measured at a high rate with high accuracy by this hybrid sensor. Input omni-directional images (resolution: 2048×1024 pixels) are shown in Figure 8. We captured the omni-directional movies along a free path as shown in Figure 9.

Figure 10 shows input images whose vertical center lines are observed along a same ray in the real world. Figure 11 shows the result of depth estimation on the line. In Figure 10, the distances between objects in an outdoor environment and camera position become far in the upper part. In the result of depth estimation as shown in Figure 11, the distances between objects and captured position are almost same as the real environment. From Figure 12, we can confirm that the distortion on the boundary was reduced by rendering coordinated with depth information. Using a PC (Pentium D 3.0GHz, memory 3.0GB), calculation cost is about 5.8 sec for one image generation. We generated binocular stereoscopic images in off-line processing when the view direction is turned by five degrees at a time. Examples of stereoscopic views (resolution: 1024×768 pixels) are given in Figure 13. We confirmed that these images can correctly show a stereoscopic view with glasses and display for a stereoscopic vision. Using generated and stored images, we can see stereoscopic view and free looking around following the indication of a user interactively were possible.

4 Conclusion

In this paper, we have proposed a method for generating novel stereoscopic view from omni-directional image sequences in wide outdoor environments. The proposed method can reduce the distortion, which is generated by conventional method in[11], by rendering coordinated with depth information that is estimated by DP matching between two images which are captured a same ray. In experiments, a user could look around a scene in outdoor environments and well perceive parallax in generated stereoscopic view. When omni-directional image sequences are captured in an outdoor environment, some moving objects such as human or vehicle are often observed. In order to generate a novel stereoscopic image, we should investigate a method for eliminating moving objects from the omni-directional image sequences.

References

1. Moezzi, S.: (ed): Special Issue on Immersive Telepresence, IEEE MultiMedia 4(1), 17–56 (1997)
2. El-Hakim, S.F., Brenner, C., Roth, G.: A Multi-sensor Approach to Creating Accurate Virtual Environments. Journal of Photogrammetry & Remote Sensing 53, 379–391 (1998)

3. Zhao, H., Shibasaki, R.: Reconstruction of Textured Urban 3D Model by Fusing Ground-Based Laser Range and CCD Images. IEICE Trans. Inf. & Syst. E-83-D(7), 1429–1440 (2000)
4. Asai, T., Kanbara, M., Yokoya, N.: 3D Modeling of Outdoor Environments by Integrating Omnidirectional Range and Color Images. In: Proc. Int. Conf. on 3-D Digital Imaging and Modeling (3DIM), pp. 447-454 (2005)
5. Adelson, E.H., Bergen, J.R.: The Plenoptic Function and the Elements of Early Vision. In: Landy, M., Movshon, J. (eds.) Computer Models of Visual Processing, pp. 3–20. MIT Press, Cambridge (1991)
6. McMillan, L., Bergen, J.: Plenoptic Modeling: An Image-Based Rendering System, In: Proc. SIGGRAPH'95, pp. 39–46 (1995)
7. Gortler, S., Grzeszczuk, R., Szeliski, R., Cohen, M.: The Lumigraph. In: Proc. SIGGRAPH'96, pp. 43–54. ACM Press, New York (1996)
8. Shum, H.Y., He, L.W.: Rendering with Concentric Mosaics. In: Proc. SIGGRAPH'99, pp. 299-306 (1999)
9. Chen, E.: QuickTime VR -An Image-Based Approach to Virtual Environment Navigation. In: Proc. SIGGRAPH'95, pp. 29–38. ACM, New York (1995)
10. Ikeda, S., Sato, T., Kanbara, M., Yokoya, N.: Immersive Telepresence System with a Locomotion Interface Using High-Resolution Omnidirectional Videos. In: Proc. IAPR Conf. on Machine Vision Applications, pp. 602–605 (2005)
11. Yamaguchi, K., Yamazawa, K., Takemura, H., Yokoya, N.: Real-Time Generation and Presentation of View-Dependent Binocular Stereo Images Using a Sequence of Omnidirectional Images. In: Proc. 15th IAPR Int. Conf. on Pattern Recognition (ICPR2000), vol. IV, pp. 589–593 (2000)
12. Ono, S., Ogawara, K., Kagesawa, M., Kawasaki, H., Onuki, M., Honda, K., Ikeuchi, K.: Driving View Simulation Synthesizing Virtual Geometry and Real Images in an Experimental Mixed-Reality Traffic Space. In: Int. Sympo. on Mixed and Augmented Reality, pp. 214–215 (2005)
13. Levoy, M., Hanrahan, P.: Light Field Rendering. In: Proc. SIGGRAPH'96, pp. 31–42. ACM, New York (1996)
14. Bellman, R.: Dynamic Programming. Princeton University Press, Princeton (1957)
15. Bellman, R., Dreyfus, S.: Applied Dynamic Programming. Princeton University Press, Princeton (1962)
16. Sakoe, H., Chida, S.: A Dynamic Programming Algorithm Optimization for Spoken Word Recognition. IEEE Tran. on Acoust. Speech and Sinal Proc. ASSP-26(1), 43–49 (1978)

Using Hidden Markov Models for Recognizing Action Primitives in Complex Actions

Volker Krüger and Daniel Grest

Computer Vision and Machine Intelligence Lab
CIT, Aalborg University
Lautrupvang 15
2750 Ballerup
Denmark
vok@cvmi.aau.dk

Abstract. There is biological evidence that human actions are composed out of action primitives, like words and sentences being composed out of phonemes. Similarly to language processing, one possibility to model and recognize complex actions is to use *grammars* with action primitives as the alphabet. A major challenge here is that the action primitives need to be recovered first from the noisy input signal before further processing with the action grammar can be done. In this paper we combine a Hidden Markov Model-based approach with a simplified version of a condensation algorithm which allows to recover the action primitives in an observed action. In our approach, the primitives may have different lengths, no clear "divider" between the primitives is necessary. The primitive detection is done online, no storing of past data is required. We verify our approach on a large database. Recognition rates are slightly lower than the rate when recognizing the singular action primitives.

1 Introduction

There is biological evidence that actions and activities are composed out of action primitives similarly to phonemes being concatenated into words[21,7,20].

In this sense, one can define a hierarchy of *action primitives* at the coarsest level, and then *actions* and *activities* as the higher abstract levels where actions are composed out of the action primitives while activities are, in turn, a composition of the set of actions [2,17][1]. If the action primivies are used as an *alphabet* one can use action grammars [12,23] to model actions and activities. It is an open problem how to define and detect these action primitives and how to define these grammars. It is reasonable to assume that these things can only be defined in context of the specific application at hand.

If an observed complex action is given and a grammar should be used for parsing and recognition, then the first necessary step is to recover the *letters* in

[1] In the following, we define the term *action* as a sequence of action primitive of arbitrary length.

B.K. Ersbøll and K.S. Pedersen (Eds.): SCIA 2007, LNCS 4522, pp. 203–212, 2007.

this observed action, i.e., the action primitives. Once the observed (continuous) sequence has been translated into a discrete set of symbols (letters), parsing based on the grammar description can be done.

In other words, if we have given an alphabet of action primitives P and if we define any *action* O to be a composition $O = a_1 a_2 a_3 \dots a_T$ of these action primitives, then our goal is to recover these primitives and their precise order. The same problem is also found in speech recognition where the goal is to find the right sequences of phonemes (see Sec. 2). Once we have recovered the sequence of action primitives in the observed sequence, we can identify the action through parsing. (In speech recognition, the sequence of detected phonemes is used to identity the corresponding word.)

The recovery of the action primitives is a non-trivial problem. Unlike phonemes (see also discussion in Sec. 2), action primitives can have a "long" durations and the variance of their execution speed may vary greatly. Also, action primitives can be heavily smeared out which complicates the distinction between them.

In this paper we deal with the recovery of the sequence of the action primitives out of an action, when a set (or alphabet) of action primitives is given.

In order to take into account possible noise and imperfect data, we base our approach on Hidden Markov Models (HMMs) [9,18] and represent our action primitives with HMMs.

Thus, given a set of action primitives P where each action primitive is represented by an HMM and given an observed sequence O of these action primitives where

1. the order of the action primitives and
2. the duration of each single action primitive and the position of their boundaries

are unknown, we would like to identify the most likely sequence of action primitives in the observation sequence O.

According to the biological findings, the representation for action recognition is closely related to the representation for action synthesis (i.e. the motor representation of the action) [21,7,20]. This motivates us to focus our considerations in this paper to actions represented in joint space. Thus, our actions are given as sequences of joint settings. A further justification for this approach is that this action representation can then be used, in future work, to bias 3D body trackers as it operates directly on the 3D parameters that are to be estimated by the 3D tracker. However, our focus on joint data is clearly without limiting generality and our technique can be applied also to other types of action representations as long as the features live in a metric space. In our on-going research we have applied the same techniques of this paper also to action recognition based on silhouettes[15].

This paper is structured as follows: In Sec. 2 will will give an overview of related work. In Sec. 3 we will discuss our approach for the HMM-based recognition of the action primitives. In Sec. 4 we present our extensive experimental results. The paper is concluded then in Sec. 5 with final comments.

2 Related Work

The recovery of phonemes in speech recognition is a closely related to our problem and thus the techniques applied there wore worthwhile to be investigated. In speech recognition, acoustic data gets sampled and quantized, followed by using Linear Predictive Coding (LPC) to compute a *cepstral* feature set. Alternatively to LPC, a Perceptual Linear Predictive (PLP) analysis [8] is often applied. In a later step, time slices are analyzed. Gaussians are often used here to compute the likelihoods of the observations of being a particular phoneme[10]. An alternative way to the Gaussians is to analyze time slices with an Artificial Neural Network[3]. Timeslices seem to work well on phonemes as the phonemes have usually a very short duration. In our case, however, the action primitives have usually a much longer duration and one would have to deal with a combinatorial problem when considering longer time slices.

In the following we will shortly review the most recent publications that consider the action recognition problem based on action primitives.

As mentioned above, the human visual system seems to relate the visual input to a sequence of motor primitives when viewing other agents performing an action [21,7,20]. These findings have gained considerable attention from the robotics community [22,6]. In *imitation learning* the goal is to develop a robot system that is able to relate perceived actions to its own motor control in order to learn and to later recognize and perform the demonstrated actions.

In [14,13], Jenkins *et al.* suggest applying a spatio-temporal non-linear dimension reduction technique on manually segmented human motion capture data. Similar segments are clustered into primitive units which are generalized into parameterized primitives by interpolating between them. In the same manner, they define action units ("behavior units") which can be generalized into actions. In [11] the problem of defining motor primitives is approached from the motor side. They define a set of nonlinear differential equations that form a control policy (CP) and quantify how well different trajectories can be fitted with these CPs. The parameters of a CP for a primitive movement are learned in a training phase. These parameters are also used to compute similarities between movements. In [5,1,4] a HMM based approach is used to learn characteristic features of repetitively demonstrated movements. They suggest to use the HMM to synthesize joint trajectories of a robot. For each joint, one HMM is used. In [5] an additional HMM is used to model end-effector movement. In these approaches, the HMM structure is heavily constrained to assure convergence to a model that can be used for synthesizing joint trajectories.

In [16], Lu *et al.* also approach the problem from a system theoretic point of view. Their goal is to segment and represent repetitive movements. For this, they model the joint data over time with a second order auto-regressive (AR) model and the segmentation problem is approached by detection significant changes of the dynamical parameters. Then, for each motion segment and for each joint, they model the motion with a damped harmonic model. In order to compare actions, a metric based on the dynamic model parameters is defined. An approach of using meaningful instants in time is proposed by Reng *et al.* [19] where key poses are found based on the curvature and covariance of the normalized trajectories.

3 Representing and Recognizing Action Primitives Using HMMs

In order to approach the action recognition problem, we model each of the action primitives $P = \{a^1, a^2, \ldots, a^N\}$ with a continuous mixture-HMM. A Hidden Markov Model (HMM) probabilistic version of a finite state machine. It is generally defined as a triplet $\lambda = (A, B, \pi)$, where A gives the transition likelihoods between states, B the observation likelihoods, conditioned on the present state of the HMM, and the starting state π (see the classics [9,18] for a further introduction). In case of the continuous mixture HMM, the observation likelihoods are given as Gaussian mixtures with $M \geq 1$ mixtures.

Our HMMs are trained on demonstrations of different individuals and the Gaussian mixtures are able to capture the variability between them. The training results into a set of HMMs $\{\lambda_i | i = 1 \ldots N\}$, one for each action primitive.

Once each action primitive is represented with an HMM, the primitives can generally simply be recognized with the classical recognition technique for HMMs, a maximum likelihood or a maximum a-posteriori classifier: Given an observation sequence O_t of an action primitive, and a set of HMMs λ_i, the maximum likelihood (ML)

$$\max_i P(O_t | \lambda_i) \tag{1}$$

identifies the most likely primitive. An alternative to the ML technique is the maximum a-posteriori (MAP) estimate that allows to take into account the likelihood of observing each action primitive:

$$\max_i P(\lambda_i | O_t) = \max_i P(O_t | \lambda_i) P(\lambda_i) \ , \tag{2}$$

where $P(\lambda_i)$ is the likelihood that the action, represented by the HMM λ_i appears.

Recognition with HMMs

In general, the likelihood of an observation for some HMM λ_i can be computed as

$$P(O|\lambda_i) = \sum_S P(O, S|\lambda_i) \tag{3}$$

$$= \sum_S P(O|S, \lambda_i) P(S|\lambda_i) \tag{4}$$

$$= \sum_S \prod_{t=0}^{T} P(O_t|S_t, \lambda_i) \prod_{t=0}^{T} P(S_t|S_{t-1}, \lambda_i) \ . \tag{5}$$

Here, one needs to marginalizes over all possible state sequences $S=\{S_0,\ldots,S_T\}$ the HMM λ_i can pass through.

To apply this technique to our problem directly is difficult in our case: In Eq. 3-5 we evaluate at the end of the observation O and select the HMM which explains this observation best. In our case, however, we are not able to identify when one primitive ends and where a new one starts. Thus, the problem is that we do not know *when* to evaluate, i.e. at what time steps t we should stop and do the maximum-likelihood estimation to find the most likely action primitive that was just now being observed.

Instead of keeping the HMMs distinct, our suggestion is to insert the "action identifier" i of the HMM λ_i as a random variable into Eq. (5) and to rewrite it as

$$P(O|a) = \sum_S \prod_{t=0}^{T} P(O_t|S_t, i_t) P(S_t, i_t|S_{t-1}, i_{t-1}). \tag{6}$$

In other words, we would like to estimate at each time step the likelihood of action i and the state S from the previously seen observations, or, respectively, the probability of λ_i being a model of the observed action:

$$P(S_T, i_T|O_{0:T}) = \prod_{t=0}^{T} P(O_t|S_t, i_t) P(S_t, i_t|S_{t-1}, i_{t-1}). \tag{7}$$

The difference in the interpretation becomes more clear when we write Eq. (7) in a recursive fashion:

$$P(S_{t+1}, i_{t+1}|O_{0:t+1}) = P(O_{t+1}|S_{t+1}, i_{t+1}) P(S_{t+1}, i_{t+1}|S_t, i_t) P(S_t, i_t|O_{0:t}) \tag{8}$$

This is the classical Bayesian propagation over time. It computes at each time step t the likelihood of observing the action i_t while having observed $O_{0:t}$. If we ignore the action identifier i_t, then Eq. (8) explains the usual efficient implementation of the forward algorithm [9]. Using the random variable i_t, Eq. (8) defines a pdf across the set of states (where the state vector S_t is the concatenation of state vectors of each individual HMM) and the set of possible actions. The effect of introducing the action i might not be obvious: using i, we do not any more estimate the likelihood of an observation, given a HMM λ_i. Instead, we compute *at each time step* the probability mass function (pmf) $P(S_t, i_t|O_{0:t})$ of each state and each identity, given the observations. By marginalizing over the states, we can compute the pmf $P(i_t|O_{0:t})$ for the action at each time step. The likelihood $P(i_t|O_{0:t})$ converges to the most likely action primitive as time progresses and more data becomes available (see Fig. 1).From Fig. 1 it is apparent that the pmf $P(i_t|O_{0:t})$ will remain constant after convergence as one action primitive will have the likelihood 1 and all other primitive likelihoods have vanished. To properly evaluate the entire observation sequence, we apply a voting scheme that counts the votes after each convergence and then restarts the HMMs. The states are initialized with the present observation likelihoods and then propagated with

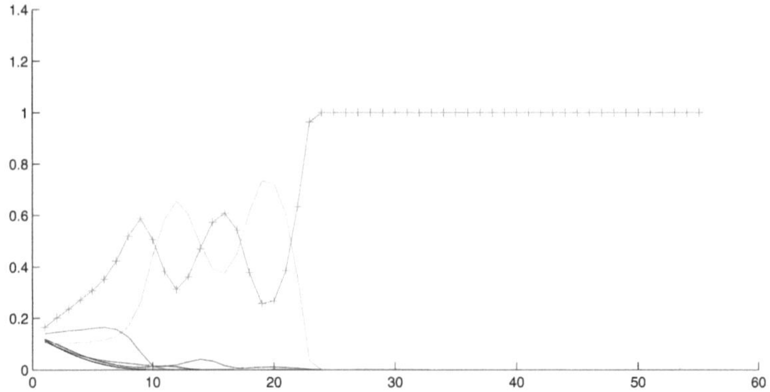

Fig. 1. shows an example for a typical behavior of the pmf $P(i_t|O_{0:t})$ for each of the actions i as time t progresses. One can see that the likelihood for one particular action (the correct one in this example, marked with "+") converges to 1 while the likelihoods for the others vanish.

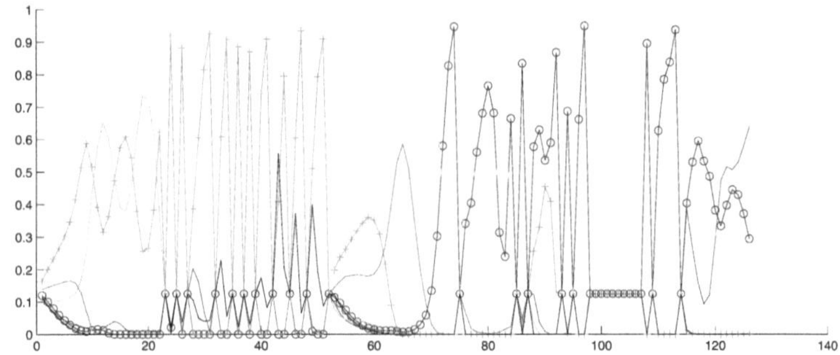

Fig. 2. shows an example for a typical behavior of the pmf $P(i_t|O_{0:t})$ as time t progresses. The input data consisted of two action primitives: first, action primitive "2", marked with "+", then, action primitive "3", marked with "o". One can see that until $\approx$ sample 52 the system converges to action "2", after sample 70, the system converges to primitive 3. The length of the first sequence is 51 samples, the length of sequence 2 is 71 samples.

the transition matrix as usual. Fig. 2 shows the repeated convergence and the restarting of the HMMs. In the example shown in Fig. 2 we have used two concatenated action primitives, denoted by the green curve with the "+" and by the blue curve with the "o", respectively. The first action primitive was in the interval between 0 and 51, while the second action primitive was from sample 52 to the end. One can see that the precise time step when primitive 1 ended and when primitive 2 started cannot be identified. But this does not pose a problem for our recovery of the primitives as for us the order matters but not their precise duration. In Fig. 1 a typical situation can be seen where the observed data did not give enough evidence for a fast recognition of the true action.

4 Experiments

For our experiments, we have used our MoPrim [19] database of human one-arm movements. The data was captured using a **FastTrack** Motion capture device with 4 electromagnetic sensors. The sensors are attached to the torso, shoulder, elbow and hand (see Fig. 3). Each sensor delivers a $6D$ vector, containing $3D$ position and $3D$ orientation thus giving a $24D$ sample vector at each time-step (4 sensors with each $6D$). The MoPrim database consists of 6 individuals, showing 9 different actions, with 20 repetitions each. The actions in the database are simple actions such as *point forward, point up, "come here", "stop!"*. Each sequence consists of $\approx$ 60-70 samples and each one starts with 5 samples of the arm in a resting position, i.e., it is simply hanging down.

Instead of using the sensor positions directly, we transform the raw $24D$ sensor data into joint angles: one elbow angle, one shoulder angle between elbow, shoulder and torso and a 3D orientation of the normal of this shoulder-elbow-torso-triangle. The orientation of the normal is given with respect to the normal of this triangle when the arm is in resting position. All angles are given in radians. No further processing of the MoPrim data was done.

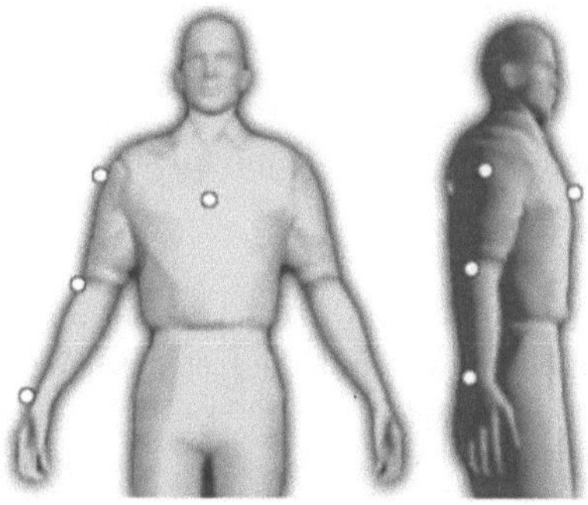

Fig. 3. marks the positions of the magnetic sensor on the human body

We have carried out several different experiments:

1. In the first test, we tested for invariance with respect to the performing human. We have trained nine HMM for nine action. Each of the HMMs was trained on 6 individuals and all the 20 repetitions of the actions. The recognition testing was then carried out on the remaining individual (leave-one-out-strategy). The HMMs we use were mixture HMMs with 10 states and 5 mixtures per state (these numbers are chosen arbitrarily, and they gave good results).
2. Here, we tested for invariance with respect to the variations within the repetitions. We have trained nine HMMs for nine actions. Each HMM was trained on all individuals but only on 19 repetitions. The test set consisted of the 20th repetition of the actions.

3. As a base line reference, we have tested how good the HMMs are able to recognize the actions primitives by testing action primitive sequences of length 1. Here, the HMMs were trained as explained under 2 above. This test reflects the recognition performance of the classical maximum-likelihood approach.
4. We have repeated the above three experiments after having added Gaussian noise with zero mean and a standard deviation of $\sigma = 0$, $\sigma = 0.3$ and $\sigma = 1$ to the training and testing data. As all angles are given in radians, thus, this noise is considerable.

To achieve a good statistic we have for each test generated 10.000 test actions of random length ≤ 100. Also, we have systematically left out each individual (action) once and trained on the remaining ones. The results below are averaged across all leave-one-out tests. In each test action, the action primitives were chosen randomly, identically and independently. Clearly, in reality there is a strong statistical dependency between action primitives so that our recognition results can be seen as a lower bound and results are likely to increase considerably when exploiting the temporal correlation by using an action grammar (e.g. another HMM).

The results are summarized in Table 1. One can see that the recognition rates of the individual action primitives is close to the general base-line of the HMMs. The recognition rates degrade with increasing noise which was to be expected, however, the degradation effect is the same for all three experiments (identities, repetition, baseline).

Table 1. summarizes the results of our various experiments. In the experiments, the training of the HMMs were done without the test data. We tested for invariance w.r.t. identity and w.r.t. the action. The *baseline* shows the recognition results when the test action was a single action primitives.

Leave-one-Out experiments		
Test	Noise σ	Recognition Result
Identities (Test 1)	0	0.9177
Repetitions (Test 2)	0	0.9097
Baseline (Test 3)	0	0.9417
Identities (Test 1)	0.5	0.8672
Repetitions (Test 2)	0.5	0.8710
Baseline (Test 3)	0.5	0.8649
Identities (Test 1)	1	0.3572
Repetitions (Test 2)	1	0.3395
Baseline (Test 3)	1	0.3548

All actions in the action database start and end in a resting pose. To assure that the resting pose does not effect the recognition results, we have repeated the above experiments on the action primitives where the rest poses were omitted. However, the recognition results did not change notably.

5 Conclusions

In this work we have presented an approach to recover the motion primitives from an action where the motion primitives are represented with a Hidden Markov Model. The approach we have taken is to consider the joint distribution of the state and the action at the same time instead of using the classical maximum likelihood approach. The experiments show that the approach is able to successfully recover the action primitives in the action with a large likelihood. It is worth pointing out that in our experiments the pairwise appearance of action primitives was statistically independent. Thus, for the recovery of the action primitives no temporal constraints between the action primitives were used or exploited. Temporal constraints between the action primitives are later introduced at a higher level though action grammars.

In future work we will use a further HMM to learn sequences of action primitives from training examples to learn such an action grammar.

Acknowledgment. This work was partially funded by PACO-PLUS (IST-FP6-IP-027657).

References

1. Billard, A., Epars, Y., Calinon, S., Schaal, S., Cheng, G.: Discovering Optimal Imitation Strategies. Robotics and Autonomous Systems 47, 69–77 (2004)
2. Bobick, A.F.: Movements, Activity, and Action: The Role of Knowledge in the Perception of Motion. In: Royal Society Workshop on Knowledge-based Vision in Man and Machine, London, England, February 1997 (1997)
3. Bourlard, H., Morgan, N.: Connectionist Speech Recognition: a Hybrid Approach. Kluwer Academic Publishers, Dordrecht (1994)
4. Calinon, S., Billard, A.: Stochastic Gesture Production and Recognition Model for a Humanoid Robot. In: International Conference on Intelligent Robots and Systems, Alberta, Canada, August 2-6, 2005 (2005)
5. Calinon, S., Guenter, F., Billard, A.: Goal-Directed Imitation in a Humanoid Robot. In: International Conference on Robotics and Automation, Barcelona, Spain, April 18-22, 2005 (2005)
6. Dariush, B.: Human Motion Analysis for Biomechanics and Biomedicine. Machine Vision and Applications 14, 202–205 (2003)
7. Giese, M., Poggio, T.: Neural Mechanisms for the Recognition of Biological Movements. Nature Reviews 4, 179–192 (2003)
8. Hermansky, H.: Perceptual linear predictive (plp) analysis of speech. Journal of Acoustical Society of America 87(4), 1725–1738 (1990)
9. Huang, X.D., Ariki, Y., Jack, M.A.: Hidden Markov Models for Speech Recognition. Edinburgh University Press (1990)
10. Huang, X.D., Jack, M.A.: Semi-continous hidden markov models for speech signals. Computer Speech and Language 3, 239–252 (1989)
11. Ijspeert, A.J., Nakanishi, J., Schaal, S.: Movement Imitation withNonlinear Dynamical Systems in Humanoid Robots. In: International Conference on Robotics and Automation, Washington DC, USA (2002)

12. Ivanov, Y., Bobick, A.: Recognition of Visual Activities and Interactions by Stochastic Parsing. IEEE Transactions on Pattern Analysis and Machine Intelligence 22(8), 852–872 (2000)
13. Jenkins, O.C., Mataric, M.: Deriving Action and Behavior Primitives from Human Motion Capture Data. In: International Conference on Robotics and Automation, Washington DC, USA (2002)
14. Jenkins, O.C., Mataric, M.J.: Deriving Action and Behavior Primitives from Human Motion Data. In: International Conference on Intelligent Robots and Systems, pp. 2551–2556, Lausanne, Switzerland, September 30 – October 4, 2002 (2002)
15. Krueger, V., Anderson, J., Prehn, T.: Probabilistic model-based background subtraction. In: Scandinavian Conference on Image Analysis, pp. 180–187, June 19-22, Joensuu, Finland (2005)
16. Lu, C., Ferrier, N.: Repetitive Motion Analysis: Segmentation and Event Classification. IEEE Transactions on Pattern Analysis and Machine Intelligence 26(2), 258–263 (2004)
17. Nagel, H.-H: From Image Sequences Towards Conceptual Descriptions. Image and Vision Computing 6(2), 59–74 (1988)
18. Rabiner, L.R., Juang, B.H.: An introduction to hidden Markov models. IEEE ASSP Magazine, pp. 4–15 (1986)
19. Reng, L., Moeslund, T.B., Granum, E.: Finding Motion Primitives in Human Body Gestures. In: Gibet, S., Courty, N., Kamp, J.-F. (eds.) GW 2005. LNCS (LNAI), vol. 3881, pp. 133–144. Springer, Heidelberg (2006)
20. Rizzolatti, G., Fogassi, L., Gallese, V.: Parietal Cortex: from Sight to Action. Current Opinion in Neurobiology 7, 562–567 (1997)
21. Rizzolatti, G., Fogassi, L., Gallese, V.: Neurophysiological Mechanisms Underlying the Understanding and Imitation of Action. Nature Reviews 2, 661–670 (2001)
22. Schaal, S.: Is Imitation Learning the Route to Humanoid Robots? Trends in Cognitive Sciences 3(6), 233–242 (1999)
23. Stolcke, A.: An Efficient Probabilistic Context-Free Parsing Algorithm That Computes Prefix Probabilities. Computational Linguistics 21(2), 165–201 (1995)

Variational Segmentation Using Dynamical Models for Rigid Motion

Jan Erik Solem and Anders Heyden

Applied Mathematics Group
School of Technology and Society
Malmö University, Sweden
{jes,heyden}@ts.mah.se

Abstract. This paper deals with the segmentation of multiple moving objects in image sequences. A method for estimating the motion of objects without the use of features is presented. This is used to predict the position and orientation in future frames of the sequence. Experiments on real data show that this estimation can be used to improve segmentation.

1 Introduction

Segmentation is the process of decomposing an image in meaningful regions, such as e.g. different objects and background. In this paper, segmentation in image sequences of objects undergoing rigid (and possibly non-rigid) motion is studied. Based on the segmentation in previous frames we are interested in obtaining estimates of the object position in the next frame. We are especially interested in regions and boundaries that are difficult to track using standard methods. Examples are textureless objects and objects having boundaries without easily identifiable points. If the scene contains textured regions where feature points can be detected and tracked through the sequence, the estimation of a rigid (or non-rigid) transformation between frames is straightforward. When the objects are featureless however, it is much harder.

A method for estimating the between-frame translation and rotation directly from the object boundaries without the use of features or landmarks is presented. Based on these estimates, dynamic models of the motion of the scene objects can be derived to estimate the position in new frames. This gives a useful online initialization algorithm. We use the Chan-Vese segmentation model [1] and a simple linear motion model to illustrate the procedure for a number of examples. The procedure is general enough to be used as initialization for curve based methods such as traditional active contour methods in the form of snakes [2], geometric active contours [3], and other similar models [4,5].

We will use a level set representation [6,7] for representing the image regions and the boundary curve of the segmentation. The proposed estimation method uses normal velocities and is therefore compatible with any representation of the boundary, explicit or implicit, cf. [8].

B.K. Ersbøll and K.S. Pedersen (Eds.): SCIA 2007, LNCS 4522, pp. 213–222, 2007.

The method in itself is based on the property that any curve evolution can be partitioned into translation, rotation and deformation. This was earlier studied in [9] and [10]. In this paper we use the partition of a particular flow and integrate over the evolution time to find the rigid component of the object displacement in the image.

When segmenting space-time data, a simple (and effective) idea is to perform the segmentation in the space-time volume using surfaces. This was studied in e.g. [11]. This enables regularization of the segmentation both within images and between images. However, a necessary requirement is that the image sequence is dense enough to make the between image difference sufficiently small. Large displacements are clearly not handled using such volumetric methods. Another (perhaps less serious drawback is that volumetric techniques are "global" since the entire sequence is used and the method can not be used to estimate object positions at consecutive frames.

The paper is organized as follows; Section 2 presents the necessary background material, Section 3 shows how translation and rotation can be estimated based on the shape of the objects in consecutive frames, Section 4 describes how this can be used to build motion models, Section 5 presents some experiments and Section 6 our conclusions.

2 Background

As a courtesy to the reader, the necessary background on the level set method, geometric gradient descent and on determining rotation and translation by orthogonal projections is briefly recalled here.

2.1 The Level Set Representation

The level set method for evolving implicit surfaces was introduced independently by [12] and [6]. The time-dependent curve $\Gamma(t)$ is represented implicitly as the zero level set of a function $\phi(\mathbf{x}, t) : \mathbf{R}^2 \times \mathbf{R} \to \mathbf{R}$ as

$$\Gamma(t) = \{\mathbf{x} \ ; \ \phi(\mathbf{x}, t) = 0\} \ . \tag{1}$$

The sets $\Omega_0 = \{\mathbf{x} \ ; \ \phi(\mathbf{x}, t) < 0\}$ and $\Omega_1 = \{\mathbf{x} \ ; \ \phi(\mathbf{x}, t) > 0\}$ are called the *interior* and the *exterior* of Γ, respectively. Using this definition the outward unit normal $\mathbf{n}$ and the mean curvature κ are given as

$$\mathbf{n} = \frac{\nabla \phi}{|\nabla \phi|} \quad \text{and} \quad \kappa = \nabla \cdot \frac{\nabla \phi}{|\nabla \phi|} \ . \tag{2}$$

To evolve the surface according to some derived normal velocity v, a PDE of the form

$$\frac{\partial \phi}{\partial t} + v|\nabla \phi| = 0 \ , \tag{3}$$

is solved. For a more thorough treatment of the level set method and implicit representations, cf. [7].

In the variational level set method, functionals are used to derive gradient descent motion PDEs of the form (3). This is done through the use of the differential (Gâteaux derivative). The differential dE is related to the L^2-gradient ∇E of a functional E as

$$dE(\Gamma)v = \langle \nabla E, v \rangle_\Gamma := \int_\Gamma \nabla E \, v \, d\sigma \ , \tag{4}$$

where v is the normal component of a perturbation of Γ, and $d\sigma$ is the curve length element. For details cf. e.g. [8,9,13]. The gradient descent flow for E is then obtained by solving the following initial value problem

$$\frac{\partial \phi}{\partial t} = \nabla E \, |\nabla \phi|, \qquad \phi(\mathbf{x}, 0) = \phi_0(\mathbf{x}) \ , \tag{5}$$

where ϕ_0 is a level set function for the initial curve Γ_0, specified by the user.

2.2 The Chan-Vese Model

Let us briefly recall the Chan-Vese model [1] which we will use to segment images in the experiments in Section 5. Let $I = I(\mathbf{x}) : D \to \mathbf{R}$ denote the image to be segmented, $D \subset \mathbf{R}^2$ being the image domain. Also, let Γ denote a simple closed curve in the image domain (or a non-overlapping union of such curves, bearing in mind that this is allowed in the level set framework). Consider the functional

$$E(\boldsymbol{\mu}, \Gamma) = \frac{1}{2} \int_{\Omega_0} |I(\mathbf{x}) - \mu_0|^2 \, d\mathbf{x} + \frac{1}{2} \int_{\Omega_1} |I(\mathbf{x}) - \mu_1|^2 \, d\mathbf{x} + \alpha |\Gamma|, \tag{6}$$

where $\boldsymbol{\mu} = (\mu_0, \mu_1) \in \mathbf{R}^2$ is a pair of parameters, $|\Gamma|$ denotes the length of the curve Γ, and $\alpha > 0$ is a fixed weight. The idea of the method presented in [1] is to find a curve Γ^* and a pair of parameters $\boldsymbol{\mu}^*$ which solves the optimization problem,

$$E(\boldsymbol{\mu}^*, \Gamma^*) = \min_{\boldsymbol{\mu}, \Gamma} E(\boldsymbol{\mu}, \Gamma). \tag{7}$$

The segmentation of the image I is defined as the partition of the image domain induced by the optimal curve Γ^*. This partition is found using gradient descent on Γ where the gradient is

$$\nabla E = \frac{1}{2}(I(\mathbf{x}) - \mu_0)^2 - \frac{1}{2}(I(\mathbf{x}) - \mu_1)^2 + \alpha\kappa \ . \tag{8}$$

It is easy to find the optimal parameters for each fixed Γ; they are simply the mean intensities of the image taken over each of the sub-domains cut out by Γ

$$\mu_i(\Gamma) = \frac{1}{|\Omega_i|} \int_{\Omega_i} I(\mathbf{x}) \, d\mathbf{x}, \qquad (i = 0, 1), \tag{9}$$

where $|\Omega_i|$ denotes the area of the set $\Omega_i \subset \mathbf{R}^2$.

2.3 Decomposition of Evolutions

Given a curve evolution, described by the normal velocity as in (3) or (5), it is possible to decompose this evolution into translation, rotation and deformation, cf. [9,10]. In this section we briefly describe this process.

The L^2-gradient ∇E, can be divided into three components $\Pi_T \nabla E$, $\Pi_R \nabla E$, and $\Pi_D \nabla E$. The two first terms are the orthogonal projections of ∇E onto the subspaces (of normal velocities at Γ) generated by translations and rotations, respectively, and the last term is the residual $\Pi_D \nabla E = \nabla E - \Pi_T \nabla E - \Pi_R \nabla E$. The residual corresponds to what is left after a rigid transformation of the curve, i.e. deformation.

The operator Π_T is the projection on the subspace L_T of $L^2(\Gamma)$ spanned by all translations. The elements of L_T are then exactly the normal velocities which come from pure translation motions. The operator can be shown to be

$$\Pi_T v = \mathbf{n}^T \mathbf{v} = \mathbf{n}^T \left[\int_\Gamma \mathbf{n}\mathbf{n}^T \, d\sigma \right]^{-1} \int_\Gamma v\mathbf{n} \, d\sigma \ , \tag{10}$$

for all normal velocities $v \in L^2(\Gamma)$, cf. [10]. The positive semi-definite matrix $\int_\Gamma \mathbf{n}\mathbf{n}^T \, d\sigma$, in the right-hand side, is called the *structure tensor* of the curve.

Let us define $\hat{\mathbf{x}} = \left[\begin{smallmatrix} 0 & -1 \\ 1 & 0 \end{smallmatrix}\right] \mathbf{x}$ as the $\pi/2$-rotation of the vector $\mathbf{x}$. The projection on the space of rotations L_R around a point $\mathbf{x}_0 \in \mathbf{R}^2$, can be shown to be

$$\Pi_R v = \frac{\mathbf{n}^T(\hat{\mathbf{x}} - \hat{\mathbf{x}}_0) \displaystyle\int_\Gamma v\mathbf{n}^T(\hat{\mathbf{x}} - \hat{\mathbf{x}}_0) \, d\sigma}{\displaystyle\int_\Gamma |\mathbf{n}^T(\hat{\mathbf{x}} - \hat{\mathbf{x}}_0)|^2 \, d\sigma} \ . \tag{11}$$

The point $\mathbf{x}_0$ in (11) is chosen such that the two subspaces L_T and L_R are orthogonal which means that $\mathbf{x}_0$ must satisfy the following vector relation $\int_\Gamma \left[\mathbf{n}^T(\hat{\mathbf{x}} - \hat{\mathbf{x}}_0)\right] \mathbf{n} \, d\sigma = 0$, hence

$$\hat{\mathbf{x}}_0 = \left[\int_\Gamma \mathbf{n}\mathbf{n}^T \, d\sigma \right]^{-1} \int_\Gamma (\mathbf{n}^T \hat{\mathbf{x}}) \mathbf{n} \, d\sigma , \tag{12}$$

where, interestingly enough, the structure tensor for Γ appears again. Since L_T and L_R are now orthogonal, it follows that the residual operator is also an orthogonal projection. For more details we refer to [10].

3 Computing Translation and Rotation

In this section we propose a method of computing translation and rotation between frames in an image sequence without the use of landmarks or features. If the boundary curve of the segmentation is represented using landmarks of some kind it is trivial to compute the translation and rotation given two curves. Since it is often difficult to reliably extract landmarks along the boundary we would like to do this directly in the implicit framework. The translation can be estimated using the centroid of the regions but the rotation component is more difficult. One choice is to use moments of the regions. This fails, however, when there are no dominant directions.

We propose to rigidly align the boundary curve Γ_{i-1} from one frame to the boundary curve Γ_i in the consecutive frame using the gradient descent of the area of symmetric difference functional, cf. [14],

$$E(\Gamma) = \frac{1}{2} \int (\chi_{\Omega_{i-1}} - \chi_{\Omega_i})^2 \, d\mathbf{x} \ . \tag{13}$$

Here $\chi_{\Omega_{i-1}}$ and χ_{Ω_i} are characteristic functions for the regions. The corresponding gradient flow for Γ_{i-1} is given by

$$\nabla E(\Gamma) = \frac{1}{2} - \chi_{\Omega_i} \ . \tag{14}$$

This evolution is projected on the space of rigid transformations using the operators in Section 2.3. After discarding the deformation part, the evolution is simply

$$\frac{\partial \phi}{\partial t} = (\Pi_T \nabla E + \Pi_R \nabla E)|\nabla \phi| \ . \tag{15}$$

At each iteration the translation vector $\mathbf{v}$ can be recovered using the relation

$$\mathbf{v} = \Big[\int_\Gamma \mathbf{n}\mathbf{n}^T \, d\sigma \Big]^{-1} \int_\Gamma v \, \mathbf{n} \, d\sigma \ ,$$

from (10). Similarly, the rotation angle ω is given by

$$\omega = \frac{\displaystyle\int_\Gamma v\mathbf{n}^T(\hat{\mathbf{x}} - \hat{\mathbf{x}}_0) \, d\sigma}{\displaystyle\int_\Gamma |\mathbf{n}^T(\hat{\mathbf{x}} - \hat{\mathbf{x}}_0)|^2 \, d\sigma} \ .$$

Now we just sum up the angles and displacements for all iterations, multiplying with the length of the time step of the rigid alignment evolution, to obtain the rotation ω_i and translation $\mathbf{v}_i$ of the object between the two images. In the next section we will use this to estimate the future values ω_{i+1} and $\mathbf{v}_{i+1}$.

4 Initialization Using Motion Models

In this section we will use the rigid body transformations obtained from the procedure above to initialize the segmentation in a given image. The goal is to estimate the position and orientation of the objects in frame $i + 1$ given the observations in previous frames $1, \ldots, i$. If the images are time-stamped one can fit polynomials, splines or curves to the previous values of $\mathbf{v}$ and ω and extrapolate to predict the values that transform the object to the next frame.

Here, for the sake of simplicity, we will use a simple linear model which states that $\mathbf{v}_{i+1} = \mathbf{v}_i$ and $\omega_{i+1} = \omega_i$. We consider the extension to other more sophisticated models to be an interesting field for future studies. Examples of interesting extensions of the simple model are:

– **Periodic motion.** Many medical examples, such as segmentation in cross sections of MR cardiac images, show periodic behavior in the sequence.
– **Physical models.** Incorporate (angular)velocity, acceleration and (angular)momentum of the objects.
– **Missing data.** Interpolate position and orientation between frames, in missing frames or when the objects are occluded.

Note that the objects in the scene are treated as a single rigid constellation if they are represented using one level set function. If there are several independent rigid parts moving non-rigidly one function for each part can be used. In this case the translation and rotation is estimated for each function at each iteration.

5 Experiments

The framework presented above is applied to some examples in this section. We show three examples of moving scenes captured by a digital camera. The scenes contain objects with very little texture and feature points that can be tracked. We show that the proposed method of estimating rigid motion and using this to predict the position of the objects in future frames makes it possible to reliably segment the objects.

In all three examples, the initial segmentation is obtained using standard level set implementation of the Chan-Vese model described in Section 2.2 and [1]. This means that the boundary curve is evolved using (8) in a gradient descent,

$$\frac{\partial \phi}{\partial t} = \left[\frac{1}{2}(I(\mathbf{x}) - \mu_0)^2 - \frac{1}{2}(I(\mathbf{x}) - \mu_1)^2 + \alpha\kappa\right] |\nabla \phi| \ ,$$

until steady state, starting from some initial curve. The boundary curve of the segmentation is represented implicitly and no points, landmarks or features are used. We are also assuming that the objects are moving rigidly when the motion is estimated and they are therefore represented using a single level set function.

In general, we find that the translation vectors obtained using our approach are very close to the translation obtained through the difference of the region centroids. The estimation of the angles appear, at least qualitatively, to be accurate.

The first sequence consists of six images of three apples in a translation-like motion. The images and the resulting segmentation are shown in Figure 1 from left to right and top down. The first two images are automatically segmented and the translation and rotation of the apples computed by solving (15) and integrating the translation vector and angle as described in Section 4. The white curve shows the initial curve obtained using a simple linear model for the motion. The

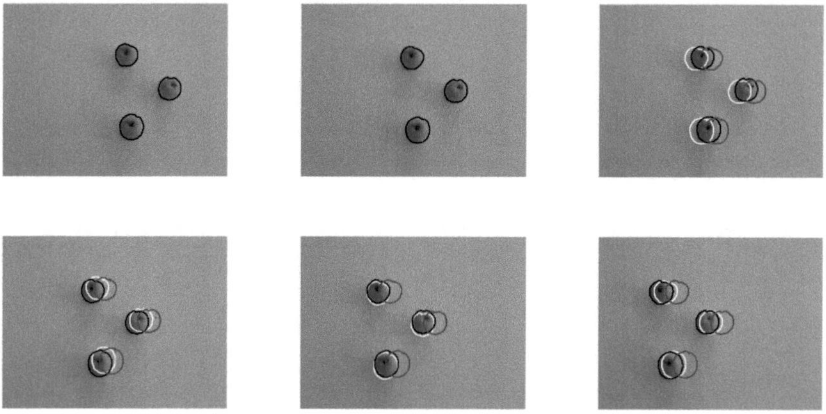

Fig. 1. Example of a translating scene (right and down). The black curve is the boundary contour of the final segmentation for each image. The white curve is the estimated position used as initial curve based on the last two frames and the red curve is the curve from the previous frame. As can be seen from the position of these curves the estimated position is better.

Fig. 2. Example of a translating scene (right and down). The blue curve is the boundary contour of the final segmentation for each image. The white curve is the estimated position used as initial curve based on the last two frames and the red curve is the curve from the previous frame. As can be seen from the position of these curves the estimated position is better.

red curve is the initial curve given by the boundary from the previous frame. It is clear that the initial segmentation is better with the motion model. In two of the frames the red curve is to far from the objects and the Chan-Vese motion then segments the background as the interior region with respect to the boundary.

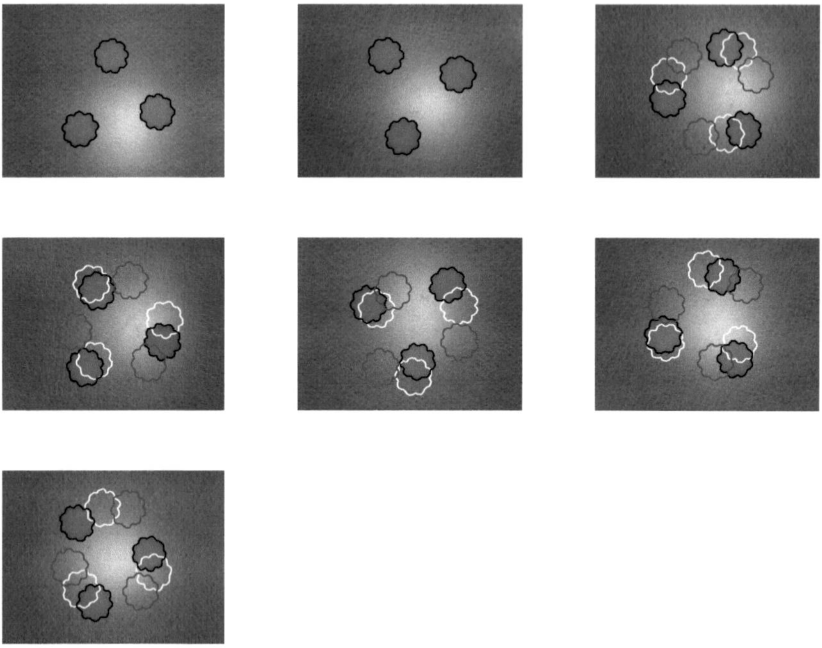

Fig. 3. Example of a rotating scene (right and down). The black curve is the boundary contour of the final segmentation for each image. The white curve is the estimated position used as initial curve based on the last two frames and the red curve is the curve from the previous frame. As can be seen from the position of these curves there are several frames where segmentation starting from the red curve would fail.

The final segmentation will of course be correct for these simple images but it takes very long time for the segmentation to converge in this case.

The second sequence shows five images of printed text on a table side. Three letters are segmented in the first and second image and the final boundary is shown in blue in Figure 2. The white curve shows the initial curve obtained using the same simple motion model as above. The red curve is the initial curve given by the boundary from the previous frame. As with the example above, it is clear that the initial segmentation is better with the motion model. The initial red curve in the last frame is too far from the letters and as a consequence the segmentation is slow and "inverted".

The third example is a sequence containing seven images of three cookies on a table undergoing a rotation-like motion. The images and the resulting segmentation is shown in Figure 3. The black curve is the boundary contour of the final segmentation for each image. The white curve is the estimated position used as initial curve based on the last two frames and the red curve is the curve from the previous frame. The in-plane rotation of the cookies is quite large between frames and there is no overlap of the regions enclosed by the red curves. The red curves are therefore a very bad initialization. A "standard" ring-like pattern, cf. [1], covering the image would be better in this case. The white curves however, are much better even if just a linear motion model is used.

6 Conclusions

This paper presented a method for estimating the motion of objects between images which can be used for featureless objects. Rotation and translation are obtained by projecting a particular warping flow on the subspace of rigid transformations and integrating over the evolution time of the flow. Using the transformation from previous frames in an image sequence makes it possible to estimate the position and orientation of the objects in future frames. Experiments show that this can be used for initialization for segmentation.

Future work and extensions include:

- More sophisticated models for the motion of the objects.
- Multiple objects moving non-rigidly using multiple functions.
- Models for the deformation.

References

1. Chan, T., Vese, L.: Active contours without edges. IEEE Trans. Image Processing 10, 266–277 (2001)
2. Kass, M., Witkin, A., Terzopoulos, D.: Snakes: Active contour models. Int. J. Computer Vision 1, 321–331 (1987)
3. Caselles, V., Kimmel, R., Sapiro, G.: Geodesic active contours. Int. Journal of Computer Vision (1997)
4. Cohen, L.D.: On active contour models and balloons. CVGIP: Image Understanding 53, 211–218 (1991)
5. Xu, C., Prince, J.L.: Snakes, shapes and gradient vector flow. IEEE Trans. on Image Processing 7, 359–369 (1998)
6. Osher, S., Sethian, J.A.: Fronts propagating with curvature-dependent speed: Algorithms based on Hamilton-Jacobi formulations. Journal of Computational Physics 79, 12–49 (1988)
7. Osher, S.J., Fedkiw, R.P.: Level Set Methods and Dynamic Implicit Surfaces. Springer Verlag, Heidelberg (2002)
8. Solem, J.E., Overgaard, N.: A geometric formulation of gradient descent for variational problems with moving surfaces. In: The 5th International Conference on Scale Space and PDE methods in Computer Vision, Scale Space 2005, Hofgeismar, Germany, pp. 419–430. Springer, Heidelberg (2005)
9. Charpiat, G., Keriven, R., Pons, J.P., Faugeras, O.: Designing spatially coherent minimizing flows for variational problems based on active contours. In: ICCV 2005. International Conference on Computer Vision, Beijing, China, pp. 1403–1408. IEEE Computer Society Press, Los Alamitos (2005)
10. Overgaard, N.C., Solem, J.E.: Separating rigid motion for continuous shape evolution. In: International Conference on Pattern Recognition, Hong Kong (2006)
11. Cremers, D., Soatto, S.: Variational space-time motion segmentation. In: International Conference on Computer Vision, Nice, France. vol. 2, pp. 886–892 (2003)

12. Dervieux, A., Thomasset, F.: A finite element method for the simulation of Rayleigh–Taylor instability. In: Tomas, G., Überhuber, C.W. (eds.) Visualization of Scientific Parallel Programs. LNCS, vol. 771, pp. 145–158. Springer, Heidelberg (1994)
13. Yezzi, A., Mennucci, A.: Conformal metrics and true gradient flows for curves. In: ICCV 2005. International Conference on Computer Vision, Beijing, China, pp. 913–919. IEEE Computer Society Press, Los Alamitos (2005)
14. Chan, T., Zhu, W.: Level set based prior segmentation. Technical Report UCLA CAM Report 03-66, University of California at Los Angeles (2003)

Context-Free Detection of Events

Benedikt Kaiser[1] and Gunther Heidemann[2]

[1] University of Karlsruhe, Institute for Process Control and Robotics
Building 40.28, Kaiserstr. 12, D-76128 Karlsruhe, Germany
[2] University of Stuttgart, Intelligent Systems Group
Universitätsstr. 38, D-70569 Stuttgart, Germany

Abstract. The detection of basic events such as turning points in object trajectories is an important low-level task of image sequence analysis. We propose extending the SUSAN algorithm to the spatio-temporal domain for a context-free detection of salient events, which can be used as a starting point for further motion analysis. While in the static 2D-case SUSAN returns a map indicating edges and corners, we obtain in a straight forward extension of SUSAN a 2D+1D saliency map indicating edges and corners in both space and time. Since the mixture of spatial and temporal structures is still unsatisfying, we propose a modification better suited for event analysis.

1 Introduction

For the analysis of static images, the detection of regions of interest or points of interest (representing a region) is an important technique to direct the focus of attention. Thus the most relevant patches for further processing can be found. Such methods serve two purposes: Making computations more efficient, and pre-selecting relevant patterns. That is, even the close-to-signal algorithms are actually part of the pattern classification. Therefore, multiple cues such as colour and texture are in use (e.g. [1]).

The benefit of attentional techniques such as segmentation and interest point (IP) detection is that they are free of context [2], i.e. not adapted to a particular domain. Thus, they serve as a purely data driven starting point for the processing cycle. But in spite of the success in the static case, image sequence analysis rarely makes use of attentional methods, at least not of such that really process the spatio-temporal data. In other words, attention is directed to regions based on isolated frames. While for segmentation there are some approaches (e.g. [3]) that exploit the 2D+1D image data (time being the additional dimension), in the field of IP-detection so far only the Harris-detector [4] has been extended to spatio-temporal data [5].

This is astonishing, since for the decomposition of static scenes into meaningful components, IPs are a standard technique to filter out and represent areas which appear relevant at the signal level. Applications are image retrieval [6], active vision [7], object recognition [8], or image compression [9]. In the present

B.K. Ersbøll and K.S. Pedersen (Eds.): SCIA 2007, LNCS 4522, pp. 223–232, 2007.

paper, we will therefore transfer the concept to the spatio-temporal domain for the detection of baisc actions and events. In comparison to the common technique which is computation of the optical flow, IP-detection is less computationally costly. Certainly, it will not be possible to cover the entire complexity of natural actions by IP-detection, but in the same way as IPs offer cues for static patterns such as symmetry [2,8], basic events like turning points, acceleration, rotation, or approach of two objects (a closing gap) can be detected.

For static imagery, IPs are points which are "salient" or "distinctive" as compared to their neighbourhood. To be more precise, an IP is not necessarily an isolated salient pixel, rather, the IP at a pixel location which stands for a salient patch of the image. Most algorithms for IP detection are aimed at the detection of corners or edges in grey value images [10,4,11,12,13,14]. Methods of this kind which detect edges or corners are particularly promising for the spatio-temporal case, since 3D-corners may indicate the turning points of 2D-corners in a temporal sequence. Thus they would indicate saliency both in the geometrical sense and in the sense of an event.

Laptev and Lindeberg [5] have shown how the concept of spatial edge- and corner-based IPs can be extended to the spatio-temporal domain for the HARRIS detector [4]. They extend the detection of IPs from the eigenvalues of the 2D autocorrelation matrix of the signal to the 3D matrix in a straight forward approach, in addition, they propose a scale space approach to deal with different spatial and temporal scales. But since the 2D- and 3D-HARRIS detector depends on the often problematic computation of grey value derivatives, we chose to extend another IP-detector to the spatio-temporal domain: The SUSAN detector proposed by Smith and Brady [12], which detects edges and corners merely from the grey values.

We will first describe the spatial SUSAN detector (section 2), then its extension to the spatio-temporal domain is described in section 3. The new 3D-SUSAN detector is tested in section 4 using artificial image sequences displaying prototypical events. But since the tests uncover shortcomings of the straight forward extension from 2D to 3D, a modification is introduced in section 5. Finally, the new approach is tested on real image sequences.

2 The SUSAN-Detector for Static Images

Smith and Brady have proposed an approach to detect edges and corners, i.e., one- and two-dimensional image features [12]. While most algorithms of this kind rely on the (first) derivatives of the image matrix, the SUSAN-detector relies on the local binarisation of grey values. To compute the edge- or corner strength of a pixel (called the "nucleus"), a circular mask A around the pixel is considered. By choice of a brightness difference threshold ϑ, an area within the mask is selected which consists of pixels similar in brightness to the nucleus. This area is called USAN ("Univalue Segment Assimilating Nucleus"). To be more precise, let $I(r)$ denote the grey value at pixel r, n the area (i.e. # pixels) of the USAN, and r_0 the nucleus. Then

$$n(r_0) = \sum_{r \in A} c(r, r_0), \qquad \text{with} \quad c(r, r_0) = \begin{cases} 1 \text{ for } |I(r) - I(r_0)| \leq \vartheta \\ 0 \text{ for } |I(r) - I(r_0)| > \vartheta \,. \end{cases} \tag{1}$$

The response of the SUSAN-detector at pixel r_0 is given by

$$R(r_0) = \begin{cases} g - n(r_0) & \text{for } n(r_0) < g \\ 0 & \text{else}\,, \end{cases} \tag{2}$$

where g is called the *geometric threshold.* For edge detection, a suitable value is $g = \frac{3}{4} n_{max}$, for corner detection $g = \frac{1}{2} n_{max}$. It can be shown the these values are optimal in certain aspects for the assumption of a particular signal-to-noise ratio.

Smith and Brady [12] obtain a saliency map which indicates edge and corner strength as the inverted USAN area for each nucleus pixel. IPs are then found as the local minima of the USAN area, thus the name SUSAN (= Smallest Univalue Segment Assimilating Nucleus).

To find the local direction of an edge and to localize corners precisely, geometrical features of the USAN have to be exploited, see [12] for details.

The SUSAN-approach is well suited for a fast detection of one- and two-dimensional basic image features with the benefit that both localization precision and the implicitly built-in noise reduction are robust to changes of the size of the circular mask.

3 Spatio-temporal Extension of SUSAN

In this section we introduce the extension of the normal 2D-SUSAN-detector to the spatio-temporal ("3D") domain [15].

The generalization of an isotropic circular mask in two dimensions is a sphere in three dimensions. But since the spatial coordinates are independent of time, a (rotationally symmetric) ellipsoid around the time axis is betters suited for event detection since it allows suitable scaling (note the same physical event may, e.g., be captured using different frame rates). However, also other 3D-shapes with a circular cross section come into question. In the following, two algorithms using different 3D-masks M_E and M_Z are investigated:

$$M_E(x, y, t) = \begin{cases} 1 \text{ if } \frac{x^2+y^2}{R_{xy}^2} + \frac{t^2}{R_t^2} < 1 \\ 0 \text{ otherwise} \end{cases} \tag{3}$$

$$M_Z(x, y, t) = \begin{cases} 1 \text{ if } \frac{x^2+y^2}{R_{xy}^2} \ \wedge \ -R_t \leq t \leq R_t \\ 0 \text{ otherwise} \end{cases} \tag{4}$$

where R_{xy} denotes the radius in the x-y-plain, and R_t the extension of the mask in on the temporal $t-$axis. In the same way as the 2D-SUSAN-detector is applied to

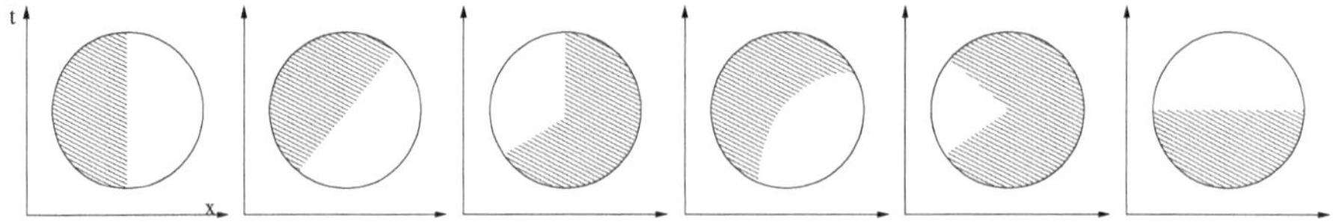

(a) (b) (c) (d) (e) (f)

Fig. 1. A selection of different kinds of motion. The USANs are shown in the $x-t$-plain: (a) Rest, (b) constant velocity, (c) stop-event, (d) acceleration, (e) turn-around, (f) sudden appearance.

each spatial point, now the mask must be applied to each spatio-temporal point of an image sequence, and the grey value of the nucleus is compared to the other pixels within the mask to obtain the USAN-volume. Instead of the original binary decision function, we use the improved version as proposed by Smith and Brady:

$$c(r, r_0) = e^{-(\frac{I(r)-I(r_0)}{t})^6} \tag{5}$$

By this means, the robustness of the algorithm is improved since now slight variations of luminance may not lead to a large variation of the output. The USAN-volume can be calculated as

$$V(x, y, z) = \sum_{x', y', z'} I(x', y', z')M(x + x', y + y', z + z'), \tag{6}$$

Fig. 1 illustrates the way SUSAN3D processes different motion events (in only one spatial dimension). The x-axis points from left to right, the t-axis upwards. The USAN is the white area within the mask. While Fig. 1(a) shows an edge element at rest, Fig. 1(b) depicts motion at constant velocity. The result of $V = 0.5$ (of the area) is the same in both cases. Figs. 1(c) and 1(d) depict a stop-event and acceleration, respectively. For these cases, values V clearly below 0.5 are to be expected. The smallest value V is to be expected in the case depicted in Fig. 1e, which shows a turning point. Fig. 1(f) shows either a sudden appearance of the object or motion at a velocity to high to be resolved. Again, the volume is $V = 0.5$.

Summarizing, by evaluating the 3D-USAN values in the manner of the conventional 2D-SUSAN-detector "salient" events can be detected, such as acceleration and turn around (values the smaller the stronger curvature). However, rest, constant motion, and sudden appearance (all three $V = 0.5$) can not be discriminated. While this is still satisfactory for rest and constant motion ($V = 0.5$ being larger than for acceleration and turn-around), sudden appearance should get the smallest value (i.e. the largest saliency output).

4 Evaluation of the Naive Spatio-temporal Algorithm

In the following, we test the SUSAN3D-detector on artificial image sequences, using as a mask a cylinder of $2R_{xy} = 7$ pixels and $2R_t = 7$ frames, yielding a total volume of 224. The brightness threshold is set to 27 as in the 2D-version.

4.1 Simulations

The first test sequence shows squares moving in different ways. Fig. 2(a) shows, from top to bottom, the SUSAN3D response maps for an accelerating square, a moving one with high constant velocity, a moving one with low constant velocity, and a square at rest. Obviously, the response of the SUSAN3D is mainly governed by geometrical features, not by dynamical features. To reduce the influence of geometrical features, the response of SUSAN3D was tested in 2(b)-(h) by a moving filled circle, i.e. an object without any corners. Fig. 2(b) is the resulting spatio-temporal map of a circle at rest. Figs. 2(c)-(e) show the results for a circle moving to the right at a constant velocity of one, two, and three pixels per frame. In 2 (f), the circle is accelerating by a constant acceleration of $a = 2\,\text{pixels/frame}^2$, in 2 (g), acceleration grows exponentially Fig. 2 (h) shows a turn-around.

Now we searched for the minimum of the USAN. To compare the USAN values achieved at a certain spot of the moving circle, in a first test we searched for the minimum not within the entire response map but only in the area of the righthand circle border on a horizontal line through the middle of the map. The rest of the map was discarded. Results are listed in table 1. Remarkably, the minimal USAN-value is always 112 — except for the turn around — which is half of the mask volume (first line of table 1). So, the expectation that acceleration expresses itself in lower USAN values did not come true in this experiment. For a better analysis of this result, the seven circular "slices" of the mask cylinder have been analysed in separation in table 1 lines "-3" ... "3" ("0" corresponds to the central slice of the mask cylinder). Obviously, the contributions of the partial volumina are different. While the distribution differs for columns $v = 0, 1, 2$ and thus reflects the fact that the object moves at a different velocity, columns $v = 3, a = 2$ and *exp* do not exhibit any difference, because velocity is too large

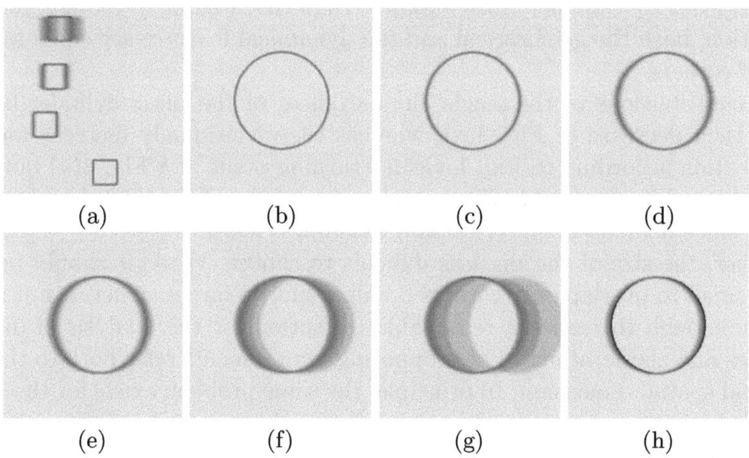

Fig. 2. Output of SUSAN3D at a single moment for different image sequences, see text

Table 1. Results of the first experiment with SUSAN3D, see Fig. 2

	$v = 0$	$v = 1$	$v = 2$	$v = 3$	$a = 2$	exp	turn
USAN	112	112	112	112	112	112	36
-3	16	0	0	0	0	0	0
-2	16	4	0	0	0	0	0
-1	16	10	4	0	0	0	10
0	16	16	16	16	16	16	16
1	16	22	28	32	32	32	10
2	16	28	32	32	32	32	0
3	16	32	32	32	32	32	0

for the chosen mask size. I.e., the object is so fast that each of the three motions is equivalent to an "appearance out of nowhere" for the detector (cf. Fig. 1(f)). In column "turn", the turn around becomes visible both in a small USAN-value and in the distribution throughout the mask cylinder (cf. 1(e)).

4.2 Discussion on the Simulations

There is still a flaw in the "naive" extension of SUSAN: Geometrical and dynamical features are coupled implicitly. Therefore, an accelerating straight edge leads to smaller output in the sequence of saliency maps than a stationary corner. For event detection, however, the first case is more relevant, so the influence of the geometrical features should be attenuated. Fig. 3 illustrates that this is a non-trivial problem: In both cases, the corners move at a constant velocity from the left to the right, the only difference being the rotation of the object. The circles are the slices of the mask cylinder corresponding to successive frames of the sequence, which exhibit different intersections with the corners. In total, case Fig. 3(a) leads to a smaller USAN-volume than case Fig. 3(b), though, regarded in isolation, both the geometrical and the dynamical features are equal for both types of corners.

The contributions of the single circular slices of the mask cylinder first increase, then decrease in Fig. 3(a), whereas they continually increase for 3(b). Classification according to Fig. 1 yields "turning point" for Fig. 3(a) but "constant velocity" for Fig. 3(b). Thus, application of the SUSAN3D-detector is not feasible since it mixes geometrical and dynamical features.

Further, the size of the mask is difficult to choose: While it should be sufficiently small to overlap only corners or edges but no larger structures, it should be large enough to realize a reasonable resolution for the analysis of different velocities and accelerations. These opposing requirements refer both to the temporal and spatial dimension. In principle, the same problem exists for the spatial SUSAN-detector and in general for any windowed function, but it becomes more difficult in the spatio-temporal domain. While in the spatial domain a given window size simply selects a certain scale, in the spatio-temporal domain the coupling between typical spatial and temporal scales has to be dealt with.

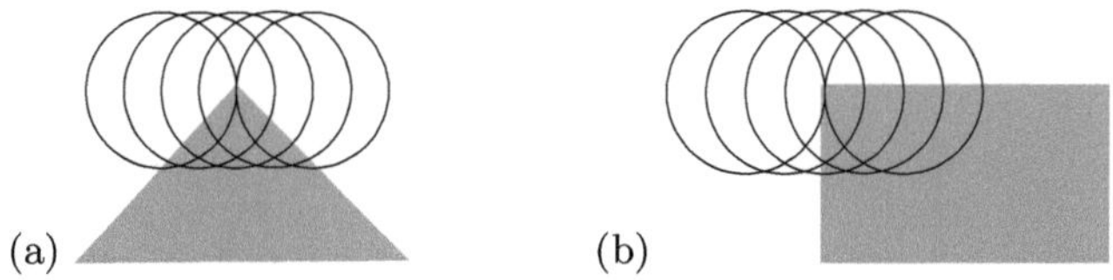

Fig. 3. A corner with constant velocity moving from left to right at different angles (a, b). The cylindrical mask of the object is indicated by its temporal "slices".

5 The SUSANinTime Detector

The discussion of the last section has shown that a straight forward extension of the SUSAN-detector to three dimensions is not satisfying. Of course, additional features could be computed to correct the detector response, but then the simple and elegant idea of using the USAN-volume as a feature for IP-detection would be more or less abandoned. Therefore, in the following we will outline an alternative approach, which applies the SUSAN principle only on the temporal dimension.

The first step is computation of the USAN-area within a cylindrical volume around the nucleus. The single x-y-slices of the cylindrical mask are evaluated to find the USAN-areas for every frame, these values are saved in a 1D-array (`areas[]`). Then the SUSAN principle is applied to the 1D-array `areas[]` in the following way: The USAN-area at the current time is considered to be a (second) nucleus value (`nucleus2`). Note the second nucleus value is an *area*, not a grey value. Then the array `areas[]` is binarised with respect to the `nucleus2` value, and the final detector response is the sum of the now binarised array.

In the pseudocode given in Fig. 4, `mask[x][y][t]` takes a value of 1 if `x,y,t` is inside the volume covered by the detector, else 0. c_1 and c_2 denote the thresholding functions for the spatial binarisation of the x-y-slices and the binarisation of the `areas[]` array, respectively.

```
SUSANinTime(x, y, t)
    nucleus <- getpixel(x,y,t)
    FOR tt FROM -R TO R DO
        areas[tt] <- 0
        FOR yy FROM -r TO r DO
            FOR xx FROM -r TO r DO
                IF mask[x][y][t] = 1 THEN
                    pixel <- getpixel(x + xx, y + yy, t + tt);
                    areas[tt] <- areas[tt] + c_1(nucleus, pixel)
    nucleus2 <- areas[0]
    value <- 0
    FOR tt FROM -R TO R DO
        value <- value + c_2(nucleus2, areas[tt])
    RETURN value
```

Fig. 4. Pseudocode of SUSANinTime, see text

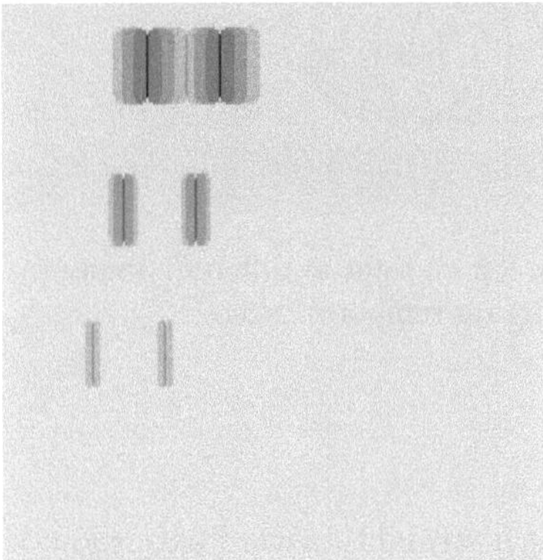

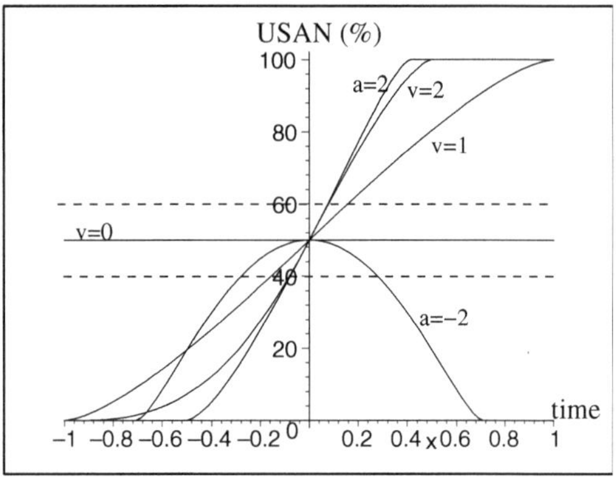

Fig. 5. Left: Output of the SUSANinTime-detector to the sequences of moving squares used in Fig. 2. Right: USAN-area as a function of time.

The idea of the SUSANinTime algorithm is to give a high response to such space-time volumes which exhibit a high activity, where "activity" is defined as a high temporal variation of the USAN-area. Fig. 5 illustrates the principle. It shows the USAN-area (as the percentage of a complete circular slice) as a function of time (to be more precise, as a function of the t-coordinate of the cylindrical mask). The nucleus value is 50% for all of the motion sequences. The SUSANinTime-detector computes for which span of time the USAN-areas are still within a surroundings the nucleus value. This time span takes a maximum for zero velocity ($v = 0$) and decreases with increasing velocity ($v = 1, v = 2$). Thus, it becomes also clear that the return value of SUSANinTime does not allow measurement of acceleration ($a = 2, a = -2$). Though the SUSANinTime-algorithm can not classify space-time events in categories velocity / acceleration, it has nevertheless highly useful properties. Fig. 5, left, shows the response of the SUSANinTime detector, where the input sequence is the one of Fig. 2(a). The detector response is approximately proportional to the velocity, for stationary regions, the detector is "blind" (here, small intensity values denote a high response). In contrast to Fig. 2(a), the corners of the squares yield no stronger response than the edges, though being geometrically more salient. So the response is determined by the dynamics, not stationary features — a major advantage, since now geometrical features detected by separate modules can be included in a final saliency map in a well-defined way.

First experiments on real world image sequences have shown that the algorithm yields robust results. Fig. 6 shows the output of the SUSANinTime detector for a sequence. At first, the hand accelerates in a movement to the right, then stops. Different intensities of the map reflect the different velocities. Note that the appearance or disappearance of (otherwise) static structures in the background is likewise detected as motion.

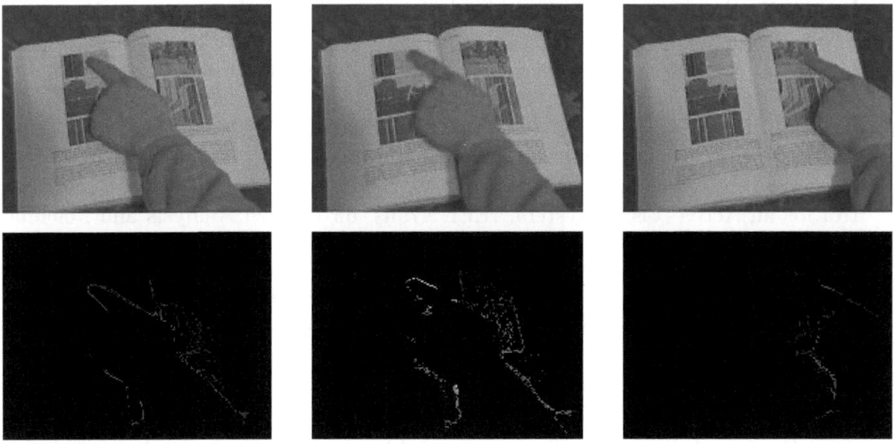

Fig. 6. Output of the SUSANinTime detector for a real image sequence. The pointing event can be clearly detected.

6 Summary and Conclusion

For the detection of generic events in image sequences such as turning points, we have introduced two extensions of the SUSAN IP-detector: A naive extension of SUSAN to a third dimension (time) is unsatisfactory, because dynamical features are less prominent in the computed sequence of saliency maps than the static edge- and corner-features. The SUSANinTime algorithms overcomes these problems both on artificial and real image sequences. Tests for human gestures and object manipulation by human hands have shown that important aspects of motion such as pointing events can be well captured.

In future work, we want to apply SUSANinTime for the classification of more complex events such as grasping an object. For this we plan to gather spatio-temporal IPs over a period of time long enough to capture the movement. Around each of the IPs local features will be extracted from the space-time volume. While in isolation features of this kind are not sufficient to characterize complex motion, we hope that a whole "cloud" of IPs provides sufficient information, which we intend to classify in a way similar to the (static) IP-based scene classification described in [16].

References

1. Martin, D.R., Fowlkes, C.C., Makik, J.: Learning to Detect Natural Image Boundaries Using Local Brightness, Color, and Texture Cues. IEEE Trans. on Pattern Analysis and Machine Intelligence vol. 26(1) (2004)
2. Reisfeld, D., Wolfson, H., Yeshurun, Y.: Context-Free Attentional Operators: The Generalized Symmetry Transform. of Computer Vision 14, 119–130 (1995)
3. Goldberger, J., Greenspan, H.: Context-Based Segmentation of Image Sequences. IEEE Trans. on Pattern Analysis and Machine Intelligence 28(3), 463–468 (2006)

4. Harris, C., Stephens, M.: A Combined Corner and Edge Detector. In: Proc. 4th Alvey Vision Conf. pp. 147–151 (1988)
5. Laptev, I., Lindeberg, T.: Space-time Interest Points. In: Proc. ICCV 2003. pp. 432–439 (2003)
6. Tian, Q., Sebe, N., Lew, M.S., Loupias, E., Huang, T.S.: Image Retrieval Using Wavelet-Based Salient Points. J. of Electronic Imaging 10(4), 835–849 (2001)
7. Backer, G., Mertsching, B., Bollmann, M.: Data- and Model-Driven Gaze Control for an Active-Vision System. IEEE Trans. on Pattern Analysis and Machine Intelligence 23(12), 1415–1429 (2001)
8. Heidemann, G.: Focus-of-Attention from Local Color Symmetries. IEEE Trans. on Pattern Analysis and Machine Intelligence 26(7), 817–830 (2004)
9. Privitera, C.M., Stark, L.W.: Algorithms for Defining Visual Regions-of-Interest: Comparison with Eye Fixations. IEEE Trans. on Pattern Analysis and Machine Intelligence 22(9), 970–982 (2000)
10. Moravec, H.P.: Towards Automatic Visual Obstacle Avoidance. In: Proc. 5th Int'l Joint Conf. on Artificial Intelligence, Cambridge, Massachusetts, USA, pp. 584–587 (1977)
11. Schmid, C., Mohr, R.: Local Grayvalue Invariants for Image Retrieval. IEEE Trans. on Pattern Analysis and Machine Intelligence 19(5), 530–535 (1997)
12. Smith, S., Brady, J.: SUSAN – A New Approach to Low Level Image Processing. of Computer Vision 23(1), 45–78 (1997)
13. Zheng, Z., Wang, H., Teoh, W.: Analysis of Gray Level Corner Detection. Pattern Recognition Letters 20, 149–162 (1999)
14. Zitová, B., Kautsky, J., Peters, G., Flusser, J.: Robust detection of significant points in multiframe images. Pattern Recognition Letters 20(2), 199–206 (1999)
15. Heidemann, G., Kaiser, B., Bax, I., Bekel, H., Ritter, H.: Spatiotemporal Events and Action Sequences. Technical report, Bielefeld Univ., Neuroinformatics Group (2005)
16. Heidemann, G.: Unsupervised image categorization. Image and Vision Computing 23, 861–876 (2005)

Supporting Structure from Motion with a 3D-Range-Camera

Birger Streckel[1], Bogumil Bartczak[1], Reinhard Koch[1], and Andreas Kolb[2]

[1] Institute of Computer Science
Christian-Albrechts-University of Kiel, 24098 Kiel, Germany
rk@informatik.uni-kiel.de
[2] Computer Graphics Group
University of Siegen, 57068 Siegen, Germany

Abstract. Tracking of a camera pose in all 6 degrees of freedom is a task with many applications in 3D-imaging as i.e. augmentation or robot navigation. Structure from motion is a well known approach for this task, with several well known restrictions. These are namely the scale ambiguity of the calculated relative pose and the need of a certain camera movement (preferably lateral) to initiate the tracking.

In the last few years time-of-flight imaging sensors were developed that allow the measuring of metric depth over a whole region with a frame rate similar to a standard CCD-camera.

In this work a camera rig consisting of a standard 2D CCD camera and a time-of-flight 3D camera is used. Structure from motion is calculated on the 2D image, aided by the depth measurement from the time-of-flight camera to overcome the restrictions named above. It is shown how the additional 3D-information can be used to improve the accuracy of the camera pose estimation.

1 Introduction

Determining the position of a single moving video camera without the help of markers is a well researched problem. One of the most promising solutions is the Structure From Motion (SfM) technique, where simultaneous to the camera pose estimation a sparse scene structure is reconstructed from prominent 2D-features, that are tracked throughout a video sequence.

A recent approach to measure the shape of a surface in a field of view (FOV) is the PMD-Camera. This camera is using the Photo Mixing Detector (PMD) technology to measure depth values in a FOV, with a frame rate comparable to a common CCD video camera. The PMD-Camera is based on the time-of-flight principle [9]. This paper is discussing how a time-of-flight camera can be used to aid SfM and which benefits and limitations can be expected.

2 Previous Work

Much work was undertaken in the field of camera tracking and SLAM [2] using a single moving 2D-camera. The main contributions were gathered by Hartley and

B.K. Ersbøll and K.S. Pedersen (Eds.): SCIA 2007, LNCS 4522, pp. 233–242, 2007.

Zissermann in [3], a profound overview was given by Lepetit and Fua in [10]. There was also much research done in the field of 3D-imaging based on the time-of-flight principle [15][7].

Using the depth images of a time-of-flight camera to improve camera tracking in 6 degrees of freedom is a fairly new idea and is based on the work of Prasad et al. [12]. There a low resolution 3D-imaging sensor is combined with a conventional high resolution 2D-CCD. The pixel correspondence is assured by an optical image multiplier which is delivering the same optical information to both measuring devices. The high resolution image information of this device enables SfM calculation aided by a known corresponding depth.

2.1 Structure from Motion

The basic idea of SfM is discussed in [11], where a single camera is moved in a static scene and simultaneously the camera position is tracked and a sparse 3D-model of the scene is generated. This method works in realtime [4] without any prior knowledge of the scene, especially without knowing the scene depth.

There are approaches where already depth information is used to aid the SfM task. Koeser et al. [5] i.e. create a scene model in an offline phase. This model is used to generate 3D-points in the online phase, where the camera pose can be estimated in realtime.

There are restrictions in the SfM approach, that limit its usability for certain applications as i.e. robot navigation. One restriction is the necessity of an initial camera movement to be able to start the tracking. A lateral movement leads to better initialization results than a forward movement. In addition the camera pose and scene structure is computed "up to scale", which means that all gathered distances are scaled with an arbitrary factor.

Based on only SfM navigation a robot would have to start moving without knowing anything about its surrounding and also would be unable to detect the real metric distance to an obstacle in its way. In addition the direction for a common robot movement is forward. This is an ill posed problem for the standard SfM approach.

2.2 Photo Mixing Detector (PMD) Technology

The time-of-flight technique measures the metric distance to an object. Modulated light is sent out, reflected by the object and received by an appropriate detector. By sampling and correlating the incoming optical signal with a reference signal it is possible to calculate the time-of-flight for the light ray. Knowing the time, it is easy to calculate the distance the light covered from sending to receiving and thus the distance of the reflecting object.

The PMD technology uses the time-of-flight principle to measure the depth over a whole FOV [15]. The scene is illuminated by LEDs sending out modulated near infrared light. The light is reflected by all visible objects and received by an image sensor, that is comparable to a CMOS chip of a common digital camera.

Also the manufacturing corresponds to the standard CMOS manufacturing process. This allows a very economic production of the device. Due to an automatic suppression of background light the sensor can be used for indoor as well as for outdoor scenes, which is a strong precondition for mobile robot navigation.

Current devices provide a resolution of 64×48 with active background light suppression, the maximum FOV is 40°. The restriction in the FOV is due to the necessity of a bright active illumination of the measured scene. The camera frame rate is 25 Hz and the modulation frequency is 20 MHz, resulting in an unambiguous range of $7.5m$.

2.3 Camera Setup and Test Sequences

The performance increase of SfM with 3D-imaging was measured on simulated image sequences as well as on real camera data.

The real image sequences were generated using a rig of a high resolution 2D camera rigidly fixed to a low resolution PMD-camera, figure 1 shows the setup. We used a PMD-camera from PMD-Tech [7], based on the technology described in section 2.2, with a resolution of 64×48 and a horizontal FOV of only 23.1°. The 2D camera was a standard CCD-camera with a resolution of 1024×768 and a horizontal FOV of 42.3°. The alignment of the cameras assured that the FOV of the PDM-camera was completely covered by the 2D camera. For calibrating the rig we used the Bouguet calibration toolbox [1] similar to Kolb et al. in [6]. With this calibration the depth data was mapped to the 2D-image, an image pair is shown in figure 1. In recent publications the accuracy of the PMD-camera was evaluated [8] and the measured depth values were calibrated [6]. The results from these publications were used to get well calibrated depth data with known uncertainty.

For the real image sequence no ground truth information is available, so it is necessary to evaluate the pose tracking performance also on synthetic data, where we are able to quantify the estimation results. The simulated 2D-images have a resolution of 1024×768 and a FOV of 80°, the depth images have a resolution of 64×48 pixel and a FOV of 40°. Image pairs from the simulated sequence are shown in figure 2. We use the results of [8] to simulate depth noise for the synthetic PMD-images.

Fig. 1. 2D/3D-camera rig with image pair, PMD-image mapped to match 2D-image

3 Using Depth Information for Camera Tracking

The usage of metric depth images for camera pose estimation has three main advantages:

- The camera poses can be estimated in a metric coordinate system.
- The camera pose estimation starts from the first frame, no initial movement is necessary (see section 3.1).
- The camera pose estimation is improved by using the additional depth information.

How these improvements are achieved is described in the following sections.

3.1 Metric Pose Estimation from the First Image

Traditional SfM camera tracking always needs two images for initialization. These images must provide an adequate baseline in order to triangulate 3D-points with the necessary accuracy. Movements parallel to the optical axis are ill conditioned for small FOV cameras, a stable triangulation of 3D points can hardly be achieved. An initial sideways camera movement is mandatory.

With additional metric depth information available it is possible to estimate metric 3D scene points and to establish a metric coordinate system for a single camera image. The coordinate system has its point of origin in the first camera center and the cameras optical axis is at zero rotation. Tracking can start from the first frame without an initial camera movement.

To facilitate tracking we estimate covariances for the 3D-points. The standard deviation of the PMD-depth data was measured by Kuhnert et al. in [8] as

$$\sigma_z = 2.734 * 10^{-3} d^2 + 2.867 * 10^{-3} d - 4.230 * 10^{-4},$$

d is the measured depth in meters. The standard deviation in x and y direction is influenced by the 2D feature tracking accuracy. In this work the KLT-corner-tracker [13] was used. Its standard deviation is given as $\sigma_{KLT} = 0.25$ pixel. This can be back-projected to the 3D-point by

$$\sigma_x = \sigma_y = \sigma_{KLT} / f_{PMD} * d$$

with d as above and f_{PMD} as the focal length of the PMD-camera in pixel.

Knowing the standard deviations along the three axes we are able to approximate the covariance matrix Σ_{3D}^{Cam} of a 3D-point X in the camera coordinate system. The transformation of Σ_{3D}^{Cam} into the global coordinate system Σ_{3D}^{World} needs the 4×4 affine transformation matrix T, with rotation matrix R and camera center C in global coordinates.

$$\Sigma_{3D}^{World} = T \Sigma_{3D}^{Cam} T^T \text{ with } \Sigma_{3D}^{Cam} = \begin{pmatrix} \sigma_x^2 & 0 & 0 \\ 0 & \sigma_y^2 & 0 \\ 0 & 0 & \sigma_z^2 \end{pmatrix} \text{ and } T = \begin{pmatrix} R & C \\ 0\ 0\ 0 & 1 \end{pmatrix}$$

3.2 Pose Estimation and Structure Update

By knowing 3D-points and a camera pose in a metric coordinate system it is now possible to estimate the camera movement for the next image in metric scale. 2D-features in the high resolution 2D-images can be tracked with high accuracy, and the 6DoF-camera pose can be estimated by a Levenberg-Marquardt-minimization from the known 2D/3D-correspondences, facilitated by a RANSAC to remove outliers [3].

Knowing the pose, again 3D-points and their covariances can be estimated in the global metric coordinate system as described above. Having two instances of the same 3D-point, these can be merged with a Kalman Filter [14]. Because the 3D-point is assumed to be static and its coordinates can be measured directly, the Kalman equations can be simplified to

$$\begin{aligned} X_{new} &= X_1 + G * (X_2 - X_1) \\ \Sigma_{X_{new}} &= (I_{3\times 3} - G) * \Sigma_{X_1} \end{aligned} \quad \text{with} \quad G = \Sigma_{X_2} * (\Sigma_{X_1} + \Sigma_{X_2})^{-1}$$

where $X_{1,2}$ are the 3D-positions of the points 1 and 2, $\Sigma_{X_{1,2}}$ are the 3×3-covariance matrices of points 1 and 2. X_{new} is the new merged point and $\Sigma_{X_{new}}$ its covariance. G is the gain matrix from the Kalman equations.

It is still possible to additionally use all SfM standard methods, i.e. triangulation of 3D-points in 2D-image parts without depth information. 3D-points can still be optimized by minimizing the back-projection error into the current 2D-image, a standard technique for SfM on 2D-images.

4 Results

The described method was evaluated on real as well as synthetic image sequences. The results are described in the following sections.

4.1 Synthetic Image Sequence

To be able to calculate correct pose estimation errors a synthetic 2D/3D-image sequence was used. For a description of the simulated camera setup see section 2.3, sample images are shown in figure 2. The image sequence is difficult to track for standard SfM, because it starts with a forward movement that is common i.e. for a moving robot but very disadvantageous for SfM.

The movement is starting in z-direction and the main movement is in the x-z-plane. It is overall covering $2.1m$ in x, $0.27m$ in y and $2.4m$ in z, the scene is at a distance of $4.5m$ to $7.5m$ for the starting image. The camera is also rotated during the movement. Figure 4 shows the camera path, the path is a closed loop of 101 images. Two sequences were generated, the first with ideal depth data despite the low spatial resolution, the second with noise added to the depth data. The σ_z of the noise was taken from [8].

Fig. 2. Synthetic 2D/3D-image pairs. Left: no noise, Right: noise added.

SfM was run on these sequences evaluating three different scenarios:

1. 3D-points were created only from triangulation of 2D-features.
2. 3D-points were created only from the 3D-depth-images.
3. 3D-points were created from depth images as well as from 2D-triangulation, in regions where no 3D-information is available. This increases the solid angle in which known 3D-points are tracked to the FOV of the 2D-camera, while still keeping the metric initialization and the 3D-point stabilization.

For scenario 1 the whole sequence was scaled for comparison. The scale factor was calculated using the known ground truth distance for the initialization movement. In scenario 2 and 3 the estimated camera poses were not scaled or modified. The average translation and rotation errors are shown in table 1, absolute in m or degree as well as relative to the overall camera movement. The error progression over the sequences is shown in figure 3 for the ideal and noised depth data. Strong improvements are visible for scenarios 2 and 3 compared to scenario 1. Scenario 3 outperforms scenario 2 only for rotation estimation. The rotation in scenario 3 is stabilized by the higher FOV while translation estimation is worse due to the large z-errors of the triangulated points. On the sequence with the noise added, certain camera poses are estimated wrong. This does not introduce a drift, the pose estimation regenerates fast.

Please notice that in scenario 2 for the first 20 images all 3D-points are located in a solid angle of 40°, since all points are created from the narrow 3D-image. With the sideways camera movement 3D-points are then moving out of the PMD-camera FOV and points in a larger solid angle are tracked.

Figure 4 is showing the ground truth and the estimated camera paths as a top view on the x-z-plane. In figure 4(a) scenario 1 is shown. The main reason for the translational error is the drift that is accumulated over the whole sequence.

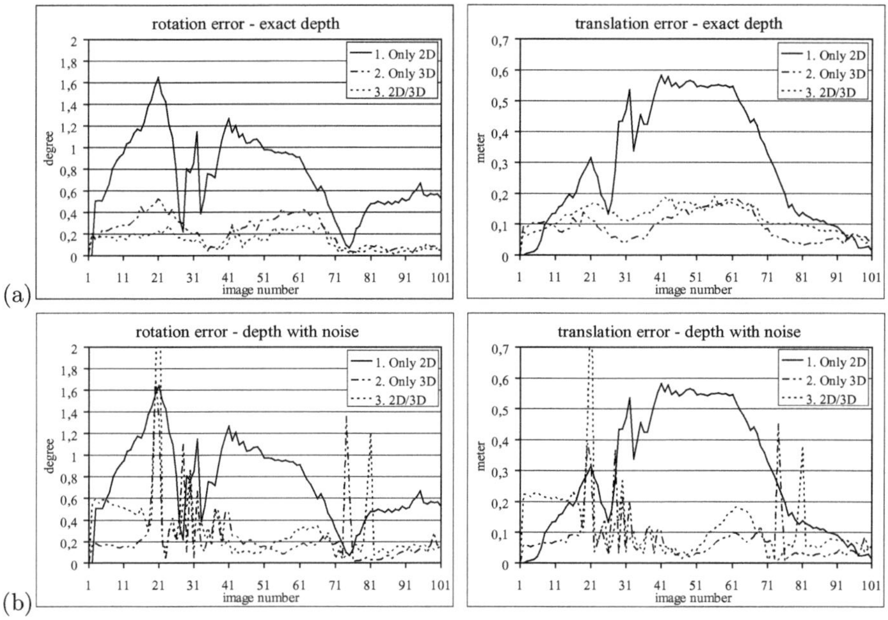

Fig. 3. Errors on synthetic sequence with (a) exact depth (b) noise on depth

Table 1. Average translation and rotation errors for synthetic sequences. The relative translation is relative to the total camera movement.

	Absolute Average Translation Error	Relative Average Translation Error	Absolute Average Rotation Error
1. Only 2D	$0.29m$	10.16%	0.77°
2. Only 3D	$0.09m$	2.63%	0.22°
3. 2D/3D	$0.12m$	3.41%	0.13°
2. Only 3D with noise	$0.08m$	3.37%	0.23°
3. 2D/3D with noise	$0.12m$	4.30%	0.32°

Its origin is the initial forward movement, that provides very bad triangulation baselines. The camera path for scenario 2 is shown in figure 4(b). Here the pose estimation for the forward movement is more accurate and thus the estimation overall better resembles the ground truth data.

4.2 Real Image Sequence

We used the 2D/3D-camera pair from section 2.3 generate an image sequence that is again starting with a forward movement of about 30 images and is very difficult for standard SfM. To be able to assess the quality on a free hand camera movement, 75 images were processed in a forward-backward-loop with a hard turn at image 75. Ideally the camera positions for forward and backward

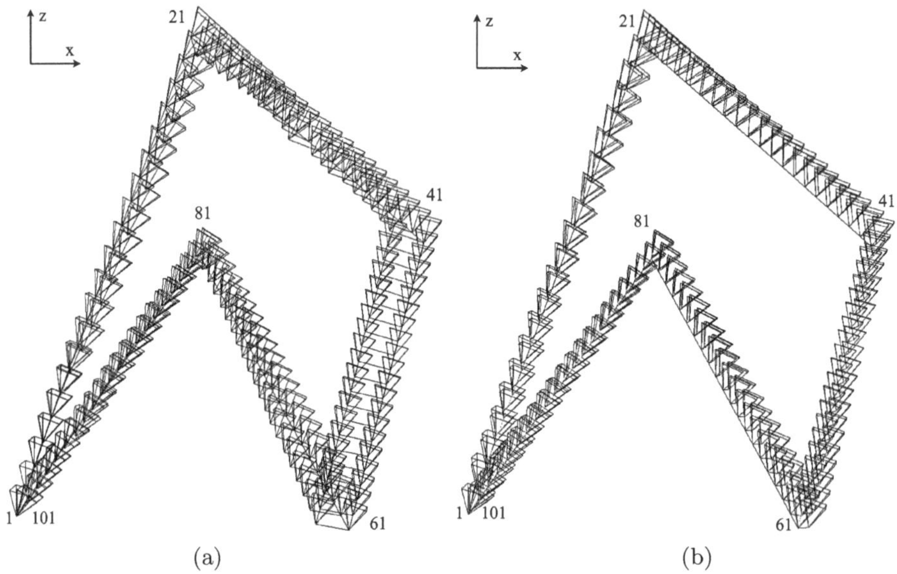

Fig. 4. Ground truth vs. estimated camera tracks on synthetic sequence without noise. (a) Scenario 1 - Only 2D (b) Scenario 2 - Only 3D.

Fig. 5. 2D/3D-pairs from the real image sequence

movement should correspond. The scene distance is between $0.6m$ and $1.1m$, the camera movement spans $0.24m$ in x, $0.1m$ in y and $0.44m$ in z direction. Some images of the processed sequence and their depth maps are shown in figure 5.

The forward movement and the small 2D-camera FOV (42.3°) make SfM on 2D-images (scenario 1) impossible. While the movement direction is estimated

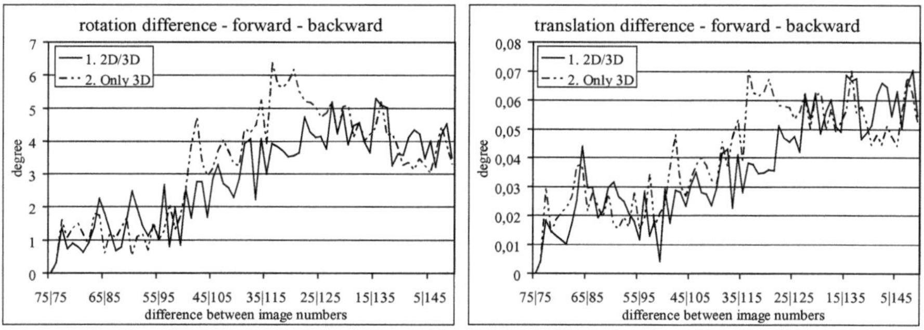

Fig. 6. Errors for the real image sequence. Differences between forward and backward track. Tracking on pure 2D data was impossible on this sequence.

correctly, the moved distance is estimated very inaccurate. The baseline for triangulating 3D-points is extremely short, thus the 3D-points have high errors in z-direction. This leads to a strong drift and a fast degradation of the camera track. No convergence could be reached.

The PMD-depth data is suitable to compensate for this distance misestimation. When using the 3D-data to aid pose tracking, SfM estimates a reasonable camera track. The average translation error between identical images in forward and backward movement for scenario 3 is $0.037m$, the average rotation error is $2.9°$. The overall performance on the real sequence is shown in figure 6. The error is small near image 75, where the forward-backward turn was just performed. Near image 1 the drift accumulated over 150 images under extremely difficult conditions. In contrast to the evaluation on synthetic data, scenario 3 slightly outperformed scenario 2 on real data, for translation as well as for rotation estimation. This was repeatable on different image sequences.

5 Conclusion

In this work it was shown how a 3D-PMD camera can be used to improve camera pose estimation with a modified SfM approach. The results show that the 3D-camera enables a valuable improvement to the camera tracking performance. Qualitatively because the estimation can be done starting from the first image in a metric coordinate system and quantitatively by improving the accuracy of the estimated pose.

Since PMD-cameras will experience a significant price drop in the next few years when productions processes are improved, such a camera provides an interesting and simple way to enhance the pose estimation of a single moving camera.

Acknowledgement. This work was supported by the German Research Foundation (DFG), KO-2044/3-1.

References

1. Bouguet, J.Y.: Camera calibration toolbox for matlab. www.vision.caltech.edu/bouguetj/calib_doc/index.html (1998)
2. Andrew J. Davison. Real-time simultaneous localisation and mapping with a single camera. In: Proc. ICCV (2003)
3. Hartley, R., Zissermann, A. (eds.): Multiple View Geometry in Computer Vision, 2nd edn. Cambridge university press, Cambridge (2004)
4. Koch, R., Koeser, K., Streckel, B., Evers-Senne, J.-F.: Markerless image-based 3d tracking for real-time augmented reality applications. In: WIAMIS 2005, Montreux, Switzerland (2005)
5. Koeser, K., Bartczak, B., Koch, R.: Drift-free pose estimation with hemispherical cameras. In: Proceedings of CVMP 2006, London (2006)
6. Kolb, A., Lindner, M.: Lateral and depth calibration of pmd-distance sensors. In: International Symposium on Visual Computing (ISVC06) (2006)
7. Kraft, H., Frey, J., Moeller, T., Albrecht, M., Grothof, M., Schink, B., Hess, H., Buxbaum, B.: 3d-camera of high 3d-frame rate, depth-resolution and background light elimination based on improved pmd (photonic mixer device)-technologies. In: 6 th Intl Conference for Optical Technologies, Optical Sensors and Measuring Techniques (OPTO 2004) (2004)
8. Kuhnert, K.-D., Stommel, M.: Fusion of stereo-camera and pmd-camera data for real- time suited precise 3d environment reconstruction. In: IEEE/RSJ International Conference on Intelligent Robots and Systems (IROS06) (2006)
9. Lange, R., Seitz, P., Biber, A., Schwarte, R.: Time-of-flight range imaging with a custom solid-state imagesensor. In: EOS/SPIE Laser Metrology and Inspection, vol. 3823 (1999)
10. Lepetit, V., Fua, P.: Monocular model-based 3d tracking of rigid objects: A survey. Foundations and Trends in Computer Graphics and Vision 1(1), 1–89 (2005)
11. Pollefeys, M., van Gool, L., Vergauwen, M., Verbiest, F., Cornelis, K., Tops, J.: Visual modeling with a hand-held camera. International Journal of Computer Vision 59(3), 207–232 (2004)
12. Prasad, T.D.A., Hartmann, K., Wolfgang, W., Ghobadi, S.E., Sluiter, A.: First steps in enhancing 3d vision technique using 2d/3d sensors. In: Chum, V., Franc, O. (eds) Computer Vision Winter Workshop 2006, pp. 82–86, Telc, Czech Republic (2006)
13. Shi, J., Tomasi, C.: Good features to track. In: Conference on Computer Vision and Pattern Recognition, Seattle, June 1994, pp. 593–600. IEEE Computer Society Press, Los Alamitos (1994)
14. Welch, G., Bishop, G.: An introduction to the kalman filter. Technical Report TR 95-041, University of North Carolina, Department of Computer Science (2001)
15. Xu, Z., Schwarte, R., Heinol, H., Buxbaum, B., Ringbeck, T.: Smart pixel - photonic mixer device (pmd). In: M2VIP '98 - International Conference on Mechatronics and Machine Vision in Practice, pp. 259 – 264 (1998)

Object Recognition Using Frequency Domain Blur Invariant Features

Ville Ojansivu and Janne Heikkilä

Machine Vision Group, Department of Electrical and Information Engineering, University of Oulu, PO Box 4500, 90014, Finland
{vpo,jth}@ee.oulu.fi

Abstract. In this paper, we propose novel blur invariant features for the recognition of objects in images. The features are computed either using the phase-only spectrum or bispectrum of the images and are invariant to centrally symmetric blur, such as linear motion or defocus blur as well as linear illumination changes. The features based on the bispectrum are also invariant to translation, and according to our knowledge they are the only combined blur-translation invariants in the frequency domain. We have compared our features to the blur invariants based on image moments in simulated and real experiments. The results show that our features can recognize blurred images better and, in a practical situation, they are faster to compute using FFT.

1 Introduction

Recognition of objects and patterns in images is a fundamental part of computer vision with numerous applications. The task is difficult as the objects rarely look exactly similar in different conditions. In real applications, images contain various artifacts such as geometrical and convolutional degradations. Image analysis systems should be able to operate also in these nonideal conditions. There has been a vast amount of research in this field of invariant pattern and object recognition [1]. However, the invariant recognition of objects degraded by blur is a much less studied topic.

In various applications, images may contain blur, which can result, for example, from atmospheric turbulence, out-of-focus, or relative motion between the camera and the scene. This degradation process can be modeled as a linear shift-invariant system in which the relation between an ideal image $f(\mathbf{x})$ and an observed image $g(\mathbf{x})$ is given by

$$g(\mathbf{x}) = f(\mathbf{x}) * h(\mathbf{x}) + n(\mathbf{x}) \ , \tag{1}$$

where $\mathbf{x}$ is a 2-D spatial coordinate vector, $h(\mathbf{x})$ the point spread function (PSF) of the system, $n(\mathbf{x})$ additive noise, and $*$ denotes 2-D convolution. The point spread function $h(\mathbf{x})$ represents blur while other degradations are captured by the noise term $n(\mathbf{x})$.

The analysis of blurred images is often carried out by first deblurring the images and then applying standard methods for further analysis. Unfortunately,

B.K. Ersbøll and K.S. Pedersen (Eds.): SCIA 2007, LNCS 4522, pp. 243–252, 2007.

image deblurring is a difficult problem. Conventional solutions include the estimation of the PSF of the blur and deconvolution of the image using that PSF. When the PSF is known, the latter ill-posed problem can be solved using approaches which use regularization [2]. In practice, the PSF is often unknown and very hard to estimate accurately. In this case, blind image restoration algorithms are used [3].

The analysis of the blurred images can also be performed without deblurring using features which are invariant to blur. Flusser and Suk proposed the first blur invariant features in [4]. These invariants are based on geometric moments, central moments (MOMIs) or the spectrum (SPEIs) of the image. Of these, the MOMIs are also invariant to translation. The MOMIs have been applied to template matching [4], recognition of defocused objects [5] and registration of X-ray images [6]. In [7], blur, rotation and scale invariants based on complex moments were proposed. In [8,9], the theory was extended to blur and affine moment invariants. A shortcoming of these blur invariant features is their sensitivity to noise, especially in the case of SPEIs [10]. Probably for this reason, the SPEIs do not have known applications. In addition, translation invariance has not been incorporated into Fourier domain blur invariants.

The features presented in this paper are invariant to centrally symmetric blur, which is exactly the same condition as with the MOMIs and SPEIs. The invariants are computed from the phase-only spectrum, (phase blur invariants, PBIs) or using phase-only bispectrum in which case we achieve blur-translation invariants (PBTIs). The computation of the invariants can be done efficiently using FFT.

2 Frequency Domain Blur Invariant Features

2.1 Features Invariant to Blur

In this section, we show how invariance to blur is obtained in the Fourier domain by using the phase-only spectrum.

If noise $n(\mathbf{x})$ is neglected, (1) can be expressed in the Fourier domain using the convolution theorem by

$$G(\mathbf{u}) = F(\mathbf{u}) \cdot H(\mathbf{u}) \ , \tag{2}$$

and in the phasor form by

$$G(\mathbf{u}) = |G(\mathbf{u})| \, e^{-i\phi_g(\mathbf{u})} \ , \tag{3}$$

where $\mathbf{u}$ is a vector in the 2-D frequency space.

If the Fourier transform $G(\mathbf{u})$ is normalized by its magnitude, only the complex exponential containing the phase remains, namely

$$\frac{G(\mathbf{u})}{|G(\mathbf{u})|} = e^{-i\phi_g(\mathbf{u})} = e^{-i[\phi_f(\mathbf{u})+\phi_h(\mathbf{u})]} \ , \tag{4}$$

where $\phi_f(\mathbf{u})$ is the phase of the original image $f(\mathbf{x})$ and $\phi_h(\mathbf{u})$ the phase of the blur PSF $h(\mathbf{x})$.

Since $h(\mathbf{x})$ is assumed to be centrally symmetric, its Fourier transform $H(\mathbf{u})$ is real and its phase $\phi_h(\mathbf{u})$ has only two possible values

$$\phi_h(\mathbf{u}) = 0 \vee \phi_h(\mathbf{u}) = \pi \ . \tag{5}$$

It follows from this and from the periodicity of the complex argument that the equality

$$\begin{aligned}[e^{-i\phi_g(\mathbf{u})}]^{2n} &= e^{-i2n\phi_g(\mathbf{u})} \\ &= e^{-i2n\phi_f(\mathbf{u})}e^{-i2n\phi_h(\mathbf{u})} \\ &= [e^{-i\phi_f(\mathbf{u})}]^{2n}\end{aligned} \tag{6}$$

holds for any integer n.

Thus, any even power of the normalized Fourier transform, i.e. $e^{-i2n\phi(\mathbf{u})}$, of the observed image is invariant to the convolution of the original image with any centrally symmetric PSF. In other words, any even multiple of the Fourier transform phase modulo 2π is also invariant. We construct the invariants in this way using value n=1, namely

$$\mathcal{B}(\mathbf{u}) = 2\phi(\mathbf{u}) \bmod 2\pi \ , \tag{7}$$

where $\phi(\mathbf{u})$ is the phase spectrum of the image. Henceforth, $\mathcal{B}(\mathbf{u})$ is called phase blur invariant (PBI). Some similarities with the derivation of the SPEIs in [4] can be observed. In this case, the invariance is obtained by using the tangent of the phase spectrum $\phi(\mathbf{u})$.

2.2 Features Invariant to Blur and Translation

In this section, we extent the theory presented in Section 2.1 to incorporate also invariance to translation.

The phase spectrum of an image, which is used to construct the invariants in Section 2.1, is not invariant to translation. The amplitude spectrum is invariant to translation, but, on the other hand, blur invariants can not be constructed from it. This is the reason why we have to turn to the higher order spectra defined by

$$\Psi_n(\mathbf{u}_1, \mathbf{u}_2, \cdots, \mathbf{u}_n) = F^*(\mathbf{s}) \prod_{i=1}^{n} F(\mathbf{u}_i) \ , \tag{8}$$

where $\mathbf{u}_i$ with $i = 1, \cdots, n$ are vectors in the 2-D frequency space, and $\mathbf{s} = \mathbf{u}_1 + \mathbf{u}_2 + \cdots + \mathbf{u}_n$. It can be easily shown that Ψ_n is shift invariant [11].

When $n = 1$ in (8) we get the power spectrum and further with the value $n = 2$ the bispectrum, namely

$$\Psi_2(\mathbf{u}_1, \mathbf{u}_2) = F(\mathbf{u}_1)F(\mathbf{u}_2)F(\mathbf{u}_2 + \mathbf{u}_2) \ . \tag{9}$$

We are interested in the bispectrum as besides its invariance with respect to translation it does not loose any essential information about the original image. This means that it also retains the phase information in contrast to the power spectrum. For these reasons, it is possible to construct a blur invariant phase-only bispectrum where the exponentials containing the phase are raised to the second power, similar to (6), namely

$$\begin{aligned} P(\mathbf{u}_1, \mathbf{u}_2) &= e^{-i2\phi(\mathbf{u}_1)} e^{-i2\phi(\mathbf{u}_2)} e^{-i2\phi(\mathbf{u}_1+\mathbf{u}_2)} \\ &= e^{-i2[\phi(\mathbf{u}_1)+\phi(\mathbf{u}_2)+\phi(\mathbf{u}_1+\mathbf{u}_2)]} \ . \end{aligned} \qquad (10)$$

By looking the equation (9), we see that the bispectrum is a function of two vector arguments, containing totally four scalar variables. Assuming that $F(\mathbf{u})$ is an N-by-N discrete Fourier transform (DFT) of an image $f(\mathbf{x})$ the bispectrum becomes a four-dimensional N-by-N-by-N-by-N matrix. Fortunately, it is not necessary to evaluate the whole bispectrum. It is possible to take only 2-D slices of the original bispectrum, which contain basically the same information. There are various ways of defining the slices [12,13]. We define the slices as

$$S_k(\mathbf{u}) = \Psi_2(\mathbf{u}, k\mathbf{u}) \qquad \forall k \in \mathcal{R} \ . \qquad (11)$$

We then form blur invariants corresponding to (10) using only one slice (11) with value $k = 1$ of the whole bispectrum (9), namely

$$\begin{aligned} P'(\mathbf{u}) &= e^{-i2[\phi(\mathbf{u})+\phi(\mathbf{u})+\phi(\mathbf{u}+\mathbf{u})]} \\ &= e^{-i2[2\phi(\mathbf{u})+\phi(2\mathbf{u})]} \ . \end{aligned} \qquad (12)$$

Finally, we construct the invariants with respect to blur and translation used throughout this paper, similarly to (7), as

$$\mathcal{T}(\mathbf{u}) = 2[2\phi(\mathbf{u}) + \phi(2\mathbf{u})] \bmod 2\pi \ . \qquad (13)$$

These invariants are called phase blur-translation invariants (PBTIs).

It would also be possible to use more slices in building up the invariants. It can be seen from (9), (10) and (11) that we need frequency samples in points $\mathbf{u}$ but also at $k\mathbf{u}$ and $(k+1)\mathbf{u}$ to compute an arbitrary slice. If we assume that the images are discrete and that the spectrum is computed using DFT, the samples in the two latter cases can be extracted from DFT by utilizing its conjugate symmetry and periodicity. In our case, we only need frequency samples from points $\mathbf{u}$ and $2\mathbf{u}$.

If the DFT size is N-by-N we have approximately $N/2$ non-redundant invariants available at once. This is opposite to the MOMIs, of which computation time depends on the number of invariants used. However, the invariants corresponding to the lower frequencies have higher signal-to-noise ratio (SNR), and there is some optimal number of invariants that should be used depending on the noise level. In the experiments, we have used L invariants of (7) or (13) for which

$\sqrt{u_1^2+u_2^2} \le r$, when $\mathbf{u} = [u_1, u_2]$, but without using the conjugate symmetric or zero frequency components. Other possibility would be to weight the invariants according to SNR.

The distance between images $f(\mathbf{x})$ and $g(\mathbf{x})$ to be classified is computed as

$$D = \sqrt{\sum_{i=1}^{L} e_i^2} \ , \tag{14}$$

where

$$e_i = \min\left\{|I_i^{(f)} - I_i^{(g)}|, |I_i^{(f)} - I_i^{(g)}| - 2\pi\right\} \ . \tag{15}$$

Here, I_i, i = 1,...,L, are the invariants of (7) or (13).

It should be noted that the composition of our frequency domain invariants is quite robust to noise in contrast to the SPEIs of [4], which are based on the discontinuous tangent of the phase. Another useful property of our invariants, which results from the normalization of the amplitude information, is the invariance to uniform illumination changes.

As mentioned earlier, our invariants, as well as the moment invariants, are invariant to centrally symmetric blur. However, this is not exactly true for arbitrary images as the blur exchanges the information across the boundaries of the image causing some error. Also the translation invariance retains only if there is no information flow across the image boundaries.

3 Experiments

In the experiments, we classified blurred and noisy images using nearest neighbor classification and compared the results of our blur invariant features to the MOMIs and SPEIs, which were implemented as proposed in [4]. For all the frequency domain invariants $r=\sqrt{5}$ or $\sqrt{10}$, and the order of the MOMIs is up to 5 or 7. This results to either L=10 or 18 invariants.

In the first experiment, we compared the features invariant only to blur. In comparison we used the invariants based on central moments (MOMIs), which are also translation invariant, as they seemed to perform better compared to the invariants based on ordinary moments. As test images, we had 40 computer generated uniformly distributed noise images, which were filtered using a Gaussian low pass filter of size 10-by-10 with the standard deviation σ=1 to get an image, as in Figure 1(a), that resemples some natural texture. We picked one image at time, blurred and added noise to it, as in Figure 1(b), and tried to classify it as one of the original 40 image categories using the invariants. The blur was circular with a radius varying from 0 to 10 pixels with steps of 2 pixels. This kind of blur is a simple model of the out-of-focus effect found in many imaging systems [2]. The image size was cropped finally to 80-by-80 containing only the valid part of the convolutions. The experiment was repeated 20 times for each blur size and for each of the 40 images.

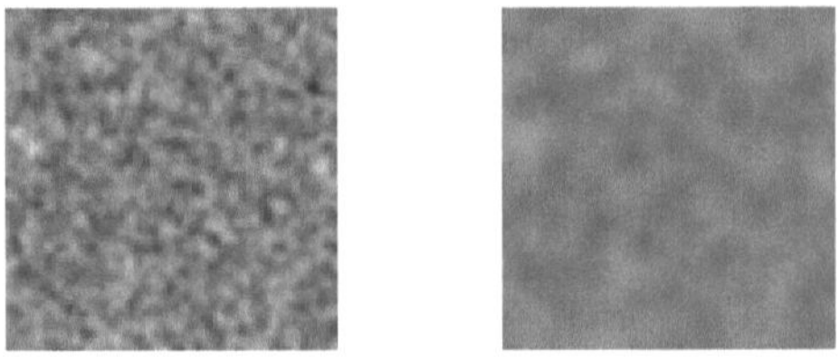

(a) (b)

Fig. 1. (a) An example of the 40 filtered noise images used in the first experiment, and (b) a degraded version of it (blur radius is 5 and PSNR 30 dB)

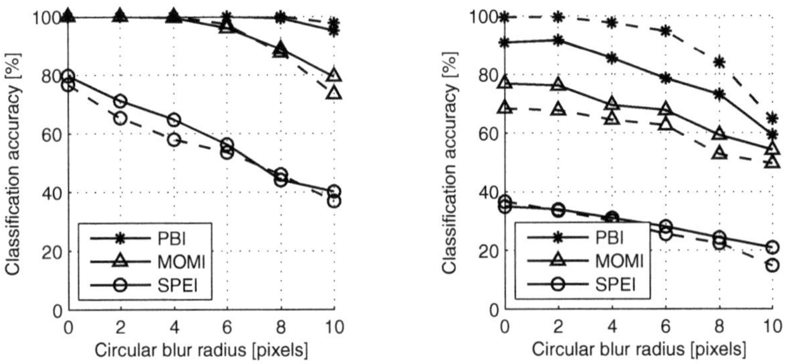

(a) (b)

Fig. 2. Classification accuracy of nearest neighbor classification of blurred and noisy images when PSNR is (a) 40 dB (b) 20 dB. The solid lines show the cases where 10 invariants are used and the dashed lines the case of 18 invariants.

The percentage of correct classification for the three methods, our PBIs, the MOMIs and the SPEIs, is shown in Figures 2(a) and 2(b) when the peak signal-to-noise ratio (PSNR) is 40 and 20 dB, respectively. The solid lines show the case of 10 invariants and the dashed lines the case in which 18 invariants are used.

In theory, all the methods are invariant to blur, but in practice, there are differences in their robustness to noise and boundary error. The results show that the SPEIs perform worst as was expected because of their unstable construction. If compared to the PBIs, also the MOMIs seem to be more sensitive at least to noise. The use of 18 invariants seems to give significantly better results compared to the use of 10 invariants only for the PBIs. In contrast, the result of the SPEIs and the MOMIs mainly degrades when 18 invariants are used. This means that the higher order MOMIs and the high frequency SPEIs are already too sensitive to perturbations in these conditions.

The second experiment mimics the situation in which there is an object to be classified on a uniform background. Here we compared the blur-translation invariants, the PBTIs and the MOMIs. We created 20 60-by-60 images, which

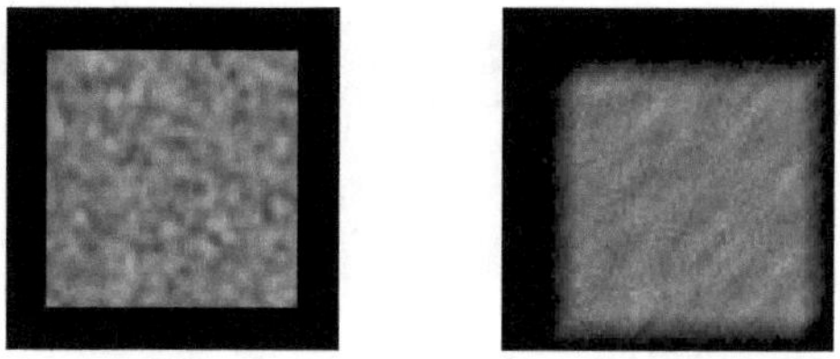

(a) (b)

Fig. 3. (a) An Example image used in the second experiment containing an artificial object (filtered noise). (b) A motion blurred and noisy version of the same image (blur length 10 and PSNR 30 dB).

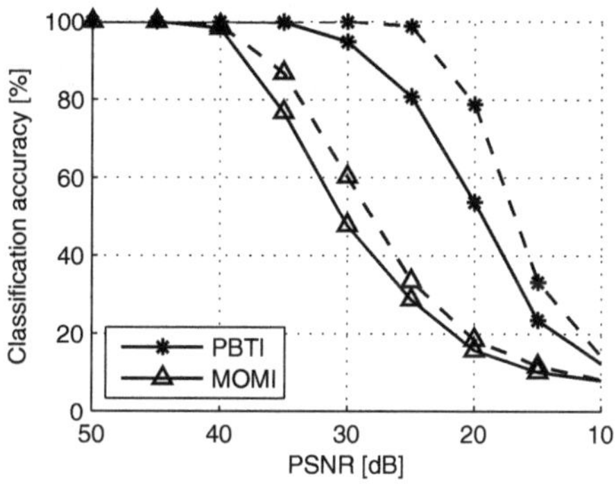

Fig. 4. The object classification accuracy of the blur-translation invariant methods. The solid lines show the cases where 10 invariants are used and the dashed lines the case of 18 invariants.

were similar to the filtered noise images of the first experiment, on a black 80-by-80 background depicting an artificial object, as shown in Figure 3(a). Distorted versions of these images were then generated, which were to be classified as one of the originals using each type of invariants and the nearest neighbor rule. The distortion included random displacement of the object in the range [-5,5] pixels using linear interpolation, motion blur with the length of 10 pixels in a random direction and additive Gaussian noise (also on the background) with PSNR varying from 50 dB to 10 dB in steps of 5 dB. Figure 3(b) shows a degraded object. The border effect does not distort the invariants now when the background is uniform.

Figure 4 presents the classification accuracy of the two methods as a function of PSNR, when the experiment is repeated 20 times for each noise level and for each test image. Similar to the previous experiment, the number of the invariants used is 10 and 18 shown by the solid and dashed lines, respectively. It is clear from the results that the MOMIs are affected much more by the noise. Especially, the noise in the background seems to be harmful. On the contrary, the PBTIs can handle also this situation and perform very well compared to the MOMIs.

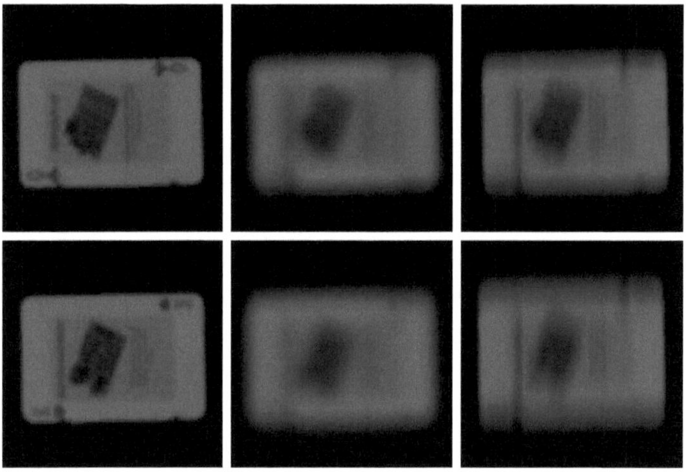

Fig. 5. Each row contains examples of one class (Queen of Clubs and King of Hearts, respectively) of the total of 10 classes of playing card images used in the third classification experiment. Cards were unusual and contained very similar figures. From left to right, there are a sharp, defocused and motion blurred image of each class.

In the final experiment, we wanted to confirm the results of the second experiment in practice. We classified real gray scale images of playing cards on the black background which were captured using a Sony DFW-X710 video camera. First, we formed 10 classes by photographing sharp images of 10 playing cards. Then, we captured three motion blurred and three out-of-focus blurred versions of these 10 cards resulting to 60 blurred card images that were to be classified as one of the 10 classes. The size of the images was 130-by-130. Motion blur was generated by panning the camera vertically in a tripod with a relatively long shutter time. Out-of-focus blur was created by defocusing the camera slightly. In Figure 5, there are sample images of two classes. In each row, from left to right, there is a sharp, defocused and motion blurred image of the same class. As can be seen, the blur is so strong that it is impossible to recognize the cards visually. Table 1 shows the classification accuracies for the two methods when 10 and 18 invariants are used. The best accuracy 95 % is achieved using 18 PBTIs. The number of correctly classified images is nearly double compared to the 18 MOMIs with the accuracy of 48 %. For example, PBTIs were able to classify all the blurred images in Figure 5, but MOMIs could only classify the blurred images on the top row were the level of blurring is lower. These results are in line with the results obtained using the artificial images.

Finally, we discuss about the computational demands of the invariants. Asymptotic complexity of the central moments, as well as the MOMIs which are build on them, is $O(N^2)$. This is less than the complexity of all types of the frequency domain invariants using radix-2 FFT, which is $O(N^2 \log_2 N)$. However, for practical image sizes the MOMIs require much more computation than the frequency domain invariants, as shown in Figure 6 where the approximate number of arithmetic

Table 1. Classification accuracies of the PBTIs and the MOMIs in the case of real degraded images of playing cards when 10 and 18 invariants are used

Method	PBTIs 18	PBTIs 10	MOMIs 18	MOMIs 10
Accuracy	95 %	73 %	48 %	45 %

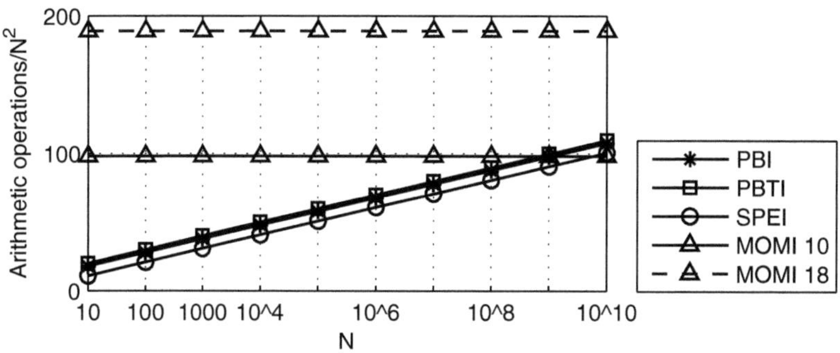

Fig. 6. The number of arithmetic operations needed to compute different types of invariants for an N-by-N image. For MOMIs the value depends on the number of invariants, which is 10 and 18.

operations per pixel is shown as the function of N when the image size is N-by-N. All operations, including complex operations of FFT and high exponents of the moments, are counted as one for simplicity. In addition, the computation load for the MOMIs also depends on the number of invariants used, while for the frequency domain methods $N/2$ non-redundant invariants are derived simultaneously. As can be seen, 10 MOMIs becomes preferable in terms of operation count at the image size 10^9-by-10^9. For 18 MOMIs this size is 10^{19}-by-10^{19}. Both of these image sizes are unrealistic. The explanation for this property is the very large number of geometric moments that have to be computed to get the corresponding MOMI features.

4 Conclusions

In this paper, we have shown how it is possible to derive new blur invariant features, PBIs and PBTIs, based on even powers of the phase-only spectrum and bispectrum of the images, respectively. The features are invariant to centrally symmetric blur, and the PBTI features based on the bispectrum are also invariant to translation. The features can be used to recognize blurred images or objects as demonstrated in the paper. Because the features are based on the normalized phase-only spectrum or bispectrum, they are also invariant to linear brightness changes. In addition, they can be computed efficiently using FFT.

We compared the PBIs and the PBTIs with the MOMI and SPEI features proposed in [4] for classification of blurred images and objects on an uniform background. In both cases, our features performed better. In practice, our features need also less computation, if we do not take into account the SPEIs which are very sensitive to noise.

A shortcoming of our features is that they do not carry further invariance to geometrical transformations such as rotation and scaling. On the other hand, there is no other frequency domain blur invariant method available that would carry the invariance to translation.

References

1. Wood, J.: Invariant pattern recognition: A review. Pattern Recognition 29, 1–17 (1996)
2. Banham, M.R., Katsaggelos, A.K.: Digital image restoration. IEEE Signal Processing Magazine 14(2), 24–41 (1997)
3. Kundur, D., Hatzinakos, D.: Blind image deconvolution. IEEE Signal Processing Magazine 13(3), 43–64 (1996)
4. Flusser, J., Suk, T.: Degraded image analysis: An invariant approach. IEEE Trans. Pattern Anal. Machine Intell. 20, 590–603 (1998)
5. Flusser, J., Suk, T., Saic, S.: Recognition of blurred images by the method of moments. IEEE Transactions on Image Processing 5(3), 533–538 (1996)
6. Bentoutou, Y., Taleb, N., Mezouar, M.C.E., Taleb, M., Jetto, L.: An invariant approach for image registration in digital subtraction angiography. Pattern Recognition 35, 2853–2865 (2002)
7. Flusser, J.: Combined invariants to linear filtering and rotation. Int. J. Pattern Recognition and Artificial Intelligence 13(8), 1123–1136 (1999)
8. Zhang, Y., Wen, C., Zhang, Y., Soh, Y.C.: Determination of blur and affine combined invariants by normalization. Pattern recognition 35(1), 211–221 (2002)
9. Suk, T., Flusser, J.: Combined blur and affine moment invariants and their use in pattern recognition. Pattern Recognition 26(12), 2895–2907 (2003)
10. Ojansivu, V.: Motion blur concealment of digital video using invariant features. In: Blanc-Talon, J., Philips, W., Popescu, D., Scheunders, P. (eds.) ACIVS 2006. LNCS, vol. 4179, Springer, Heidelberg (2006)
11. Chandran, V., Carswell, B., Boashash, B., Elgar, S.: Pattern recognition using invariants defined from higher order spectra: 2-d image inputs. IEEE Transactions on Image Processing 6(5), 703–712 (1997)
12. Dianat, S.A., Rao, R.M.: Fast algorithms for phase and magnitude reconstruction from bispectra. Optical Engineering 29(5), 504–512 (1990)
13. Petropulu, A.P., Pozidis, H.: Phase reconstruction from bispectrum slices. IEEE Transactions on Signal Processing 46(2), 527–530 (1998)

Regularized Neighborhood Component Analysis*

Zhirong Yang and Jorma Laaksonen

Laboratory of Computer and Information Science
Helsinki University of Technology
P.O. Box 5400, FI-02015 TKK, Espoo, Finland
{zhirong.yang,jorma.laaksonen}@tkk.fi

Abstract. Discriminative feature extraction is one of the fundamental problems in pattern recognition and signal processing. It was recently proposed that maximizing the class prediction by neighboring samples in the transformed space is an effective objective for learning a low-dimensional linear embedding of labeled data. The associated methods, Neighborhood Component Analysis (NCA) and Relevant Component Analysis (RCA), have been proven to be useful preprocessing techniques for discriminative information visualization and classification. We point out here that NCA and RCA are prone to overfitting and therefore regularization is required. NCA and RCA's failure for high-dimensional data is demonstrated in this paper by experiments in facial image processing. We also propose to incorporate a Gaussian prior into the NCA objective and obtain the Regularized Neighborhood Component Analysis (RNCA). The empirical results show that the generalization can be significantly enhanced by using the proposed regularization method.

1 Introduction

Discriminant Analysis (DA) is one of the central problems in pattern recognition and signal processing. The supervised training data set consists of pairs $(\mathbf{x}_j, c_j)$, $j = 1, \ldots, n$, where $\mathbf{x}_j \in \mathbb{R}^m$ is the primary data, and the auxiliary data c_j takes categorical values. DA seeks for a transformation $\mathbf{f} : \mathbb{R}^m \rightarrow \mathbb{R}^r$ (usually $r \ll m$) such that $\mathbf{y}_j = \mathbf{f}(\mathbf{x}_j)$ encodes only the relevant information with respect to c_j. Here relevance or discrimination can be measured by the expectation of predictive probability $E\{p(c|\mathbf{y})\}$. Because in real applications only finite data are available, one has to estimate $p(c|\mathbf{y})$ according to a certain density model, which leads to different DA algorithms.

Fisher's *Linear Discriminant Analysis* (LDA) [3] is a classical method for this task. LDA maximizes the trace quotient of between-class scatter against within-class scatter and can be solved by *Singular Value Decomposition* (SVD). LDA is attractive for its simplicity. Nevertheless, each class in LDA is modeled by a single Gaussian distribution and all classes share a same covariance, which

* Supported by the Academy of Finland in the projects *Neural methods in information retrieval based on automatic content analysis and relevance feedback* and *Finnish Centre of Excellence in Adaptive Informatics Research.*

B.K. Ersbøll and K.S. Pedersen (Eds.): SCIA 2007, LNCS 4522, pp. 253–262, 2007.

restricts the applicability of LDA. For example, the output dimensions of LDA cannot be more than the number of classes minus one. Furthermore, LDA is prone to overfitting because no complexity control is involved.

Recently Goldberger et al. [6] presented an algorithm that learns a Mahalanobis distance measure. Their method, called *Neighborhood Component Analysis* (NCA), maximizes the approximated discrimination $\sum_j p(c_j|\mathbf{y}_j)$. NCA is able to learn a transformation matrix such that the *Nearest-Neighbor* (NN) classifier performs well in the linear transformed space. Peltonen and Kaski [8] proposed a tightly connected technique, *Relevant Component Analysis* (RCA), where the objective is to maximize $\sum_j \log p(c_j|\mathbf{y}_j)$ and the transformation matrix is constrained to be orthonormal. Both NCA and RCA can handle complicated class distributions and output arbitrary number of dimensions. They have been applied to several data sets, where both methods seem to generalize well [6,8].

However, we argue that these two methods are not free of overfitting in very high-dimensional spaces and complexity control is therefore required for the transformation matrix. To improve the generalization, we propose to incorporate a Gaussian prior to the NCA objective and obtain a novel method called *Regularized Neighborhood Component Analysis* (RNCA). In this paper, several empirical examples in facial image processing are presented, where the dimensionality of data is much higher than those used in [6,8]. It turns out that both NCA and RCA behave poorly in our generalization tests, but the overfitting problems can be significantly overcome by using RNCA.

The remaining of the paper is organized as follows. We briefly review the principles of NCA and RCA in Section 2. Next we describe the motivation of regularizing the transformation matrix and present the Regularized NCA in Section 3. In Section 4 we demonstrate the experimental results, both qualitative and quantitative, on facial image processing. Finally the conclusions are drawn in Section 5.

2 Neighborhood Component Analysis

Neighborhood Component Analysis (NCA) [6] learns an $r \times m$ matrix $\mathbf{A}$ by which the primary data $\mathbf{x}_i$ are transformed into a lower-dimensional space. The objective is to maximize the *Leave-One-Out* (LOO) performance of nearest neighbor classification. NCA measures the performance based on "soft" neighbor assignments in the transformed space:

$$p_{ij} = \frac{\exp\left(-\|\mathbf{A}\mathbf{x}_i - \mathbf{A}\mathbf{x}_j\|^2\right)}{\sum_{k \neq i} \exp\left(-\|\mathbf{A}\mathbf{x}_i - \mathbf{A}\mathbf{x}_k\|^2\right)}, \quad p_{ii} = 0. \tag{1}$$

Denote n the number of samples and

$$p_i = \sum_{j: c_i = c_j} p_{ij}. \tag{2}$$

The objective of NCA can then be expressed as maximization of

$$\mathcal{J}_{\text{NCA}}(\mathbf{A}) = \sum_{i=1}^{n} \sum_{j:c_i=c_j} p_{ij} = \sum_{i=1}^{n} p_i. \tag{3}$$

The NCA learning algorithm is based on the gradient

$$\frac{\partial \mathcal{J}_{\text{NCA}}(\mathbf{A})}{\partial \mathbf{A}} = 2\mathbf{A} \sum_{i=1}^{n} \left(p_i \sum_{k=1}^{n} p_{ik} \mathbf{x}_{ik} \mathbf{x}_{ik}^T - \sum_{j:c_i=c_j} p_{ij} \mathbf{x}_{ij} \mathbf{x}_{ij}^T \right) \tag{4}$$

and employs an optimizer such as conjugate gradients.

A method closely related to NCA is the *Relevant Component Analysis* (RCA) proposed by Peltonen and Kaski [8]. RCA maximizes the sum of log-probability, i.e. the discriminative information

$$\mathcal{J}_{\text{RCA}}(\mathbf{W}) = \sum_{i=1}^{n} \log \left(\sum_{j:c_i=c_j} p_{ij}^{\text{RCA}} \right) = \sum_{i=1}^{n} \log p_i^{\text{RCA}}. \tag{5}$$

Different from NCA, the transformation matrix $\mathbf{W}$ in RCA is restricted to be orthogonal. That is, the rows of $\mathbf{W}$ form an orthonormal basis. Due to this constraint, RCA requires an additional parameter $\beta > 0$ to control the smoothness of the "soft" assignment:

$$p_{ij}^{\text{RCA}} = \frac{\exp\left(-\beta \|\mathbf{W}\mathbf{x}_i - \mathbf{W}\mathbf{x}_j\|^2\right)}{\sum_{k \neq i} \exp\left(-\beta \|\mathbf{W}\mathbf{x}_i - \mathbf{W}\mathbf{x}_k\|^2\right)}, \quad p_{ii}^{\text{RCA}} = 0. \tag{6}$$

The optimization of RCA is based on a stochastic gradient of (5) with respect to the parameters in Givens rotation [7] of $\mathbf{W}$.

3 Regularized NCA

Goldberger et al. claimed that they had never observed an "overtraining" effect when using NCA [6]. In the presented empirical results [6,8], it seems that NCA and RCA generalize well on several different data sets. On the other hand, the objectives of these two methods do not involve any explicit complexity control on the matrix elements. One is therefore tempted to ask whether they really are free of overlearning.

NCA and RCA do have some implicit complexity control mechanisms on the "soft" assignments or density estimation in order to avoid overfitting. The formulation of p_{ij} leaves out the i-th training sample, which improves the generalization of NCA (although it might cause divide-by-zero errors for outliers). The β parameter in RCA also controls the smoothness of the predictive probability estimation. Smaller β usually leads to coarser estimation but better generalization. All in all, these two strategies seem to work well for low-dimensional data sets.

However, as shown in the next section by experiments on high-dimensional facial images, NCA and RCA behave poorly beyond the training sets. Actually, given a training data set, the number of parameters to be learned by a DA algorithm is r times of that in one-dimensional projection methods such as the Support Vector Machines (SVMs) [2]. That is, the DA problem is more ill-posed and regularization is hence necessary. This motivates us to impose regularization on DA algorithms.

For many parameterized machine learning algorithms, techniques of regularization [5,1] have demonstrated great success for improving the generalization performance in ill-posed problems. The overlearning effect caused by a small training data set and a large number of parameters to be learned can be significantly reduced by attaching a penalty term behind the original objective. Viewed from Bayesian inference, regularization actually incorporates a certain prior on the parameters themselves, and the learning maximizes the posterior

$$p(\mathbf{A}|\{\mathbf{x}_j, c_j\}) \propto p(\{\mathbf{x}_j, c_j\}|\mathbf{A})p(\mathbf{A}). \tag{7}$$

We choose the decomposable Gaussian prior

$$p(\mathbf{A}) = \prod_{k=1}^{r}\prod_{l=1}^{m} p(A_{kl}) = \prod_{k=1}^{r}\prod_{l=1}^{m} \frac{\sqrt{\lambda}}{\pi} \exp(-\lambda A_{kl}^2) \tag{8}$$

for regularization because the first-order derivative of its logarithm is simple and exists throughout the parameter sapce, which facilitates gradient-based optimization. If we model

$$p(\{\mathbf{x}_j, c_j\}|\mathbf{A}) \propto \exp\left(\sum_{i=1}^{n} p_i\right) \tag{9}$$

the regularized NCA objective becomes:

$$\begin{aligned} \text{Maximize} \quad \mathcal{J}_{\text{RNCA}}(\mathbf{A}) &= \log p(\mathbf{A}|\{\mathbf{x}_j, c_j\}) && (10) \\ &= \log p(\{\mathbf{x}_j, c_j\}|\mathbf{A}) + \log p(\mathbf{A}) + \text{constant} && (11) \\ &= \sum_{i=1}^{n} p_i - \lambda \|\mathbf{A}\|_F^2 + \text{constant}. && (12) \end{aligned}$$

Here λ acts as a non-negative trade-off parameter and $\|\mathbf{A}\|_F^2$ notates the Frobenius matrix norm $\sum_{k=1}^{r}\sum_{l=1}^{m} A_{kl}^2$. For one-dimensional subspace analysis, $y = \mathbf{a}^T\mathbf{x}$, the matrix norm reduces to $\mathbf{a}^T\mathbf{a}$, i.e. the large margin regularizer used in SVMs. We call the new algorithm *Regularized Neighborhood Component Analysis* (RNCA). It is equivalent to NCA when $\lambda = 0$. Note that the Gaussian prior is not applicable to RCA optimization because the Frobenius norm of an orthonormal matrix is a constant r. The estimation of a suitable λ in (12) is a further problem that could presumably be solved by Bayesian methods. In this paper, we simply try different values for λ empirically.

There exist other priors, e.g. the Laplacian prior [11], on linear transformation parameters. Some of them are reported to produce better results for particular learning problems, but here it is difficult to obtain convenient optimization algorithms when combining these priors with the NCA or RCA objective.

4 Experiments

4.1 Data

Several empirical results of NCA and RCA have been provided in [6,8]. However, the data used in these experiments have been low-dimensional relative to the number of training samples. The highest dimensionality is 560 in [6] and 76 in [8]. In addition, both training and testing data in their visualization and classification experiments have been selected from the same database. Here we present the learning results of NCA, RCA and RNCA on much higher-dimensional data. Furthermore, we performed the tests on a database obtained from another source, which will certainly demonstrate the generalization powers of the compared methods better.

We have used the FERET database of facial images [9] as the training data set. After face segmentation, 2,409 frontal images (poses “fa” and “fb”) of 867 subjects were stored in the database for the experiments. We obtained the coordinates of the eyes from the ground truth data of FERET collection, with which we calibrated the head rotation so that all faces are upright. Afterwards, all face boxes were normalized to the size of 32×32, with fixed locations for the left eye (26,9) and the right eye (7,9). Each image were reshaped to a 1024-dimensional vector by column-wise concatenation. The testing data set we used is the UND database (collection B) [4], which contains 33,247 frontal facial images of 491 subjects. We applied the same preprocessing procedure to the UND images as to the FERET database.

We compared the DA methods in two problems where the extracted features were used to discriminate the *gender* of a subject and whether she or he is wearing *glasses*. Table 1 shows the statistics of the classes.

Table 1. Images (subjects) of the experimented classes

	gender		*glasses*	
	Male	Female	Yes	No
FERET	1495 (501)	914 (366)	262 (126)	2147 (834)
UND	2524 (63)	855 (19)	2601 (149)	30538 (482)

4.2 Visualizing the Transformation Matrix

It is intuitive to inspect the elements in the trained transformation matrix **A** before performing quantitative evaluation. Each row of the transformation matrix acts as a linear filter and can be displayed like a *filter image*. If the transformation matrix works well for a given DA problem, it is expected to find some

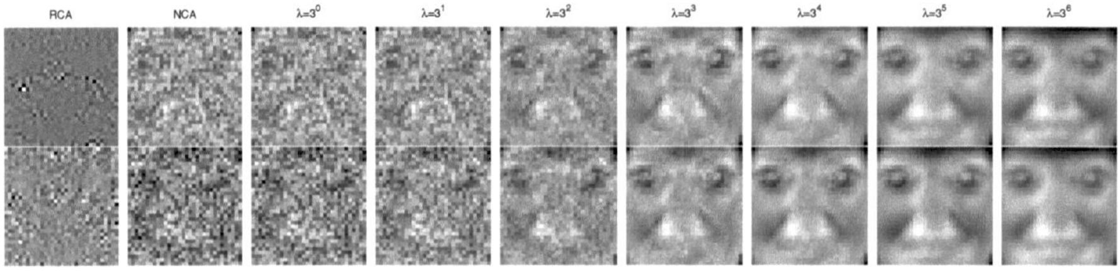

(a) gender

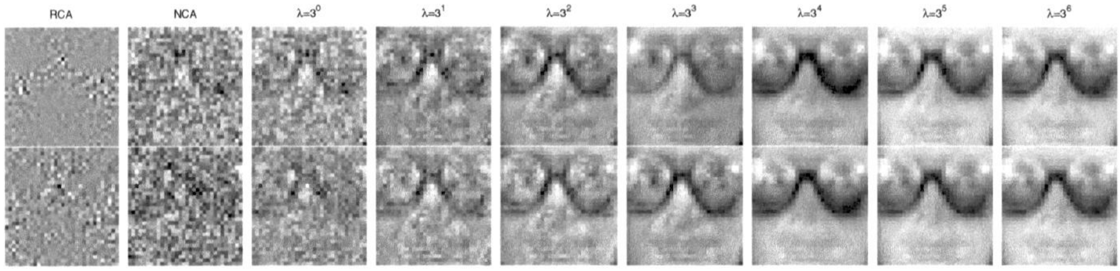

(b) glasses

Fig. 1. The rows of transformation matrices are plotted as filter images

semantic connections between the filter images and our common prior knowledge of the considered classes.

Figure 1 shows the resulting filter images for the *glasses* and *gender* DA problems with a two-dimensional transformed space, i.e. $r = 2$. In both cases, RCA has stuck in a local optimum, where some pixel pairs of high contrast can be found in the plotted images. These pixel pairs look like Gabor wavelets, but they are too small to represent any semantically relevant visual patterns of gender or glasses. Such overfitting effect could be relieved by reducing the image size, as employed in [6], but much useful visual information would be lost during downsampling. The filter images trained by NCA are composed of nearly random pixels, from which it is difficult to perceive any visual patterns.

NCA is a special case of RNCA with $\lambda = 0$ in (12). Next we increased λ in the power scale of three and run RNCA with these different values of λ. The results are shown in the third to ninth columns in Figure 1. We can see that the facial patterns become clearer with higher λ values. However, too large λ's cause underfitting—all filter vectors lie in the straight line passing the two class means and thus the filter images look identical. A proper tradeoff value for λ should therefore occur in between. In the following experiments we chose to use $\lambda = 3^3$ for the *gender* case and $\lambda = 3^2$ for *glasses*. Careful readers can in these cases perceive the small differences between the displayed filter images.

4.3 Visualizing the Transformed Distributions

NCA, as well as RCA, is able to extract more than one discriminative component for two-class DA problems, which allows plotting the 2-D transformed data. RNCA inherits the same property from NCA and can hence also be used for visualization. In this section we illustrate a qualitative comparison of NCA (3),

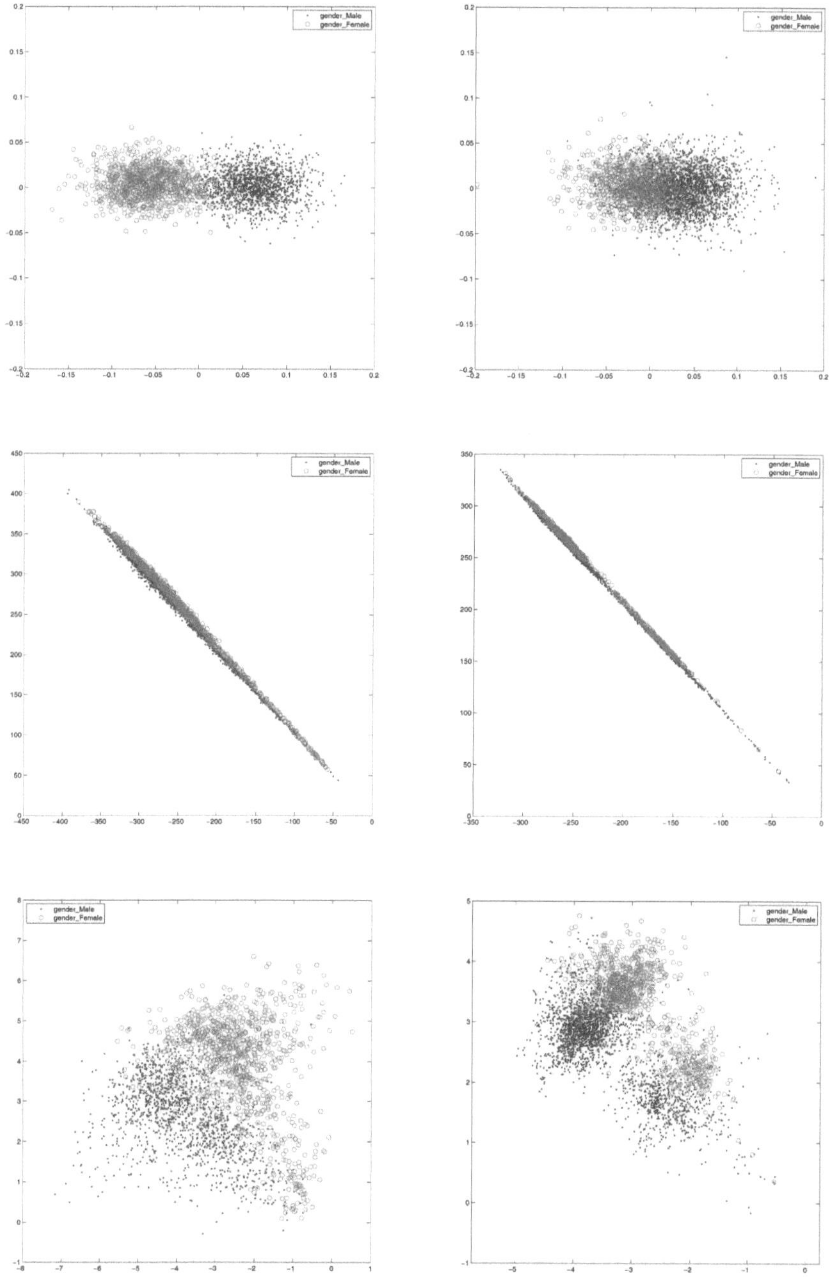

Fig. 2. 2-D transformed *gender* training data (left) and testing data (right). Rows from top to bottom are results of RCA, NCA and RNCA ($\lambda = 3^3$), respectively.

Table 2. Numbers of classification errors (false positives, false negatives) using $r = 2$ components

	gender		*glasses*	
	training	testing	training	testing
RCA	46, 49	380, 376	2, 2	2140, 2101
NCA	84, 94	249, 248	15, 1	2113, 2141
RNCA	124, 121	201, 211	64, 67	2002, 2051

RCA (5) and RNCA (12). Due to the space limit, only the plots of *gender* are displayed. Similar results can be obtained for the *glasses* case.

The training and testing results are shown in Figure 2. RCA starts from an orthonormalized LDA result and tries to improve it [8], but in this case the initial solution already separates data well and hence RCA does only little change. On the other hand, it can be seen in the right column that the *gender* classes are heavily mixed for the testing data.

NCA learns a transformed space in which both training data and testing data mainly distribute around a straight line, representing the *boundary direction*. The two classes are slightly separated in a *discriminative direction* nearly orthogonal to the boundary for the training data. However, such discrimination can barely be seen from the transformed testing data, where the two classes are heavily overlapped. Moreover, the 2-D Euclidean metric would not perform properly in the transformed space because of the presence of a dominant direction.

By contrast, one can easily see the almost separated classes in both training and testing cases with RNCA. Although the separation is not as clear as that of RCA for the training data, it is relatively better preserved for the testing data. That is, the overfitting effect is much alleviated by employing our regularization technique. The neighborhood based on the 2-D Euclidean metric should be more suitable for the RNCA results because the scales in the boundary and discriminative directions have become comparable.

4.4 Classification Results

One of the most important applications of discriminative features is for classification. The classification results therefore serve as a quantitative comparison measure of different DA methods. Following the conventional terms in binary classification, we specify the prediction of *gender_Male* and *glasses_Yes* as positive and their counterparts as negative.

Table 2 shows the Nearest-Neighbor classification error counts when using the DA results with $r = 2$. A pair of numbers are shown in each table entry, the first for false positives and the second for false negatives. Although RNCA is not as good as RCA and NCA in classifying the training data, it performs best for the testing data. This conforms to the qualitative results demonstrated in the previous section.

The classification accuracy of RNCA can be further improved by increasing the number of components. Figure 3 illustrates the classification error rates on

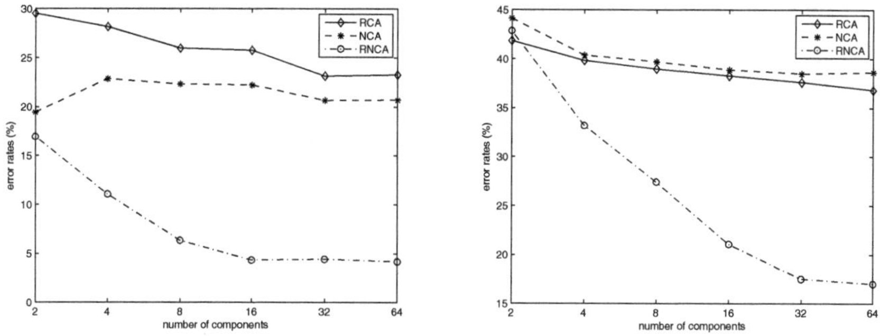

Fig. 3. Nearest-Neighbor classification error rates of the compared methods for *gender* (left) and *glasses* (right) with different numbers of r components

the testing data with different values of r. Here we have taken the average of the false positive and false negative error rates. RNCA achieves its best performance for classifying *gender* when $r \geq 16$. More components for the *glasses* case are required because there exist more different eyeglasses styles in the UND database than those in FERET [4,9]. By contrast, RCA and NCA benefit only little from the additional components. Although high dimensionality of the transformed subspace brings more expressive power, RCA and NCA suffer from severe overfitting without regularization on the additional parameters.

The nearest neighbor classifier based on the RNCA results can even outperform the well-known Support Vector Machines (SVMs) [2]. The best classification accuracies we obtained with RNCA+NN are 95.3% for *gender* and 82.5% for *glasses*, while SVMs with linear kernel achieve only 90.8% and 78.2%, respectively. All the DA methods discussed in this paper, as well as SVM, can be generalized to non-linear cases by adopting the kernel trick [10]. However, more efforts are required to tune the additional parameter involved in the kernel, which is beyond the scope of this paper.

5 Conclusions

Two existing discriminant analysis methods, NCA and RCA, have gained success with low-dimensional data. In this paper we have pointed out that they are prone to overfitting with high-dimensional facial image data. We also proposed regularizing the neighborhood component analysis by imposing a Gaussian prior on the transformation matrix. Experimental results confirm our statement and show that the Regularized NCA becomes more robust in extracting discriminative features.

Moreover, we demonstrated that more than one component exists for two-class discriminant analysis problems. Unlike SVM and other algorithms dedicated for classification, our RNCA method can be applied to many other applications, for instance, preprocessing of discriminative feature visualization and creation of Discriminative Self-Organizing Maps [12].

Similar to other linear subspace methods, RNCA is readily extendable to non-linear versions. The nonlinear discriminative components can be obtained by mapping the primary data to a higher-dimensional space with appropriate kernels. This will be a topic of our future work.

References

1. Chen, Z., Haykin, S.: On different facets of regularization theory. Neural Computation 14, 2791–2846 (2002)
2. Cristianini, N., Shawe-Taylor, J.: An Introduction to Support Vector Machines. Cambridge University Press, Cambridge (2000)
3. Fisher, R.A.: The use of multiple measurements in taxonomic problems. Annals of Eugenics vol. 7 (1963)
4. Flynn, P.J., Bowyer, K.W., Phillips, P.J.: Assessment of time dependency in face recognition: An initial study. Audio- and Video-Based Biometric Person Authentication, pp. 44–51 (2003)
5. Girosi, F., Jones, M., Poggio, T.: Regularization theory and neural networks architectures. Neural Computation 7(2), 219–269 (1995)
6. Goldberger, J., Roweis, S.T., Hinton, G.E., Salakhutdinov, R.: Neighbourhood components analysis. In: NIPS 2004 (2004)
7. Golub, G.H., van Loan, C.F.: Gene H, 2nd edn. The Johns Hopkins University Press, Baltimore (1989)
8. Peltonen, J., Kaski, S.: Discriminative components of data. IEEE Transactions on Neural Networks 16(1), 68–83 (2005)
9. Phillips, P.J., Moon, H., Rizvi, S.A., Rauss, P.J.: The FERET evaluation methodology for face recognition algorithms. IEEE Trans. Pattern Analysis and Machine Intelligence 22, 1090–1104 (2000)
10. Schölkopf, B., Smola, A.: Learning with Kernels. MIT Press, Cambridge, MA (2002)
11. Tibshirani, R.: Regression shrinkage and selection via the lasso. J. Royal. Statist. Soc. B. 58(1), 267–288 (1996)
12. Yang, Z., Laaksonen, J.: Zhirong Yang and Jorma Laaksonen. In: Singh, S., Singh, M., Apte, C., Perner, P. (eds.) ICAPR 2005. LNCS, vol. 3687, pp. 216–225. Springer, Heidelberg (2005)

Finding the Minimum-Cost Path Without Cutting Corners

R. Joop van Heekeren, Frank G.A. Faas, and Lucas J. van Vliet

Quantitative Imaging Group, Delft University of Technology, The Netherlands
L.J.vanVliet@tudelft.nl

Abstract. Applying a minimum-cost path algorithm to find the path through the bottom of a curvilinear valley yields a biased path through the inside of a corner. DNA molecules, blood vessels, and neurite tracks are examples of string-like (network) structures, whose minimum-cost path is cutting through corners and is less flexible than the underlying centerline. Hence, the path is too short and its shape too stiff, which hampers quantitative analysis. We developed a method which solves this problem and results in a path whose distance to the true centerline is more than an order of magnitude smaller in areas of high curvature. We first compute an initial path. The principle behind our iterative algorithm is to deform the image space, using the current path in such a way that curved string-like objects are straightened before calculating a new path. A damping term in the deformation is needed to guarantee convergence of the method.

1 Introduction

Algorithms for computing the minimum-cost path have played an important role in various fields of science and engineering. These algorithms try to find the path connecting a selected start and end point that minimizes the integrated costs. In optics, light rays travel along a minimum-cost path from source to destination. A wave front of light propagates with a speed that depends inversely proportional on the refractive index of the medium. A space varying velocity map suggests that the path with the shortest arrival time will in general be longer than the Euclidean distance between the start and end points. If you consider the local cost as the inverse of the local speed, then calculating the minimum integrated cost corresponds to calculating the smallest possible arrival time from a start point to all points in the domain.

In many fields of science and engineering we encounter images of string-like structures in which the centerline conveys important information about the underlying objects. Examples are DNA-strands (cf. Fig. 1), blood vessels, or neurite tracks. The tracking results as depicted in [5] display exactly the problems that we are addressing in this paper. The minimum-cost path does not follow the curvilinear valley of the cost function, but is biased towards the inside of corners. In general, the minimum-cost path is cutting corners, and is therefore shorter and stiffer than the underlying centerline of the cost valley. Quantification of the

B.K. Ersbøll and K.S. Pedersen (Eds.): SCIA 2007, LNCS 4522, pp. 263–272, 2007.

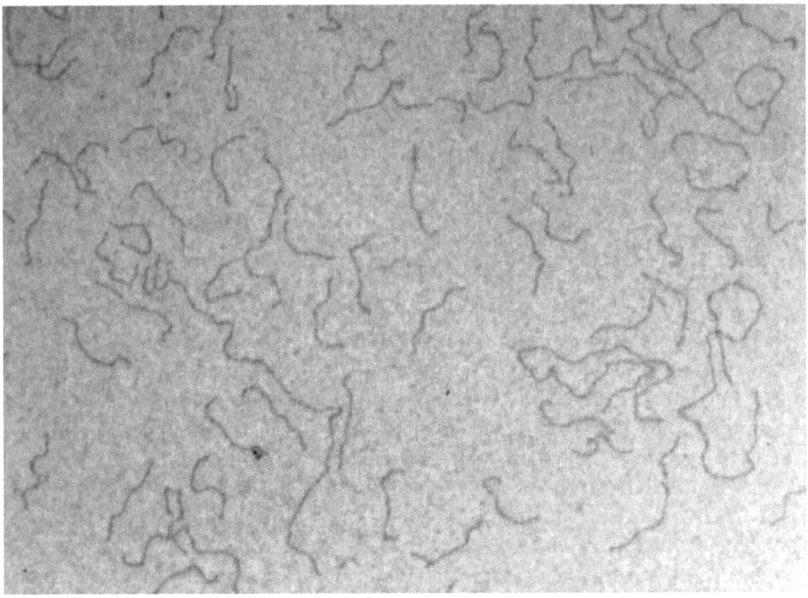

Fig. 1. DNA molecules labeled with uranyl acetate and visualized by transmission electron microscopy. The images are kindly provided by Dr. Dmitry Cherny, PhD, Dr. Sc.

bending energy of DNA plays a key role in understanding cellular processes. To verify the competing models and measure the so-called persistence length, an accurate path through these structures is required. The key to quantifying the length and shape of these objects is to find the centerline of these objects. A minimum-cost path guarantees a connected path that approximates this centerline even if the curvilinear object contains small gaps and is corrupted by noise (cf. Fig. 1). Solving the bias problem of minimum-cost path algorithms will be of utmost importance in many fields of science and engineering.
A typical implementation of such a standard minimum-cost path approach consists of the following steps:

- Define one start point in the image domain. Having an end point is not mandatory, but may assist in defining an early stopping criterion.
- Compute the minimum integrated cost from the starting point to all points in the domain, or until the pre-defined end point has been reached.
- Descend along the opposite gradient direction of the integrated cost image from the end point to the start point. Due to the smoothness of the integrated cost images one can obtain sub-pixel accuracy in the location of the minimum-cost path.

The minimum integrated cost T is given as the minimum cumulative cost along all possible paths $\boldsymbol{P}$ connecting the start point S with any end point E in our domain. Or mathematically:

$$T = \min_{\forall \boldsymbol{P}_{SE}} \int_S^E I(\boldsymbol{P}_{SE}(l))dl \tag{1}$$

with $\boldsymbol{P}_{SE}(l) = (x(l), y(l))$. This is equivalent to solving the Eikonal equation [2]

$$|\nabla u(x,y)| = I(x,y) \tag{2}$$

in which $I(x,y)$ denotes the local cost function and $u(x,y)$ the local arrival time. For uniform costs, the solution of the Eikonal equation is identical to the result given by the (domain constrained) Euclidean distance transform. For space variant costs we have the gray-weighted distance transform (GDT) [9,7,6,4] and fast marching (FM) algorithms [1,3,8]. Both methods are based on wave front propagation. The GDT constructs a path using a superposition of cost-weighted basis vectors, thereby quantizing the local path direction to the directions of a set of basis vectors in a 3x3 (or 5x5) neighborhood. The FM algorithm models the wave front by a straight front, which does not restrict the propagation direction to a limited set of discrete directions. Both methods produce an image containing the minimum integrated cost from a starting point to all points in the domain.

The minimum-cost path can be obtained by a steepest descend (from the end point back to the start point) along the opposite gradient direction of the integrated cost map created by the aforementioned methods. Since the cost function is usually smooth, the integrated cost function is even smoother. This permits sub-pixel accuracy in computing the steps taken during the steepest descend. Due to the finite step size and approximation errors in the aforementioned algorithms, the path will not end exactly at the starting point but in very close (sub-pixel) proximity.

In section 2 we quantify the cutting corner problem for circular arcs with a Gaussian line profile and present our iterative algorithm to solve it. In section 3 we present quantitative results on the displacement error as a function of the number of iterations and qualitative results on TEM images of uranyl acetate labeled DNA. Section 4 presents the conclusions of our work.

2 Method

A correct implementation of a minimum-cost path algorithm applied to curved linear structures will always result in a path that is shorter and stiffer (less bending energy) than the centerline of the underlying linear structure. Especially in highly curved areas the minimum-cost path is cutting corners. The minimum-cost path does not follow the path through the minimum of the cost function in curved areas. To illustrate this we consider a circular path with a Gaussian cross-section

$$I(r, \sigma, c) = 1 + c(1 - exp(-\frac{(r - R_c)^2}{2\sigma^2})) \tag{3}$$

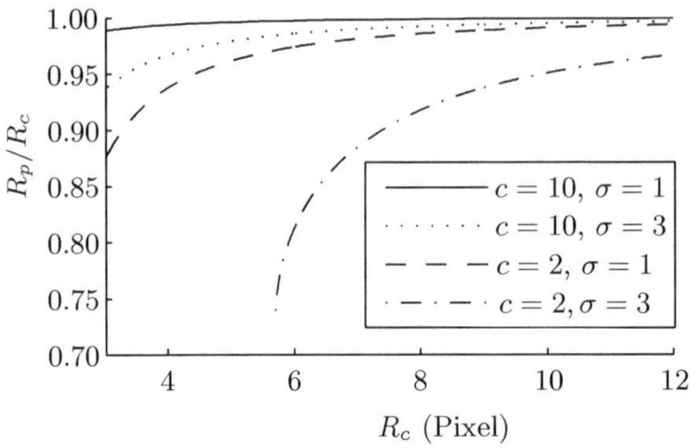

Fig. 2. The relative radius (R_p/R_c) of the minimum-cost path for the Gaussian profile as a function of centerline radius R_c

with the cost contrast $c = \frac{c_b - c_p}{c_p}$, in which c_b and c_p are respectively the cost values of the background and the path, and R_c the radius of curvature of the centerline. The integrated cost of a circular path with radius r around such a circle is

$$T(r) = 2\pi r\, I(r, \sigma, c) = 2\pi r\, (1 + c\,(1 - exp(\frac{-(r - R_c)^2}{2\sigma^2}))) \tag{4}$$

To find the minimum-cost path, we calculated the radius R_p for which $T(r)$ is minimized, $R_p = argmin\, T(r)$. Fig. 2 shows the relative radius R_p/R_c of the minimum-cost paths for different values of line width σ and contrast c. The results suggest that increasing the contrast or decreasing the line width (for example by scaling the cost function: $I(\boldsymbol{r}) \rightarrow I^{\alpha}(\boldsymbol{r}), \alpha > 1)$) of the cost function will reduce the bias in the minimum-cost path. In practice the bias will be reduced by these measures to some extent, but will never produce the desired smooth centerline path. This is shown by considering the limit ($c \rightarrow \infty$ or $\alpha \rightarrow \infty$), this will reduce the problem to a binary problem discarding all the gray value information and therefore produce a rough, binary skeleton type path instead of smooth centerline path. This skeleton path will also be hampered and possibly even interrupted due to the presence of noise in the original image. We claim to have developed an algorithm not based on increasing the contrast or decreasing the line width which solves the bias problem and still finds a smooth path, approximating the true centerline, through this class of objects.

Algorithm

Our method is based on the idea that a standard minimum-cost path algorithm such as FM will only give the correct centerline path for straight string-like objects (assuming the start and end points are located on the centerline). Hence, the principle behind our algorithm is to deform the image space in such a way

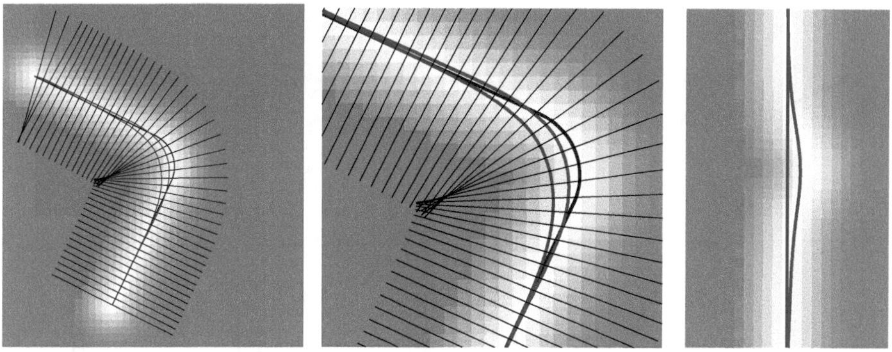

(a) (b) (c)

Fig. 3. (a) The red line is the path extracted by a classical minimum-cost path algorithm, the green line is the path after the first iteration and the blue line is the true centerline of the object. (b) Zoomed in version of (a). (c) Deformed image obtained by equi-distant sampling perpendicular to the initial (red) path. The green path is the minimum cost path formed in the deformed image.

that a curvilinear object becomes straight. After an initial path through the object is extracted using a standard minimum-cost path algorithm, two cubic splines are defined through the data points found by a steepest descend; one for the x-values and one for y-values of the data points, using the distance from the end points along the path as the running variable. Using cubic splines guarantees that the path is continuous up to the second derivative. As shown in Fig. 3(a-b), lines perpendicular to the splines separated by a distance of one pixel are defined. A new image Fig. 3(c) is sampled using cubic interpolation on equi-distant points along these perpendicular lines. A new minimum-cost path is calculated in the deformed space (the green line in Fig. 3(c)). This new path is again represented by two splines. Next, the perpendicular distance between the centerline of the deformed image and the splines is calculated. By defining points on the perpendicular lines in the original image with the same distance from the original path, the new path is transferred back to the original image space. Two new splines are fitted through the coordinates of these points to produce a new path. As shown in Fig. 3(a), this path is already much closer to the desired centerline. Repeating the process described above yields a path that converges to the true centerline of the object.

3 Results

We first tested our algorithm on synthetic data, allowing us to measure its performance by comparing the results with a ground truth. Later we used images of DNA-strands made using an electron microscope to examine its real world performance qualitatively.

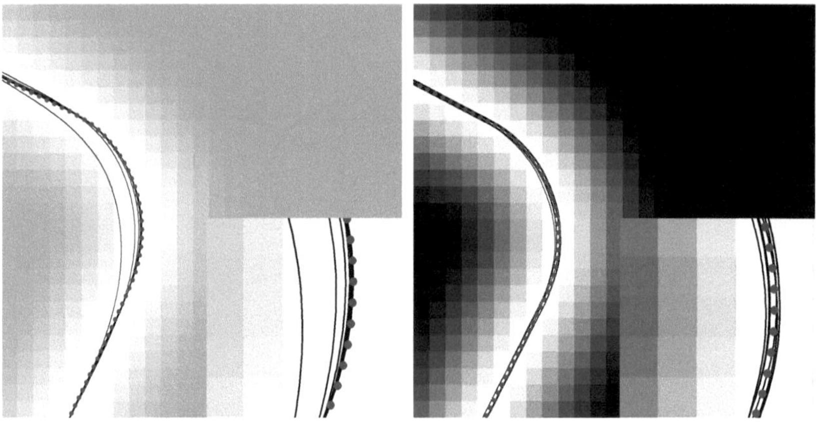

(a) (b)

Fig. 4. (a) A low contrast ($c = 1$) image depicting the path converging to the true centerline of the object. (b) A High contrast ($c = 10$) image depicting the oscillation effect causing the path to lie alternately on either side of the true centerline of the object. In both images the true centerline is denoted by the red dotted line.

3.1 Synthetic Data

As synthetic data we used images of curved Gaussian line profiles, with a ninety degrees change in orientation and a curvature radius R_c (Fig. 3(a)). The cross section of this profile is defined as in Eq. 3. The algorithm was tested for different centerline radii R_c, noise levels and contrast ratios c. As shown in Fig. 4(a) the first iteration already results in a path which is significantly closer to the centerline. We measured the performance by looking at the distance between the centerline of the object and the path found using our algorithm. We computed the root-mean-square (RMS) of the perpendicular distance between the path and the ground truth at ten points separated by a pixel in the middle of the curve.

Initially this RMS error decreases for all the settings. However, after a number of iterations (one to three for the high contrast images and about six for the low contrast images) it starts to increase for certain values of radius R_c and contrast level c. This is due to an oscillation effect, which results in the paths lying alternately on either side of the centerline of the object between successive iterations. The effect is depicted in the close up of Fig. 4(b). We suspect it originates from the two fundamentally different ways to cut a corner. In Fig. 5 the two possible ways are shown. On the left side the radius of the path R_p which cuts the corner is larger than the radius of the centerline R_c of the object. In contrast to the situation on the right where the radius of the path is smaller. Due to this sharper bend, we overcompensate for the bending and hence change sign of the curvature in the deformed space. In cases with a bend which is less sharp than the true centerline, we only partially compensate and hence do not change the sign of the curvature. Therefore, the oscillating effect is only observed when

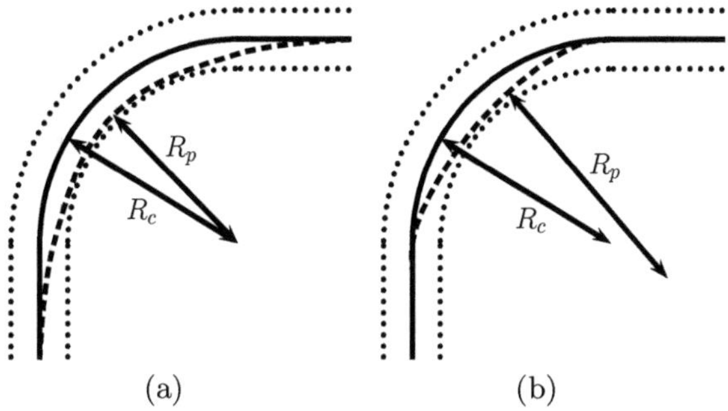

Fig. 5. (a) The first way to cut a corner. Notice the radius of the path (R_p) being larger than the radius of the centerlineÂ (R_c). (b) The second way to cut a corner. Notice the radius of the path (R_p) being smaller than the radius of the centerline (R_c).

an intermediate path has a sharper bend than the true centerline of the object and significantly cuts the corner.

To counteract this effect we introduced a damping term at the transition of the path from the deformed space to the original image space. The damping is being reduced exponentially. After N iterations the distance between the last and the new path in the nth iteration is multiplied by a damping factor $D^{(n-N)}$ ($D < 1$). This damping assures stable results. Elaborate testing has showed us that $D = 0.7$ is either the optimal or near optimal over a wide range of values for c and R_c. Only very low contrast settings require less damping to allow the path to reach the centerline.

In Fig. 6 the mean of the RMS distance of twenty realizations is plotted as a function of the number of iterations with low ($c = 1$) and high ($c = 10$) contrast settings and no noise added. The plots show that the RMS error decreases dramatically in comparison with standard minimum-cost path algorithms (the 0th iteration) for all radii and contrast levels. The damping is switched on after the fourth iteration. We observed that the damping decreased the RMS distance on all high contrast images, but on the low contrast images only for the curves with a large radius.

Fig. 7 shows the mean RMS distance for images with 5 percent Gaussian noise added. The RMS distance slightly increases after a number of iterations. This is not due to oscillating behavior but caused by the fact that the path also adapt to the noise pattern. For medium to high SNR's the path corrections in the first iterations are dominated by the signal. The iterative procedure should stop when the noise becomes the dominant factor.

3.2 Real Data

The proposed algorithm has been extensively tested on transmission electron microscope images of DNA-strands labeled with uranyl acetate to quantify their shape. Empirically we deduced that twenty-five iterations were sufficient to reach a stable result on all of the images. Because no oscillating behavior was observed,

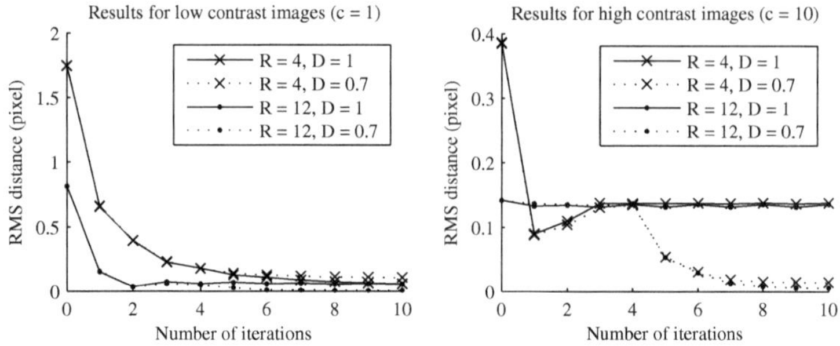

Fig. 6. The mean of the RMS distance for twenty realizations as a function of the number of iterations using low ($c = 1$) and high ($c = 10$) contrast and line width $\sigma = 2.5$. For the cases with $D = 0.7$ the damping is switched on after four iterations.

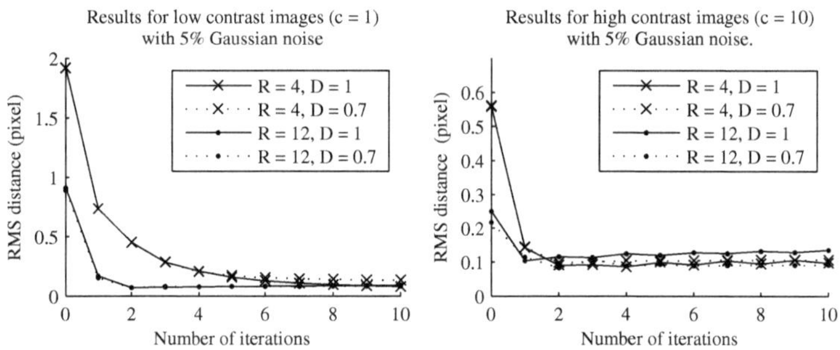

Fig. 7. The mean of the RMS distance for twenty realizations as a function of the number of iterations using low ($c = 1$) and high ($c = 10$) contrast and line width $\sigma = 2.5$. 5% Gaussian noise is added to the images. For the cases with $D = 0.7$ the damping is switched on after four iterations.

no exponential damping was used. Fig. 8 shows four typical results from the more than thousand molecules that were processed. The red line is depicting the path found by the fast marching algorithm, the blue indicates the final result after twenty-five iterations, the green lines in between are the results after respectively one and four iterations. Note that the final results describe the centerline of the object much better, especially in regions with high curvature. The blue line follows the local minimum of the cost function without cutting corners. This work permits the computation of the persistence length of DNA with much greater accuracy, especially over small distances. Earlier results always overestimated the persistence length in this regime due to the stiffness of the minimum-cost path.

3.3 Computational Speed

The time needed for one iteration is comparable to the time needed to calculate a classical minimal-cost path. Therefore, it is evident that the amount of

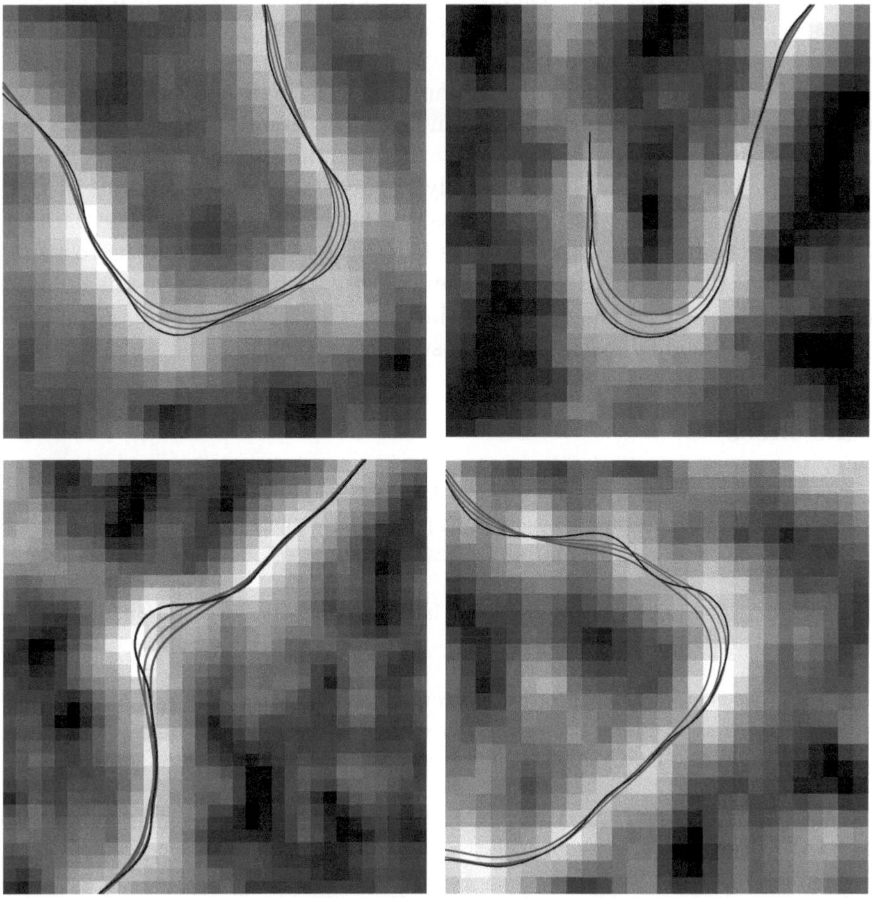

Fig. 8. Four TEM images of DNA-strands. The red line depicts the path found by a standard minimum-cost path algorithm (FM). The green lines indicate the results of our method after one and four iterations, the blue line after 25 iterations.

computation needed increases approximately linearly with the number of iterations. Note that one often can limit the amount of image space to be evaluated after the first iteration, hence reducing the computation time in subsequent iterations.

4 Conclusion

In this paper we present an improvement on minimum-cost path algorithms, which significantly boosts their performance in describing the centerline of string-like objects. The method can be incorporated in any minimum-cost path algorithm. We have demonstrated that our algorithm results in a path which corresponds much better to the centerline of both simulated and real-world string-like objects. The RMS displacement error decreases more then a factor of ten, especially in highly curved areas. Displacement errors of several pixels can be repaired. The behavior depends on conditions such as contrast, noise level and line width. Under certain conditions, such as high contrast, the method only converges after incorporating a damping term. Ten to twenty-five iterations are needed, using an exponentially reducing damping term after several iterations. The method has been successfully applied to several thousands of DNA molecules in high-resolution images obtained by TEM and AFM. The paths we obtained on the images of DNA-strands follow the valley through the cost function without cutting corners. Hence the length measurement remains unbiased and the curvature is no longer underestimated. This is of utmost importance for measuring the bending energy of DNA-strands on a nanometer scale.

References

1. Adalsteinsson, D., Sethian, J.A.: A fast level set method for propagating interfaces. Journal of Computational Physics 118(2), 269–277 (1995)
2. Born, M., Wolf, E.: Principles of Optics, 6th edn. 1977 Pergamon Press, London (1980)
3. Danielsson, P.-E., Lin, Q.: A modified fast marching method. In: Bigun, J., Gustavsson, T. (eds.) SCIA 2003. LNCS, vol. 2749, pp. 1154–1161. Springer, Heidelberg (2003)
4. Fouard, C., Gedda, M.: An objective comparison between gray weighted distance transforms and distance transforms on curved spaces. In: Kuba, A., Nyúl, L.G., Palágyi, K. (eds.) DGCI 2006. LNCS, vol. 4245, pp. 259–270. Springer, Heidelberg (2006)
5. Meijering, E., Jacob, M., Sarria, J.-C.F., Steiner, P., Hirling, H., Unser, M.: Design and validation of a tool for neurite tracing and analysis in fluorescence microscopy images. Cytometry 58A(2), 167–176 (2004)
6. Saha, P.K., Wehrli, F.W., Gomberg, B.R.: Fuzzy distance transform: Theory, algorithms, and applications. Computer Vision and Image Understanding 86(3), 171–190 (2002)
7. Toivanen, P.J.: New geodesic distance transforms for gray-scale images. Pattern Recognition Letters 17(5), 437–450 (1996)
8. Tsitsiklis, J.N.: Efficient algorithms for globally optimal trajectories. IEEE Transactions on Automatic Control 40(9), 1528–1538 (1995)
9. Verbeek, P.W., Verwer, B.J.H.: Shading from shape, the eikonal equation solved by grey-weighted distance transform. Pattern Recognition Letters 11, 681–690 (1990)

Object Class Detection Using Local Image Features and Point Pattern Matching Constellation Search

Alexander Drobchenko[1], Jarmo Ilonen[1], Joni-Kristian Kamarainen[2], Albert Sadovnikov[1], Heikki Kälviäinen[1], and Miroslav Hamouz[2]

[1] Machine Vision and Pattern Recognition Research Group
Lappeenranta University of Technology
[2] Centre for Vision, Speech and Signal Processing
University of Surrey

Abstract. Several novel methods based on locally extracted image features and spatial constellation models have recently been introduced for invariant object class detection and recognition. The accuracy and reliability of the methods depend on the success of both tasks: image feature extraction and spatial constellation model search. In this study a novel method for object class detection is introduced. It combines supervised Gabor-based confidence-ranked image features and affine invariant point pattern matching. The method is able to deal with occlusions and its potential is demonstrated on a standard face database.

1 Introduction

Object class (category) recognition has recently become a popular research topic in computer vision. The popularity probably originates from the problem of face detection where the faces establish an object class. The traditional detection methods can be considered as image or window based approaches where a scene is exhaustively scanned with a window and delivered as an input to a classifier system (e.g., template matching or support vector machine). Lately, the image based approaches have faced competition in the form of feature based methods (e.g., [1,2,3]). These methods utilize locally detected features which along with their spatial configuration are combined using "a constellation model" to establish a complete representation of an object.

Feature based methods yield certain advantages over image based methods, but hitherto most of them have been based on simple key points (e.g., [4]). The advantages of the key points are their generality (shared by many classes) and their semi-supervised nature, that is, objects must be only segmented, named and aligned. The main disadvantages are their incapability to generalize over varying presences of the same feature (e.g., human eye) and to specialize for a specific object class. Since the advantage of semi-supervised training is quite artificial an alternative approach can be utilized by labeling image features and training them in a supervised manner. That would enable a more representative

B.K. Ersbøll and K.S. Pedersen (Eds.): SCIA 2007, LNCS 4522, pp. 273–282, 2007.

set of features allowing a computationally lighter realization of the spatial model (e.g., [1]); The detection load is shared by the both, the image feature detection and the constellation model.

In this study, we propose an object class detection and localisation method, which utilizes the supervised feature detection in [5] and an affine transformation based point pattern matching constellation model.

2 Related Research

Partitioning an object to more easily detectable local patches and combining the patches using a spatial constellation model is not a new approach but originally introduced by Fishler and Elschlager in 1973 [6]. Since then a well-known graph matching method utilizing a similar structure was proposed by Lades et al. [7], but it cannot be used as a general object class detection method since it requires a sufficient initial guess of the object pose. Modern and currently state-of-the-art spatial constellation models appeared recently along with efficient methods for key point detection, e.g., by Lowe [8,3] and Burl and Perona et al. [9,1].

Lowe uses an approach resembling Hough transform for object detection based on SIFT features [3,8]. SIFT features with similar scale, orientation and translation (relative to the model) are grouped in bins. Then bins are sorted according to the number of hits and each bin is verified using an approximated affine mapping of the model onto features in the bin. Outliers are determined by a threshold on the difference in scale, rotation and translation from the parameters obtained in the affine model mapping.

Simultaneously, a probabilistic constellation model was developed by the Perona's group. The core of the system is the estimation of how normal noise is distorted by affine and projective transforms. A breadth first search (reviewing most probable models first) is then used for locating the most likely affine correspondence of the model and extracted image features.

The main difference of these state-of-the-art methods is in their use of the unsupervised key points by Lowe [8] or Kadir [10]. In this study we propose to use supervised image features and a similar spatial search method as proposed by Hamouz et al. [2], with the difference that the spatial constellation model is replaced by a direct affine point pattern matching. The method provides the global optimum over the given image features and is capable to estimate locations of missing features.

3 Supervised Image Feature Extraction

The extraction method was introduced by the authors in [5] and is based on simple Gabor features [11] and feature ranking based on confidence information derived from Gaussian mixture model pdf's [12]. In the following section the method will be shortly reviewed.

3.1 Simple Gabor Features

The simple Gabor feature space and its properties have been originally introduced by the authors in [11]. The features are based on responses of complex Gabor filters on multiple scales and orientations, thus forming a multi-resolution Gabor frame structure.

Responses of Gabor filters, $\psi(x, y; f, \theta)$, over the whole image $\xi(x, y)$,

$$\begin{aligned} r_\xi(x, y; f, \theta) &= \psi(x, y; f, \theta) * \xi(x, y) \\ &= \iint_{-\infty}^{\infty} \psi(x - x_\tau, y - y_\tau; f, \theta)\xi(x_\tau, y_\tau)dx_\tau dy_\tau \ , \end{aligned} \tag{1}$$

are calculated for several frequencies f_k and orientations θ_l and arranged into a matrix form as $\mathbf{G} =$

$$\begin{pmatrix} r(x_0,y_0;f_0,\theta_0) & r(x_0,y_0;f_0,\theta_1) & \cdots & r(x_0,y_0;f_0,\theta_{n-1}) \\ r(x_0,y_0;f_1,\theta_0) & r(x_0,y_0;f_1,\theta_1) & \cdots & r(x_0,y_0;f_1,\theta_{n-1}) \\ \vdots & \vdots & \ddots & \vdots \\ r(x_0,y_0;f_{m-1},\theta_0) & r(x_0,y_0;f_{m-1},\theta_1) & \cdots & r(x_0,y_0;f_{m-1},\theta_{n-1}) \end{pmatrix} \tag{2}$$

where rows correspond to responses on the same frequency and columns correspond to responses on the same orientation. The first row is the highest frequency and the first column is typically the angle 0°. This kind of feature structure is capable to accurately represent local image patches and facilitates invariance operations for image feature search over arbitrary rotations, scale and translation and by normalization also achieves illumination invariance [13,11].

3.2 Classification and Ranking of Features

In general, any classifier can be used to learn and to classify features into image feature classes. However, certain advantages advocate the use of statistical methods. Most importantly, not only class labels for observed features are desired but also it should be possible to rank features in a scene and to sort them in the best matching order returning only a fixed number of the best candidates.

In order to apply statistical classification and ranking it is necessary to estimate class conditional pdf's for every feature class. However, a single Gaussian cannot represent class categories, such as eyes, since they may contain inherited sub-classes, such as closed eye, open eye, Caucasian eye, Asian eye, eye with glasses, and so on. Inside a category there are instances from several sub-classes which can be distinct in the feature space. In this sense Gaussian mixture model is an effective principal distribution to represent the statistical behaviour of simple Gabor features.

There are several methods to estimate parameters of Gaussian mixture models (GMMs) and some of them can automatically estimate the number of components in a GMM [12]. Pdf's are estimated separately for different image feature types from the complex vectors of Gabor feature matrix in (2) as

$$\boldsymbol{g} = [r(x_0, y_0; f_0, \theta_0) \ r(x_0, y_0; f_0, \theta_1) \ldots r(x_0, y_0; f_{m-1}, \theta_{n-1})] \ . \tag{3}$$

Using estimated pdfs it is possible to assign a class for features extracted at any location of an image by simply applying the Bayes decision making. However, as posteriors do not act as inter-class measures but as between-class measures for a single observation, class-conditional probability (likelihood) is a preferred choice to act as a ranking confidence score [12]; it is a measure of how reliable the class assignment is. Ranking facilitates an efficient search, image features with the highest confidences can be processed first.

The algorithms utilizing simple Gabor features and Gaussian mixture model feature ranking have been given in [5].

4 Affine Transform Based Spatial Constellation Model

In this as well as in the aforementioned studies ([9,1,6,2,3]) the detection is applied to real 3-D objects spanning real 3-D surfaces, but for simplicity, the objects have been treated as planar, that is, they can be uniquely represented in two dimensions. It is well known that for example frontal human faces with a low degree of in-depth rotation can be accurately detected using 2-D image processing techniques not utilizing the 3-D shape of faces. Extracted image features are 2-D projections of planar point sets in 3-D vector space, that is, we consider 2-D projective geometry invariant to affine transforms. The properties of affine space will be considered next and then a spatial search method utilizing them will be introduced.

4.1 Affine Transform

Affine transformation in N-dimensional vector space $\mathbb{R}^N$ can be represented as a matrix multiplication in homogenous coordinates as $T : \mathbb{R}^N \rightarrow \mathbb{R}^N,\ T(\boldsymbol{x}) = A\boldsymbol{x}$ where, in case of 2-D coordinates, the transform matrix becomes

$$A = \begin{pmatrix} a & b & c \\ d & e & f \\ 0 & 0 & 1 \end{pmatrix} . \tag{4}$$

It is difficult to interpret the parameters in this form. There are various possible decompositions into a set of geometrically meaningful parameters. Parameters such as rotation, scale, shear, squeeze, scaling along first and second axis (dilation) may be involved. In this study we apply the following decomposition which is one of the easiest to derive and interpret

$$A = \underbrace{\begin{pmatrix} 1 & 0 & c \\ 0 & 1 & f \\ 0 & 0 & 1 \end{pmatrix}}_{\text{translation}} \underbrace{\begin{pmatrix} \cos(\phi) & \sin(\phi) & 0 \\ -\sin(\phi) & \cos(\phi) & 0 \\ 0 & 0 & 1 \end{pmatrix}}_{\text{rotation}} \underbrace{\begin{pmatrix} p & 0 & 0 \\ 0 & q & 0 \\ 0 & 0 & 1 \end{pmatrix}}_{\text{scaling/squeeze}} \underbrace{\begin{pmatrix} 1 & n & 0 \\ 0 & 1 & 0 \\ 0 & 0 & 1 \end{pmatrix}}_{\text{shear}} \tag{5}$$

Since it is often computationally most convenient to operate on the original transform matrix A, the motivation is to utilize the transform parameters a, b, c, d, e, and f in (4) as functions of the decomposed parameters ϕ, p, q and n in (5).

4.2 Model Representation and Training

Local image features can be detected by the method proposed in Sec. 3, but in addition their configuration topology must be restricted in order to apply spatial constellation model search. In the following we assume planar objects and their 3-D projections on 2-D image plane. This assumption provides a sufficient framework for analysis where the spatial relationship of image features, the constellation, remains affinely almost fixed, e.g., for every two facial images there exists an affine transform which maps the features from one image to the corresponding features in another. The suitability of affine mapping for roughly frontal facial images is demonstrated in Fig. 1 where 10 image features are represented for two different fixed coordinates and for an affinely mapped space where the smallest variance is achieved.

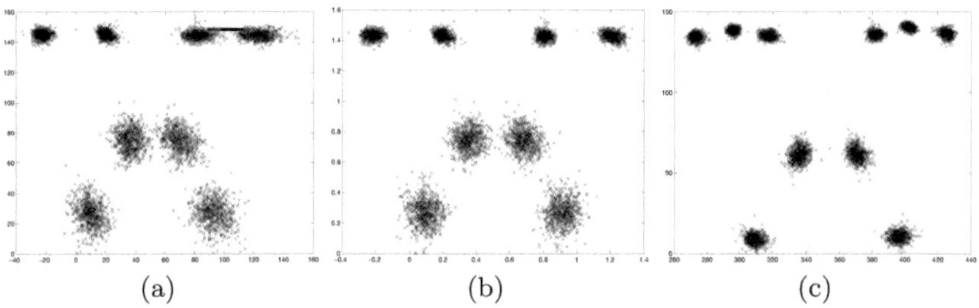

Fig. 1. Image features from 600 training images in the XM2VTS database: (a) left eye center and eyes' angle fixed; (b) left and right eye fixed; (c) affine mapped (LSQ) to features of one face

A spatial model can be generated by storing the feature configuration from a single image. Once having such a model any other object can be mapped to the model, for example by the least mean square fit (LSQ), and accepted as a detection if the deviation from the model is within acceptable limits. However, in practice spatial variability of face features is not completely affine and selection of the proper representative for "ideal face" affects the model's performance. An iterative scheme presented in Algorithm 1. can be used to generate the model from a training set.

4.3 Spatial Constellation Search

In this section we solve a specific task of Point Pattern Matching (PPM). The task of PPM is to find a transform of a given type that best fits one set of points onto another in terms of a given metric. The role of the metric is important and different fits are provided depending on the selected metric (including for example subset fitting).

In our case we are fitting image feature coordinates with the affine model representative. In the simplest case the metric (or distance between the two point sets) can be defined as a sum of squares of distances between points with

Algorithm 1. Spatial constellation model training

```
Assign the model representation to the first element in the training set: M_1 = X_1
for n = 1...N {Over all training set} do
    for m = 1...n do
        Find an affine transform mapping F_m : X_m → Y_m, for which || Y_m − M_n ||^2
        achieves its minimum (RSS)
    end for
    Map first n elements to the current model M_n using the corresponding mappings
    F_n
    Update the current model as the mean of the mappings: M_{n+1} = (Σ_{i=1}^n Y_i)/n, where
    summations and divisions by scalar are done elementwise
end for
Return the current model, M_{N+1}, as the result
```

the same label in different sets. In a such method the inputs would be: 1) a point set containing labeled candidate image feature locations, $\boldsymbol{S}_{i,j}$, such that $\boldsymbol{S}_{i,j}$ is the location (x,y) of the j-th most probable candidate for i-th image feature, and 2) the model M, obtained from the training step. The output of the method is an object hypothesis index vector, $\boldsymbol{I}_i$, denoting numbers of candidate locations used for object hypotheses generation, $\boldsymbol{S}_{i,I_i}$, $\forall\ i:\ I_i \neq 0$. Zero values in the index denote omitted features (omission handling described later).

Object search is based on yet another assumption, which allows to reduce the search complexity greatly. Once images for three points (triplet) in the model have been selected, the affine transform is uniquely defined and other points from hypothesis can be selected as the closest corresponding candidate locations. We are assuming that if we try all possible triplets of model points, we would not miss the best possible, globally optimal, hypothesis.

For reducing computation time it might be useful to check only a subset of all the possible triplets, since time complexity is linearly dependent on the number of mapping triplets. The number of triplets to be checked can thus be from 1 to $\frac{n!}{(n-3)!n!}$ where n is the number of image features. The amount of triplets to check depends on the time constraints and desired omission resistance. The search is described in Algorithm 2..

Handling missing image features. We call an image feature omitted if its correct coordinates on the image are not well enough described by one of the candidate locations for the corresponding feature. The reason for the features being omitted can be for example partial occlusion or a feature detection failure.

Although massive omissions decrease method performance, it is possible to recover the correct hypothesis if enough features are remaining. A crucial point for the correct hypothesis recovering is that at least one matching triplet exists.

A simple omission detection approach exists: if there are no extracted image features which contribute to the overall error (RSS) for less than a given threshold, the point is considered to be omitted. Another parameter is the omission penalty the amount which is added to the overall error for each point

Algorithm 2. Spatial constellation search

for all selected triplets o, p, q **do**
  for all possible label values i, j, k **do**
    Find affine mapping F of triangle $[M(o)M(p)M(q)]$ on triangle $[S(o,i)S(o,j)S(q,k)]$ (equivalent to solving system of 6 linear equations).
    Create image from the model using F: $M^F = FM$.
    Select points closest to model points with equal labels for each label and store their indices in $I_{o,p,q,i,j,k}$
    Calculate the sum of squared distances (or Residual Square Sum): $RSS_{o,p,q,i,j,k} = \sum_t \| M^F(t) - S(t, I_{o,p,q,i,j,k}) \|$
  end for
end for
Sort values in $RSS_{o,p,q,i,j,k}$ and return corresponding $I_{o,p,q,i,j,k}$.

considered as omitted. The introduction of these two parameters is justified by two reasons: 1) the distance from the predicted feature location to the actual local feature response should be discarded if it is much greater than the feature size – the found feature is most likely not related to the current model mapping; 2) It is useful to be able to control the balance between hypotheses with a few omitted features, high error and a substantial number of omissions or lower error for the rest of the image features.

After the points are checked for omission, algorithm proceeds in the same way, with the only difference that omission penalty is added to the final hypotheses error instead of a squared distance of each omitted image feature.

Applying affine transformation restrictions. A model of the frontal human face will not represent only frontal faces if it is subject to an unrestricted affine transform. A transformation between two frontal images of the same face is a similarity transform (a combination of shift, rotation and scaling) which is only a custom case of affine transform. Affine transform is a similarity transform if both shear and squeeze parameters in its decomposition are fixed to zero. Since in the real world applications faces cannot be absolutely frontal, some degree of shearing and squeezing should be still allowed in the transformation model for better performance. Another example is restricting possible rotations - often getting an upside-down image is something completely not acceptable and thus means false detection with a very high probability.

Restrictions are implemented as multipliers for the final hypothesis residual square sum, based on the parameters of transform used for model mapping in current hypothesis generation. The final coefficient was combined of four functions: $P_{transform} = P_{shear} \times P_{squeeze} \times P_{scale} \times P_{rotation}$. Each of these functions is an inverse of the estimated probability density for corresponding transform parameters, with a small added value limiting maximum penalty: $P_{param}(t) = \frac{1}{pdf_{param}(t) + \Delta}$.

5 Experiments

The XM2VTS facial image database used in the experiments is a publicly available database for benchmarking face detection and recognition methods [14]. The frontal part of the database contains 600 training images and 560 test images of size 720 × 576 (width × height) pixels. 10 marked image feature detectors were trained and searched as described in [5]. The localisation accuracy was measured by d_{eye}, which is defined as maximum over distances between detected features and groundtruth normalized by groundtruth eye distance [15]. The eye-distance normalization makes d_{eye} scale independent. d_{eye} is standard and recommended face localisation error measure [16].

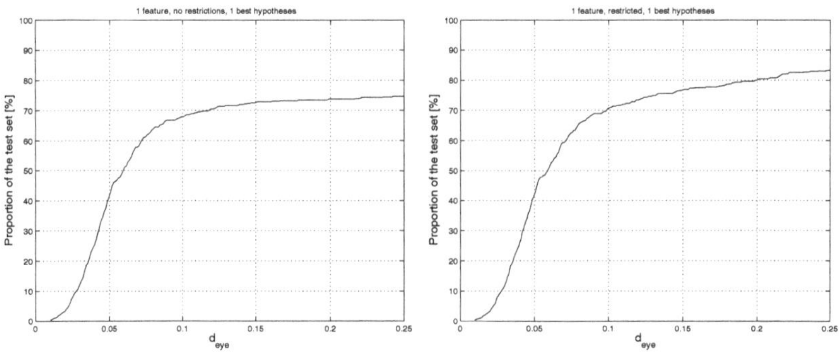

Fig. 2. Results for 1 best hypothesis using 1 best image feature (for each 10 feature classes)

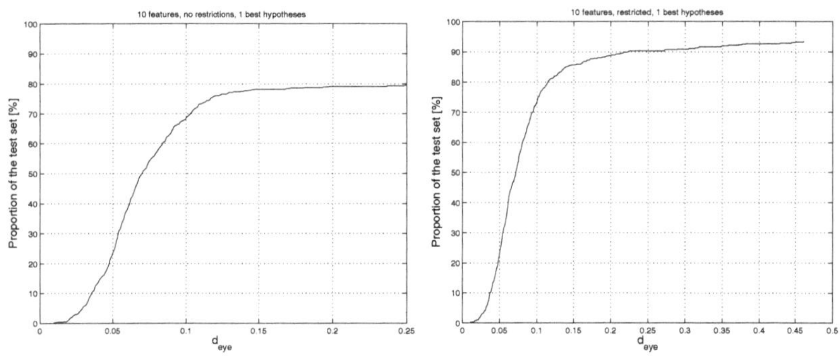

Fig. 3. Results for 1 best hypothesis using 10 best image features

Fig. 2 shows the results from an experiment where only one feature per class was extracted and only the best face hypothesis was accepted. The experiment was performed with and without affine restrictions estimated from the training set images. The results merely represent the accuracy and reliability of detected image features and in 70% of images already the feature detector provides the correct face ($d_{eye} = 0.1$).

In the second experiment, results shown in Fig. 3, the number of image features was increased to 10 (total of 100 for 10 different classes) which improved the results, but also revealed a problem; the best hypothesis is often misaligned with respect to the ground truth.

Hypotheses close to the misaligned best hypothesis were included by allowing detection of 10 best hypotheses, which significantly increased the detection accuracy to 90% (Fig. 4).

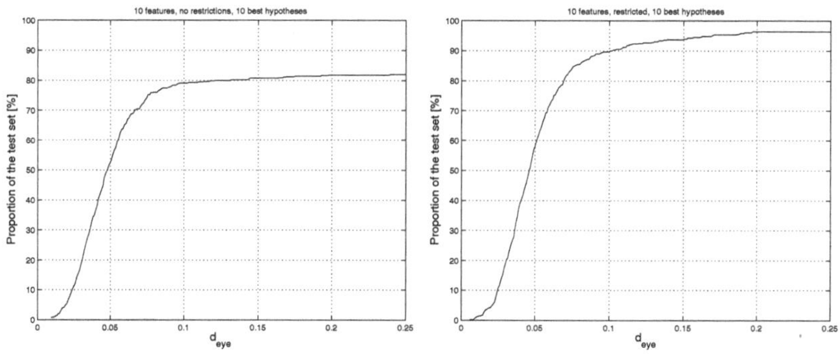

Fig. 4. Results for 10 best hypotheses using 10 best image features

It is noteworthy that the results of this simple method are comparable to the much more complicated but state-of-the-art method reported in [2].

6 Conclusions

In this study we proposed a new feature based method for the detection of object classes in gray-level still images. The proposed method follows state-of-the-art approaches by separating the process to image feature extraction and spatial constellation search. The image feature extraction is based on simple Gabor features and their statistical ranking providing very accurate and reliable results. The spatial search was formulated as a point pattern matching problem over affine invariant point sets. The method finds the globally optimal constellation of extracted image features, is robust to outlier features, and provides estimated location of missing image features.

Acknowledgements

This work was supported by EPSRC project "2D + 3D = ID" (GR/S98528/01), with contributions from EU Project BIOSECURE, and Academy of Finland (204708).

References

1. Fergus, R., Perona, P., Zisserman, A.: Object class recognition by unsupervised scale-invariant learning. In: Proc. of the IEEE Computer Society Conference on Computer Vision and Pattern Recognition (2003)
2. Hamouz, M., Kittler, J., Kamarainen, J., Paalanen, P., Kalviainen, H., Matas, J.: Feature-based affine-invariant localization of faces. IEEE Trans. on Pattern Analysis and Machine Intelligence 27, 1490–1495 (2005)
3. Helmer, S., Lowe, D.: Object recognition with many local features. In: Workshop on Generative Model Based Vision. (2004)
4. Mikolajczyk, K., Schmid, C.: A performance evaluation of local descriptors. IEEE Trans. on PAMI vol. 27 (2005)
5. Kamarainen, J.K., Ilonen, J., Paalanen, P., Hamouz, H., Kälviäinen, H., Kittler, J.: Object evidence extraction using simple gabor features and statistical ranking. In: Proc. of the 14th Scandinavian Conf. of Image Processing, Joensuu, Finland pp. 119–129 (2005)
6. Fischler, M., Elschlager, R.: The representation and matching of pictorial structures. IEEE Trans. on Computers 22, 67–92 (1973)
7. Lades, M., Vorbrüggen, J., Buhmann, J., Lange, J., von der Malsburg, C., Würtz, R., Konen, W.: Distortion invariant object recognition in the dynamic link architecture. IEEE Transactions on Computers 42, 300–311 (1993)
8. Lowe, D.G.: Object recognition from local scale-invariant features. In: Proc. of the International Conference on Computer Vision, Corfu, Greece pp. 1150–1157 (1999)
9. Burl, M.C.: Recognition of Visual Object Classes. PhD thesis, California Institute of Technology (1997)
10. Kadir, T.: Scale, Saliency and Scene Description. PhD thesis, Oxford University (2002)
11. Kyrki, V., Kamarainen, J.K., Kälviäinen, H.: Simple Gabor feature space for invariant object recognition. Pattern Recognition Letters 25, 311–318 (2003)
12. Paalanen, P., Kamarainen, J.K., Ilonen, J., Kälviäinen, H.: Feature representation and discrimination based on Gaussian mixture model probability densities - practices and algorithms. Pattern Recognition 39, 1346–1358 (2006)
13. Kamarainen, J.K., Kyrki, V., Kälviäinen, H.: Invariance properties of Gabor filter based features - overview and applications. IEEE Trans. on Image Processing 15, 1088–1099 (2006)
14. Messer, K., Matas, J., Kittler, J., Luettin, J., Maitre, G.: XM2VTSDB: The extended M2VTS Database. In: Chellapa, R. (ed.) Proc. of Second Int. Conf. on Audio and Video-based Biometric Person Authentication. pp. 72–77 (1999)
15. Jesorsky, O., Kirchberg, K., Frischholz, R.: Robust face detection using the hausdorff distance. In: Proc. of 3rd Int. Conf. on Audio- and Video-based Biometric Person Authentication. pp. 90–95 (2001)
16. Rodriguez, Y., Cardinaux, F., Bengio, S., Mariéthoz, J.: Measuring the performance of face localization systems. Image and Vision Computing 24, 882–893 (2006)

Image Segmentation with Context

Anders P. Eriksson, Carl Olsson, and Fredrik Kahl

Centre for Mathematical Sciences
Lund University, Lund, Sweden
anderspe, calle, fredrik@maths.lth.se

Abstract. We present a technique for simultaneous segmentation and classification of image partitions using combinatorial optimization techniques. By combining existing image segmentation approaches with simple learning techniques we show how prior knowledge can be incorporated into the visual grouping process through the formulation of a quadratic binary optimization problem. We further show how such to efficiently solve such problems through relaxation techniques and trust region methods. This has resulted in an method that partitions images into a number of disjoint regions based on previously learned example segmentations. Preliminary experimental results are also presented in support of our suggested approach.

1 Introduction

Image segmentation is normally defined as the task of distinguishing objects from background in unseen images. This visual grouping process is typically based on low-level cues such as intensity, homogeneity or image contours. Popular approaches include thresholding techniques, edge based methods and region-based methods. Regardless of the method, the difficulty lies in formulating and describing the perception of what constitutes foreground and background in an arbitrary image. Furthermore, such a grouping is also highly contextually driven, certain image regions may be labeled differently depending on the task at hand - are we looking for people, buildings or trees? If one also allows for more labels than only foreground and background, the problem becomes increasingly harder and requires a much higher level of scene understanding. And even then, what constitutes visually relevant regions is not always obvious.

In this paper we make an attempt at addressing the problem of contextually based multiclass image segmentation. By combining image segmentation approaches with standard learning techniques we seek to include prior knowledge into the visual grouping process. Our wish is to segment an image into any number of parts based on previously seen and manually annotated examples.

The approach taken here is based on graph cut techniques from combinatorial optimization. This choice was motivated by the proven success of these methods and that it allowed for a straightforward incorporation of prior knowledge into its

B.K. Ersbøll and K.S. Pedersen (Eds.): SCIA 2007, LNCS 4522, pp. 283–292, 2007.

formulation. We show how this problem can be stated as a large-scale quadratic 0-1 optimization program with linear constraints. Typically, such proven NP-complete problems are solved by relaxing or simply dropping some constraints and rewriting the problem as one that can be solved efficiently, see for example [1] for semidefinite relaxation techniques. Another popular approach is the energy minimization method of [2] for optimizing objective functions that are submodular; such discrete problems can be solved exactly. Unfortunately, multiclass labeling does not belong to this set of objective functions. Nevertheless, in [3] the authors suggest that these types of problems can be approximated by solving a number of subproblems exactly. We argue that this level of accuracy is not required for image segmentation problems, since the problem itself is vaguely formulated striving for such exactness may be wasteful.

Instead we propose the use of efficient relaxation techniques capable of handling large problem instances. We have examined two different such methods. Spectral relaxation is a standard relaxation technique, based on eigenvalue computations it is well suited for large-scale problems. In [4], a relaxation technique for solving non-submodular large-scale quadratic combinatorial optimization problems is applied to image restoration, binary partitioning and registration. In addition to the theoretical developments in this paper, results on applying these two methods to the problem of multiclass segmentation with prior information will be given.

2 Combinatorial Optimization and Image Segmentation

A graph cut is the process of partitioning a directed or undirected graph into disjoint sets. The concept of optimality of such cuts is usually introduced by associating an energy to each cut. Problems of this kind have been well studied within the field of graph theory but can for graphs with more than only a few nodes be notoriously difficult (that is, NP-hard). Nevertheless, ever since it became apparent that many low-level vision problems can be posed as finding cuts in graphs, these techniques have received a lot of attention in the computer vision community. Graph cut methods have been successfully applied to stereo, image restoration, texture synthesis and image segmentation, for example [5,6,7]. Below we give a brief overview of graph cuts for image segmentation as well as an introduction to some basic definitions.

2.1 Graph Cuts

Given a graph $G = \{V, E, W\}$, where V denotes its nodes, E its edges and W the affinity matrix, which associates a weight to each edge in E. A cut on a graph is a partition of V into k subsets $A_1, ...A_k$ such that $\bigcup A_i = V, \quad A_i \bigcap A_j = \emptyset, \ i \neq j$. Perhaps the simplest and best known graph cut method is the minimal cut formulation. The min-cut of a graph is the cut that partitions G into disjoint segments such that the sum of the weights associated with edges between the different segments are minimized. That is, the partition that minimizes

$$C_{min}(\{A_i\}_{i=1}^{k}) = \sum_{i=1}^{k} \sum_{u \in A_i, v \notin A_i} w_{uv}. \tag{1}$$

However, as this is in most cases an NP-hard combinatorial optimization problem, the task of finding the solution can be a formidable one. Instead we attempt to find such cuts by rewriting the original formulation and by relaxing some of the constraints and thus arriving at a problem that can be efficiently solved. First we will describe how our initial problem of segmenting images is connected to graphs cuts.

2.2 Graph Representations of Images

The general approach of constructing an undirected graph from an image is shown in fig. 2.2. Basically each pixel in the image is viewed as a node in a graph.

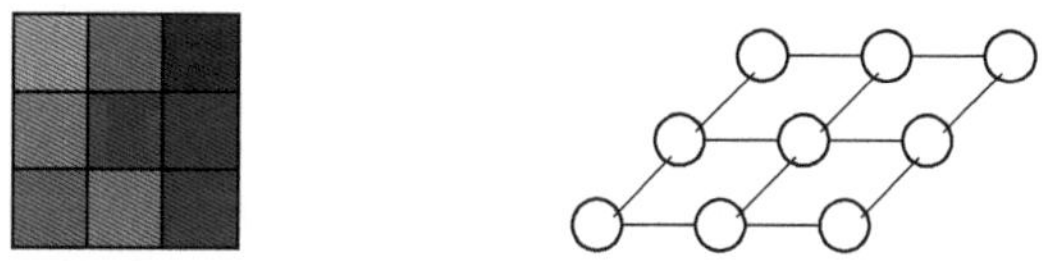

Fig. 1. Graph representation of a 3×3 image

Edges are formed between nodes with weights corresponding to how alike two pixels are, given some measure of similarity, as well as the distance between them. In an attempt to reduce the number of edges in the graph, only pixels within a small, predetermined neighborhood $\mathcal{N}$ of each other are considered. Cuts made in such a graph will then correspond to a segmentation of the underlying image. Owing to the definition of image-pixel resemblance this segmentation should then be a partition such that pixels close to each other with a high degree of intensity similarity will end up in the same partition. Any spatial structure in the image will hopefully be preserved.

2.3 Including Prior Information

In order to be able to include prior information into the visual grouping process we modify the construction of the graphs in the following way. To the graph G we add k artificial nodes. These nodes do not correspond to any pixels in the image, instead they are meant to represent the k different classes the image is to be partitioned into. The contextual information that we wish to incorporate is modeled by a simple statistical model. Edges between the class nodes and the images nodes are added, with weights proportional to how likely a particular pixel is to a certain class. With the labeling of the k class nodes fixed, a

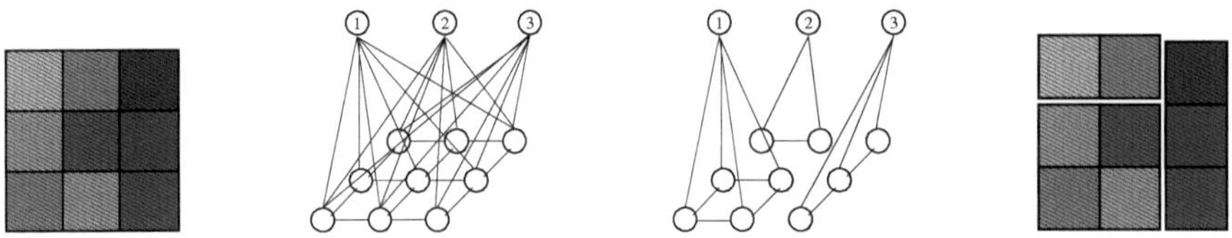

(a) Original image (b) Corresponding graph (c) Multiclass min-cut (d) Resulting segmentation

Fig. 2. A graph representation of an image and an example three-class segmentation. Unnumbered nodes corresponds to pixels and numbered ones to the artificial class nodes.

minimal cut on such a graph should group together pixels according to their class likelihood and still preserving the spatial structure, see fig. 2.

2.4 Combinatorial Optimization

In this section we derive the optimization problem related to (1). Let $Z = [\, z_1, \ldots, z_k \,] \in \{-1, 1\}^{n \times k}$ denote the $n \times k$ assignment matrix for all the n nodes. A 1 in row i of column j signifies that pixel i of the image belongs to class j, and of course -1 in the same position signifies the opposite. If we let W contain the inter-pixel affinities, the min-cut (without pixel class probabilities) can then be written

$$C_{min} = \inf_Z \sum_{i=1}^{k} \sum_{\substack{u \in A_i \\ v \notin A_i}} w_{uv} = \inf_Z \sum_{i,j,l} w_{jl}(z_{ij} - z_{il})^2 = \inf_Z \sum_{i=1}^{k} z_i^T (D - W) z_i. \quad (2)$$

Here D denotes diag($W\mathbf{1}$). The assignment matrix Z must satisfy $Z\mathbf{1} = (2 - k)\mathbf{1}$. In addition, if the pixel/class-node affinities $P = [\, p_1, \ldots, p_k \,]$ (that is, the probabilities of a single pixel belonging to a certain class) are included and also the labels of the class-nodes are fixated, we get

$$C_{min} = \inf_{\substack{Z \in \{-1,1\}^{n \times k} \\ Z\mathbf{1} = (2-k)\mathbf{1}}} \sum_{i=1}^{k} z_i^T \underbrace{(D - W)}_{L} z_i - 2p_i^T z_i = \inf_{\substack{Z \in \{-1,1\}^{n \times k} \\ Z\mathbf{1} = (2-k)\mathbf{1}}} tr\left(Z^T L Z\right) +$$

$$+2 \underbrace{[\, -p_1^T, \ldots, -p_k^T \,]}_{b^T} \underbrace{\begin{bmatrix} z_1 \\ \vdots \\ z_k \end{bmatrix}}_{z} = \inf_{\substack{z \in \{-1,1\}^{nk} \\ Z\mathbf{1} = (2-k)\mathbf{1}}} z^T \underbrace{\begin{bmatrix} L & 0 & \ldots & 0 \\ 0 & L & \ldots & 0 \\ 0 & \vdots & \ddots & 0 \\ 0 & \vdots & 0 & L \end{bmatrix}}_{A} z + 2b^T z. (3)$$

As $z \in \{-1,1\}^{nk} \Leftrightarrow z_i^2 = 1$, we arrive at the quadratically constrained quadratic program

$$\mu = \inf_z z^T A z + 2b^T z \tag{4}$$
$$\text{s.t.} \quad z_i^2 = 1 \tag{5}$$
$$Z\mathbf{1} = (2-k)\mathbf{1}. \tag{6}$$

Since the constraint (5) implies $z^t z = nk$ this redundant equality can be added to (4) without changing the problem. The above problem is still equivalent to the original min-cut formulation (1). For efficiently solving this problem we here turn our attention to two relaxations that are tractable from a computational perspective. When dropping the $z_i^2 = 1$ and $Z\mathbf{1} = (2-k)\mathbf{1}$ constraints we obtain the problem

$$\mu_{tr} = \inf_{||z||^2 = nk} z^T A z + 2b^T z. \tag{7}$$

A common approach for solving this problem is to homogenize (7). That is, we add an extra variable z_{nk+1} and solve

$$\mu_{spec} = \inf_{||z||^2 + z_{nk+1}^2 = nk+1} \begin{pmatrix} z \\ z_{nk+1} \end{pmatrix}^T \begin{pmatrix} A & b \\ b^T & 0 \end{pmatrix} \begin{pmatrix} z \\ z_{nk+1} \end{pmatrix}. \tag{8}$$

Note that if we add the constraint $z_{nk+1} = 1$ to (8) we obtain (7). It is therefore clear that we always have $\mu_{spec} \leq \mu_{tr}$. Equality will only occur if z_{nk+1} happens to be ± 1. The reason for solving (8) instead of (7) is that (8) is easily solved by computing the eigenvector corresponding to the smallest eigenvalue of the matrix

$$H = \begin{pmatrix} A & b \\ b^T & 0 \end{pmatrix}. \tag{9}$$

The downside is of course that, in practice z_{nk+1} is often quite far away ± 1, resulting in poor relaxations. In our case it is easy to see that an incorrect value of z_{nk+1} changes the balance between the effects of the quadratic smoothing therm and the linear term containing the prior information. To remedy this problem we propose to solve (7), using a method borrowed from the trust region problem.

2.5 The Trust Region Subproblem

In this section we review how to solve (7) (see also [4], [8] and [9]). A problem closely related to (7) is

$$\inf_{||z||^2 \leq nk} z^T A z + 2b^T z. \tag{10}$$

This problem is usually referred to as the trust region subproblem. Solving the problem is one step in a general optimization scheme for descent minimization and it is known as the trust region method [10]. Instead of minimizing a general function, one approximates it with a second order polynomial $z^T Az + 2b^T z + c$. A constraint of the type $||z||^2 \leq m$ then specifies the set in which the approximation is believed to be good (the trust region).

The trust region subproblem have been studied extensively in the optimization literature ([8,9,11,12,13]). A remarkable fact is that it is a non convex problem with no duality gap (see [14]). This is always the case when we have quadratic objective function and only one quadratic constraint. The dual problem of (10) is

$$\sup_{\lambda \leq 0} \inf_{z} z^T Az + 2b^T z + \lambda(nk - z^T z). \tag{11}$$

In [12] is shown that z^* is the global optimum of (10) if and only if (z^*, λ^*) is feasible in (11) and fulfills the following system of equations:

$$(A - \lambda^* I)z^* = -b \tag{12}$$

$$\lambda^*(nk - z^T z) = 0 \tag{13}$$

$$A - \lambda^* I \succeq 0. \tag{14}$$

The first two equations are the KKT conditions for a local minimum, while the third determines the global minimum. From equation (14) it is easy to see that if A is not positive semidefinite, then λ^* will not be zero. Equation (13) then tells us that $||z||^2 = nk$. This shows that for an A that is not positive semidefinite problems (7) and (10) are equivalent. Note that we may always assume that A is not positive semidefinite in (7). This is because we may always subtract mI form A since we have the constant norm condition. Thus replacing A with $A - mI$ for sufficiently large m gives us an equivalent problem with A not positive definite.

A number of methods for solving this problem has been proposed. In [13] semidefinite programming is used to optimize the function $nk(\lambda_{\min}(H(t)) - t)$, where

$$H(t) = \begin{pmatrix} A & b \\ b^T & t \end{pmatrix}, \tag{15}$$

and $\lambda_{\min}$ is the algebraically smallest eigenvalue. In [15] the authors solve $\frac{1}{\psi(\lambda)} - \frac{1}{\sqrt{nk}} = 0$ where $\psi(\lambda) = ||(A - \lambda I)^{-1} b||$. This is a rational function with poles at the eigenvalues of A. To ensure that that $A - \lambda I$ is positive semidefinite a Cholesky factorization is computed. If one can afford this, Cholesky factorization is the preferred choice of method. However, the LSTRS-algorithm developed in [8] and [9] is more efficient for large scale problems. LSTRS works by solving a parameterized eigenvalue problem. It searches for a t such that the eigenvalue problem

$$\begin{pmatrix} A & b \\ b^T & t \end{pmatrix} \begin{pmatrix} y \\ 1 \end{pmatrix} = \lambda_{min} \begin{pmatrix} y \\ 1 \end{pmatrix} \tag{16}$$

or equivalently

$$\begin{aligned} (A - \lambda_{min} I)y &= -b \\ t - \lambda_{min} &= -b^T y \end{aligned} \tag{17}$$

has a solution. Finding this t is done by determining a λ such that $\phi'(\lambda) = nk$, where ϕ is defined by

$$\phi(\lambda) = b^T(A - \lambda I)^\dagger b = -b^T y. \tag{18}$$

It can be shown that λ gives a solution to (17). Since ϕ is a rational function with poles at the eigenvalues of A, it can therefore be expensive to compute. Instead rational interpolation is used to efficiently determine λ. For further details see [8] and [9].

Regardless of relaxation the solution will be a vector with continuous entries that will most likely not fulfil constraints (5) and (6). Obtaining a discrete solution that fulfills all the constraints of the original problem, from the relaxed optima - known as rounding - is hence necessary. From the optima of either relaxation, the vector z^*, a $n \times k$ matrix Z^* is formed and the discrete solution Z is found through non-maximum supression of the rows of Z^*. That is, the largest value in each row of Z^* is set to 1 and the others to -1 thus ensuring that both $Z \in \{-1,1\}^{n\times k}$ and $Z\mathbf{1} = (2-k)\mathbf{1}$ holds.

3 Experimental Results

As mentioned in the previous section prior knowledge is incorporated into the graph cut framework through the k artificial nodes. For this purpose we need a way to describe each pixel as well as model the probability of that pixel belonging to a certain class.

The image descriptor in the current implementation is based on color alone. Each pixel is simply represented by their three RGB color channels. The probability distribution for these descriptors are modeled using a Gaussian Mixture Model (GMM).

$$p(v|\Sigma,\mu) = \sum_{i=1}^{k} \frac{1}{\sqrt{2\pi|\Sigma_i|}} e^{(-\frac{1}{2}(v-\mu_i)^T \Sigma_i^{-1}(v-\mu_i))} \tag{19}$$

From a number of manually annotated training images the GMM parameters are then fitted through Expectation Maximization, [16]. This fitting is only carried out once and can be viewed as the learning phase of our proposed method.

The edge weight between pixel i and j and the weights between pixel i and the different class-nodes are given by

$$w_{ij} = e^{(-\frac{r(i,j)}{\sigma_R})} e^{(-\frac{||s(i)-s(j)||^2}{\sigma_W})} \tag{20}$$

$$p_{ki} = \alpha \frac{p(w(i)|i\in k)}{\sum_j p(w(i)|i\in j)}. \tag{21}$$

Here $||\cdot||$ denotes the euclidian norm, $r(i,j)$ the distance between pixel i and j and λ, σ_R and σ_W are tuning parameters weighing the importance of the different features. Hence, w_{ij} contains the inter-pixel similarity, that ensures that the segmentation more coherent. p_i describes how likely a pixel is to belong

Fig. 3. Sample training images

to class k. α is a parameter weighting the importance of spatial structure vs. class probability.

Preliminary tests of the suggested approach were carried out on a limited number of images. We chose to segment the images into four simple classes, sky, grass, brick and background. Gaussian mixture model for each of these classes was firstly acquired from a handful of training images manually chosen as being representative of such image regions, see fig. 3. For an unseen image the pixel affinity matrix W and class probabilities were computed according to (20) and (21). The resulting optimization program was then solved using both the spectral relaxation and the trust region subproblem method. The outcome can be seen in fig. 4. Parameters used in these experiments were $\sigma_R = 1$, $\sigma_W = 1$, $\alpha = 10$ and $\mathcal{N}$ a 9×9 neighborhood structure.

Both relaxations produce visually relevant segmentations, based on very limited training data our proposed approach does appear to use the prior information in a meaningful way. Taking a closer look at the solutions supplied by the trust region method and the spectral relaxation for these two examples does however reveal one substantial difference. The spectral relaxation (8) was reached by ignoring the constraint on the homogenized coordinate $z_{nk+1} = 1$. The solutions to the examples in fig. 4 produces an homogeneous coordinate value of $z_{nk+1} \approx 120$, in both cases. As the class probabilities of the pixels are represented by the linear part of eq. 4, the spectral relaxation, in these two cases, thus yields an image partition that that weights prior information much higher than spatial coherence. Any spatial structure of an image will thus not be preserved, the spectral relaxation is basically just a maximum-likelihood classification of each pixel individually.

We conclude that the trust region formulation seems provide the degree of accuracy required for these types of problems and that spectral relaxation does not.

Original images.

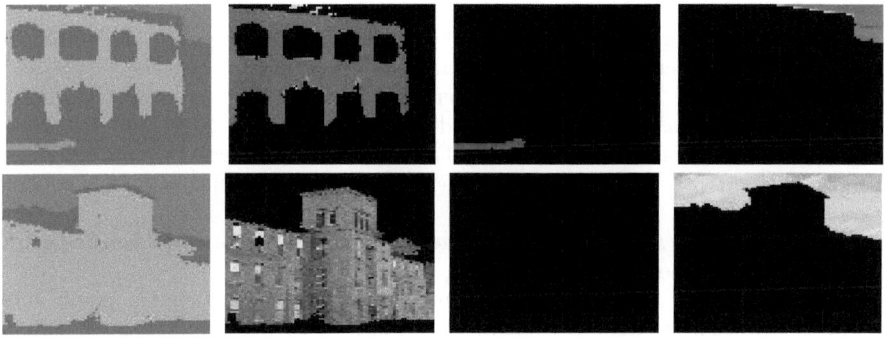

(TSP) Resulting class labelling.

(SR) Resulting class labelling.

Fig. 4. Example segmentation/classification of an image using both Trust Region Subproblem (TSP) formulation and Spectral Relaxation (SR)

4 Summary and Conclusions

In this paper we have proposed a method for multiclass image segmentation with context. We describes how prior information can be brought into a graph cut framework through the use of terminal node weights and learning techniques. In particular, an efficient implementation that brings forward the trust region subproblem formulation as an alternative to existing approaches for finding these image partitions is presented. We also give some promising results on a number of color images.

References

1. Wolkowicz, H., Saigal, R., Vandenberghe, L. (eds.): Handbook of Semidefinite Programming. Kluwer Academic Publishers, Dordrecht (2000)
2. Kolmogorov, V., Zabih, R.: What energy functions can be minimized via graph cuts? IEEE Trans. Pattern Analysis and Machine Intelligence 26(2), 147–159 (2004)
3. Boykov, Y., Veksler, O., Zabih, R.: Fast approximate energy minimization via graph cuts. IEEE Trans. Pattern Analysis and Machine Intelligence 23(11), 1222–1239 (2001)
4. Olsson, C., Kahl, A.E., Solving, F.: large scale binary quadratic problems: Spectral methods vs. semidefinite programming. CVPR (2007)
5. Kolmogorov, V., Zabih, R.: Multi-camera scene reconstruction via graph cuts. In: Heyden, A., Sparr, G., Nielsen, M., Johansen, P. (eds.) ECCV 2002. LNCS, vol. 2352, pp. 82–96. Springer, Heidelberg (2002)
6. Shi, J., Malik, J.: Normalized cuts and image segmentation. IEEE Trans. Pattern Analysis and Machine Intelligence 22(8), 888–905 (2000)
7. Rother, C., Kolmogorov, V., Blake, A.: Grabcut: interactive foreground extraction using iterated graph cuts. In: ACM Transactions on Graphics. pp. 309–314 (2004)
8. Rojas, M., Santos, S., Sorensen, D.: A new matrix-free algorithm for the large-scale trust-region subproblem. SIAM Journal on optimization 11(3), 611–646 (2000)
9. Rojas, M., Santos, S., Sorensen, D.: Lstrs: Matlab software for large-scale trust-region subproblems and regularization. Technical Report 2003-4, Department of Mathematics, Wake Forest University (2003)
10. Fletcher, R.: Practical Methods of Optimization. John WIley & Sons, New York (1987)
11. Sorensen, D.: Minimization of a large-scale quadratic fuction subject to a spherical constraint. SIAM J. Optim. 7(1), 141–161 (1997)
12. Sorensen, D.: Newton's method with a model trust region modification. SIAM Journal on Nomerical Analysis 19(2), 409–426 (1982)
13. Rendl, F., Wolkowicz, H.: A semidefinite framework for trust region subproblems with applications to large scale minimization. Math. Prog. 77(2 Series B), 273–299 (1997)
14. Boyd, S., Vandenberghe, L.: Convex Optimization. Cambridge University Press, Cambridge (2004)
15. Moré, J., Sorensen, D.: Computing a trust region step. SIAM J. Sci. Stat. Comput. 4(3), 553–572 (1983)
16. Dempster, A., Rubin, M.L., Maximum, D.: likelihood from incomplete data via the em algorithm. J. R. Stat. Soc (1977)

Improving Hyperspectral Classifiers: The Difference Between Reducing Data Dimensionality and Reducing Classifier Parameter Complexity

Asbjørn Berge and Anne Schistad Solberg

Department of Informatics
University of Oslo, Norway
asbjorb@ifi.uio.no

Abstract. Hyperspectral data is usually high dimensional, and there is often a scarcity of available ground truth pixels . Thus the task of applying even a simple classifier such as the Gaussian Maximum Likelihood (GML) classifier usually forces the analyst to reduce the complexity of the implicit parameter estimation task. For decades, the common perception in the literature has been that the solution to this has been to reduce data dimensionality. However, as can be seen from a result by Cover [1], reducing dimensionality increases the risk of making the classification problem more complex.Using the simple GML classifier we compare state of the art dimensionality reduction strategies with a recently proposed strategy for sparsing of parameter estimates in full dimension [2]. Results show that reducing parameter estimation complexity by fitting sparse models in full dimension have a slight edge on the common approaches.

1 Introduction

Hyperspectral imaging, an increasingly common tool in remote sensing, is sampling of the spectrum of reflected sunlight in wavelengths from ultraviolet to infrared. As a natural extension of the multispectral sensors, hyperspectral sensors sample reflected sunlight in 50 to several hundred contiguous narrow bands. Thus more information can be extracted from a single pixel compared to a multispectral image, however the high dimension of the resulting feature space makes classification of pixels a complex problem. Features also usually exhibit high correlation, adding a redundancy to the data that in some cases may obscure the information important for classification. When the number of training samples is low compared to data dimensionality the so called curse of dimensionality impacts the generalization capability of the classifiers designed.

The common approach for dealing with the curse of dimensionality in the literature is to reduce the dimensionality, and thus indirectly reducing the number of parameters to estimate. Contrary to this approach, it is possible to reduce the number of parameters to estimate by choosing to fit simpler models in full

B.K. Ersbøll and K.S. Pedersen (Eds.): SCIA 2007, LNCS 4522, pp. 293–302, 2007.

dimension. We will discuss the simple classifier resulting from Bayes rule when assuming that classes are distributed as Gaussians. The main contribution of this paper is to present results and a discussion comparing indirect (dimensionality reduction) and direct (parameter sparsing) simplifications of such classifiers. The motivation for this comparison can be found in Covers theorem [1]. We want to ascertain whether dimensionality reduction can be seen to make the classification problem more complex.

In section 2 and section 3 we present the classifier, and point out the effect of Covers theorem. Section 4 discusses the contrary approach of reducing the number of parameters to estimate by fitting a sparse model. In section 5 we briefly review some of the dimensionality reduction strategies proposed in the literature on classification on remotely sensed hyperspectral data. Section 6 presents and discuss the results of several experiments on four different hyperspectral images. Section 7 concludes this paper.

2 The Classification Task

Consider a classification problem with k classes, assuming class conditional distributions to be Gaussian with mean μ_k and class-wise covariance matrices Σ_k. It is well known that this reduces to comparing the k quadratic discriminant functions $g_k(x) = -\frac{1}{2}\log|\Sigma_k| - \frac{1}{2}(x-\mu_k)'\Sigma_k^{-1}(x-\mu_k) + \log\pi_k$, where π_k is the a priori probability for class k. The parameters of these distributions are usually calculated from the maximum likelihood estimates and plugged into the above rule. This decision rule is commonly referred to a Gaussian Maximum Likelihood(GML) classifier. The common problem with the GML classifier is that the number of parameters to estimate grows quadratically with the dimensionality of the feature space. Clearly this means that we quickly run out of samples to reliably estimate these parameters.

3 The Separability of Patterns as a Function of Dimensionality

Our classification problem may be an intrinsically non-linear problem, or may *become* a non-linear problem after dimensionality reduction according to Cover's theorem [1] regarding the separability of patterns. For a set of N samples, and a classifier represented by a surface with d degrees of freedom, where any labeling is equally probable, the probability of randomly picking a class labeling of samples that can be perfectly separated by the chosen classifier is [1]

$$P(N,d) = (\frac{1}{2})^{(N-1)} \sum_{k=0}^{d-1} \binom{N-1}{k}.$$

Plainly what this states is that by increasing the degrees of freedom of a surface intended to separate the classes, the probability of being able to separate the classes approaches one. This is a very common justification for the efficacy of support vector machines and other kernel methods, where you map the data up into a much higher dimensional space and solve a linear classification problem. A linear decision boundary between two classes in a space with m features has $d = m+1$ degrees of freedom corresponding to the dimensionality m of its normal vector, plus one allowing for an arbitrary intercept. Likewise, fitting a quadratic decision boundary corresponds to $d = \binom{m+2}{2}$ degrees of freedom, i.e., finding the coefficients and intercept of a quadratic function in the feature space. When introducing a dimensionality reduction on the feature space, this would be the same as only allowing decision boundaries in some subspace, so consequently the degrees of freedom will be reduced as a function of the reduced dimensionality. Thus, for a fixed number of samples, reducing the dimensionality by any feature extraction or selection method reduces the probability of randomly picking a labeling that can be separated - and for a fixed number of samples this probability is dependent on the degrees of freedom of the classifier. Consequently, it might be harder in reduced dimensions to find a linear classifier that separates the data than a more complex classifier. In section 6 we compare the classification performance of a linear and quadratic classifier as a function of dimensionality to observe this phenomenon with real data.

4 Parameter Sparsing in Full Dimension

Crude models for reducing the number of parameters to estimate in a GML classifier are well known. Constraints such as the assumption that features are uncorrelated, well known as a naïve bayes classifier, reduces the number of parameters to estimate for each distribution down to the dimensionality. We recently [2] proposed an approach for reducing the number of parameters needed to estimate when designing classifiers in high dimensional feature spaces, sparse cholesky triangle inverse covariance (STIC) estimates. The method is based on time series theory regarding the Cholesky decomposition of the inverse covariance matrix, $\Sigma_k^{-1} = L_k D_k L_k^T$. where L_i is a lower triangular matrix with ones on the diagonal and D a diagonal matrix. (See Fig. 1a) If we were to consider the features of each sample as a time-series, the elements in L can be seen row-wise as parameters in autoregressive processes of the same order as the row r. Several authors in the time series literature have noted this [3], [4]. We will use this fact to transform the task of approximating covariance matrices into a sequence of regressions. For each row, r, one could then "predict" the next feature based on the $r-1$ preceding features. Assuming zero mean for readability, this can be expressed as: $x_r = \sum_{j=1}^{r-1} \alpha_{r,j} x_j + \varepsilon_r$ where the rth diagonal entry of $D_{r,r} = \text{var}(\varepsilon_r)$. This parametrization has the effect that the resulting covariance matrix will still be positive definite, as long as the diagonal elements of D are positive.

The general idea is to start by approximating the covariance matrices with the simplest possible models, i.e.,diagonal matrices, and add parameters to the

$$L = \begin{bmatrix} 1 & & & & \\ -\alpha_{2,1} & 1 & & & \\ -\alpha_{3,1} & -\alpha_{3,2} & 1 & & \\ \vdots & & & \ddots & \\ -\alpha_{p,1} & & -\alpha_{p,2} & -\alpha_{p,p-1} & 1 \end{bmatrix}$$

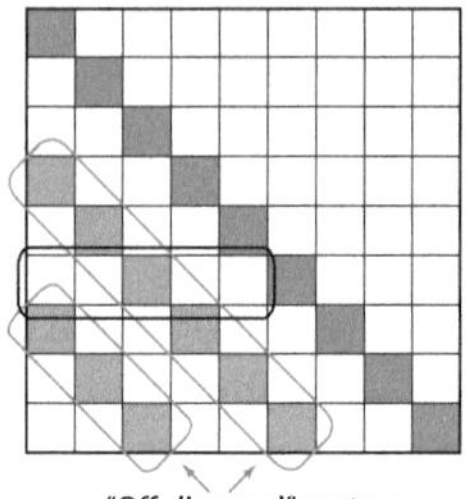

Fig. 1. Illustration of a matrix of correlations, L, for the inverse covariance matrix. The matrix is lower triangular, with 1 on the diagonal. Sparsity in the covariance estimate is obtained by only estimating the matrix elements in *some* off-diagonal vectors. The matrix is estimated by a sequence of regressions, one for each row in the matrix L, indicated with black rectangle.

approximation until the classification performance of the model no longer improves. A very simple heuristic, illustrated in Fig. 1b is to gradually add off-diagonal elements one diagonal vector at a time in a sequential forward fashion. One specific diagonal vector corresponds to the set of correlations with some specific distance between features. The search is guided by ten-fold cross-validation (10-CV) as a performance measure. Further details of this approach can be found in [2].

5 Dimensionality Reduction

In the literature on classification of remotely sensed hyperspectral data several linear feature extraction techniques has been proposed with the goal of improving the performance of the classification. In our comparison we include two classic approaches and two approaches designed for hyperspectral data, decision boundary feature extraction (DBFE) [5] and non-parametric weighted feature extraction (NWFE) [6].

By far the most common dimensionality reduction method in multivariate statistics is the well known principal component analysis (PCA) which eigenanalyze the scatter (or equivalently covariance) matrix of the dataset, corresponding to projecting the data onto a set of orthogonal vectors where the scatter (variance) of the data is highest. Another classical feature extraction technique is Fisher's linear discriminant (LDA).

A method for eigenanalyzing the "scatter" of the decision boundary, called decision boundary feature extraction (DBFE), was proposed in [5], with applications on hyperspectral data in mind. DBFE has been relatively popular in the literature on classification of hyperspectral images. The general idea is to find a set of virtual samples as intersection points between the decision boundary and lines between samples of different classes. The principal components of these virtual samples can thus be argued to describe the most important vectors in the decision boundary and possibly discriminative features. The heuristic used for finding these virtual samples is to classify the dataset in full dimension and then intersecting the estimated decision boundary with lines between the closest samples in opposite classes.

Another popular feature extraction method developed for hyperspectral images is the nonparametric weighted feature extraction(NWFE) proposed in [6], which is a nonparametric extension of LDA by redefining the scatter according to distance. By weighing the influence of samples according to the distance to samples in opposite classes, samples near the decision boundary are considered more important. In LDA, the between-class scatter is of deficit rank. In NWFE this is overcome by redefining this scatter to represent the scatter of between samples and a distance-weighted mean in an opposite class.

In the literature, several feature selection approaches have been proposed, tailored for specific hyperspectral classification tasks. However, for our purposes it is reasonable to compare with feature selection methods that does not need any initialization of the number of features wanted, and keeps the features in the set after selection. Reasonably, when a modest number of data is available for training, the optimal number of features might be low. Sequential forward search (SFS) is the simplest possible algorithm in such cases, adding features sequentially ranked by some criterion, the experiments presented here use the Mahalanobis distance.

6 Experimental Results and Discussion

Four hyperspectral datasets are analyzed in this work. The first, *Fontainebleu*, is from an airborne sensor (ROSIS), containing forest pixels from Fontainebleu south of Paris. It is divided into three classes, have 81 bands and a pixel size of 5.6m. The second dataset *Pavia*[7] is also from an airborne sensor (DAIS), depicting urban landcover pixels over Pavia, Italy. The dataset has 71 bands and 2.6m pixels. The third dataset contains a wetland vegetation scene acquired over the Okavango delta in *Botswana*[8] acquired by the Hyperion sensor aboard the EO-1 satellite, with a total of 145 bands after removal of uncalibrated and noisy bands. The image has 30m pixels. The last image we use is the from an AVIRIS airborne sensor, over Kennedy Space Center (*KSC*)[8], is a vegetation dataset, with 18m pixels and 176 bands. For all the datasets, the average number of training pixels per class is 700, 100, 196, and 115 in presented order. For all datasets we designed (as far as it was possible) spatially separate datasets for training and testing to avoid fitting the classifiers to the similarities between neighboring pixels due to spatial correlation. The same set of 10-fold cross-validation rotation on the training data was used for model choice in all methods, i.e., guiding the number of features or the number of nonzero parameters. The reported performance is average overall correct classification rate. We compare classification performance for the models chosen by cross-validation of the different dimensionality reduction methods: principal component analysis (PCA), Fisher's linear discriminant (LDA), non-parametric weighted feature extraction (NWFE), decision boundary feature extraction (DBFE), sequential forward feature selection (SFS) and the parameter sparsing approach using sparse cholesky triangle covariance estimates (STIC).

6.1 Results

Table 1 reports the test set classification performance and the number of parameters estimated (as a fraction of a full dimensional GML classifier). The reported results are included as a supplement to the results given in the figures. One notes that STIC has a slight edge on all dimension reduction strategies, however, an interesting result is that we commonly use more degrees of freedom to estimate sparse models in full dimension.

In Fig. 2a-2d the performance of a linear classifier versus a quadratic classifier as a function of the number of features (extracted by PCA) is given. The solid black line in these plots indicate the performance of the chosen STIC model. Several conclusions can be made from the presented results. We can observe

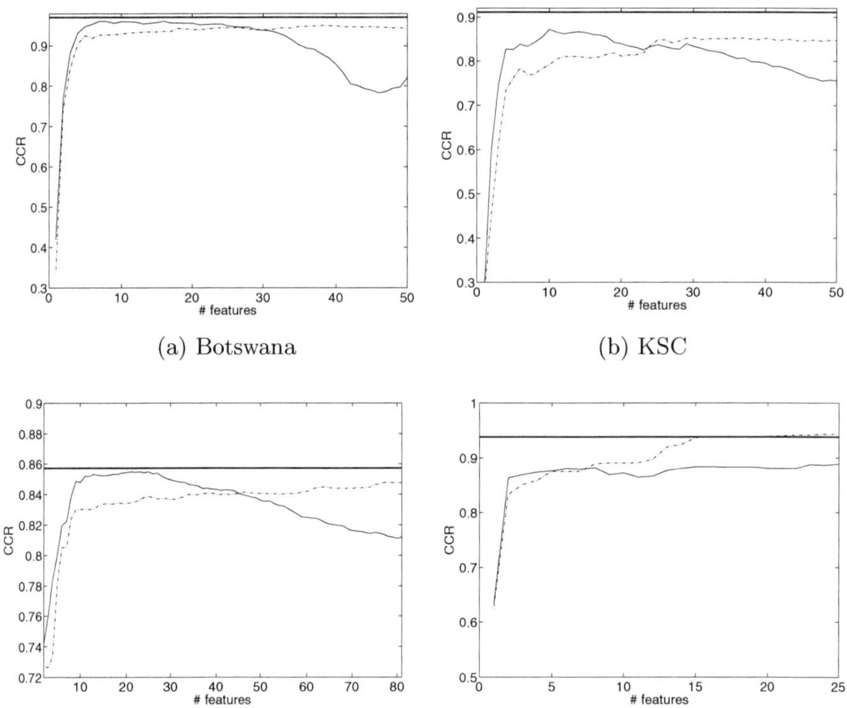

(a) Botswana (b) KSC

(c) Fountainebleu (d) Pavia

Fig. 2. Comparison of correct classification on test data to the number of features retained using PCA as dimension reduction. The performance of a linear classifier is represented by a stippled line and a quadratic classifier by a solid line. The thick solid line indicates the performance of the STIC model chosen by cross-validation. For visualization purposes, only the 50 first features is shown for the KSC and Botswana images and the first 25 for the Pavia image.

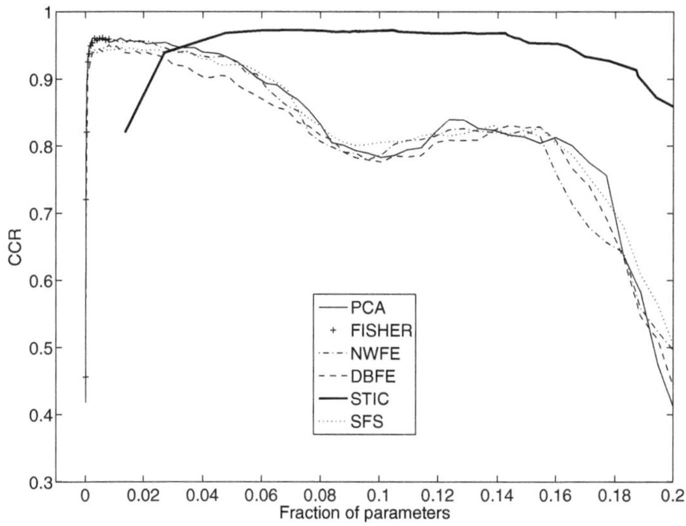

(a) Botswana

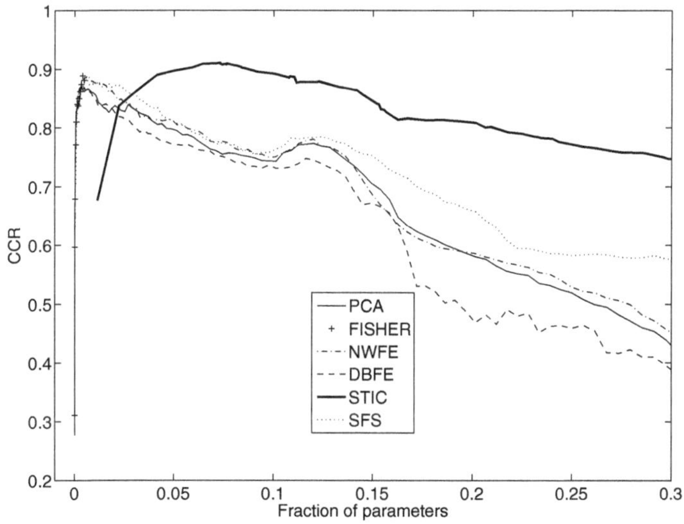

(b) KSC

Fig. 3. Correct classification rates on test data compared to the fraction of covariance parameters for a full model. All methods decay quite rapidly for cases using more than 30% of the parameters of a full model.

similarities in Fig. 2a-2d. All datasets have reasonably low amounts of data available for training. Thus when dimensionality increases, the number of degrees of freedom of the quadratic classifier grows so fast that the estimates become unstable, due to the curse of dimensionality. Not unexpectedly the simpler linear classifier decays slower, and overtakes the quadratic classifier at some point.

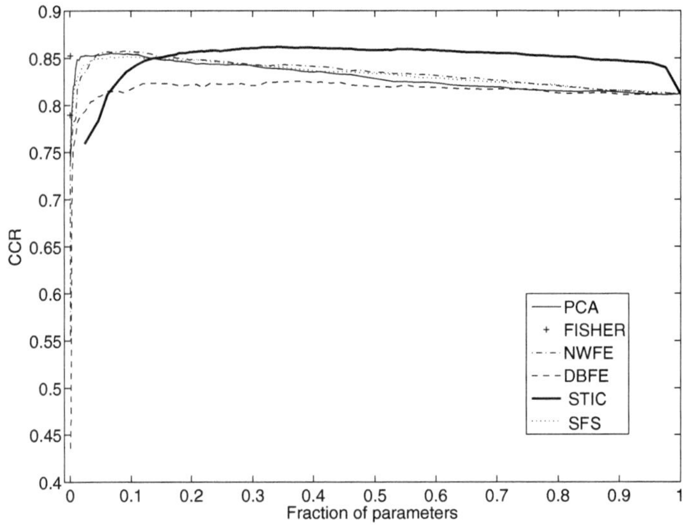

(a) Fountainebleu

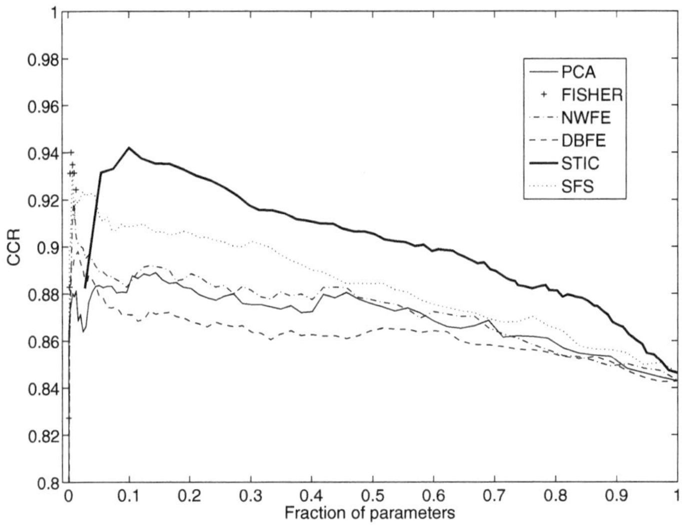

(b) Pavia

Fig. 4. Correct classification rates on test data compared to the fraction of covariance parameters for a full model

These results can also be viewed as support for the conclusions drawn from Cover's theorem, arguing that when dimension is low, the decision boundary tends to be less linear than it is in full dimension. This indicates that parameter sparsing by STIC, i.e., a search for simple classifier models in full dimension is reasonable.

Table 1. Average overall test set performance for the models chosen by cross-validation for the different datasets. In parentheses the number of parameters estimated as the fraction of a full model. All measures in percent.

Dataset	Fontainebleu	Pavia	Botswana	KSC
PCA	85.2(4.8)	87.5(41.8)	95.8(1.1)	86.7(0.6)
LDA	85.2(0.06)	93.4(0.85)	95.8(0.7)	88.9(0.4)
NWFE	85.7(7.7)	88.4(35.8)	95.4(1.4)	88.6(0.5)
DBFE	82.5(52.3)	87.1(13.2)	93.8(1.9)	86.7(0.4)
SFS	85.1(6)	90.7(10.9)	94.7(1.2)	87.5(1.05)
STIC	85.7(23)	93.8(17)	97.1(9.6)	91.1(8.5)

Fig. 3a-4b illustrates the performance for the different dimensionality reduction strategies as a function of the number of parameters estimated compared to a full model. In these plots, the performance of the parameter sparsing strategy, STIC, is also given. The general conclusion that can be drawn from Fig. 3a-4b is that fitting sparse models in full dimensional space is fairly effective over a wide range of parameter sizes. Even so, the best model found for the STIC, especially in the case of the high dimensional images see Fig. 3a and 3b, estimates a lot more parameters than the correspondingly optimal models using dimensionality reduction. (See Table 1.) One can note that the dimensionality reduction results from these images are fairly similar for all feature extraction and selection methods. One possible explanation for this might be that since the features are so highly correlated, any feature reduction will cover mostly the same discriminative information, regardless of approach. From Fig. 4a and 4b we can observe typical performance of full dimensional sparsed models over the entire range. As can be seen, when dimensionality is fairly low, and the amount of training data available is high, as with the Fontainebleu image, little is gained by using sparse models compared with feature extraction. The classes in this dataset is known to be overlapping even in the full dimensional space, and two of the three classes are extremely similar, so this dataset might be complex to classify even in full dimension. Considering class separability, the Pavia image is an example of the opposite - classes can be reasonably classified using a linear classifier in full dimension. This can be seen in the fairly high performance of LDA as a feature extractor.

7 Conclusion

We have discussed the difference between reducing classifier complexity using dimension reduction versus parameter reduction. Theoretical results [1], and supporting experimental results indicate the soundness of fitting simple models in full dimensional space compared to using more complex classifiers after reducing dimensionality. Specifically, our previously proposed strategy, STIC, seems to have a slight edge on dimensionality reduction. However, STIC is still more of a proof of concept than a fully developed method. The heuristic used for selection of non-zero parameters is a bit crude, and as can be seen in Table 1, we usually use fairly many degrees of freedom for describing the classifier.

References

1. Cover, T.M.: Geometrical and statistical properties of systems of linear inequalities with applications in pattern recognition. IEEE Transactions on Electronic Computers (3), 326–334 (1965)
2. Berge, A., Jensen, A.C., Solberg, A.S.: Sparse inverse covariance estimates for hyperspectral image classification. IEEE Trans. Geosci. Remote Sensing, Accepted for publication (2007)
3. Smith, M., Kohn, R.: Parsimonius covariance matrix estimation for longitudinal data. Journal of the American Statistical Association 97(460), 1141–1153 (2002)
4. Pouhramadi, M.: Foundations of Time Series Analysis and Prediction Theory. Wiley, Chichester (2001)
5. Lee, C., Landgrebe, D.: Feature extraction based on desicion boundaries. IEEE Trans. Pattern Anal. Machine Intell. 15(15), 388–400 (1993)
6. Kuo, B.C., Landgrebe, D.: A robust classification procedure based on mixture classifiers and nonparametric weighted feature extraction. Remote Sensing 40(11), 2486–2494 (2002)
7. Gamba, P.: A collection of data for urban area characterization. In: Proc. IEEE Geoscience and Remote Sensing Symposium (IGARSS'04) (2004)
8. Ham, J., Chen, Y., Crawford, M.M., Ghosh, J.: Investiagtion of the random forest framework for classification of hyperspectral data. Remote Sensing 43(3), 492–501 (2005)

A Hierarchical Texture Model for Unsupervised Segmentation of Remotely Sensed Images

Giuseppe Scarpa[1,2], Michal Haindl[2], and Josiane Zerubia[1]

[1] ARIANA Research Group, INRIA/I3S, Sophia Antipolis, France
[2] Pattern Recognition Dep., ÚTIA, Academy of Sciences, Prague, Czech Republic
giscarpa@unina.it,haindl@utia.cas.cz,zerubia@sophia.inria.fr

Abstract. In this work a novel texture model particularly suited for unsupervised image segmentation is proposed. Any texture is represented at region level by means of a finite-state hierarchical model resulting from the superposition of several Markov chains, each associated with a different spatial direction. Corresponding to such a modeling, an optimization scheme, referred to as Texture Fragmentation and Reconstruction (TFR) algorithm, has been introduced.

The TFR addresses the model estimation problem in two sequential layers: the former "fragmentation" step allows to find the terminal states of the model, while the latter "reconstruction" step is aimed at estimating the relationships among the states which provide the optimal hierarchical structure to associate with the model. The latter step is based on a probabilistic measure, i.e, the region gain, which accounts for both the region scale and the inter-region interaction.

The proposed segmentation algorithm was tested on a segmentation benchmark and applied to high resolution remote-sensing forest images as well.

Keywords: Segmentation, texture model, Markov chain, remote sensing, forest classification.

1 Introduction

Image segmentation [1,2,3,4] is a low-level processing which is of critical importance for many applications in several domains, like medical imaging, remote sensing, source coding, and so on. Although segmentation has been widely studied in the last decades in many cases it remains still open, as for textured images, where the spatial interactions may cover long ranges asking for high order complex modeling. In this work we focus on a remote sensing application, which is the segmentation of forest images [5,6] that represents a basic step for land cover classification and monitoring.

There are a large number of approaches to segmentation, but due to space limitations, here we confine ourselves to reviewing only those that have been tested using the same benchmarking system [7] as we use, and which therefore serve as points of comparison. In [8] image blocks are modeled by means of local

B.K. Ersbøll and K.S. Pedersen (Eds.): SCIA 2007, LNCS 4522, pp. 303–312, 2007.

Gauss Markov Random Fields (GMRF) and the segmentation is performed in the parameter space by assuming an underlying Gaussian Mixture. Similar to the previous, but with an auto-regressive 3-D model (AR3D) in place of the Gauss MRF, is the method presented in [3]. In [9] an approach, namely the JSEG, is presented where segmentation is achieved in two steps: a color quantization followed by a processing of the label map which accounts for spatial interaction. Another method taken in consideration is the segmentation algorithm underlying the content-based image retrieval system *Blobworld* [1]. Here a Gaussian Mixture model is assumed in a feature space, where contrast, anisotropy and polarity are the salient texture descriptors, and the EM algorithm carries out the clustering. Finally, the algorithm presented in [10] (EDISON) combines a region-based approach with a contour-based one, hence balancing the global evidence which characterizes a region-based model with the local information typically dominant in the contour modeling.

In this work we present a method based on a hierarchical finite-state probabilistic texture modeling. The model is coupled with an optimization scheme, namely the Texture Fragmentation and Reconstruction (TFR) algorithm, which first estimates the states at the finest level (fragmentation), and then relates them hierarchically (reconstruction) as to provide the desired hierarchical segmentation.

In order to assess the accuracy of the proposed method, we have used the Prague Texture Segmentation Data Generator Benchmark [7] where all the algorithms mentioned above were tested as well. Furthermore, we provide a few results obtained by the TFR in the case of high resolution remotely sensed images portraiting wooded areas.

2 Hierarchical Texture Model

In this work we present a *hierarchical*, *discrete* and *region-based* probabilistic model for texture representation, which is particularly suited for *unsupervised* image segmentation. In order to apply the model, an early processing is then needed to provide a discrete image that roughly represents the original data. In general this processing may be any known pixel-wise texture feature extraction followed by a clustering, but in practice we reduce it to a simple color-based segmentation, since the textural information will be handled in the discrete space. Obviously, this first operation is associated with an information loss which reduces the description capability of the model. However, while this could be a rather serious limit in a synthesis framework, it is not that critical in an analysis problem like segmentation, and especially in an unsupervised setting where robustness, rather than precision, is quite often the most relevant issue.

To introduce the model, let us consider the example in Fig.1, where a textile pattern (a) is associated with some graphical representations. Imagine first a simple 3-level discrete approximation of the data (say, the color-states *blue*, *black* and *red*), and consider its partition in uniform connected regions. A Region Adjacency Graph (RAG) representation of this partition is shown in (b).

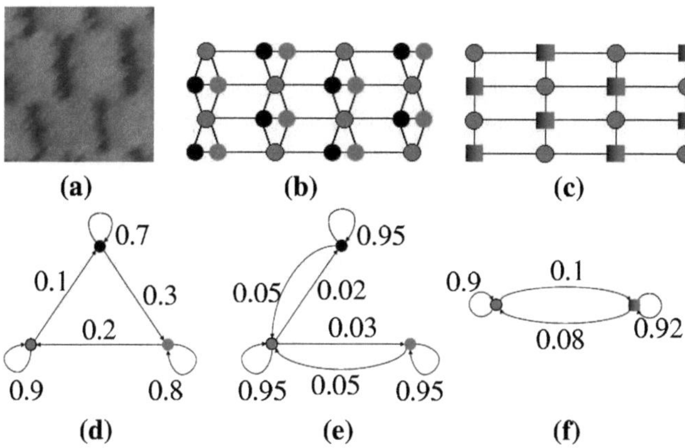

Fig. 1. Hierarchical texture model. Textile pattern (a); H-RAG: 3- and 2-state RAG, (b) and (c) respectively; 3-state chain models for east and south directions, (d) and (e) respectively; and 2-state chain for east direction (f).

Likewise, in case of a 2-state partition (for example, let *black* and *red* collapse in a single state) we would get a RAG like that depicted in (c). Notice that, by merging state *black* with *red* without involving the *blue* one, we established a clear relationship between the two graphs, which form together a Hierarchical RAG (H-RAG) [11]. In this toy example the H-RAG has only two layers because we have considered only two nested partitions, but in practice it has usually more layers as we start from much finer segmentations.

Now, let us observe how the textural properties are reflected in the adjacency graphs (b) and (c) as cyclic occurrence (strictly periodic in the specific example) of subgraphs of three and two nodes, respectively for (b) and (c). Such phenomenon can be synthetically represented for any given spatial direction by means of state diagrams, as in (d) and (e) for directions east and south respectively, when three color states are considered (b), and in (f) for east direction if we have only two states (c). As well as the RAGs, and for the same reasons, these diagrams are hierarchically related for any given direction, (see for example (d) and (f)). The example also clearly shows that, for a fixed periodical texture component, the coarser the scale of the RAG representation, the lesser the order at which it is revealed on the graph. In other words, the multiscale representation allow us to represent simultaneously both micro- and macro-textural features with the same (low) order but in different layers of the hierarchical model.

As can be seen, the compact representation (d)-(f) not only accounts for the adjacency among states but also for their directionality (mutual positioning) and relevance, through the specification of transition probabilities (TP) on a pixel-by-pixel step basis. Approximated TPs are indicated on the graphs just to give an idea of their relationship with the visual appearance of the texture. In particular, observe that intra-region TPs account for the shape of the texture components. As an example, consider the blue patches that regularly occur in the sample. Due to their rectangular shape, the associated intra-region TP in the vertical direction (e) is larger than the horizontal one (d). The remaining, inter-region, TPs accounts instead for the spatial context, that is, the relative occurrence and positioning of the neighboring regions.

More precisely our texture model refers to the graphical representations introduced above and is basically a simultaneous hierarchical finite-state Markov model that for a given texture is completely defined by the triple $(\Omega, \mathcal{T}, \mathcal{P})$, where Ω is the set of states of the finest, but discrete, version of the texture, $\mathcal{T}$ is a tree structure representing the hierarchical relationships among the states[1] and, finally, $\mathcal{P} = \{\mathbf{P}_\omega\}_{\omega \in \Omega}$ is the set of TP matrices (TPMs) for the terminal states. TPMs are given by

$$\mathbf{P}_\omega(\omega', j) = \frac{|\mathcal{S}_{\omega \xrightarrow{j} \omega'}|}{|\mathcal{S}_\omega|} \qquad \forall \omega' \in \Omega, \;\; 1 \leq j \leq 8, \tag{1}$$

where $\mathcal{S}_\omega$ is the set of pixels with state ω and $\mathcal{S}_{\omega \xrightarrow{j} \omega'}$ is the restriction of $\mathcal{S}_\omega$ to its sites whose neighbor in position (direction) j belongs to state ω'. While the TPMs defined above describe globally a texture, a single connected region element n of a given state ω has itself an own TPM, $\mathbf{P}^n_\omega$, computed through the same formula but restricted to the region $\mathcal{S}^n_\omega \subseteq \mathcal{S}_\omega$.

Observe that at coarser level representations the states are completely defined by combination of related offspring states according to the given structure $\mathcal{T}$, which means that their TPMs are derived by simple weighted averages. Moreover, notice that in general a color may occur in a texture according to different configurations, hence increasing the number of states which do not necessarily represent different colors.

3 Texture Fragmentation and Reconstruction (TFR) Algorithm

Let us consider now the application of the above modeling in the particular case of unsupervised segmentation. The image to be segmented is then a composition of an unknown number of different textures whose corresponding models are unknown as well and need to be estimated during the process of texture identification. The model fitting consists in estimating the states (with related TPMs) at the finest scale and the hierarchical tree which univocally defines each intermediate state.

The determination of the number of textures of a given image, classically referred to as *cluster validation problem*, is strictly related to the spatial scale (hence to the hierarchical structure) at which we are interpreting the image. When the scale is not fixed somehow, the cluster validation becomes an ill-posed problem. To give an example, the same texture of Fig.1 may be interpreted as a composition of three different textures if we refer to a finer scale.

As a consequence we aim at solving this problem simultaneously with the estimation of the internal structures, according to the model defined above. In

[1] Hence, the states of Ω are associated with the terminal nodes, while the root represents the whole image.

practice, this means that we fit the image with only one hierarchical model which (when correctly derived) includes as non-overlapped substructures the marginal models associated with the single textures. Then, by specifying a spatial scale, we automatically get the proper pruning of the structure which provides us with the marginal models and the associated image partition.

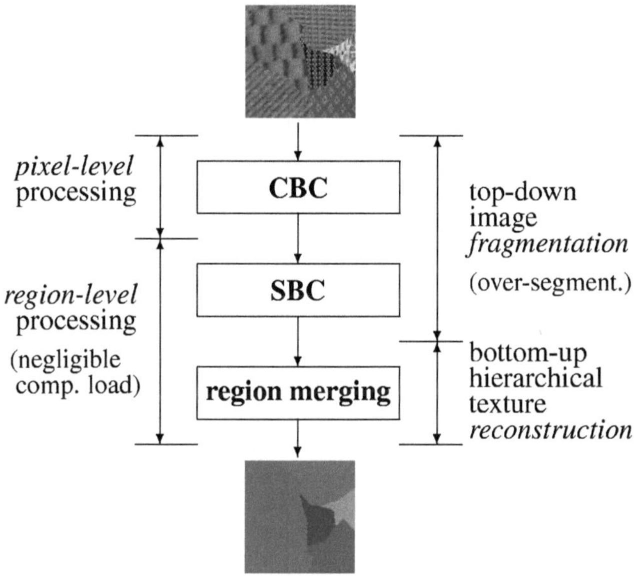

Fig. 2. TFR algorithm flow chart

In order to estimate this overall model we implemented the optimization scheme shown in Fig.2, namely the Texture Fragmentation and Reconstruction (TFR) algorithm, which first extracts a proper number of terminal states through the top-down fragmentation step, composed of blocks CBC (Color-Based Clustering) and SBC (Spatial-Based Clustering), and then relates them by means of a recursive bottom-up merging step, as to reconstruct the whole hierarchical structure.

The estimation of the states is performed in two steps, the former (CBC) dealing with color information, hence working at the pixel level, the latter (SBC) focused on the spatial information at the region level in the TPM space. In principle, CBC may be any color quantization process, but in our implementation we preferred the use of the TS-MRF (tree-structured Markov random field) segmentation algorithm [2], since it avoids the generation of punctiform regions (which are not reliably characterized in terms of TPM) due to regularization of the MRF. Furthermore, the tree-structured formulation ensures a quick processing and allows to balance the energy among the discrete color states.

Once the color segmentation has been obtained, we switch to a region-based representation, by taking connected regions with uniform color as basic elements characterized by TPMs. Since the color of a region only partially defines its state[2], the SBC applies to each set of elements with common color, as to split it in subgroups which are homogeneous also in terms of TPM, that is providing the desired states. The split is realized by means of a k-means algorithm [12] applied in the feature space resulting from a PCA (Principal Component Analysis) [12] on the TPM space. The PCA was necessary because of the large dimensionality of the full feature space w.r.t. the number of elements which does not allow a reliable characterization.

The steps described above, CBC and SBC, realize the "fragmentation" whose goal is the estimation of the terminal states of the hierarchical model. The "reconstruction", that is the estimation of the hierarchy structure, is realized by means of the region (or state) merging process, which is nothing but a sequential binary combination of the states driven by a specific parameter, namely the *region gain* which accounts for the mutual spatial relationships among the corresponding regions. Indeed the merging selection process is not symmetric, as the gain is a measure of the scale of the region weighted by an additional term which quantifies the attraction operated by the other regions (candidates for the merging). The scale factor allows to always privilege the merging of small regions so that the final hierarchy is such that micro-textural features are represented at the bottom, while the macro ones will appear at upper levels, and finally inter-texture mergings will be placed at the top of the structure, in order to keep separate the marginal sub-models corresponding to the different textures.

In this work we compare two different region gains. The former, already proposed in [13], is defined as

$$\mathcal{G}^i \triangleq \frac{p(s\in\mathcal{R}_i)}{\max_{j\neq i} p(r\in\mathcal{R}_j|s\in\mathcal{R}_i)} = p(s \in \mathcal{R}_i) \cdot \frac{1}{p(r\notin\mathcal{R}_i|s\in\mathcal{R}_i)} \cdot \frac{p(r\notin\mathcal{R}_i|s\in\mathcal{R}_i)}{\max_{j\neq i} p(r\in\mathcal{R}_j|s\in\mathcal{R}_i)}$$

where $\mathcal{R}_i$ is the region of interest, s is an image site and r is any of the eight neighbors of s. The first two factors represent the scale, since one is proportional to the area of the region and the other quantifies its compactness. The third term accounts for the relative occurrence of the nearest neighbour region (context).

The latter, introduced here, is a modification of the former where the contextual term has been reinforced by means of the Kullback-Leibler Divergence (KLD), $D(q_i\|q_j)$, between the region spatial distributions, that is

$$\log \mathcal{G}^i_{\mathrm{KLD}} \triangleq \min_{j\neq i} \left\{ \log \frac{p(s \in \mathcal{R}_i)}{p(r \in \mathcal{R}_j|s \in \mathcal{R}_i)} + D(q_i\|q_j) \right\}, \tag{2}$$

where q_i and q_j are normals (see details about KLD for Gaussians in [14]).

[2] More states may correspond to the same color, because either it appears in different configurations in a texture or it occurs in different textures.

4 Application to the Prague Benchmark and Numerical Evaluation

The proposed algorithm, that is the TFR or the TFR+ (when the gain includes the KLD term), is compared with other algorithms which were tested on the same benchmark system [7] and are briefly recalled in the introduction. The system provides a comparison w.r.t. a large number of indicators, some of which are region-based, some others are pixel-wise accuracy indicators, and a few of them give a measure of consistency. A complete description of all the parameters can be found on the system website [7].

Table 1. Up arrows indicate that larger values of the parameters are better; down arrows, the opposite. Benchmark criteria: CS, correct segmentation; OS, over-segmentation; US, under-segmentation; ME, missed error; NE, noise error; O, omission error; C, commission error; CA, class accuracy; CO, recall - correct assignment; CC, precision - object accuracy; I., type I error; II., type II error; EA, mean class accuracy estimate; MS, mapping score; RM, root mean square proportion estimation error; CI, comparison index; GCE (LCE), Global (Local) Consistency Error.

	Benchmark – Colour						
	TFR+	TFR	AR3D	GMRF	JSEG	Blobworld	EDISON
↑ *CS*	51.25	46.13	37.42	31.93	27.47	21.01	12.68
↓ *OS*	5.84	2.37	59.53	53.27	38.62	7.33	86.91
↓ *US*	7.16	23.99	8.86	11.24	5.04	9.30	0.00
↓ *ME*	31.64	26.70	12.54	14.97	35.00	59.55	2.48
↓ *NE*	31.38	25.23	13.14	16.91	35.50	61.68	4.68
↓ *O*	23.60	27.00	35.19	36.49	38.19	43.96	68.45
↓ *C*	22.42	26.47	11.85	12.18	13.35	31.38	0.86
↑ *CA*	67.45	61.32	59.46	57.91	55.29	46.23	31.19
↑ *CO*	76.40	73.00	64.81	63.51	61.81	56.04	31.55
↑ *CC*	81.12	68.91	91.79	89.26	87.70	73.62	98.09
↓ *I.*	23.60	27.00	35.19	36.49	38.19	43.96	68.45
↓ *II.*	4.09	8.56	3.39	3.14	3.66	6.72	0.24
↑ *EA*	75.80	68.62	69.60	68.41	66.74	58.37	41.29
↑ *MS*	65.19	59.76	58.89	57.42	55.14	40.36	31.13
↓ *RM*	6.87	7.57	4.66	4.56	4.62	7.52	3.09
↑ *CI*	77.21	69.73	73.15	71.80	70.27	61.31	50.29
↓ *GCE*	20.35	15.52	12.13	16.03	18.45	31.16	3.55
↓ *LCE*	14.36	12.03	6.69	7.31	11.64	23.19	3.44

For the sake of brevity we do not show here the segmentation maps, which can be found on the benchmark web site [7] as well, but just the numerical results summarized in Tab.1. The interpretation of these indicators may seem quite ambiguous since (of course) no algorithm outperforms uniformly all the other ones. However it can be easily recognized that the two versions of TFR seem to outperform the other ones w.r.t. many indicators, with TFR+ being generally better than TFR. In particular, the main drawback of the reference methods is the tendency to over-segment while, on the contrary, only the TFR has a tendency to under-segment. In this regard, the best trade-off is reached by the TFR+, which outperforms TFR and can be considered as the best one.

5 Application to Remotely Sensed Data

Finally, in this section an application of the proposed method to remote-sensing data is presented. We worked on high resolution (50cm) aerial images covering wooded areas, which match well with the proposed modeling since they present different relevant texture patterns with acceptable stationarity. Such images are curtesy of the "French Forest Inventory" (IFN).

We present two experiments. The former, see Fig.3, refers to an area composed of several classes of trees plus no tree lands and shadows. Since we have no ground-truth related to these data, we build up the latter experiment where a mosaic image was obtained which is composed of four square subimages, see Fig.4. Three of them represent different quasi stationary tree textures, while the last one (bottom-left) is a mixing of an urban class and one of the other (bottom-right) tree textures.

We experimented only the case of TFR+, since it has been shown to be better than TFR in the previous section. Also no comparative algorithms have yet been tested on these data, and eventually we can only make conjectures about the performances of TFR+. A comparison with another method currently under development could be made later.

Fig. 3. Left: Forest image, south of Bourgogne, France. ©IFN. Right: Segmentation map obtained by the TFR+ algorithm (5 classes: two kinds of poplars, oaks, no trees, and shadows).

The 1024×1024 forest image and the associated 5-class TFR+'s segmentation are shown in Fig.3. One class represents just the shadows, one is associated with low vegetation areas, the remaining three classes correspond to different tree patterns. The segmentation seems to be quite promising according to a visual inspection. Indeed, in order to obtain such good result, a slight modification of the TFR+ algorithm was necessary. In fact, the proposed optimization schemes (meaning both TFR and TFR+) are sensitive to the presence of continuous regions, like background colors, because these are typically large and, hence, work

Fig. 4. Left: Mosaic of different kinds of remotely sensed forest patterns, south of Bourgogne ©IFN. Right: Segmentation map obtained by the TFR+ (4 classes).

as collectors of other regions. This becomes a critical problem when different textures have the same background color and share a long contour, where we can found many of such regions which cross the border and, therefore, link the textures forcing a merging. Unfortunately this was the case of the shadow regions present in the image (see Fig.3, left hand side). For this reason we decided to simply detect the background regions (just the shadows, in this case) after the CBC step, and ignore them in the subsequent steps (SBC and region merging).

Instead, in the latter experiment such modification was not necessary. The results are encouraging in this case as well. In particular, from the segmentation shown in Fig.4, we can see that the three different tree patterns have been detected with satisfactory precision. As for the mixed urban-tree area (bottom-left), the urban elements are assigned with a fourth class, while the trees are largely assigned with the correct tree class (that at bottom-right).

6 Conclusion

In this work we have presented a novel texture model which is particularly suited for the task of image segmentation in an unsupervised framework. The model aims at describing each texture at multiple scales through a region-based hierarchical approach which allows a very simple, but effective, segmentation scheme (the TFR) which processes color and spatial information in two independent steps, as to obtain an image decomposition in texture states. Finally a region merging procedure allows us to properly relate the states hierarchically, and single out the textured regions.

Numerical results proved the superior performance of the proposed method w.r.t. to other algorithms on the Prague benchmark data. Encouraging results have been obtained as well on satellite images. Future research will be focused on the replacement of k-means at SBC layer with a more effective clustering method.

Acknowledgments. This work was carried out during the tenure of an ERCIM fellowship (Scarpa's postdoc), and supported by EU MUSCLE project (e-team: shape modelling), FP6-507752, and partially by the project 1ET400750407. The authors would also like to thank the "French Forest Inventory" (IFN) for providing the remotely sensed data covering the forest areas.

References

1. Carson, C., Thomas, M., Belongie, S., Hellerstein, J.M., Malik, J.: Blobworld: A system for region based image indexing and retrieval. In: 3th ICVIS, Amsterdam, The Netherlands, pp. 509–516. Springer, Heidelberg (1999)
2. D'Elia, C., Poggi, G., Scarpa, G.: A Tree-Structured Markov random field model for Bayesian image segmentation. IEEE Transactions on Image Processing 12(10), 1259–1273 (2003)
3. Haindl, M., Mikeš, S.: Colour texture segmentation using modelling approach. In: Singh, S., Singh, M., Apte, C., Perner, P. (eds.) ICAPR 2005. LNCS, vol. 3687, pp. 484–491. Springer, Heidelberg (2005)
4. Poggi, G., Scarpa, G., Zerubia, J.: Supervised segmentation of remote-sensing images based on a tree-structured MRF model. IEEE Transactions on Geoscience and Remote Sensing 43(8), 1901–1911 (2005)
5. Beaulieu, J.-M., Touzi, R.: Segmentation of textured polarimetric SAR scenes by likelihood approximation. IEEE Transactions on Geoscience and Remote Sensing 42(10), 2063–2072 (2004)
6. S.Hese. Segmentation of forest stands in very high resolution stereo data. In Proc. IGARSS'01, Sydney (AUS), July 2001, vol.4, pp.1654–1656 (2001)
7. Mikeš, S., Haindl, M.: Prague texture segmentation data generator and benchmark. ERCIM News 64, 67–68 (2006) `http://mosaic.utia.cas.cz`.
8. Haindl, M., Mikeš, S.: Model-based texture segmentation. In: Campilho, A., Kamel, M. (eds.) ICIAR 2004. LNCS, vol. 3212, pp. 306–313. Springer, Heidelberg (2004)
9. Deng, Y., Manjunath, B.S.: Unsupervised Segmentation of Color-Texture Regions in Images and Video. IEEE Transactions on Pattern Analysis Machine Intelligence 23(8), 800–810 (2001)
10. Christoudias, C.M., Georgescu, B., Meer, P.: Synergism in Low Level Vision. In: proc. 16th ICPR, August 2002, vol.4, pp.150–155, Los Alamitos (2002)
11. Fischer, B., Thies, C.J., Guld, M.O., Lehmann, T.M.: Content-based image retrieval by matching hierarchical attributed region adjacency graphs. In: Proc. SPIE, vol. 5370, pp. 598–606, San Diego, CA (USA) (2004)
12. Duda, R.O., Hart, P.E., Stork, D.G.: Pattern Classification, 2nd edn. Wiley, Chichester (2000)
13. Scarpa, G., Haindl, M.: Unsupervised Texture Segmentation by Spectral-Spatial-Independent Clustering. In: Proc. 18th ICPR, Hong Kong (China), August 2006, vol.2, pp.151–154 (2006)
14. Penny, W.D.: Kullback-Leibler divergences of normal, Gamma, Dirichlet and Wishart densities. Technical report, Wellcome Department of Imaging Neuroscience, University College London (UK) (2001)

A Framework for Multiclass Reject in ECOC Classification Systems

Claudio Marrocco, Paolo Simeone, and Francesco Tortorella

DAEIMI, Università degli Studi di Cassino
Via G. Di Biasio 43, 03043 Cassino (FR), Italia
{c.marrocco,paolo.simeone,tortorella}@unicas.it

Abstract. ECOC is a diffused and successful technique to implement a multiclass classification system by decomposing the original problem in several two-class problems. In this paper we propose ECOC systems with a reject option carried out through two different schemes. The first one estimates the reliability of the output of the ECOC system and does not require any change in its structure. The second scheme, instead, estimates the reliability of the internal dichotomizers and implies a slight modification in the decoding stage. A final investigation is done on the sequential combination of both methods.

Keywords: ECOC, reject option, multiple classifiers systems.

1 Introduction

A diffused technique to face a classification problem with many possible classes is to decompose it into a set of two class problems. The rationale of this approach rely on the stronger theoretical roots and better comprehension characterizing two class classifiers (dichotomizers) such as Perceptrons or Support Vector Machines that, with this method, become employable in multiclass problems.

In this framework, Error Correcting Output Coding (ECOC) has emerged as a well established technique for many applications in the field of Pattern Recognition and Data Mining, mainly for its good generalization capabilities. In short, ECOC decomposition labels each class with a bit string (*codeword*) of length L, higher than the number of classes. The codewords are arranged as rows of a *coding matrix*, whose columns define each a two class problem; thus, for each problem, the set of the original classes parts into two complementary superclasses. On such problems induced by the coding matrix, L dichotomizers have to be trained in the learning phase. In the operating phase, the dichotomizers will provide a string of L outputs for each sample to be classified. The Hamming distance of such string from each of the codewords of the coding matrix is then evaluated and the class that corresponds to the nearest codeword is chosen. Usually, the codewords are chosen so as to have a high Hamming distance between each other; in this way, ECOC is robust to potential errors made by the dichotomizers. The reasons for the classification efficiency exhibited by ECOC seem to be the reduction of both bias and variance [1] and the achievement of

B.K. Ersbøll and K.S. Pedersen (Eds.): SCIA 2007, LNCS 4522, pp. 313–323, 2007.

a large margin [2]. After the seminal paper by Dietterich and Bakiri [3], many studies have been proposed which have analyzed several aspects of ECOC such as the factors affecting the effectiveness of ECOC classifiers [4], techniques for designing codes from data [5], evaluations of coding and decoding strategies [2].

A very common point in many applications in which the ECOC approach is used is that a classification error could have serious consequences, usually expressed by means of an error cost. In some cases, such cost can be so high that it is convenient to reject the sample (i.e. to suspend the decision and call for a further test) instead of risking a wrong decision. Obviously, also this choice involves a not negligible cost given by the charge of employing a more powerful system or requiring the decision of a human expert. Thus a rule is needed to find the optimal trade off between errors and rejects for the application at hand.

This paper proposes the introduction of a reject option for ECOC systems accomplished through two different schemes. The first one works on the output of the whole classification system and the reject is accomplished by considering the Hamming distance among the output codeword and the rows of the coding matrix. In the second scheme the reject option is performed on the base classifiers output by taking into account the confidence degrees provided by the dichotomizers. Such scheme makes use of a particular decoding technique for the erased bit in the codeword corresponding to rejects. To generalize the reject option, the cascade of the two approaches has been considered too.

In the rest of the paper we present, after a short description of the ECOC approach, the two schemes performing the reject option and the cascade of them. The successive section describes the results obtained from experiments performed on some UCI repository data sets. Some conclusions and future developments are drawn in the last section.

2 The ECOC Approach

The Error Correcting Output Coding has been introduced to decompose a multiclass problem into a set of complementary binary problems. Each class label is represented by a bit string of length L, called *codeword*, with the only requirement that distinct classes are represented by distinct codewords. If n is the number of the original classes, a code is a $n \times L$ matrix $\mathbf{C} = \{c_{hk}\}$ where $c_{hk} \in \{0, 1\}$. Each row of $\mathbf{C}$ corresponds to a codeword for a class, while each column corresponds to a binary problem. In this way, the multiclass problem is reduced to L binary problems on which L dichotomizers have to be trained. An example of coding matrix with $n = 5$ and $L = 12$ is shown in table 1. In the training phase, each dichotomizer is learned from a finite set of samples. In the operating phase, the sample $\mathbf{x}$ to be classified is fed to all the dichotomizers and each of them produces a binary value: all such values are collected to make a vector of binary decisions (*output vector*) to be compared with the codewords of the coding matrix. It is possible that some dichotomizer makes a wrong prediction, but this does not necessarily lead to an irrecoverable error in the multiclass problem since the code matrix is built by n distinct codewords of length $L > n$,

Table 1. An example of a coding matrix for a 5 classes problem

classes	codewords											
	f_1	f_2	f_3	f_4	f_5	f_6	f_7	f_8	f_9	f_{10}	f_{11}	f_{12}
1	0	0	1	1	1	1	0	1	0	1	1	0
2	1	0	0	1	0	0	0	1	1	1	1	0
3	0	0	1	1	0	1	1	1	0	0	0	0
4	1	1	1	0	1	0	1	1	0	0	1	0
5	0	1	0	0	1	1	0	1	1	1	0	0

so as to make the Hamming distance between every pair of codewords as large as possible. The Hamming distance D_H between two words is given by the number of position where the bit patterns of the two words differ.

The minimum Hamming distance (MHD) $d = \min_{i,j} D_H(\mathbf{c}_i, \mathbf{c}_j)$ between any pair of codewords is a measure of the quality of the code. In particular it is possible to correct codewords which contains no more than $\lfloor (d-1)/2 \rfloor$ single bit errors. In this way, a single bit error does not influence the result, as it can happen when using the usual one-per-class coding, where the Hamming distance between each pair of strings is 2. To pass from the binary to the multiclass problem, the most common approach consists in evaluating the Hamming distances between the output vector $\mathbf{o}$ and the codewords of the matrix and choose for the nearest codeword, i.e. for the codeword exhibiting the minimum Hamming distance from the output vector. Therefore, the decision for the k-th class ω_k corresponding to the $\mathbf{c}_k$ codeword is taken according to:

$$\omega_k = \arg\min_j (D_H(\mathbf{c}_j, \mathbf{o})), \qquad (1)$$

In particular, if the dichotomizers have soft output (e.g. they provide a confidence degree which is a real value ranging from 0 to 1), it is necessary to threshold their responses to obtain the value of the bits in the corresponding positions of the codeword.

3 The Multiclass Reject Option

The goal of this paper is to introduce a reject option for a multiclass problem in order to decrease the total classification cost by turning as many errors as possible into rejects. In fact, for a realistic problem, the error cost should be higher than a reject cost and thus an effective reject option is advantageous for the original multiclass classification problem. In general, a reject option is accomplished on a classifier by evaluating in some way the reliability of the decision taken by the classifier and rejecting the decision if the reliability is lower than a given threshold. In the case of an ECOC-based classification system, there are actually two places in which a decision is taken: the first place is the decoding stage, where the final multiclass decision is taken on the basis of the MHD. The second place is given by all the dichotomizers, each taking a two-class decision.

As a consequence, two different strategies are possible. The first one affects the decoding stage and evaluates the reliability of the multiclass decision on the basis of the MHD obtained; we will define *external* such scheme since it works at the output of the whole classification system. The second scheme (*internal* scheme), instead, evaluates the reliability of the outputs coming from the dichotomizers and rejects the decisions not sufficiently reliable. This approach affects the structure of the output vector since, in this case, it will contain, besides the usual values of 0 and 1, another symbol (let us call it r) which indicates that for the corresponding dichotomizer a reject has been taken. The decoding algorithm has to be consequently modified in order to handle the 3-value output vector. Obviously, this makes the second approach quite less general since, besides the change on the decoding stage, it puts some requirements on the characteristics of the dichotomizers to be employed. The two schemes are described in the following sections together with the cascade of them.

3.1 The External Reject Option

In literature, the decision in the ECOC approach has been principally based on the minimization of the Hamming distance among the codewords of the coding matrix and the output vector produced by the dichotomizers. Every employed dichotomizer gives an output that can be thresholded and combined to determine the final output vector: $\mathbf{o} = (o_1, o_2, \ldots, o_L)$.

Let us consider two codewords $\mathbf{c}_h$ and $\mathbf{c}_k$ that differs on d bits. If the number of erroneous bits is lower than $d/2$ we can correctly decode the word by using the MHD rule. When the number of errors is higher than $d/2$ it is not possible to recover the right codeword, i.e., the final decoding will be erroneous. This means that the greater is the Hamming distance between the output vector and the correct codeword the greater is the probability of an erroneous decision. In this situation it is possible to consider a reject rule based on the Hamming distance that introduces a reject region between the two codewords. This allows us to avoid to take a decision when the distance between the output vector and its nearest codeword is too high. If t_e is the reject threshold and ω_k is the class chosen according to eq. (1) the reject rule is:

$$r(\mathbf{o}, t_e) = \begin{cases} \omega_k & \text{if } D_H(\mathbf{c}_k, \mathbf{o}) < t_e, \\ reject & \text{if } D_H(\mathbf{c}_k, \mathbf{o}) \geq t_e. \end{cases} \tag{2}$$

Fig. 1 shows an example for such a problem. In fig. 1.a two samples belonging to the class ω_p produce two output vectors $\mathbf{o}_1$ and $\mathbf{o}_2$. In the first case a correct decision is taken while $\mathbf{o}_2$ will be assigned to the wrong class ω_q. Introducing the reject rule, a decision for the vector $\mathbf{o}_2$ will not be taken so avoiding an error (see fig. 1.b). It is worth noting that the lowest Hamming distance is zero while the highest one depends on the codewords of the matrix $\mathbf{C}$. If L is the maximum distance that we can have between two codewords (i.e., in the coding matrix there are two complementary rows) an upper bound of the maximum distance allowable for the reject threshold is $L/2$. It is worth noting that such scheme

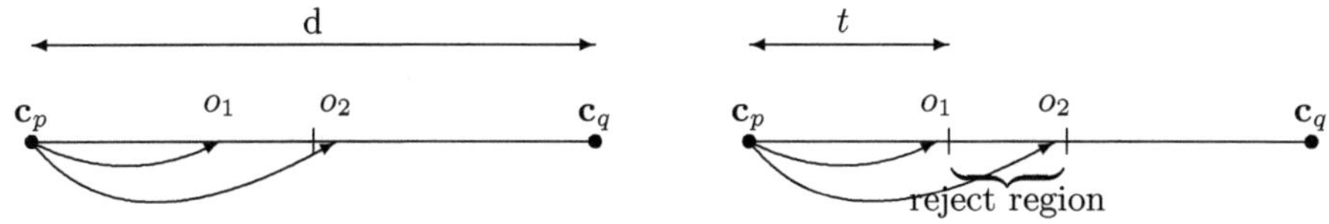

Fig. 1. Example of the decoding method based on the MHD in the standard approach (a) and with an external reject option (b)

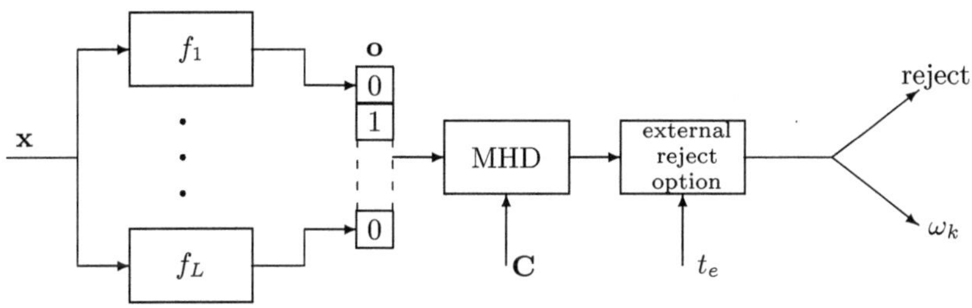

Fig. 2. The block diagram for the external reject rule

does not require any assumption neither on the dichotomizers nor on the coding matrix. The whole scheme is described in fig. 2.

3.2 The Internal Reject Option

Let us now suppose that we can estimate the reliability of the output of each dichotomizer in the ECOC system. For example, let us consider a model for the dichotomizer which provides a soft value ranging from 0 to 1. In this case, we should threshold the soft output to have a crisp response with a typical threshold value of 0.5. However, it is easy to see that a value for the soft output falling near the threshold will be much less reliable than a value near 0 or near 1. As a consequence, we can adopt a reject rule for each dichotomizers as:

$$o_j(\mathbf{x}, t_i) = \begin{cases} 1 & \text{if } f_j(\mathbf{x}) > 0.5 + t_i \\ 0 & \text{if } f_j(\mathbf{x}) < 0.5 - t_i \\ r & \text{otherwise} \end{cases} \tag{3}$$

Since in this case the output vector can also contain rejected bits, i.e. $c_i \in \{0, 1, r\}$, we have to focus on a decoding rule able to handle the 3 values. To this aim, it is possible to analyze the effect of an erasure (i.e. a reject) on the ECOC system. If μ is the number of erasures, the minimum distance between codewords (evaluated on the unerased bits) becomes $d - \mu$ and the error correcting capability of the code decreases to $\lfloor (d - \mu - 1)/2 \rfloor$. Therefore, to have a correct decision the number of errors and erasures should verify the following condition:

$$2\nu + \mu < d \tag{4}$$

where ν is the number of errors. This means that is twice difficult to correct an error than to correct an erasure. To show how the internal reject can be advantageous for the final decision, let us consider an output vector affected by ν errors. Without internal reject a correct decision will be taken if $2\nu < d$. Applying the internal reject rule we turn some erroneous bits (say μ_1) into erasures while the remaining erasures (say $\mu_2 = \mu - \mu_1$) come from correct decisions. In this case, the correct decision will be taken if $2\nu_1 + \mu_1 + \mu_2 < d$ where $\nu_1 = \nu - \mu_1$. Therefore, we will take advantage from the internal reject if $\mu_1 < \mu_2$ that is if at least half of the erasures comes from erroneous bits.

In order to take a decision, an erasure filling method called *erasure decoding* [6] is adopted in the decoding stage. To understand its rationale, let us suppose to replace all the erased bits by 0 and decode the obtained vector. If no more than half of the erasures should have been ones and eq. (4) is satisfied, then the number of errors is still less than half of d and the decoding will be correct. On the other hand, if more than half of the erasures should have been ones then we are introducing other bit errors and the decision will be erroneous. In this case, if we fill all the erased bits with 1 the decision will be successful. Therefore, the erasure filling procedure consists in decoding twice and choose the codeword that is closer to the output vector in terms of Hamming distance. The resulting procedure can be summarized as follows:

1. Place zeros in all erased position and decode to the closer codeword (in Hamming distance terms) $\mathbf{c}^{(0)}$;
2. Place ones in all erased position and decode to the closer codeword (in Hamming distance terms) $\mathbf{c}^{(1)}$;
3. Choose the closest $\mathbf{c}^{(j)}$ to the received codeword in the unerased positions, where $j = 0, 1$.

The first two steps of the algorithm are meant to solve the rejects/erasures while the last one exploits the error correction capability of the code.

However, it could happen that the output vector falls (according to the erasure decoding) on the halfway between two different codewords. In this case, the decision can not be reliably taken and thus a reject is produced. The complete rule can be described as:

$$r(\mathbf{C}, \mathbf{x}) = \begin{cases} \omega_k^{(0)} & \text{if } D_H^*(\mathbf{c}_k, \mathbf{c}^{(0)}) < D_H^*(\mathbf{c}_k, \mathbf{c}^{(1)}), \\ \omega_k^{(1)} & \text{if } D_H^*(\mathbf{c}_k, \mathbf{c}^{(1)}) < D_H^*(\mathbf{c}_k, \mathbf{c}^{(0)}), \\ \mathit{reject} & \text{if } D_H^*(\mathbf{c}_k, \mathbf{c}^{(0)}) = D_H^*(\mathbf{c}_k, \mathbf{c}^{(1)}). \end{cases} \tag{5}$$

where D_H^* is the Hamming distance on the unerased bits. The resulting system is shown in fig. 3.

3.3 The Cascade Reject Option

It is worth noting that the output of the ECOC system provided with the internal reject option is still based on the MHD criterion. Therefore, it is possible to implement a cascade of the two procedures before described using the output

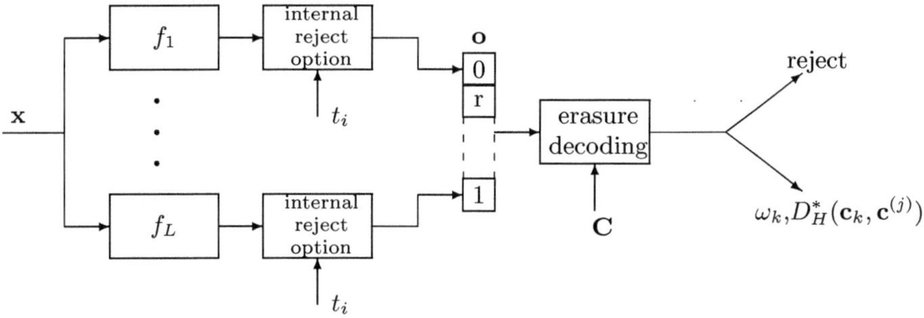

Fig. 3. The block diagram for the internal reject rule

of the internal rule as input for the external reject option. In such a case the Hamming distance between the codewords and the output vector (and then the threshold for the external reject option) is evaluated only on the unerased bits. The goal of the cascade of the two methods is to reduce the number of erroneous decision that we obtain after the internal rule. It is worth noting that in this case we have to choose two different thresholds. A block scheme of this approach (that we called *cascade reject rule*) is reported in fig. 4.

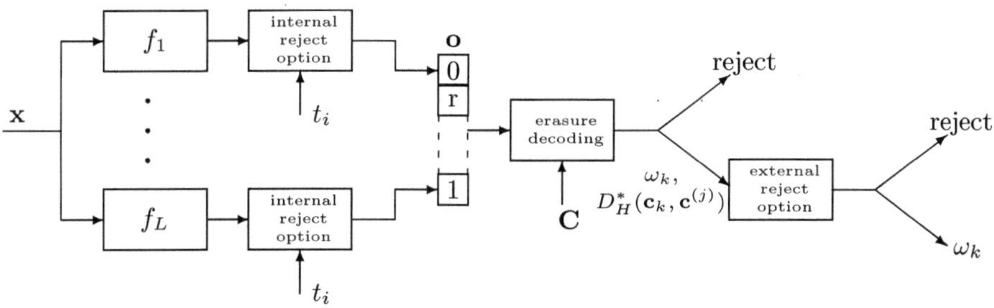

Fig. 4. The block diagram for the cascade reject rule

4 Experiments

In order to evaluate the performance of the proposed methods, experiments were made on some data sets publicly available at the UCI Machine Learning Repository [7]; all of them have numerical input features and a variable number of

Table 2. Data sets and coding matrices used in the experiments

Data Sets	# Classes	# Features	Coding Matrix	Length (L)	# Samples
Glass	6	9	Exhaustive	31	214
SatImage	6	36	Exhaustive	31	6435
Yeast	10	8	BCH 31-21	31	1484
Vowel	11	10	14-11	14	435

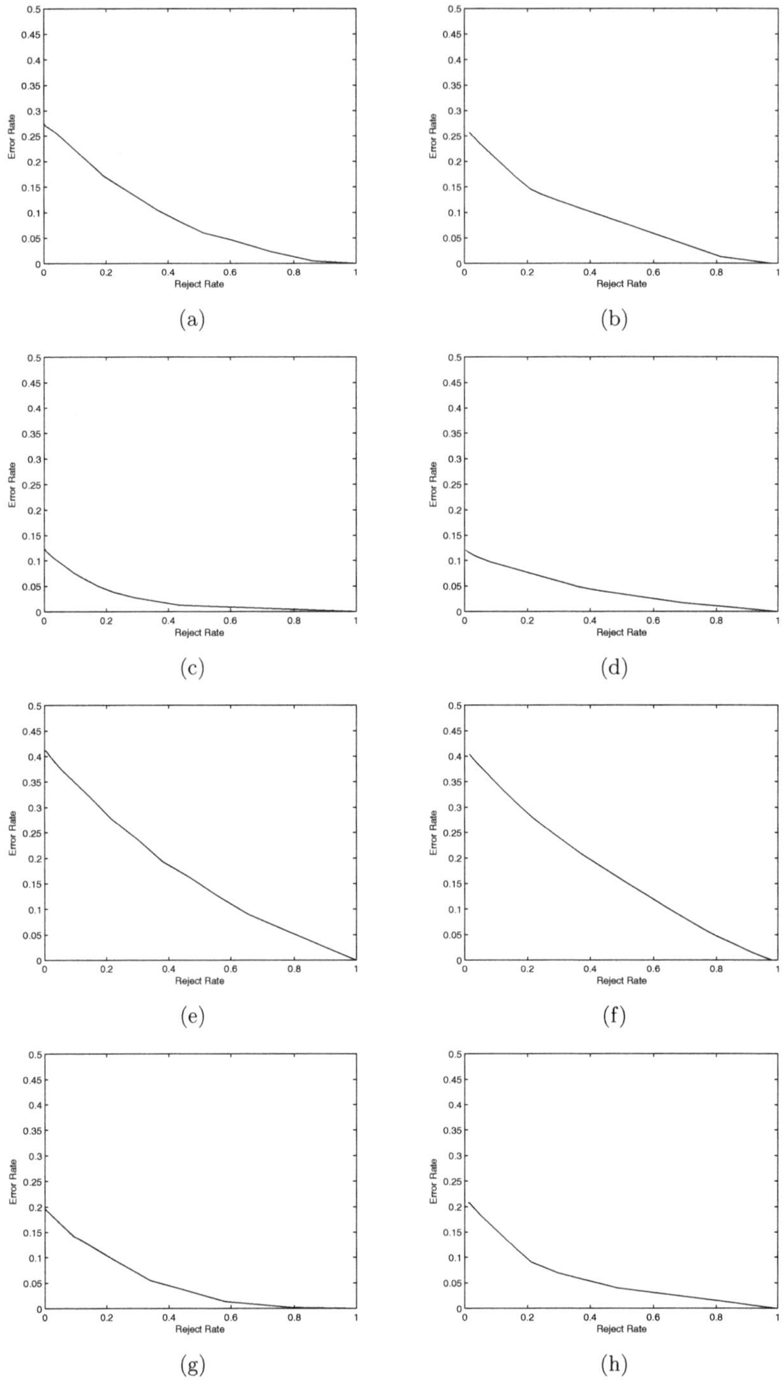

Fig. 5. Comparison between external (left side) and internal (right side) reject option on the different data sets: (a-b) Glass, (c-d) SatImage, (e-f) Yeast, (g-h) Vowel

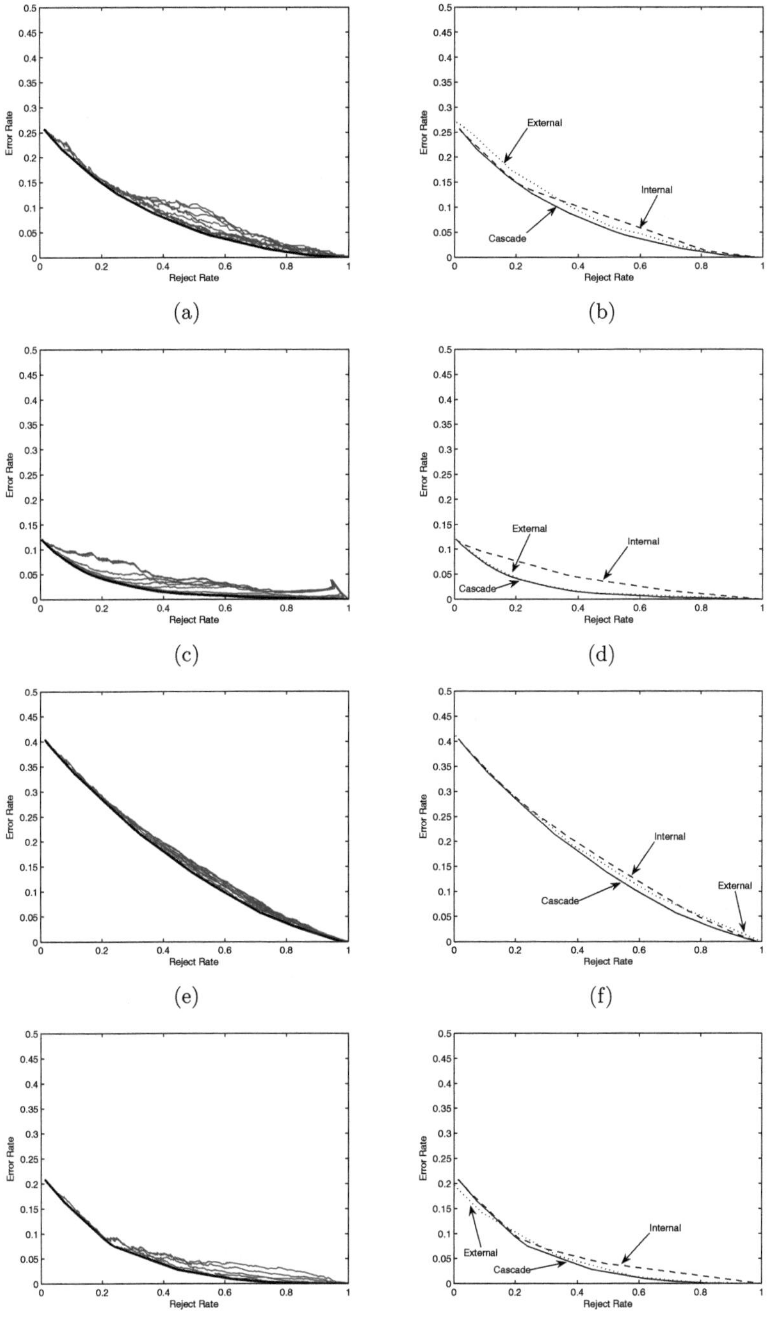

Fig. 6. The results obtained with the cascade option (left side) and the comparison between the three methods (right side) on the different data sets: (a-b) Glass, (c-d) SatImage, (e-f) Yeast, (g-h) Vowel

classes. More details for the data sets are given in table 2. The table provides also the type of ECOC matrix employed for each data set. As in [3] we have chosen an exhaustive code for the sets that have a number of classes lower than 8 and a BCH code for those having a number of classes greater than 8. In particular, for Vowel data set we used a matrix (named 14-11) with a reduced number of columns available at `http://web.engr.oregonstate.edu/~tgd/software/ecoc-codes.tar.gz` As base dichotomizers in the ECOC framework Modest AdaBoost [8] has been used using a simple decision tree as weak learner with a randomized number of splits in every run. To avoid any bias in the comparison, 12 runs of a multiple hold out procedure have been performed on all the data sets. In each run, the data set has been split in three subsets: a training set (containing the 70% of the samples of each class) to train the base classifiers, a validation set and a test set (each containing the 15% of the samples of each class) used respectively to normalize the outputs into the range $[0, 1]$ and to evaluate the performance for the multiclass classification.

To compare the different methods a useful representation to evaluate the benefits of a reject option is the error-reject curve that has been built varying the opportune thresholds t_i and t_e for all the data sets. In fig. 5 we report the results of the comparison between the external and internal schemes. The number of reject thresholds for the two cases are different: the external approach considers values ranging between $[0, L/2]$ as discussed in section 3.1 while the internal rule considers all the possible normalized output values observed in the range $[0, 0.5]$. It should be also noted that since for the internal option we fix a multiclass reject rule (see eq. 5) we obtain a reject rate always greater than zero since we can have a reject even if $t_i = 0$. Experimental results does not show better performance of one of these strategies on the other but they are practically equivalent. In fig. 6 we show (on the left side) the results obtained on each data set with the cascade approach. In each graph the error-reject curves varying the internal threshold for a fixed external threshold are reported. For the sake of comparison the convex hull of all these curves has been evaluated and compared with the two previous methods in the right side of fig. 6. The cascade option presents always a lower error-reject curve on all the data sets with only one exception on Vowel data set (see fig. 6) where for the range $[0, 0.18]$ the curve of the external reject rule exhibits lower error probabilities.

5 Conclusions

In this paper we have proposed two schemes to provide an ECOC classification system with a reject option. The experiments have shown that the two methods give similar results, even though they are effective on different situations. In fact, when both are activated in a cascade scheme, the results obtained are clearly better. The future work will focus on the analysis of particular codes more suitable for erasure decoding.

References

1. Kong, E.B., Dietterich, T.G.: Error-Correcting Output Coding Corrects Bias and Variance In: International Conference on Machine Learning, pp.313–321 (1995)
2. Allwein, E.L., Schapire, R.E., Singer, Y.: Reducing multiclass to binary: A unifying approach for margin classifiers. In: Proceedings of the Seventeenth International Conference on Machine Learning, pp. 9–16 (2000)
3. Dietterich, T.G., Bakiri, G.: Solving multiclass learning problems via error-correcting output codes. Journal of Artificial Intelligence Research 2, 263–286 (1995)
4. Masulli, F., Valentini, G.: An experimental analysis of the dependence among codeword bit errors in ECOC learning machines. Neurocomputing 57, 189–214 (2004)
5. Pujol, O., Radeva, P., Vitrià, J.: Discriminant ECOC: A heuristic method for application dependent design of error correcting output codes. IEEE Trans. On Pattern Analysis And Machine Intelligence 28(6), 1001–1007 (2006)
6. Morelos-Zaragoza, R.H.: The Art of Error Correcting Coding. Wiley & Sons, Chichester (2002)
7. Blake, C., Keogh, E., Merz, C.J.: UCI Repository of Machine Learning Databases, www.ics.uci.edu/~mlearn/MLRepository.html (1998)
8. Vezhnevets, A., Vezhnevets,V.: Modest AdaBoost - Teaching AdaBoost to Generalize Better. Graphicon-2005, Novosibirsk Akademgorodok, Russia, http://graphics.cs.msu.su/en/research/boosting/index.html (2005)

Scale-Space Texture Classification Using Combined Classifiers

Mehrdad J. Gangeh[1,2], Bart M. ter Haar Romeny[2], and C. Eswaran[3]

[1] Multimedia University, Faculty of Engineering, Cyberjaya, Malaysia
mehrdad@mmu.edu.my
[2] Eindhoven University of Technology, Department of Biomedical Engineering, Biomedical Image Analysis, Eindhoven, The Netherlands
{M.Gangeh,B.M.terhaarRomeny}@tue.nl
[3] Multimedia University, Faculty of Information Technology, Cyberjaya, Malaysia
eswaran@mmu.edu.my

Abstract. Since texture is scale dependent, multi-scale techniques are quite useful for texture classification. Scale-space theory introduces multi-scale differential operators. In this paper, the N-jet of derivatives up to the second order at different scales is calculated for the textures in Brodatz album to generate the textures in multiple scales. After some preprocessing and feature extraction using principal component analysis (PCA), instead of combining features obtained from different scales/derivatives to construct a combined feature space, the features are fed into a *two-stage combined classifier* for classification. The learning curves are used to evaluate the performance of the proposed texture classification system. The results show that this new approach can significantly improve the performance of the classification especially for small training set size. Further, comparison between combined feature space and combined classifiers shows the superiority of the latter in terms of performance and computation complexity.

Keywords: Scale-space, multi-scale, texture classification, combined classifiers, Gaussian derivatives.

1 Introduction

There is a vast literature on texture analysis, as can be judged from the innumerable applications of texture analysis in various fields [1].

In recent years, multi-scale and multiresolution techniques have been recognized as vital tools for texture analysis. This is because texture displays a multi-scale property. Whatever may be the representation, it is applicable in different scales.

Some of multiresolution techniques on texture analysis in the literature are: multiresolution histograms [2, 3] including locally orderless images [4], techniques based on multi-scale local autocorrelation features [5], multi-scale local binary patterns [6], multiresolution Markov random fields [7], wavelets [8, 9], Gabor filters [8, 10, 11], Gabor wavelet filters [12], Markov models in the wavelet domain [13], and techniques based on scale-space theory [4, 14].

B.K. Ersbøll and K.S. Pedersen (Eds.): SCIA 2007, LNCS 4522, pp. 324–333, 2007.

In almost all papers related to multiresolution texture classification, the features obtained from different scales are concatenated to construct a combined feature space. These combined features are then fed to a classifier for the purpose of texture classification. However, fusion of features obtained from different scales produces a high dimensional feature space. Working in this high dimensional feature space usually imposes problems as we need more data samples for training the classifier. This phenomenon is called the 'curse of dimensionality'. It may cause the peaking phenomena in classifier design [15].

This problem is addressed in some of above papers by using classifiers that behave better in high dimensional feature space, e.g. support vector machines (SVMs) [3, 11]. The rest of the papers try to solve this problem by applying severe feature reduction by using feature selection/extraction techniques.

On the other hand, the multiresolution papers in texture classification only report the classification error for a single specific training set size (usually a large training set size). This keeps the behavior of the classifier unrevealed in small training set sizes that might be important in some applications especially those where obtaining a large training set size is difficult, costly or even impossible. This is particularly the case in texture classification applications on medical images: obtaining medical images for some specific diseases is cumbersome especially as standardized protocols for image acquisition need to be followed [16].

In this paper, we address these two issues. Firstly, we use combined classifiers instead of combining features. In combined classifiers, features produced in each scale are applied to a base classifier and hence feature fusion is no longer required. The outputs of these base classifiers are then combined using a combining rule for a final decision on the class of each texture. Secondly, the learning curves for training the classifiers are constructed using different training set sizes. This clearly shows how the training of the classifier evolves as we increase the training set size. As the results show, this leads to an important conclusion: the classifier performance is improved especially in small training set size which is important in applications mentioned above.

Scale-space theory in the context of multi-scale texture classification is presented in section 2. In section 3, the experiments are explained followed by the results in section 4. Eventually, the effectiveness of the method especially in small training set sizes is discussed in section 5.

2 Scale-Space Texture Classification

A texture classification system typically consists of several stages such as preprocessing, feature extraction and classification. Each stage is explained below in the context of scale-space texture classification.

2.1 Scale-Space Theory

Texture is usually considered as a repetitive pattern and the basic repetitive structure is of varying size in different textures. This inspires us to apply multi-scale techniques in texture analysis. Here, scale-space theory, which is biologically motivated by the models of early stages of human vision [14], is used for multi-scale texture classification.

The key notion in scale-space theory is that images are observed through a finite aperture device (CCD element, rod/cone etc.). The size is a free parameter (scale). A linear scale-space is a stack of Gaussian blurred images. The generating equation of the images at different scales is the linear isotropic diffusion equation as given in (1)

$$\Delta_{\mathbf{x}} L(\mathbf{x};\sigma) = \frac{\partial L(\mathbf{x};\sigma)}{\partial t} , \tag{1}$$

where Δ is the Laplacian, $t = \sigma^2/2$ is the variance, and σ is the scale. Diffusion of intensity is equivalent to convolving the image with Gaussian kernels of variance σ^2 given in (2), as the Gaussian kernel is the Green's function of (1).

$$G_{2D}(\mathbf{x};\sigma) = \frac{1}{2\pi\sigma^2} e^{-\frac{\mathbf{x}^2}{2\sigma^2}} . \tag{2}$$

This convolution blurs the image and depending on the scale of the Gaussian kernel will emphasize coarser structures in the image, which is desirable in texture analysis as textures are scale dependent.

Derivatives are additional information about textures. Derivatives of discrete data are ill-posed. According to scale-space theory, observed data is a Gaussian convolution with the image

$$L(\mathbf{x};\sigma) = L(\mathbf{x}) \otimes G_{2D}(\mathbf{x};\sigma) , \tag{3}$$

where $\otimes$ is the convolution operator. If we take the derivative of both sides we obtain

$$\frac{\partial}{\partial \mathbf{x}} L(\mathbf{x};\sigma) = \frac{\partial}{\partial \mathbf{x}} [L(\mathbf{x}) \otimes G_{2D}(\mathbf{x};\sigma)] \cdot \tag{4}$$

Both convolution and differentiation are linear operators, so we can exchange the order of these two operators

$$\frac{\partial}{\partial \mathbf{x}} L(\mathbf{x};\sigma) = L(\mathbf{x}) \otimes \frac{\partial}{\partial \mathbf{x}} G_{2D}(\mathbf{x};\sigma) , \tag{5}$$

which means that the derivative of an observed image can be calculated by convolving the image with the derivative of the Gaussian kernel. The order of the derivative determines the type of structure extracted from the image, e.g. the 1st order derivative emphasizes on the edges, the 2nd order on ridges and corners and so on.

To construct multi-scale texture images we simply convolve the Gaussian derivatives including the zeroth order derivative (the Gaussian kernel itself) with the texture image. However, there are two main questions in this multi-scale texture construction. First, up to what order do we need to consider the derivatives? Second, in each order, in how many different orientations the derivatives should be taken? The answer to the first question is application dependent. Practically, we only consider the derivatives up to the second order to prevent excessive computational load. The answer to the second question is based on the steerability property of Gaussian derivatives [14]. The nth order derivative at any given orientation can be constructed from n+1 independent orientations. E.g. if n = 1, from only L_x and L_y the derivatives in all other orientations can be calculated using

$$L_{\mathbf{x}}^{\theta}(x,y) = Cos(\theta)L_x(x,y) + Sin(\theta)L_y(x,y) \ . \tag{6}$$

where $L_{\mathbf{x}}^{\theta}$ indicates the derivative of image $L(\mathbf{x})$ in orientation θ.

To discriminate two or more textures, we use the additional information provided in different scales/derivatives to achieve a better performance (comparing to single scale). To this end, the patches are extracted from the textures in the original and derivative scale spaces. The size of the patch is scale dependent. As we go to higher scales, i.e. lower resolutions, more emphasis is given to coarser structures and (in similar fashion as our visual system) we need to look at these structures through larger windows. Thus, we increase the size of patches when we go to higher scales. These patches can be subsampled to reduce the computation load.

The main goal in this multi-scale texture classification system is to identify to what extent the additional information provided by other scales/derivatives can improve the performance. Hence, the patches should be taken from the same spatial locations in the scale-space. This is shown graphically in Fig. 1 for one patch extracted from different scales/derivatives for a typical texture image.

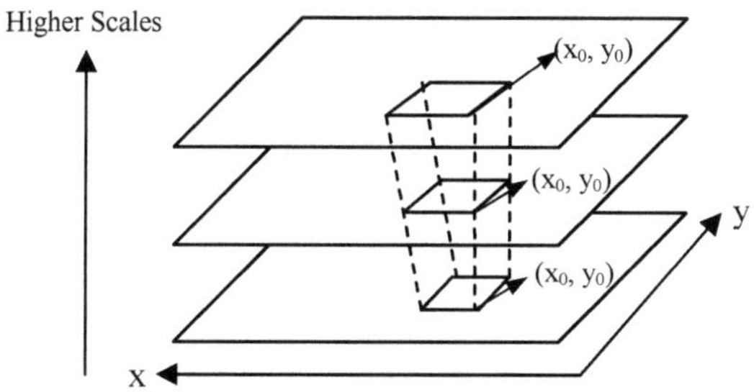

Fig. 1. Illustration of the method for selection of a typical patch from three different scales of the same texture. The patches are taken from the same spatial locations in different scales. However, the size of patch increases as we go to higher scales. The spatial location should be also the same in different derivatives of the same texture.

2.2 Feature Extraction

The patch size affects the dimensionality of the feature space as larger patches generate more features. This may impose problems with respect to the computation speed and the necessary training set size. Increasing the feature space dimension usually degrades the performance of the classifier unless we dramatically increase the number of data samples for training (curse of dimensionality) and may cause the peaking phenomena in classifier design. There are two solutions to this problem: to increase the training set size, and to reduce the feature space dimension using feature selection/extraction techniques, e.g. Principal Component Analysis (PCA).

2.3 Combined Classifier

As mentioned in the introduction, the common practice in multiresolution texture classification is concatenating the features obtained from different resolutions to construct a combined feature space. This combined feature space is fed to a classifier. This produces a high dimensional feature space and therefore severe feature reduction is required using feature selection/extraction techniques. The alternative method proposed in this paper is using *combined classifiers.*

Combined classifiers are used in multiple classifier source applications like different feature spaces, different training sets, different classifiers applied for example to the same feature space, and different parameter values for the classifiers, e.g. k in k-nearest neighbor (k-NN) classifiers.

In this approach, the features obtained from each scale are applied to a base classifier. As the features spaces generated from different scales/derivatives are different in our method, parallel combined classifiers seem to be natural choice. Use of combined classifiers lifts the requirement for feature space fusion. Moreover, each base classifier can be examined separately, e.g. by drawing the corresponding learning curve, to determine the contribution of each scale/derivative towards the overall performance of the classification algorithm as will be shown in section 4.

Here, we propose a two-stage parallel combined classifier with the structure shown in Fig. 2. As can be seen in this figure, in the first stage, the feature space from each scale/derivative is applied to a base classifier. At the output of this stage a fixed combining rule is applied to combine the outputs of different scales for one particular order of derivative. For example scales s1, s2 and s3 for the L_x are combined using a fixed combining rule. In the second stage, the outputs of the first stage, i.e. all different derivatives, are combined to produce the overall decision on the texture classes.

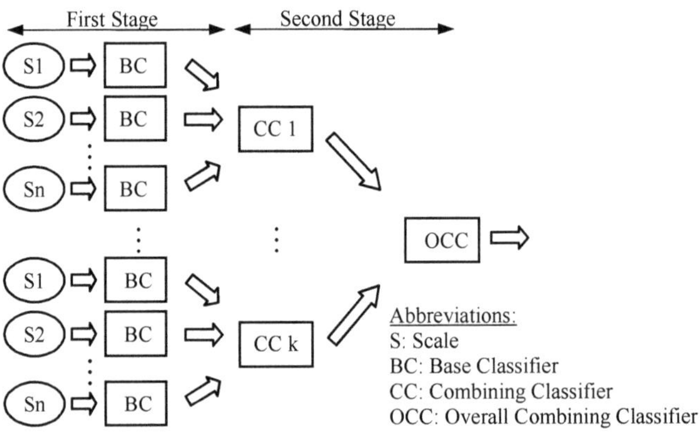

Fig. 2. The structure of two-stage combined classifier proposed in this paper

This two-stage parallel combined classifier provides a versatile tool to examine the contribution of each derivative in the overall performance of the classifier. However, it must be noticed that if two combining rules are the same, e.g. if both are fixed 'mean combining' rule, this two stage combined classifier is not different from one stage where all the feature spaces are applied to the base classifiers and the outputs are combined using the same combining rule.

Two parameters have to be selected in this combined classifier: the type of the base classifier and the combining rule. The selection of these two options is discussed in section 4.

2.4 Evaluation

One of the main shortcomings of the papers in multiresolution texture classification is reporting the performance of the algorithm for one single training set size. This causes that the behavior of the algorithm for different training set sizes remains unrevealed.

In this paper, to overcome this problem and to investigate the performance of the classifier, the learning curves are drawn by calculating the error for different training set sizes.

Since the train and test sets are generated from the same image in each scale/derivative, they may spatially overlap in the image domain. Hence, to assure that the train and test sets are separate, we split the texture images spatially into two halves and generate the train and test sets from each half.

As we have constructed the train set and test sets from two different halves of the image we can use a fixed test set size for calculating the error of the combined classifier trained using different train set sizes. This may improve the accuracy of error computation as we can increase the test set size independent from the train set size at the cost of higher computational load.

3 Experiments

To evaluate the effectiveness of the proposed method, some experiments are performed on a supervised classification of some test images. The test images are from the Brodatz album as shown in Fig. 3.

Construction of Multi-scale Texture Images. The N-jet of scaled derivatives up to the second order is chosen to construct the multi-scale texture images from each texture. Based on steerability of Gaussian derivatives, we have used the zeroth order derivative, i.e. the Gaussian kernel itself, L_x, L_y, L_{xx}, L_{xy}, L_{yy}, and L_{xx+} L_{yy} where the last one is the Laplacian. For each derivative (including the zeroth order) three scales are computed. The variances (σ^2) of the Gaussian derivatives in scales s1, s2 and s3 are 1, 4 and 7 respectively. In this way, for each texture image 7×3 texture images are computed.

Preprocessing. To make sure that for all textures the full dynamic range of the gray level is used contrast stretching is performed on all textures in different scales. Also, to make the textures indiscriminable to mean or variance of the gray level, DC cancellation and variance normalization are performed.

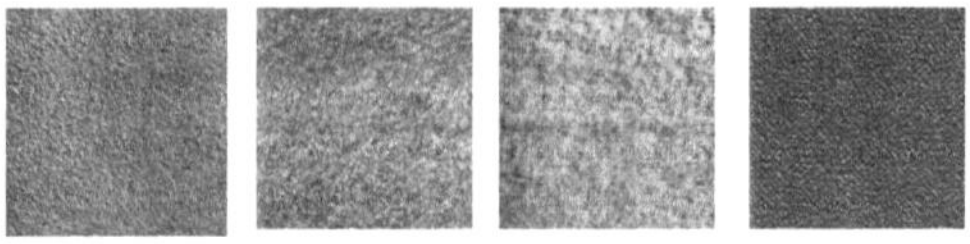

Fig. 3. Textures D4, D9, D19 and D57 from Brodatz album used in the experiments

Construction of Train and Test Sets. To ensure that the train and test sets are completely separate, they are extracted from the upper and lower half of each image respectively. For one texture, the patches are with sizes of 18×18, 24×24, and 30×30 in scales s1, s2 and s3 respectively. For the time being no subsampling is performed in higher scales but this might be considered in future for optimization of the algorithm in terms of speed and memory usage. Test set size is fixed at 900. Train set size increases from 10 to 1500 to construct the learning curves.

Feature Extraction. PCA is used for feature extraction. The number of components used for dimension reduction is chosen to preserve 95% of the original variance in the transformed (reduced) space.

Combined Classifier. A two-stage parallel combined classifier is used. A variety of base classifiers and combining rules are tested to find the best one. Among the base classifiers tested are some normal-based density classifiers like quadratic discriminant classifier (qdc), linear discriminant classifier (ldc), and nearest mean classifier (nmc). The Parzen classifier was also tested as a representative of non-parametric based density classifier. Twenty one base classifiers, one for each scale in one derivative order are used. The mean, product and voting fixed combining rules as well as the nearest mean trainable combining rule are tested for comparison.

Evaluation. The performance of the texture classification system is evaluated by drawing the learning curves for training set sizes up to 1500. The error is measured 5 times for each training set size and the results are averaged.

4 Results

A variety of tests are performed using different parameters as explained in the previous section. Among the tested base classifiers qdc performed the best. This can be understood as PCA is a linear dimension reduction that performs integration in the feature space. Consequently, the features tend to be normally distributed based on the central limit theorem. Of the tested combining rules 'mean combiner' performs best. Hence, the results are shown here based on experiments using qdc as base classifier and mean combining rule for both stages.

The left graph in Fig. 4 displays the learning curves obtained at the output of stage 1 (combined scales for each derivative order) and stage 2 (combine all derivatives in all scales) of the two-stage combined classifier. As can be seen, the overall performance of the classifier is improved comparing to each derivative order in different scales. This is especially significant in small training set sizes. However, for very small training set sizes, i.e. below 200, the output of the combined classifier is degraded significantly. This is because of peaking phenomena as the dimension of feature space in scale 1 is rather high even after applying PCA.

Combined Classifier with Regularization of Base Classifiers in Scale 1. To overcome the size problem, we applied regularization to the qdc base classifier in scale 1 for all derivative orders (right graph in Fig. 4). The performance improved significantly in very small training set sizes, i.e. below 200. One of the major drawbacks of regularization is that it ruins the results for large training set sizes. However, as we have only applied the regularization to scale 1, the information from other scales prevents deteriorating the performance in large training set sizes as can be judged from the graph.

Combined Classifier with the Same Patch Size in All Scales. The left graph in Fig. 5 illustrates the importance of increasing patch size in higher scales. In this figure, the patch sizes extracted from the multi-scale texture images in different scales are the same (18×18). It is apparent that the performance of the classifiers in higher scales is degraded compared to Fig. 4.

Combined Feature Space versus Combined Classifier. To demonstrate the superiority of our method to the common trend in the literature on multiresolution texture classification, we compare the results of two classification tests: combining the features obtained from different scales in the zeroth order derivative with the case that we use a combined classifier for the same derivative order in different scales. The results are shown in the right graph of Fig. 5 and the superiority of using a combined classifier to combined feature space is obvious. Combining all features generated from different scales/derivatives produces very high dimensional feature space that leads to extremely poor performance in combined feature space technique. Hence, the comparison has been made in only one derivative order. It should be also highlighted here that combined classifier approach is computationally much less expensive than combined feature space technique as applying fused feature space to one base classifier needs huge amount of memory space simultaneously and runs slowly compared to the combined classifiers.

5 Discussion and Conclusion

Scale-space theory, PCA and combined classifiers are integrated into a texture classification system. The algorithm is very efficient especially in small training set size. The system can significantly improve the performance of classification in comparison with single scale and/or to single derivative order based on the information provided in multiple scales and multiple derivative orders.

The two-stage combined classifier along with the learning curves proposed in this paper provides a versatile means to investigate the significance of different scales/derivatives in overall performance of the classifier. Regularization of qdc base classifier in scale 1 further improves the performance especially in small and very small training set sizes. This, however, does not deteriorate the results in large

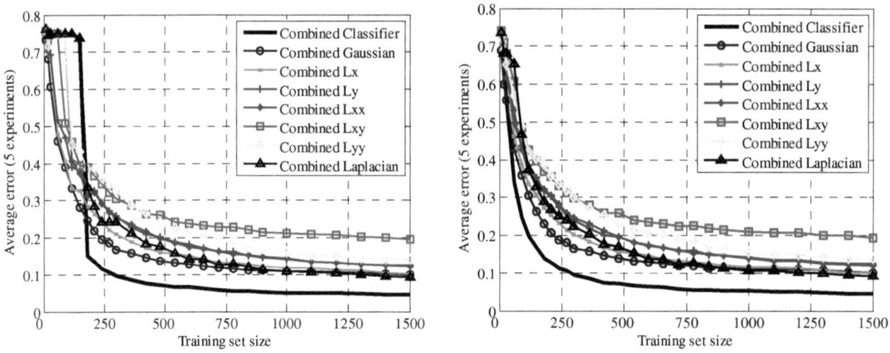

Fig. 4. Learning curves for the classification of 4 textures from the Brodatz album at multiple scales of single and multiple derivative orders without regularization (*left graph*) and with regularization (*right graph*) of base classifiers at scale 1

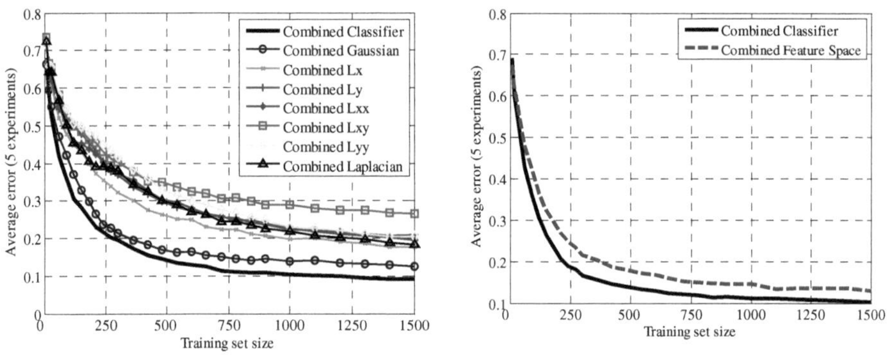

Fig. 5. Learning curves for the classification of 4 textures from the Brodatz album with regularization of base classifiers at scale 1 using the same patch size for all scales (*Left graph*). Learning curves obtained using the combined classifier and the combined feature space methods for the zeroth order derivative (*Right graph*).

training set sizes thanks to the information provided from other scales. This approach is superior to the combined feature space in terms of performance and computation load.

This algorithm is effective in situations where gathering many data samples for the train/test sets is difficult, costly or impossible. E.g. in ultrasound tissue characterization image acquisition has to be standardized and the ground truth for diffused liver diseases is biopsy, which is invasive, and collection of data is very time consuming. In these situations this algorithm can improve the performance of the classification over classical approaches as the train set size is small.

In future work, we want to apply this method to distinguish normal from abnormal lung tissue in high resolution CT chest scans.

Acknowledgement. The authors would like to thank R.P.W. Duin from Delft University of Technology for useful discussions throughout this research.

References

1. Materka, A., Strzelecki, M.: Texture Analysis Methods-A Review. Technical University of Lodz, Institute of Electronics, COSTB 11 report, Brussels (1998)
2. Hadjidemetriou, E., Grossberg, M.D., Nayar, S.K.: Multiresolution Histograms and Their Use for Recognition. IEEE Trans. on PAMI 26, 831–847 (2004)
3. Andra, S., Wu, Y.J.: Multiresolution Histograms for SVM-Based Texture Classification. In: Kamel, M., Campilho, A. (eds.) ICIAR 2005. LNCS, vol. 3656, pp. 754–761. Springer, Heidelberg (2005)
4. van Ginneken, B., ter Haar Romeny, B.M.: Multi-scale Texture Classification from Generalized Locally Orderless Images. Pattern Recognition, Vol. 36 pp. 899–911 (2003)
5. Kang, Y., Morooka, K., Nagahashi, H.: Scale Invariant Texture Analysis Using Multi-Scale Local Autocorrelation Features. In: Kimmel, R., Sochen, N.A., Weickert, J. (eds.) Scale-Space 2005. LNCS, vol. 3459, pp. 363–373. Springer, Heidelberg (2005)
6. Ojala, T., Pietikainen, M., Maenpaa, T.: Multiresolution Gray-Scale and Rotation Invariant Texture Classification with Local Binary Patterns. IEEE Trans. on PAMI 24(7), 971–987 (2002)
7. Wang, L., Liu, J.: Texture Classification Using Multiresolution Markov Random Field Models. Pattern Recognition Letters 20, 171–182 (1999)
8. Randen, T., Husoy, J.H.: Filtering for Texture Classification: A Comparative Study. IEEE Trans. on PAMI 21, 291–310 (1999)
9. Li, S.T., Shawe-Taylor, J.: Comparison and Fusion of Multiresolution Features for Texture Classification. Pattern Recognition Letters 26, 633–638 (2005)
10. Jain, A.K., Farrokhnia, F.: Unsupervised Texture Segmentation Using Gabor Filters. Pattern Recognition 24, 1167–1186 (1991)
11. Kim, K.I., Jung, K., Park, S.H., Kim, H.J.: Support Vector Machines for Texture Classification. IEEE Trans. on PAMI 24(11), 1542–1550 (2000)
12. Haley, G.M., Manjunath, B.S.: Rotation-Invariant Texture Classification Using a Complete Space-Frequency Model. IEEE Trans. on Image Processing 8(2), 255–269 (1999)
13. Choi, H., Baraniuk, R.G.: Multi-scale Image Segmentation Using Wavelet-Domain Hidden Markov Models. IEEE Trans. on Image Processing 10(1), 1309–1321 (2001)
14. ter Haar Romeny, B.M.: Front-End Vision and Multi-scale Image Analysis: Multi-scale Computer Vision Theory and Applications (Written in Mathematica). Kluwer Academic Publishers, Dordrecht, the Netherlands (2003)
15. Jain, A.J., Duin, R.P.W., Mao, J.: Statistical Pattern Recognition: A Review. IEEE Trans. on PAMI 22(1), 4–37 (2000)
16. Kadah, Y.M., Farag, A., Zurad, J.M., Badawi, A.M., Youssef, A.B.M.: Classification Algorithms for Quantitative Tissue Characterization of Diffuse Liver Disease from Ultrasound Images. IEEE Trans. on Medical Imaging 15, 466–478 (1996)

Multiresolution Approach in Computing NTF

Arto Kaarna[1,2], Alexey Andriyashin[3], Shigeki Nakauchi[2], and Jussi Parkkinen[3]

[1] Lappeenranta University of Technology, Department of Information Technology
P.O. Box 20, FIN-53851 Lappeenranta, Finland
[2] Toyohashi University of Technology
Department of Information and Computer Sciences
1-1 Hibarigaoka, Tenpaku-cho, Toyohashi, 441-8580, Japan
[3] University of Joensuu, Laboratory of Computer Science
P.O. Box 111, FI-80101 Joensuu, Finland

Abstract. The computation of non-negative tensor factorization may become very time-consuming when large datasets are used. This study shows how to accelerate NTF using multiresolution approach. The large dataset is preprocessed with an integer wavelet transform and NTF results from the low resolution dataset are utilized in the higher resolution dataset. The experiments show that the multiresolution based speed-up for NTF computation varies in general from 2 to 10 depending on the dataset size and on the number of required basis functions.

1 Introduction

Non-negative basis for data description is useful for two reasons. First, the approach is natural since many measuring devices output only non-negative values. Secondly, non-negative filters can be physically implemented. Thus, many application possibilities exist for non-negative bases. They include feature extraction in image databases, band selection in spectral imaging, and even image compression.

In general, the basis is a low-dimensional mapping of a high-dimensional data. The traditional approaches are the principal component analysis, the vector quantization, and the singular value composition [1]. Their outputs are either representatives for groups of samples or eigenimages, which are then mixed to get the reconstruction. The mixing coefficients may be both positive, negative, or they may have zero values.

Two approaches to find a non-negative factorization of a data set V have arisen recently. The non-negative matrix factorization (NMF) generates the basis functions W and their multipliers H for a composition $V_r = WH$ [1]. The number of columns in W and the number of rows in H is the rank k. If the data set contains images, then these images must be vectorized to apply NMF. Thus, the approach is not able to utilize the structural features of images. In the non-negative tensor factorization (NTF) the original shape of the data is maintained [2]. The factorization outputs the reconstruction as $V_r = \sum u \otimes v \otimes w$, where u, v, and w are the factors for each domain, and the sum is over the rank k.

B.K. Ersbøll and K.S. Pedersen (Eds.): SCIA 2007, LNCS 4522, pp. 334–343, 2007.

In NTF, the factorization in practical applications results in a unique solution, while in NMF that may vary depending on the initial values for the solution process.

Both in NMF and NTF the factorization is obtained through an iterative process. The target is to minimize the distance between the original data and the reconstructed data. Depending on the application, the distance may be measured as the energy norm, the entropy, or as the Kullback-Liebler divergence.

The iteration is a time-consuming process. Typically, hours of computation time is needed to find the factorization for the data set. In most cases the basis functions are only needed and in the literature, these results are mostly reported, compared, and evaluated [1,2,3,4,5,6,7].

In this study, our hypothesis is that the number of iterations can be limited with a selection of relevant initial values. Instead of random initial values, a starting point generated from the dataset can speed up the iteration. In the proposed approach, the multiresolution of the data set is used to enhance the computation of the non-negative tensor factorization. The multiresolution is obtained through the integer wavelet transform. For a low resolution image the NTF is computed and then the components u, v, and w for that resolution are interpolated for the next higher resolution level.

The structure of the report is as follows. In Chapter 2 we introduce the NTF process. In Chapter 3 the give the background for the multiresolution approach based on the integer wavelet transform. In Chapter 4 the consider the computational complexity of the multiresolution approach for NTF computation. In Chapter 5 we show the results form the experiments. The discussion and the conclusions are in Chapter 6.

2 Non-negative Tensor Factorization

Recently, new approaches have emerged to define non-negative bases for datasets. Two of the methods are the non-negative matrix factorization (NMF) [1], and the non-negative tensor factorization (NTF) [2]. The basic approach for both of these is to find a solution for the problem

$$\min_{V_o, V_r \geq 0} ||V_o - V_r|| \tag{1}$$

where V_o is the original data and V_r is the reconstructed data. In composing V_r, all the components or substructures required in composition are non-negative.

In NTF [2] the reconstruction V_r is obtained as a sum of tensor products

$$V_r = \sum_{j=1}^{k} u^j \otimes v^j \otimes w^j \tag{2}$$

where u^j are bases for the first domain, v^j are bases for the second domain and w^j are bases for the third domain. k is the rank, a normal requirement is $(r + s + t)k < rst$, where r, s, and t are the number of samples in each domain.

Every element i in u^j, v^j, and w^j is non-negative. The number of domains naturally depends on the dataset. Three domains, u, v, and w are needed for example for the analysis of grayscale facial image databases, where the first two domains are the spatial domains of the images and the third domain comes from the stack of several facial images [2], [3] [4], [5].

We have used the well-known iterative process [2], that minimizes the reconstruction error in energy sense. Now the iteration steps for u_i^j, v_i^j, and w_i^j are defined, respectively, as

$$u_i^j \leftarrow \frac{u_i^j \sum_{s,t} G_{i,s,t} v_s^j w_t^j}{\sum_{m=1}^k u_i^m < v^m, v^j >< w^m, w^j >} \tag{3}$$

$$v_i^j \leftarrow \frac{v_i^j \sum_{r,t} G_{r,i,t} u_r^j w_t^j}{\sum_{m=1}^k v_i^m < u^m, u^j >< w^m, w^j >} \tag{4}$$

$$w_i^j \leftarrow \frac{w_i^j \sum_{r,s} G_{r,s,i} u_r^j v_s^j}{\sum_{m=1}^k w_i^m < u^m, u^j >< v^m, v^j >} \tag{5}$$

where $< .,. >$ refers to the inner product, matrix G contains the values from V_0.

3 Integer Wavelet Transform

For the multiresolution approach an approximation of the original data is required. This can be performed in may ways. A simple approach would be to subsample the data, i.e. select every second value from each domain. This approach is not suitable for basis function generation, since the bases are typically required to represent the low-frequency properties of the data. At least with synthetic data this approach would lead to problems since some features may be only one pixel wide [2].

The wavelet transform performs the appropriate approximation of the data. The original data is transformed to the approximative component and to the detail component [9]. In the inverse wavelet transform these two components are used to reconstruct the data. The wavelet transform carries the perfect reconstruction property. In Fig. 1, a), b), the principle of multiresolution is illustrated. The lower level approximation is received as values a_{j+1} etc. from the original values a_j. In practice the transform is performed using convolution with low-pass filter h and high-pass filter g. In definition of the filters different requirements can be set [9].

The wavelet transform is one-dimensional in nature. In the two-dimensional case, the one-dimensional transform is applied to the rows and columns of the image. In the three-dimensional case, the one-dimensional transform is applied to the spatial and spectral domains separately. In Fig. 1, c), the principle of the three-dimensional, separable transform is shown.

The datasets in imaging normally contain integer data. Thus, an integer version of the wavelet transform suits well to our case. Similarly to the floating case, there exists different integer wavelet transforms [10,11,13]. The integer wavelet

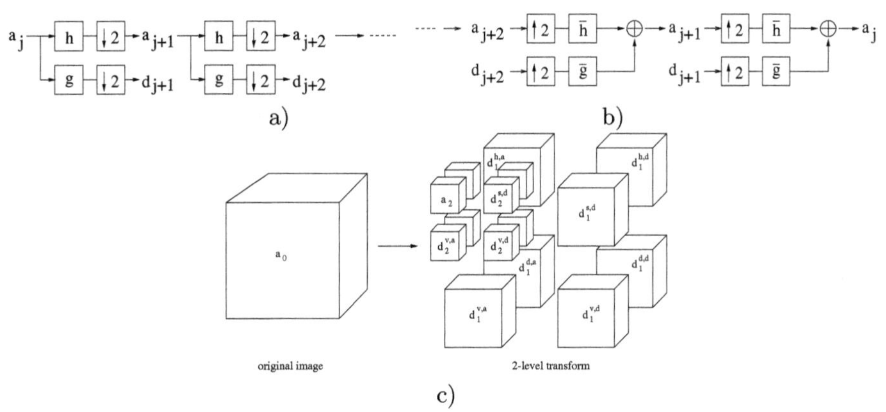

Fig. 1. Wavelet transform. a) Forward transform. b) Inverse transform, c) Separable three-dimensional wavelet transform applid twice.

transform is based on the lifting scheme: different filters are derived by combining the prediction step with the update step [11].

The basic form S of the integer wavelet transform subtracts the even samples from the odd samples to get the difference d_1 and the new approximation a_1 as

$$d_{1,l} = a_{0,2l+1} - a_{0,2l}, \quad a_{1,l} = a_{0,2l} + \lfloor d_{1,l}/2 \rfloor \tag{6}$$

where the original data is stored in $a_{0,\cdot}$. The second subscript refers to the index in the sample vector. The exact reconstruction comes from calculating the values in reverse order as

$$a_{0,2l} = a_{1,l} - \lfloor d_{1,l}/2 \rfloor, \quad a_{0,2l+1} = a_{0,2l} + d_{1,l} \tag{7}$$

In general, the integer wavelet transform consists of prediction and of update based on the lifting where the number of vanishing moments is increased. In [11], [12], [13], [10] several integer wavelet transforms are defined. We have implemented the following transforms: TS-transform, $S+P$-transform, $(2+2,2)$-transform, and 5/3-transform. In Eqs. 8, 9, 10, and 11 the forward transforms are given, respectively.

$$TS \quad \begin{cases} d_{1,l} = a_{0,2l+1} - a_{0,2l}, \quad a_{1,l} = a_{0,2l} + \lfloor d_{1,l}/2 \rfloor \\ d_{1,l} = d_{1,l} + \lfloor 1/4(a_{1,l-1} - a_{1,l}) + 1/4(a_{1,l} - a_{1,l+1}) \rfloor \end{cases} \tag{8}$$

$$S+P \quad \begin{cases} d_{1,l} = a_{0,2l+1} - a_{0,2l}, \quad a_{1,l} = a_{0,2l} + \lfloor d_{1,l}/2 \rfloor \\ d_{1,l} = d_{1,l} + \lfloor 2/8(a_{1,l-1} - a_{1,l}) + \\ \qquad 3/8(a_{1,l} - a_{1,l+1}) + 2/8 d_{1,l+1} \rfloor \end{cases} \tag{9}$$

$$(2+2,2) \quad \begin{cases} d_{1,l} = a_{0,2l+1} - \lfloor 1/2(a_{0,2l} + a_{0,2l+2}) + 1/2 \rfloor \\ a_{1,l} = a_{0,2l} + \lfloor 1/4(d_{1,l-1} + d_{1,l}) + 1/2 \rfloor \\ d_{1,l} = d_{1,l} - \lfloor 1/8(-1/2 a_{1,l-1} + a_{1,l} - 1/2 a_{1,l+1}) + \\ \qquad 1/8(-1/2 a_{1,l} + a_{1,l+1} - 1/2 a_{1,l+2}) + 1/2 \rfloor \end{cases} \tag{10}$$

$$5/3 \qquad \begin{cases} d_{1,l} = a_{0,2l+1} - \lfloor 1/2(a_{0,2l+2} + a_{0,2l}) \rfloor \\ a_{1,l} = a_{0,2l} + \lfloor 1/4(d_{1,l} + d_{1,l-1} + 1/2) \rfloor \end{cases} \tag{11}$$

The integer wavelet transform outputs non-negative values for the approximation coefficients a if the original data is non-negative. When details d are added in the inverse transform, the output is still non-negative. Thus, the transform does not violate the requirement of non-negativeness of the data for NTF.

4 Definition and Computational Complexity of the Proposed Algorithm

Non-negative matrix factorization outputs k vectors for each domain. The number of samples for each domain are r, s, and t, if a three-dimensional dataset is used. This is the normal case, when NTF is used with a grayscale facial image dataset or with a spectral image. For the spectral dataset, one domain, like v, can be neglegted, since the data is only two-dimensional, the first domain is the spectral domain and the second domain consists of the large number of spectra.

4.1 Definition of the Algorithm

The algorithm for the multiresolution approach for computing the NTF consists of the integer wavelet transform (IWT) and of NTF computaion. The details are given in Algorithm 1.

Algorithm 1

1. Compute the lowest resolution transform using IWT for the original data set.
2. Compute u, v, and w for this lowest level in multiresolution.
3. Interpolate u, v, and w for the next higher level in multiresolution.
4. Use inverse IWT to compute the next higher level in multiresolution.
5. Compute u, v, and w for the current multiresolution level.
6. If u, v, and w are computed for the highest level in multiresolution, then Stop. Otherwise, goto Step 3.

4.2 Computational Complexity

The one-dimensional wavelet transform is of order $O(n)$, where n is the number of samples. In three-dimensional case the number of samples n is $n = rst$. At each step to lower level in resolution, the number of samples is divided by eight, so one half of the samples in each domain is transfered to the next level. Each IWT described in Eqs. 8, 9, 10, and 11, require from 4 (S-transform, Eq. 6) to 24 ($(2+2, 2)$-transform, Eq. 10) operations to get the two new values in the lower resolution level. Normally, a low number of levels are needed in the transform, like from 3 to 5 levels.

In NTF, u, v, and w are obtained through an iterative process. In each iteration step j, see Eqs. 3, 4, 5, the computational load is proportional to $O(rst)$.

The number of iterations depends on the data set used in the application. Typically, hundreds or even thousands of iterations are needed for the process to converge [8].

In the implementation, we selected various number of iterations for each resolution level. At low resolution, it was possible to select a large number of iterations with only a nominal effect to the whole computational time. The final target was to minimize the required number of iterations at the highest resolution, since at the highest resolution the iteration was most time-consuming in finding the final components u, v and w for the data set.

5 Experiments

Three phases were performed in the expriments. The first phase was to select a suitable integer wavelet filter for the multiresolution analysis. The second phase considered the number of levels in the multiresolution. In the last phase NTF within the multiresolution approach was performed.

5.1 Experimental Data

Three data set were used in the experiments. The first set is a two-dimensional set containing $t = 1269$ color spectra. Each spectrum had $r = 384$ samples. Typical representatives of the spectra set are shown in Fig. 2 [14].

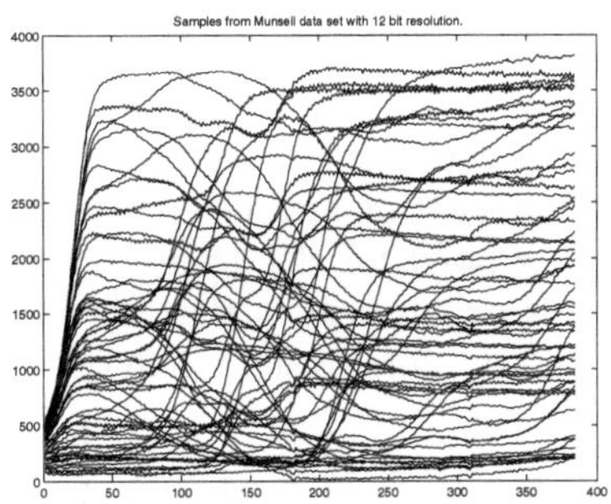

Fig. 2. Test set 1: Samples from Munsell color spectra set

The second data set was constructed from facial images from CBCL data set [15]. The number of images was $t = 192$, and each image was of size $rs = 96 \times 96$ pixels. Samples of the data set are shown in Fig. 3.

The third data set is a spectral image of size $rst = 256 \times 256 \times 224$. The image is part of Moffet Field remote sensing image captured using AVIRIS equipment [16].

Fig. 3. Test set 2: Facial image dataset, samples from the set

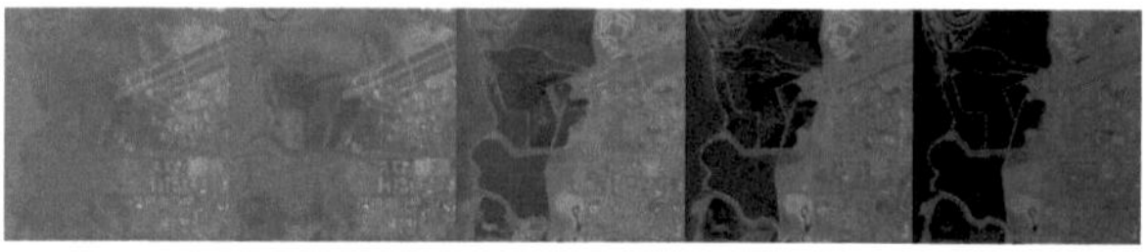

Fig. 4. Test set 3: Moffet Field spectral image, bands 11, 21, 41, 61, and 91

5.2 Selection of the Integer Filter and the Number of Levels in Multiresolution

In [10], the best lossless compression results for gray-scale images were obtained with the 5/3-transform. This means that the filter can capture essential features from the data set to a low number of coefficients. Our target is different. The purpose is to describe the data set such that features from a lower resolution to a higher resolution remain similar. Thus, the first task is to select a suitable IWT filter.

We computed the algorithm with $k = 3$ for the third data set with various IWT filters. The results are collected to Table 1. The quality is computed as the signal-to-noise ratio and it is expressed in decibels (dB). For filter S the results for the whole multiresolution solution are shown. For other filters, only the final result is shown. The starting point for the iteration was the same for all filters.

The conclusion from this experiment is that all filters act similarly, except filter 5/3, which outputs slightly worse results than the others. The final selection

Table 1. IWT filter selection, $k = 3$

Filter name	level	# of iterations	Quality (dB)	Relative time
S	4	2000	18.453	1.080
	3	600	15.975	2.190
	2	300	14.333	27.910
	1	50	13.119	74.540
	0	1	12.408	24.730
TS	0	1	12.408	31.780
SP	0	1	12.408	42.830
$(2+2,2)$	0	1	12.405	41.070
5/3	0	1	12.393	31.730

criterion is the computational time which is lowest for the filter S. Thus, all the experiments were performed with filter S.

Next, we wanted to find out, how many steps in the multiresolution ladder are needed. The experiment was performed with filter S, with the spectral image data set. The results are shown in Table 2.

Table 2. Selection of the number of levels in the multiresolution. IWT filter S, $k = 3$.

level	# of iterations	Quality (dB)	Relative time
3	600	15.974	2.020
2	300	14.332	24.430
1	50	13.119	70.050
0	1	12.408	24.300
2	300	14.320	32.230
1	50	13.112	77.640
0	1	12.402	24.400
1	50	12.627	67.740
0	1	12.020	22.630

From Tables 1 and 2 we can conclude that at least three levels of multiresolution are required to achieve the good quality in reconstruction. In the next experiments, four levels in multiresolution were applied, since the computational cost in the lowest level is very small compared to the whole process.

5.3 Computational Results for NTF

The last experiment considerers the whole process described in Alg. 1. The three datasets were used, in IWT the filter was filter S, and four levels in the multiresolution ladder were used. The results from the experiment are shown in Figs. 5 a), b), and c). On the horizontal axis there is the relative computational time with a logaritmic scale, and on the vertical axis there is the reconstruction quality in dB.

Each subfigure in Fig. 5 contains ranks $k = 1, 2, 3, 4, 6, 8, 16, 32$. For each k without multiresolution, the number of iterations were $2, 4, 8, 16, 32, 64, 128, 256, 512, 1024$. The qualities for the latest iterations are shown for each k. For the multiresolution case, only the final result is shown as a single box ($\Box$) for each k.

6 Conclusions

In this study, we have enhanced the computation of u, v, and w for the non-negative tensor factorization with a multi-resolution approach. The multiresolution was computed using the integer wavelet transform.

The following conclusions can be made from the experiments. In general, the proposed approach is from 2 to 10 times faster than the original computation. Especially, for a data set with large values for r and s, the approach is very good. When the rank k is large, the original solution requires time that is fifty-fold

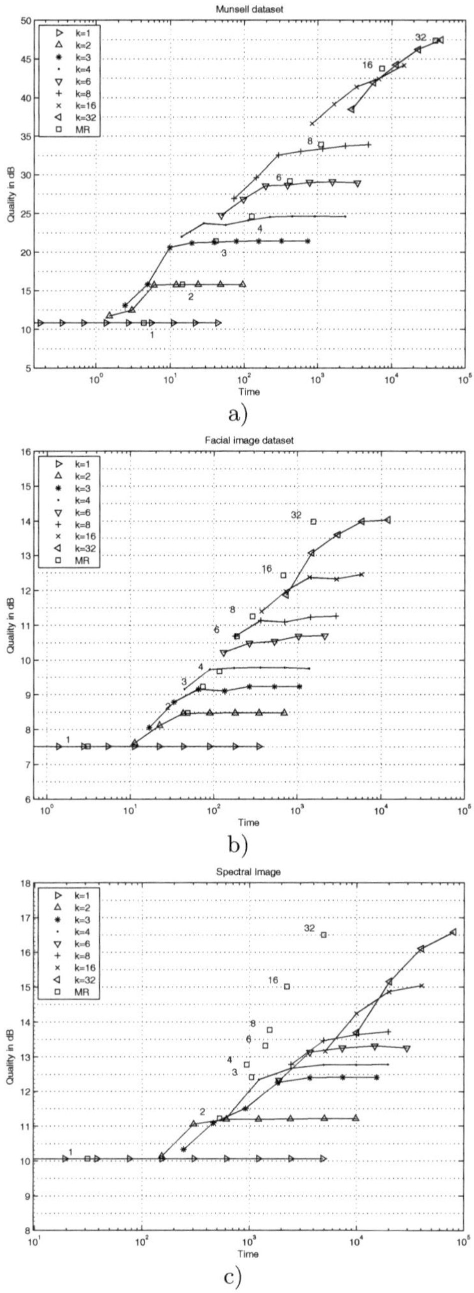

Fig. 5. Computational time vs. reconstruction quality, a) Munsell data set. b) Facial image data set. c) Moffet Field spectral image. For each run 2,4,8,16,32,64,128,256, and 1024 iteraton steps were used. For larger rank k, only last steps are marked. For each k, the computational time for the highest quality of the proposed method is marked by □.

compared to the proposed approach, see Fig. 5, c). The gain from the proposed approach is from 2 to 10 fold. For facial image data set, the same conclusions can be drawn, see Fig 5, b).

For the Munsell data, the approach provides clear gain when k is larger, like $k = 4, 6, 8, 16, 32$. With $k = 1$ the proposed approach is not usable, there is the extra load of the IWT compared to the original solution, see Figs. 5 a), b), c). In practice, a small number of iterations (even 2 iterations) results in the converged solution.

References

1. Lee, D.D., Seung, N.S.: Learning the Parts of Objects by Non-negative Matrix Factorization. Nature 401, 788–791 (1999)
2. Hazan, T., Polak, S., Shashua, A.: Sparse Image Coding using a 3D Non-negative Tensor Factorization. IEEE International Conference on Computer Vision (ICCV'05) (2005)
3. Hoyer, P.O.: Non-negative Matrix Factorization with Sparseness Constraints. Journal of Machine Learning Research 5, 1457–1469 (2004)
4. Li, S., Hou, X.W., Zhang, H.J., Cheng, Q.S: Learning Spatially Localized, Part-Based Representation. In: Proceedings of IEEE Conference on Computer Vision and Pattern Recognition (CVPR'01), Hawaii, USA 1, 207–212 (2001)
5. Wild, S., Curry, J., Dougherty, A.: Improving Non-negative Matrix Factorizations through Structured Initialization. Pattern Recognition 37, 2217–2232 (2004)
6. Shashua, A., Hazan, T.: Non-negative Tensor Factorization with Applications to Statistics and Computer Vision. In: Proceedigs of the 22nd International Conference on Machine Learning, Bonn, Germany, pp. 792–799 (2005)
7. Heiler, M., Schnörr, C.: Controlling Sparseness in Non-negative Tensor Factorization. In: Leonardis, A., Bischof, H., Pinz, A. (eds.) ECCV 2006. LNCS, vol. 3951, pp. 56–67. Springer, Heidelberg (2006)
8. Yuan, Z., Oja, E.: Projective Nonnegative Matrix Factorization for Image Compression and Feature Extraction. In: Kalviainen, H., Parkkinen, J., Kaarna, A. (eds.) SCIA 2005. LNCS, vol. 3540, pp. 333–342. Springer, Heidelberg (2005)
9. Daubechies, I.: Ten Lectures on Wavelets, CBMS-NSF. Regional Conference Series in Applied Mathematics, 61, SIAM, USA (1992)
10. Adams, M.D., Kossentini, F.: Reversible integer-to-integer wavelet transforms for image compression: performance evaluation and analysis. IEEE Transactions on Image Processing 9(6), 1010–1024 (2000)
11. Calderbank, A.R., Daubechies, I., Sweldens, W., Yeo, B-L.: Wavelet transforms that map integers to integers. Applied and Computational Harmonic Analysis 5(3), 332–369 (1998)
12. Calderbank, A.R., Daubechies, I., Sweldens, W., Yeo, B-L.: Lossless Image Compression using Integer to Integer Wavelet Transforms. IEEE International Conference on Image Processing (ICIP'97) 1, 596–599 (1997)
13. Daubechies, I.: Recent results in wavelet applications. Journal of Electronic Imaging 7(4), 719–724 (1998)
14. Spectral Database. University of Joensuu Color Group, Accessed: October 26, 2006 Available: `http://spectral.joensuu.fi/`
15. Face Recognition Database. MIT-CBCL, accessed: November 10, 2006, availabe: `http://cbcl.mit.edu/software-datasets/heisele/facerecognition-database.html`
16. Homepage, AVIRIS: accessed: November 10, 2006 available: `http://aviris.jpl.nasa.gov/`

Generation and Empirical Investigation of hv-Convex Discrete Sets

Péter Balázs*

Department of Computer Graphics and Image Processing
University of Szeged
Árpád tér 2, H-6720 Szeged, Hungary
pbalazs@inf.u-szeged.hu

Abstract. One of the basic problems in discrete tomography is the reconstruction of discrete sets from few projections. Assuming that the set to be reconstructed fulfils some geometrical properties is a commonly used technique to reduce the number of possibly many different solutions of the same reconstruction problem. Since the reconstruction from two projections in the class of so-called hv-convex sets is NP-hard this class is suitable to test the efficiency of newly developed reconstruction algorithms. However, until now no method was known to generate sets of this class from uniform random distribution and thus only ad hoc comparison of several reconstruction techniques was possible. In this paper we first describe a method to generate some special hv-convex discrete sets from uniform random distribution. Moreover, we show that the developed generation technique can easily be adapted to other classes of discrete sets, even for the whole class of hv-convexes. Several statistics are also presented which are of great importance in the analysis of algorithms for reconstructing hv-convex sets.

Keywords: discrete tomography; hv-convex discrete set; decomposable configuration; random generation; analysis of algorithms.

1 Introduction

The reconstruction of two-dimensional discrete sets from their projections plays a central role in discrete tomography and it has several applications in pattern recognition, image processing, electron microscopy, angiography, radiology, non-destructive testing, and so on [11,12]. Since taking projections of an object can be expensive or time-consuming the number of projections used in the reconstruction is usually small (in most cases two or four). This can yield extremely many solutions with the same projections or/and NP-hard reconstruction causing the developed reconstruction algorithm hardly applicable in practice. One way to get rid of these problems is to suppose that the discrete set to be reconstructed belongs to a certain class described by some geometrical properties

* This work was supported by OTKA grant T48476.

B.K. Ersbøll and K.S. Pedersen (Eds.): SCIA 2007, LNCS 4522, pp. 344–353, 2007.

such as convexity, connectedness, etc. One of the first such approaches was presented in [14] where the author gave a reconstruction heuristic for the class of horizontally and vertically convex (shortly, *hv*-convex) discrete sets using only two projections. Later, it was shown that this reconstruction task is NP-hard [18]. However, by this time it was known that assuming that the set to be reconstructed is also connected makes polynomial-time reconstruction possible [4,7]. Thus, researchers began to study what makes the reconstruction in the general class of *hv*-convexes so difficult. For certain subclasses it was found that the reconstruction can be done in polynomial time [1,6]. Surprisingly, it was also shown that the reconstruction is no longer intractable if absorption in the projections is present (at least for certain absorption coefficients) [15]. Therefore, during the last few years the class of *hv*-convex discrete sets became one of the indicators of newly developed exact or heuristic reconstruction algorithms from the viewpoint of effectiveness [5,8]. Unfortunately, all the developed techniques for solving the reconstruction problem in the class of *hv*-convexes had to face the problem that no method was known to generate sets of this class from uniform random distribution and thus no exact comparison of the techniques was possible. In this paper we outline algorithms for generating certain *hv*-convex discrete sets from uniform random distributions and study properties of randomly generated *hv*-convex sets from several point of view. The structure of the contribution is the following. First, the necessary definitions are introduced in Section 2. In Section 3 we describe the generation method for a subclass of *hv*-convexes. Then, in Section 4 we investigate some properties of randomly generated *hv*-convex discrete sets of the above class that can affect the complexity of several reconstruction algorithms. In Section 5 we discuss our results and show how the presented generation technique can be adapted to other classes of discrete sets, in particular, even for the whole class of *hv*-convexes.

2 Preliminaries

Discrete tomography aims to reconstruct a discrete set (a finite subset of the two-dimensional integer lattice defined up to translation) from its line integrals along several (usually horizontal, vertical, diagonal, and antidiagonal) directions. Discrete sets can be represented by binary pictures or binary matrices (see Fig. 1) and thus the above problem is equivalent to the task of reconstructing a binary matrix from its row, column (and sometimes also diagonal and antidiagonal) sums. In the following we will use both terms discrete set and binary matrix depending on technical convenience. To stay consistent, without loss of generality we will assume that the vertical axis of the 2D integer lattice is directed top-down and the upper left corner of the smallest containing rectangle of a discrete set is the position $(1, 1)$. Clearly, definitions given for discrete sets always have natural counterparts in matrix theory. Vice versa, a definition described in matrix theoretical form can be expressed in the language of discrete sets, too.

A discrete set F is *4-connected* (with an other term *polyomino*), if for any two positions $P \in F$ and $Q \in F$ of the set there exist a sequence of distinct positions

$(i_0, j_0) = P, \ldots, (i_k, j_k) = Q$ such that $(i_l, j_l) \in F$ and $|i_l - i_{l+1}| + |j_l - j_{l+1}| = 1$ for each $l = 0, \ldots, k-1$. A discrete set is called *hv-convex* if all the rows and columns of the set are 4-connected, i.e., the 1s of the corresponding representing matrix are consecutive in each row and column. For example, the discrete set in Fig. 1 is hv-convex.

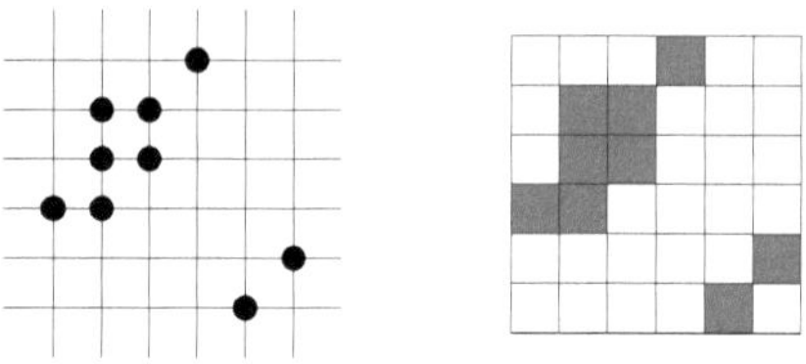

$$\begin{bmatrix} 0&0&0&1&0&0\\ 0&1&1&0&0&0\\ 0&1&1&0&0&0\\ 1&1&0&0&0&0\\ 0&0&0&0&0&1\\ 0&0&0&0&1&0 \end{bmatrix}$$

Fig. 1. A discrete set represented by its elements (*left*), a binary picture (*center*), and a binary matrix (*right*)

The maximal 4-connected subsets of a discrete set F are called the *components of F*. For, example the discrete set in Fig. 1 has four components: $\{(1,4)\}$, $\{(2,2),(2,3),(3,2),(3,3),(4,1),(4,2)\}$, $\{(5,6)\}$, and $\{(6,5)\}$.

3 Generation of Special *hv*-Convex Binary Matrices

The first result towards the uniform random generation of hv-convex binary matrices was given by Delest and Viennot [9] who proved that the number P_n of hv-convex polyominoes with perimeter $2n+8$ is

$$P_n = (2n+11)4^n - 4(2n+1)\binom{2n}{n} . \tag{1}$$

Later, based on the above result in [10] it was shown that the number $P_{m+1,n+1}$ of hv-convex polyominoes with a minimal bounding rectangle of size $(m+1)\times(n+1)$ is

$$P_{m+1,n+1} = \frac{m+n+mn}{m+n}\binom{2m+2n}{2m} - \frac{2mn}{m+n}\binom{m+n}{m}^2 . \tag{2}$$

We first will consider a special class of hv-convex matrices (denoted by $\mathcal{S}$), namely where the components' bounding rectangles are connected to each other with their bottom right and upper left corners and there are no rows or columns with zero-sums. Especially, every hv-convex polyomino belongs to this class, too, as they have only one component. Then, a binary matrix $F \in \mathcal{S}$ of size $m \times n$ is either an hv-convex polyomino or it contains a polyomino of size $k \times l$ (where $k < m$ and $l < n$) as a submatrix in the upper left corner and the remaining part of F is a binary matrix of size $(m-k)\times(n-l)$ which also belongs to the class $\mathcal{S}$. Denoting the number of binary matrices of $\mathcal{S}$ of size $m \times n$ with $Q_{m,n}$ this observation can be expressed by the following recursive formula

$$Q_{m,n} = P_{m,n} + \sum_{k<m,\ l<n} P_{k,l} \cdot Q_{m-k,n-l} . \tag{3}$$

Using Equation (2) and the initial values $Q_{1,j} = P_{1,j} = 1$ $(j = 1, \ldots, n)$ and $Q_{i,1} = P_{i,1} = 1$ $(i = 1, \ldots, m)$ $Q_{m,n}$ can be calculated by a dynamic programming approach in $O(m^2n^2)$ time with $O(mn)$ memory requirement. Based on this we now can describe the algorithm for generating *hv*-convex binary matrices of $\mathcal{S}$ from uniform random distribution.

Algorithm 1. for generating matrices of $\mathcal{S}$ from uniform random distribution
Input: The integers m and n.
Output: A binary matrix $F \in \mathcal{S}$ of size $m \times n$.

Step 1 calculate $Q_{m,n}$;
Step 2 generate a number $r \in [1, Q_{m,n}]$ from uniform random distribution;
Step 3 if ($r > P_{m,n}$)
{ $r = r - P_{m,n}$;
for $k = 1$ **to** $m - 1$
for $l = 1$ **to** $n - 1$
{ **if** ($r > P_{k,l} \cdot Q_{m-k,n-l}$) $r = r - P_{k,l} \cdot Q_{m-k,n-l}$;
else call Algorithm 1 with parameters $m - k$ and $n - l$; }
}
Step 4 generate the components from uniform random distribution;

This algorithm works as follows. First, in Step 1 it calculates $Q_{m,n} = P_{m,n} + P_{1,1} \cdot Q_{m-1,n-1} + P_{1,2} \cdot Q_{m-1,n-2} + P_{2,1} \cdot Q_{m-2,n-1} + \ldots + P_{m-1,n-1} \cdot Q_{1,1}$. Choosing a number randomly in the interval $[1, Q_{m,n}]$ (Step 2) it can be decided whether it is in the interval $[1, P_{m,n}]$, $[P_{m,n} + 1, P_{m,n} + P_{1,1} \cdot Q_{m-1,n-1}]$, $[P_{m,n} + P_{1,1} \cdot Q_{m-1,n-1} + 1, P_{m,n} + P_{1,1} \cdot Q_{m-1,n-1} + P_{1,2} \cdot Q_{m-1,n-2}]$, etc. Thus, the size of the upper left component can be identified, and this method can be repeated for the remaining set, too (Step 3). Now, we only have to generate the components themselves from uniform random distribution knowing their bounding rectangles which is possible with the algorithm given in [13] (Step 4).

The above method can be extended to *hv*-convex binary matrices possibly having zero row or/and column sums, too (but still having the same configuration of the components as in the class $\mathcal{S}$). This class will be denoted by $\mathcal{S}'$. Clearly, $\mathcal{S} \subset \mathcal{S}'$. In fact, a binary matrix $F \in \mathcal{S}'$ of size $m \times n$ is either an *hv*-convex polyomino or it contains a polyomino of size $k \times l$ (where $k < m$ and $l < n$) as a submatrix in the upper left corner and the remaining part of F is a binary matrix of size $(m - k) \times (n - l)$ such that it possibly has some zero rows, or/and columns in the upper left corner and the remaining part belongs to the class $\mathcal{S}'$. Denoting the number of binary matrices of $\mathcal{S}'$ of size $m \times n$ with $Q'_{m,n}$ we get a formula similar to Equation (3)

$$Q'_{m,n} = P_{m,n} + \sum_{k<m,\ l<n} P_{k,l} \cdot \Big(\sum_{i \le m-k,\ j \le n-l} Q'_{i,j} \Big) . \qquad (4)$$

Again, on the basis of Equation (2) and the initial values $Q'_{1,j} = P_{1,j} = 1$ $(j = 1, \ldots, n)$ and $Q'_{i,1} = P_{i,1} = 1$ $(i = 1, \ldots, m)$ the above formula can be evaluated by a dynamic programming approach in $O(m^3n^3)$ time with $O(mn)$ memory requirement. Then, an algorithm similar to Algorithm 1 can be given to generate hv-convex binary matrices of $\mathcal{S}'$ from uniform random distribution.

4 Statistics on *hv*-Convex Matrices

In order to test some important properties of hv-convex binary matrices we have generated test data sets with Algorithm 1 (and its modified version in the case of matrices with possible zero rows or/and columns). Each set of test data consisted of 1000 hv-convex matrix with the same size generated from uniform random distribution from the classes $\mathcal{S}$ and $\mathcal{S}'$. The algorithms were implemented in C++ and the long integer functions of library NTL-5.4 [16] were used. The test run on a PC with Intel Pentium 4 processor of 3.2 GHz and 1 GB RAM under Debian GNU/Linux 3.1, Kernel 2.6.17.13.

4.1 The Number of Special *hv*-Convex Discrete Sets

Our first simple investigation focuses on the number of special hv-convex discrete sets. Table 1 shows the number of hv-convex polyominoes and hv-convex sets from the classes $\mathcal{S}$ and $\mathcal{S}'$ with bounding rectangles of semi-perimeter n for the first 15 values of n denoted by $P(n)$, $Q(n)$, and $Q'(n)$, respectively. The first column can also be calculated by formula (1) and this is Sequence A005436 in [17]. For $n = 5$ the corresponding binary pictures of all three classes are shown in Fig. 2.

Table 1. The number $P(n)$, $Q(n)$, and $Q'(n)$ of hv-convex polyominoes, and hv-convex sets from the classes $\mathcal{S}$ and $\mathcal{S}'$, respectively, depending on the semi-perimeter n of the bounding rectangle

n	$P(n)$	$Q(n)$	$Q'(n)$
2	1	1	1
3	2	2	2
4	7	8	8
5	28	32	34
6	120	139	150
7	528	618	674
8	2344	2779	3056
9	10416	12528	13898
10	46160	56404	63178
11	203680	253152	286570
12	894312	1131849	1296008
13	3907056	5040412	5842442
14	16986352	22359981	26255254
15	73512288	98837102	117642282

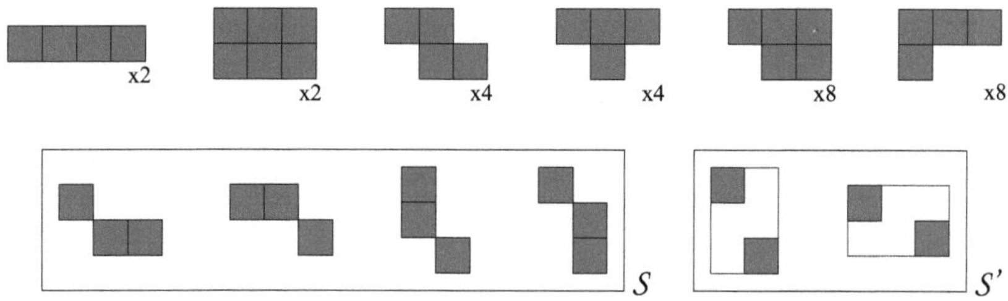

Fig. 2. All the *hv*-convex polyominoes (*first row*) and *hv*-convex sets from the classes $\mathcal{S}$ and $\mathcal{S}'$ (*second row*) with bounding rectangles of semi-perimeter 5. Small numbers in the first row indicate other solutions that can be get by mirroring or/and rotating the given polyomino.

4.2 The Number of Components

The second experiment treats the number of components of special *hv*-convex sets. It is important information when reconstructing such kind of sets. For example, as it was mentioned in Section 1 if the set consists of a single component then the reconstruction can be executed in polynomial-time. Table 2 shows the number of components of the generated sets depending on size when the sets do not have empty rows and columns (top half of the table) and when empty rows and columns are also permitted (bottom half of the table). Note that the sum of the elements in the last two rows are less than 1000. Due to space considerations we omitted 2 sets of size 80×80 and 22 sets of size 100×100 that have more than 15 components. The numerical investigation shows that generating *hv*-convex sets from the classes $\mathcal{S}$ and $\mathcal{S}'$ from uniform random distribution there is a great possibility that the set consists of a single component if the size of the set is small (namely, less than or equal to 20×20) but there is almost no chance to apply the well-known polynomial-time algorithms for reconstructing *hv*-convex polyominoes for sets of greater sizes.

It is interesting and could be quite useful in the reconstruction that the number of components can be estimated in advance knowing only the size of the set. Let $E(n)$ and $D^2(n)$ denote the expected number of components and its variance, respectively, for a set of size $n \times n$ generated from uniform random distribution from the class $\mathcal{S}$ or $\mathcal{S}'$. If $n \leq 100$ then the estimated values of $E(n)$ and $D^2(n)$ can be calculated directly from Table 2. For larger sets a good estimation can be given using the following equations

$$E(n) \approx 0.075n \quad \text{and} \quad D^2(n) \approx 0.04n \tag{5}$$

in the class $\mathcal{S}$, and

$$E(n) \approx 0.100n \quad \text{and} \quad D^2(n) \approx 0.06n \tag{6}$$

in the class $\mathcal{S}'$.

Moreover, for each size of sets the number of components follows a normal-like distribution with expected value $E(n)$ and with variance $D^2(n)$. In order

Table 2. The number of components of 1000 hv-convex discrete sets with bounding rectangle of size $n \times n$ generated from uniform random distribution from the classes $\mathcal{S}$ (*top half of the table*) and $\mathcal{S}'$ (*bottom half of the table*)

Size	1	2	3	4	5	6	7	8	9	10	11	12	13	14	15
5×5	785	191	23	1											
7×7	746	225	27	2											
10×10	659	272	60	9											
20×20	314	403	196	73	12	2									
40×40	29	189	318	257	135	54	8	6	3	1					
60×60	2	36	118	240	260	171	106	40	19	6	1	1			
80×80		3	34	100	183	216	186	138	85	36	12	6	1		
100×100		1	9	30	69	160	189	190	149	92	55	32	17	6	1
5×5	725	241	32	2											
7×7	656	280	54	9	1										
10×10	518	335	120	22	5										
20×20	175	326	315	123	51	10									
40×40	9	72	206	271	205	132	57	33	11	1	3				
60×60		10	33	102	169	224	192	126	87	28	12	5	1	1	
80×80			5	24	55	106	148	196	178	122	75	49	21	14	5
100×100				2	11	30	98	123	147	169	146	94	77	52	29

to check this we have generated two more test sets consisting of 1000 uniformly chosen discrete sets of sizes 200×200 and 500×500 with nonempty rows and columns (the generation of this latter set took about half a day). Figure 3 shows the differences between the test results and the normal distributions with the estimated parameters.

4.3 The Number of Decomposable Configurations

Given an ordered pair of binary matrices (C, D) we say that we get the binary matrix F by *NorthWest gluing* (or shortly, NW-gluing) C to D if

$$F = \begin{pmatrix} C & \mathbf{0} \\ \mathbf{0} & D \end{pmatrix} .$$

NE-, SE- and SW-gluings are defined similarly. Then, given a binary matrix F consisting of $k \geq 2$ components we say that the components are in a *decomposable configuration* if the following properties hold

- the sets of the row and column indices of the components are disjoint, and
- if $k > 2$ then we get F by gluing a single component to a binary matrix having a decomposable configuration consisting of $k - 1$ components using one of the four gluing operators.

For example, the four components of the matrix depicted in Fig. 1 are in a non-decomposable configuration. Decomposability was introduced in [1] as a new

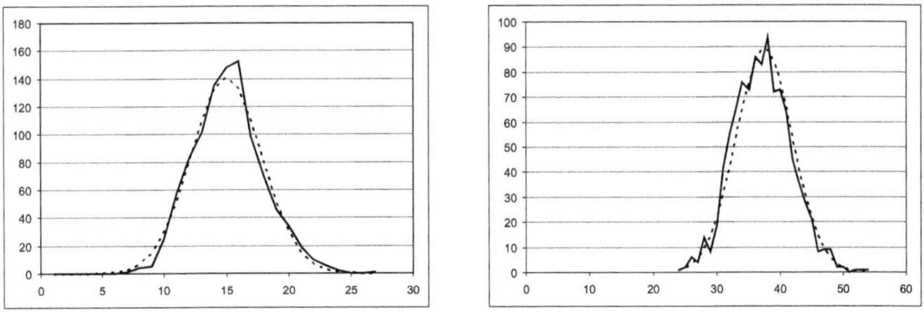

Fig. 3. The number of sets (*vertical axis*) having given number of components (*horizontal axis*) in the test data (*solid lines*) and the corresponding normal distribution (*dashed lines*) for sets of sizes 200×200 (*left*) and 500×500 (*right*)

property that can guarantee polynomial-time reconstruction from four projections. Later, in [2] it was also shown how the knowledge that the components of an hv-convex discrete set form a decomposable configuration can facilitate the reconstruction in the case of four projections. In our third experiment we tested whether decomposable configurations often occur in the class of hv-convex sets. In order to test this we also generated a uniformly chosen random permutation π of order k for each previously generated test data if the set consisted of $k \geq 2$ components. Then, we permuted the set of the column indices of the components in that test data according to the generated permutation π. Although in this way we did not get a uniform distribution in the whole class of hv-convexes a useful estimation could be made about the number of decomposable configurations. Table 3 represents the number of decomposable configurations for each set of test data having size up to 100×100 if empty rows and columns are not present ($N(dec)$) and if empty rows and columns are permitted ($N'(dec)$).

Table 3. The number of decomposable configurations from 1000 hv-convex sets of the given size $n \times n$

Size $n \times n$	2	3	4	5	7	10	20	40	60	80	100
$N(dec)$	163	166	160	214	254	340	669	826	599	344	160
$N'(dec)$	176	208	230	275	343	477	777	673	354	124	45

From this study we can see that decomposition reconstruction algorithms (see [1,2]) have the best chance to succeed on the generated test sets if the size of the set is between about 10×10 and 80×80 if the set has no empty rows and columns, and 7×7 and 60×60 if empty rows and columns are possibly present. From the definition it follows that every discrete set consisting of a single component is non-decomposable. Moreover, if a discrete set has two or three components then they necessarily form a decomposable configuration. Finally, the more component the discrete set has the less likely that the components are in a decomposable configuration. Thus, our statistic corresponds to the previous one about the expected number of components.

5 Conclusions and Discussion

We have developed a technique for generating special hv-convex binary matrices from uniform random distribution. Then, several statistics are given that can be useful in the complexity analysis of reconstruction algorithms. In a forthcoming contribution we plan to do an empirical investigation of the effectiveness of several reconstruction algorithms for hv-convex discrete sets that could be valuable in designing further more efficient reconstruction algorithms.

The main advantage of the developed method is that it can be applied for any class of matrices having disjoint components if the components themselves can be generated from uniform random distribution knowing their bounding rectangles and if it is possible to enumerate them. For a simple example let us assume that the discrete set to be generated does not have empty rows and columns and all the components are rectangles. However, we do not assume that the components are in a special configuration, we only suppose that the sets of their row and column indices are disjoint. Let $R^{(t)}_{m,n}$ denote the number of such discrete sets of size $m \times n$ having exactly t components. For each i and j there exists exactly one discrete rectangle of size $i \times j$, i.e., $R^{(1)}_{i,j} = 1$ for each $i = 1, \ldots, m$ and $j = 1, \ldots, n$. Moreover, $R^{(t)}_{i,j} = 0$ if $i < t$ or $j < t$, and for $t > 1$ the following recursive formula holds

$$R^{(t)}_{m,n} = \sum_{i<m,\ j<n} R^{(t-1)}_{i,j} \cdot t \tag{7}$$

where the factor t represents the proper weights for describing the possible permutations of the column sets of the t components. Then, the total number of such discrete sets of size $m \times n$ is $\sum_{t=1}^{\min\{m,n\}} R^{(t)}_{m,n}$ and the given algorithm can be modified in a straightforward way (see [3] for further details).

In particular, the above method can also be extended to the whole class of hv-convexes, too (see again [3]). Let $Q^{(t)}_{m,n}$ denote the number of arbitrary hv-convex discrete sets with minimal bounding rectangle of size $m \times n$ with nonempty rows and columns and having exactly t components. Then $Q^{(t)}_{i,j} = 0$ if $i < t$ or $j < t$, and $Q^{(1)}_{i,j} = P_{i,j}$ for each $i = 1, \ldots, m$ and $j = 1, \ldots, n$. Finally, for $t > 1$ the following recursive formula holds

$$Q^{(t)}_{m,n} = \sum_{k<m,\ l<n} P_{k,l} \cdot Q^{(t-1)}_{m-k,n-l} \cdot t\ . \tag{8}$$

Then, we get that the total number of arbitrary hv-convex discrete sets of size $m \times n$ with nonempty rows and columns is $\sum_{t=1}^{\min\{m,n\}} Q^{(t)}_{m,n}$. However, due to its huge computational complexity this generation method is applicable for discrete sets of moderate sizes only. Although several useful statistics could be done even in the classes $\mathcal{S}$ and $\mathcal{S}'$ it is an important open question whether more sophisticated and more efficient generation techniques for the whole class of *hv*-convexes can be developed.

References

1. Balázs, P.: A decomposition technique for reconstructing discrete sets from four projections. Image and Vision Computing, accepted.
2. Balázs, P.: On the ambiguity of reconstructing *hv*-convex binary matrices with decomposable configurations. Acta Cybernetica, submitted.
3. Balázs, P.: Reconstruction of discrete sets from their projections using geometrical priors. Doctoral Dissertation at the University of Szeged (in preparation)
4. Barcucci, E., Del Lungo, A., Nivat, M., Pinzani, R.: Reconstructing convex polyominoes from horizontal and vertical projections. Theor. Comput Sci. 155, 321–347 (1996)
5. Batenburg, K.J.: An evolutionary approach for discrete tomography. Discrete Applied Mathematics 151, 36–54 (2005)
6. Brunetti, S., Del Lungo, A., Del Ristoro, F., Kuba, A., Nivat, M.: Reconstruction of 4- and 8-connected convex discrete sets from row and column projections. Lin. Algebra Appl. 339, 37–57 (2001)
7. Chrobak, M., Dürr, C.: Reconstructing hv-convex polyominoes from orthogonal projections. Inform. Process. Lett. 69, 283–289 (1999)
8. Dahl, G., Flatberg, T.: Optimization and reconstruction of *hv*-convex $(0,1)$-matrices. Discrete Appl. Math. 151, 93–105 (2005)
9. Delest, M.P., Viennot, G.: Algebraic languages and polyominoes enumeration. Theor. Comput. Sci. 34, 169–206 (1984)
10. Gessel, I.: On the number of convex polyominoes. Ann. Sci. Math. Québec 24, 63–66 (2000)
11. Herman, G.T., Kuba, A. (eds.): Discrete Tomography: Foundations, Algorithms and Applications, Birkhäuser, Boston (1999)
12. Herman, G.T., Kuba, A. (eds.): Advances in Discrete Tomography and Its Applications, Birkhäuser, Boston, to appear in (2007)
13. Hochstättler, W., Loebl, M., Moll, C.: Generating convex polyominoes at random. Discrete Math. 153, 165–176 (1996)
14. Kuba, A.: The reconstruction of two-directionally connected binary patterns from their two orthogonal projections. Comp. Vision, Graphics, and Image Proc. 27, 249–265 (1984)
15. Kuba, A., Nagy, A., Balogh, E.: Reconstruction of *hv*-convex binary matrices from their absorbed projections. Discrete Applied Mathematics 139, 137–148 (2004)
16. Shoup, V.: NTL: A library for doing number theory, `http://www.shoup.net/ntl`
17. Sloane, N.J.A.: The on-line encyclopedia of integer sequences, `http://www.research.att.com/~njas/sequences/`
18. Woeginger, G.W.: The reconstruction of polyominoes from their orthogonal projections. Inform. Process. Lett. 77, 225–229 (2001)

The Statistical Properties of Local Log-Contrast in Natural Images

Jussi T. Lindgren, Jarmo Hurri, and Aapo Hyvärinen

Helsinki Insititute for Information Technology
Department of Computer Science
University of Helsinki
firstname.lastname@cs.helsinki.fi

Abstract. The study of natural image statistics considers the statistical properties of large collections of images from natural scenes, and has applications in image processing, computer vision, and visual computational neuroscience. In the past, a major focus in the field of natural image statistics have been the statistics of outputs of linear filters. Recently, attention has been turning to nonlinear models. The contribution of this paper is the empirical analysis of the statistical properties of a central nonlinear property of natural scenes: the local log-contrast. To this end, we have studied both second-order and higher-order statistics of local log-contrast. Second-order statistics can be observed from the average amplitude spectrum. To examine higher-order statistics, we applied a higher-order-statistics-based model called independent component analysis to images of local log-contrast. Our results on second-order statistics show that the local log-contrast has a power-law-like average amplitude spectrum, similarly as the original luminance data. As for the higher-order statistics, we show that they can be utilized to learn intuitively meaningful spatial local-contrast patterns, such as contrast edges and bars. In addition to shedding light on the fundamental statistical properties of natural images, our results have important consequences for the analysis and design of multilayer statistical models of natural image data. In particular, our results show that in the case of local log-contrast, oriented and localized second-layer linear operators can be learned from the higher-order statistics of the nonlinearly mapped output of the first layer.

1 Introduction

The study of natural image statistics considers the statistical properties of large collections of images and their transformations (for a review see [1]). The most important application areas of natural image statistics are image processing and computer vision (e.g., [2]), and visual computational neuroscience (e.g., [3,4]). In the case of image transformations, a majority of the research on natural image statistics has focused on the statistical properties of outputs of linear filters (e.g., [3,5]). Recently, attention has been shifting to nonlinear models to account for higher-order visual information such as contours [6] and cue invariance [7].

B.K. Ersbøll and K.S. Pedersen (Eds.): SCIA 2007, LNCS 4522, pp. 354–363, 2007.

One important nonlinear property of images is *local luminance contrast*, which in general refers to the relationship between the luminance of an object and its immediate surrounding. (From here on, the term 'local contrast' refers to local luminance contrast.) Spatial variations in local contrast can be viewed as a *second-layer*[1] visuel cue, because a model with two layers of filtering – with a nonlinearity in between – can be used to detect lines and edges formed by such cues (see, e.g., [8]). Such models are called *filter-rectify-filter* (FRF) models [8], where *rectification* refers to the nonlinear layer.

The contribution of this paper is an empirical study of the statistical properties of local contrast in natural images. While second-layer cues have received some attention in the natural image statistics research (e.g., [9,10]; see also [7]), to our knowledge the statistical properties of local contrast have not been studied in detail in previous research. The importance of the study of the statistics of local contrast has further increased due to recent research, which suggests that local luminance and local contrast have small statistical dependencies in natural scenes [10]; this suggests the applicability of a particularly practical and simple (factorizable) statistical model, in which the cues are first modeled separately and then combined in a straightforward manner.

Here we have analyzed both the second-order (correlation-based) statistics of local contrast, and some higher-order statistics. Second-order statistics are revealed by the average amplitude spectrum. While the spectra of rectified signals have been derived in some special cases (e.g., [11]), the richness of image data makes it unfeasible to derive an analytical expression for filtered and rectified images. This intractability also holds for higher-order statistics, such as higher-order cumulants. Therefore, we have approached the problem by analyzing these statistics empirically in an ensemble of images.

The results of our empirical analysis suggest that in natural images, *local log-contrast* has important statistical properties. Here 'log' refers to the use of a logarithmic function as a nonnegative, compressive function in the computation of the contrast; in our model, this nonlinearity is closely related to the rectification in an FRF model (see below). Our results on amplitude spectra show that local log-contrast has similar second-order statistics as luminance: the average amplitude spectra of both fall of as f^{-a}, with $a > 0$. This suggests that local log-contrast is scale invariant [12]. As for the higher-order statistics, in this work they were probed with independent component analysis (or, equivalently, sparse coding), which has been a very influential higher-order-statistics-based method in the analysis of statistical properties of linear transformations of natural images [3,5]. In the analysis of higher-order statistics, we were particularly interested in what kind of spatial patterns can be learned, because these patterns may form the second layer in FRF models. Our results show that meaningful spatial patterns of local contrast – such as contrast edges and bars – do emerge from higher-order statistics. This suggests that higher-order-statistics-based methods are able to discover

[1] A commonly used term for these cues is *second-order*, but we will use the term second-layer here to avoid confusion with second-order (correlation-based) *statistics*.

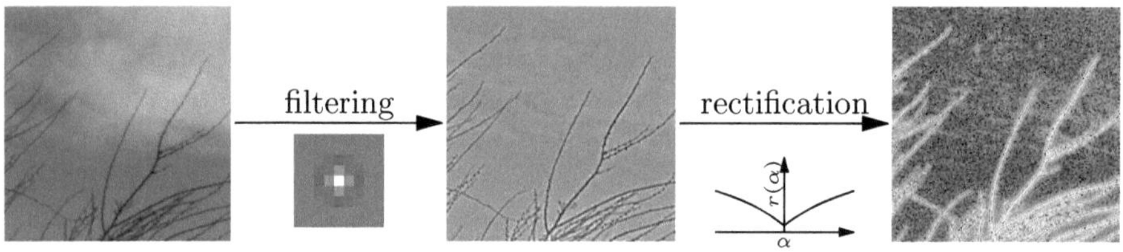

Fig. 1. The measure of local contrast employed in this paper is computed by a cascade of linear filtering and a pointwise logarithmic nonlinearity. The linear filter is a center-surround filter, and has been derived from the statistics of natural images (see text for details). This filter responds strongly (either positively or negatively) when the luminance difference between the center location and its surrounding is large, that is, when there is intuitively a large luminance contrast at the center location. The pointwise nonlinear logarithmic function (the *rectification* function) is $r(\alpha) = \ln(|\alpha| + d)$, $d \geq 1$, whose output is a nonnegative measure of local log-contrast; this function also compresses the large range of contrasts present in natural images. Note that in this figure, the images and the filter are represented with a highly different scale: the example images are of size 200×200, while the size of the filter is 9×9 pixels.

nonlinear image properties by learning second-layer linear operators in an FRF cascade.

The rest of this paper is organized as follows. In Section 2 we describe the measure of local contrast employed in this work. The natural image data is described shortly in Section 3. Second-order statistics are reported in Section 4, and results on higher-order statistics in Section 5. Finally, this paper ends with conclusions in Section 6.

2 Measuring Local Contrast

Luminance contrast refers in general to the relationship between the luminance of an object and its immediate surrounding. There is no unique definition of contrast in a natural image [13]. Two commonly employed types of measures of local luminance contrast utilize either center-surround filtering followed by a rectifying (nonnegative) nonlinearity, or localized measures of variance, possibly normalized by local luminance (see, e.g., [13,14,10]). While there are many different ways of computing local contrast, in this paper we have chosen to apply the center-surround filtering approach [13,15] (illustrated in Figure 1) due to its simplicity and relationship with other research on natural image statistics: it offers us a chance to specify the linear part of the computation of the local contrast – including the exact degree of localization – in terms of natural image statistics. This approach also gives us a view of the effect of the nonlinear part in the computation of contrast.

To specify the exact structure of the center-surround filter, we utilize a well-known statistical criterion (e.g., [16,17]): we specify that the filter will *whiten* the data – that is, the output of the filter will on the average have a flat power

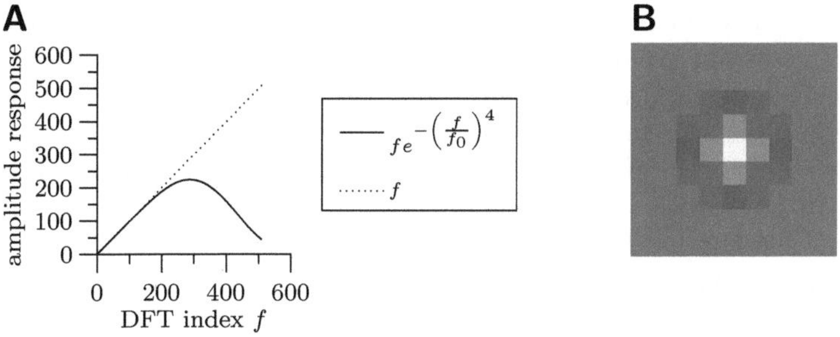

Fig. 2. The center-surround filter. (**A**) The specification of the filter in the Fourier domain. The amplitude response of the filter follows the linear whitening filter at low frequencies, but drops at high frequencies in order to not amplify high-frequency noise [17]. For comparions, the plot also shows the amplitude response of the whitening filter with no noise control (dashed). The amplitude responses are plotted as functions of discrete Fourier transform (DFT) index. In the whitening filter, the cutoff frequency $f_0 = 0.4N = 407.2$, where $N = 1018$ is the size of the images. (**B**) The central part of the resulting filter in the spatial domain.

spectrum – with the constraint that the filter should not amplify unduly the low-power high-frequency part of the spectrum, because a significant part of the highest frequencies consist of noise and sampling artifacts (see below). Because the average power spectrum is also the Fourier transform of the autocorrelation, the whitening condition is equivalent to spatial uncorrelatedness of filter output. The whitening criterion leads to the specification of the filter in the Fourier domain, where whitening determines the amplitude response of the filter, and the symmetrical structure of the filter results from setting the phase response to zero for all frequencies.

It is well-known that when averaged over a set of natural images, the amplitude spectrum falls off essentially as f^{-a}, where f is the frequency, and $a > 0$ and is typically between 0.7 and 1.5 (see, e.g., [14]). Therefore, for natural images, a whitening filter should have an amplitude response proportional to f^a; typically, the "average" value $a = 1$ is used in the specification of a whitening filter. However, in whitening a precaution should be taken to avoid amplifying high-frequency noise and sampling artifacts [16,17]. Because of the f^{-a} amplitude spectrum of natural images, the high frequencies would need to be strongly amplified in whitening. Unfortunately this would magnify the high-frequency noise and sampling artifacts resulting from the use of a rectangular sampling grid. Therefore, a typical whitening filter drops off at high frequencies. Here we specify the exact form of the whitening filter as in [17]. Let $|G(f)|$ be the one-dimensional Fourier amplitude response of the (spherically symmetric) filter g; then

$$|G(f)| = fe^{-(f/f_0)^4}, \tag{1}$$

where f_0 is the frequency cut-off; here $f_0 = 0.4N$, when $N \times N$ is the size of our natural images. The amplitude response of the filter is plotted in Figure 2A. The filter has the effect of removing the DC (constant component), dampening the low frequencies that are dominant in natural images, and attenuating the highest frequencies which have the worst signal-to-noise ratio. The two-dimensional whitening filter is a spherically symmetric 2D version of equation (1); Figure 2B shows the resulting filter in the spatial domain. For details on the used filter, see [17]. To improve computational efficiency, convolution with the filter was implemented in this work by pointwise multiplication in the Fourier space.

After filtering with the center-surround filter, the value of local contrast was obtained by applying a *rectifying* (nonnegative, nonlinear) function $r(\cdot)$ to the filtering output. Let $*$ denote convolution, $g(x, y)$ the center-surround filter, and $I(x, y)$ an $N \times N$ image; then the local contrast $c(x, y)$ is obtained by

$$c(x, y) = r\left(g * I\right). \tag{2}$$

The exact form of the rectifying function $r(\cdot)$ is related to contrast gain control – a way of compressing the large range of contrasts in images – which is a general property of the biological visual system [18] and is also used in image processing to facilitate the interpretation of images with high contrast (e.g., [19]). In this paper we examine in particular the local log-contrast, that is, the case where

$$r(\alpha) = \ln\left(|\alpha| + d\right); \tag{3}$$

where a relatively small constant d is added to the absolute value to make the rectification function nonnegative and well-behaved (here d is 10% of the mean of the absolute values of the center-surround-filtered images) . In psychophysics, the logarithmic nonlinearity is related to the relationship between stimulus contrast and the perceived contrast (e.g., [20]).

3 Natural Image Data

In the experiments we used the natural image dataset provided by van Hateren and van der Schaaf [4]. This dataset contains over 4000 grayscale images of natural scenes, each image having a resolution of 1024×1536 pixels. Out of the two versions of these images available, we used the deblurred versions that have been compensated for the point spread function of the camera lens. To avoid border artifacts present in the data and to be able to apply the (square) whitening filter, we cropped each image to a resolution 1018×1018 pixels.

4 Second-Order Statistics

We examined the second-order statistics of the (original) luminance images, center-surround-filtered (whitened) images and local-log-contrast images. For this purpose, 200 natural images were sampled at random from the natural image collection, and the local-log-contrast computation scheme described in

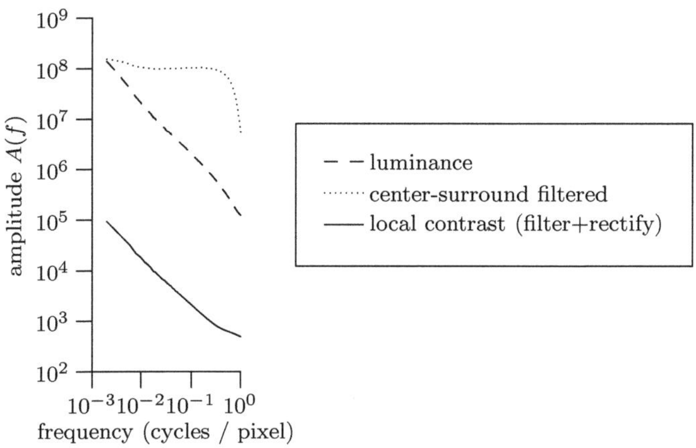

Fig. 3. Local log-contrast has similar second-order statistics as luminance: the average amplitude spectrum of local-log-contrast images exhibits similar f^{-a} behaviour as the average amplitude spectrum of the luminance images. This log-log plot shows the amplitude spectra of luminance images, filtered images, and local-log-contrast images (filtering + rectification), averaged over 200 natural images and all orientations. In a log-log plot, a f^{-a} characteristic corresponds to an affine plot with slope $-a$; the intercept reflects a scaling of the data and is irrelevant. Least-squares estimation of the slope and the intercept gives a slope $a_{\mathrm{lum}} \approx 1.2$ for the luminance data and $a_{\mathrm{con}} \approx 0.72$ for the local contrast data. Note the role of the rectification: before the rectification, the average amplitude spectrum of the filtered images is approximately flat (except for the highest frequencies), a consequence of the whitening principle.

Section 2 was applied to these images. Amplitude spectra averaged over the 200 images and all orientations were computed from the luminance images, center-surround-filtered images and local-log-contrast images.

The results are shown in Figure 3; please note that this is a log-log plot. As can be seen, the average amplitude spectrum of the luminance images follows the familiar f^{-a} curve, which corresponds to an affine curve in a log-log plot; least-squares estimation of an affine model $\mathrm{Est}\left\{\log_{10} A_{\mathrm{lum}}\right\}(f) = b_{\mathrm{lum}} - a_{\mathrm{lum}} \log_{10} f$, where $\mathrm{Est}\left\{\cdot\right\}$ denotes an estimator, gives a slope $a_{\mathrm{lum}} \approx 1.2$. The intercept b_{lum} corresponds to a global scaling of the data and is irrelevant here. The amplitude spectra of the center-surround images are approximately flat, except for the highest frequencies, as can be expected from the construction of the center-surround filter from the whitening principle (see Section 2).

The local contrast has a similar power-law-like f^{-a} form as the luminance; least-squares estimation of an affine model yields $a_{\mathrm{con}} \approx 0.72$ for the local contrast image data. Note in particular the effect of the rectifying nonlinearity, evident in the difference between the amplitude spectra of the center-surround-filtered images and the rectified contrast images. This implies that in natural images, local log-contrast enjoys similar scale-invariance as luminance [12].

5 Higher-Order Statistics

In this work, the higher-order statistics were probed with independent component analysis (ICA), which has been an influential model in the analysis of higher-order statistics of natural images [4,5] and is closely related [17,5] to another influential model called sparse coding [3]. In particular, we were interested in what kind of local-log-contrast patterns would emerge from higher-order statistics; it is well known that the luminance patterns that emerge from the application of ICA or sparse coding to luminance data are localized, oriented and bandpass and resemble – depending on the viewpoint – Gabor functions, edge and line detectors, and the receptive fields of simple cells in the primary visual cortex [17,4].

In its basic form, ICA assumes that the observed data $\boldsymbol{x}$, which is a random vector of dimension n, has been generated by a linear generative model

$$\boldsymbol{x} = \boldsymbol{A}\boldsymbol{s}, \tag{4}$$

where $\boldsymbol{A}$ is a constant *mixing matrix* (to be estimated) and $\boldsymbol{s}$ a random vector of unknown *independent components*. The dimension of $\boldsymbol{s}$ is assumed to be equal to the dimension of $\boldsymbol{x}$, possibly after the observed data $\boldsymbol{x}$ have been reduced to a smaller dimension by principal component analysis [5]. ICA tries to estimate both $\boldsymbol{s}$ and the parameter matrix $\boldsymbol{A}$ from the observed data $\boldsymbol{x}$; when the model holds, this can be done with very few assumptions (up to a scaling and ordering of the independent components and columns of matrix $\boldsymbol{A}$). In the absence of data following the generative model (4), many ICA algorithms – including the FastICA algorithm we use [5] – can still be interpreted as minimization of higher-order statistical dependencies between the components of the estimated random vector $\hat{\boldsymbol{s}} = \hat{\boldsymbol{A}}^{-1}\boldsymbol{x}$ [5]. In addition, when statistical dependencies are reduced by searching for maximally non-Gaussian projections – as in FastICA – the projections often turn out to be sparse, and the method can be viewed as computing a sparse coding basis for the observed data.

We computed the ICA decomposition of localized local-log-contrast data; to be more precise, of local-log-contrast image patches of size 19×19. To compute this decomposition, we sampled a set of $100,000$ patches of size 19×19 pixels from local-log-contrast images. Samples were taken from all of the over 4000 images. After sampling, the local DC component (the mean value in the patch, typically considered an "uninteresting" projection direction) was removed from each image patch. The patches were then vectorized; the resulting vectors formed a sample of the observed data $\boldsymbol{x}$ used in the ICA estimation. We used the FastICA algorithm [5] with $\tanh(\cdot)$ nonlinearity and symmetric (simultaneous) estimation of all components. The input dimension was $19 \times 19 = 361$, and principal component analysis was used to reduce the dimension for ICA to $16\times16 = 256$, retaining over 85% of the variance (dimensionality reduction is a standard procedure in ICA to reduce the effect of noise). This resulted in the estimation of 256 independent components, or maximally non-Gaussian directions.

The resulting set of ICA basis vectors (columns of matrix $\boldsymbol{A}$ in equation (4)) is shown in Figure 4A. As can be seen, these basis vectors of local-log-contrast

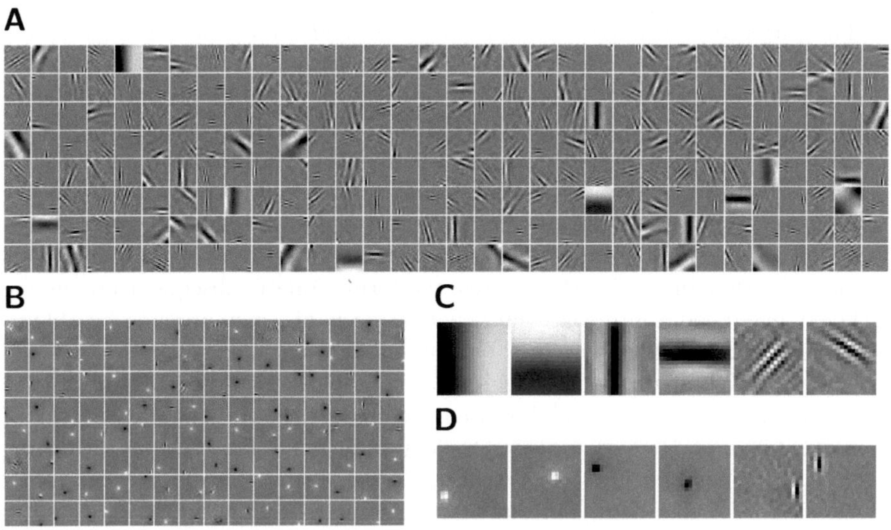

Fig. 4. Important contrast patterns emerge when independent component analysis (ICA) is applied to local-log-contrast data. (**A**) A set of 256 ICA basis vectors of size 19×19 pixels, computed from the local-log-contrast data. (**B**) For comparison, the ICA basis vectors obtained in a control experiment with a noncompressive rectification function $r(\alpha) = |\alpha|$. The basis vectors in (B) are the same pixel size as in (A), but are shown at a smaller scale. (**C**) Magnified examples of different types of local-log-contrast patterns in the basis vector set (A): contrast edges, contrast bars, and high-frequency localized spatial contrast patterns. (**D**) Magnified examples of different types of patterns from the control experiment results in (B).

data correspond to oriented spatial contrast patterns in many different scales (frequencies). These patterns correspond to spatial variations in contrast such as contrast edges, contrast bars and spatially localized frequency patterns; examples of these are shown in Figure 4C. These basis vectors could form the starting point of a second-layer statistical image representation. To our knowledge, learning such second-layer (FRF) representations of local contrast has not been demostrated in previous research. Our results are therefore an important step in the development of multilayer statistical models of natural image data. For example, when learning multilayer models of images, one approach is to learn the layers one by one. Here we have demonstrated that meaningful second-layer patterns can be learned by employing ICA on top of a rectified first-layer output after the first-layer filter has been learned from the whitening principle.

It should be noted that in our model, nonlinear detectors of contrast edges are not simply linear luminance edge detectors with rectified outputs. For example, whereas a rectified linear edge detector would respond positively to an edge regardless of the polarity (sign) of the edge, it would *not* signal the existence of a contrast edge, because there is no luminance difference across a contrast edge.

To verify that the results shown in Figure 4A are not trivial or an artifact, we performed two control experiments.

1. The purpose of this control experiment was to study the role of the rectification ofunction. We repeated the computation of the ICA basis, but this time used the rectification function $r(\alpha) = |\alpha|$. The results are shown in Figure 4B, with close-ups in Figure 4D. As can be seen, in this case only a minority of the patterns are oriented, and these are of a single scale (high frequency); this suggests that a linear, noncompressive measure of contrast does not possess the statistical properties enjoyed by local log-contrast, which are needed to learn oriented spatial patterns in multiple scales. This indicates that in FRF architectures, the formation of subsequent processing layers may need appropriate gain control mechanisms in order to be able to perceive certain structures in the incoming data as salient.
2. The objective of this control experiment was to verify that our results are not an artifact of our model, but do indeed reflect the statistical properties of natural images. Here we repeated the computation of the ICA basis for three different types of white (uncorrelated) noise, each with different marginal distributions: Gaussian, uniform, and the same as in the original image data. For all of these noise models, the ICA estimation failed, suggesting that the filtered and rectified noise had a Gaussian distribution [5]. Our main results do therefore reflect the statistical structure of natural images.

6 Conclusions

We have shown in this paper that local log-contrast has important statistical properties in natural images. An examination of the average amplitude spectrum – which is equivalent to an examination of the second-order statistics – reveals a power-law form of f^{-a}, implying scale invariance. The ICA basis vectors – which reflect higher-order statistics – correspond to spatial local-log-contrast patterns such as contrast edges and bars. This suggests that higher-order statistics can be employed in learning multilayer statistical models of natural images.

The current work presents the first steps in the examination of the statistical properties of local log-contrast in natural images. For a more complete understanding of multilayer contrast processing, further theoretical and empirical research is under way. In particular, the effects caused by the interaction of the input data, the first-layer linear filter, and the used rectification function must be understood in more detail. One possibility is that the optimal parameterizations of the used mechanisms depend strongly on the requirements of subsequent processing stages.

Acknowledgments. This work was supported by the IST Programme of the European Community, under the PASCAL Network of Excellence, IST-2002-506778 (J.L., A.H.), by the FDK Center of Excellence of the Academy of Finland (J.L.), and by the Academy of Finland project #116962 (J.H.). This publication only reflects the authors' views.

References

1. Simoncelli, E.P.: Statistical modeling of photographic images. In: Bovik, A. (ed.) Handbook of Image & Video Processing, 2nd edn. pp. 431–441. Academic Press, San Diego (2005)
2. Portilla, J., Strela, V., Wainwright, M.J., Simoncelli, E.P.: Image denoising using scale mixtures of Gaussians in the wavelet domain. IEEE Transactions on Image Processing 12(11), 1338–1351 (2003)
3. Olshausen, B.A., Field, D.: Emergence of simple-cell receptive field properties by learning a sparse code for natural images. Nature 381(6583), 607–609 (1996)
4. van Hateren, J.H., van der Schaaf, A.: Independent component filters of natural images compared with simple cells in primary visual cortex. Proceedings of the Royal Society of London B. 265(1394), 359–366 (1998)
5. Hyvärinen, A., Karhunen, J., Oja, E.: Independent Component Analysis. John Wiley & Sons, Chichester (2001)
6. Hoyer, P.O., Hyvärinen, A.: A multi-layer sparse coding network learns contour coding from natural images. Vision Research 42(12), 1593–1605 (2002)
7. Hurri, J.: Learning cue-invariant visual responses. In: Weiss, Y., Schölkopf, B., Platt, J. (eds.) Advances in Neural Information Processing Systems, vol. 18, pp. 539–546. The MIT Press, Cambridge (2006)
8. Landy, M.S., Graham, N.: Visual perception of texture. In: Chalupa, L.M., Werner, J.S. (eds.) The Visual Neurosciences, vol. 2, pp. 1106–1118. MIT Press, Cambridge (2004)
9. Johnson, A.P., Baker Jr., C.: First- and second-order information in natural images: a filter-based approach to image statistics. Journal of the Optical Society of America A. 21(6), 913–925 (2004)
10. Frazor, R.A., Geisler, W.S.: Local luminance and contrast in natural images. Vision Research 46(10), 1585–1598 (2006)
11. Regan, M.P.: Half-wave linear rectification of a frequency modulated sinusoid. Applied Mathematics and Computation 79(2–3), 137–162 (1996)
12. Ruderman, D.L.: Origins of scaling in natural images. Vision Research 37, 3358–3398 (1997)
13. Tadmor, Y., Tolhurst, D.J.: Calculating the contrasts that retinal ganglion cells and LGN neurones encounter in natural scenes. Vision Research 40(22), 3145–3157 (2000)
14. Bex, P.J., Makous, W.: Spatial frequency, phase, and the contrast of natural images. Journal of the Optical Society of America A. 19(6), 1096–1106 (2002)
15. Itti, L., Koch, C., Niebur, E.: A model of saliency-based visual attention for rapid scene analysis. IEEE Transactions on Pattern Analysis and Machine Intelligence 20(11), 1254–1259 (1998)
16. Atick, J.J., Redlich, A.N.: What does the retina know about natural scenes? Neural Computation 4(2), 196–210 (1992)
17. Olshausen, B.A., Field, D.: Sparse coding with an overcomplete basis set: A strategy employed by V1? Vision Research 37(23), 3311–3325 (1997)
18. Schwartz, O., Simoncelli, E.P.: Natural signal statistics and sensory gain control. Nature Neuroscience 4(8), 819–825 (2001)
19. Tumblin, J., Hodgins, J.K., Guenter, B.K.: Two methods for display of high contrast images. ACM Transactions on Graphics 18(1), 56–94 (1999)
20. Fiorentini, A., Maffei, L.: Contrast perception and electrophysiological correlates. Journal of Physiology 231(1), 61–69 (1973)

A Novel Parameter Decomposition Approach for Recovering Poses of Distal Locking Holes from Single Calibrated Fluoroscopic Image*

Guoyan Zheng and Xuan Zhang

MEM Research Center-ISTB, University of Bern, CH-3014, Bern, Switzerland
guoyan.zheng@ieee.org

Abstract. One of the most difficult steps of intramedullary nailing of femoral shaft fractures is distal locking - the insertion of distal transverse interlocking screws, for which it is necessary to know the position and orientation of the distal locking holes of the intramedullary nail. This paper presents a novel parameter decomposition approach for solving this problem using single calibrated X-ray image. The problem is formulated as a model-based optimal fitting process, where the to-be-optimized parameters are decomposed into two sets: (a) the angle between the nail axis and its projection on the imaging plane, and (b) the translation and rotation of the geometrical models of the distal locking holes around the nail axis. By using a hybrid optimization technique coupling an evolutionary strategy and a local search algorithm to find the optimal values of the latter set of parameters for any given value of the former one, we reduce the multiple-dimensional model-based optimal fitting problem to a one-dimensional search along a finite interval. We report the $in-vitro$ experimental results, which demonstrate that the accuracy of our approach is adequate for successful distal locking of intramedullary nails.

Keywords: distal locking, fluoroscopy, pose estimation, parameter decomposition, hybrid optimization, model-based method.

1 Introduction

It has been recognized that one of the most difficult steps of intramedullary nailing of femoral shaft fractures is distal locking - the insertion of distal interlocking screws, for which it is necessary to know the positions and orientations of the distal locking holes of the intramedullary nail [1]. Complicating the process of locating and inserting the distal interlocking screw is the nail deformation with insertion. It has been reported that deformation occurs in several planes due to medial-lateral and anterior-posterior flexion of the distal nail after it has been inserted. Deformation analysis of solid 9 mm femoral nails using a magnetic tracking system in a cadaveric study has shown lateral translations of 18.1 ±

* This work was supported in part by Swiss National Science Foundation through project NCCR CO-ME.

B.K. Ersbøll and K.S. Pedersen (Eds.): SCIA 2007, LNCS 4522, pp. 364–373, 2007.

10.0 mm, dorsal translations of -3.1 $\pm$ 4.3 mm, and rotational deformation of -0.1 $\pm$ 0.2 degrees for the center of the distal transverse locking holes [1]. The reason for the wide variations of the insertion-related femoral nail deformation is due to the fact that the nail has to deform to the shape of the medullary canal upon insertion. The shape of the canal varies widely from person to person. Therefore, it is very difficult, to determine what the resultant locations and orientations of the distal locking holes will be relative to their initial position before it is deformed. The surgeon depends heavily on intra-operative X-ray means in a conventional surgical procedure for providing precise locations and orientations of the distal locking holes. It requires positioning the axis of the fluoroscope perpendicular to the locking holes so that these holes appear perfectly circular in the images. This is achieved through a trial-and-error method and requires long time X-ray exposure for both the surgeon and patient. It has been reported that the surgeon's direct exposure to radiation for each conventional surgical procedure was 3 - 30 min, of which 31% - 51% was used for distal locking [2].

The desire to target accurately with as little as possible X-ray exposure has led to various attempts to develop image-based methods for recovering the positions and orientations of the distal locking holes [3][4][5]. These methods require either multiple calibrated images or single image but with perfectly circular holes in the image, which normally requires the X-ray technician to use a try-and-move method several times to achieve.

This paper presents a novel parameter decomposition approach for solving this problem using single calibrated fluoroscopic image. We do not ask for an image with perfectly circular holes but we do put a constraint on its acquisition, i.e., the reduced patient shaft should be roughly parallel to the image intensifier of the fluoroscopy machine, which is much easier to be achieved intraoperatively. We then formulate the pose recovery as a model-based fitting problem and decompose the to-be-optimized parameters into two sets: (a) the angle between the nail axis and its projection on the imaging plane, and (b) the translation and rotation of the geometrical models of the locking holes around the nail axis. By using a hybrid optimization technique coupling an evolutionary strategy and a local search algorithm to find the optimal values of the latter set of parameters for each give value of the former one, we reduce the multiple-dimensional optimal fitting problem to a one-dimensional search along a finite interval.

The paper is organized as follows. Section 2 describes image calibration, geometrical models, and preprocessing. In Section 3, we describe the proposed approach in details. Section 4 presents our in-vitro experimental results, followed by conclusions in Section 5.

2 Image Calibration, Geometrical Models, and Preprocessing

(1) Image Calibration: In reality, the proximal fragment, the distal fragment, and the nail may be treated as three rigid bodies and registered independently. The rigid transformations between these three rigid bodies can be trivially obtained from a navigator such as an optoelectronic tracker, a magnetic tracker, or even a medical robot. As this is not our focus in this paper, here we assume that the fractured femur has already been reduced and the proximal fragment and distal fragment are kept fixed relative to each other at the time of image acquisition. We also assume that the nail has been inserted till the distal end of the femur and has been locked proximally by screw so that the complete femur and the nail can be treated as one rigid body. A local coordinate system is established on this rigid body through a so-called dynamic reference base technique [6]. In the following description, let's denote this patient coordinate system as $A - COS$. All computations are done in this reference coordinate sytem.

To relate a pixel in the two-dimensional (2D) projection image to $A - COS$, the acquired image has to be calibrated for physical projection properties and be corrected for various types of distortion. We have chosen a weak-perspective pin-hole camera model for modeling the C-arm projection [7]. Using such a camera model, a 2D pixel V_{I} is related to a three-dimensional (3D) point V_{A} by following equations:

$$S_{\mathrm{A}} = \frac{(V_{\mathrm{A}} - f_{\mathrm{A}})}{||V_{\mathrm{A}} - f_{\mathrm{A}}||}; \; and \; \begin{bmatrix} V_{\mathrm{I},x} \\ V_{\mathrm{I},y} \\ 1 \end{bmatrix} = \begin{bmatrix} c_{\mathrm{A},x} & c_{\mathrm{A},y} & c_{\mathrm{A},z} & p_{\mathrm{I},x} \\ \mathrm{r}_{\mathrm{A},x} & \mathrm{r}_{\mathrm{A},y} & \mathrm{r}_{\mathrm{A},z} & p_{\mathrm{I},y} \\ 0 & 0 & 0 & 1 \end{bmatrix} \begin{bmatrix} S_{\mathrm{A},x} \\ S_{\mathrm{A},y} \\ S_{\mathrm{A},z} \\ 1 \end{bmatrix} \quad (1)$$

where $||\cdot||$ means to calculate the length of a vector and the vectors f_{A}, r_{A}, c_{A}, p_{I} represent the position of focal point, the vector along image row increasing direction, the vector along image column increasing direction, and the 2D position of piercing point, respectively. They are projection parameters used to describe the projection properties of the C-arm and can be calibrated preoperatively or intraoperatively.

Eq. (1) can be used for both forward and backward projections. For example, if we want to calculate the direction S_{A} of the forward projection ray of an image point V_{I}, an additional constraint $||S_{\mathrm{A}}|| = 1$ can be used together with Eq. (1) to solve for it. The projection ray of point V_{I} is defined by the focal point and the direction S_{A}.

The position of the imaging plane in $A - COS$ and the focal length in our camera model is implicitly determined using the calibrated focal point f_{A} and the vectors r_{A} and c_{A}. Any 2D image point V_{I} corresponds to a 3D spatial point V_{A} in this imaging plane, which is the intersection between its forward projection ray and this plane.

(2) Geometrical models: The distal part of an intramedullary nail containing the two distal locking holes, which is what we are interested in, is modeled as a cylinder (Fig. 1, left). The distance L between the centers of the two distal locking holes can be accurately extracted from its product information. The geometrical model of each locking hole is represented by two circles as shown by Fig. 1, right, and is used later to simulate X-ray projections.

To obtain the coordinates of those points (visualized as red dots in Fig. 1, right) used to describe the model of the lcoking hole, a local coordinate system

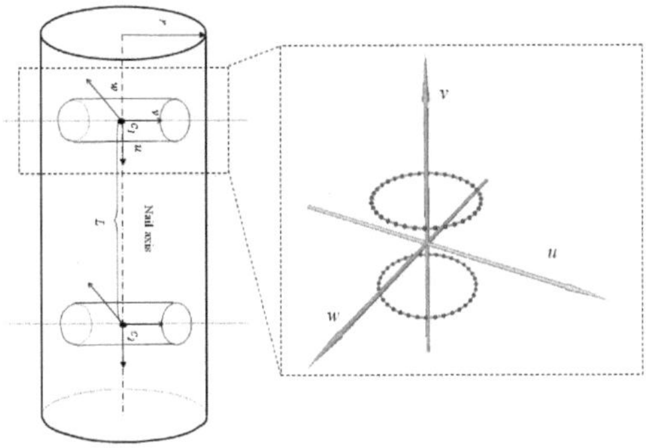

Fig. 1. The geometrical model of the distal part of the intramedullary nail (left) and the geometrical model of the distal locking hole (right)

C'^{uvw} is established by taking the intersection point C (it is also called the center of the locking hole) between the axis of the hole and the axis of the nail as the origin, the axis of the nail as the u axis, and the axis of the locking hole as the v axis (see Fig. 1 for details). The coordinates of those points expressed in this local coordinate system can be directly measured from the nail using a caliper, thanks to the symmetrical property of the locking hole; or extracted from the engineering drawings of the nail, if they are available.

(3) Preprocessing: The task of the preprocessing is to determine the projections of the distal locking holes. To extract these feature points from the image, Hough transform [8] is used to find the two mostly parallel edge lines of the projection of the distal part of the nail after applying a Canny edge detector to the image. The projection of the axis of the nail is considered as the middle line between these two mostly parallel edge lines. To determine those edge pixels belonging to the locking holes, the method reported in [5] is modified for our purpose. A parallelpiped window, whose sizes are equal to the distance between the detected edge lines, is swept along the middle line to find two locations which contain the maximum number of edge pixels and whose distance is greater than a pre-selected distance threshold τ (e.g. the width of the window). The centroids of the detected edge pixels in both locations are then calculated. The projection point of the center of each locking hole is then determined by finding the closest point on the middle line to the associated centroid. A preprocessing example is shown in Fig. 2.

3 The Proposed Approach

3.1 Model-Based Fitting for Pose Recovery

Using above detected feature points, we can find their corresponding spatial points on the imaging plane. Let's denote them as d_1 corresponding to the projection point of the center C_1 of the distal hole (the hole that is closer to the

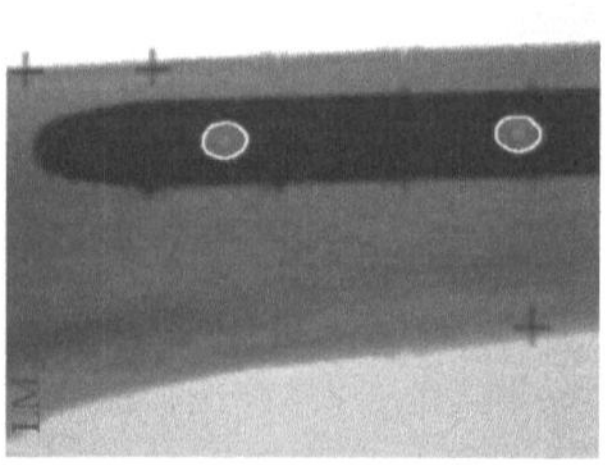

Fig. 2. A preprocessing example. The detected projection centers are displayed together with the extracted edge pixels of the distal locking holes.

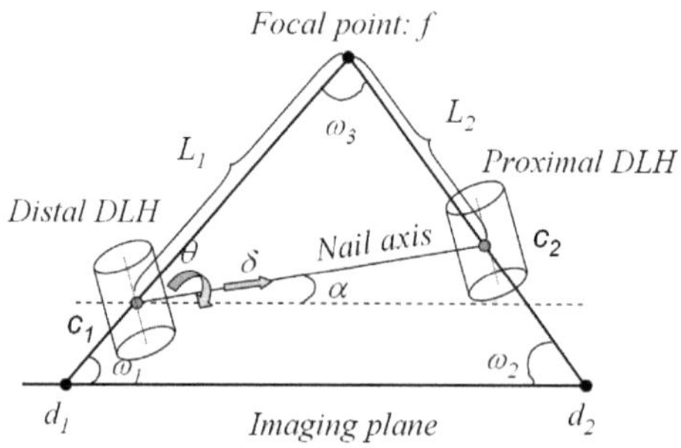

Fig. 3. Schematic view of the model-based fitting for pose recovery

nail tip), and d_2 corresponding to the projection point of the center C_2 of the proximal hole, respectively, as shown in Fig. 3. These two points define a line in $A-COS$. This line together with the focal point f defines a plane where the axis of the distal part of the nail should fall in. As we know the coordinates for point f, d_1, and d_2, we can calculate three internal angles ω_1, ω_2, ω_3 of the triangle fd_1d_2. Assume the angle between the nail axis and its projection in the imaging plane is α, then the coordinates of the centers of both locking holes are calculated as following:

$$
\begin{array}{l}
C_1 = f + L_1 \cdot \frac{(d_1-f)}{||d_1-f||}\,;\, C_2 = f + L_2 \cdot \frac{(d_2-f)}{||d_2-f||} \\
L_1 = L \cdot \frac{\sin(\omega_2+\alpha)}{\sin(\omega_3)}\,;\, L_2 = L \cdot \frac{\sin(\omega_2+\alpha)\cdot\cos(\omega_3)}{\sin(\omega_3)} + L\cdot\cos(\omega_2+\alpha) \\
where\ \alpha \in (-\pi/2, \pi/2)
\end{array}
\qquad (2)
$$

where L is the distance between the centers of two holes. It can be measured or extracted from the product information.

Accoding to Eq. (2), the coordinates of both centers only depends on the parameter α, so as the direction of the nail axis $[n_x, n_y, n_z]^T$.

Assuming that the coordinates of the center C of one of the locking holes is denoted as $[C_x, C_y, C_z]^T$, the problem to estimate the pose of the locking hole in $A-COS$ is now changed to find the rotation angle α of the nail axis, the rotation angle θ and the translation distance δ of the geometrical model of the distal locking hole along the nail axis so that the simulated projection can be fitted to its real X-ray projection (see Fig. 3 for details). This constrained transformation around the parameterized nail axis could be described by a 3×3 rotation matrix $rot(\alpha, \theta, \delta)$:

$$\begin{bmatrix} n_x^2 + SS_{yz}\cos(\theta) & M_{xy}\cdot CC - n_z\sin(\theta) & M_{zx}\cdot CC + n_y\sin(\theta) \\ M_{xy}\cdot CC + n_z\sin(\theta) & n_y^2 + SS_{zx}\cos(\theta) & M_{yz}\cdot CC - n_x\sin(\theta) \\ M_{zx}\cdot CC - n_y\sin(\theta) & M_{yz}\cdot CC + n_x\sin(\theta) & n_z^2 + SS_{xy}\cos(\theta) \end{bmatrix} \tag{3}$$
$$where\ SS_{xy} = n_x^2 + n_y^2;\ SS_{yz} = n_y^2 + n_z^2;\ SS_{zx} = n_z^2 + n_x^2;$$
$$and\ M_{xy} = n_x n_y;\ M_{yz} = n_y n_z;\ M_{zx} = n_z n_x;\ CC = 1 - \cos(\theta)$$

and a translational vector $trans(\alpha, \theta, \delta) = [t_x, t_y, t_z]^T$:

$$\begin{cases} t_x = (C_x + \delta\cdot n_x)\cdot(n_y^2 + n_z^2) - n_x\cdot((C_y + \delta\cdot n_y)\cdot n_y + (C_z + \delta\cdot n_z)\cdot n_z) + \\ \quad (n_x\cdot((C_y + \delta\cdot n_y)\cdot n_y + (C_z + \delta\cdot n_z)\cdot n_z) - (C_x + \delta\cdot n_x) \\ \quad \cdot(n_y^2 + n_z^2))\cos(\theta) + ((C_y + \delta\cdot n_y)\cdot n_z - (C_z + \delta\cdot n_z)\cdot n_y)\sin(\theta) \\ t_y = (C_y + \delta\cdot n_y)\cdot(n_x^2 + n_z^2) - n_y\cdot((C_x + \delta\cdot n_x)\cdot n_x + (C_z + \delta\cdot n_z)\cdot n_z) + \\ \quad (n_y\cdot((C_x + \delta\cdot n_x)\cdot n_x + (C_z + \delta\cdot n_z)\cdot n_z) - (C_y + \delta\cdot n_y) \\ \quad \cdot(n_x^2 + n_z^2))\cos(\theta) + ((C_z + \delta\cdot n_z)\cdot n_x - (C_x + \delta\cdot n_x)\cdot n_z)\sin(\theta) \\ t_z = (C_z + \delta\cdot n_z)\cdot(n_x^2 + n_y^2) - n_z\cdot((C_x + \delta\cdot n_x)\cdot n_x + (C_y + \delta\cdot n_y)\cdot n_y) + \\ \quad (n_z\cdot((C_x + \delta\cdot n_x)\cdot n_x + (C_y + \delta\cdot n_y)\cdot n_y) - (C_z + \delta\cdot n_z) \\ \quad \cdot(n_x^2 + n_y^2))\cos(\theta) + ((C_x + \delta\cdot n_x)\cdot n_y - (C_{yz} + \delta\cdot n_y)\cdot n_x)\sin(\theta) \end{cases} \tag{4}$$

The pose recovery problem can then be formulated as an optimal model-based fitting:

$$\underset{\{\alpha^*, \theta^*, \delta^*\}}{\arg\min} \sum_i ||e_{j=CP(i)} - Pr(rot(\alpha, \theta, \delta)\cdot m_i + trans(\alpha, \theta, \delta))||^2 \tag{5}$$

where $\{e_i\}$ are the detected edge pixels of the locking holes; $\{m_i\}$ are the points used to describe their geometrical models; $Pr(\cdot)$ denotes the projection operator; $CP(\cdot)$ denotes the action of finding the closest edge pixel of the simulated projection point into the image of a model point.

3.2 Parameter Estimation

Various techniques have been proposed for estimating parameters for model-based fitting. Lowe [9] suggests to minimize the non-linear error function on image domain, where the perpendicular distance between projected model line and extracted edge point will be minimized. The correspondence between the model projection to image edge is found by selecting the one who has the shortest perpendicular distance. This strategy can lead to some ambiguity in fitting process when part of the model line has been occluded by structure of the model itself. This problem was solved by Fua [10] through applying hidden algorithm to avoid this pitfall. All these algorithms suffer from the facts that they are easily to be trapped by a local minimum and that the interpretation and initialization of model parameter values have to be done by the operator, which is not desirable for an intra-operative application in a sterilized environment.

Parameter decomposition approach is a powerful optimization method that tries to decompose a high-dimensional problem into small, low-dimensional components and estimate the parameters for each component separately, thus reducing the computational complexity. The general idea of model decomposition for parameter estimation has bee successfully applied in many domains, e.g., geometrical curve fitting [11] and Bayesian model learning [12].

According to our observation that the size of the geometrical models of the distal locking holes (around 10 mm in each dimension) is relatively small compared to the focal length of the X-ray image (around 1000mm), we decompose the control parameters in Eq. (5) into two sets: (a) the angle α between the nail axis and its projection in the imaging plane; and (b) the rotation and translation distance of the geometrical models of the locking holes along the nail axis (θ, δ). Now the original optimization problem can be re-formulated as:

$$\arg\min_{\alpha^*}[(\arg\min_{\{\theta^*,\delta^*\}} \sum_i ||e_{j=CP(i)} - Pr(rot(\alpha,\theta,\delta) \cdot m_i + trans(\alpha,\theta,\delta))||^2)] \quad (6)$$

Where the term in the square brackets simply means the minimum sum of distance for a fixed α and all possibilities of (θ, δ). The advantage of such decomposition lies in the fact that the latter set of parameters can be calculated by using a hybrid optimization technique coupling an evolutionary strategy and an iterative closest projection point algorithm (ICPP) as described below, which then reduces the original multiple-dimensional optimization problem to a one-dimensional search in a finite interval.

A. Initialization: Given a fixed α, we can estimate the positions of both centers of the locking holes and the orientation of the nail axis. Then, the initial transformation between the local coordinate system of the geometrical model of the locking holes and $A - COS$ can be obtained by taking the estimated center as the origin, the estimated nail axis as the u axis, and the normal of the imaging plane as the v axis. All points defined in the local coordinate system of the geometrical model can then be transformed to $A - COS$ using this transformation. The optimal values of the rotation θ and the translation δ of the models along the nail axis can be optimally estimated by fitting the geometrical models of the locking holes to the image as by a hybrid optimization technique as described below.

B. The Iterative Closest Projection Point (ICPP) Algorithm: Let us denote E be a set of N_E detected 2D edge pixels $\{e_1, e_2, ..., e_{N_E}\}$ of the locking hole projection. Further denote M^{t-1} be a set of N_M model point $\{m_0^{t-1}, m_1^{t-1}, ..., m_{N_M}^{t-1}\}$ at iteration step $t-1$. Now in the iteration step t, we perform following steps:

Simulating X-ray projection: In this step, we simulate the X-ray projection of the geometrical models of the locking holes to remove invisible points. Let P^{t-1} be a set of N_P 2D projection points $\{p_1^{t-1}, p_2^{t-1}, \cdots, p_{N_P}^{t-1}\}$ obtained by simulating X-ray projection of the 3D models into the image. Normally $N_P << N_M$. Thus, for each 2D projection point p_i^{t-1}, we know its associated 3D model point m_i^{t-1}.

Find closest projection point: In this step, we try to find the closest neighbor edge pixel e_i of each 2D model projection point p_i^{t-1}.

Establishing 3D-2D correspondence: For each 2D matched pairs (e_i, p_i^{t-1}), calculate the forward projection ray BP_i of the 2D edge pixel e_i. Then for the ray BP_i, calculate a 3D point pair $PP_i^{t-1} = (be_i^{t-1}, m_i^{t-1})$, where be_i^{t-1} is a point on the line BP_i that is closest to the 3D model point m_i^{t-1} of the model projection point p_i^{t-1}.

Estimating pose: For all calculated 3D point pairs $PPS^{(t-1)} = \{PP_i^{t-1}\}$, find an optimal local solution of all pose parameters by minimizing following cost function:

$$\begin{aligned} &\arg\min_{\{\theta^{(t-1)*},\delta^{(t-1)*}\}} \quad S(\theta^{(t-1)}, \delta^{(t-1)}); \; where \; S(\theta^{(t-1)}, \delta^{(t-1)}) = \\ &\sum_i ||be_i^{t-1} - (rot(\theta^{(t-1)}, \delta^{(t-1)}) \cdot m_i^{t-1} + trans(\theta^{(t-1)}, \delta^{(t-1)})||^2 \end{aligned} \quad (7)$$

where we drop the symbol α from the expressions, as its value is fixed.

These steps are repeated until all pose parameters are converged.

C. The Evolutionary Strategy: The ICPP algorithm can be regarded as a local minimum search algorithm but we are trying to find the global minimum of the disparity function that may be well hidden among many poorer local minima, especially when a poor initialization is used. In our approach, this is handled by combining a conventional genetic algorithm [13] with the ICPP algorithm. The genetic algorithm acts as a random generator for possible parameter sets that solve the minimization problem. All generated individual parameter set is then fed through the ICPP algorithm before being rated using the disparity function. Five best ones (with the lowest values of $S(\theta^{(t-1)}, \delta^{(t-1)})$) become the parents of next generation. The algorithm stops when the differences of the disparity function values of all five best ones are smaller than a pre-selected threshold.

D. Optimization of parameter α: We now convert a multiple-dimensional optimization problem to a one-dimensional one, where the parameter α can be optimized by a search along a finite interval $[-30^o, +30^o]$ (due to the acquisition constraint that the nail should be put roughly parallel to the imaging plane). A typical optimization space of this parameter is shown in Fig. 4. It has a symmetrical shape and a clear global optimum around the ground truth $\alpha = 10.4^o$. We could separate the optimization space into two sub-intervals, i.e. $[-30^o, 0)$ and $[0, 30^o]$. In each sub-interval, the optimum of that sub-interval could be easily found by a local search algorithm starting from any initialization value. Actually, in all experiments, we have simply initialized α by the middle value of each sub-interval. The global minimum is then found by taking the better one of the two optima

4 Experimental Results

We design and conducted in vitro experiments to analyze the accuracy and robustness of the proposed approach. A SYNTHES® (STRATEC Medical,

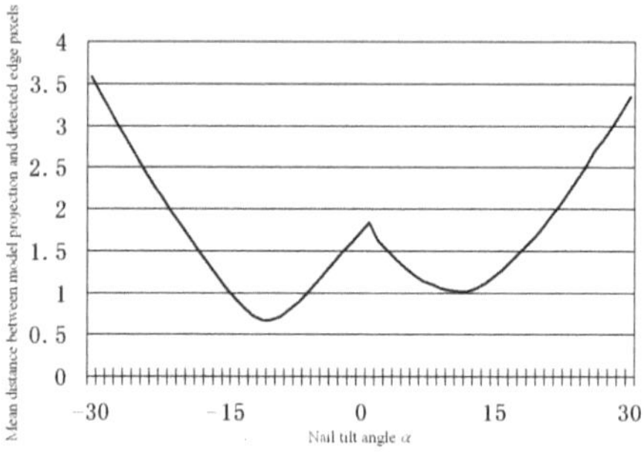

Fig. 4. Optimization space of the parameter α

Oberdorf, Switzerland) 9 mm solid titanium femoral nail was used in our study. It was inserted into a cadaveric human femur and was locked proximally. A Siemens ISO-C C-arm (Siemens AG, Erlangen, Germany) was used to acquire fluoroscopic images for our experiments. The ground truth of the positions of the locking holes was obtained after image acquisition by inserting a custom-made steel rod through the hole and then digitizing both top and bottom centers of the rod using an optically trackable sharp pointer (OPTOTRAK 3020, Northern Digital Inc, Waterloo, Canada).

Three images acquired from different directions were used in our experiments, as shown in Fig. 5. For each image, we applied the proposed approach ten times to estimate the poses of the distal locking holes. The estimated results were compared to the ground truth to compute the errors for each hole, which were defined as the angular difference between the estimated hole axis and the one obtained through pointer-based digitization and the positional difference of the entry point and its ground truth along the plane perpendicular to the hole axis, because the positional difference along the hole axis is not important for the task for insertion of distal locking screws.

In all studies, the poses of the distal locking holes could be automatically recovered. The angular and positional errors are shown in Table 1. Compared to ground truths, the average angular error was found to be 1.0^o (std=0.4^o) and the average positional error along the plane perpendicular to the hole axis was found to be 0.6 mm (std=0.4 mm).

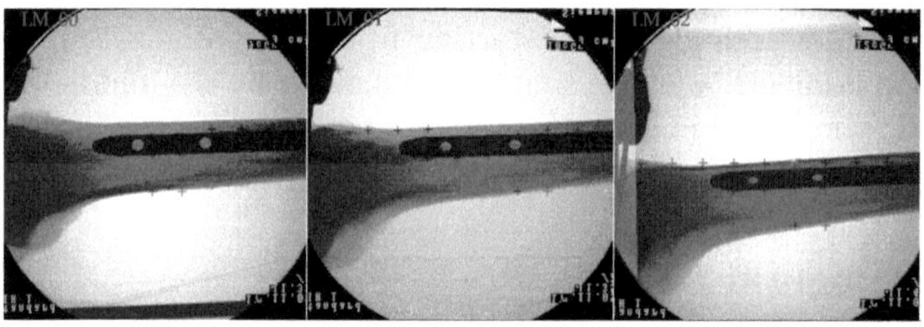

Fig. 5. Three images used in the *in − vitro* experiment: LM_00, LM_01, LM_02

Table 1. *In − vitro* experimetal results

Image	Angular differences (^{o})	Positional differences (mm)
LM_00	0.7 ± 0.3	0.2 ± 0.0
LM_01	0.9 ± 0.2	0.4 ± 0.1
LM_01	1.5 ± 0.2	1.1 ± 0.1
Overall	1.0 ± 0.4	0.6 ± 0.4

5 Conclusions

We have presented a novel parameter decomposition approach for automatic pose recovery of distal locking holes from single calibrated fluoroscopic image. Unlike previously introduced method [5], our approach does not ask for an image with perfectly circular holes. Our in-vitro experimental results demonstrate that the accuracy of our approach is adequate for successful distal locking of intramedullary nails.

References

1. Krettek, C., Mann, J., Miclau, T. et al.: Deformation of femoral nails with intramedullary insertion. J. Orthop. Res. 16, 572–675 (1998)
2. Skjeldal, S., Backe, S.: Interlocking medullary nails - radiation doses in distal targeting. Arch Orthopaedic Trauma Surg. 106, 179–181 (1987)
3. Zhu, Y., Phillips, R., Griffiths, J.G., et al.: Recovery of distal hole axis in intramedullary nail trajectory planning. In: Proc Inst Mech Eng [H], vol. 216, pp. 323–332 (2002)
4. Leloup, T., Schuind, F., Warzee, N.: Process for the acquisition of information intended for the insertion of a locking screw into an orifice of an endomedullary device. European Patent Application Number: 04447153.0 (2004)
5. Yaniv, Z., Joskowicz, L.: Precise robot-assisted guide positioning for distal locking of intramedullary nails. IEEE Trans Med. Imaging 24, 624–635 (2005)
6. Nolte, L.-P., Visarius, H., Arm, E. et al.: Computer-aided fixation of spinal implants. J. Image Guid Surg. 1, 88–93 (1995)
7. Gremban, K.D., Thorpe, C.E., Kanade, T.: Geometric camera calibration using systems of linear equations. In: Proceedings of IEEE conference on robotics and automation, pp. 562–567(1988)
8. Jain, R., Kasturi, R., Schunk, B.G.: Machine Vision. McGraw-Hill, New York (1995)
9. Lowe, F.G.: Fitting parameterized three-dimensional models to images. IEEE Trans Pattern Anal Mach Intell 13, 441–450 (1991)
10. Fua, P.: Model-based optimization: accurate and consistent site modeling. In: Proceedings of the 18th Congress, International Society for Photogrammetry and remote sensing, pp. 222–223 (1996)
11. Jiang, X., Cheng, D.C.: A novel parameter decomposition approach to faithful fitting of quadric surfaces. In: Kropatsch, W.G., Sablatnig, R., Hanbury, A. (eds.) Pattern Recognition. LNCS, vol. 3663, pp. 168–175. Springer, Heidelberg (2005)
12. Neapolitan, R.E.: Learning Bayesian Networks (1st Ed). Prentice Hall
13. Goldberg, D.E.: Genetic algorithms in search, optimization, and machine learning, Reading, MA. Addison-Wesley, London (1989)

Covariance Estimation for SAD Block Matching

Johan Skoglund and Michael Felsberg

Computer Vision Laboratory
Department of Electrical Engineering
Linköping University
SE-581 83 Linköping, Sweden
{skoglund,mfe}@isy.liu.se

Abstract. The estimation of a patch position in an image is a long established but still relevant topic with many applications, e.g. pose estimation and tracking in image sequences. In most systems the position estimate needs to be fused with other estimates, and hence, covariance information is required to weight the different estimates in the right way. In this paper we address the issue with covariance estimation in the case of sum of absolute difference (SAD) block matching. First, we derive the theory for covariance estimation in the case of SAD matching. Second, we evaluate the suggested method in a virtual 3D patch tracking scenario in order to verify the performance in real-world scenarios.

1 Introduction

Motion information from images is useful for many applications. In this paper we focus on one type of method for doing this, called block matching. Typically a patch is given and the algorithm finds the best match in an image.

A common problem is merging of different measurements. Most common is to average a number of similar measurements to reduce average error but the measurements might also come from different sensors. Optimal merging of the measurements, requires knowledge about the accuracy for each measurement. Accuracy of these measurements and information whether the error in the measurements are dependent or not is stored in the "covariance matrix".

In this paper we present both the theory and also an evaluation of a method for estimating the covariance for block matching using *sum of absolute difference* (SAD), which is to our knowledge not yet an established topic.

1.1 Block Matching

Block matching is a common name for all algorithms that try to find the position of a patch p, in an image b. This can be done in a number of different ways but the goal is to find the minimum of an error function

$$\min_{\gamma} e(p, T(b, \gamma)) \ . \tag{1}$$

B.K. Ersbøll and K.S. Pedersen (Eds.): SCIA 2007, LNCS 4522, pp. 374–382, 2007.

The function e measures the difference between the patch and the image. A number of parameters γ, containing the patch position and possibly other interesting parameters like rotation of the patch or other shape information are estimated. To fully utilize the information both the pose γ and information about the accuracy, $\mathrm{cov}(\gamma)$ are needed. In this paper a dc-invariant error function based on the L_1 norm is used.

$$e_1 = \sum_{x,y} |(p(x,y) - \overline{p}) - (b(x,y) - \overline{b})| \tag{2}$$

This method, combined with fast subpixel interpolation is described in [1].

2 Covariance

When several measurements are combined, an assumption of the covariance of the different measurements is required. This assumption might be implicit, might assume that each measurement has the same covariance or explicit like Kalman filters. As an example we can look at a weighted linear least square problem:

$$\arg\min_x ||Ax = b||_2 \tag{3}$$

Which solution is

$$x = (A^T w^{-1} A)^{-1} A^T w^{-1} b \tag{4}$$

Where w is the covariance matrix for b. Solving (3) can be classified into four groups depending on the assumption about the covariance matrix, from equal weight of each measurement to the full covariance:

1. Assume that all measurements are independent and with the same accuracy. w is the unit matrix.
2. Different accuracy for different measurements but that the measurements are independent. w is a diagonal matrix where each element that comes from one measurement is the same.
3. Assume that the measurements can be divided into groups where each measurement might influence all other in that group. Different groups are assumed to be independent. w is a block diagonal matrix with one block from each measurement.
4. One full covariance matrix for all measurements. This makes it possible to have different accuracy in different directions, and also that measurements are dependent. w is an arbitrarily positive definite matrix.

2.1 Covariance Derivation

Details of definition and computational laws for covariances can be found in many text books about statistics [2]. This section is therefore only a short summary containing the most important formulas needed for the rest of this paper. Practically, the covariance can be seen as a measurement of the spread of a stochastic vector x. The covariance contains both a measurement of the spread of different components in x and a measurement of the dependencies between the different parts. The covariance is defined as:

$$\mathrm{cov}[x] = E[(x - \overline{x})^T(x - \overline{x})] \tag{5}$$

The covariance matrix is a symmetric positive semidefinite matrix. Besides the definition, the rule for covariance propagation is needed. Covariance propagation is the calculation of the covariance of the output from a function based on the covariance of the input, estimating $cov[f(x)]$ from $cov[x]$. The exact solution of this problem can easily be found if f is a linear function, $f(x) = Ax + b$. In this case the covariance is:

$$\mathrm{cov}[f(x)] = A\mathrm{cov}[x]A^T \tag{6}$$

Finding the exact solution for an arbitrarily function is usually not possible and a common approximation, sometimes called the *Gauss approximation formula* [3], is to use a linear model of the function and approximate A with the Jacobian:

$$\mathrm{cov}[f(x)] \approx [\frac{d}{dx}f(E[x])]\mathrm{cov}[x][\frac{d}{dx}f(E[x])]^T \tag{7}$$

3 Covariance Estimation

For each patch used in the tracking an estimate of the covariance is needed. The most basic form of confidence measure is the SAD distance between the patch and the image. However, this measurement has three important limitations:

- The measurement depends on intensity scaling in an unsuitable way, e.g. if the patch and image are scaled by 2 the error is scaled by 2.
- This gives an isotrop covariance. The certainty of the result might be different in different directions, i.e. we need a anisotropic measurement.
- SAD measures the distance between the patch and the image and gives a measurement of the similarity. Most applications do however need a measurement of the position accuracy, not the similarity.

Therefore, a more advanced covariance estimate is needed, an estimate which is able to model anisotrop covariances. This representation, makes it is also possible to represent the certainty for 1-D features like lines and edges. Figure 1 shows two examples of covariances estimated with the method proposed later. We can see that the accuracy is much higher perpendicular to the edge than along the edge.

The covariance can be estimated in two different ways [4]:

- Estimation from the structure of the error function around the minimum
- Estimation from the influence of each pixel in the patch

Fig. 1. Shape of uncertainties estimated for two patches

3.1 Covariance from Each Pixel

The most obvious way for estimating the covariance is probably to use covariance propagation (7). This requires knowledge about the covariance for each pixel and to differentiate the position of the minimum of the error wrt each pixel. One way to do this is explained in [5]. This paper shows that the covariance is estimated as:

$$\text{cov}(\gamma) \approx [-\frac{de^2}{d\gamma^2}]^{-1}[\frac{de^2}{d\gamma dx}]\text{cov}(x)[\frac{de^2}{d\gamma dx}]^T[-\frac{de^2}{d\gamma^2}]^{-T} \tag{8}$$

This method has been successfully used for error function like the L_2 norm [4]. It is however not possible to use this for the L_1 norm because the differentiation of the error function (2) wrt each pixel gives:

$$\frac{de}{dx} = sign(p - b) \ . \tag{9}$$

Differentiating (9) wrt γ gives 0 and that the whole covariance is 0. Therefore, the covariance needs to be estimated from the error function instead.

3.2 Covariance from the Error Function

The covariance can also be estimated from the structure of the error function around the minimum. We propose to apply (7) on the error function (2).

$$\text{Var}[e(\gamma)] \approx [\frac{de}{d\gamma}]\text{cov}(\gamma)[\frac{de}{d\gamma}]^T \ . \tag{10}$$

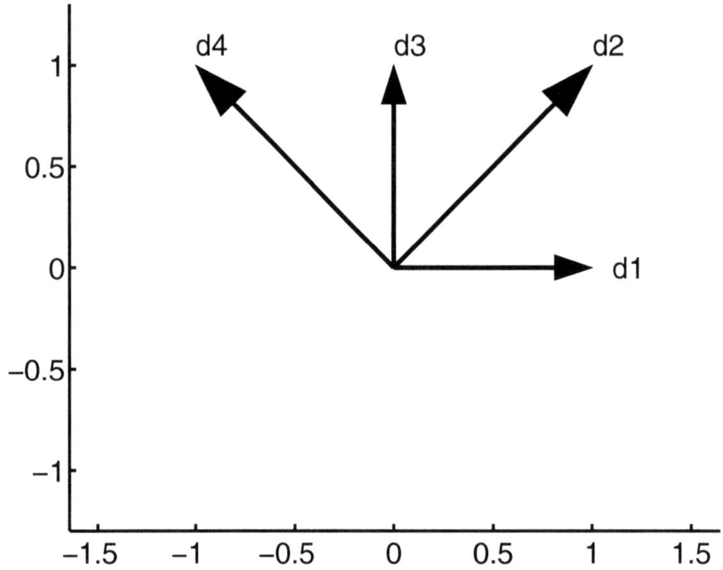

Fig. 2. Directions for the four derivatives d_1 ... d_4

Cov(γ) is a symmetric covariance matrix, hence a real-valued eigensystem decomposition exists and is given as

$$\mathrm{cov}[\gamma] = \lambda_1 e_1 e_1^T + \lambda_2 e_2 e_2^T \; . \tag{11}$$

Plugging this into (10) results in

$$\mathrm{Var}[e(\gamma)] \approx \; [\tfrac{df}{d\gamma}]^T(\lambda_1 e_1 e_1^T + \lambda_2 e_2 e_2^T)[\tfrac{df}{d\gamma}] \tag{12}$$

$$= \lambda_1([\tfrac{df}{d\gamma}]^T e_1)^2 + \lambda_2([\tfrac{df}{d\gamma}]^T e_2)^2 \; . \tag{13}$$

Rewriting (13) using the Frobenius product $\langle\rangle_F$ [6]gives

$$\mathrm{Var}[e(\gamma)] \approx \lambda_1\langle[\tfrac{df}{d\gamma}][\tfrac{df}{d\gamma}]^T|e_1 e_1^T\rangle_F + \lambda_2\langle[\tfrac{df}{d\gamma}][\tfrac{df}{d\gamma}]^T|e_2 e_2^T\rangle_F \tag{14}$$

$$= \quad \langle[\tfrac{df}{d\gamma}][\tfrac{df}{d\gamma}]^T|\mathrm{Cov}[\gamma]\rangle_F \; . \tag{15}$$

To be able to estimate the full covariance matrix, at least three different derivatives are needed [7]. Using derivatives in four directions is however useful since this makes it easy to sample the derivatives regularly. If the derivatives d_1 to d_4 are estimated in the x,y and the diagonal directions according to figure 2, these four responses correspond to the frame tensors

$$B_1 = \begin{bmatrix} 1 & 0 \\ 0 & 0 \end{bmatrix} \qquad B_2 = \begin{bmatrix} 1 & 1 \\ 1 & 1 \end{bmatrix} \tag{16}$$

$$B_3 = \begin{bmatrix} 0 & 0 \\ 0 & 1 \end{bmatrix} \qquad B_4 = \begin{bmatrix} 1 & -1 \\ -1 & 1 \end{bmatrix} \; . \tag{17}$$

For the tensor computation we need the dual frame with minimum norm given as [7]

$$\tilde{B_1} = \begin{bmatrix} 0.6 & 0 \\ 0 & -0.4 \end{bmatrix} \qquad \tilde{B_2} = \begin{bmatrix} 0.2 & 0.25 \\ 0.25 & 0.2 \end{bmatrix} \tag{18}$$

$$\tilde{B_3} = \begin{bmatrix} -0.4 & 0 \\ 0 & 0.6 \end{bmatrix} \qquad \tilde{B_4} = \begin{bmatrix} 0.2 & -0.25 \\ -0.25 & 0.2 \end{bmatrix} . \tag{19}$$

Frame theory gives that the least square solution of (15) is [7]:

$$\mathrm{Cov}[\gamma] = \mathrm{Var}[e(\gamma)] \sum_{i=1}^{4} \tilde{B}_i |d_i|^{-2} . \tag{20}$$

To simplify the notation, we define the tensor

$$T = \sum_{i=1}^{4} \tilde{B}_i |d_i|^{-2} . \tag{21}$$

For the next step an assumption about the distribution of the errors $e(\gamma)$ is needed, how $Var[e(\gamma)]$ should be approximated from the minimum of the error function. In general, the estimate of the variance is:

$$\mathrm{Var}[e(\gamma)] = c\mathrm{E_{min}}^n . \tag{22}$$

where C and n depends on the assumed distribution. C and n are found by analyzing

$$\mathrm{Var}[e(\gamma)] = cE[e(\gamma)]^n \tag{23}$$

for the given distribution. Examples of c and n for different distributions can be found in table 1. Combining (20) and (22) gives that the covariance of γ can be estimated as:

$$\mathrm{Cov}[\gamma] = c\mathrm{E_{min}}^n T^{-2} \tag{24}$$

Table 1. C and n for different distributions

Distribution	C	n
Positive uniform	4/3	2
Abs of normal	$\pi/2$	2
χ^2	2	1
Poisson	1	1

4 Evaluation

This evaluation has been done within the *MATRIS*[1] project. The goal of this project is to develop a real time system for pose estimation of cameras. One

[1] http://www.ist-matris.org

central part of this system is efficient algorithms for patch tracking in video sequences. Within the system, the result from the patch tracking is merged with data from an *Inertial Measurement unit* (IMU). To be able to combine the information in an efficient way a covariance estimate is needed for each patch.

There is a number of problems with an evaluation of tracking algorithms. The most important problem is probably to generate the ground truth without bias. In this evaluation, we solve this problem by using a synthetic image sequence generated from real textures. With this method it is possible to generate the ground truth and to simulate illumination changes.

To generate the test images a number of tools used or developed in the *MATRIS* project was used. The test data was created using this procedure:

1. Create a textured 3d-model consisting of planar patches from a number of real images.
2. Render a sequence of images showing the model from different poses. Save the 2D center position for the planar patches together with the image.
3. Start to create patches and predict the position for patches in all images. This is done with software developed in the project.
4. Estimate the start pose for the camera, the position of the camera for the first image using the image content.
5. Warp all patches that are visible from the estimated pose using a homography and save these patches.
6. Track the position for all visible patches and use this to improve the pose, combining the 3D-model and patch positions.
7. Iterate step 5-6 for all images.

This gives a number of images together with almost correctly transformed patches and their correct position. These images are then modified to simulate noise. The data has previously been used for evaluation of tracking algorithms [1].

The purpose of the evaluation is to compare the suggested covariance estimation method with an L_1 based tracking algorithm. The evaluation uses the dc-invariant tracking algorithm with subpixel accuracy [1]. To do this one condition for covariances has been derived from the definition (5), that

$$E[(x-\overline{x})^T C^{-1}(x-\overline{x})] = dim(x) \; . \tag{25}$$

This condition is used because it is simple to evaluate. Showing that a method satisfies this condition is probably good enough for practical situations, we should however note that this is not a proof that the estimated covariance is correct.

To simulate noise in the camera a number of different models are available. For this evaluation, additive Gaussian noise was used. The pixels were in the interval [0:255] and Gaussian noise with σ between 0 and 20 was added. The most interesting part with the evaluation is to compare the result for (25) with different amount of noise. Accuracy of the tracking decreases when the amount of noise increases. The experiment evaluates whether the covariance estimate will increase with the same speed.

5 Results

The covariance estimation (24) has two parameters, C and n. Most important of these parameters is n, which significantly influences the whole result whereas C "only" scales the result. The evaluation showed that $n = 1$ gave best result and is therefore used for the following results. A scaling using $C = 1$ is used, which corresponds to an assumption that the error has a Poisson distribution.

Figure 3 shows the RMS error from the tracking and the square root of (25) which is the error scaled by the estimated covariance. Most important is the shape of the curves and the RMS error measured in pixels is therefore scaled to simplify the evaluation. Originally the RMS tracking error started at ≈ 0.15. The figure shows that the normalized average error is closer to a constant function than the original tracking error, especially for small amounts of noise, $\sigma < 10$. The figure shows a trend, the estimated covariance underestimates the increase of the error in the tracking with high noise levels. Whether the covariance estimate is too optimistic or performance of the block matching could be improved with high noise levels is still an open question.

We can also see that the scaling of the covariance is slightly wrong, the graph does not start at 2, which is the dimension of the estimated parameter. To be able to use the covariance in combination with other sensors, this scaling has to be manually adjusted.

The results show that the suggested method is significantly better than using no covariance at all and the estimation of the covariance is very fast.

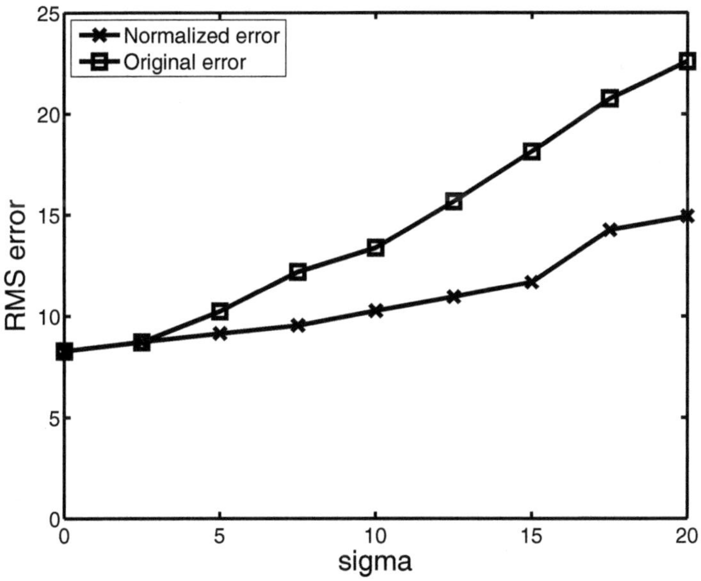

Fig. 3. Normalized error vs (scaled) original error

6 Conclusions

In this paper we proposed a method for estimating the covariance matrix of SAD block matching. An algorithm for computing the covariance using dual frames has been formulated. This provides an efficient method to calculate the covariance.

In the second part of the paper, an evaluation of the proposed method for covariance estimation has been performed. In the evaluation a dc-invariant SAD block matching method with subpixel interpolation was used. The evaluation showed that the suggested method for covariance estimation was significantly better than assuming that each patch has the same error. The calculational complexity of the suggested method is low and it can therefore be applied, with almost unchanged computational complexity.

Acknowledgment

This work has been supported by EC Grant IST-2002-002013 MATRIS.

References

1. Skoglund, J., Felsberg, M.: Evaluation of subpixel tracking algorithms. In: ISVC (2), pp. 374–382 (2006)
2. Dougherty, E.R.: Random Processes for Image and Signal Processing. SPIE press (1999)
3. Ljung, L.: Sytem Identifiaction. Prentice hall, Englewood Cliffs (1999)
4. Kanazawa, Y., Kanatani, K.: Do we really have to consider covariance matrices for image features? ICCV 02, 301 (2001)
5. Fessler, J.A.: Mean and variance of implicitly defined biased estimators (such as penalized maximum likelihood): applications to tomography. IEEE Tr. Im. Proc. 5(3), 493–506 (1996)
6. Sun, Q., DeJong, G.: Feature kernel functions: Improving SVMs using high-level knowledge. In: CVPR (2). pp. 177–183 (2005)
7. Granlund, G., Knutsson, H.: Signal Processing for Computer Vision. Kluwer Academic Publishers, Dordrecht (1995)

Infrared-Visual Image Registration Based on Corners and Hausdorff Distance*

Tomislav Hrkać, Zoran Kalafatić, and Josip Krapac

University of Zagreb, Faculty of Electrical Engineering and Computing
Unska 3, 10000 Zagreb, Croatia
tomislav.hrkac@fer.hr

Abstract. The paper presents an approach to multimodal image registration. The method is developed for aligning infrared (IR) and visual (RGB) images of facades. It is based on mapping clouds of points extracted by a corner detector applied to both images. The experiments show that corners are suitable features for our application. In the alignment process a number of transformation hypotheses is generated and evaluated. The evaluation is performed by measuring similarity between the RGB corners and the transformed corners from IR image. Directed partial Hausdorff distance is used as a robust similarity measure. The implemented system has been tested on various IR-RGB pairs of images of buildings. The results show that the method can be used for image registration, but also expose some typical problems.

1 Introduction

Registration is a fundamental problem in computer vision. The main task is to align two images taken at different times, from different viewpoints, or by different sensors. There are several comprehensive reviews of the subject [1,2].

When developing an image registration algorithm, several key issues have to be addressed:

- appropriate image transformation class for aligning one image to another;
- features to be used as landmarks guiding the registration procedure;
- strategy for hypothesis generation;
- similarity measure used for hypothesis evaluation.

In the case of multisensor registration, the images to be aligned are not necessarily similar, so that appearance-based or correlation-based registration methods cannot be used. It is necessary to use features that are stable with respect to sensor, i.e., the same physical artifact produces features in both images. Usual features used for multisensor registration are edges, line segments [3], or virtual line intersections [4,5]. Corners can also be used, but it seems that they are

* This work has been supported by the Croatian Ministry of Science, Education and Sports, as a part of the TEST (technological R&D) programme, administrative number #4046 (2004).

B.K. Ersbøll and K.S. Pedersen (Eds.): SCIA 2007, LNCS 4522, pp. 383–392, 2007.

not very popular due to their sensitivity to scale, skew, rotation, illumination changes, etc. [4].

This work is aimed to aligning infrared (IR) and visual (RGB) images of facades in an application for thermal isolation inspection. Experiments with various features extracted from IR and RGB images of buildings have shown that corners have the desired property of being stable in both images, due to the typical scene structure containing windows, doors, balconies, and similar rectangular areas (Fig. 1). As a typical scene contains mostly planar surfaces and both images (IR and RGB) are taken from approximately the same position and frontally to the facade, a simple similarity transform can be used for aligning the images. The strategy for hypothesis generation is based on establishing correspondences between pairs of corners from IR image and corner pairs in RGB image. Each hypothesis is evaluated by computing the corresponding similarity measure based on partial Hausdorff distance [6] between the transformed corners from IR image and the corners in RGB image. This similarity measure is not based on the correlation between the detected corners' neighbourhoods [7], but on the constellation of the corner positions.

Fig. 1. A typical RGB-IR image pair

The rest of the paper is organized as follows. The specific assumptions of the proposed approach to the multi-sensor image registration are described in Section 2. In Section 3 the registration procedure is presented, together with some implementation details. Section 4 presents and discusses the results of applying the implemented method. Some concluding remarks and a discussion of future research are given in Section 5.

2 Registration Assumptions

The proposed registration algorithm is tailored to the registration of IR and RGB images of facades. The nature of the application enables us to impose several constraints and simplify the algorithm design.

The first constraint is due to the fact that the images are usually taken frontally to the facade and the photographer tries to take the pictures of the same part of the building. Since both images are taken from viewpoints that are not far apart, the general transform for registration of planar scenes (planar homography) can be approximated by much simpler similarity transform. The transformation is defined by four parameters: scaling (s), rotation (α), and translation (t_x, t_y):

$$\begin{bmatrix} x_{RGB} \\ y_{RGB} \end{bmatrix} = s \cdot \begin{bmatrix} \cos\alpha & -\sin\alpha \\ \sin\alpha & \cos\alpha \end{bmatrix} \cdot \begin{bmatrix} x_{IR} \\ y_{IR} \end{bmatrix} + \begin{bmatrix} t_x \\ t_y \end{bmatrix} . \tag{1}$$

Assuming that the geometric transformation between the two images is global, two pairs of corresponding points are enough to compute the transformation parameters. As the structure of the images is such that it promotes corners in both the IR and RGB images, hypotheses are generated based on points obtained by a corner detector. The hypothesis generation procedure is discussed in Sec. 3.

Two corner detectors were tried out: SUSAN [8] and Harris [9]. In the conducted experiments, Harris corner detector has given significantly better results. That is consistent with [10]. An example of corners detected by Harris corner detector is shown in Fig. 2.

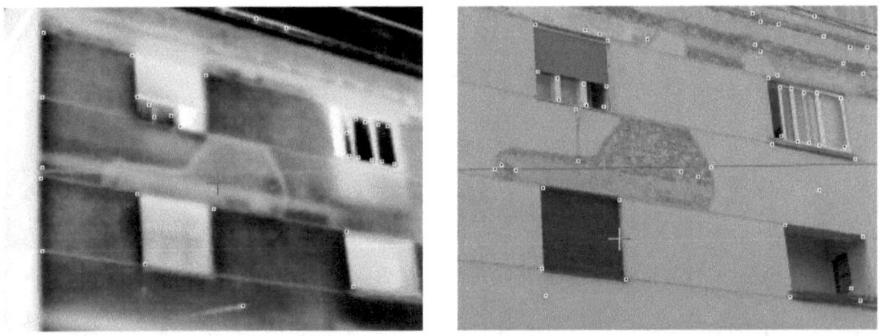

Fig. 2. Corners detected in IR (left) and RGB (right) images by using Harris corner detector

The Harris corner detector [9] determines the matrix $\mathbf{C}$ describing the intensity structure of the local neighborhood W for each image pixel (Eq. 2):

$$\mathbf{C} = \begin{bmatrix} \sum_W (I_x(x,y))^2 & \sum_W (I_x(x,y) \cdot I_y(x,y)) \\ \sum_W (I_x(x,y) \cdot I_y(x,y)) & \sum_W (I_y(x,y))^2 \end{bmatrix} , \tag{2}$$

where $I_x(x,y) = \partial I(x,y)/\partial x$ and $I_y(x,y) = \partial I(x,y)/\partial y$. The detector output (corner "strength") is computed as $s = \det(\mathbf{C}) - \kappa \cdot \mathrm{trace}^2(\mathbf{C})$. Local maxima of these corner "strengths" indicate potential corner positions. In the experiments the parameter κ was set to 0.04. In order to prune the list of corners, local maxima having $s < 0.01 \cdot s_{max}$ were filtered out, where s_{max} denotes the strength of the global maximum. Since the IR images had significantly lower contrast than the RGB images, the contrast of IR images was enhanced by histogram equalization.

3 Hypothesis Generation and Evaluation

In order to compute the four transformation parameters (s, α , t_x, t_y), a pair of points in the IR image and the corresponding pair of points in the RGB image must be known, providing a set of four equations. If the two corner points in the IR image are denoted by $A_1(x_{A1}, y_{A1})$ and $A_2(x_{A2}, y_{A2})$, and the two corresponding corner points in the RGB image by $B_1(x_{B1}, y_{B1})$ and $B_2(x_{B2}, y_{B2})$ (where point B_1 in the RGB image corresponds to the point A_1 in the IR image, and point B_2 corresponds to the point A_2), we get the following set of equations:

$$\begin{aligned} x_{B1} &= s \cdot (\cos\alpha \cdot x_{A1} - \sin\alpha \cdot y_{A1}) + t_x, \\ y_{B1} &= s \cdot (\sin\alpha \cdot x_{A1} + \cos\alpha \cdot y_{A1}) + t_y, \\ x_{B2} &= s \cdot (\cos\alpha \cdot x_{A2} - \sin\alpha \cdot y_{A2}) + t_x, \\ y_{B2} &= s \cdot (\sin\alpha \cdot x_{A2} + \cos\alpha \cdot y_{A2}) + t_y. \end{aligned} \tag{3}$$

By solving the above set of equations, the parameters of the transformation can be found as:

$$\begin{aligned} \alpha &= \mathrm{arctg}\left(\frac{\Delta x_A \cdot \Delta y_B - \Delta x_B \cdot \Delta y_A}{\Delta y_A \cdot \Delta y_B + \Delta x_B \cdot \Delta x_A}\right), \\ s &= \frac{\Delta x_B}{\cos\alpha \cdot \Delta x_A - \sin\alpha \cdot \Delta y_A}, \\ t_x &= x_{B1} - s \cdot (\cos\alpha \cdot x_{A1} - \sin\alpha \cdot y_{A1}), \\ t_y &= y_{B1} - s \cdot (\sin\alpha \cdot x_{A1} + \cos\alpha \cdot y_{A1}), \end{aligned} \tag{4}$$

where $\Delta x_A = x_{A2} - x_{A1}$, $\Delta y_A = y_{A2} - y_{A1}$, $\Delta x_B = x_{B2} - x_{B1}$, and $\Delta y_B = y_{B2} - y_{B1}$.

The basic idea is to use all combinations of corner pairs from IR and RGB image as *match hypotheses*. For each match hypothesis, parameters of similarity transform are calculated according to equations (4), and each hypothesis is evaluated according to the criterion described later in this section. The transformation obtaining the best similarity measure is then selected as the final solution.

However, the described simple scenario has an obvious drawback: if *all* combinations of corner pairs were used, a very large number of hypotheses would be generated and evaluated. Many of those hypotheses are obviously false, which results in an unacceptably large computational load. Therefore, to achieve better computational times, the number of generated hypotheses has to be significantly reduced. Two strategies to reduce the number of hypotheses were employed.

First, because the images were taken from approximately the same viewpoint, the corresponding corners in the IR and RGB image should not be too far away from each other. Therefore, if the corner A_1 is selected in the IR image, not all of the RGB corners are considered as its potential matching corners, but only those within a circular area of radius R ($R = 50$ pixels was used in the experiments) around the RGB point A_1' with the same coordinates as A_1 (Fig. 3).

The second strategy is based on the fact that, due to the noise and imperfectness of corner detectors, the positions of the detected corners are often imprecise. This imprecision is usually not greater than a few pixels. However, if two relatively close corner points from the same image are used for hypothesis generation, the parameters of the resulting hypothesis can be significantly distorted due to such imprecision in corner positions (for example, the length and inclination of a very short line vary significantly with small displacement of its end points, resulting in a significant variation of transformation parameters s and α). In order to avoid this problem and to further reduce the number of generated hypotheses, not all pairs of corners in one image are considered to generate a hypothesis, but only those pairs that are at least d pixels away form each other ($d = 50$ pixels was used in the experiments) (Fig. 3).

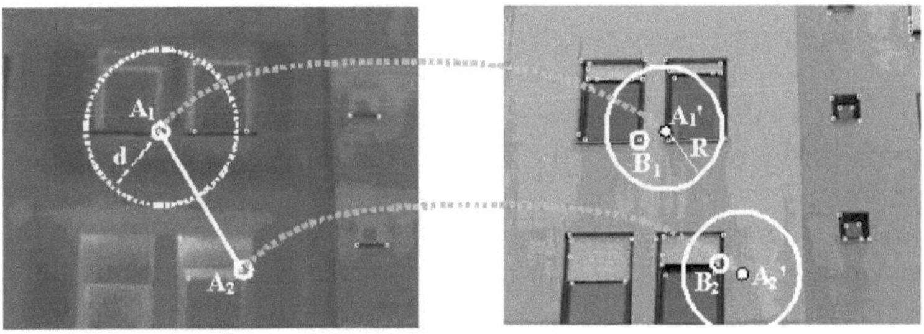

Fig. 3. Reducing the number of hypotheses

To evaluate a hypothesis, all corner points from the IR image are transformed with a hypothesized transformation and a similarity measure between the transformed points and the corner points in RGB image is computed. In multimodal images, a lot of outliers (i.e. feature points that are present only in one but not in another image) can be present. Therefore, the similarity measure has to be robust in order to reduce negative impact of these outliers. A well-known robust measure of similarity called directed partial Hausdorff distance [6,11] was used. In order to determine the directed partial Hausdorff distance between two sets of points, for each point in one set (A), the nearest point in other set (B) has to be found and the Euclidian distance between these two points has to be calculated. The partial Hausdorff distance H_k can be defined as the k-th smallest distance in the set of distances:

$$H_k(A, B) = K^{th}_{a \in A} \min_{b \in B} \text{dist}(a, b), \quad (5)$$

where K^{th} is an operator returning the k-th smallest element of the set [11]. The parameter k can also be expressed in terms of *quantile* – percentage of total number of points in the set.

Usually, the median distance is used, i.e., k is set to 50% of the number of points in the set A, assuming that there is no more than 50% of outliers in the set A (i.e., corners in the IR image). The number of outliers in the set B

(i.e. corners in the RGB image) can be greater than 50% without degrading the method performance.

In order to reduce the computational complexity of calculating Hausdorff distance for each hypothesis to be evaluated, the distance transform is used [6]. The distance transform is applied to the image containing the corners detected in the RGB image. The result is a gray-level image expressing for every pixel the distance to its nearest RGB corner. The transformation is computed only once for each image pair to be registered and can be used to determine the partial Hausdorff distance efficiently.

4 Experimental Results and Evaluation

The method was evaluated on an image database containing 40 pairs of IR and RGB images of facades taken roughly from the same viewpoint and under similar viewing angle. Despite the effort to maximize the overlap between the images of each pair, later analysis had shown that it was much worse than expected. The IR images have the resolution 320×240 and the RGB images were scaled to the same resolution.

The first evaluation was based on (subjective) visual inspection of synthetic images composed of edges found in the IR images and the corresponding RGB images. The IR edges detected by an edge detector [12] were mapped to the RGB image by the transformation found by the registration algorithm (Fig. 4).

Fig. 4. Subjective evaluation of the registration. The rightmost images are obtained by laying the edges from IR images (leftmost) over the RGB images (middle).

The registration result was subjectively evaluated either as success or failure. In total, 34 out of 40 test pairs were successfully registered by proposed method. It is worth mentioning that no parameter tuning was necessary except for three of those image pairs. The only parameter is the percentage of point distances used for partial Hausdorff distance computation. It is by default set to 50%, but the method is successful over large variations of the parameter. For images with a large number of outliers the results improve by setting the parameter to 30%. However, six test image pairs could not be correctly registered even under extensive parameter tuning. Analysis of the detected corners in both images have shown that the reason for the failure was the large number of outliers – different corners appearing in the images (e.g., graffiti produce many corners in RGB image and none in the corresponding IR image; a tree in front of the building produces many unstable corners that overwhelm the "useful" corners corresponding to building artifacts – Fig. 5).

Fig. 5. An example of very problematic image. The corners in IR (left) and RGB (middle) images are also shown. The "registration" is obviously worthless.

Inspired by [13], we tried to gain insight into distribution of hypotheses by clustering hypotheses for "problematic" image pairs. Each hypothesis is represented by a vector of similarity transform parameters. Two approaches were taken: representation of hypotheses distribution by a histogram, and clustering hypotheses using mean-shift algorithm [14]. Unfortunately, the results were disappointing. In fact, the image pairs for which the hypotheses formed significant clusters were also successfully registered by the previously described method. On the contrary, the hypotheses generated by "problematic" image pairs don't tend to form distinct clusters. Moreover, it turned out that some of the successfully registered image pairs didn't generate clustered hypotheses. Therefore, we were eager to get some insight about the quality of the generated hypotheses. Also, an objective evaluation of transform quality was needed. In order to compare and evaluate the hypotheses, a rough approximation of *ground truth* information was prepared by manually marking 10 pairs of corresponding points for each image pair.

For each image pair the registration algorithm has been executed and every generated hypothesis has been recorded. Each record contained the transformation parameters and the values of partial Hausdorff distances with various quantiles (10 – 90% in the increments of 10%). Then, each hypothesis has been evaluated by measuring the average distance of *ground truth* corners from IR image, transformed by the hypothesized parameters, to the corresponding (ground truth) RGB corners. This measure will be denoted by *DGT – distance to the*

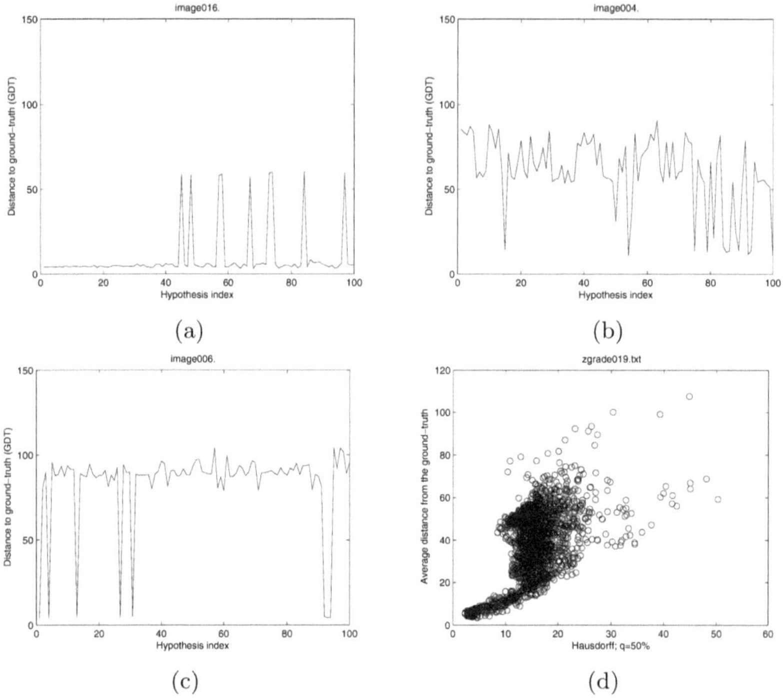

Fig. 6. (a–c) DGT graphs showing the sorted Hausdorff distances (HD) and the corresponding distances to the ground truth (DGT). (d) Scatter diagram of HD vs. DGT distances for the generated hypotheses.

ground truth. The smaller the DGT distance, the better the transformation. All hypotheses can be sorted by their Hausdorff distances (for a given quantile) and their corresponding DGT can be plotted. It is interesting that just by analyzing those graphs the success of the registration can be infered. The successfully registered image pairs have one or more best ranked hypotheses with small DGT (Fig. 6a). If the best ranked hypothesis has a large DGT, obviously the registration procedure will fail (Fig. 6b). On the other hand, if there are many highly ranked hypotheses with small DGT, a clustering approach could filter out the possibly best ranked false hypotheses. Also, the graphs revealed some image pairs that were successfully registered due to the luck of the draw. Their graphs show that there are some generated hypotheses that are completely wrong (have large DGT) while having almost equally small Hausdorff distance (HD) as the best ranked one (Fig. 6c). The same data can be plotted as scatter diagram showing the generated hypotheses as points with coordinates (HD, DGT). Image pairs with scatter diagrams similar to the one shown in Fig. 6d are easily registered by described method because the hypotheses with small HD have small DGT (the points close to the graph origin) and the hypotheses with large DGT have large HD.

However, the diagram shown in Fig. 7a reveals that although there are many hypotheses with small HD and DGT values, many misleading hypotheses are also generated (the points with small HD and large DGT). Those hypotheses have good Hausdorff measure although they are far away from the ground truth. Closer analysis of the corresponding image pair shows that the reason is its periodic structure (the windows from IR image can be registered with the windows in RGB image using two different horizontal translations, but only one is considered as correct – Fig. 7b). Subjective evaluation of the result considers both registrations as correct. Even humans have problems with such registration and can resolve the ambiguity only by identifying some details that can be recognized in both (multisensor) images (e.g., a broken window) – i.e., by using higher level knowledge.

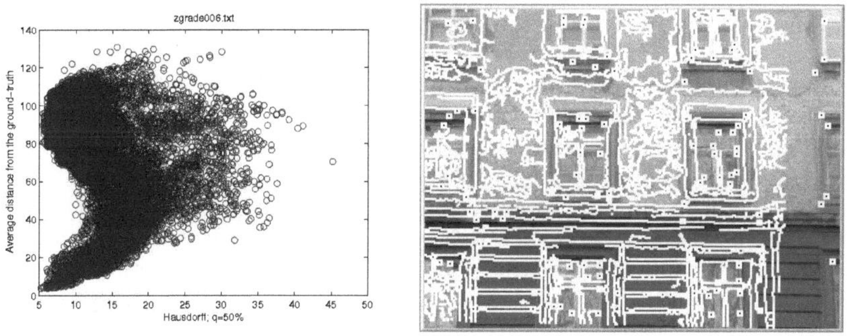

(a) (b)

Fig. 7. (a) Scatter diagram for image pair with periodic structure. (b) The corresponding registered image.

In spite of previously discussed problems, the method has proven to be robust and reliable because in most cases (85%), the correct registration was obtained. The method is also relatively fast, although it depends on the number of detected corners, i.e. the number of generated hypotheses. Typical time required for registration of an image pair is about one second on a computer running at 1.8 GHz, with 512 MB RAM.

5 Conclusion

A simple yet effective method for infrared-visual image registration has been presented. It has been applied for aligning IR images of facades to the corresponding RGB images taken from similar viewpoint. It is intended to be a part of a system for thermal isolation inspection.

The advantage of the proposed method is that it uses small number of parameters and with same parameter setting performs well on wide range of images. The only parameter whose change influenced the results was quantile percentage. Future work would include automatic determination of quantile percentage from Euclidean distances of transformed points. We presume another major source of misregistration is insufficient characterization of corner points. Corners could be additionally characterized e.g., using cornerity measure and corner orientation which can be calculated from matrix (2). Other usual approaches to corner characterization through corner neigborhood could fail due to different modalities of neighborhoods. The method assumes simple global similarity transformation between the images. Although this assumption does not always hold, the results show that it can be used for obtaining rough alignment of the images. Residual DGT could be made smaller using better geometric models of image transformation, e.g., refining similarity transform to affine transform or planar homography.

References

1. Brown, L.G.: A survey of image registration techniques. ACM Computing Surveys 24(4), 325–376 (1992)
2. Zitova, B., Flusser, J.: Image registration methods: a survey. Image and Vision Computing 21(11), 977–1000 (2003)
3. Krüger, W.: Robust and efficient map-to-image registration with line segments. Machine Vision and Applications 13(1), 38–50 (2001)
4. Coiras, E., Santamaria, J., Miravet, C.: A segment-based registration technique for visual-IR images. Optical Engineering 39(1), 282–289 (2000)
5. Segvic, S.: A multimodal image registration technique for structured polygonal scenes. In: Proc of 4th Symp. on Image and Signal Processing and Analysis, Zagreb, Croatia pp. 500–505 (2005)
6. Huttenlocher, D.P., Klauderman, G.A., Rucklidge, W.J.: Comparing images using the Hausdorff-distance. IEEE Trans. on PAMI 15, 850–863 (1993)
7. Irani, M., Anandan, P.: Robust multi-sensor image alignment. In: International Conference on Computer Vision, Bombay, India, pp. 959–966 (1998)
8. Smith, S., Brady, J.: Susan - a new approach to low level image processing. Int. Journal of Computer Vision 23(1), 45–78 (1997)
9. Harris, C.J., Stephens, M.: A combined corner and edge detector. In: Proc. 4th Alvey Vision Conferences, pp. 147–151 (1988)
10. Vincent, E., Laganiere, R.: An empirical study of some feature matching strategies. In: Proc. Conf. Vision Interface, Calgary, Canada, pp. 139–145 (2002)
11. Mount, D.M., Netanyahu, N.S., Moigne, J.L.: Efficient algorithms for robust feature matching. Pattern Recognition 32, 17–28 (1998)
12. Canny, J.: A computational approach to edge detection. IEEE Trans. PAMI 8, 679–714 (1986)
13. Stockman, G., Kopstein, S., Benett, S.: Matching images to models for registration and object detection via clustering. IEEE Trans. PAMI 4(3), 229–241 (1982)
14. Comaniciu, D., Meer, P.: Mean shift: A robust approach toward feature space analysis. IEEE Trans. PAMI 24(5), 603–619 (2002)

Watertight Multi-view Reconstruction Based on Volumetric Graph-Cuts

Mario Sormann[1], Christopher Zach[1], Joachim Bauer[1], Konrad Karner[1], and Horst Bishof[2]

[1] VRVis Research Center,
Inffeldgasse 16, 8010 Graz, Austria
sormann@vrvis.at
[2] Institute for Computer Graphics and Vision, Graz University of Technology
Inffeldgasse 16, 8010 Graz, Austria
bischof@icg.tu-graz.ac.at

Abstract. This paper proposes a fast 3D reconstruction approach for efficiently generating watertight 3D models from multiple short baseline views. Our method is based on the combination of a GPU-based plane-sweep approach, to compute individual dense depth maps and a subsequent robust volumetric depth map integration technique. Basically, the dense depth map values are transformed to a volumetric grid, which are further embedded in a graph structure. The edge weights of the graph are derived from the dense depth map values and if available, from sparse 3D information. The final optimized surface is obtained as a min-cut/max-flow solution of the weighted graph. We demonstrate the robustness and accuracy of our proposed approach on several real world data sets.

Keywords: volumetric 3D reconstruction, graph-cut, dense depth maps, virtual 3D models.

1 Introduction

In our approach we consider the problem of creating virtual 3D models solely from a set of digital input images, which is still a challenging problem in computer vision. The principal reason for utilizing digital images as input source, is the independancy of the 3D reconstruction process from the size of the objects to be modeled.

Current state of the art approaches for multi-view reconstruction are divided in two main categories: one pass (or directed) methods versus two pass (or indirect) methods. Direct methods, recently proposed by Vogiatzis et al. [19] or Hornung and Kobbelt [9] process all available input images from different viewpoints simultaneously. Their methods are based on finding a minimum cut in a graph structure, which is embedded in a volumetric grid. One of the main benefits of these methods is that they generate watertight surfaces such that the final 3D model does not contain any disturbing holes. Clearly, a drawback is that these approaches still rely on existing object silhouettes to consider only voxels which are close to the visual hull. But, the extraction of visual hull information, especially for complex environments, can be a tedious and time consuming process. Therefore we introduce an indirect, two pass method which extracts in a

B.K. Ersbøll and K.S. Pedersen (Eds.): SCIA 2007, LNCS 4522, pp. 393–402, 2007.

first pass a set of dense depth maps, whereas the second pass enforces a robust integration of the depth maps to create proper and watertight 3D models. Thus, we are able to provide intermediate results and can bring, if necessary, a human operator into the reconstruction loop. Especially for large data sets, an user assisted visual evaluation of intermediate results can be very helpful to detect errors at the earliest possible point in the 3D reconstruction pipeline. Therefore, we try to combine the main benefits of both, direct and indirect approaches. Consequently, the main contributions of our approach are the following:

1. Our approach avoids the incorporation of visual hull information, because the extraction of visual hull information is a tedious and time consuming process.
2. As a side effect of our indirect reconstruction process, we can easily bring a human operator into the reconstruction loop for quality assessment.
3. The proposed method is able to reconstruct 3D models even from dense depth maps containing outliers.
4. Due to the fact that our method utilizes global minimization techniques we can guarantee a watertight and global optimized surface.
5. Our algorithm can deal with high volumetric resolutions as well as a large number of input images.

2 Related Work

The automatic 3D reconstruction of complex objects is still an active research field within the computer vision community. There are two major approaches to the problem of 3D real world modeling: range-based modeling and image-based modeling. Range-based modeling is based on laser scanners. A very well known approach in this field is The Digital Michelangelo Project carried out by M. Levoy et.al. [15].

In this work we focus on image-based modeling, which represents the 3D reconstruction of real world objects from a dense set of photographs. A comparative evaluation of image-based and range-based methods can be found in El-Hakim and Beraldin [5]. Image-based modeling techniques utilize in general widely available hardware and developed systems can be used for a wide range of different objects and scenes. Furthermore such algorithms produce realistic models with an increasing level of automation.

All range-based methods as well as most of the image-based modeling methods generate 2.5D heightfields. In order to generate true 3D models, a robust fusion of this set of heightfields into a single 3D surface is necessary. The fundamentals of robust depth image fusion in the context of laser scanned data was proposed by Curless and Levoy [3]. The basic idea of volumetric range image integration is the conversion of depth maps to corresponding 3D distance fields and a subsequent robust averaging of these distance fields. The resolution and the accuracy of the final model are determined by the quality of the source images and the resolution of the target volume. Recently, Zach et. al. [23] introduced a fast GPU-based method, based on the original work of Curless and Levoy [3]. Since this method is a pure local method, the final 3D model can still contain many holes.

In contrast shape from silhouette methods try to overcome these restrictions. They recover the shape of the objects from their contours, known as visual hull, and no depth map information is used. A practical system to generate 3D models from its profiles was introduced by Wong and Cipolla [20]. This approach uses only the silhouettes of a sculpture for both motion estimation and model reconstruction, and neither corner detection nor matching is necessary. The method is robust and fast, but as drawback they are limited to simple shaped objects. Therefore, recent developed methods combine the visual hull information with a photo-consistency function, which is further embedded in a graph structure. A general approach combining multi-camera stereo reconstruction with graph-cuts was presented by Kolmogorov et. al. [13]. A comparison of energy functional types, which can be minimized using graph-cuts is given by Kolmogorov and Zabih in [12]. Several applications of graph-cut based energy minimization for volumetric reconstruction were presented in Vogiatzis et. al. [19] and Hornung and Kobbelt [9]. In these approaches, individual voxels correspond to nodes in the graph, used to determine the maximum flow. These techniques still rely on existing object silhouettes in order to consider only voxels close to the visual hull. Additionally, visibility information is mainly introduced from the visual hull to find occluded views for each voxel.

The inspiration for the method presented in this paper is given by a number of above mentioned volumetric 3D reconstruction approaches and efficient energy minimization techniques utilizing graph-cuts. More precisely, most of the above mentioned methods have in common that it is in general difficult to generate watertight and global optimized 3D models from dense depth maps, which is the standard output of most image-based modeling techniques. Furthermore, discussed methods either perform the 3D reconstruction in two passes, but then can not guarantee a watertight and global optimized surface, or in one pass, but then bypass the dense depth maps and extract the 3D model directly. Consequently, there is still a need to combine the ideas and benefits of both schema.

3 Dense Depth Map Estimation

Our work targets the reconstruction of objects from arbitrary image sequences taken with a calibrated digital consumer camera. The process of camera calibration and pose estimation, which are not the topics of this paper, are well studied problems in computer vision and determine the internal and external parameters of a camera [8].

The set of images with known calibration and orientation is used to generate a 3D model of the object in a fully automated manner. For dense depth map estimation a fast reconstruction method suitable for small-baseline settings is applied for every view. Basically, we utilize a plane-sweep approach [21] to create the set of dense depth maps, using up to 5 images simultaneously for matching (one key image in the middle and one or two neighboring reference images on each side). For each depth value, the reference images are projected onto the key image plane, located at the given depth and a correlation measure with respect to the key image is calculated. Occlusion handling is addressed by the best half-sequence strategy. The set of slices filled with correlation values comprise a data structure similar to the disparity space image. A final matching algorithm (e.g. scanline optimization [17]) establishes the dense depth map from the disparity space image. Depending on the resolution, plane-sweep matching requires 5.5 seconds for each reference image at an resolution of 1024x1024 pixels. More details of our developed GPU-based plane-sweeping technique can be found in Zach et. al. [23].

4 Graph-Cut Based Volumetric Depth Map Integration

In this section we give an overview of the basic ideas, datastructures and processing steps of our approach. All different steps are discussed in more detail in the following subsections. An overview is given in Figure 1.

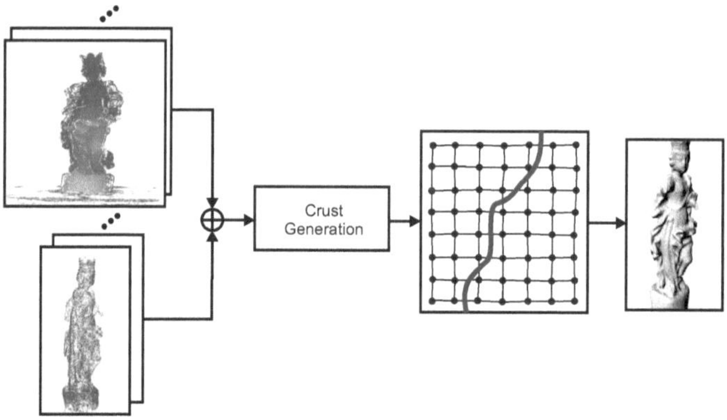

Fig. 1. Overview of our graph-cut based volumetric depth map integration pipeline: First, dense deph map values and if available sparse 3D information is transformed to a volumetric grid. Then, we direct extract the surface confidence in the vicinity of the dense depth maps. Finally, a min-cut/max-flow algorithm is performed to determine a watertight global optimized surface on the selected volumetric resolution. Note, that all available 3D information are in same coordinate system.

The required input for the volumetric integration approach is a set of dense depth maps and, but not necessarily, sparse 3D information obtained beforehand as proposed by Bauer et. al. [1]. Instead of utilizing visual hull information as proposed by Vogiatzis et. al. [19], we derive a so called crust band directly in the vicinity of the dense depth maps. This crust band can be interpreted as a confidence map, which represents the probability that the final unknown surface passes through. The confidence values are computed as an unsigned distance function ϕ over the underlying volumetric grid, which is described in more detail in section 4.1.

As soon as the confidence values are computed, we determine a global optimized surface S_{opt}, which approximates the true but unknown surface, with respect to the used energy functional. Previous work already have shown that such problems can be efficiently solved by a min-cut/max-flow algorithm [19]. Additionally, we incorporate sparse 3D information into our energy functional, which further enhance the obtained 3D reconstruction results.

Finally, the voxel based representation is transformed into a triangular mesh based on a standard marching cube algorithm introduced by Lorenson and Cline [16].

4.1 Crust Generation

The first step in our approach is the determination of crust voxels lying on both sides of the true surface. The generated crust should be as small as possible in order to obtain the maximal computational efficiency. On the other hand, the crust must be able to reflect potential concavities arising in the true model geometry. Consequently, the generation of the proper crust is non-trivial, and recently proposed strategies include incorporation of the visual hull [19] and coarse-to-fine approaches [10].

We select a different path by employing the initial, still noisy 3D result of our efficient depth map integration scheme as the primary indicator of crust voxels. Our volumetric depth image integration method [23] robustly averages the set of approximated signed distance fields induced by the depth maps. For every voxel a statistic is accumulated, which is based on the signed distance of the voxel to the approximately closest surface point indicated by the current depth map. Finally, a voting scheme determines the final signed distance value of a voxel, which can be used to extract the isosurface. We utilize the accumulated signed distance field to determine the initial set of crust voxels by including voxels close to the isosurface (with respect to a user-specified distance threshold). This set is enhanced by a number of dilation steps d to achieve a watertight separation of interior and exterior regions. In all our experiments we generally set $d = 2$.

Since isosurfaces generated from signed distance tend to have unnecessary high genus, positive surface confidences $\phi(v)$ are employed in the extraction procedure instead of signed distance, using a similar approach to [10]. Voxels crossed by the isosurface as well as voxels which are filled from sparse 3D information have confidence value zero (indicating high certainty), and the confidences of all other crust voxels are initialized with 1. The confidence map ϕ is subsequently smoothed using a homogeneous diffusion scheme.

Figure 2 illustrates all intermediate results of our initial crust generation process.

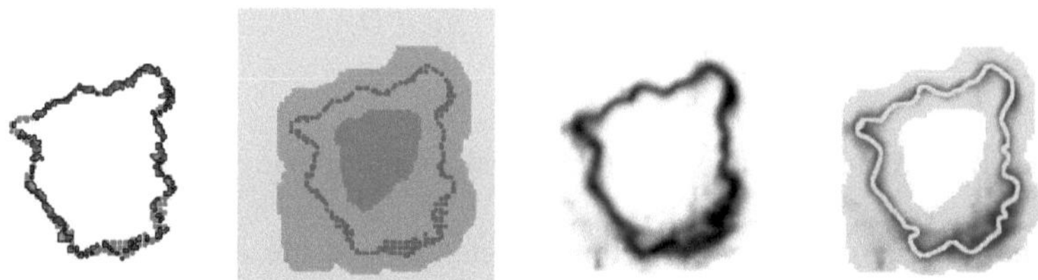

(a) (b) (c) (d)

Fig. 2. This image illustrates all intermediate results showing one slice of the volumetric grid of the St. Barbara data set. **(a)** Dense depth map values (light grey) and sparse 3D information (dark grey). **(b)** Obtained voxel crust (green), exterior (light grey) and interior (dark grey) component **(c)** Confidence band derived from dense depth map values, where darker values correspond to higher confidence. **(d)** Optimal surface (blue) extracted by a min-cut/max-flow algorithm.

4.2 Surface Reconstruction

This section is dedicated to discuss our graph-cut based surface reconstruction procedure. Since, our goal is to extract an optimal as well as watertight surface S_{opt}, we transform the volumetric grid to a graph based structure and solve the optimization problem by performing a min-cut/max-flow algorithm. Similar to other approaches we minimize $E(D) = \sum_{v \in D} \omega(v)$ where D is a weighted sum of dense depth map values.

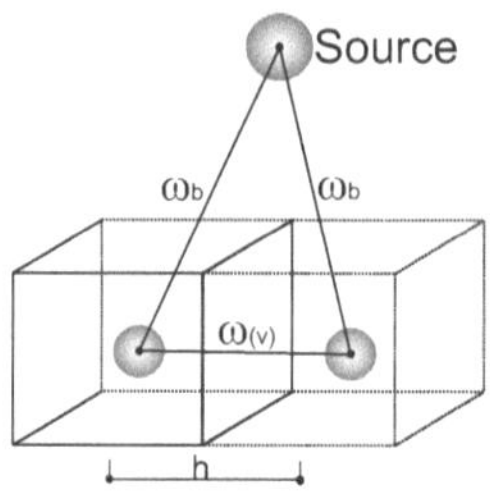

Fig. 3. Correspondence between crust voxels and nodes in the graph

The geometric configuration of the correspondence between crust voxels and nodes in the graph is shown in Figure 3.

The corresponding graph, consisting of all crust voxels, is connected over a regular six-neighborhood. The edge weight $\omega(v)$ is assigned to all edges of the embedded graph and can be derived from the unsigned distance value $\phi(v)$. Basically, $\omega(v)$ is defined as $\omega(v) = (\phi(v))^s$, where s can be interpreted as some kind of smoothness factor. Additionally, as discussed in Vogiatzis et. al. [19] we add a ballooning force ω_b, which connects every crust voxel to the source node with a constant weight of $\omega_b = \lambda h^3$, where λ is a weight parameter and and h represents the quantized size of a voxel. The ballooning force avoids a cut across thin structures of the object. As usual, interior voxels V_{int} are connected to the source node and exterior voxels V_{ext} are connected to the sink node. As stated, for exploiting sparse 3D information we extend $\omega(v)$ in the following way:

$$\omega(v) = \begin{cases} (\phi(v))^s & \forall v \in D_V \\ 0.0 & \forall v \in E_V \end{cases} \tag{1}$$

where D_V and E_V represents dense depth map values and sparse 3D information respectively.

After the min-cut/max-flow algorithm has determined the optimal surface voxels S_{opt}, a standard marching cube algorithm converts the voxel based surface into a triangular mesh for possible further processing.

To summarize, our approach reconstructs watertight 3D surface models, even from non-outlier free dense depth maps. In contrast to related approaches and due to the fact that our approach do not rely on visual hull information we avoid the complex, time consuming and tedious task of acquiring such information. In addition, we do not need any hole filling algorithm, since large gaps are effectively closed due to the embedded energy functional. And finally, incorporated sparse 3D information enhances the quality of our final 3D models.

5 Results

This section is dedicated to discuss the visual and quantitative results of our approach. We applied our depth map integration method to several real-world data sets. All experiments were performed on 4 GHZ PC with 2GB main memory and a GeForce 7800 GT with 256MB graphics memory. The images were taken with a calibrated digital consumer camera at a geometric resolution of 4064x2704 pixels. After pose estimation, the source views are resized to 1024x1024 pixels and the obtained dense depth map have the same resolution (unless noted otherwise).

Table 1 demonstrates quantitative results and compares the number of input images, target resolution, mesh complexity as well as the timing for each of our data sets. The reconstruction time includes the dense depth map estimation as well as the volumetric depth map integration, which is the less dominant computational factor.

Table 1. Illustration of time and space complexity for each of our data sets. The obtained reconstruction time can be separated into a dense depth map estimation part and a volumetric depth map integration part, which is the more dominant computational factor.

Dataset	Images	Resolution	Triangles	Time [min.]
Barbara	46	256x384x256	704446	9.5
Pedestal	74	256x256x384	890358	14.5
Temple	47	256x256x384	790186	7.5

The first data set depicted in Figure 4 shows the lime stone statue of St. Barbara, which was reconstructed from 46 images. The statue is 55cm tall with a diameter of 13cm at the pedestal. The final 3D model was reconstructed in less than 10 minutes and consists of approximately 700k triangles.

(a) (b) (c) (d) (e)

Fig. 4. 3D Reconstruction of the statue of St. Barbara from 46 images. **(a)** One input image of the data set. **(b)** Obtained dense depth map. **(c)-(d)** Two viewpoints of the reconstructed 3D model consisting of approximately 700k triangles. **(e)** Textured version of our obtained 3D reconstruction.

The second experiment (Figure 5) illustrates a pedestal (3x2x1.5 meters) of a statue located in front of the Austrian National Library in Vienna. We obtained the final 3D model in about 15 minutes at an geometric resolution of 900k triangles. For the reconstruction of the pedestal we used 76 images. We are able to obtain a visually appealing as well as watertight 3D model, even in textureless regions around the fresco's.

Finally, Figure 6 illustrates the well known temple data set from the multi-view stereo evaluation page [18] consisting of 47 images. The dense depth maps were obtained

(a) (b) (c) (d) (e)

Fig. 5. 3D Reconstruction of a pedestal, located in front of the Austrian National Library from 76 images. (**a**) One image of the data set. (**b**) Obtained dense depth map. (**c**)-(**d**) Two viewpoints of the final 3D reconstruction consisting of approximately 900k triangles. (**e**) Textured 3D model.

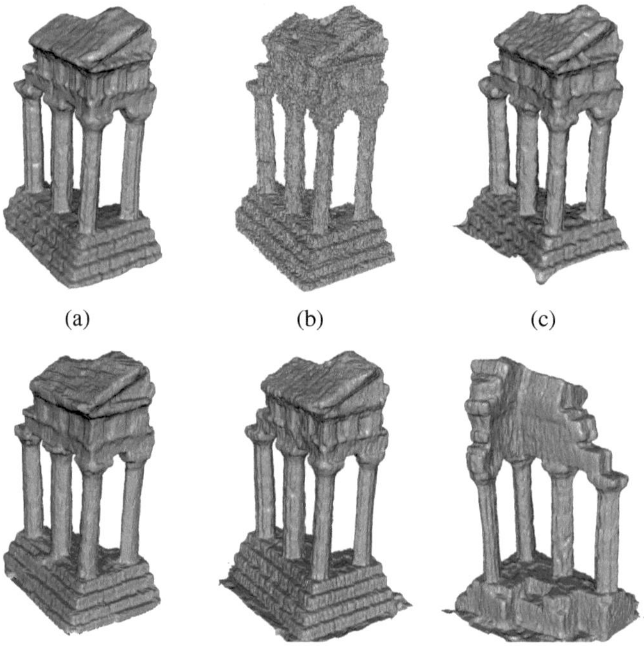

(a) (b) (c)

(d) (e) (f)

Fig. 6. 3D reconstruction of the well known temple data set from the multi-view stereo evaluation page [18] consisting of 47 input images. (**a**) 3D reconstruction proposed by Furukawa and Ponce [6]. (**b**) 3D reconstruction of Kolmogorov and Zabih [11]. (**c**) Obtained 3D model of Vogiatzis et. al. [19]. (**d**) 3D reconstruction of Hornung and Kobbelt [9]. (**e-f**) Two views of our achieved 3D reconstruction consisting of approximately 800k triangles.

at a resolution of 640x480 pixels. The visual comparison of our results against four other related multi-view reconstruction methods is shown in Figure 6(a-d). Figure 6(e-f) illustrates two views of our 3D reconstruction consisting approximately 800k triangles. Note, that all presented results, except the one shown in Figure 6(a), are utilizing graph-cuts for global optimization. Of course, the quantitative as well as qualitative evaluation of our results is given at the multi-view stereo evaluation page [18].

6 Conclusion

In this paper we demonstrated a fast and robust method for the 3D reconstruction of proper 3D models, even from non-outlier free dense depth maps. The achieved quality of our 3D models mainly depends on the grade of the dense depth maps as well as the selected target resolution. One main advantage of the proposed method is, that there is no need for some kind of visual hull information during the 3D reconstruction process. Due to a min-cut/max-flow optimization we can guarantee a watertight and global optimized surface.

Though the results are very promising there are several improvements that can made to our approach. Further work needs to include the already generated error map, which provides a confidence measurement for each dense depth map value, into the cost functional of the min-cut/max-flow algorithm. Finally, we plan to evaluate and compare several edge weight functions.

Acknowledgments

This work is partly funded by the VRVis Research Center, Graz and Vienna/Austria (`http://www.vrvis.at`). We would also like to thank the Vienna Science and Technology Fund (WWTF).

References

1. Bauer, J., Zach, C., Karner, K., Bischof, H.: Efficient sparse 3d reconstruction by space sweeping. In: 3DPVT (Chapel Hill, USA, June 2006). CD Proceedings (2006)
2. Collins, R.T.: A space-sweep approach to true multi-image matching. In: IEEE Computer Society Conference on Computer Vision and Pattern Recognition (San Francisco, USA, June 1996), pp. 358–363 (1996)
3. Curless, B., Levoy, M.: A volumetric method for building complex models from range images. In: ACM SIGGRAPH (New Orleans, USA, August 1996), vol. 1, pp. 303–312 (1996)
4. Debevec, P.E., Taylor, J., Malik, J.: Modeling and rendering architecture from photographs. In: ACM SIGGRAPH (New Orleans, USA, August 1996), pp. 11–20 (1996)
5. El-Hakim, S., Beraldin, J.: Configuration analysis for sensor integration. In: Proceedings of SPIE (Philadelphia, USA, October 1995), vol. 2, pp. 274–285 (1995)
6. Furukawa, Y., Ponce, J.: High-fidelity image-based modeling. Tech. rep. UIUC (2006)
7. Goesele, M., Curless, B., Seitz, S.: Multi-view stereo revisited. In: IEEE Computer Society Conference on Computer Vision and Pattern Recognition (New York, USA, June 2006), vol. 1, pp. 2402–2409 (2006)
8. Heikkilä, J.: Geometric camera calibration using cicular control points. IEEE Transactions on Pattern Analysis and Machine Intelligence 22, 10 (October 2000), pp. 1066–1077 (2000)
9. Hornung, A., Kobbelt, L.: Hierarchical volumetric multi-view stereo reconstruction of manifold surfaces based on dual graph embedding. In: IEEE Computer Society Conference on Computer Vision and Pattern Recognition (New York, USA, June 2006), vol. 1, pp. 503–510 (2006)
10. Hornung, A., Kobbelt, L.: Robust reconstruction of watertight 3d models from non-uniformly sampled point clouds without normal information. In: Eurographics Symposium on Geometry Processing (Sardinia, Italy, June 2006), vol. 1, pp. 41–50 (2006)

11. Kolmogorov, V., Zabih, R.: Multi-camera scene reconstruction via graph cuts. In: European Conference on Computer Vision (Copenhagen, Denmark, May 2002), vol. 3, pp. 82–96 (2002)
12. Kolmogorov, V., Zabih, R.: What energy functions can be minimized via graph cuts? IEEE Transactions on Pattern Analysis and Machine Intelligence 26, 2 (February 2004), 147–159 (2004)
13. Kolmogorov, V., Zabih, R., Gortler, S.J.: Generalized multi-camera scene reconstruction using graph cuts. In: IEEE Computer Society Conference on Computer Vision and Pattern Recognition (Madison, USA, June 2003), vol. 1, pp. 501–516 (2003)
14. Kutulakos, K.N., Seitz, S.M.: A theory of shape by space carving. International Journal of Computer Vision 38 3, 199–218 (2000)
15. Levoy, M., Pulli, K., Curless, B., Rusinkiewicz, S., Koller, D., Pereira, L., Ginzton, M., Anderson, S., Davis, J., Ginsberg, J., Shade, J., Fulk, D.: The digital michelangelo project: 3d scanning of large statues. In: ACM SIGGRAPH (New Orleans, USA, July 2000), vol. 1, pp. 131–144 (2000)
16. Lorenson, W., Cline, H.: A high resolution 3d surface reconstruction algorithm. In: ACM SIGGRAPH (Anaheim, USA, July 1987), vol. 1, pp. 163–170 (1987)
17. Scharstein, D., Szeliski, R.: A taxonomy and evaluation of dense two-frame stereo correspondence algorithms. International Journal of Computer Vision 47(1-3), 7–42 (2002)
18. Seitz, S., Curless, B., Diebel, J., Scharstein, D., Szeliski, R.: A comparison and evaluation of multi-view stereo reconstruction algorithms. In: IEEE Computer Society Conference on Computer Vision and Pattern Recognition (New York, USA, June 2006), vol. 1, pp. 519–526 (2006)
19. Vogiatzis, G., Torr, P.H.S., Cipolla, R.: Multi-view stereo via volumetric graph-cuts. In: IEEE Computer Society Conference on Computer Vision and Pattern Recognition (Washington D.C., USA, June 2005), vol. 1, pp. 391–398 (2005)
20. Wong, K.-Y., Cipolla, R.: Reconstruction of outdoor sculptures from silhouettes under approximate circular motion of an uncalibrated hand-held camera. IEEE Transactions on Information and Systems 87(1), 27–33 (2004)
21. Yang, R., Welch, G., Bishop, G.: Real-time consensus based scene reconstruction using commodity graphics hardware. In: Proceedings of Pacific Graphics (Beijing, China, October 2002), pp. 358–363 (2002)
22. Yezzi, A., Soatto, S.: Stereoscopic segmentation. International Journal of Computer Vision 53(1), 31–43 (2003)
23. Zach, C., Sormann, M., Karner, K.: High performance multi-view reconstruction. In: 3DPVT (Chapel Hill, USA, June 2006). CD Proceedings (2006)

Grain Size Measurement of Crystalline Products Using Maximum Difference Method

Leena Lepistö[1], Iivari Kunttu[1], Matti Lähdeniemi[1], Tero Tähti[2], and Juha Nurmi[2]

[1] Satakunta University of Applied Sciences
Faculty of Technology and Maritime Management
Tekniikantie 2, FI-28600 Pori, Finland
{Leena.Lepisto, Iivari.Kunttu}@samk.fi
http://www.samk.fi

[2] Danisco Texturants & Sweeteners, Innovation & Technology,
Sokeritehtaantie 20, FI-02460 Kantvik, Finland
{Tero.Tahti, Juha.Nurmi}@danisco.com
http://www.danisco.com

Abstract. Texture analysis methods are widely used in various monitoring and measurement tasks in machine vision solutions. In this paper we present a novel method for the determination of grain size distributions in the manufacturing processes of crystalline products. Our method, *maximum difference histogram* (MDH), is based on statistical gray level differences in the texture images. Using this method, it is possible to estimate the grain size distributions in the images. It is also possible to monitor the average grain sizes in the image series acquired during the crystallization process. This is carried out by determining the center of gravity (CoG) of the distribution represented by MDH. Experimental results obtained from images acquired from a carbohydrate crystallization process reveal that the proposed method is useful in in-line grain size measurement tasks.

1 Introduction

The use of image information in material characterization has been in strong growth during recent years. Materials in various industrial processes can be effectively characterized and their visual properties can be measured using image analysis and machine vision methods. In the process industry, significant amounts of information on the process can be acquired using machine vision. This information is utilized in process monitoring and control tasks.

The inspection and measurements of granular products is a fundamental problem in several industrial processes. Traditionally, the characterization of the granular products has been carried out by manual inspection. That is, the grain properties have been inspected by sieving or microscope analysis. This kind of approach is time consuming and hence unable to provide on-line information from the granulation process. For this reason, image analysis methods have been adopted to inspection of grain properties in several industrial processes. In the field of pharmaceutical sciences, the characterization of grain size distributions of powders has become an actively studied topic. Laitinen et al. [2],[3] for example, have used image analysis methods for the

B.K. Ersbøll and K.S. Pedersen (Eds.): SCIA 2007, LNCS 4522, pp. 403–410, 2007.

inspection of pharmaceutical powder distributions. In the field of crystallization, Qu et al. [7] have inspected crystal growth using image analysis applied to image material acquired using a video microscope.

The application field of this paper is related to the manufacturing process of crystalline products. In this process the average size and size distribution of the product play a constitutive role. The properties of the crystal population are important as well for the end use functionality as for the ease of downstream processing. In industrial processes fluctuations in the processing conditions can occur for several reasons. Such fluctuations must be compensated by the operators in order to keep the product quality stable. Visual observation of the process has remained an important tool in evaluation of the expected product quality in terms of crystal size. Up to now quantitative size measurements from the in-line process data have been scarce due to difficulties in receiving reliable information on the crystal size from very dense suspensions present in industrial processes.

Texture analysis is common in various industrial machine vision applications. Texture analysis is used to e.g. estimate different properties of surfaces. Typical industrial applications of texture analysis include the inspection of different surface materials such as paper, metal or textiles [6]. In these applications, the classification and recognition of different surfaces is based on texture properties. These properties are for example roughness, granular size or directionality of texture. According to Rao and Lohse [8], the most essential texture properties in human perception are repetitiveness, directionality, granularity, and complexity. The directionality of non-homogenous natural textures has been discussed in our earlier work [4]. The analysis of grain properties in rock images can be found in [5].

When image analysis is utilized in the analysis of grain properties of crystalline products, one needs to choose between two kinds of approaches. The traditional approach is to extract the grains from the image by using some image segmentation method. The grain properties, such as size or shape, are then measured from the segmented image. However, there are several difficulties with this kind of approach. Firstly, the image segmentation algorithms are often sensitive to illumination changes in the process. In addition, the color (or gray level) distributions of the grains are not always homogenous. On the contrary, the colors of the grains to be extracted from the images may vary significantly. These factors may cause difficulties in the segmentation process. Secondly, the segmentation causes computational load that may be critical in the case of on-line analysis and inspection applications. An alternative for the extraction of the grains from the image background is the employment of texture analysis in the granularity measurements. It has been found that texture analysis methods can be applied to the analysis and inspection of granular products. Using texture analysis tools, it is often advantageous to inspect the particle populations as larger surfaces, not by extracting single grains from the image. Compared to the traditional approach, texture analysis does not have the problems related to image segmentation process. In addition, the whole image can be used in the granularity inspection, not only the single grains.

In this paper, we apply texture analysis methods for granularity analysis of crystalline products. The testing material used in this study is image data obtained from the crystallization process in carbohydrate manufacturing. The rest of this paper is organized as follows. Chapter 2 describes statistical texture analysis methods and particularly the proposed method, *Maximum difference histogram*. In Chapter 3, we present two kinds of experiments in which the proposed texture analysis method is used to estimate granular sizes of the carbohydrate crystals. The experimental results are discussed in Chapter 4.

2 Statistical Texture Analysis

Numerous techniques have been proposed for texture description. Tüceryan and Jain [9] have divided texture description methods into four main categories: statistical, geometric, model-based, and signal processing methods. In this study, we concentrate on the statistical texture analysis methods.

Statistical techniques are based on the description of the spatial organization of the image grey levels. On the basis of the grey level distribution, it is possible to calculate several kinds of simple statistical features. Grey level co-occurrence matrix developed by Haralick [1] has been a popular tool in texture analysis and classification. Co-occurrence matrix estimates the second order joint probability density functions $g(i,j \mid d, \Theta)$. Each $g(i,j \mid d, \Theta)$ is the probability of going from grey level i to grey level j, when the relative position between the gray levels is d and the direction is Θ. These probabilities create the co-occurrence matrix $\mathbf{M}(i,j \mid d, \Theta)$. It is possible to extract textural features from the matrix [1]. The most commonly used textural features extracted from the matrix are contrast, entropy, and energy. Preliminary experiments have revealed the contrast is the most sensitive of these features for granularity.

In addition to the co-occurrence matrix, several other statistical texture analysis methods have been presented. One example of them is grey level difference method originally presented by Weszka et al. [10]. In this method, the histograms formed by absolute differences between pairs of gray levels are used. The grey level difference calculation is based on pixel pairs whose relative position is defined by displacement vector $\boldsymbol{d}=(d_x,d_y)$.

2.1 Maximum Difference Method

In the statistical methods presented in the previous Chapter, the relative position or distance between the gray level pairs, d, is fixed. Consequently it is necessary to define constant spatial distance between the gray levels used in the statistical measurement for the whole image set to be inspected. This can be problematic because in certain texture types the grain sizes are varying. For this reason, the use of fixed distance d is not practical, especially in the cases in which texture analysis is employed to measure the granular sizes of texture. For this reason, fixed distance is not the best alternative for granularity measurements.

In our approach, we use adaptive relative position d in the texture statistics. In the proposed statistical texture measure, *maximum difference histogram,* a histogram formed by relative positions that produce maximum difference between gray levels of the pixel pairs is formed. The method can be presented as follows.

1. The texture image is inspected pixel by pixel.
2. For each pixel (gray level) i, select the following n pixels in direction Θ.
3. Find the highest absolute difference between gray level i and the gray levels within the following n pixels. If computational efficiency is required, the n should be selected small.
4. The gray level j that has the highest absolute difference to gray level i is selected.
5. The spatial distance d between gray levels i and j is determined.
6. This procedure is repeated for each gray level of the texture. The values of d are presented as a histogram.

The resulting histogram is called *maximum difference histogram* (MDH), and it can be employed in the estimation of the grain size distribution. If only the average of the grain sizes in the population is interesting, it is possible to determine it based on the histogram. This can be done by determining the center of gravity (CoG) of the histogram.

3 Experiments

In the experimental part of this paper, we present two kinds of test results obtained from the granularity inspection of carbohydrate crystals. In the first one, we use the proposed *maximum difference histogram* method for the estimation of granular size distributions of images acquired from the carbohydrate crystallization process. In the second experiment, an in-line measurement of the crystallization process is presented. Before these experiments, is given a short description of the carbohydrate manufacturing process.

3.1 Carbohydrate Manufacturing Process

The crystallizing substance used in this study is a carbohydrate product. The crystallization process of the carbohydrate is based on evaporation of the solvent in a constant temperature. The resulting supersaturation in the carbohydrate-solution acts as a driving force for the phase change operation from liquid to solid. During the evaporation process the final crystal population is created through growth of existing crystals and formation of new crystals, starting from an initially given seed crystal population. The target is to obtain a pre-defined average crystal size and a possibly narrow crystal size distribution. This is achieved by balancing the processes of crystal growth and crystal formation by adjustment of the process parameters.

3.2 Determination of Grain Size Distribution

In the first experiment, we estimated grain size distributions of some selected microscope images obtained from the carbohydrate crystallization process. We used the *maximum difference method* to make a histogram that estimates the grain size distributions of the images. Figure 1 presents two examples of the grain images and their maximum difference histograms. The purpose is to estimate the grain size distribution. The histograms express normalized granular size distributions of the images in pixels. In all the histograms, the distance d is determined in horizontal direction.

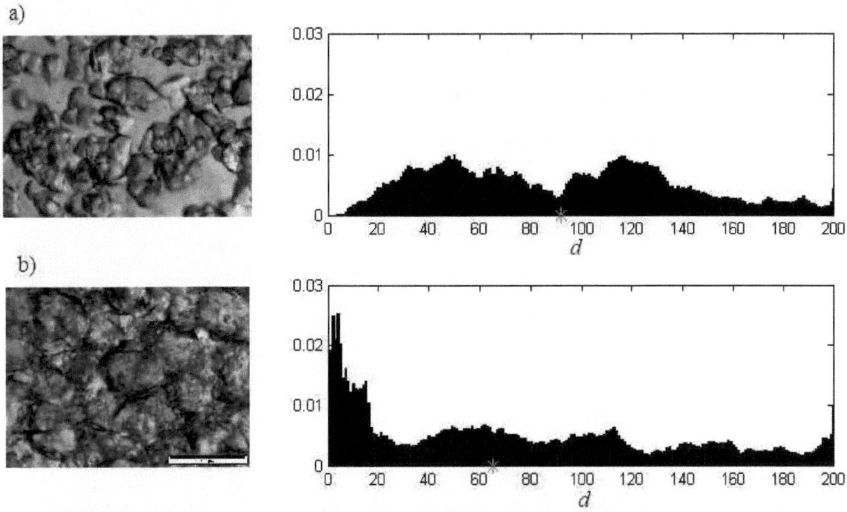

Fig. 1. Grain images and their respective *maximum difference histograms*. The center of gravity (CoG) is marked with green star.

The results presented in figure 1 show that the *Maximum difference histogram* is able to estimate the grain size distributions with reasonable accuracy. In subfigure (b), the effect of small grains is visible in the histogram. On the other hand, in subfigure (a), these small particles are not present. So, the method can be used to track the formation of small crystals during the process. However, just by looking at the obtained average value the growth into larger crystals visible in subfigure (b) can not be detected. The histograms shown in figure 1 represent number based distributions of the crystal sizes. If the small particles are not considered interesting, the growth of larger crystals can be followed easier by representing the histograms in a volume based (d^3) form. The grain image presented in subfigure (a) has also certain smooth regions which do not contain any grains. These kinds of regions can be problematic for the proposed grain size estimation method and they may cause some errors to the distributions. Also noise and other distortions may cause small variations. However, the method seems to estimate the mean grain size (marked with green star in figure 1) sufficiently good. Another topic to be considered is the transform from pixels to metric system, because distribution expresses the grain sizes in pixels. This can be done experimentally.

3.3 Measurement of Mean Grain Level in the Crystallization Process

In the practical carbohydrate manufacturing, one needs to monitor and control the growth of the grain size during the crystallization process. In the beginning of the process, the grains are small sized and their amount is also small. However, the grains start growing very soon after the beginning of the process and also their number grows fast. After that, the grain sizes tend to stabilize on certain level although their shapes get smoother. To monitor the grain size growth, one needs some image descriptor that is capable of expressing the mean grain size in each image.

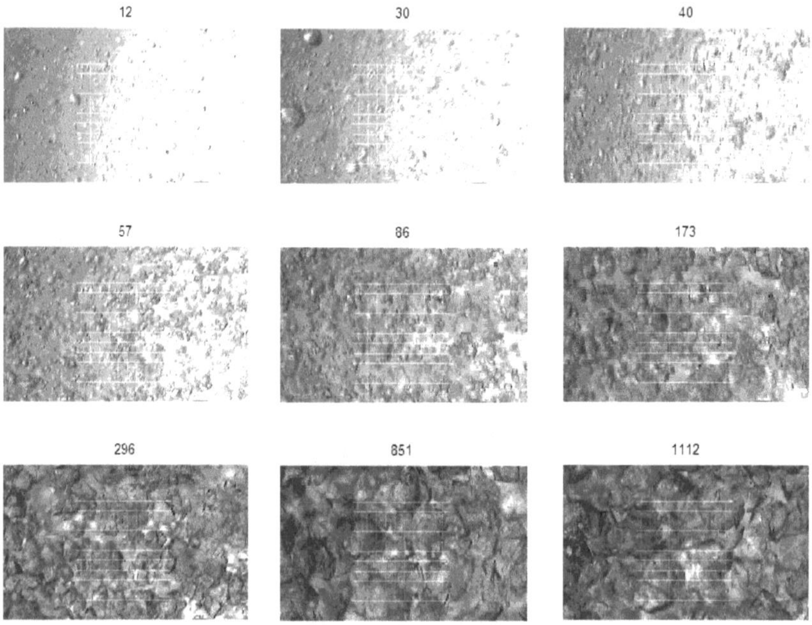

Fig. 2. Example images from the test set acquired from the carbohydrate crystallization process. The images are numbered from 1 to 1200.

In order to test the procedure for monitoring an industrial crystallization process using textural methods, images were acquired through a common process control microscope in a real production process. Images were taken using a process microscope and a process surveillance camera attached on the side of the evaporation vessel. The camera was connected to a computer with a frame grabber card. Images were saved on computer using Image-Pro Plus image analysis software. Additional in-house software was used for automation and timing of the image capture procedure. The image acquisition was started at the time of seed crystal addition and stopped at the end of the process. That means that in addition to crystal size and crystal size distribution, also the solids fraction in the images changes remarkably. The resulting set of images contained approximately 1200 images. Image size was 768x460 pixels. Figure 2 shows some example images acquired in different phases of the crystallization process.

We used *maximum difference method* to estimate the growth of the grain size. In this experiment, only the average size of the grains was determined. This was carried out by calculating the center of gravity (CoG) for each MDH. The MDH was calculated using value 50 for parameter n in horizontal direction. The CoG was calculated for each image in the sequence. Figure 3 presents the CoG during the process.

The results presented in figure 3 show that the graph representing the center of gravity follows the real crystal growth process. The graph rises very fast between samples 1 and 150. After that, it stabilizes on certain level. This kind of growth can be seen also in the image set presented in figure 2. Based on manual inspection, this kind

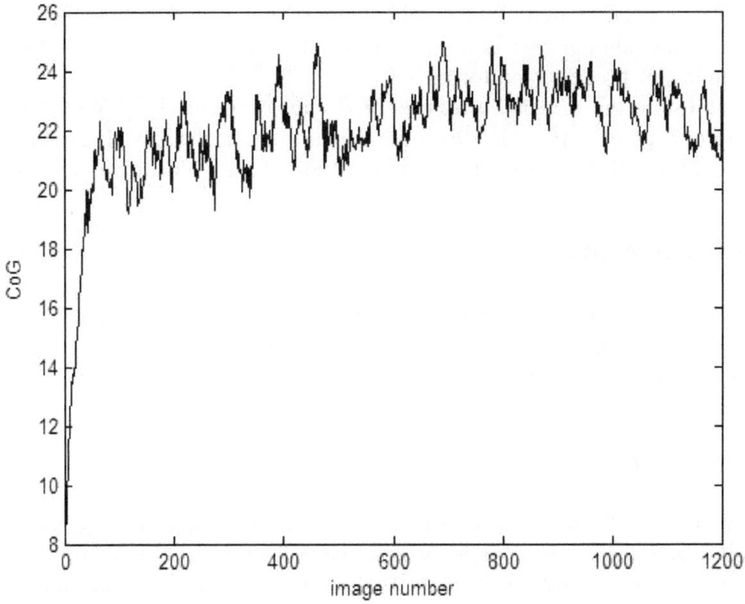

Fig. 3. Mean grain sizes in the sequence of 1200 crystal images acquired from a carbohydrate crystallization process

of graph seems to estimate the real development in the crystal population during the process reasonably well. Formation of new crystals and attrition of existing large crystals sets into a dynamic balance, while the number of crystals keeps rising until the end of the process.

4 Discussion

In this paper we have presented a novel texture description method, *Maximum difference histogram* for the estimation of grain size distributions in the images. The experimental results reveal that this method is able to relatively well estimate the mean grain size in the population. The distribution is somewhat sensitive to noise and other distortions in the image. Also areas without grains may cause problems. For this reason, the shape of the distribution may have small differences. These difficulties can be overcome by using e.g. image preprocessing, edge detection and thresholding, or other image enhancement before the calculation of the histogram. This could be subject of further investigations.

On the other hand, in its current form the *Maximum difference histogram* method has proved to be able to approximate the mean grain size in the grain population. We employed the method in the monitoring of carbohydrate production. The experimental results show that the method is able to estimate the grain growth in the crystallization process.

The computational cost of the proposed method is at the same level as with other statistical texture analysis methods. The computation can be made lighter by e.g.

limiting the inspection area in the pixel neighborhoods (selecting parameter n). In any case, the computational cost is not a significant problem in the use of the proposed method in in-line process monitoring.

In conclusion, the *Maximum difference histogram* method is able to estimate the mean grain sizes of a particle distribution relatively accurately. It is suitable to be used in real process monitoring tasks in industrial machine vision applications.

Acknowledgments

The authors wish to thank Danisco Sweeteners & Pharma for assistance with the image acquisition in the plant, and Mr. Esa Wainio from Picomega Oy for invaluable help with the image acquisition tools.

References

1. Haralick, R.M., Shanmugam, K., Dinstein, L.: Textural Features for Image Classification. IEEE Transactions on Systems, Man, and Cybernetics 3, 610–621 (1973)
2. Laitinen, N., Antikainen, O., Yliruusi, J.: Does a powder surface contain all necessary information for particle size distribution analysis? European Journal of Pharmaceutical Sciences 17, 217–227 (2002)
3. Laitinen, N., Rantanen, J., Antikainen, O., Yliruusi, J.: New perspectives for visual characterization of pharmaceutical solids. Journal of Pharmaceutical Sciences 93, 165–176 (2004)
4. Lepistö, L., Kunttu, I., Autio, J., Visa, A.: Classification Method for Colored Natural Textures Using Gabor Filtering. In: Proceedings of 12th International Conference on Image Analysis and Processing, pp. 397-401 (2003)
5. Lepistö, L., Kunttu, I., Autio, J., Visa, A.: Rock image retrieval and classification based on granularity. In: Proceedings of 5th International Workshop on Image Analysis for Multimedia Interactive Services (2004)
6. Pietikäinen, M., Ojala, T., Silven, O.: Approaches to texture-based classification, segmentation and surface inspection. In: Chen, C.H., Pau, L.F., Wang, P.S.P. (eds.) Handbook of Pattern Recognition and Computer Vision, 2nd edn. pp. 711–736. World Scientific Publishing Company, Singapore (1998)
7. Qu, H., Louhi-Kultanen, M., Kallas, J.: In-line image analysis on the effects of additives in batch cooling crystallization. Journal of Crystal Growth 289, 286–294 (2006)
8. Rao, A.R., Lohse, G.L.: Towards a texture naming system: identifying relevant dimensions of texture. In: Proceedings of IEEE Conference on Visualization, San Jose, California, pp. 270-227 (1993)
9. Tüceryan, M., Jain, A.K.: Texture Analysis. In: Chen, C.H., Pau, L.F., Wang, P.S.P. (eds.) Handbook of Pattern Recognition and Computer Vision, pp. 235–276. World Scientific, Singapore (1993)
10. Weszka, J.S., Dyer, C.R., Rosenfeld, A.: A Comparative Study of Texture Measures for Terrain Classification. IEEE Transactions on Systems, Man, and Cybernetics 6, 269–285 (1976)

Robust Boundary Delineation Using Random-Phase-Shift Active Contours

Astrit Rexhepi and Farzin Mokhtarian

Centre for Vision, Speech, and Signal Processing
School of Electronics and Physical Sciences
University of Surrey
Guildford GU2 7XH
United Kingdom
a.rexhepi@surrey.ac.uk,
f.mokhtarian@surrey.ac.uk
http://www.surrey.ac.uk

Abstract. When an active contour is applied to a noisy image, the contour is sometimes attracted to a local energy minimum, since the noise gives rise to high rates of change of the image gray levels. In this paper we will describe a novel method of overcoming this problem by using a sparse set of points to represent the active contour C and randomly varying the positions of these points.

1 Introduction

Active contours, also known as "snakes" or deformable models, have proven to be an effective method of boundary delineation. Since the original work by Kass, Witkin, and Terzopoulos (KWT) in 1988 [1], extensive research has been carried out on such models [2]. An active contour locates a boundary in an image by minimizing an energy function. This function includes "internal" terms that depend on the length and curvature of the contour; these terms are small when the contour is short and smooth. It also includes "external" terms that depend on the image gray levels at or near the points of the contour. For example, if the inverse rate of change of the image gray level is used as the external term, it will be small when the contour lies close to a strong boundary in the image.

In the KWT active contour model, the energy function is defined by an expression of the form

$$E(C) = \int_C [w_1|v_s|^2 + w_2|v_{ss}|^2 + \xi(v)]ds$$

where C is a curve in the image plane; s is a parameter representing a point on C; $v = (x, y)$ is the position of the point in the plane; v_s and v_{ss} are the first and second derivatives of v with respect to s; w_1 and w_2 are (possibly variable) weights; and $\xi(v)$ is a function of the image values in a neighborhood of v.

In this expression for $E(C)$, the first two terms are called "internal" energy terms because they depend only on the geometry of C itself, while the third

B.K. Ersbøll and K.S. Pedersen (Eds.): SCIA 2007, LNCS 4522, pp. 411–420, 2007.

term is called an "external" energy term because it depends on the image gray levels at or near C. For example, suppose the value of ξ at a point (x, y) is the inverse rate of change of the image gray level at (x, y); then ξ has low values on or near image boundaries and high values elsewhere.

It is well known that at a minimum of $E(C)$, the coordinates x, y of the points of C must satisfy the Euler equations

$$e_x \equiv w_1 x_{ss} + w_2 x_{ssss} + (\partial \xi / \partial x) = 0$$

$$e_y \equiv w_1 y_{ss} + w_2 y_{ssss} + (\partial \xi / \partial y) = 0$$

In the KWT model these equations are solved by an iterative process in which C is approximated by a discrete set of points, and at each iteration, the positions of the points are adjusted so as to reduce e_x and e_y.

It has long been realized that this basic active contour model must be modified in order to enable it to detect a distant object boundary, avoid local energy minima due to noise, and conform to the details of a boundary's shape. Since $\xi(x, y)$ depends only on the image gray levels near (x, y), it has an influence on $E(C)$ only near image features. For example, the inverse rate of change of the image gray level is low only near an object boundary; elsewhere, the boundary has no effect on $E(C)$, and C has no tendency to get closer to the boundary. On the other hand, since noise gives rise to high rates of change of the image gray level, the inverse is low at noise points, and $E(C)$ may have a local minimum when C passes through or near noise points. If C does succeed in approaching an object boundary, it may have difficulty conforming to parts of the boundary that have high curvature, since the internal energy of C is high on such parts. Methods of overcoming these difficulties will be discussed in the next sections.

In the remainder of this section we review selected publications on active contours, emphasizing papers that made contributions to the representation of the contour; the definitions of the internal and external energy terms; the methods of initialization and energy minimization; and the methods of handling noise.

To improve the performance of active contours, alternative representations for the contour have been proposed. Kass *et al.* [1] represented an active contour by a digital curve. Delagnes *et al.* [3] defined adjustable polygons: sets of active line segments that can approximate any object shape; this representation gives good results if the object to be delineated is noise-free. Wang *et al.* [4] used a spline representation; this resulted in some improvement in accuracy and convergence speed over the KWT model. Wong *et al.* [5] proposed a segmented snake model; this converted the problem of global optimization of a closed curve into local optimization of a number of open arcs. Their approach was able to locate convex, concave and high-curvature parts of an object boundary; its performance was similar to that of Wang's spline representation. Chesnaud *et al.* [6] proposed a region snake model rather than using a boundary-based representation. Their model was based on Maximum Likelihood estimation of the statistics of the inner and outer regions defined by the boundary. This approach works well if we can use a priori assumptions about the statistics of the regions, and if these statistics are invariant or at least easy to classify. Ray *et al.* [7] proposed a multiresolution approach in which the snake algorithm is applied at an initial scale, and after the snake stabilizes, a higher resolution is used to adjust it. A potential disadvantage of this model is that high-curvature parts of the boundary and thick curves may be eliminated at the coarser resolution. Velasco and Marroquin [8] proposed Sandwich Snakes, which can detect contours that have complex shapes and reject false minima (up to some level) due to noise. Their model consists of two snakes, one inside and the other outside the boundary; it requires a one-to-one correspondence between the two snakes.

The performance of active contour models can be improved by modifying the definition of their internal energy. Cohen [9] proposed the use of a pressure force which inflates the active contour in the normal direction until it conforms to the boundary of the object. This model gives good results in noise-free images, but improper selection of the pressure force yields poor results. Davatzikos and Prince [10] proposed spatio-temporal variation of the internal energy terms as functions of position in the image and the number of iterations. This allows the model to handle high-curvature parts of the boundary more effectively than fixed-parameter algorithms. Xu *et al.* [11] proposed a method of compensating for the normal internal force so as to make it independent of shape. The resulting model works well, with no need to fine-tune internal parameters, and can conform to high-curvature parts of a boundary such as corners; however, its ability to overcome noise is reduced. Wang *et al.* [4] divided the energy minimization process into multiple stages. The first stage was designed to optimize the convergence speed in order to allow the snake to quickly approach the minimum-energy state. The second stage was devoted to snake refinement and local minimization of the energy, thereby driving the snake to a quasi-minimum-energy state. Finally, the third stage used the Bellman optimality principle to fine-tune the snake to a global minimum-energy state. Metaxas and Kakadiaris [12] presented a technique for the automatic adaptation of a deformable model's elastic parameters in a Kalman filter framework. The parameters are initialized and are subsequently modified, depending on the data and the noise distribution, until the contour conforms to the boundary; this works well if the spatial variations of the data are smooth. Mokhtarian and Mohanna [20] proposed an active contour model in which the smoothness internal evergy term is replaced by the output of a Curvature Scale Space filtering process.

Other authors have modified the definition of the external energy to increase the capture range of a snake and thus make the snake robust to noise. Kass *et al.* [1] used a scale space approach to guide a snake toward the boundary of an object. Xu and Prince [13] proposed a new external energy term called the "Gradient Vector Flow Field" computed by diffusion of the gradient vectors of a gray-level or binary edge map derived from the image. This force field is insensitive to initialization of the snake and allows the snake to move into concave boundary regions in noise-free images. However, using the diffusion of gradient vectors to develop this field may increase the effect of noise. Peterfreund [15] used spatio-velocity space (a combination of optical flow and image forces) to track boundaries of nonrigid objects on cluttered backgrounds.

An active contour model can be semi- or fully automatic, depending on how it is initialized. Kass *et al.* [1] initialized the snake near a boundary. Cohen [9] initialized the snake inside or outside an object and used a pressure force to push the contour outward or inward until it reached the boundary. Neuenschwander *et al.* [16] presented a model in which the user has to specify only the two endpoints of the contour rather than a polygonal approximation. The snake converges from this minimal initialization by propagating image information along the contour from both endpoints. The Gradient Vector Flow Field used by Xu and Prince [13] made the snake insensitive to initialization.

Alternative algorithms for minimizing the energy of an active contour have also been used. Kass *et al.* [1] minimized the energy by solving the Euler equations. Amini *et al.* [17] used dynamic programming to optimize an active contour; their approach was more stable than the original KWT approach, but it was time-consuming. Williams and Shah [18] proposed a greedy algorithm, which gave results comparable to those of Amini's method but was much faster. A common disadvantage of both methods was that they are local and hence are relatively sensitive to noise. Caselles *et al.* [19] proposed Geodesic Active Contours, which combined a geometric contour model with energy function minimization. The performance of this approach is comparable to that of conventional active contour models up to a constant that depends on the initial parameterization of the contour. The Geodesic Active Contour model combined with level set methods can be used to delineate the boundaries of multiple objects. This model has advantages over the original active contour model, but it has the drawback of being nonlinear.

The convergence speed and accuracy of active contours depends greatly on the level of noise in the image. Filtering techniques can be used to reduce the noise to some degree, but it is almost impossible to eliminate it completely. As a result, a snake may get stuck at energy minima caused by noise before it reaches a boundary. To avoid this situation, Davatzikos and Prince [10] proposed an algorithm in which the internal energy varies spatially. By giving high weight to the internal energy in noisy parts of the image they were able to overcome local minima. His model worked well when the object to be delineated had smooth boundaries. Delagnes *et al.* [3] proposed a new energy function based on textural characteristics of objects to resolve conflict situations when tracking objects on a cluttered background. Their method worked well when the textures were easily distinguishable. Other active contour models that were designed to overcome noise include those proposed by Metaxas and Kakadiaris [12], Chesnaud *et al.* [6], and Velasco and Marroquin [8]; these models were described above.

The organization of this paper is as follows: In Section 2 we show how local minima in the energy of an active contour, due to noise in the image, can be avoided by perturbing the contour representation during the energy minimization process. Section 3 and Section 4 describes our methods. Section 5 describes an experiment in which an active contour is used to delineate the boundary of a moving hand in an image sequence. Section 6 summarizes our methods.

2 Overcoming Local Energy Minima Due to Noise

When an active contour is applied to a noisy image, the contour sometimes is attracted to a local energy minimum, since the noise gives rise to high rates of change of the image gray levels. In this section we will describe a method of overcoming this problem by using a sparse set of points to represent the active contour C and randomly varying the positions of these points.

The number of points chosen to represent C must be a fraction of the total number of pixels on the (digital) contour, so there will be room to vary the position of the points. (We can roughly estimate the number of points on the contour by examining the output of the boundary extraction process [21]) On the other hand, the fraction cannot be too small, since it must be possible to closely approximate the shape of the contour by interpolating between the points. In the experiments described in this section, C was several hundred pixels long, and we represented it by about 60 points. Note that when we use scale-space methods to detect object boundaries at a distance, we are reducing (and afterwards increasing) the number of pixels on the contour, and the number of points used to represent the contour must be reduced or increased correspondingly.

In an active contour algorithm that uses a sparse representation, the contour C is represented by (say) n points. At each iteration of the algorithm, the points shift slightly; the new n points represent a new contour C'. This process is repeated at each iteration.

We have investigated a method of incorporating random variation into the points that represent the contour. In our method the number n of points remains constant. At each iteration of the active contour algorithm, we interpolate a smooth digital curve C' through the points. We choose n equally spaced points of C', one of which coincides with one of the original points on C'. We then randomly shift the new points along C' by an amount less than the spacing between the points. The points all shift together; their spacing remains the same. We refer to this method as "phase perturbation".

If many of the points that represent the contour coincide with noise points in the image, the external energy of the contour will be low, since external energy is inversely related to the gradient of the image gray level, which is high at noise points. Thus a contour configuration in which many of the points coincide with noise points may give rise to a local minimum in the energy, and the active contour algorithm may not be able to leave this minimum. However, if we perturb the points that represent the contour using phase perturbation, the perturbed points will no longer coincide with the noise points, so the contour has a chance of escaping the local energy minimum.

When phase perturbation is used, the number n of points that represent the contour remains constant, so the expected number of coincidences between these points and noise points is also constant. Thus the random displacement of the n points can be chosen from a uniform distribution over an interval.

We will now verify experimentally that using a uniform distribution is preferable in phase perturbation.

Figure 1(a) shows a 100 × 100 image that contains a blurred hollow square on a noisy background; the outer boundary of the square is represented by a solid curve S, and an initial active contour C surrounding S is overlaid on the image. We chose n equally spaced points to initially represent C, and performed 500 iterations of a sparse version of the KWT algorithm. (Randomness can be introduced into any active contour algorithm, but in our experiments we used the KWT algorithm.) After each iteration we also had the option of performing a phase perturbation by shifting the n points by an integer number of steps along the contour, where the number never exceeded the spacing s between the points. We performed six versions of this experiment:

a) No shift.
b-e) A shift chosen randomly in the range [0,2], [0,3], [0,4], or [0,s].
f) The shift that resulted in the highest external energy of C' (to maximize the likelihood of C escaping from the local energy minimum).

where (a to f) stands for (top-down). Figure 1(b) shows plots of the area (in pixel units) contained between C and S as a function of iteration number, averaged over 500 instances of the noise. We see that in all cases, the area at first decreases rapidly from its initial value of about 5000 as C shrinks toward S, but it then levels off. When no shifts were used (version (a)) the area levels off at about 2700. When random shifts were used (versions (b-e)), the area continues to drop; the larger the range of the shifts, the greater the drop, because there are more possibilities for increasing the energy. For shifts in the range [0,s] the area drops

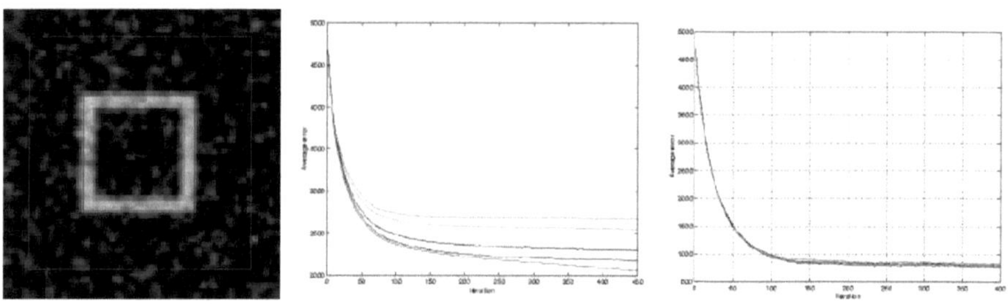

(a) (b) (c)

Fig. 1. (a)- A square in a noisy image. The initial active contour and the boundary of the square are overlaid on the image. (b)- Phase perturbation. Each curve shows the area between the active contour and the boundary of the square as a function of iteration number. The curves are averages over 500 instances of the image noise. Top to bottom: (a) No perturbation. (b-e) Shifts chosen randomly in the ranges [0,2], [0,3], [0,4], and [0,s], where s is the spacing between the points that represent the active contour. (f) The shifts that resulted in the highest external energy of the active contour. (c)- Comparison of two algorithms on a non-noisy version of Figure 1(a): a sparse version of the KWT algorithm, and algorithm that incorporated phase perturbation.

to less than 2200; this is nearly as good as when we use the shift that results in the highest external energy of C (version (f)).

Our algorithms represent a contour by a discrete set of points, and interpolate a smooth digital curve on these points each time a perturbation is applied. It might be thought that this repeated interpolation would result in a contour evolution process, but in fact this did not happen. To demonstrate this, we applied two algorithms to a version of Figure 1(a) that contained no noise: a sparse version of the KWT algorithm, and versions that incorporated phase perturbation. As we see in Figure 1(C), the plots of the area between C and S are virtually indistinguishable for both algorithms. This demonstrates that our use of repeated interpolation did not result in contour evolution.

In the experiments described in this section, the active contour C never penetrated the boundary of the square S; we could therefore use the area contained between C and S as an error measure. In the real examples described in the next section it is possible for C to penetrate the object boundary B. We will therefore use a more general error measure: the area of the symmetric difference between the regions surrounded by C and B.

3 Delineating an Object Boundary

Active contour performance can be improved by dividing the energy minimization process into stages [4] and allowing the energy function to vary during the process [12,13]. In Section 4 we will describe how such an adaptive active contour algorithm can be used to detect an object boundary at a distance and then locate details of the boundary's shape.

An active contour can be used to track the boundary of a moving object in an image sequence. This is usually done by locating the boundary (by minimizing the energy of the contour) in each frame of the sequence, and then using the result to initialize the contour in the next frame. In Section 5 we will use an active contour to locate the boundary of a moving object in an image sequence, using the moving boundary extraction process described in [21]. We will describe an experiment in which an active contour is used to locate the boundary of a hand moving against a complex background.

4 Detecting and Conforming to the Boundary

Since the external energy ξ of C depends only on the image values in the vicinity of C, distant object boundaries have no effect on ξ. Thus if C is initialized far from an image boundary, minimization of $E(C)$ does not attract C toward the boundary. This problem can be overcome by blurring the image before initializing C; but blurring the image may destroy details of the shapes of object boundaries. To achieve both detection at a distance and accurate location of shape details, we can vary the amount of image blur during the minimization process [1,12,13]. The blur remains high until ξ becomes low, indicating that C is approaching a boundary; the blur can then be gradually reduced so that C

can accurately conform to the boundary shape. The spacing of the points that we use to represent the contour does not exceed the amount of the blur.

The need to conform to boundary parts that have high curvatures introduces another problem: the internal energy of C is high when its curvature is high. This problem can be overcome [13] by gradually reducing the weight given to the curvature term of $E(C)$ as C approaches the boundary.

5 An Application: Delineating the Boundary of a Moving Hand in an Image Sequence

In this section we use an active contour to delineate the boundary of a moving object; the boundary is initially extracted by the method described in [21].

Figure 2(a) shows part of a frame of a 20-frame video sequence of a hand and arm moving in an indoor scene. A number of final boundaries located by the active contour after energy minimization is overlaid on the image. Figure 2(b) shows the moving boundary points extracted from that frame by the method described in [21]. Most of these points are concentrated near the hand and arm boundaries.

C was initialized on a large square which was close to the image border. Gaussian blur was initially applied to the moving boundaries that were extracted

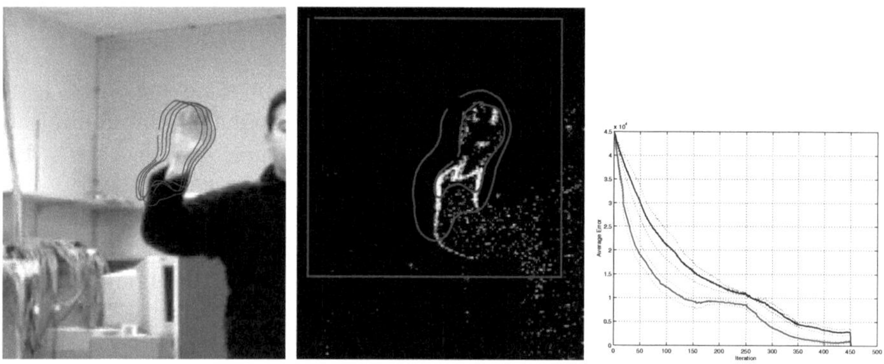

(a) (b) (c)

Fig. 2. Delineation of a moving hand in a video. (a) One frame of the video, with the final contours overlaid. (b) Initial contour, final Stage 1 contour, and final Stage 2 contour overlaid on the output of the moving boundary points that were extracted from the image. (c) Area (in pixels) of the symmetric difference between the hand and the interior of the contour, as a function of iteration number, averaged over 20 frames. The solid curve is the mean; the dotted curves are one standard deviation above and below the mean. Upper curves: Algorithm without perturbation. Lower curves: Algorithm with perturbation. Note the shoulder in the curves at iteration 250, when the first stage ended.

from that image (Figure 2(b)). The rate of change of the gray level in this blurred image was used as the external energy term of C. This rate of change is a maximum at the inflections of the Gaussian; hence its inverse is a minimum. Minimization of $E(C)$ thus causes C to shrink and to approach the hand and arm boundary until it reaches the inflections. In Figure 2(b) the location of C after 250 iterations overlaid on the image.

By this time the external energy of C was quite low. A second stage of energy minimization was then initiated, in which the amount of image blur and the weight given to curvature in the internal energy of C were both progressively decreased, as described in Section 4. This allowed C to approach the boundary closely and to conform to its shape. The location of C after 200 iterations of the second-stage process is also overlaid on the Figure 2(b).

To reach its final location, C must cross noisy parts of the image background. As discussed in Section 2, it is possible for C to be trapped by a local energy minimum caused by the noise, but this can be avoided by applying phase perturbation to the points that define C.

To study how this perturbation improves performance, we applied two versions of our active contour algorithm to the 20 frames of our image sequence; the first version perturbed the points that represent the contour and the second version did not. Figure 2(c) compares the average performance of the two versions on the 20 frames; the upper curves are for the second version and the lower curves for the first version. In each frame, we computed the area of the symmetric difference between the hand/arm region and the interior of the active contour, as a function of iteration number. (The solid curve is the 20-frame average; the dotted curves are one standard deviation above and below the average.) We see that the first version of the algorithm converged more quickly and approximated the boundary more accurately. The lower curve comes close to a minimum after 150 iterations in the first stage of the process, and after 450 iterations it is less than 20% as high as the upper curve.

6 Concluding Remarks

This paper has made the following contributions: First; an active contour can be trapped by local energy minima when too many of its points are influenced by image noise. We have shown that this situation can be prevented by incorporating randomness to the points that represents the contour. Second; to speed up the process of convergence and to conform to boundary shape we developed a two stage algorithm. Our model is a mixture of deterministic and random components that made it robust with respect to noise. We have shown that our model performs equally well as other sparse models when there is no noise, the robustness becomes visible only when the image contains noise. Using this active contour model, we were able to locate and track a moving boundary in a sequence of images in the presence of noise.

References

1. Kass, M., Witkin, A.P., Terzopoulos, D.: Snakes: Active Contour Models. International Journal of Computer Vision 1, 321–331 (1988)
2. Blake, A., Isard, M.: Active Contours. Springer, Heidelberg (1998)
3. Delanges, P., Benois, J., Barba, D.: Active Contours Approach to Object Tracking in Image Sequences with Complex Background. Pattern Recognition Letters 16, 171–178 (1995)
4. Wang, M., Evans, J., Hassebrook, L., Knapp, C., Multistage, A.: Optimal Active Contour Model. IEEE Trans. on Image Processing 5, 1586–1591 (1996)
5. Wong, Y.Y., Yuen, P.C., Tong, C.S.: Segmented Snake for Contour Detection. Pattern Recognition 31, 1669–1679 (1998)
6. Chesnaud, C., Refregier, P., Boulet, V.: Statistical Region Snake-Based Segmentation Adapted to Different Physical Noise Models. IEEE Trans. on Pattern Analysis and Machine Intelligence 21, 1145–1157 (1999)
7. Ray, N., Chanda, B., Das, J.: A Fast and Flexible Multiresolution Snake with a Definite Termination Criterion. Pattern Recognition 34, 1483–1490 (2001)
8. Velasco, F.A., Marroquin, J.L.: Robust Parametric Active Contours: The Sandwich Snakes. Machine Vision and Applications 12, 238–242 (2001)
9. Cohen, L.D.: On Active Contour Models and Balloons. Computer Vision, Graphics, and Image Processing 53, 211–218 (1991)
10. Davatzikos, C., Prince, J.L.: Adaptive Active Contour Algorithms for Extracting and Mapping Thick Curves. Computer Vision and Pattern Recognition, pp. 524-529 (1993)
11. Xu, G., Segawa, E., Tsuji, S.: Robust Active Contours with Insensitive Parameters. Pattern Recognition 27, 879–884 (1994)
12. Metaxas, D., Kakadiaris, I.A.: Elastically Adaptive Deformable Models, European Conference on Computer Vision, pp. 550–559 (1996)
13. Xu, C., Prince, J.L.: Gradient Vector Flow: A New External Force for Snakes. In: Conference on Computer Vision and Pattern Recognition, pp. 66–71 (1997)
14. Davatzikos, C., Prince, J.L.: Convexity Analysis of Active Contour Problems. Image and Vision Computing 17, 27–36 (1999)
15. Peterfreund, N.: The Velocity Snake: Deformable Contour for Tracking in Spatio-Velocity Space. Computer Vision and Image Understanding 73, 346–356 (1999)
16. Neuenschwander, W., Fua, P., Szekely, G., Kubler, O.: Making Snakes Converge from Minimal Initialization. International Conference on Pattern Recognition, pp. 613–615 (1994)
17. Amini, A., Tehrani, S., Weymouth, T.E.: Using Dynamic Programming for Minimizing the Energy of Active Contours in the Presence of Hard Constraints. In: International Conference on Computer Vision, pp. 95–99 (1988)
18. Williams, D.J., Shah, M.: A Fast Algorithm for Active Contours. In: International Conference on Computer Vision, pp. 592–598 (1990)
19. Caselles, V., Kimmel, R., Sapiro, G.: Geodesic Active Contours. International Journal on Computer Vision 22, 61–79 (1997)
20. Mokhtarian, F., Mohanna, F.: Fast Active Contour Convergence through Curvature Scale Space Filtering. Image and Vision Computing, pp. 157–162 (2001)
21. Rexhepi, A., Rosenfeld, A., Mokhtarian, F.: Extracting Boundaries from Images by Comparing Cooccurrence Matrices. Digital Image Computing Techniques and Applications DICTA2003 (2003)

Accurate Spatial Neighborhood Relationships for Arbitrarily-Shaped Objects Using Hamilton-Jacobi GVD*

Sumit K. Nath, Kannappan Palaniappan, and Filiz Bunyak

MCVL, Department of Computer Science, University of Missouri-Columbia, MO, USA
{naths, palaniappank, bunyak}@missouri.edu

Abstract. Many image segmentation approaches rely upon or are enhanced by using spatial relationship information between image regions and their object correspondences. Spatial relationships are usually captured in terms of relative neighborhood graphs such as the Delaunay graph. Neighborhood graphs capture information about which objects are close to each other in the plane or in space but may not capture complete spatial relationships such as containment or holes. Additionally, the typical approach used to compute the Delaunay graph (or its dual, the Voronoi polytopes) is based on using only the point-based (i.e., centroid) representation of each object. This can lead to incorrect spatial neighborhood graphs for sized objects with complex topology, eventually resulting in poor segmentation. This paper proposes a new algorithm for efficiently, and accurately extracting accurate neighborhood graphs in linear time by computing the Hamilton-Jacobi generalized Voronoi diagram (GVD) using the exact Euclidean-distance transform with Laplacian-of-Gaussian, and morphological operators. The algorithm is validated using synthetic, and real biological imagery of epithelial cells.

1 Introduction

Spatial neighborhood relationships among objects is an important characteristic in many image analysis, computer vision and robotics applications. One common approach is to compute Delaunay graphs from an ordinary Voronoi diagram (OVD), using information from centroids of objects [1]. In the context of biological image analysis, the OVD has been used for accurate segmentation and analysis of confluent migrating cells [2, 3], tissue architecture characterization [4], or endothelial cell classification [5]. Our application is primarily focused on accurate segmentation and tracking of cells in biomedical video sequences that undergo complex shape changes like mitosis and apoptosis.

An OVD using points is insensitive to object properties like size, shape, orientation or containment. Thus, neighborhood graphs derived from point-based centroid representations of arbitrarily-shaped objects often lead to incorrect neighborhood relationships as shown in Figs. 1(b) and (e). Applications that depend on accurate spatial neighborhood relationships would consequently fail or lead to unpredictable behavior. For example, incorrect neighborhood relationships may lead to false merges of neighboring cells in the segmentation algorithm described in [2]. In other applications, such as robot path

* This work was supported by a U.S National Institute of Health NIBIB award R33 EB00573.

B.K. Ersbøll and K.S. Pedersen (Eds.): SCIA 2007, LNCS 4522, pp. 421–431, 2007.

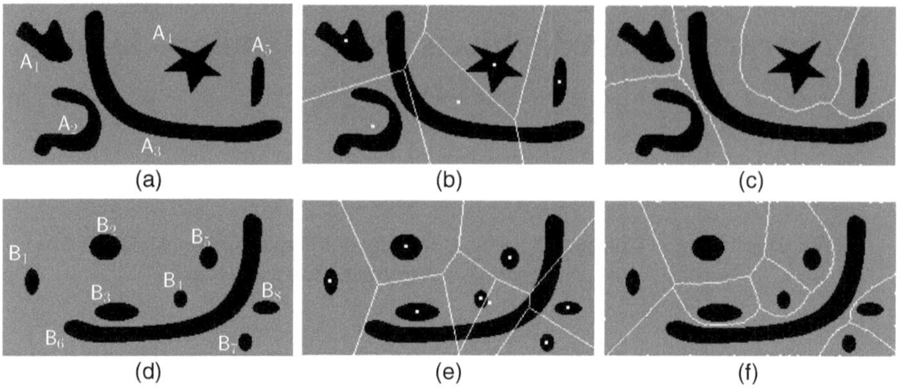

Fig. 1. [**a** and **d**]: Synthetic images showing arbitrarily-shaped objects. [**b** and **e**]: It is evident that an ordinary Voronoi diagram (OVD), computed from centroid points of objects (shown as white squares) leads to incorrect neighborhood relationships. [**c** and **f**]: However, using a generalized Voronoi diagram (GVD) leads to correct boundaries and neighborhood relationships. Corresponding neighborhood adjacency graphs are shown in Tables 1 and 2.

planning, inaccurate neighborhoods obtained from OVD's may impede the movement of the robot or lead to weak navigation performance [6]. An alternative to the OVD is to compute the generalized Voronoi diagram (GVD) that takes into account the size, shape, orientation, and placement of objects when computing neighborhood relationships. As seen from Figs. 1(c) and 1(f) the GVD accurately identifies the neighborhoods of complex-shaped objects (e.g., the thin long non-convex worm-like object).

The GVD in any dimension can be precisely defined using point to object distance measures [1, p 280]. Let $\mathbf{A} = \{A_1, A_2, \ldots, A_N\}$ be a set of arbitrarily shaped objects in a $d-$dimensional space $\mathbb{R}^d$. Now, for any point $\mathbf{p} \in \mathbb{R}^d$, let $\mathsf{D}(\mathbf{p}, A_i)$ denote a distance measure representing how far the point $\mathbf{p}$ is from the object A_i which is typically the minimum distance from $\mathbf{p}$ to any point in object A_i. The dominance region (also known as influence-zone) of A_i, is then defined as

$$\text{Dom}(A_i, A_j) = \Big\{\mathbf{p}\,|\mathsf{D}(\mathbf{p}, A_i) \leq \mathsf{D}(\mathbf{p}, A_j),\ \forall j, j \neq i\Big\} \tag{1}$$

A generalized Voronoi boundary, between A_i and Aj, can then be defined as the loci of equidistant points between both objects, $\mathcal{L}(A_i, A_j)$, where

$$\mathcal{L}(A_i, A_j) = \Big\{\mathbf{p}\,|\mathsf{D}(\mathbf{p}, A_i) = \mathsf{D}(\mathbf{p}, A_j)\Big\}, \tag{2}$$

and the corresponding influence zone for A_i, $V(A_i)$, is the set intersection

$$V(A_i) = \bigcap_{i \neq j} \text{Dom}(A_i, A_j) \tag{3}$$

Hence, the generalized Voronoi diagram of $\mathbf{A}$, GVD($\mathbf{A}$), is given by the union of of such generalized Voronoi regions, as

$$\begin{aligned}\mathrm{GVD}(\mathbf{A}) &= \bigcup_i V(A_i) \\ &= \bigcup_i \bigcap_{i \neq j} \left\{ \mathbf{p} \,|\, \mathsf{D}(\mathbf{p}, A_i) \leq \mathsf{D}(\mathbf{p}, A_j),\ \forall j, j \neq i \right\}\end{aligned} \tag{4}$$

When $\mathbf{A}$ is a collection of points rather than sized objects, GVD($\mathbf{A}$) reduces to an ordinary Voronoi diagram, OVD($\mathbf{A}$). Note that OVD boundaries are always straight lines or hyperplanes, whereas GVD boundaries can be complex curves or surfaces. Fig. 1 shows examples of the OVD and GVD for objects in a plane (i.e., $d = 2$). For those interested in properties of the OVD for point objects, we direct them to the book by Okabe *et al.* [1] and the survey paper by Aurenhammer [7].

The GVD representation of a set of objects has a number of useful properties: (i) it is a thin set that partitions a space into connected regions (ii) it is homotopic to the number of objects, (iii) it is invariant under transformations applied to all objects, and (iv) each region of the GVD is guaranteed to contain the entire object.

Sugihara presents an algorithm to construct an approximate GVD by reducing an object to a collection of points [8]. A different class of algorithms to construct GVD's is based on morphological operators and label propagation. This consists of labeling connected components (objects) in an image, and simultaneously growing them using dilation operators. The loci of points at which these regions stop growing determine the influence zone of each object. In the literature, this algorithm is referred to as *skeletons by influence zone* (SKIZ) and is described in detail by Vincent [9,5]. Lu and Tan have presented a variation of SKIZ by approximating connected components as polygons and expanding the regions using Freeman codes for document image analysis [10]. Hoff *et al.* have reported a fast algorithm for GVD construction using graphics hardware [11].

Recently, Siddiqi *et al.* proposed a new class of algorithms to compute object skeletons using the average outward flux of the gradient of a distance transform [12]. Homotopy preserving properties of this algorithm makes it a strong alternative to other algorithms that use the Euclidean distance transform (EDT) to compute object skeletons. A Hamilton-Jacobi formulation for *shock tracking*, combined with homotopy preserving thinning leads to a robust and low-complexity implementation. As an original contribution, we propose using the Hamilton-Jacobi formulation to compute GVD's. The focus of this paper is on efficiently extracting exact neighborhood relationships of arbitrarily shaped objects (e.g., biological cells) using the GVD as the basic underlying framework, based on a fast EDT. It should be noted that even though our algorithm aims at solving a problem in biological image analysis, it can be applied to other applications in computer vision such as robot navigation, remote sensing of urban areas or content-based image retrieval.

The paper is organized as follows. In Sec. 2, we summarize our proposed algorithm and explain its key features. Comparative results of using OVD versus GVD for computing cell neighborhood relationships are shown in Sec. 3, and conclusions in Sec. 4.

Algorithm 1. `Compute a 2D Neighborhood Adj. Graph`

Input : $\mathbf{P}$, a 2D mask with N labeled objects,
T_{LD}, threshold to detect ridges,
T_{HS}, threshold for max. hole size, and
σ, to control smoothing.

Output : $\mathcal{N}(\mathbf{P})$, the adjacency graph of $\mathbf{P}$

1: Remove labeled 8-connected pixels in $\mathbf{P}$ that are adjacent to one or more different labels.
2: Convert the processed mask into a binary image $\mathcal{B}$.
3: Compute the Euclidean distance transform (EDT), $\mathcal{D}$, of $\mathcal{B}$ using the FH-EDT algorithm [13].
4: Compute $\mathsf{E} = \nabla^2 G_\sigma \circledast \mathcal{D}$, the Laplacian of the smoothed EDT.
5: Obtain a binary image, E_{thr}, from E using a threshold value T_{LD}.
6: Fill holes using T_{HS}, the hole-size threshold.
7: Apply a suitable thinning algorithm (e.g., [14, 15]) on E_{thr} to obtain an image with 1-pixel thick GVD boundaries, E_{thr}^{thn}.
8: Apply any homotopy-preserving algorithm [16] to prune *branches* from the generalized Voronoi diagram, E_{thr}^{thn}.
9: Assign $\mathcal{Q} \leftarrow (\mathsf{E}_{thr}^{thn})^c$; the complementary image of E_{thr}^{thn}
10: Using 4-connectivity, label the connected components of $\mathcal{Q}$.
11: Update the neighborhood relationship map $\mathcal{N}(\mathbf{P})$ by checking a 3×3 neighborhood of each background pixel (i.e., boundary pixels of connected components) in $\mathcal{Q}$.

2 Neighborhood Adjacency Graphs Using GVD

The proposed algorithm to compute a neighborhood adjacency graph $\mathcal{N}(\mathbf{P})$ for an image $\mathbf{P}$ containing $N-$arbitrarily shaped objects, using GVD in $\mathbb{R}^2$ is shown in Algorithm 1., and described in detail in the following paragraphs.

In order to compute reliable GVD boundaries touching objects need to be separated by at least a one-pixel gap. In Step 1, labeled pixels are (temporarily) removed from the image if they are adjacent to one or more different labeled pixels, without any gap. In Step 2, we convert the modified multi-labeled mask into a binary image with non-zero pixels representing N distinct connected components.

Siddiqi *et al.* have reported using a Borgefors distance transform (BDT) in their skeletonization algorithm [12]. However, the BDT is an approximation of the Euclidean distance transform (EDT). Hence, in Step 3, we compute the exact EDT using a "separable algorithm" proposed by Felsenzwalb and Huttenlocher (FH-EDT) that is fast (linear time), and efficient to implement [13].

Let $\mathcal{G}_1 = \{0, 1, \ldots, n-1\}$ be a 1D grid, and $f : \mathcal{G}_1 \rightarrow \mathbb{R}$ an arbitrary function on the grid. The one-dimensional FH-EDT of f is defined as

$$\mathcal{D}_f(p) = \min_{q \in \mathcal{G}_1} \Big((p-q)^2 + f(q) \Big) \tag{5}$$

with the added constraint that for each point $q \in \mathcal{G}_1$, the distance transform of f is bounded by a parabola rooted at $(q, f(q))$. The distance transform at point p is the height of the lower envelope of all such parabolas [13, Fig. 1]. The FH-EDT algorithm

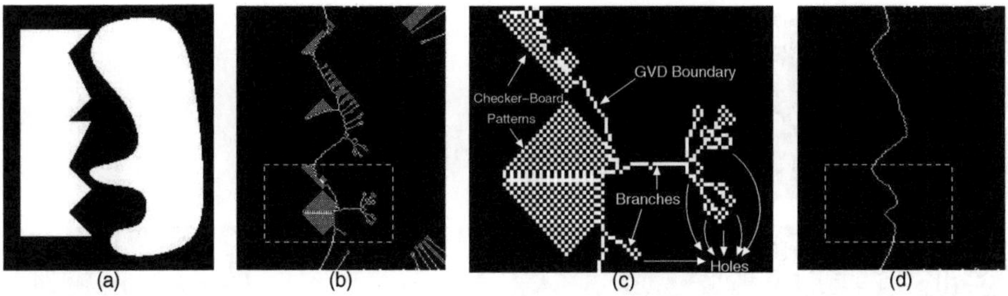

Fig. 2. Examples of isolated holes and checker-board pattern holes that are formed when a large number of single pixel width ridges appear very close to each other. Non-pruned branches may affect the performance of such applications as robot navigation. A relatively higher value of T_{LD} can reduce such occurrences, with the possibility of breaking actual GVD boundaries. This figure is related to problems that are solved in Steps 6 - 8 in Algorithm 1.. The actual GVD boundary is shown in (d).

computes the distance transform in $O(n)$ time. The efficiency of this algorithm is evident by considering a two-dimensional grid $\mathcal{G}_2 = \{0, 1, \ldots, n-1\} \times \{0, 1, \ldots, m-1\}$, and $f : \mathcal{G}_2 \rightarrow \mathbb{R}$ an arbitrary function on the grid. The two-dimensional distance transform of f is given by

$$\begin{aligned} \mathcal{D}_f(x, y) &= \min_{x', y'} \Big((x - x')^2 + (y - y')^2 + f(x', y'), \Big) \\ &= \min_{x'} \Big((x - x')^2 + \min_{y'} \Big((y - y')^2 + f(x', y') \Big) \Big), \\ &= \min_{x'} \Big((x - x')^2 + \mathcal{D}_{f|_{x'}}(y) \Big), \end{aligned} \tag{6}$$

where $\mathcal{D}_{f|_{x'}}(y)$ is the 1D distance transform of f restricted to the column indexed by x'. Hence, the 2D distance transform can be computed separably in linear time.

In order to detect points of singularities (or *shock points*), Siddiqi *et al.* propose to compute the average outward flux at every point in a vector field $\dot{\mathbf{q}}$ (derived from the distance transform) using a Hamilton-Jacobi formulation [12]. Using the divergence theorem, a relationship between the divergence of the vector field $\operatorname{div}(\dot{\mathbf{q}})$, and the average outward flux is given by [12]

$$\operatorname{div}(\dot{\mathbf{q}}) \equiv \lim_{\Delta a \rightarrow 0} \frac{\int_{\delta R} < \dot{\mathbf{q}}, \mathcal{N}_s > \, ds}{\Delta a}, \tag{7}$$

where δR is the bounding contour of the region R, $\mathcal{N}_s$ is the outward normal at each point of the contour, and ds is the element of integration. The divergence $\operatorname{div}(\dot{\mathbf{q}})$ can be equivalently written as the sum of partial derivatives with respect to each of the vector field's component directions. However, the vector field (i.e., distance field) is differentiable at all points except at *singular or shock points*. This is the justification provided by Siddiqi *et al.* for using Eq. 7, and a limit approximation, to locate singularities in $\dot{\mathbf{q}}$. As an alternative, in Step 4, we propose using a 2D Laplacian-of-Gaussian

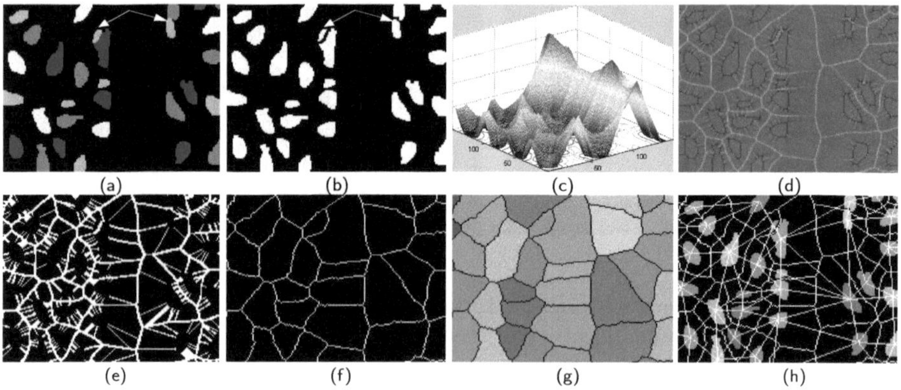

Fig. 3. A flow diagram that describes Algorithm 1.. (a) A section of the original mask with four unique foreground colors obtained from Nath *et al.* algorithm [2]. Cells that are touching each other are marked with arrows. (b) A binary image $\mathcal{B}$ is obtained as per rules outlined in as per Step 1 of Algorithm 1.. (c) 3D view of the distance transform ($\mathcal{D}$), that shows the difficulty in isolating ridges (i.e., Voronoi boundaries). (d) $\nabla^2 \mathcal{D}_\sigma$. (e) A thresholded version of Fig. 3(d) after removal of small holes. (f) A pruning step removes *branches* from the generalized Voronoi diagram. (g) Connected-component labeling, followed by generation of $\mathcal{N}(\mathbf{P})$, is implemented on a complement of the image, obtained in Fig 3(f), as per Steps 10-11 of Algorithm 1.. (h) The final generalized Voronoi diagram and neighbors of cells are shown in white and yellow, while cells are shown in colors used previously for labeling Voronoi cells in Fig. 3(g).

$\nabla^2 G_\sigma \circledast \mathcal{D}$ operator on the distance transform, $\mathcal{D}$, in order to detect regions of local maxima (or minima, depending on how the Laplacian operator is applied), i.e., ridge points. The Gaussian operator G_σ smooths the distance transform prior to applying the Laplacian operator insuring differentiability at shock points. Smoothing, however, does not guarantee homotopy preservation of GVD boundary points. Hence, to satisfy both constraints, the regularization parameter σ is set to a small value.

In Step 5, we threshold $\mathsf{E} = \nabla^2 G_\sigma \circledast \mathcal{D})$ to obtain the binary image, E_{thr},

$$\mathsf{E}_{thr} = \begin{cases} 1 & \mathsf{E} > T_{LD}, \\ 0 & \text{otherwise}, \end{cases}$$

before computing the GVD, A suitable choice of the threshold value, T_{LD}, is critical in homotopy preservation of GVD boundaries. A low threshold value results in larger number of spurious features (such as branches and associated holes), while a larger threshold significantly reduces these features at the cost of breaking real object boundaries. We set $T_{LD} = 0$ by default.

After binarization of E, the background *should* normally be segmented into N connected generalized Voronoi regions, corresponding to N input objects. However, when computing the Laplacian of the EDT, regions of local maxima, i.e., ridges, may appear very close to each other and interact to produce "*holes*" that are small connected background components (shown in Fig. 2(b) and (c)). In our algorithm, each influence zone (i.e., $V(A_i)$) corresponds to a unique object (A_i) in the image. Hence, in Step 6, such

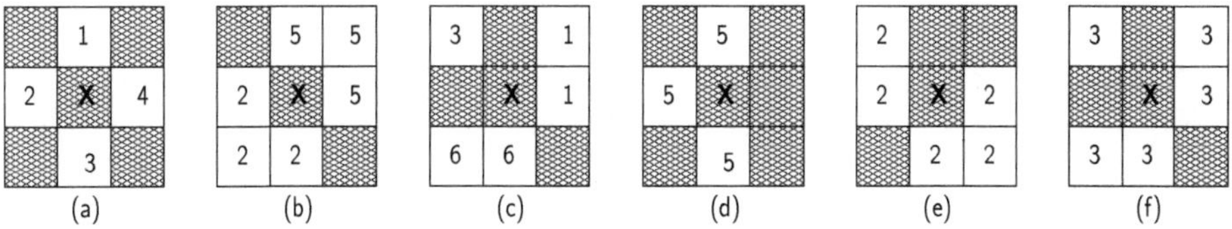

Fig. 4. Examples of neighborhood relationships between connected components when centering a 3×3 neighborhood on a boundary pixel (marked with a X). Shaded regions indicate one pixel thick boundaries of connected components in $\mathcal{Q}$. Valid generalized Voronoi boundary pixels separate different influence zones, resulting in *at least* two different sets of foreground labels in the neighborhood (Figs. 4(a) - 4(c)). On the other hand, spurs/branches are contained within a single influence zone, thus resulting in a single set of foreground labels in the neighborhood. As a result, no changes need to be made in the adjacency graph $\mathcal{N}(\mathbf{P})$ (Figs. 4(d) - 4(f)).

holes are removed using a threshold parameter T_{HS}, prior to computing the GVD. Non-removal of such holes prevents further removal of ridges that are attached to such holes, termed as *branches*. Hole removal is effected by size-constrained connected component analysis. The binarized image, obtained in Step 5 of Algorithm 1. is inverted followed by a connected component analysis. All connected components below a certain size are classified as part of the background which results in "hole-filling".

In Step 8, a thinning algorithm (c.f. [15]) is applied to the hole-filled, binarized image in order to reduce ridge boundaries to single pixel thickness. This step is necessary in order to simplify the search for neighborhood adjacency relationships along boundaries. A key component of any thinning algorithm is the preservation of end points. Thus, after thinning, spurious ridges, without holes, remain attached to actual GVD boundaries. We term such ridges as *spurs* (see Fig. 5(f) for example). Hence, in Step 9, we remove such spurs by applying a pruning algorithm having the same features as standard thinning algorithms (e.g., [15]) but enforcing the constraint of non-preservation of end points. Let this thinned (and optionally pruned) image be represented as E_{thr}^{thn}.

After obtaining one-pixel thick GVD boundaries, we invert E_{thr}^{thn} in Step 9 as $\mathcal{Q} = (\mathsf{E}_{thr}^{thn})^c$. This is followed, in Step 10, by a connected component analysis on $\mathcal{P}$ and assigning unique labels to each GVD influence zone, i.e., $\mathbf{Q} = \bigcup_i \mathcal{Q}(V_i)$, where $\mathcal{Q}(V_i)$ is the ith connected component formed from the corresponding generalized Voronoi influence zone. Finally, in Step 11, a 3×3 window positioned at each boundary pixel (i.e., pixels not part of any connected component) is analyzed, from which a neighborhood relationship map $\mathcal{N}(\mathbf{P})$ is constructed (see Fig. 4 for some examples). To complement the discussion in previous paragraphs, key steps of our algorithm are shown in Fig. 3.

3 Results and Discussion

The Hamilton-Jacobi GVD algorithm for determining accurate neighborhood graphs was applied to a biomedical application involving cell segmentation and tracking [2]. Time-lapse phase contrast microscopy of epithelial cells moving in a monolayer sheet

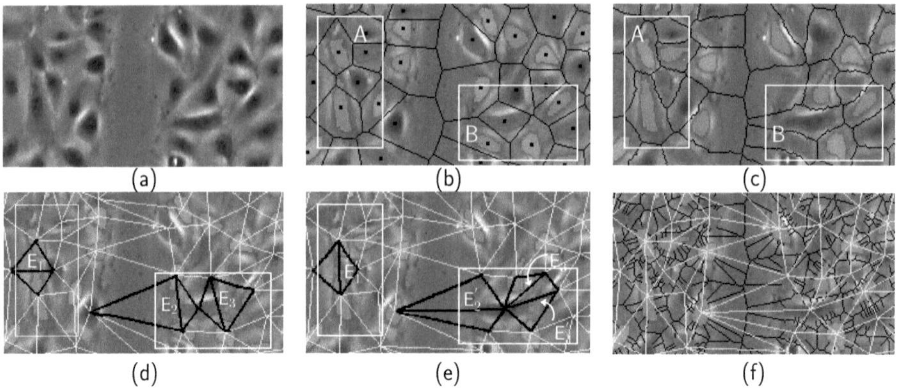

Fig. 5. (a) A representative region from the original image (Frame 46). **(b)** OVD boundaries superimposed with four unique colors. The centroids are represented as small squares. **(c)** GVD boundaries superimposed with four unique colors. **(d)** Neighborhood relationships in two representative regions A and B, when using the OVD from (b). The OVD leads to incorrect neighborhood relationships, as shown by edges E_1, E_2 and E_3. This leads to neighboring cells being assigned the same color during graph-vertex coloring [2]. **(e)** Neighborhood relationships using GVD with correct assignments indicated by $\mathsf{E}_1^{'}$, $\mathsf{E}_2^{'}$ and $\mathsf{E}_3^{'}$ with an additional neighborhood relationship $\mathsf{E}_4^{'}$ that is detected when. **(f)** Neighborhood relationships without pruning branches of the GVD does not affect neighborhood relationships between cells. This feature will be addressed in a different paper. Parameters used in computing the Hamilton-Jacobi GVD are: $\sigma = 0.5$, 9-tap Laplacian kernel with a center weight of 8, $T_{LD} = 0.0$, and $T_{HS} = 5$.

are imaged at $0.13\mu m$ resolution, and appear as a clustering of dark colored nuclei with indistinct boundaries (Fig. 5(a)) [3, 2, 17].

The OVD regions, and associated Delaunay graph based on centroids of cell nuclei in Fig. 5(a) are shown Figs. 5(b) and (d), respectively. Edges E_1, E_2, and E_3 show incorrect object adjacency relationships based on OVD regions A, and B. The object colors are based on graph-vertex coloring and used to implement a fast 4-color level set-based cell segmentation algorithm incorporating spatial coupling constraints [2]. The main feature of the 4-color level set algorithm is to assign different colors to neighboring cells, in order to prevent false merges. From Fig. 5(b) it can be observed from region A that the two green-colored cells are neighbors of each other, yet they are not marked as neighbors when using an OVD. However, in Fig. 5(c) and 5(e), these cells are correctly classified as neighbors when using our proposed GVD algorithm (the cells have been recolored).

The neighborhood adjacency graphs for the synthetic images shown in Figs. 1(a) and 1(d) using the OVD and GVD are shown in Tables 1 and 2, respectively. It is clearly evident that the Hamilton-Jacobi GVD algorithm correctly identifies neighbors of objects in both images, while errors are evident when using OVD to compute the spatial adjacencies of objects. For example, the long thing worm-like object, B_6, is adjacent to smaller elliptical objects B_1, B_3, B_4, B_5, B_7, and, B_8. It does not overlap any other object and has a worm-like influence zone based on the GVD, as seen in Fig. 1(f).

Table 1. Neighborhood map of Fig. 1(a)

$\mathcal{Q}(V_i)$	OVD	GVD
A_1	A_2, A_3	A_2, A_3, A_4
A_2	A_1, A_3	A_1, A_3
A_3	A_1, A_2, A_4, A_5	A_1, A_2, A_4, A_5
A_4	A_3, A_5	A_1, A_3, A_5
A_5	A_3, A_4	A_3, A_4

Table 2. Neighborhood map of Fig. 1(d)

$\mathcal{Q}(V_i)$	OVD	GVD
B_1	B_2, B_3	B_2, B_3, B_6
B_2	B_1, B_3, B_4, B_5	B_1, B_3, B_4, B_5
B_3	B_1, B_2, B_4, B_6	B_1, B_2, B_4, B_6
B_4	B_2, B_3, B_5, B_6	B_2, B_3, B_5, B_6
B_5	B_2, B_4, B_6, B_8	B_2, B_4, B_6
B_6	B_4, B_5, B_7, B_8	$B_1, B_3, B_4, B_5, B_7, B_8$
B_7	B_6, B_8	B_6, B_8
B_8	B_5, B_6, B_7	B_6, B_7

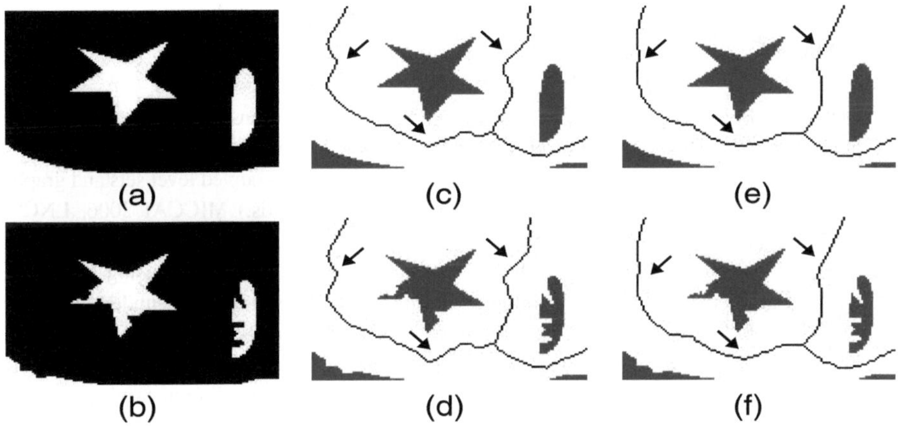

Fig. 6. [**a**]: Subset of objects from Fig. 1(a) showing objects A_4 and A_5. [**b**]: Same objects with perturbed boundaries. [**c** and **d**]: SKIZ-based implementation of GVD with 8-connected label propagation. [**e** and **f**]: Hamilton-Jacobi GVD.

We compare the robustness of the Hamilton-Jacobi GVD algorithm with a watershed-based, fast implementation of SKIZ [18, pg. 170-173] in MATLAB. Figs. 6 (c) and (d), and Figs. 6 (e) and (f) show that the SKIZ-based GVD, and the Hamilton-Jacobi GVD are both relatively insensitive to perturbations in object boundaries as indicated by the arrows. However, Hamilton-Jacobi GVD boundaries are more accurate (same arrows), since the exact EDT is used.

4 Conclusion

In this paper, we have presented a novel algorithm for computing Hamilton-Jacobi based GVD's to build accurate spatial neighborhood adjacency graphs for arbitrarily-shaped objects. Our algorithm extends the Hamilton-Jacobi skeletonization algorithm of Siddiqi *et al.* [12], and is coupled with morphological-based operators to remove spurious regions from the initial GVD boundaries. A fast Laplacian-of-Gaussian (LoG) filter is used to detect potential GVD boundary locations (i.e., shock points). Useful features of the LoG filter, like the guarantee of closed contours, continuity of ridges, and non-formation of new ridges with an increase in scale (smoothing) makes it appealing for our algorithm. We compare the performance of our Hamilton-Jacobi GVD algorithm, with a previously developed OVD framework for cell segmentation in [2] on real biological, as well as synthetic images. In all instances, we demonstrate the superiority of our GVD algorithm.

As a future work, we would like to present a comparison of our algorithm with other state-of-the-art algorithms described in the literature. Due to the separable nature of the FH-EDT algorithm [13], we can obtain neighborhood relationships between objects in higher dimensions. Hence, we would like to extend our algorithm to $\mathbb{R}^d,\ d > 2$.

References

1. Okabe, A., Boots, B., Sugihara, K., Chiu, S.N.: Spatial Tesselations: Concepts and Applications of Voronoi Diagrams. 2nd edn. John Wiley & Sons Ltd. West Sussex, UK (2000)
2. Nath, S., Palaniappan, K., Bunyak, F.: Cell segmentation using coupled level sets and graph-vertex coloring. In: Larsen, R., Nielsen, M., Sporring, J. (eds.) MICCAI 2006. LNCS, vol. 4190, pp. 101–108. Springer, Heidelberg (2006)
3. Nath, S., Bunyak, F., Palaniappan, K.: Robust tracking of migrating cells using four-color level set segmentation. In: Blanc-Talon, J., Philips, W., Popescu, D., Scheunders, P. (eds.) ACIVS 2006. LNCS, vol. 4179, pp. 4179–4920. Springer, Heidelberg (2006)
4. Keenan, S., Diamond, J., McCluggage, W., Bharucha, H., Thompson, D., Bartels, P., Hamilton, P.: An automated machine vision system for the histological grading of cervical intraepithelial neoplasia (CIN). J. Pathol. 192, 351–362 (2000)
5. Vincent, L., Masters, B.: Morphological image processing and network analysis of cornea endothelial cell images. In: Proc. SPIE - Image Algebra Morph. Image Proc. III, San Diego, CA, vol. 1769 pp. 212–226 (1992)
6. Choset, H., Walker, S., Ard, K., Burdick, J.: Sensor-based exploration: Incremental construction of the hierarchical generalized voronoi graph. I. Journ. Robotics Res. 19, 126–148 (2000)
7. Aurenhammer, F.: Voronoi diagrams - A survey of a fundamental geometric data structure. ACM Comp. Surveys 23, 345–405 (1991)
8. Sugihara, K.: Approximation of generalized Voronoi diagrams by ordinary Voronoi diagrams. Com. Vis. Graph. Image Process 55, 522–531 (1993)
9. Vincent, L.: Graphs and mathematical morphology. Sig. Proc. 16, 365–388 (1989)
10. Lu, Y., Tan, C.: Constructing area Voronoi diagram in document images. In: Proc. 88th IEEE Int. Conf. Doc. Anal. Recog. IEEE Comp. Soc. pp. 342–346 (2005)
11. Hoff III, K., Keyser, J., Lin, M., Manocha, D., Culver, T.: Fast computation of generalized voronoi diagrams using graphics hardware. In: SIGGRAPH-99. 26th Ann. Conf. Comp. Graphics Inter. Tech, pp. 277–286. ACM Press/Addison-Wesley Publications, New York (1999)
12. Siddiqi, K., Bouix, S., Tannenbaum, A., Zucker, S.: Hamilton-Jacobi skeletons. Int. J. Comput. Vis. 48, 215–231 (2002)
13. Felzenswalb, P., Huttenlocher, D.: Distance transforms of sampled functions. Technical Report TR2004-1963. Dept. of Comp. Sci. Cornell University, Ithaca, NY (2004)

14. Rosenfeld, A.: A characterization of parallel thinning algorithms. Inform. Control 29, 286–291 (1975)
15. Cychosz, J.: Efficient binary image thinning using neighborhood maps. In: Graphics gems IV, pp. 465–473. Academic Press Professional, Inc, San Diego, CA, USA (1994)
16. Lam, L., Lee, S., Suen, C.Y.: Thinning methodologies-A comprehensive survey. IEEE Trans. Patt. Anal. Machine Intel. 14, 869–885 (1992)
17. Bunyak, F., Palaniappan, K., Nath, S.K., Baskin, T.I., Dong, G.: Quantitive cell motility for in vitro wound healing using level set-based active contour tracking. In: Proc. 3rd IEEE Int. Symp. Biomed. Imaging (ISBI), Arlington, VA, pp. 1040–1043. IEEE Computer Society Press, Los Alamitos (2006)
18. P.Soille: Morphological Image Analysis. 2 edn. Springer, New York, USA, ISBN 3-540-429883-0 (2004)

FyFont: Find-your-Font in Large Font Databases

Martin Solli and Reiner Lenz

Dept. Sci & Tech., Linköping University, 601 74 Norrköping, Sweden
marso@itn.liu.se reile@itn.liu.se

Abstract. A search engine for font recognition in very large font databases is presented and evaluated. The search engine analyzes an image of a text line, and responds with the name of the font used when writing the text. After segmenting the input image into single characters, the recognition is mainly based on eigenimages calculated from edge filtered character images. We evaluate the system with printed and scanned text lines and character images. The database used contains 2763 different fonts from the English alphabet. Our evaluation shows that for 99.8 % of the queries, the correct font name is one of the five best matches. Apart from finding fonts in large databases, the search engine can also be used as a pre-processor for Optical Character Recognition.

1 Introduction

When selecting a font for a text one often has an idea about the desired appearance of the font, but the font name is unknown. Sometimes we may have a text written in a similar but unknown font. In that case we would like to find out if that font is contained in a given database. Examples of databases can be the collection on our own personal computer, a commercial database belonging to a company, or a set with free fonts. In this paper we describe a new font recognition approach called Eigenfonts, based on eigenvectors and eigenvalues of images. The approach is closely related to Eigenfaces, used for face recognition, see Turk and Pentland [1]. We also introduce some improvements, mainly in the pre-processing step. In our experiments we use three different databases. First the original database containing character images created directly from font files. A few examples of images from this collection can be seen in Fig. 1. The second database contains images from the original database that are printed and scanned. The third contains images from unknown fonts, also printed and scanned. Database two and three are used for evaluation. In the following chapters when we refer to 'testdb1' or 'testdb2', we use collections of images from the second database. More about the font databases can be found in section 4.2. The search engine can also be utilized as a pre-processor for Optical Character Recognition, or as a font browser if retrieved images are used as queries.

The paper is organized as follows: In the next chapter some previous attempts in font recognition are described. In chapter 3 we present the design of the search engine FyFont (**F**ind-**y**our-**Font**), a publicly available search engine for free fonts[1]. This includes a description of the Eigenfonts method together

[1] http://media-vibrance.itn.liu.se/fyfont/

B.K. Ersbøll and K.S. Pedersen (Eds.): SCIA 2007, LNCS 4522, pp. 432–441, 2007.

Fig. 1. Examples of character a

with the design parameters. In chapter 4, we summarize the search engine and evaluate the overall performance. This will also include a description of the font databases, and the dependence on properties of the search image. Then we draw conclusions in chapter 5.

2 Background

There are two major application areas for font recognition or classification; as a tool for font selection, or as a pre-processor for OCR. A difference between these areas is the typical size of the font database. In a font selection task, we can have several hundreds or thousands of fonts, whereas in OCR systems it is usually sufficient to distinguish between less than 50 different fonts.

To our knowledge, only one search engine is available for font selection: WhatTheFont. The engine is commercially operated by MyFonts.com. An early description of the method can be found in [2]. Fonts are identified by comparing features obtained from a hierarchical abstraction of binary character images at different resolutions. Their database consisted of 1300 characters from 50 different fonts rendered at 100 pts/72 dpi. The best performance was achieved with a non-weighted kd-tree metric at 91% accuracy for a perfect hit. We are not aware of newer, publicly available, descriptions of improvements that are probably included in the commercial system.

Others used font classification as a pre-processor for OCR systems. There are mainly two approaches, a local approach based on features for individual characters or words, and a global approach using features for blocks of text. An example of the local approach describing font clustering and cluster identification in document images can be found in [3]. Four different methods (bitmaps, DCT coefficients, eigencharacters, and Fourier descriptors) are evaluated on 65 fonts. All result in adequate clustering performance, but the eigenfeatures result in the most compact representation. Another contribution to font recognition is [4], considering classification of typefaces using spectral signatures. A classifier capable of recognizing 100 typefaces is described in [5], resulting in significant improvements in OCR-systems. In [6] font attributes such as "serifness" and "boldness" where estimated in an OCR-system.

A technique using typographical attributes such as ascenders, descenders and serifs from word images is presented in [7]. These attributes are used as input to a neural network classifier for classifying seven different typefaces with different sizes commonly used in English documents. Non-negative matrix factorization (NMF) for font classification was proposed in [8]. 48 fonts were used in a hierarchical clustering algorithm with the Earth Mover Distance (EMD) as distance metric. In [9] clusters of words are generated from document images and then matched to a database of function words such as "and", "the" and "to". Font

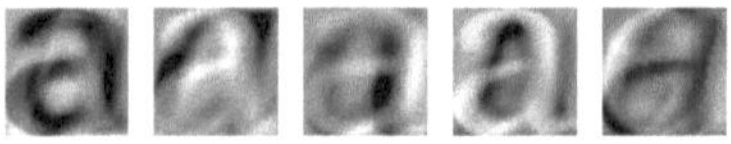

Fig. 2. First five eigenimages for character a (with normalized intensity values)

families are recognized in [10] through global page properties such as histograms and stroke slopes together with information from graph matching results of recognized short words such as "a", "it" and "of".

In a global approach [11], text blocks are considered as images containing specific textures, and Gabor filters are used for texture recognition. A recognition rate above 99% is achieved for 24 frequently used Chinese fonts and 32 frequently used English fonts. The authors conclude that their method is able to identify more global font attributes, such as weight and slope, but less appropriate to distinguish finer typographical attributes. Similar approaches with Gabor filters can be found in [12] and [13]. Also [14] describes an approach based on global texture analysis. Features are extracted using third and fourth order moments. 32 commonly used fonts in Spanish texts are investigated in the experiments.

Here we describe a new local approach based on the processing of character images.

3 Search Engine Design

In this chapter we introduce the eigenfonts method. We discuss important design parameters, like alignment and edge filtering of character images, and end with fine tuning the parameters of the system.

3.1 Eigenfonts Basics

We denote by $I(char, k)$ the kth gray value image (font) of character *char*. For instance, $I(a, 100)$ is the 100th font image of character 'a'. Images of characters from different fonts are quite similar in general; therefore images can be described in a lower dimensional subspace. The principal component analysis (or Karhunen-Loeve expansion) reduces the number of dimensions, leaving dimensions with highest variance. Eigenvectors and eigenvalues are computed from the covariance matrix of each character in the original database. The eigenvectors corresponding to the K highest eigenvalues describe a low-dimensional subspace on which the original images are projected. The coordinates in this low-dimensional space are stored as the new descriptors. The first five eigenimages for character 'a' can be seen in Fig. 2.

The method works as follows. Using the given images $I(char, k)$, for each character we calculate the mean image from different font images

$$m(char) = \frac{1}{N} \sum_{n=1}^{N} I(char, n) \tag{1}$$

After reshaping images to one column vectors, sets of images will be described by the matrix $I(char) = (I(char, 1), ..., I(char, N))$. From each set we subtract the mean image and get $\widehat{I}(char) = I(char) - m(char)$. The covariance matrix is then given by $C(char) = \widehat{I}(char)\widehat{I}(char)'/N$, and the eigenvectors u_k, corresponding to the K largest eigenvalues λ_k, are computed. The corresponding coefficients are used to describe the image. A query image is reshaped to a vector Q, and for $k = 1, ..., K$ its weights $\omega_1, ..., \omega_K$ are computed as: $\omega_k = u_k'(Q - m)$. The weights form a vector that describes the representation of the query image in the eigenfont basis. Using the eigenfonts approach requires that all images of a certain character are of the same size and have the same orientation. We also assume that they have the same color (black letters on a white paper background is the most obvious choice). We therefore apply the following pre-processing steps before we compute the eigenfont coefficients: 1) Grey value adjustments: If the character image is a color image, color channels are merged, and then gray values are scaled to fit a pre-defined range. 2) Orientation and segmentation: If character images are extracted from a text line, the text line is rotated to a horizontal position prior to character segmentation. 3) Scaling: Character images are scaled to the same size.

3.2 Character Alignment and Edge Filtering

Since we are using the eigenfonts method, images must have the same size, but the location of the character within each image can vary. We consider two choices: characters are scaled to fit the image frame exactly, or characters are aligned according to their centroid value, leaving space at image borders. The later requires larger eigenfont images since the centroid value varies between characters, which will increase the computational cost. Experiments showed that frame alignment gives significantly better retrieval accuracy than centroid alignment.

Most of the information about the shape of a character can be found in the contour, especially in this case when we are dealing with black text on a white paper background. Based on this assumption, character images were filtered with different edge filters before calculating eigenimages. The images used are rather small and therefore we use only small filter kernels (max 3×3 pixels). Extensive experiments with different filter kernels (reported elsewhere) resulted in the following filters used in the final experiments (in Matlab notation): H=[1 2 1;0 0 0;-1 -2 -1]; V=[1 0 -1;2 0 -2;1 0 -1]; D1=[2 1 0;1 0 -1;0 -1 -2]; D2=[0 1 2;-1 0 1;-2 -1 0]; D3=[-1 0;0 1]; D4=[0 -1;1 0]; Four diagonal filters, one horizontal and one vertical edge filter. Retrieval results for character 'a', filtered with different filters can be seen in Table 1. PERFECT corresponds to the percentage of when the correct font is returned as the best match, and TOP5 when the correct font can be found within the five best matches. The same notation will be used in the rest of this paper. When several filters were used, we filtered the image with each filter separately, and then added the resulting images.

The table shows that the combination of one horizontal and two diagonal filters gives the best result. Some experiments with varying image sizes showed that images of size 25×25 pixels seem to be a good choice. Sizes below 15×15

Table 1. Retrieval accuracy for different filter combinations. (Character 'a' from testdb1, image size 40 × 40 pixels).

Filter	PERFECT	TOP5	Filter comb.	PERFECT	TOP5
H	73	95	H+D1+D2	**77**	**97**
V	57	83	H+D1	75	94
D1	69	97	H+D2	63	96
D2	66	94	H+V	73	98
D3	66	86			
D4	56	87			

Table 2. Retrieval accuracy for different filter combinations, for character 'd', 'j', 'l', 'o', 'q' and 's' from testdb2. (P=PERFECT, T5=TOP5).

	d		j		l		o		q		s	
Filter	P	T5	P	T5	P	T5	P	T5	P	T5	P	T5
H	88	100	**86**	**99**	72	82	79	97	91	100	**92**	**100**
V	86	98	70	94	58	78	81	99	87	99	**91**	**100**
D1	**90**	**100**	82	98	64	85	81	97	91	100	**92**	**100**
D2	**91**	**100**	80	98	66	84	**84**	**99**	92	100	**91**	**100**
H+V	89	100	82	98	68	85	**85**	**99**	**95**	**100**	**91**	**100**
H+V+D1+D2	88	100	80	99	66	85	**82**	**98**	**93**	**100**	**91**	**100**
H+D1+D2	**90**	**100**	**85**	**98**	**72**	**88**	**83**	**98**	**93**	**100**	**91**	**100**
V+D1+D2	88	99	79	94	59	82	**84**	**98**	89	100	**91**	**100**
D1+D2	89	100	79	97	65	84	**85**	**98**	**93**	**100**	**92**	**100**
H+D1	89	100	**86**	**99**	**75**	**89**	80	97	**93**	**100**	**91**	**100**
H+D2	**90**	**100**	**85**	**99**	**72**	**88**	83	97	91	100	90	100

pixels decreased the retrieval accuracy significantly. To verify the result from character 'a', a second test was carried out with characters 'd', 'j', 'l', 'o', 'q' and 's', from testdb2. The result for different combinations of filters can be seen in Table 2. The retrieval results vary slightly between different characters, but usually filter combinations "H+D1+D2" and "H+D1" perform well. We choose the first combination, a horizontal Sobel filter together with two diagonal filters. The vertical filter does not improve the result, probably because many characters contain almost the same vertical lines.

3.3 Selection of Eigenimages

The selection of eigenimages is a tradeoff between accuracy and processing time and is critical for search performance. Increasing the number of eigenimages used leads first to an increased performance but the contribution of eigenimages with low eigenvalues is negligible. The number of eigenimages and retrieval performance for scanned and printed versions of character 'a' (size 24 × 24 pixels, edge filtered) can be seen in Fig. 3. The figure shows that 30 to 40 eigenimages are

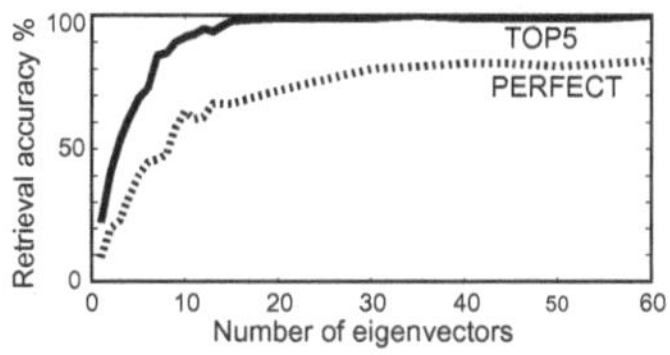

Fig. 3. Retrieval performance for different number of eigenvectors (eigenimages)

appropriate for character 'a'. Preliminary tests were carried out with other image sizes, and other characters, and most of them show that using 40 eigenimages is sufficient.

3.4 Additional Search Engine Components

We continue by analyzing the influence of image size and interpolation methods. As mentioned in chapter 3.2, a suitable image size for character 'a' is 24×24 pixels. However, quadratic images is not suitable for characters like 'l' and 'i', therefore we use rectangular eigenimages for "rectangular characters". As an example, the perfect match accuracy for character 'b' (considered as a "rectangular character") increased from 77 to 81 % when the image size changed from 24×24 to 24×20 pixels (rows×columns). Scaling requires interpolation, and the influence of the interpolation method on the search performance was evaluated for three common interpolation techniques: Nearest neighbor, bilinear and bicubic interpolation. The result showed that the interpolation method is of minor importance as long as something more advanced than nearest neighbor is used.

We also evaluated if features not derived from eigenimages could improve retrieval performance. We evaluated the influence of the ratio between character height and width (before scaling), the ratio between the area of the character and the area of the surrounding box, and center of gravity (centroid) values. The only extra feature that resulted in significant improvements is the ratio between height and width. However, it is important that the value is weighted properly.

Similarity between feature vectors is calculated with the L_2 norm, or Euclidean distance. We also tested other distance measures (L_1, and Mahalanobis with different covariance matrices), but the L_2 norm gave the best result. This might be related to the fact that eigenimages are calculated with Principal Component Analysis, which is defined as the minimizer of the L_2 approximation error.

4 Result

In this chapter the overall results are presented. We also describe our experiments investigating the effects of different pre-processing methods and noise sensitivity.

4.1 Final Configuration of the Search Engine

The components of the search engine and its internal parameters are as follows:

* **Image scaling:** Square images, size 24 × 24 pixels, for "square characters" like a, e, and o. Rectangular images, for example 24 × 20 pixels, for "rectangular characters" like l, i, and t. Align image borders for characters instead of center of gravity. Scaling with bilinear interpolation.
* **Edge filtering:** Three Sobel filters, one horizontal and two diagonal.
* **Number of eigenimages:** 40
* **Extra features:** Ratio between character height and width before scaling.
* **Distance metric:** L_2 norm (Euclidean distance)

4.2 Font Databases

The original database contains 2763 different fonts. Single characters are represented by images, typically around 100 pixels high. For evaluation, three test databases were created. The first contains images from the original database that are printed in 400 dpi with an ordinary office laser printer, and then scanned in at 300 dpi with an ordinary desktop scanner (HP Scanjet 5590, default settings). As a first step characters 'a' and 'b' from 100 randomly selected fonts were scanned (testdb1). For the second database (testdb2) the same 100 fonts were used, but this time all small characters were scanned, giving totally 2600 images. These images are used in the evaluation of the final search engine. The third test database also adopts the print and a scan procedure, as mentioned above, but with fonts that are not in the database. Only seven fonts were used, all of them downloaded from dafont (www.dafont.com). Both fonts with an "ordinary look", and fonts with unusual shapes were used.

4.3 Overall Results

The search performance for different characters from testdb2 is shown in Table 3. The mean values for a perfect match and a top 5 result is 88,2 and 99,1 %. Characters like 'l' and 'i' that contain relatively few lines and details that can be used for distinguishing fonts from each other decrease the mean values rather significantly. Without 'l' and 'i', the mean values increase to 89,2 and 99,8 %. Since the input to the search engine is a text line, tricky characters like 'l' can be removed or weighted down, and usually the remaining characters will be sufficient for producing an accurate result.

4.4 Image Resolution and Rotation Influence

Experiments based on image resolution/scaling, JPEG compression and character rotation are presented in this section. Fig. 4 shows the relationship between query image size and search performance. In the figure we observe that a query image height below 40 pixels will decrease the retrieval performance significantly. We also evaluated the correlation between different JPEG compression rates and

Table 3. Search performance for different characters. (PE=PERFECT, T5=TOP5).

Character	PE	T5	Character	PE	T5	Character	PE	T5	Character	PE	T5
a	94	100	h	88	100	o	83	98	v	89	99
b	90	100	i	82	94	p	88	100	w	91	99
c	89	99	j	85	99	q	93	100	x	90	100
d	90	100	k	86	100	r	86	100	y	87	100
e	91	100	l	71	88	s	91	100	z	90	100
f	89	100	m	91	100	t	90	100			
g	88	100	n	91	100	u	91	100			
									MEAN	88.2	99.1

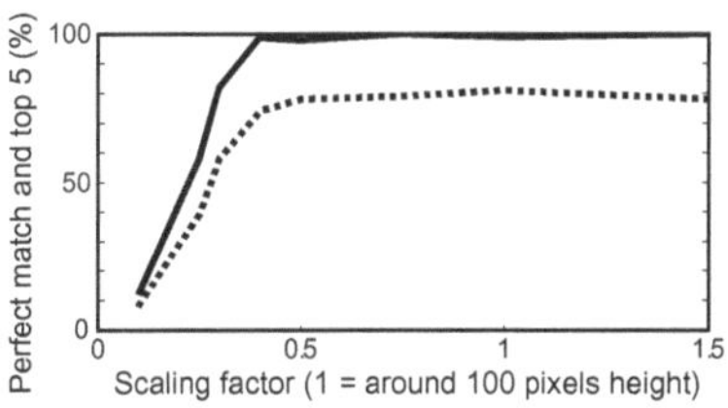

(a) The relationship between query image size and search performance.

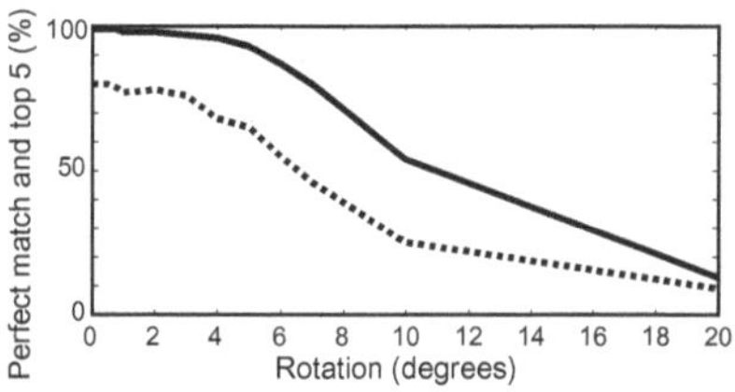

(b) The relationship between query image rotation and search performance.

Fig. 4. Image resolution and rotation influence. Dashed line corresponds to perfect match, solid line corresponds to a top 5 result. (Character 'a' from testdb1).

search performance. In JPEG compression, the quality can be set between 0 and 100, where 100 corresponds to the best quality (lowest compression). The result showed that only when the quality is below 50 the retrieval performance is affected. Finally, the relationship between query image orientation (rotation) and search performance can be seen in Fig. 4. When input images are rotated more than 5 degrees, the performance declines sharply. However, an angle around or over 5 degrees is rarely encountered in real samples. In our experience rotation angles are below 1 degree after pre-processing.

4.5 An Example of a Complete Search

This section illustrates a complete search from input image to the font name output. An example of an input image can be seen in the top left part of Fig. 5. First edge filtering and the Hough transform are used to automatically rotate

Fig. 5. Top left: Example of an input image. Bottom left: Input image after rotation. Right: 12 first sub images after segmentation.

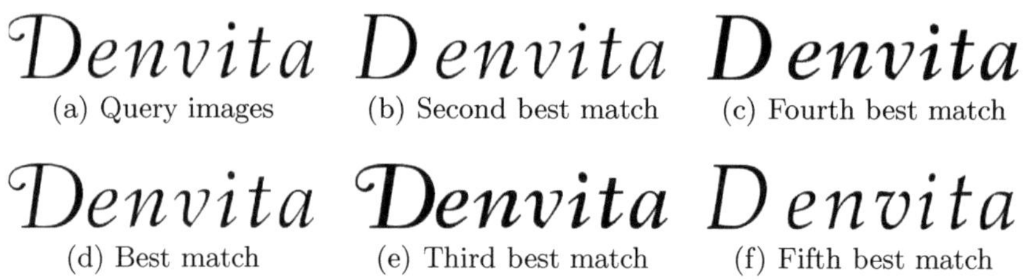

Fig. 6. (a) Query images (b)-(f) The five most similar fonts in the database

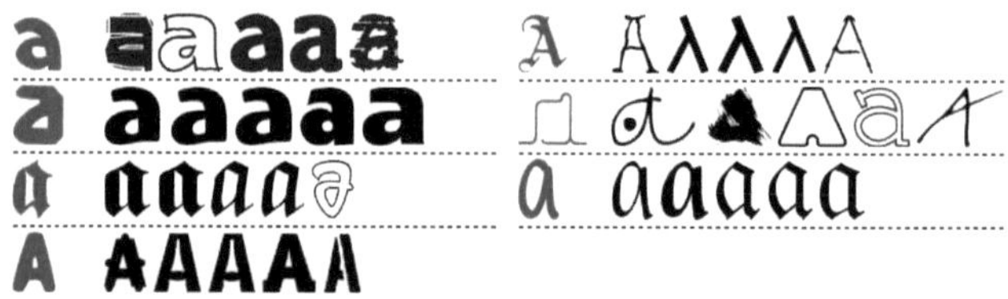

Fig. 7. Search result for character 'a' from seven fonts not present in the database. The search image to the left, followed by the five best matches.

the text line to a horizontal position. The result can be seen in the bottom left part of Fig. 5. Then the text line is segmented into sub images. The first 12 sub images are shown in the right side of Fig. 5. The user will then assign letters to sub images that are to be used in the database search. Since the segmentation algorithm did not manage to segment 'k' and 'j', the user should not assign a letter to this sub image. The character segmentation can be improved, but since we are primarily interested in font recognition we did not spend time on developing a more sophisticated segmentation method. See [15] for a survey of character segmentation. Assigned images will be used as input to the search engine. Results from individual characters are weighted and combined to a final result, presenting the most similar fonts in the database. Fig. 6 shows the five most similar fonts for the query image. In this case the first seven characters were assigned letters and used as input.

Ending this section we describe experiments with query images created with fonts not in the database. For these images the retrieval accuracy can't be measured, the result must be evaluated visually. Fig. 7 shows seven query images from fonts not present in the database, together with the five best matches.

5 Conclusions

A search engine for very large font databases has been presented and evaluated. The input is an image with a text from which the search engine retrieve the name of the font used to writing the text. Apart from font retrieval, the search engine can also be used as a pre-processor to OCR systems. In an OCR context, the method would be classified as a local approach since features are calculated for individual characters. The recognition is mainly based on eigenimages calculated from character images filtered with three edge filters. Even for the very large font database, containing 2763 fonts, the retrieval accuracy is very high. For individual characters, the mean accuracy for a perfect match is 89,2 %, and the probability to find the correct font name within the five best matches is 99,8 %. In practice, the overall accuracy will increase since the search engine will work with text lines, giving the opportunity to combine the result from many characters. To resemble a real life situation, retrieval experiments were made with printed and scanned text lines and character images from the original database.

References

1. Turk, M., Pentland, A.: Eigenfaces for recognition. Journal of Cognitive Neuroscience 3(1), 71–86 (1991)
2. Sexton, A., Todman, A., Woodward, K.: Font recognition using shape-based quad-tree and kd-tree decomposition. In: JCIS 2000, pp. 2: 212–215 (2000)
3. Östürk, S., Sankur, B., Abak, A.T.: Font clustering and cluster identification in document images. Journal of Electronic Imaging 10(2), 418–430 (2001)
4. Morris, R.A.: Classification of digital typefaces using spectral signatures. Pattern Recognition 25(8), 869–876 (1992)
5. Baird, H.S., Nagy, G.: Self-correcting 100-font classifier. In: Proceedings of SPIE, vol. 2181, pp. 106–115 (1994)
6. Cooperman, R.: Producing good font attribute determination using error-prone information. In: SPIE, vol. 3027, pp. 50–57 (1997)
7. Jung, M., Shin, Y., Srihari, S.: Multifont classification using typographical attributes. In: ICDAR '99, p. 353. IEEE Computer Society Press, Los Alamitos (1999)
8. Lee, C.W., Jung, K.C.: NMF-based approach to font classification to printed english alphabets for document image understanding. Modeling Decisions for Artificial Intelligence, Proceedings 3558, 354–364 (2005)
9. Khoubyari, S., Hull, J.J.: Font and function word identification in document recognition. Computer Vision and Image Understanding 63(1), 66–74 (1996)
10. Shi, H., Pavlidis, T.: Font recognition and contextual processing for more accurate text recognition. In: ICDAR '97, pp. 39–44. IEEE Computer Society, Los Alamitos (1997)
11. Zhu, Y., Tan, T.N., Wang, Y.H.: Font recognition based on global texture analysis. IEEE T-PAMI 23(10), 1192–1200 (2001)
12. Ha, M.H., Tian, X.D., Zhang, Z.R.: Optical font recognition based on Gabor filter. In: Proc. 4th Int. Conf. on Machine Learning and Cybernetics (2005)
13. Yang, F., Tian, X.D., Guo, B.L: An improved font recognition method based on texture analysis. In: Proc. 1st Int. Conf. on Machine Learning and Cybernetics (2002)
14. Avilés-Cruz, C., Rangel-Kuoppa, R., Reyes-Ayala, M., Andrade-Gonzalez, A., Escarela-Perez, R.: High-order statistical texture analysisfont recognition applied. Pattern Recognition Letters 26(2), 135–145 (2005)
15. Casey, R.G., Lecolinet, E.: A survey of methods and strategies in character segmentation. IEEE T-PAMI 18(7), 690–706 (1996)

Efficiently Capturing Object Contours for Non-Photorealistic Rendering

Jiyoung Park[1] and Juneho Yi[2]

[1] Computer Graphics Research Team, Digital Content Research Division, Electronics and Telecommunications Research Institute, Daejeon 305-700, Korea
jiyp@etri.re.kr
[2] School of Information and Communication Engineering, Sungkyunkwan University, Suwon 446-740, Korea
jhyi@ece.skku.ac.kr

Abstract. Non-photorealistic rendering (NPR) techniques aim to outline the shape of objects and reduce visual clutter such as shadows and inner texture edges. As the first phase result of our entire research, this work is concerned with a structured light based approach that efficiently detects depth edges in real world scenes. Depth edges directly represent object contours. We exploit distortion of the light pattern in the structured light image along depth discontinuities to reliably detect depth edges. However, in reality, distortion along depth discontinuities may not occur or be large enough to detect depending on the distance from the camera or projector. For practical application of the proposed approach, we have presented a novel method that guarantees the occurrence of the distortion along depth discontinuities for a continuous range of object location. Experimental results show a great promise that the technique can successfully provide object contours to be used for non-photorealistic rendering.

Keywords: depth edges, structured light, non-photorealistic rendering.

1 Introduction

Depth edges directly outline shape contours of objects [2-5]. Unfortunately, there have been reported few research results that only provide depth discontinuities without computing 3D information at every pixel in the input image of a scene. On the other hand, most effort has been devoted to stereo vision problems in order to obtain depth information. In fact, stereo methods for 3D reconstruction would fail in textureless regions and along occluding edges with low intensity variation [6, 7]. Recently, the use of structured light was reported to compute 3D coordinates at every pixel in the input image [8, 9]. However, the fact that this approach needs a number of structured light images makes it hard to be applicable in realtime.

One notable technique was reported recently for non-photorealistic rendering [10]. They capture a sequence of images in which different light sources illuminate the scene from various positions. Then they use shadows in each image to assemble a depth edge map. This technique was applied to stylized rendering highlighting boundaries between

B.K. Ersbøll and K.S. Pedersen (Eds.): SCIA 2007, LNCS 4522, pp. 442–451, 2007.

geometric shapes. Although very attractive, it only works where shadows can be reliably created. In contrast, our method is shadow free. In addition, by a slight modification of the imaging system so that it can capture white and structured images at the same time, it can be easily applied to dynamic scenes where the camera moves.

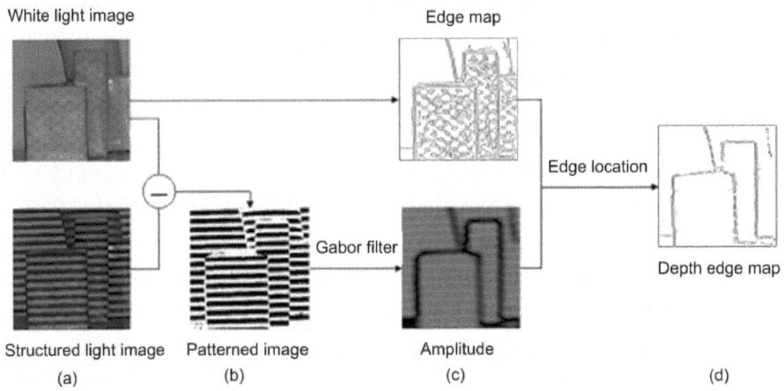

Fig. 1. Illustration of the basic idea to compute a depth edge map: (a) capture of a white light image and structured light image, (b) patterned image, (c) detection of depth edges by applying a Gabor filter to the patterned image with edge information from the white light image, (d) final depth edge map

The eventual goal of this research is to produce a depth edge map that conveys object contours to be used for non-photorealistic rendering. This work describes the first stage research result that is concerned with reliably capturing depth edges. We present a structured light based framework for reliably capturing depth edges in real world scenes without dense 3D reconstruction.

We have illustrated in Fig. 1 the basic idea that detects depth edges. First, as can be seen in Fig. 1 (a), we project a white light and structured light in a row onto a scene where depth edges are to be detected. The structured light contains a special light pattern. In this work, we have used simple black and white horizontal stripes with the same width. Vertical stripes can be used with the same analysis applied to horizontal stripes. We capture the white light image and structured light image. Second, we extract horizontal patterns simply by differencing the white light and structured light images. We call this difference image 'patterned image'. Refer to Fig. 1 (b). Third, we identify depth edges in the patterned image guided by edge information from the white light image. We exploit distortion of light pattern in the structured light image along depth edges. Since the horizontal pattern can be considered a periodic signal with specific frequency, we can easily detect candidate locations for depth edges by applying a Gabor filter to the patterned image [11]. The amplitude response of Gabor filter is very low where distortion of light pattern occurs. Fig. 1 (c) illustrates this process. Last, we accurately locate depth edges using edge information from the white light image, yielding a final depth edge map as in Fig. 1 (d).

However, distortion along depth discontinuities may not occur or be sufficient to detect depending the distance from the camera or projector. For practical application of the proposed approach, it is essential to have a solution that guarantees the occurrence of the distortion along depth discontinuities irrespective of object location. Fig. 2 shows an example situation. Along the depth edges between objects A and B, C and D, the distortion of pattern almost disappears. This makes it not feasible to detect these depth edges using a Gabor filter.

We propose a method based on a single projector-camera system that guarantees the occurrence of the distortion for a continuous range of object location. Based on a modeled imaging geometry of camera, projector, object, and its mathematical analysis, we first compute the exact ranges of object location where detection of distortion is not feasible. We simply use several structural light images with different width of horizontal stripes. We have used a general purpose LCD projector, however, an infrared projector can be employed with the same analysis in order to apply the method to humans. Experimental results have confirmed that the proposed method works very well for shapes of human hand and body as well as general objects.

The remaining of this paper is organized as follows. In section 2, we describe the application of Gabor filter to detect depth edges in a patterned image. Section 3 present our method that guarantees the occurrence of the distortion along depth discontinuities for a continuous range of object location. We report our experimental results in section 4. Finally, conclusions and future work are discussed in section 5.

(a) (b) (c)

Fig. 2. Problem of disappearance of distortion along depth edges depending on the distance of an object from the camera and projector: (a) white light image (b) patterned image (c) amplitude response of Gabor filter. Along the depth edges between objects A and B, C and D, in the pattered image (b), the distortion of pattern almost disappears. This makes it not feasible to detect these depth edges using a Gabor filter.

2 Detecting Depth Edges

We detect depth edges by projecting structured light onto a scene and exploit distortion of light pattern in the structured light image along depth discontinuities. In order to exploit distortion of light pattern, we use 2D Gabor filtering that is known to be useful in segregating textural regions [1, 12]. We first find candidate depth edges by applying a Gabor filter to the patterned image. We then accurately locate depth edges using edge information from white light image.

2.1 The Use of Gabor Filter

Since a horizontal pattern can be considered a spatially periodic signal with specific frequency, we can easily detect candidate locations for depth edges by applying a

Gabor filter to the patterned image. A 2-D Gabor filter is an oriented complex sinusoidal grating modulated by a 2-D Gaussian function, which is given by

$$G_{\sigma,\phi,\theta}(x,y)=g_{\sigma}(x,y)\cdot\exp[2\pi j\phi(x\cos\theta+y\sin\theta)] \tag{1}$$

$$\text{where} \quad g_{\sigma}=\frac{1}{2\pi\sigma^{2}}\exp\left[-\left(x^{2}+y^{2}\right)/2\sigma^{2}\right].$$

The frequency of the span-limited sinusoidal grating is given by ϕ and its orientation is specified as θ. $g_{\sigma}(x,y)$ is the Gaussian function with scale parameter σ. Decomposing $G_{\sigma\phi\theta}(x,y)$ into real and imaginary parts gives

$$G_{\sigma,\phi,\theta}(x,y)=R_{\sigma,\phi,\theta}(x,y)+jI_{\sigma,\phi,\theta}(x,y) \tag{2}$$

$$\text{where} \quad \begin{aligned} R_{\sigma,\phi,\theta}(x,y)&=g_{\sigma}(x.y)\cdot\cos[2\pi\phi(x\cos\theta+y\sin\theta)] \\ I_{\sigma,\phi,\theta}(x,y)&=g_{\sigma}(x.y)\cdot\sin[2\pi\phi(x\cos\theta+y\sin\theta)]. \end{aligned}$$

The Gabor filtered output of an image $f(x,y)$ is obtained by the convolution of the image with the Gabor function $G_{\sigma\phi\theta}(x,y)$. Thus, its amplitude response can be computed as follows:

$$E_{\sigma,\phi,\theta}(x,y)=\sqrt{[R_{\sigma,\phi,\theta}(x,y)*f(x,y)]^{2}+[I_{\sigma,\phi,\theta}(x,y)*f(x,y)]^{2}} \tag{3}$$

2.2 Referring to Edges in White Light Image

It is possible to accurately locate depth edges by combining the Gabor filter output and edge information from the white light image. In this work, we use a gradient based technique to detect edges in the white light image, although other methods could also be applied. Fig. 3 (b) represents the gradient magnitude of the intensity along the line in the white light image in Fig. 3(a). We only take the gradient magnitude where Gabor amplitude has low values. Fig. 3 (c) illustrates the regions of low Gabor amplitude. The accurate locations of depth edges are obtained by finding peak points of the gradient magnitude in these regions (see Fig. 3 (d)).

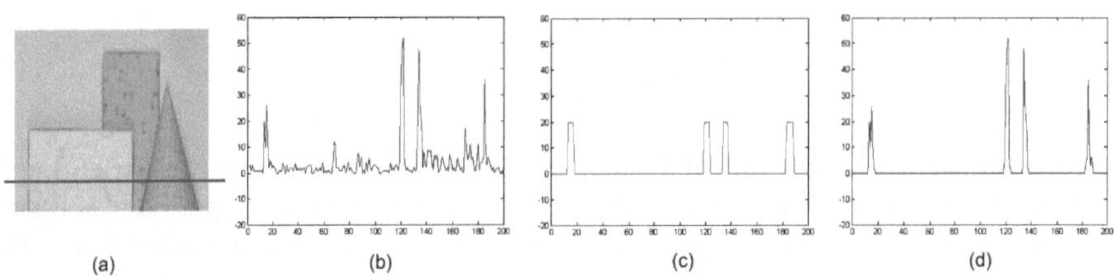

Fig. 3. Location of depth edges using edge information from the white light image: (a) white light image, (b) gradient magnitude along the line in (a), (c) regions of low Gabor amplitude (d) location of depth edges

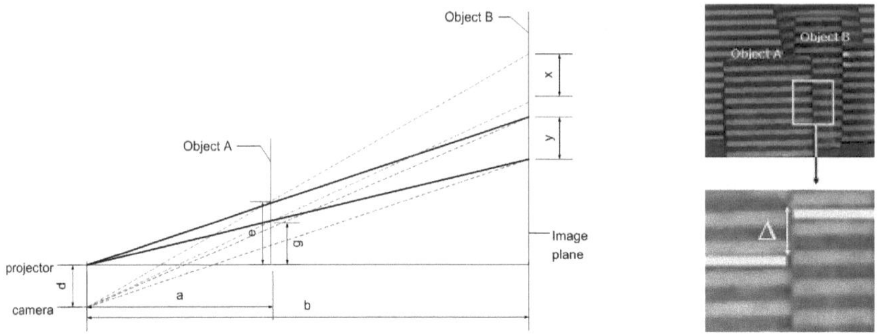

(a) (b)

Fig. 4. Imaging geometry and the amount of distortion: (a) spatial relation of camera, projector and two object points viewed from the side, : disparity in the image plane of the same horizontal stripe projected onto different object points, (b) the magnitude of pattern distortion, , in a real image

3 Detectable Range of Depth Edges

We have described that we can easily detect depth edges by exploiting the distortion along depth discontinuities in the patterned image. However, as previously mentioned, the distortion may not occur or be sufficient to detect depending the distance of depth edges from the camera or projector. In this section, we present a method to guarantee the occurrence of the distortion for a continuous range of object location.

3.1 Reliably Detectable Distortion

In order to compute the exact range where depth edges can be detectable, we have modeled imaging geometry of camera, projector and object as illustrated in Fig. 4. The solid line represents a light ray from the projector. When structured light is projected onto object points A and B, they are imaged at different locations in the image plane due to different depth values. That is, distortion of horizontal pattern occurs along the depth discontinuity. The amount of distortion is denoted by Δ . From this model, we can derive the following equation using similar triangles:

$$\Delta = fd\left(\frac{1}{a} - \frac{1}{b}\right) = \left(\frac{fdr}{a(a+r)}\right) \tag{4}$$

However, the exact amount of Δ may not be measurable because we use simple black and white stripes with equal width and the amount of offset is periodic as it gets large. In order for a depth edge to be detectable by applying a Gabor filter, the measurable Δ , in the image plane should be above a certain amount. We have confirmed through experiments that 2/3 of the width of a horizontal stripe used is necessary for reliable detection of the distortion. Thus, the range of Δ for reliable detection of pattern distortion can be written as in equation (5).

$$2wk + \frac{2w}{3} \le \Delta \le 2wk + \frac{4w}{3}, \quad k = 0,1,\cdots. \tag{5}$$

where w denotes the width of horizontal stripes. From equation (5), given the distance, r, between two object points and the separation, d, between the camera and the projector, we can compute the exact range of the foreground object point, A, from the camera where reliable detection of distortion is guaranteed. For practical application of the proposed approach, we need to guarantee the occurrence of distortion for a continuous range. Note that the width of horizontal stripes projected onto object locations A and B are the same in the image plane although they have different depth values. This is because the perspective effect of the camera and projector are canceled each other out.

3.2 Extending the Detectable Range of Depth Edges

We use several structured light images with different width of horizontal stripes to extend the range where detection of distortion can be guaranteed. Fig. 5(a) depicts the relationship between Δ and a for any k and $k+1$. The marked regions in the horizontal axis, a, represent the ranges of the foreground object point, A, from the camera that correspond to reliably detectable distortion Δ in the vertical axis. We can see that there exist ranges where we cannot detect depth edges due to the lack of distortion depending the distance of a depth edge from the camera or projector. Therefore, to extend the range of detectable distortion, Δ, we use additional structured light whose spatial period is halved such as $w_2=2w_1$, $w_3=2w_2$, $w_4=2w_3$, $\cdots$, as shown in Fig. 5(b). When n such structured light images are used, the range of detectable distortion, Δ, is expressed as follows.

$$\frac{2}{3}w_1 < \Delta < \frac{(n^2+3n)}{3}w_1 \tag{6}$$

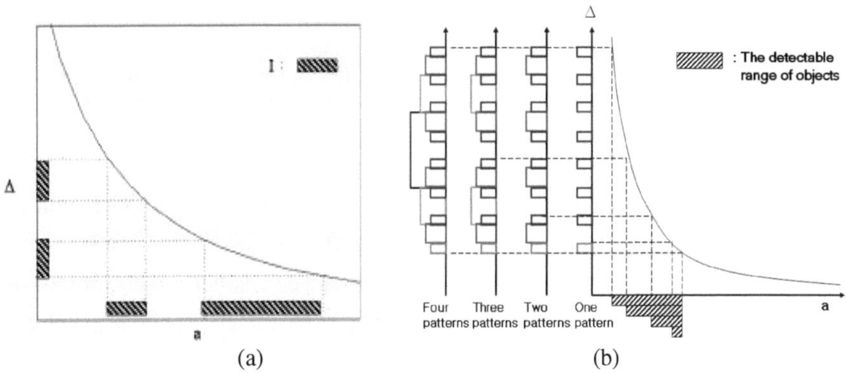

Fig. 5. The detectable range of depth edges (a) can be increased like (b) by projecting additional structured light with different width of stripes

3.3 Computation of the Detectable Range of Depth Edges

As shown in Fig. 6 (a), the detectable range of depth edges $[a_{\min}, a_{\max}]$ is computed in the following two steps: (1) Setting the maximum distance of the detectable range, $a_{\max}$, and the minimum distance between object points, r_{min}, determines the width of stripes, w, in structured light image. r_{min} can be set to different values depending on applications. (2) This w gives the minimum distance of the detectable range, a_{min}, resulting in the detectable range of depth edges, $[a_{min}, a_{max}]$.

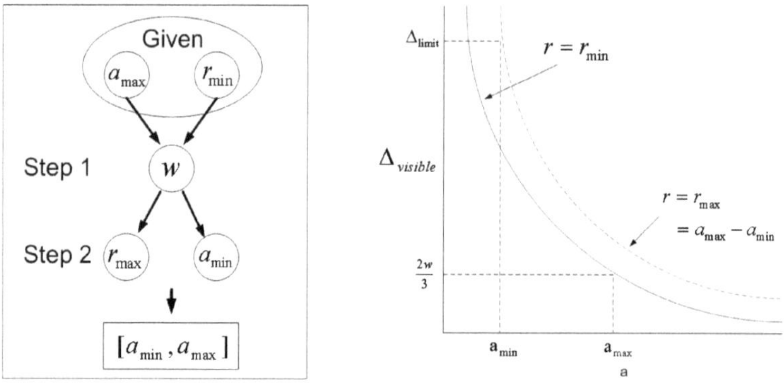

(a) (b)

Fig. 6. Computation of the detectable range of depth edges: (a) computation process of $[a_{min}, a_{max}]$, (b) computation of a_{min}

Step 1: Determination of the width of a stripe, w, in the structured light
First, we set a_{max} to the distance from the camera to the farthest background. Given r_{min}, w can be computed by equation (7) which is derived from equation (4).

$$w = \frac{3fd_1 r_{min}}{2a_{max}(a_{max} + r_{min})} \tag{7}$$

Thus, given a_{max} and r_{min}, we can compute the optimal width of stripes of the structured light.

Step 2: The minimum of the detectable range, a_{min}
Given w from step 1, we can compute a_{min} that corresponds to the upper limit of Δ, Δ_{limit}, as shown in Fig. 6 (b). After determining Δ_{limit} and r_{max}, a_{min} can be computed by equation (9). r_{max} denotes the maximum distance between object points in the range $[a_{min}, a_{max}]$ that guarantees the occurrence of the distortion along depth discontinuities. Clearly, the distance between any two object points is bounded by $(a_{max}-a_{min})$, i. e., $r_{max}= a_{max}-a_{min}$. Thus Δ_{limit} becomes:

$$\Delta_{limit} = \frac{fdr_{max}}{a_{min}(a_{min} + r_{max})} = \frac{fd(a_{max} - a_{min})}{a_{max} a_{min}}. \tag{8}$$

On the other hand, $\Delta_{\text{limit}} = \frac{n^2+3n}{3} w_1$ from equation (6). Solving for $a_{\min}$ gives:

$$a_{\min} = \frac{fd_1 a_{\max}}{fd_1 + \frac{(n^2+3n)w_1}{3} a_{\max}}. \tag{9}$$

This way, we can employ structured light of optimal spatial resolution so that we are guaranteed to detect depth edges of all object points located in the range $[a_{\min}, a_{\max}]$, and apart from each other no less than $r_{\min}$ and no more than $r_{\max}$.

4 Experimental Results

For capturing structured light images, we have used a HP xb31 DLP projector and Cannon IXY 500 digital camera. Fig. 7 shows the result of depth edge detection using three structured light images with different width of horizontal stripes. Fig. 7 (a) and (b) display the front and side views of the scene, respectively. All the objects are located within the range of 2.4m ~ 3m from the camera. Setting f=3m, d=0.173m, $a_{\max}$ =3m and $r_{\min}$ =0.1m, w_1 and $a_{\min}$ are determined as 0.0084m and 2.325m, respectively. That is, the detectable range of depth edges becomes [2.325m, 3m] and the length of the range is 0.675m. Thus, the widths of stripes of the three structured light that guarantee the detection of depth edges in this range are w_1, $2w_1$ and $4w_1$.

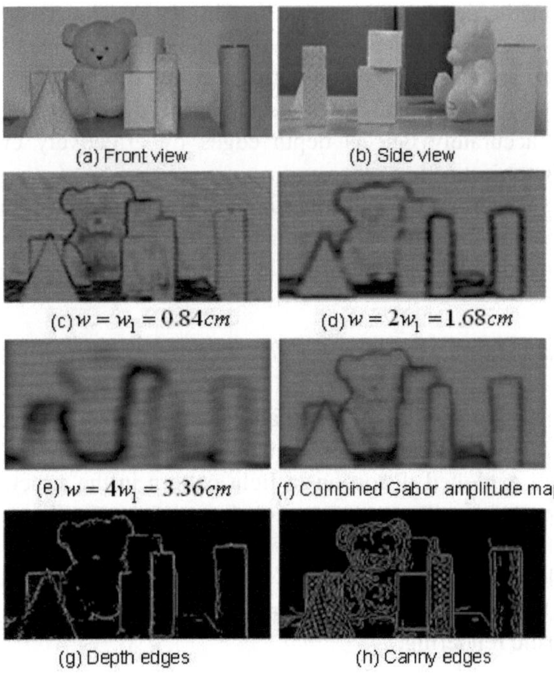

Fig. 7. Detecting depth edges using a single camera and projector

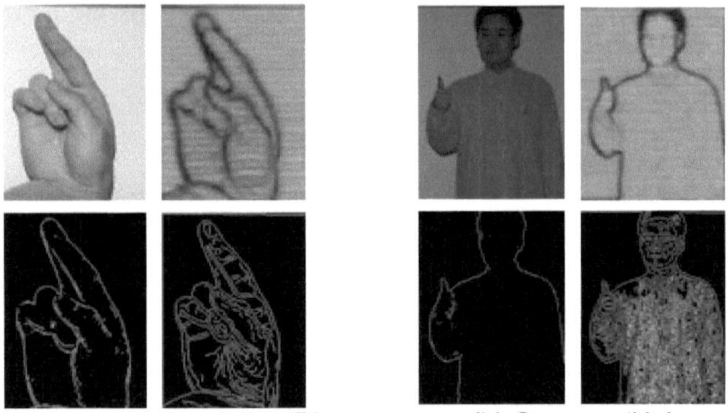

Fig. 8. (a) Detection of depth edges in the case of hand configurations (b) Detecting human body contours. Clockwise from top left: white light image, Gabor amplitude map, depth edges and canny edges.

Fig. 7 (c)~(e) show Gabor amplitude maps in the three cases. Each Gabor amplitude map shows that we cannot detect all the depth edges in the scene using a single structured light image. However, combining the results from the three cases, we can obtain the final Gabor amplitude map as in Fig. 7 (f) where distortion for detection is guaranteed to appear along depth discontinuities in the range of [2.325m, 3m]. Finally, we can get the depth edge map as in Fig. 7 (g). The result shows that this method is capable of detecting depth edges of all the objects located in the detectable range. We have also compared the result with the output of the traditional Canny edge detector (Fig. 7 (h)) where cluttered inner texture edges are also detected. The proposed method accurately detects depth edges by effectively eliminating inner texture edges of the objects.

Fig. 8 (a) shows detection of depth edges in the case of hand configuration. We have also applied our method to human body scenes. Fig. 8 (b) shows the result of detecting of human body contours. Our method effectively suppress inner texture details.

5 Conclusions

We have presented a structured light based approach for effectively suppressing inner texture details with only depth edges contained. We have strategically projected structured light and exploited distortion of light pattern in the structured light image along depth discontinuities. Through a modeled imaging geometry and mathematical analysis, we have also described a method that can guarantee the occurrence of the distortion along depth discontinuities for a continuous range of object location. The second phase of the research using the proposed method is currently under progress for non-photorealistic rendering.

Acknowledgement

This work was supported in part by the Korea Science and Engineering Foundation (KOSEF) through the Biometrics Engineering Research Center (BERC) at Yonsei University.

References

1. Bovik, A.C., Clark, M., Geisler, W.S.: Multichannel Texture Analysis Using Localized Spatial Filters. IEEE Trans. on Pattern Analysis and Machine Intelligence, pp. 55-73 (1990)
2. Cass, T.A.: Robust Affine Structure Matching for 3D Object Recognition. IEEE Trans. on Pattern Analysis and Machine Intelligence, pp. 1264-1265 (1998)
3. Chen, Y., Medioni, G.: Object Modelling by Registration of Multiple Range Image. Image and Vision Computing, pp. 145-155 (1992)
4. Loncaric, S.: A survey of shape analysis techniques. Pattern Recognition, pp. 983-1001 (1998)
5. Weiss, I., Ray, M.: Model-based recognition of 3D object from single vision. IEEE Trans. on Pattern Analysis and Machine Intelligence, pp. 116-128 (2001)
6. Frohlinghaus, T., Buhmann, J.M.: Regularizing phase-based stereo. In: Proc. of 13th International Conference on Pattern Recognition, pp. 451-455 (1996)
7. Hoff, W., Ahuja, N.: Surfaces from Stereo: Integrating Feature Matching, Disparity Estimation, and Contour Detection. IEEE Trans. on Pattern Analysis and Machine Intelligence 11(2), 121–136 (1989)
8. Lee, S., Choi, J., Kim, D., Jung, B., Na, J., Kim, H.: An Active 3D Robot Camera for Home Environment. In: Proc. of 4th IEEE Sensors Conference (2004)
9. Scharstein, D., Szeliski, R.: High-Accuracy Stereo Depth Maps Using Structured Light. IEEE Computer Society Conference on Computer Vision and Pattern Recognition 1, 195–202 (2003)
10. Raskar, R., Tan, K.H., Feris, R., Yu, J., Turk, M.: Non-photorealistic Camera: Depth Edge Detection and Stylized Rendering Using Multi-Flash Imaging. In: Proc. of ACM SIGGRAPH Conference, Vol. 23, pp. 679–688 (2004)
11. Feris, R., Turk, M., Raskar, R., Tan, K., Ohashi, G.: Exploiting Depth Discontinuities for Vision-based Fingerspelling Recognition. IEEE Workshop on Real-Time Vision for Human-Computer Interaction (2004)
12. Ma, W., Manjunath, B.S.: EdgeFlow: a technique for boundary detection and image segmentation. IEEE Trans. on Image Processing 9, 1375–1388 (2000)

Weighted Distances Based on Neighbourhood Sequences in Non-standard Three-Dimensional Grids

Robin Strand

Centre for Image Analysis, Uppsala University,
Box 337, SE-75105 Uppsala, Sweden
robin@cb.uu.se

Abstract. By combining weighted distances and distances based on neighbourhood sequences, a new family of distance functions with potentially low rotational dependency is obtained. The basic theory for these distance functions, including functional form of the distance between two points, is presented for the face-centered cubic grid and the body-centered cubic grid. By minimizing an error function, the optimal combination of weights and neighbourhood sequence is derived.

1 Introduction

When using non-standard grids such as the face-centered cubic (fcc) grid and the body-centered cubic (bcc) grid for 3D images, less samples are needed to obtain the same representation/reconstruction quality compared to the cubic grid [1]. This is one reason for the increasing interest in using these grids in e.g. image acquisition [1], image processing [2,3,4], and image visualization [5].

Measuring distances is of great importance in many applications. Because of its low rotational dependency, the Euclidean distance is often used as distance function. There are, however, applications where other distance functions are better suited. For example, when minimal cost-paths are computed, a distance function defined as the shortest path between any two points is better suited, see e.g. [6], where the constrained distance transform is computed using the Euclidean distance resulting in a complex algorithm. The corresponding algorithm using a path-based approach is simple, fast, and easy to generalize to higher dimensions [7]. Examples of path-based distances are weighted distances, where weights define the cost (distance) between neighbouring grid points [8,3,2], and distances based on neighbourhood sequences (ns-distances), where the cost is fixed but the adjacency relation is allowed to vary along the path [9,4]. These path-based distance functions are generalizations of the well-known city-block and chessboard distance function defined for the square grid in [10]. The rotational dependency of weighted distances and ns-distances can be minimized by optimizing the weights [8,3,2] and neighbourhood sequences [9,4], respectively.

In this paper, weighted distances and ns-distances for the fcc and bcc grids, presented in [2,3] and [4] respectively, are combined by defining the distance

B.K. Ersbøll and K.S. Pedersen (Eds.): SCIA 2007, LNCS 4522, pp. 452–461, 2007.

as the shortest path between two grid points using both weights for the local distance between neighbouring grid points *and* allowing the adjacency relation to vary along the path. Functional forms are given of the distance functions for grid points in the fcc and bcc grids. Moreover, the rotational dependency is minimized by finding the weights and neighbourhood sequence that minimize an error function. The results are compared with results for weighted distances and ns-distances, which are both special cases of the proposed distance function.

2 Basic Notions

The following definitions of the fcc and bcc grids are used:

$$\mathbb{F} = \{(x, y, z) : x, y, z \in \mathbb{Z} \text{ and } x + y + z \equiv 0 \pmod 2\}. \tag{1}$$

$$\mathbb{B} = \{(x, y, z) : x, y, z \in \mathbb{Z} \text{ and } x \equiv y \equiv z \pmod 2\}. \tag{2}$$

When the result is valid for both $\mathbb{F}$ and $\mathbb{B}$, the notation $\mathbb{G}$ is used. Two grid points $\mathbf{p}_1 = (x_1, y_1, z_1), \mathbf{p}_2 = (x_2, y_2, z_2) \in \mathbb{G}$ are ρ-neighbours, $1 \leq \rho \leq 2$, if

1. $|x_1 - x_2| + |y_1 - y_2| + |z_1 - z_2| \leq 3$ and
2. $\max\{|x_1 - x_2|, |y_1 - y_2|, |z_1 - z_2|\} \leq \rho$

The points $\mathbf{p}_1, \mathbf{p}_2$ are *adjacent* if $\mathbf{p}_1$ and $\mathbf{p}_2$ are ρ-neighbours for some ρ. The ρ-neighbours which are not $(\rho - 1)$-neighbours are called *strict* ρ-neighbours. The neighbourhood relations are visualized in Figure 1 by showing the Voronoi regions (the voxels) corresponding to some adjacent grid points.

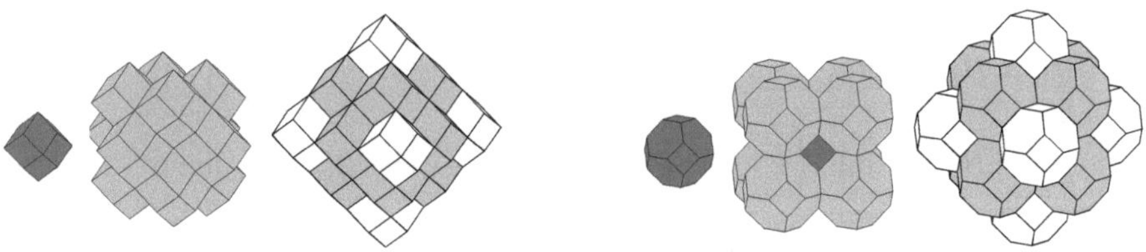

Fig. 1. The grid points corresponding to the dark and the light grey voxels are 1-neighbours. The grid points corresponding to the dark grey and white voxels are (strict) 2-neighbours. Left: fcc, right: bcc.

A ns B is a sequence $B = (b(i))_{i=1}^{\infty}$, where each $b(i)$ denotes a neighbourhood relation in $\mathbb{G}$. If B is periodic, i.e., if for some fixed strictly positive $l \in \mathbb{Z}_+$, $b(i) = b(i + l)$ is valid for all $i \in \mathbb{Z}_+$, then we write $B = (b(1), b(2), \ldots, b(l))$. A *path*, denoted $\mathcal{P}$, in a grid is a sequence $\mathbf{p}_0, \mathbf{p}_1, \ldots, \mathbf{p}_n$ of adjacent grid points. A path is a *B-path* of length n if, for all $i \in \{1, 2, \ldots, n\}$, $\mathbf{p}_{i-1}$ and $\mathbf{p}_i$ are $b(i)$-neighbours. The notation 1- and (strict) 2-steps will be used for a step to a 1-neighbour and step to a (strict) 2-neighbour, respectively. The number of 1-steps and strict 2-steps in a given path $\mathcal{P}$ is denoted $\mathbf{1}_{\mathcal{P}}$ and $\mathbf{2}_{\mathcal{P}}$, respectively.

Definition 1. *Given the ns B, the ns-distance $d(\mathbf{p}_0, \mathbf{p}_n; B)$ between the points $\mathbf{p}_0$ and $\mathbf{p}_n$ is the length of (one of) the shortest B-path(s) between the points.*

Let the real numbers α and β (the *weights*) and a path $\mathcal{P}$ of length n, where exactly l ($l \leq n$) adjacent grid points in the path are strict 2-neighbours, be given. The *length of the (α, β)-weighted B-path $\mathcal{P}$* is $(n-l)\alpha + l\beta$. The B-path $\mathcal{P}$ between the points $\mathbf{p}_0$ and $\mathbf{p}_n$ is a *shortest (α, β)-weighted B-path between the points $\mathbf{p}_0$ and $\mathbf{p}_n$* if no other (α, β)-weighted B-path between the points is shorter than the length of the (α, β)-weighted B-path $\mathcal{P}$.

Definition 2. *Given the ns B and the weights α, β, the weighted ns-distance $d_{\alpha,\beta}(\mathbf{p}_0, \mathbf{p}_n; B)$ is the length of (one of) the shortest (α, β)-weighted B-path(s) between the points.*

The following notation is used:

$$\mathbf{1}_B^k = |\{i : b(i) = 1, 1 \leq i \leq k\}| \text{ and}$$
$$\mathbf{2}_B^k = |\{i : b(i) = 2, 1 \leq i \leq k\}|.$$

3 Distance Function in Discrete Space

We now state a functional form of the distance between two grid points $(0,0,0)$ and (x,y,z), where $x \geq y \geq z \geq 0$. We remark that by translation-invariance and symmetry, the distance between any two grid points is given by the formula below. The formulas in Lemma 1 are proved (as Theorem 2 and 5) in [4].

Lemma 1. *Let the ns B and the point $\mathbf{p} = (x,y,z) \in \mathbb{G}$, where $x \geq y \geq z \geq 0$ be given. The ns-distance between $\mathbf{0}$ and $\mathbf{p}$ is given by*

$$d(\mathbf{0}, \mathbf{p}; B) = \min\left\{k \in \mathbb{N} : k \geq \max\left(\frac{x+y+z}{2}, x - \mathbf{2}_B^k\right)\right\} \text{ for } \mathbf{p} \in \mathbb{F}$$

$$d(\mathbf{0}, \mathbf{p}; B) = \min\left\{k \in \mathbb{N} : k \geq \max\left(\frac{x+y}{2}, x - \mathbf{2}_B^k\right)\right\} \text{ for } \mathbf{p} \in \mathbb{B}$$

Given a shortest B-path $\mathcal{P}$ of length $\hat{k}$ (obtained by the formula in Lemma 1), Lemma 2 and 3 give the number of 1-steps and 2-steps in $\mathcal{P}$, i.e. $\mathbf{1}_\mathcal{P}$ and $\mathbf{2}_\mathcal{P}$, respectively.

Lemma 2. *Let the point $(x,y,z) \in \mathbb{F}$, where $x \geq y \geq z \geq 0$, and the value of $\hat{k} = \min\left\{k : k \geq \max\left(\frac{x+y+z}{2}, x - \mathbf{2}_B^k\right)\right\}$ be given. If only the steps $(1,1,0)$, $(1,-1,0)$, $(0,1,1)$, $(1,0,1)$, and $(2,0,0)$ are used for a shortest B-path between $\mathbf{0}$ and (x,y,z), then*

$$\mathbf{1}_\mathcal{P} = \begin{cases} \hat{k} & \text{if } x \leq y+z \\ 2\hat{k} - x & \text{otherwise.} \end{cases}$$

and

$$\mathbf{2}_\mathcal{P} = \begin{cases} 0 & \text{if } x \leq y+z \\ x - \hat{k} & \text{otherwise.} \end{cases}$$

Proof. First of all, in the proof of Theorem 2 in [4] it is shown that there is a shortest B-path with only the steps $(1,1,0)$, $(1,-1,0)$, $(0,1,1)$, $(1,0,1)$, and $(2,0,0)$ when $x \geq y \geq z \geq 0$.

For the case $x \leq y+z$, the length of the shortest B-path is independent of B and there is a shortest B-path consisting of only 1-steps, see Theorem 2 in [4]. Hence, the number of 1-steps is $\hat{k}$ and the number of 2-steps is 0.

In the proof of Theorem 2 in [4], it is shown that only the steps $(2,0,0)$, $(1,1,0)$, $(1,-1,0)$, and $(1,0,1)$ are needed to find a shortest B-path between $(0,0,0)$ and $(x,y,z) \in \mathbb{F}$, where $x \geq y \geq z \geq 0$ and $x > y+z$. Thus, $(x,y,z) = (2,0,0)\mathbf{a}_1 + (1,1,0)\mathbf{a}_2 + (1,-1,0)\mathbf{a}_3 + (1,0,1)\mathbf{a}_4$ for some $\mathbf{a}_1, \mathbf{a}_2, \mathbf{a}_3, \mathbf{a}_4 \in \mathbb{N}$. Obviously, the length $\hat{k}$ of the path is $\mathbf{a}_1 + \mathbf{a}_2 + \mathbf{a}_3 + \mathbf{a}_4$, $\mathbf{1}_{\mathcal{P}} = \mathbf{a}_2 + \mathbf{a}_3 + \mathbf{a}_4$, and $\mathbf{2}_{\mathcal{P}} = \mathbf{a}_1$. We get:

$$\begin{aligned} \hat{k} &= \mathbf{a}_1 + \mathbf{a}_2 + \mathbf{a}_3 + \mathbf{a}_4 \\ x &= 2\mathbf{a}_1 + \mathbf{a}_2 + \mathbf{a}_3 + \mathbf{a}_4. \end{aligned}$$

Thus, $\mathbf{1}_{\mathcal{P}} = \mathbf{a}_2 + \mathbf{a}_3 + \mathbf{a}_4 = 2\hat{k} - x$ and $\mathbf{2}_{\mathcal{P}} = \mathbf{a}_1 = x - \hat{k}$. □

Lemma 3. *Let the point $(x,y,z) \in \mathbb{B}$, where $x \geq y \geq z \geq 0$, and the value of $\hat{k} = \min\left\{k : k \geq \max\left(\frac{x+y}{2}, x - \mathbf{2}_B^k\right)\right\}$ be given. If only the steps $(1,1,1)$, $(1,1,-1)$, $(1,-1,-1)$, and $(2,0,0)$ are used for a shortest B-path between $\mathbf{0}$ and (x,y,z), then*

$$\mathbf{1}_{\mathcal{P}} = 2\hat{k} - x$$

and

$$\mathbf{2}_{\mathcal{P}} = x - \hat{k}.$$

Proof. In the proof of Theorem 5 in [4], it is shown that only the steps $(2,0,0)$, $(1,1,1)$, $(1,1,-1)$, and $(1,-1,-1)$ are needed to find a shortest B-path between $(0,0,0)$ and $(x,y,z) \in \mathbb{B}$, where $x \geq y \geq z \geq 0$. Thus, $(x,y,z) = (2,0,0)\mathbf{a}_1 + (1,1,1)\mathbf{a}_2 + (1,1,-1)\mathbf{a}_3 + (1,-1-1)\mathbf{a}_4$ for some $\mathbf{a}_1, \mathbf{a}_2, \mathbf{a}_3, \mathbf{a}_4 \in \mathbb{N}$. Obviously, $\mathbf{1}_{\mathcal{P}} = \mathbf{a}_2 + \mathbf{a}_3 + \mathbf{a}_4$ and $\mathbf{2}_{\mathcal{P}} = \mathbf{a}_1$. Using the same technique as in the proof of Lemma 2, the result follows. □

Lemma 4. *Let the ns B and the point $(x,y,z) \in \mathbb{G}$, where $x \geq y \geq z \geq 0$ be given. If $\mathcal{P}$ is a shortest B-path between $\mathbf{0}$ and (x,y,z) consisting of only the steps*

$$(2,0,0),\ (1,1,0),\ (1,-1,0),\ (0,1,1),\ \textit{and}\ (1,0,1)\quad \textit{(fcc)}$$
$$(2,0,0),\ (1,1,1),\ (1,1,-1),\ \textit{and}\ (1,-1,-1)\quad \textit{(bcc)}$$

such that the weights $\alpha, \beta \in \mathbb{R}$ are such that $0 < \alpha \leq \beta \leq 2\alpha$, then $\mathcal{P}$ is also a shortest (α,β)-weighted B-path between $\mathbf{0}$ and (x,y,z).

Proof. From [4], we know that a shortest B-path is obtained by using these steps when $x \geq y \geq z \geq 0$, cf. Lemma 2 and 3.

Let n be the length of $\mathcal{P}$ and let l be the number of 2-steps in $\mathcal{P}$. The length of $\mathcal{P}$ can then be written $n = (n-l)1 + l1$. The length of the (α, β)-weighted B-path is $(n-l)\alpha + l\beta$. Assume that the shortest (α, β)-weighted B-path $\mathcal{P}'$ is a shorter (α, β)-weighted B-path between $\mathbf{0}$ and (x, y, z) than $\mathcal{P}$. Let the length of $\mathcal{P}'$ ((1, 1)-weighted) be n' ($n' \geq n$ since $\mathcal{P}$ is a shortest B-path) and the number of 2-steps be l'. The length of the (α, β)-weighted B-path $\mathcal{P}'$ is $(n'-l')\alpha + l'\beta$. By assumption,

$$(n'-l')\alpha + l'\beta < (n-l)\alpha + l\beta. \tag{3}$$

We get the following cases:

i $l' < l$
Since $\mathcal{P}'$ is a path between $\mathbf{0}$ and (x, y, z) and the only 2-step that is used is $(2, 0, 0)$, there are at least $2(l-l')$ more 1-steps in $\mathcal{P}'$ compared to $\mathcal{P}$. We get $(n'-l') \geq (n-l) + 2(l-l') \Rightarrow (n'-l') - (n-l) \geq 2(l-l') \Rightarrow ((n'-l') - (n-l))\,\alpha \geq 2(l-l')\alpha$. By the assumption (3), we have $((n'-l') - (n-l))\,\alpha < (l-l')\beta$. Thus, $2(l-l')\alpha \leq ((n'-l') - (n-l))\,\alpha < (l-l')\beta \Rightarrow 2\alpha < \beta$, which is a contradiction.

ii $l' > l$

$$\begin{aligned} (n'-l')\alpha + l'\beta &< (n-l)\alpha + l\beta && \text{by (3)} \\ (n-l')\alpha + l'\beta &< (n-l)\alpha + l\beta && \text{since } n' \geq n \\ (l'-l)\beta &< (l'-l)\alpha && \\ \beta &< \alpha && \text{since } l' > l, \end{aligned}$$

which is a contradiction.

This proves that the assumption (3) implies $l' = l$. Rewriting (3) with $l' = l$ gives $n' < n$, which contradicts the fact that $\mathcal{P}$ is a shortest B-path (i.e. that $n' \geq n$). Therefore (3) is false, so

$$(n'-l')\alpha + l'\beta \geq (n-l)\alpha + l\beta.$$

It follows that $\mathcal{P}$ is a shortest (α, β)-weighted B-path. □

The following theorems (Theorem 1 and 2) are obtained by summing up the results from Lemma 1–4. Given a shortest B-path consisting of as many 1-steps as possible between $\mathbf{0}$ and $(x, y, z) \in \mathbb{G}$ with $x \geq y \geq z \geq 0$, if $0 < \alpha \leq \beta \leq 2\alpha$, we know by Lemma 4 that the path is also a shortest (α, β)-weighted B-path. Using Lemma 2 and 3, where the number of 1-steps and 2-steps in the path are given, the formulas in Lemma 1 can be rewritten such that the length of the path is given by summing the number of 1-steps and 2-steps. By multiplying these numbers with the corresponding weights, we get the following functional forms of the weighted distance based on neighbourhood sequences between two points in the fcc and bcc grids.

Theorem 1. *Let the ns B, the weights α, β s.t. $0 < \alpha \leq \beta \leq 2\alpha$, and the point $(x, y, z) \in \mathbb{F}$, where $x \geq y \geq z \geq 0$, be given. The weighted ns-distance between $\mathbf{0}$ and (x, y, z) is given by*

$$d_{\alpha,\beta}\left(\mathbf{0},(x,y,z);B\right)=\begin{cases} k\cdot\alpha & \textit{if } x\le y+z\\ (2k-x)\cdot\alpha+(x-k)\cdot\beta & \textit{otherwise,}\end{cases}$$
$$\textit{where } k=\min_k : k\ge\max\left(\frac{x+y+z}{2}, x-\mathbf{2}_B^k\right).$$

Theorem 2. *Let the ns B, the weights α, β s.t. $0<\alpha\le\beta\le 2\alpha$, and the point $(x,y,z)\in\mathbb{B}$, where $x\ge y\ge z\ge 0$, be given. The weighted ns-distance between* $\mathbf{0}$ *and (x,y,z) is given by*

$$d_{\alpha,\beta}\left(\mathbf{0},(x,y,z);B\right)=(2k-x)\cdot\alpha+(x-k)\cdot\beta$$
$$\textit{where } k=\min_k : k\ge\max\left(\frac{x+y}{2}, x-\mathbf{2}_B^k\right).$$

To illustrate the discrete distance functions, balls of radius 20 in the fcc grid with $\alpha=1$, $\beta=1.4862$, and $B=(1,2)$ and the bcc grid with $\alpha=1$, $\beta=1.2199$, and $B=(1,2)$ are shown in Figure 2.

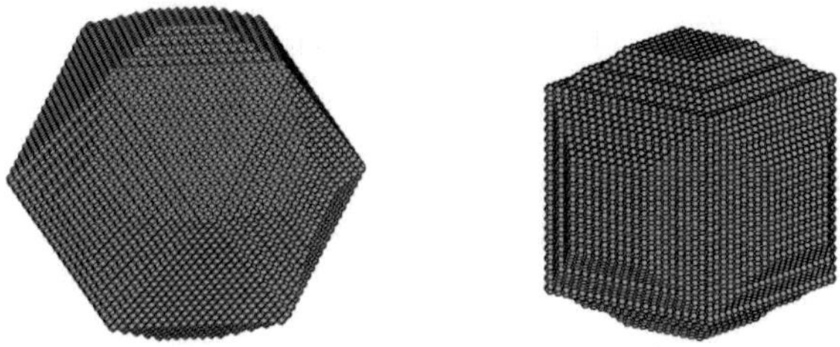

(a) (b)

Fig. 2. Balls of radius 20 in the fcc grid with $\alpha=1$, $\beta=1.4862$, and $B=(1,2)$ (a) and the bcc grid with $\alpha=1$, $\beta=1.2199$, and $B=(1,2)$ (b)

4 Distance Function in Continuous Space

The optimization is carried out in $\mathbb{R}^3$ by finding the best shape of polyhedra corresponding to balls of constant radii using the proposed distance functions. To do this, the distance functions presented for the fcc and bcc grids in the previous section are stated in a more general form valid for all points $(x,y,z)\in\mathbb{R}^3$, where $x\ge y\ge z\ge 0$. The following distance functions are considered:

$$d_{\alpha,\beta}^{fcc}\left(\mathbf{0},(x,y,z);\gamma\right)=\begin{cases} k\cdot\alpha & \text{if } x\le y+z\\ (2k-x)\cdot\alpha+(x-k)\cdot\beta & \text{otherwise,}\end{cases}$$
$$\text{where } k=\min_k : k\ge\max\left(\frac{x+y+z}{2}, x-(1-\gamma)k\right)$$

and

$$d_{\alpha,\beta}^{bcc}\left(\mathbf{0},(x,y,z);\gamma\right)=(2k-x)\cdot\alpha+(x-k)\cdot\beta$$
$$\text{where } k=\min_k : k\ge\max\left(\frac{x+y}{2}, x-(1-\gamma)k\right),$$

where $k \in \mathbb{R}$ and $\gamma \in \mathbb{R}$, $0 \leq \gamma \leq 1$ is the fraction of the steps where 2-steps are *not* allowed (so $\mathbf{1}_B^k$ and $\mathbf{2}_B^k$ corresponds to γk and $(1-\gamma)k$, respectively). In this way we obtain a generalization of the distance functions in discrete space $\mathbb{G}$ valid for all points (x, y, z) where $x \geq y \geq z \geq 0$ in continuous space $\mathbb{R}^3$. By considering

$$d_{\alpha,\beta}^{fcc}\left(\mathbf{0},(x,y,z);\gamma\right) = r \text{ and } d_{\alpha,\beta}^{bcc}\left(\mathbf{0},(x,y,z);\gamma\right) = r, \tag{4}$$

the points on a sphere of constant radius are found.

Remark 1. For a fixed point (x, y, z), $\frac{x+y+z}{2}$ and $\frac{x+y}{2}$ are constant and $x-(1-\gamma)k$ is decreasing w.r.t. k and when $k = 0$, $0 \leq \max\left(\frac{x+y}{2}, x\right) \leq \max\left(\frac{x+y+z}{2}, x\right)$. Therefore both $k = \max\left(\frac{x+y+z}{2}, x-(1-\gamma)k\right)$ and $k = \max\left(\frac{x+y}{2}, x-(1-\gamma)k\right)$ has a solution $k \in \mathbb{R}$. Thus, when $k \in \mathbb{R}$,

$$k = \min_k : k \geq \max\left(\frac{x+y+z}{2}, x-(1-\gamma)k\right) \Leftrightarrow k = \max\left(\frac{x+y+z}{2}, x-(1-\gamma)k\right)$$
$$k = \min_k : k \geq \max\left(\frac{x+y}{2}, x-(1-\gamma)k\right) \Leftrightarrow k = \max\left(\frac{x+y}{2}, x-(1-\gamma)k\right)$$

4.1 Reformulation of d^{fcc}

Using Remark 1, the expression for $d_{\alpha,\beta}^{fcc}$ is rewritten:

$$d_{\alpha,\beta}^{fcc}\left(\mathbf{0},(x,y,z);\gamma\right) = \begin{cases} k \cdot \alpha & \text{if } x \leq y+z \\ (2k-x)\cdot\alpha + (x-k)\cdot\beta & \text{otherwise,} \end{cases}$$
$$\text{where } k = \max\left(\frac{x+y+z}{2}, x-(1-\gamma)k\right).$$

We get two cases:

i) $\frac{x+y+z}{2} \geq x-(1-\gamma)k$
$k = \frac{x+y+z}{2} \Rightarrow$

$$d_{\alpha,\beta}^{fcc}\left(\mathbf{0},(x,y,z);\gamma\right) = \begin{cases} \frac{x+y+z}{2}\cdot\alpha & \text{if } x \leq y+z \\ (y+z)\cdot\alpha + \left(\frac{x-(y+z)}{2}\right)\cdot\beta & \text{otherwise.} \end{cases}$$

ii) $\frac{x+y+z}{2} < x-(1-\gamma)k$
$k = x-(1-\gamma)k \Rightarrow k = \frac{x}{2-\gamma}$

$$d_{\alpha,\beta}^{fcc}\left(\mathbf{0},(x,y,z);\gamma\right) = \begin{cases} \frac{x}{2-\gamma}\cdot\alpha & \text{if } x \leq y+z \star \\ \left(2\frac{x}{2-\gamma}-x\right)\cdot\alpha + \left(x-\frac{x}{2-\gamma}\right)\cdot\beta & \text{otherwise.} \end{cases}$$

Observe that when $x \leq y+z$, $\frac{x+y+z}{2} \geq x \geq x-(1-\gamma)k$ and thus the case $\star$ will not occur.

4.2 Reformulation of d^{bcc}

Using Remark 1, the expression is rewritten also for $d^{bcc}_{\alpha,\beta}$:

$$d^{bcc}_{\alpha,\beta}\left(\mathbf{0},(x,y,z);\gamma\right) = (2k-x)\cdot\alpha + (x-k)\cdot\beta$$
$$\text{where } k = \max\left(\frac{x+y}{2}, x-(1-\gamma)k\right).$$

Again, there are two cases:

i) $\frac{x+y}{2} \geq x-(1-\gamma)k$
$k = \frac{x+y}{2} \Rightarrow$

$$d^{bcc}_{\alpha,\beta}\left(\mathbf{0},(x,y,z);\gamma\right) = (y)\cdot\alpha + \left(\frac{x-y}{2}\right)\cdot\beta$$

ii) $\frac{x+y}{2} < x-(1-\gamma)k$
$k = x-(1-\gamma)k \Rightarrow k = \frac{x}{2-\gamma}$

$$d^{bcc}_{\alpha,\beta}\left(\mathbf{0},(x,y,z);\gamma\right) = \left(2\frac{x}{2-\gamma}-x\right)\cdot\alpha + \left(x-\frac{x}{2-\gamma}\right)\cdot\beta$$

Together with (4), this describes the portions of polyhedra satisfying $x \geq y \geq z \geq 0$. By using symmetry, the entire polyhedra are described. There polyhedra are spheres with the proposed distance functions.

4.3 Optimization of the Parameters

For any triplet α,β,γ ($\alpha,\beta > 0$ and $0 \leq \gamma \leq 1$), (4) defines a polyhedron in $\mathbb{R}^3$. The shape of the polyhedra obtained for some values of α,β,γ are shown in Figure 3. Let A be the surface area and V the volume of the (region enclosed by the) polyhedron. The values of A and V are determined by the vertices of the polyhedra which are derived using (4) together with the expressions for $d^{fcc}_{\alpha,\beta}$ and $d^{bcc}_{\alpha,\beta}$ derived in Section 4.1 and 4.2, respectively. The vertices satisfying $x \geq y \geq z \geq 0$ are

$$\left(\frac{2-\gamma}{\gamma\alpha+\beta-\beta\gamma}, \frac{\gamma}{\gamma\alpha+\beta-\beta\gamma}, 0\right) \text{ and } \left(\frac{1}{\alpha}, \frac{1}{\alpha}, 0\right) \text{ for } d^{fcc} \text{ and}$$
$$\left(\frac{2-\gamma}{\gamma\alpha+\beta-\beta\gamma}, \frac{\gamma}{\gamma\alpha+\beta-\beta\gamma}, \frac{\gamma}{\gamma\alpha+\beta-\beta\gamma}\right) \text{ and } \left(\frac{1}{\alpha}, \frac{1}{\alpha}, \frac{1}{\alpha}\right) \text{ for } d^{bcc}$$

The following error function (often called the compactness ratio) is used

$$E = \frac{\frac{A^3}{V^2} - 36\pi}{36\pi},$$

which is equal to zero if and only if A is the surface area and V is the volume of a Euclidean ball.

The values of α, β, and γ that minimize E are computed. The optimal values are found in Table 1 and visualized by the shape of the corresponding polyhedra in Figure 4.

Fig. 3. Shapes of balls for $d^{fcc}_{\alpha,\beta}$ (left 3×3 set of figures) and $d^{bcc}_{\alpha,\beta}$ (right 3×3 set of figures) for $\alpha = 1$, $r = 1$ when $\beta = 1$ (top row), $\beta = 1.5$ (middle row), $\beta = 2$ (bottom row) and $\gamma = 0$ (left column), $\gamma = 0.5$ (middle column), and $\gamma = 1$ (right column) in each block

Table 1. Performance of weighted distance based on neighbourhood sequences (wns), weighted distance (w), and distance based on neighbourhood sequences (ns) using the error function E. The value of E is attained whenever t is a strictly positive real number. The values shown in bold are fixed in the optimization.

	fcc				bcc			
Name	α	β	γ	E	α	β	γ	E
w	t	$1.5302t$	**0**	0.1367	t	$1.2808t$	**0**	0.2808
ns	**1**	**1**	0.8453	0.2794	**1**	**1**	0.5857	0.2147
wns	t	$1.4862t$	0.4868	0.1276	t	$1.2199t$	0.4525	0.1578

5 Conclusions

A new distance function is defined for the fcc and bcc grids, namely the weighted distance based on neighbourhood sequences. The value of the error function is lower for this distance function compared to the weighted distance and ns-distance. The weighted distances and ns-distances are special cases of the proposed distance function and the results in Figure 4 and Table 1 are similar (since another error function is used) to the weights obtained in [2,3] and equal to the neighbourhood sequences obtained in [4]. When $\gamma = 0$ (the left columns in Figure 3), weighted distances are obtained and when $\alpha = 1$ and $\beta = 1$ (the top row in Figure 3), distances based on neighbourhood sequences are obtained.

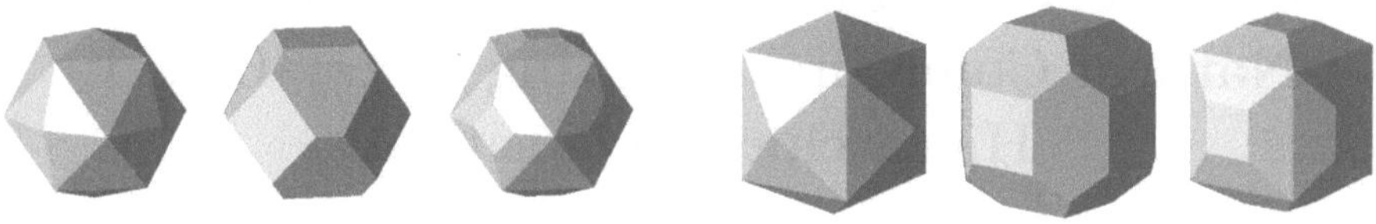

Fig. 4. Shapes of balls using values of α, β, γ that minimize E, see Table 1. (a) and (d): Weighted distance ($\gamma \equiv 0$). (b) and (e): Distance based on neighbourhood sequences ($\alpha \equiv 1, \beta \equiv 1$). (c) and (f): Proposed distance function. (a)–(c): fcc. (d)–(f): bcc.

For the balls in Figure 2, the neighbourhood sequence $B = (1, 2)$ is used. This corresponds to $\gamma = 0.5$, which approximates the optimal values quite good.

Since the distance function is path-based it will be a good choice in applications where a path-based distance function with low rotational dependency should be considered.

References

1. Matej, S., Lewitt, R.M.: Efficient 3D grids for image reconstruction using spherically-symmetric volume elements. IEEE Transactions on Nuclear Science 42(4), 1361–1370 (1995)
2. Strand, R., Borgefors, G.: Distance transforms for three-dimensional grids with non-cubic voxels. Computer Vision and Image Understanding 100(3), 294–311 (2005)
3. Fouard, C., Strand, R., Borgefors, G.: Weighted distance transforms generalized to modules and their computation on point lattices. Accepted for publication in Pattern Recognition. Available online, `www.sciencedirect.com` (2007)
4. Strand, R., Nagy, B.: Distances based on neighbourhood sequences in non-standard three-dimensional grids. Discrete Applied Mathematics 155(4), 548–557 (2006)
5. Carr, H., Theussl, T., Möller, T.: Isosurfaces on optimal regular samples. In: Bonneau, G.-P., Hahmann, S., C.D., H., (eds.) Proceedings of the symposium on Data visualisation 2003, Eurographics Association, pp. 39–48 (2003)
6. Coeurjolly, D., Miguet, S., Tougne, L.: 2D and 3D visibility in discrete geometry: an application to discrete geodesic paths. Pattern Recognition Letters 25(5), 561–570 (2004)
7. Verwer, B.J.H., Verbeek, P.W., Dekker, S.T.: An efficient uniform cost algorithm applied to distance transforms. IEEE Transactions on Pattern Analysis and Machine Intelligence 11(4), 425–429 (1989)
8. Borgefors, G.: Distance transformations in digital images. Computer Vision, Graphics, and Image Processing 34, 344–371 (1986)
9. Rosenfeld, A., Pfaltz, J.L.: Distance functions on digital pictures. Pattern Recognition 1, 33–61 (1968)
10. Rosenfeld, A., Pfaltz, J.L.: Sequential operations in digital picture processing. J. ACM 13(4), 471–494 (1966)

Unsupervised Perceptual Segmentation of Natural Color Images Using Fuzzy-Based Hierarchical Algorithm

Junji Maeda, Akimitsu Kawano, Sato Saga, and Yukinori Suzuki

Muroran Institute of Technology, Muroran 050-8585, Japan
junji@csse.muroran-it.ac.jp

Abstract. This paper proposes unsupervised perceptual segmentation of natural color images using a fuzzy-based hierarchical algorithm. $L^\star a^\star b^\star$ color space is used to represent color features and statistical geometrical features are adopted as texture features. A fuzzy-based homogeneity measure makes a fusion of color features and texture features. Proposed hierarchical segmentation method is performed in four stages: simple splitting, local merging, global merging and boundary refinement. Experiments on segmentation of natural color images are presented to verify the effectiveness of the proposed method in obtaining perceptual segmentation.

1 Introduction

Image segmentation is a process to partition an image into meaningful regions and is an important step before an image recognition process. In this paper, we are concerned with unsupervised perceptual segmentation of natural color images. Perceptual segmentation or rough segmentation is defined to obtain segmentation that produces a small number of segmented regions, and each region should represent a main object or a meaningful part of an object without paying much attention to region interiors. For example, perceptual segmentation could regard a tree that has many branches and leaves as one object in contrast to the conventional detailed segmentation.

Since natural color images contain various textures with different properties and all kinds of man-made objects, perceptual segmentation of natural color scenes is still a significantly difficult problem. Therefore, an effective segmentation method based on a set of texture features having good discriminating capability is essential in order to segment natural color images.

Though there is an extensive literature on image segmentation [1]-[4], the papers on perceptual segmentation are limited. Among them, Mirmehdi and Petrou [5] proposed the method based on the multiscale perceptual tower and the probabilistic relaxization method. Shi and Malik [6] proposed the perceptual grouping method based on graph theory, and Ma and Manjunath [7] proposed the technique based on Gabor filter and edge flow. Recently, Chen *et al.* [8] proposed the approach based on the adaptive clustering algorithm and steerable

B.K. Ersbøll and K.S. Pedersen (Eds.): SCIA 2007, LNCS 4522, pp. 462–471, 2007.

filter decomposition. Although these methods perform well, their algorithms are complicated and difficult to implement.

Searching for simpler algorithm, we proposed to use local fractal dimension and a fuzzy region-growing algorithm for rough segmentation [9]-[11]. However, our approach were not able to reduce the number of segmented regions sufficiently.

Ojala and Pietikäinen [12] proposed a texture segmentation method that has the advantages of a simple algorithm and easy implementation. Their method is a hierarchical segmentation algorithm that uses local binary pattern and contrast (LBP/C) features as texture measures and executes segmentation in hierarchical splitting, agglomerative merging and pixelwise classification. Though the method performs well for perceptually uniform segmentation of texture images, it is insufficient to segment natural color images for the following reasons.

1) It is difficult to use a minimum block size smaller than 16 due to the unstable histogram distribution of LBP/C features. Since the method is split-and-merge technique, it is preferable to use smaller block size at the split stage.
2) The method only treats gray-scale images and has to be adapted to color images.
3) The algorithm has a heavy computational burden due to the agglomerative merging stage since all possible pairs must be searched at each step just to merge two adjacent regions.

In this paper we propose a new technique, a fuzzy-based hierarchical algorithm, to perform unsupervised perceptual segmentation of natural color images based on the method by Ojala and Pietikäinen. We improve their algorithm in the following four points.

1) We adopt Statistical Geometrical Features (SGF) [13] as texture measures because it enables the algorithm to set a small minimum block size of 4 and the SGF can discriminate various types of textures remarkably.
2) We adopt fuzzy reasoning [14] to incorporate color features as well as texture features that enables the algorithm to treat color images.
3) We change the stage of hierarchical splitting into simple splitting to reduce the computational cost and the number of parameters.
4) We introduce a new stage of local merging that merges adjacent regions locally in order to drastically reduce the number of regions to be used at the stage of global merging. It is expected that the new stage will considerably reduce the total computational cost.

As a result of these improvement, the proposed algorithm has the capability of unsupervised perceptual segmentation of natural color images by the simple algorithm with easy implementation that has four hierarchical stages: simple splitting, local merging, global merging and boundary refinement. During the latter three stages, we measure the similarity of any adjacent regions by fuzzy homogeneity which combines the similarity of color features and texture features with different weights of importance. We use the $L^\star a^\star b^\star$ color space to represent color features and the SGF as texture features.

The adoption of fuzzy-based homogeneity measure simplifies the complex mechanism of integrating different features by using symbolic representations. It also reduces the difficulty in choosing the many threshold values inherent in segmentation methods, though the tuning of the fuzzy membership functions is still required for each image.

In this paper we also propose to introduce a new type of parameter that is the number of segmented regions N instead of the threshold value. Since the expected number of segmented regions is very small in perceptual segmentation, it is possible for a user to presuppose a rough estimate of a desirable number of segmented regions in advance depending on the contents of each image and user's intention. We consider around 5-15 number of segmented regions as optimal range of N in the sense of perceptual segmentation. In the practical implementation, the proposed algorithm has the significant ability to produce the segmentation results by reducing the number of segmented regions one by one at each step. Thus, the user is able to determine the optimal result with an appropriate roughness by observing the several segmented results.

Several experiments are made to confirm the effectiveness of the proposed method in obtaining unsupervised perceptual segmentation of natural color images.

2 Color and Texture Features

2.1 $L^\star a^\star b^\star$ Color Features

The $L^\star a^\star b^\star$ color space is a perceptually uniform color space, where $L^\star$ represents brightness and $a^\star$ and $b^\star$ represent chromatic information. We obtain the $L^\star a^\star b^\star$ color space from the RGB color space, then the three components are normalized and used as three color features.

2.2 SGF Texture Features

The SGF [13] are a set of texture features based on the statistics of geometrical properties of connected regions in a sequence of binary images obtained from an original image. The extraction of the SGF starts by thresholding the each component of a color image $C(x,y)$ (where $C = L^\star$, $a^\star$ and $b^\star$) with a threshold value α that produces the binary image C_b defined as

$$C_b(x,y;\alpha) = \begin{cases} 1 & C(x,y) \geq \alpha \\ 0 & otherwise, \end{cases} \tag{1}$$

where $1 \leq \alpha \leq n_l - 1$ and n_l is the maximum gray level of each component.

Each binary image $C_b(x,y;\alpha)$ comprises a several connected regions. The number of connected regions of 1-valued pixels and that of 0-valued pixels give two geometrical measures, $NOC_1(\alpha)$ and $NOC_0(\alpha)$, respectively. Next a measure of irregularity (or a measure of non-circularity) is defined to each of the

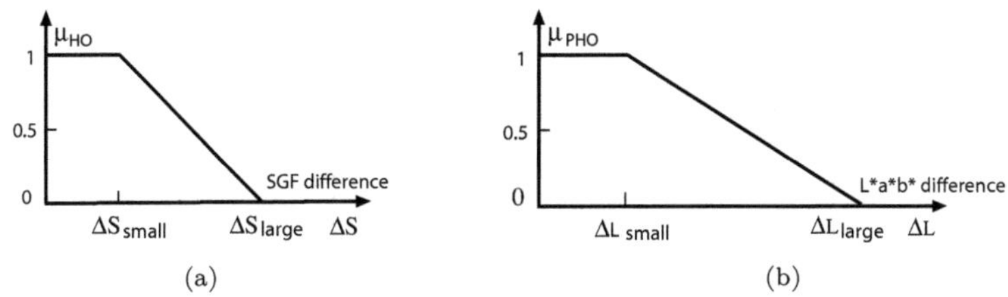

Fig. 1. Fuzzy membership functions of (a) SGF difference (ΔS) and (b) $L^{\star}a^{\star}b^{\star}$ difference (ΔL)

connected regions. The average of irregularity measure of 1-valued pixels in the binary image $C_b(x, y; \alpha)$ and that of 0-valued pixels are represented by $\overline{IRGL_1}(\alpha)$ and $\overline{IRGL_0}(\alpha)$, respectively.

Each of these four functions is further characterized using the three statistics over an entire image: the average value, the sample mean and the sample standard deviation. This gives a total of 36 texture features. At the implementation, a sliding overlapping window of size 5×5 is used to calculate the SGF of each pixel of an original image.

3 Fuzzy-Based Homogeneity Measure

Homogeneity is a measure to test similarity of two regions under consideration during segmentation procedure. We adopt a fuzzy-based homogeneity measure to integrate the different features: the $L^{\star}a^{\star}b^{\star}$ color features and the SGF texture features. We use the following fuzzy rules where each rule has a corresponding membership function.

1) Rule 1: if SGF difference is SMALL, then HOMOGENEOUSE (HO); else NOT HOMOGENEOUSE (NHO).
2) Rule 2: if $L^{\star}a^{\star}b^{\star}$ difference is SMALL, then PROBABLY HOMOGENEOUSE (PHO); else PROBABLY NOT HOMOGENEOUSE (PNHO).

These fuzzy rules give the SGF texture features a higher priority than the $L^{\star}a^{\star}b^{\star}$ color features because we consider that the texture features provide more important information for textured color images. In these fuzzy rules, a SGF difference is the Euclidean distance of 36 SGF between two regions under consideration, and a $L^{\star}a^{\star}b^{\star}$ difference is the Euclidean distance of the three $L^{\star}a^{\star}b^{\star}$ components between them. Four conditions HO, NHO, PHO and $PNHO$ represent different grades of homogeneity between two regions. Their homogeneity values μ_{HO}, μ_{NHO}(=1-μ_{HO}), μ_{PHO} and μ_{PNHO}(=1-μ_{PHO}) can be obtained from fuzzy membership functions of the SGF difference and the $L^{\star}a^{\star}b^{\star}$ difference as shown in Fig. 1. Here ΔS represents the SGF difference and ΔL represents the $L^{\star}a^{\star}b^{\star}$ difference. The values of ΔS_{small}, ΔS_{large}, ΔL_{small} and ΔL_{large} in Fig. 1 have to be tuned empirically.

A final homogeneity measure H is inferred by min-max inference by using the fuzzy set as shown in Fig. 2 and the centroid defuzzification method [14]. Suppose

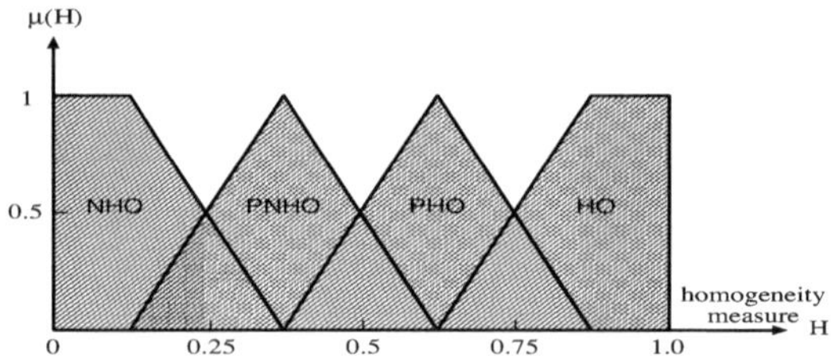

Fig. 2. The fuzzy set used for homogeneity inference

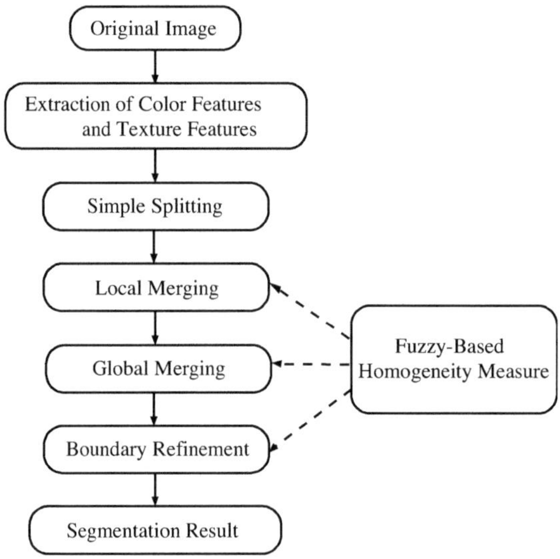

Fig. 3. Block diagram of the segmentation procedure

the homogeneity limit is set 0.5, then if the inferred homogeneity measure is over 0.5, the two regions being concerned are regarded as homogeneous and they are merged. We use the value of the homogeneity measure H in the proposed segmentation algorithm.

4 Segmentation Algorithm

The proposed fuzzy-based hierarchical segmentation procedure is shown in Fig. 3. We first obtain the $L^\star a^\star b^\star$ color features and the SGF texture features for each pixel of an original image. We then execute the segmentation in four stages: simple splitting, local merging, global merging and boundary refinement. During the latter three stages, the fuzzy-based homogeneity measure H is used as similarity measure. In the following, we will demonstrate the progress of segmentation on a 256×256 natural color image as shown in Fig. 4(a).

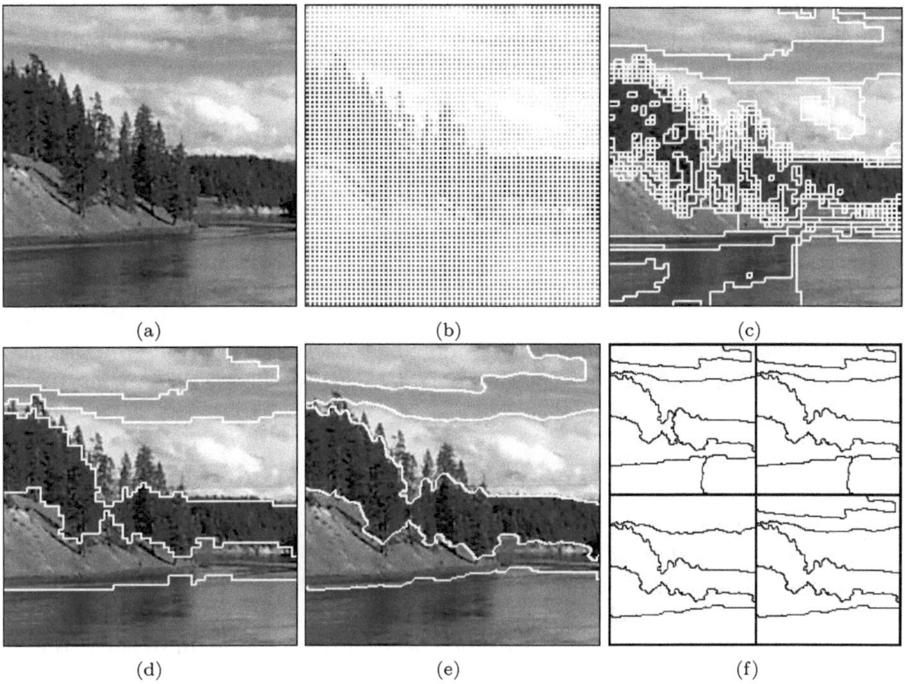

Fig. 4. Segmentation of a natural color image: (a) original image; (b) result of simple splitting; (c) result of local merging; (d) result of global merging with $N = 6$; (e) result of boundary refinement; (f) results of segmentation with N = 8, 7, 6 and 5 shown clockwise from the upper left

4.1 Simple Splitting

Ojala and Pietikäinen used the hierarchical splitting algorithm that recursively splits an original image into square blocks of varying size. Since their algorithm requires computational cost and the threshold value, we changed the stage of hierarchical splitting into simple splitting to reduce the computational cost and the number of parameters. In simple splitting, the image is divided into rectangular subblocks of size 4×4 as shown in Fig. 4(b). It is noted the adoption of the SGF texture features and the incorporation of local merging in our algorithm enable the use of simple splitting.

4.2 Local Merging

Local merging is a newly proposed stage by us to merge adjacent regions locally for drastically reducing the number of regions to be used at the stage of global merging. The SGF of each 4×4 subblock are obtained by averaging the texture features of all pixels within the subblock, so does the $L^{\star}a^{\star}b^{\star}$ color features of each subblock.

The homogeneity between any current region and its neighboring adjacent region is measured individually. Then the two adjacent regions having the largest homogeneity measure H_{max} are regarded as similar and merged to become one region if the value of H_{max} is higher than a threshold 0.5. The process is continued until all regions are scanned. We set the threshold 0.5 to avoid over merging in this stage. The result of local merging is shown in Fig. 4(c).

4.3 Global Merging

Global merging is a stage to merge similar adjacent regions globally. A pair of adjacent regions with the smallest merger importance value among all possible mergers in the entire image will be merged at each step. Merger importance MI is defined as the ratio of the number of pixels in the smaller region to its homogeneity measure of adjacent regions

$$MI = \frac{P_{small}}{H}. \tag{2}$$

The procedure finds the best possible pair of adjacent regions globally whose merging introduces the smallest change in the segmented image. Since global merging reduces the number of segmented regions one by one at each step and it removes unimportant regions first, the essential regions remain to the end and thus perceptual segmentation is achieved.

It is also easy to stop the algorithm when the number of segmented regions reaches the specified number of segmented regions N. Fig. 4(d) shows the result of global merging when we set $N = 6$ to obtain perceptual segmentation.

4.4 Boundary Refinement

Boundary refinement is finally performed to improve the localization of boundaries. If an image pixel is on the boundary of at least two distinct regions, a discrete disk with radius 3 will be placed on it. Then the homogeneity measure H between the disk and its neighboring region is calculated individually to decide if the pixel needs to be relabeled. The next scan will check the neighborhoods of the relabeled pixels until no pixels are relabeled. The final segmentation after boundary refinement is shown in Fig. 4(e).

In the practical implementation, the user can easily choose the desirable optimal result with an appropriate roughness from among the several-segmented results. The results of segmentation when $N = 8$, 7, 6 and 5 are shown clockwise from the upper left in Fig. 4(f). This figure demonstrates how the number of segmented regions decreases in the proposed algorithm.

5 Experimental Results

In this section, we present experimental results to assess the performance of the proposed segmentation method. For comparison, we show the results by

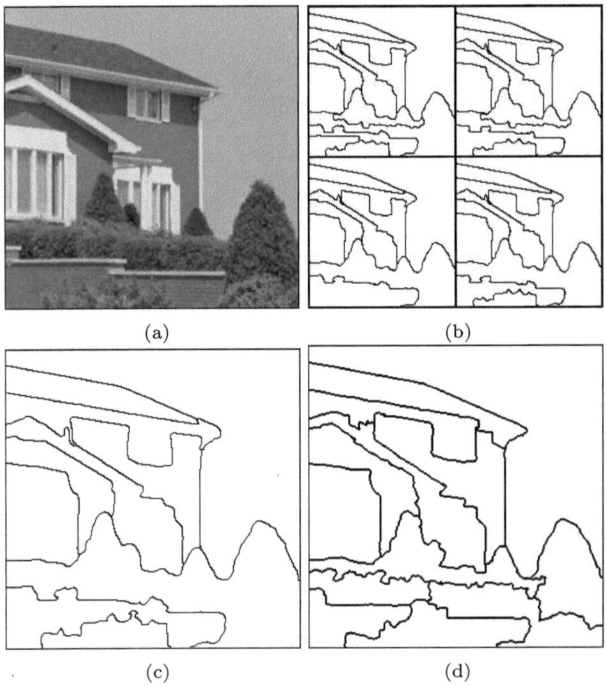

Fig. 5. Segmentation of a natural color image containing man-made objects: (a) original image; (b) results of segmentation with $N = 12$, 11, 10 and 9 shown clockwise from the upper left; (c) result by proposed method with $N = 10$; (d) result by EdgeFlow method

the EdgeFlow method using the algorithm made by the authors [15]. Since the EdgeFlow algorithm cannot produce the specified number of segmented regions, we chose the best result with nearest number of segmented regions to our result by tuning the parameters. Since we set $\Delta S_{small} = 0$ and $\Delta L_{small} = 0$ in Fig. 2, we have to tune two parameters ΔS_{large} and ΔL_{large} of the fuzzy membership functions differently according to each image and they were decided empirically.

We apply the proposed method to a 300×300 natural color image containing man-made objects shown in Fig. 5(a). The image is composed of the sky, a house, trees and a wall. The boundaries within the house further divide it into the main parts of objects such as roofs and windows. The perceptual segmentation is rather difficult because it is necessary to obtain accurate boundaries as well as uniform texture regions. Fig. 5(b) is the results of perceptual segmentation by the proposed algorithm when $N = 12$, 11, 10 and 9. The user can easily decide the optimal result by observing these segmented results. The selected optimal result with $N = 10$ is shown in Fig. 5(c) and the result by the EdgeFlow method is shown in Fig. 5(d), respectively. We only show the boundaries of the segmented regions for clarity. Although two algorithms show the same degree of rough segmentation, the proposed algorithm represents a slightly better result than

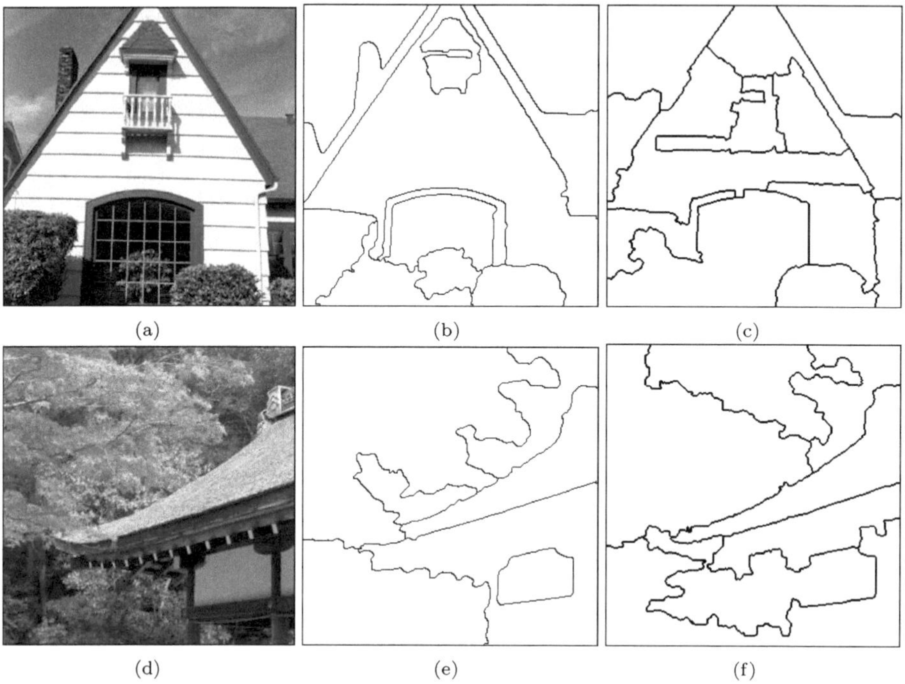

Fig. 6. Segmentation of a natural color image containing man-made objects: (a) and (d) original images; (b) and (e) results by proposed method with $N = 12$ and $N = 7$; (c) and (f) results by EdgeFlow method

the EdgeFlow method. In Fig. 5(c), the trees and the main parts of the house are well maintained as complete regions, and thus the segmented regions are more uniform and the boundaries of each region are maintained spatially more accurate than the result by the EdgeFlow method. The comparison demonstrates the effectiveness of the proposed method.

We then apply the proposed method to 300×300 natural color images shown in Fig. 6(a) and (d). The results of perceptual segmentation by the proposed algorithm when $N = 12$ and $N = 7$ are shown in Fig. 6(b) and (e), respectively. The results of segmentation by the EdgeFlow method are shown in Fig. 6(c) and (f). These results also represent that the proposed algorithm produces perceptual segmentation more accurately than the EdgeFlow method. However, further investigations are necessary to precisely compare the proposed algorithm with other methods.

In order to assess the time cost by the newly introduced local merging stage, we compared the processing time with and without this stage. As a result, the processing time with this stage became $12 - 3\%$ of the time without this stage. Thus, the introduction of local merging was confirmed to be effective to reduce the total computational cost.

6 Conclusions

In this paper, we presented unsupervised perceptual segmentation of natural color images using the proposed fuzzy-based hierarchical algorithm that makes a reliable fusion of the $L^{\star}a^{\star}b^{\star}$ color features and the SGF texture features. The hierarchical segmentation using the fuzzy-based homogeneity measure is effective in obtaining perceptual segmentation that maintains uniform texture regions and accurate boundaries. The proposed algorithm has the prospective advantage of the capability to determine the desirable optimal result with an appropriate roughness, since it can produce the segmentation results by reducing the number of segmented regions one by one at each step.

References

1. Fu, K.S., Mu, J.K.: A survey on image segmentation. Pattern Recognition 13(1), 3–16 (1981)
2. Haralick, R.M., Shapiro, L.G.: Image segmentation techniques. Comput. Vision Graphics Image Processing 29, 100–132 (1985)
3. Pal, N.R., Pal, S.K.: A review on image segmentation techniques. Pattern Recognition 26(9), 1277–1294 (1993)
4. Reed, T.R., du Buf, J.M.H.: A review of recent texture segmentation and feature extraction techniques. CVGIP: Image Understanding 57, 359–372 (1993)
5. Mirmehdi, M., Petrou, M.: Segmentation of color textures. IEEE Trans. Pattern Anal. Mach. Intell. 22(2), 142–159 (2000)
6. Shi, J., Malik, J.: Normalized cuts and image segmentation. IEEE Trans. Pattern Anal. Mach. Intell. 22(8), 888–905 (2000)
7. Ma, W.Y., Manjunath, B.S.: EdgeFlow: A technique for boundary detection and image segmentation. IEEE Trans. Image Processing 9(8), 1375–1388 (2000)
8. Chen, J., Pappas, T.N., Mojsilovic, A., Rogowitz, B.E.: Adaptive perceptual color-texture image segmentation. IEEE Trans. Image Processing 14(10), 1524–1536 (2005)
9. Maeda, J., Novianto, S., Saga, S., Suzuki, Y., Anh, V.V.: Rough and accurate segmentation of natural images using fuzzy region-growing algorithm. In: Proc. IEEE Int. Conf. on Image Processing, vol. 3, pp. 227–231 (1999)
10. Maeda, J., Ishikawa, C., Novianto, S., Tadehara, N., Suzuki, Y.: Rough and accurate segmentation of natural color images using fuzzy region-growing algorithm. In: Proc. 15th Int. Conf. on Pattern Recognition, vol. 3, pp. 642–645 (2000)
11. Novianto, S., Suzuki, Y., Maeda, J.: Near optimum estimation of local fractal dimension for image segmentation. Pattern Recognition Letters 24(1-3), 365–374 (2003)
12. Ojala, T., Pietikäinen, M.: Unsupervised texture segmentation using feature distributions. Pattern Recognition 32(3), 477–486 (1999)
13. Chen, Y.Q., Nixon, M.S., Thomas, D.W.: Statistical geometrical features for texture classification. Pattern Recognition 28(4), 537–552 (1995)
14. Zadeh, L.A.: Fuzzy sets. Inform. Control 8, 338–353 (1965)
15. `http://vision.ece.ucsb.edu/segmentation/edgeflow/software/`

Line-Stepping for Shell Meshes

Kenny Erleben and Jon Sporring

Department of Computer Science, University of Copenhagen, Denmark
{kenny,sporring}@diku.dk

Abstract. This paper presents a new method for creating a thick shell tetrahedral mesh from a triangular surface mesh. Our main goal is to create the thickest possible shell mesh with the lowest possible number of tetrahedrons.

Low count tetrahedral meshes is desirable for animating deformable objects where accuracy is less important and to produce shell maps and signed distance fields. In this work we propose to improve convergence rate of past work.

1 Introduction

Many graphical models of solid objects are given as surface meshes [1,2], since this is an economical representation for visualization, and easily obtainable by laser scanning of real objects or by hand-modeling using a 3 dimensional drawing tool. However, animating deformations of solid objects requires a notion of inner structure which is surprisingly difficult to obtain. Existing algorithms such as [3] are difficult to implement and do not use the natural, intrinsic representation of shape by symmetry sets [4,5].

Shell meshes are attractive since they give a volume representation of a surface mesh with a very low tetrahedral count, which is desirable for animation or similar purposes, where speed is preferred over accuracy of deformation. The shell mesh finds applications in animating solid objects, for shell maps [6], and for the calculation of signed distance field [7,8].

In this paper we will present an extension of [9,10], and our main goal is to create the thickest possible shell mesh with the lowest possible tetrahedral count. I.e. given a polygonal surface mesh, we create a tetrahedral volume mesh representing a thick version of the surface mesh, a shell mesh. The thickness shells are the most challenging to produce, and all thinner shells are a subset of these, hence we will in this article focus on producing the thickest possible shells. In past work vertices of the polygonal surface mesh are displaced inward, thereby creating a new version of the surface mesh. This operation is in the literature termed inward extrusion or just simply extrusion, although intrusion seems a better term. Following an inward extrusion the original surface mesh is used to generate the outside of the shell and the extruded surface mesh is used to generate the inside of the shell mesh. The two meshes is then used to create a triangle prism shell mesh. Finally, the triangle prism mesh are converted into a consistent tetrahedral mesh, also known as tetrahedral tessellation.

B.K. Ersbøll and K.S. Pedersen (Eds.): SCIA 2007, LNCS 4522, pp. 472–481, 2007.

A linear randomized tessellation algorithm, the ripple tessellation, was developed in [9]. The ripple method suffers from several problems. It is not deterministic, but relies on picking random ripple directions to fix inconsistencies. Further, no proof has been given on existence of a consistent tetrahedral tessellation of the triangle prism shell. A safe conservative, upper extrusion length limit is used in [9] and later improved in [10]. Nevertheless, both algorithms are very slow due to bad and unpredictable convergence of the bisection search method. In present article we present a new extrusion approach that seek to solve the bad and unpredictable convergence of the bisection search method.

2 Adaptive Signed Distance Field Extrusion

The signed distance field for a closed surface mesh is a scalar field, ϕ, who's magnitude is the distance to the closest point on the surface, and where values are positive outside and negative inside the surface mesh.In the neighborhood of the surface mesh, the gradient of the signed distance field, $\nabla\phi$, has length 1 and a direction parallel to the surface normal, $\boldsymbol{n}$, but further away from the mesh, the distance field may experience singular points, exactly where two or more points on the surface are closest. At these points, the signed distance field is first order discontinuous. As an example consider the signed distance map of a circle, which looks like a cone, passing through the circle and whose apex is the center of the circle, since the center is equally close to every point on the circle. The points that are closest to two or more points on a shape are known as the symmetry set [4] and a subset is the skeleton or the Medial surface representation [5].

We perform an inward extrusion of a triangle on the surface, thus producing a prism. This is unproblematic as long as the extrusion is still in the neighborhood of the surface, however, when the extrusion extends beyond the singular points of the distance field, then prisms will overlap with the singular points, and therefore also overlap with an extrusion from another part of the object. Since we are only concerned with inward extrusions, we only need to consider the medial surface part of the symmetry sets, since the medial surface is the locus of the singular points of the signed distance field inside the object. Thus, given the medial surface, the maximum inward extrusion lengths avoiding overlapping prisms would be easily obtained by intersection of the triangle normals with the medial surface. Algorithms do exist to compute the medial surfaces of polygonal models [11], but these are often not applicable to general polygonal models made by artist. In practice one seeks approximations [12,13,14]. We propose to use the signed distance field directly, since it contains the medial surface implicitly. Thus, we avoid the computational burden of the approximation, and gain precision, since the signed distance field is numerically more accurate than an approximation based hereof. In the following we will describe our algorithm.

We attack the problem of detecting the singularities in the signed distance map. Hence, given a signed distance map, ϕ, for a surface, its singularities may be found by a binary search from the surface inwardly along the surface normal, which is the same as in the signed direction of the gradient of the in to find the medial surface point in the opposite direction of a distance field gradient. Hence, in a neighborhood of the a surface point p with corresponding normal $\boldsymbol{n}$, we seek the largest value of ε fulfilling the criteria

$$\varepsilon = -\phi(\boldsymbol{p} - \varepsilon \boldsymbol{n}), \quad \text{and} \quad \varepsilon \leq \varepsilon_{\text{user}}, \tag{1}$$

where we for practical purposes allow for a user-specified, maximum size, $\varepsilon_{\text{user}}$. Thus prior to overshooting the medial surface, we must have

$$\varepsilon = -\phi, \tag{2}$$

and immediately after overshooting we must have

$$\varepsilon \neq -\phi. \tag{3}$$

Hence, a root searching algorithm is easily devised, unfortunately the bisection approach works rather badly in practice, since discretization errors and interpolation can be quite large. This implies that

1) The gradient direction is not very accurate.
2) The distance value at any given position might be a little off.

Thus, given a surface point, $\boldsymbol{p}$, we cannot expect (1) to be very accurate, and an effective stopping criteria for the bisection method is not easily designed. In our tests we used a simple iteration limit of 1000, and a quite large threshold value in the comparison of (1). Figure 1(a) shows a 2D result. Notice that some surface points are extruded quite poorly, and this approach results in too poor results at too high computational cost. In some cases, where we step along symmetry lines, we can not even rely on the assumption in (1). Any root-search method will therefore fail.

An alternative approach is line-stepping, which is based on the idea of stepping along the extrusion line with a fixed increment of $\Delta\varepsilon$. In each step the extrusion length is updated by

$$\varepsilon_{i+1} = \varepsilon_i + \Delta\varepsilon, \tag{4}$$

and the current extrusion point, $\boldsymbol{q}(\varepsilon)$, is found by

$$\boldsymbol{q}(\varepsilon) = \boldsymbol{p} - \varepsilon \nabla\phi(\boldsymbol{p}). \tag{5}$$

The stepping is performed as long as $\nabla\phi(\boldsymbol{q})$ points in the same direction as $\nabla\phi(\boldsymbol{p})$. If the cell size of the regular sampled signed distance field be given by Δx, Δy, and Δz then the increment is chosen as

$$\Delta\varepsilon = \frac{\min(\Delta x, \Delta y, \Delta z)}{2}. \tag{6}$$

This ensures that each step along the extrusion line is not faster than the information changes in the signed distance field. This works due to spatial coherence of the values in the signed distance field, since the value at a neighboring grid

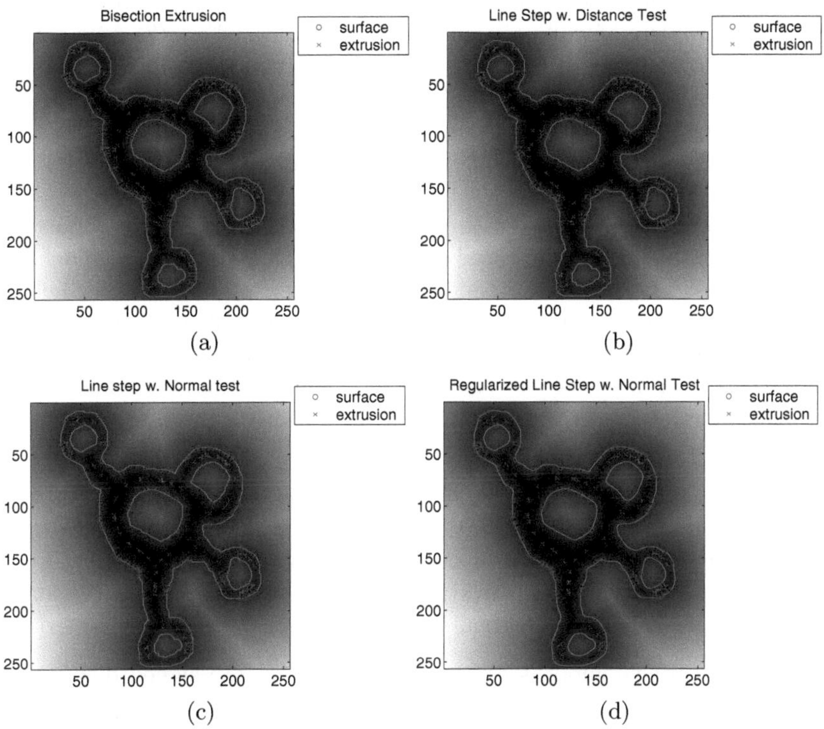

Fig. 1. 2D examples of inward extrusions on random points on the surface. (a) The bisection method gives poor extrusion lengths. (b) The line stepping using distance testing as stopping criteria results in some points extruded badly. (c) The line stepping using normal testing as stopping criteria, improves the extrusion lengths over (b). (d) The regularized line stepping using normal testing as stopping criteria has the diminished the crossing of extrusion lines.

node in the signed distance field differs by at most $\sqrt{\Delta x^2 + \Delta y^2 + \Delta z^2}$. The stopping criteria we use is to keep on increasing ε, while

$$\nabla\phi(\boldsymbol{p}) \cdot \nabla\phi(\boldsymbol{q}) > \rho, \tag{7}$$

where $\rho > 0$ is a user specified threshold to control accuracy. We call this the normal-test. The intuition behind the normal-test is: if we pass the medial surface from one side of an object to the opposite side, then the gradients in the signed distance field will flip directions. That is, the sign of the normal-test will flip from positive to negative. Later we will give a more rigorous explanation of this stopping criteria. One should think that a better stopping criteria would be to only step along an extrusion line while

$$|\varepsilon - \phi(\boldsymbol{q})| < \rho. \tag{8}$$

Unfortunately in practice this distance test is far to sensitive to discretization and interpolation errors in the signed distance field and requires large values of ρ.

Leading to a questionable approximation of the medial surface. For now we will overlook the problem of walking along symmetry lines, where (8) does not hold. Figure 1(b) shows a test result using the distance test, while Figure 1(c) shows the same test using the normal test. Comparing Figure 1(c) with Figure 1(a) and Figure 1(b) we see that extrusions are much larger and more evenly distributed for the line-stepping approach using normal testing, and therefore we favor this method.

The line-stepping approach with the normal test works quite robustly, although some problems remain: Firstly, the gradient, $\nabla\phi(\boldsymbol{p})$, is not well defined at surface vertices. This is normally not noticeable, since signed distance fields are typically sampled on a regular grid, implying a regularization of the signed distance field gradient. Secondly, typical finite difference approximation schemes for the gradient operator are not accurate enough, which results in surface normals, where the direction of the extrusion lines might cause a swallow tail problem. To circumvent these problems, two solutions may be adopted:

1) Instead of using $\nabla\phi(\boldsymbol{p})$, we use the angle weighted pseudo-normals [15]. These can be computed directly from the input surface mesh at high accuracy.
2) We can regularize the signed distance field by a curvature flow (or Gaussian convolution etc.), this seems to straighten out the normal directions at small scale features, where the normal direction is poor due to sampling artifacts.

The second approach smooths errors, but it destroys the signed distance field property, this could be recovered by reinitialization [8]. However, in our experience it does not cause major changes in the overall direction of the gradient. Figure 1(d) shows the result using regularization. Comparing Figure 1(d) with Figure 1(c) it is seen that the extrusion lines tends to cross much less, when using regularization.

Smoothing the signed distance field by curvature flow works well in two dimensions, but is costly in three dimensions, where we prefer angle weighted surface normal solution, to be described below.

It is worth noting that left-hand-side of the normal-test in (7) is in fact the directional derivative at position $\boldsymbol{q}$. The sign of the directional derivative therefore tell us, how ϕ changes, as we move in the opposite direction of the surface normal vector $\boldsymbol{n}$. In our specific case the following rules applies:

$$\nabla\phi(\boldsymbol{q})\cdot\boldsymbol{n}\begin{cases}<0 & \phi \text{ is increasing}\\ >0 & \phi \text{ is decreasing}\ .\\ =0 & \phi \text{ is constant}\end{cases} \tag{9}$$

The actual value of the directional derivative tells us how fast ϕ changes in the normal direction. For our inward extrusions, we want to extrude as long as ϕ is decreasing. However, this is not quite enough: as illustrated in Figure 2(a), an extrusion line can cross over the symmetry set, if the surface angle a is less than the accepted angle difference between $\nabla\phi(\boldsymbol{q})$ and $\boldsymbol{n}_{\boldsymbol{p}}$, which is undesirable. In this case, the directional derivative, $\nabla\phi(\boldsymbol{q})\cdot\boldsymbol{n}$, is 1 until the symmetry line is hit, but beyond the symmetry line, the distance value continues to decrease. To stop

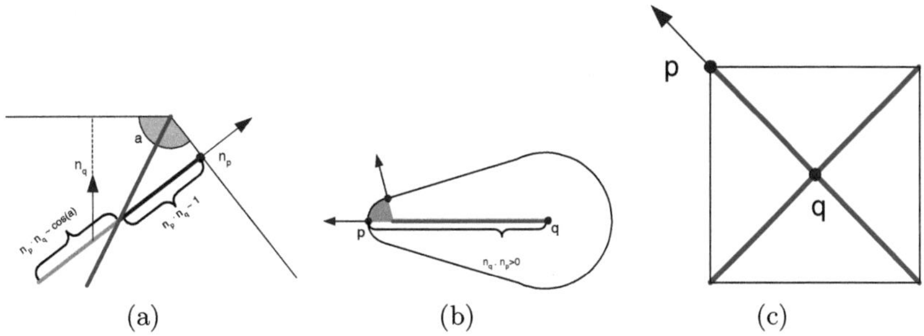

Fig. 2. (a) An inward extrusion line may cross over the symmetry line. (b) A tentacle is created along the symmetry axis. (c) Not always do we wish to walk along a symmetry line.

at the symmetry line, we must detect the change in the value of the directional derivative. Another approach would be to require that the gradient of the signed distance field does not differ from the surface normal by more than some small user specified angle, i.e.

$$\nabla \phi(\boldsymbol{q}) \cdot \boldsymbol{n_p} > \rho \tag{10}$$

where ρ is the cosine of the accepted angle difference, α.

$$\rho = \cos(\alpha) \tag{11}$$

In our test examples we used $\alpha = 0.4363$ radians, which means $\rho \approx 0.9$. We experience that this value greatly reduces the number of cross-overs. Depending on the resolution and accuracy of the signed distance field (10) can be extremely sensitive to numerical errors. In such cases setting ρ too tight would result in almost no extrusion.

In some rare cases a surface normal can be aligned with the symmetry set in such a way that the signed distance value along the extrusion line keeps on decreasing, while stepping along the symmetry set. That is, the directional derivative is still positive while stepping along inside the symmetry set. This is illustrated in Figure 2(b). The directional derivative, $\nabla \phi(\boldsymbol{q}) \cdot \boldsymbol{n}$, is 1 until the symmetry axis is hit. However, the distance value is decreasing while stepping along the symmetry axis. The distance value will not increase before we reach the point $\boldsymbol{q}$ at the end of the symmetry line. This creates a tentacle for the green prism. The problem is very unlikely in practice, and we have not encountered it in our test-runs, most likely due to noise caused by approximation errors in $\nabla \phi(\boldsymbol{q})$) and interpolation errors in $\phi(\boldsymbol{q})$. This observation yields another hint at how to minimize the chance of this problem to occur. An initial slight tangential perturbation of the surface mesh vertices, will destroy any special alignment with the symmetry set.

It should be noted that in some cases we actually do want to walk along a symmetry line. This is illustrated in Figure 2(c). In this case we want to step along the symmetry line, until we hit the internal junction point. At the junction point we have a singularity of $\nabla\phi(\boldsymbol{q})$. However, along the symmetry line we have a constant value of the directional derivative. If we stepped beyond the junction point, we would see a sign flip of the directional derivative. One possible resolution to both the cross-over and the tentacle problems may be to extend the normal test with an upper extrusion length limit such as the adaptive thin-shell limit [10]. There are some disadvantages of doing this, which we will discuss later on. Another possibility may be to detect the change in the slope of $\phi(\boldsymbol{q}(\varepsilon))$. However, it is numerically very sensitive to make a good estimate of the slope, and we must ask our self the question of how big a numerical error can be accepted?

Other artifacts can occur as well, we attribute these to sampling artifacts, since they are completely dependent on the original placement of the input surface mesh vertices. The first sampling artifact, illustrated in Figure 3(a), is the creation of void regions due to too coarse a sampling. Thus, after extrusion the extruded prisms provide a poor fit to the medial axis leading to an empty void region inside the object. If we add more surface mesh vertices the extruded prisms

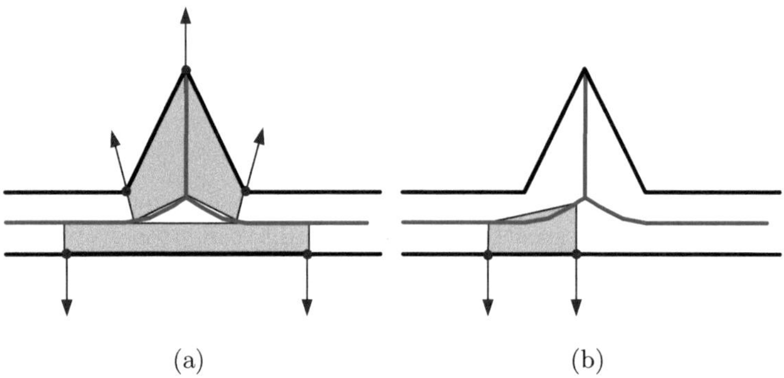

Fig. 3. (a) The coarse sampling of the blue surface mesh vertices is not enough to capture the higher order curve of the medial axis, shown in red. (b) The coarse sampling of the blue surface mesh vertices combined with unfortunate placement causes the extruded prism to pass over the medial axis creating a potential overlap with prisms extruded from the opposite side.

will come closer and closer to the medial axis. The coarse sampling combined with unfortunate placement of sampling points may even cause overlapping regions as illustrated in Figure 3(b). Both the creation of void and overlapping regions are due to sampling artifacts, thus one way to improve upon these problems is to re-sample the input surface mesh. Either by detecting good places to insert sample points or simple brute subdivision of mesh faces. Neither of these two solutions have been used for our presented test results using signed distance fields.

3 Conclusion on Signed Distance Field Extrusions

In [7] the tetrahedral GPU scan conversion method was presented, and we used this method for generating the signed distance fields in our computations. The narrow-band size was determined by talking half of the diagonal length of a tightly enclosing axis aligned bounding box around each surface mesh.

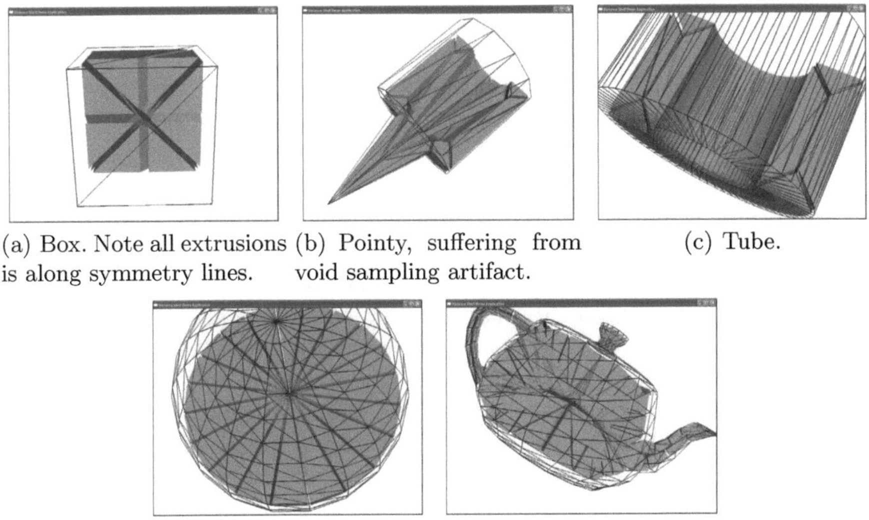

(a) Box. Note all extrusions is along symmetry lines. (b) Pointy, suffering from void sampling artifact. (c) Tube.

(d) Sphere. (e) Teapot, suffering from cross-over problem.

Fig. 4. Results of the adaptive distance shell creation method. Using signed distance field of resolution 256^3, and a normal-test with $\rho = 0.9$. Observe the overlapping shell mesh of the teapot.

In Figure 4 we have shown planar cross-intersections of a few tetrahedral shell meshes generated using our adaptive distance shell creation method. The more complex teapot shape does suffer from the cross-over problem explained above. However, aside from this it is clear that the presented extrusion method is capable of filling the internal void inside the surface meshes (shown as black wire-frame).

The signed distance field resolution is very important for the quality of the extrusion in thin regions of objects. We suggest the following rule of thumb: the higher the resolution the better the quality. This phenomena is rather trivial because the fixed extrusion length increment $\Delta\varepsilon$ is determined by the resolution of the signed distance field. The higher the resolution the smaller the increment.

To illustrate how the cross-over problem could be handled by combining with an upper extrusion length bound, we have combined the adaptive distance shell method with the extrusion limit from the previous adaptive shell method [10].

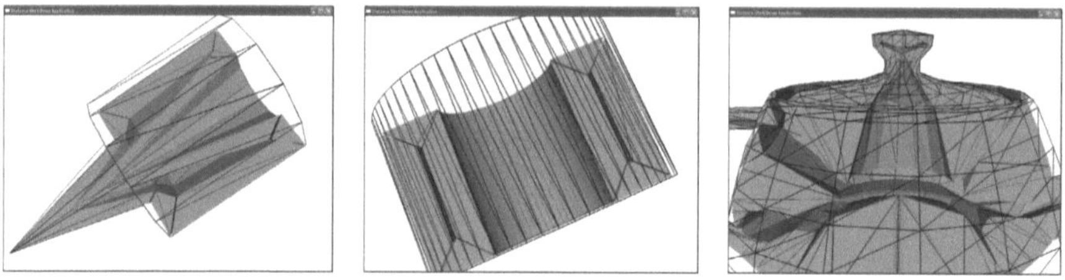

(a) Pointy, suffering from void sampling artifact. (b) Tube. (c) Teapot.

Fig. 5. Results of combining the adaptive distance shell method with upper bounds from the adaptive thin shell. Observe that overlaps have disappeared, but the artifacts of the extrusion length computation method is present.

Our results is shown in Figure 5. It is clear that the cross-over problem have been resolved. On the other hand the quality is somewhat inferior. This is because the upper limit used is not the maximum upper limit. It is though a safe limit, and it is very dependent on the local shape of the surface mesh faces.

The time-complexity of the line-stepping method is govern by two parameters, the signed distance field resolution, N and the number of vertices, V. Each extrusion line is treated independently. Thus, the algorithm scales linear in the number of vertices. The number of steps that can be taken a long an extrusion line, is bounded above by the maximum number of grid nodes encountered on the diagonal of the regular grid. Thus,

$$O(V\sqrt{3\tfrac{1}{2}N^2}) \approx O(VN). \tag{12}$$

This is extremely fast, since the operations done during each step have very low constants dominated by the computation of the gradient of a scalar field sampled on a regular grid. The major performance cost is the computation of the signed distance fields. We refer to the paper [7] for performance details hereof.

The proposed method for computing extrusion points have resulted in many discoveries: line-stepping for finding the "center-position" is the best solution over any root search method. In terms of numerical robustness normal testing is a far better stopping criteria than distance testing. Finally, thresholding on normal testing is a necessary evil due to fix-precision floating point arithmetic. This leads to possible overlapping interior regions. In practice this calls for parameter tuning. The implication is that it is not always possible to find an acceptable bound on the normal test which restrict internal overlaps and allow for the most aggressive extrusion lengths.

In conclusion, line-stepping has been identified as an algorithm that works exceptionally well on many surface meshes, but there are some cases, in which improvements are needed. The main issue is how to deal with line stepping along symmetry lines, which will be our future point of research.

References

1. Mantyla, M.: Introduction to Solid Modeling. W. H. Freeman & Co, New York, USA (1988)
2. Botsch, M., Steinberg, S., Bischoff, S., Kobbelt, L.: Openmesh–a generic and efficient polygon mesh data structure. In: Proc. Open SG Symposium (2002)
3. Molino, N., Bridson, R., Teran, J., Fedkiw, R.: Adaptive physics based tetrahedral mesh generation using level sets. (in review) (2004)
4. Diatta, A., Giblin, P.: Pre-symmetry sets of 3D shapes. In: Olsen, O.F., Florack, L.M.J., Kuijper, A. (eds.) DSSCV 2005. LNCS, vol. 3753, pp. 36–49. Springer, Heidelberg (2005)
5. Pizer, S.M., Fletcher, P.T., Joshi, S., Thall, A., Chen, J.Z., Fridman, Y., Fritsch, D.S., Gash, A.G., Glotzer, J.M., Jiroutek, M.R., Lu, C., Muller, K.E., Tracton, G., Yushkevich, P., Chaney, E.L.: Deformable m-reps for 3d medical image segmentation. International Journal of Computer Vision 55(2/3), 85–106 (2003)
6. Porumbescu, S.D., Budge, B., Feng, L., Joy, K.I.: Shell maps. ACM Trans. Graph. 24(3), 626–633 (2005)
7. Erleben, K., Dohlmann, H.: Scan conversion of signed distance fields. In: Olsen, S.I., (ed.) Proceedings of DSAGM. pp. 81–91 (2006)
8. Sethian, J.A.: Level Set Methods and Fast Marching Methods. In: Evolving Interfaces in Computational Geometry, Fluid Mechanics, Computer Vision, and Materials Science. Cambridge Monograph on Applied and Computational Mathematics, Cambridge University Press, Cambridge (1999)
9. Erleben, K., Dohlmann, H.: The thin shell tetrahedral mesh. In: Olsen, S.I., (ed.) Proceedings of DSAGM, pp. 94–102 (2004)
10. Erleben, K., Dohlmann, H., Sporring, J.: The adaptive thin shell tetrahedral mesh. Journal of WSCG, pp. 17–24 (2005)
11. Culver, T., Keyser, J., Manocha, D.: Accurate computation of the medial axis of a polyhedron. Technical Report TR98-034, University of North Carolina, Chapel Hill (1998)
12. Bradshaw, G., O'Sullivan, C.: Adaptive medial-axis approximation for sphere-tree construction. ACM Transactions on Graphics vol. 23(1) (2004)
13. Bouix, S., Siddiqi, K.: Divergence-based medial surfaces. In: ECCV '00: Proceedings of the 6th European Conference on Computer Vision-Part I, London, UK, pp. 603–618. Springer, Heidelberg (2000)
14. Dimitrov, P., Damon, J.N., Siddiqi, K.: Flux invariants for shape. In: CVPR (1), pp. 835–841 (2003)
15. Baerentzen, J.A.: Signed distance computation using the angle weighted pseudonormal. IEEE Transactions on Visualization and Computer Graphics, Member-Henrik Aanaes 11(3), 243–253 (2005)

Nonlinear Functionals in the Construction of Multiscale Affine Invariants

Esa Rahtu[1], Mikko Salo[2], and Janne Heikkilä[1]

[1] Machine Vision Group
Department of Electrical and Information Engineering
P.O. Box 4500, 90014 University of Oulu, Finland
{erahtu,jth}@ee.oulu.fi
[2] Department of Mathematics and Statistics / RNI
P.O. Box 68, 00014 University of Helsinki, Finland
mikko.salo@helsinki.fi

Abstract. In this paper we introduce affine invariants based on a multiscale framework combined with nonlinear comparison operations. The resulting descriptors are histograms, which are computed from a set of comparison results using binary coding. The new constructions are analogous to other multiscale affine invariants, but the use of highly nonlinear operations yields clear advantages in discriminability. This is also demonstrated by the experiments, where comparable recognition rates are achieved with only a fraction of the computational load. The new methods are straightforward to implement and fast to evaluate from given image patches.

1 Introduction

In computer vision, invariant features have provided an elegant way of identifying objects under geometric transformations. The appropriate transformations depend heavily on the application, but in many cases the affine transformation provides a reasonable model. Several affine invariant features have been considered. The first method, the affine invariant moments [1,2] was introduced already in 1962. Since then affine invariant spectral signatures [3], crossweighted moments [4], and the trace transform [5] have been presented. Unfortunately these methods often suffer from sensitivity to nonaffine distortions, implementational difficulties, and the lack of discriminating features.

One recent approach to affine invariants is the multiscale framework. The idea in this framework is to extend the given image to a set of affine covariant versions, each carrying slightly different information, and then to extract some known invariant characteristics from each of them separately. The construction of the affine covariant set is the key part of the approach, and it is done by combining several scaled representations of the original image. The advantage is the possibility for variations, which is also demonstrated by the amount of methods created using this framework: multiscale autoconvolution [6], spatial

B.K. Ersbøll and K.S. Pedersen (Eds.): SCIA 2007, LNCS 4522, pp. 482–491, 2007.

multiscale affine invariants [7], generalized affine invariant moments, multiscale autoconvolution histograms [8], and ridgelet-based affine invariants.

Hitherto, the invariants introduced using the multiscale framework apply simple pointwise products and convolution for creating the affine covariant sets from scaled representations. While these linear operations offer robustness to noise and other distortions, they can also easily compromise the discriminability of the features. In this paper we introduce new multiscale affine invariants, which apply nonlinear comparison operations to the combinations of the scaled representations. We also introduce a way for combining several comparisons together using a binary code construction. These nonlinear operations and the binary code construction have not been used in previous multiscale invariants. The experiments performed demonstrate that the use of comparison operations has a clear impact on the performance, and the new methods achieve comparable or better results with only a fraction of the computational load of the earlier methods.

2 Multiscale Approach

We begin with a description of the multiscale approach in constructing affine invariants. First we recall the definition of an affine transformation.

Definition 1. *A spatial affine transformation* $\mathscr{A}$ *of coordinates* $x \in \mathbf{R}^2$ *is given by* $\mathscr{A}(x) = Tx + t$*, where* $t \in \mathbf{R}^2$ *and* T *is a* 2×2 *nonsingular matrix with real entries. Further, let* $f\colon \mathbf{R}^2 \to \mathbf{R}$*,* $f \geq 0$*, be an image function. The affine transformed version* $f_{\mathscr{A}}$ *of* f *is given by*

$$f_{\mathscr{A}}(x) = f \circ \mathscr{A}^{-1}(x) = f(T^{-1}x - T^{-1}t).$$

The construction of multiscale affine invariants can be done in three steps.

1. The image f is represented in n different scales $f(\alpha_1 x), \ldots, f(\alpha_n x)$.
2. The scaled images are combined to a new image $Gf(x)$. The combination is required to be affine covariant, which means that for any affine transformation $\mathscr{A}$ one has
$$G(f \circ \mathscr{A}^{-1})(x) = (Gf)(\mathscr{A}^{-1}(x)).$$
3. An affine invariant operation is applied to f and Gf to obtain the full invariant If.

The procedure is illustrated in Figure 1. The advantage of the method is that by varying the scales α_i, the combinations G, and the affine invariant operations, it is possible to create a great variety of different features for many purposes.

The first step, scaling of images, is straightforward. The third step can also be quite simple. Possible choices for the affine invariant operation include the normalized integration

$$If = \frac{1}{\|f\|_{L^1}} \int_{\mathbf{R}^2} Gf(x)\, dx, \tag{1}$$

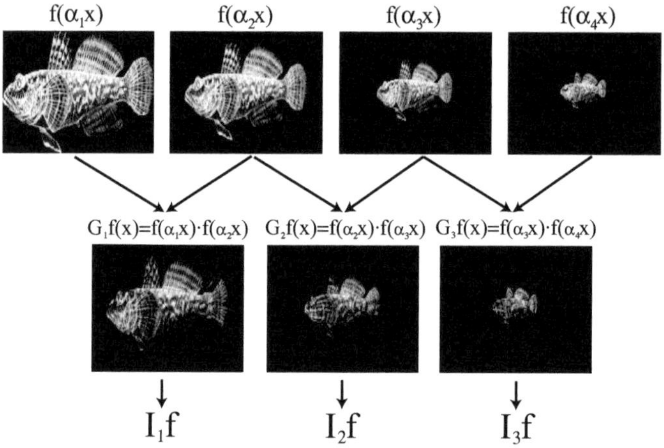

Fig. 1. Example of multiscale scheme, where two scaled representations are combined together with pointwise product. The final invariants are computed from the resulting images $G_i f$.

or more generally, any affine invariant moment [2] can be used. Also, one may apply a histogramming operation to Gf.

The second step is often the most complicated one in this approach. One needs to take the scaled images $f(\alpha_i x)$ and combine them to a new image $Gf(x)$, so that Gf and $G(f \circ \mathscr{A}^{-1})$ are related by the affine transformation $\mathscr{A}$. The reason for using scaled images is that scaling commutes with matrix products, i.e. $\alpha T = T\alpha$. The translation causes some problems. It may either be normalized away by computing the image centroid, or one may choose G more carefully so that translation invariance is obtained without finding the centroid.

We will discuss a few examples for G. First we consider the spatial multiscale invariants (SMA) [7], given here in a slightly modified form. We choose Gf to be a product of the original f and two scaled representations of it, $f(\alpha x)$, and $f(\beta x)$. In this formulation, the translation component must be normalized and it is done by computing the image centroid $\mu(f)$. The operator G is given by

$$Gf(x) = f(x)f(\alpha x + (1-\alpha)\mu(f))f(\beta x + (1-\beta)\mu(f)), \tag{2}$$

where $\alpha, \beta \in \mathbf{R}$. We then construct the invariant feature I by using the normalized integration (1). The result is the SMA transform, given by

$$Sf(\alpha,\beta) = \frac{1}{\|f\|_{L^1}} \int_{\mathbf{R}^2} f(x)f(\alpha x + (1-\alpha)\mu(f))f(\beta x + (1-\beta)\mu(f))\, dx. \tag{3}$$

Due to its simplicity, $Sf(\alpha, \beta)$ is very fast to evaluate, and the possibility of varying the scales results in an infinite number of different descriptors.

Another example is the multiscale autoconvolution (MSA) [6], where one uses a combination of convolutions and products to form Gf. One advantage of this descriptor is the fact that the translation component does not have to be considered separately. Define

$$Gf(x) = \frac{1}{\|f\|_{L^1}^2} f(x)(f_\alpha * f_\beta * f_\gamma)(x), \tag{4}$$

where $\alpha, \beta, \gamma \in \mathbf{R}$, $\alpha+\beta+\gamma = 1$, $f_a(x) = a^{-2} f(x/a)$, and $*$ denotes convolution. The actual invariant features are again constructed by normalized integration, which gives the MSA transform

$$Mf(\alpha, \beta) = \frac{1}{\|f\|_{L^1}^3} \int_{\mathbf{R}^2} f(x)(f_\alpha * f_\beta * f_\gamma)(x)\, dx. \tag{5}$$

This formulation is not computationally appealing, but fortunately Mf can be computed using the Fourier transform $\hat{f}$ as

$$Mf(\alpha, \beta) = \frac{1}{\hat{f}(0)^3} \int_{\mathbf{R}^2} \hat{f}(-\xi)\hat{f}(\alpha\xi)\hat{f}(\beta\xi)\hat{f}(\gamma\xi)\, d\xi.$$

These two examples illustrate the application of the multiscale framework in constructing affine invariants. Further examples could include generalized affine invariant moments, multiscale autoconvolution histograms, and ridgelet-based affine invariants. The basic idea of applying the multiscale framework is similar in all the examples, although the preprocessing and the computation of the actual invariant characteristics may differ.

3 The New Approach

All the multiscale affine invariants presented above perform the combination Gf of scaled images by using convolutions or pointwise products. These operations, which have a linear character, seem to behave robustly under noise, but in many applications they can compromise the discriminability of the methods. For this reason, better performance might be achieved by using other, nonlinear functionals in the combination of the scaled images.

We propose here to replace the products in the earlier constructions by pointwise comparison operations. This approach is motivated by the excellent performance of the local binary patterns (LBP) [9], where similar comparison operations were used to construct highly discriminative texture descriptors. Another motivation is the fact that comparison operations perform well under illumination distortions, which are very common in real applications. We demonstrate the new approach in two cases, based on similar formulations as in SMA and MSA.

3.1 Invariant Based on Comparison of Scaled Images

The first new invariant is analogous to SMA. However, instead of two scales we take only one, and we replace the product by the comparison operation

$$Gf(x) = G_\alpha f(x) = X(f(x), f(\alpha x + (1-\alpha)\mu(f))), \tag{6}$$

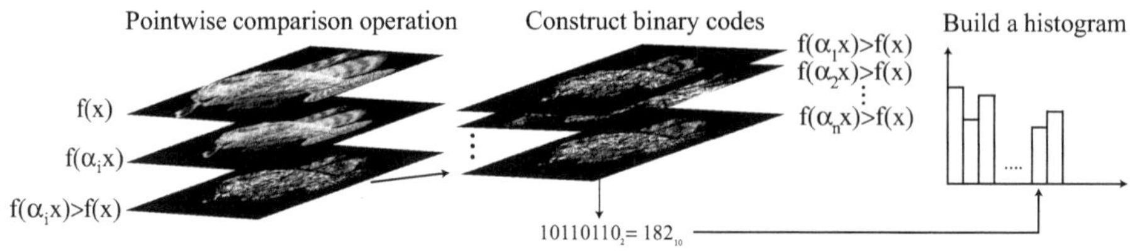

Fig. 2. Illustration of the process for generating the invariant histograms

where $\alpha \in \mathbf{R}$ and $\mu(f)$ is the image centroid, and

$$X(a,b) = \begin{cases} 1 & \text{if } a > b, \\ 0 & \text{otherwise.} \end{cases} \tag{7}$$

It is easy to show that $G(f \circ \mathscr{A}^{-1})(x) = Gf(\mathscr{A}^{-1}(x))$, which was required in the multiscale framework. Now one could simply compute the normalized integral from the binary image Gf to form the affine invariant. However, this would destroy the spatial information in Gf, and one would need to use many different α values to achieve desired discriminability. Instead, we propose to compute $G_\alpha f(x)$ with only a few different α values, say $\alpha_1, \alpha_2, \ldots, \alpha_n$, and then to use a binary code construction as in LBP to combine all the information to a new function

$$Bf(x) = B_{\alpha_1,\ldots,\alpha_n} f(x) = G_{\alpha_1} f(x) + G_{\alpha_2} f(x) \cdot 2 + \ldots + G_{\alpha_n} f(x) \cdot 2^{n-1}. \tag{8}$$

The function Bf has integer values from 0 to $2^n - 1$, and it completely encodes the information in the functions $G_{\alpha_1} f, \ldots, G_{\alpha_n} f$. It follows immediately that $B(f \circ \mathscr{A}^{-1})(x) = Bf(\mathscr{A}^{-1}(x))$, so also Bf fulfills the requirement of affine relationship in the multiscale framework. Thus, to compute the final affine invariant, one could just evaluate the normalized integral of Bf. This approach, however, does not make sense, since the different combinations would have an uneven impact to the resulting invariant value. Instead we construct a histogram $HSf(k)$, with $2^n - 2$ bins, from nonzero values of Bf, and we normalize the histogram so that the sum over all bins is equal to one. Compared to the direct integration of $G_\alpha f$, this makes it possible to preserve the relative spatial arrangements in the functions $G_{\alpha_1} f, \ldots, G_{\alpha_n} f$. The construction of the invariant is illustrated in Figure 2.

3.2 Invariant Based on Comparison of Image and Its Scaled Autoconvolution

The previous construction has the disadvantage that one needs to eliminate the translation by computing the image centroid. Also, it shares the same incompleteness issues as SMA [7]. For these reasons, we base the next construction on a formulation which is similar to MSA. Consider the convolution of two scaled representations of f as

$$Cf(x) = C_\alpha f(x) = \frac{1}{\|f\|_{L^1}}(f_\alpha * f_{1-\alpha})(x).$$

Here $\alpha \in \mathbf{R}$ and $f_a(x) = a^{-2}f(x/a)$. It is easy to see that $C(f \circ \mathscr{A}^{-1})(x) = Cf(\mathscr{A}^{-1}(x))$, and Cf is an affine covariant operator.

We take the functions $C_\alpha f$ as a basis for the new invariant, and use comparison operations for combining them. If X is the comparison operator in (7), we define

$$G_\alpha f(x) = X(f(x), C_\alpha f(x)). \tag{9}$$

It immediately follows that $G_\alpha(f \circ \mathscr{A}^{-1})(x) = G_\alpha f(\mathscr{A}^{-1}(x))$, and the requirement in the multiscale approach is satisfied. With each α value we get a binary image $G_\alpha f$, and we use these binary images in the expression (8) to get Bf. One may then form a normalized histogram of Bf as in the preceding invariant. The resulting histogram is denoted by $HMf(k)$.

4 Implementational Issues

Evaluating HS can be done similarly as SMA [7], with a few straightforward modifications. Basically the only differences are that instead of a product we use a comparison operation, which is then followed by the binary coding, and the histogram operation instead of the sum. It is also possible to apply a similar interpolation scheme as in SMA, where the interpolation grid is designed so that all the required samples are computed at once. This gave a significant speed advantage in SMA. Due to the very similar implementations, HS is almost as efficient to evaluate as SMA. The only difference is the histogramming, which is slightly slower than summing. It is easy to show that the asymptotical complexity for an $N \times N$ image is $O(N^2)$ for both HS and SMA.

The computation of HM is similar to MSA [6]. However, the product in MSA can be handled in Fourier domain, which is not possible for the comparison operation. The convolution involved in (9) can still be evaluated using the Fourier transform, and we also have the advantage that if $0 < \alpha < 1$ the Fourier transform does not have to be zero padded for accurate evaluation. In addition to this, we need to select a way to perform the scaling to produce $f(x/\alpha)$ and $f(x/(1-\alpha))$. We could do this also in Fourier domain, but as in MSA, better results are achieved by scaling in the spatial domain before taking the Fourier transform. The final asymptotical complexity of the method is the same as in discrete Fourier transform, i.e. $O(N^2 \log N)$.

The Matlab programs which were used to evaluate both HS and HM are available at the website: `http://www.ee.oulu.fi/research/imag/cmp_inv/`.

5 Experiments

In this section we assess the two new invariants HS and HM in classification experiments and compare their performance to MSA, MSA histograms

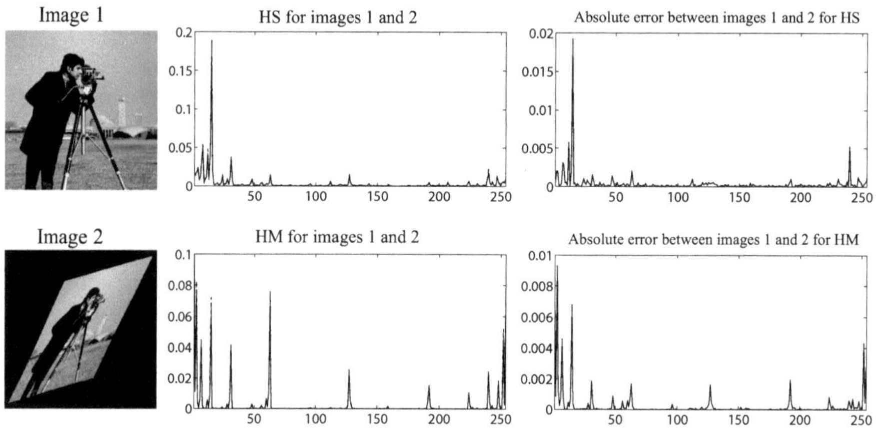

Fig. 3. Examples of the proposed affine invariant histograms. The first row has histograms HS for images 1 and 2, and the absolute difference between these. The second row has the same data for the histograms HM.

[8], SMA, and affine invariant moments (AMI) [2]. We start by giving some examples of the invariant histograms HS and HM in Figure 3. The scaling parameters used with HS were $\{0.6, 0.7, 0.8, 0.9, 1.1, 1.2, 1.3, 1.4\}$ and with HM $\{0.02, 0.055, 0.09, 0.11, 0.16, 0.205, 0.35, 0.45\}$. The same parameters are also used in the classification experiments later. Notice the slight differences between histograms for images 1 and 2, which are due to nonaffine discretization errors.

In the classification experiments, we selected the parameters for comparison methods so that they have approximately the same computational load as the corresponding new methods. In the case of MSA we used 5 invariants with $(\alpha, \beta) = \{(-0.1,\ 0.1), (-0.1, 0.3), (-0.2, 0.2), (-0.2, 0.4), (-0.3, 0.4)\}$, and in the case of SMA 10 invariants with $(\alpha, \beta) = \{(-1, -1), (-1, -0.25), (-1, 0.75), (-0.75, -0.5), (-0.75, 0.5), (-0.5, -0.5), (-0.5, 0.5), (-0.25, -0.25), (-0.25, 0.75), (0.25, 0.5)\}$. In addition to these we also computed the MSA and SMA with a larger amount of features, namely 19 for MSA and 36 for SMA. For other comparison methods we selected the MSA histogram with $(\alpha, \beta) = (-0.1, 0.3)$ and AMI with 60 independent invariants. We will refer to these methods as MSA5, SMA10, MSA19, SMA36, MSAhist, and AMI, respectively. Since the methods except for HS are not illumination invariant, we normalized the images so that they have mean 128 and standard deviation 30. The classification was performed using a simple nearest neighbor classifier, where the distance measure was the histogram intersection for the histogram approaches and the Euclidean distance for the others. The calculation of Euclidean distance was preceded by PCA decorrelation and dimension reductions in order to enhance the classification performance.

In the first experiment we classified 256×256 gray-scale images of postcards, obtained from photographs taken from different viewing angles. The training set included one image from each of the 50 different postcards and the test set

Fig. 4. Some examples of training images (first row) and angle distorted images (second row)

included 175 images of the same postcards taken from various angles in natural light. The camera we used was Canon EOS 10D with EF 17-40mm F/4L USM lens. The resulting classification problem is not easy, since in addition to the affine deformations, the test images are subjected to many other distortions due to real photographing conditions. Some examples of the training and test images are shown in Figure 4. The achieved classification results are given in Table 1, along with the approximate computation times per one 256 × 256 image. The results indicate that *HS* has the clearly the best performance if we take into account the computation time and the classification accuracy. The difference is especially large compared to SMA. Also *HM* performs well, outperforming MSA5 and MSA histograms in classification. With 19 features for MSA we were able to achieve better recognition, but with a significantly larger computational load.

Table 1. Classification error percents under real view angle distortions

	HS	HM	MSA5	MSA19	MSA hist.	SMA10	SMA36	AMI
Classification error	3.4 %	5.7 %	22.9 %	3.4 %	27.4 %	36.6 %	26.3 %	65.7 %
Execution time	1.08 s	1.89 s	2.42 s	11.40 s	0.19 s	2.62 s	5.09 s	0.29 s

Table 2. Classification error percents under illumination distortions

	HS	HM	MSA19	SMA36
Underexposure by 1.5 apertures	2.4 %	2.4 %	1.6 %	10.4 %
Underexposure by 3 apertures	4.0 %	5.6 %	0.0 %	11.2 %

Table 3. Classification error percents under heavy illumination and noise distortions

	HS	HM	MSA19	SMA36
Underexposure by 2 and ISO 3200	24.0 %	30.0 %	2.0 %	4.0 %

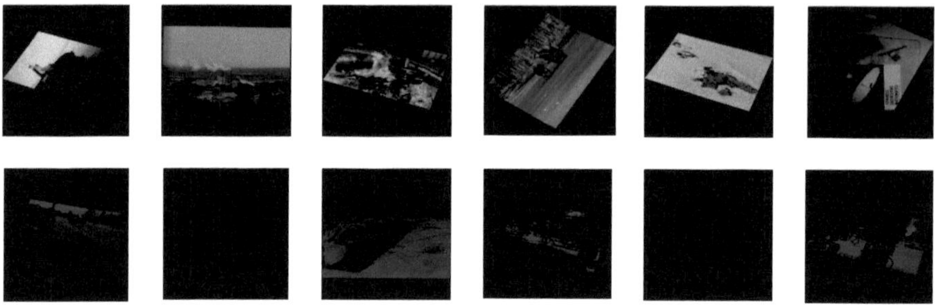

Fig. 5. Examples of illumination distorted images. With the images on first row the underexposure is 1.5 apertures and on the second row 3 apertures.

We continued by introducing more distortions to the image. First we added illumination distortions by taking a new set of 125 test photos, underexposing them first by 1.5, and then by 3 aperture steps of the camera. The changes in the viewing angle were kept quite small in this experiment. Some examples of the new test sets are shown in Figure 5. The results of this experiment are illustrated in Table 2. We omitted the MSA5, MSA histogram, SMA10, and AMI methods from this experiment due to their high error rates in the first experiment. It can be observed that all the tested methods, except SMA36, are very tolerant to illumination changes. As expected, MSA with 19 invariants works quite robustly, but similar results are achieved with HS and HM with only a small fraction of the required computational load. In the case of HS there was also no need to perform illumination normalization, which simplifies the overall procedure.

As a final illustration we created one more test set of 50 images, by underexposing by 2 aperture steps and increasing the sensitivity (ISO) value of the camera to 3200. The resulting images were severely distorted by noise. The results of this experiment are shown in Table 3. The new methods, which are based on comparison operations, react to the substantial changes in gray-scale values more strongly than MSA and SMA which use product operations. In many cases, reactivity is a desirable feature of the method, but in this experiment robustness leads to better results. The experiment clearly illustrates the trade-off between robustness and discriminability.

6 Conclusions

In this paper we introduced a novel way of creating affine invariants from the multiscale framework by applying comparison operations and binary coding. The application of these nonlinear operations offered a new way to increase the discriminability of the invariants. The simplicity of the new operations made the proposed methods efficient to evaluate. The experiments performed indicated that using already a few scales in the construction of the invariants can outperform similar approaches with linear functionals. The amount of features in the traditional methods had to be drastically increased to achieve even comparable results. In addition to the two examples provided here, we expect that similar nonlinear constructions can be applied to other multiscale invariants.

Acknowledgments

This work was supported by the Academy of Finland (project no. 110751). The authors would like to thank Mr. Janne Kenttälä for imparting to us his superior expertise in photographic matters, and Mr. Andrew Kenttälä for his valiant efforts in making this possible.

References

1. Hu, M.: Visual pattern recognition by moment invariants. IEEE Trans. Information Theory 8, 179–187 (1962)
2. Flusser, J., Suk, T.: Pattern recognition by affine moment invariants. Pattern Recognition 26(1), 167–174 (1993)
3. Ben-Arie, J., Wang, Z.: Pictorial recognition of objects employing affine invariance in the frequency domain. IEEE Trans. Pattern Analysis and Machine Intelligence 20(6), 604–618 (1998)
4. Yang, Z., Cohen, F.S.: Cross-weighted moments and affine invariants for image registration and matching. IEEE Trans. Pattern Analysis and Machine Intelligence 21(8), 804–814 (1999)
5. Petrou, M., Kadyrov, A.: Affine invariant features from the trace transform. IEEE Trans. Pattern Analysis and Machine Intelligence 26(1), 30–44 (2004)
6. Rahtu, E., Salo, M., Heikkilä, J.: Affine invariant pattern recognition using multiscale autoconvolution. IEEE Transactions on Pattern Analysis and Machine Intelligence 27(6), 908–918 (2005)
7. Rahtu, E., Salo, M., Heikkilä, J.: A new efficient method for producing global affine invariants. In: Proc. International Conference on Image Analysis and Processing, 407–414, Cagliari, Italy (2005)
8. Rahtu, E., Salo, M., Heikkilä, J.: Multiscale autoconvolution histograms for affine invariant pattern recognition. In: Proc. the 16th British Machine Vision Conference, Edinburgh, UK, vol. 3 pp. 1059–1068 (2006)
9. Ojala, T., Pietikäinen, M., Mäenpää, T.: Multiresolution gray-scale and rotation invariant texture classification with Local Binary Patterns. IEEE Transactions on Pattern Analysis and Machine Intelligence 24(7), 971–987 (2002)

A New Fuzzy Impulse Noise Detection Method for Colour Images

Samuel Morillas[1,⋆,⋆⋆], Stefan Schulte[2,⋆], Etienne E. Kerre[2,⋆⋆⋆], and Guillermo Peris-Fajarnés[1]

[1] Technical University of Valencia, E.P.S. de Gandia, Carretera Nazaret-Oliva s/n, 46730 Grao de Gandia, Spain
smorillas@ieee.org

[2] Ghent University, Department of Applied Mathematics and Computer Science, Krijgslaan 281 - S9, 9000 Gent, Belgium
Stefan.Schulte@UGent.be

Abstract. This paper focuses on fuzzy image denoising techniques. In particular, we develop a new fuzzy impulse noise detection method. The main difference between the proposed method and other state-of-the-art methods is the usage of the colour components for the impulse noise detection method that are used in a more appropriate manner. The idea is to detect all noisy colour components by observing the similarity between (i) the neighbours in the same colour band and (ii) the colour components of the two other colour bands. Numerical and visual results illustrate that the proposed detection method can be used for an effective noise reduction method.

1 Introduction

Reduction of noise in digital images is one of the most basic image processing operations. Recently a lot of fuzzy based methods have shown to provide efficient image filtering [1,2,3,4,5,6,7]. These fuzzy filters are mainly developed for images corrupted with fat-tailed noise like impulse noise. Although these filters are especially developed for greyscale images, they can be used to filter colour images by applying them on each colour component separately. This approach generally introduces many colour artefacts mainly on edge and texture elements. To overcome these problems several nonlinear vector-based approaches were successfully introduced [8,9,10,11,12,13,14,15,16,17,18,19]. Nevertheless all these vector-based methods have the same major drawbacks, i.e. (i) the higher the noise level is the lower the noise reduction capability is in comparison to the component-wise approaches and (ii) they tend to cluster the noise into a

⋆ Corresponding author.

⋆⋆ S. Morillas acknowledges the support of Spanish Ministry of Education, Science under program "Becas de Formación de Profesorado Universitario FPU".

⋆⋆⋆ S. Schulte and E.E. Kerre acknowledge the support of Ghent University under the GOA-project 12.0515.03.

B.K. Ersbøll and K.S. Pedersen (Eds.): SCIA 2007, LNCS 4522, pp. 492–501, 2007.

larger array which makes it even more difficult to reduce. The reason for these disadvantages is that the vector-based approaches consider each pixel as a whole unit, while the noise can appear in only one of the three components.

In this paper another colour filtering method is proposed. As in most other applications we use the RGB colour space. The main difference between the proposed method and other state-of-the-art methods is the usage of the colour components for the impulse noise detection. The idea behind this detection phase is to detect all colour components which are dissimilar (i) to the neighbours in the same colour band and (ii) to the colour components of the two other colour bands. The proposed method illustrates the advantage of using the colour information in a more appropriate way to improve the noise reduction method. This work should also stimulate more research in the field of colour processing for image denoising.

The paper is organized as follows: in section 2 the new colour based impulse noise detection method is explained. A noise reduction method that uses the performed detection is described in section 3. Section 4 illustrates the performance of the proposed method in comparison to other state-of-the-art methods and the conclusions are finally drawn in section 5.

2 Fuzzy Impulse Noise Detection

In this section a novel fuzzy impulse noise detection method for colour images is presented. In comparison to the vector-based approaches the proposed fuzzy noise detection method is performed in each colour component separately. This implies that a fuzzy membership degree (within $[0, 1]$) in the fuzzy set *noise-free* will be assigned to each colour component of each pixel. When processing a colour, the proposed detection method examines two different relations between the central colour and its neighbouring colours to perform the detection: it is checked both (i) whether each colour component value is similar to the neighbours in the same colour band and (ii) whether the value differences in each colour band corresponds to the value differences in the other bands. In the following, the method is described in more detail.

Since we are using the RGB colour-space, the colour of the image pixel at position i is denoted as the vector $\mathbf{F}_i$ which comprises its red (R), green (G), and blue (B) component, so $\mathbf{F}_i = (F_i^R, F_i^G, F_i^B)$. Let us consider the use of a sliding filter window of size $n \times n$, with $n = 2c + 1$ and $c \in \mathbb{N}$, which should be centered at the pixel under processing. The colours within the filter window are indexed according to the scheme shown in Figure 1 for the 3×3 case. For larger window sizes the indexing will be performed in an analogous way. The colour pixel under processing is always represented by $\mathbf{F}_0 = (F_0^R, F_0^G, F_0^B)$.

First, we compute the absolute value differences between the central pixel $\mathbf{F}_0$ and each colour neighbour as follows:

$$\Delta F_k^R = |F_0^R - F_k^R|, \ \Delta F_k^G = |F_0^G - F_k^G|, \ \Delta F_k^B = |F_0^B - F_k^B| \tag{1}$$

1	2	3
4	0	5
6	7	8

Fig. 1. Vector index in the filter window

where $k = 1, \ldots, n^2 - 1$ and $\Delta F_k^R, \Delta F_k^G, \Delta F_k^B$ denote the value difference with the colour at position k in the R, G and B component, respectively. Now, we want to check if these differences can be considered as small. Since small is a linguistic term, it can be represented as a fuzzy set [20]. Fuzzy sets in turn can be represented by a membership function. We compute the membership degree in the fuzzy set *small*$_1$ using the $1-S$-membership function [20] over the computed differences. This function is defined as follows

$$1 - S(x) = \begin{cases} 1 & \text{if } x < \alpha_1 \\ 1 - 2\left(\frac{x-\gamma_1}{\gamma_1-\alpha_1}\right)^2 & \text{if } \alpha_1 < x < \frac{\alpha_1+\gamma_1}{2} \\ 2\left(\frac{x-\alpha_1}{\gamma_1-\alpha_1}\right)^2 & \text{if } \frac{\alpha_1+\gamma_1}{2} < x < \gamma_1 \\ 0 & \text{if } x > \gamma_1 \end{cases} \tag{2}$$

where we have experimentally found that $\alpha_1 = 10$ and $\gamma_1 = 50$ receive satisfying results in terms of noise detection. In this case we denote $1 - S$ by S_1, so that $S_1(\Delta F_k^R)$, $S_1(\Delta F_k^G)$, $S_1(\Delta F_k^B)$ denote the membership degrees in the fuzzy set *small*$_1$ of the computed differences with respect to the colour at position k. Now, we use the values $S_1(\Delta F_k^R)$, $S_1(\Delta F_k^G)$, $S_1(\Delta F_k^B)$ for $k = 1, \ldots, n^2 - 1$ to decide whether the values F_0^R, F_0^G and F_0^B are similar to its component neighbours. The calculation of the membership degree in the fuzzy set *noise-free* is illustrated for the R component only but is performed in an analogous way for the G and B component. Because of the noise some of the neighbours could be corrupted with noise and therefore the values of $S_1(\Delta F_k^R)$ for $k = 1, ..., n^2 - 1$ are sorted in descending order so that only the most relevant differences are considered. The value occupying the j-th position in the ordering is denoted by $S_1(\Delta F_{(j)}^R)$. Next, the similarity to the neighbour values is determined by checking that the value difference should be *small* with respect to, at least, a certain number K of neighbours. The number K of considered neighbours will be a parameter of the filter and its importance is discussed in section 4. So, we apply a fuzzy conjunction operator (fuzzy AND operation represented here by the triangular product t-norm [21,22]) among the first K ordered membership degrees in the fuzzy set *small*$_1$. The conjunction is calculated as follows:

$$\mu^R = \prod_{j=1}^{K} S_1(\Delta F_{(j)}^R), \tag{3}$$

where μ^R denotes the degree of similarity of F_0^R with respect to K of its neighbours in the most favourable case. Notice that in the case that F_0^R is noisy a low similarity degree μ^R should be expected.

The next step of the detection process is to determine whether the observed differences in the R component of the processed colour corresponds to the same observations in the G and B component. We want to check if these differences agree at least for a certain number K of neighbours. Then, for each neighbour we compute the absolute value of the difference between the membership degrees in the fuzzy set *small*$_1$ for the red and the green and for the red and the blue components, i.e. $|S_1(\Delta F_k^R) - S_1(\Delta F_k^G)|$ and $|S_1(\Delta F_k^R) - S_1(\Delta F_k^B)|$, where $k = 1, \ldots, n^2 - 1$, respectively. Now, in order to see if the computed differences are *small* we compute their fuzzy membership degrees in the fuzzy set *small*$_2$. A $1-S$-membership function is also used but now we used $\alpha_2 = 0.10$ and $\gamma_2 = 0.25$, which also have been determined experimentally. In this case we denote the membership function as S_2. So we calculate

$$\begin{aligned} \mu_k^{RG} &= S_2(|S_1(\Delta F_k^R) - S_1(\Delta F_k^G)|), \\ \mu_k^{RB} &= S_2(|S_1(\Delta F_k^R) - S_1(\Delta F_k^B)|), \end{aligned} \tag{4}$$

where μ_k^{RG} and μ_k^{RB} denote the degree in which the observed difference in the red component is similar to the observed difference in the green and blue components with respect to the colour located at position k, respectively. Now, since we want to require that the differences are similar with respect to at least K neighbours, the values of μ_k^{RG} and μ_k^{RB} are also sorted in descending order, where $\mu_{(j)}^{RG}$ and $\mu_{(j)}^{RB}$ denote the values ranked at the j-th position. Consequently, the joint similarity with respect to K neighbours is computed as

$$\mu^{RG} = \prod_{j=1}^{K} \mu_{(j)}^{RG}, \qquad \mu^{RB} = \prod_{j=1}^{K} \mu_{(j)}^{RB}, \tag{5}$$

where μ^{RG} and μ^{GB} denote the degree in which the observed differences for the red component are similar to the observed differences in the green and blue components, respectively. Notice that if F_0^R is noisy and F_0^G and F_0^B are not, then the observed differences can hardly be similar and therefore, low values of μ^{RG} and μ^{RB} are expected.

Finally, the membership degree in the fuzzy set *noise-free* for F_0^R is computed using the following fuzzy rule 1

Fuzzy Rule 1. *Defining the membership degree $NF_{F_0^R}$ for the red component F_0^R in the fuzzy set* noise-free*:*

IF μ^R *is* large *AND* μ^{RG} *is* large *AND* μ^G *is* large OR
μ^R *is* large *AND* μ^{RB} *is* large *AND* μ^B *is* large
THEN *the noise-free degree of* F_0^R *is* large

A colour component is considered as noise-free if (i) it is similar to some of its neighbour values (μ^R) and (ii) the observed differences with respect to some of its neighbours are similar to the observed differences in some of the other colour components (μ^{RG} and μ^{GB}). In addition, the degrees of similarity of the

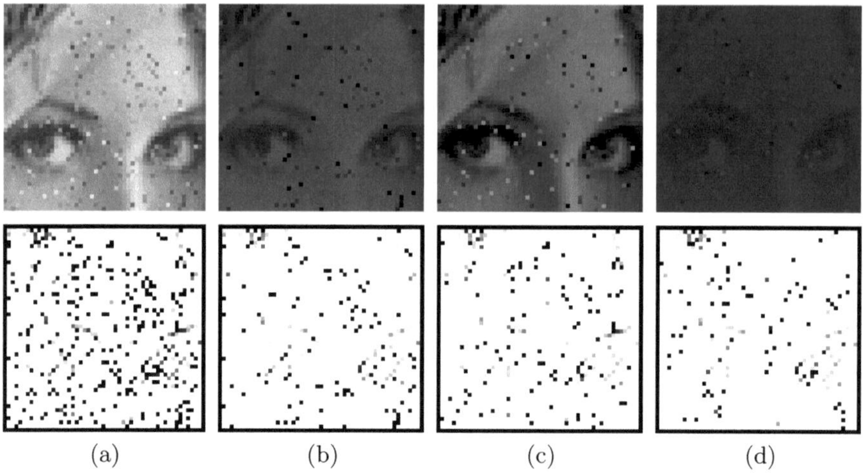

Fig. 2. An example of the proposed noise detection performance, with (a)-(d) Detail of the "Lena" image contaminated with 10% random-value impulse noise in each colour channel, and the computed noise-free degrees. Dark/white points indicate high/low noise-degree, respectively.

other components values with respect to their neighbour values, i.e. μ^G and μ^B, are included so that a probably noisy component (with a low μ^G or μ^B value) can not be taken as a reference for the similarity between the observed differences. The fuzzy rule 1 contains four conjunctions and one disjunction. In fuzzy logic triangular norms and co-norms are used to represent conjunctions and disjunctions [21,22], respectively. Since we use the product triangular norm to represent the fuzzy AND (conjunction) operator and the probabilistic sum co-norm to represent the fuzzy OR (disjunction) operator the noise-free degree of F_0^R which we denote as $NF_{F_0^R}$ is computed as follows:

$$NF_{F_0^R} = \mu^R \mu^{RG} \mu^G + \mu^R \mu^{RB} \mu^B - \mu^R \mu^{RG} \mu^G \mu^R \mu^{RB} \mu^B. \quad (6)$$

Notice that all the variables in the antecedent of the fuzzy rule 1 are already appropriate fuzzy values, so that no *fuzzyfication* is needed. Moreover, since we aim at computing a fuzzy noise-free degree, any *defuzzyfication* is neither needed.

Analogously to the calculation of the noise-free degree for the red component described above, we obtain the noise-free degrees of F_0^G and F_0^B denoted as $NF_{F_0^G}$ and $NF_{F_0^B}$ as follows

$$\begin{aligned} NF_{F_0^G} &= \mu^G \mu^{RG} \mu^R + \mu^G \mu^{GB} \mu^B - \mu^G \mu^{RG} \mu^R \mu^G \mu^{GB} \mu^B, \\ NF_{F_0^B} &= \mu^B \mu^{RB} \mu^R + \mu^B \mu^{GB} \mu^G - \mu^B \mu^{RB} \mu^R \mu^B \mu^{GB} \mu^G. \end{aligned} \quad (7)$$

In fuzzy logic, involutive negators are commonly used to represent negations. We use the standard negator $N_s(x) = 1 - x$, with $x \in [0, 1]$ [21]. By using this negation we can also derive the membership degree in the fuzzy set *noise* for each colour component, i.e. $N_{F_0^R} = 1 - NF_{F_0^R}$, where N denotes the membership degree in the fuzzy set *noise*. An example of the proposed detection method is shown in Figure 2. Note that first/last rows/columns have not been processed in this example.

3 Image Denoising Method

Now we briefly explain an image denoising method that uses the fuzzy detection in section 2. The image is denoised so that (i) each colour component is smoothed according to its noisy degree and (ii) the colour information is used to estimate the output values. We propose to compute a weight for each colour component in order to calculate a weighted averaging to obtain the output. Now we illustrate the case of the R component but it is done in an analogous way for the G and B components. The denoised R component is obtained as follows

$$\hat{F}_0^R = \frac{\sum_{k=0}^{n^2-1} W_{F_k^R} F_k^R}{\sum_{k=0}^{n^2-1} W_{F_k^R}} \qquad (8)$$

where $\hat{F}_0^R$ denotes the estimated value for the R component, $F_k^R, k = 0, ..., n^2 - 1$ denote the R component values in the filter window and $W_{F_k^R}$ are their respective weights. The weight of the component being processed $W_{F_0^R}$ is set proportionally to its noise-free degree $NF_{F_0^R}$ so that it will be less weighted, and therefore more smoothed, if its noise-free degree is lower. The weight of the neighbour components is set inversely proportional to the noise-free degree of the component being denoised $NF_{F_0^R}$. Therefore, the neighbours are more weighted as $NF_{F_0^R}$ is lower. In addition, in order to take into account the colour information, we will weigh more those components F_k^R for which it can be observed that F_k^G is similar to F_0^G or that F_k^B is similar to F_0^B. The underlying reasoning is that if two colours have similar G or B components then it is observed that the R component is also similar. Notice that in a extremely noisy situation it may happen that $W_{F_k^R} = 0, \forall k$ and then the weighted average cannot be performed. In such situations we perform a weighted vector median (WVM) operation, instead. In the WVM the weight of each vector should be set according to the vector noise-free degree which is computed as the conjunction of the noise-free degree of its RGB components.

4 Parameter Setting and Experimental Results

In this section we evaluate the performance of the proposed method and compare it with the performance of other methods. We use the Peak Signal-to-Noise Ratio (PSNR) [9] as objective measure to evaluate the quality of the filtered images.

In order to set the K parameter of the filter we have taken different images and we have contaminated them with random-value impulse noise varying its percentage from 1% to 50% in each colour component. We have computed the performance (PSNR) achieved by the proposed filter using a 3×3 filter window for all possible values of $K \in \{1, \dots, 8\}$. The obtained results seem to indicate that the most appropriate values of the K parameter are $K = 2, 3$. When the images are contaminated with low-medium percentages of noise, setting $K = 2, 3$ makes the filter able to properly detect and reduce impulse noise while preserving noise-free image areas, specially edges and textures. However, when the percentage of noise is high it is observed that some clusters of similar noisy pixels may occasionally appear in the noisy images. Using a value of $K = 2, 3$ may not be able to reduce clusters of noisy pixels larger than or equal to 3 or 4 pixels. This problem can be solved by using a larger value for K (maybe $K = 4, 5$), but in this case the performance for low densities of noise would be far from optimal because the detail-preserving ability is not so good as it is for lower values of K. Instead of this, we propose to perform a filtering based on a two-step approach. In the first step the noisy image is filtered using a 3×3 window and $K = 2$. In this step, isolated noisy pixels are reduced while uncorrupted edges and details are properly preserved. In the second step, the image is now filtered using a 5×5 window and $K = 5$. This step is intended to remove possible clusters of noisy pixels that may still remain in the image.

In the following the performance of the proposed filtering procedure, which we will entitle as *impulse noise reduction* method (INR), is compared to the performance of other state-of-the-art filters. The set of filters chosen for the comparison includes some filters for grayscale images applied in a component wise way (UF [19] and FRINRM [6]) and some colour image filters (VMF [11], PGSF [15], FISF [16] and FIDRMC [7]). Notice that some of the mentioned filters are also based on fuzzy concepts (FRINRM, FISF and FIDRMC). We have used the three well-known images Baboon, Boats and Parrots for the tests. These images have been corrupted with different percentages of random-value impulse noise in each colour channel. We have used the following percentages: 5%, 10%, 15%, 20%, 25%, 30%, 40%.

Since the proposed method uses a two-step procedure we have also filtered the test-images with the proposed filters using an analogous two-step design. The first step uses a 3×3 filter window where we used the (optimal) parameter setting suggested in the corresponding works. After the first step we have performed a second step where we use a 5×5 window size and where the corresponding (optimal) parameters are changed accordingly to the number of pixels in the window. In Tables 1-2 we have illustrated the PSNR performance achieved by all filters. The performance of the state-of-the-art methods included in the tables corresponds with the best performance achieved by the first or second step. Numbers followed by a $*$ indicate that the best performance is achieved in the first step. If no $*$ is used then the best performance is achieved by the second step. Some outputs of the filters for visual comparison are included in Figure 3 using a detail of the Parrots image corrupted by 15% of noise in each colour channel. From these results we can make the following conclusions:

Table 1. Some experimental results for comparison in terms of PSNR using the Baboon image corrupted with different densities of random-value impulse noise

Filter	5%	10%	15%	20%	25%	30%	40%
	PSNR	PSNR	PSNR	PSNR	PSNR	PSNR	PSNR
None	21.98	18.95	17.18	15.95	15.01	14.18	12.97
VMF	22.95*	22.68*	22.35*	21.93*	21.70	21.46	20.77
PGSF	25.22*	24.00*	22.83*	22.04	21.51	20.95	19.66
FISF	25.29*	24.08*	23.33*	22.96*	22.39*	21.75*	20.74
FIDRMC	26.09*	25.48*	24.72*	24.02*	23.30*	22.86	21.85
UF	24.17*	23.94*	23.65*	23.37*	23.07*	22.72*	21.95
FRINRM	29.12*	26.85*	25.25	24.55	23.75	22.82	20.29
INR	**30.64***	**28.88***	**27.03**	**25.99**	**25.09**	**24.24**	**22.61**

Table 2. Some experimental results for comparison in terms of PSNR using the Boat image corrupted with different densities of random-value impulse noise

Filter	5%	10%	15%	20%	25%	30%	40%
	PSNR	PSNR	PSNR	PSNR	PSNR	PSNR	PSNR
None	21.75	18.78	16.99	15.79	14.82	13.99	12.73
VMF	30.28*	29.42*	28.20*	26.70*	26.11	25.46	23.81
PGSF	33.42*	30.30*	28.45	27.24	26.08	24.64	22.08
FISF	31.63*	30.14*	29.01*	27.80*	26.73*	25.16*	23.87
FIDRMC	34.25*	32.41*	31.00	29.79	29.05	27.95	25.80
UF	33.08*	32.13*	31.32*	30.46*	29.65*	28.67*	**26.79**
FRINRM	36.80*	32.38	31.28	30.10	28.88	27.14	23.07
INR	**38.48***	**34.77***	**32.84**	**31.36**	**30.31**	**29.10**	26.37

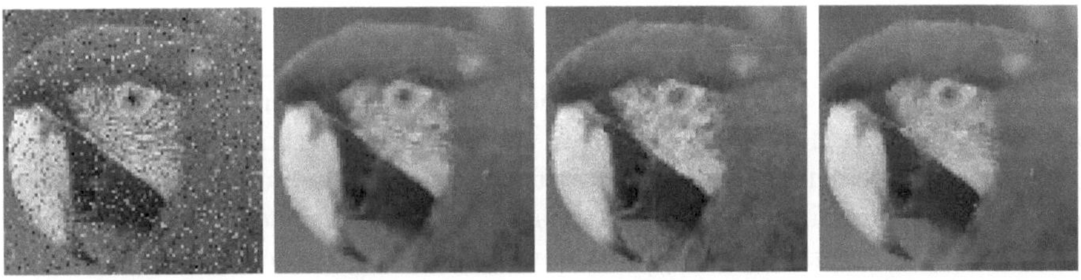

(a) PSNR = 16.50 (b) PSNR = 26.38* (c) PSNR = 26.39 (d) PSNR = 27.47*

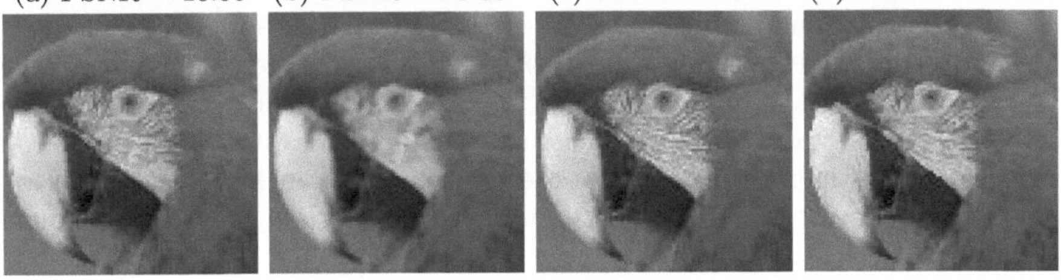

(e) PSNR = 29.36 (f) PSNR = 28.01* (g) PSNR = 29.99 (h) PSNR = 30.80

Fig. 3. Visual comparison of filters performance. (a) Detail of Parrots image with 15% of random-value impulse noise in each colour channel and outputs from (b) VMF, (c) PGSF, (d) FISF, (e) FIDRMC, (f) UF, (g) FRINRM and (h) INR.

- The proposed method generally receives the best PSNR values, which indicates that the proposed method receives the best filtering capability. Other filters such as the UF, may eventually receive slightly better PSNR values however, from the visual results we illustrate that the other methods have some important disadvantages in comparison to the proposed method.
- From the images we observe the main problem of the filtering algorithms that are applied component-wise, i.e. they even introduce (impulse like) colour artefacts in heterogeneous areas like edges or fine texture areas. By processing each component separately it often happens that colour component differences were destroyed.
- The vector based approaches do not introduce artefacts but tend to cluster the noise in larger areas, as in the case of PGSF. This makes it much more difficult to reduce the remaining noise. Additionally we observe that the results from the vector based approaches tend to make the images much blurrier (smoother) than the other methods so that important image structures are destroyed.
- The best visual results were obtained by the proposed method. We observe that the proposed method reduces the noise very well, while preserving the colour information and the important image features like edges and textures.

From both the numerical and visual results we can conclude that the proposed method can be advised for reducing random-value impulse noise in colour images since it generally outperforms other state-of-the-art methods.

5 Conclusion

In this paper a new fuzzy filter for impulse noise reduction in colour images is presented. The main difference between the proposed method (denoted as INR) and other classical noise reduction methods is that the colour information is taken into account in a more appropriate way (i) to develop a better impulse noise reduction method and (ii) to develop a noise reduction method which reduces the noise effectively. The advantages of the proposed method are (i) it reduces random-value impulse noise (for low and high noise levels) effectively, (ii) it preserves edge sharpness and (iii) it doesn't introduce blurring artefacts or new colours artefacts in comparison to other state-of-the art methods. This method also illustrates that colour images should be treated differently than grayscale images in order to increase the visual performance. Also, this method could possibly be extended to process multispectral images produced by so many satellites.

References

1. Wang, J.H., Liu, W.J., Lin, L.D.: Histogram-Based Fuzzy Filter for Image Restoration. IEEE Transactions on Systems man and cybernetics part B.-cybernetics 32(2), 230–238 (2002)

2. Schulte, S., Nachtegael, M., De Witte, V., Van der Weken, D., Kerre, E.E.: A Fuzzy Impulse Noise Detection and Reduction Method. IEEE Transactions on Image Processing 15(5), 1153–1162 (2006)
3. Farbiz, F., Menhaj, M.B.: A fuzzy logic control based approch for image filtering. In: Kerre, E.E., Nachtegael, M. (eds.) Fuzzy Techniques in Image Processing, vol. 52, pp. 194–221. Springer Physica Verlag, Berlin Heidelberg New York (2000)
4. Xu, H., Zhu, G., Peng, H., Wang, D.: Adaptive fuzzy switching filter for images corrupted by impulse noise. Pattern Recognition Letters 25, 1657–1663 (2004)
5. Kalaykov, I., Tolt, G.: Real-time image noise cancellation based on fuzzy similarity. In: Nachtegael, M., Van der Weken, D., Van De Ville, D., Kerre, E.E. (eds.) Fuzzy Filters for Image Processing, vol. 122, pp. 54–71. Springer Physica Verlag, Berlin Heidelberg New York (2003)
6. Schulte, S., De Witte, V., Nachtegael, M., Van der Weken, D., Kerre, E.E.: Fuzzy random impulse noise reduction method. Fuzzy Sets and Systems. In press (2007)
7. Schulte, S., De Witte, V., Nachtegael, M., Van der Weken, D., Kerre, E.E.: Fuzzy two-step filter for impulse noise reduction from color images. IEEE Transactions on Image Processing 15(11), 3567–3578 (2006)
8. David, H.A., Nagaraja, H.N.: Order Statistics, 3rd edn. Wiley, New York (2003)
9. Plataniotis, K.N., Venetsanopoulos, A.N.: Color Image Processing and Applications. Springer, Heidelberg (1998)
10. Lukac, R.: Adaptive vector median filtering. Pattern Recognition Letters 24(12), 1889–1899 (2003)
11. Astola, J., Haavisto, P., Neuvo, Y.: Vector Median Filters. IEEE Proceedings 78(4), 678–689 (1990)
12. Barni, M., Cappellini, V., Mecocci, A.: Fast vector median filter based on Euclidean norm approximation. IEEE Signal Processing Letters 1(6), 92–94 (1994)
13. Lukac, R., Plataniotis, K.N., Venetsanopoulos, A.N., Smolka, B.: A statistically-switched adaptive vector median filter. Journal of Intelligent and Robotic Systems 42(4), 361–391 (2005)
14. Camacho, J., Morillas, S., Latorre, P.: Efficient Impulse Noise suppression based on Statistical Confidence Limits. Journal of Imaging Science and Technology 50(5), 427–436 (2006)
15. Smolka, B., Chydzinski, A.: Fast detection and impulsive noise removal in color images. Real-Time Imaging 11(5-6), 389–402 (2005)
16. Hore, S., Qiu, B., Wu, H.R.: Improved vector filtering for color images using fuzzy noise detection. Optical Engineering 42(6), 1656–1664 (2003)
17. Morillas, S., Gregori, V., Peris-Fajarnés, G., Latorre, P.: A fast impulsive noise color image filter using fuzzy metrics. Real-Time Imaging 11(5-6), 417–428 (2005)
18. Lukac, R., Smolka, B., Martin, K., Plataniotis, K.N., Venetsanopoulos, A.N.: Vector filtering for color imaging. IEEE Signal Processing Magazine 22(1), 74–86 (2005)
19. Garnett, R., Huegerich, T., Chui, C., He, W.: A universal noise removal algorithm with an impulse detector. IEEE Transactions on Image Processing 14(11), 1747–1754 (2005)
20. Kerre, E.E.: Fuzzy sets and approximate Reasoning. Xian Jiaotong University Press (1998)
21. Lee, C.C.: Fuzzy logic in control systems: fuzzy logic controller-parts 1 and 2. IEEE Transactions on Systems, Man, and Cybernetics 20(2) 404-435
22. Fodor, J.: A new look at fuzzy-connectives. Fuzzy sets and Systems 57(2), 141–148 (1993)

On Reasoning over Tracking Events

Daniel Rowe[1], Jordi Gonzàlez[2], Ivan Huerta[1], and Juan J. Villanueva[1]

[1] Computer Vision Centre / Computer Science Department, UAB, Barcelona, Spain
[2] Institut de Robòtica i Informàtica Industrial, UPC, Barcelona, Spain

Abstract. High-level understanding of motion events is a critical task in any system which aims to analyse dynamic human-populated scenes. However, current tracking techniques still do not address complex interaction events among multiple targets. In this paper, a principled event-management framework is proposed, and it is included in a hierarchical and modular tracking architecture. Multiple-target interaction events, and a proper scheme for tracker instantiation and removal according to scene events, are considered. Multiple-target group management allows the system to switch among different operation modes. Robust and accurate tracking results have been obtained in both indoor and outdoor scenarios, without considering a-priori knowledge about either the scene or the targets based on a previous training period.

1 Introduction

High-level event understanding is a complex and essential task in any Image Sequence Evaluation (ISE) system [10,3]. This transforms image-sequence data into semantic descriptions; subsequently, these descriptions are processed, and the system reacts in terms of signal triggers or conceptual terms. Such a system could perform a smart video surveillance, an intelligent gestural user-computer interface, or any other application in orthopedics and athlete performance analysis, natural-language scene description, or computer animation [1,5,7].

A robust and accurate multiple-people tracking is a crucial component of any ISE system. However, a proper event detection and management is critical for tracking success. Further, this provides a valuable knowledge to achieve scene understanding. Thus, complex event management requires (i) considering simultaneously multiple target interactions, specially when no assumption is made with respect to the targets' trajectories; and (ii), since in an open-world scenario targets can enter and exit the scene, a procedure has to be implemented to reliably perform tracker instantiation and removal.

Despite this interest and the increasing number of proposed algorithms which deal with multiple interacting targets in open-world scenarios, this still constitutes an open problem which is far from been solved. Yang et al. [14] proposed a system with some similarities to ours, albeit grouped targets are not independently tracked and no complex situation —for instance, those in which a group of more-than-two members split— can satisfactorily be tackled. The cues and models used are essentially different. Wu et al. [13] address occlusions events within

B.K. Ersbøll and K.S. Pedersen (Eds.): SCIA 2007, LNCS 4522, pp. 502–511, 2007.

a Particle Filter (PF) framework by implementing a Dynamic Bayesian Network (DBN) with an extra hidden process for occlusion handling. BraMBLe [6] is an interesting approach to multiple-blob tracking which models both background and foreground using Mixtures of Gaussians (MoG). However, no model update is performed, there is a common foreground model for all targets, and suffers for the curse of dimensionality, as all PF-based methods which tackle multiple-target tracking combining information about all targets in every sample. Alternatively, several approaches take advantage of 3D information by making use of a known camera model and assuming that agents move on a known ground plane. These and other assumptions relative to a known Sun position or constrained standing postures allow the system presented in [15] to initialise trackers on people who do not enter the scene isolated.

Simultaneous tracking of numerous target has been just recently considered [9]. This forces tracking systems to tackle more complex interacting events than before. In this paper, a principled event-management framework is proposed, and it is included in a hierarchical and modular tracking architecture. Multiple-target interaction events are handled by means of a state machine, which consider all possible grouping configuration. This is crucial in order to achieve successful performances, by allowing the system to switch among different tracking approaches depending on the current event [8]. Further, a proper scheme for tracker instantiation and removal is proposed, which is basic in open-world applications.

The remainder of this paper is organized as follows. Section 2 outlines the system architecture. Section 3 details the event management approach. Section 4 shows some experimental results obtained from well-known databases, and finally, section 5 summarises the conclusions, and proposes future-work lines.

2 Tracking Framework

Due to the inherent complexity involved in non-supervised multiple-human tracking, a structured framework is proposed to accomplish this task. We take advantage of the modular and hierarchically-organised system published in preliminary works [4,12]. This is based on a set of co-operating modules distributed in three levels. These levels are defined according to the different functionalities to be performed, namely target detection, low-level tracking, and high-level tracking, see Fig. 1. A remarkable characteristic of this architecture is that the tracking task is split into two levels: a lower level based on a short-term blob tracker, and a long-term high-level appearance tracker. The latter automatically builds and tunes multiple appearance models, manages the events in which the target is involved, and selects the most appropriate tracking approach according to these.

In general, reliable target segmentation is critical in order to achieve an accurate feature extraction without considering any prior knowledge about potential targets, specially in dynamic scenes. However, complex interacting agents who move through cluttered environments require high-level reasoning. Thus, our proposal combines in a principled architecture both bottom-up and top-down approaches: the former provides the system with initialisation, error-recovering

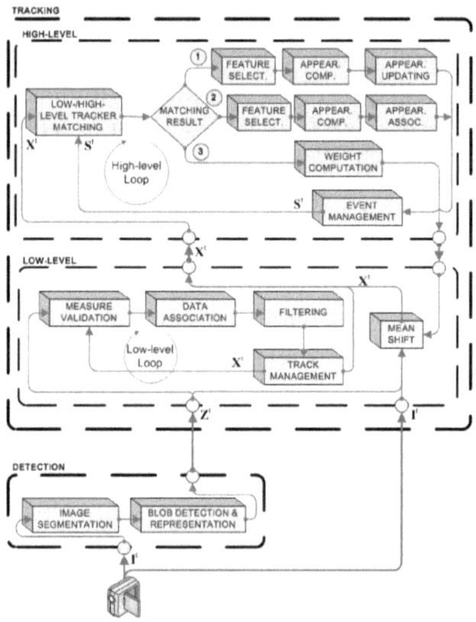

Fig. 1. System architecture. $\mathbf{I}^t$ represents the current frame, $\mathbf{Z}^t$ represents the observations, $\mathbf{X}^t$ the target's low-level state, and $\mathbf{S}^t$ the target's high level state. Matching results are explained in the text.

and simultaneous modelling and tracking capabilities, while the latter builds the models according to a high-level event interpretation, and allows the system to switch among different operation modes.

The first level performs target detection. First, the segmentation task is accomplished following a statistical colour background-subtraction approach. Next, the obtained image masks are filtered, and object blobs are extracted. Each blob is labelled, their contours are computed, and they are parametrically represented. Consequently, the spurious structural changes that they may undergo are constrained. These include target fragmentation due to camouflage, or the inclusion of shadows and reflections. Moreover, this representation can be handled by the low-level tracker, thereby filtering the target state and reducing also these effects. An ellipse representation —which keeps the blob first and second order moments— is chosen [2,11]. Thus, the j-observed blob at time t is given by the vector $\mathbf{z}_j^t = \left(x_j^t, y_j^t, h_j^t, w_j^t, \theta_j^t\right)$, where x_j^t, y_j^t represent the ellipse centroid, h_j^t, w_j^t are the major and minor axes, respectively, and the θ_j^t gives the angle between the abscissa axis and the ellipse major one. Low-level trackers establish coherent target relations between frames by setting correspondences between observations and trackers, and by estimating new target states according to the associated observations using a bank of Kalman filters. Finally, the *track-management* module (i) initiates tentative tracks for those observations which are not associated; (ii) confirms tracks with enough supporting observations; and (iii) removes low-quality ones. Results are forwarded to high-level trackers, and fed back to the measure-validation module. See [4] for details.

A high-level tracker is instantiated whenever a low-level track is first confirmed. Hence, tracking events can be managed. This allows target tracking even when image segmentation is not feasible, and low-level trackers are removed, such as during long-duration occlusions or grouping. As a result of the tracker matching, three cases are considered: (i) if the track is stable, the target appearance is computed and updated, see matching result (1) in Fig. 1; (ii) those high-level trackers which remain orphans are processed to obtain an appearance-based data association, thereby establishing correspondences between lost high-level trackers and new ones, see matching result (2). The details of this procedure can be found in [12]; and, (iii) those targets which have no correspondence are tracked in a top-down process using appearance-based trackers, see matching result (3). An *event* module determines what is happening within the scene, such as a target is grouping or it is entering into the scene. These results are fed back, thereby allowing low-level and high-level tracker matching. The aim of this paper is to propose an approach for event management.

3 Event Management

Multiple-people tracking requires considering potential target interactions among them, specially when no assumption is made with respect to their trajectories. These interactions will be referred in the following as *interaction events*. Further, in open worlds targets can enter and exit the scene, or a Region Of Interest (ROI) defined on it. These events will be referred as *scene events*, and they have an important role in matching low-level and high-level trackers, and in managing the latters. Both types of events will be managed as follows.

3.1 On Interaction Events

A proper detection of these events is crucial to achieve successful performances, since a different tracking approach must be used in each case. On the one hand, whenever a detected blob clusters more than one target, tracking by motion detection is no longer feasible, and no accurate target position can be obtained. On the other hand, appearance-based trackers suffer from a poor target localisation, and therefore they are not the optimal choice when an appropriate detection can be performed. Thus, by detecting these events, several operation modes could be introduced and properly selected. Further, this represents a significant knowledge which can be used for scene understanding.

Two targets are said to be *in-collision* when their *safety areas* superpose themselves. These areas are defined according to the targets' sizes. Thus, the following states are defined: (i) a target is considered as *single* if it does not collide with any other target within the scene; (ii) targets are said to be *grouping* if they do collide, but no group is being tracked in their area; (iii) targets are considered as *grouped* if they collide, they are over a group tracker area, and the group tracker is currently associated with an observation; (iv) finally, trackers are said to be *splitting* once the group has no longer an observation, but they

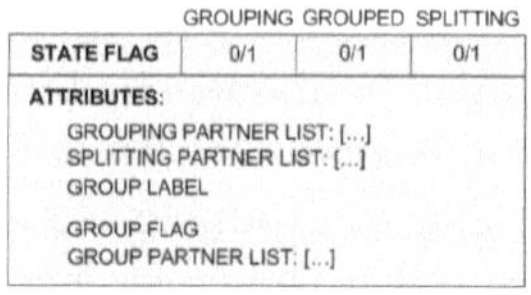

Fig. 2. Target state coding

do still collide. The frame rate is supposed to be high enough so that a target cannot change from grouped to single without ever being splitting.

Unfortunately, the above-presented classification does not suffice in complex scenarios where clusters of more than one target may be formed; for instance, one target could be grouping with a second one at the same time as splitting from a third one. Hence, the aforementioned scheme should be generalised by taking into account multiple and different target interactions.

The interaction state is coded using a three-bit vector, where each bit point outs whether the target is grouping, grouped or splitting. When every bit is set to zero, the target's state is single. Otherwise, the state could be a mixture of the previously defined situations. Secondly, several attributes are associated with each state. These point out relevant information to solve queries about current interaction events: which targets are interacting, which ones are simultaneously grouping and splitting, with which targets are they grouping, etc. Two cases are distinguished, depending on whether the tracker tracks a target or a group of them. In the first case, two lists of grouping and splitting partners are kept. Further, the group label, if this exists, is stored. In the second one, a flag pointing out that the tracker tracks a group is defined. In addition, a list of grouped targets is also kept. Thus, the eight possible states include all potential tracking situation, and these, along with the associated attributes, constitute all the necessary knowledge to solve any query relative to target interaction, see Fig 2.

Next, several events must be taken into account in order to define state transitions. These include issues such as target collision with another target, or with a group, whether the group has an associated observation or not, if there are new partners in collision, or whether old ones are no longer partners.

Thus, once all targets' positions and sizes are estimated, a collision map is computed. The collision map is also used to determine whether a new-born tracker represents a group: in this case, it is instantiated over a collision zone. Then, when two single targets are colliding, and none of them is a new target, their states change into grouping. If they also collide with a group tracker with an associated observation, their states are set to grouped. Once the group tracker has no longer an associated observation, but they still collide, their states change into splitting. More complex situations can be taken into account by considering the previous and current partner list. Finally, a tracker that stop colliding becomes single again. As an example of complex interaction, consider a target whose state is grouped; the following events occur: (i) it is colliding with some other targets, (ii) the group has no associated observation, and (iii) new partners are also colliding. As a result, it changes its state into *grouping and splitting.*

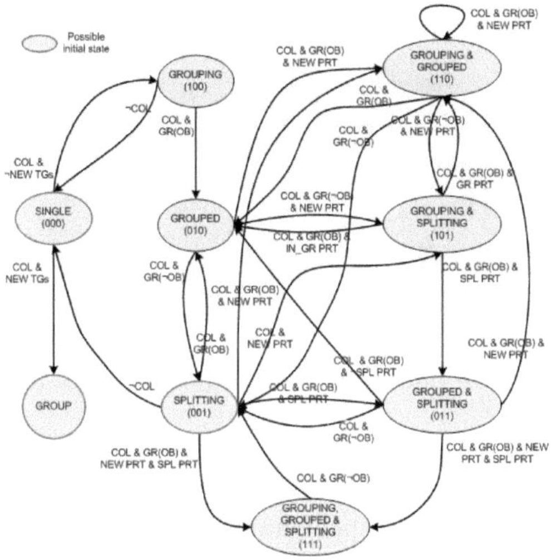

Fig. 3. Group management. Eight possible target states, and a state for group trackers, are defined (represented by ellipses). Interaction events are denoted by arrows. Notice that some of the less frequent transitions are not drawn for the sake of clarity.

The state machine that models the group management is defined by eight plus one states. The formers are defined for target trackers, and the latter for group trackers. Thus, there are 56 potential transitions between target states, although a fraction of them are not feasible according to the aforementioned assumptions. For instance, grouped targets cannot become single, since they have to split before. It is possible to perform changes in the attributes without this meaning a state transition. This is the case when several targets are already grouping, and a new one joins them. The state machine is show in Fig. 3. It should be remarked that it is not possible to add new partners to a group without first removing the group and then creating a new one. This happens because new observations won't be assigned to the former group since both position are shape would have undergone important changes. This is however a desirable effect since the new group would have a different number of partners, and therefore it is actually a different group. Although the current proposal do not allow yet to initially track people who do not enter into the scene isolated, it do detect them as they split and stable trackers are instantiated over the group region.

3.2 On Scene Events

A proper handling of scene events is essential in order to achieve successful system performances in open-world applications. In these, the number of targets within the scene is not a-priori known, and it may vary as new targets enter the scene, or other ones exit it. By defining a Region of Interest (ROI) within the scene boundaries three aims can be achieve: (i) it is not necessary to fully process the whole image, and therefore this favours accomplishing

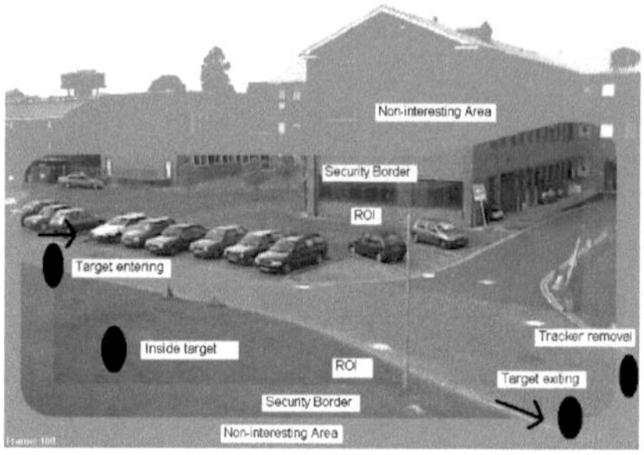

Fig. 4. Scene regions. The three regions defined on an image from PETS database.

real-time performances; (ii) the number of false positives can be effectively reduced, by avoiding detections in non-plausible or non-interesting areas, like the sky in a pedestrian-surveillance application; and (iii) targets can be completely segmented.

Three regions are here defined: a ROI, a security border, and non-interesting areas. These are used to determine where targets can be detected, where low-level and high-level trackers can be instantiated, and when they can be removed. The security border prevents the system from creating and removing trackers following the same target placed on the ROI frontier.

Thus, pixel segmentation is carried out in the whole image, since targets' sizes are not a-priori known. However, targets are only detected if the centroid of the corresponding blob lies within the ROI or the security border. For each detected target, a low-level tracker is instantiated. Once a low-level tracker is confirmed, a high-level tracker can be instantiated. This requires that the tracker has an associated observation, which implies that the target centroid is within the aforementioned area, and that the target is at least partially within the ROI. High-level trackers are instantiated as *entering*, except when they come from a group that have split. This status last until they completely lie within the ROI. When a part of the target is partially outside the ROI and the security border, the target is marked as *exiting*. The target can now either return to the ROI, or lie completely outside the area defined by the ROI and the security border. The latter implies the tracker removal. Trackers are also removed if they are partially in the outer zone and they are being tracked by a low-confidence appearance tracker, thereby avoiding a senseless gradient-based search when the target has actually exited. An example is shown in Fig. 4.

4 Experimental Results

The performance of the system has been tested using sequences taken from two well-known data-sets: the CAVIAR database[1], and PETS 2001 Test Case Scenario[2]. The former corresponds to indoor sequences which have been recorded

[1] http://homepages.inf.ed.ac.uk/rbf/CAVIAR
[2] http://peipa.essex.ac.uk/ipa/pix/pets

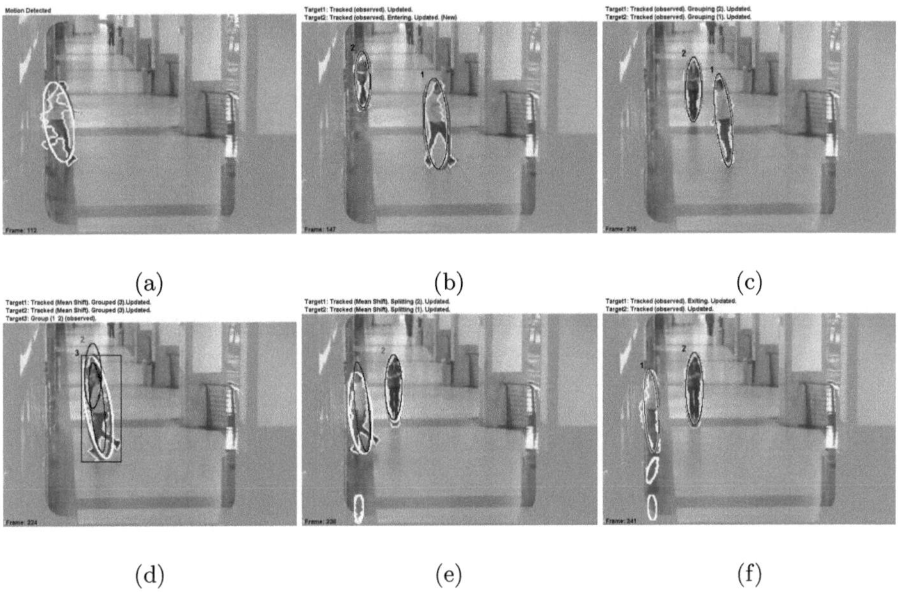

Fig. 5. Tracking results on a indoor sequence

in a mall centre, whereas the latter contains outdoor sequences taken in a scene which includes roads, parking places, green areas, and several buildings.

In the sequence *OneLeaveShopReenter1cor* (CAVIAR database, 389 frames at 25 fps, 384 x 288 pixels), two targets are tracked simultaneously, despite their being articulated and deformable objects whose dynamics are highly non-linear, and that move through an environment which locally mimics the target colour appearance. The first target performs a rotation and heads towards the second one, eventually occluding it. The background colour distribution is so similar to the target one that it constitutes a strong source of clutter. Furthermore, several oriented lighting sources are present, dramatically affecting the target appearance depending on its position and orientation (notice the bluish effect on the floor on the right of the corridor, and the reddish one on the floor of the left of the corridor). Thus, significant speed, size, shape and appearance changes can be observed, jointly with events such as people grouping, partial occlusions and group splitting. The sequence *DATASET1_TESTING_CAMERA1* (PETS database, 2688 frames at 29.97 fps, 768 x 576 pixels) presents a high variety of targets entering into the scene: three isolated people, two groups of people, three cars, and a person who exits from a parked car. These cause multiple tracking events in which several targets are involved in different grouping, grouped, and splitting situations simultaneously.

Targets are accurately tracked along both sequences. All events are correctly detected. Fig. 5 shows a sequence successful event detections for both targets. Blobs in motion are detected and low-level trackers are created. Once they enter the scene, high-level trackers are instantiated and associated to the stable low-level ones. A grouping event is correctly detected, and the operation mode is changed to

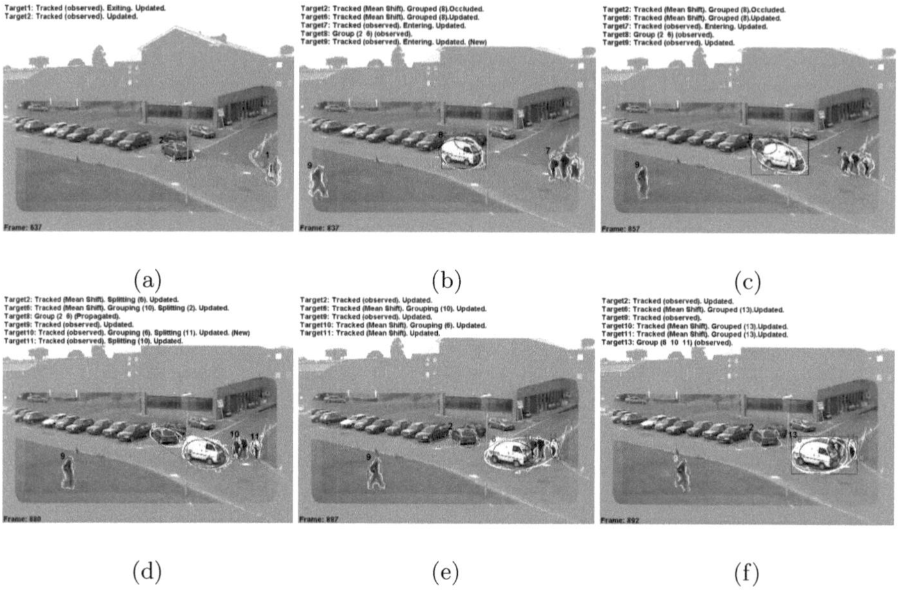

Fig. 6. Tracking results on an outdoor sequence

appearance tracking. Despite the strong occlusion of target 2, both targets are accurately tracked while they are grouped. Finally, the split event is detected and the operation mode is again changed to tracking by motion. Fig. 6 shows a more complex sequence of interaction events. A group enter the scene together, see Fig. 6.(b), but an independent tracker have been associated to one person as they momentarily split. In Fig. 6.(d) targets 2 and 6 are tracked using appearance-based methods, while targets 9, 10 and 11 are tracked by motion detection. In this frame, target 2 is splitting from 6, which is also grouping with target 10. The latter is in fact a group of two people who are grouping with target 6 while splitting from target 11. In Fig. 6.(f), targets 6, 10 and 11 have conformed a stable group and all of then are being tracked by means of appearance tracking.

5 Concluding Remarks

A principled event-management framework is proposed, and it is included in a structured multiple-target tracking framework. No a priori knowledge about either the scene or the targets, based on a previous training period, is used. A remarkable characteristic of the system is its ability to manage multiple interactions among several targets. This provides a valuable knowledge in order to obtain high-level scene descriptions, while allowing the system to switch among different operation modes. The latter is crucial to achieve successful performances, since non-supervised multiple-human tracking is a complex task which demands different approaches according to different situations.

Experiments on complex indoor and outdoor scenarios have been successfully carried out, thereby demonstrating the system ability to deal with difficult situations in unconstrained and dynamic scenes. Future work will focus on segmenting groups of people who do not enter the scene isolated, thereby allowing a robust and independent target tracking. In addition, targets will be classified by distinguishing among people, vehicles and other objects in motion.

Acknowledgements. This work was supported by the Catalan Research Agency (AGAUR), the EC grants IST-027110 for the HERMES project and IST-045547 for the VIDI-Video project, and the Spanish MEC under projects TIN2006-14606 and DPI-2004-5414. Jordi Gonzàlez also acknowledges the support of a Juan de la Cierva Postdoctoral fellowship from the Spanish MEC.

References

1. Collins, R., Lipton, A., Kanade, T.: A System for Video Surveillance and Monitoring. In: 8th ITMRRS, Pittsburgh, USA, pp. 1–15. ANS (1999)
2. Comaniciu, D., Ramesh, V., Meer, P.: Kernel-based Object Tracking. PAMI 25(5), 564–577 (2003)
3. González, J.: Human Sequence Evaluation: The Key-frame Approach. PhD thesis, UAB, Spain (2004)
4. González, J., Rowe, D., Andrade, J., Villanueva, J.J.: Efficient Management of Multiple Agent Tracking through Observation Handling. In: 6th VIIP, Mallorca, Spain, pp. 585–590. IASTED/ACTA PRESS (2006)
5. Haritaoglu, I., Harwood, D., Davis, L.: W4: real-time surveillance of people and their activities. PAMI 22(8), 809–830 (2000)
6. Isard, M., MacCormick, J.: BraMBLe: A Bayesian Multiple-Blob Tracker. In: 8th ICCV, Vancouver, Canada, vol. 2, pp. 34–41. IEEE Computer Society Press, Los Alamitos (2001)
7. Kahn, R., Swain, M., Prokopowicz, P., Firby, R.: Gesture recognition using the perseus architecture. In: CVPR, San Francisco, USA, pp. 734–741. IEEE Computer Society Press, Los Alamitos (1996)
8. Matsuyama, T., Hwang, V.: SIGMA A Knowledge Based Aerial Image Understanding System. Plenum Press, New York (1990)
9. Moeslund, T., Hilton, A., Krüger, V.: A Survey of Advances in Vision-Based Human Motion Capture and Analysis. CVIU 104, 90–126 (2006)
10. Nagel, H.: Image Sequence Evaluation: 30 years and still going strong. In: 15th ICPR, Barcelona, Spain, vol. 1, pp. 149–158. IEEE Computer Society Press, Los Alamitos (2000)
11. Nummiaro, K., Koller-Meier, E., Van Gool, L.: An Adaptive Color-Based Particle Filter. IVC 21(1), 99–110 (2003)
12. Rowe, D., Reid, I., Gonzàlez, J., Villanueva, J.: Unconstrained Multiple-people Tracking. In: 28th DAGM, Berlin, Germany. LNCS, pp. 505–514. Springer, Heidelberg (2006)
13. Wu, Y., Yu, T., Hua, G.: Tracking Appearances with Occlusions. In: CVPR, Wisconsin, USA, vol. 1, pp. 789–795. IEEE Computer Society Press, Los Alamitos (2003)
14. Yang, T., Li, S., Pan, Q., Li, J.: Real-time Multiple Object Tracking with Occlusion Handling in Dynamic Scenes. In: CVPR, San Diego, USA, vol. 1, pp. 970–975. IEEE Computer Society Press, Los Alamitos (2005)
15. Zhao, T., Nevatia, R.: Tracking Multiple Humans in Complex Situations. PAMI 26(9), 1208–1221 (2004)

FPGA Implementation of kNN Classifier Based on Wavelet Transform and Partial Distance Search

Yao-Jung Yeh, Hui-Ya Li, Wen-Jyi Hwang*, and Chiung-Yao Fang

Graduate Institute of Computer Science and Information Engineering,
National Taiwan Normal University, Taipei, 117, Taiwan
spmark@ice.ntnu.edu.tw, f3807391@ms8.url.com.tw, whwang@ntnu.edu.tw,
violet@ice.ntnu.edu.tw

Abstract. A novel algorithm for field programmable gate array (FPGA) realization of kNN classifier is presented in this paper. The algorithm identifies first k closest vectors in the design set of a kNN classifier for each input vector by performing the partial distance search (PDS) in the wavelet domain. It employs subspace search, bitplane reduction and multiple-coefficient accumulation techniques for the effective reduction of the area complexity and computation latency. The proposed implementation has been embedded in a softcore CPU for physical performance measurement. Experimental results show that the implementation provides a cost-effective solution to the FPGA realization of kNN classification systems where both high throughput and low area cost are desired.

Keywords: FPGA Implementation, Partial Distance Search, Image Processing, Pattern Recognition, Nonparametric Classification.

1 Introduction

Nonparametric classification techniques such as kNN [3,5,7] have been shown to be effective for applications of statistical pattern recognition. These techniques can achieve a high classification accuracy for problems of unknown distributions. Besides, when dealing with problems of nonnormal distributions, these techniques enjoy a lower classification error rate than that achieved by the commonly used parametric classification approaches, such as linear classifiers and quadratic classifiers. However, the nonparametric classification has a major drawback that a large amount of design vectors are required in classifiers. Large design sets, which result in high computation and storage complexities for classifiers, are necessary since the nonparametric methods require adequate statistical information. Therefore, it might be difficult to apply nonparametric classification techniques for the pattern recognition applications where real-time processing is desired.

---

* To whom all correspondence should be sent.

B.K. Ersbøll and K.S. Pedersen (Eds.): SCIA 2007, LNCS 4522, pp. 512–521, 2007.

One way to reduce the computational complexity is based on partial distance search (PDS) [2,3,4] in the original or transform domains. The PDS algorithms remove undesired vectors with less number of multiplications by calculating only the partial distance. An undesired vector is identified when its partial distance is larger than the full distance of the *current optimal vector*. These algorithms have been found to be effective for accelerating the encoding process of vector quantizers without increasing the distortion. Moreover, when applies to the kNN classifier, these algorithms do not increase the classification error rate. However, because the PDS is usually implemented by software, only a moderate acceleration can be achieved.

The objective of this paper is to present a novel hardware implementation for kNN classifier using FPGA technique. In addition to the employment of FPGAs, a subspace PDS with bitplane reduction and multiple coefficient accumulation is adopted for attaining higher throughput, higher flexibility, and lower area complexity. The PDS in the transform domain may have superior performance over the PDS in the original domain provided that the transform is orthogonal, and is able to compact the energy to few coefficients in the transform domain. In our implementation, a simple orthogonal wavelet, the Haar wavelet, is applied to each input test pattern because of its simplicity for VLSI realization. Coefficients in some highpass subbands will be truncated before fast search because they contain little energy of the input pattern. In addition, the bitplanes of the least significant bits (LSBs) of remaining coefficients can be removed because they have only a limited impact on results of the PDS. The partial distance can also be accumulated multiple coefficients at a time to further accelerate the throughput. The combination of subspace search, bitplane removal and multiple coefficient accumulation effectively enhances the search speed and reduce the codebook storage size at the expense of a slight degradation in performance.

The proposed implementation has been adopted as a custom logic block in the arithmetic logic unit (ALU) of the softcore NIOS processor running on 50 MHz. Custom instructions are also derived for accessing the custom logic block. The CPU time of the NIOS processor executing the PDS program with custom instructions is measured. Experiment results show that the CPU time is lower than that of 3.0-GHz Pentium processors executing the PDS programs without the support of custom hardware. In pratical applications, the proposed circuit may be beneficial for the implementation of smart camera for low network bandwidth consumption, which directly produces the classification results instead of delivering image raw data to main host computers for classification.

2 PDS for kNN Classifier

First we briefly introduce some basic facts of the wavelet transform [6]. Let $\mathbf{X}$ be the n-stage discrete wavelet transform (DWT) of a $2^n \times 2^n$ pixel vector $\mathbf{x}$. The DWT $\mathbf{X}$ is also a $2^n \times 2^n$ pixel block containing blocks $\mathbf{x}_{L0}$ and $\mathbf{x}_{Vi}, \mathbf{x}_{Hi}, \mathbf{x}_{Di}, i = 0, .., n-1$. In the DWT, the blocks $\mathbf{x}_{L(m-1)}$ (lowpass blocks), and $\mathbf{x}_{V(m-1)}, \mathbf{x}_{H(m-1)}, \mathbf{x}_{D(m-1)}$ (vertical, horizontal, and diagonal orientation selective highpass blocks), $m = 1, .., n$, are obtained recursively from $\mathbf{x}_{Lm}$ with $\mathbf{x}_{Ln} = \mathbf{x}$, where the blocks $\mathbf{x}_{Lm}, \mathbf{x}_{Vm}, \mathbf{x}_{Hm}$ and $\mathbf{x}_{Dm}, m = 0, .., n-1$ are with dimension $2^m \times 2^m$ pixels, respectively.

In the kNN classifier, suppose there are N classes: $\omega_1, ..., \omega_N$. Each class ω_i is associated with a design set $\mathcal{S}_i$. Let $\mathcal{S} = \{\mathbf{y}^j, j = 1, ..., t\} = \cup_{i=1}^{N} \mathcal{S}_i$ be the union of the design sets for kNN classification, where t is the total number of vectors in the union. In addition, let $\mathbf{x}$ be the input vector to be classified. Both $\mathbf{x}$ and $\mathbf{y}^j$ have the same dimension $2^n \times 2^n$ pixels. In the original kNN classifier, the first k closest vectors to $\mathbf{x}$ among the vectors in $\mathcal{S}$ are first identified. Let $k_i, 1 \leq i \leq N$ be the number of vectors that belong to $\mathcal{S}_i$ (i.e. the class ω_i) in these k closest vectors. The kNN algorithm classifies the input vector $\mathbf{x}$ to class ω_p iff $p = \arg\max_i k_i$.

Let $\mathbf{Y}^j$ be the DWT of $\mathbf{y}^j$. In addition, let $D(\mathbf{u}, \mathbf{v}) = \sum_i (u_i - v_i)^2$ be the squared distance between $\mathbf{u}$ and $\mathbf{v}$. It can be shown that, for an orthogonal DWT,

$$D(\mathbf{x}, \mathbf{y}^j) = D(\mathbf{X}, \mathbf{Y}^j) \tag{1}$$

Let $\mathcal{F}_k = \{f_1, f_2, ..., f_k\}$ be the indices of the first k closest vectors to $\mathbf{x}$ among the vectors in $\mathcal{S}$, where $D(\mathbf{X}, \mathbf{Y}^{f_1}) \leq D(\mathbf{X}, \mathbf{Y}^{f_2}) \leq ... \leq D(\mathbf{X}, \mathbf{Y}^{f_k})$.

The objective of the PDS algorithm is to reduce the computation time for finding $\mathcal{F}_k$ for kNN classification. Let X_i and Y_i^j be the i-th coefficient of $\mathbf{X}$ and $\mathbf{Y}^j$, respectively, where the coefficients in the wavelet domain are indexed in the zig-zag order. Moreover, let $D^q(\mathbf{X}, \mathbf{Y}^j) = \sum_{i=1}^{q} (X_i - Y_i^j)^2$ be the partial distance between $\mathbf{X}$ and $\mathbf{Y}^j$. Since $D(\mathbf{X}, \mathbf{Y}^j) > D^q(\mathbf{X}, \mathbf{Y}^j)$, it follows that

$$D(\mathbf{x}, \mathbf{y}^j) > D^q(\mathbf{X}, \mathbf{Y}^j). \tag{2}$$

Let $D_i, i = 1, ..., k$, be the squared distance between $\mathbf{x}$ and *current* $\mathbf{y}^{f_i}$ during PDS process. Before PDS is started, we set the initial *current* $f_i = 0$, $i = 1, .., k$, and the corresponding initial D_i is set to be $D_i = \infty$.

Starting from $\mathbf{Y}^1$, we check each vector (except initial *current* $\mathbf{Y}^{f_1}$ itself) until $\mathbf{Y}^t$ is reached. For each vector $\mathbf{Y}^j$ to be searched, we compute $|Y_1^j - X_1|$. Suppose $|Y_1^j - X_1| > \sqrt{D_k}$, then $\sqrt{D(\mathbf{x}, \mathbf{y}^j)} > \sqrt{D_k}$. Hence, j does not belong to $\mathcal{F}_k$ and $\mathbf{Y}_j$ can be rejected. If $|Y_1^j - X_1| < \sqrt{D_k}$, then we employ the following fast search process. Beginning with $q = 2$, for each value of q, $q = 2, .., 2^n \times 2^n$, we evaluate $D^q(\mathbf{X}, \mathbf{Y}^j)$. Suppose $D^q(\mathbf{X}, \mathbf{Y}^j) > D_k$, then $D(\mathbf{x}, \mathbf{y}^j) > D_k$ and $\mathbf{Y}^j$ can be rejected. Otherwise, we go to the next value of q and repeat the same process. Note that the partial distance $D^q(\mathbf{X}, \mathbf{Y}^j)$ can be expressed as

$$D^q(\mathbf{X}, \mathbf{Y}^j) = D^{q-1}(\mathbf{X}, \mathbf{Y}^j) + (X_q - Y_q^j)^2. \tag{3}$$

Therefore, the partial distance of the new q can use the partial distance of the previous q, and only the computation of $(X_q - Y_q^j)^2$ is necessary.

This PDS process is continued until $\mathbf{Y}^j$ is rejected or q reaches $2^n \times 2^n$. If $q = 2^n \times 2^n$, then we compare $D(\mathbf{x}, \mathbf{y}^j)$ with D_k. If $D(\mathbf{x}, \mathbf{y}^j) < D_k$, then we remove f_k from $\mathcal{F}_k$, insert j into $\mathcal{F}_k$, and re-sort elements in $\mathcal{F}_k$. After the final $\mathcal{F}_k$ is found, we then compute k_i for each class ω_i. The class having highest k_i

Table 1. The classification error rate and area complexity (LEs) of kNN design set for various l values with or without subspace search

Subspace	No	Yes	Yes
l	0	0	6
Classification error rate	5.41%	6.67%	7.08%
Design Set LEs	26706	15235	3912

is then identified as the class that $\mathbf{x}$ belongs to. This completes the PDS-based kNN classification.

3 The Architecture

3.1 PDS for Hardware Realization

The PDS adopted here for hardware realization features the subspace search, the bitplane reduction, and multiple-coefficient partial distance accumulation.

Subspace Search. Because the DWT is able to compact the energy of a vector to a lowpass subband, the PDS in the wavelet domain can be accelerated further by scanning only the coefficients in the lowpass subband. The kNN design set C for subspace search contains only $\mathbf{Y}^j_{Lm}$(i.e., the DWT coefficients of $\mathbf{y}^j_{Lm}$), $j = 1, ..., t$. Therefore, the subspace search is also beneficial for the VLSI realization of a kNN classifier since it significantly reduce the storage size of the design set.

Although the $\mathbf{Y}^j_{Lm}, j = 1, ..., t$, can be obtained offline, the $\mathbf{X}_{Lm}$ should be obtained online from a source vector $\mathbf{x}$. Therefore, the DWT computation is still necessary in the subspace PDS. We use the Haar wavelet for the DWT hardware realization because it has simple lowpass filter (i.e., impulse response $= \{\frac{1}{2}, \frac{1}{2}\}$) and highpass filer (i.e., impulse response $= \{-\frac{1}{2}, \frac{1}{2}\}$). Consequently, no multiplication is necessary for the DWT implementation.

Bitplane Reduction. In addition to its simplicity, another advantage of the Haar wavelet is that the coefficients are of finite precision. All coefficients can be precisely stored in the ROM for PDS operations. On the other hand, it may not be necessary to store all bitplanes in $\mathbf{X}_{Lm}$ since the removal of LSB bitplanes have limited impact on the PDS performance. Table 1 shows the classification error rate of kNN classifier based on subspace PDS with/without bitplane reduction. Area complexities of the ROM are also included for comparison purpose. The classification error rate is defined as the number of test vectors that are misclassified divided by the total number of test vectors. The area complexity is defined as the number of logic elements (LEs) required for the FPGA implementation of the ROM containing the design sets for kNN classification. The FPGA used for the measurement of the storage size is the Altera Stratix [8]. The bitplane reduction level, denoted by l, is defined as the number of LSB bitplanes removed from the subspace PDS.

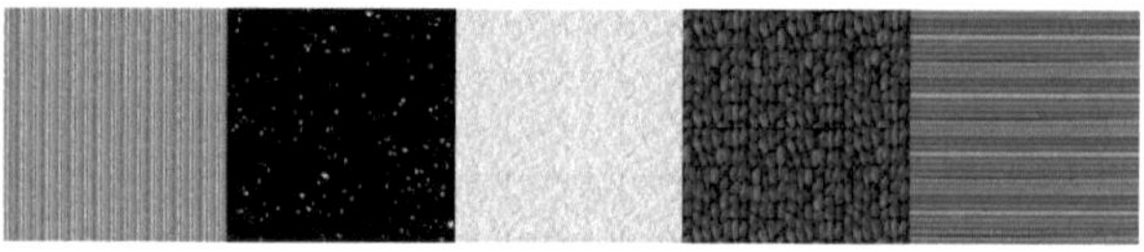

Fig. 1. Five textures for classification

Table 2. The area complexity (LEs) of VSDC unit for various l and ä values

	l	0	2	4	6
	1	46	40	34	28
	2	123	107	92	75
δ	**4**	278	242	206	170
	8	589	513	437	361
	16	1212	1056	900	744

In this experiment, there are 5 classes (i.e., $N = 5$). Each class represents a texture as shown in Fig. 1. The design set C_i associated with each class contains 256 vectors. Therefore, there are 1280 vectors ($t = 1280$) in the design set C. The dimension of the vectors is 8×8 pixels (i.e., $n = 3$). For the subspace search, the dimension of $\mathbf{X}_{Lm}$ and $\mathbf{Y}^j_{Lm}$ is 4×4 pixels (i.e., $m = 2$).

As compared with the basic PDS process, the classification error rate is only slightly increased when using subspace PDS and removing 6 LSB bitplanes ($l =$ 6). However, the deduction in ROM area complexity is quite substantial. In fact, the classification error rate is only increased by 1.67% (from 5.41% to 7.08%), but the deduction in area complexity is 85.35% (from 26706 LEs to 3912 LEs).

Multiple-Coefficient Partial Distance Accumulation. From eq.(3), it follows that the partial distance is accumulated one coefficient at a time in the basic PDS. Therefore, the hardware realization of the basic PDS requires only one multiplier. The speed of partial distance computation can be accelerated for enhancing the throughput by accumulating δ coefficients at a time. The partial distance is then computed by

$$D^q(\mathbf{X}_{Lm}, \mathbf{Y}^j_{Lm}) = D^{q-\delta}(\mathbf{X}_{Lm}, \mathbf{Y}^j_{Lm}) + \sum_{i=q-\delta+1}^{q} (X_i - Y^j_i)^2, \qquad (4)$$

where X_i and Y^j_i are the i-th coefficient of $\mathbf{X}_{Lm}$ and $\mathbf{Y}^j_{Lm}$, respectively. The size of the lowpass subband $2^m \times 2^m$ pixels should be a multiple of δ.

Table 2 shows the area complexity of the vector squared distance computation (VSDC) units with various δ and l values. The FPGA used for the area complexity measurement is also the Altera Stratix. It can be observed from the table that, although VSDC units with larger δ values have higher encoding throughput, their area complexity is also very large. Therefore, the employment

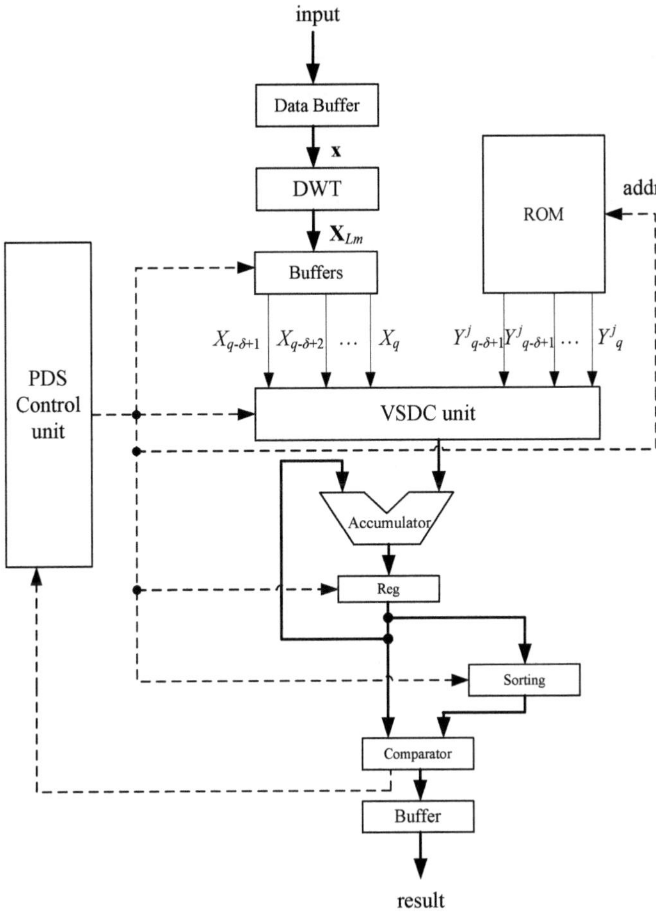

Fig. 2. The VLSI architecture of the proposed PDS algorithm

of VSDC with very large δ values may be difficult for some applications where the cost is an important concern. On the contrary, smaller δ values (e.g., $\delta = 4$) requires significantly lower area complexity while improving the throughput.

3.2 FPGA Implementation

The basic VLSI architecture for realizing the proposed PDS is shown in Fig. 2, which contains a ROM, a DWT unit, a VSDC unit, an accumulator, a comparator, a control unit, a sorting circuit, and a number of registers storing the values of intermediate results. The ROM contains $\mathbf{Y}^1_{Lm}, ..., \mathbf{Y}^t_{Lm}$ for the fast search. The DWT unit is used for computing $\mathbf{X}_{Lm}$ of an input vector $\mathbf{x}$. The VSDC and accumulator are used to compute and store the partial distance $D^q(\mathbf{X}_{Lm}, \mathbf{Y}^j_{Lm})$ for $q = \delta, 2\delta, ..., 2^m \times 2^m$. The comparator then compares $D^q(\mathbf{X}_{Lm}, \mathbf{Y}^j_{Lm})$ with D_k. The comparison results are reported to the control unit, which determines

Table 3. The latency of the kNN classifier for various ä values

δ	1	2	4	8	16
$\mathcal{L}$	1.4836	1.2096	1.0864	1.0307	1

whether the reset of the accumulator and the update of $\mathcal{F}_k$ are necessary. When $\mathcal{F}_k$ is required to be updated, the sorting circuit will be activated. The circuit first replaces current f_k by j and updates corresponding D_k. After that, the circuit sorts the elements of $\mathcal{F}_k$ in accordance with $D_i, i = 1, ..., k$. Each of the new $f_i, i = 1, ..., k$, and its corresponding D_i after sorting operations will be stored in registers for subsequent PDS comparisons.

Table 3 shows the latency of the VLSI architecture for various δ values. Let $\mathcal{T}(\mathbf{x})$ be the number of clock cycles required for the completion of the subspace PDS given an input vector $\mathbf{x}$. The latency $\mathcal{L}$ is then defined as

$$\mathcal{L} = \frac{1}{tW} \sum_{j=1}^{W} \mathcal{T}(\mathbf{x}^j), \tag{5}$$

where W is the total number of input vectors and t is the total number of codewords. For each input vector, the latency $\mathcal{L}$ then represents the average number of clock cycles required for the computation of each input vector. The design sets used in this experiment are identical to those in Table 1. Therefore, $N = 5$, $t = 1280$, $n = 3$, and $m = 2$.

It can be observed from the Table 3 that, the average latency of the PDS with a relatively small $\delta > 1$ for detecting an undesired codeword is close to 1. In addition, it also follows from Tables 1 and 2 that the subspace PDS with small δ and large l can lower the area complexity. Therefore, with a lower area complexity and a faster clock rate, the proposed PDS is able to achieve an average latency close 1 for squared distance calculation and undesired codeword detection.

3.3 The PDS Architecture as a Custom User Logic in a Softcore CPU

To physically measure the performance of the proposed architecture, the proposed PDS architecture has been implemented as a user logic in the ALU of the softcore NIOS CPU, which is a general purpose, pipelined, single-issued reduced instruction set computer (RISC) processor. The major distinction between normal microprocessors and softcore processors is that softcore processors have a re-configurable architecture. Therefore, custom VLSI architectures can be embedded in the softcore processors for physical performance measurement.

Fig. 3 shows the position of the kNN architecture in the ALU of the NIOS CPU. It can be observed from the figure that the kNN architecture and the basic ALU unit share the same input and output buses. The kNN architecture can be accessed by the custom instructions associated with the architecture.

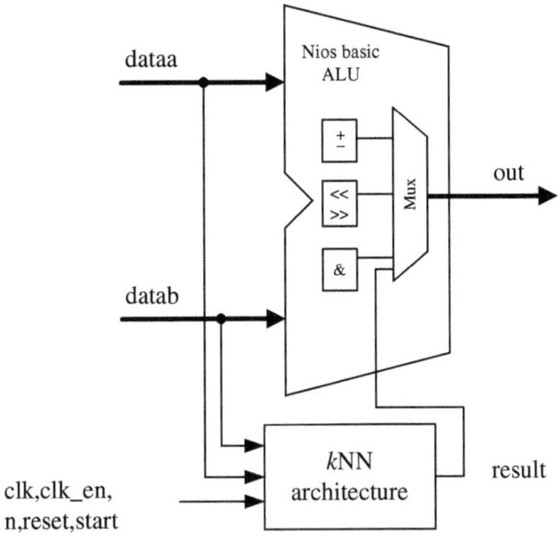

Fig. 3. The position of the kNN architecture in the ALU of the NIOS CPU

The custom instructions accessing the buffers and status register of the kNN architecture can be called using macros in C/C++. Therefore, the C program for kNN classification running on the NIOS CPU involves only the input vector loading and kNN results collection operations. The C code implementation is straightforward, and requires few computations. This can be beneficial for embedded systems where the CPU has limited computational capacity for high level language execution.

4 Experimental Results and Conclusions

This section presents some physical performance measurements of the proposed FPGA implementation. In our experiments, the dimension of vectors in the design set is 8×8 pixels (i.e., $n = 3$). Only $\mathbf{Y}^1_{Lm}, ..., \mathbf{Y}^t_{Lm}$ are stored for PDS, where the total number of vectors t in the design set is 1280. We set $m = 2$ so that the dimension of $\mathbf{Y}^j_{Lm}$ is 4×4 pixels. The bitplane reduction level is $l = 6$. The VSDCs are implemented with $\delta = 4$. There are $N = 5$ textures for classification as shown in Fig. 1. We also set $k = 5$ for the kNN classification.

The whole kNN system consisting of the NIOS CPU embedded by the proposed kNN architecture is implemented by the Altera Stratix 1P1S40 FPGA, which has maximum number of 41250 LEs [8]. The operating frequency of the system is 50 MHz, and the system consumes only 6391 LEs for the implementation of both NIOS CPU and subspace PDS circuit. Therefore, the area complexity of the proposed circuit is significantly less than the maximum capacity of the target FPGA device.

Table 4 compares the execution time of NIOS with that of the Pentium IV for various PDS operations, where the execution time is defined as the average

Table 4. The CPU time of various kNN systems

CPU type	PentiumIV 3.0 GHz		PentiumIV 1.8 GHz		NIOS 50 MHz
algorithm	*k*NN	Subspace PDS	*k*NN	Subspace PDS	Subspace
Implementation	Software	Software	Software	Software	Hardware/ Software Codesign
CPU time (μs)	16079.2	262.5	24912.5	497.5	137.77

Table 5. Comparison between kNN and HVQ

FPGA Implementation	FPGA Device	Number of codewords	Vector Dimension	Number of LEs	Maximum Clock Rate
Subspace *k*NN (*k*=5)	Altera Stratix EP1S40	1280	64	2415	50 MHz
Hierarchical VQ[1]	Xilinx Vertex 400	256	64	6276	16MHz

CPU time (in μs) required for identifying the final class to each input vector. The NIOS CPU is running with the support of the proposed hardware; whereas, the Pentium IV CPU is executing solely on C codes. Note that, for the NIOS systems with software/hardware codesigns, the measurements in fact cover the complete execution process for codeword search including the memory accesses (the fetching of NIOS instructions and source vectors), buffer filling, executions of NIOS instructions, executions of PDS subspace hardware, and the retrieval of encoding results.

It can be observed from Table 4 that the average CPU time of the NIOS and Pentium IV 3.0 GHz are 137.77 μs and 262.5 μs, respectively. Therefore, although the operating frequency of the NIOS CPU is only 50 MHz, its average execution time is still lower than that of the Pentium IV CPU operating at 3.0 GHz. Table 4 also includes the CPU time of the basic kNN algorithm executed in Pentium IV without PDS. In this case, the CPU time is 16079.2 μs for Pentium IV 3.0 GHz CPU, which is 116.71 times longer then the NIOS CPU with hardware acceleration.

To further access the performance of the proposed implementation, Table 5 gives a comparison of our subspace PDS design with the FPGA implementation of the vector quantizer (VQ) presented in [1], which aims to reduce the computational cost for VQ encoding. We can view the VQ encoding circuit as a special kNN classification circuit with $k = 1$. The softcore CPU is not adopted as test bed in [1]. Therefore, only the kNN classfication circuits are considered for comparison. Consequently, in Table 5, the area complexity of the subspace PDS implementation is only 2415 LEs. Although our circuit contains larger number of vectors in the design set, we can see from the table that our subspace PDS architecture requires significantly lower area complexity. The FPGA implementation in [1] has higher area cost because it is based on the hierarchical VQ (HVQ), which needs extra intermediate vectors/nodes to build a tree structure accelerating the VQ encoding process. The tree structure requires large area overhead for the FPGA implementation. By contrast, our design achieves fast computation while reducing the area complexity by the employment of subspace search and bitplane reduction. All these facts demonstrate the effectiveness of the proposed architecture and implementation.

References

1. Bracco, M., Ridella, S., Zunino, R.: Digital Implementation of Hierarchical Vector Quantization. IEEE Trans. Neural Networks 14, 1072–1084 (2003)
2. Hwang, W.J., Jeng, S.S., Chen, B.Y.: Fast Codeword Search Algorithm Using Wavelet Transform and Partial Distance Search Techniques. Electronic Letters. 33, 365–366 (1997)
3. Hwang, W.J., Wen, K.W.: Fast kNN Classification Algorithm Based on Partial Distance Search. Electronics letters. 34, 2062–2063 (1998)
4. Mcnames, J.: Rotated Partial Distance Search for Faster Vector Quantization Encoding. IEEE Signal Processing Letters pp. 244–246 (2000)
5. Ridella, S., Rovetta, S., Zunino, R.: K-Winner Machines for Pattern Classification. IEEE Trans. Neural Networks 12, 371–385 (2001)
6. Vetterli, M., Kovacevic, J.: Wavelets and Subband Coding. Prentice Hall, New Jersy (1995)
7. Xie, A., Laszlo, C.A., Ward, R.K.: Vector Quantization Technique for Nonparametric Classifier Design. IEEE Trans. Pattern Anal. Machine Intell 15, 1326–1330 (1993)
8. Stratix Device Handbook `http://www.altera.com/literature/lit-stx.jsp` (2005)
9. Custom Instructions for NIOS Embedded Processors, Application Notes 188 (2002) `http://www.altera.com/literature/lit-nio.jsp`

Affine Illumination Compensation for Multispectral Images

Pedro Latorre Carmona[1], Reiner Lenz[2], Filiberto Pla[1], and Jose M. Sotoca[1]

[1] Depto. Lenguajes y Sistemas Informáticos, Universidad Jaume I
Campus del Riu Sec s/n, 12071, Castellón de la Plana, Spain
{latorre,pla,sotoca}@lsi.uji.es
[2] Department of Science and Technology, Linköping University, Campus Norrköping
Norrköping, Sweden
reile@itn.liu.se

Abstract. We apply a general form of affine transformation model to compensate illumination variations in a series of multispectral images of a static scene and compare it to a particular affine and a diagonal transformation models. These models operate in the original multispectral space or in a lower-dimensional space obtained by Singular Value Decomposition (*SVD*) of the set of images. We use a system consisting of a multispectral camera and a light dome that allows the measurement of multispectral data under carefully controlled illumination conditions to generate a series of multispectral images of a static scene under varying illumination conditions. We evaluate the compensation performance using the *CIELAB* colour difference between images. The experiments show that the first 2 models perform satisfactorily in the original and lower dimensional spaces.

1 Introduction

Colour image processing based on data acquired by a camera follows a complex image formation process involving the properties of the camera, the reflection properties of the object points, the spectral characteristics of the illumination source and the geometric relation between all these components. In many applications we are however only interested in one of the parts of this process. For instance, in industrial inspection, remote sensing, or automatic colour correction, just to cite a few.

In this paper we propose the application of a general affine transformation model and analyze its performance in relation to other 2 models for the description of illumination changes. For this purpose we measure a static scene by a multichannel camera with 33 channels in the visible range of the spectrum. The scene illumination is provided by 120 lamps arranged on a semi-dome to provide a homogeneous illumination environment. The simplest model to describe color changes is given by a diagonal transform of the colour space [1,4,5]. This model, which corresponds to the so-called *von-Kries adaptation* in human colour vision [16], may be generalized by the introduction of an *offset* in the transformation model (a translation vector) without changing the diagonal nature of the transformation matrix [6], or by considering a particular class of affine transformations of the distribution of the colour content in the colour or spectral space [7,12]. In the cases where these affine models provide sufficiently accurate descriptions

B.K. Ersbøll and K.S. Pedersen (Eds.): SCIA 2007, LNCS 4522, pp. 522–531, 2007.

of the colour changes they can be used to develop invariant features. Such features can be computed from RGB images obtained by conventional commercial cameras, and from multispectral images. Typical application areas are illumination-invariant recognition of objects, or robust content-based retrieval of satellite images obtained under different illumination conditions and acquisition geometry [7,8,9]. In this paper we will analyze the applicability of a general affine transformation model with the help of series of multispectral images of static scenes under carefully controlled illumination changes.

The organization of the paper is as follows: in Section 2 we introduce the mathematical framework for the estimation of the parameters of the models. In Section 3 we describe the experimental device. In Section 4 we analyze the main features of the illumination changes, and compare the compensation models, both in the original and in a lower-dimensional spaces. Conclusions can be found in Section 5.

2 Affine Transformation Estimation for Illumination Changes

We use a vector $x \in \mathbb{R}_{+}^{D}$ to denote a measurement from a D bands multispectral camera of an object point under some illumination condition. Under a change in the illumination characteristics this vector will undergo a change which can be described by the transformation $x \rightarrow \widetilde{x}$. Assuming a specific model of light-camera interaction, Healey *et al* [7,8,9] consider the following Equation to describe this change:

$$\widetilde{x} = \mathbf{A} \cdot x, \tag{1}$$

where $\mathbf{A}$ is a $D \times D$ matrix. On the other hand, a *general affine transformation model* in the form of Eq. 2 [2] could be motivated by the inclusion of effects like noise in the camera, or others:

$$\widetilde{x} = \mathbf{B} \cdot x + t, \tag{2}$$

where $\mathbf{B}$ is also a $D \times D$ matrix. The estimation of the $\mathbf{B}$ matrix and the t vector follows the description by Heikkila *et al* in [10], who applied the *model* to the movement of rigid objects in grey-scale images. We consider the two point sets $\mathbf{X}$ and $\widetilde{\mathbf{X}}$ as $N \times D$ matrices with N the number of points in the set, and $\mathbf{C}$ and $\widetilde{\mathbf{C}}$ their *covariance matrices*. Let us introduce the *Cholesky Factorization*:

$$\begin{aligned} \mathbf{C} &= \mathbf{F} \cdot \mathbf{F}^t \\ \widetilde{\mathbf{C}} &= \widetilde{\mathbf{F}} \cdot \widetilde{\mathbf{F}}^t \end{aligned} \tag{3}$$

where $\mathbf{F}^t$ and $\widetilde{\mathbf{F}}^t$ are the transpose matrices of $\mathbf{F}$ and $\widetilde{\mathbf{F}}$ respectively. Points in the set are first whitened (only shown for the first group), i. e.

$$y = \mathbf{F}^{-1} \cdot \overline{x}, \tag{4}$$

where $\overline{x} = x - \mathbb{E}\{x\}$. Taking into account Eq. 2 and Eq. 4 we have $\widetilde{\mathbf{F}} \cdot \widetilde{y} = \mathbf{B} \cdot \mathbf{F} \cdot y$, and creating a quadratic form of this last expression, we get

$$\widetilde{\mathbf{F}} \cdot \widetilde{\mathbf{F}}^t = \mathbf{B} \cdot \mathbf{F} \cdot \mathbf{F}^t \cdot \mathbf{B}^t, \tag{5}$$

In [15] Sprinzak *et al* proved that an equation of the form $\mathbf{T}\cdot\mathbf{T}^t = \mathbf{S}\cdot\mathbf{S}^t$ has a solution of the form $\mathbf{T} = \mathbf{S}\cdot\mathbf{M}$, where $\mathbf{M}$ is an orthonormal matrix. This will help find the final relation $y \rightarrow \widetilde{y}$. Applying this to Eq. 5 and solving for $\mathbf{B}$, we have

$$\mathbf{B} = \widetilde{\mathbf{F}}\cdot\mathbf{M}^t\cdot\mathbf{F}^{-1}, \tag{6}$$

Substituting Eq. 6 in $\widetilde{\mathbf{F}}\cdot\widetilde{y} = \mathbf{B}\cdot\mathbf{F}\cdot y$, yields:

$$\widetilde{y} = \mathbf{M}^t\cdot y, \tag{7}$$

The assessment of the $\mathbf{M}$ matrix is known as the *Orthogonal Procrustes problem* (see [13] for details). The solution matrix is $\mathbf{M} = \mathbf{V}\cdot\mathbf{W}^t$, where $\mathbf{V}\cdot\mathbf{D}\cdot\mathbf{W}^t$ is the *Singular Value Decomposition* of $(\mathbf{Y}^t\cdot\widetilde{\mathbf{Y}})$. $\mathbf{Y}$ and $\widetilde{\mathbf{Y}}$ are $N\times D$ matrices formed by the vectors y and $\widetilde{y}$ of the point sets. We obtain $\mathbf{B}$ replacing $\mathbf{M}$ in Eq. 6. Applying the *Expectation Operator* to $\widetilde{x} = \mathbf{B}\cdot x + t$, we get $t = \mathbb{E}\{\widetilde{x}\} - \mathbf{B}\cdot\mathbb{E}\{x\}$.

The matrix $\mathbf{A}$ in the particular affine transformation model can also be obtained using the definition of the *Moore-Penrose* inverse. Following [4], who applied this *model* to illumination changes in *RGB* images, consider a $D\times N$ matrix $\mathbf{X}^t$ of points under some reference illumination condition. Denote by $\widetilde{\mathbf{X}^t}$ the corresponding matrix when there is an illumination change. The matrix $\mathbf{A}$ that accomplishes:

$$\widetilde{\mathbf{X}^t} \approx \mathbf{A}\cdot\mathbf{X}^t, \tag{8}$$

is:

$$\mathbf{A} = \widetilde{\mathbf{X}^t}\cdot[\mathbf{X}^t]^+, \tag{9}$$

$[\mathbf{X}^t]^+$ is the *Moore-Penrose* inverse of matrix $\mathbf{X}^t$ (i. e., $[\mathbf{X}^t]^+ = \mathbf{X}(\mathbf{X}^t\cdot\mathbf{X})^{-1}$). The diagonal transform matrix $\mathbf{A}^d$ can be obtained from Eq. 9. Considering [4]:

$$A^d_{ii} = \widetilde{X}^t_i\cdot[X^t_i]^+ = \frac{\widetilde{X}^t_i\cdot X_i}{X^t_i\cdot X_i}, \tag{10}$$

where the single subscript *i* denotes the *ith* matrix row and the double subscript *ii* denotes matrix element at row *i* column *i*.

3 Experimental Set-Up

In our experiments we use a multichannel camera built around the CCD *QImaging* Retiga EX camera (12-bit, Monochrome Cooled camera without IR Filter). The sensor resolution is 1036×1360, down 516×676 pixels. Connected to the camera is a *Liquid Crystal Tunable Filter* (LCTF). The spectral sampling is 10 nm in the range from 400 to 720 nm, resulting in 33 channels. The tunable filter is fixed in front of the camera and the camera/filter combination is mounted on top of the hemisphere, looking down towards its center.

The illumination chamber is shaped as a hemisphere with a diameter of 60 cm and contains 120 halogen lights of 10 Watts each. The lamps are powered by three-phase AC 12 volts adjustable power supply. Each group powers 40 lamps. The even spatial

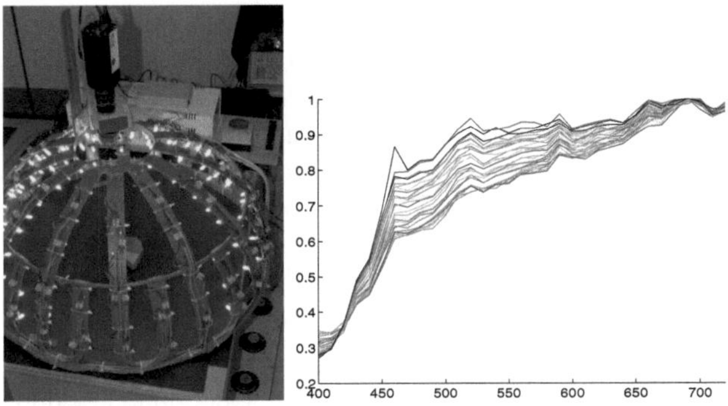

(a) Experimental set-up (b) Normalized Illumination spectra

Fig. 1. (a) Experimental set-up with hemispherical illumination and 3 phase potenciometers, (b) Spectra of the 26 illumination levels

distribution of the lamps inside of the hemisphere avoids flickering effects, due to the AC current. By adjusting the voltage level of the illumination chamber power supply we can change the spectral power distribution of the illuminant. The experimental set-up is shown in Figure 1(a).

4 Illumination Transformation Estimation

4.1 Change in Illumination of the Experimental Set-Up

We change the power supply of the illumination uniformly over the whole range to get 26 different illumination levels and for each illuminant we capture an image of a perfect reflectance diffuser object (*spectralon*) to serve as a spectral descriptor of the light source. In Figure 1(b) the spectral power distribution of the illuminants is shown. They are normalized to 1 dividing each one by its own maximum.

We first characterize the properties of light sources. This is done using the following methods:

Colour Temperature: It is defined as the temperature in *Kelvins* at which a heated *black-body radiator* matches the *hue* of a lamp [16]

Colour Rendering: Defined as a value in the interval $[0 - 100]$ which measures the effect of a light source on the colour appearance of objects in comparison with their colour appearance under a reference illuminant (see [3]).

Colour Difference in the *CIELAB* and *CIELUV* Colour Spaces: These Coordinate systems are derived from properties of human color vision in which euclidean distance corresponds to perceptual difference (for a description, see [16]).

Figure 2 shows the changes in *Colour Temperature*, *Colour Rendering*, the *chromaticity* measured as (a^*, b^*) vectors in the *CIELAB* system (see Section 4.2 for more details on *CIELAB* conversion) , and the change in *Lightness* vs the *chromaticity* C^*_{ab} for the 26 illuminants. The $26th$ illuminant (with highest L^* value) is taken as reference. For the calculation of the *CIELAB* colour difference, the procedure to manage the *white point* explained in section 4.2 is used. From Figures 2(a) and (b) we can see that the illuminants correspond to different *Colour Temperatures* and that they are similar to the corresponding *Black Body Radiators* (i.e., *Colour Rendering* values are very close to 100). Figures 2 (c) and (d) show significant variation both in L^* and *Chroma* values. Figure 3 shows an important colour difference measured in *CIELAB* and *CIELUV* colour spaces for the illumination changes.

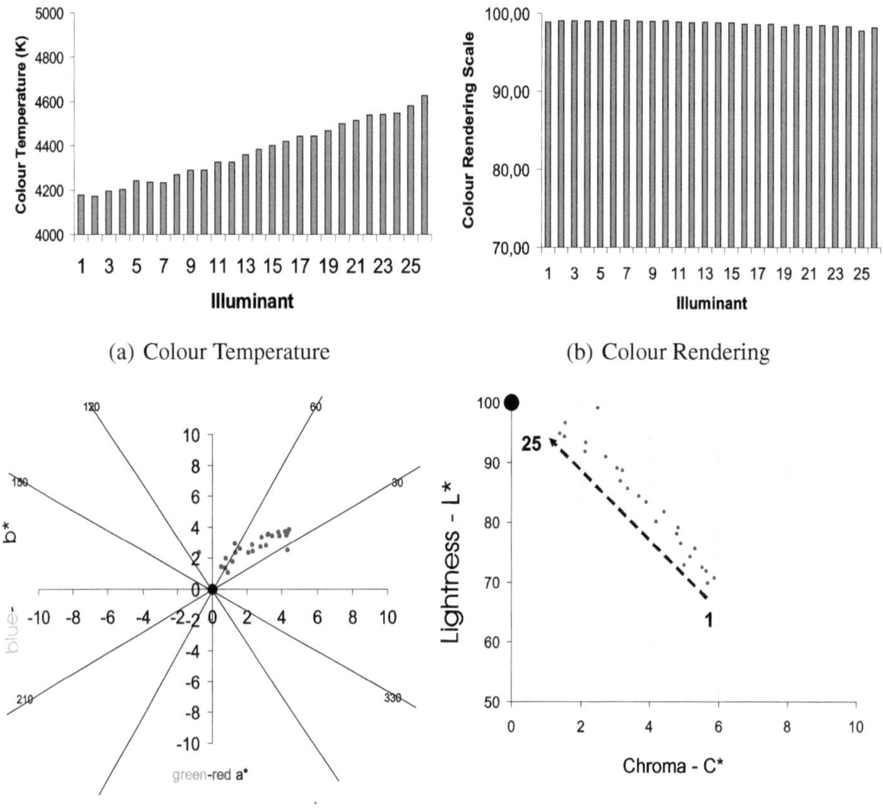

(a) Colour Temperature

(b) Colour Rendering

(c) a^* vs b^*

(d) *Lightness* vs C^*_{ab}

Fig. 2. Analysis of the 26 Illumination spectra

4.2 Assessment of the Compensation

We take a series of 26 images of 4 wooden geometric objects (denoted as Image 1 to Image 26), together with the illuminants of Section 4.1, and apply the affine models to Image k, $k = 2, \ldots, 26$ so they are as similar as possible to Image 1 (the reference image) after the affine transformation (where the ordering in k is such that Image 26 corresponds to the illumination in Section 4.1 with highest L^* value). We use the *CIELAB* colour space [11] to assess their compensation performance, converting the

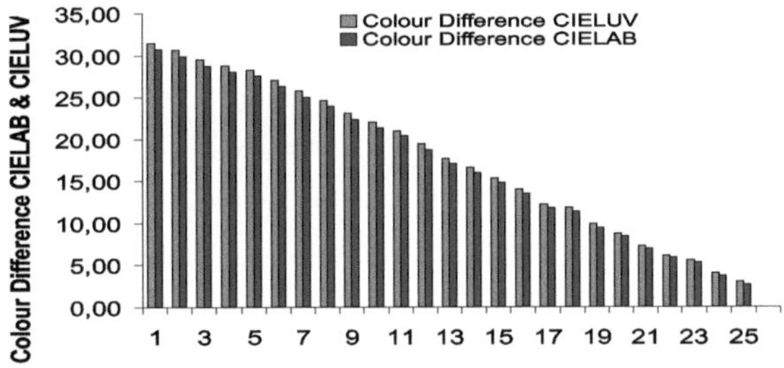

Fig. 3. ΔE_{ab} and ΔE_{uv} colour difference values between Illuminant k ($k = 1, \ldots, 26$) and Illuminant 26, taken as reference

spectral curves of each image of the geometrical objects using the conventional formulae to change from spectra to the XYZ colour space [11]:

$$X \propto \sum_{\lambda} P(\lambda)R(\lambda)\bar{x}(\lambda), \quad Y \propto \sum_{\lambda} P(\lambda)R(\lambda)\bar{y}(\lambda), \quad Z \propto \sum_{\lambda} P(\lambda)R(\lambda)\bar{z}(\lambda) \tag{11}$$

where $P(\lambda)$ is the Power Distribution of the Illuminant, $R(\lambda)$ is the reflectance spectrum of the object and the product $P(\lambda) \cdot R(\lambda)$ is the signal arriving at the camera. $[\bar{x}(\lambda), \bar{y}(\lambda), \bar{z}(\lambda)]$ are the *Colour Matching Functions* of the 10° 1964 *CIE Supplementary Standard Observer* [11]. The *white point* change between illuminations (X_{ni}, Y_{ni}, Z_{ni}), $i = 1, ...26$, is obtained evaluating the XYZ coordinates of the images of the spectralon for each illuminant, and normalizing the Y value of the highest illuminant to 100. The rest of the (X_{ni}, Y_{ni}, Z_{ni}) values are changed accordingly:

$$X_{ni} \leftarrow \frac{X_{ni}}{Y_{n26}} \cdot 100, \quad Y_{ni} \leftarrow \frac{Y_{ni}}{Y_{n26}} \cdot 100, \quad Z_{ni} \leftarrow \frac{Z_{ni}}{Y_{n26}} \cdot 100. \tag{12}$$

This normalization constant is used for any *CIELAB* colour difference.

4.3 Model Evaluation in the Original Space

In Figure 4 we can see the mean ΔE_{ab} Colour Difference for the comparison between Image 1 and the rest for the 3 models (general, particular and diagonal). The application of the first 2 models decreases ΔE_{ab} substantially, but this is not the case for the diagonal model.

In Figure 5 we show that the general affine transformation model gives better ΔE_{ab} results than the particular affine model for low illumination differences between images. As the difference gets higher, both Models tend to give similar results, though the particular model tends to *oscillate* more. We also generate simulated RGB images before and after compensation using the 33×3 transformation matrix of a commercial camera obtained following the procedure described in [14]. In Figure 6 we see the *Simulated Images* 1, 26, and the change from 26 to 1 using the general affine model.

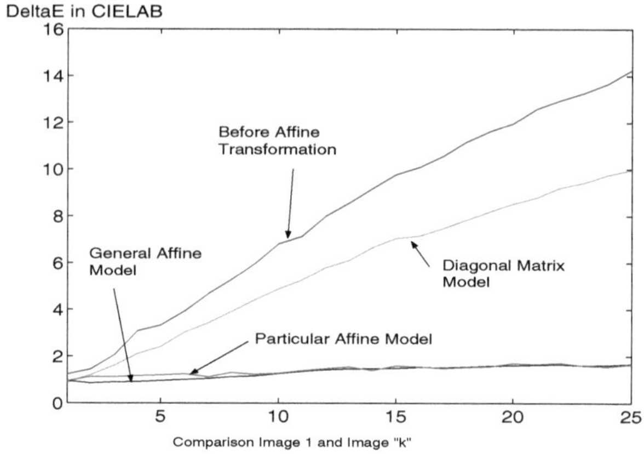

Fig. 4. Comparison mean ΔE_{ab} value before and after compensation using the general, particular and diagonal affine transformation models in the 33 dimensional space between Image 1 and Image *k* with $k = 2, \ldots, 26$

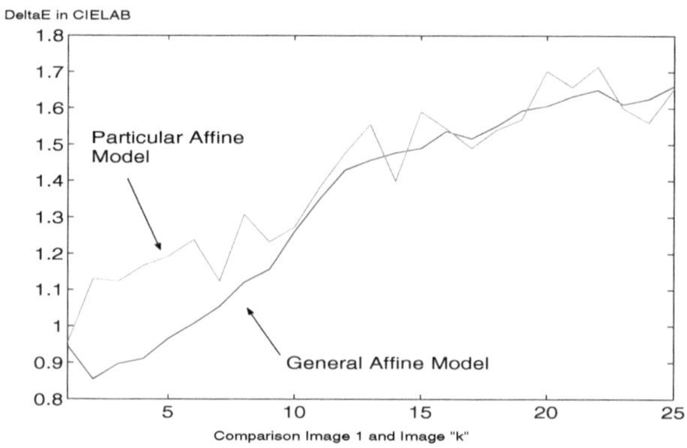

Fig. 5. Zoomed version Comparison mean ΔE_{ab} value before and after compensation using both affine transformation models in the 33 dimensional space

(a) $RGB\,1$ (b) $RGB\,26$ (c) $RGB\,26 \rightarrow 1$

Fig. 6. Simulated RGB Images created with the sensitivity curves of a Nikon camera

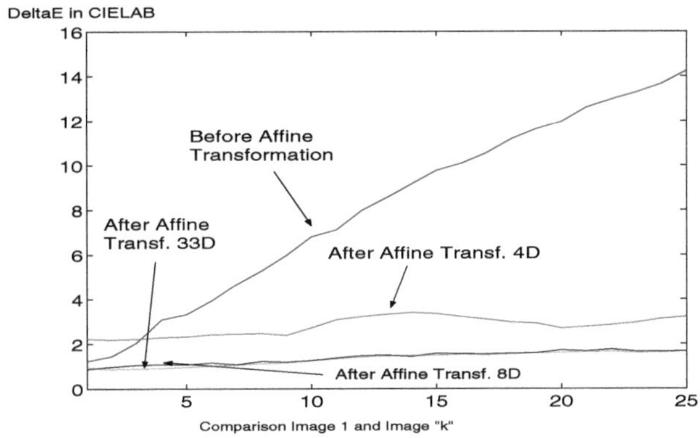

Fig. 7. Mean ΔE_{ab} value using the general affine transformation model for a 4, 8 and 33 dimensional space

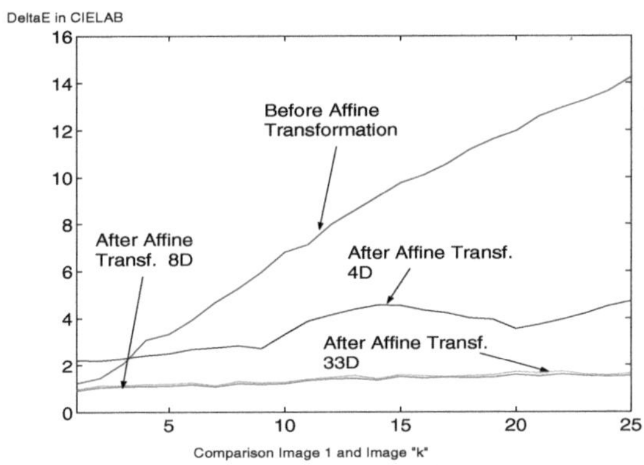

Fig. 8. Mean ΔE_{ab} value using the particular affine transformation model for a 4, 8 and 33 dimensional space

4.4 Model Evaluation in a Low-Dimensional Space

We focus on the first two types of models. We apply the *Singular Value Decomposition* transform to a random selection of the 5% of the whole amount of pixels of the 26 multispectral images acquired under different illuminations. We make a selection of the first 4 and 8 eigenvectors ordered by their corresponding eigenvalues, and project the original images in the 33 dimensional space to these lower dimensional spaces using the following matrix projection formula:

$$\mathbf{X}_{jD} = \mathbf{X}_{33D} \cdot \mathbf{P}_{33 \rightarrow j}, \quad (13)$$

where $j = 4, 8$. We apply both affine transformation models in this lower dimensional space, and recover then the transformed data in the original space using:

$$\mathbf{X}_{ch,33D} = \mathbf{X}_{ch,6D} \cdot \mathbf{P}^t_{33 \rightarrow j}, \quad (14)$$

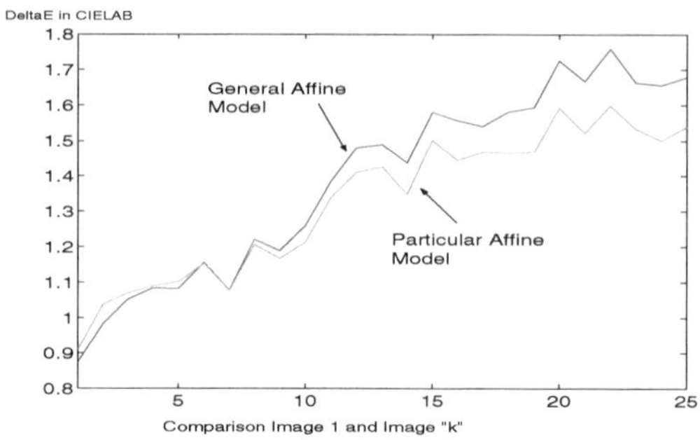

Fig. 9. Zoomed version comparison mean ΔE_{ab} value for the general and particular affine transformation models in an 8 dimensional space

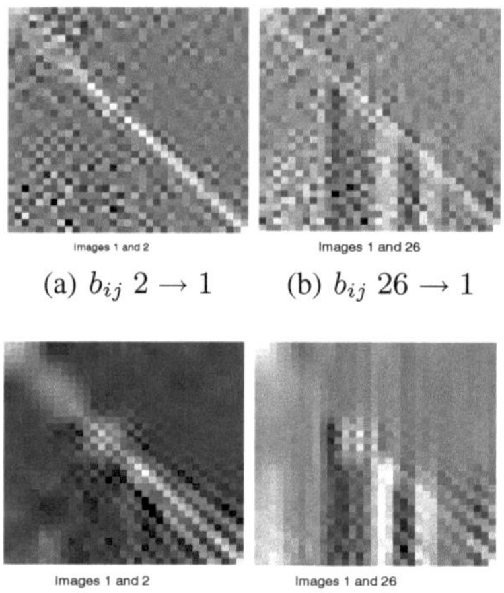

(a) b_{ij} $2 \rightarrow 1$ (b) b_{ij} $26 \rightarrow 1$

(c) a_{ij} $2 \rightarrow 1$ (d) a_{ij} $26 \rightarrow 1$

Fig. 10. Grey-scale image general and particular models matrix elements in the original space

We evaluate the models in the same way as in Section 4.3. Figures 7 and 8 show that we can work in an 8 dimensional space and obtain the same compensation performance results as in the original 33 dimensional space, for both models. In Figure 9 we show that for low illumination changes the general affine model gives better results than the particular affine model in an 8 dimensional space. Finally, in Figure 10 we present 2 examples of the matrix elements values of the models to analyze their distribution in the 33 dimensional space. The main diagonal and other parts of the matrices contribute. There is then an interaction among channels due to some internal device processes.

5 Conclusion

In this paper we showed that the affine model used by Heikkila *et al* in [10] for the analysis of rigid body movements can be applied to the compensation of illumination changes. This model gives better results in general than the particular affine model in the original 33 dimensional space. The analysis of illumination changes as affine transformations opens the door to the development of invariant representations of images.

Acknowledgments

This work has been partially funded by the Ministry of Education and Science of the Spanish Government through the *DATASAT* project ($ESP-2005-00724-C05-C05$). Pedro Latorre Carmona is a *Juan de la Cierva* Programme researcher (Ministry of Education and Science). The authors also thank Prof. Verdu for many useful discussions, and for the variability analysis of the illumination source.

References

1. Barnard, K., Finlayson, G., Funt, B.: Color constancy for scenes with varying illumination. Computer Vision and Image Understanding 65, 311–321 (1997)
2. Begelfor, E., Werman, M.: Affine invariance revisited. In: IEEE Conf. on Computer Vision and Pat. Rec. vol. 2, pp. 2087–2094 (2006)
3. Method of measuring and specifying colour rendering properties of light sources. CIE-Technical Report 13.3 (1995)
4. Finlayson, G.D., Drew, M.S., Funt, B.V.: Spectral sharpening: sensor transformations for improved color constancy. Journal of the Opt. Soc. of America, A. 11, 1553–1563 (1994)
5. Finlayson, G., Chatterjee, S.S., Funt, B.V.: Color Angular Indexing. In: Buxton, B.F., Cipolla, R. (eds.) ECCV 1996. LNCS, vol. 1065, pp. 16–27. Springer, Heidelberg (1996)
6. Finlayson, G.D., Hordley, S.D., Xu, R.: Convex programming colour constancy with a diagonal-offset model. In: IEEE Int. Conf. on Image Processing vol. 3, pp. 948–951 (2005)
7. Healey, G., Slater, D.: Global color constancy: recognition of objects by use of illumination-invariant properties of color distributions. Journal of the Opt. Soc. of America, A. 11, 3003–3010 (1994)
8. Healey, G., Slater, D.: Computing illumination-invariant descriptors of spatially filtered color image regions. IEEE Trans. on Image Proc. 6, 1002–1013 (1997)
9. Healey, G., Jain, A.: Retrieving multispectral satellite images using physics-based invariant representations. IEEE Trans. on Pat. Analysis and Mach. Intel. 18, 842–848 (1996)
10. Heikkilä, J.: Pattern Matching with Affine Moment Descriptors. Pattern Recognition 37, 1825–1834 (2004)
11. Hunt, R.W.G.: Measuring Colour, Fountain Press (1998)
12. Lenz, R., Tran, L.V., Meer, P.: Moment based normalization of color images, IEEE 3rd Workshop on Multimedia Signal Processing, pp. 103-108 (1998)
13. Schonemann, P.H.: A generalized solution of the orthogonal Procrustes problem. Psychometrika 31, 1–10 (1966)
14. Solli, M., Andersson, M., Lenz, R., Kruse, B.: Color measurements with a consumer digital camera using spectral estimation techniques. In: Kalviainen, H., Parkkinen, J., Kaarna, A. (eds.) SCIA 2005. LNCS, vol. 3540, pp. 105–114. Springer, Heidelberg (2005)
15. Sprinzak, J., Werman, M.: Affine Point Matching. Pat. Rec. Letters 15, 337–339 (1994)
16. Wyszecki, G., Stiles, W.S.: Color Science: concepts and methods, quantitative data and formulae, Wiley Classics Library (2000)

GPU-Based Edge-Directed Image Interpolation

Martin Kraus[1], Mike Eissele[2], and Magnus Strengert[2]

[1] Computer Graphics and Visualization Group,
Informatik 15, Technische Universität München,
Boltzmannstraße 3, 85748 Garching, Germany
krausma@in.tum.de
[2] Visualization and Interactive Systems Group,
Institut VIS, Universität Stuttgart,
Universitätsstraße 38, 70569 Stuttgart, Germany
mike.eissele@informatik.uni-stuttgart.de,
magnus.strengert@informatik.uni-stuttgart.de

Abstract. The rendering of lower resolution image data on higher resolution displays has become a very common task, in particular because of the increasing popularity of webcams, camera phones, and low-bandwidth video streaming. Thus, there is a strong demand for real-time, high-quality image magnification. In this work, we suggest to exploit the high performance of programmable graphics processing units (GPUs) for an adaptive image magnification method. To this end, we propose a GPU-friendly algorithm for image up-sampling by edge-directed image interpolation, which avoids ringing artifacts, excessive blurring, and staircasing of oblique edges. At the same time it features gray-scale invariance, is applicable to color images, and allows for real-time processing of full-screen images on today's GPUs.

1 Introduction

Digital image magnification is by no means a trivial technical task—in fact, many real-life scenarios include perceptional issues which are not covered by the theory of signal processing. In particular, human subjects often perceive theoretically optimal magnifications of images as less sharp and more blurred than images magnified with algorithms that heuristically add high frequencies to the sampled signal. This is due to the fact that humans often have a more precise model of the physical signal than the sampled image data can provide. For example, we often expect regions of uniform colors with sharp edges in images although the finite sampling of a digital image cannot provide this information. Thus, improving image magnification algorithms for subjectively sharper results is—in a technical sense—an ill-posed problem. Nonetheless, the challenge exists and became more important in recent years due to the increasing popularity of higher resolution display devices such as PC screens, video beamers, and HDTVs, while video DVDs and TV signals still provide lower resolutions. There are also new popular image sources, e.g., webcams, camera phones, and internet video streams, which

B.K. Ersbøll and K.S. Pedersen (Eds.): SCIA 2007, LNCS 4522, pp. 532–541, 2007.

often provide even lower resolutions. Thus, it is common to up-sample images before rendering them.

In many of the conceivable scenarios, programmable graphics processing units (GPUs) are employed to render the image data. Therefore, we propose to use these GPUs for advanced image magnification techniques—in particular because off-line preprocessing of image data is often not possible due to the lack of communication bandwidth, available memory, or the requirement to avoid latencies; for example, in interactive applications. The image magnification algorithm presented in this work is particularly well suited for implementations on GPUs.

Additional requirements and related work are discussed in Section 2. In Section 3, a one-dimensional edge model is presented and a method for magnifying this ideal edge without blurring nor ringing artifacts is derived. The adaptation of this method for GPU-based image magnification is discussed in Section 4 while Section 5 presents experiments and results.

2 Requirements and Related Work

Several previously published concepts and ideas are crucial in the design of our method. In this section, we discuss the most important design requirements and related publications.

2.1 Pyramidal Magnification

Many image magnification methods are restricted to a magnification factor of 2. Factors equal to powers of 2 are implemented by the corresponding number of magnification operations while arbitrary factors are implemented by up-sampling to the smallest power of 2 that is greater than the requested factor and a minification step that employs, for example, bilinear interpolation. This pyramidal approach provides the optimal (linear) time complexity while considerably simplifying the up-sampling algorithm and its implementation—in particular if the method is implemented in hardware. Figure 1a illustrates an image pyramid, which consists of the coarse image (1×1 pixel) at the top and two finer image levels (2×2 pixels and 4×4 pixels), which are synthesized from the original image by expanding the image data by a factor of two in each magnification step.

There are two different schemes for the positioning of the new samples, which are called primal and dual scheme (or face-split and vertex-split scheme) in the literature on subdivision surfaces. The primal scheme inserts new samples between old samples while keeping the old samples. This is the more traditional interpolation scheme in image magnification methods since it guarantees an interpolation of the original colors by preserving them, and also avoids most computations for all the old samples, i.e., one quater of the pixels of the magnified image.

The dual scheme places all new samples symmetrically between old samples and discards the old samples as illustrated in Fig. 1b. This approach has been

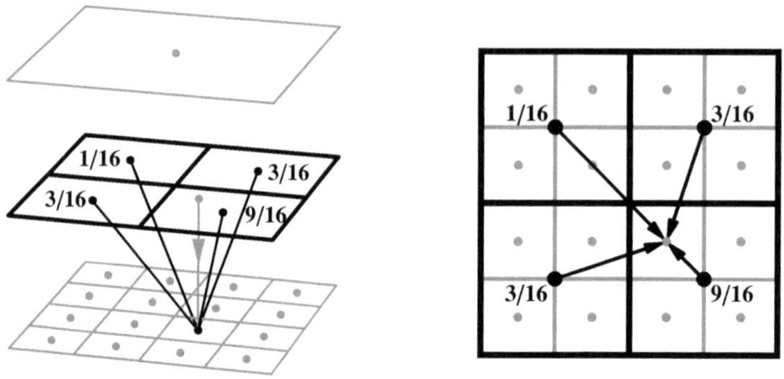

(a) (b)

Fig. 1. (a) Image pyramid consisting of three levels. (b) Dual subdivision scheme.

employed by Zhao and de Haan [1] and Strengert et al. [2] as it is more appropriate for a single-pass implementation and/or a GPU-based implementation. Moreover, the uniform interpolation of samples according to the dual scheme with the weights depicted in Fig. 1 results in a C^1-continuous biquadratic B-spline filtering in the limit of infinitely many up-sampling steps [2].

2.2 Edge Preservation

In practice, image data is usually undersampled. Therefore, reconstructing images with the ideal sinc filter would result in ringing artifacts unless an additional low-pass filter is applied. This low-pass filtering, however, is often perceived as a blurring of the image. As mentioned in the introduction, human subjects estimate the optimal sharpness of edges in images not only based on the displayed data but also based on their knowledge about the depicted objects. Thus, human subjects often correctly assume that the physical signal featured sharper edges than the sampled image data. Therefore, model-based image up-sampling methods have been proposed that attempt to solve the inverse problem of determining the physical signal, which led to the given image data under the assumption of a certain observation model [3].

A somewhat more modest goal is realized by adaptive interpolation methods, which detect strong edges in images and adapt the interpolation weights accordingly to avoid an interpolation across these edges. Therefore, this approach is also called edge-directed image interpolation. In terms of a model-based approach, this corresponds to identifying those edges which have not been sampled at a sufficiently high frequency and, therefore, appear blurred in a technically correct reconstruction of the finite resolution image data. Figures 2c and 2f on page 537 illustrate the sharpening effect of edge-directed interpolation in a one-dimensional example. Since the physical signal of the edge is assumed to feature an infinitely sharp edge as depicted in Fig. 2a, the slope of the up-sampled edge signal in Fig. 2f is increased (in comparison to the original edge signal in Fig. 2c) to approximate the sampling of this edge at a higher resolution. It should be emphasized that excessive sharpening of edges has to be avoided since the lack of any blurring due to a sampling process is very noticeable. Moreover, excessive sharpening of two-dimensional images results in staircasing artifacts of oblique edges similar to aliasing artifacts.

Techniques for edge detection in adaptive edge-directed interpolation methods are usually based on pixel correlation [1,4], local pixel classification [5,6,7], or local gradients [8,9]. In this work, we adapt the boundary model proposed by Kindlmann and Durkin [10], which is related to the edge detectors by Canny [11] and by Marr and Hildreth [12].

2.3 Gray-Scale Invariance

Adaptive interpolation methods are nonlinear mappings of images because the interpolation weights depend on the image data. In general, this can result in undesirable dependencies on the overall intensity or on the local illumination. These disadvantages can be avoided if the adaptive interpolation method is required to be homogeneous; i.e., the up-sampled image should show a multiplicative scaling behaviour for scaled input data. In the context of edge-directed interpolation, this requires an edge detection method that works independently of the absolute scale of edges. Of course, this assumes a signal-to-noise ratio, which is also scale-invariant. As quantized image data already violates this assumption, gray-scale invariance cannot be achieved perfectly. Nonetheless, it is an important design requirement for adaptive interpolation methods as noted, for example, by Pietikäinen [7].

2.4 Color Interpolation

Applicability to color images is an obvious requirement for general image magnification methods. Applying the adaptive interpolation separately to all color components will usually result in unpleasant color shifts. Therefore, edge-directed interpolation methods (including our approach) are usually designed to detect edges in luminance images and adapt the weights for an interpolation of color vectors.

2.5 GPU-Based Interpolation

Some of the requirements for high-performance GPU-based interpolation methods can be identified in the design of the linear, non-adaptive image zooming method proposed by Strengert et al. [2]. Specifically, this method also applies pyramidal magnification with the dual sampling scheme. Even more important is the consistent use of bilinear image interpolation supported by OpenGL graphics hardware. In the context of adaptive interpolation, this amounts to offsetting the sampling coordinates of a (dependent) bilinear image interpolation such that the weights of the GPU-based bilinear interpolation are adapted automatically by the GPU. This indirect adaptation of interpolation weights is also applicable to color images if the offsets to the sampling coordinates are computed from luminance data while the dependent bilinear image lookup interpolates color image data.

3 Ideal Edge Characterization

Analogously to the work by Kindlmann and Durkin [10], we first choose a one-dimensional edge model and develop an exact magnification method for this continuous model. The method does not suffer from ringing artifacts but preserves the ideal edge without any blurring. Moreover, it is gray-scale invariant and can be applied to color images.

For an ideal edge the observed physical signal is assumed to feature an arbitrarily sharp, discontinuous change from a value $y_{\min}$ to $y_{\max}$ at position x_0 as depicted in Fig. 2a. Therefore, this physical signal is modeled by a parameterized step function:

$$y_{\min} + (y_{\max} - y_{\min})\,\Theta(x - x_0). \tag{1}$$

Due to the measurement, however, the sampled signal is blurred. For the sampled edge signal $f(x)$, this observation process is modeled by a convolution with a normal distribution with standard deviation σ (Fig. 2b):

$$f(x) = (y_{\min} + (y_{\max} - y_{\min})\,\Theta(x - x_0)) \otimes \frac{1}{\sqrt{2\pi}\sigma} \exp\left(-\frac{x^2}{2\sigma^2}\right). \tag{2}$$

The resulting function $f(x)$ is a parameterized error function plotted in Fig. 2c and serves as our model of the sampled signal for an ideal edge:

$$f(x) = \frac{y_{\min} + y_{\max}}{2} + \frac{y_{\max} - y_{\min}}{2} \operatorname{erf}\left(\frac{x - x_0}{\sqrt{2}\sigma}\right). \tag{3}$$

The position x_0 of the ideal edge is characterized by the maximum of $f'(x)$ and the zero crossing of $f''(x)$ as illustrated in Fig. 2d. These criteria are exploited by the edge detectors by Canny [11] and by Marr and Hildreth [12], respectively. However, $f'(x)$ and $f''(x)$ are less appropriate for the characterization of the transition region of an ideal edge as their scale depends on $y_{\min}$ and $y_{\max}$, which are not known a priori. Therefore, we form a scale-invariant expression from $f'(x)$ and $f''(x)$:

$$d(x) \stackrel{\text{def}}{=} \frac{-\sigma^2 f''(x)}{f'(x)}. \tag{4}$$

For our model of the ideal edge, $d(x)$ is equal to $x - x_0$ as plotted in Fig. 2e; thus, a zero crossing of $d(x)$ corresponds to the position x_0 of an (ideal) edge. Moreover, the definition of $d(x)$ is independent of $y_{\min}$, $y_{\max}$, and x_0; thus, it can be employed for a scale-invariant characterization of transition regions in an edge magnification method as explained next.

We consider the magnification of the ideal edge signal depicted in Fig. 2c by a factor of 2. At twice the resolution, the edge signal should be twice as

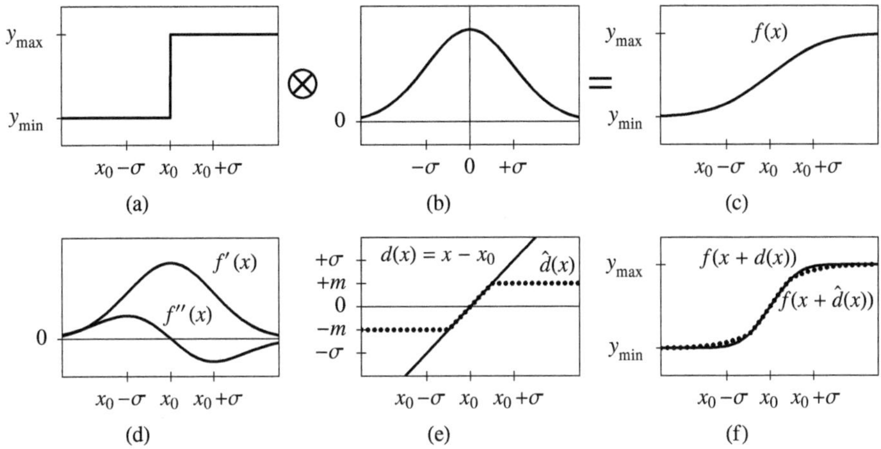

Fig. 2. Detection and sharpening of an ideal edge: (a) parameterized step function, (b) normal distribution filter modeling the measurement process, (c) sampled ideal edge signal, (d) first and second derivative of the sampled signal, (e) offset $d(x)$ for resampling (*solid line*) and clamped offset $\hat{d}(x)$ (*dotted line*), (f) resampled (and sharpened) signal $f(x + d(x))$ (*solid line*) and $f(x + \hat{d}(x))$ (*dotted line*)

sharp, i.e., the convolution in (2) should be computed with a normal distribution with half the standard deviation $\sigma/2$. Since the scaling in x direction is only dependent on σ, it is also possible to compute the magnified edge signal by rescaling the distance $x - x_0$, i.e., the resulting sharper edge signal is given by $f(2(x - x_0) + x_0)$ as depicted in Fig. 2f. This expression, however, allows us to employ our scale-invariant edge characterization:

$$f(2(x - x_0) + x_0) = f(x + (x - x_0)) = f(x + d(x)) . \tag{5}$$

Thus, we can achieve the sharpening of an ideal edge due to a magnification by a factor of 2 simply by resampling the signal with the positional offset $d(x) = -\sigma^2 f''(x)/f'(x)$. It should be emphasized that this resampling of the data with a positional offset maps very well to the GPU-based bilinear image interpolation discussed in Section 2.5.

While this method works for the ideal edge signal, it is obivously not very useful for arbitrary signals since the numerical computation of the offset $d(x)$ is unstable. However, this problem can be easily avoided by clamping the offset between symmetrical bounds $-m$ and $+m$:

$$\hat{d}(x) \stackrel{\text{def}}{=} \max(-m, \min(+m, d(x))). \tag{6}$$

As illustrated by the dotted curves in Figs. 2e and 2f, the clamping has only limited effect in the case of the ideal edge signal if $f(x)$ is already close to y_{min} for $x < x_0 - m$ and close to y_{max} for $x > x_0 + m$. On the other hand, clamping the offset stabilizes the resampling process even for $f'(x) \to 0$ since any offset $\hat{d}(x) \in [-m, +m]$ will lead to approximately the same resampling result $f(x + \hat{d}(x)) \approx f(x)$ for $f'(x) \approx 0$.

4 Proposed Method for GPU-Based Up-Sampling

First we present our method for a gray-scale image $f(\mathbf{x})$ with pixel data defined at integer coordinates of $\mathbf{x}$. As mentioned in Section 2 our method employs a dual up-sampling scheme, i.e., the coordinates of the two new pixel positions between the integer coordinates n and $n+1$ are $n+\frac{1}{4}$ and $n+\frac{3}{4}$ as illustrated in Fig. 1b. The resampling of $f(\mathbf{x})$ at these positions employs a bilinear interpolation; therefore, it corresponds to the subdivision scheme of biquadratic B-splines [2]. Analogously to the one-dimensional magnification method discussed in Section 3, we offset the resampling positions $\mathbf{x}$ by a vector $\hat{\mathbf{d}}(\mathbf{x})$ to sharpen undersampled edges by resampling at $f(\mathbf{x} + \hat{\mathbf{d}}(\mathbf{x}))$.

For the two-dimensional generalization of the one-dimensional offset $d(x) = -\sigma^2 f''(x) \;/\; f'(x)$ from Section 3, it is necessary to compute derivatives across edges, i.e., in the direction of the gradient $\nabla f(\mathbf{x})$. Approximating the second derivative across an edge by the Laplacian $\Delta f(\mathbf{x})$ [10] and choosing the normalized gradient $\nabla f(\mathbf{x})/|\nabla f(\mathbf{x})|$ for its direction, the offset $\mathbf{d}(\mathbf{x})$ becomes:

$$\mathbf{d}(\mathbf{x}) \stackrel{\text{def}}{=} \frac{-\sigma^2 \Delta f(\mathbf{x})}{|\nabla f(\mathbf{x})|^2} \nabla f(\mathbf{x}). \tag{7}$$

Additionally, the absolute value of the offset $\mathbf{d}(\mathbf{x})$ should be clamped, say to m:

$$\hat{\mathbf{d}}(\mathbf{x}) \stackrel{\text{def}}{=} \min(m, |\mathbf{d}(\mathbf{x})|) \frac{\mathbf{d}(\mathbf{x})}{|\mathbf{d}(\mathbf{x})|}. \tag{8}$$

Alternatively, each coordinate of $\mathbf{d}(\mathbf{x})$ could be clamped between $-m$ and $+m$ separately. Actual values of m should be approximately 0.25 because of the sampling scheme depicted in Fig. 1b: For larger values of m, the translated sampling positions might no longer be well separated, which leads to a susceptibilty to noise. If m is significantly smaller, the sharpening will become ineffective.

The second free parameter of our method is the standard deviation σ of the Gaussian filter simulating the blurring due to the observation process. This parameter controls the maximum scale of edges that are sharpened; i.e., the smaller σ, the fewer (harder) edges are sharpened. For $\sigma = 0$ no edges are sharpened and our method degenerates to the biquadratic B-spline filtering proposed by Strengert et al. [2]. On the other hand, the larger σ, the more (softer) edges are sharpened. It should be noted that for $\sigma \gtrsim 1$, anti-aliased and other intentionally soft edges might be sharpened. This should be avoided because it is likely to result in aliasing artifacts such as staircasing of oblique edges.

In general, an optimal value of σ does not exist; thus, it is preferable to let users adjust σ between 0 and 1 according to their preferences. Another alternative might be to determine an appropriate σ from statistics about edges detected at various scales: If no soft edges are detected, a large value of σ is less likely to result in artifacts. If, however, no hard edges are detected, even a rather small value of σ can result in artifacts due to too strong sharpening.

The approximations of the terms $\nabla f(\mathbf{x})$ and $\Delta f(\mathbf{x})$ in our proposed computation of $\mathbf{d}(\mathbf{x})$ are based on a low-pass filtered version of the image $f(\mathbf{x})$, which is denoted by $\tilde{f}(\mathbf{x})$. Similarly to the technique presented by Strengert et al. [2], the employed 3×3 Bartlett filter can be implemented by a sequence of two convolutions with 2×2 box filters, each of which can be implemented by a single bilinear image interpolation:

$$\tilde{f} \stackrel{\text{def}}{=} \frac{1}{16}\begin{bmatrix}1&2&1\\2&4&2\\1&2&1\end{bmatrix} \otimes f = \frac{1}{4}\begin{bmatrix}1&1&0\\1&1&0\\0&0&0\end{bmatrix} \otimes \frac{1}{4}\begin{bmatrix}0&0&0\\0&1&1\\0&1&1\end{bmatrix} \otimes f. \tag{9}$$

This smoothing is necessary for a numerical robust computation of $\nabla f(\mathbf{x})$ by central differences at a new pixel position $\mathbf{x}$:

$$\nabla f(\mathbf{x}) \approx \frac{1}{2}\begin{pmatrix}\tilde{f}\left(\mathbf{x}+(1\ 0)^\top\right) - \tilde{f}\left(\mathbf{x}-(1\ 0)^\top\right)\\ \tilde{f}\left(\mathbf{x}+(0\ 1)^\top\right) - \tilde{f}\left(\mathbf{x}-(0\ 1)^\top\right)\end{pmatrix}. \tag{10}$$

The Laplacian operator for the computation of $\Delta f(\mathbf{x})$ is approximated by a particular filter applied to $\tilde{f}(\mathbf{x})$, which can be evaluated by means of a second convolution with a Bartlett filter:

$$\Delta f(\mathbf{x}) \approx \frac{1}{4}\begin{bmatrix}1&2&1\\2&-12&2\\1&2&1\end{bmatrix} \otimes \tilde{f}(\mathbf{x}) = 4\left(\frac{1}{16}\begin{bmatrix}1&2&1\\2&4&2\\1&2&1\end{bmatrix} \otimes \tilde{f}(\mathbf{x}) - \tilde{f}(\mathbf{x})\right). \tag{11}$$

In summary, the computation of $\mathbf{d}(\mathbf{x})$ requires 7 (non-dependent) bilinear image interpolations per new, resampled pixel (2 for $\tilde{f}$ per pixel of the original image, i.e., 0.5 per pixel of the magnified image; 4 for $\nabla f(\mathbf{x})$; 0.5 for the Bartlett-filtered $\tilde{f}$; and 2 for $\Delta f(\mathbf{x})$). The clamped offset vector $\hat{\mathbf{d}}(\mathbf{x})$ is then employed for a dependent bilinear image interpolation $f(\mathbf{x}+\hat{\mathbf{d}}(\mathbf{x}))$, which determines the pixel data of the magnified image. It is important that this last dependent image interpolation accesses the unfiltered data f (instead of $\tilde{f}$); thus, the low-pass filtering, which is necessary to compute smooth derivatives, does not lead to any blurring of the magnified image.

As discussed in Section 2.4, the extension of this method to color images is straightforward if luminance edges are detected and sharpened. For each pixel of a color image f_c, the luminance is computed and stored in a gray-scale image f. This image is used to compute $\hat{\mathbf{d}}(\mathbf{x})$ as described above. However, the dependent image interpolation accesses the original color image f_c in order to adaptively up-sample this image. Results of a prototypical implementation of the proposed magnification method are presented in the next section.

5 Experiments and Results

Figure 3 presents some results of our adaptive image up-sampling algorithm. Figure 3c shows a magnification by factor 4 and Figure 3f by factor 8, both for the parameter settings $\sigma = 0.7$ and $m = 0.25$. For comparison, Figs. 3a and 3d depict the original pixels with constant pixel colors, while Figs. 3b and 3e show the magnification with biquadratic B-spline filtering [2], which corresponds to our method for $\sigma = 0$.

We have implemented our method for GPUs that support the OpenGL extensions `GL_ARB_fragment_program` and `GL_EXT_framebuffer_object` using four 16 bit floating-point RGBA buffers. For zoom factors of 2, 4, 8, and 64 with a target image size of one megapixel, our implementation for static color images achieves frame rates of 256, 186, 172, and 165 frames per second (i.e., 3.9, 5.4, 5.8, and 6.1 milliseconds per frame) on an NVIDIA GeForce 6800 GT, which was released in 2004. For the same problem, the more recent NVIDIA GeForce 8800 GTX performed at 1437, 1055, 961, and 876 frames per second (0.70, 0.95, 1.04, and 1.14 milliseconds).

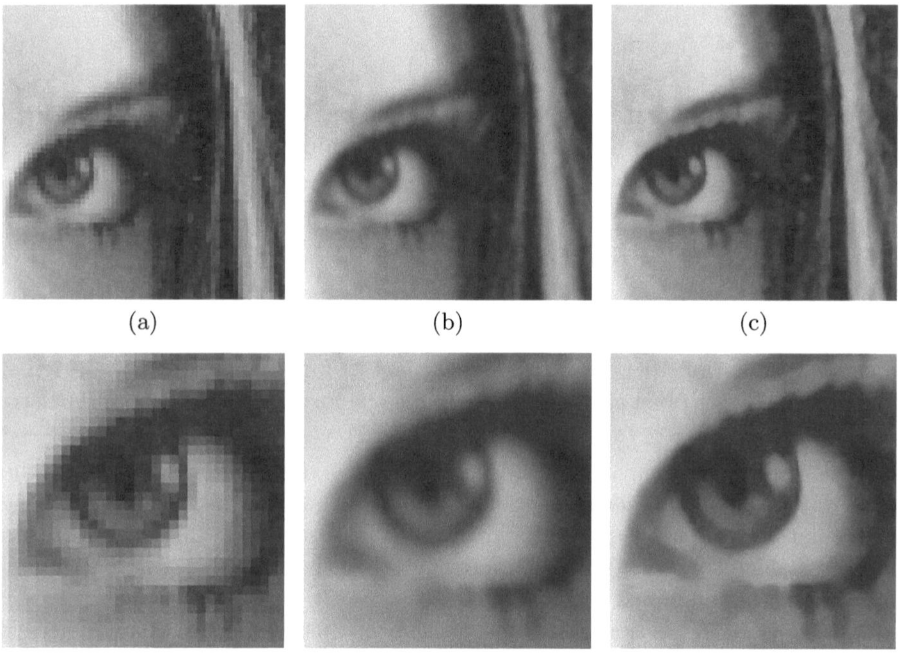

(a) (b) (c)

(d) (e) (f)

Fig. 3. Comparison of three image magnification methods for the 512×512 Lena image with magnification factor 4 in the top row and factor 8 in the bottom row: (a) and (d) sample-and-hold, (b) and (e) biquadratic B-spline filtering (a special case of our method for $\sigma = 0$), (c) and (f) our method for $\sigma = 0.7$ and $m = 0.25$

6 Conclusion

We have identified several important requirements for a GPU-based, adaptive image magnification algorithm in order to design and implement an appropriate method on programmable GPUs. In particular, our algorithm features gray-scale invariance, is applicable to color images, and adapts interpolation weights for an edge-directed image interpolation to avoid the blurring of edges. The method is designed to exploit GPU-supported dependent image interpolation for the adaptation of bilinear interpolation weights; therefore, our implementation makes good use of the rasterization performance offered by modern GPUs and provides full-screen image zooming in real time for almost all application scenarios that include a GPU.

Apart from improving the proposed method and its parameters, long-term future work should include research on GPU-based implementations of alternative image magnification algorithms, e.g., adaptive interpolation based on pixel correlation [1,4] and model-based image magnification methods [3].

References

1. Zhao, M., de Haan, G.: Content adaptive video up-scaling. In: Proceedings ASCI 2003. pp. 151–156 (2003)
2. Strengert, M., Kraus, M., Ertl, T.: Pyramid methods in gpu-based image processing. In: Proceedings Vision, Modeling, and Visualization 2006. pp. 169–176 (2006)
3. Aly, H.A., Dubois, E.: Image up-sampling using total-variation regularization with a new observation model. IEEE Transactions on Image Processing 14(10), 1647–1659 (2005)
4. Li, X., Orchard, M.T.: New edge-directed interpolation. IEEE Transactions on Image Processing 10(10), 1521–1527 (2001)
5. Atkins, C., Bouman, C., Allebach, J.: Optimal image scaling using pixel classification. In: International Conference on Image Processing 2001. vol. 3. pp. 864–867 (2001)
6. Kondo, T., Node, Y., Fujiwara, T., Okumura, Y.: Picture conversion apparatus, picture conversion method, learning apparatus and learning method. US-patent 6,323,905 (2001)
7. Pietikäinen, M.: Image analysis with local binary patterns. In: Image Analysis, pp. 115–118. Springer, Heidelberg (2005)
8. Hwang, J.W., Lee, H.S.: Adaptive image interpolation based on local gradient features. IEEE Signal Processing Letters 11(3), 359–362 (2004)
9. Wang, Q., Ward, R.K.: A new edge-directed image expansion scheme. In: ICIP (3). pp. 899–902 (2001)
10. Kindlmann, G., Durkin, J.W.: Semi-automatic generation of transfer functions for direct volume rendering. In: Proceedings 1998 IEEE Symposium on Volume Visualization, pp. 79–86. IEEE Computer Society Press, Los Alamitos (1998)
11. Canny, J.F. (ed.): A computational approach to edge detection, pp. 184–203. Morgan Kaufmann Publishers, San Francisco (1987)
12. Marr, D., Hildreth, E.: Theory of edge detection. In: Proceedings of the Royal Society of London, Series 207(1167) pp. 187–217 (1980)

Graph-Based Range Image Registration Combining Geometric and Photometric Features

Ikuko Shimizu[1], Akihiro Sugimoto[2], and Radim Šára[3]

[1] Tokyo University of Agriculture and Technology, Japan
[2] National Institute of Informatics, Japan
[3] Czech Technical University, Czech Republic
ikuko@cc.tuat.ac.jp, sugimoto@nii.ac.jp, sara@cmp.felk.cvut.cz

Abstract. We propose a coarse registration method of range images using both geometric and photometric features. The framework of existing methods using multiple features first defines a single similarity distance summing up each feature based evaluations, and then minimizes the distance between range images for registration. In contrast, we formulate registration as a graph-based optimization problem, where we independently evaluate geometric feature and photometric feature and consider only the order of point-to-point matching quality. We then find as large consistent matching as possible in the sense of the matching-quality order. This is solved as one global combinatorial optimization problem. Our method thus does not require any good initial estimation and, at the same time, guarantees that the global solution is achieved.

1 Introduction

Automatic 3D model acquisition of the real-world object is important for many applications such as CAD/CAM or CG. A range sensor, which is a sensing device directly measuring 3D information of an object surface, is a useful tool in modeling 3D objects. An image of an object captured by a range sensor is called a range image and it provides a partial shape of the object in terms of the 3D coordinates of surface points in which the coordinate system is defined by the position and orientation of the range sensor. To obtain the full shape of an object, therefore, we have to align range images captured from different viewpoints. This alignment, i.e., finding the rigid transformation between coordinate systems that aligns given range images, is called range image registration.

Widely used methods for range image registration are the iterative closest point (ICP) method proposed by [1] and its extensions [2,8,14,16]. These methods iterate two steps: Each point in one range image is transformed by a given transformation to find the closest point in the other range image. These point correspondences are then used to estimate the transformation minimizing matching errors. In order to robustly[1] realize range image registration, some features

[1] The terminology "robust" in this paper means that the possibility of successful registration is enhanced; registration is more successful.

B.K. Ersbøll and K.S. Pedersen (Eds.): SCIA 2007, LNCS 4522, pp. 542–552, 2007.

reducing matching ambiguity are proposed in addition to simply computed geometric features [4,5,6,7,12,13]. They are, for example, color attributes [5], chromaticity [7], normal vectors [4], curvatures themselves and their features [6,12], and attributes representing overlapping areas of planes [13]. Combining different kinds of features enhances robustness for registration; nevertheless, defining one common meaningful metric for similarity using different kinds of features is still even difficult.

On the other hand, a method using a graph-based optimization algorithm for range image registration is proposed [11]. The method formalizes the matching problem as a discrete optimization problem in an oriented graph so that optimal matching becomes equivalent with the uniquely existing maximum strict sub-kernel (SSK) of the graph. As a result, this method does not require any good initial estimation and, at the same time, guarantees that the global solution is achieved. In addition, it also has an advantage that a part of data is rejected rather than forcefully interpreted if evidence of correspondence is insufficient in the data or if it is ambiguous. The method, however, deals with geometric features only and fails in finding matching for data of an object having insufficient shape features.

In this paper, we extend the graph-based method [11] so that it does work even for the case of data with insufficient shape features. We incorporate the combination of geometric and photometric features into the framework to enhance the robustness of registration. Existing methods [4,5,6,7,12,13] combining such features define a single metric by adding or multiplying similarity criteria computed from each feature to find point matches. In contrast, our proposed method first evaluates each point match independently using each feature, and then determines the order of matching quality among all possible matches. To be more concrete, for two point-matches, if similarity of one match is greater than the other over all features, we regard that the former is strictly superior to the latter. Otherwise, we leave the order between the two matches undetermined. This is because both geometric and photometric features should be consistently similar with each other for a correct match. Introducing this partial order on matching quality to the graph-based method for range image registration allows us to find as large consistent matching with given data among all possible matches. The maximum SSK algorithm enables us to uniquely determine the largest consistent matching of points with guaranteeing the global solution. This indicates that our proposed method is useful for coarse registration.

2 Multiple Features for Reducing Matching Ambiguity

2.1 3D Point Matching Problem

A range image is defined as a set of discretely measured 3D points of an object surface where each point is represented by the coordinate system depending on a viewpoint and its orientation. Let $\boldsymbol{x}_k^i$ be the coordinates of the k-th point in the i-th image ($i = 1, 2$). We assume in this paper that RGB values $\boldsymbol{r}_k^i$ of the point (with coordinates $\boldsymbol{x}_k^i$) is also measured.

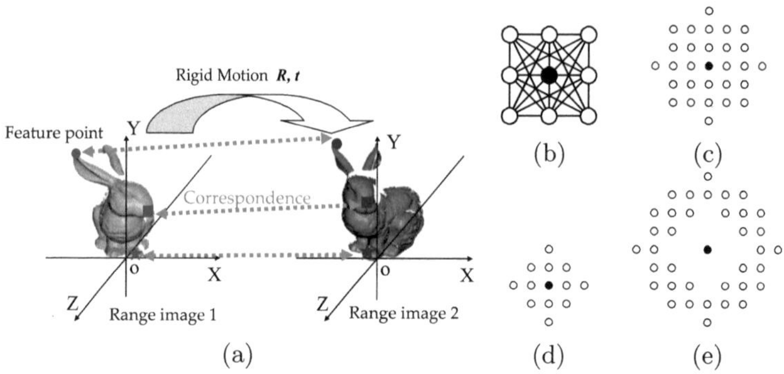

Fig. 1. Point-based registration of two range images (a) and augmented triangular mesh over 3×3 vertex neighborhood. (b) 24 elementary triangles sharing the central vertex. (c) local surface vertices neighborhood used to estimate a local normal vector from 332 triangles. (d) neighborhood for computing the triple and photometric features (52 vertices). (e) neighborhood for computing the triple feature (604 triangles).

Two coordinate systems representing two given range images are related with each other by a rigid transformation $(\boldsymbol{R}, \boldsymbol{t})$, where $\boldsymbol{R}$ is a rotation matrix and $\boldsymbol{t}$ is a translation vector. If two measured points, $\boldsymbol{x}_k^1$ and $\boldsymbol{x}_{k'}^2$, are the same point (namely, corresponding), then $\boldsymbol{x}_{k'}^2 = \boldsymbol{R}\boldsymbol{x}_k^1 + \boldsymbol{t}$. The range image registration is to find $(\boldsymbol{R}, \boldsymbol{t})$ using the corresponding points (Fig.1 (a)).

Searching for corresponding points is realized by comparing features between measured points. If two points are corresponding, then invariant features against rigid transformations should be equivalent with each other. In addition, geometric concordance should be preserved over all corresponding points, which can be evaluated by covariant features.

Some cases exist where an object shape is too smooth to discriminate measured points and thus geometric features alone do not work for reducing ambiguity in finding matching. We, therefore, employ photometric features in addition to geometric features to achieve robust registration.

2.2 Employed Features for Registration

The features we will use are computed from the augmented triangular mesh [11] which includes all possible triangles among triples of vertices in a small vertex neighborhood (Fig.1 (b)). We have chosen four local features, three of which are geometric and the other is photometric: (A) oriented surface normal, (B) structure matrix, (C) triple feature, and (D) chromaticity. We note that (A) and (B) are covariant features whereas (C) and (D) are invariant.

(A) Oriented surface normal. For each measured point $\boldsymbol{x}_k^i$, we compute its oriented surface normal $\boldsymbol{n}_k^i$ as the average over the oriented surface normals of neighboring triangles. In our experiments, we used the augmented triangular mesh over 7×7 neighborhood as shown in Fig.1 (c). We remark that these computed $\boldsymbol{n}_k^i$'s are used for computing structure matrix $\boldsymbol{S}_k^i$ and triple feature F_k^i.

(B) Structure matrix. A set of surface normals $\boldsymbol{n}_j$'s gives 3×3 structure matrix $\boldsymbol{S} = \sum_j \boldsymbol{n}_j \boldsymbol{n}_j^\top$ [11]. In our experiments, we used the augmented triangular mesh over 7×7 neighborhood as shown in Fig.1 (c).

When $\boldsymbol{x}_k^1$ and $\boldsymbol{x}_{k'}^2$ are corresponding, their structure matrices, $\boldsymbol{S}_k^1$ and $\boldsymbol{S}_{k'}^2$, satisfy $\boldsymbol{S}_{k'}^2 = \boldsymbol{R}\boldsymbol{S}_k^1\boldsymbol{R}^\top$. Letting their SVD be $\boldsymbol{S}_k^1 = \boldsymbol{U}\boldsymbol{D}\boldsymbol{U}^\top$ and $\boldsymbol{S}_{k'}^2 = \boldsymbol{U}'\boldsymbol{D}'(\boldsymbol{U}')^\top$, we have two conditions:

$$\boldsymbol{U}'\boldsymbol{P} = \boldsymbol{R}\boldsymbol{U}, \boldsymbol{D}' = \boldsymbol{D}, \tag{1}$$

where $\boldsymbol{P}$ is the 3×3 diagonal matrix $\boldsymbol{P} = \mathrm{diag}(s_1, s_2, s_1 \cdot s_2)$ $(|s_1| = |s_2| = 1)$ representing ambiguity in signs. We use this relationship to evaluate geometric concordance of transformations over corresponding points.

(C) Triple feature. Given a surface as an augmented triangular mesh, the triple feature $F_k^i = \{f_k^i(\ell), \ell = 1, 2, \ldots, t\}$ at point $\boldsymbol{x}_k^i$ represents its neighboring convexity/concavity and is defined by

$$f_k^i(\ell) = \frac{\det[\boldsymbol{n}_k^i, \boldsymbol{n}^i_{\Delta_1^k(\ell)}, \boldsymbol{n}^i_{\Delta_2^k(\ell)}]}{\|(\boldsymbol{x}^i_{\Delta_1^k(\ell)} - \boldsymbol{x}_k^i) \times (\boldsymbol{x}^i_{\Delta_2^k(\ell)} - \boldsymbol{x}_k^i)\|}, \tag{2}$$

where three vertices of the ℓ-th triangle are $\boldsymbol{x}_k^i$, $\boldsymbol{x}^i_{\Delta_1^k(\ell)}$, $\boldsymbol{x}^i_{\Delta_2^k(\ell)}$ and their oriented normal vectors are $\boldsymbol{n}_k^i$, $\boldsymbol{n}^i_{\Delta_1^k(\ell)}$, $\boldsymbol{n}^i_{\Delta_2^k(\ell)}$.

In our experiments, we computed $F_k^i[j]$ $(j = 1, 2)$ using two augmented triangular meshes (see Fig.1 (d), (e)). In our case, $t = 52$ for $j = 1$, while $t = 604$ for $j = 2$.

(D) Chromaticity. Photometric features are useful for robust registration. In particular, when an object has smooth surfaces or similar surfaces in shape, geometric features are not sufficiently discriminative while photometric features are sometimes discriminative.

In our method, as a photometric feature, we consider color distribution over neighboring points. Since RGB values themselves are sensitive to illumination conditions, we employ chromaticity which eliminates the luminance from color information.

Letting $\boldsymbol{r}_k^i$, $\boldsymbol{r}^i_{\Delta_1^k(\ell)}$, $\boldsymbol{r}^i_{\Delta_2^k(\ell)}$ be RGB values respectively at measured points $\boldsymbol{x}_k^i$, $\boldsymbol{x}^i_{\Delta_1^k(\ell)}$, $\boldsymbol{x}^i_{\Delta_2^k(\ell)}$, and $\bar{\boldsymbol{r}} = \frac{\boldsymbol{r}}{\|\boldsymbol{r}\|}$, we compute, for a measured point $\boldsymbol{x}_k^i$,

$$c_k^i(\ell)[j] = \frac{\bar{\boldsymbol{r}}_k^i[j] + \bar{\boldsymbol{r}}^i_{\Delta_1^k(\ell)}[j] + \bar{\boldsymbol{r}}^i_{\Delta_2^k(\ell)}[j]}{3}, \tag{3}$$

where $\boldsymbol{r}_k^i[j]$ is the j-th entry of $\boldsymbol{r}_k^i$ $(j = 1, 2, 3)$. We then define chromaticity distribution over neighborhood $C_k^i[j] = \{c_k^i(\ell)[j], \ell = 1, \ldots, t\}$ $(j = 1, 2, 3)$.

In our experiments, we used the augmented triangular mesh of Fig.1 (d) and thus $t = 52$ in this case.

2.3 Distribution Based Similarity Evaluation

In our method, our employed triple feature and chromaticity are computed over neighboring points and, therefore, they are defined as collections of computed values. The Kolmogorov-Smirnov distance (KS distance) [3] enables us to compute the similarity between two collections[2]. For given triple features $F_k^1[j]$ and $F_\ell^2[j]$, the similarity $c_F(\boldsymbol{x}_k^1, \boldsymbol{x}_\ell^2)$ of the triple feature between them is defined by

$$c_F(\boldsymbol{x}_k^1, \boldsymbol{x}_\ell^2) = \prod_{j=1}^{2} \left(1 - \mathrm{KS}\left(F_k^1[j],\, F_\ell^2[j]\right)\right), \tag{4}$$

where $\mathrm{KS}\left(F_k^1[j],\, F_\ell^2[j]\right)$ represents the KS distance between $F_k^1[j]$ and $F_\ell^2[j]$. In the same way, we define the similarity $c_C(\boldsymbol{x}_k^1, \boldsymbol{x}_\ell^2)$ of chromaticity by

$$c_C(\boldsymbol{x}_k^1, \boldsymbol{x}_\ell^2) = \prod_{j=1}^{3} \left(1 - \mathrm{KS}\left(C_k^1[j],\, C_\ell^2[j]\right)\right). \tag{5}$$

3 Graph-Based Registration Method Using Multiple Features

We now extend the graph-based method [11] so that it can handle multiple features within the same framework. The graph-based matching method [11] selects as many consistent matches in best agreement with data as possible among all possible matches. In line with this idea, we formalize the range image registration problem using both geometric and photometric features in a graph. We note that our method is distinguished from existing methods in the sense that each employed feature is independently evaluated only to determine the matching-quality order and that the obtained order allows us to combinatorially determine the best matching.

3.1 Generating an Unoriented Graph $\mathcal{G}$

Along with [11], we first create an unoriented graph $\mathcal{G}$ representing uniqueness constraint of matching and geometric concordance constraint of the rigid transformation. In evaluating geometric concordance constraint, we use covariant features.

The vertex set P of $\mathcal{G}$ is defined as all putative correspondences $p = (\boldsymbol{x}_k^1,\, \boldsymbol{x}_\ell^2)$. We remark that, in the case where a search range of rigid transformations is known in advance, we can restrict putative correspondences further using our covariant feature evaluation.

The edge set E represents uniqueness of matching and geometric concordance of transformations. Namely, two vertices (i.e., two pairs of matches) are joined if they cannot occur in a matching simultaneously or if no rigid transformation exists that realizes the two pairs of matches simultaneously.

[2] The choice of the KS distance as a similarity measure in fact allows us to combine our triple feature and chromaticity by computing the product of c_F and c_C in Eqs. (4) and (5). It should be stressed, however, that our approach is general and does work for any other similarity measures and their combination.

3.2 Generating an Oriented Graph $\mathcal{D}$

It should be clear that if $M \subset P$ is a solution of the matching problem, no pair of entries in M should be connected by any edge in $\mathcal{G}$. In other words, every possible matching M is an independent vertex subset of $\mathcal{G}$. Since many independent vertex subsets exist in $\mathcal{G}$, we, therefore, select the one that is in best agreement with data.

To do so, we here add orientations to the edges of $\mathcal{G}$ to create oriented graph $\mathcal{D}$ where we evaluate invariant features to determine the orientation of an edge. In this evaluation, [11] uses a single feature alone while our method employs multiple features.

For a putative correspondence $p = (\boldsymbol{x}_k^1, \boldsymbol{x}_\ell^2) \in P$, we denote by $\boldsymbol{c}(p)$ the 2D vector whose entries are the similarities of the triple feature and chromaticity, respectively. Based on $\boldsymbol{c}(\cdot)$, we give the orientation to each edge in $\mathcal{G}$ to define $\mathcal{D} = (P, A \cup A^*)$. Here, A and A^* represent bidirectional edges and unidirectional edges, respectively. We note that $A \cap A^* = \emptyset$.

For two pairs of matches, $p, q \in P$, if $\boldsymbol{c}(p) - \boldsymbol{c}(q)$ is positive for all entries of $\boldsymbol{c}(\cdot)$, then let $(q, p) \in A^*$ (i.e., an oriented edge from q to p). Inversely, if $\boldsymbol{c}(p) - \boldsymbol{c}(q)$ is negative for all entries of $\boldsymbol{c}(\cdot)$, then let $(p, q) \in A^*$. Otherwise, we define $(p, q), (q, p) \in A$. We remark here that we can incorporate robustness further by testing if $\boldsymbol{c}(p) - \boldsymbol{c}(q) > t$, where t is a small positive constant.

3.3 Strict Sub-kernel of $\mathcal{D}$

The maximum matching we are looking for is identical with the maximum strict sub-kernel (SSK in short) [10] of oriented graph $\mathcal{D} = (P,\ A \cup A^*)$ defined above. We remind that the SSK, $K \subseteq P$, of $\mathcal{D}$ is an independent vertex subset in $\mathcal{D}$ and that, for any $p \in K$, the existence of $r \in K$ is ensured such that $(q, r) \in A^*$ for every $(p, q) \in A \cup A^*$. Uniqueness of the SSK in $\mathcal{D}$ is guaranteed and an polynomial algorithm for finding the SSK is known [9,10,11].

Finally we summarize the characteristics of our approach. First, for each feature, similarity between two points is independently evaluated. In other words, for two pairs of matches, each feature independently gives the matching-quality order only and our method focuses on the combination of matches based on this order. Differently from other existing methods, our method does not either define any single metric for similarity using multiple features or minimize any cost function derived from employed features. Secondly, employing geometric features as well as photometric features in the graph-based method enhances robustness in registration. If an object has insufficiently discriminative surfaces in shape, the SSK using geometric features alone may find incorrect matching because two points locally having similar shapes happen to generate the SSK. In contrast, incorporating evaluation of photometric features as well leads to excluding the possibility of generating the SSK that includes such points. Accordingly, ambiguities in matching are reduced and robust registration is achieved. This can be also understood from the fact that points included in the SSK have to be superior in both geometric and photometric similarities to all competing points.

4 Range Image Registration Using SSK

4.1 Interest Point Detection

We detect interest points using triple features among measured points. For each measured point $\boldsymbol{x}_k^i$, we compute the standard deviation of triple features over its neighborhood[3]: $L_k^i = \mathrm{std} \bigcup_{j=1}^{r} F_k^i(j)$. L_k^i becomes large for a point whose neighboring surface shape is not uniform. We thus detect points with local maxima of L_k^i. We call them interest points in this paper.

Then, we use two sets of interest points, each of which is independently detected from one of two given range images, and generate a table for all possible matches. In generating the table, we eliminate matches that do not satisfy a given search range of rigid transformations. To be more concrete, for a given corresponding pair of points, we compute their structure matrices and then decompose them using SVD to find the rotation relating the pair (cf. Eq. (1)). Next, we eliminate the pair from the table if the rotation is not admissible.

4.2 Maximum SSK and Matching

Based on the generated table, we consider all possible matches and then define the vertex set of unoriented graph $\mathcal{G}$. We then define edges along with uniqueness of matching and geometric concordance. Next we give the orientation to the edges in $\mathcal{G}$ using Eqs. (4) and (5).

As a result, we obtain oriented graph $\mathcal{D}$ representing our problem. The SSK algorithm uniquely finds the best matching in $\mathcal{D}$ [11].

5 Experiments

To demonstrate the potential applicability of the proposed method, we applied our method to synthetic range images.

We used a horse model provided by [17] and attached randomly generated texture to it in order to generate its range images (Fig. 2). The body of the horse model was scanned at 20 degree rotation steps with respect to the Y-coordinate and 18 range images with the size of 200×200 pixels were obtained. We then perturbed the Z-coordinate of each point in the range images by adding Gaussian noise with zero mean and standard deviation of $\sigma = 0.1$. This implies that if the height of the horse body is about 60cm, the added noise is about 1mm.

We applied our method to all adjacent pairs in the range image sequence above. We set the search range of ration angles be $\pm 15°$ different from the ground truths just for reducing computational cost. We remark that we did not have any assumption about the rotation axis to be found. To see the effectiveness of our method, we also applied a method using geometric features alone to the same data. The registration results are shown in Fig. 3 and Table 1. Fig. 3 presents selected interest points and their matches obtained by the two methods.

[3] $F_k^i(1)$ and $F_k^i(2)$ are computed with Eq. (2) over 52 and 604 triangles respectively, as shown in Fig. 1 (d), (e). L_k^i is standard deviation over them.

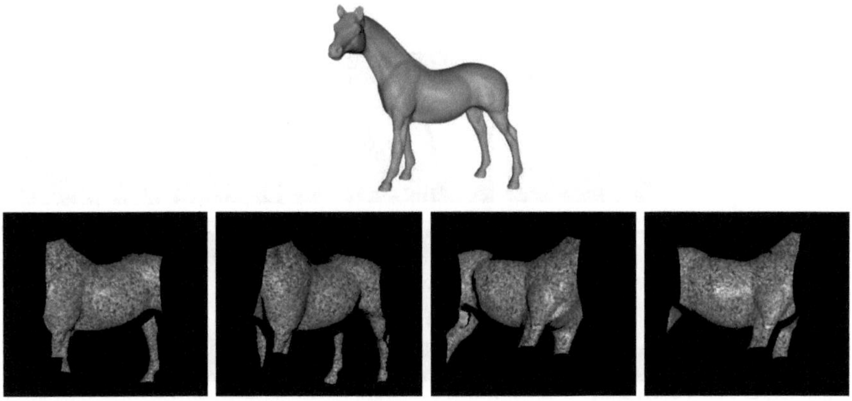

Fig. 2. Horse model and its synthetic range images

Table 1. Evaluation of registration results ("–" means failure in estimation)

	i	1	2	3	4	5	6	7	8	9
	points	11616	10888	9913	9374	9442	9778	10503	11589	12118
	IPs of i-th image	173	241	210	173	197	210	260	178	172
	IPs of $(i+1)$-th image	237	229	172	193	269	243	190	171	172
shape and color	matches	11	5	7	7	8	11	9	7	9
	estimated rotation [°]	20.1	21.1	16.6	19.9	20.1	19.9	20.1	20.2	20.0
	rotation error [°]	0.2	1.0	5.5	0.3	0.8	0.2	1.2	0.3	0.3
	translation error	0.2	1.5	4.4	0.1	0.2	0.1	0.4	0.2	0.2
shape only	matches	3	13	11	11	9	13	10	17	4
	estimated rotation [°]	46.9	19.7	35.2	24.8	15.8	21.1	20.4	19.9	-168.9
	rotation error [°]	63.1	1.7	14.6	19.6	17.2	6.8	11.6	0.2	71.5
	translation error	29.2	0.4	20.2	7.3	21.9	1.4	1.4	0.2	21.2

	i	10	11	12	13	14	15	16	17	18
	Points	11929	11464	10735	9779	8957	9198	10105	11228	11725
	IPs of i-th image	179	256	302	331	250	178	145	166	135
	IPs of $(i+1)$-th image	225	285	328	307	154	140	165	153	156
shape and color	matches	3	2	7	12	2	3	8	5	10
	estimated rotation [°]	19.9	-	19.6	19.8	-	19.9	19.9	20.0	20.1
	rotation error [°]	1.1	-	1.2	0.3	-	2.2	0.8	1.0	0.1
	translation error	0.1	-	0.6	0.3	-	0.8	0.1	0.6	0.0
shape only	matches	2	3	15	10	12	5	17	4	21
	estimated rotation[°]	-	16.2	15.8	43.4	24.5	11.9	19.5	23.4	19.4
	rotation error [°]	-	71.5	18.7	67.4	8.4	38.1	2.1	30.9	3.6
	translation error	-	38.9	7.1	40.8	8.5	19.9	0.6	10.7	1.3

In Table 1, the i-th column corresponds to the registration result of the i-th and $(i+1)$-th images. The number of measured points, the number of detected interest points (IPs), the number of obtained matches (the number of vertices in the obtained SSK), the estimated rotation angle, error of the estimated rotation axes, and translation error are presented there. Errors of the estimated rotation axes were evaluated by the difference from the ground truth while translation errors were by the difference between norms. Since the rigid transformation was estimated using the 3D coordinates of matched points [15], we need at least three matches. "–" was used in the case of less than three matches, which means failure in estimation.

Table 1 shows that over all the cases, the registration accuracy of our method is not only significantly higher but also numerically more stable, compared with the method using geometric features alone. In fact, in our method, errors of

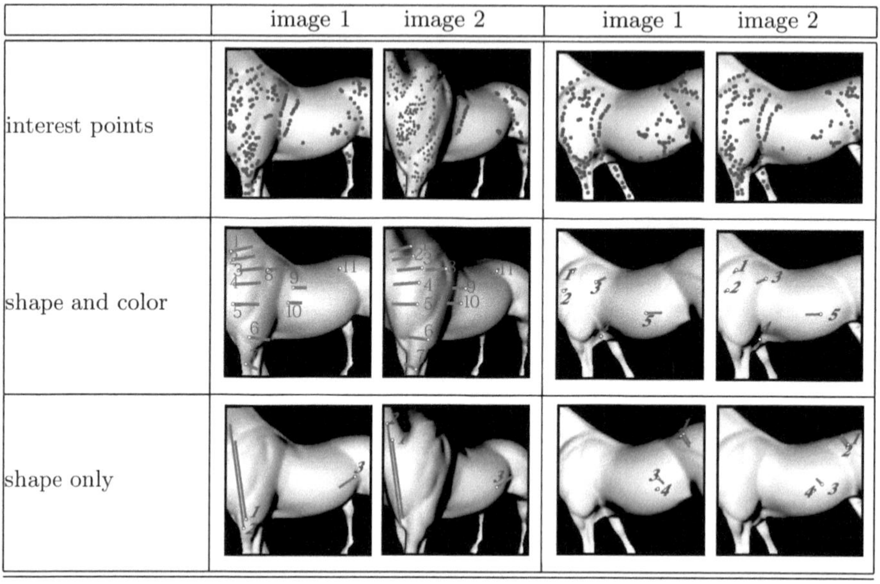

Fig. 3. Examples of registration results

estimated rotations were within ±1 degree and translation errors were within 1.0 except for two cases failing in estimation. These observations can be understood by the fact that discriminative geometric features are not expected due to smoothness of the shape of the horse body while photometric features are discriminative even for such shapes in this case. Fig. 3 certificates this because vectors connecting matched feature points have the uniform direction in our method while they do not in the case of geometric features alone. Similarity derived from our photometric feature reduces matching ambiguity using geometric features alone and matching-quality of incorrect matches as well, which prevents such matches from being included in the SSK. Here we remark again that we did not assume any rotation axis to be found; this suffices to show that our proposed method, differently from existing methods, does not require any good initial estimation. We can thus conclude that our method achieves sufficiently accurate registration without any good initial estimation.

6 Conclusion

We extended a graph-based range image registration method so that it can handle both geometric and photometric features simultaneously. Namely, we formulated registration as a graph-based optimization problem where we independently evaluate geometric feature and photometric feature and then consider only the order of point-to-point matching quality. We then find as large consistent matching as possible in the sense of the matching-quality order. This is solved as one global combinatorial optimization problem of polynomial complexity. The advantage of our method is that each match is independently evaluated by each employed feature and the order of matching-quality is only concerned. Differently from existing methods, our proposed method need not define any single metric of similarity for evaluating matching. Our experimental results demonstrate the effectiveness of our method for coarse registration.

The proposed method will reduce the possibility of finding an incorrect matching but cannot be expected to increase the number of matches significantly. This follows from the fact that both the two similarity criteria have to be consistent. In principle, it is also possible to combine the two criteria in such a way that when one of them strictly favors the match of q to p and the other is at least indifferent between p and q, the edge joining p and q becomes unidirectional. Such definition requires using a different matching algorithm from the one used in this paper. This research direction is our ongoing work.

Acknowledgments. A part of this work was done under the framework of MOU between the Czech Technical University and National Institute of Informactics. This work is in part supported by the Czech Academy of Sciences under project 1ET101210406 and by the EC project MRTN-CT-2004-005439.

References

1. Besl, P.J., McKay, N.D.: A Method for Registration of 3-D Shapes. IEEE Trans. on PAMI 14(2), 239–256 (1992)
2. Chen, Y., Medioni, G.: Object Modeling by Registration of Multiple Range Images. IVC 10(3), 145–155 (1992)
3. Duda, R.O., Hart, P.E., Stork, D.G.: Pattern Classification, 2nd edn. John Willey and Sons, Inc. New York (2001)
4. Feldmar, J., Ayache, N., Berring, F.: Rigid, Affine and Locally Affine Registration of Free-Form Surfaces. IJCV 18(2), 99–119 (1996)
5. Godin, G., Laurendeau, D., Bergevin, R.: A Method for the Registration of Attributed Range Images, Proc. of 3DIM, pp. 179–186 (2001)
6. Guest, E., Berry, E., Baldock, R.A., Fidrich, M., Smith, M.A.: Robust Point Correspondence Applied to Two and Three-Dimensional Image Registration. IEEE Trans. on PAMI 23(2), 165–179 (2001)
7. Okatani, I.S., Sugimoto, A.: Registration of Range Images that Preserves Local Surface Structures and Color, Proc. 3DPVT, pp. 786–796 (2004)
8. Rusinkiewicz, S., Levoy, M.: Efficient Variants of the ICP Algorithm, Proc. of 3DIM, pp. 145–152 (2001)
9. Šára, R.: Finding the largest unambiguous component of stereo matching. In: Heyden, A., Sparr, G., Nielsen, M., Johansen, P. (eds.) ECCV 2002. LNCS, vol. 2352, pp. 900–914. Springer, Heidelberg (2002)
10. Šára, R.: Robust Correspondence Recognition for Computer Vision. In: Proc. of 17th ERS-IASC Symposium on Computational Statistics, pp. 119–131 (2006)
11. Šára, R., Okatani, I.S., Sugimoto, A.: Globally Convergent Range Image Registration by Graph Kernel Algorithm, Proc. of 3DIM, pp. 377–384 (2005)
12. Sharp, G.C., Lee, S.W., Wehe, D.K.: ICP Registration Using Invariant Features. IEEE Trans. on PAMI 24(1), 90–102 (2002)
13. Silva, L., Bellon, O.R.P., Boyer, K.L.: Enhanced, Robust Genetic Algorithms for Multiview Range Image Registration, Proc. of 3DIM, pp. 268–275 (2003)

14. Turk, G., Levoy, M.: Zipped Polygon Meshes from Range Images, ACM SIGGRAPH Computer Graphics, pp. 311–318 (1994)
15. Umeyama, S.: Least-Square Estimation of Transformation Parameters Between Two Point Patterns. IEEE Trans. on PAMI 13(4), 376–380 (1991)
16. Zhang, Z.: Iterative Point Matching for Registration of Free-Form Curves and Surfaces. IJCV 13(2), 119–152 (1994)
17. Georgia Institute of Technology Large Geometric Models Archive `http://www-static.cc.gatech.edu/projects/large_models/`

Automatic Identification and Validation of Tie Points on Hyperspectral Satellite Images from CHRIS/PROBA

André R.S. Marçal

Faculdade de Ciências, Universidade do Porto
DMA, Rua do Campo Alegre, 687, 4169-007 Porto, Portugal

Abstract. There is great interest in the development of automatic geometric correction systems for satellite images. A fully automatic system, based exclusively on the identification of tie points (image to image control points) by image matching needs to use efficient selection and validation methods. Four Tie Point Suitability Indices (TPSI) are proposed to select the most suitable areas in an image to search for tie points. Three tie point validation parameters are also proposed. The validation parameters make use of the various spectral bands available in hyperspectral and multispectral satellite images. The proposed TPSIs and validation parameters were tested with hyperspectral high-resolution satellite images from the CHRIS/PROBA sensor.

1 Introduction

The manual registration and geometric correction of satellite images is a laborious and time-consuming task. The increasingly wider access to satellite images prompts an interest in the development of algorithms for the automatic or semi-automatic geometric correction of these images. As long as the images have certain similarities, it is possible to implement a fully automatic geometric correction system based on the identification of tie points (image to image control points) by image matching. The tie points are used to establish a transformation function between the input and the base or reference image, which once applied performs the geometric correction of the input image. In order for this process to be effective, a significant overlap (over 50%) between the two images is required. However, even two images of the same place acquired by the same sensor, on different dates or with different viewing conditions, often look very different. This causes difficulties for a fully automatic geometric correction system based on automatic identification of tie points. One possible way to tackle this problem is to separate the image transformation process in two parts. Initially, a set of candidate tie points is searched for, but only a subset of trustable tie points are used to establish a first order (affine) transformation function. The second step consists of searching for a new set of tie points, but limiting the search to a small window centred on the locations predicted by the affine transformation.

B.K. Ersbøll and K.S. Pedersen (Eds.): SCIA 2007, LNCS 4522, pp. 553–561, 2007.

This methodology is strongly dependent on the ability to select the right candidate tie points that are used to establish the affine transformation function. The purpose of this work is to propose a methodology to select and validate tie points identified on hyperspectral satellite images. The method was developed for hyperspectral satellite images from CHRIS/PROBA.

CHRIS/PROBA is the first hyperspectral satellite sensor with pointing capabilities and high spatial resolution [1]. A CHRIS/PROBA scene is composed of 5 images with different viewing angles, with fly-by Zenith Angles (ZA) of 55, 36, 0, -36 and 36 degrees [2]. In the most common operational mode, the sensor acquires data with a nominal spatial resolution of 17 meters over the full swath (13 km), with 18 spectral bands from 400 to 1050 nm [3].

2 Method

In this section a method to establish a set of candidate tie points is described, and a number of parameters that can be used for a validation criteria proposed.

2.1 Tie Point Selection by Image Matching

The tie point selection process is based on image matching by normalized two-dimensional cross-correlation in the spatial domain. A target matrix T (of size t, usually small) is established in the reference image and a search window S (of size s, larger than t) is examined in the input image. The convolution between T and all sub-window of S (of size t) is performed, resulting is a set of correlation coefficients, from -1 to 1. The best match will be the pixel of highest correlation in the search window. The MATLAB implementation of the normalized two-dimensional cross-correlation function was used in this work [4].

2.2 Tie Point Suitability Indices

The purpose of a Tie Point Suitability Index (TPSI) is to identify the best locations on an image to search for tie points. These locations should have distinct features in order to maximise the chances of a correct selection on the image matching process. For a given target matrix of size t (for example 3×3 for $t = 3$) a TPSI value is attributed to each pixel of the image, thus producing a TPSI image, where the highest values should correspond to the most promising locations to search for tie points. Four TPSI are proposed: Basic (B), Composed (C), Ratio (R) and Prewitt (P).

For a sub-section of an image of size t (t by t pixels), Hh is the highest possible sum of pixel values, or Digital Numbers (DNs), from two horizontally adjacent pixels, Hv is the highest sum of DNs from two vertically adjacent pixels, and Lh and Lv are the lowest DN sums for two pixels horizontally and vertically adjacent. For example, the value of Hh of Figure 1's section (a) is 0.74, as the highest pair is formed by the pixels with DN values of 0.36 and 0.38 on the top right corner. The values for the other parameters for this section are: $Lh = 0.51$,

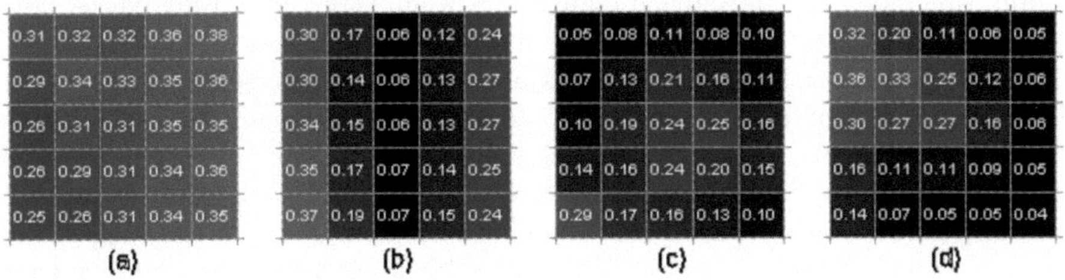

Fig. 1. Example of 5x5 test sites

Table 1. TPSI values for the 5x5 test sites of figure 1 (band 14 as reference)

SITE	Hh	Lh	Hv	Lv	BASIC TPSI	COMPOSED TPSI	RATIO TPSI	PREWITT TPSI
A	0.74	0.51	0.74	0.51	0.1903	0.0349	0.0057	0.0086
B	0.56	0.18	0.72	0.12	0.7188	0.3605	0.0420	0.0028
C	0.49	0.13	0.48	0.12	0.6028	0.3539	0.1122	0.0117
D	0.69	0.09	0.68	0.09	0.7710	0.5849	0.1791	0.1311

$Hv = 0.74$ and $Lv = 0.51$. The values for the other 3 sites of Figure 1 are presented in Table 1.

The Basic TPSI (B) uses the highest and lowest pairs, regardless of the orientation, to compute an index from 0 to 1, using (1), where $H = Max\{Hh, Hv\}$ and $L = Max\{Lh, Lv\}$. The Basic TPSI (B) has high values when the difference between the highest DN pair and lowest DN pair is high. In homogeneous areas the values of B will be low. The Composed TPSI (C), computed by (2), has some similarities with the Basic TPSI. However, the information about horizontal and vertical variability within the matrix is used separately. This index only reaches high values when there is a large difference between the highest and lowest DN pairs both horizontally and vertically.

$$B = \frac{H - L}{H + L} \tag{1}$$

$$C = \frac{(Hv - Lv) \times (Hh - Lh)}{(Hv + Lv) \times (Hh + Lh)} \tag{2}$$

The Ratio TPSI (R) is computed in a different way from B and C, searching for edges locally. For each pair of horizontally adjacent pixels in the sub-section of the image being tested, the ratio between the absolute difference and the sum of their DNs is calculated. The maximum for all pairs is selected as Rh. The same process is done for all vertical pairs, resulting in the maximum ratio Rv. The Ratio TPSI is obtained by multiplying Rh and Rv. The index only has high values when there are both strong horizontal and vertical edges.

The final TPSI proposed is based on a standard edge detector - the Prewitt operator [6]. The Prewitt TPSI is obtained by the multiplication of the results of 3x3 horizontal and vertical Prewitt operators applied to the input image.

Again, this index will only reach high values when there are strong edges, both horizontally and vertically, within the image sub-section tested.

As an illustration, four 5x5 sections of an image are presented in Figure 1. These are 5 by 5 pixel sections used to compute the TPSI for the central pixel (t=5). The examples in Figure 1 illustrate different cases: (a) a homogeneous region, (b) a strong vertical edge, (c) a weak and small peak, and (d) an area with two clearly distinct zones, providing both horizontal and vertical edges. The site (d) should definitely be the best choice for a candidate tie point and the second best should be site (c). Sites (a) and (b) are clearly inadequate to find a tie point. The values of the TPSIs for these four image sections are presented in Table 1. All indices rated site (d) as the best choice, but the TPSIs B and C failed to identify site (c) as the second best choice. These two indices ranked highly the strong vertical edge in site (b).

2.3 Tie Point Validation

Initially a spectral band is used to produce a TPSI. A criterion will select which pixels will be used as candidate tie points. For example, the image can be divided into sectors, and the pixel with highest TPSI in each sector selected.

For each pixel selected as candidate tie point in the base image, the matching process will provide a conjugate pair in the input image. This will be the location in the search window where the convolution between the target window and the search sub-window is maximum. However, this will not necessarily be a suitable match, as the presence of clouds, noise, or other similar locations elsewhere might result on the selection of the wrong location. It is important to have a consistent criterion to reject these bad matches. The hyperspectral characteristic of the images can provide additional information to properly identify the correct matches.

A convolution between the target window and the search sub-window, centred on the location selected by the image matching process, is performed in several spectral bands. In this work only 9 bands were used (bands 1, 4, 6, 8, 10, 12, 14, 16 and 18) but more bands could be easily used. Three parameters are considered: (i) the number of bands with a correlation coefficient (r) above 0.95 ($N1$), (ii) the number of bands with $r > 0.90$ ($N2$) and (iii) the average of the 3 highest r values ($R123$). An adequate match will hopefully score high in all 3 parameters while a wrong match should score low in at least one of them.

3 Results

Three CHRIS/PROBA scenes of the same location were acquired to test the performance of the TPSIs and the validation parameters. The image centre target was a point in Arcos de Valdevez, northwest Portugal, with longitude -8.42 and latitude 41.8. Each image scene includes 5 images (766 by 748 pixels) with different viewing angles, each with 18 spectral bands. Due to the uncertainty in CHRIS/PROBA pointing, the centre might be displaced by as much as

7 km, which is more than half the image size (image swath of 13 km) [1]. This is true both for images acquired on different dates, and for different viewing angle images of the same scene.

3.1 Testing Strategy

Five pairs of CHRIS/PROBA images were selected for testing. Table 2 indicates the main characteristics of the images used (ZA–Zenith Angle). The image pairs tested were the pairs formed by the three vertical views (IM1 to IM2, IM1 to IM3 and IM2 to IM3), and the pairs formed between the vertical view and two oblique views for image 1 (IM1 to IM1B and IM1 to IM1C). Figure 2 shows an example of a vertical view image (IM3), the near infrared band 14 (781 nm) [2]. In this image it is easy to distinguish the noticeable feature of the river that crosses the image nearly horizontally, which given its irregularity should provide good locations to search for tie points.

Table 2. CHRIS/PROBA images used in the TPSI test (ZA–Zenith Angle)

Label	Aquisition date	Fly-by ZA	Observation ZA	Solar ZA
IM1	11 04 2006	0	4.7	35.0
IM1B	11 04 2006	+36	28.3	35.0
IM1C	11 04 2006	-36	33.4	35.0
IM2	28 05 2006	0	3.6	30.0
IM3	24 05 2006	0	6.2	23.0

For each pair of images, one was selected as base and the other as input image. Between 9 and 15 control points were identified manually for each image pair, which were used to establish first order and second order polynomial transformation functions. The TPSIs B, C, R and P were computed for selected reference bands of the base images. Only bands 1, 6, 10, 14 and 18 were used as reference. As an example, Figure 2 (centre) shows the Composed TPSI (C) produced for IM3 with reference band 14 (the image displayed is $C \times 2$ as the original image has low contrast). The next step is to select a reasonable number of widespread tie point candidates. This goal is achieved by establishing a 5 by 5 grid of non-adjacent sectors. An illustration of this grid is displayed over a TPSI image in Figure 2 (right). For each TPSI image, the pixel with maximum value on each sector is selected as a candidate tie point. The image-matching algorithm is used to select the conjugate pixel on the input image, with a search window size of 251 by 251 pixels. The central pixel of the sub-window with maximum correlation coefficient is selected as the tie point conjugate pair. The correlation between the base and input image for a 5×5 window is computed for this pair for all 9 bands. The 3 validation parameters ($N1$, $N2$ and $R123$) are also computed.

A tie point provided by the image matching process is evaluated according to the root mean square difference between its coordinates in the input image and those predicted by the second order polynomial transformation function

Fig. 2. IM3 Band 14 (left), Composed TPSI (centre), Ratio TPSI with sectors and points selected (right)

established for the image pair. If the difference was less than 2 pixels the point was labelled as correct, if it was above 5 pixels it was labelled as incorrect, and those pixels with a difference between 2 and 5 pixels were labelled as doubtful. The pixels that the polynomial transformation function projected to a location outside the input image were considered invalid, as no successful matching was possible. Those points were discarded from subsequent analysis.

3.2 Reference Band Test

The first test was to evaluate the importance of the spectral band used as reference. The procedure described in the previous section was applied to the image pair IM1 (base) and IM2 (input) using bands 1, 6, 10, 14 and 18 as reference bands. The results are summarised in Table 3. Only the correct (good) and the incorrect (bad) matches are displayed in the table. For each of the four TPSI the variables tested were the number of tie points, and the validation parameters $N1$, $N2$ and $R123$. Ideally one would have a large number of correct tie points and few incorrect tie points. The validation criteria should have high values for correct points and low values for incorrect points. The values underlined in Table 3 correspond to the best reference band for each TPSI, and in bold is the best overall for each parameter tested. For example, the best combination in terms of the number of valid (good) tie points was to use the Ratio TPSI with band 10 as reference. This combination resulted in 16 correct and only 5 incorrect tie points (2 were doubtful and the remaining 2 invalid).

The best scenario for the validation parameters is to have a large difference between the values of good and bad tie points. This is a requirement to make the parameter able to properly discriminate the good and bad (valid or invalid) tie points. The values underlined in Table 3 correspond to the reference bands that proved more efficient in this context. The Ratio TPSI performed best in all 4 parameters tested. The reference band 14 seems to be the most efficient in terms of tie point validation, while band 10 or band 1 are good choices in terms of getting a high number of valid tie points. This might be partially justified by the spectral location of these bands. Band 1 is in the blue part of the visible spectrum (442 nm), Band 10 in the red edge (703 nm) and Band 14 in the near infrared

Table 3. Summary of the TPSI performance using 5 different reference bands (image pair IM1–IM2)

TPSI:	BASIC		COMPOSED		RATIO		PREWITT	
REFERENCE / TP type:	Good	Bad	Good	Bad	Good	Bad	Good	Bad
No. TPs								
BAND 1	14	7	14	7	14	7	12	9
BAND 6	10	10	9	10	10	11	10	12
BAND 10	11	8	11	8	**16**	**5**	8	10
BAND 14	11	7	9	9	10	6	8	10
BAND 18	8	11	12	6	9	8	11	8
N1								
BAND 1	5.00	2.21	5.71	2.86	5.57	0.93	4.33	0.83
BAND 6	6.80	2.50	6.67	2.30	3.40	1.73	5.90	1.42
BAND 10	6.55	1.88	5.73	2.25	4.13	0.80	6.00	0.50
BAND 14	5.45	4.00	5.56	4.00	**6.50**	**0.50**	5.38	3.10
BAND 18	6.25	3.18	6.42	2.67	6.22	0.88	5.73	2.50
N2								
BAND 1	6.57	4.07	7.14	4.43	7.57	2.57	7.67	3.25
BAND 6	7.70	4.60	7.89	4.40	6.90	2.55	7.10	3.17
BAND 10	7.64	4.38	7.55	3.25	7.06	3.20	8.00	2.90
BAND 14	6.91	6.00	7.11	5.56	**7.90**	**2.00**	6.50	4.40
BAND 18	7.75	5.27	7.42	5.17	7.56	2.63	6.55	4.25
R123								
BAND 1	0.973	0.947	0.976	0.955	0.968	0.907	0.963	0.925
BAND 6	0.978	0.964	0.981	0.954	0.958	0.923	0.972	0.918
BAND 10	0.978	0.944	0.971	0.923	0.960	0.848	0.974	0.890
BAND 14	0.921	0.821	0.946	0.663	**0.951**	**0.357**	0.922	0.640
BAND 18	0.963	0.838	0.965	0.755	0.946	0.480	0.929	0.769

(781nm) [2]. There is considerable scattering by the Earth's atmosphere for low visible wavelengths (particularly blue), which tends to reduce the variability of the surface reflectances in Band 1. On the contrary, the spectral signatures of water and vegetation are clearly distinguishable in the near infrared. However, using near infrared reference bands, it is very likely that some vegetation features are selected as candidate tie points, which will often fail to produce a suitable match due to vegetation change between the two image acquisition dates.

Overall, the performance of band 10 as reference band can be considered rather good, except for the validation parameter $R123$, which has a small difference between good and bad tie points.

3.3 TPSI Test

Another test of the TPSI performance was done using the 5 image pairs available. The reference band used was band 10, as this seems to offer a good compromise between the various parameters of interest. The results are summarised in Table 4, again with only the correct (good) and incorrect (bad) tie points

Table 4. Summary of the TPSI performance using 5 images pairs (reference band 10)

TPSI:	Basic		Composed		Ratio		Prewitt	
Image pair / TP type:	Good	Bad	Good	Bad	Good	Bad	Good	Bad
No. TPs								
IM1 to IM2	11	8	11	8	**16**	**5**	8	10
IM1 to IM3	8	8	7	7	5	12	4	13
IM2 to IM3	12	8	12	8	13	7	10	11
IM1 to IM1B	7	14	7	15	6	19	4	19
IM1 to IM1C	6	16	5	18	7	17	6	17
N1								
IM1 to IM2	6.5	1.9	5.7	2.3	4.1	0.8	6.0	0.5
IM1 to IM3	6.0	2.3	5.7	2.4	5.8	0.9	**6.8**	**1.2**
IM2 to IM3	7.5	3.0	7.7	3.1	5.5	1.4	5.8	1.5
IM1 to IM1B	7.7	2.7	7.9	2.7	7.3	2.7	6.0	2.2
IM1 to IM1C	7.7	4.3	6.6	3.8	3.4	1.5	5.2	1.6
N2								
IM1 to IM2	7.6	4.4	7.5	3.3	7.1	3.2	**8.0**	**2.9**
IM1 to IM3	7.4	3.6	7.1	3.9	6.8	2.5	7.3	3.3
IM2 to IM3	8.7	5.0	8.7	4.4	7.8	4.0	7.5	3.1
IM1 to IM1B	8.9	4.9	8.9	4.4	8.5	4.3	8.3	4.0
IM1 to IM1C	8.0	6.0	7.8	5.6	7.4	3.5	7.2	4.4
R123								
IM1 to IM2	0.978	0.944	0.971	0.923	**0.960**	**0.848**	0.974	0.890
IM1 to IM3	0.980	0.908	0.980	0.932	0.973	0.903	0.987	0.924
IM2 to IM3	0.981	0.957	0.981	0.940	0.970	0.923	0.975	0.912
IM1 to IM1B	0.985	0.911	0.986	0.904	0.969	0.908	0.972	0.929
IM1 to IM1C	0.982	0.944	0.982	0.928	0.941	0.882	0.969	0.918

displayed. The underlined values in this Table correspond to the best TPSI performer, for each image pair and parameter. The bold indicates the best performance overall for a parameter. Out of the 20 image pairs and parameters tested, the number of wins was 3 for the Basic TPSI (B), 4 for C, 7 for R and 6 for P. The R and P indices got 2 best overall rates each. Although the results do not seem to indicate a clear favourite, the Ratio and Prewitt indices performed better.

The results in Tables 3 and 4 suggest that a suitable discrimination criteria based on the $N2$ value can be established. The results seem to indicate that a threshold of 5 could perhaps be effective to distinguish between good and bad tie points. However, the values presented in these tables are average values, calculated for all tie point candidates labelled as good or bad. A detailed analysis was carried out to investigate if this could be an appropriate discrimination criterion. The results are summarised in Table 5. The accuracy of this criterion depends on the TPSI index, reference band and the limiting threshold (l) considered, but are generally around 80%.

Table 5. Evaluation of a discriminative criteria based on $N2$

Label	Ratio 10	Prewitt 10	Ratio 14	Prewitt 14
No. Points	107	101	104	100
Success rate (l=6)	79.1%	84.1%	83.9%	76.5%
Success rate (l=5)	78.1%	80.4%	81.0%	74.2%

4 Conclusions

The successful implementation of a fully automatic geometric correction system, based on the identification of tie points (image to image control points) by image matching, is dependent both on the ability to identify suitable areas to search for tie points, and to validate the candidate tie points. Four Tie Point Suitability Indices (TPSI) were proposed, which aim to select the most suitable areas to be used as candidate tie points in an image. Three tie point validation parameters were also proposed, which can be used with hyperspectral or multispectral images. The validation parameters make use of the fact that there is a high correlation between the neighbourhoods of correctly matched tie point for a large number of spectral bands. The proposed TPSIs and validation parameters were tested with 5 CHRIS/PROBA hyperspectral high-resolution satellite images. A criterion to distinguish between correct and incorrect candidate tie points was tested, with an accuracy of about 80%. The results are promising but further research is still required in order to establish the most effective TPSI and validation criteria.

Acknoledgments

This work was done with the support of Centro de Investigação em Ciências Geo-Espaciais, Faculdade de Ciências da Universidade do Porto, financed by the Portuguese National Science Foundation (Fundação para a Ciência e a Tecnologia - FCT).

References

1. Alonso, L., Moreno, J.: Advances and limitations in a parametric geometric correction of CHRIS/PROBA data. In: Proc. of the 3rd ESA CHRIS/Proba Workshop, March 21-23, ESRIN, Frascati, Italy ESA SP–593 (2005)
2. Cutter M. A.: CHRIS data format, revision 4.2 SIRA, Kent, UK (2005)
3. Barnsley, M.J., Settle, J.J., Cuter, M.A., Lobb, D.R., Teston, F.: The PROBA/CHRIS Mission: A Low-Cost Smallsat for Hyperspectral Multiangle Observations of the Earth Surface and Atmosphere. IEEE Transactions on Geoscience and Remote Sensing 45, 1512–1520 (2004)
4. Using Matlab, Version 6.5. The MathWorks, Inc. Natick. MA (2002)
5. Gonzalez, R.C., Woods, R.E.: Digital Image Processing. Prentice-Hall, Englewood Cliffs (2002)
6. Gonzalez, R.C., Woods, R.E., Eddins, S.L.: Digital Image Processing using MATLAB. Prentice-Hall, Englewood Cliffs (2004)

Boneless Pose Editing and Animation

J. Andreas Bærentzen[1], Kristian Evers Hansen[2], and Kenny Erleben[3]

[1] Informatics and Mathematical Modelling
Technical University of Denmark
[2] 3Shape
[3] Computer Science, University of Copenhagen
jab@imm.dtu.dk, evers@exsite.dk, kenny@diku.dk

Abstract. In this paper, we propose a pose editing and animation method for triangulated surfaces based on a user controlled partitioning of the model into deformable parts and rigid parts which are denoted handles. In our pose editing system, the user can sculpt a set of poses simply by transforming the handles for each pose. Using Laplacian editing, the deformable parts are deformed to match the handles. In our animation system the user can constrain one or several handles in order to define a new pose. New poses are interpolated from the examples poses, by solving a small non-linear optimization problem in order to obtain the interpolation weights. While the system can be used simply for building poses, it is also an animation system. The user can specify a path for a given constraint and the model is animated correspondingly.

1 Introduction

Almost all modern animation systems are based on the concept of *skeletons* and *skinning*. Skinning is basically the process whereby a trained professional relates a triangle mesh (the skin) to a bone structure (the skeleton). When this work is done, the mesh may be animated by transforming the bones in the skeletal structure. The transformations of the bones, in turn, can be computed using *keyframes*, *inverse kinematics*, *motion capture*, or some other method.

While skeleton based animation is well established and used in numerous commercial systems, it is by no means an ideal solution. In particular, it is difficult to create a good correspondence between the skeleton and the mesh (skin). One goal of this paper is to convince the reader that we can do without skeletons. Instead of a skeleton, we propose to allow the user to paint rigid handles on the model and animate by transforming these handles.

1.1 Contributions and Related Work

Traditional inverse kinematics [1] combined with linear blend skinning [2] is used for character animation in packages such as Maya®and 3DStudio Max®. This combination offers the advantage of direct control and manipulation of the skin and skeletons. However, there are certain issues. Linear blend skinning is

B.K. Ersbøll and K.S. Pedersen (Eds.): SCIA 2007, LNCS 4522, pp. 562–571, 2007.

notoriously known for joint collapse and "candy wrap" artifacts. Many people have tried to fix these problems, for instance by replacing the interpolation method [3], or improving on the skin weights using statistical analysis and adding correction terms to the linear blending [4]. Also advanced methods exist for assigning skin weights, e.g. [5]. Nevertheless, animators still spend a considerable amount of time tweaking skin weights. In fact, traditional skeleton and skinning animation is known for requiring tweaking of skin-weights on a per motion basis, i.e. skin-weights that work well for a jumping motion may work badly for a running motion, and skinning for a lower arm twist probably won't do for wrist flexing.

Skeleton based animation also falls a bit short with regard to what types of animations that can be achieved. Typically, animators work in one of two ways: Either the model is skinned and animated by rotation of bones (skeletal animation) or individual poses are sculpted simply by moving individual vertices to specific positions in specific poses (interpolation based animation). The former is more useful for overall character animation while the latter is needed for facial expressions. Ideally, and in order to fix the problems with bones based animation, they should be combined, and in [6] Lewis et al. present a system where the user can manipulate vertices directly for a given pose and at the same time use a skeleton driven deformation. Arguably, we obtain the same advantage in a simpler fashion.

Using our method, only one weight is associated with each vertex, and it has a very clear significance, namely the rigidity of a vertex. Using a simple selection tool, the user marks clusters of rigid vertices which become the *handles* of the model. Once a set of handles has been defined, the user can sculpt poses by rigidly transforming the handles. The deformable parts of the model are subsequently transformed using *Laplacian editing* which is discussed briefly in Section 2. (C.f. [7] for more details). Handle selection and pose editing are described in section 3. At least two benefits are gained from our approach

- **Simplicity:** The user only needs to paint "rigidity" and not "degree of association with any bone in a set of bones". Moreover, there are no bones which need to be aligned with the model in its rest position.
- **Genericity:** Our method is based on rigid motion (translation or rotation) of handles followed by a smooth deformation of the remaining parts of the mesh. This technique can be used to achieve the same effects as both skeletal and interpolation based animation.

Inverse kinematics problems are usually underdetermined in the sense that different motions will lead to the same goal [8]. In order to tackle this issue, recent authors have considered reconstructing plausible motion from examples. In [9] Grochow et al. represent a dense set of poses as feature vectors in a low dimensional space and given a set of constraints finds an interpolated pose close to one of the examples. In Mesh based inverse kinematics [10] Sumner et al. also finds a model in pose space which best fits a set of constraints. Our approach is conceptually similar, but the problem to be solved is simpler since the pose space is a space of handle transformations and not vertex transformations. The technique used for animation is described in detail in section 4. Finally, in section 5, we discuss our findings, draw conclusions and point to future work.

2 Laplacian Editing

The fundamental idea in Laplacian editing is to represent mesh vertices in terms of *differential coordinates*. The differential coordinates capture the position of a vertex relative to its neighbours, and by nailing down the translational freedom, we can reconstruct the mesh directly from this representation. Moreover, we can impose several constraints and reconstruct the remaining vertices from the differential coordinates in the least squares sense which is what allows us to perform deformations.

To be more specific the differential coordinates of a vertex i are

$$\boldsymbol{\delta}_{\mathbf{i}} = \frac{1}{|N_i|} \sum_{j \in N_i} \mathbf{v}_j - \mathbf{v}_i \tag{1}$$

where j is a vertex in the neighbourhood, N_i, of i, and $\mathbf{v}_i$ and $\mathbf{v}_j$ are the positions of vertex i and its neighbour j, respectively. This operator, known as the umbrella operator, (1) constitutes a mesh Laplace operator given a simple local parameterization [7]. It is clearly possible to compute the mesh Laplacian simultaneously for all vertices if we write it as a matrix $\mathbf{L}$ where $\mathbf{L}_{ii} = -1$ and $\mathbf{L}_{ij} = \frac{1}{|N_i|}$ if there is an edge from vertex i to vertex j and zero otherwise. Provided we have all the vertices and a column vector $\mathbf{P}$, we can now compute the differentical coordinates for all vertices $\mathbf{D} = \mathbf{LP}$ The matrix does not have full rank [7] but adding just a single constrained vertex allows us to reconstruct the positions $\mathbf{P}$ from the $\mathbf{D}$ vector. In general, we would add multiple constraints and then recompute the vertex positions by solving for $\mathbf{D}$ in the least squares sense. If we wish to constrain a vertex i, we add an equation of the form $w_i \mathbf{v}_i = w_i \mathbf{c}_i$, where w_i is the weight of our constraint and $\mathbf{c}_i$ is the position of the constrained vertex. This equation is added as a row to the system $\mathbf{D} = \mathbf{LP}$. Let m be the number of constraints. In this case the system becomes

$$[\mathbf{D}|w_1\mathbf{c}_1 \; w_2\mathbf{c}_2 \; ... \; w_m\mathbf{c}_m]^T = \left[\frac{\mathbf{L}}{I_{m \times m}|0}\right] \mathbf{P} \ , \tag{2}$$

where we have assumed that the first m vertices are constrained. Of course, the vertex numbering is arbitrary and the constraints can be imposed on any m vertices.

To simplify notation, let the extended $\mathbf{D}$ vector of Laplacians be denoted $\tilde{\mathbf{D}}$ and the extended $\mathbf{L}$ matrix be denoted $\tilde{\mathbf{L}}$. In this case, the above system is $\tilde{\mathbf{D}} = \tilde{\mathbf{L}}\mathbf{P}$, and to reconstruct our vertices, we need to compute the least squares solution

$$\mathbf{P} = (\tilde{\mathbf{L}}^T \tilde{\mathbf{L}})^{-1} \tilde{\mathbf{L}}^T \tilde{\mathbf{D}} \tag{3}$$

Unfortunately, the differential coordinates $\mathbf{D}$ are not invariant with respect to rotation or scaling. In animation, the first problem is particularly severe. Several elegant solutions exist, e.g. [11,12,13]. Inspired by the work of Lipman et al. [13], we choose the approach we find to be the simplest. The idea is to have a smooth version of the mesh. For each vertex of the smoothed mesh, we compute a frame and represent the differential coordinates in terms of this frame. When the surface is deformed by moving the constrained vertices, we compute a new smoothed version of the mesh and corresponding frames for each vertex. The differential coordinates are then rotated simply by transferring them to the new frame.

What lends efficiency to this scheme is that the smooth solution is simply obtained by solving (3) with $\mathbf{D} = \mathbf{0}$ since this solution minimizes the membrane energy [7]. Intuitively, minimizing the Laplacians corresponds to placing each unconstrained vertex at the barycenter of its neighbours. Clearly, the constraints are also important in this case since an unconstrained mesh would simply collapse.

For a large mesh solving (3) at interactive rates is not feasible unless attention is paid to the structure of the problem. Fortunately, we are looking for a least squares solution which means that we are dealing with a positive definite matrix. Moreover, $(\tilde{\mathbf{L}}^T\tilde{\mathbf{L}})$, is fairly sparse, and a good way of solving such a linear system is by Cholesky factorization [14]. We do this using the Taucs library which is sparse matrix library known for its speed.

3 Pose Editing

In the following, we describe how an animator might define a set of handles and edit the model using these handles. The discussion is based on our test system which works well, but clearly vertex selection and handle rotation or translation can be done in a number of ways. We have chosen simple methods that work well with mouse and keyboard. First of all, It is necessary to have a relatively flexible mechanism for selecting the vertices which belong to a given handle since a single handle can be an arbitrary and not necessarily connected group of vertices. In

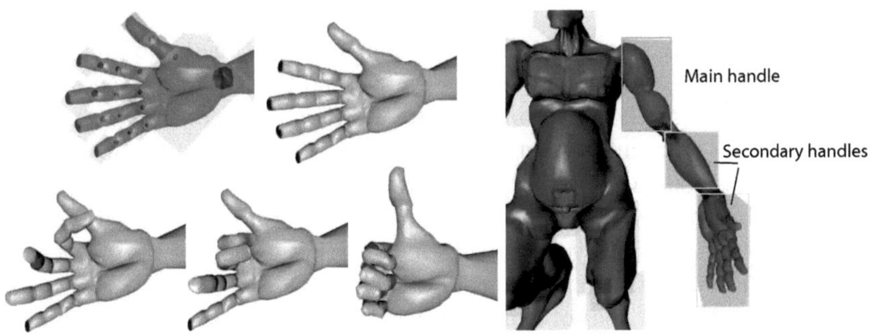

Fig. 1. Handles painted onto a hand model, and some sculpted poses (left). When we need to transform an arm it is crucial to be able to select several handles and rotate using the center of rotation for just one of these (right).

our system, handles can be defined using a paint metaphor where the user sprays the regions that are to be defined as handles. It is also possible to make a screen space box selection and the user can choose to select only the visible vertices which fall within the box or to select only one connected component. An example of handles on a hand model and corresponding sculpted poses is shown in Fig. 1 (left). As a part of defining the handles, the animator also needs to indicate the center of rotation for each handle.

Having selected the handles and rotation centers, the user proceeds to sculpting the pose. This is basically done by moving and rotating handles. A handle is moved or rotated simply by selecting and dragging with the mouse. Since we use a mouse as interface all translation is parallel to the view plane and rotation is around an axis perpendicular to the view plane. However, the user can choose arbitrary views.

Clearly, we sometimes wish to transform several handles in one operation (See Fig. 1 right). In order to do so, the user specifies a set of handles to transform (typically rotate) and then the handles are transformed using the first selected handle's center of rotation. Selecting several handles imposes some structure of the handles reminiscent of a skeletal structure, but the structure itself is not stored, only the transformations which are applied to each handle in the selection.

It is also important to state that when a handle is transformed, the actual transformation is stored (e.g. a rotation axis and the angle in degrees) and not simply a transformation matrix or, worse, the new vertex positions, since knowing the precise transformation allows us to smoothly go from rest position of the handles to the deformed state by using only a specific fraction of each transformation.

Apart from simply selecting the handles, the user also has the option of specifying that parts of the handle are less rigid than others (using the paint metaphor). This can have a large impact as shown in Fig. 2 bottom left.

4 Pose Blending and Animation

The poses which have been defined using the method outlined above can be used in a number of ways. In some computer games, models are animated simply by interpolating between the positions of corresponding vertices in two (keyframe) poses. That tends to be a better strategy than skeletal based animation for things like facial expressions and it works well if rotations are not too big, e.g. if we have a dense set of keyframe poses. Our pose blending and animation system could indeed be used simply as a tool for constructing keyframe meshes to be used in conjunction with linear interpolation.

However, the second goal of this project was to develop an animation system with a direct coupling to the pose editing system just described.

In Lewis et al. [6] a pose space is defined as the space spanned by the variations of the controls needed to specify the various poses. In the current context, our controls are simply the transformations imposed on the handles.

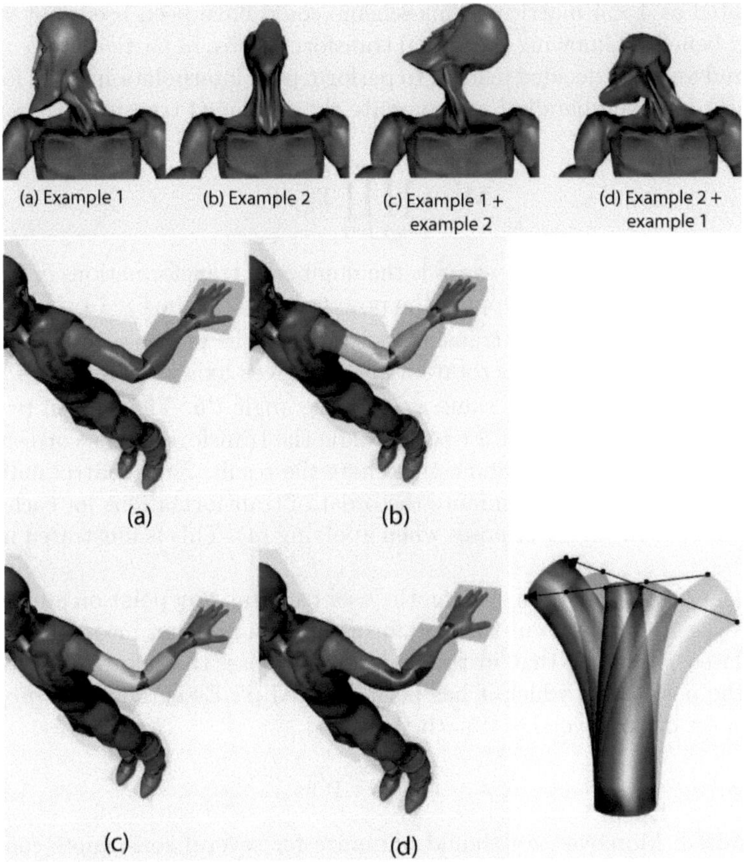

Fig. 2. This figure shows that examples do not commute (top row), and that if we use semirigid vertices we obtain different results. Here weights of 1, 2/3, 1/3, 0 were used (bottom left). Finally, simply interpolating linearly between examples does not suffice (bottom right).

Our goal is not only to interpolate between poses but also to find the best pose to match a given constraint. In this respect our work is similar to that of Sumner et al. [10]. However, the strategy is different. Sumner et al. are not concerned with how the pose is obtained whereas we employ the transformations used to define the pose when interpolating in pose space. Moreover, they constrain individual vertices whereas we constrain handles.

An issue confronted when interpolating in pose space is the fact that for large rotations linear interpolation simply does not work as shown in Fig. 2 bottom right. This means that we cannot just linearly interpolate between the vertex positions. Nor can we perform linear interpolation between transformation matrices in a meaningful fashion. This last problem has actually been addressed by Alexa [15] who proposed a scheme for meaningful commutative addition and scalar multiplication of transformations (rotations, scalings, and translations) represented as 4×4 matrices. This scheme could have been used, but since we have the benefit of knowing the actual transformations, in particular the rotation angles and axes, we decided instead to perform pose interpolation in the following fashion. For a given handle h we compute the combined transformation matrix, M^h, as follows

$$\mathbf{M}^h = \prod_{i=1}^{P} \prod_{j=1}^{N} \mathbf{T}_{i,j}^h(t^i) \tag{4}$$

where P is the number of poses, N is the number of transformations of handle h, and $T_{i,j}^h(t^i)$ is a function mapping the pose weight, t^i, to a 4×4 transformation matrix representing the j'th transformation of the the i'th pose scaled by t^i. For instance, if $T_{i,j}^h(1)$ represents rotation about a given axis with angle θ, $T_{i,j}^h(t^i)$ is simply a rotation about the same axis by the angle $t^i\theta$. Thus, given two poses with weight 0.5 we construct for each handle the transformations corresponding to half the pose transforms and concatenate the result. Since matrix multiplication does not, in general, commute the order of transformations for each pose is significant as is the order of poses when applying (4). This is illustrated in Fig. 2 top.

In our pose interpolation system, the user can grab any point on a handle and simply drag it. The system then performs an optimization in order to find an interpolated pose such that in the transformed pose, the selected point, $\mathbf{p}$ will match the position to which it has been dragged $\mathbf{p}'$. Essentially, the problem is to find a set of pose weights t^i such that

$$\|M_h\mathbf{p} - \mathbf{p}'\| \tag{5}$$

is minimized. Moreover, we should minimize for several constraints contemporaneously.

4.1 Optimization

There are two steps to this minimization, namely picking the best order of the poses and picking the best weights. Assume we are given an order, we simply need to minimize the target distances as discussed above. Unfortunately, (5) does not suffice in itself since there is no penalty for a pose which does not contribute to reaching the target. In other words, if the foot is constrained, adding in a pose which raises an arm is not going to increase the energy. Hence, we have experimented with a number of energy terms to add to (5). The most useful are 1: Sum of weights, 2: One minus sum of weights, 3: Distance to example pose, 4: Distance from last interpolated pose

1 often gives good results but effectively keeps us close to the rest pose which is often not desired. Thus, we recommend 2 which keeps the sum of weights close to a division of unity. If the pose space is dense and we wish to remain close to the examples, 3 is useful. Finally, 4 basically constrains the solution to be close to the previous solution. If we are performing an animation, this term avoids discontinuities.

A number of strategies were considered for performing the actual optimization. Bearing in mind that the energy functional is relatively complex and consists of heterogeneous terms, we decided to use a simple method (inspired by Hooke and Jeeves [16]) which only requires evaluation of the energy for various combinations of the weights: From the initial value, the weight of the first pose is iteratively increased halving the step length if no improvement is made, until the step size is below a given threshold. If increasing did not work, decreasing is attempted. Once a best weight has been found, the next pose is tried. This can be seen as a simple hill climbing method which we found to work well in practice.

Only one iteration is used to ensure program responsiveness. However, when the constraint position is moved, we rerun the optimization procedure starting from the previously found weights since they are almost always a good starting guess.

The final ingredient is the ordering of poses. We choose the simple greedy strategy of ordering the poses by the distance of the pose (by itself) to the target constraints. Note that in a polished system based on our method, the user would be exposed to few, if any, of the choices mentioned above.

5 Results and Discussion

In this paper, we have proposed an effective method for pose editing and animation which is entirely based on painting rigidity (i.e. handles) and placing centers of rotation. This is arguably much simpler than traditional skinning. Performancewise, the system can handle fairly large models at interactive frame rates. In Table 5 (left) we have listed the precomputation time and the time it takes to actually carry out the Laplacian editing for a range of model sizes. Note that even for a mesh of 45000 vertices, it takes less than a second total to compute the deformed pose given handle positions. These numbers seem to be about the same as what Sumner et al. report for their MeshIK system, but notice that the numbers cited there is for one iteration of their linear solver, and six iterations are generally needed for the solution to the nonlinear problem [10].

A particularly strong point of our method is the fact that it combines, in a homogeneous fashion, rotational motion (which is simple in bones based systems) with translational deformations. In Fig. 3 top, a bent leg is shown. On the left,

Table 1. These tables show performance numbers for deformation and the frame rate for an animation

Deformation

Model	Vertices	Precomp. (s)	Per frame (s)
Boba Fett (small)	2138	0.09	0.04
Boba Fett (medium)	10600	0.52	0.18
Armadillo	17300	1.04	0.30
Boba Fett (large)	45000	3.45	0.83

Animation

	Example poses		
handles	8	12	20
18	17.4	15.4	12.4
24	17.2	14.9	11.7
	fps		

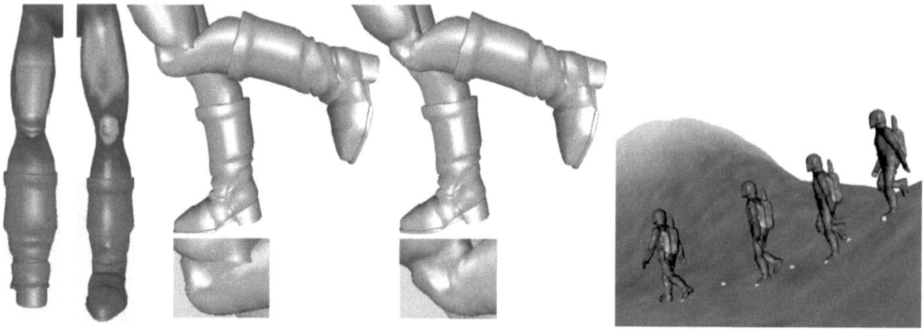

Fig. 3. Top: Using bones it is tricky to achieve an artifact free bent knee. On the left, we have shown the handles. The middle figure shows what happens if we simply rotate the lower leg. On the right is the result after some tweaking of the knee and uper leg handles. Bottom: A walk animation.

the leg is simply bent by rotating the lower leg. This leads to self intersections and an unnatural looking knee. Through minor adjustments (mostly translations) a much more natural pose is obtained. It is not clear how these adjustments could have been made in a purely bones based system.

Table 5 right contains the frame rates for animation of the small Boba Fett model (2138 vertices) which is shown walking down a grassy slope in Fig. 3 bottom. The animation is done simply by placing constraints on the feet of the model and then alternately moving the left and right constraint position. The balls indicate constraint positions. Note that for each frame, the optimal pose is found; the handles are transformed accordingly, and finally Laplacian editing is used to deform the model.

Our method leads to a homogeneous paradigm for animation and a simple workflow for the animator. Unsurprisingly, these advantages do not come for free. Unlike the simple matrix blending used in bones systems, which is easily implemented on a GPU, we need to solve a (possibly large) linear system for each frame. For this reason it may not be directly feasible to use this form of animation in a game, but it would be possible to sculpt keyposes and use the pose interpolation to create a dense set of keyframes which would be very suitable for real-time animation. One might also envision our system being used for actual animation in a non-real time setting our in a real time setting in a few years if there is a sufficient advance in the speed at which we can solve linear systems. For instance, one direction of future investigation would be to offload the solution of the linear system to graphics hardware.

References

1. Wellman, C.: Inverse kinematics and geometric constraints for articulated figure manipulation. Master's thesis, Simon Fraser University (1993)
2. Fernando, R., Kilgard, M.J.: The Cg Tutorial, The Definitive Guide to Prgrammable Real-Time Graphics (Chapter 6, Section 9.1). Addison-Wesley, London (2003)

3. Kavan, L., Zara, J.: Spherical blend skinning: a real-time deformation of articulated models. In: SI3D '05: Proceedings of the 2005 symposium on Interactive 3D graphics and games, New York, USA, pp. 9–16. ACM Press, New York (2005)
4. Kry, P.G., James, D.L., Pai, D.K.: Eigenskin: real time large deformation character skinning in hardware. In: SCA '02: Proceedings of the 2002 ACM SIGGRAPH/Eurographics symposium on Computer animation, pp. 153–159. ACM Press, New York (2002)
5. Mohr, A., Gleicher, M.: Building efficient, accurate character skins from examples. ACM Trans. Graph. 22(3), 562–568 (2003)
6. Lewis, J.P., Cordner, M., Fong, N.: Pose space deformation: a unified approach to shape interpolation and skeleton-driven deformation. In: SIGGRAPH '00. 27th annual conference on Computer graphics and interactive techniques, New York, USA, pp. 165–172. ACM Press/Addison-Wesley, New York (2000)
7. Sorkine, O.: Differential representations for mesh processing. Computer Graphics Forum vol. 25(4) (2006)
8. Maciejewski, A.A.: Motion simulation: Dealing with the ill-conditioned equations of motion for articulated figures. IEEE Comput. Graph. Appl. 10(3), 63–71 (1990)
9. Grochow, K., Martin, S.L., Hertzmann, A., Popović, Z.: Style-based inverse kinematics. ACM Trans. Graph. 23(3), 522–531 (2004)
10. Sumner, R.W., Zwicker, M., Gotsman, C., Popović, J.: Mesh-based inverse kinematics. ACM Trans. Graph. 24(3), 488–495 (2005)
11. Sorkine, O., Lipman, Y., Cohen-Or, D., Alexa, M., Rössl, C., Seidel, H.P.: Laplacian surface editing. In: Proceedings of the Eurographics/ACM SIGGRAPH Symposium on Geometry Processing, Eurographics Association, pp. 179–188. ACM Press, New York (2004)
12. Lipman, Y., Sorkine, O., Levin, D., Cohen-Or, D.: Linear rotation-invariant coordinates for meshes. In: P.o.A.S. (ed.) Proceedings of ACM SIGGRAPH 2005, pp. 479–487. ACM Press, New York (2005)
13. Lipman, Y., Sorkine, O., Cohen-Or, D., Levin, D., Rössl, C., Seidel, H.P.: Differential coordinates for interactive mesh editing. In: Shape Modeling International 2004. (2004)
14. Botsch, M., Bommes, D., Kobbelt, L.: Efficient linear system solvers for mesh processing. In: Martin, R., Bez, H., Sabin, M.A. (eds.) Mathematics of Surfaces XI. LNCS, vol. 3604, pp. 62–83. Springer, Heidelberg (2005)
15. Alexa, M.: Linear combination of transformations. ACM Trans. Graph. 21(3), 380–387 (2002)
16. Hooke, R., Jeeves, T.A.: Direct search solution of numerical and statistical problems. Journal of the ACM 8, 212–229 (1961)

Text Driven Face-Video Synthesis Using GMM and Spatial Correlation

Dereje Teferi, Maycel I. Faraj, and Josef Bigun

School of Information Science, Computer, and Electrical Engineering (IDE)
Halmstad University
P.O.Box 823, SE-301 18
Halmstad, Sweden
{Dereje.Teferi,Maycel.Faraj,Josef.Bigun}@ide.hh.se

Abstract. Liveness detection is increasingly planned to be incorporated into biometric systems to reduce the risk of spoofing and impersonation. Some of the techniques used include detection of motion of the head while posing/speaking, iris size in varying illumination, fingerprint sweat, text-prompted speech, speech-to-lip motion synchronization etc. In this paper, we propose to build a biometric signal to test attack resilience of biometric systems by creating a text-driven video synthesis of faces. We synthesize new realistic looking video sequences from real image sequences representing utterance of digits. We determine the image sequences for each digit by using a GMM based speech recognizer. Then, depending on system prompt (sequence of digits) our method regenerates a video signal to test attack resilience of a biometric system that asks for random digit utterances to prevent play-back of pre-recorded data representing both audio and images. The discontinuities in the new image sequence, created at the connection of each digit, are removed by using a frame prediction algorithm that makes use of the well known block matching algorithm. Other uses of our results include web-based video communication for electronic commerce and frame interpolation for low frame rate video.

1 Introduction

People have to be uniquely identified to get access to an increasing number of services; their office, their bank account, their computer, their mail, even to enter a country etc. To this effect, person identification is done in many ways, one of which is biometrics.

Biometrics is the study of automated methods for uniquely recognizing humans based upon one or more intrinsic physiological or behavioral traits. Biometric systems are in use in many applications such as security, finance, banking etc [1], [2], [3]. In spite of their high level of accuracy, biometric systems have not been used massively in the aforementioned areas. One of the main drawbacks is that biometric data of a person (such as face, speech, etc) are not secret and cannot be replaced anytime the user wants to or whenever they are compromised

B.K. Ersbøll and K.S. Pedersen (Eds.): SCIA 2007, LNCS 4522, pp. 572–580, 2007.

by a third party for spoofing. This problem is minimal if the authentication system works with the help of a human supervisor as in border control where the presented trait can be visually checked to see if it is genuine or fake. However, this risk is high for remotely controlled biometric applications such as those that use the internet [4]. The risk of spoofing on biometric systems can be reduced by combining multiple traits into the system and incorporating liveness detection.

Recent technological advances such as those in audio-video capture and processing have enabled researchers to develop sophisticated biometric systems. As such it has also given spoofers the chance to become more vicious system attackers. Some spoofers can impersonate clients even in multimodal biometric applications. Impersonation is done, for example, by audio-video playback and application of image processing techniques without the presence of the real client.

The actual presence of the client can be assured to a certain degree by liveness detection systems. Liveness detection is an anti-spoofing mechanism to protect biometric systems [5], [6], [7]. It is performed by, for example, face part detection and optical flow of lines to determine liveness score [8], analysis of fourier spectra of the face image [9], lip-motion to speech synchronization[10], body temperature, on the spot random queries such as pronouncing of random sequences of digits, iris size scanning under varying illumination etc.

Random text-prompted liveness detection or audio-video based recognition systems use random digits as their text prompts. This is because random numbers are easier to read for the client than random texts and also easier to synthesize. Accordingly, the digit speaking image sequences of the XM2VTS database is used in this work for the experiment[11]. In our approach we develop a text prompted face-video synthesis system for testing the performance of liveness detection as well as audio-visual recognition systems.

We use GMM based speech recognizer to identify the locations of each digit spoken by the subject. The pre-recorded image sequence is then reshuffled according to system prompt (or digits entered). However, the process of shuffling creates discontinuities between digits in the new image sequence. This discontinuity between the last frame of a digit and the first frame of the next digit is compensated by estimating the frames in between. The GMM based speech recognizer is used to identify the locations of the digits in a video of new training data.

A number of motion estimation techniques are discussed in [12],[13],[14], and [15]. The method applied here for motion estimation is the well known block matching which uses the Schwartz inequality.

2 Speech Recognition

Here we present a speech recognition system using the well known Gaussian Mixture Model(GMM).

Figure 1, illustrates text-dependent speech recognition of a digit for a specific person. The speech analysis is represented by Mel-Frequency Cepstral feature analysis where the extracted feature set is put into Gaussian Mixture Model system.

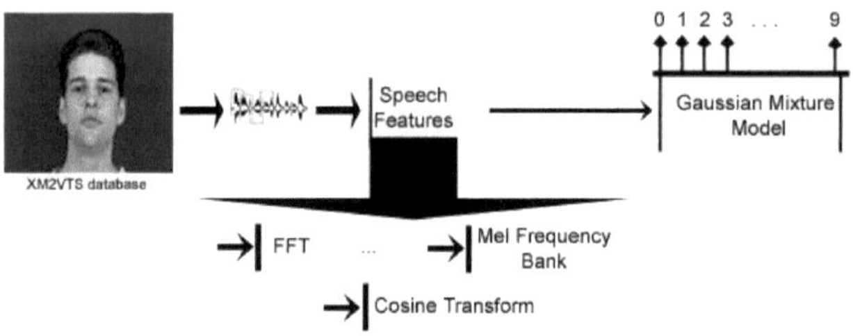

Fig. 1. Speech Recognition System

2.1 Speech Features

The vocal tract structure of a person influences the speech spectrum significantly which in turn can be used to distinguish a user from other users. Spectral representations of a person's speech can be extracted with several methods, and in this paper we implemented the commonly used Mel-Frequency Cepstral Coefficients(MFCC). Our acoustic features were constructed by MFCC according to [16]. The sampled wave files in the XM2VTS database were processed by using the HTK(Hidden Markov Model Toolkit)[17],[18].

The sampled waveform is converted into a sequence of acoustic parameter blocks, where the output sampling period is set to 10 ms and the window size is set to 25 ms, [10]. From each block we extract a 39 dimensional feature vector, consisting of 12 cepstral coefficients with normalized log-energy, 13 delta coefficients and delta-delta coefficients. The delta and delta-delta coefficients correspond to the first and second order time derivatives of extracted MFCC, also known as velocity and acceleration respectively.

2.2 Gaussian Mixture Model

A Gaussian Mixture Model(GMM) can be represented as a weighted sum of multivariate Gaussian distributions [16]. Here we use the GMM to model the person specific features [16]. In this paper, we developed the GMM using the HTK toolbox[17]. Each person is described with Gaussian model parameters that are learned from a training set, that are the mean and variances of a number of Gaussian distributions, as well as their associated weights which are used to linearly combine the individual Normal components to finally yield a multidimensional distribution for the features. More details about the set-up can be found in [10].

3 Motion Estimation

We use motion estimation technique to predict frames to reduce the discontinuity between frames. The discontinuity occurs due to the rearrangement of sequence of frames according to the prompt of the biometric system.

Motion estimation is a common technique used in video compression. A video is a sequence of frames and there is always some redundant data between adjacent frames within a certain period of time $(t_1 - t_0)$. The redundancy is small for fast paced and complex motion of objects and background in the video and high otherwise. This redundancy is exploited to compress the video. That is a reference frame (sometimes known as the independent frame) is taken from a sequence every n frame apart. Then the middle frames are predicted from these frames as well as from previously predicted frames by the codec. The codec actually uses the Motion vector and prediction error to reconstruct the *dependent* frames. Forward prediction, where the new frames are predicted from previous frames, or backward prediction, where the frames are predicted from future frames, or both can be used for estimation of the new frames in the middle. Many codecs use this technique for video compression.

The natural motion of the head while speaking is minimal. Moreover, it is not too difficult to acquire video of an arbitrary person uttering the 10 digits. Given such a sequence an attacker could proceed as discussed below.

Assuming the video is captured with a stationary camera, the background will be near-constant. Therefore, little information is lost or added between adjacent frames, such as teeth, mouth and eyes. That is, there is a high probability that part of a frame exist in another frame although translated to a different location. First the points in motion are extracted using absolute difference between the two frames. These two frames are extracted from the image sequences of the last frame of a digit and the first frame of the succeeding digit.

$$AD = |\mathbf{F}(k,l) - \tilde{\mathbf{F}}(k,l)| \tag{1}$$

Now that we know the points/blocks in motion, the motion vector (MV) is calculated only for these points. For each point or block in motion on frame $\mathbf{F}$, we look for its parallel pattern in frame $\tilde{\mathbf{F}}$ within a local neighborhood by using block matching algorithm.

3.1 Block Matching Algorithm

Block matching is a standard video compression techniques to encode motion in video sequences [14]. A review of block matching algorithms is given in [15], [14], [13], [19].

In our approach, equal sized non-overlapping blocks are created over the frame $\mathbf{F}$ (*the frame at time* t_0). Then, for those blocks on $\mathbf{F}$ containing points in motion, a search area is defined on frame $\tilde{\mathbf{F}}$ (*the frame at time* t_1). The search area is larger than the size of the block by *expected* displacement of the object in motion e. Then we apply Schwartz inequality to find the most parallel pattern for the block in frame $\mathbf{F}$ from the search area in frame $\tilde{\mathbf{F}}$.

Let $\mathbf{f}$ and $\tilde{\mathbf{f}}$ be vector representations of patterns from frame $\mathbf{F}$ and $\tilde{\mathbf{F}}$ and $<,>$ be the scalar product defined over the vector space. We then have

$$| < \mathbf{f}, \tilde{\mathbf{f}} > | \leq \|\mathbf{f}\| \|\tilde{\mathbf{f}}\|$$

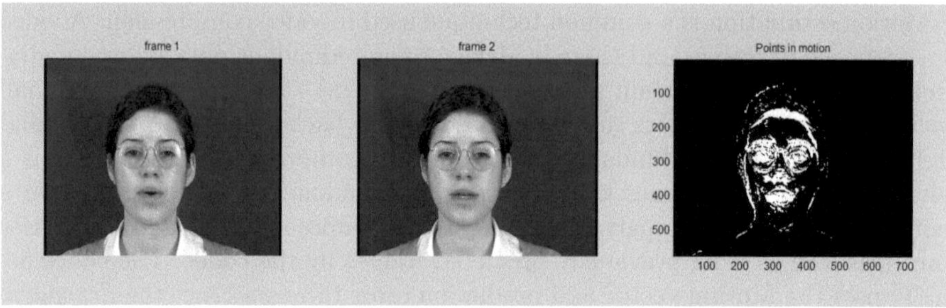

Fig. 2. Points in Motion between frame $\mathbf{F}$ and $\tilde{\mathbf{F}}$

$$cos(\theta) = \frac{|<\mathbf{f}, \tilde{\mathbf{f}}>|}{||\mathbf{f}||\,||\tilde{\mathbf{f}}||} = \frac{|\mathbf{f}^T\tilde{\mathbf{f}}|}{||\mathbf{f}||\,||\tilde{\mathbf{f}}||} \leq 1 \tag{2}$$

where $\mathbf{f} = (f_1, f_2, ..., f_{k-1}, f_k, f_{k+1}, ...)$ and $\tilde{\mathbf{f}} = (\tilde{f}_1, \tilde{f}_2, ..., \tilde{f}_{k-1}, \tilde{f}_k, \tilde{f}_{k+1}, ...)$ are vector forms of the 2D pattern $\mathbf{f}$ and $\tilde{\mathbf{f}}$ from frames $\mathbf{F}$ and $\tilde{\mathbf{F}}$ respectively and $cos(\theta) \in [0, 1]$ is the similarity measure between the patterns.

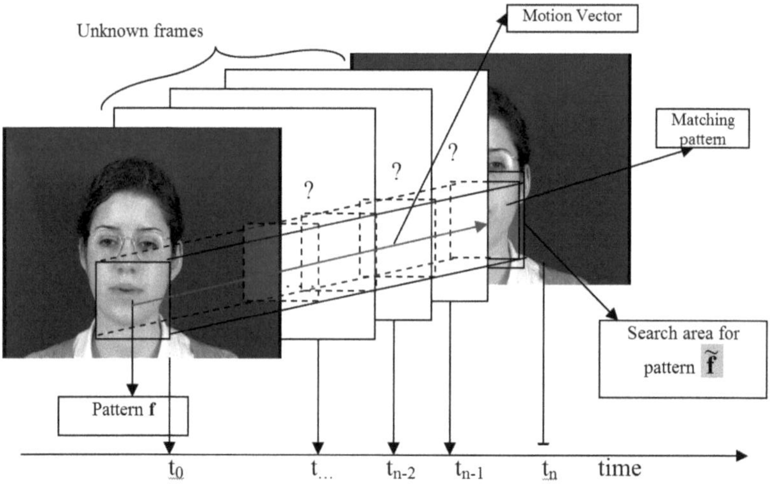

Fig. 3. Motion Vector and unknown frames in a sequence

The most parallel pattern $\tilde{\mathbf{f}}$ in the frame $\tilde{\mathbf{F}}$ is found by maximizing $cos(\theta)$. This can be done by repetitive scalar products. That is the pattern $\mathbf{f}$ from frame $\mathbf{F}$ is glided over an *expected* parallel local neighborhood of $\tilde{\mathbf{f}}$ in the frame $\tilde{\mathbf{F}}$ and the most similar pattern is selected as a match.

The motion vector for the point at the center of pattern $\mathbf{f}$ is calculated as the displacement between pattern $\mathbf{f}$ and pattern $\tilde{\mathbf{f}}$ (Fig. 3). That is:

$$MV(k, l) = x + iy \tag{3}$$

Where x and y are the horizontal and vertical displacements respectively of the block/pattern **f**, *(k,l)* is the index for the center of pattern **f** and $i = \sqrt{-1}$.

3.2 Frame Prediction

Finally, the motion vector is used to predict the unknown frames between **F** and $\tilde{\mathbf{F}}$ (Fig. 3) if the image sequence is to be perceived realistic with minimum amount of discontinuity. The number of frames to be predicted depends on the norm of the motion vector and is determined at run-time. For a length 2 time units, the actual frame prediction is done by dividing the motion vector at point (k,l) by 2 and moving the block in frame $\tilde{\mathbf{F}}$ centered at $(k+x,l+y)$ to the middle frame at $(k+x/2,l+y/2)$. Then the block at the new location in frame $\tilde{\mathbf{F}}$ is moved back the same distance. That is, let **F** be the frame at t_0, $\tilde{\mathbf{F}}$ the frame at t_1 and $\mathbf{F}^{'}$ be the frame at $\frac{(t_1+t_0)}{2}$, then

$$\mathbf{F}^{'}(k+x/2,l+y/2) = \tilde{\mathbf{F}}(k+x,l+y) \tag{4}$$

$$\mathbf{F}^{'}(k,l) = \tilde{\mathbf{F}}(k+x/2,l+y/2) \tag{5}$$

where x and y are the real and imaginary parts of the motion vector MV at (k,l).

To avoid overlaps in the interpolation process, those blocks that are already been interpolated are flagged. Consecutive predictions are made in analogous manner. The motion vector is adjusted and a new frame is created as necessary

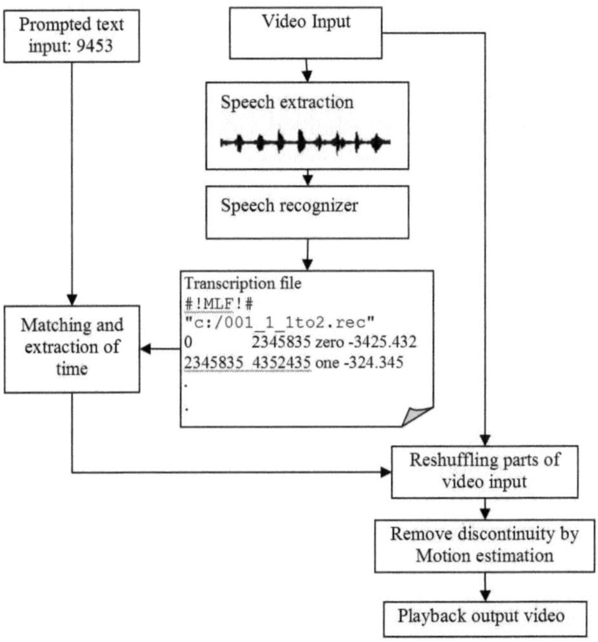

Fig. 4. Process flowchart

between frames $\mathbf{F}$ and $\mathbf{F}'$ as well as between $\mathbf{F}'$ and $\tilde{\mathbf{F}}$. This process continues until all the necessary frames are created.

4 Video Synthesis

Audio signal is extracted from the input audio-video signal and forwarded to the speech recognizer. The recognizer uses the GMM models to return transcription file containing the start and end time of each digit spoken in the audio signal. A search for the prompted text is done against the transcription file and the time gap of each prompted digits within the video signal is captured (Fig. 4).

The discontinuity between the image sequences of each digit is compensated by predicting the unknown frames (Fig. 3) using the motion estimation technique summarized in section 3. The new predicted frames are attached to a silence sound and are inserted to their proper locations in the video signal to decrease the discontinuity of utterances. Finally, the video is played to represent the digit sequence prompted by the biometric system.

5 Experiment

Th experiments are conducted on all the digit speaking face videos of the XM2VTS database (295 persons). The accuracy of the reshuffled video signal is mainly dependent on the accuracy of the speech recognition system. The accuracy of our GMM based speech recognition system is 96%. The frame prediction algorithm works well when the block size is set to 3x3 pixels. When the block size is larger some visible deformations appear on the predicted frame mainly due to rotational effects. Therefore, we used 3x3 blocks to predict the unknown frames. The discontinuity of the reshuffled video signal is reduced significantly as evaluated by the human eye, the authors.

Biometric authentication and liveness detection systems that make use of motion information of zoomed in face, head, lip and text prompted audio-video are easy targets of such system attacks.

6 Conclusion

The risk of spoofing is forcing biometric systems to incorporate liveness detection. Assuring liveness especially on remotely controlled systems is a challenging task. The proposed method shows a way to produce play-back attacks against text-prompted systems using audio and video. The result shows that assuring liveness remotely by methods that rely on apparent motion can be bypassed. Our results suggest the need to increase the sophistication level of biometric systems to stand up against advanced play-back attacks.

Acknowledgment

This work has been sponsored by the Swedish International Development Agency (SIDA).

References

1. Jain, A., Ross, A., Prebhakar, S.: An introduction to biometric recognition. IEEE Transactions on Circuits and Systems for Video Technology, Special Issue on Image- and Video-Based Biometrics, January 2004, vol. 14(1) (2004)
2. Jain, A., Pankanti, S., Prabhakar, S., Hong, L., Ross, A.: Biometrics: A grand challenge. In: Proceedings of the 17th International Conference on Pattern Recognition, ICPR 2004. vol. 2. pp. 935–942 (2004)
3. Ortega-Garcia, J., Bigun, J., Reynolds, D., Gonzalez-Rodriguez, J.: Authentication gets personal with biometrics. IEEE Signal Processing Magazine 21(2), 50–62 (2004)
4. Faundez-Zanuy, M.: Biometric security technology. IEEE Aerospace and Electronic Systems Magazine 21(6), 15–26 (2006)
5. Bigun, J., Fronthaler, H., Kollreider, K.: Assuring liveness in biometric identity authentication by real-time face tracking. In: IEEE international Conference on Computational Intelligence for Homeland Security and Personal Safety. Venice, Italy, July 2004 (2004)
6. Stephanie, A., Schukers, C.: Spoofing and anti-spoofing measures. Information Security Technical Report (2002)
7. Ratha, N.K., Connell, J.H., Bolle, R.M.: Enhancing security and privacy in biometrics-based authentication systems. IBM Systems Journal 40(2), 614–634 (2001)
8. Kollreider, K., Fronthaller, H., J., B.: Evaluating liveness by face images and the structure tensor. In: AutoID 2005: Fourth Workshop on Automatic Identification Advanced Technologies, October 2005, pp. 75–80. IEEE Computer Society Press, Los Alamitos (2005)
9. Li, J., Wang, Y., Tan, T., Jain, A.K.: Live face detection based on the analysis of fourier spectra. In: Jain, A.K., Ratha, N.K., (eds.) Biometric Technology for Human Identification. Proceedings of the SPIE, August 2004, vol. 5404. pp. 296–303(2004)
10. Faraj, M., Bigun, J.: Person verification by lip-motion. In: Computer Vision and Pattern Recognition Workshop (CVPRW), June 2006, pp. 37–45 (2006)
11. Messer, K., Matas, J., Kitler, J., Luettin, J., Maitre, G.: XM2VTSDB: The extended M2VTS database. In: 2nd International Conference on Audio and Video-based Biometric Person Authentication (AVBPA'99). pp. 72–77 (1999)
12. Bigun, J.: Vision with Direction:A Systematic Introduction to Image Processing and Computer Vision. Springer-Verlag, Berlin Heidlberg (2006)
13. Jain, J., Jain, A.K.: Displacement measurement and its application in interframe image coding. IEEE Transactions on Communication COM 29, December 1981, pp. 1799–1808 (1981)
14. Gyaourova, A., Kamath, C., Cheung, S.C.: Block matching for object tracking. Technical report UCRL-TR-200271. Laurence Livermore Technical Laboratory (Occtober 2003)

15. Cheng, K.W., Chan, S.C.: Fast block matching algorithms for motion estimation. In: ICASSP-96: IEEE International Conference on Acoustic Speech and Signal Processing. Vol. 4(1) 2311–2314 (May 1996)
16. Reynolds, D., Rose, R.: Robust text independent speaker identification using gaussian mixture models. IEEE Transactions on Speech and Audio Processing 3(1), 72–83 (1995)
17. Young, S., Evermann, G., Gales, M., Hein, T., Kershaw, D., Moore, G., Odell, J., Ollason, D., Povey, D., Valtchev, V., Woodland, P.: The htk book. for version 3.3. http://htk.eng.cam.ac.uk/docs/docs.shtml (April 2005)
18. Veeravalli, A.G., Pan, W., Adhami, R., Cox, P.G.: A tutorial on using hidden markov models for phoneme recognition. In: Thirty-Seventh Southeastern Symposiumon System Theory, SSST 2005 (2005)
19. Aly, S., Youssef, A.: Real-time motion based frame estimation in video lossy transmission. In: Symposium on Applications and the Internet, January 2001 pp. 139–146 (2001)

Accurate 3D Left-Right Brain Hemisphere Segmentation in MR Images Based on Shape Bottlenecks and Partial Volume Estimation

Lu Zhao, Jussi Tohka, and Ulla Ruotsalainen

Institute of Signal Processing, Tampere University of Technology, P.O.Box 553, FIN-33101, Finland
lu.zhao@tut.fi, jussi.tohka@tut.fi, ulla.ruotsalainen@tut.fi

Abstract. Current automatic methods based on mid-sagittal plane to segment left and right human brain hemispheres in 3D magnetic resonance (MR) images simply use a planar surface. However, the two brain hemispheres, in fact, can not be separated by just a simple plane properly. A novel automatic method to segment left and right brain hemispheres in MR images is proposed in this paper, which is based on an extended shape bottlenecks algorithm and a fast and robust partial volume estimation approach. In this method, brain tissues firstly are extracted from the MR image of human head. Then the information potential map is generated, according to which a brain hemisphere mask with the same size of the original image is created. 10 simulated and 5 real T1-weighted MR images were used to evaluate this method, and much more accurate segmentation of human brain hemispheres was achieved comparing with the segmentation with mid-sagittal plane.

Keywords: Brain asymmetry, Mid-sagittal plane, Stereotaxic registration.

1 Introduction

The hemisphere segmentation is required by the study of the interhemispheric human brain asymmetry that can reveal the evolutionary, hereditary, developmental and pathological information of human brain [21]. Because the left and right hemispheres of a healthy human brain build a roughly bilaterally symmetric structure with respect to the mid-sagittal plane, namely the longitudinal median plane bisecting the brain, the mid-sagittal plane has been employed as a popular tool to segment human brain hemispheres in various neuroimages. In the existing methods using mid-sagittal plane, the searched plane can be defined as either the plane best matching the cerebral interhemispheric fissure [3,16], or the plane maximizing the bilateral symmetry [1,13,14,17,19,20]. In addition, the stereotaxic registration [2], i.e. transforming the images for different subjects to match a common brain template, presently is applied widely for human brain asymmetry study [10]. It produces the mid-sagittal plane appearing as the middle line in the transverse and coronal views of the stereotaxic space. Although

B.K. Ersbøll and K.S. Pedersen (Eds.): SCIA 2007, LNCS 4522, pp. 581–590, 2007.

acceptable results have been achieved considering the rough brain symmetry, the performance of these conventional techniques is always limited by the fact that human brains are never perfectly symmetric [5,6,9]. Even for normal brains, the interhemispheric boundary actually is a curved surface but not a plane due to the conspicuous anatomical asymmetry [21]. Fig.1 shows examples of brain hemisphere segmentation with mid-sagittal plane generated by stereotaxic registration for simulated and real MR images. In both examples, the errors can be spotted clearly in the regions highlighted with circles, despite the simulated MR data was designed to match the stereotaxic template perfectly. Therefore, the mid-sagittal plane based techniques are not sufficient enough when more precise asymmetry analysis is required.

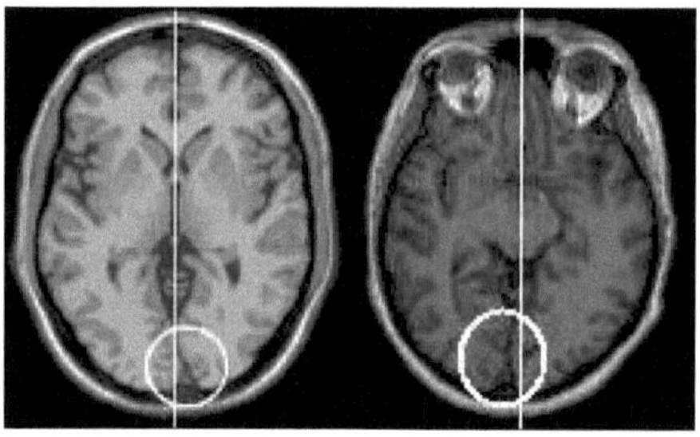

Fig. 1. Examples of the transverse view of brain hemisphere segmentation in simulated (left) and real (right) MR images using mid-sagittal plane generated by stereotaxic registration

In this work, a novel method segmenting left and right brain hemispheres in 3D T1-weighted MR images more accurately than the segmentation using mid-sagittal plane has been developed. This is built on an extended shape bottlenecks algorithm and a fast and robust partial volume estimation approach [22]. The shape bottlenecks algorithm [15], simulating the steady state of an information transmission process between two parts of a complex object, has been proved to be useful to detect the corpus callosum, anterior commissure and brain stem in white matter (WM). However the original algorithm was only designed for WM and concerned no other brain tissues. In [11] an application on whole brain has been done to identify the anterior and posterior commissures, but the partial volume effect (PVE) problem, i.e. the phenomenon that a voxel can represent more than one tissue types, still was not concerned like in the original one. Here, because of our different aim, we extend the original algorithm to the whole brain, including both WM and grey matter (GM), to generate the information potential map (IMP) of the brain. Moreover, prior to implementing the shape bottlenecks algorithm, an effective partial volume estimation approach [22] is implemented to improve the accuracy of the brain extraction. This shape bottlenecks and partial volume estimation based (SB-PVE) method was tested with a set of 10 simulated T1-weighted MR images [4,12] that had been registered into stereotaxic space, and the results were evaluated with the manual segmentation template [23]. In addition, the method was also applied to 5 real MR images. Compared with the segmentation with mid-sagittal plane from stereotaxic registration, our SB-PVE method achieved considerably superior performance with regard to the error rate. The error rates were calculated with respect to the whole brain (integrated hemispheres) and certain regions of interest (ROI).

2 Method

2.1 Segmentation Strategy

In this section, a novel automatic algorithm to segment the left and right hemispheres of human brain in MR images is proposed, and we begin by giving a general strategy of it. Firstly, the brain is extracted from the 3D MR image of human head. Then the extended shape bottlenecks algorithm is applied on the brain to acquire its information potential map (IPM). Next, a mask of brain hemispheres is produced by classifying the voxels in IPM into two clusters with respect to their information potential values (IPV). Finally, the left and right brain hemispheres can be simply identified in the original image with this mask.

2.2 Brain Extraction

Skull-Stripping and Non-Uniformity Correction. To extract the brain, the skull, scalp and other extraneous tissues, except the cerebrospinal fluid (CSF), need to be removed initially. This is completed using the Brain Surface Extractor (BSE) [18]. Moreover, because the intensity non-uniformity, also described as bias field, is commonly seen in MR images and can significantly degrade automatic segmentation and prevent quantitative analysis, the Bias Field Corrector (BFC) [18] is utilized to compensate this artifact in the skull-stripped volume.

Partial Volume Estimation. The partial volume effect (PVE) is an inevitable problem in the MR image analysis due to the finite spatial resolution of imaging devices, and its existence demotes the quality of automatic tissue classification. Taking this issue into account, the output volume from BSE and BFC contains not only the cerebrospinal fluid (CSF) and the brain tissues [grey matter (GM) and white matter (WM)], but also the partial volume combinations of them (WM/GM, GM/CSF and CSF/background). Thus, the precision of brain extraction would be improved by, simultaneously with the CSF removal, removing an appropriate amount of partial volume voxels relating to CSF according to the information provided by partial volume estimation. Here, partial volume estimation refers to the estimation of the amount of each tissue type in each voxel.

Although a fast partial volume estimation technique [18] is available together with the BSE and BFC, another approach [22] is employed here to obtain more accurate estimation. In this method, the MR image of brain with PVE is modeled as a statistical mixel model allowing distinct variances for different tissue types. Then the tissue types contained in each voxel are identified, and following the proportion of each tissue type in each voxel is obtained by solving a simplified maximum likelihood estimation problem. The superiority of this technique stems mainly from more accurate estimates of the tissue type parameters (mean and variance) in the mixel PVE model by the trimmed minimum covariance determinant (TMCD) method. The TMCD method combines an outlier detection scheme to the robust minimum covariance determinant (MCD) estimator to estimate the tissue type parameters based on a rough tissue classification which here is based on the tissue classifier available in BrainSuite [18]. The accuracy of the tissue parameter estimates is of fundamental importance for the partial volume estimation as note [22].

From this partial volume estimation, three images are produced for the three tissue types (CSF, GM or WM) respectively, whose elements reflect the proportion of the corresponding tissue type in each voxel.

CSF Removal. The CSF removal involves both the pure CSF voxels and the partial volume voxels containing certain amount of CSF. This is completed by, in one hand, discarding all the partial volume voxels of CSF/background from the skull-stripped volume; in another hand, removing the voxels of CSF/GM in which the percentage of CSF is greater than a threshold value. In this work, after assessing the influence of the amount of removed CSF voxels on the following IMP generation, the threshold value is set to 30%, i.e. partial volume voxels of CSF/GM, where the amount of GM is higher than 70%, will be retained as GM voxels. Till now, all the non-brain tissues are eliminated from the MR image of human head.

2.3 Information Potential Map Generation

Modeling. The IMP of the tested brain is generated by using an extended shape bottlenecks algorithm. The basic idea is to use partial differential equations to simulate a steady state of information transmission process between left and right hemispheres. Because the information sources are supposed to be only on the outmost layer of the cortical surface, the information transmission process has a conservative flow inside brain. This means that the following condition is always fulfilled throughout the brain domain Ω:

$$\int_{\partial\Omega} \nabla i \cdot \boldsymbol{n} \, \mathrm{d}(\partial\Omega) = 0, \tag{1}$$

where ∇ is the gradient operator, $i(x,y,z)$ is the information potential at point (x,y,z), and $\boldsymbol{n}$ denotes the normal oriented towards Ω exterior. Applying the Green formula to Eq.1, the well-known Laplace equation is obtained:

$$\triangle i = 0, \tag{2}$$

where $\triangle$ refers to the Laplace operator. Thus, the information potential value (IPV) of each brain voxel is obtained by solving the Laplace equation. Using the standard discretization of Laplace operator, the usual consistent second order discrete Laplace operation with assumption of isotropic voxels is given for the interior of Ω as

$$i(x, y, z) = \frac{1}{6} \sum_{\mathbf{m} \in N_6(x,y,z)} i(\mathbf{m}), \tag{3}$$

where $N_6(x, y, z)$ denotes the set of 6 neighbors of point (x,y,z).

In our case the Dirichlet-Neumann boundary condition is applicable. Let Φ denote the whole boundary. The information is supposed to be propagated from a boundary subset $H \subset \Phi$ with high IPV h towards another subset $L \subset \Phi$ with low IPV l, where h and l are constant values with $h > l$. Additionally, the value of $\partial i / \partial \boldsymbol{n}$ is set to be 0 on $\Phi - (H + L)$. Subsequently, the discrete Dirichlet boundary condition, existing on $H + L$, is straightforward:

$$\forall \mathbf{m} \in H \quad i(\mathbf{m}) = h; \quad \forall \mathbf{m} \in L \quad i(\mathbf{m}) = l. \tag{4}$$

However, the Neumann boundary condition, occurring on $\Phi - (H + L)$, is more complicated to discretize. Set $\mathbf{m}$ to be a boundary point, i.e. one of its 6 neighborhood points is not contained in Ω. The number of grid directions, for which there is at least one of the 6 neighbors of $\mathbf{m}$ belonging to Ω, is denoted as d. Finally, let $\mathcal{N}$ be the set of directions (x, y or z) with one point $\mathbf{p}$ of the 6 neighbors included in Ω, and $\mathcal{T}$ be the set of directions with two neighbors $\mathbf{t}_1$ and $\mathbf{t}_2$ in Ω. Then the second order discrete version of Neumann boundary condition (Eq.5) can be produced by substituting normal second order partial derivatives with tangential second order partial derivatives [8]:

$$\forall \mathbf{m} \in (\Phi - (H + L)) \quad i(\mathbf{m}) = \frac{1}{d} \sum_{\mathcal{N}} i(\mathbf{p}) + \frac{1}{2d} \sum_{\mathcal{T}} (i(\mathbf{t}_1) + i(\mathbf{t}_2)). \tag{5}$$

Numerical Implementation. An iterative scheme is employed to solve the linear system built with Eq.3, 4 and 5. Initially, the voxels on H and L are defined with Eq.4, and the ones on $\Omega - (H + L)$ are set as:

$$\forall \mathbf{m} \in (\Omega - (H + L)), \quad i^0(\mathbf{m}) = \frac{h + l}{2}. \tag{6}$$

The iterative process can be described as

$$\begin{gathered} \forall \mathbf{m} \in (\Omega - (H + L)), \quad \forall k \geq 0 \\ i^{(k+1)}(\mathbf{m}) = (1 - \omega)\, i^k(\mathbf{m}) + \omega \sum_{\mathbf{p} \in N_6(\mathbf{m})} \alpha_{\mathbf{p}}(\mathbf{m})\, i^k(\mathbf{p}). \end{gathered} \tag{7}$$

where $1 < \omega < 2$, and $\alpha_{\mathbf{p}}(\mathbf{m})$ is a coefficient given by Eq.3 or 5 or zero when $\mathbf{p} \notin \Omega$, namely if $\mathbf{m} \in (\Omega - \Phi)$, Eq.3 is used; or Eq.5 if $\mathbf{m} \in (\Phi - (H + L))$.

In our experiments, boundary subsets H and L are found out as the leftmost and rightmost parts of the whole boundary, which are determined with two longitudinal planes in the left and right hemispheres respectively. h=5000 on H

and l=1000 on L. The IPVs of the voxels in the other regions are initialized with Eq.6. Then the iterative process described in Eq.7 is applied with $\omega = 1.5$. The number of iterations is set to 1000 to guarantee the convergence. Here, all the relative parameters are defined with the instruction from [15]. The left part in Fig.2 shows an example of the obtained IPM, where the bottlenecks of brain can be identified as several high information flows between left and right hemispheres.

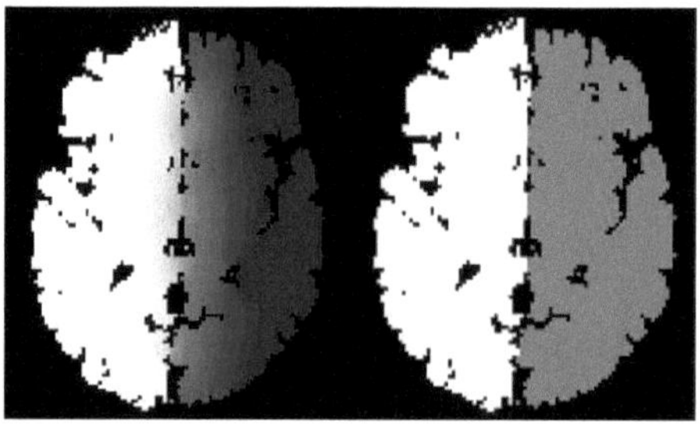

Fig. 2. Transverse slices of the information potential map (left) and the corresponding hemisphere mask (right) of a tested brain image

2.4 Hemisphere Mask

In the IPM shown in Fig.2, it can be seen that most of the voxels in different hemisphere present different intensity level, namely different IPV level. Thus a mask for identifying left and right hemispheres for the processed brain can be produced by implementing the k-means clustering [7] to classify the voxels of IPM into two clusters according to their IPVs. An example of the generated mask of brain hemispheres is presented in the right part of Fig.2.

3 Qualitative and Quantitative Evaluation

3.1 Qualitative Evaluation

The BrainWeb Simulated Brain Database of Montreal Neurological Institute [4,12] (http://www.bic.mni.mcgill.ca/brainweb) was employed to test the method proposed in this paper. Ten distinct T1-weighted MR images ($181 \times 217 \times 181$ voxels) were used for testing, which had the same voxel size $1 \times 1 \times 1\,mm^3$, but different noise and intensity non-uniformity (INU) levels. Furthermore, all these images had been registered to the stereotaxic, Talairach-based brain space used by BrainWeb, i.e. the left and right hemispheres of the brains shown in them were segmented already with the mid-sagittal plane appearing as the middle line in the coronal or transverse view. From the visualization of the experimental results, excellent segmentation performance was achieved for all these data. Fig.3 gives one example of the similar experimental results for different simulated MR images.

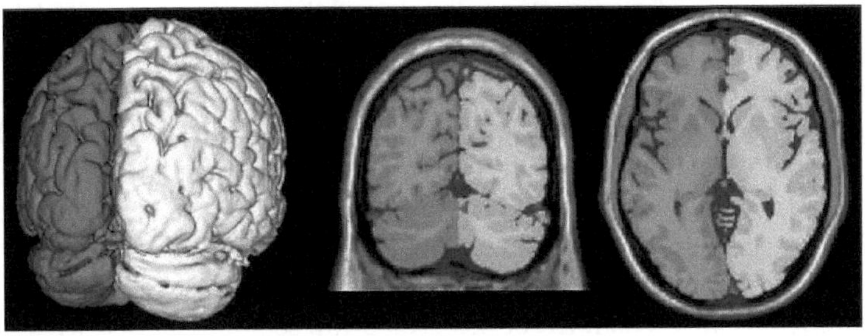

Fig. 3. Example of the segmentation result for one of the 10 simulated T1-weighted MR images. Left is the corresponding hemisphere mask in 3D that is shown from the back view of brain where the curved interhemispheric boundary is more obvious. Middle and right parts are the coronal and transverse slices of the segmentation result respectively. The transverse slice is the counterpart of the one shown in the left side of Fig.1.

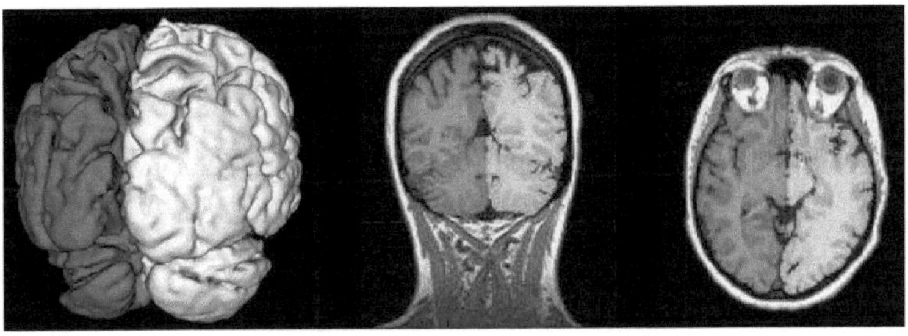

Fig. 4. Example of the segmentation result for one of the 5 real T1-weighted MR images. Left is the corresponding hemisphere mask in 3D that is shown from the back view of brain where the curved interhemispheric boundary is more obvious. Middle and right parts are the coronal and transverse slices of the segmentation result respectively. The transverse slice is the counterpart of the one shown in the right side of Fig.1.

The SB-PVE method was also applied on five real T1-weighted MR images of healthy volunteers that contained $128 \times 256 \times 256$ voxels of size $1.5 \times 1.5 \times 1.5\,mm^3$ and did not undertake any posterior image processing, e.g. stereotaxic registration. Because of the similarity of the experimental results, only one example is given here (Fig.4). It can be seen that the left and right brain hemispheres were segmented appropriately, even though extra asymmetry might be caused by the abnormal position of the object's head. From the two examples illustrated in Fig.3 and 4 it is demonstrated once more that the brain hemispheres can not be segmented properly with just a planar surface as the interhemispheric boundary is actually a curved surface.

3.2 Quantitative Evaluation

The quantitative evaluation is based on the experimental results of simulated MR data. Both of the segmentation results obtained by using the SB-PVE method and mid-sagittal plane from stereotaxic registration were assessed with the Automated Anatomical Labeling (AAL) template [23] that was, in fact, achieved by manually segmenting the same simulated data. Then the error rates, namely the percentage of misclassified voxels, of the new and conventional approaches were compared in the whole brain (Table.1) and a series of specific ROIs defined with AAL (Table.2). From the comparison, the global error rate was abated from more than 0.37% for the mid-sagittal plane segmentation to lower than 0.10% by our SB-PVE method, although it increased slightly when the level of noise or intensity non-uniformity increased. In the selected brain regions (named with *Calcarine fissure and surrounding cortex*, *Cuneus*, *Superior frontal gyrus, medial* and *Supplementary motor area* by AAL), the improvement of the segmentation accuracy was more remarkable, due to the curved boundary between the left and right parts of these areas (See the regions highlighted with circles in Fig.1). Particularly in the *Medial superior frontal gyrus* the error rate was reduced by about 95% averagely.

Table 1. Error rates (ER) in whole brain (1479969 voxels) for the SB-PVE method and mid-sagittal plane method (MSP). The first and second figures in the name of each image represent the levels of noise and intensity non-uniformity respectively.

Images	1_20	1_40	3_20	3_40	5_20	5_40	7_20	7_40	9_20	9_40
ER of SB-PVE (%)	0.080	0.080	0.079	0.085	0.082	0.088	0.087	0.090	0.099	0.095
ER of MSP (%)	0.370	0.371	0.369	0.384	0.365	0.386	0.371	0.379	0.371	0.376

Table 2. Error rates (ER) in specific ROIs for the SB-PVE method and mid-sagittal plane method (MSP). The first and second figures in the name of each image represent the levels of noise and intensity non-uniformity respectively.

Images	1_20	1_40	3_20	3_40	5_20	5_40	7_20	7_40	9_20	9_40
ROI	*Calcarine fissure and surrounding cortex* (33042 voxels)									
ER of SB-PVE (%)	0.978	0.948	0.978	0.952	1.048	0.954	0.993	0.966	1.092	0.938
ER of MSP (%)	5.555	5.450	5.578	5.413	5.554	5.459	5.405	5.272	5.286	5.113
ROI	*Cuneus* (23456 voxels)									
ER of SB-PVE (%)	0.538	0.519	0.483	0.533	0.502	0.528	0.577	0.557	0.602	0.528
ER of MSP (%)	4.364	4.339	4.269	4.224	4.253	4.213	4.426	4.167	4.315	4.110
ROI	*Superior frontal gyrus, medial* (40831 voxels)									
ER of SB-PVE (%)	0.064	0.081	0.086	0.130	0.086	0.115	0.107	0.140	0.125	0.172
ER of MSP (%)	1.968	2.093	1.984	2.326	1.994	2.315	1.989	2.306	1.999	2.295
ROI	*Supplementary motor area* (36167 voxels)									
ER of SB-PVE (%)	0.075	0.067	0.069	0.084	0.060	0.088	0.094	0.103	0.106	0.139
ER of MSP (%)	0.644	0.632	0.594	0.808	0.610	0.807	0.650	0.769	0.689	0.800

4 Conclusion

The segmentation of human brain hemispheres in MR images using mid-sagittal plane is problematic because the interhemispheric boundary is actually a curved surface. A new method to solve this problem is introduced, which is based on an extended shape bottlenecks algorithm and a fast and robust partial volume estimation approach. This method is iterative and fully automated. The qualitative evaluation was done using real and simulated MR images, and excellent brain hemisphere segmentation performance was presented from the visualization of the experimental results of every tested image. A manually segmented brain template was used to evaluate the error rates for both the proposed method and the bmid-sagittal plane from stereotaxic registration. Our method acquired much higher accuracy compared with the mid-sagittal plane method.

Acknowledgment

This work was supported by the Academy of Finland under the grants 108517, 104834, and 213462 (Finnish Centre of Excellence program (2006-2011)). The MR images were provided by the Turku PET Centre.

References

1. Ardekani, B.A., Kershaw, J., Braun, M., Kanno, I.: Automatic detection of the mid-sagittal plane in 3D brain images. IEEE Transactions on Medical Imaging 16(6), 947–952 (1997)
2. Brett, M., Johnsrude, I.S., Owen, A.M.: The problem of functional localization in the human brain. Nature Reviews Neuroscience 3(3), 243–249 (2002)
3. Brummer, M.E.: Hough transform detection of the longitudinal fissure in tomographic head images. IEEE Transactions on Medical Imaging 10, 74–81 (1991)
4. Collins, D.L., Zijdenbos, A.P., Kollokian, v., Sled, J.G., Kabani, N.J., Holmes, C.J., Evans, A.C.: Design and construction of a realistic digital brain phantom. IEEE Trans. Med. Imaging 17(3), 463–468 (1998)
5. Crow, T.J.: Schizophrenia as an anomaly of cerebral asymmetry. In: Imaging of the Brian in Psychiatry and Related Fields, Berlin, Germany, pp. 1–17. Springer, Heidelberg (1993)
6. Davidson, R.J., Hugdahl, K.: Brain Asymmetry. MIT Press/Bradford Books, Cambridge, MA (1996)
7. Duda, R.O., Hart, P.E., Stork, D.G.: Pattern Classification, 2nd edn. Wiley-Interscience, New York (2000)
8. Euvrard, D.: Resolution numerique des equations aux derivees partielles. Masson, Paris, France (1988)
9. Geschwind, N., Levitsky, W.: Human brain: Left–right asymmetries in temporal speech region. Science 161, 186–187 (1968)
10. Kovalev, V.A., Kruggel, F., von Cramon, D.Y.: Gender and age effects in structural brain asymmetry as measured by MRI texture analysis. NeuroImage 19(3), 895–905 (2003)
11. Kruggel, F., von Cramon, D.Y: Alignment of magnetic-resonance brain datasets with the stereotactical coordinate system. Medical image analysis, vol. 3(2) (1999)
12. Kwan, R.-S., Evans, A.C., Pike, G.B.: MRI simulation based evaluation and classifications methods. IEEE Trans. Med. Imaging 18(11), 1085–1097 (1999)

13. Liu, Y., Collins, R.T., Rothfus, W.E.: Automatic bilateral symmetry (midsagittal) plane extraction from pathological 3D neuroradiological images. In: SPIE International Symposium on Medical Imaging, 1998. Proceedings 3338-161 (1998)
14. Liu, Y., Collins, R.T., Rothfus, W.E.: Robust midsagittal plane extraction from normal and pathological 3D neuroradiology images. IEEE Transactions on Medical Imaging 20(3), 175–192 (2001)
15. Mangin, J.-F., Régis, J., Frouin, V.: Shape bottlenecks and conservative flow systems. In: IEEE Work. MMBIA, pp. 319–328, San Francisco, CA (1996)
16. Marais, P., Guillemaud, R., Sakuma, M., Zisserman, A., Brady, M.: Visualising cerebral asymmetry. In: Höhne, K.H., Kikinis, R. (eds.) VBC 1996. LNCS, vol. 1131, pp. 411–416. Springer-Verlag, Heidelberg (1996)
17. Prima, S., Ourselin, S., Ayache, N.: Computation of the mid-sagittal plane in 3D brain images. IEEE Transactions on Medical Imaging 21(2), 122–138 (2002)
18. Shattuck, D.W., Sandor-Leahy, S.R., Schaper, K.A., Rottenberg, D.A., Leahy, R.M.: Magnetic resonance image tissue classification using a partial volume model. NeuroImage 13(5), 856–876 (2001)
19. Smith, S., Jenkinson, M.: Accurate robust symmetry estimation. In: Taylor, C., Colchester, A. (eds.) MICCAI '99. LNCS, vol. 1679, pp. 308–317. Springer-Verlag, Heidelberg (1999)
20. Sun, C., Sherrah, J.: 3D symmetry detection using the extended Gaussian image. IEEE Trans. Pattern Anal. Mach. Intell. 19(2), 164–168 (1997)
21. Toga, W., Thompson, P.M.: Mapping brain asymmetry. Nature Reviews Neuroscience 4(1), 37–48 (2003)
22. Tohka, J., Zijdenbos, A., Evans, A.C.: Fast and robust parameter estimation for statistical partial volume models in brain MRI. NeuroImage 23(1), 84–97 (2004)
23. Tzourio-Mazoyer, N., Landeau, B., Papathanassiou, D., Crivello, F., Etard, O., Delcroix, N., Mazoyer, B., Joliot, M.: Automated anatomical labeling of activations in SPM using a macrosopic anatomical parcellation of the MNI MRI single-subject brain. NeuroImage 15(1), 273–289 (2002)

Image Inpainting by Cooling and Heating

David Gustavsson[1], Kim S. Pedersen[2], and Mads Nielsen[2]

[1] IT University of Copenhagen
Rued Langgaards Vej 7, DK-2300 Copenhagen S,Denmark
davidgsson@itu.dk
[2] DIKU, University of Copenhagen
Universitetsparken 1, DK-2100 Copenhagen Ø, Denmark
{kimstp,madsn}@diku.dk

Abstract. We discuss a method suitable for inpainting both large scale geometric structures and stochastic texture components. We use the well-known FRAME model for inpainting. We introduce a temperature term in the learnt FRAME Gibbs distribution. By using a fast cooling scheme a MAP-like solution is found that can reconstruct the geometric structure. In a second step a heating scheme is used that reconstruct the stochastic texture. Both steps in the reconstruction process are necessary, and contribute in two very different ways to the appearance of the reconstruction.

Keywords: Inpainting, FRAME, ICM, MAP, Simulated Annealing.

1 Introduction

Image inpainting concerns the problem of reconstruction of the image contents inside a region Ω with unknown or damaged contents. We assume that Ω is a subset of the image domain $D \subseteq \mathbb{R}^2$, $\Omega \subset D$ and we will for this paper assume that D form a discrete lattice. The reconstruction is based on the available surrounding image content. Some algorithms have reported excellent performance for pure geometric structures (see e.g. [1] for a review of such methods), while others have reported excellent performance for pure textures (e.g. [2,3,4]), but only few methods [5] achieve good results on both types of structures.

The variational approaches have been shown to be very successful for geometric structures but have a tendency to produce a too smooth solution without fine scale texture (See [1] for a review). Bertalmio et al [5] propose a combined method in which the image is decomposed into a structure part and a texture part, and different methods are used for filling the different parts. The structure part is reconstructed using a variational method and the texture part is reconstructed by image patch pasting.

Synthesis of a texture and inpainting of a texture seem to be, more or less, identical problems, however they are not. In [6] we propose a two step method for inpainting based on Zhu, Wu and Mumford's stochastic FRAME model (Filters, Random fields and Maximum Entropy) [7,8]. Using FRAME naively for

B.K. Ersbøll and K.S. Pedersen (Eds.): SCIA 2007, LNCS 4522, pp. 591–600, 2007.

inpainting does not produce good results and more sophisticated strategies are needed and in [6] we propose such a strategy. By adding a temperature term T to the learnt Gibbs distribution and sampling from it using two different temperatures, both the geometric and the texture component can be reconstructed. In a first step, the geometric structure is reconstructed by sampling using a cooled - i.e. using a small fixed T - distribution. In a second step, the stochastic texture component is added by sampling from a heated - i.e. using a large fixed T - distribution.

Ideally we want to use the MAP solution of the FRAME model to reconstruct geometric structure of the damaged region Ω. In [6] we use a fixed low temperature to find a MAP-Like solution in order to reconstruct the geometric structure. To find the exact MAP-solution one must use the time consuming simulated annealing approach such as described by Geman and Geman [9]. However to reconstruct the missing contents of the region Ω, the true MAP solution may not be needed. Instead a solution which is close to the MAP solution may provide visually good enough results. In this paper we propose a fast cooling scheme that reconstruct the geometric structure and approaches the MAP solution. Another approach is to use the solution produced by the Iterated Conditional Modes (ICM) algorithm (see e.g. [10]) for reconstruction of the geometric structure. Finding the ICM solution is much faster than our fast cooling scheme, however it often fails to reconstruct the geometric structure. This is among other things caused by the ICM solutions strong dependence on the initialisation of the algorithm. We compare experimentally the fast cooling solution with the ICM solution.

To reconstruct the stochastic texture component the Gibbs distribution is heated. By heating the Gibbs distribution more stochastic texture structures will be reconstructed without destroying the geometric structure that was reconstructed in the cooling step. In [6] we use a fixed temperature to find a solution including the texture component. Here we introduce a gradual heating scheme.

The paper has the following structure. In section 2 FRAME is reviewed, in section 2.1 filter selection is discussed and in section 2.2 we explain how FRAME is used for reconstruction. Inpainting using FRAME is treated in section 3. In section 3.1 a temperature term is added to the Gibbs distribution, the ICM solution and fast cooling solution is discussed in sections 3.2 and 3.3. Adding the texture component by heating the distribution is discussed in section 3.4. In section 4 experimental results are presented and in section 5 conclusion are drawn and future work is discussed.

2 Review of FRAME

FRAME is a well known method for analysing and reproducing textures [8,7]. FRAME can also be thought of as a general image model under the assumptions that the image distribution is stationary. FRAME constructs a probability distribution $p(I)$ for a texture from observed sample images.

Given a set of filters $F^\alpha(I)$ one computes the histogram of the filter responses H^α with respect to the filter α. The filter histograms are estimates of marginal distributions of the full probability distribution $p(I)$. Given the marginal distributions for the sample images one wants to find all distributions that have the same expected marginal distributions, and among those find the distribution with maximum entropy, i.e. by applying the maximum entropy principle. This distribution is the least committed distribution fulfilling the constraints given by the marginal distributions. This is a constrained optimisation problem that can be solved using Lagrange multipliers. The solution is

$$p(I) = \frac{1}{Z(\Lambda)} \exp\{-\sum_i \sum_\alpha \lambda_i^\alpha H_i^\alpha\} \tag{1}$$

Here i is the number of histogram bins in H^α for the filter α and $\Lambda = \{\lambda_i^\alpha\}$ are the Lagrange multipliers which gives information on how the different values for the filter α should be distributed. The relation between λ^α:s for different filters F^α gives information on how the filters are weighted relative to each other.

An Algorithm for finding the distribution and Λ can be found in [7]. FRAME is a generative model and given the distribution $p(I)$ for a texture it can be used for inference (analysis) and synthesis.

2.1 The Choice of Filter Bank

We have used three types of filters in our experiments: The delta filter, the power of Gabor filters and Scale Space derivative filters. The delta, Scale Space derivative and Gabor filters are linear filters, hence $F^\alpha(I) = I * F^\alpha$, where $*$ denotes convolution. The power of the Gabor filter is the squared magnitude applied to the linear Gabor filter.

The Filters F^α are:

- Delta filter - given by the Dirac delta $\delta(x)$ which simply returns the intensity at the filter position.
- the power of Gabor filters - defined by $\mid I * G_\sigma e^{-i\omega x} \mid^2$, where $i^2 = -1$. Here we use 8 orientations, $\omega = 0, \frac{\pi}{4}, \frac{\pi}{2}, \frac{3\pi}{4}, \pi, \frac{5\pi}{4}, \frac{3\pi}{2}, \frac{7\pi}{4}$ and 2 scales $\sigma = 1, 4$, in total 16 Gabor filters have been used.
- Scale space derivatives - using 3 scales $\sigma = 0.1, 1, 3$ and 6 derivatives G_σ, $\frac{\partial G_\sigma}{\partial x}$, $\frac{\partial G_\sigma}{\partial y}$, $\frac{\partial^2 G_\sigma}{\partial x^2}$, $\frac{\partial^2 G_\sigma}{\partial y^2}$, $\frac{\partial^2 G_\sigma}{\partial x \partial y}$.

For both the Gabor and scale space derivative filters the Gaussian aperture function G_σ with standard deviation σ defining the spatial scale is used,

$$G_\sigma(x, y) = \frac{1}{2\pi\sigma^2} \exp\left(-\frac{x^2 + y^2}{2\sigma^2}\right) .$$

Which and how many filters should be used have a large influence on the type of image that can be modelled. The filters must catch the important visual appearance of the image at different scales. The support of the filters determines a Markov neighbourhood. Small filters add fine scale properties of the image, while large filters add coarse scale properties of the image. Hence to model properties at different scales, different filter sizes must be used. The drawback of using large filters is that the computation time increases with the filter size. On the other hand large filters must be used to catch coarse scale dependencies in the image.

Gabor filters are orientation sensitive and have been used for analysing textures in a number of papers and are in general suitable for textures (e.g. [11,12]). By carefully selecting the orientation ω and the scale σ, structures with different orientations and scales will be captured.

It is well known from scale space theory that scale space derivative filters capture structures at different scales. By increasing σ in the Gaussian kernel, finer details are suppressed, while coarse structures are enhanced. By using the full scale-space both fine and coarse scale structures will be captured [13].

2.2 Sampling

Once the distribution $p(I)$ is learnt, it is possible to use a Gibbs sampler to synthesise images from $p(I)$. I is initialised randomly (or in some other way based on prior knowledge). Then a site $(x,y)_i \in D$ is randomly picked and the intensity $I_i = I((x,y)_i)$ at $(x,y)_i$ is updated according to the conditional distribution [14,10]

$$p(I_i|I_{-i}) \tag{2}$$

where the notation I_{-i} denotes the set of intensities at the set of sites $\{(x,y)_{-i}\} = D\backslash(x,y)_i$. Hence $p(I_i|I_{-i})$ is the probability for the different intensities in site $(x,y)_i$ given the intensities in the rest of the image. Because of the equivalence between Gibbs distributions and Markov Random Fields given a neighbourhood system N (the Hammersley-Clifford theorem, see e.g. [10]), we can make the simplification

$$p(I_i|I_{-i}) = p(I_i|I_{N_i}) \tag{3}$$

where $N_i \subset D\backslash(x,y)_i$ is the neighbourhood of $(x,y)_i$. In the FRAME model, the neighbourhood system N is defined by the extend of the filters F^α.

By sampling from the conditional distribution in (3), I will be a sample from the distribution $p(I)$.

3 Using FRAME for Inpainting

We can use FRAME for inpainting by first constructing a model $p(I)$ of the image, e.g. by learning from the non-damaged part of the image, $D\backslash\Omega$. We then use the learnt model $p(I)$ to sample new content inside the damaged region Ω. This is done by only updating sites in Ω. A site $(x,y)_i \in \Omega$ is randomly picked and updated by sampling from the conditional distribution given in (3). If the site $(x, y)_i$ is close (in terms of filter size) to the boundary $\partial\Omega$ of the damaged region, then the filters get support from both sites inside and outside Ω. The sites outside Ω are known and fixed, and are boundary conditions for the inpainting. We therefore include a small band region around Ω in the computation of the histograms H^α. Another option would have been to use the whole image I to compute the histogram H^α, however this has the downside that the effect of updates inside Ω on the histograms are dependent on the the relative size ratio between ω and D, causing a slow convergence rate for small Ω.

3.1 Adding a Temperature Term $\beta = \frac{1}{T}$

Sampling from the distribution $p(I)$ using a Gibbs sampler does not easily enforce the large scale geometric structure in the image. By using the Gibbs sampler one will get a sample from the distribution, this includes both the stochastic and the geometric structure of the image, however the stochastic structure will dominate the result.

Adding an inverse temperature term $\beta = \frac{1}{T}$ to the distribution gives

$$p(I) = \frac{1}{Z(\Lambda)} \exp\{-\beta \sum_\alpha \sum_i \lambda_i^\alpha H_i^\alpha\} . \tag{4}$$

In [6] we proposed a two step method to reconstruct both the geometric and stochastic part of the missing region Ω:

1. Cooling: By sampling from (4) using a fixed small temperature T value, structures with high probability will be reconstructed, while structures with low probability will be suppressed. In this step large geometric structures will be reconstructed based on the model $p(I)$.
2. Heating: By sampling from (4) using a fixed temperature $T \approx 1$, the texture component of the image will be reconstructed based on the model $p(I)$.

In the first step the geometric structure is reconstructed by finding a smooth MAP-like solution and in the second step the texture component is reconstructed by adding it to the large scale geometry.

In this paper we propose a novel variation of the above discussed method. We consider two cooling schemes and a gradual heating scheme which can be considered as the inverse of simulated annealing.

3.2 Cooling - The ICM Solution

Finding the MAP solution by simulated annealing is very time consuming. One alternative method is the Iterated Conditional Modes (ICM) algorithm. By letting $T \to 0$ (or equivalently letting $\beta \to \infty$) the conditional distribution (3) will become a point distribution. In each step of the Gibbs sampling one will set the new intensity for a site $(x, y)_i$ to

$$I_i^{\text{new}} = \arg\max_{I_i} p(I_i \mid I_{N_i}) . \tag{5}$$

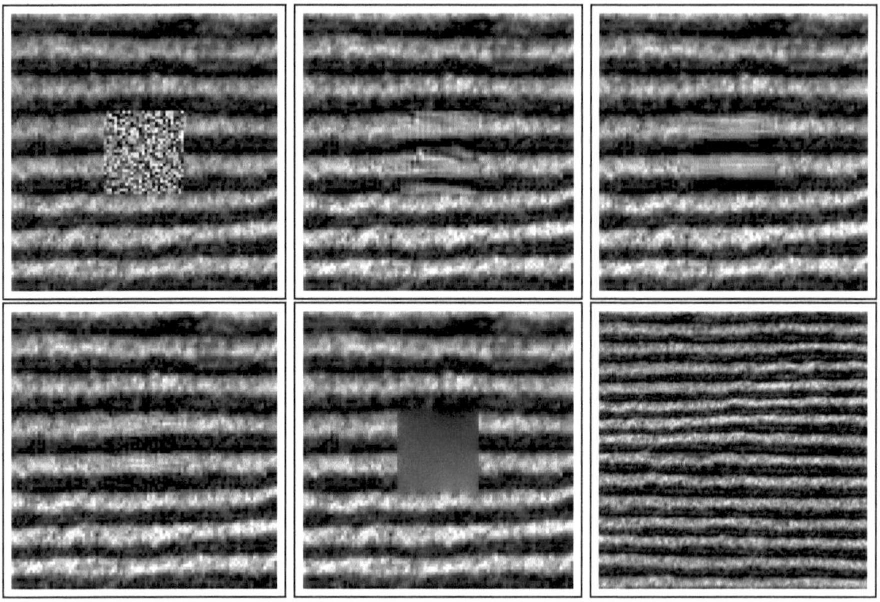

Fig. 1. From top left to bottom right: a) the image containing a damaged region b) the ICM solution c) the fast cooling solution d) adding texture on top of the fast cooling solution by heating the distribution e) total variation (TV) solution and f) the reconstructed region in context (can you find it?)

This is a site-wise MAP solution (i.e. in each site and in each step the most likely intensity will be selected). This site-wise greedy strategy is not guaranteed to find the global MAP solution for the full image. The ICM solution is similar but not identical to the high β sampling step described in [6]. The ICM solution depends on initialisation of the unknown region Ω. Here we initialise by sampling pixel values identically and independent from a uniform distribution on the intensity range.

3.3 Cooling - Fast Cooling Solution

The MAP solution for the inpainting is the most likely reconstruction given the known part of the image $D \backslash \Omega$,

$$I^{\text{MAP}} = \arg \max_{I_i \forall (x,y)_i \in \Omega} p(I \mid I(D \backslash \Omega), \Lambda) \ . \tag{6}$$

Simulated annealing can be used for finding the MAP solution. Replacing β in (4) with an increasing (decreasing) sequence β_n called a cooling (heating) scheme. Using simulated annealing one starts to sample using a high temperature T and slowly cooling down the distribution (4) by letting $T \to 0$. If β_n

is increasing slowly enough and letting $n \to \infty$ then simulated annealing will find the MAP solution (see e.g. [9,10,14]). Unfortunately simulated annealing is very time consuming.

To reconstruct Ω, the true MAP solution may not be needed, instead a solution which is close to the MAP solution may be enough. We therefore adopt a fast cooling scheme, that does not guarantee the MAP solution. The goal is to reconstruct the geometric structure of the image and suppress the stochastic texture.

The fast cooling scheme used in this paper is defined as (in terms of β)

$$\beta_{n+1} = C^{+} \cdot \beta_n \tag{7}$$

where $C^{+} > 1.0$ and $\beta_0 = 0.5$.

3.4 Heating - Adding Texture

The geometric structures of the image will be reconstructed by sampling using the cooling scheme. Unfortunately the visual appearance will be too smooth, and the stochastic part of the image needs to be added.

The stochastic part should be added in such a way that it does not destroy the large scale geometric part reconstructed in the previous step. This is done by sampling from the distribution (4) using a heating scheme similar to the cooling scheme presented in previous section and using the solution from the cooling scheme as initialisation.

The heating scheme in this paper is

$$\beta_{n+1} = C^{-} \cdot \beta_n \tag{8}$$

where $C^{-} < 1.0$ and $\beta_0 = 25$.

By using a decreasing β_n, value finer details in the texture will be reproduced, while coarser details in the texture will be suppressed.

4 Results

Learning the FRAME model $p(I)$ is computational expensive, therefore only small image patches have been used. Even for small image patches the optimisation times are at least a few days. After the FRAME model has been learnt, inpainting can be done relatively fast if Ω is not to large.

The dynamic range of the images have been decreased to 11 intensity levels for computational reasons. The images that have been selected includes both large scale geometric structures as well as texture.

The delta filter, 16 Gabor filters and 18 scale space derivative filters have been used in all experiments and 11 histogram bins have been used for all filters (see section 2.1 for a discussion).

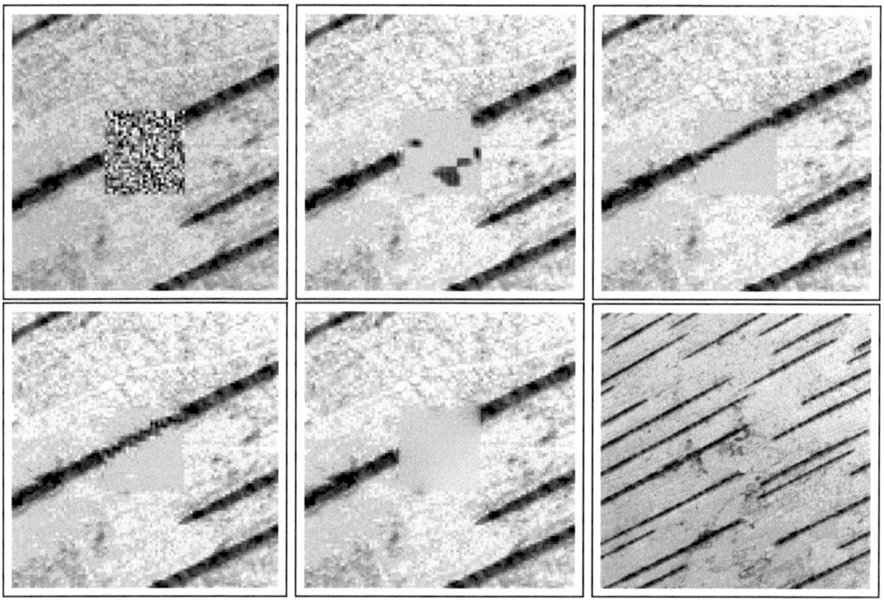

Fig. 2. From top left to bottom right: a) the image containing a damaged region b) the ICM solution c) the fast cooling solution d) adding texture on top of the fast cooling solution by heating the distribution e) total variation (TV) solution and f) the reconstructed region in context (can you find it?)

In the cooling scheme (7), we use $\beta_0 = 0.5$, $C^+ = 1.2$ and the stopping criterion $\beta_n > 25$ in all experiments. In the heating scheme (8), we use $\beta_0 = 25$,$C^- = 0.8$ and the stopping criterion $\beta_n < 1.0$.

Each figure contains an unknown region Ω of size 30×30 that should be reconstructed. Figure 1 contains corduroy images, figure 2 contains birch bark images and figure 3 wood images. Each figure contains the original image with the damaged region Ω with initial noise, the ICM and fast cooling solutions and the solution of a total variation (TV) based approach [1] for comparison.

The ICM solution reconstruct the geometric structure in the corduroy, but fails to reconstruct the geometric structure in both the birch and the wood images. This is due to the local update strategy of ICM, which makes it very sensitive to initial conditions. If ICM starts to produce wrong large scale geometric structures it will never recover.

The fast cooling solution on the other hand seem to reconstruct the geometric structure in all examples and does an even better job than the ICM solution for the corduroy image. The fast cooling solutions are smooth and have suppressed the stochastic textures. Because of the failure of ICM we only include results on heating based on the fast cooling solution.

The results - image d) - after the heating are less smooth Ω's, but it is still smoother than $I \backslash \Omega$. The total variation (TV) approach produce a too smooth solution even if strong geometric structures are present in all example.

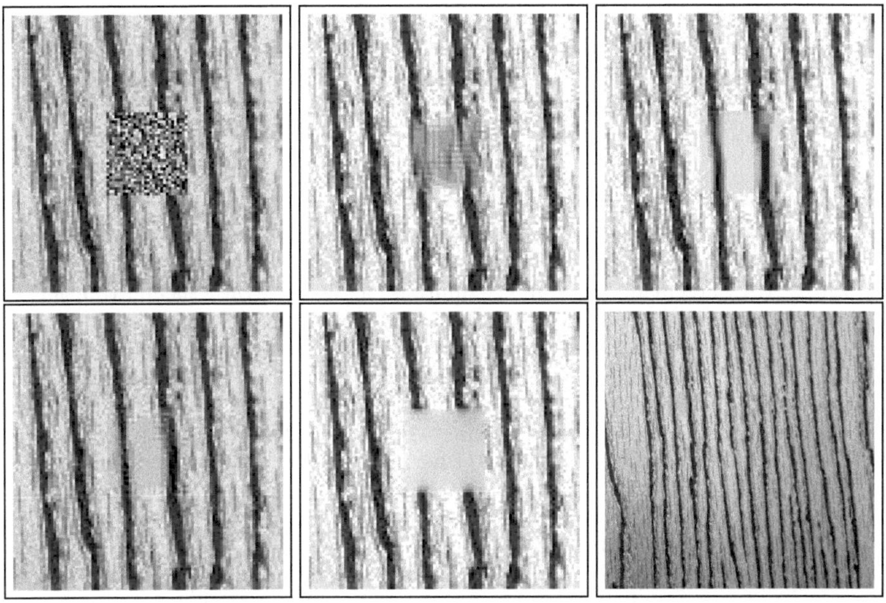

Fig. 3. From top left to bottom right: a) the image containing a damaged region b) the ICM solution c) the fast cooling solution d) adding texture on top of the fast cooling solution by heating the distribution e) total variation (TV) solution and f) the reconstructed region in context (can you find it?)

5 Conclusion

Using FRAME to learn a probability distribution for a type of images gives a Gibbs distribution. The boundary condition makes it hard to use the learnt Gibbs distribution as it is for inpainting; it does not enforce large scale geometric structures strongly enough. By using a fast cooling scheme a MAP-like solution is found that reconstruct the geometric structure. Unfortunately this solution is too smooth and does not contain the stochastic texture. The stochastic texture component can be reproduced by sampling using a heating scheme. The heating scheme adds the stochastic texture component to the reconstruction and decrease the smoothness of the reconstruction based on the fast cooling solution.

A possible continuation of this approach is to replace the MAP-like step with a partial differential equation based method and a natural choice is the Gibbs Reaction And Diffusion Equations (GRADE) [15,16], which are build on the FRAME model.

We decompose an image into a geometric component and a stochastic component and use the decomposition for inpainting. This is related to Meyer's [17,18] image decomposition into a smooth component and a oscillating component (belonging to different function spaces). We find it interesting to explore this theoretic connection with variational approaches.

Acknowledgements

This work was supported by the Marie Curie Research Training Network: Visiontrain (MRTN-CT-2004-005439).

References

1. Chan, T.F., Shen, J.: Variational image inpainting. Communications on Pure and Applied Mathematics vol. 58 (2005)
2. Efros, A.A., Freeman, W.T.: Image quilting for texture synthesis and transfer. In: Proceedings of SIGGRAPH '01, Los Angeles, California, USA (2001)
3. Efros, A.A., Leung, T.K.: Texture synthesis by non-parametric sampling. In: IEEE International Conference on Computer Vision, Corfu, Greece, pp. 1033–1038 (1999)
4. Bonet, J.S.D.: Multiresolution sampling procedure for analysis and synthesis of texture images. In: Computer Graphics, ACM SIGGRAPH, pp. 361–368 (1997)
5. Bertalmio, M., Vese, L., Sapiro, G., Osher, S.: Simultaneous structure and texture image inpainting. IEEE Transcations On. Image Processing 12(8), 882–889 (2003)
6. Gustavsson, D., Pedersen, K.S., Nielsen, M.: Geometric and texture inpainting by gibbs-sampling. In: SSBA07 (2007)
7. Zhu, S.C., Wu, Y.N., Mumford, D.: Filters, random fields and maximum entropy (frame): To a unified theory for texture modeling. International Journal of Computer Vision 27(2), 107–126 (1998)
8. Zhu, S.C., Wu, Y.N., Mumford, D.: Minimax entropy principle and its application to texture modelling. Neural Computation 9(8), 1627–1660 (1997)
9. Geman, S., Geman, D.: Stochastic relaxation, gibbs distribution, and the bayesian restoration of images. IEEE Transaction PAMI 6, 721–741 (1984)
10. Winkler, G.: Image Analysis, Random Fields, and Markov Chain Monte Carlo Methods. In: Number 27 in Stochastic Modelling and Applied Probability, Springer, Heidelberg (2006)
11. Bigun, J.: Vision with Direction - A Systematic Introduction to Image Processing and Computer Vision. Springer, Heidelberg (2006)
12. Jain, A.K., Farrokhnia, F.: Unsupervised texture segmentation using gabor filters. Pattern Recogn. 24(12), 1167–1186 (1991)
13. ter Haar Romeny, B.M.: Front-End Vision and Multi-Scale Image Analysis: Multi-Scale Computer Vision Theory and Applications, written in Mathematica. Computional Imaging and Vision, vol. 27. Kluwer Academic Publishers, Dordrecht (2003)
14. Liu, J.S.: Monte Carlo Strategies in Scientific Computing. Springer Series in Statistics. Springer, Heidelberg (2004)
15. Zhu, S.C., Mumford, D.: Prior learning and gibbs reaction-diffusion. IEEE Transaction on Pattern Analysis and Machine Intelligence 19(11), 1236–1250 (1997)
16. Zhu, S.C., Mumford, D.: Grade: Gibbs reaction and diffusion equation - a framework for pattern synthesis, denoising, image enhancement, and clutter removal. IEEE Trans. PAMI 19(11), 1627–1660 (1997)
17. Meyer, Y.: Oscillating Patterns in Image Processing and Nonlinear Evolution Equations: The Fifteenth Dean Jacqueline B. Lewis Memorial Lectures. American Mathematical Society (AMS), Boston, MA, USA (2001)
18. Aujol, J.F., Gilboa, G., Chan, T., Osher, S.: Structure-texture image decomposition — modeling, algorithms, and parameter selection. International Journal of Computer Vision 67(1), 111–136 (2006)

Evaluating a General Class of Filters for Image Denoising

Luis Pizarro, Stephan Didas, Frank Bauer, and Joachim Weickert

Mathematical Image Analysis Group,
Faculty of Mathematics and Computer Science,
Building E1 1, Saarland University, 66041 Saarbrücken, Germany
{pizarro,didas,bauer,weickert}@mia.uni-saarland.de
http://www.mia.uni-saarland.de

Abstract. Recently, an energy-based unified framework for image denoising was proposed by Mrázek et al. [10], from which existing nonlinear filters such as M-smoothers, bilateral filtering, diffusion filtering and regularisation approaches, are obtained as special cases. Such a model offers several degrees of freedom (DOF) for tuning a desired filter. In this paper, we explore the generality of this filtering framework in combining nonlocal tonal and spatial kernels. We show that Bayesian analysis provides suitable foundations for restricting the parametric space in a noise-dependent way. We also point out the relations among the distinct DOF in order to guide the selection of a combined model, which itself leads to hybrid filters with better performance than the previously mentioned special cases. Moreover, we show that the existing trade-off between the parameters controlling similarity and smoothness leads to similar results under different settings.

1 Introduction

Many denoising techniques have been proposed in literature, many of them are application-dependent. However, there are only few strategies that combine different approaches and allow further generalizations. One of them is the energy-based approach proposed by Mrázek et al. [10], which combines the well-known M-smoothers [5] with bilateral filtering [17]. By extending the spatial influence that neighbouring pixels have on a central pixel, they end up with a general nonlocal filtering framework that rewards similarity to the input image and penalises large deviations from smoothness. Many other existing nonlinear filters are obtained as special cases of this framework. In this paper we explore the relations among the different degrees of freedom available for tuning a specific filter. Particularly, we focus on degradation processes governed by Gaussian and impulse perturbations. Moreover, in Bayesian analysis we find suitable foundations to instantiate probabilistically this model. Finally, we point out that the use of combined (nonlocal) *data* and (nonlocal) *smoothness* constraints leads to better denoising results than considering those filters obtained as special cases of this general approach. Our paper is organised as follows: Section 2 presents

B.K. Ersbøll and K.S. Pedersen (Eds.): SCIA 2007, LNCS 4522, pp. 601–610, 2007.

the *Nonlocal Data and Smoothness* (NDS) filtering framework. Section 3 provides brief insides into Bayesian analysis that will help us in tuning the model parameters in Section 4. Finally, some conclusions are outlined in Section 5.

2 The NDS Framework

Let $f \in \mathbb{R}^n$ be a given degraded version of the unknown 1-D or 2-D image $u \in \mathbb{R}^n$, and let i, j be pixel indices running along the set $\Omega = \{1, \ldots, n\}$. In [10] a unifying variational approach to restore the original image u was proposed. It minimises the energy function

$$\begin{aligned} E(u) = \quad & \alpha \sum_{i,j \in \Omega} \Psi_D(|u_i - f_j|^2)\, w_D(|x_i - x_j|^2) \\ & + (1-\alpha) \sum_{i,j \in \Omega} \Psi_S(|u_i - u_j|^2)\, w_S(|x_i - x_j|^2). \end{aligned} \tag{1}$$

This energy-based approach comprises a linear combination of two constraints, i. e. $E(u) = \alpha\, E_D(u) + (1-\alpha)\, E_S(u)$, where the constant $\alpha \in [0,1]$ determines the relative importance of both assumptions. The *data term* $E_D(u)$ rewards similarity to the input image f, while the *smoothness term* $E_S(u)$ penalises deviations from (piecewise) homogeneity in the restored u. On one hand, the kernels $\Psi(\cdot)$ are increasing functions that penalise large *tonal distances* s^2; that is, the distance between two pixel grey values s_i and s_j. On the other hand, the kernels $w(\cdot)$ are nonnegative and (possibly) nonlocal windows that ponder the influence of distant pixels; the Euclidean distance between two pixel locations x_i and x_j is called *spatial distance* x^2. See Table 1 and Table 2 for a non-exhaustive list of kernels $\Psi(\cdot)$ and $w(\cdot)$ proposed in literature.

Table 1. Popular choices for tonal weights Ψ

$\Psi(s^2)$		$\Psi'(s^2)$	known in the context of
s^2	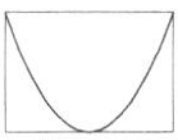	1	Tikhonov regularisation [16]
$2\left(\sqrt{s^2+\varepsilon^2}-\varepsilon\right)$	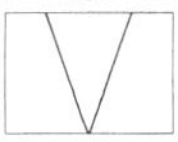	$\left(s^2+\varepsilon^2\right)^{-\frac{1}{2}}$	regularised total variation (TV) [15]
$2\lambda^2\left(\sqrt{1+\frac{s^2}{\lambda^2}}-1\right)$	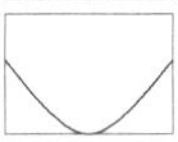	$\left(1+\frac{s^2}{\lambda^2}\right)^{-\frac{1}{2}}$	nonlinear regularisation, Charbonnier *et al.* [4]
$\lambda^2 \log\left(1+\frac{s^2}{\lambda^2}\right)$	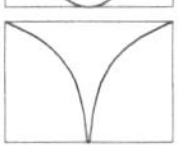	$\left(1+\frac{s^2}{\lambda^2}\right)^{-1}$	nonlinear diffusion, Perona-Malik 1 [14]
$2\lambda^2\left(1-\exp\left(-\frac{s^2}{2\lambda^2}\right)\right)$	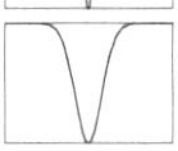	$\exp\left(-\frac{s^2}{2\lambda^2}\right)$	nonlinear diffusion, Perona-Malik 2 [14]
$\min(s^2, \lambda^2)$	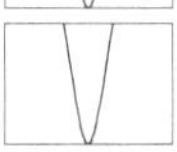	$\begin{cases} 1 & \lvert s\rvert < \lambda \\ 0 & \text{else} \end{cases}$	segmentation, Mumford and Shah [11]

Table 2. Possible choices for spatial weights w

$w(s^2)$			known in the context of
$\begin{cases} 1 & \lvert s \rvert < \theta \\ 0 & \text{else} \end{cases}$	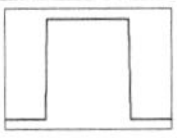	hard window	locally orderless images, Koenderink and van Doorn [8]
$\exp\left(-\frac{s^2}{\theta^2}\right)$	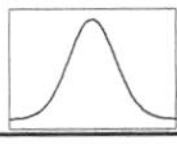	soft window	Chu *et al.* [5]

Table 3. Nonlinear filters derived as special cases of the NDS framework (1)

	Filter	Model
(a)	histogram operations	$\sum_{i,j} \Psi(\lvert u_i - f_j \rvert^2)$
(b)	M-smoothing	$\sum_{i,j} \Psi_D(\lvert u_i - f_j \rvert^2)\, w_D(\lvert x_i - x_j \rvert^2)$
(c)	bilateral filtering	$\sum_{i,j} \Psi_S(\lvert u_i - u_j \rvert^2)\, w_S(\lvert x_i - x_j \rvert^2)$
(d)	Bayesian/regularisation/diffusion	$\int \left(\lvert u - f \rvert^2 + \alpha \Psi_S(\lvert \nabla u \rvert^2)\right) dx$

The NDS function (1) presented above exhibits wide generality in choosing the parameters Ψ_D, Ψ_S, w_D, w_S and α to derive a desired filter. In particular, Table 3 shows some well-known nonlinear filters obtained as special cases by tuning the different degrees of freedom.

To minimise the expression (1) we make use of the Jacobi method. See [6] for a comparison study among different minimisation methods for the NDS functional.

3 Statistical Background

The most common degradation model is given by $f = u + \eta$, where u is the true image, η represents a zero-mean additive noise with standard deviation σ, and f is the recorded image. In the following we consider these quantities as realisations of the random variables U, η, and F, denoting as p_U, p_η, and p_F their probability density function (pdf), respectively. In Bayesian analysis [18], the *maximum a posteriori* (MAP) estimator

$$\begin{aligned} \hat{u}_{\text{MAP}} &= \arg\max_u \log p_{U|F}(u|f) \\ &= \arg\min_u \left(-\log p_{F|U}(f|u) - \log p_U(u)\right) \end{aligned} \tag{2}$$

yields the most likely image u given f. The conditional distribution $p_{F|U}$, also called *likelihood*, models the degradation process of U to F and is therefore considered as the noise distribution, i. e. $p_{F|U}(f|u) = p_\eta(\eta) = \prod p_\eta(f_i - u_i)$, assuming that the noise is independent and identically distributed (i.i.d.). We will focus on Gaussian and impulse noise. These types of noise are well modelled by the *Gaussian* and *Laplacian* distributions, being respectively their pdf's

$$p_{\eta_G} = \left(\frac{1}{\sqrt{2\pi}\sigma}\right)^n \exp\left(-\frac{1}{2\sigma^2}\sum_{i\in\Omega}|\eta_i|^2\right), \tag{3}$$

$$p_{\eta_I} = \left(\frac{1}{2\sigma}\right)^n \exp\left(-\frac{1}{\sigma}\sum_{i\in\Omega}|\eta_i|\right). \tag{4}$$

Pluging these noise models into the MAP estimator (2), and observing the structural resemblance to our NDS function (1), suggests that the penalisers for the data term can be instantiated as $\Psi_D(s^2) = s^2$ for Gaussian noise, and as $\Psi_D(s^2) = |s|$ for impulse noise, see Table 1.

The previous noise distributions are special cases of a more general probabilistic law: the *generalized Gaussian* distribution [7], with parameters mean μ, variance σ^2, and $\nu > 0$ (Gaussian case $\nu = 2$, Laplacian case $\nu = 1$); and pdf

$$p_Z(z) = \frac{\nu\Gamma(3/\nu)^{1/2}}{2\sigma\Gamma(1/\nu)^{3/2}} \exp\left(-\frac{|z-\mu|^\nu}{\sigma^\nu}\left(\frac{\Gamma(3/\nu)}{\Gamma(1/\nu)}\right)^{\nu/2}\right), \tag{5}$$

where $\Gamma(\cdot)$ is the Euler Gamma function. This distribution has been also utilized for modelling probabilistic *prior* knowledge about the signal u to recover. In (2), this information is represented in terms of the *prior* distribution p_U of the grey values of U. Besag [2] proposed the Laplacian law as model for p_U, which was later extended by Bouman & Sauer with their *generalized Gaussian Markov Random Field* [3] based on the distribution (5) for $\nu \in [1, 2]$. Since choosing a particular model for the prior distribution is essentially equivalent to specify the penaliser Ψ_S for the smoothness term in our NDS framework, we can instantiate such tonal kernel as $\Psi_S(s^2) = |s|^\nu$. This function is nonconvex for $0 < \nu < 1$, what may give rise to local minima in (1). However, nonconvex penalisers can allow almost exact restoration quality [9,12,13].

In summary, the Bayesian framework provides a founded basis for choosing appropriate tonal kernels $\Psi(\cdot)$ for the data and smoothness terms in (1). Studying other types of noise and the properties of the signal to recover, will lead to different criteria for selecting the penalisers.

4 Tuning the Model Parameters

4.1 Linear Combination of Kernels

The problem of determining α for the image simplification approach described in the Section 2 is crucial to obtain an optimal combination of similarity and smoothness. We intend to justify the use of such framework for $\alpha \notin \{0, 1\}$, i. e. for a wider spectrum of filters than those special cases outlined in Table 3.

A function $\varphi : [0, 1] \to \mathbb{R}$ is called *unimodal* on $[0, 1]$ if it contains a single minimum in that interval. Then, we obtain an estimate of the true image u as

$$\hat{u} = \arg\min_{\alpha} \varphi(\alpha), \tag{6}$$

assuming that we deal with the unimodal function $\varphi(\alpha) := \|u - u_\alpha\|_1$, where u_α is the solution image for a specific value of α from (1). Exploiting the empirical unimodality[1] of φ on $[0, 1]$, we employ the *Fibonacci Search* method to find an optimal value for α that solves (6). This line-search strategy ensures fast convergence. For *multimodal* functions it is better to utilize the *Simulated Annealing* technique as minimization strategy, which guarantees finding a global minimum in finite time. See [1] for the implementation details.

Another important ingredient to achieve an appropriate mixture of similarity and smoothness is the determination of the support of the spatial kernels $w(\cdot)$ in both terms of (1). Moreover, we want to find out if there is a certain interrelation between both supports. In the following, we use as spatial kernels the disk-shaped hard window function of Table 2, denoted by $\mathcal{B}(\cdot)$ with supporting radius θ, i. e. with diametrical support $(2\theta + 1)$.

Let us first consider the case of Gaussian noise. As suggested by the statistical framework, we select the penaliser $\Psi_D(s^2) = s^2$ for the data term, and the function $\Psi_S(s^2) = |s|^\nu$ for the smoothness term, focusing on $\nu \in \{1, 2\}$. Thus, our designed nonlocal filter for Gaussian noise reads

$$E(u) = \alpha \sum_{i \in \Omega, j \in \mathcal{B}(i)} |u_i - f_j|^2 + (1 - \alpha) \sum_{i \in \Omega, j \in \mathcal{B}(i)} |u_i - u_j|^\nu . \tag{7}$$

Let us apply this model to reconstruct the Gaussian noise signal depicted in *top left* of Fig. 1. All model parameters were optimised and the best five results for both values of ν are shown in the first two sections of the left-hand side of Table 4. In the third section we present the performance of those filters from Table 3(d), also optimising their parameters. Mean and median stand as representatives of M-smoothers, and four instantiations of Ψ_S for regularisation were included. Without exceptions, our designed model outperforms all the well known filters obtained as particular cases of the unifying NDS filtering framework.

Now, if we have data contaminated by impulse noise, for instance salt-and-pepper noise, we just need to modify from our previous model the penaliser in the data term by $\Psi_D(s^2) = |s|$, as it is proposed in the Bayesian framework. Our new model reads

$$E(u) = \alpha \sum_{i \in \Omega, j \in \mathcal{B}(i)} |u_i - f_j| + (1 - \alpha) \sum_{i \in \Omega, j \in \mathcal{B}(i)} |u_i - u_j|^\nu . \tag{8}$$

We use this filter for denoising the salt-and-pepper noise signal plotted on *top right* of Fig. 1. We conduct the same comparative analysis as before, and conclude again that our model beats all the other particular filters.

Exploiting the NDS image denoising framework for $\alpha \notin \{0, 1\}$ and nonlocal window functions leads to hybrid methods for image simplification. It is important to mention that we have deliberately chosen penalisers that do not require any gradient threshold parameter. This keeps our model simple and efficient.

[1] Even though we can guarantee neither the continuous dependence of φ with respect to α nor a unique solution, we have observed this behavior in most of our experiments.

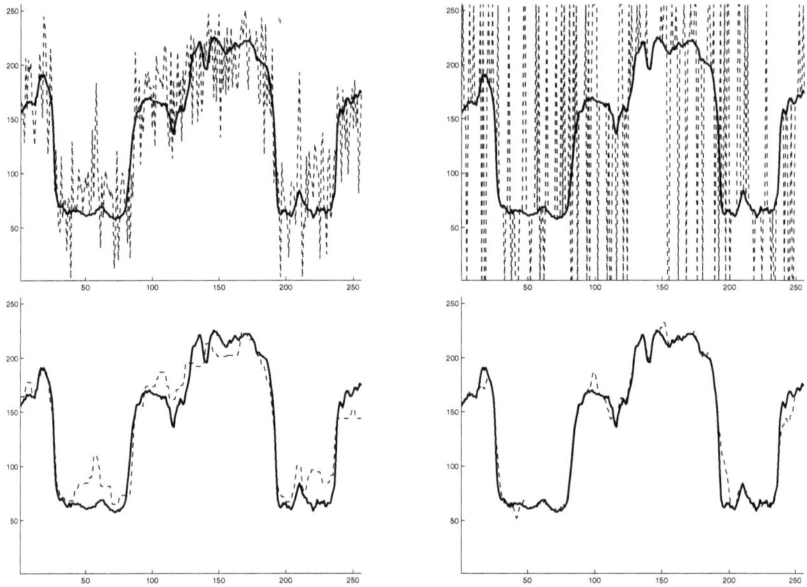

Fig. 1. Examples of signal denoising using the NDS function. Original signal in solid line. *Top left*: Noisy signal perturbed by zero-mean Gaussian noise with $\sigma = 40$ in dashed line, $\ell^1 = 27.30$. *Bottom left*: Recovered version of the Gaussian noise signal in dashed line, $\ell^1 = 13.83$. *Top right*: Noisy signal perturbed by 40% of a zero-mean salt-and-pepper noise in dashed line, $\ell^1 = 48.04$. *Bottom right*: Recovered version of the salt-and-pepper noise signal in dashed line, $\ell^1 = 4.61$.

4.2 Smoothing Effects

Trade-off Between α and the Radius of w_D. If we consider a functional which only consists of a data term, we notice that increasing the support of the spatial window leads to smoothing. On the other hand, if we leave the spatial window of the data term small and add a smoothness term, this has visually almost the same effect. In this experiment we want to quantify the difference more accurately and search for α corresponding to a certain support. To this end we consider the two following functions. The first function

$$E_D(u) = \sum_{i,j \in \Omega} (u_i - f_j)^2 w_D\left(|x_i - x_j|^2\right) \tag{9}$$

consists only of a data term, but allows for a larger window given by the disc-shaped hard window function w_D with supporting radius θ_D. The second function has a local data term and a smoothness term which only takes the direct neighbours $\mathcal{N}(i)$ of pixel i into consideration

$$E_C(u) = \alpha \sum_{i \in \Omega} (u_i - f_i)^2 + (1 - \alpha) \sum_{i \in \Omega, j \in \mathcal{N}(i)} (u_i - u_j)^2 \tag{10}$$

Table 4. Numerical comparison of different filters. *Left:* Denoising results of the Gaussian noise signal of Fig. 1. *Right:* Denoising results of the salt-and-pepper noise signal of Fig. 1. The best results are written in bold letters and plotted in Fig. 1.

Filter	θ_D	θ_S	α	ℓ^1
	3	1	0.223	14.18
	3	2	0.572	14.24
model (7), $\nu = 2$	2	1	0.208	14.26
	2	2	0.542	14.30
	3	3	0.793	14.32
	2	**2**	**0.072**	**13.83**
	2	3	0.133	13.83
model (7), $\nu = 1$	2	4	0.178	13.85
	2	5	0.223	13.93
	3	2	0.098	14.00
mean	4	-	1.000	14.93
median	4	-	1.000	14.90
Tikhonov	0	1	0.329	14.57
TV	0	1	0.001	15.62
Perona-Malik 1	0	1	0.298	14.47
Charbonnier	0	1	0.314	14.53

Filter	θ_D	θ_S	α	ℓ^1
	0	**1**	**0.903**	**4.61**
	3	1	0.810	4.67
model (8), $\nu = 2$	3	2	0.936	4.80
	4	1	0.793	4.90
	2	1	0.757	4.91
	3	8	0.895	5.20
	3	9	0.910	5.27
model (8), $\nu = 1$	3	10	0.921	5.38
	4	9	0.918	5.59
	3	7	0.892	5.51
mean	6	-	1.000	23.95
median	6	-	1.000	6.98
Tikhonov	0	1	0.096	23.22
TV	0	1	0.001	35.04
Perona-Malik 1	0	1	0.095	23.21
Charbonnier	0	1	0.096	23.21

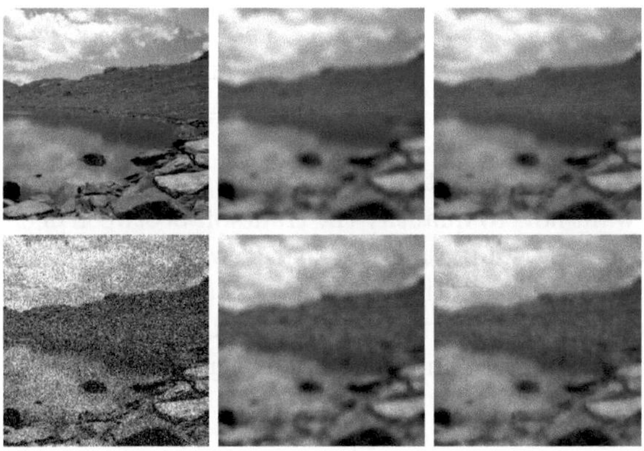

Fig. 2. Example of the trade-off between different parameters. *Top left:* Original image (256×256 pixels). *Top middle:* Smoothed version with E_D, radius $\theta_D = 5$. *Top right:* Smoothing with E_C, $\alpha = 0.503$. *Bottom left:* Image with additive Gaussian noise, standard deviation $\sigma = 50$. *Bottom middle:* Denoising with E_D, radius $\theta_D = 5$. *Bottom right:* Denoising with E_C, $\alpha = 0.424$.

Here, only changing the value α is used to steer the amount of smoothness. Fig. 2 shows two examples of the trade-off between these parameters. We see that in both cases with and without noise it is possible to obtain very similar results with both functionals. We are interested in knowing how far away the results

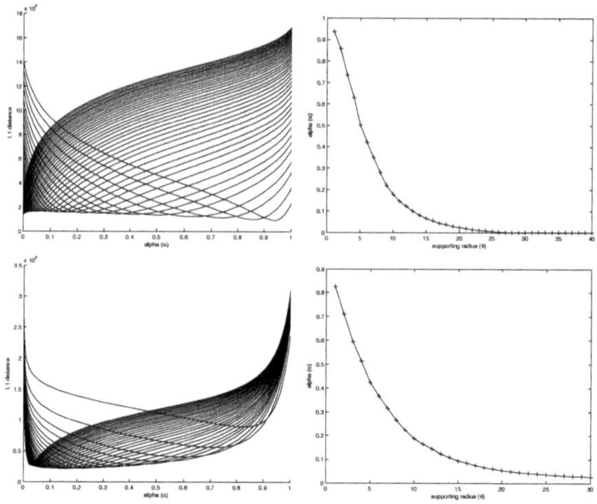

Fig. 3. Trade-off between large kernels and small α. *Top left:* ℓ^1-error between E_D and E_C depending on α, for noise-free image. *Top right:* Optimal value of α depending on the radial support θ_D. *Bottom:* Same plots for the noisy input image.

obtained with the functions E_D and E_C are from each other, and how they approach each other by means of tuning θ_D and α, respectively. In Fig. 3, we display some measurements to quantify the trade-off between these parameters. In the left column, each curve stands for a certain radius size, and there is one value of α that minimizes the ℓ^1 distance between their estimates. The minimum distance achieved for every pair (θ_D, α) is displayed in the right column.

Trade-off Between α and the Radius of w_S. Similarly to the previous experiment, we want now to quantify the trade-off beetween decreasing the value of α and increasing the support of the spatial window in the smoothness term; both procedures lead to smoothing. Let us assume a level $\sigma = 20$ of Gaussian noise. From our experiments in Section 4.1, we know that the following function produces satisfactory results under Gaussian perturbations:

$$E(u) = \alpha \sum_{i \in \Omega, j \in \mathcal{B}(i)} |u_i - f_j|^2 + (1-\alpha) \sum_{i \in \Omega, j \in \mathcal{B}(i)} |u_i - u_j|. \qquad (11)$$

Keeping fixed $\theta_D = 1$, Fig. 4 shows the ℓ^1 distance between the original and the restored images for α ranging in $[0, 1]$ and different radial support θ_S. One can see that slightly better results are attained when both α and θ_S are small. Fig. 5 shows an example where similar restoration quality is achieved under different parameterization.

These experiments show that it is possible to interchange certain smoothing approaches by some others. This is important when one is searching for fast and efficient denoising algorithms. By using the NDS framework one does not has necessarily to give up quality for speed of computation.

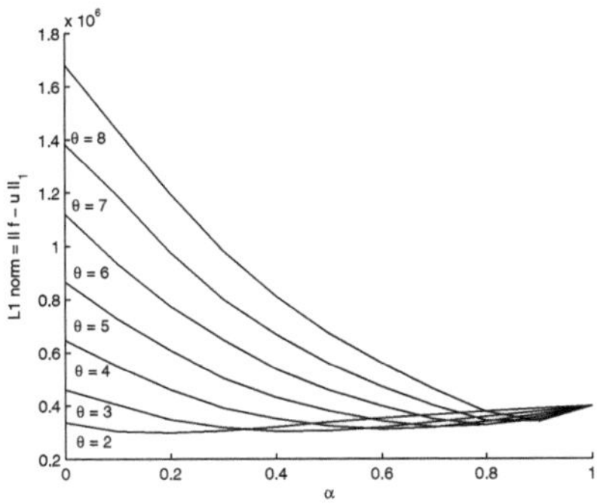

Fig. 4. Denoising properties of the functional (11). The plot outlines the ℓ^1 distance between the original and the denoised images for different values of the smoothing parameters α and θ_S. Original image in Fig. 5 *left*, and noisy image in Fig. 5 *middle left*.

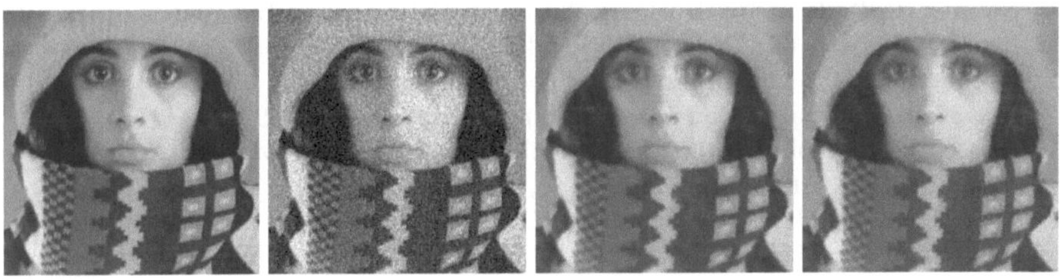

Fig. 5. Denoising properties of the functional (11). *Left*: original image. *Middle left*: disturbed image with Gaussian noise $\sigma = 20$, $\ell^1 = 16.02$. *Middle right*: restored image with $\alpha = 0.2$, $\theta_S = 2$, $\ell^1 = 4.88$. *Right*: restored image with $\alpha = 0.8$, $\theta_S = 6$, $\ell^1 = 5.18$.

5 Conclusions

We have shown the capabilities of the NDS framework as unifying filtering approach. We saw that excellent results can be obtained when terms that reward fidelity to the observations and penalise smoothness in the solution are nonlocally combined. By tuning its different degrees of freedom it is possible to design hybrid filters that outperform the performance of classical filters. The NDS functional possesses such a versatility that it is even possible to attain very similar results by tuning the parameters with different criteria and directions, what is particularly useful in looking for alternative ways to solve a denoising problem.

Acknowledgements. We gratefully acknowledge partly funding by *Deutscher Akademischer Austauschdienst (DAAD)* and by the priority programme SPP1114 of the *Deutsche Forschungsgemeinschaft (DFG)*, project WE 2602/2-3.

References

1. Bard, J.: Practical Bilevel Optimization. Algorithms and Applications. Springer, Heidelberg (1999)
2. Besag, J.: Toward Bayesian image analysis. Journal of Applied Statistics 16(3), 395–407 (1989)
3. Bouman, C., Sauer, K.: A generalized gaussian image model for edge-preserving MAP estimation. IEEE Transactions on Image Processing 2(3), 296–310 (1993)
4. Charbonnier, P., Blanc-Féraud, L., Aubert, G., Barlaud, M.: Two deterministic half-quadratic regularization algorithms for computed imaging. In: Proc. IEEE International Conference on Image Processing (ICIP-94, Austin, November. 13-16, 1994), vol.2, pp. 168–172 (1994)
5. Chu, C.K., Glad, I.K., Godtliebsen, F., Marron, J.S.: Edge-preserving smoothers for image processing. Journal of the American Statistical Association 93(442), 526–541 (1998)
6. Didas, S., Mrazek, P., Weickert, J.: Energy-based image simplification with non-local data and smoothness terms. In: Iske, A., Levesley, J. (eds.) Algorithms for Approximation, pp. 51–60. Springer, Heidelberg (2006)
7. Herman, G.T., Hurwitz, H., Lent, A., Lung, H-P.: On the Bayesian Approach to Image Reconstruction. Information and Control 42, 60–71 (1979)
8. Koenderink, J.J., Van Doorn, A.L.: The structure of locally orderless images. International Journal of Computer Vision 31(2/3), 159–168 (1999)
9. Künsch, H.R.: Robust priors for smoothing and image restoration. Annals of the Institute of Statistical Mathematics 46, 1–19 (1994)
10. Mrázek, P., Weickert, J., Bruhn, A.: On robust estimation and smoothing with spatial and tonal kernels. In: Klette, R., Kozera, R., Noakes, L., Weickert, J. (eds.) Geometric Properties for Incomplete Data, November 2005. Computational Imaging and Vision, vol. 31, Springer, Heidelberg (2005)
11. Mumford, D., Shah, J.: Optimal approximation of piecewise smooth functions and associated variational problems. Communications on Pure and Applied Mathematics 42, 577–685 (1989)
12. Nikolova, M.: Minimizers of cost-functions involving nonsmooth data-fidelity terms. Application to the processing of outliers. SIAM Journal on Numerical Analysis 40(3), 965–994 (2002)
13. Nikolova, M.: Analysis of the recovery of edges in images and signals by minimizing nonconvex regularized least-squares. Multiscale Modeling & Simulation 4(3), 960–991 (2005)
14. Perona, P., Malik, J.: Scale space and edge detection using anisotropic diffusion. IEEE Transactions on Pattern Analysis and Machine Intelligence 12, 629–639 (1990)
15. Rudin, L.I., Osher, S., Fatemi, E.: Nonlinear total variation based noise removal algorithms. Physica D. 60, 259–268 (1992)
16. Tikhonov, A.N.: Solution of incorrectly formulated problems and the regularization method. Soviet Mathematics Doklady 4(2), 1035–1038 (1963)
17. Tomasi, C., Manduchi, R.: Bilateral filtering for gray and colour images. In: Proc. of the 1998 IEEE International Conference on Computer Vision, Bombay, India, January 1998. pp. 839–846, Narosa Publishing House (1998)
18. Winkler, G.: Image Analysis, Random Fields and Markov Chain Monte Carlo Methods: A Mathematical Introduction. Springer, Heidelberg (2006)

Efficient Feature Extraction for Fast Segmentation of MR Brain Images

László Szilágyi[1,2], Sándor M. Szilágyi[2], and Zoltán Benyó[1]

[1] Budapest University of Technology and Economics,
Dept. of Control Engineering and Information Technology, Budapest, Hungary
[2] Sapientia - Hungarian Science University of Transylvania,
Faculty of Technical and Human Science, Târgu-Mureş, Romania
lalo@ms.sapientia.ro

Abstract. Automated brain MR image segmentation is a challenging problem and received significant attention lately. Various techniques have been proposed, several improvements have been made to the standard fuzzy c-means (FCM) algorithm, in order to reduce its sensitivity to Gaussian, impulse, and intensity non-uniformity noises. In this paper we present a modified FCM algorithm, which aims at accurate segmentation in case of mixed noises, and performs at a high processing speed. As a first step, a scalar feature value is extracted from the neighborhood of each pixel, using a filtering technique that deals with both spatial and gray level distances. These features are clustered afterwards using the histogram-based approach of the enhanced FCM algorithm. The experiments using 2-D synthetic phantoms and real MR images show, that the proposed method provides better results compared to other reported FCM-based techniques. The produced segmentation and fuzzy membership values can serve as excellent support for level set based cortical surface reconstruction techniques.

Keywords: image segmentation, fuzzy c-means algorithm, feature extraction, noise elimination, magnetic resonance imaging.

1 Introduction

By definition, image segmentation represents the partitioning of a digital image into non-overlapping, consistent pixel sets, which appear to be homogeneous with respect to some criteria concerning gray level intensity and/or texture.

The fuzzy c-means (FCM) algorithm is one of the most widely studied and applied methods for data clustering, and probably also for brain image segmentation [2,6]. However, in this latter case, standard FCM is not efficient by itself, as it is unable to deal with that relevant property of images, that neighbor pixels are strongly correlated. Ignoring this specificity leads to strong noise sensitivity and several other imaging artifacts.

Recently, several solutions were given to improve the performance of segmentation. Most of them involve using local spatial information: the own gray

B.K. Ersbøll and K.S. Pedersen (Eds.): SCIA 2007, LNCS 4522, pp. 611–620, 2007.

level of a pixel is not the only information that contributes to its assignment to the chosen cluster. Its neighbors also have their influence while getting a label. Pham and Prince [8] modified the FCM objective function by including a spatial penalty, enabling the iterative algorithm to estimate spatially smooth membership functions. Ahmed et al. [1] introduced a neighborhood averaging additive term into the objective function of FCM, calling the algorithm bias corrected FCM (BCFCM). This approach has its own merits in bias field estimation, but it computes the neighborhood term in every iteration step, giving the algorithm a serious computational load. Moreover, the zero gradient condition at the estimation of the bias term produces a significant amount of misclassifications [10]. Chuang et al. [5] proposed averaging the fuzzy membership function values and reassigning them according to a tradeoff between the original and averaged membership values. This approach can produce accurate clustering if the tradeoff is well adjusted empirically, but it is enormously time consuming.

In order to reduce the execution time, Szilágyi et al. [11], and Chen and Zhang [4] proposed to evaluate the neighborhoods of each pixel as a pre-filtering step, and perform FCM afterwards. The averaging and median filters, followed by FCM clustering, are referred to as FCM_S1 and FCM_S2, respectively [4]. Paper [11] also pointed out, that once having the neighbors evaluated, and thus for each pixel having extracted a scalar feature, FCM can be performed on the basis of the gray level histogram, clustering the gray levels instead of the pixels, which significantly reduces the computational load, as the number of gray levels is generally smaller by orders of magnitude. This latter quick approach, combined with an averaging pre-filter, is referred to as enhanced FCM (EnFCM) [3,11]. All BCFCM, FCM_S1, and EnFCM suffer from the presence of a parameter denoted by α, which controls the strength of the averaging effect, balances between the original and averaged image, and whose ideal value unfortunately can be found only experimentally. Another drawback is the fact, that averaging and median filtering, besides eliminating salt-and-pepper noises, also blurs relevant edges. Due to these shortcomings, Cai et al. [3] introduced a new local similarity measure, combining spatial and gray level distances, and applied it as an alternative pre-filtering to EnFCM, having this approach named fast generalized FCM (FGFCM). This approach is able to extract local information causing less blur than the averaging or median filter, but still has an experimentally adjusted parameter λ_g, which controls the effect of gray level differences.

In this paper we propose a novel method for MR brain image segmentation that simultaneously aims at high accuracy in image segmentation, low noise sensitivity, and high processing speed.

2 Methods

2.1 Standard Fuzzy C-Means Algorithm

The fuzzy c-means algorithm has successful applications in a wide variety of clustering problems. The traditional FCM partitions a set of object data into a number of c clusters based on the minimization of a quadratic objective function.

When applied to segment gray level images, FCM clusters the intensity level of all pixels (x_k, $k = 1 \ldots n$), which are scalar values. The objective function to be minimized is:

$$J_{\mathrm{FCM}} = \sum_{i=1}^{c} \sum_{k=1}^{n} u_{ik}^{m} (x_k - v_i)^2 \ , \tag{1}$$

where $m > 1$ is the fuzzyfication parameter, v_i represents the prototype value of cluster i, $u_{ik} \in [0, 1]$ is the fuzzy membership function showing the degree to which pixel k belongs to cluster i. According to the definition of fuzzy sets, for any pixel k, we have $\sum_{i=1}^{c} u_{ik} = 1$. The minimization of the objective function is reached by alternately applying the optimization of J_{FCM} over $\{u_{ik}\}$ with v_i fixed, $i = 1 \ldots c$, and the optimization of J_{FCM} over $\{v_i\}$ with u_{ik} fixed, $i = 1 \ldots c$, $k = 1 \ldots n$ [6]. During each cycle, the optimal values are computed from the zero gradient conditions, and obtained as follows:

$$u_{ik}^{\star} = \frac{(v_i - x_k)^{-2/(m-1)}}{\sum_{j=1}^{c} (v_j - x_k)^{-2/(m-1)}} \quad \forall i = 1 \ldots c, \forall k = 1 \ldots n \ , \tag{2}$$

$$v_i^{\star} = \frac{\sum_{k=1}^{n} u_{ik}^{m} x_k}{\sum_{k=1}^{n} u_{ik}^{m}} \quad \forall i = 1 \ldots c \ . \tag{3}$$

After adequate initialization of centroids v_i, (2) and (3) are applied alternately until the norm of the variation of vector $\mathbf{v}$ is less than a previously set small value ε. FCM has invaluable merits in making optimal clusters, but in image processing it has severe deficiencies. The most important one is the fact, that it fails to take into consideration the position of pixels, which is also relevant information in image segmentation. This drawback led to introduction of spatial constraints into fuzzy clustering.

2.2 Fuzzy Clustering Using Spatial Constraints

Ahmed et al. [1] proposed a modification to the objective function of FCM, in order to allow the labeling of a pixel to be influenced by its immediate neighbors. This neighboring effect acts like a regularizer that biases the solution to a piecewise homogeneous labeling [1]. The objective function of BCFCM is:

$$J_{\mathrm{BCFCM}} = \sum_{i=1}^{c} \sum_{k=1}^{n} \left[u_{ik}^{m} (x_k - v_i)^2 + \frac{\alpha}{n_k} \sum_{r \in N_k} u_{ik}^{m} (x_r - v_i)^2 \right] \ , \tag{4}$$

where x_r represents the gray level of pixels situated in the neighborhood N_k of pixel k, and n_k is the cardinality of N_k. The parameter α controls the intensity of the neighboring effect, and unfortunately its optimal value can be found only experimentally. Having the neighbors computed in every computation cycle, this iterative algorithm performs extremely slowly.

Chen and Zhang [4] reduced the time complexity of BCFCM, by previously computing the neighboring averaging term or replacing it by a median filtered term, calling these algorithms FCM_S1 and FCM_S2, respectively. These algorithms outperformed BCFCM, at least from the point of view of time complexity.

Szilágyi et al. [11] proposed a regrouping of the processing steps of BCFCM. In their approach, an averaging filter is applied first, similarly to the neighboring effect of Ahmed et al. [1]:

$$\xi_k = \frac{1}{1+\alpha}\left(x_k + \frac{\alpha}{n_k}\sum_{r\in N_k} x_r\right) . \tag{5}$$

This filtering is followed by an accelerated version of FCM clustering. The acceleration is based on the idea, that the number of gray levels is generally much smaller than the number of pixels. In this order, the histogram of the filtered image is computed, and not the pixels, but the gray levels are clustered [11], by minimizing the following objective function:

$$J_{\mathrm{EnFCM}} = \sum_{i=1}^{c}\sum_{l=1}^{q} h_l u_{il}^m (l - v_i)^2 , \tag{6}$$

where h_l denotes the number of pixels with gray level equaling l, and q is the number of gray levels. The optimization formulas in this case will be:

$$u_{il}^\star = \frac{(v_i - l)^{-2/(m-1)}}{\sum_{j=1}^{c}(v_j - l)^{-2/(m-1)}} \qquad \forall i = 1 \dots c, \forall l = 1 \dots q , \tag{7}$$

$$v_i^\star = \frac{\sum_{l=1}^{q} h_l u_{il}^m l}{\sum_{l=1}^{q} h_l u_{il}^m} \qquad \forall i = 1 \dots c . \tag{8}$$

EnFCM drastically reduces the computation complexity of BCFCM and its relatives [3,11]. If the averaging pre-filter is replaced by a median filter, the segmentation accuracy also improves significantly [3,12].

2.3 Fuzzy Clustering Using Spatial and Gray Level Constraints

Based on the disadvantages of the aforementioned methods, but inspired of their merits, Cai et al. [3] introduced a local (spatial and gray) similarity measure that they used to compute weighting coefficients for an averaging pre-filter. The filtered image is then subject to EnFCM-like histogram-based fast clustering. The similarity between pixels k and r is given by the following formula:

$$S_{kr} = \begin{cases} s_{kr}^{(s)} \cdot s_{kr}^{(g)} & \text{if } r \in N_k \setminus \{k\} \\ 0 & \text{if } r = k \end{cases} . \tag{9}$$

where $s_{kr}^{(s)}$ and $s_{kr}^{(g)}$ are the spatial and gray level components, respectively. The spatial term $s_{kr}^{(s)}$ is defined as the L_∞-norm of the distance between pixels k and r. The gray level term is computed as $s_{kr}^{(g)} = \exp(-(x_k - x_r)^2/(\lambda_g \sigma_k^2))$, where σ_k denotes the average quadratic gray level distance between pixel k and its neighbors. Segmentation results are reported more accurate than in any previously presented case [3].

2.4 The Proposed Method

Probably the most relevant problem of all techniques presented above, BCFCM, EnFCM, FCM_S1, and FGFCM, is the fact that they depend on at least one parameter, whose value has to be found experimentally. The parameter α balances the effect of neighboring in case of the former three, while λ_g controls the tradeoff between spatial and gray level components in FGFCM.

The zero value in the second row of (9) implies, that in FGFCM, the filtered gray level of any pixel is computed as a weighted average of its neighbors. Having renounced to the original intensity of the current pixel, even if it is a reliable, noise-free value, unavoidably produces some extra blur into the filtered image. Accurate segmentation requires this kind of effects to be minimized [9].

In the followings we propose a set of modifications to EnFCM/FGFCM, in order to improve the accuracy of segmentation, without renouncing to the speed of histogram-based clustering. In other words, we need to define a complex filter that can extract relevant feature information from the image while applied as a pre-filtering step, so that the filtered image can be clustered fast afterwards based on its histogram. The proposed method consists of the following steps:

1. As we are looking for the filtered value of pixel k, we need to define a small square or diamond-shape neighborhood N_k around it. Square windows of size 3×3 were used throughout this study.

2. We search for the minimum, maximum, and median gray value within the neighborhood N_k, and we denote them by min_k, max_k and med_k, respectively.

3. We replace the gray level of the maximum and minimum valued pixel with the median value (if there are more than one maxima or minima, replace them all), unless they are situated in the middle pixel k. In this latter case, pixel k remains unchanged, just labeled as unreliable value.

4. Compute the average quadratic gray level difference of the pixels within the neighborhood N_k, using the formula

$$\sigma_k = \sqrt{\frac{\sum_{r \in N_k \setminus \{k\}} (x_r - x_k)^2}{n_k - 1}} \ . \tag{10}$$

5. The filter coefficients will be defined as:

$$C_{kr} = \begin{cases} c_{kr}^{(s)} \cdot c_{kr}^{(g)} & \text{if } r \in N_k \setminus \{k\} \\ 1 & \text{if } r = k \wedge x_k \notin \{max_k, min_k\} \\ 0 & \text{if } r = k \wedge x_k \in \{max_k, min_k\} \end{cases} \ . \tag{11}$$

The central pixel k will be ignored if its value was found unreliable, otherwise it gets unitary weight. All other neighbor pixels will have coefficients $C_{kr} \in [0, 1]$, depending on their space distance and gray level difference from the central pixel. In case of both terms, higher distance values push the coefficients towards 0.

6. The spatial component $c_{kr}^{(s)}$ is a negative exponential of the Euclidean distance between the two pixels k and r: $c_{kr}^{(s)} = \exp(-L_2(k, r))$. The gray level term depends on the difference $|x_r - x_k|$, according to a bell-shaped function defined as follows:

$$c_{kr}^{(g)} = \begin{cases} \left[\cos\left(\pi \frac{x_r - x_k}{8\sigma_k}\right)\right]^2 & \text{if } |x_r - x_k| \leq 4\sigma_k \\ 0 & \text{if } |x_r - x_k| > 4\sigma_k \end{cases} . \tag{12}$$

7. The extracted feature value for pixel k, representing its filtered intensity value, is obtained as a weighted average of its neighbors:

$$\xi_k = \frac{\sum_{r \in N_k} C_{kr} x_r}{\sum_{r \in N_k} C_{kr}} . \tag{13}$$

Algorithm. We can summarize the proposed method as follows:

1. Pre-filtering step: for each pixel k of the input image, compute the filtered gray level value ξ_k, using (10), (11), (12), and (13).
2. Compute the histogram h_l of the pre-filtered image, $l = 1 \ldots q$.
3. Initialize v_i with valid gray level values, differing from each other.
4. Compute new fuzzy memberships u_{il}, $i = 1 \ldots c$, $l = 1 \ldots q$, using (7).
5. Compute new cluster prototypes v_i, $i = 1 \ldots c$, using (8).
6. If there is relevant change in the v_i values, go back to step 4. This is tested by comparing any norm of the difference between the new and the old vector $\mathbf{v}$ with a preset small constant ε.

The algorithm converges quickly, however, the number of necessary iterations depends on ε and on the initial cluster prototype values.

3 Results and Discussion

In this section we test and compare the accuracy of four algorithms: BCFCM, EnFCM, FGFCM, and the proposed method, on several synthetic and real images. All the following experiments used 3×3 window size for all kinds of filtering.

The proposed filtering technique uses a convolution mask whose coefficients are context dependent, and thus computed for the neighborhood of each pixel. Figure 1 presents the obtained coefficients for two particular cases. Figure 1(a) shows the case, when the central pixel is not significantly noisy, but some pixels in the neighborhood might be noisy or might belong to a different cluster. Under such circumstances, the upper three pixels having distant gray level compared to the value of the central pixel, receive small weights and this way they hardly contribute to the filtered value. Figure 1(b) presents the case of an isolated noisy pixel situated in the middle of a relatively homogeneous window. Even though all computed coefficients are low, the noise is eliminated, resulting a convenient filtered value 111. The migration of weights from the local maximum and minimum towards the median valued pixel, indicated by the arrows, is relevant in the second case and useful in the first.

The noise removal performances were compared using a 256×256-pixel synthetic test image taken from IBSR [7], having a high degree of mixed noise. Results are summarized in Fig. 2. Visually, the proposed method achieves best results, slightly over FGFCM, and significantly over all others.

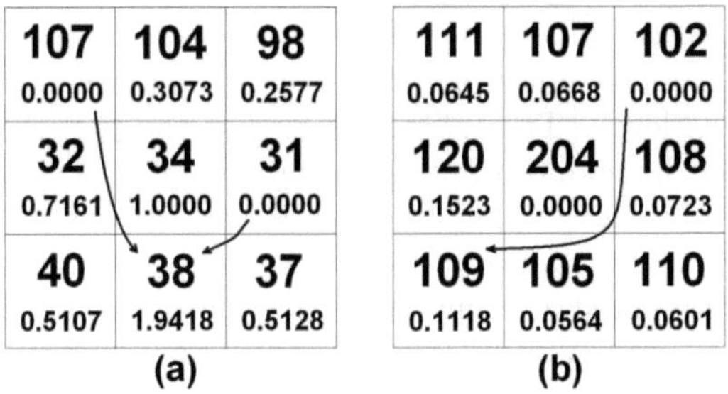

Fig. 1. Filter mask coefficients in case of a reliable pixel intensity value (a), and a noisy one (b). The upper number in each cell is the intensity value, while the lower number represents the obtained coefficient. The arrows show, that the coefficients of extreme intensities are transferred to the median valued pixel.

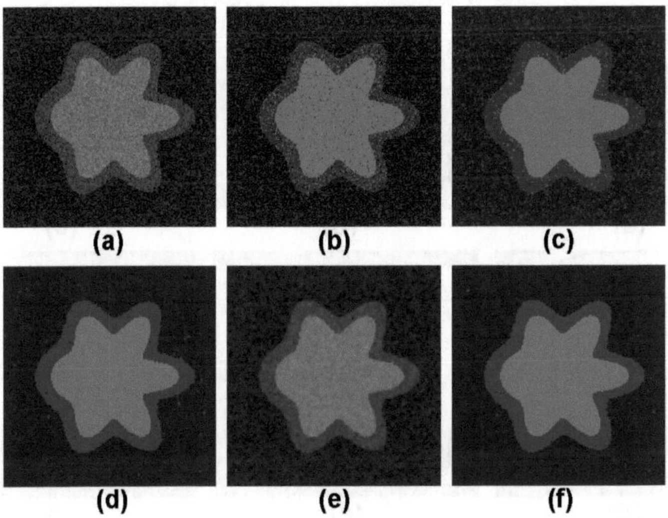

Fig. 2. Segmentation results on phantom images: (a) original, (b) segmented with traditional FCM, (c) segmented using BCFCM, (d) segmented using FGFCM, (e) filtered using the proposed pre-filtering, (f) result of the proposed segmentation

Table 1. Segmentation results on synthetic test images: misclassification rates at different noise types and levels

Noise type	BCFCM	EnFCM	FGFCM	Proposed
Gaussian 1	0.232 %	0.192 %	0.102 %	0.090 %
Gaussian 2	7.597 %	0.972 %	0.405 %	0.330 %
Gaussian 3	18.545 %	4.647 %	2.975 %	2.155 %
Impulse	0.250 %	0.192 %	0.130 %	0.120 %
Mixed 1	0.345 %	0.220 %	0.130 %	0.110 %
Mixed 2	5.542 %	1.025 %	0.675 %	0.547 %

Table 1 gives a statistical analysis of the synthetic images contaminated with different noises (Gaussian noise, salt-and-pepper impulse noise, and mixtures of these) at different levels. The table reveals that the proposed filter performs best at removing all these kinds of noises. Consequently, the proposed method is suitable for segmenting images corrupted with unknown noises, and in all cases it performs at least as well as his ancestors.

We applied the presented filtering and segmentation techniques to several T1-weighted real MR images. A detailed view, containing numerous segmentations, is presented in Fig. 3.

The original slice in Fig. 3 (a) is taken from IBSR. We produced several noisy versions of this slice, by adding salt-and-pepper impulse noise and/or Gaussian

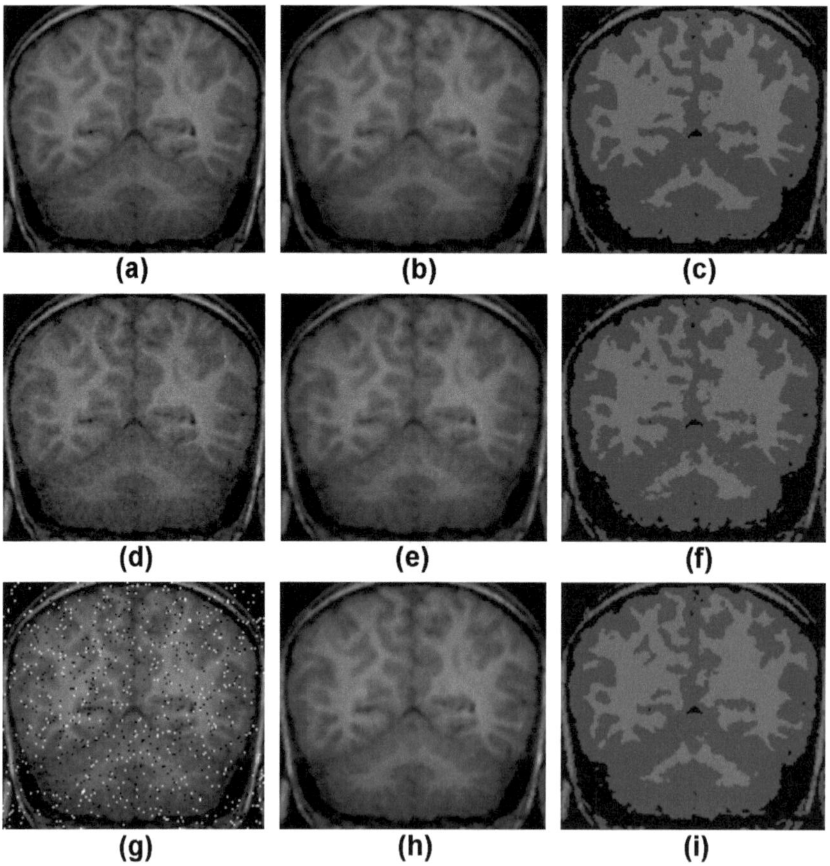

(a) (b) (c)
(d) (e) (f)
(g) (h) (i)

Fig. 3. Filtering and segmentation results on real T1-weighted MR brain images, corrupted with different kinds and levels of artificial noise. Each row contains an original or noise-corrupted brain slice on the left side, the filtered version (using the proposed method) in the middle, and the segmented version on the right side. Row (a)-(c) comes from record number 1320_2_43 of IBSR [7], row (d)-(f) is corrupted with 10 % Gaussian noise, while row (g)-(i) contains mixed noise of 5 % impulse + 5 % Gaussian.

noise, at different intensities. Some of these noisy versions are visible in Fig. 3 (d), (g). The filtered versions of the three above mentioned slices are presented in the middle column of Fig. 3. The segmentation results of the chosen slices are shown in Fig. 3 (c), (f), (i). From the segmented images we can conclude, that the proposed filtering technique is efficient enough to make proper segmentation of any likely-to-be-real MRI images in clinical practice, at least from the point of view of Gaussian and impulse noises.

Table 2 takes into account the behavior of three mentioned segmentation techniques, in case of different noise types and intensities, computed by averaging the misclassifications on 12 different T1-weighted real MR brain slices, all taken from IBSR. The proposed algorithm has lowest misclassification rates in most of the cases.

Table 2. Misclassification rates in case of real brain MR image segmentation

Noise type	EnFCM	FGFCM	Proposed
Original, no extra noise	0.767 %	0.685 %	0.685 %
Gaussian 4 %	1.324 %	1.131 %	1.080 %
Gaussian 8 %	3.180 %	2.518 %	1.489 %
Gaussian 12 %	4.701 %	2.983 %	2.654 %
Impulse 1 %	0.836 %	0.717 %	0.726 %
Impulse 3 %	1.383 %	0.864 %	0.823 %
Impulse 5 %	1.916 %	1.227 %	0.942 %
Impulse 10 %	3.782 %	1.268 %	1.002 %
Impulse 5 % + Gaussian 4 %	2.560 %	1.480 %	1.374 %
Impulse 5 % + Gaussian 8 %	3.626 %	2.013 %	1.967 %
Impulse 5 % + Gaussian 12 %	6.650 %	4.219 %	4.150 %

Further tests also revealed, that the proposed method performs well in case of T2-weighted MR brain images, too. We applied the proposed segmentation method to several complete head MR scans in IBSR. The dimensions of the image stacks were 256 × 256 × 64 voxels. The average total processing time for one stack was around 10 seconds on a 2.4 GHz Pentium 4.

4 Conclusions

We have developed a modified FCM algorithm for automatic segmentation of MR brain images. The algorithm was presented as a combination of a complex pre-filtering technique and an accelerated FCM clustering performed over the histogram of the filtered image. The pre-filter uses both spatial and gray level criteria, in order to achieve efficient removal of Gaussian and impulse noises without significantly blurring the real edges. Experiments with synthetic phantoms and real MR images show, that our proposed technique accurately segments the different tissue classes under serious noise contamination. We compared our results with other recently reported methods. Test results revealed that our approach performed better than these methods in many aspects, especially in the accuracy of segmentation and processing time.

Although the proposed method segments 2-D MR brain slices, it gives a relevant contribution to the accurate volumetric segmentation of the brain, because the segmented images and the obtained fuzzy memberships can serve as excellent input data to any level set method that constructs 3-D cortical surfaces.

Further works aim to reduce the sensitivity of the proposed technique to intensity inhomogeneity noises, and to introduce adaptive determination of the optimal number of clusters.

Acknowledgements. This research was supported by the Sapientia Institute for Research Programmes, Domus Hungarica Scientiarium et Artium, the Communitas Foundation, and the Pro Progressio Foundation.

References

1. Ahmed, M.N., Yamany, S.M., Mohamed, N., Farag, A.A., Moriarty, T.: A modified fuzzy c-means algorithm for bias field estimation and segmentation of MRI data. IEEE Trans. Med. Imag. 21, 193–199 (2002)
2. Bezdek, J.C., Pal, S.K.: Fuzzy models for pattern recognition. IEEE Press, Piscataway, NJ (1991)
3. Cai, W., Chen, S., Zhang, D.Q.: Fast and robust fuzzy c-means algorithms incorporating local information for image segmentation. Patt. Recogn. 40, 825–838 (2007)
4. Chen, S., Zhang, D.Q.: Robust image segmentation using FCM with spatial constraints based on new kernel-induced distance measure. IEEE Trans. Syst. Man. Cybern. Part. B. 34, 1907–1916 (2004)
5. Chuang, K.S., Tzeng, H.L., Chen, S., Wu, J., Chen, T.J.: Fuzzy c-means clustering with spatial information for image segmentation. Comp. Med. Imag. Graph. 30, 9–15 (2006)
6. Hathaway, R.J., Bezdek, J.C., Hu, Y.: Generalized fuzzy c-means clustering strategies using L_p norm distances. IEEE Trans. Fuzzy Syst. 8, 576–582 (2000)
7. Internet Brain Segmentation Repository, at `http://www.cma.mgh.harvard.edu/ibsr`
8. Pham, D.L., Prince, J.L.: Adaptive fuzzy segmentation of magnetic resonance images. IEEE Trans. Med. Imag. 18, 737–752 (1999)
9. Pham, D.L.: Unsupervised tissue classification in medical images using edge-adaptive clustering. Proc. Ann. Int. Conf. IEEE EMBS 25, 634–637 (2003)
10. Siyal, M.Y., Yu, L.: An intelligent modified fuzzy c-means based algorithm for bias field estimation and segmentation of brain MRI. Patt. Recogn. Lett. 26, 2052–2062 (2005)
11. Szilágyi, L., Benyó, Z., Szilágyi, S.M., Adam, H.S.: MR brain image segmentation using an enhanced fuzzy C-means algorithm. Proc. Ann. Int. Conf. IEEE EMBS 25, 724–726 (2003)
12. Szilágyi, L.: Medical image processing methods for the development of a virtual endoscope. Period. Polytech. Ser. Electr. Eng. 50(1–2), 69–78 (2006)

Automated Mottling Assessment of Colored Printed Areas

Albert Sadovnikov, Lasse Lensu, and Heikki Kälviäinen

Machine Vision and Pattern Recognition Group
Laboratory of Information Processing
Department of Information Technology
Lappeenranta University of Technology
P.O.Box 20, 53851 Lappeenranta, Finland
Firstname.Lastname@lut.fi
http://www.it.lut.fi/ip/research/mvpr/

Abstract. Mottling is one of the most significant defects in modern offset printing using coated papers. Mottling can be defined as undesired unevenness in perceived print density. Previous research in the field considered only gray scale prints. In our work, we extend current methodology to color prints. Our goal was to study the characteristics of the human visual system, perform psychometric experiments and develop methods which can be used at industrial level applications. We developed a method for color prints and extensively tested it with a number of experts and laymen. Suggested approach based on pattern-color perception separability proved to correlate with the human evaluation well.

1 Introduction

Paper printability and quality of prints are important attributes of modern printing applications. These issues are of equal significance for all parties involved in printed media production, from paper mills and print houses to end-consumers. High print quality is a basic requirement in printed products containing images. There are several undesired effects in prints because of non-ideal interactions of paper and ink in high-speed printing processes. One of these effects is mottling which is related to density and gloss of print. It is the uneven appearance of solid printed areas, and it depends on the printing ink, paper type, and printing process. Depending on the phenomenon causing this unevenness, there exist three types of mottling: back-trap mottle (uneven ink absorption in the paper), water-interface mottle (insufficient and uneven water absorption of the paper causing uneven ink absorption), and ink-trap mottle (wet or dry; incorrect trapping of the ink because of tack. In the task of automated mottling measurement the main cause of mottling is often overlooked, namely the human perception. Truly, if a person could not perceive printed area unevenness then the attention to mottling would be a lot less.

B.K. Ersbøll and K.S. Pedersen (Eds.): SCIA 2007, LNCS 4522, pp. 621–630, 2007.

Several methods to evaluate mottling by an automatic machine vision system have been proposed. The ISO 13660 standard includes a method for monochrome images. It is based on calculating the standard deviation of small tiles within sufficiently large area [1]. In the standard, the size of the tiles is set to a fixed value, which is a known limitation [2]. The first improvement to the standard method was to use tiles of variable sizes [3]. Other methods relying on clustering, statistics, and wavelets have also been proposed [4,5,6]. Other approaches to evaluate gray scale mottling have their basis in frequency-domain filtering [7] and frequency analysis [8], which were thoroughly studied in [9]. All of the before-mentioned methods are designed for binary or gray scale images. If color prints were assessed, the correlation of the methods to human assessments would be severely limited. Also the grounds for the methods do not arise from any models for the phenomena causing mottling, nor vision science.

Mottling can be physically defined, but it becomes problematic when a print is perceived. If a person looking at a solid print perceives unevenness, mottling is a problem. Thus, the properties and limits of the human visual system must be taken into account when proper methods to assess mottling are designed. This is especially very important in the assessment of color images. When perception of image noise is of concern, visual sensitivity to contrast and spatial frequencies of the human visual system (HVS) is independent of luminance within common luminance levels [10]. However, contrast sensitivity depends on spatial frequency [11], thus, mottles of different sizes are perceived differently. The peak sensitivity of the HVS is approximately at 3 cycles/degree, and the maximum detected frequency is from 40 cycles/degree (sinusoidal gratings) [12] to over 100 cycles/degree (single cycle) [13].

The purpose of this work was to design the artificial method for a human assessment of color mottling samples. In our study, we sought proper background for the methodological selections based on vision science. We implemented the method based on gray scale approach, and modified for color prints, as needed to accommodate appropriate knowledge concerning the HVS.

2 Background

Method presented in the paper is based on the current approaches to gray scale mottling evaluation and pattern-color separability hypothesis. It is worth to give a short introduction to those.

2.1 Gray Scale Mottling Evaluation

The human perception of mottling relies on the spatial frequency theory, which is based on an atomistic assumption: the representation of any image, no matter how complex, is an assemblage of many primitive spatial "atoms". The primitives of spatial frequency theory are spatially extended patterns called sinusoidal gratings: two dimensional patters whose luminance varies according to a sine wave over one spatial dimension and is constant over the perpendicular dimension [14].

The idea for this method comes from the bandpass method [7] and later extended in [9]. The value of the perceived mottling can be computed using the following formula:

$$M = \frac{1}{R} \int_{u \in U,\, \phi \in \Phi} |F(u, \phi)| C(u)\, du d\phi \tag{1}$$

where $F(u, \phi)$ image in frequency space, $C(u)$ contrast sensitivity function (CSF), R is a mean image reflectance. The implementation of the method uses Mannos CSF defined as [15]:

$$C(u) = 2.6(0.0192 + 0.114u)e^{(-(0.114u)^{1.1})} \tag{2}$$

Mannos CSF formulation is in cycles per degree (cpd) units and has a maximum at 8 cpd (see Fig. 1).

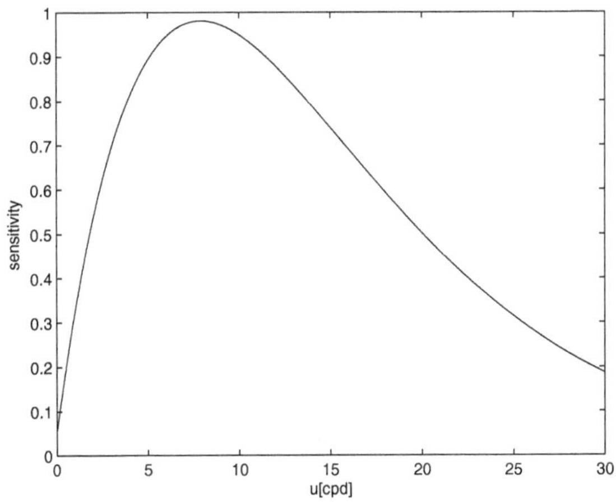

Fig. 1. Mannos contrast sensitivity function model

The method can be defined as a sum of the image energy filtered with CSF in the spectrum that falls within the visual area, which is $U = [0.04, 30]$ cpd in the current implementation.

The effect of orientation(ϕ) sensitivity of the has also been studied HVS [12]. It is known that human sensitivity is lowest around 45 and 135 and highest at vertical and horizonal directions. However, experiments showed low significance of introducing the orientational dependent filtering. This can also be understood based on the stochastic nature of mottling.

2.2 Pattern-Color Separability

A mechanism of color mottling stimuli perception is pattern-color separable when [16]:

- Its relative pattern sensitivity is invariant as the test stimulus' spectral composition is changed.
- Its relative wavelength selectivity is invariant when stimulus' spatial pattern is changed.

The absolute level of the pattern and color sensitivity can vary, but the relative pattern and color sensitivities must not. From this definition it is evident that pattern-color separability is required before it is possible to say that a mechanism has a unique pattern or wavelength sensitivity function. The pattern-color separable model provides a framework for thinking about how different visual pathways might contribute to visual sensitivity [16].

If the previous ideas are right, then color mottling perception depends separately on the spatial characteristics of an image (gray scale mottling) and color characteristics (color range). The following section will describe how it is possible to decompose an image in the pattern and color sense.

3 Pattern-Color Separable Method

Let us consider a gray scale image $f(x, y)$ with a given mottling index M. The question arises, how the mottling perception will change if the image will be colored with the following coloring procedure: red component $f_r(x, y) = c_r\, f(x, y) + m_r$, green component $f_g(x, y) = c_g\, f(x, y) + m_g$ and blue component $f_b(x, y) = c_b\, f(x, y) + m_b$. It is clear that perceived mottling value will depend on the pattern $f(x, y)$ and on the visual color differences along the line $(m_r, m_g, m_b) + k\,(c_r, c_g, c_b)$. This also holds for the gray scale perception model (see Eq. 1), where mottling value depends on the mean reflectance R (like (m_r, m_g, m_b)), but since the color orientation in gray level images is constant $(1, 1, 1)$, it is not used.

Basically, the $1/R$ factor in Eq. 1 reflects simplified correspondence between lightness and luminance, making color difference along the line $(1, 1, 1)$ locally constant. This fact leads to an idea to use a perceptually uniform color system for computing mottling index, where the role of the mean sample color and the color orientation will be reduced to a minimum. For the method design 1976 $CIE\ L^*a^*b^*$ color space was used.

Color mottling evaluation can be summarized in Algorithm 1.

Algorithm 1. *Pattern-color separable method*

1: *Transform an image from the input color space to* 1976 $CIE\ L^*a^*b^*$.
2: *Compute at each plane* (L^*, a^*, b^*) *mottling value with Eq.1, omitting* $1/R$ *factor.*
3: *Compute the final mottling index as* $M = \sqrt{M_L^2 + M_a^2 + M_b^2}$.

4 Human Evaluation

4.1 Pairwise Comparison Test

The basic method of paired comparisons [17] consists of sequentially presenting pairs of samples to an observer and asking the observer which one of the pair has the greatest amount of mottling.

For this test we used a set of 62 black at 70% (K70) samples, with previously known and tested mottling values [18], covering a wide range of mottling. It is obviously impossible to evaluate a perception model for mottling using only K70 (gray scale samples). To extend the sample set, the following coloring procedure was used:

$$(f_r(x,y), f_g(x,y), f_b(x,y)) = (c_r f(x,y)+m_r, c_g f(x,y)+m_g, c_b f(x,y)+m_b) \quad (3)$$

where $f(x,y)$ is an input K70 sample, f_r, f_g, f_b - new image color planes, m_r, m_g, m_b - new image mean color value, and vector (c_r, c_g, c_b) describes image color range orientation in red-green-blue (RGB) color space and also includes scaling effect, i.e., vector (c_r, c_g, c_b) is not normalized.

Since it is not feasible to cover all the visible color space, the following areas of interest were used: cyan at 70% (C70), magenta at 70% (M70), yellow at 70% (Y70), K70. These inks are usually used as printing primaries, and therefore are of more interest.

Pairwise comparison test can be summarized in the Algorithm 2.

Algorithm 2. *Pairwise comparison test*

1: *Select randomly two samples from K70 set.*
2: *Select mean color m_r, m_g, m_b, i.e. C70, M70, Y70 or K70.*
3: *Randomly generate two color orientation vectors of the form (c_r, c_g, c_b).*
4: *Color each sample with previously defined mean color and orientations.*
5: *Present a pair for evaluation.*

When presented a pair (see the test layout in Fig. 2) an observer is asked to give a mark for the difference between the samples, ranged in -3, -2, -1, 0, 1, 2, 3. The test is organized in the way which allows to avoid color adaptation, i.e. saturated samples (C70, M70 and Y70) are alternated with unsaturated (K70).

The observers were divided into two groups: experts and laymen. Experts were individuals with vast mottling evaluation experience(10 persons), and others were defined as laymen(15 persons). Each pair evaluation was limited by a certain time (15 seconds), to avoid such artifacts as pattern/color adaptation and guessing.

4.2 Reference Set

Current mottling evaluation practise is based on experts opinion. Given a mottling sample, an expert is asked to put it into a certain predefined category,

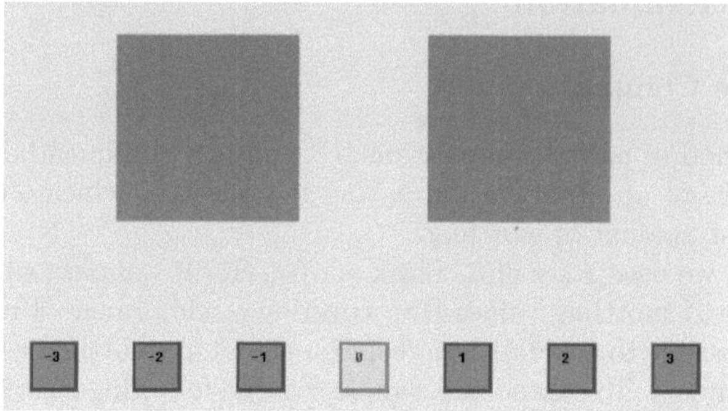

Fig. 2. Pairwise comparison test layout

based on the perceived mottling characteristics. The number of categories vary from a party involved (paper manufacturers, printing houses).

Grading a mottling sample requires a reference set, i.e., a set of samples with different levels of mottling. This set is constructed basing on several experts' experience and makes the evaluation procedure straightforward: an evaluated sample is placed among the reference samples and given a category of the closest reference sample.

For the performance testing of our method we used 10 reference sets, 5 for cyan at 50% (C50) and 5 for black at 50% (K50). C50 reference set consisted of 7 different classes of mottling, ranged from 0 to 6, and K50 sample set ranged from 1 to 6, making it 35 C50 samples and 30 K50 samples. Smaller values correspond to less uneven images.

Given reference samples were used in correlation computation, in method performance evaluation, and in separability checks.

5 Parameter Estimation

The architecture of the method presented in Section 3 is relatively simple. However it contains a number of parameters that need to be estimated carefully for better performance.

For a feasible and relatively simple analysis we will use so called correlation images [19]. Consider a pairwise test, described in Section 4.1. Denote i^{th} pair of images as $f_i(x, y)$ and $g_i(x, y)$. Corresponding pair of Fourier magnitudes will be $|F_i(u, v)|$ and $|G_i(u, v)|$. Difference in mottling values for this, marked by an observer will be H_i. If our model is consistent, then (simplified case)

$$\begin{aligned} \int_{u,v} |F_i(u,v)| C(u,v) dudv - \int_{u,v} |G_i(u,v)| C(u,v) dudv \sim H_i \\ \int_{u,v} (|F_i(u,v)| - |G_i(u,v)|) C(u,v) dudv \sim H_i \end{aligned} \tag{4}$$

and in the discrete case:

$$\sum_{u,v} (|F_i(u,v)| - |G_i(u,v)|) C(u,v) \sim H_i \tag{5}$$

If we make an assumption about the cross frequency independence, i.e., certain frequency component of an image can not be estimated with other frequency components, then the amount of information introduced by each frequency point to the perceived mottling should correspond to the contrast sensitivity at this point. We measure this information by using correlation image $R_H(u,v)$ for the perceived difference H (N stands for the number of pairs):

$$R_H(u,v) = corr(\{|F_i(u,v)| - |G_i(u,v)|\}_1^N, \{H_i\}_1^N) \tag{6}$$

In our case, since the amount of information introduced by each of the planes L^*, a^* and b^* to the perception model is assumed to be equal (color space uniformity), we assume the following:

$$\begin{array}{c} R_H(u_L, v_L) \sim C(u_L, v_L) \\ R_H(u_a, v_a) \sim C(u_a, v_a) \\ R_H(u_b, v_b) \sim C(u_b, v_b) \end{array} \tag{7}$$

which means that correlation images of each color plane should model the correlation function to some extent. In Fig. 3 it can be noticed that correlation planes yield similar information in terms of correlation of certain frequency bands with perceived mottling. In can be also seen that L^* correlation plane has more

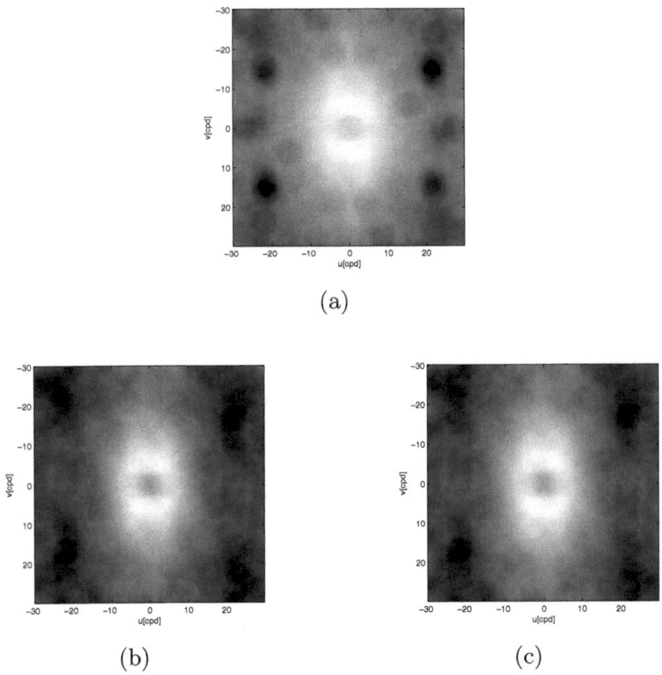

(a) (b) (c)

Fig. 3. Correlation planes: (a) L^* plane; (b) a^* plane; (c) b^* plane

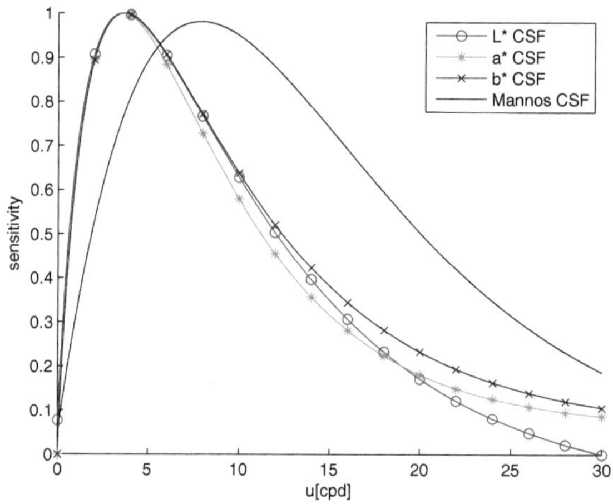

Fig. 4. Contrast sensitivity function models

information and this is probably due to linear nature of random stimuli generation. However the relative energy distribution is similar to those in a^* and b^* planes.

Consequently, if we modify Mannos CSF to follow the correlation planes, we can derive for each plane separate CSF by fitting the curve parameters. The derived functions are presented in Fig.4. It can be concluded, that the pattern-color separable model works well in this case of stimuli, because the derived sensitivity curves are similar.

6 Results and Conclusion

We have computed correlations of the methods' performances in the case of pairwise test and reference set. The numbers are comprised in the Table 1. It is clear that pattern-color separable method outperforms the gray scale one because of the simple reason: it includes information from the a^* and b^* planes in a way relevant to HVS.

Extremely good performance on the reference sets can be explained by a big relative mottling scale distance between samples, and a 100% agreement between experts on a samples' classification. However, from Fig. 5 it can be seen that in case of gray scale method, mottling classes 3 and 4 are not quite clearly separated.

The overall performance of the pattern-color separable method can be described as extremely promising, though there is still a room for improvements. Used CSF formulations can be replaced by the most recently developed ones [10]. Also, the perceptual uniformity of the 1976 *CIE* $L^*a^*b^*$ color space is questionable [20].

Table 1. Mottling assessment correlations

Methods	Laymen (1100 pairs)	Experts (300 pairs)	Overall (1400 pairs)	Reference
Gray scale	0.73	0.77	0.75	0.98
Pattern-Color	0.82	0.87	0.85	0.99

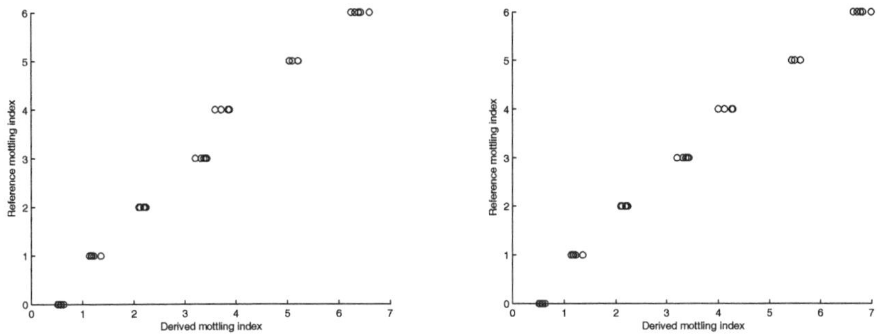

(a) (b)

Fig. 5. Reference mottling indexes versus derived with: (a) Gray scale method; (b) Pattern-Color separable method

Acknowledgments

This work was done as a part of Papvision project funded by European Union, National Technology Agency of Finland (TEKES Projects No. 70049/03, 70056/04, 40483/05), Academy of Finland (Project No. 204708), and Research Foundation of Lappeenranta University of Technology.

References

1. ISO/IEC: 13660:2001(e) standard. information technology - office equipment - measurement of image quality attributes for hardcopy output - binary monochrome text and graphic images. ISO/IEC (2001)
2. Briggs, J., Forrest, D., Klein, A., Tse, M.K.: Living with ISO-13660: Pleasures and perils. In: IS&Ts NIP 15: 1999 International Conference on Digital Printing Technologies, IS&T, Springfield VA, pp. 421–425 (1999)
3. Wolin, D.: Enhanced mottle measurement. In: PICS 2002: IS&T's PICS conference, IS&T pp. 148–151 (2002)
4. Armel, D., Wise, J.: An analytic method for quantifying mottle - part 1. Flexo pp. 70–79 December 1998 (1998)
5. Armel, D., Wise, J.: An analytic method for quantifying mottle - part 2. Flexo, January 1999, pp. 38–43 (1999)
6. Streckel, B., Steuernagel, B., Falkenhagen, E., Jung, E.: Objective print quality measurements using a scanner and a digital camera. In: DPP 2003: IS&T International Conference on Digital Production Printing and Industrial Applications. pp. 145–147 (2003)

7. Johansson, P.Å.: Optical Homogeniety of Prints. PhD thesis, Kunglika Tekniska Högskolan, Stockholm (1999)
8. Rosenberger, R.R.: Stochastic frequency distribution analysis as applied to ink jet print mottle measurement. In: IS&Ts NIP 17: 2001 International Conference on Digital Printing Technologies, IS&T, Springfield VA, pp. 808–812 (2001)
9. Sadovnikov, A., Lensu, L., Kamarainen, J., Kalviainen, H.: Quantifed and perceived unevenness of solid printed areas. In: Xth Ibero-American Congress on Pattern Recognition, Havana, Cuba, pp. 710–719 (2005)
10. Barten, P.: Contrast Sensitivity of the Human Eye and its Effects on Image Quality. SPIE (1999)
11. Schade, O.H.: Optical and photoelectric analog of the eye. Journal of the Optical Society of America 46, 721–739 (1956)
12. Kang, H.R.: Digital Color Halftoning. SPIE IEEE Press, New York (1999)
13. Campbell, F.W., Carpenter, R.H.S., Levinson, J.Z.: Visibility of aperiodic patterns compared with that of sinusoidal gratings. Journal of Physiology pp. 204 283–298
14. Palmer, S.: Vision Science: Photons to Phenomenology, 1st edn. MIT Press, Cambridge (1999)
15. Mannos, J., Sakrison, D.: The effects of a visual fidelity criterion on the encoding of images. IEEE Transactions on Information Theory 20(4), 525–536 (1974)
16. Poirson, A., Wandell, B.: Pattern-color separable pathways predict sensitivity to simple colored patterns. Vision Research 36, 515–526 (1996)
17. Engeldrum, P.G.: Psychometric Scaling: A Toolkit For Imaging Systems Development. 1st edn. Imcotec Press (2000)
18. Sadovnikov, A., Salmela, P., Lensu, L., Kamarainen, J., Kalviainen, H.: Mottling assessment of solid printed areas and its correlation to perceived uniformity. In: 14th Scandinavian Conference of Image Processing, Joensuu, Finland, pp. 411–418 (2005)
19. Sadovnikov, A., Lensu, L., Kämäräinen, J., Kälviäinen, H.: Model based generation of random stimuli and parameter analysis for print unevenness evaluation. In: Proceedings of the 29th European Conference on Visual Perception, St. Petersburg, Russia, pp. 49–50 (2006)
20. Johnson, G., Fairchild, M.A.: A top down description of S-CIELAB and CIEDE2000. Color Research and Application 28(6), 425–435 (2003)

Image Based Measurements of Single Cell mtDNA Mutation Load

Amin Allalou[1], Frans M. van de Rijke[2], Roos Jahangir Tafrechi[2], Anton K. Raap[2], and Carolina Wählby[1]

[1] Centre for Image Analysis, Uppsala University, Sweden
[2] Department of Molecular Cell Biology, Leiden University Medical Center, The Netherlands

Abstract. Cell cultures as well as cells in tissue always display a certain degree of variability, and measurements based on cell averages will miss important information contained in a heterogeneous population. This paper presents automated methods for image based measurements of mitochondiral DNA (mtDNA) mutations in individual cells. The mitochondria are present in the cell's cytoplasm, and each cytoplasm has to be delineated. Three different methods for segmentation of cytoplasms are compared and it is shown that automated cytoplasmic delineation can be performed 30 times faster than manual delineation, with an accuracy as high as 87%. The final image based measurements of mitochondrial mutation load are also compared to, and show high agreement with, measurements made using biochemical techniques.

Keywords: single cell analysis, cytoplasm segmentation, mitochondrial DNA, image cytometry.

1 Introduction

Great improvements in microscopy hardware have made it possible to produce thousands of high resolution cell images in a short period of time. It has led to a great demand for high-throughput automated cell image analysis.

Image based analysis of microscopy data and cell segmentation is not new [1]. The interest for high- throughput image based techniques is however growing fast, an it has been shown in a recent study [9] that relatively simple methods for nuclear and cytoplasmic segmentation combined with specific stains for a number of target molecules can reveal dose-dependent phenotypic effects of drugs in human cell cultures, providing information useful for discovering the mechanisms and predicting the toxicity of new drugs.

An image of a cell culture often contains different cells that all possess different characteristics. Taking the average over such an image will not reveal the biologically important differences and variations between the cells. Single cell analysis is clearly the only option to observe the dissimilarities between the cells.

To be able to assign a signal to a particular cell it has to be delineated. A common approach is to use a cytoplasm staining as a guide in delineation of the cell

B.K. Ersbøll and K.S. Pedersen (Eds.): SCIA 2007, LNCS 4522, pp. 631–640, 2007.

at its cytoplasmic borders [6], [13]. Another approach is to use a membrane stain, which binds to the cytoplasmic surface. In combination with nuclear staining individual cells can then be delineated by a combination of gradient curvature flow techniques and seeded watershed segmentation [7]. In many studies a blue stain is used for the nucleus, and red and green stain for molecular detection. Due to fluorescence spectral overlap this may limit the possibility of using a unique color for a stain that helps segmenting the cytoplasm. Also, a cytoplasmic stain may not be compatible with the molecular stain of interest.

Mitochondrial DNA (mtDNA) is a small extra-nuclear genome, present in 100s to 1000s of copies per mammalian cell. The genetic information contained in the ~16kbp human mtDNA is essential for a major energy-generating process of the cell called oxidative phosphorylation. All DNA mutates and so does mtDNA. When the mutation is pathogenic it needs to accumulate to relative large amounts (>80% of all mtDNAs) for the cell's energy provision to become so subverted that cell functions are lost and cells die. Such mutation accumulation leads to devastating diseases if the mutation is inherited from the mother or to normal aging phenomena if it is acquired somatically. A major factor in determining cellular mutation loads is the process of mitotic segregation. To understand mtDNA segregation and with it mtDNA mutation accumulation, our research focuses on mtDNA segregation patterns in *in vitro* cultured cells. Experimentally, this requires the determination of the mutation load in hundreds of individual cells in multiple serial cell culture passages of a cloned heteroplasmic founder cell (i.e., a single cell carrying mutant and wildtype-mtDNA molecules). *In situ* genotyping mtDNA with the padlock/rolling circle method [5] provides an elegant approach for detection of mtDNA sequences variants at the (sub-) cellular level. However, in our experience thusfar no cytoplasmic or membrane staining proved compatible with the padlock/rolling circle method. One way to approximate the outline of the cytoplasm is using a fixed radius for each cell [3]. A fixed radius may not always be the best choice since cells are often not spherical. Thus, to analyze thousands of cells, the challenge is to develop for this application a cell segmentation in absence of a cytoplasmic stain. Here we describe development of such an automated image cytometric procedure for fully automated measurements of mtDNA mutation loads of single cells. Preliminary results show that it greatly facilitates the determination of the mutant mtDNA fraction of heteroplasmic cells stained for the wild-type and mutant locus at position 3243 of human mtDNA.

2 Materials and Methods

2.1 Cell Preparation and Image Acquisition

The nuclei are stained with DAPI (blue). Padlock probes for mutated mtDNA are detected with Cy5 stain (red) and padlock probes for wild-type DNA are detected with FITC stain (green). For visualization of the cytoplasm tubulin is detected with mouse anti-tubulin antibodies and two different secondary antibodies, rabbit anti-mouse FITC (green) and goat anti-mouse Alexa 594 (red). Cytoplasmic stains can not be used together with padlock probes detected with the same color due to spectral overlap. Images are acquired using an epiflourescent microscope (Leica, Leica Microsystems GmbH, Wetzlar, Germany) equipped with a cooled monochrome CCD camera (Quantix, Photometrix, Melbourne, Australia).

2.2 Delineation of Nuclei

The cell segmentation is initiated by a segmentation of the image channel representing the nuclear stain (Fig. 1A). Otsu's method of thresholding, which minimizes the variance of the foreground and the background, separates the nuclei from the background [8] (Fig. 1B). Sometimes, dark areas inside the nuclei appear as holes after the threshold. These holes are filled using a flood fill algorithm.

Clustered nuclei are separated by watershed segmentation [4]. The watershed segmentation can be understood as seeing the image as a landscape. The gray level intensity represents the differences in elevation. Water enters through the local minima and starts to rise. A lake around a local minimum is created and referred to as catchment basin. The rising of the water stops when it reaches a pixel at the same geodesic distance from two different catchment basins. As two catchment basins meet they form a dam or watershed that separates the two objects. The implementation of the watershed can be done with sorted pixel lists, therefore the segmentation can be done very fast [12]. For our version of the watershed segmentation water raises from the maxima, i.e., the local maxima are the seeds. Water rises and floods until two catchment basins meet and generate a watershed. Given that flooding only starts from the seeds, every seed will produce one object. Incorporated into the watershed is an area count of each object label. This count is used for removing objects that are smaller than a user defined area minimum.

The binary image representing the nuclei is transformed to a landscape-like image using distance transformation (Fig. 1C). The distance image is produced using the 5-7-11-chamfer distance transform on the binary image [2]. The chamfer distance transform was preferred over the Euclidian distance transform due to lower computational cost and yet sufficient result.

Seeds that represent the different nuclei are needed in order to separate clustered nuclei into different objects. Due to imperfect circularity of the nuclei distance transform may lead to multiple seeding points or local maxima, for the same nucleus. This will result in over-segmentation. The h-maxima transform is able to suppress maxima whose depth is smaller than a given threshold t[10]. A low value of t in contrast to a high t value will result in more seed points, hence more over segmentation. By suppressing all small maxima several adjacent local maxima are merged into one regional maximum, i.e., one seed point for each nucleus is achieved. The value t is directly proportional to the radius of an object and can therefore also be used to remove objects that are too small to be true cell nuclei, and thus have a radius less than a specified value. The result of the watershed segmentation is shown in Fig. 1D.

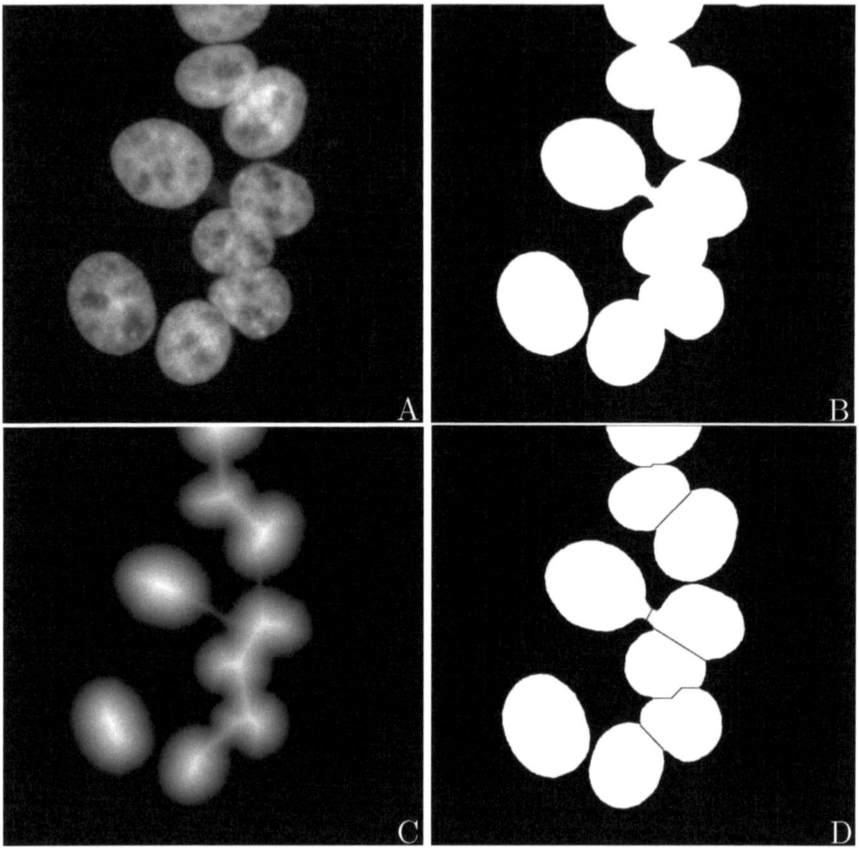

Fig. 1. **A:** Nuclear stain. **B:** Resulting binary image after thresholding. **C:** Distance transformation of **B**. **D:** Final segmentation result.

2.3 Delineation of Cytoplasms

Approach 1: no cytoplasmic stain (*NCS*). If no cytoplasm staining is present the delineation of the cytoplasm is purely based on a fixed distances from the nucleus. A distance transform is applied to the background of the binary image of the nuclei. This results in an image that represents the distance to the nearest nuclei for each pixel. A user defined threshold, corresponding to the maximum radius of the cytoplasm, is applied to the distance transformed background. A watershed is again used to define the borders of the objects. In order to be able to use the same watershed as previously, i.e., with water rising from image maxima, the distance transformed image is inverted. Water rises from the maxima in the image and rises until water from two catchment basins meet and a watershed line, separating two cytoplasms, is formed, see result in Fig. 2A.

Approach 2: with cytoplasmic stain (*CS*). The second approach to delineation of the cytoplasm makes use of a cytoplasm staining (tubulin stain). Tubulin is present throughout the whole cytoplasm and can therefore be used as a marker for the cytoplasm. A variance filter is applied to the channel representing the tubulin and areas of high intensity variation (tubulin areas) are enhanced. Thereafter, an average filter is applied to smooth the variance. The smoothed image is thresholded by Otsu‘s method, but to include all of the cytoplasm the threshold is adjusted by multiplying it with 0.25. This may be avoided by using a different thresholding method. A watershed transformation seeded by the nuclei and restricted to the binary image of the cytoplasm is there after applied, see result in Fig. 2B.

Approach 3: manual delineation (*O*). The third method for delineation of the cytoplasm is a manual segmentation. Here, two observers used the software Visiopharm Integrator System (VIS, Visiopharm, Hørsholm, Denmark) to outline the tubulin stained images manually.

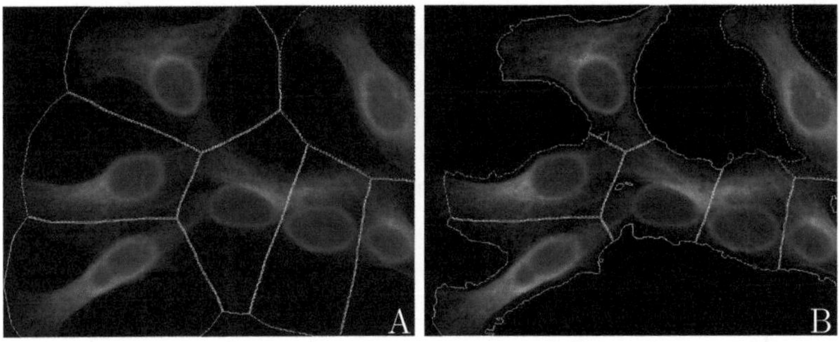

Fig. 2. A: Result of cytoplasmic segmentation not making use of the cytoplasmic stain (*NCS*). **B:** Result when cytoplasmic stain is included (*CS*).

2.4 Localization of Padlock Signals

The image channels containing the padlock signals are filtered with a 5x5 kernel that enhances areas of local maxima. A varying background can be a problem in some of the images, and background reduction by subtraction of an average filtered image is applied prior to signal detection. The average filtration is done three times with a 7x7 mean filter. The signals are separated from the image background by a user defined threshold set to a default value that localizes the major proportion of the signals. However, a lower threshold may be used in images with less background noise. The same threshold was used for all images, and in order to evaluate the influence of the threshold on the final measure of mutation load the thresholds were increased or decrease by 20%. The variation caused by these changes is shown as error bars in Fig. 4-Left. The binary image representing the signals is further reduced to single pixels by distance transformation and detection of local maxima. Thus, each signal event is represented by a single pixel.

2.5 Comparison of Segmentation Methods

In the evaluation of the different methods for cytoplasm segmentation, accuracy (agreement with truth), precision (reproducibility) and efficiency (time) are considered, based on the ideas by Udupa et. al. [11]. Define S as being the result of the segmentation method being compared to S_t, the true segmentation. The accuracy makes use of three definitions; False Negative Area Fraction (FNAF), False Positive Area Fraction (FPAF) and True Positive Area Fraction (TPAF). FNAF is the fraction of S_t that was missed by S. FPAF denotes the area that is falsely identified by S as a fraction S_t. In the current case, the parts of the S that overlap with the image background, as defined by S_t, are not counted as falsely identified because the background does not give rise to any signals and will not affect the calculation of signals per cell. TPAF describes the total amount of cytoplasm defined by S that coincides with S_t as a fraction of S_t.

Precision is the ability to reproduce the same result. Naturally, a fully automated method will always reproduce the same result. At manual delineation, the result will most likely not be fully reproducible, we will have inter- and intra-observer variation.

Two factors must be considered when comparing the efficiency of a segmentation method; the computational time and the human operator time required to complete the segmentation.

3 Results

The results consist of two parts; the first part is a comparison between three different methods of delineating cytoplasms. In the second part, image based measurements of single cell mutation load are compared to measurements based on single cell PCR-RFLP (Polymer Chain Reaction-Restriction Fragment Length Polymorphism), a biochemical method that measures mutation load in single cells by quantifying DNA-fragment length variation This comparison was performed to validate the image based method of analysis.

3.1 Comparison of Segmentation Methods

For a full comparison of segmentation methods, accuracy, precision, and efficiency should be considered. The comparative study of methods for cytoplasm segmentation was performed on 9 images containing a total of 56 cells. Two fully automated image based segmentation methods, one using information from a cytoplasmic stain (referred to CS), and which does not make use of a cytoplasmic stain (referred to as NCS) were compared to each other and to manual segmentation (referred to as O) of the same cytoplasms. Both automated methods are seeded by the same image of the cell nuclei, and the same threshold t for the h-maxima transform (the only input parameter) was used in all images. As no gold standard or ground truth is possible to produce, it is assumed that the manual segmentation method (O) results in the true delineation, defined as S_t^O. Manual segmentation was performed three times by two different persons to provide measurements of precision (reproducibility) in terms of inter- and intra-observer variability (referred to as O_{1a}, O_{1b} and O_2). The results can be seen in Table 1.

First of all, considering the accuracy, *NCS* and *CS* is significantly (α=0.05) less accurate than O. Between *CS* and *NCS* no significant (α= 0.05) difference can be seen in terms of accuracy. Furthermore, method O has noticeably lower precision than the other methods, as the computer based methods will reproduce the same result if re-run on the same image data, i.e., 100% precision, while manual segmentation varies both between observers (inter-observer precision is 79%) and for the same observer assessing the data at different times (intra-observer precision is 84%). Finally, the efficiency of *NCS* and *CS* is approximately 30 times higher than that of the manual segmentation O when using a 2.53 GHz Intel Pentium 4 processor.

Table 1. Comparison of segmentation methods

Method	Accuracy				Precision	Efficiency
	vs.	TPAF	FNAF	FPAF	(%)	Cells/min
NCS	O_{1a}	0.87±0.03	0.14±0.03	0.12±0.04	100	30
CS	O_{1a}	0.85±0.03	0.16±0.03	0.11±0.03	100	30
O_2	O_{1a}	0.84±0.02	0.16±0.02	0.02±0.01	79	1
O_{1b}	O_{1a}	0.90±0.02	0.10±0.02	0.03±0.01	84	1

3.2 Image Based Measurement of Mutation Load vs. PCR-RFLP

Mutation load is the proportion of mutated mtDNA (number of red padlock signals) compared to wild type mtDNA (number of green padlock signals) per cell. The image based analysis was first performed on a padlock probed co-culture, meaning that cells with 100% wild type mtDNA were mixed and cultured together with cells having 100% mutated mtDNA, see Fig. 3. This data set consisted of 29 images containing a total of 178 cells. A histogram of mutation load per cell measured from image data is shown in Fig. 4,left. The data from the co-culture shows distinct distributions at the extremes, i.e., cells with 100% and 0% mutation load. The automated analysis performed very well considering that hardly any intermediate levels were found. A large amount of intermediate levels would have been an indication of a high degree of error in the analysis method. Second, an analysis of a padlock probed culture (G55) of cells with a ~50% mtDNA mutation load was made on 66 cells in 10 images. As predicted, the analysis from G55 has a clear peak close to 50% mutation load (Fig. 4,left).

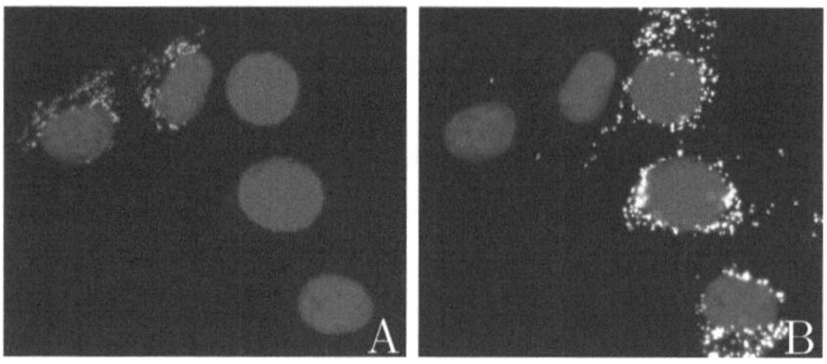

Fig. 3. Two views of the same image of co-cultured cells where padlock probes are seen as small spots and cell nuclei are shown in darker gray. **A:** Image channel R and B, showing padlock probes against mutated DNA. **B:** Image channel G and B, showing padlock probes against wild type DNA. In this data set, cells should either be 100% mutant or 100% wild type.

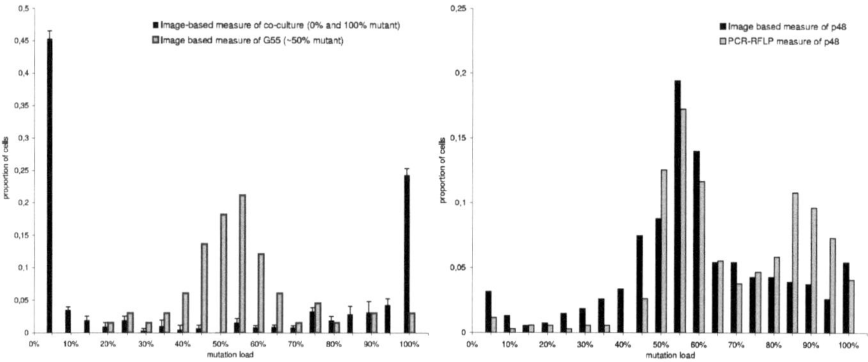

Fig. 4. Left: Histogram of proportion of cell population against mutation load as achieved by image based measures. Error bars show variation caused when varying the threshold for signal detection. **Right:** Histogram of image and PCR-RFLP based measure of mutation load in cells from passage 48 of a clone known to be heteroplasmic for the A3243G mtDNA mutation.

To study the segregation of mtDNA (originating from a heteroplasmic founder cell) different passages in the progression must be analyzed. Mutation loads of single cells from passage 48 of a heteroplasmic clone were measured by the padlock probed image based analysis and compared with a PCR-RFLP based analysis of the same passages. The image analysis was done on 536 cells in 58 images.

4 Conclusions

Comparison of methods for cytoplasmic segmentation show that presence of a cytoplasmic stain does not result in a significant increase in accuracy. This may seem strange as a cytoplasmic stain will guide the segmentation mask to the true edges of the cytoplasm. However, in the presented analysis, inclusion of parts of the image background does not affect the measurement of mutation load, as no signals are present in the background. Therefore, we have chosen not to count the inclusion of background as part of the false positive area fraction (FPAF). For other applications, e.g., if cytoplasmic area is to be measured, a segmentation method making use of the information from a cytoplasmic stain may be necessary. It is also worth mentioning that the agreement between manual cytoplasm segmentation and either of the fully automated methods is about the same as the agreement between manual cytoplasm segmentation performed by two different persons.

The automated method not including a cytoplasmic stain turned out to be a sufficiently accurate and fast method for analysis of single cell mutation load. The fact that no cytoplsmic stain was included also allows the use of two different colors for mutant and wild type mtDNA without problems with overlapping fluorescence spectra.

The image based measurements of mutation load show good agreement with measurement made by PCR-RFLP. In combination with automated image acquisition and batch processing of image files, the presented methods opens the possibility of high-throughput analysis of large numbers of cells with little human interaction or observer bias. This is also the next step to take within this project. Compared to other methods for single cell analysis, such as single cell PCR or flow cytometry, image based cytometry always has the option of returning to the original image data for visual inspection in cases of suspicious outliers.

Acknowledgments

This project was funded by the EU-Strep project ENLIGHT (ENhanced LIGase based Histochemical Techniques). The authors would also like to thank Chatarina Larsson and Mats Nilsson at the Department of Genetics and Pathology, Uppsala University, Sweden, for advice and help with the padlock probing techniques, George Janssen and Marchien van de Sande at the Department of Molecular Cell Biology of the Leiden University Medicine Center for their intellectual and experimental contributions in segregation analysis, and the staff at Visiopharm, Hørsholm, Denmark, for providing a free users license and help with integration of new functionality in VIS.

References

1. Bengtsson, E., Nordin, B., Gombrich, P., Domanik, R.: Mapping the cellular contents of pap smears. Analytical and Quantitative Cytology and Histology 18(1), 49–50 (1996)
2. Borgefors, G.: Distance transformations in digital images. Computer Vision, Graphics and Image Processing 34, 344–371 (1986)
3. Carpenter, A.E., Jones, T.R., Lamprecht, M.R., Clarke, C., Kang, I.H., Friman, O., Guertin, D.A., Chang, J.H., Lindquist, R.A., Moffat, J., Golland, P., Sabatini, D.M.: Cellprofiler: image analysis software for identifying and quantifying cell phenotypes. Genome Biology, 7(10) (2006)

4. Lantuéjoul, C., Beucher, S.: On the use of geodesic metric in image analysis. Journal of Microscopy 121, 39–49 (1981)
5. Larsson, C., Koch, J., Nygren, A., Janssen, G., Raap, A.K., Landegren, U., Nilsson, M.: In situ genotyping individual DNA molecules by target- primed rolling- circle amplification of padlock probes. Nature Methods 1, 227–232 (2004)
6. Lindblad, J., Wählby, C., Bengtsson, E., Zaltsman, A.: Image analysis for automatic segmentation of cytoplasms and classification of Rac1 activation. Cytometry 57(1), 22–33 (2004)
7. de Solorzano, C.O., Malladi, R., Lelievre, S.A., Lockett, S.J.: Segmentation of nuclei and cells using membrane related protein markers. Journal of Microscopy 201(3), 404–415 (2001)
8. Otsu, N.: A threshold selection method from gray-level histograms. IEEE Trans. on System Man. and Cybernetics 9(1), 62–69 (1979)
9. Perlman, Z.E., Slack, M.D., Feng, Y., Mitchison, T.J., Wu, L.F., Altschuler, S.J.: Multidimensional drug profiling by automated microscopy. Science 306, 1194–1198 (2004)
10. Soille, P.: Morphological Image Analysis: Principles and Applications. Springer, Heidelberg (1999)
11. Udupa, J.K., LeBlanc, V.R., Zhuge, Y., Imielinska, C., Schmidt, H., Currie, L.M., Hirsch, B.E., Woodburn, J.: A framework for evaluating image segmentation algorithms. Computerized Medical Imaging and Graphics 30, 75–87 (2006)
12. Vincent, L., Soille, P.: Watersheds in digital spaces: An efficient algorithm based on immersion simulations. IEEE Transactions on Pattern Analysis and Machine Intelligence 13(6), 583–597 (1991)
13. Wählby, C., Lindblad, J., Vondrus, M., Bengtsson, E., Björkesten, L.: Algorithms for cytoplasm segmentation of fluorescence labelled cells. Analytical Cellular Pathology 24, 101–111 (2002)

A PCA-Based Technique to Detect Moving Objects

Nicolas Verbeke and Nicole Vincent

Laboratoire CRIP5-SIP, Université René Descartes Paris 5, 45 rue des Saints-Pères, 75270 Paris Cedex 06, France
{nicolas.verbeke,nicole.vincent}@math-info.univ-paris5.fr
http://www.sip-crip5.org

Abstract. Moving objects detection is a crucial step for video surveillance systems. The segmentation performed by motion detection algorithms is often noisy, which makes it hard to distinguish between relevant motion and noise motion. This article describes a new approach to make such a distinction using principal component analysis (PCA), a technique not commonly used in this domain. We consider a ten-frame subsequence, where each frame is associated with one dimension of the feature space, and we apply PCA to map data in a lower-dimensional space where points picturing coherent motion are close to each other. Frames are then split into blocks that we project in this new space. Inertia ellipsoids of the projected blocks allow us to qualify the motion occurring within the blocks. The results obtained are encouraging since we get very few false positives and a satisfying number of connected components in comparison to other tested algorithms.

Keywords: Data analysis, motion detection, principal component analysis, video sequence analysis, video surveillance.

1 Introduction

Digital video cameras available on the market are less and less expensive and more and more compact. At the same time, nowadays the computation power of computers enables us to consider real-time processing of video sequences in a serious way. That is why industrialists now tend to choose vision-based solutions to solve problems, for which, a few years back, they would have chosen another type of solution, such as human surveillance or more mechanical sensors. Video sequences thus obtained are three-dimensional data (two spatial dimensions and one temporal dimension) and may be considered as 2D+T volumes. Various issues are encountered, but the first task of a video sequence analysis system is always motion detection, and if possible, moving objects detection (segmentation). This task can be more or less difficult depending on the light conditions and the expected processing speed and accuracy. A detailed list of problems related to light conditions and to the scene content can be found in [1]. In this paper we will address the case of a static video camera.

B.K. Ersbøll and K.S. Pedersen (Eds.): SCIA 2007, LNCS 4522, pp. 641–650, 2007.

Most motion detection algorithms in the literature are described as background subtraction methods. A survey of recent advances can be found in [2]. The process is to build a background model, then to apply a decision function over each new frame in order to label each point as background or foreground. In other words, video data is not considered as a 2D+T volume but as a series of two-dimensional image pairs.

When the video is a set of frames indexed by time, the simplest background model is to consider frame $t-1$ as the background in frame t and to classify every pixel that changed significantly between t and $t-1$ as belonging to the foreground. In other words, motion detection is achieved by temporal derivation of the sequence. This derivative can be computed very quickly, but it is also very unstable because of its sensitiveness to any kind of noise. In addition, only short term past is considered so that slow or arrhythmic motions are ill-detected. Thus, temporal derivative is almost always post-processed, as in [3] where it is regularized with respect to the optical flow history at each point.

Instead of using frame $t-1$ as a background model, one may use a "reference image". Unfortunately such an image is not always available, and if it is, it becomes quickly out-of-date, especially in outdoor environment. That is why the authors who use this technique always offer a function to update the reference image. This method is called "background maintenance", as in [4] where the reference image is continuously updated when temporal derivative is negligible.

The background model is often a statistical model built to estimate the probability of a gray value or a color at a given point. If this probability is high the pixel is considered as belonging to the background, otherwise it must belong to an object. Sometimes the model is the parameter set of a distribution from a known family, as in [5] where the color values are modeled as Gaussian distributions. In other cases, no prior assumption is made about the distribution to be estimated, and a non-parametric estimation is achieved, as in [6]. The background model is then a set of past measurements used to punctually estimate probability densities thanks to a Parzen window or a kernel function. Background subtraction may also be viewed as a prediction problem. The most commonly used technique is Wiener [1] or Kalman [7] filtering.

Thus, most methods aim at estimating a background model to detect moving objects. Nonetheless, other works can be mentioned such as [8] where moving areas are those where spatiotemporal entropy of the sequence reaches a maximum. Unlike foregoing techniques, temporal dimension is fully used by a local analysis algorithm through the 2D+T video volume. In [9], this approach is slightly modified to compute the temporal derivative's entropy instead of the input sequence's entropy, in order to prevent the detection of spatial edges as moving areas. Our study lies in the same category: we aim at detecting moving object by analyzing the video volume with no explicit background model. We will first introduce the chosen feature space and then we will explain the criteria used to select moving objects. At last we will analyze the results of our experiments and evaluate them.

2 Feature Space

Initially we consider video data represented by a function defined in a three-dimensional space: two spatial dimensions (x, y) and a temporal one (t). With each point of this space is associated a gray value at point (x, y) and time t. So the semantic entities (background, moving objects) are subsets of points from this space. In order to identify them, we have to aggregate them into classes with respect to shared features. In such a space, the amount of points to consider is huge, that is why the background model-based approach is so commonly used: the only points to consider are those from current frame, while the background model is supposed to sum up all past observations. We would prefer to keep a knowledge of the past less synthetic because relevant information we have to extract is not always the same. Thus we need a feature space that fits better to the sequence itself, rather than to each frame, and that enables us to consider motion without modifying input data. With each point (x, y) in the image space, is associated a vector containing gray values at location (x, y) along the considered time interval. Moreover, in the prospect of using data analysis techniques, the sequence is not regarded as a function anymore but as a set of individuals, which are the pixels we observe when we watch the sequence. During this step, spatial relationships between pixels are therefore ignored. To avoid doing a fine analysis, we do not track objects anymore but we focus on a fixed location within the image. We will keep the evolution of gray values for each pixel along time. A dozen values may be kept (let p be that number), and each pixel becomes an individual characterized by a set of parameters. Individuals have p coordinates. As our algorithm processes p frames at a time, we can allow the computation to be p times slower than if we would have processed each frame individually; so, we can use more time-consuming techniques. Nevertheless, for the process to be fast enough, we have to reduce the amount of information. There are many dimension reduction techniques, such as principal component analysis (PCA), factor analysis (FA), the whole family of independent component analysis (ICA) methods, or neural network-based algorithms as Kohonen self-organizing maps. For an extended survey of dimension reduction techniques, one may refer to [10]. As PCA is known to be the best linear dimension reduction technique in terms of mean squared error, we chose this method to avoid losing information that best discriminates points.

PCA was developed in the early 19th century to analyze data from human science. It is a statistical technique to simplify a dataset representation by expressing it in a new coordinate system so that the greatest variance comes to lie on the first axis. Thus we can reduce the search space dimensionality by keeping the first few coordinates in the new frame of reference. A basis of this space consists of the eigenvectors of the covariance matrix of the data sorted in the decreasing order of the corresponding eigenvalues magnitude.

The coordinates of background pixels are likely to be more or less equal to each other, while the coordinates of moving objects pixels should vary. We want to detect this variation, it is therefore interesting to find the axis, that is, the good basis in the p-dimensional space where the factor's variance is maximum.

In the case of a video sequence, the data matrix $\mathbf{X}$ contains all the features of the points (x, y, t) to consider. From now on, let n be the number of rows of $\mathbf{X}$, that is the number of pixels in the image, and let p be its number of columns, that is the number of features associated with each pixel. The two first coordinates (x, y) can get a finite number of values; let $\mathcal{D}_P$ be the pixel domain. On the other hand, the time domain is assumed to be infinite. So we have to choose a range that should contain all relevant information. We decide to use $\mathcal{D}_t = \{t - \Delta t, \ldots, t\}$ as time domain, where t is current time. Each row of $\mathbf{X}$ is a data element corresponding to a pixel $(x, y) \in \mathcal{D}_P$, and each column of $\mathbf{X}$ is an observed variable, that is a gray value at time $\tau \in \mathcal{D}_t$. The new basis of the feature space is then associated with the eigenvectors of the data covariance matrix $\mathbf{C}$.

$$\mathbf{C} = \bar{\mathbf{X}}^T \cdot \mathbf{D}_p \cdot \bar{\mathbf{X}}, \tag{1}$$

where $\bar{\mathbf{X}}$ is the centered data matrix, and $\mathbf{D}_p = \frac{1}{p}\mathbf{I}_p$ ($\mathbf{I}_p$ being the p-order identity matrix.)

The method must be as invariant as possible towards the various acquisition conditions, therefore we will focus on gray variations rather than on gray values. Thus we can remove one dimension from the feature space by filling $\mathbf{X}$ with the temporal derivative rather than with the raw grayscale images. We have then a $(p-1)$-dimensional space. Let $\mathbf{Y}$ be the data matrix in this space.

Let us consider a 10-frame sub-sequence with 288 rows and 720 columns. Figure 1(a) pictures the first frame from this sub-sequence. Matrix $\mathbf{X}$ has therefore 288×720 rows and 10 columns, while $\mathbf{Y}$, over which we will perform PCA, has 288×720 rows and 9 columns. Figure 1 pictures the nine projections of $\mathbf{Y}$ on the principal axes given by the PCA algorithm. More precisely, we consider the image domain and we build an image where gray values are proportional to the values of the feature vectors projected on a given principal axis.

According to Fig. 1, moving areas are clearly shown when $\mathbf{Y}$ is projected on the two first principal axes. The difference between a static area and a moving area is emphasized on these axes. This observation is confirmed by the histogram of variance explained by the factors (Fig. 2). The variance explained by an axis is defined as the ratio between the eigenvalue associated with this axis and the sum of all eigenvalues of the covariance matrix.

Thus, if we choose to keep only the two first principal axes, 20% of the initial amount of data is enough to preserve 80% of the observed variance. This first experimentation confirms our approach, which remains very global. In next section, we will use this feature space. The process is then to compute a PCA for every set of ten successive frames. To achieve greater consistency, we will now consider a more local approach relying on this first global study.

3 Detection of Coherent Moving Areas

The data representation as shown on Fig. 1(b) enables to detect local motion (pixel motion) more easily. Indeed, selecting the pixels whose absolute

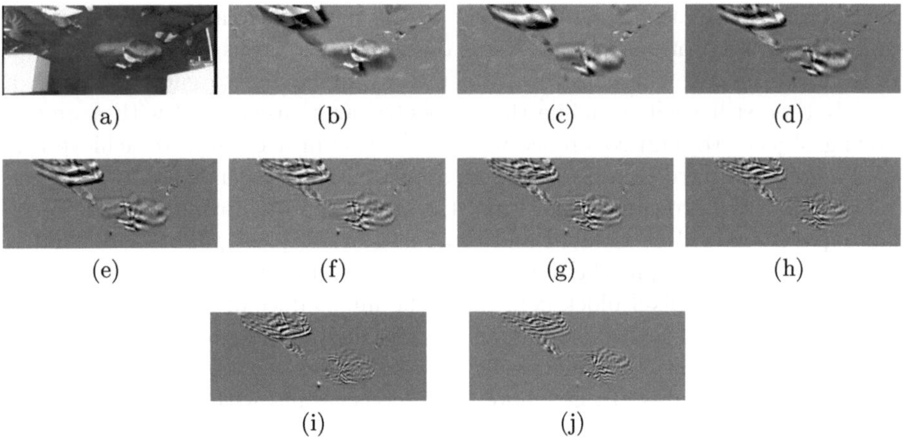

Fig. 1. (a) Input sequence. (b)—(j) Projections of $\mathbf{Y}$ on each of the nine principal axes outlined by PCA. Subfigures are ordered with respect to the rank of the associated factor.

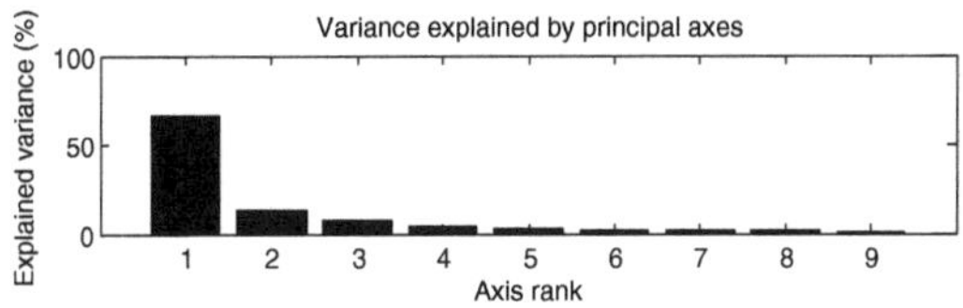

Fig. 2. Variance explained by the principal axes

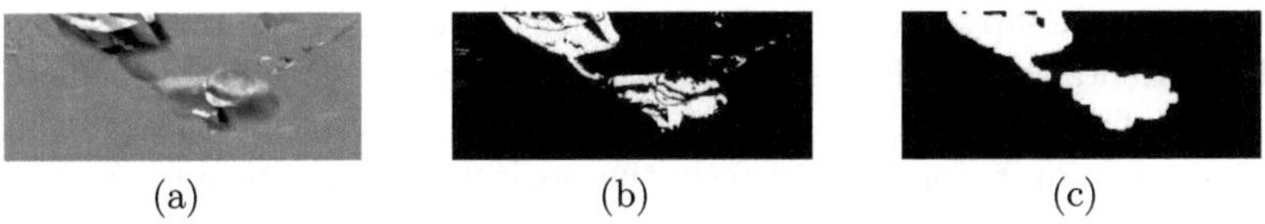

Fig. 3. (a) Projection of $\mathbf{Y}$ on the first principal axis, (b) background segmentation achieved from (a), (c) segmentation improved by morphological operations

value is high (the darkest and the lightest ones) is enough to get a moving objects/background segmentation. Figure 3(b) shows the automatic segmentation of the projection of $\mathbf{Y}$ on the first principal axis.

We find ourselves in the same situation as most methods present in the literature; such connected components of a binary picture would be labeled to obtain a moving object detection. Like in most cases, in Fig. 3, image post-processing would be necessary to remove the false positives and restore the connectivity of the objects (Fig. 3(c)). Such an approach provides an accurate segmentation, but choosing the morphological operations to carry out is often a delicate job. A mistake in choosing a structuring element could erase an important object, connect two different objects, validate a false positive, etc. A learning phase is necessary to adapt the general method to the particular case of the sequence studied. The need of post-processing is due to the global aspect of the method all over the image. Therefore, we prefer to avoid having to carry out such a step, but we still need to define the connected areas associated with a unique moving object. To gain coherence we have to lose in accuracy. To achieve this we are going to introduce a more local approach that nevertheless is based on the results of the previous study. From the global representation studied above, sub-populations will be isolated and compared in the pixel population. That is why we will start to split the data ($\mathbf{Y}$) into many subsets. Each subset is associated with a $b \times b$ pixel block defined by the nine values of the data matrix $\mathbf{Y}$ which constitute the values of the factors revealed in the global study. On the initial video volume three-dimensional blocks of size $b \times b \times 10$ are thus studied through 9 new features. To get more continuous results without increasing the computation time too much, we chose blocks that overlay in half along the space dimensions.

The subsets thus obtained are represented in a $(p-1)$-dimensional space. We will study the relative locations of those subsets. To simplify the computation, we will represent each subset by its inertia ellipsoid. Projections of the ellipsoids will be compared in the plane formed by the first two factors of the global representation.

4 Comparison of the Detected Areas

Figure 4 shows a set of inertia ellipsoids projected on the first factorial plane of the global image. Each corresponds to a spatiotemporal block as described in Sect. 3.

The observed ellipses differ by way of their location in the plane, their area and their orientation. As far as this study is concerned, we will not focus on the orientation of the ellipses. The data being centered, the frame of Fig. 4 has for origin the mean of $\mathbf{Y}$ (or more exactly the projection of the mean.) As a result, an ellipse that is far from the origin represents a block of which many points are in motion. The area of the ellipses gives a hint about the variability of the points in the block it represents. Thus, we can distinguish four cases:

1. A small ellipse close to the origin represents a block in which no motion occurs.
2. A large ellipse close to the origin represents a block in which the different points have dissimilar motions, but in which the mean of the motions is close to zero. In other words, we can speak of noise.
3. A small ellipse far from the origin represents a block in which the mean motion is important, and whose points have almost the same motion. They are blocks fully included in a moving object.
4. A large ellipse far from the origin represents a block in which the mean motion is important and whose points show various motions. They are blocks that could be found for example on the edge of a moving object.

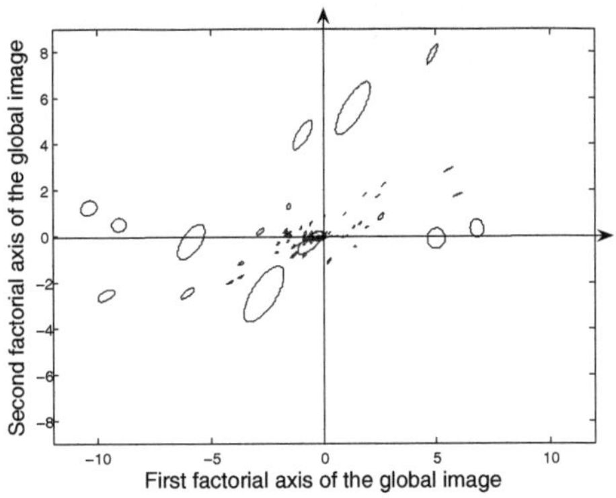

Fig. 4. Each three-dimensional block is modeled by the inertia ellipsoid of the points which constitute it, and each ellipsoid is projected on the plane formed by the two first factors from the global study (PCA)

To detect moving objects, the most interesting blocks are those corresponding to cases 3 and 4, in other words the ellipses far from the origin. Therefore, it is necessary to threshold with respect to the distance from the origin to the center of the ellipses, that is to say the mean or the sum of the points belonging to the corresponding blocks.

5 Results

The size of spatiotemporal blocks introduced in section 3 is still to be discussed. The edges of the detected moving objects could be not accurate enough if the blocks are too large, while blocks too small would imply greater computation time, and impair the objects connectivity. Figure 5 pictures the results obtained for the same ten-frame sequence, with blocks of size $b \times b \times 9$, where b equals successively 16, 32, 64 and 128. Besides we measured the computation time necessary to get these results, as well as their accuracy. Accuracy is given by the number of false positives and false negatives observed when the result is compared to an ideal segmentation. These measurements are put down in Table 1.

We notice that the computation time does not depend a lot on the blocks size. As a consequence it can be chosen depending only on the sequence to be analyzed, without worrying about computation time. In this case, the image size is 720×288 pixels and the blocks size providing the best results is 32×32.

To evaluate our algorithm, we use five video sequences which differ in the issue raised by the application and/or the intrinsic difficulties of the sequence. The first sequence represents a people counting application, where most people stand still for a long time in the field of vision before passing the monitored gate. It is then necessary that the algorithm does not detect insignificant motion. The second sequence is another people counting application, where people tend to move in connected groups. Therefore, the algorithm has to be accurate enough to

(a) (b) (c) (d) (e)

Fig. 5. Results obtained from one sequence by only changing the spatiotemporal blocks size. (a) Input sequence. (b) $b = 16$. (c) $b = 32$. (d) $b = 64$. (e) $b = 128$.

Table 1. Algorithm performances achieved with different block sizes

	$b = 16$	$b = 32$	$b = 64$	$b = 128$
Computation time (%)	100	97.4	87.8	63.9
False positives (pixels)	975	3375	10,372	28,445
False negatives (pixels)	10,851	6874	2149	506

discern the different members of each group. The third one is a vehicle monitoring application. The images are very noisy, due to the sun passing through the trees on the left side of the picture and the vehicles moving in the background. The two last sequences are classical benchmark sequences used in numerous articles[1]. They are used to facilitate the comparison of our results with other methods.

In Fig. 6 we can see the foreground motion detections achieved on those five video sequences thanks to four different algorithms. Row 2 shows the results obtained with the algorithm presented in this article; row 3 is the smoothed temporal derivative of the sequence (with a fixed smoothing coefficient); row 4 shows a non-parametric background subtraction [6]; row 5 is the difference-based spatiotemporal entropy of the sequence [9]. As methods 3 to 5 usually require a post-processing step before labeling connected components, we applied a morphological closing by a 5-pixel-diameter disk followed by an opening by the same structuring element to obtain results shown in rows 3 to 5.

With the smoothed temporal derivation method, the smoothing coefficient is a critical issue and has to be carefully chosen depending on the sequence to be analyzed. A coefficient too high produces a "ghost effect" (row 3, column 1), while a coefficient too low tends to detect only the edges and ignore the inside (row 3, columns 4 and 5).

The non-parametric background modeling algorithm (row 4) detects accurately the edges of the moving objects. Still, this method is very noise-sensitive and unless the post-processing is chosen very specifically contingent on the sequence, the results obtained do not constitute a good segmentation of moving objects.

Our method (row 2), as well as spatiotemporal entropy (row 5), both sacrifice the edges' accuracy for greater robustness. However, there is more noise with entropy than with the method presented here.

[1] They come from the EC funded CAVIAR project/IST 2001 37540, found at URL: `http://homepages.inf.ed.ac.uk/rbf/CAVIAR/`

(1)
(2)
(3)
(4)
(5)

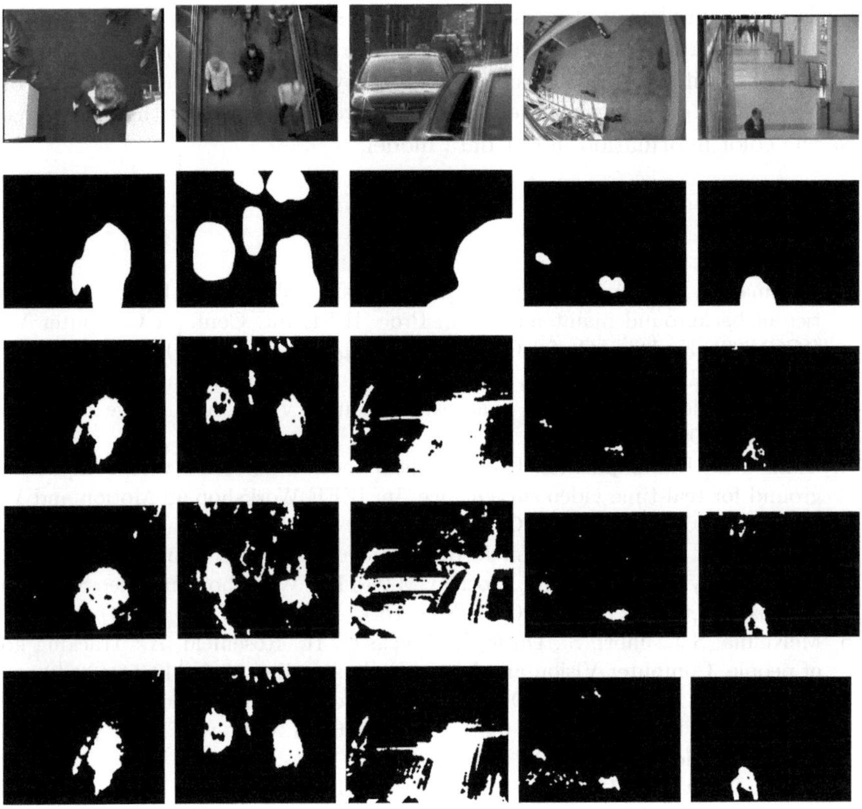

Fig. 6. Results obtained on 5 sequences with 4 algorithms: (1) Input sequence, (2) our algorithm, (3) smoothed temporal derivative, (4) non-parametric modeling, (5) difference-based spatiotemporal entropy

6 Conclusion

In this paper, we have presented a new coherent motion detection technique in a video sequence. Unlike most of the methods present in the literature, we do not aim at modeling the background of the scene to detect objects, but rather to select significant information and to express it in a lower-dimensional space, in which classifying motion areas and still areas is easier. To get this space, we carry out a principal components analysis on the input data, and we only keep the first two principal factors. Then the sequence is split into spatiotemporal blocks which are classified with respect to the location of their respective inertia ellipse in the first factorial plane. The results are satisfactory if we consider that the number of connected components matches the expected number of objects. However, the edges of the objects are less accurately detected than with statistical background modeling algorithms. Nevertheless, in the context of an industrial use of the method, edge accuracy is not a capital issue. It is far more important to know precisely the number of objects present in the scene as well as their approximate location and area. If very accurate edges are required, one may use an active contour [11] initialized on the contour given by our algorithm. Besides, in order to take better advantage of the input data, we plan to study the way we could use the color information in our data model.

References

1. Toyama, K., Krumm, J., Brummit, B., Meyers, B.: Wallflower: Principles and practice of background maintenance. In: Proc. IEEE Int. Conf. on Computer Vision (ICCV'99). Kerkyra, Corfu, Greece, vol. 1, pp. 255–261 (1999)
2. Moeslund, T.B., Hilton, A., Krüger, V.: A survey of advances in vision-based human motion capture and analysis. Computer Vision and Image Understanding 104(2), 90–126 (2006)
3. Tian, Y.L., Hampapur, A.: Robust salient motion detection with complex background for real-time video surveillance. In: IEEE Workshop on Motion and Video Computing. Breckenridge, CO, vol. II, pp. 30–35 (2005)
4. Yang, T., Li, S.Z., Pan, Q., Li, J.: Real-time and accurate segmentation of moving objects in dynamic scene. In: Proc. ACM 2nd Int. Workshop on Video Surveillance & Sensor Networks (VSSN 2004), New York, NY pp. 136–143 (2004)
5. McKenna, S.J., Jabri, S., Duric, Z., Wechsler, H., Rosenfeld, A.: Tracking goups of people. Computer Vision and Image Understanding 80(1), 42–56 (2000)
6. Elgammal, A.M., Harwood, D., Davis, L.S.: Non-parametric model for background subtraction. In: Vernon, D. (ed.) ECCV 2000. LNCS, vol. 1842, pp. 751–767. Springer, Heidelberg (2000)
7. Koller, D., Weber, J., Malik, J.: Robust multiple car tracking with occlusion reasoning. Technical Report UCB/CSD-93-780, University of California at Berkeley, EECS Department, Berkeley, CA (1993)
8. Ma, Y.F., Zhang, H.J.: Detecting motion object by spatio-temporal entropy. In: Proc. IEEE Int. Conf. on Multimedia and Expo (ICME 2001), Tokyo, Japan pp. 265–268 (2001)
9. Guo, J., Chng, E.S., Rajan, D.: Foreground motion detection by difference-based spatial temporal entropy image. In: Proc. IEEE Region 10 Conf. (TenCon 2004), Chiang Mai, Thailand pp. 379–382 (2004)
10. Fodor, I.K.: A survey of dimension reduction techniques. Report UCRL-ID-148494, Lawrence Livermore National Laboratory, Livermore, CA (2002)
11. Kass, M., Witkin, A., Terzopoulos, D.: Snakes: Active contour models. International Journal of Computer Vision 1(4), 321–331 (1988)

Page Frame Detection for Marginal Noise Removal from Scanned Documents

Faisal Shafait[1], Joost van Beusekom[2], Daniel Keysers[1], and Thomas M. Breuel[2]

[1] Image Understanding and Pattern Recognition (IUPR) research group
German Research Center for Artificial Intelligence (DFKI) GmbH
D-67663 Kaiserslautern, Germany
faisal@iupr.dfki.de, keysers@iupr.dfki.de

[2] Department of Computer Science, Technical University of Kaiserslautern
D-67663 Kaiserslautern, Germany
joost@iupr.dfki.de, tmb@informatik.uni-kl.de

Abstract. We describe and evaluate a method to robustly detect the page frame in document images, locating the actual page contents area and removing textual and non-textual noise along the page borders. We use a geometric matching algorithm to find the optimal page frame, which has the advantages of not assuming the existence of whitespace between noisy borders and actual page contents, and of giving a practical solution to the page frame detection problem without the need for parameter tuning. We define suitable performance measures and evaluate the algorithm on the UW-III database. The results show that the error rates are below 4% for each of the performance measures used. In addition, we demonstrate that the use of page frame detection reduces the optical character recognition (OCR) error rate by removing textual noise. Experiments using a commercial OCR system show that the error rate due to elements outside the page frame is reduced from 4.3% to 1.7% on the UW-III dataset.

1 Introduction

For a clean document, the page frame is defined as the rectangular region enclosing all the foreground pixels in the document image. When a page of a book is scanned or photocopied, textual noise (extraneous symbols from the neighboring page) and/or non-textual noise (black borders, speckles, ...) appears along the border of the document. The goal of page frame detection is to find the actual page, ignoring the noise along the page border. The importance of page frame detection in document image analysis is often underestimated, although a good page frame detection algorithm can help to improve the performance considerably. Since the state-of-the-art page segmentation algorithms report textual noise regions as text-zones [1], the OCR accuracy decreases in the presence of textual noise, because the OCR system usually outputs several extra characters in these regions. Including page frame detection as a document preprocessing step can thus help to increase OCR accuracy.

B.K. Ersbøll and K.S. Pedersen (Eds.): SCIA 2007, LNCS 4522, pp. 651–660, 2007.

The most common approach to eliminate marginal noise is to perform document cleaning by filtering out connected components based on their size and aspect ratio [2,3,4]. However, when characters from the adjacent page are also present, they usually cannot be filtered out using these features alone. Some approaches try to perform document cleaning as a part of layout analysis [5]. These approaches remove black borders and speckles resulting from photocopy effects with high accuracy but report a number of false alarms [1].

Instead of removing individual components, researchers have also tried to explicitly detect and remove the marginal noise. Le et al. [6] have proposed a rule-based algorithm using several heuristics for detecting the page borders. The algorithm relies upon the classification of blank/textual/non-textual rows and columns, object segmentation, and an analysis of projection profiles and crossing counts to detect the page frame. Their approach is based on the assumption that the page borders are very close to edges of images and borders are separated from image contents by a white space, i.e. the borders do not overlap the edges of an image content area. However, this assumption is often violated when pages from a thick book are scanned or photocopied. Avila et al. [7] and Fan et al. [8] propose methods for removing non-textual noise overlapping the page content area, but do not consider textual noise removal. Cinque et al. [9] propose a method for removing both textual and non-textual noise from greyscale images based on image statistics like horizontal/vertical difference vectors and row luminosities. However, their method is not suitable for binary images.

Here, we propose an algorithm to detect the page frame that can be used to remove both textual and non-textual noise from binary document images. The method does not assume the existence of whitespace between noisy borders and actual page contents, and can locate the page contents region even if the noise overlaps some regions of the page content area. Instead of trying to detect and remove the noisy borders, we focus on using geometric matching methods to detect the page frame in a document image. The use of geometric matching for solving such a problem has several advantages. Instead of devising carefully crafted rules, the problem is solved in a more general framework, thus allowing higher performance on a more diverse collection of documents.

2 Geometric Matching for Page Frame Detection

Connected components, textlines, and zones form different levels of page segmentation. We use a fast labeling algorithm to extract connected components from the document image. In recent comparative studies for the performance evaluation of page segmentation algorithms [1,10], it is shown that the constrained textline finding algorithm [3] has the lowest error rates among the compared algorithms for textline extraction, and the Voronoi-diagram based algorithm [5] has the lowest error rates for extracting zone-level information. Therefore, we use these two algorithms for extracting textlines and zones from the document image, respectively. After extracting connected components, textlines, and zones, the next step is to extract the page frame from the document image.

Page Frame Detection. Given the sets of connected components C, textlines L, and zones Z, we are interested in finding the best matching geometric primitives for the page frame with respect to the sets C, L, and Z. Since the bounding box of the page frame is an axis-aligned rectangle, it can be described by four parameters $\vartheta = \{l, t, r, b\}$ representing the left, top, right, and bottom coordinates respectively. We compute the best matching parameters for the page frame by finding the maximizing set of parameters

$$\hat{\vartheta}(C, L, Z) := \arg\max_{\vartheta \in T} Q(\vartheta, C, L, Z) \tag{1}$$

where $Q(\vartheta, C, L, Z)$ is the total quality for a given parameter set, and T is the total parameter space.

Design of the Quality Function. The design of an appropriate quality function is not trivial. We may define the page frame as a rectangle that touches many character bounding boxes on its four sides. The character bounding boxes are obtained from C by filtering out noise and non-text components based on their area and aspect ratio. However, this approach has some limitations:

1. The top and bottom lines do not necessarily touch more characters than other lines in the page (especially when there is only a page number in the header or footer). Also in some cases, there can be non-text zones (images, graphics, ...) at the top or bottom of the page. Hence the parameters t and b can not be reliably estimated using character level information.
2. Changes in text alignment (justified, left-aligned, etc) of a page may result in arbitrary changes in the estimated l and r parameters.

Instead of using connected component level information, we can use textlines. We may define the page frame as a rectangle that touches many line bounding boxes on its two sides, besides containing most of the textlines in the page. In order to search for optimal parameters, we decompose the parameters into two parts: $\vartheta_h = \{l, r\}$ and $\vartheta_v = \{t, b\}$. Although ϑ_h and ϑ_v are not independent, such a decomposition can still be used because of the nature of the problem. We first set parameters ϑ_v to their extreme values ($t = 0, b = H$ where H is the page height) and then search for optimal ϑ_h. This ensures that we do not lose any candidate textlines based on their vertical position in the image. The decomposition not only helps in reducing the dimensionality of the searched parameter space from four to two, but also prior estimates for ϑ_h make the estimation of ϑ_v a trivial task, as we will discuss below. Hence the optimization problem of Equation (1) is reduced to

$$\hat{\vartheta_h}(L) := \arg\max_{\vartheta_h \in T} Q(\vartheta_h, L) \tag{2}$$

We employ the RAST technique [11] to perform the maximization in Equation (2). RAST is a branch-and-bound algorithm that guarantees to find the globally optimal parameter set by recursively subdividing the parameter space and processing the resulting parameter hyper-rectangles in the order given by an upper bound on the total quality. The total upper bound of the quality Q can be written as the sum of local quality functions

$$Q(\vartheta_h, L) := \sum_j q(\vartheta_h, L_j) \tag{3}$$

We then compute an upper and lower bound for the local quality function q. Given a line bounding box (x_0, y_0, x_1, y_1), we determine intervals $d(l, x_i)$ and $d(r, x_i)$ of possible distances of the x_i from the parameter intervals l and r, respectively. The local quality function for a given line and a parameter range ϑ_h can then be defined as

$$q_1(\vartheta_h, (x_0, x_1)) = \max\left(0, 1 - \frac{d^2(l, x_0)}{\epsilon^2}\right) + \max\left(0, 1 - \frac{d^2(r, x_1)}{\epsilon^2}\right) \tag{4}$$

Where ϵ defines the distance up to which a line can contribute to the page frame. Textlines may have variations in their starting and ending positions within a text column depending on text alignment, paragraph indentation, etc. We use $\epsilon = 150$ pixels in order to cope with such variations. This quality function alone already works well for single column documents, but for multi-column documents it may report a single column (with the highest number of textlines) as the optimal solution. In order to discourage such solutions, we introduce a negative weighting for textlines on the 'wrong' side of the page frame in the form of the quality function

$$q_2(\vartheta_h, (x_0, x_1)) = -\max\left(0, 1 - \frac{d^2(l, x_1)}{(2\epsilon)^2}\right) - \max\left(0, 1 - \frac{d^2(r, x_0)}{(2\epsilon)^2}\right) \tag{5}$$

the overall local quality function is then defined as $q = q_1 + q_2$ It can be seen that this quality function will yield the optimal parameters for ϑ_h even if there are intermediate text-columns with larger number of textlines. However, if the first or last column contain very few textlines, the column is possibly ignored.

Parameter Refinement. After obtaining the optimal parameters for ϑ_h in terms of mean square error, we refine the parameters to adjust the page frame according to different text alignments and formatting styles as shown in Figure 1. Hence we obtain an initial estimate for the parameters ϑ_v by inspecting all lines that contribute positively to the quality function $Q(\vartheta_h, L)$ and setting t to the minimum of all occurring y_0-values and b to the maximum of all occurring y_1-values. A page frame detected in this way is shown in Figure 1. This approach gives the correct result for most of the documents, but fails in these cases:

1. If there is a non-text zone (images, graphics, ...) at the top or bottom of the page, it is missed by the page frame.
2. If there is an isolated page number at the top or bottom of the page, and it is missed by the textline detection, it will not be included in the page frame.

An example illustrating these problems is shown in Figure 2. In order to estimate the final values for $\vartheta_v = \{t, b\}$, we use zone level information as given by

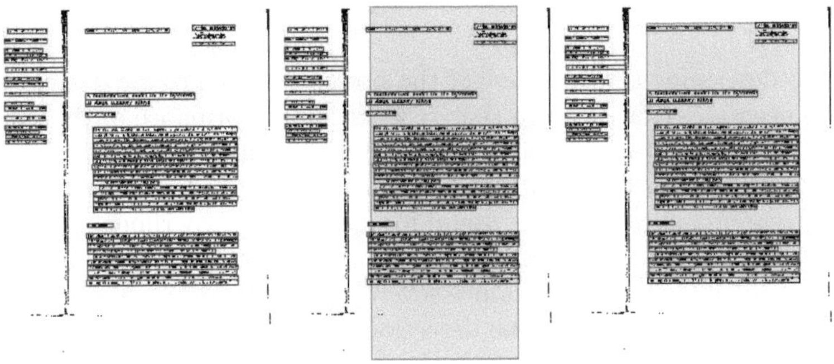

Fig. 1. Example demonstrating refinement of the parameters ϑ_h to adapt to text alignment. The detected textlines are shown in the left image, one paragraph being indented more than the remaining lines. A page frame corresponding to the optimal parameters with respect to Equation 3 is shown in the center. The image on the right shows the initial page frame after adjusting for text alignment.

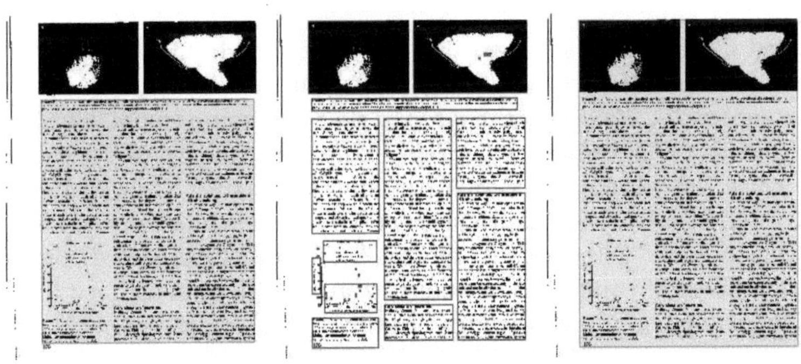

Fig. 2. Example image to demonstrate inclusion of non-text boxes into the page frame. The initially detected page frame based only on the textlines is shown on the left. Note that the images on the top and the page number at the bottom are not part of the page frame. The middle image shows the zones detected by the Voronoi algorithm. The right image shows the final page frame obtained using zone-level information.

the Voronoi algorithm [5]. We perform filtering on the zones obtained by the Voronoi algorithm, such that all the zones that lie completely inside, or do not overlap horizontally with the detected page frame are removed. Then, we consider including all possible combinations of the remaining zones into the page frame and calculate the aspect ratio of the resulting page frames. We finally select the page frame for which aspect ratio is closest to a target value. This target value can be chosen depending on the class of documents under consideration. For a typical journal article in A4 or letter size, the aspect ratio of the page frame usually lies in the interval [1.4, 1.5]. An example result is shown in Figure 2.

3 Error Measures

In order to determine the accuracy of the presented page frame detection algorithm, we need an error measure that reflects the performance of the evaluated algorithm. Previous approaches for marginal noise removal [6,7,8,9] use manual inspection to decide whether noise regions have been completely removed or not. While these approaches might be useful for small scale experiments, we need an automated way of evaluating border noise removal for evaluation on a large sized dataset. In the following, we introduce several performance measures to evaluate different aspects of our page frame detection algorithm.

Area Overlap. Let F_d be the detected page frame and F_g be the ground-truth page frame. Then the area overlap between the two regions can be defined as $A = (2|F_d \cap F_g|)/(|F_d| + |F_g|)$. However, the area overlap A does not give any hints about the errors made. Secondly, small errors like including a noise zone near the top or bottom of the page into the page frame may result in large errors in terms of area overlap.

Connected Components Classification. The page frame partitions the connected components into two sets: the set of document components and the set of noise components. Based on this property, and defining components detected to be within the page frame as 'positive', the performance of page frame detection can be measured in terms of the four quantities 'true positive', 'false positive', 'true negative', and 'false negative'. The error rate can then be defined as the ratio of 'false' detections to the total number of connected components. This classification of connected components gives equal importance to all components, which may not be desired. For instance, if the page number is missed by the algorithm, the error rate is still very low but we loose important information about the document. Considering the page number as an independent zone, a performance measure based on detection of ground-truth zones is introduced.

Ground-Truth Zone Detection. For the zone-based performance measure, three different values are determined:

- Totally In: Ground-truth zones completely within the computed page frame
- Partially In: Ground-truth zones partially inside the computed page frame
- Totally Out: Ground-truth zones totally outside the computed page frame

Using this performance measure, we analyze the 'false negative' detections in more detail. As the page numbers are considered independent zones, losing page numbers will have a higher impact on the error rates in this measure.

OCR Accuracy. In order to demonstrate the usefulness of page frame detection in reducing OCR error rates by eliminating false alarms, we chose to use Omnipage 14 - a commercial OCR system. We use the edit distance [12] between the OCR output and the ground-truth text as the error measure. Edit distance is the minimum number of point mutations (insertion, deletions, and substitutions) required to convert a given string into a target string. The ground-truth text provided with the UW-III dataset has several limitations when used to evaluate an OCR system. First, there is no text given for tables. Secondly, the formating of the documents is coded as latex commands. When an OCR system is tested on this ground-truth using error measures like the Edit distance, the error rate is unjustly too high. Also, our emphasis in this work is on the improvement of OCR errors by using page frame detection, and not on the actual errors made by the OCR system. Hence, we first cleaned the UW-III documents using the ground-truth page frame, an then used the output of Omnipage on the cleaned images as the ground-truth text. This type of ground-truth gives us an upper limit of the performance of a page frame detection algorithm, and if the algorithm works perfectly, it should give 0% error rate, independent of the actual error rate of the OCR engine itself.

4 Experiments and Results

The evaluation of the page frame detection algorithm was done on the University of Washington III (UW-III) database [13]. The database consists of 1600 skew corrected English document images with manually edited ground-truth of entity bounding boxes. These bounding boxes enclose page frame, text and non-text zones, textlines, and words. The database contains a large number of scanned photocopies presenting copy borders and parts of text from the adjacent page, making it suitable for the evaluation of our page frame detection algorithm. The documents in the UW-III database can be classified into different categories based on their degradation type (see [13] for more details):

- Direct Scans (Scans): the original document has been scanned directly.
- First Generation Photocopies (1Gen): the original has been copied and this copy has been scanned.
- Nth Generation Photocopies (NGen): the Nth generation photocopy of the original has been scanned ($N > 1$).

The dataset was divided into 160 training images and 1440 test images. In order to make the results replicable, we included every 10th image (in alphabetical order) from the dataset into the training set. Hence our training set consists of images A00A, A00K, ..., W1UA. The evaluation was done on the remaining 1440 images. Some examples of page frame detection for documents from the collection are shown in Figure 3. The rightmost image in Figure 3 shows an example where the marginal noise overlaps with some textlines at the bottom of the page. The use of page frame detection successfully detects the page contents region and removes the border noise while keeping the page contents intact.

When the page frame detection was evaluated on the basis of overlapping area, the overall mean overlap was 91%. However, the UW3 ground-truth page frame has a white border of unspecified size around the rectangle covering all page zones. In order to eliminate this border, we modified the ground-truth page

Fig. 3. Some example images showing the detected page frame in yellow color

Table 1. Results for the connected component based evaluation. The number in brackets gives the number of documents of that class. Error rates in [%].

Document Type	True Positive	False Negative	True Negative	False Positive
Scans (392)	99.84	0.16	76.6	23.4
1Gen (1029)	99.78	0.22	74.0	26.0
*N*Gen (19)	99.93	0.07	42.8	57.2
all (1440)	99.80	0.20	73.5	26.5
total (absolute)	4,399,718	8,753	187,446	67,605

frame by determining the smallest rectangle containing all of the ground-truth zones as page frame. Testing with these page frames as ground-truth gave an overall mean area overlap of 96%. In the following, when mentioning the ground-truth page frame, we refer to this corrected ground-truth page frame.

The results for the connected component based metric are given in Table 1. The high percentage of true positives shows that mostly, the page frame includes all the ground-truth components. The percentage of true negatives is about 73.5%, which means that a large part of noise components are removed. The total error rate defined as the ratio of 'false' detections to the total number of connected components is 1.6%. The results for the zone based metric are given in Table 2. Compared to the number of missed connected components, one can see that the percentage of missed zones is slightly higher than the corresponding percentage of false negatives on the connected component level. One conclusion that can be drawn from this observation is that the zones missed do not contain a lot of components, which is typically true for page numbers, headers, and footers of documents. In some cases, the textline finding merges the textlines consisting of textual noise to those in the page frame. In such cases, a large portion of the textual noise is also included in the page frame.

The use of page frame detection in an OCR system showed significant improvement in the OCR results. First, we ran the OCR on the original images and computed the Edit distance to the estimated ground-truth text (cf. Sec. 3). Then, we used the computed page frame to remove marginal noise from the documents, and re-ran the experiments. The results (Table 3) show that the use

Table 2. Results for the zone based evaluation. Error rates in [%].

Document Type	Totally In	Partially In	Totally Out
Scans (392)	97.6	0.7	1.7
1Gen (1029)	97.1	1.0	1.9
NGen (19)	97.5	0.0	2.5
all (1440)	97.2	0.9	1.9

Table 3. Results for the OCR based evaluation with page frame detection (PFD) and without page frame detection

	Total Characters	Deletions	Substitutions	Insertions	Total Errors	Error Rate
Without PFD	4831618	34966	29756	140700	205422	4.3%
With PFD	4831618	19544	9828	53610	82982	1.7%

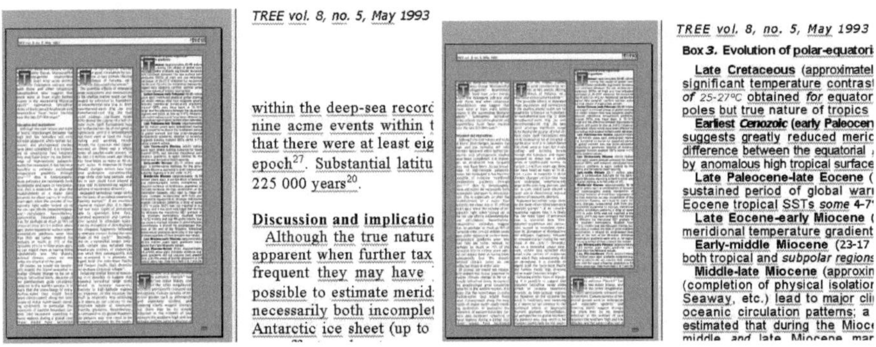

Fig. 4. Screenshot of Omnipage 14 showing the recognized text of the original document (left) and the document cleaned using page frame detection (right). Note that the reading order of the text has changed, probably due to the slightly changed geometry.

of page frame detection for marginal noise removal reduced the error rate from 4.3% to 1.7%. The insertion errors are reduced by a factor of 2.6, which is a clear indication that the page frame detection helped in removing a lot of extraneous symbols that were treated previously as part of the document text. There are also some deletion errors, which are the result of changes in the OCR software's reading order determination. One example is shown in Figure 4, for which the reading order changed after document cleaning.

5 Conclusion

We presented an approach for page frame detection using geometric matching methods. Our approach does not assume the existence of whitespace between marginal noise and the page frame and can detect the page frame even if the noise overlaps some regions of the page content area. We defined several error measures based on area overlap, connected component classification, and ground-truth zone detection accuracy. It was shown that the algorithm performs well on all three performance measures with error rates below 4% in each case. The major source of errors was missing isolated page numbers. Locating the page numbers as a separate process and including them in the detected page frame may further decrease the error rates. The benefits of the page frame detection in practical applications were highlighted by using it with an OCR system, where we showed that the OCR error rates were significantly reduced.

Acknowledgments. This work was partially funded by the BMBF (German Federal Ministry of Education and Research), project IPeT (01 IW D03).

References

1. Shafait, F., Keysers, D., Breuel, T.M.: Pixel-accurate representation and evaluation of page segmentation in document images. In: 18th Int. Conf. on Pattern Recognition, Hong Kong, China pp. 872–875 (August 2006)
2. Baird, H.S.: Background structure in document images. In: Bunke, H.e.a. (ed.) Document Image Analysis, pp. 17–34. World Scientific, Singapore (1994)
3. Breuel, T.M.: Two geometric algorithms for layout analysis. In: Document Analysis Systems, pp. 188–199. Princeton, New York (August 2002)
4. Gorman, L.O.: The document spectrum for page layout analysis. IEEE Trans. on Pattern Analysis and Machine Intelligence 15(11), 1162–1173 (1993)
5. Kise, K., Sato, A., Iwata, M.: Segmentation of page images using the area Voronoi diagram. Computer Vision and Image Understanding 70(3), 370–382 (1998)
6. Le, D.X., Thoma, G.R., Wechsler, H.: Automated borders detection and adaptive segmentation for binary document images. In: 13th Int. Conf. Patt. Recog. Vienna, Austria pp. 737–741 (August 1996)
7. Avila, B.T., Lins, R.D.: Efficient removal of noisy borders from monochromatic documents. In: Int. Conf. on Image Analysis and Recognition, Porto, Portugal pp. 249–256 (September 2004)
8. Fan, K.C., Wang, Y.K., Lay, T.R.: Marginal noise removal of document images. Patt. Recog. 35, 2593–2611 (2002)
9. Cinque, L., Levialdi, S., Lombardi, L., Tanimoto, S.: Segmentation of page images having artifacts of photocopying and scanning. Patt. Recog. 35, 1167–1177 (2002)
10. Shafait, F., Keysers, D., Breuel, T.M.: Performance comparison of six algorithms for page segmentation. In: 7th IAPR Workshop on Document Analysis Systems, Nelson, New Zealand pp. 368–379 (February 2006)
11. Breuel, T.M.: A practical, globally optimal algorithm for geometric matching under uncertainty. Electr. Notes Theor. Comput. Sci. 46, 1–15 (2001)
12. Levenshtein, V.I.: Binary codes capable of correcting deletions, insertions and reversals. Soviet Physics Doklady 10(8), 707–710 (1966)
13. Phillips, I.T.: User's reference manual for the UW english/technical document image database III. Technical report, Seattle University, Washington (1996)

Representing Pairs of Orientations in the Plane

Magnus Herberthson[1], Anders Brun[2], and Hans Knutsson[2]

[1] Department of Mathematics, Linköping University, Sweden
[2] Department of Biomedical Engineering, Linköping University, Sweden

Abstract. In this article we present a way of representing pairs of orientations in the plane. This is an extension of the familiar way of representing single orientations in the plane. Using this framework, pairs of lines can be added, scaled and averaged over in a sense which is to be described. In particular, single lines can be incorporated and handled simultaneously.

1 Introduction

Consider the two different rectangles in fig. 1. In each rectangle, two regions with different linear structure meet. If the dominant orientation is estimated, the (classical)

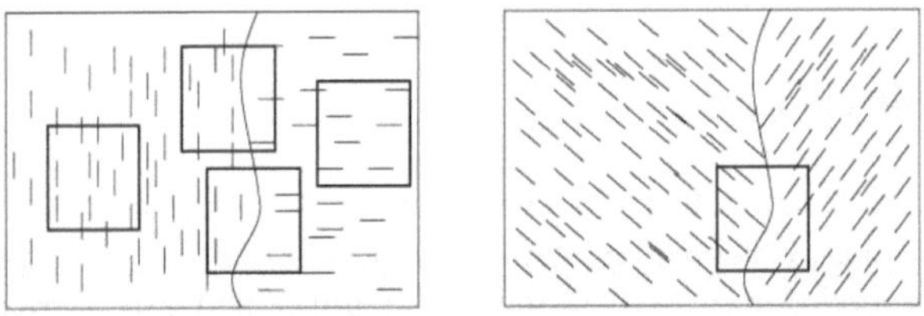

Fig. 1. When estimating a single orientation, it is hard to distinguish the border region in the rectangle to the left from the border region to the right. This can be remedied by allowing for pairs of orientations.

estimates formed near the border between the regions will be poor. The two middle 'windows' to the left and the window to the right will all give more or less the same information, namely that the average orientation is zero or isotropic. With the method presented here, these two cases can be distinguished.

2 The Problem

Let us consider the problem of adding, or making weighted averages of, *pairs* of indistinguishable orientations in the plane $\mathbf{P} \sim \mathbf{R}^2$, each direction being represented by a line through the origin. A pair of lines in $\mathbf{P}$, both containing the origin point, can be represented in several ways, for instance by the triple $(\hat{v}, \hat{u}, \alpha)$, where $\hat{v}$ and $\hat{u}$ are units vectors along the lines, and where $\alpha \in \mathbf{R}$ is a weight. It is noticed that $\pm\hat{v}$ and $\pm\hat{u}$ represent the same object, and any calculation must be invariant to these changes of

B.K. Ersbøll and K.S. Pedersen (Eds.): SCIA 2007, LNCS 4522, pp. 661–670, 2007.

Fig. 2. A number of pairs of orientations in the plane. The thickness indicates strength and/or certainty. What is the dominating pair of orientations? How certain is it? How to you measure the 'spread'?

sign. Also, since the lines are indistinguishable, the representation should be invariant under the change $\hat{v} \longleftrightarrow \hat{u}$.

The situation is illustrated in figure 2, where we note that both lines in a pair of orientations have the same strength (or certainty). This is a condition which can be relaxed, i.e., the two orientations within a pair can indeed have different strengths, but this situation is most conveniently considered at a later stage. The problem is now the following:

Given a family of line pairs $\{(\hat{v}_i, \hat{u}_i, \alpha_i)\}_i$, how do we determine the (weighted) formal sum (or average) $\sum_{i=1}^{n}(\hat{v}_i, \hat{u}_i, \alpha_i)$?

3 Sums of Single Lines

In order to settle the notation, let us review the familiar corresponding process for a family of single lines, i.e., we consider, symbolically, $\sum_{i=1}^{n}(\hat{v}_i, \alpha_i)$. For a single line, the weight α can be encoded in $\overline{v} = \alpha\hat{v}$, so that we should consider the formal sum $\sum_{i=1}^{n}\overline{v}_i$ (or formal average $\frac{1}{n}\sum_{i=1}^{n}\overline{v}_i$), which again should be unaffected by any change(s) $\overline{v}_i \to -\overline{v}_i$. It is customary to form the sum of the outer products $A = \sum \overline{v}_i\overline{v}_i^t$, which gives a symmetric mapping $\mathbf{P} \to \mathbf{P}$ (or $\mathbf{R}^n \to \mathbf{R}^n$ if $v_i \in \mathbf{R}^n$). A can of course also be considered as a quadratic form. One then takes the largest eigenvalue and let the corresponding eigenline represent the average. See figure 3.

Example 1. In standard coordinates, put $\hat{f}_1 = \begin{pmatrix}1\\0\end{pmatrix}$, $\hat{f}_2 = \begin{pmatrix}0\\1\end{pmatrix}$, $\hat{h}_1 = \frac{1}{\sqrt{2}}\begin{pmatrix}1\\1\end{pmatrix}$, $\hat{h}_2 = \frac{1}{\sqrt{2}}\begin{pmatrix}1\\-1\end{pmatrix}$. Then $\hat{f}_1\hat{f}_1^t + \hat{f}_2\hat{f}_2^t = \begin{pmatrix}1\,0\\0\,1\end{pmatrix}$. In both cases, the sum is isotropic, and no particular orientation can be distinguished, see figure 4. In addition, the information

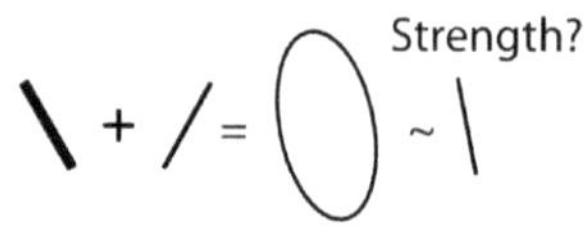

Fig. 3. The 'sum' of two lines is given by the major axis of the ellipse corresponding to the symmetric mapping formed from the lines. What is the strength?

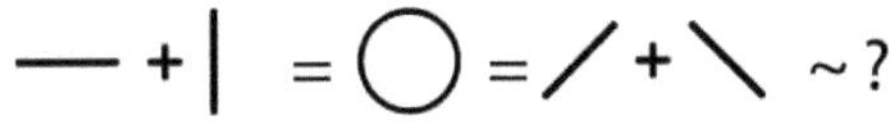

Fig. 4. Two different ways of adding up to a isotropic sum. No direction is singled out.

that the lines forming the sum were either horizontal/vertical or had slopes $\pm 45°$ is completely lost. Cf. figures 9 and 10. **Remark**: In the abstract index notation, [2], $A = \sum \overline{v}_i \overline{v}_i^t$ would be written $A^a{}_b = \sum (v_i)^a (v_i)_b$, omitting the bars.

4 Sums of Pair of Lines

The triple $(\hat{v}_i, \hat{u}_i, \alpha_i)$ can be represented, non-uniquely, by the pair $(\overline{v}_i, \overline{u}_i) \sim (v^a, u^a)$, where $\alpha_i = |\overline{v}_i| \cdot |\overline{u}_i|$. Using this, we make the following definition.

Definition 1. *Let a pair of orientations in the plane be given by the the pair* $(v^a, u^a) \equiv (\overline{v}, \overline{u})$ *where* $\pm\hat{v}, \pm\hat{u}$ *are directed along the orientations and where* $|\overline{v}| \cdot |\overline{u}|$ *gives the strength. We then represent this pair of orientations by the symmetric tensor* $v^{(a}u^{b)} := \frac{1}{2}(v^a u^b + v^b u^a)$.

In terms of column vectors, where $v^a \sim \overline{x}$, $u^a \sim \overline{y}$, $v^{(a}u^{b)} := \frac{1}{2}(\overline{x}\,\overline{y}^t + \overline{y}\,\overline{x}^t)$, i.e. the symmetrized outer product. The primitive representations $v^{(a}u^{b)}$ specifies v^a, u^a up to relative scaling and is *not* invariant under change of sign. To form outer products of such elements, they must live in a vector space. We provide $\mathbf{P}$ with the standard scalar product, and also introduce standard cartesian coordinates (X, Y) so that $\hat{X} = \hat{f}_1$, $\hat{Y} = \hat{f}_2$.

Definition 2. *We let* $\mathbf{P}^{(ab)}$ *denote the vector space of symmetric rank* $\binom{2}{0}$ *tensors over* $\mathbf{P}$.

Elements in $\mathbf{P}^{(ab)}$ can be written either $v^{(ab)}$ or $v^{(a}u^{b)}$ where in the latter case it is understood that $v^a \in \mathbf{P}^a, u^a \in \mathbf{P}^a$. In terms of coordinates, elements in $\mathbf{P}^{(ab)}$ are represented by symmetric 2 by 2 matrices, in particular

Lemma 1. *dim(*$\mathbf{P}^{(ab)}$*)=3*

Note however, that the set $\{v^{(a}u^{b)}, v^a \in \mathbf{P}^a, u^a \in \mathbf{P}^a\}$ does not constitute a vector space. As can easily be checked explicitly, cf. lemma 3, the symmetric 2 by 2 matrix $\hat{X}^{(a}\hat{X}^{b)} + \hat{Y}^{(a}\hat{Y}^{b)} \sim \begin{pmatrix} 1 & 0 \\ 0 & 0 \end{pmatrix} + \begin{pmatrix} 0 & 0 \\ 0 & 1 \end{pmatrix} = \begin{pmatrix} 1 & 0 \\ 0 & 1 \end{pmatrix}$ can not be written as $v^{(a}u^{b)}$ for any v^a, u^b. We now turn to the scalar product on $\mathbf{P}^{(ab)}$. This scalar product will be induced by the scalar product on $\mathbf{P}$, which we denote by g_{ab}. Thus $g_{ab} : \mathbf{P} \times \mathbf{P} \to \mathbf{R}$, so that $\overline{v} \cdot \overline{u} = g(\overline{v}, \overline{u}) = g_{ab} v^a u^b = v^a u_a$. Again, the subscripts ab in g_{ab} indicates the covariant nature of the scalar product g_{ab} as explained in [2].

Definition 3. *The scalar product on* $\mathbf{P}^{(ab)} : \mathbf{P}^{(ab)} \times \mathbf{P}^{(cd)} \to \mathbf{R}$ *is given by* $(v^{(a}u^{b)}, w^{(c}z^{d)}) \to g_{ac} g_{bd} v^{(a}u^{b)} w^{(c}z^{d)} = \frac{1}{2}(g_{ac} v^a w^c g_{bd} u^b z^d + g_{ac} v^a z^c g_{bd} u^b w^d) = \frac{1}{2}((\overline{v} \cdot \overline{w})(\overline{u} \cdot \overline{z}) + (\overline{v} \cdot \overline{z})(\overline{u} \cdot \overline{w}))$.

In particular, the squared norm of the pair $\hat{v}^{(a}\hat{u}^{b)}$ is $||\hat{v}^{(a}\hat{u}^{b)}||^2 = \frac{1}{2}[1 + (\hat{v} \cdot \hat{u})^2] = \frac{1}{2}[1 + \cos^2\phi]$, where ϕ is the angle between $\hat{v}$ and $\hat{u}$. One easily checks the following lemma.

Lemma 2. *Let $v^{(a}u^{b)}$ and $w^{(c}z^{d)}$ be represented by the symmetric matrices A and B. The scalar product is then simply given by $v^{(a}u^{b)} \cdot w^{(c}z^{d)} = tr(AB)$.*

It was noted above that the not every element of $\mathbf{P}^{(ab)}$ is of the primitive form $v^{(a}u^{b)}$. We therefore ask the question *what elements in $\mathbf{P}^{(ab)}$ can be written as $v^{(a}u^{b)}$, with $v^a \in \mathbf{P}^a, u^a \in \mathbf{P}^a$?* To illustrate this, we introduce suitable cartesian coordinates x, y, z on $\mathbf{P}^{(ab)} \cong \mathbf{R}^3$.

In terms of standard coordinates, it is easy to see that an ON-basis is given by

$$\hat{e}_1 = \frac{1}{\sqrt{2}}\begin{pmatrix} 1 & 0 \\ 0 & -1 \end{pmatrix}, \hat{e}_2 = \frac{1}{\sqrt{2}}\begin{pmatrix} 0 & 1 \\ 1 & 0 \end{pmatrix}, \hat{e}_3 = \frac{1}{\sqrt{2}}\begin{pmatrix} 1 & 0 \\ 0 & 1 \end{pmatrix}$$

We can now express each $v^{(ab)} \in \mathbf{P}^{(ab)}$ as $v^{(ab)} = x\hat{e}_1 + y\hat{e}_2 + z\hat{e}_3 = \alpha^1\hat{e}_1 + \alpha^2\hat{e}_2 + \alpha^3\hat{e}_3 = \alpha^j\hat{e}_j$ or simply by the coordinates $[v^{(ab)}] = \begin{pmatrix} x \\ y \\ z \end{pmatrix}$. One can easily check that $\hat{e}_1$ and $\hat{e}_2$, but not $\hat{e}_3$ can be decomposed. This is the content of the following lemma.

Lemma 3. *Suppose $[w^{(ab)}] = \begin{pmatrix} x \\ y \\ z \end{pmatrix}$ is given. Then $w^{(ab)} \in \mathbf{P}^{(ab)}$ is decomposable as $w^{(ab)} = v^{(a}u^{b)}$ for some $v^a, u^a \in V^a$ if and only if $z^2 \leq x^2 + y^2$.*

The proof is fairly straight forward, using the ansatz $\hat{v} = \begin{pmatrix} \cos\phi \\ \sin\phi \end{pmatrix}$, $\hat{u} = \begin{pmatrix} \cos\gamma \\ \sin\gamma \end{pmatrix}$.

4.1 Illustrations of Pair of Orientations in $\mathbf{P}^{(ab)}$

We have introduced cartesian coordinates x, y, z in $\mathbf{P}^{(ab)}$, and know that that decomposable primitive objects corresponds to vectors whose coordinates satisfies $z^2 \leq x^2 + y^2$. We also have the ambiguity that $v^{(ab)}$ and $-v^{(ab)}$ represent the same pair of orientations (and strength) in the plane $\mathbf{P}$.

Apart from the strength, each pair of orientations can be represented by the symmetrized outer product of $\hat{v}_\phi$ and $\hat{u}_\gamma$ where ϕ and γ indicates the angle between the unit vector in question and a reference direction, e.g. the positive x-axis. Since both unit vectors are interchangeable, and since $\hat{v}_\phi$ and $\hat{v}_{\phi+\pi}$ represent the same orientation, many different pairs of angles (ϕ, γ) will represent the same pair of orientation. This is illustrated in figure 5 where the gray triangle gives an example of a region containing all possible values of $\hat{v}_\phi^{(a}\hat{u}_\gamma^{b)}$. In this triangle, each object $\hat{v}_\phi^{(a}\hat{u}_\gamma^{b)}$ is represented by a single point, except for the boundary points, which are to be identified in a certain way. For instance, in figure 5, the three small squares all represent the same pair of orientations. Similarly, the two small circles are to identified. As can be anticipated, the topology of this triangle with appropriate boundary points identified is non-trivial. (In fact, the topology is the Moebuis band.)

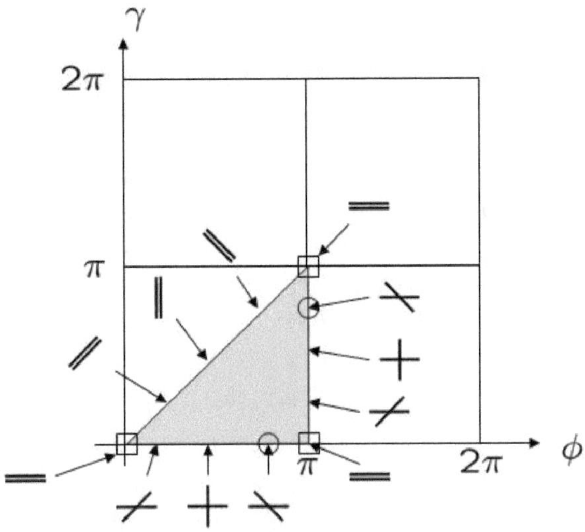

Fig. 5. The gray triangle is an example of a region where each element $v^{(ab)} \in \mathbf{P}^{(ab)}$ i uniquely represented by one point, except for the boundary of the triangle where certain points are to be identified. This identification gives a non-trivial topology.

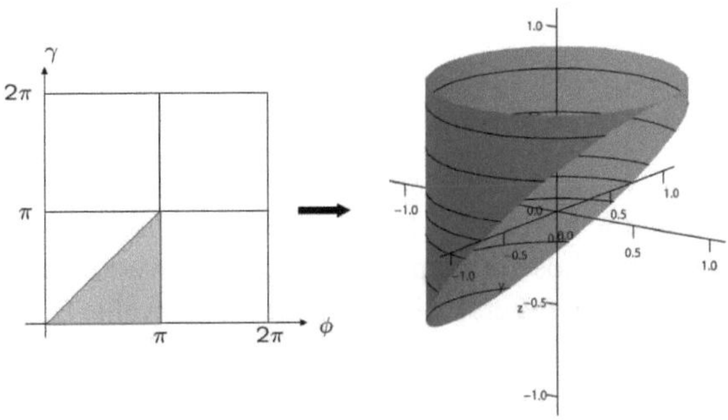

Fig. 6. The triangle is mapped in an almost one-to-one fashion to the surface to the right. Antipodal points on the ellipse are to be identified.

Each element $\hat{v}^{(a}_{\phi}\hat{u}^{b)}_{\gamma}$ is mapped in to $\mathbf{P}^{(ab)}$, and consequently, the set $M = \{\hat{v}^{(a}_{\phi}\hat{u}^{b)}_{\gamma} \mid 0 \leq \phi \leq 2\pi, 0 \leq \gamma \leq 2\pi\}$ is represented by a surface in $\mathbf{P}^{(ab)}$. In terms of the coordinates used in lemma 3, the surface M occupies the cylinder $\{(x, y, z) \mid x^2 + y^2 = 1, -1 \leq z \leq 1\}$. However, the image of M is not injective, since the elements $\hat{v}^{(a}_{\phi}\hat{u}^{b)}_{\gamma}$ and $\hat{v}^{(a}_{\phi}\hat{u}^{b)}_{\gamma+\pi}$ represent the same pair of orientations, but differ by a sign when represented in $\mathbf{P}^{(ab)}$, antipodal points in $\mathbf{P}^{(ab)}$ must be identified. In order to get a (almost) one-to-one mapping between elements in a part of M and a surface in $\mathbf{P}^{(ab)}$, we look at figure 6. Apart from the boundary points of the indicated triangle, each element represents a pair of orientation uniquely. The triangle is mapped to the cylindrical surface to the right, where now only antipodal points on the shown ellipse are to be identified.

4.2 Sums or Averages of Pairs of Orientations

To each pair of orientations in the plane $\mathbf{P}$, we have the associated element $v^{(a}u^{b)} \in \mathbf{P}^{(ab)}$. This representation is not invariant under the inversion $v \to -v$ or $u \to -u$, which is reflected by the property that antipodal elements in $\mathbf{P}^{(ab)}$ are to be identified. To handle this sign ambiguity, we proceed as in the single line case, i.e., we note that the outer product of $v^{(a}u^{b)}v_{(c}u_{d)}$ is invariant under $v \to -v$ or $u \to -u$. Thus, repeating the steps for the single line case, the procedure is as follows.

Suppose that we are given n elements $v_i^{(a}u_i{}^{b)} \in \mathbf{P}^{(ab)}$. To calculate their formal 'sum', we form the sum of the tensor products

$$\sum_{i=1}^{n} v_i^{(a}u_i{}^{b)}(v_i)_{(c}(u_i)_{d)}$$

This gives a symmetric mapping $\mathbf{P}^{(ab)} \to \mathbf{P}^{(ab)}$. We then let the eigenvectors corresponding to the largest eigenvalue of this mapping correspond to the 'sum', or dominating pair of orientations, *if these eigenvectors* (in $\mathbf{P}^{(ab)}$) *are decomposable*.

Again we remark that the sum should be regarded is the symmetric mapping as a whole, and that if the eigenvalues of this mapping are of the same order, it may be highly missleading to focus on the largest eigenvalue.

In this case, we also have the complication that the eigenvectors in $\mathbf{P}^{(ab)}$ corresponding to the largest eigenvalue, may not be decomposable as a pair of lines in $\mathbf{P}$. However, this is not a real problem as this situation corresponds to an isotropic situation, where the two lines are replaced by a ellipsoid (or circle).

Example 2. In figure 7 we have started with four pairs of lines in the plane, drawn to the left. By representing them in $\mathbf{P}^{(ab)}$, forming the outer products and summing, we get a symmetric mapping, whose principal ellipsoid is drawn to the right. The pair of lines corresponding to the eigenvector with the largest eigenvalue is also shown.

4.3 Some Properties of Pair Orientation Averages

In this section we mention some of the properties of the presented representation.

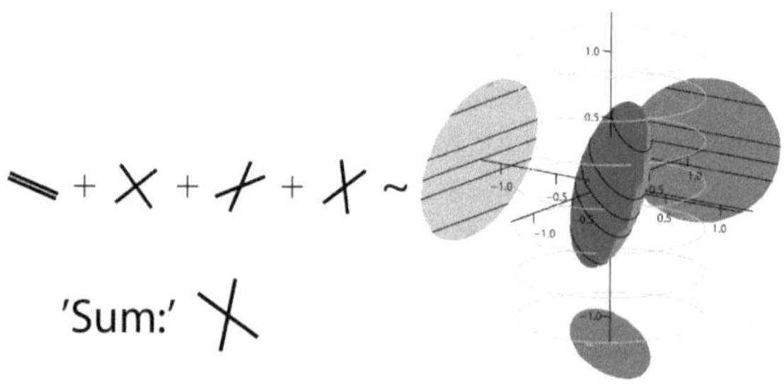

Fig. 7. The sum of four pairs of lines. The resulting ellipsoid is shown to the left, together with its projections on the coordinate planes.

Sums Containing a Common Orientation. A natural situation is to consider sums of pair of orientations $\sum_i v^{(a}u_i{}^{b)}$, i.e., we have an orientation v^a which is common to all pairs. The natural question is to ask then whether the resulting (dominant) pair of orientations always will contain the orientation represented by $\hat{v}$ or not. For consistency reasons, we expect the answer to be yes, and that this is the case is content of the following theorem.

Theorem 1. *Suppose that we are given n elements $v^{(a}u_i{}^{b)}$, so that all elements contain a common orientation (given by v). If we form outer products of these elements regarded as vectors in $\mathbf{P}^{(ab)}$, and sum them to a symmetric mapping $\mathbf{P}^{(ab)} \to \mathbf{P}^{(ab)}$, then any eigenvector corresponding to the largest eigenvalue will contain v^a, i.e., it can be written $v^{(a}u^{b)}$ for some $u^b \in \mathbf{P}^b$.*

This is illustrated in figure 8, where all pairs contain a common direction. The corresponding ellipsoid lies in a plane, which means that one eigenvalue is zero. Again, the 'dominating' pair is also shown.

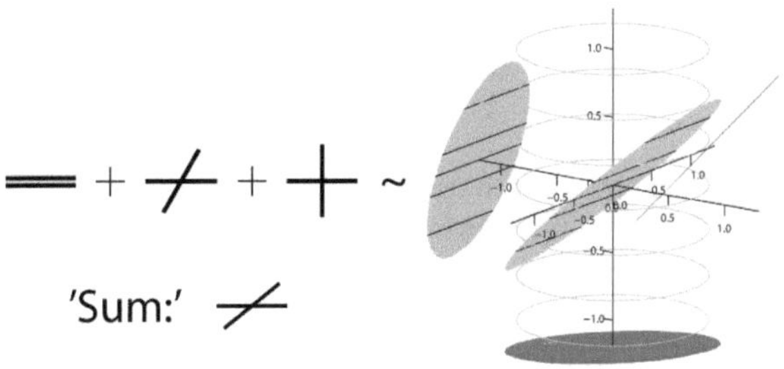

Fig. 8. The sum of three pairs of lines. This time all pairs have one common orientation. As a result, the corresponding ellipsoid lies in a plane, i.e. one eigenvalue is 0.

Before we start the proof of theorem 1, we introduce a notation and a useful lemma.

Definition 4. *Suppose that $v \in \mathbf{P}$ is given. $v_\perp \in \mathbf{P}$ is then defined as the vector obtained by rotating v $\pi/2$ counter-clockwise. Furthermore, by $v_\perp^{(ab)} \in \mathbf{P}^{(ab)}$ we mean $v_\perp^{(ab)} = v_\perp^{(a}v_\perp^{b)}$*

By construction, v and $v_\perp$ are orthogonal, i.e., $v_\perp^a v_a = 0$. As regards $v_\perp^{(ab)}$, we have the following lemma.

Lemma 4. *Let $v^a \in V^a$ be given. Then $v_\perp^{(ab)}u_{(ab)} = 0 \Leftrightarrow u^{(ab)} = v^{(a}u^{b)}$ for some $u^a \in \mathbf{P}^a$.*

Proof of lemma 4
$\Leftarrow$: Suppose $u^{(ab)} = v^{(a}u^{b)}$ for some $u^a \in \mathbf{P}^a$. Then $v_\perp^{(ab)}u_{(ab)} = v_\perp^{(a}v_\perp^{b)}v_{(a}u_{b)} = 0$.
$\Rightarrow$: It is easy to see that for a given (non-zero) $v^a \in \mathbf{P}^a$, the set $\Pi = \{v^{(a}u^{b)} : u^a \in V^a\}$ is a two-dimensional vector space. Also, Π lies in the orthogonal complement to $v_\perp^{(ab)} : [v_\perp^{(ab)}]^\perp$. But since $v_\perp^{(ab)}$ is a non-zero vector in the three-dimensional vector space $\mathbf{P}^{(ab)}$, its orthogonal complement $[v_\perp^{(ab)}]^\perp$ is two-dimensional, i.e., it is Π.

Proof of theorem 1
Put $\Pi = [v_{\perp}^{(ab)}]^{\perp}$. All elements $v^{(a}{u_i}^{b)}$ lie in Π. Therefore, to the (positive semidefinite) symmetric mapping formed by the sum of the outer products of these elements, all eigenvectors corresponding to nonzero eigenvalues must lie in Π. This is perhaps most easily seen as a consequence of the fact that $v_{\perp}^{(ab)}$ is an eigenvector with eigenvalue 0, and that eigenvectors corresponding to different eigenvalues are orthogonal.

Representing Single Orientations. As we saw in example 1 and figure 4, the sum of two lines may be totally isotropic, which in particular means that all information of the

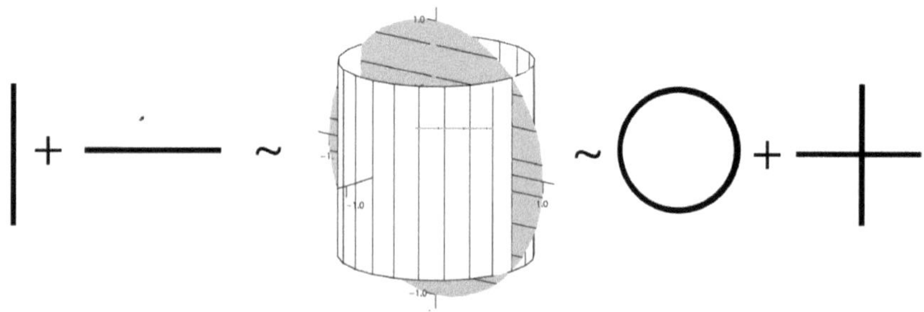

Fig. 9. The two lines to the left are each represented by pair of identical orientations. Their sum is represented by the disk in the figure, which can equally well be interpreted as the sum of the two objects to the right. Cf. figure 10.

individual lines are lost. Clearly, it could be advantageous to find the isotropic sum, but still be able to distinguish between the two cases (and others) in figure 4. This can be done by representing the orientation of one single line as a pair of parallel coinciding lines in the manner described in this paper. In both cases, there is a common eigenvector which is vertical in the figures 9 and 10. This eigenvector is not decomposable, i.e. can not be written as $v^{(a}u^{b)}$, and corresponds to the isotropic part to the right. However, the eigenvectors orthogonal to the common vertical eigenvector differs in the cases displayed in the figures 9 and 10. Each of these vectors is decomposable and corresponds to pairs of orthogonal lines, as shown in the figures.

4.4 Pair of Lines with Different Strength

Although explicitly stated earlier that the orientations we have represented should be indistinguishable, this statement can be slightly modified. Uptil now, we have not assigned any individual strength to v^a or u^a in the composite object $v^{(a}u^{b)}$, i.e., from the composite object, the constituent parts are only determined up to a scaling. However, the crucial property is rather that when summing pairs of orientations, $v^{(a}u^{b)}$ with $w^{(a}z^{b)}$ say, there should not be any rule connecting v^a with either w^a or z^a. Apart from that, one could very well consider cases where the lines/orientations *have* different strength. This is modeled in such a way that the orientations given by v^a and u^a, where the line given by v^a is 'stronger' than the one given by u^a is modeled as a sum of $v^{(a}v^{b)}$ and $v^{(a}u^{b)}$.

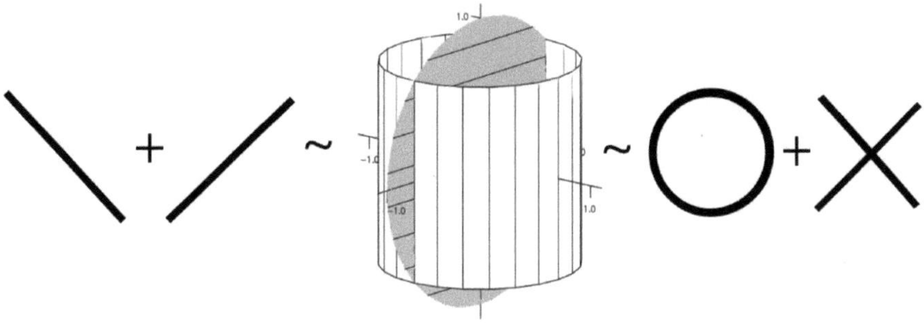

Fig. 10. The two lines to the left are each represented by pair of identical orientations. Their sum is represented by the disk in the figure, which can equally well be interpreted as the sum of the two objects to the right. Compared to figure 9, both disks have a common vertical (eigen) vector. This corresponds to the common isotropic part. The horisontal vectors contained in the discs in this and the previous figure differ, which shows up by the different orientations of the orthogonal line pairs.

4.5 Generalizations

The concept presented here has obvious generalisations . For instance, one might consider pairs of lines in a vector space V with $\dim(V) = 3$ instead of 2. The same construction is useful, i.e., one consider elements $v^{(a}u^{b)} \in V^{(ab)}$ where $v, u \in V$. Most earlier calculations have obvious counterpart, the most important difference is perhaps the correspondence to lemma 3. This time, we can regard elements in $V^{(ab)}$ as symmetric 3×3 matrices. However, elements decomposable as $v^{(a}u^{b)}$ will, among other conditions, correspond to a surface (rather than a volume) in $V^{(ab)}$, since the corresponding symmetric matrices have rank 2. On the other hand, one can equally well consider three orientations in a three-dimensional space, i.e., elements of the type $v^{(a}u^{b}w^{c)}$. These elements lie in vector space of dimension 10. To proceed, one can consider n lines (through the origin) in a m-dimensional vector space V, by forming the symmetrized outher products $v_1^{(a}v_2^{b}\ldots v_n^{c)}$.

4.6 Example/Applications

Let us return to the situation in figure 1. There we considered the problem of estimating the dominating orientation of linear structures. Handled in the traditional way, these estimates normally fail to describe the situation satisfactory near 'borders' i.e., where two areas with different orientations meet. See also figures 4, 9 and 10. Using the approach desribed in this work, the situation becomes different. This is illustrated in figure 11, where, to the left, two regions with different orientations meet. Using a kernel with 5x5 elements, averages of neighbouring orientations are formed, and these averages are then presented as pairs of lines. Away from the border these pairs are dominated by a single orientations, while in the transit region, the averages show up as crosses, reflecting both ingoing orientations. Among others, it is possible to discriminate between the different sitations described in figure 4. Further postprocessing and presentation is of course application dependent.

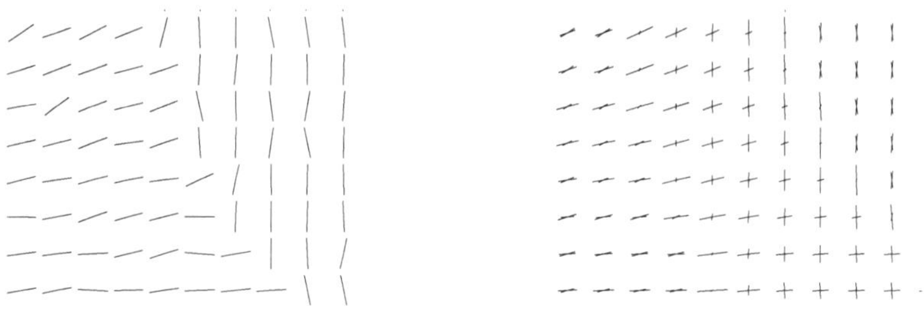

Fig. 11. To the left: two regins with different orientations meet. To the right: Estimates of the local orientations are given. By using the framework presented in this work, the estimates near the border between the regions are pairs of lines rather than isotropic circles or near isotropic ellipses.

References

1. Knutsson, H.: Representing Local Structure Using Tensors. The 6th Scandinavian Conference on Image Analysis, Oulu, Finland (1989)
2. Wald, R.M.: General Relativity, University of Chicago Press (1984)

Improved Chamfer Matching Using Interpolated Chamfer Distance and Subpixel Search

Tai-Hoon Cho

School of Information Technology, Korea University of Technology and Education,
307 Gajun-ri, Byungchun-myun, Chonan, Choongnam, Korea
thcho@kut.ac.kr

Abstract. Chamfer matching is an edge based matching technique that has been used in many applications. The matching process is to minimize the distance between transformed model edges and image edges. This distance is usually computed at the pixel resolution using a distance transform, thus reducing accuracy of the matching. In this paper, an improved approach for accurate chamfer matching is presented that uses interpolation in the distance calculation for subpixel distance evaluation. Also, instead of estimating the optimal position in subpixel using a neighborhood of the pixel position with the minimum distance, for more accurate matching, we use the Powell's optimization to find the distance minimum through actual distance evaluations in subpixel. Experimental results are presented to show the validity of our approach.

1 Introduction

Object matching, finding objects belonging to one image in another image, is a very important problem in computer vision and image analysis. It is particularly useful for industrial applications, where a model of an object must be aligned with an image of the object. The transformation or pose obtained by this object matching process can be used for various tasks, e.g., automatic pose adjustment in pick and place operations. In most cases, the model of the object is generated from an image of the object.

There are roughly two approaches in object matching: image-based matching and feature-based matching. Feature-based matching uses features, e.g., edges or corners, extracted from the image and the model, instead of using gray values directly. Perhaps edges are the most important low-level features. The edge based matching is more robust than the image based matching under non-uniform illumination conditions that can occur in typical industrial applications. Also, since edges in an image are a more compact representation than the image itself, the edge-based matching is computationally more efficient than the image-based matching in general.

A notable approach among edge-based matching approaches is chamfer matching, which was first proposed by Barrow et al. [1], and further improved in [2]. Since then it has been extensively used in many applications, e.g., [3][4][5]. The matching process is minimization of the distance between transformed model edges and image edges. One critical problem of this approach is that calculation of the distance between the transformed model edges and the image edges is usually done at the pixel

B.K. Ersbøll and K.S. Pedersen (Eds.): SCIA 2007, LNCS 4522, pp. 671–678, 2007.

resolution using a distance transform, thus reducing accuracy of the matching. With pixel-level distance computation, even use of a subpixel algorithm for the matching will not improve much the accuracy due to insensitivity to subpixel variation.

In this work, an improved approach for accurate chamfer matching is presented that uses interpolation in the distance calculation for subpixel distance evaluation. For more accurate matching, instead of estimating the optimal position in subpixel using the neighborhood of the pixel position with minimum distance, we use an optimization method like Powell's [6] that can find the minimum of a function through actual function evaluations in subpixel. Experimental results are presented to show the validity of our approach.

2 Chamfer Matching

2.1 Overview

Chamfer matching [2] is a technique that can recognize an object via 2-dimensional edge contour matching. The edge contour image of a known object is called "prepolygon image", and the edge contour image of the image, within which the object is searched, is called "predistance image". The edge contour image is a binary image consisting of edges extracted by an edge detection operator.

In the predistance image, each non-edge pixel is given a distance value from nearest edge pixel. Computation of the true Euclidean distance requires excessive time and memory resources, therefore an approximation is desirable. The operation converting a binary image to an approximate distance image is called a distance transformation (DT). The DT used in the matching algorithm should be a reasonably good approximation of the Euclidean distance; otherwise the discriminating power of the matching measure, computed from the distance values, becomes poor. It is also desirable that global distances in the image are approximated by propagating local distances (distances between neighboring pixels) over the image.

The 3-4 DT, a common choice for the DT, uses 3 for the distance between horizontal/vertical neighbors and 4 for the distance between diagonal neighbors in the 3x3 neighborhood. The 3-4 DT is known to have maximum difference of 8% compared with the Euclidean distance, thus being a good approximation of the Euclidean distance [2]. Since edge points are influenced by noise, computing the exact Euclidean distance is usually not necessary. For a more accurate DT than the 3-4 DT, the 5-7-11 DT [11] based on the 5x5 neighborhood could be used, although its computation should be more complex.

The distance image based on the DT can be obtained very efficiently by a sequential DT algorithm [2]. In the binary edge image, each edge pixel is first set to zero, and each non-edge pixel is set to infinity. Then, two passes are made over the image, first "forward" in the raster scan order (from left to right and from top to bottom), and then "backward" in the reverse raster scan.

In the prepolygon image edge pixels are extracted and converted to a list of (x,y) coordinate pairs, where x and y represents column and row number, respectively.

From this list the edge points that are actually used for matching are chosen. The list of selected points is called the "polygon".

When the polygon is superimposed on the distance image, an average of the pixel values of the distance image at the polygon coordinates is the measure of correspondence between the edges, called the "edge distance". A perfect fit between the two edges will yield edge distance zero since each polygon point corresponds to a pixel of distance value zero in the distance image. The actual matching is minimizing this edge distance. For the matching measure, one can choose median, arithmetic average, root mean square average (r.m.s.), or maximum.

An edge distance is computed for each position of the polygon, determined by the transformation equations. The position with the minimal edge distance is defined as the position with the best fit. The transformation equations that change the position of the polygon points should be parametric. Let (x,y) be the polygon coordinates and (X,Y) the position in the distance image. For example, for translation and rotation, the transformation equations become $X = c_x + x \cos\Theta - y \sin\Theta$ and $Y = c_y + x \sin\Theta + y \cos\Theta$, where Θ is the rotation angle, and c_x and c_y are the translation parameters in the X- and Y-directions, respectively. Since the (X,Y) coordinates are not usually integers, they are frequently rounded to the nearest integer values.

Finding the optimal polygon position is equal to finding the global minimum of a multidimensional function. Thus, the minimization must be started very close to the optimal position to avoid falling in false local minima. For efficient search, a hierarchical matching algorithm is frequently used. If the minimum edge distance found is below a threshold (typically, 1 – 1.5), it is considered that there is the model in the search image.

2.2 Implementation

Although there exist many methods for edge detection including the Canny edge detector [7], the Sobel edge operator [8] was used here for computational efficiency and good performance to extract the edge magnitude image.

Thresholding the edge gradient magnitude image obtained by applying the Sobel edge operator usually contains many thick edges. As the number of edges is increased, the matching speed gets slower. Thus, to reduce the number of edges, a non-maxima suppression technique [9] was used for thinning edges, which suppresses all values not being the local maximum along the line of the gradient direction in the edge gradient magnitude image.

Fig. 1 shows an example of a model image and a search image, within which the model is searched. The search image was obtained by rotating by 12 degrees counterclockwise an image, from which Fig. 1(a) was extracted. Fig. 2 shows the edges extracted by applying the Sobel edge operator followed by the non-maxima suppression. Fig. 3 shows a distance image obtained by applying the 3-4 DT to Fig. 2(b). Edge pixels are completely black, and pixels are getting brighter as they get more distant from the edge pixels.

(a) (b)

Fig. 1. (a) A model image. (b) A search image, within which the model is searched.

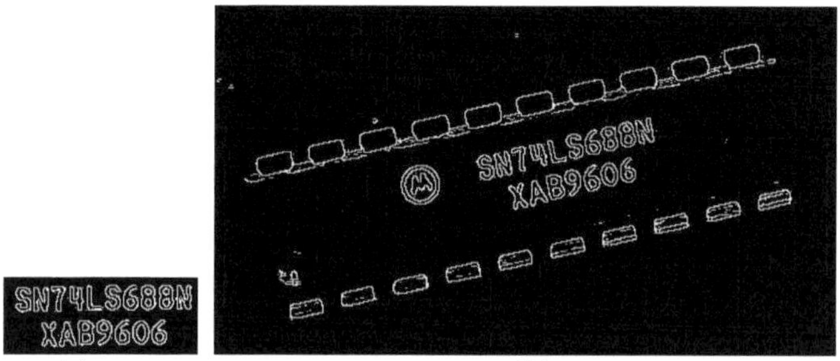

(a) (b)

Fig. 2. (a) Edges of Fig. 1(a). (b) Edges of Fig. 1(b).

Fig. 3. A distance image of Fig. 2(b)

For the matching measure, among median, arithmetic average, root mean square average (r.m.s.), and maximum, r.m.s. was used here since it is known to give significantly fewer false minima than others [2]. The r.m.s. is given by

$$d_{rms} = \frac{1}{C}\sqrt{\frac{1}{n}\sum_{i=1}^{n} d_i^2},$$

where d_i is the distance value of the i-th point among n points in the polygon, and C is a constant to compensate the unit distance in the DT. (C=3 for the 3-4 DT, and C=5 for the 5-7-11 DT.)

Start positions for chamfer matching are very important since they should be close to the optimal matching position to avoid falling in false local minima. Thus, the start positions were provided as outputs of a generalized Hough transform [10] for approximate matching positions. Here, three pose transformation parameters, translation and rotation, were assumed. Since in many industrial applications the appearance of the object to be found has limited degrees of freedom, rigid transformation (translation and rotation) is often sufficient.

Fig. 4 shows the results of matching the model of Fig. 2(a) with the edges of Fig 2(b); model edges are accurately overlaid on the test image, demonstrating that the matching position and angle were correctly found. (The positions of edges are denoted in black, and the white cross represents the reference point of the model, which is set to the center of the model image.)

Fig. 4. The results of matching using the model of Fig. 2(a)

2.3 An Improved Algorithm

In computing the edge distance for each position of the polygon, determined by the transformation equation from the polygon coordinates (x,y) to the position (X,Y) in the distance image, the transformed coordinates (X,Y) are not usually integers, thus has been usually rounded to the nearest integer values.

However, for more accurate matching, we propose that the distance value at (X,Y) should be interpolated using adjacent pixels around (X,Y) of the distance image. Bi-linear or bi-cubic interpolation can be used for this purpose. If bi-linear interpolation is used in the distance image D, the interpolated distance D_{xy} at (X,Y) can be computed by

$$\begin{aligned} D_{XY} &= (1-\alpha)(1-\beta)D(\lfloor X \rfloor, \lfloor Y \rfloor) + (1-\alpha)\beta D(\lfloor X \rfloor, \lfloor Y \rfloor + 1) \\ &\quad + \alpha(1-\beta)D(\lfloor X \rfloor + 1, \lfloor Y \rfloor) + \alpha\beta D(\lfloor X \rfloor + 1, \lfloor Y \rfloor + 1) \end{aligned}$$

where $\alpha = X - \lfloor X \rfloor$, $\beta = Y - \lfloor Y \rfloor$, and $\lfloor X \rfloor$ represents the truncated integer of X.

Table 1 shows the results of applying chamfer matching at the pixel level with/without bilinear interpolation in the distance image. We can notice that the distance value is considerably decreased by using the distance interpolation.

Table 1. The results of applying chamfer matching

	Position [pi xel]	Angle [degree]	Distance value
Without distance interpolation	(292, 130)	12	0.537
With distance interpolation	(292, 130)	12	0.444

To find more accurate matching parameters, a subpixel algorithm like a simple 1-dimensional parabolic interpolation method can be used. Suppose some evaluation function f(x) has the maximum value at discrete pixel value x=H, and let $f(H-1)=f_l$ and $f(H+1)=f_r$, then the locally maximum subpixel position H_s can be estimated by the following parabolic interpolation:

$$H_s = H + \frac{f_r - f_l}{2(2f_m - f_r - f_l)}$$

If the evaluation function has three parameters x, y, and angle, the 1-d interpolation method is applied independently for each parameter to estimate parameter values in subpixel. However, this method does estimate the matching position using interpolation without actual evaluation of the distance value in subpixel. Thus, a more accurate matching can be achieved by an optimization method with the interpolated distance evaluation for the minimum chamfer distance search. To this end, the Powell's method [6] may be used since it needs only the function evaluation without requiring function derivative computation.

Table 2 shows the results of applying the subpixel algorithms to Fig. 4 without the distance interpolation, and Table 3 shows the results of applying the subpixel

Table 2. The results of applying chamfer matching without the distance interpolation followed by the subpixel algorithm

	Position[pixel]	Angle[degree]	Distance value
Before subpixel algorithm	(292, 130)	12	0.537
After subpixel algorithm (parabolic interpolation)	(291.82, 129.80)	12.08	0.478
After subpixel algorithm (Powell)	(291.76, 129.73)	12.00	0.463

Table 3. The results of applying chamfer matching with the distance interpolation followed by the subpixel algorithm

	Position[pixel]	Angle[degree]	Distance value
Before subpixel algorithm	(292, 130)	12	0.444
After subpixel algorithm (parabolic interpolation)	(291.82, 129.79)	12.08	0.386
After subpixel algorithm (Powell)	(291.72, 129.70)	12.00	0.378

algorithms to Fig. 4 with the distance interpolation. We can notice that the distance value is further reduced by the subpixel algorithm. For the subpixel algorithm, the Powell's method seems to yield better results than the parabolic interpolation estimation.

3 Accuracy Test of the Improved Algorithm

The accuracy of the improved algorithm implemented was measured. First, to measure x, y accuracy, an image containing Fig. 1(a) as a subimage, was translated in x and y axis direction in the range of -0.8 ~ 0.8 pixel by step size of 0.2 pixel, thus yielding 80 images in total. Using these images, the algorithm was run to obtain the translational accuracy (standard deviation and maximum error). Also, the test image was rotated within the range of +/-30 degrees by step size of 1 degree, and 60 images in total were obtained. Our algorithm was also run for these images to estimate the error of the rotation angle.

These results are shown in Table 4 and Table 5. (The model image used is Fig. 1(a), and bilinear interpolation was utilized in the translation and rotation of the image.) We can notice that the subpixel algorithm based on the parabolic interpolation yields x, y accuracy of 0.14 pixel and angle accuracy of 0.15 degree in the worst case, while the Powell's method yields x accuracy of 0.06 pixel, y accuracy of 0.08 pixel, and angle accuracy of 0.04 degree in the worst case. Without the distance interpolation, the Powell's method for subpixel search gives poor results. The reason for this is that distance evaluation without interpolation should not be accurate due to insensitiveness to very small variations of x, y, and angle, thus being easily fallen into false local minima. Thus, the Powell's subpixel algorithm with the chamfer distance interpolation is found to be the most accurate.

The 3-4 DT was used throughout the above test, but we obtained nearly the same results with the 5-7-11 DT, known to be more accurate than the 3-4 DT. However, the 5-7-11 DT may be preferable to the 3-4 DT depending on the application.

Table 4. Accuracy of position and angle estimated without the chamfer distance interpolation

Subpixel method		x [pixel]	y [pixel]	angle [degree]
Parabolic interpolation	Standard dev.	0.09	0.09	0.07
	Max. error	0.14	0.12	0.19
Powell's optimization	Standard dev.	0.30	0.32	0.06
	Max. error	0.42	0.71	0.31

Table 5. Accuracy of position and angle estimated with the chamfer distance interpolation

Subpixel method		x [pixel]	y [pixel]	angle [degree]
Parabolic interpolation	Standard dev.	0.09	0.09	0.07
	Max. error	0.14	0.12	0.15
Powell's optimization	Standard dev.	0.03	0.03	0.02
	Max. error	0.06	0.08	0.04

4 Conclusion

In this paper, an improved approach for accurate chamfer matching was presented that uses interpolation in the distance calculation for subpixel distance evaluation. For more accurate matching, instead of estimating the optimal position in subpixel using neighborhood of the pixel position with the minimum distance, we used the Powell's optimization to find the distance minimum through actual distance evaluations in subpixel. Experimental results demonstrated the effectiveness of our approach.

References

1. Barrow, H.G., Tenenbaum, J.M., Bolles, R.C., Wolf, H.C.: Parametric correspondence and chamfer matching: Two new techniques for image matching. In: Proc. 5th Int. Joint Conf. Artificial Intelligence, Cambridge, MA, pp. 659–663 (1977)
2. Borgefors, G.: Hierarchical chamfer matching: a parametric edge matching algorithm. IEEE Trans. Pattern Analysis and Machine Intelligence 10(6), 849–865 (1988)
3. Chetverikov, D., Khenokh, Y.: Matching for Shape Defect Detection. In: Solina, F., Leonardis, A. (eds.) CAIP 1999. LNCS, vol. 1689, pp. 367–374. Springer, Heidelberg (1999)
4. Gavila, D.: Pedestrian detection from a moving vehicle. In: Vernon, D. (ed.) ECCV 2000. LNCS, vol. 1842, pp. 37–49. Springer, Heidelberg (2000)
5. Thayananthan, A., Stenger, B., Torr, P.H.S., Cipolla, R.: Shape context and chamfer matching in cluttered scenes. In: Proc. CVPR 2003, Madison, Wisconsin, pp. 127–135 (2003)
6. Press, W.H., Flannery, B.P., Teukolsky, S.A., Vetterling, W.T.: Numerical Recipes in C, 2nd edn. Cambridge University Press, Cambridge (1992)
7. Canny, J.: A computational approach to edge detection. IEEE Trans. Pattern Analy. and Mach. Intelli. 8(6), 679–698 (1986)
8. Davies, E.R.: Machine Vision, 3rd edn. Morgan Kaufmann, San Francisco (2005)
9. Jain, R., Kasturi, R., Schunck, B.G.: Machine Vision. McGraw-Hill, New York (1995)
10. Ballard, D.H.: Generalizing Hough transform to detect arbitrary shapes. Pattern Recognition 13(2), 111–122 (1981)
11. Borgefos, G.: Distance transformations in digital images. Computer Vision, Graphics, and Image Processing 34(3), 344–371 (1986)

Automatic Segmentation of Fibroglandular Tissue

Christina Olsén and Aamir Mukhdoomi

Department of Computing Science,
Umeå University, SE-901 87 Umeå, Sweden
{colsen, c01ami}@cs.umu.se

Abstract. In this paper, a segmentation algorithm is proposed which extracts regions depicting fibroglandular tissue in a mammogram. There has been an increasing need for such algorithms due to several reasons, the majority of which are related to the development of techniques for Computer Aided Diagnosis of breast cancer from mammograms. The proposed algorithm consists of a major phase and a post-processing phase. The purpose of the major phase is to calculate the threshold value that yields a segmentation of glandular tissue which is achievable by thresholding. The method by which we calculate this threshold value is based on the principle of minimizing the cross-entropy between two images. The resulting segmentation is then post-processed to remove artifacts such as noise and other unwanted regions. The algorithm has been implemented and evaluated with promising results. In particular, its performance seems to match that of medical professionals specialized in mammography.

1 Introduction

Breast cancer in females often occur in fibroglandular part of the breast tissue. Due to several reasons, the majority of which are related to the development of techniques for Computer Aided Diagnosis of breast cancer from mammograms, there has been an increasing need for develop an automated segmentation algorithm for extracting the glandular tissue. A first step in that process might be to locate and extract the glandular tissue disc. The next step would be to analyze the different structures within that disc. The location and delineation of the glandular tissue disc is also important, as a quality assurance, in order to determine if the entire glandular tissue disc is depicted in the mammogram. To make this procedure fully automatic is a highly difficult segmentation task since many different tissue types within the tree-dimensional breast superimpose as projected down to a two-dimensional image plane.

Several methods have been proposed previously to perform the segmentation of glandular tissue in digital mammograms. Saha et al. [13] described an automatic method to segment dense tissue regions from fat within breasts using scale-based fuzzy connectivity methods. Bovis and Singh [1] and Karssemeijer [4] have devised segmentation algorithms based on classification methods through

B.K. Ersbøll and K.S. Pedersen (Eds.): SCIA 2007, LNCS 4522, pp. 679–688, 2007.

feed-forward Artificial Neural Network and kNN classifier respectively. Various studies found in the literature [12], are also based on the texture analysis of the tissue in mammograms. Different tools are used for analyzing texture such as for instance fractal dimensions and image filters [5].

In addition to the kind of methods mentioned above, a method based on a simple concept of analyzing the histogram of an image is also found in the literature. From this method an optimal threshold value is calculated that can be used to create a distinction between two regions of the image that have different intensity compositions. In this case, the required optimal threshold should create a clear distinction between the fibroglandular tissue and the remaining breast tissue. This method of finding the optimal threshold is called *Minimum Cross-Entropy*, and was used by Masek [8] to segment different regions in digital mammograms. His method however requires manual setting of many parameters for the different types of tissue. The aim of this paper is to develop an objective and fully automated segmentation algorithm for extracting the glandular tissue disc from mammograms. To reach this objective the idea by Masek [8] is used, because of its simplicity, robustness and the fact that the segmentation is independent of the exact location of the glandular tissue disc in a mammogram.

2 The Method

In this paper, a collection of 200 mammograms were randomly chosen from the two databases that are commonly used in image analysis projects related to digital mammography, i.e. *Mammographic Image Analysis Society's* (MIAS) digital mammography database and the *Digital Database for Screening Mammography* (DDSM). These 200 mammograms are part of a bigger research project, where the randomly chosen images were manually examined by five medical experts [10]. Among other tasks, these experts were asked to mark what they perceived as anatomical landmarks in each mammogram. This dataset was divided into 40 randomly chosen training images used during the development and 160 test images used for evaluation of the proposed algorithm. All the images used have dimension 1024×1024 and are 8-bit grayscale images.

2.1 Cross-Entropy Based Thresholding

It has already been established that the region of interest (ROI) for this algorithm is the region of the mammogram that depicts glandular tissue. In the process of extracting the glandular tissue disc it is assumed that the position of the nipple, the breast boundary, the pectoralis muscle and other unwanted regions such as for instance labels are extracted by Olsén's algorithms [9].

In order to use thresholding based segmentation, it is important to examine the images in question to determine if their intensity composition allows segmentation of ROIs by thresholding. However, in these images it can be observed that there are well defined peaks that correspond to the glandular tissue in the histogram of each mammogram. Such a peak has been found to exist in all the mammograms in varying shapes and sizes. Therefore a robust algorithm is needed that can detect this peak and calculate an acceptable threshold intensity with consistent performance independent of the shape and size of the peak.

Cross-Entropy. The concept of cross-entropy was first introduced by Kullback [6], and has proven to be very useful in different applications as it provides a robust method to measure how good one probability distribution approximates another probability distribution by quantifying the difference between the two.

If P and Q are two discrete probability distributions, such that $P = p_1, \ldots, p_n$ and $Q = q_1, \ldots, q_n$, then the cross-entropy distance between P and Q is given as:

$$HCE(Q, P) = \sum_{i=1}^{n} q_i log \frac{q_i}{p_i} \tag{1}$$

Minimum Cross-Entropy and the LLBP Approach. If a digital image substitutes P in Equation (1) and its thresholded version substitutes Q, then the cross-entropy measure explains how good the thresholded image estimates the original image. Following this line of thought, the principle of minimum cross-entropy states that in order to find the solution (i.e. thresholded image) that best estimates the original image, one should minimize the difference (cross-entropy) between them. In our case this will practically involve minimizing cross-entropy values between an image and its thresholded version over a certain range of threshold values.

Li and Lee [7] and Brink and Pendock [2] have designed image thresholding algorithms based on minimum cross-entropy as well. These two algorithms, along with two other approaches [15,11], have been tested and evaluated on synthetic images as well as actual mammograms by Masek [8]. Masek [8] has combined them into a single method designated as *LLBP approach* that has been shown to give best result when used to segment components of a mammogram. The LLBP approach is the basic idea behind the algorithm proposed below. Formally stated, the main principle in LLBP approach is that, given an image with histogram P, one strives to find a thresholded histogram Q such that distance (cross-entropy) between P and Q is minimized.

2.2 The Complete Segmentation Algorithm

The image shown in Fig. 1(a) will be used as an example of an input image, I_{input}, to explain the different steps of this algorithm. Fig. 1(b) shows the histogram h_{breast} of this image. Black background can and should be exempted from any calculations since this part will not include any tissue depicted in the image, therefore the histogram is calculated on the breast tissue part, I_{breast}, of the image only. This algorithm is divided into two phases. First, an optimal thresholded image is acquired and then the image is post processed (including several image processing operations) to compensate for incorrect segmentations.

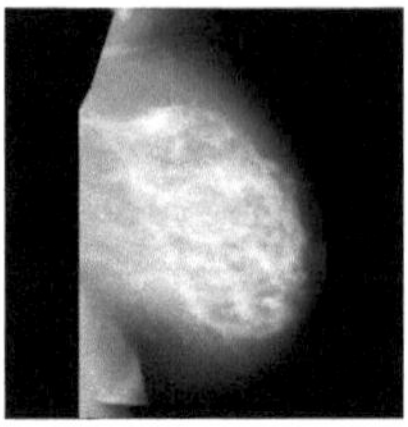

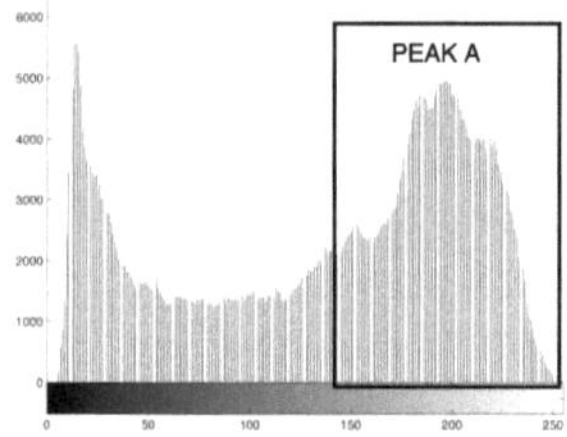

Fig. 1. a) Image mdb032 from MIAS database. b) Image histogram of the pixels belonging to breast tissue of mammogram.

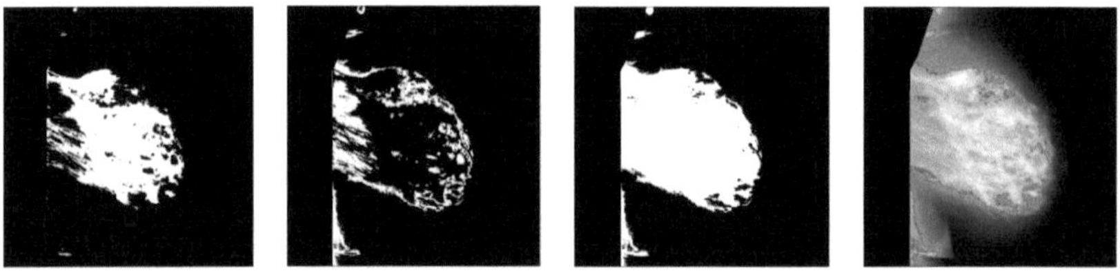

(a) (b) (c) (d)

Fig. 2. a) Segmentation produced by thresholding in stage 1. b) Segmentation produced by thresholding in stage 2. c) The combined segmentation from stage 1 and stage 2, obtained as the union of the segmentations in a) and b). d) Segmentation from c) marked in the input image from Fig. 1(a).

Phase I. Let $CEM(f(.), t_1, t_2)$ denote a function that performs cross-entropy minimizations on the image as will be explained below. The minimization is performed over a range of intensity values, in which each intensity value is chosen as a threshold and cross-entropy is calculated between the original and the thresholded image until the minimum cross-entropy is reached. Here, $f(.)$ is the image to be thresholded and the parameters t_1 och t_2 define the respective lower and upper bound of this range, as explained later.

Through series of tests it has been observed that the choice of an appropriate range of intensity values, especially the lower bound, for the minimization process is important to produce an optimal result. Experimentation shows that peak A corresponds to the pixels belonging to the glandular tissue (Fig. 1(b)). Empirical tests show that a threshold value near the left end of peak A as an initial value often gives the desired optimal threshold. A reliable and automated method to calculate such an initial value has been devised here that uses the cumulative histogram to detect the ascents and descents of the major peaks in an image histogram. The bends in the cumulative histogram curve correspond to ascends and descends of peak A in the image histogram. To find these bends a line segment, joining the end points of the cumulative histogram curve, is used to calculate the distance between the points on the curve and the line segment.

The point farthest away from the line has shown to be a good initial value for the cross-entropy minimization [3].

Depending on the size and structure of glandular tissue disc, the form of peak A can vary. Kurtosis is a measure useful for describing the histograms relative flatness. Calculations show that if the tip of this peak is flatter in form compared to the tip of a normal distribution i.e. if $kurtosis(h_{breast}) < a$, then the above mentioned method gives a good initial value. However, for images that give $kurtosis(h_{breast}) \geq a$, such an initial value causes the minimization to converge too fast towards a local minimum giving a too low threshold value. In these cases, the intensity value with highest count in the image histogram has proven to be a better initial value. This step of using the maximum intensity count as the initial value instead of using the cumulative histogram is called *the max step*.

With a suitable initial value, cross-entropy minimization is performed to receive a suitable threshold used to create the thresholded image. This is referred to as *stage 1 thresholding* and is illustrated in Fig. 2(a). Let T_{s1} be the threshold value from stage 1 and I_{s1} denote the set of image pixels from I_{breast} that have intensity higher than T_{s1}. Then I_{s1} denotes the glandular tissue disc according to stage 1 thresholding.

In many cases, stage 1 leaves out parts of glandular tissue disc due to too high threshold value. Therefore thresholding through the function CEM is repeated in stage 2. The input image is denoted $I_{input_{s2}}$ to make it clear that it is the input image to stage 2. $I_{input_{s2}}$ consists of all the pixels that belong to the breast tissue except those pixels that were segmented in stage 1 thresholding. It has been observed that the set of pixels of $I_{input_{s2}}$ does not require the *max step*, since the minimization does not converge to local minima as in stage 1. Therefore the initial value is chosen using the cumulative histogram method only. The resulting thresholded image from this stage will be referred to as $I_{output_{s2}}$, see Fig. 2(b). The segmentations from stage 1 and stage 2 are combined to create a single segmented image, denoted $I_{output_{s1 \vee s2}}$, as shown in Fig. 2(c).

Phase II. During the training of the proposed algorithm it was observed that in some cases the resulting extraction of the glandular tissue disc did not agree with the markings of the glandular tissue disc done by five experts. The reason was that in these cases regions that have a high probability of not belonging to the actual glandular tissue disc was not properly removed by the steps in Phase I. Therefore, in this phase, the final segmentation from phase I is post-processed in order to remove these regions. These regions are a result of image noise and/or simply the fact that threshold based segmentation always has its limitations. To remove unwanted regions and improve the segmentation, a sequence of image processing operations has been devised, that can be divided into several procedures.

First we want to remove unwanted regions that might belong to fat or muscle tissue. The area of all connected components (clusters of white pixels) in the $I_{output_{s1 \vee s2}}$ is calculate, i.e the number of pixels for each cluster. A limit is estimated such as that A_{limit} is the fraction between the largest area and

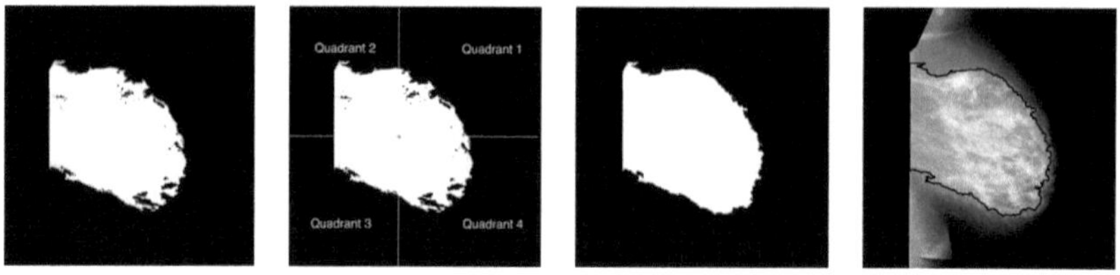

(a) (b) (c) (d)

Fig. 3. a) Binary image BW_{large}. b) Division of BW_{large} into four quadrants with the centroid of the largest region as point of the origin. c) The final binary segmentation. d) The final segmentation c) marked in the original input image (Fig. 1(a)).

a constant k, $A_{limit} = \frac{A_{largest}}{k}$. Based on these calculations a new binary image, $BW_{largeinitial}$, is created that only contains regions from $I_{output_{s1 \vee s2}}$ that have area greater than or equal to A_{limit}. However, for cases were a elongation representing the muscle is connected to a large cluster in $BW_{largeinitial}$, further post-processing is necessary. This region is detected by analyzing relatively abrupt changes in width of the regions. For this purpose, $BW_{largeinitial}$ is sliced horizontally into l rectangular small images of equal heights. In each sliced image, width of the middle row is calculated to represent the width of the whole slice. If the difference in width between the set of two consecutive slices is greater than a certain limit, m then the change in width is too sharp and thus revealing the presence of a elongated upper region. When this limit is reached the segment is cut off to exclude the unwanted elongation. However, it is important that the limit of differences, m, is calculated such that it has the same relativity to the widest part of the segmentation. To ensure this, the segmented region is resized so that the widest part has width of n pixels. Ratio between height and width of the segmented region is preserved in all cases. The final segmentation of this post-processing procedure, thus containing only the regions with large pixel count, is denoted BW_{large} (Fig. 3(a)).

A shortcoming of removing connected elongation is that too many small regions might be excluded resulting is underdetermined glandular tissue disc. It is the anatomical fact that glandular tissue gets more concentrated and compact as one moves closer to the nipple from inside of the breast. That is, the probability of a small region belonging to glandular tissue disc increases. In addition, all regions that are located in the same quadrant as the nipple are assumed to belong to the glandular tissue. Therefore, BW_{large} is divided into 4 quadrants like a Cartesian coordinate system such that the centroid of the largest region acts as the point of the origin; see Fig. 3(b). Since the position of the nipple is already known, its quadrant is determined. Let BW_{extra} be, $BW_{extra} = I_{output_{s1 \vee s2}} \setminus BW_{large}$, in other words, all the smaller regions that were excluded in the post-processing procedure resulting in the binary image BW_{large}. Centroid for all these regions is calculated and a new binary image is created that contains small regions whose centroid is located in the same quadrant as the nipple. This new binary image is denoted BW_{mam}.

The small regions BW_{small}, is defined as $BW_{small} = BW_{extra} \setminus BW_{mam}$, and is processed separately. For each quadrant a rectangle and a circular sector is calculated. Let x_{max} and y_{max} be respective maximum horizontal and vertical distances achieved by segmented pixels in BW_{large}, then x_{max} and y_{max} are used as width and height of the rectangle. The maximum radial distance from the origin, achieved by segmented pixels in BW_{large} is used as a radius to calculate the sector. All of the small regions, whose centroid lies within the intersection of the rectangle and the sector, are accepted as part of the glandular tissue disc. This procedure is repeated for all the quadrants except the quadrant of the nipple.

Finally, the union of all the binary images, obtained in previous post-processing procedures, is returned as the final segmentation. Morphological closing is done to smooth the boundary of this segmentation. As a last step, any holes present in this segmentation are also filled to insure a closed boundary of the glandular disc. The resultant binary- and intensity images respectively are shown in Fig. 3(c) and Fig. 3(d).

3 Results

3.1 Empirical Evaluation of the Different Parameters

The results from the training data show that the desired result of the cross-entropy minimization procedure in stage 1, explained in Phase I Sect. 2.2, is reached if performed on pixels within intensity range $[max(I_{min}, t_1), min(t_2, I_{max})]$. Here, I_{min} and I_{min} are the minimum and maximum intensity values in the input image respectively and $t_1 = 20$ and $t_2 = 200$. For stage 2, where the cross-entropy minimization procedure is iterated, the parameters t_1 and t_2, should be 20 and $max(I_{input_{s2}})$ respectively.

Furthermore, for the post-processing procedures, in Phase II Sect. 2.2, empirical tests on the 40 training images show that an optimal post-processing performance is reached if the parameters have certain defined values. For the kurtosis analysis $a = 3$ is shown to provide successful results for estimating the degree of flatness of the image histogram. The optimal constant, k for calculating the area limit, is found to be $k = 5$. Here, the area limit was useful for determining which clusters where large enough to be assumed to represent the depicted glandular tissue disc in the original image.

We saw that for some cases where a elongation representing the muscle is connected to a large cluster it is desirable to remove this part. Therefore the current binary image at that stage was sliced horizontally into a number of rectangular small images of equal heights. The number of rectangular small images needed for this post-processing procedure is $l = 10$. This elongation was automatically detected by a difference analysis of the width between the set of two consecutive rectangular small images and the limit for abrupt changes was empirical found to be $m = 140$. It is also necessary for this part that $n = 500$.

Table 1. Result from an evaluation method proposed by Olsén and Georgsson, [10]. Rank 1 corresponds to the best agreement and rank 6 to the least agreement with the rest of the markings in the ensemble ($A_1 - A_6$). The level of agreement is calculated on a leave one out basis.

	Rank sum	Expert
1	359	A_4
2	365	A_2
3	434	A_5
4	436	A_3
5	480	A_6
6	488	A_1

3.2 The Performance of the Algorithm

A common method of measuring the performance, of an automated segmentation algorithm, is to compare the results with a ground-truth. It is impossible to establish true segmentation of the glandular tissue disc. Hence, for segmentation evaluation, we need to choose a surrogate of true segmentation. In the field of medical image analysis, a ground-truth is often obtained from markings manually created by human-experts. Segmentation evaluation based on ground truth obtained by a group of experts has been studied for a long time. However, the most appropriate way to compare computer generated segmentation to segmentations created by a group of experts is so far unclear [14]. One of the principal reasons for absence of a standard evaluation method is that the inter-expert variations among manually created segmentations are very high. This makes it difficult to create an objective ground-truth.

Fig. 4 shows an example of the inter-expert variations between the manually segmented regions of the glandular tissue disc, in the same mammogram, provided by five experts in radiology. By observing Fig. 4, it can be concluded that an automated segmentation algorithm cannot be tested against a ground-truth from just one expert, because the evaluation would be subjective. Olsén and Georgsson [10] discuss problems involved in using manually created ground-truths and propose a method striving to create an objective performance measure, based on markings provided by experts. The authors have looked at the usage of ensembles of domain experts when assessing the performance of a segmentation algorithm. Olsén and Georgsson [10] estimate a function that calculates the degree of agreement of a given segmentation with the estimated ground-truth. All the five expert markings along with the result from the proposed segmentation algorithm (referred to as the *system segmentation*) are individually ranked by this function. A rank sum is then defined which expresses the overall agreement of expert markings and the system segmentation for the whole set of test images. The rank sum calculated for the test set of 160 images is shown in Table 1, where ranks are arranged in descending order.

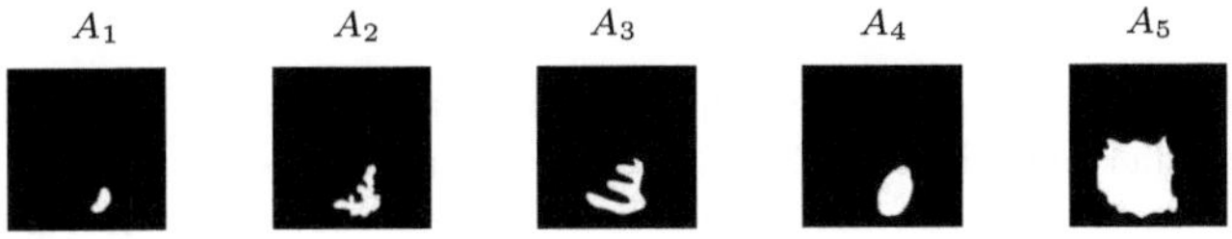

Fig. 4. The glandular tissue disc manually outlined in mammogram no. 138 (MIAS) by a panel of five experts ($A_1 - A_5$) in mammography. The inter-expert variations between the manually segmented regions of the same glandular tissue disc in the same mammogram is, as visible, large.

The five expert markings used to estimate the ground-truth are denoted as $A_{1...5}$ and A_6 denotes the system segmentation. This result indicates that proposed algorithm mixes in with the human experts. In other words, the performance of the proposed algorithm seem to match the human professionals in mammography.

4 Discussion

In Sect. 3, inter-expert variations were explained as one of the primary reasons to why it is difficult to make an objective assessment of segmentation algorithms, for fibroglandular tissue in mammograms. Furthermore, an example of evaluation was presented by showing results from a newly developed method [10]. Since this evaluation was based on markings from a certain group of experts, it is possible to get different ranks for a different expert panel. Therefore the results presented in Table 1 can not be considered as exact performance evaluation for the algorithm developed in this paper, but rather an estimation of how good segmentation this algorithm produces on average.

5 Conclusion

A novel approach for extracting the glandular tissue disc has been presented in this paper. This approach was based on global histogram analysis of the glandular regions in mammograms. Based on the arguments in Sect. 3, it can be concluded that this algorithm is able to calculate a good approximation of the location and delineations of the glandular tissue disc. The only truths we can use for segmentation evaluation is medical experts in the field. However, since the most appropriate way to compare computer generated segmentation to segmentations created by a group of experts is so far unclear, evaluation of segmentation algorithms for mammograms is very complex. In fact, in fields were only surrogate truths are available formation of an objective evaluation method is as important as construction of the segmentation routines that are being evaluated.

References

1. Bovis, K., Singh, S.: Classification of mammographic breast density using a combined classifier paradigm. PANN Research, Department of Computer Science, University of Exeter, Exeter, UK. In: 4th International Workshop on Digital Mammography, pp. 177–180 (2002)
2. Brink, A.D., Pendock, N.E.: Minimum Cross-Entropy Threshold Selection. Pattern Recognition 29(1), 179–188 (1996)
3. Chandrasekhar, R., Attikiouzel, Y.: Automatic breast border segmentation by background modeling and subtraction. In: M. J. Yaffe, editor, Digital Mammography, Computational Imaging and Vision, Medical Physics Publishing, Madison, Wisconsin pp. 560–565 (2000)
4. Karssemeijer, N.: Automate classification of parenchymal patterns in mammograms. Physics in Medicine and Biology 43, 365–378 (1998)
5. Kasparis, T., Charalampidis, D., Georgiopoulos, M., Rolland, J.: Segmentation of textured images based on fractals and image filtering. The Journal of Pattern Recognition Society, 34 (2001)
6. Kullback, S.: Information Theory and Statistics. John Wiley, New York (1959)
7. Li, C.H., Lee, C.K.: Minimum Cross-Entropy Thresholding. Pattern Recognition 26(4), 617–625 (1993)
8. Masek, M.: Hierarchical Segmentation Of Mammograms Based On Pixel Intensity. PhD thesis, Centre for Intelligent Information Processing Systems, School of Electrical, Electronic, and Computer Engineering. University of Western Australia, Crawley, WA, 6009 (February 2004)
9. Olsén, C.: Automatic Assessment of Mammogram Adequacy. Licentitate Thesis, Umeå University, Department of Computing Science (2005)
10. Olsén, C., Georgsson, F.: Assessing ground truth of glandular tissue. In: Astley, S.M., Brady, M., Rose, C., Zwiggelaar, R. (eds.) IWDM 2006. LNCS, vol. 4046, pp. 10–17. Springer, Heidelberg (2006)
11. Pal, N.R.: On Minimum Cross-Entropy Thresholding. Pattern Recognition 29(4), 575–580 (1996)
12. Rangayyan, R.M.: Biomedical Image Analysis. In: Biomedical Engineering Series, CRC Press LLC, Boca Raton (2005)
13. Saha, P.K., Udupa, J.K., Conant, E.F., Chakraborty, D.P., Sullivan, D.: Breast tissue density quantification via digitized mammograms. IEEE Transactions On. Medical Imaging 20(792–803), 575–580 (2001)
14. Warfield, S.K., Zou, K.H., Wells, W.M.: Simultaneous truth and performance level estimation (staple): An algorithm for the validation of image segmentation. IEEE Transaction on Medical Imaging 23, 903–921 (2004)
15. Xue, Y., Zhang, Y., Lin, X.: Threshold selection using cross-entropy and fuzzy divergence. Proceedings of the SPIE 3561, 152–162 (1998)

Temporal Bayesian Networks for Scenario Recognition

Ahmed Ziani and Cina Motamed

Laboratoire LASL EA 2600
Université du Littoral Côte d'Opale
Bat 2, 50 Rue F.Buisson
62228 Calais France
ziani@lasl.univ-littoral.fr, motamed@lasl.univ-littoral.fr

Abstract. This work presents an automatic scenario recognition system for video sequence interpretation. The recognition algorithm is based on a Bayesian Networks approach. The model of scenario contains two main layers. The first one enables to highlight atemporal events from the observed visual features. The second layer is focused on the temporal reasoning stage. The temporal layer integrates an event based approach in the framework of the Bayesian Networks. The temporal Bayesian network tracks lifespan of relevant events highlighted from the first layer. Then it estimates qualitative and quantitative relations between temporal events helpful for the recognition task. The global recognition algorithm is illustrated over real indoor images sequences for an abandoned baggage scenario.

Keywords: Visual-surveillance, scenario recognition, image sequence analysis, Bayesian Network.

1 Introduction

The ability to automatically resume and index video content is an important challenge of the vision and data base communities. The motion interpretation of video is an important area of investigation. Indeed in MPEG-7 standard, a set of general motion descriptors has been defined: camera motion, object motion trajectory, parametric object motion and motion activity.

The purpose of this work is the recognition of specific scenarios of human activity in visual surveillance applications [1]. For such dynamic scenes, scenarios are based on a combination of spatial, temporal and interacting events. This automatic recognition is firstly helpful for video-surveillance operator for an on-line alarm generation by highlighting abnormal situation. The second utility concerns the off-line retrieval of specific behaviours from a stored image sequence. This capability becomes naturally more powerful when the monitoring implies several cameras working simultaneously.

From a video indexing point of view, the specificity of this application with respect to general video sequences is that camera parameters and the majority of the observed background remain fixed. These favourable conditions permit to focus the attention on the object motion.

B.K. Ersbøll and K.S. Pedersen (Eds.): SCIA 2007, LNCS 4522, pp. 689–698, 2007.

A video surveillance system generally contains three main hierarchical stages. The motion detection, the tracking stage, and finally, the high level motion interpretation. The main objectives are to observe the scene, index activities and recognise the modelled scenarios.

Scenarios concern specific spatio-temporal trajectory, interaction between static or non static objects, or also a combination of both of them.

2 Motion Detection and Tracking

The detection of moving objects is then made by comparing the three RGB components:

$$\begin{aligned} &if \ \ (\max_{c=R,G,B} \left| I_c^k(P) - R_c^{k-1}(P) \right| > \omega) \\ &\qquad \rightarrow D^k(P) = 1 \\ &else \ \ \rightarrow D^k(P) = 0 \end{aligned} \tag{1}$$

D^k represents the detection decision (1: moving object, 0: background) and R^k. the reference images. The raw motion detection result generally presents many artefacts. Two kinds of cleaning procedures are applied. Firstly, we use standard morphological operations of erosion and dilatation for reducing noise in the foreground. Then, too small uninteresting image regions are removed. In our algorithm this size threshold is defined globally in an empirical manner for the entire image. The Figure 1 illustrates a motion detection result for a real sequence of human activity.

Fig. 1. Detection process: I^k and D^k

The tracking process matches the detected regions from a temporal sequence to another and takes into account the splitting and merging phenomenon during object motion. One important component of our tracking algorithm is the integration of a belief revision mechanism associated with each detected object, reflecting an instantaneous quality of its tracking [2]. We consider that the belief can increase when:

- The tracked object follows a continuous path;
- The object size is stable;
- No ambiguity appears (other target meeting: target splitting, loss of target…);

Otherwise the belief decreases.

3 Scenario Recognition

A Scenario is composed of a set of elementary events linked with constraints. These links represent the ordering of the events or temporal constraints. Several kinds of events can be used, and depend directly on the observed scene and on the objective of the surveillance task.

The difficulty of human activity scenarios is their variabilities. This variability can be both spatial, and behavioural. So the scenario recognition process must be able to handle such uncertainties.

The logical structure linking events is generally based on graph. Many approaches are used: Petri-net [3] Bayesian network [4], [5], Network of time constraints also brings an efficient way to represent temporal structures of scenarios [6]. HHM are a popular state based probabilistic approach for representing dynamic systems, they have been initially used in speech recognition and have been successfully applied over gesture recognition [7]. An interesting feature of HMM is its time scale invariance enabling scenarios in various speeds. The Couple HHM has also been developed in order to efficiently represent interaction between objects [4]. But the major disadvantage of the HMM approaches is their complexities and the need of a long learning step in order to adjust model parameter. In this work, the model of scenarios is also obtained with the Bayesian networks formalism.

3.1 Bayesian Networks, Definitions

A Bayesian network is a directed acyclic graph whose:

- Nodes represent variables;
- Arcs represent dependence relations between the variables and local probability distributions for each variable given values of its parents.

Nodes have the possibility to represent any kind of random or determinist variables. They can be attached with a measured parameter or a latent variable.

Bayesian Networks generally operate on discrete values of nodes but can also deals with continuous representation.

If there is an arc from node A to another node B, then variable B depends directly on variable A, and A is called a parent of B. If for each variable X_i, i= 1 to n, the set of parent variables is denoted by parents X_i) then the joint distribution of the variables is product of the local distributions

$$P(X_1,\ldots,X_n) = \prod_{i=1}^{n} P(X_i \mid parents(X_i)) \tag{2}$$

One important advantage of the Bayesian network is its ability to encode qualitative and quantitative contextual knowledge and their dependence and it brings an efficient inference structure for real time application [8]. The topology of the network represents a qualitative causal relation between variable and the joint probability represents the quantitative part of the network.

The joint probability is represented by the expression

$$P(X_j, \mathrm{H}_i) = P(X_j | \mathrm{H}_i) P(\mathrm{H}_i) \quad (3)$$

Then the objective is to estimate hypotheses H_i (i=1,..,N) based on evidence X_j

$$P(\mathrm{H}_i | X_j) = \frac{P(X_j, \mathrm{H}_i)}{P(X_j)} \quad (4)$$

One of the major difficulties of using Bayesian Networks approach concerns the design of the network structure. In fact every dependency must be taken into account in order to make a good representation of the problem.

3.2 Bayesian Networks and Visual Scenario Recognition

Bayesian networks have been widely used in computer vision community for event or scenario recognition. The system described by [5] delivers textual descriptions for dynamic activities occurring in a dynamic, containing vehicles and pedestrians. [9], used Bayesian Networks to recognize several activities in an American football game. [10], suggested that for events with logical and temporal relationships between them, Allen's relations can be used to describe temporal relations between sub-events. A particular form of dynamic Bayesian networks, Recurrent Bayesian Networks (RBNs), have been used for the recognition of human behaviours [11]. Such networks have the advantage to have some time scale independence.

In order to model efficiently general scenarios, the system has to model temporal constraint between elementary events. The time aspect in Bayesian Network has led to several approaches.

The first common category is known as "time slice" approach. The main technique is the Dynamic Bayes networks [12], [13]. It assumes a Markov property by considering that a single snapshot in the past is sufficient for predicting the future. The structure of a static BN is generated for a specific instant and repeated with the same structure over the time. Temporal arc are added between nodes belonging to different time instants.

The second category, less explored, represents the "event based" approach which permits to integrate explicitly temporal nodes (Temporal Nodes Bayesian Networks (TNBN) [14] and Net of Irreversible Events in Discrete Time (NIEDT) [15]. This second category of approaches has several advantages. Firstly, it is particularly adapted when the system has to manipulate the notion of time at several temporal granularities. Secondly, the event based approach can easily establish quantitative temporal relation between events as Allen interval [16]. In time slice approach there is no way to represent naturally concept as an event e_1 appears before another event e_2. The HMM approach can be seen as a particular case of DBN and consequently shares the same limitations.

3.3 Proposed Model

In this work we propose a scenario recognition algorithm based on an "event based" Bayesian Network approach. The global structure of our proposed network uses the concept of Hierarchical Bayes networks that have been proposed as a technique of incorporating a Bayesian Network inside a node of a higher level Network. Such encapsulation facilitates the hierarchical decomposition of the problem into sub-problem.

The model of the scenario is composed by two layers: the atemporal event layer and the temporal reasoning layer. Is such organisation the first layer contains low level networks representing each atemporal event.

Each atemporal event is represented a hierarchical Bayesian network. The lower level is linked with observed data X_j and the upper level represents an hypothesis H_i (i=1,..,N) representing latent nodes. In our problem, observed nodes are measured visual features X and the hypotheses H represent decision of the elementary events recognition. Visual features used at this layer are: positions, speed, direction and size of tracked objects.

The temporal reasoning layer is based on a set of Bayesian networks which contain nodes associated with temporal information. Such nodes generally use the result obtained by estimating the lifespan of hypotheses delivered by the atemporal event layer. First of all, the objective of the temporal layer is to evaluate qualitative or quantitative temporal constraints for each event and also to estimate relations between related events. Constraints represent the duration or the time of occurrence (relative or absolute) of an event. Relations between events represent their mutual temporal dependency (before, after, during etc..). The low level temporal nodes have to detect the "Start" and "End" time of their events of interest. We have chosen to propagate explicitly the uncertainty of these last estimations over the scenario recognition task. For this, each date is approximated by a normal distribution with respect to its mean and variance values. Then, Allen relationships between events are implemented by using specific nodes in order to evaluate temporal relations between events. This network contains nodes which compute some hypothesis linked with the start and end times of events (Table 1). The Ds_1s_2 is the distance of the start time of event 1 and the start time of the event 2. The Ds_1e_2 is the distance of the start time of event 1 and the end time of the event 2. The use of Bayesian network at this layer is efficient for tacking into account uncertainty of estimated dates (start and end time of event). The general structure of the Bayesian network which has to verify the temporal relations is defined in the (Fig.2).

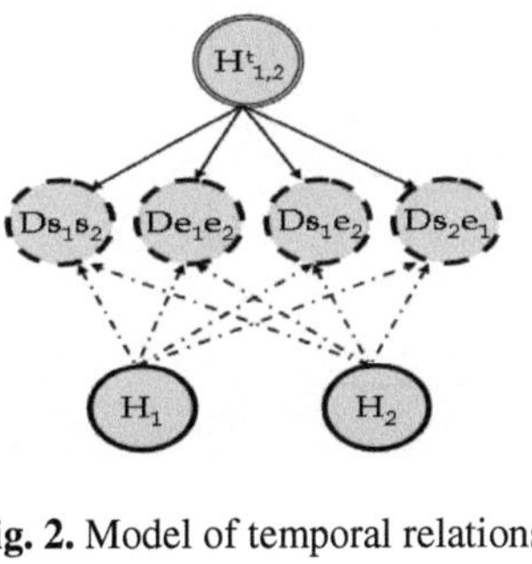

Fig. 2. Model of temporal relations

Table 1. Example of the relation 'Equal' of allen

H_1 equal H_2
$D_{s_1s_2}$ *true* $if(s_2 - s_1 = 0)$
$D_{e_1e_2}$ *true* $if(e_2 - e_1 = 0)$
$D_{s_1e_2}$ *true* $if(e_2 - s_1 > 0)$
$D_{s_2e_1}$ *true* $if(e_1 - s_2 > 0)$

The integration of the temporal constraints associated with an event is also performed with this approach, by linking observed events and with constraints which are represented by specific nodes as time point or duration (relative or absolute time-base). Generally the constraints defined from the contextual information of the scenario are defined by a relative time base with respect to other previous events. The figure 3 illustrates the general structure of the suggested scenario model.

Table 2. Various situations of the relationships of allen

Relations	Ds_1s_2	De_1e_2	Ds_1e_2	Ds_2e_1
H_1 starts H_2	$Ds_1s_2=0$	$De_1e_2<0$	$Ds_1e_2>0$	$Ds_2e_1>0$
H_1 finishes H_2	$Ds_1s_2>0$	$De_1e_2=0$	$Ds_1e_2>0$	$Ds_2e_1>0$
H_1 during H_2	$Ds_1s_2>0$	$De_1e_2<0$	$Ds_1e_2>0$	$Ds_2e_1>0$
H_1 equal H_2	$Ds_1s_2=0$	$De_1e_2=0$	$Ds_1e_2>0$	$Ds_2e_1>0$
H_1 meets H_2 H_1 before H_2	$Ds_1s_2>0$	$De_1e_2>0$	$Ds_1e_2>0$	$Ds_2e_1=0$
H_1 overlaps H_2	$Ds_1s_2>0$	$De_1e_2>0$	$Ds_1e_2>0$	$Ds_2e_1>0$

For illustration, in (Fig. 4), we present the model of a scenario of an abandoned baggage by using the proposed approach. H^t_{12} represents temporal node verifying relation between events H_1 and H_2. The node H^t_3 in the temporal layer permits to verify a temporal duration of the H_3 with a constraint of a defined duration (Δt).

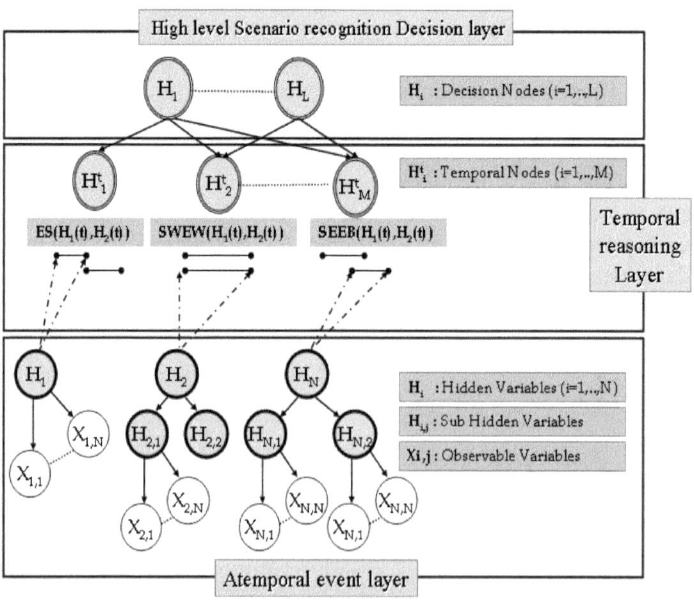

Fig. 3. General model of a scenario

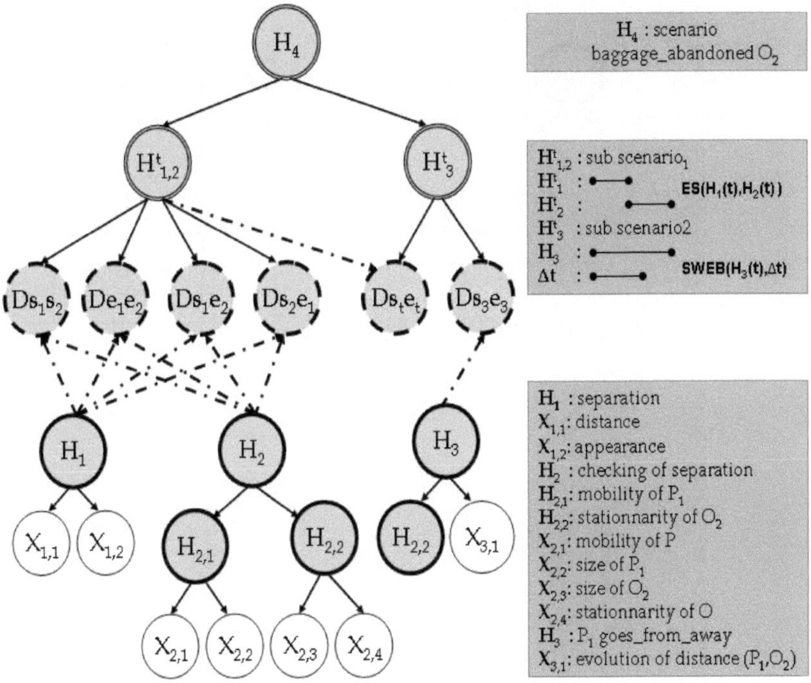

Fig. 4. Model of the scenario for an abandoned baggage

3.4 Parameters Learning

Generally, it is possible to learn the network parameters, from the experimental data, in particular the conditional probabilities tables.

$$P(X_i = x_k | parent(X_i) = c_j) = \theta_{i,j,k} = \frac{n_{i,j,k}}{\sum_K n_{i,j,k}} \tag{5}$$

Where $n_{i,j,k}$ is the number of the events in the database for which variable X_i is in the state x_k and his parents are in the configuration c_j.

In the context of visual surveillance, it is not always realistic to perform the standard learning procedure. Because, some times, it's not possible to have sufficient occurrences for each scenario. Another difficulty is that learning process has to deal with uncertain inputs. Parameters for a fixed network from incomplete data, in the presence of missing values or hidden variables can be estimated by the EM Expectation-Maximisation algorithm.

We illustrate the learning process by using the EM over an example from the abandoned baggage scenario (Fig. 5). The incomplete data are presented in Table 3.

The EM algorithm is initialized with $P^{(0)}(H_{2,1}=1)=0.5$ and $P^{(0)}(H_{2,1}=0)=0.5$. Convergence is obtained after 13th iterations: $P^{(13)}(X_{2,1}=1|H_{2,1}=1)=0.857$ and $P^{(13)}(X_{2,2}=1|H_{2,1}=0)=0.985$. The figure 6 shows the evolution of these probabilities.

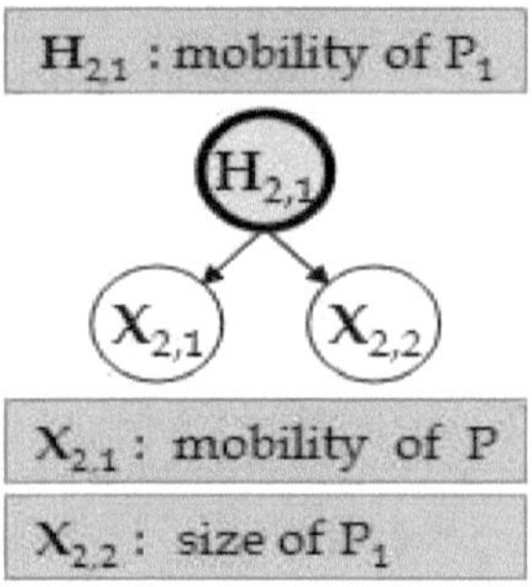

Fig. 5. Learning of a naive bayesian network, an example

Table 3. Table of occurrence

$X_{2,1}$	$X_{2,2}$	$H_{2,1}$	Nb
1	1	1	18
1	1	0	1
0	1	0	3
1	0	0	2
?	1	1	1
1	?	1	2
0	0	0	14
0	?	0	3

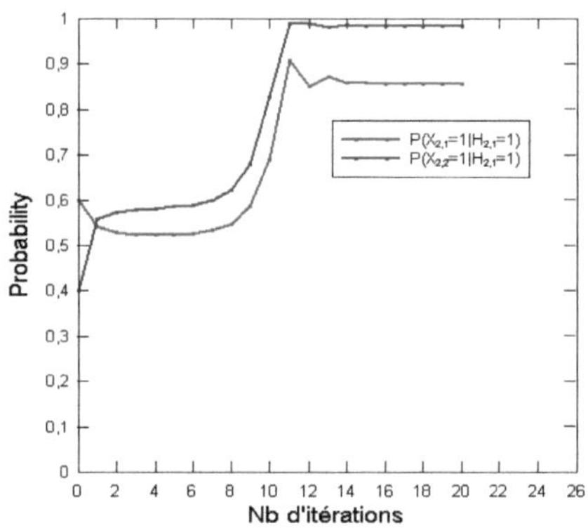

Fig. 6. Evolution of the estimated probabilities

4 Results

Two experimentations based on real indoor sequences are presented in this section. Both sequences represent a scenario of abandoned baggage. The first sequence is obtained from the scene that we have used for learning of the scenario model (Fig. 7). The second sequence comes from the PETS'04 dataset (Fig. 8). The results illustrate the evolution of the main nodes of the scenario recognition model.

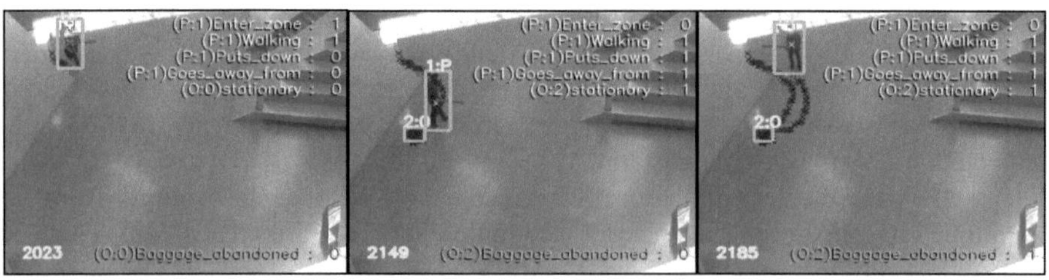

Fig. 7. First sequence

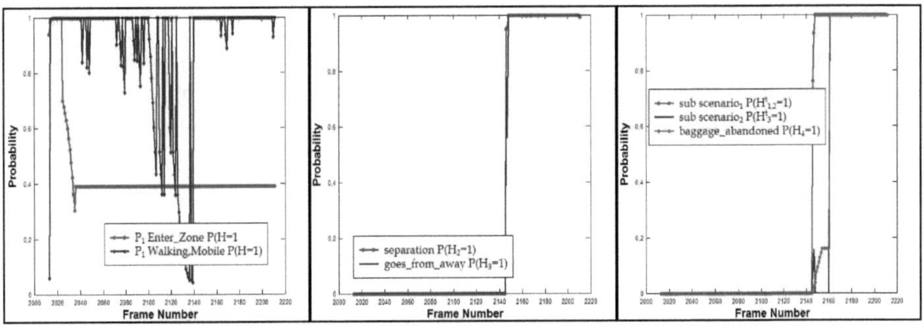

Fig. 7. *(Continued)*

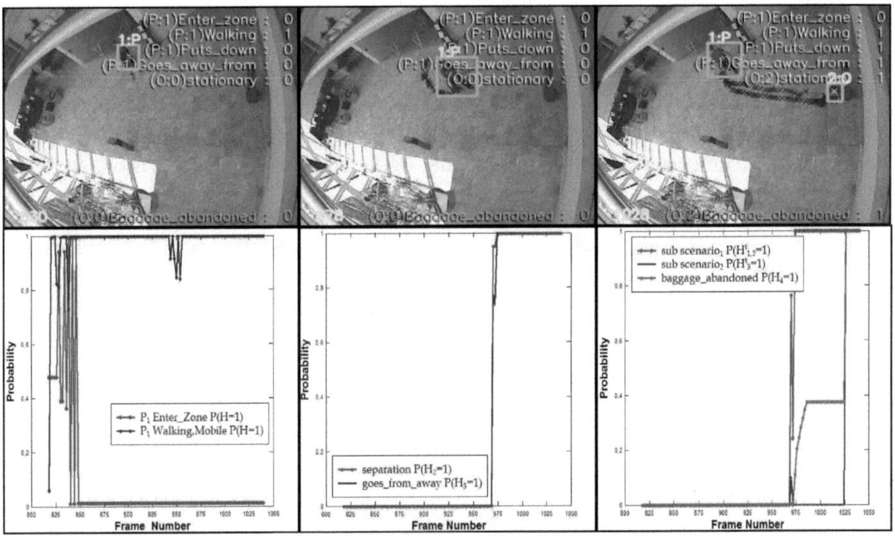

Fig. 8. Sequences from PETS'04 Workshop

5 Conclusions

The proposed scenario recognition algorithm is based on the Bayesian network. It permits to deal efficiently with uncertainty due to the low level visual extraction stage and also the variability of scenario models. The model contains two main layers. The first one permits to highlight events from the observed visual features. The second layer is focused on the temporal reasoning stage by using specific nodes based on the temporal validity of the first layer's events. The temporal reasoning permits firstly to estimate the relation between events and also to add quantitative and qualitative on the high level recognition decision. The use of node containing temporal information under the framework of the Bayesian Networks permits to have a flexible temporal reasoning capability over the recognition task. The separation of the scenario model by two complementary layers brings many advantages as modularity and clarity. The specificity of our temporal layer with respect to other event based approaches is that it can deal easily with all kind of constraints or relation (qualitative, quantitative, relative and absolute). In fact, the majority of existing event based approaches use fixed quantitative time interval. The global recognition algorithm is tested and validated on real indoor images sequences.

References

1. Buxton, H.: Learning and Understanding Dynamic Scene Activity: A Review. Image and Vision Computing 21, 125–136 (2003)
2. Motamed, C.: Motion detection and tracking using belief indicators for an automatic visual- surveillance system. Image and Vision Computing 24(11), 1192–1201 (2006)
3. Castel, C., Chaudron, L., Tessier, C.: What Is Going On? A High Level Interpretation of Sequences of Images. 4th European conference on computer vision, Workshop on conceptual descriptions from images,Cambridge UK (1996)
4. Oliver, N., Rosario, B., Pentland, A.: A Bayesian Computer Vision System for Modeling Human In-teractions. In: Christensen, H.I. (ed.) ICVS 1999. LNCS, vol. 1542, Springer, Heidelberg (1998)
5. Remagnino, P., Tan, T., Baker, K.: Agent orientated annotation in model based visual surveillance, Proceedings of the International Conference on Computer Vision, pp. 857–862 (1998)
6. Eude, V., Bouchon-Meunier, B., Collain, E.: Reconnaissance d'activités à l'aide de graphes temporels flous. In: Proc LFA'97 Logique Floue et Applications, Lyon France pp. 91–98 (1997)
7. Starner, T., Pentland, A.: Real-time American Sign Language Recognition from Video Using Hidden Markov Models. In: Proceedings of International Symposium on Computer Vision, pp. 265–270 (1995)
8. Pearl, J.: Probabilistic reasoning in Intelligent Systems. Morgan Kaufman, San Francisco (1988)
9. Intille, S.S., Bobick, A.F.: Visual Recognition of multi-agent Action using Binary Temporal Relations. In: IEEE Proceedings of Computer Vision and Pattern Recognition, Fort Collins, CO (1999)
10. Hongeng, S., Nevatia, R.: Multi-Agent Event Recognition, ICCV '01, pp. 84–93 (2001)
11. Moenne-Loccoz, N., Bremond, F., Thonnat, M.: Recurrent Bayesian Network for the Recognition of Human Behaviors from Video. In: Crowley, J.L., Piater, J.H., Vincze, M., Paletta, L. (eds.) ICVS 2003. LNCS, vol. 2626, pp. 68–77. Springer, Heidelberg (2003)
12. Dean, T., Kanazawa, K.: A model for reasoning about persistence and causation. Computational Intelligence 5(3), 142–150 (1989)
13. Nicholson, A., J.M., B.: The data association problem when monitoring robot vehicles using dynamic belief networks. In: ECAI 92: 10th European Conference on Artificial Intelligence (1992)
14. Arroyo-Figueroa, G., Sucar, L.E.: A Temporal Bayesian Network for Diagnosis and Prediction. Uncertainty in artificial intelligence. In: Proc15th Conf Uncertainty Artif Intell pp. 13–20 (1999)
15. Galan, S.F., Dıez, F.J.: Modeling dynamic causal interactions with bayesian networks: temporal noisy gates. In: CaNew', the 2nd International Workshop on Causal Networks held in conjunction with ECAI 2000, Berlin, Germany, pp.1–5 (August 2000)
16. Allen, J.F.: An interval based representation of temporal knowledge. International Joint Conference on Artificial Intelligence 81, 221–226 (1981)
17. Koller, D., Pfeiffer, A.: Object-oriented Bayesian networks. In: Koller, D., Pfeiffer, A. (eds.) Uncertainty in Artificial Intelligence: Proceedings of the Thirteenth Conference (UAI-1997), pp. 302–313. Morgan Kaufmann Publishers, San Francisco (1997)

Comparison of Combining Methods of Correlation Kernels in kPCA and kCCA for Texture Classification with Kansei Information

Yo Horikawa[1] and Yujiro Ohnishi[2]

[1] Faculty of Engineering, Kagawa University, Takamatsu, 761-0396 Japan
[2] Ryobi Systems Corporation, Okayama, 700-8504 Japan
horikawa@eng.kagawa-u.ac.jp
http://www.eng.kagawa-u.ac.jp/~horikawa/

Abstract. The authors consider combining correlations of different orders in kernel principal component analysis (kPCA) and kernel canonical correlation analysis (kCCA) with the correlation kernels. We apply combining methods, e.g., the sums of the correlation kernels, Cartesian spaces of the principal components or the canonical variates and the voting of kPCAs and kCCAs output and compare their performance in the classification of texture images. Further, we apply Kansei information on the images obtained through questionnaires to the public to kCCA and evaluate its effectiveness.

Keywords: kernel method, principal component analysis, canonical correlation analysis, correlation kernel, combining classifiers, Kansei information, texture classification.

1 Introduction

Kernel principal component analysis (kPCA) [1] and kernel canonical correlation analysis (kCCA) [2], [3], [4] are kernelized versions of PCA and CCA in multivariate statistical analysis. In PCA, the linear projections which allow to reconstruct original feature vectors are obtained with minimal quadratic errors. It is used to reduce the dimensionality of the original data retaining most existing structure in the data. In CCA, linear transformations that yield maximum correlation between two kinds of features vectors of objects, e.g., their images and sounds, are obtained. It is also applied to dimensionality reduction or feature extraction. In the kernel methods, the inner products of the feature vectors are replaced to nonlinear kernel functions [5], [6]. Nonlinear mappings of the feature vectors to high-dimensional spaces are then performed in implicit manners. Then nonlinear characteristics of the original data can be extracted with them.

Correlation kernels were recently proposed as kernel functions [7], [8]. They are inner products of the autocorrelation functions of the feature vectors. The idea was shown about forty years ago and their characteristics are that higher-order correlation kernels are effectively calculated [9]. They are suitable to image data, which have strong spatial correlations and support vector machines (SVM), kPCA and kCCA with the correlation kernels were applied to invariant texture classification [10], [11], [12].

B.K. Ersbøll and K.S. Pedersen (Eds.): SCIA 2007, LNCS 4522, pp. 699–708, 2007.

In this study we consider combining correlation kernels of different orders in kPCA and kCCA in the classification of texture images. Combining classifiers have been of wide interest in pattern recognition [13], [14] and they can show higher classification performance than any single classifiers. Some combining methods: use of the sums of the correlation kernels as kernel functions, use of Cartesian spaces of the principal components or the canonical variates as feature vectors and the voting of output of the classifiers with kPCA or kCCA are employed and compared in their classification performance. Further, we try to use Kansei information, the perceptual and cognitive ability to feel objects, e.g., impressions to images. Kansei information was applied to image retrieval systems with CCA [15] and it can be adopted in kCCA as the second feature vectors.

In Sect. 2, related theoretical background: kPCA, kCCA, correlation kernels, combining methods and Kansei information are mentioned. In Sect. 3, Method and results of texture classification experiment with kPCA and kCCA are shown. Discussions are given in Sect. 4.

2 Theoretical Background

2.1 Kernel Principal Component Analysis (kPCA)

The feature vectors $\boldsymbol{x}_i$ $(1 \le i \le n)$ of sample objects are transformed to $\boldsymbol{\varphi}(\boldsymbol{x}_i)$ in another feature spaces with an implicit nonlinear mapping $\boldsymbol{h}$. We assume that the mean of the transformed features are zero, i.e., $\sum_{i=1}^{n} \boldsymbol{h}(\boldsymbol{x}_i) = \boldsymbol{0}$, for simplicity. The mean centering can be done in the calculation of kernel functions [1]. The kernel version of PCA is PCA for $\boldsymbol{h}(\boldsymbol{x}_i)$ and the principal components are obtained through the eigenproblem

$$\boldsymbol{\Phi v} = \lambda \boldsymbol{v} \tag{1}$$

where $\boldsymbol{\Phi}$ is a kernel matrix and its elements are calculated with a kernel function $\Phi_{ij} = \boldsymbol{h}(\boldsymbol{x}_i)\cdot\boldsymbol{h}(\boldsymbol{x}_j) = \varphi(\boldsymbol{x}_i, \boldsymbol{x}_j)$. Let $\boldsymbol{v}_r = (v_{r1}, \ldots, v_{rn})^{\mathrm{T}}$ $(1 \le r \le R\ (\le n))$ be the eigenvectors in the non-increasing order of the corresponding non-zero eigenvalues λ_r, which are normalized as $\lambda_r \boldsymbol{v}_r \cdot^{\mathrm{T}} \boldsymbol{v}_r = 1$. The rth principal component u_r for a new data $\boldsymbol{x}$ is then obtained by

$$u_r = \sum_{i=1}^{n} v_{ri} \boldsymbol{h}(\boldsymbol{x}_i)\cdot\boldsymbol{h}(\boldsymbol{x}) = \sum_{i=1}^{n} v_{ri} \varphi(\boldsymbol{x}_i, \boldsymbol{x}) \tag{2}$$

Classification methods, e.g., the nearest-neighbor method, the discriminant analysis and SVMs can be applied in the principal component space $(u_1, \cdots, u_R)$.

2.2 Kernel Canonical Correlation Analysis (kCCA)

The kernel version of CCA is as follows [2], [3], [4]. Let $(\boldsymbol{x}_i, \boldsymbol{y}_i)$, $(1 \le i \le n)$ be pairs of the feature vectors of n sample objects, which describe different aspects of the objects, e.g., sounds and images. Define kernel matrices $\boldsymbol{\Phi}$ and $\boldsymbol{\Theta}$ by $\Phi_{ij} = \varphi(\boldsymbol{x}_i, \boldsymbol{x}_j)$ and $\Theta_{ij} = \theta(\boldsymbol{y}_i, \boldsymbol{y}_j)$, $(1 \le i, j \le n)$, which correspond to the inner products of implicit functions of $\boldsymbol{x}$ and $\boldsymbol{y}$, respectively. Then we obtain the eigenvectors $(\boldsymbol{f}^{\mathrm{T}}, \boldsymbol{g}^{\mathrm{T}})^{\mathrm{T}}$ of the generalized eigenproblem:

$$\begin{pmatrix} \mathbf{0} & \mathbf{\Phi\Theta} \\ \mathbf{\Theta\Phi} & \mathbf{0} \end{pmatrix} \begin{pmatrix} \boldsymbol{f} \\ \boldsymbol{g} \end{pmatrix} = \lambda \begin{pmatrix} \mathbf{\Phi}^2+\gamma_x\boldsymbol{I} & \mathbf{0} \\ \mathbf{0} & \mathbf{\Theta}^2+\gamma_y\boldsymbol{I} \end{pmatrix} \begin{pmatrix} \boldsymbol{f} \\ \boldsymbol{g} \end{pmatrix} \tag{3}$$

Small multiples of the identity matrix $\gamma_x\boldsymbol{I}$ and $\gamma_y\boldsymbol{I}$ are added for the regularization. The canonical variates u and v of $(\boldsymbol{x}, \boldsymbol{y})$ of a object are linear projections of the implicit functions of $\boldsymbol{x}$ and $\boldsymbol{y}$ which maximize correlation between them. They are obtained with the eigen vectors $\boldsymbol{f}$ and $\boldsymbol{g}$ by

$$\begin{aligned} u &= \textstyle\sum_{i=1}^{n} f_i\varphi(\boldsymbol{x}_i, \boldsymbol{x}) \\ v &= \textstyle\sum_{i=1}^{n} g_i\theta(\boldsymbol{y}_i, \boldsymbol{y}) \end{aligned} \tag{4}$$

An indicator vector is used as the second feature vector $\boldsymbol{y}$ for classification problems [2]. When an object $\boldsymbol{x}$ is categorized into one of C classes, the indicator vector corresponding to $\boldsymbol{x}$ is defined by

$$\begin{aligned} &\boldsymbol{y} = (y_1, \cdots, y_C)^{\mathrm{T}} \\ &y_c = 1 \quad \text{if } \boldsymbol{x} \text{ belongs to class } c \\ &y_c = 0 \quad \text{otherwise} \qquad (1 \le c \le C) \end{aligned} \tag{5}$$

Then the linear inner product $\boldsymbol{y}_i^{\mathrm{T}}\boldsymbol{y}_j$ is used as the kernel function $\theta(\boldsymbol{y}_i, \boldsymbol{y}_j)$. The canonical variates u_i $(1 \le i \le C\text{-}1)$ are obtained corresponding to non-zero eigenvalues of Eq. (3). This is equivalent to Fisher's discriminant analysis in two class problems. Standard classification methods are also applied in the canonical variate space.

2.3 Correlation Kernels and Their Modification

The autocorrelation of the original feature vector $\boldsymbol{x}$ is used in the correlation kernels [7], [8]. In the following, we consider 2-dimensional image data $x(l, m)$, $(1 \le l \le L, 1 \le m \le M)$ as the feature vector $\boldsymbol{x}$. The kth-order autocorrelation $r_x(l_1, l_2, \cdots, l_{k-1}, m_1, m_2, \cdots, m_{k-1})$ of $x(l, m)$ is defined by

$$\begin{aligned} &r_x(l_1, l_2, \cdots, l_{k-1}, m_1, m_2, \cdots, m_{k-1}) \\ &= \textstyle\sum_l \sum_m x(l, m)x(l+l_1, m+m_1) \cdots x(l+l_{k-1}, m+m_{k-1}) \end{aligned} \tag{6}$$

The inner product of the autocorrelations r_{xi} and r_{xj} of image data $x_i(l, m)$ and $x_j(l, m)$ is calculated by the sum of the kth power of the cross-correlation $cc_{xi,xj}(l_1, m_1)$ of the image data [9]

$$r_{xi}\cdot r_{xj}(k) = \textstyle\sum_{l1=0}^{L-1}\sum_{m1=0}^{M-1}\{cc_{xi,xj}(l_1, m_1)\}^k/(LM) \tag{7}$$

$$cc_{xi,xj}(l_1, m_1) = \textstyle\sum_{l=1}^{L-l1}\sum_{m=1}^{M-m1} x_i(l,m)x_j(l+l_1, m+m_1)/(LM) \tag{8}$$

Computational costs are reduced in the practical order even for high-order k of the correlation and large data size L and M since the calculation of the explicit values of the autocorrelations are avoided. Equation (7) is employed as the kth-order correlation kernel function $k(\boldsymbol{x}_i, \boldsymbol{x}_j)$ and thus $\mathbf{\Phi}$ in Eq. (3).

Since the performance of the correlation kernels of odd or higher-orders is degraded, the following modified versions of the correlation kernels have been proposed [12] and are used in this study.

L_p norm kernel (P)

$$r_{xi} \cdot r_{xj} = \mathrm{sgn}(cc_{xi,xj}(l_1, m_1)) |\sum_{l1,m1} \{cc_{xi,xj}(l_1, m_1)\}^k|^{1/k} \quad (9)$$

Absolute correlation kernel (A)

$$r_{xi} \cdot r_{xj} = \sum_{l1,m1} |cc_{xi,xj}(l_1, m_1)|^k \quad (10)$$

Absolute L_p norm kernel (AP)

$$r_{xi} \cdot r_{xj} = | \sum_{l1,m1} \{cc_{xi,xj}(l_1, m_1)\}^k |^{1/k} \quad (11)$$

Absolute L_p norm absolute kernel (APA)

$$r_{xi} \cdot r_{xj} = |\sum_{l1,m1} |cc_{xi,xj}(l_1, m_1)|^k |^{1/k} \quad (12)$$

Max norm kernel (Max)

$$r_{xi} \cdot r_{xj} = \max_{l1,m1} cc_{xi,xj}(l_1, m_1) \quad (13)$$

Max norm absolute kernel (MaxA)

$$r_{xi} \cdot r_{xj} = \max_{l1,m1} |cc_{xi,xj}(l_1, m_1)| \quad (14)$$

The L_p norm kernel (P) and the absolute correlation kernel (A) take the kth roots and absolute values, respectively, of the original ones. The max norm kernel (MAX) is regarded as the L_p norm kernel in the limit of $k \to \infty$. The absolute L_p norm kernel (AP), the absolute L_p norm absolute kernel (APA) and the Max norm absolute kernel (MaxA) are combinations of them.

2.4 Combining Correlation Kernels

The following combining methods in three levels are employed for combining autocorrelation features of different orders.

First is combination in a kernel level. The sums of the correlation kernels of different orders are used as kernel functions, which are used in [7], [8].

$$\varphi(\boldsymbol{x}_i, \boldsymbol{x}_j) = \sum_{k=1}^{km} r_{xi} \cdot r_{xj}(k) \quad (15)$$

Second is combination in a feature level. Cartesian spaces of the principal components in kPCA or the canonical variates in kCCA with the correlation kernels of different orders are used as combined feature spaces: $(u_1, \cdots, u_{nc}, u'_1, \cdots, u'_{nc'})$ $(0 \leq n_c, n'_c \leq C\text{-}1)$ obtained from two sets of the canonical variates $(u_1, \cdots, u_{c\text{-}1})$ and $(u', \cdots, u'_{C\text{-}1})$, for instance. Third is combination of classifiers output level. Classification methods are applied to the principal component spaces or the canonical variate spaces with different correlation kernels and their output (classification results) are combined. In this study, the simple nearest-neighbor classifier and the majority vote output are used as the classification method and combining classifiers, respectively.

2.5 Kansei Information

Kansei information can be obtained from human experts or from the public. A standard way is that we prepare impression words for objects, e.g., {beautiful, dark, clear} for images and ask the persons to choose the impression words which they think most matched to the objects. We then use the vectors the elements of which are the numbers of the votes of the impression words to the objects as Kansei information. Questionnaire systems can be employed to collect them.

The obtained vectors, which we call the impression vectors, are used as the second feature vectors $\boldsymbol{y}$ instead of the indicator vectors in Eq. (5) in kCCA. The canonical variates obtained with them retain Kansei information and then consist of the features differ from those without them. These features can contribute to increases in classification performance.

3 Texture Classification Experiment

Classification experiment with 30 texture images arbitrarily taken from the Brodatz album [16] were done using kPCA and kCCA with combining correlation kernels of different orders as well as Kansei information.

3.1 Collection of Kansei Information

We made a questionnaire system on Web to collect Kansei information from Japanese undergraduate student in our faculty. Java Server Pages (JSP) was used for the client-server system. The Web page shows 30 texture images in the Brodatz album in Fig. 1 and asks people to choose one image which matches each Japanese impression word. Twenty impression words for images [17] are used: delicate, beautiful, bold, sharp, decorative, fine, sophisticated, simple, soft, deep, impressive, quiet, elegant, chic, natural, hard, grave, silent, solid, rural. We sent requests for the questionnaire by e-mail to about 200 students and obtained answers from 50 students.

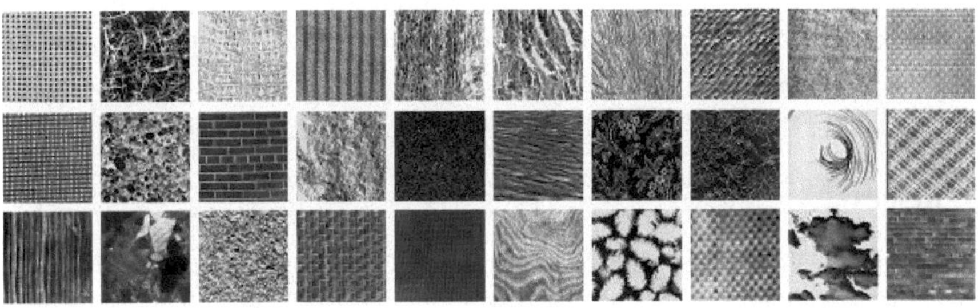

Fig. 1. Texture images from the Brodatz album

Kansei information of each texture image is a 20-dimensional vector the element of which is the number of the vote of the impression word to the image (the impression vector). For instance, the impression vector for the top-left texture image (Brodatz D101), top-left in Fig. 1, was

$$y_1 = (1,1,0,0,7,1,1,25,1,0,7,2,1,0,1,0,0,3,0,2) \tag{16}$$

which indicates that many answerers thought the image simple. We use the impression vectors as the second feature vectors $\boldsymbol{y}$ in kCCA.

3.2 Method

Thirty 8bit image data of 640×640 pixels in the Brodatz album are obtained from AMOVIP-DB [18]. Ten subimages of 10×10 pixels are taken from each original image without overlap, 300 images in total, are used as sample data and one hundred subimages for each, 3000 images in total, are used as test data as well.

As the second feature vector $\boldsymbol{y}$ in kCCA , we use the indicator vectors (Eq. (5)) or the impression vectors, in which Kansei information is taken into account. The values of regularization parameters γ_x and γ_y in kCCA are set to be $0.1n$. The principal components with 299 non-zero elements in kPCA and the canonical variates with 29 elements in kCCA with the indicator vectors and 19 elements in kCCA with the impression vectors are calculated with the sample data. In kPCA, the first 50 elements of the principal components are used for classification.

A simple nearest neighbor classifier (1-NN) in the feature space (the principal component spaces, the canonical variate spaces and their combination) in each kPCA and kCCA is used for the classification of the test data. For the combining methods, the sums of the correlation kernels of different orders in a kernel level, Cartesian spaces of the principal components and the canonical variates with the correlation kernels of different orders in a feature level, and the majority vote of the 1-NN classifiers output with kPCA and kCCA with the correlation kernels are employed as explained in Sect. 2.4. Further, the impression vectors (Kansei information) as well as the indicator vectors are adopted for the second feature vectors $\boldsymbol{y}$ in kCCA.

To express the set (M, φ) of the kernel method and the kernel function φ, the following symbols are used.

kPCA: P, kCCA with the indicator vector: I, kCCA with the impression vector K

the kth-order correlation kernel: Ck, the kth-order L_p norm kernel: Pk, the max norm kernel: Max, etc., the symbols in which are the same as Eqs. (9)-(14), for the kernel function φ

For instance, (P, C2) indicates kPCA with the 2nd-order correlation kernel.

3.3 Results

Figure 2 shows the correct classification rates (CCRs) of the single classifiers. The correct classification rates are calculated with the first r principal components (u_1, ···, u_r) ($1 \le r \le 50$)) for all r in kPCA and with the first i canonical variates (u_1, ···, u_i) ($1 \le i \le 29$ or 19) for all i in kCCA. The maximum values of CCRs are usually obtained with parts of the principal components or the canonical variates. The highest CCRs are obtained with the max norm kernel (Max) in all cases and their values are: 0.2237 (P, r = 50), 0.2250 (P (max), r = 32) in kPCA; 0.2347 (I, i =29) with the indicator vector (I), in kCCA; 0.2237 (K, i = 19), 0.2250 (K (max), i = 15) with the impression vector (K). In most cases kCCAs with the indicator vector (I) show the highest CCRs, as is expected since kCCA is superior to kPCA and the indicator vector is most suitable for classification tasks.

Figure 3 shows CCRs with the combined classifiers, the sum of the kernels (a), Cartesian spaces of the principal components and the canonical variates (b), and the voting of the 1-NN classifiers output (c). In each case, the kernels or the classifiers are added in descending order of CCRs of the single classifiers and CCRs with all elements of the principal components and the canonical variates. The highest CCRs are obtained with the combination of small numbers (1 – 14) of the classifiers, though the maximum 96 classifiers can be combined. Table 1 shows the highest CCRs and the combination of the classifiers. The maximum value 0.2560 is obtained with Cartesian subspace of canonical variates of the three correlation kernels ((I, Max), (I, AP7), (I, APA3)), and the second value 0.2553 is with Cartesian space of the six correlation kernels ((I, Max), (K, Max), (I, AP7), (I, APA3), (I, AP9), (I, A5)), both in kCCA. While the optimal combination consists of the kernels with the indicate vectors, Kansei information contributes to the second one. In the combination with the voting, the highest CCR is obtained with the combination including both kPCA and kCCA.

Since the combining with Cartesian space performs well, CCRs of all combination of two and three kernels are calculated and the optimal combination of the correlation kernels with Cartesian spaces is shown in Table 2. Using the all elements of the canonical variates, CCR increases to 0.2567 with 2 classifiers ((I, Max), (I, A3)) and 0.2684 with 3 classifiers ((I, Max), (I, AP7), (I, A7)). Using the optimal elements of the canonical variates in descending order of the corresponding eigenvalues, CCR increases to 0.2623 with 2 classifiers ((I, Max), (K, P4)) and 0.2787 with 3 classifiers ((I, Max), (I, A5), (K, C3)). They exceed the maximum CCR 0.2560 in Table 1. Note that the combination with the classifiers with low CCR in themselves, e.g., (I, C6), (I, C8), shown in Fig. 2, shows high CCRs, though the best single classifiers (I, Max) is included in most cases. The second-best single classifier (P, Max) does not appear in Table 2.

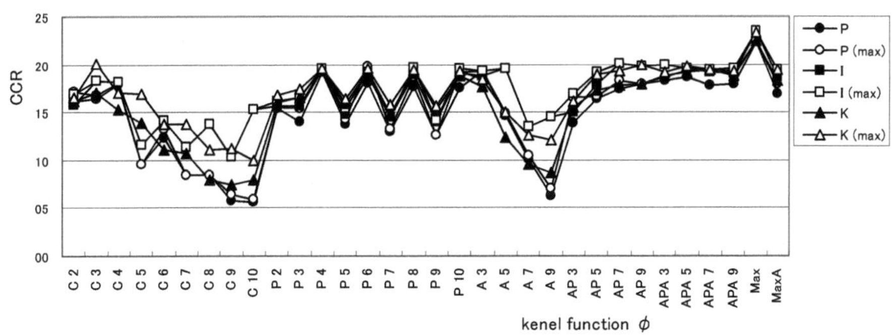

Fig. 2. Correct classification rates (CCRs) of the single classifiers

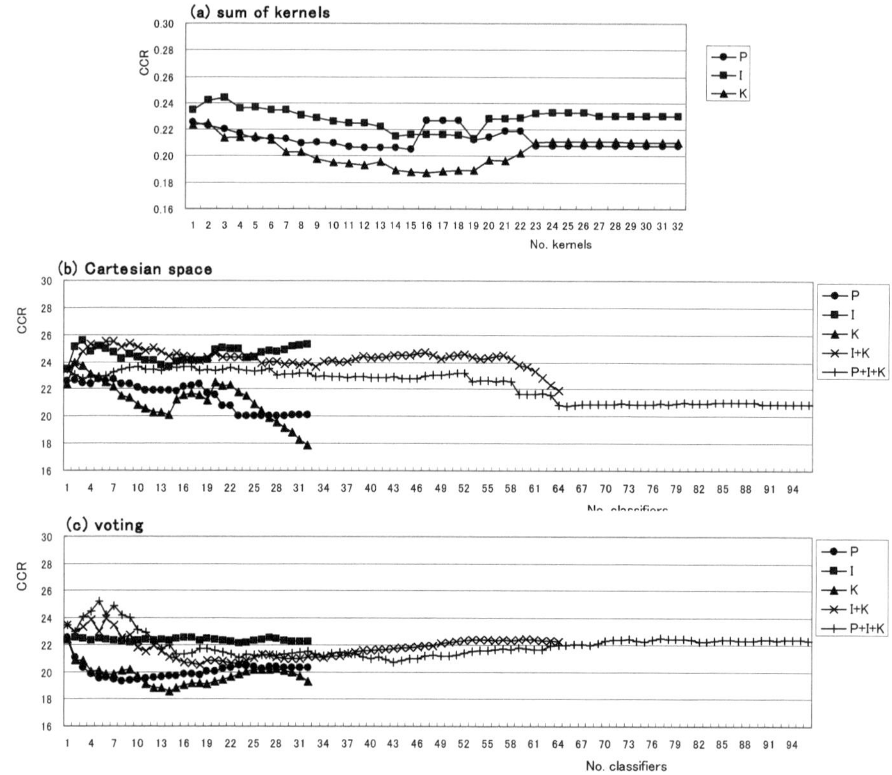

Fig. 3. CCRs of combined classifiers. The sum of the correlation kernels (a), Cartesian space of the feature vectors (b), the voting of the classifiers output (c).

Table 1. Highest CCRs of the combining classifiers in Fig3

	Highest CCR (classifiers)		
	(a) Sum	(b) Cartesian	(c) Voting
P	0.2333 (P, Max)	0.2267 (P, Max), (P, P4), (P, APA7), (P, P6), (P, P8)	0.2253 (P, Max)
I	0.2440 (I, Max), (I, AP7), (I, APA3)	0.2560 (I, Max), (I, AP7), (I, APA3)	0.2347 (I, Max)
K	0.2247 (K, Max), (K, A3)	0.2397 (K, Max), (K, A3)	0.2237 (K, Max)
I+K		0.2553 (I, Max), (K, Max), (I, AP7), (I, APA3), (I, AP9), (I, A5)	0.2390 (I, Max), (K, Max), (I, AP7), (I, APA3)
P+I+K		0.2367 (I, Max), (P, Max), (K, Max), (I, AP7), (I, APA3), (I, AP9), (I, A5), (I,APA5), (P,P4), (I, P4), (P, APA7), (P, P6), (P, P8), (K, A3)	0.2521(I, Max), (P, Max), (K, Max), (I, AP7), (I, APA3)

Table 2. Highest CCRs of the combining classifiers in Cartesian spaces of the feature vectors

Highest CCRs with Cartesian space							
2 classifiers				3 classifiers			
all elements		optimal elements		all elements		optimal elements	
0.2567	(I, Max), (I, A3)	0.2623	(I, Max) 5, (K, P4) 6	0.2683	(I, Max), (I, AP7), (I, A7)	0.2787	(I, Max) 29, (I, A5) 29, (K, C3) 9
0.2560	(I, A7), (K, Max)	0.2617	(I, Max) 21, (I, C6) 29	0.2677	(I, Max), (I, AP7), (I, C8)	0.2767	(I, Max) 29, (I, A7) 29, (K, APA7) 6
0.2557	(I, Max), (I, A7)	0.2610	(I, Max) 8, (I, A9) 29	0.2643	(I, Max), (I, P4), (I, A7)	0.2753	(I, Max) 12, (I, C6) 29, (I, APA3) 6
0.2550	(I, Max), (I, C6)	0.2603	(I, Max) 18, (I, A7) 29	0.2643	(I, Max), (I, AP7), (I, A9)	0.2753	(I, Max) 29, (I, A7) 29, (K, P6) 4
0.2550	(I, Max), (I, A9)	0.2600	(I, Max) 8, (K, C2) 29	0.2640	(I, Max), (I, C8), (K, P4)	0.2750	(I, Max) 29, (I, A7) 29, (K, P4) 5

4 Discussion

The combination of the correlation kernels of different orders and Kansei information were employed in kPCA and kCCA, and their performance is compared in the experiment of texture classification. The combining with Cartesian space of the feature elements (principal components and the canonical variates) obtained with different kernels in kPCA and kCCA performed better than the use of the sum of the correlation kernels and the voting of the output of the 1-NN classifiers with multiple kPCAs and kCCAs.

As can be seen from Fig. 2, the highest CCRs were obtained with the combination of only a few classifiers. This is ascribed to the correlations between the feature elements of the correlation kernels, and then not so many classifiers can contribute to increases in CCR. However, after the CCRs once drop to the minimum values, they increase again as the numbers of the classifiers increase in most cases. This implies that the classifiers with low CCRs in themselves have a potential ability to improve classifiers performance through combining. In fact, high CCRs were obtained with the combination of the classifiers of the highest CCR and rather low CCR in Table 2. Choosing the optimal combination of the classifiers as well as the feature elements with brute-force search is impractical as their number increases. Even the calculation of CCRs for the all combination of the feature elements of three classifiers in 96 classifiers was a formidable task. It is expected that applying feature selection methods to the feature elements obtained with multiple kPCAs and kCCAs works and it is a future problem.

References

1. Schölkopf, B. et al.: Nonlinear component analysis as a kernel eigenvalue problem. Neural Computation, 10, 1299–1319 (1998)
2. Kuss, M., Graepel, T.: The geometry of kernel canonical correlation analysis, Technical Report 108, Max Plank Institute for Biological Cybernetics (2002)
3. Melzer, T. et al.: Appearance models based on kernel canonical correlation analysis. Pattern Recognition, 36, 1961–1971 (2003)
4. Hardoon, D.R. et al.: Canonical correlation analysis; an overview with application to learning methods, Technical Report CSD-TR-03-02, Dept. of Computer Science, University of London (2003)
5. Ruiz, A., López-de-Teruel, P.E.: Nonlinear kernel-based statistical pattern analysis. IEEE Trans. Neural Networks 12, 16–32 (2001)
6. Müller, K.-R. et al.: An introduction to kernel-based learning algorithms. IEEE Trans. Neural Networks 12, 181–201 (2001)
7. Popovici, V., Thiran, J.-P.: Higher order autocorrelations for pattern classification, In: Proc. IEEE 2001 International Conference on Image Processing (ICIP2001), pp. 724–727 (2001)
8. Popovici, V., Thiran, J.-P.: Pattern recognition using higher-order local autocorrelation coefficients. Pattern Recognition Letters 25, 1107–1113 (2004)
9. McLaughlin, J.A., Raviv, J.: Nth-order autocorrelations in pattern recognition. Information and Control 12, 121–142 (1968)

10. Horikawa, Y.: Comparison of support vector machines with autocorrelation kernels for invariant texture classification, In: Proc. 17th International Conference on Pattern Recognition (ICPR 2004), vol.1, 3P.Moiii-4 (2004)
11. Horikawa, Y.: Use of autocorrelation kernels in kernel canonical correlation analysis for texture classification. In: Pal, N.R., Kasabov, N., Mudi, R.K., Pal, S., Parui, S.K. (eds.) ICONIP 2004. LNCS, vol. 3316, pp. 1235–1240. Springer, Heidelberg (2004)
12. Horikawa, Y.: Modification of correlation kernels in SVM, KPCA and KCCA in texture classification, In: Proc. 2005 International Joint Conference on Neural Networks (IJCNN 2005), P2-A2fac, 2042, pp. 2006–2011 (2005)
13. Kittler, J.K. et al.: On combining classifiers. IEEE Trans. Patt. Anal. Machine Intell. 20, 226–239 (1998)
14. Jain, A.K. et al.: Statistical pattern recognition: A review. IEEE Trans. Patt. Anal. Machine Intell. 22, 4–37 (2000)
15. Kurita, T., Kato, T.: Learning of personal visual impression for image database systems, In: Proc. the Second International Conference on Document Analysis and Recognition, pp. 547–552 (1993)
16. Brodatz, P.: Textures: A Photographic Album for Artists and Designers. Dover, New York (1966)
17. Chijiiwa, H.: Color Science (in Japanese), Fukumura, Japan, (J.Suzuki's Website, http://www.teu.ac.jp/chiit/~jsuzuki/kanseiwr.html) (1983)
18. AMOVIP-DB, http://www.iv.optica.csic.es/projects/database.html

A Visual System for Hand Gesture Recognition in Human-Computer Interaction

Matti-Antero Okkonen, Vili Kellokumpu, Matti Pietikäinen, and Janne Heikkilä

Department of Electrical and Information Engineering, P.O. Box 4500, FI-90014, University of Oulu, Finland
okm@ee.oulu.fi
http://www.ee.oulu.fi/mvg

Abstract. Visual hand gestures offer an interesting modality for Human-Computer-Interaction (HCI) applications. Gesture recognition and hand tracking, however, are not trivial tasks and real environments set a lot of challenges to algorithms performing such activities. In this paper, a novel combination of techniques is presented for tracking and recognition of hand gestures in real, cluttered environments. In addition to combining existing techniques, a method for locating a hand and segmenting it from an arm in binary silhouettes and a foreground model for color segmentation is proposed. A single hand is tracked with a single camera and the trajectory information is extracted along with recognition of five different gestures. This information is exploited for replacing the operations of a normal computer mouse. The silhouette of the hand is extracted as a combination of different segmentation methods: An adaptive colour model based segmentation is combined with intensity and chromaticity based background subtraction techniques to achieve robust performance in cluttered scenes. An affine-invariant Fourier-descriptor is derived from the silhouette, which is then classified to a hand shape class with support vector machines (SVM). Gestures are recognized as changes in the hand shape with a finite state machine (FSM).

1 Introduction

Hand tracking and gesture recognition have evolved during the past two decades from cumbersome data gloves and tracking aids to purely visual-based recognition systems. Visual interpretation of hand gestures offers a non-contact HCI-modality, but it has a lot of challenges to be tackled. The human hand is a complex object to be modeled and tracked: It consist of 27 bones and is capable of performing variegated and non-rigid motion. An accurate representation of a hand's 3D orientation requires high dimensional models, which often leads to problems with the real-time requirements of HCI-applications. In addition, complex models often contain irrelevant information from the application point of view. Besides the human hand, the real environments also contain factors that complicates the recognition process, such as background clutter and unideal lighting conditions.

B.K. Ersbøll and K.S. Pedersen (Eds.): SCIA 2007, LNCS 4522, pp. 709–718, 2007.

Skin color has a unique chromaticity value, which makes it a popular cue for hand area detection. Parametric approaches usually present the skin colour distribution with Gaussian functions [1,2], whereas non-parametric methods use simple look-up-tables or histograms [3]. The last-mentioned have the ability to present arbitrary distributions, albeit they do not share the compactness of the parametric techniques. Another discriminative characteristic among colour models is their adaptivity. Some methods, like the system of Pantrigo et al., use predefined models that remain fixed throughout the processing [4]. These models are prone to fail under environments where lighting conditions differ from the samples that the models are built on. More advanced models adapt themselves to the current conditions and thus specify the skin area more robustly [1]. Another problem with skin area segmentation is the specification of the palm and the fingers from other skin coloured areas. For example, if a user wears a t-shirt, the arm should be somehow separated from the palm. In many studies the users wear long sleeved shirts, as in the work of Zhai et al., when this problem is not addressed at all [3].

In vision-based HCI applications, there are usually two kinds of information extracted from the user: gestural and spatial signals. To replace a mouse, for example, both of these are needed. Some of the earlier approaches extract only pointing information[5,3,1], whereas other approaches analyze only the static hand poses [6]. Also methods that exploit both spatial and gestural information have been developed [4,7,8,9,2]. Fingertips are the most typical choices to be tracked when spatial information is needed. Hardenberg and Bérard also used the number of fingertips to recognize different hand poses [9]. In addition, some techniques track the area of the hand and extract the trajectory of the hand from its bounding box [3,4]. More detailed information about previous work in hand gesture recognition can be found in the references [10,11].

In this paper, a new approach for hand gesture recognition for HCI purposes is presented. The proposed method combines background subtraction techniques and colour segmentation for a robust skin area segmentation. In addition, it distinguishes palm and fingers from the arm and recognizes five different hand gestures. The rest of the paper is organized as follows: An overview of the system is given in section 2, section 3 explains the techniques for hand segmentation and tracking and section 4 presents the gesture recognition process in detail. Section 5 reports the experimental results and finally, discussion and conclusions are given in section 6.

2 Overview

At the top level, the system is controlled by a state machine. In the idle state, the system is using background subtraction to detect an appearing object. The state changes from idle to initialization, when an object bigger than a predefined threshold appears in the input stream. The colour-model and it's constraints are defined in the initialization state. After that, the system starts the actual algorithm. A flowchart of the system architecture is given in Fig. 1(a). In the

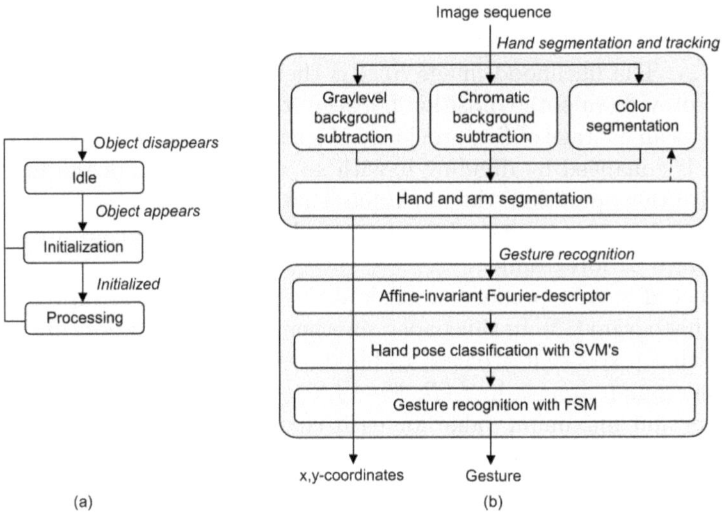

Fig. 1. (a) Top level architecture of the system. (b) Overview of the actual algorithm.

processing stage, each frame is analyzed and a possible gesture is recognized. In addition, spatial coordinates are extracted from the hand silhouette. An overview of the process is illustrated in Fig. 1(b).

3 Hand Segmentation and Tracking

Robust hand segmentation in silhouette-based systems is essential, because the rest of the processing is highly dependent on it. In the method presented here, the silhouette of the hand is extracted by combining background subtraction in gray-level space and in Normalized Color Coordinates (NCC) with adaptive color model based segmentation.

3.1 Background Subtraction and Color Segmentation

In the background subtraction method used here, the background image B is composed as an average of the first N images in a sequence, containing only the background. A binary silhouette S_G is obtained simply by subtracting each frame from the background image in gray-level values and thresholding the result. The second silhouette S_{NCC} is calculated on normalized R- channel in the same way as S_G. Now both gray-scale and a chromatic difference between the foreground object and the background have been exploited. A combined silhouette $S_G \cup S_{NCC}$ is used in object detection and colour model initialization in the *idle* and *initialization* stages of the system.

The color-based segmentation is achieved by *histogram backprojection* [12]. For a frame I_t, an r,g-histogram H_t of the foreground object is created using the segmentation result as a mask. For the following frame I_{t+1}, a skin likelihood image S_{t+1} is built using the H_t: For each pixel in S_{t+1}, a value $H_t(r', g')$ is assigned, where r' and g' are the quantized r, g-values of the corresponding pixel in I_{t+1}. The likelihood image S_{t+1} is then thresholded to a binary image S_{HBP}. To provide smooth transition between frames, the colour histogram H_t is calculated as an average of five previous object histograms. The specificity of the histogram is enhanced by dividing it with an r, g-histogram of the background image. With this operation, the r, g-values that appear often in the background are diminished in the resulting histogram and background pixels are less likely to be classified as foreground.

Soriano et al. used a predefined mask in the r,g-space for selecting the pixels for object histogram [13]. In this paper, another type of constraints are proposed: To ensure that pixels that are used to build the object histogram do not belong to the background, a mask in a $(R,\ R-G,\ G-B)$-colour space is defined with a minimum and maximum value for each channel. Minimum values α^j_{min} are constructed from normalized R-, $R-G$- and $G-B$-histograms H_j as

$$\alpha^j_{min} = \beta^j_{max} - C \tag{1}$$

by finding the maximum values β^j_{max} that satisfy the constraints

$$\sum_{k=0}^{\beta^j_{max}} H_j(k) < \gamma, \tag{2}$$

where γ is a coefficient that defines what portion of the histogram is located at bins bigger than β^j_{max}. Coefficient C is an offset value and together with γ it can be used for adjusting the looseness of the limits. The maximum values

$$\alpha^j_{max} = \beta^j_{min} + C \tag{3}$$

are defined in a similar way by finding the minimum values β^j_{min} that satisfy the constraints

$$\sum_{k=0}^{\beta^j_{min}} H_j(k) > 1 - \gamma. \tag{4}$$

In our experiments, we used the values $\gamma = 0.05$ and C $= 7$. Instead of predefined samples, the limits are established during the initialization stage of the algorithm. This assures that the limits correspond particularly to the current environment.

The three binary images, S_G, S_{NCC} and S_{HBP}, are combined as one with a bitwise logical OR operation to produce a robust segmentation of the hand.

3.2 Hand and Arm Segmentation

So far the segmentation method offers a binary silhouette of the foreground object, i.e. the user's hand. Since the interest lays on the palm and the fingers, they have to be segmented from the arm. In this approach the hand area is segmented based on it's geometrical features.

It is assumed that there is only the user's hand in the image. In this case, the biggest blob on the foreground object is the palm of the hand. The binary image is convolved with a Gaussian element, when the palm center is positioned at the maximum of the convolution. The operation is illustrated in Fig. 2.

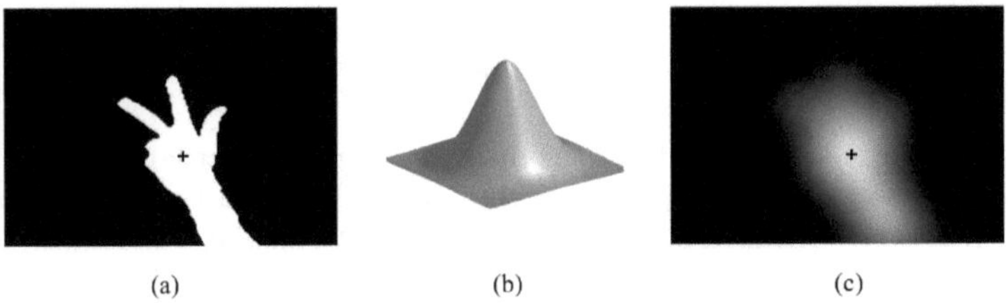

Fig. 2. Example of palm center positioning. (a) A binary hand silhouette (240x320), (b) Gaussian convolution element (100x100) and (c) Convolution of (a) and (b). The crosses in (a) and (c) mark the maximum of the convolution.

The value of the convolution maximum is dependent of the number of pixels valued one in the binary image that fall in the range of the convolution mask. Thus, if the palm is the biggest object in the binary image and the palm is considered to be disk shaped, it's radius can be estimated from the maximum of the convolution. In our approach, we used an empirically inferred polynomial function $R(\hat{I}_C)$ to model the radius as a function of the convolution maximum. Now, when the radius is known, the palm and the fingers can be segmented from hand with a mask that consist of the bounding box of the hand and an ellipse

$$\frac{(y - y_{max})^2}{\epsilon_1 R} + \frac{(x - x_{max})^2}{\epsilon_2 R} = 1, \tag{5}$$

where y_{max} and x_{max} are the coordinates of the convolution maximum and ϵ's are constants adjusting the axis lengths. In our implementation we used the values $\epsilon_1 = 1.0$ and $\epsilon_2 = 1.1$. The segmentation is illustrated in Fig. 3. The result contains the palm and fingers of the hand, and it is used for both gesture recognition and as a mask for the color model adaptation.

In addition to the segmentation, the convolution maximum also offers a stable key point for controlling a pointer in a HCI-application. Other key points, like fingertips and silhouette's center of gravity, are greatly dependent on the segmentation result and in unideal scenes the trajectories of them may involve excessive noise. The convolution, however, is insensitive to small errors in the silhouette and does not react, for example, to finger movements in the silhouette.

4 Gesture Recognition

The gesture recognition is based on the classified hand shapes. In this approach, the hand shapes are classified with SVM's and the gestures are recognized as changes from a hand shape to another.

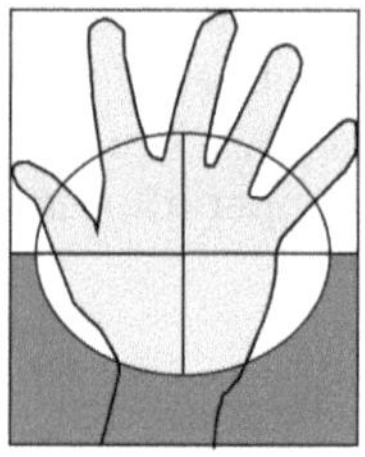

Fig. 3. The hand is segmented from the arm with a mask that consists of an ellipse positioned at the center of the palm and the bounding box of the whole object. The dark gray values are excluded from the binary silhouette of the hand.

4.1 Hand Contour Description and Posture Classification

The support vector machine is a learning machine that constructs a hyperplane based on a subset of the training data. These data points are called support vectors. A single SVM classifies data to two classes, but connecting them with some topology a multi-class classifier can be constituted. A more profound introduction is given in [14].

In our method, the extracted hand contour is re-sampled to an evenly spaced sequence with a constant length. An affine-invariant Fourier-descriptor is calculated from the resampled contour, and the pose is classified with SVM's using only the lowest components of the Fourier-descriptor. Along with the affine-invariance, another advantage of the descriptor is that small disturbances in the contour alter only the higher frequency components of the Fourier descriptor and thus do not affect the hand shape recognition. A detailed explanation of the descriptor can be found in [15]. similar approach for human body pose estimation has been used in [16]. In our implementation, we used 200 samples for the contour and the first fourteen components of the Fourier-descriptor (not counting the first one which always scales to unity). The classifier was trained altogether with 233 manually labeled samples from five different hand shape classes to provide a proper amount of within-class variance. All the hand shapes were gathered from a single person. As a kernel function we used the linear type.

4.2 Gesture Recognition with a Finite State Machine

The most common approach for dynamic gesture recognition are the Hidden Markov Models (HMM). They are well suited for time series analysis, such as recognizing hand written letters based on trajectory information. however, in our approach a gesture is simply a change from a hand pose to another with no complicated temporal behavior. Therefore there is no need for advanced time series analysis.

Here, a change of the hand's posture from on to another is interpreted as a gesture. To recognize the changes, a finite state machine (FSM) was used, which consisted of states corresponding to the shape classes and all possible transitions between the states. A transition from a state i to j happens when the last k poses are classified as j's. Through this condition, the recognition is invariant to transient shape misclassifications of duration $\tau < k$. To control the mouse in our implementation, each of the states in the FSM defines a state of the mouse, as explained in Fig. 4.

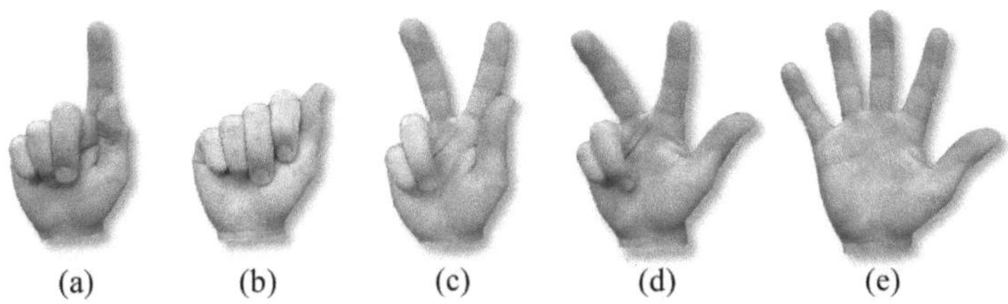

Fig. 4. (a) The mouse cursor is controlled by the palm center and none of the mouse buttons are down. (b) Left button is down. (c) Right button is down. (d) The roll of the mouse is used by extracting only the vertical coordinate of the palm center. (e) Mouse control is disabled.

5 Experiments

The system was implemented on a normal PC with 1 GB of memory and a 3.0 GHz Pentium 4 processor. The implementation processed about 18 frames per second, and the speed could be increased further with optimization. The performance of the system was tested against different backgrounds: Three test sequences were made by different persons, where a hand, positioned like in Fig. 3, moved and changed it's pose continuously. The backgrounds of the sequences are given in Fig. 5. The person used for the training of the system did not participate in the test sequences. The lengths of the sequences were approximately 55, 53 and 62 seconds long, containing altogether 131 gestures (i.e. hand shape changes). To the authors knowledge, there is no databases which could be used to evaluate the presented system.

To evaluate the system's performance, false positive and false negative ratios of the segmentation were calculated: For each sequence five randomly selected frames were manually segmented and compared to the result of the algorithm. The result are given in Table 1.

Fig. 5. Backgrounds of the test sequences one, two and three containing different levels of clutter and skin coloured objects

Table 1. False positive and false negative ratios of the segmentation

Measure	Test sequence 1	2	3
False positives	2.1%	2.6%	2.3%
False negatives	4.8%	8.7%	11.2%

The results of the gesture recognition are given in the form of a confusion matrix in Table 2, where all the gestures ending in the same hand shape are treated as one to avoid a sparse representation. From the results we can deduce the overall recognition rate of 94.6%, defined as the ratio of the correct detections to all the gestures. In addition, the results gives us a recognition accuracy of 89.2%, defined as the ratio of the correct detections to all the detections.

Table 2. Confusion matrix of the gesture recognition. The rows and columns represent the detected gestures and the ground truths, respectively. The names of the gestures are adopted from Fig. 4, where (x) means any shape.

	(x) → (a)	(x) → (b)	(x) → (c)	(x) → (d)	(x) → (e)	No gesture
(x) → (a)	27					1
(x) → (b)		25				1
(x) → (c)			30	1	1	2
(x) → (d)				24		4
(x) → (e)					18	5
No detection					5	

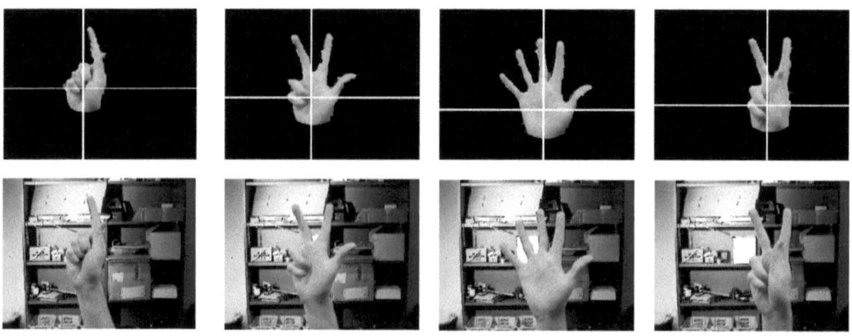

Fig. 6. Example frames and hand segmentations from tests sequence number three. The crosshairs mark the estimated palm centers.

As most of the gesture recognition systems, this approach contains some restrictions concerning the user and the environment. The camera view should be static and comprise only of a single hand with no occlusions. In addition, the hand should not rotate too much about the image coordinate axis and it should be in somewhat parallel with the image plane. As an example, some frames from test sequence number three are given in Fig. 6

6 Conclusions

In this paper, a novel method for real-time hand tracking and gesture recognition in cluttered environments is presented. This approach uses a new way to achieve robust segmentation of a users hand from video sequences. Unlike many algorithms, this method uses a colour-model that auto-initializes and adapts itself during processing. Further, the method uses the hand's geometric properties and affine-invariant features to recognize the hand poses. In our experiments, the system achieved a recognition rate of 95% and an accuracy of 89%, even when the background was cluttered and contained skin coloured objects. Altogether, the method operates both in real-time and real environments, and hence offers potential techniques for HCI-applications. For future research, the system could be possibly enhanced by using an adaptive background model instead of a static one.

References

1. Kurata, T., Okuma, T., Kourogi, M., Sakaue, K.: The hand mouse: Gmm hand-color classification and mean shift tracking. In: Proc. of the IEEE ICCV Workshop on Recognition, Analysis, and Tracking of Faces and Gestures in Real-Time Systems (RATFG-RTS'01), Vancouver, Canada, pp.119–124 (July 2001)
2. Wagner, S., Alefs, B., Picus, C.: Framework for a portable gesture interface. In: FGR '06: 7th International Conference on Automatic Face and Gesture Recognition, pp. 275–280 (2006)
3. Zhai, H., Wu, X., Han, H.: Research of a real-time hand tracking algorithm. In: Proc. of the International Conference on Neural Networks and Brain (ICNN&B'05), Beijing, China, pp. 1233–1235 (October 2005)
4. Pantrigo, J., Montemayor, A., Sanchez, A.: Local search particle filter applied to human-computer interaction. In: Pan, Y., Chen, D.-x., Guo, M., Cao, J., Dongarra, J.J. (eds.) ISPA 2005. LNCS, vol. 3758, pp. 279–284. Springer, Heidelberg (2005)
5. Quek, F., Mysliwiec, T., Zhao, M.: Fingermouse: A freehand computer pointing interface. In: Proc. of Int'l Conf. on Automatic Face and Gesture Recognition, pp. 372–377 (1995)
6. Horimoto, S., Arita, D., Taniguchi, R.-i.: Real-time hand shape recognition for human interface. In: Proc. of the 12th International Conference on Image Analysis and Processing (ICIAP '03), Mantova, Italy, pp. 20–25 (September 2003)
7. Liu, Y., Jia, Y.: A robust hand tracking for gesture-based interaction of wearable computers. In: McIlraith, S.A., Plexousakis, D., van Harmelen, F. (eds.) ISWC 2004. LNCS, vol. 3298, pp. 22–29. Springer, Heidelberg (2004)

8. Malik, S., Laszlo, J.: Visual touchpad: a two-handed gestural input device. In: Proc. of the 6th international conference on Multimodal interfaces (ICMI'04), State College, PA, USA, pp. 289–296 (October 2004)
9. von Hardenberg, C., Bérard, F.: Bare-hand human-computer interaction. In: Proceedings of the 2001 workshop on Perceptive user interfaces (PUI'01), Orlando, FL, USA, pp. 1–8 (November 2001)
10. Pavlovic, V.I., Sharma, R., Huang, T.S.: Visual interpretation of hand gestures for human-computer interaction: A review. IEEE Trans. Pattern Anal. Mach. Intell. 19(7), 677–695 (1997)
11. Ong, S.C.W., Ranganath, S.: Automatic sign language analysis: A survey and the future beyond lexical meaning. IEEE Trans. Pattern Anal. Mach. Intell. 27(6), 873–891 (2005)
12. Swain, M.J., Ballard, D.H.: Color indexing. Int. J. Comput. Vision 7(1), 11–32 (1991)
13. Soriano, M., Martinkauppi, B., Huovinen, S., Laaksonen, M.: Adaptive skin color modeling using the skin locus for selecting training pixels. Pattern Recognition 36(3), 681–690 (2003)
14. Burges, C.J.C.: A tutorial on support vector machines for pattern recognition. Data Mining and Knowledge Discovery 2(2), 121–167 (1998)
15. Arbter, K., Snyder, W.E., Burhardt, H., Hirzinger, G.: Application of affine-invariant fourier descriptors to recognition of 3-d objects. IEEE Trans. Pattern Anal. Mach. Intell. 12(7), 640–647 (1990)
16. Kellokumpu, V., Pietikäinen, M., Heikkilä, J.: Human activity recognition using sequences of postures. In: Proceedings of the IAPR Conference on Machine Vision Applications (MVA 2005), pp. 570–573 (2005)

Single View Motion Tracking by Depth and Silhouette Information

Daniel Grest[1], Volker Krüger[1], and Reinhard Koch[2]

[1] Aalborg University Copenhagen, Denmark
Aalborg Media Lab
[2] Christian-Albrechts-University Kiel, Germany
Multimedia Information Processing

Abstract. In this work[1] a combination of depth and silhouette information is presented to track the motion of a human from a single view. Depth data is acquired from a Photonic Mixer Device (PMD), which measures the time-of-flight of light. Correspondences between the silhouette of the projected model and the real image are established in a novel way, that can handle cluttered non-static backgrounds. Pose is estimated by Nonlinear Least Squares, which handles the underlying dynamics of the kinematic chain directly. Analytic Jacobians allow pose estimation with 5 FPS.

Keywords: optical motion capture, articulated objects, pose estimation, cue-integration.

1 Introduction

Valid tracking of human motion from a single view is an important aspect in robotics, where research aims at motion recognition from data, that is collected from the robot's measuring devices. Additionally, the processing time should be at least near-to-real-time to make human-robot interaction possible. Both aspects are addressed in this work.

Motion capture and body pose estimation are also applied in motion analysis for sports and medical purposes. Motion capture products used in the film industry or for computer games are usually marker based to achieve high quality and fast processing. While the accuracy of markerless approaches is comparable to marker based systems [13,4], the segmentation step makes strong restrictions to the capture environment, because these systems rely on segmentation of the person in the foreground, e.g. homogenous clothing and background, constant lighting, camera setups that cover a complete circular view on the person etc.

Our approach doesn't need explicit segmentation or homogenous clothing and gives reliable results even with non-static cluttered background. Additionally,

[1] Acknowledgment This work was partially funded by PACO-PLUS (IST-FP6-IP-027657) and by German Science Foundation project DFG-3DPoseMap.

B.K. Ersbøll and K.S. Pedersen (Eds.): SCIA 2007, LNCS 4522, pp. 719–729, 2007.

motion can be accurately tracked even from a single view, because the underlying motion and body model is directly incorporated in the image processing step. We present here a combination of depth data and silhouette information, which extends the motion estimation from stereo data [7] with additional information from silhouette correspondences. Results are given for depth data from a novel measuring technique, called *Photonic Mixer Device (PMD)*, which gives a 64×48 depth image in real-time with 25FPS. The results show, that the characteristic of this depth data is not sufficient alone for valid tracking. However in combination with silhouette information, the accuracy is increased and motion can be successfully tracked over longer sequences.

Capturing human motion by pose estimation of an articulated object is done in many approaches and is motivated from inverse kinematic problems in robotics. Solving the estimation problem by optimization of an objective function is also very common [13,8,11]. Silhouette information is usually part of this function, that tries to minimize the difference between the model silhouette and the silhouette of the real person either by background segmentation [13,11] or image gradient [12,5]. In [2] a scaled orthographic projection approximates the full perspective camera model and in [13] the minimization of 2D image point distances is approximated by 3D-line-3D-point distances. A recent extensive survey on vision-based motion capture can be found in [9].

While some kind of template body model is common in most approaches, adaption of body part sizes of the template during the motion estimation is also possible [12], where depth and silhouette information were combined to estimate the size and pose of the upper body. In contrast to their approach, we estimate pose in near-to-real-time and minimize silhouette differences in the image plane rather than in 3D, which makes the estimation more accurate. The image processing with color histograms allows us to establish valid silhouette correspondences even with moving background, which in turn allows moving cameras. By combination with depth data from a PMD device motion can be tracked, which is not trackable from a single view with only one of these data types. Our method minimizes errors, where they are observed and makes no approximations to the motion or projection model. Additionally, it allows analytical derivations of the optimization function. This speeds up the calculation by more accuracy and less function evaluations than numerical derivatives. Therefore the approach is fast enough for real-time applications in the near future as we process images already with 5 frames per second on a standard PC Pentium IV 3 GHz.

2 Body and Motion Model

Depending on the kind of work different body models are used for the estimation process. The models range from simple stick figures over models consisting of scalable spheres (meta-balls) [12] to linear blend skinned models [1]. We use models with motion capabilities as defined in the MPEG4 standard, with up to

180 DOF, an example model is shown in figure (1). The MPEG4 description allows to exchange body models easily and to reanimate other models with the captured motion data. The model for a specific person is obtained by silhouette fitting of a template model as described in [6].

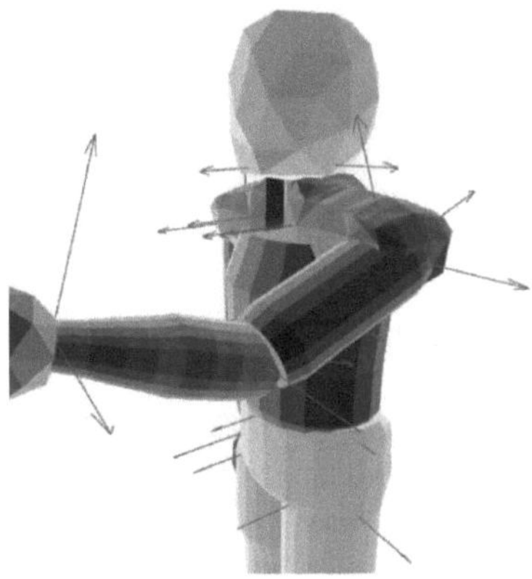

Fig. 1. The body model with rotation axes shown as arrows

The MPEG4 body model is a combination of kinematic chains. The motion of a point, e.g. on the hand, may therefore be expressed as a concatenation of rotations [7]. As the rotation axes are known, e.g. the flexion of the elbow, the rotation has only one degree of freedom (DOF), i.e. the angle around that axis. In addition to the joint angles there are 6 DOF for the position and orientation of the object within the global world coordinate frame. For an articulated object with p joints the transformation may be written according to [7] as:

$$\boldsymbol{f}(\boldsymbol{\theta},\boldsymbol{x})=(\theta_x,\theta_y,\theta_z)^T+(R_x(\theta_\alpha)\circ R_y(\theta_\beta)\circ R_z(\theta_\gamma)\circ R_{\boldsymbol{\omega},\boldsymbol{q}}(\theta_1)\circ\cdots\circ R_{\boldsymbol{\omega},\boldsymbol{q}}(\theta_p))(\boldsymbol{x})$$

where $(\theta_x,\theta_y,\theta_z)^T$ is the global translation, R_x, R_y, R_z are the rotations around the global x, y, z-axes with Euler angles α, β, γ and $R_{\boldsymbol{\omega},\boldsymbol{q}}(\theta_i)$,$i \in \{1..p\}$ denotes the rotation around the known axis with angle θ_i. The axis is described by the normal vector $\boldsymbol{\omega_i}$ and the point $\boldsymbol{q_i}$ on the axis with closest distance to the origin.

The equation above gives the position of a point $\boldsymbol{x}$ on a specific segment of the body (e.g. the hand) with respect to joint angles $\boldsymbol{\theta}$ and an initial body pose.

The first derivatives of $\boldsymbol{f}(\boldsymbol{\theta},\boldsymbol{x})$ with respect to $\boldsymbol{\theta}$ give the Jacobian matrix $J_{ki} = \frac{\partial f_k}{\partial \theta_i}$. The Jacobian for the motion of the point $\boldsymbol{x}$ on an articulated object is

$$J = \begin{bmatrix} 1\,0\,0 & & & & & & \\ 0\,1\,0 & \frac{\partial f}{\partial \theta_\alpha} & \frac{\partial f}{\partial \theta_\beta} & \frac{\partial f}{\partial \theta_\gamma} & \frac{\partial f}{\partial \theta_1} & \cdots & \frac{\partial f}{\partial \theta_p} \\ 0\,0\,1 & & & & & & \end{bmatrix}, \tag{1}$$

with the simplified derivative at zero:

$$\left.\frac{\partial f}{\partial \theta_i}\right|_0 = \left.\frac{\partial R_{\boldsymbol{\omega},\boldsymbol{q}}(\theta_i)}{\partial \theta_i}\right|_0 = \boldsymbol{\omega}_i \times (\boldsymbol{x} - \boldsymbol{q}_i) = \boldsymbol{\omega}_i \times \boldsymbol{x} - \boldsymbol{\omega}_i \times \boldsymbol{p}_i \ . \tag{2}$$

Here $\boldsymbol{p}_i$ is an arbitrary point on the rotation axis. The term $\boldsymbol{\omega}_i \times \boldsymbol{p}_i$ is also called the momentum. The simplified derivative at zero is valid, if relative transforms in each iteration step of the *Nonlinear Least Squares* are calculated and if all axes and corresponding point pairs are given in world coordinates.

2.1 Projection

If the point $\boldsymbol{x} = (x_x, x_y, x_z)^T$ is observed by a pin-hole camera and the camera coordinate system is in alignment with the world coordinate system, the camera projection may be written as:

$$p(\boldsymbol{x}) = \begin{pmatrix} s_x \frac{x_x}{x_z} + c_x \\ s_y \frac{x_y}{x_z} + c_y \end{pmatrix} \tag{3}$$

where s_x, s_y are the pixel scale (focal length) of the camera in x- and y-direction, and $(c_x, c_y)^T$ is the center of projection in camera coordinates.

We now combine $\boldsymbol{f}(\boldsymbol{\theta}, \boldsymbol{x})$ and $p(\boldsymbol{x})$ by writing $\boldsymbol{g}(s_x, s_y, c_x, c_y, \boldsymbol{\theta}, \boldsymbol{x}) = p(\boldsymbol{f}(\boldsymbol{\theta}, \boldsymbol{x}))$. The partial derivatives of g can now be easily computed using the chain rule. The resulting Jacobian reads as follows:

$$\begin{aligned} J &= \left[\begin{matrix} \frac{\partial g}{\partial s_x} & \frac{\partial g}{\partial s_y} & \frac{\partial g}{\partial c_x} & \frac{\partial g}{\partial c_y} & \frac{\partial g}{\partial \theta_x} & \frac{\partial g}{\partial \theta_y} & \frac{\partial g}{\partial \theta_z} & \frac{\partial g}{\partial \theta_\alpha} \cdots \frac{\partial g}{\partial \theta_p} \end{matrix} \right] \\ &= \left[\begin{matrix} \frac{\boldsymbol{f}(\boldsymbol{\theta})_x}{\boldsymbol{f}(\boldsymbol{\theta})_z} & 0 & 1 & 0 & \frac{s_x}{\boldsymbol{f}(\boldsymbol{\theta})_z} & 0 & s_x \frac{-\boldsymbol{f}(\boldsymbol{\theta})_x}{(\boldsymbol{f}(\boldsymbol{\theta})_z)^2} & \frac{\partial g_x}{\partial \theta_\alpha} \cdots \frac{\partial g_x}{\partial \theta_p} \\ 0 & \frac{\boldsymbol{f}(\boldsymbol{\theta})_y}{\boldsymbol{f}(\boldsymbol{\theta})_z} & 0 & 1 & 0 & \frac{s_y}{\boldsymbol{f}(\boldsymbol{\theta})_z} & s_y \frac{-\boldsymbol{f}(\boldsymbol{\theta})_y}{(\boldsymbol{f}(\boldsymbol{\theta})_z)^2} & \frac{\partial g_y}{\partial \theta_\alpha} \cdots \frac{\partial g_y}{\partial \theta_p} \end{matrix} \right] \end{aligned} \tag{4}$$

and

$$\frac{\partial g}{\partial \theta_i} = \begin{pmatrix} \frac{\partial \left(s_x \frac{f_x}{f_z} \right)}{\partial \theta_i} \\ \\ \frac{\partial \left(s_y \frac{f_y}{f_z} \right)}{\partial \theta_i} \end{pmatrix} = \begin{pmatrix} \frac{s_x \left(\frac{\partial f_x}{\partial \theta_i} \boldsymbol{f}(\boldsymbol{\theta})_z - \boldsymbol{f}(\boldsymbol{\theta})_x \frac{\partial f_z}{\partial \theta_i} \right)}{(\boldsymbol{f}(\boldsymbol{\theta})_z)^2} \\ \\ \frac{s_y \left(\frac{\partial f_y}{\partial \theta_i} \boldsymbol{f}(\boldsymbol{\theta})_z - \boldsymbol{f}(\boldsymbol{\theta})_y \frac{\partial f_z}{\partial \theta_i} \right)}{(\boldsymbol{f}(\boldsymbol{\theta})_z)^2} \end{pmatrix} \tag{5}$$

The partial derivatives $\frac{\partial f}{\partial \theta_i}, i \in \{\alpha, \beta, \gamma, 1, .., p\}$ are given in equation (1) and $\boldsymbol{f}(\boldsymbol{\theta}) = (f_x, f_y, f_z)^T$ is short for $\boldsymbol{f}(\boldsymbol{\theta}, \boldsymbol{x})$. Note that $\boldsymbol{f}(\boldsymbol{\theta})$ simplifies to $\boldsymbol{x}$, if $\boldsymbol{\theta}$ is zero.

These Jacobian allows full camera calibration from (at best) five 2D-3D correspondences or pose from 3 correspondences. An implementation of it with an extension to the *Levenberg-Marquardt* algorithm[3], which ensures an error decrease with each iteration, is available for public in our open-source C++ library[2].

3 Correspondences by Silhouette

To compensate the drift we add silhouette information to our estimation. This is achieved by calculating additional 2D-3D correspondences for the model silhouette and the silhouette of the real person. In contrast to [13] we don't utilize explicit segmentation of the images in fore- and background, but use the predicted model silhouette to search for corresponding points on the real silhouette. Previous work like [8] already took this approach by searching for a maximum grey value gradient in the image in the vicinity of the model silhouette. However we experienced that the gray value gradient alone gives often erroneous correspondences, especially if the background is heavily cluttered and the person wears textured clothes.

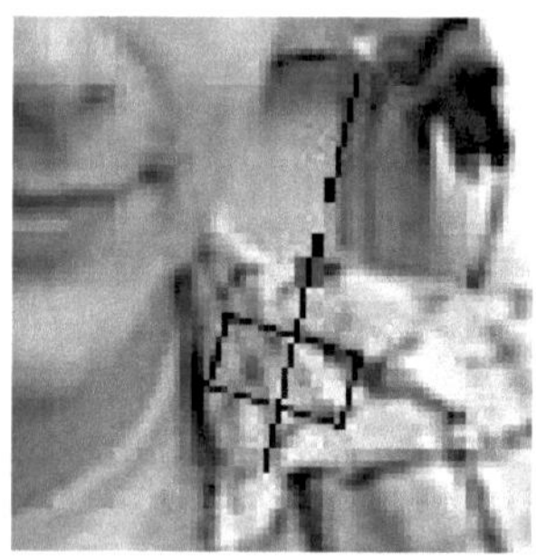

Fig. 2. Correspondence search along the normal

[2] www.mip.informatik.uni-kiel.de/Software/software.html

Therefore we also take color information into account. As the initial pose is known, it is possible to calculate a color histogram for each body segment. We use the HSL color space to get more brightness invariance. This reference histogram is then compared with a histogram calculated over a small window on the searched normal. In figure 2 the normal is shown and the rectangular window, that are used for histogram and gradient calculation. The expectation is, that the histogram difference changes most rapidly on the point on the normal of the correct correspondence, where the border between person and background is. The type of combination function was chosen by analyzing the developing of gradient and histogram values over 15 normals in different images. The actual values of the combination were then evaluated experimentally by trying different values and counting the number of correct correspondences manually for about 100 silhouette points in 4 different images.

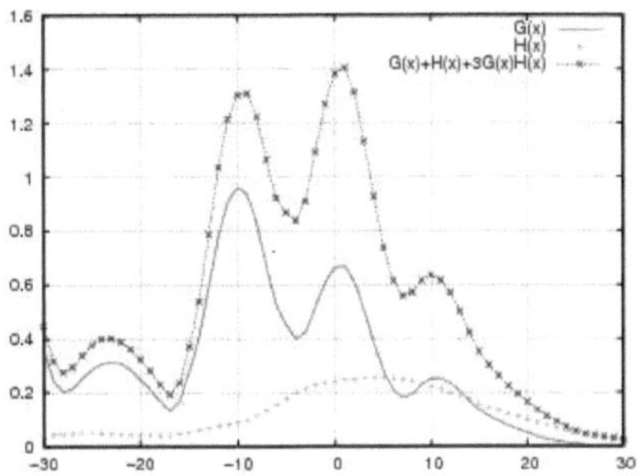

Fig. 3. The gradient (G(x)) and histogram (H(x)) values along the normal. Correct correspondence at 0.

A rather difficult case is shown in figure 3, which shows a plot of the maximum search along the normal of figure 2. The grey value gradient $G(x)$ is shown as a solid line, the gradient of the histogram differences $H(x)$ as points and the combination with lines and points. As visible, the grey value gradient alone would give a wrong correspondence, while the combination yields the correct maximum at zero.

For parallel lines it isn't possible to measure the displacement in the direction of the lines (aperture problem). Therefore we use a formulation that minimizes the distance between the tangent at the model silhouette and the target silhouette point (normal displacement), resulting in a 3D-point-2D-line correspondence as visible in figure 4. For a single correspondence the minimization is

$$\min_{\boldsymbol{\theta}} \left[(\boldsymbol{g}(\boldsymbol{\theta}, \boldsymbol{x}) - \boldsymbol{x}')^T \boldsymbol{n} - d\right]^2 \tag{6}$$

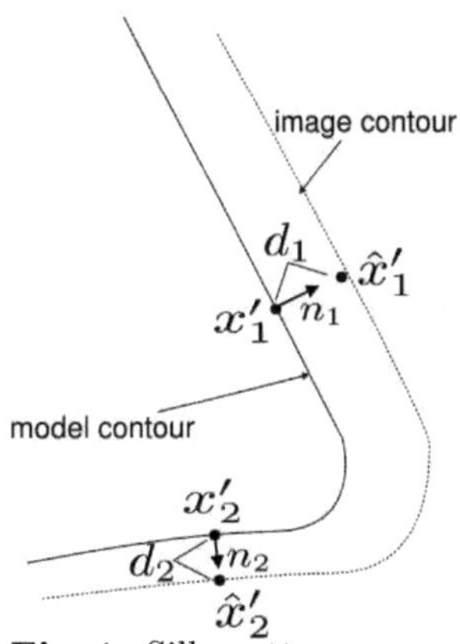

Fig. 4. Silhouette correspondences

where $\boldsymbol{n}$ is the normal vector on the tangent line and d is the distance between both silhouettes. We compute d as $d = (\hat{\boldsymbol{x}}' - \boldsymbol{x}')\boldsymbol{n}$. The point on the image silhouette $\hat{\boldsymbol{x}}'$ is the closest point to $\boldsymbol{x}'$ in direction of the normal. In this formulation a motion of the point perpendicular to the normal will not change the error. We calculate the normal vector as the projected face normal of the triangle, which belongs to the point $\boldsymbol{x}'$.

For a set $\boldsymbol{X}$ with N points and projected image points $\boldsymbol{X}'$ the optimal solution is:

$$\hat{\boldsymbol{\theta}} = \arg\min_{\boldsymbol{\theta}} \sum_{i=1}^{N} \left[(\boldsymbol{g}(\boldsymbol{\theta}, \boldsymbol{x_i}) - \boldsymbol{x}'_i)^T \boldsymbol{n_i} - d_i \right]^2$$

This problem is known as *Nonlinear Least Squares* and can be solved by *Newton's Method* [3]. We use the *Gauss-Newton Method* [3], which doesn't require the second derivatives of $\boldsymbol{g}(\boldsymbol{\theta}, \boldsymbol{x_i})$. The necessary Jacobian is given as:

$$J_{ik} = \left(\frac{\partial \boldsymbol{g}(\boldsymbol{\theta}, \boldsymbol{x_i})}{\partial \theta_k}\right)^T \boldsymbol{n_i} \tag{7}$$

Note that each of these correspondences gives one row in the Jacobian.

The solution is found by iteratively solving the following equation:

$$\boldsymbol{\theta_{t+1}} = \boldsymbol{\theta_t} - (J^T J)^{-1} J^T \left(G(\boldsymbol{\theta_t}, \boldsymbol{X}) - \hat{\boldsymbol{X}}' \right) \tag{8}$$

Here the Jacobian matrix J consists of all partial derivatives for all N points. The Jacobian for a single point is given in equation (4). In case of convergence the final solution $\hat{\boldsymbol{\theta}}$ is found.

4 Combining Multiple Cues

Integration of different vision cues into our parameter estimation problem is non trivial. Different cues like tracked edges or points give different information about the model parameters. Additionally, the measurement noise of different cues can vary dramatically.

In [5] both aspects are addressed by modeling different image cues, which are defined by regions. These regions are then propagated through the estimation by affine arithmetic. For example, tracked edges have a region that is elongated along the edge and less elongated perpendicular to it. These regions are combined into a generalized image force for each cue. The resulting region in parameter space is approximated by a Gaussian distribution. The Gaussians from each cue are then combined by a *Maximum Likelihood* Estimator and the result is integrated in a classical Euler integration procedure. The defined image regions are supposed to set hard limits on the possible displacements, however due to Gaussian approximation of the resulting parameter region the limits are softened. Therefore the approach becomes similar to a covariance based approach, where each image cue has an associated covariance matrix.

The approach taken in this work is different. The silhouette information is integrated by changing the objective function, such that the distance of the projected 3D-point to the 2D line is minimized. This is equivalent to a point-point distance with a covariance infinitely extended in direction along the edge. The different measurement noise of different cues is integrated in the estimation here by weighting each correspondence with a scalar. Weighting with a covariance matrix would be possible as well. However, for the different cues in this work the measurement noise is not exactly known and therefore covariance matrices are assumed to be diagonal and extended the same in all directions. Additionally it is assumed, that the measurement noise is the same for all measurements of one cue, resulting in one single scalar weight for each cue. In addition to the measurement noise, the weights reflect the different units of measurements, e.g. the measurement unit of 3D point positions from stereo images is meter, while the 2D measurement unit is pixels.

Let $X = \{\boldsymbol{x_0}, \boldsymbol{x_1}, .., \boldsymbol{x_k}\}$ be the set of model points and $Y = \{\boldsymbol{y}_1, .., \boldsymbol{y}_k\}$ be their corresponding 3D-points from the PMD camera found by nearest neighbor [7]. The correspondences for the silhouette information are built by the 3D-points $X = \{\boldsymbol{x_{k+1}}, .., \boldsymbol{x_{k+l}}\}$ and corresponding points on the image silhouette $X' = \{\boldsymbol{x}'_0, \boldsymbol{x}'_1, .., \boldsymbol{x}'_l\}$. Additionally assume that the pose of the person is known at that time, such that the projected body model aligns with the observed image as in the first image of figure 8. If the person now moves a little and an image I_{t+1} is taken, it is possible to capture the motion by estimating the relative joint angles of the body between the frames I_t and I_{t+1}. The pose estimation problem is to find the parameters $\hat{\boldsymbol{\theta}}$ that best fit the transformed and projected model points to the $k + l$ correspondences. This can be formulated as follows:

$$\hat{\boldsymbol{\theta}} = \arg\min_{\boldsymbol{\theta}} \sum_{i=1}^{k} w_i \; |\boldsymbol{f}(\boldsymbol{\theta}, \boldsymbol{x_i}) - \boldsymbol{y_i}|^2 + \sum_{j=1}^{l} w_j \; \left[(\boldsymbol{g}(\boldsymbol{\theta}, \boldsymbol{x_{k+j}}) - \boldsymbol{x}'_{\boldsymbol{i}})^T \boldsymbol{n_j} - d_i\right]^2 \tag{9}$$

This problem is again a *Nonlinear Least Squares* and is solved with the *Gauss-Newton method* [3]. The necessary Jacobian is a row-wise combination of the Jacobians from equation (1) and (7).

To get the initial pose, the user has to position the model manually in a near vicinity to the correct image position. After a few ICP iterations, the initial correct pose is found.

4.1 Arm Tracking from Silhouette and PMD-Data

A *Photonic Mixer Device* (PMD) is able to measure the distance to scene objects in its field of view. Similar to laser range scanners it is based on the time-of-flight of light. In contrast to the rather expensive laser range scanners, which usually give only one line of distances at a time. A PMD device gives distance values for a complete volume at a time. The construction and working principle is similar to conventional cameras. The time of flight is measured by phase differences between modulated emitted light and received light. To become more invariant to scene illumination and less disturbing, infrared light is used. More details can be found in [10].

Fig. 5. Setup used for the arm tracking. PMD camera on the top with IR-LEDs next to it.

In figure 5 the setup used in the experiments is shown. On the top one sees the PMD camera with Infrared-LEDs next to it. On the bottom is a conventional cameras installed. The PMD-depth image is best visualized with a view on the

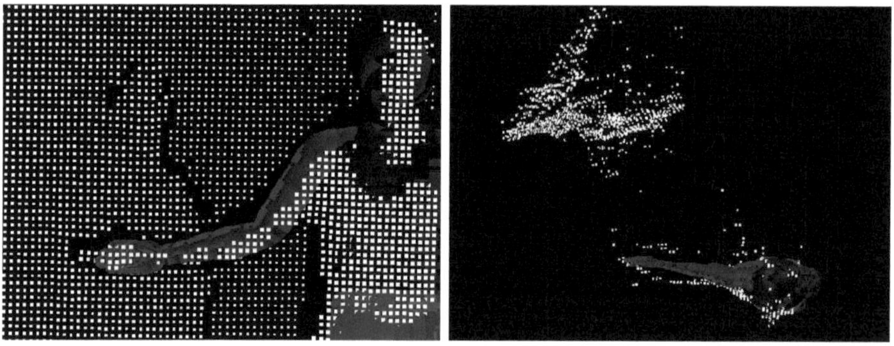

Fig. 6. Two views on the depth points, variance filtered

resulting 3D-scene points as shown in figure 6. Where one sees two views on the same point cloud from different angles. The depth image has been altered to eliminate erroneous depth values between fore- and background. To reduce the influence of in-between-points and that of outliers, a variance filter is run on the depth image, that calculates the variance within a 3×3 window and sets all pixels with a deviation larger than a threshold to zero. Typical values are in between 0.1m and 0.25m.

5 Results

The field of view of the PMD with 20 degrees is rather small and could not be exchanged with other lenses, because the lens has a special daylight filter. Therefore the compromise between a large visible scene volume and low outlier rate is taken, which is at approx. 3m distance to the camera. In this distance the motion

Fig. 7. Silhouette correspondences are accurate though the background is very dynamic

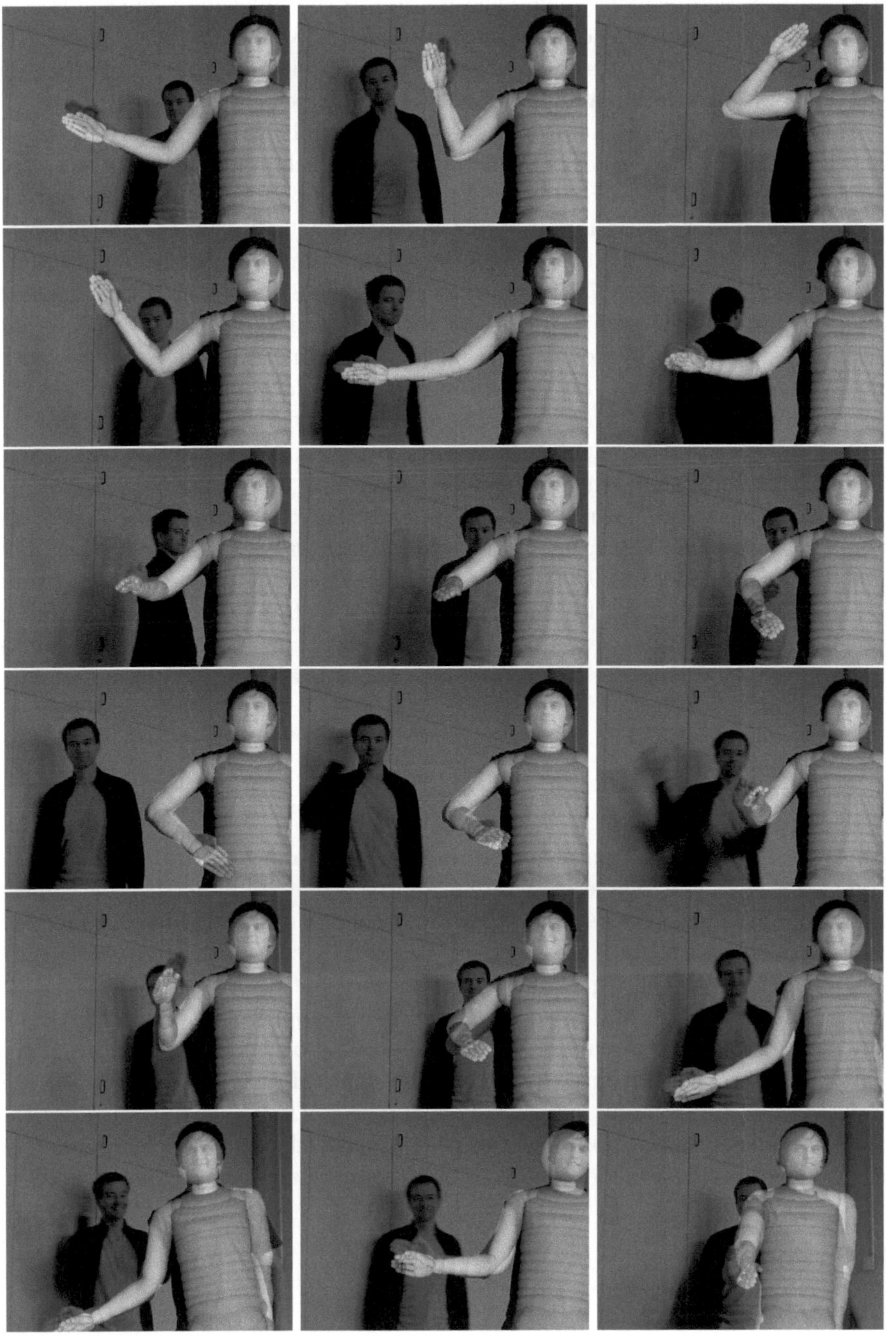

Fig. 8. Result sequence with dynamic background. The estimated model pose is overlayed on the original camera image. Ten DOF were estimated.

of one arm is completely visible. The motion in the following sequence is estimated from 3D-point-3D-point correspondences and 3D-2D-line correspondences established from silhouette information. The motion of shoulder and elbow as well as global translation and rotation were estimated, all together 10 DOF.

The motion of the right arm could be successfully tracked over the whole length of different sequences. Even though the background is non-static and cluttered (a person is walking around in the background) silhouette correspondences are accurate as visible in figure 7. This is achieved by the combination of grey value gradient and color histograms. An example sequence of 670 frames which was recorded with 7 FPS is shown in figure 8. Depicted is the image of the conventional camera superimposed with the estimated model pose. When the arms are moving in front of the body, there is not enough silhouette visible for a valid single view tracking from silhouette data alone. In that case the tracking relies on the depth data and becomes less accurate. The accuracy of the estimation is limited by the accuracy of the fitted model, which does not reveal the exact person's shape in the shoulder region.

Experiments with depth data alone showed, that the estimation is less accurate and during the 670 frame sequence tracking was lost for 50 frames. The depicted body model has 90000 points and 86000 triangles and processing time was about 1.5 seconds per frame. For this type of motion however a less detailed model is sufficient. In our experiments with a model consisting of a 10000 points and approx. 3500 triangles the processing time was about 5FPS on a standard PC Pentium IV 3GHz.

6 Conclusions

We showed how estimation of human motion can be derived from point transformations of an articulated object. Our approach uses a full perspective camera model and minimizes errors where they are observed, i.e. in the image plane. The combination of depth and silhouette information by color histograms and gradients allows to establish correct correspondences in spite of non-static background and people wearing normal clothing. Therefore the approach allows moving cameras as well. Ongoing research analyzes the quality of depth information of the PMD and stereo algorithms. We expect the depth data from stereo to be less accurate, but also exhibit less outliers than the PMD. Open problems are the necessary known initial pose and the need of a fitted body model, because the accuracy of the fitted model is a lower bound on the accuracy of the estimation.

References

1. Bray, M., Koller-Meier, E., Mueller, P., Van Gool, L., Schraudolph, N.N.: 3D Hand Tracking by Rapid Stochastic Gradient Descent Using a Skinning Model. In: CVMP. IEE (March 2004)
2. Bregler, C., Malik, J.: Tracking people with twists and exponential maps. In: Proceeding IEEE CVPR, pp. 8–15 (1998)

3. Chong, E.K.P., Zak, S.H.: An Introduction to Optimization (chapter 9), 2nd edn. Wiley, Chichester (2001)
4. Mündermann, L. et al.: Validation Of A Markerless Motion Capture System For The Calculation Of Lower Extremity Kinematics. In: Proc. American Society of Biomechanics, Cleveland, USA (2005)
5. Goldenstein, S., Vogler, C., Metaxas, D.: Statistical Cue Integration in DAG Deformable Models. PAMI, 25(7), 801–813 (2003)
6. Grest, D., Herzog, D., Koch, R.: Human Model Fitting from Monocular Posture Images. In: Proc. of VMV (November 2005)
7. Grest, D., Woetzel, J., Koch, R.: Nonlinear Body Pose Estimation from Depth Images. In: Proc. of DAGM, Vienna (September 2005)
8. Kakadiaris, I., Metaxas, D.: Model-Based Estimation of 3D Human Motion. IEEE Trans. on Pattern Analysis and Machine Intelligence (PAMI), 22(12) (2000)
9. Moeslund, T.B., Hilton, A., Krüger, V.: A Survey of Advances in Vision-Based Human Motion Capture Analysis. Journal of CVIU (2006)
10. Möller, T., Kraft, H., Frey, J., Albrecht, M., Lange, R.: Robust 3D Measurement with PMD Sensors. In: IEEE PacRim (2005)
11. Niskanen, M., Boyer, E., Horaud, R.: Articulated motion capture from 3-D points and normals. In: CVMP, London (2005)
12. Plaenkers, R., Fua, P.: Model-Based Silhouette Extraction for Accurate People Tracking. In: Proc. of ECCV, pp. 325–339. Springer-Verlag, Berlin Heidelberg New York (2002)
13. Rosenhahn, B., Kersting, U., Smith, D., Gurney, J., Brox, T., Klette, R.: A System for Marker-Less Human Motion Estimation. In: Kropatsch, W. (ed.) DAGM, Wien, Austria (September 2005)

Face Recognition with Irregular Region Spin Images

Yang Li, William A.P. Smith, and Edwin R. Hancock

Department of Computer Science, University of York, York, YO10 5DD, UK

Abstract. This paper explores how spin images can be constructed using shape-from-shading information and used for the purpose of face recognition. We commence by extracting needle maps from gray-scale images of faces, using a mean needle map to enforce the correct pattern of facial convexity and concavity. Spin images [6] are estimated from the needle maps using local spherical geometry to approximate the facial surface. Our representation is based on spin image histograms for an arrangement of image patches. Comparing to our previous spin image approach, the current one has two basic difference: Euclidean distance is replaced by geodesic distance; Irregular face region is applied to better fit face contour. We demonstrate how this representation can be used to perform face recognition across different subjects and illumination conditions. Experiments show the method to be reliable and accurate, and the recognition precision reaches 93% on CMU PIE sub-database.

1 Introduction

Face recognition is an active research area that has been approached in many ways. Roughly speaking the existing methods can be divided in two categories. The first is the feature-based method, while the second is the model-based method. Recently, it is the model-based methods that have attracted the greatest attention [2]. Here one of the most important recent developments is the work of Blanz and Vetter [3]. In this work a 3D morphable model is matched to face data using correspondences delivered by optic flow information. The method gives recognition rates of about 80% when profiles are used to recognise frontal poses. However, the construction of the model requires manual marking feature points, which is labour intensive. Hence, the automatic construction of models remains an imperative in face recognition. There are related feature-based approaches which are based on the assumption that face images are the result of Lambertian reflectance. Under this assumption 3D linear subspaces can be constructed that account for facial appearance under fixed viewpoint but under different illumination [12,1,10].

In this paper we aim to develop a feature-based method for face recognition that can be used to recognise faces using surface shape information inferred from image brightness using a Lambertian shape-from-shading scheme. Shape-from-shading is not widely accepted as a technique for face recognition. The reason is that surface normal is commonly believed to be noisy and is unstable under changes in illumination direction or change of pose. However, recently it has been shown that shape-from-shading can be used to extract useful features from real world face images [11].

One of the problems that hinders the extraction of reliable facial topography using shape-from-shading is the concave/convex inversions that arise due to the bas-relief

B.K. Ersbøll and K.S. Pedersen (Eds.): SCIA 2007, LNCS 4522, pp. 730–739, 2007.

ambiguity. A recent paper [11] have shown how this problem can be overcome using a statistical model for admissible surface normal variations trained on range data. Here we use a simplified version of this algorithm. The surface normals are constrained to fall on the Lambertian reflectance with axis in the light source direction and apex angle given by the inverse cosine of the normalised image brightness. The position of the surface normal on the cone is decided to minimize its distance to the corresponding mean surface normal direction.

To construct a surface representation from the surface normals, we turn to the spin image first developed by Johnson and Hebert [6]. A spin image is a group of histograms constructed from the polar coordinates of arbitrary reference points on a surface. The representation can capture fine topographic surface details. Unfortunately, the computational overheads associated with the method are high, since a histogram needs to be generated for each surface location. Moreover, the original spin image representation was developed for range images and hence relies on surface height rather than surface normal information. We demonstrate how these two problems can be overcome by computing local spin images on image patches using surface normal information.

2 Mean Needle Map Alignment

The shape-from-shading algorithm used to extract needle maps from brightness images is as follows. We follow the work in [13] and place the surface normal on a cone whose axis is the light source direction and whose opening angle is the inverse cosine of the normalised image brightness.

This initial surface normals typically contains errors, and in particular locations where the pattern of convexity or concavity is reversed. To overcome this problem we draw on a model that accounts for the distribution of surface normals across ground-truth facial surfaces. To construct this model we use a sample of range images of human faces. From the gradients of surface height data, we make estimates of surface normal direction. The resulting fields of surface normals are adjusted so that faces have the same overall centering, scale and orientation. At each location we compute the mean surface normal direction according to the available set of ground-truth surface normals. Here we use the Max-Planck database which has 200 sample images of male and female subjects.

We use the mean facial needle map to adjust the positions of the surface normals on the reflectance cones. Each initial surface normal is rotated on its cone so that it minimises the angle subtended with the mean surface normal at the corresponding image location.

$$f(x, y) = argmin(|\theta_r(x, y) - \theta_{mean}(x, y)|) \tag{1}$$

where θ_r and θ_{mean} are the azimuth angles of the aligned surface normal n_r and the mean surface normal n_{mean} on the surface point (x, y).

The simplest way to satisfy Eqn. 1 is to make sure the aligned surface normal n_r, the mean surface normal n_{mean}, and the illumination cone axis n_x are on the same plane.

In Fig. 3 we illustrate the improvements gained using this simple shape-from-shading procedure. In the top row of the figure we show the input images of a single subject with the light source in different directions. In the second row we show the initial estimates of the surface normal directions. Here we have visualised the needle maps by taking the inner product of the surface normal with the light-source vector perpendicular to the image plane. This is equivalent to re-illuminating the surface normals with frontal Lambertian reflectance. From the images in the second row it is clear that there are significant concave/convex inversions in the proximity of the nose and lips when the face is illuminated obliquely. In the third row of the figure we show the surface normals that result from the adjustment procedure described above. The re-illuminations reveal that the inversions are removed and the quality of the recovered facial topography is improved (Fig. 4 illustrates the solution of this inversion problem).

3 The Original Spin Image Approach

The spin image of Johnson and Hebert [6] aims to construct an object-centered representation. The representation consists of a group of histograms and is constructed in the following manner: Commence by selecting an arbitrary point on the surface as the reference point O, and $\vec{n_o}$ is the surface normal at the point O. Then select a second arbitrary point P on the surface, and $\vec{n_p}$ is the surface normal at the point P . Assume the object resides in a 3D coordinate system with the surface normal $\vec{n_o}$ as z axis and the xy plane perpendicular to $\vec{n_o}$. The Euclidean distance $\gamma = \mid \vec{OP} \mid$ can be projected onto the xy plane as α and the z axis as β respectively. After the distances α and β of all the surface points are calculated, we can use them to construct a histogram. The above procedure is performed after each point on the surface has been taken as the point P so that a single histogram is constructed, and then a group of histograms are constructed using the above steps and taking each point on the surface as the reference point O. Figure 1 illustrates the spin image construction.

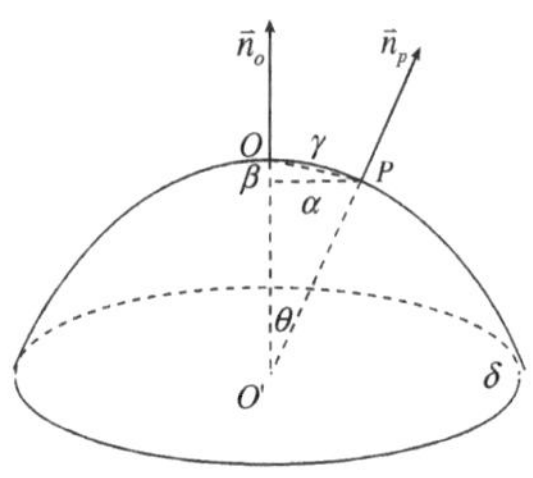

Fig. 1. Illustration of spin image construction

This object-centered representation is invariant to scale, translation and rotation since the spin image is calculated using only relative distances between object surface points.

The spin image representation is based on the availability of surface height data and can not be applied directly to fields of surface normals or needle maps. Moreover, spin image histograms need to be constructed at each image location, and this is demanding in both computation time and storage. In order to obtain this object-centered representation for an object with n surface points/image pixels, the computation cost will be $O(n^2)$.

4 Patch-Based Spin Images Adaptation

We have adapted a patch-based approach to spin image representation. We segment surface into patches and for each patch we use only the geometric center point O as reference point to construct spin image, rather than use every point of this surface as in the original spin image approach. Our histograms are constructed on a patch-by-patch basis.

4.1 Region Segmentation

As we mentioned before, we will develop patch-based spin image representation here. In our previous approach [7], a primitive fixed rectangle segmentation strategy was employed to obtain face patches. From Fig. 2, we can clearly see that the major problem in this strategy is the patches dos not represent natural face components, which makes different faces less distinguishable. Also some patches include the unnecessary background, which might involve non-face information if the face is with clutter background.

In this paper, we will propose a more sophisticate way for patch segmentation, which employs Active Appearance Models(AAM) [4]. It is a statistical model of shape and grey-level appearance. After proper training, it can locate the face feature points, and we can use this information to construct our irregular patches.

We first divide face and background into rectangular patches using our previous fixed rectangle segmentation strategy [7]. We manually preset our feature points on distinguishing face contour points, e.g. eye corners, brow outline points, etc. Then we train AAM to obtain feature points on all subject faces. Search for the overlapped parts of the rectangular patches and the polygonal area within the face contour, we can effectively exclude the background. Because the surface normal works poorly on the face parts of eyes/eyebrows/hair, we also need to exclude those face parts from the face region.

From Fig. 2, we can see that after combing the rectangular patch segmentation and the feature point location, the patches only contain face regions but not the background. Also the unnecessary parts are excluded from the patch. So the spin image can be constructed on more distinguishable face patches.

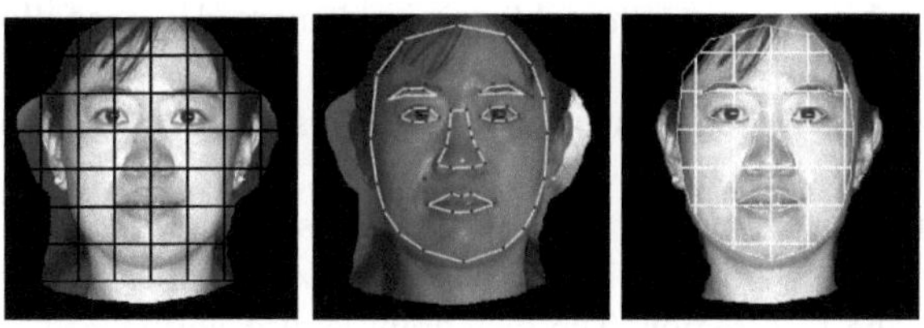

Fig. 2. This figure illustrates the original rectangular patch segmentation, the feature point location generated by AAM and the irregular patch segmentation

4.2 From Surface Normal to Spin Image Histogram

From the GGFI [9] we can obtain a surface height map Φ_{HM} over the surface, and the relative height β_{OP} between any two surface points O and P in the viewing direction $\vec{n_v}$ can be obtained from the height map Φ_{HM}.

We used to use the Euclidean distance α_{OP} between O and P perpendicular to the viewing direction $\vec{n_v}$ [7], but since the mesh-modeling can represent small regions of face surface more naturally than smooth planes or sphere regions, now we switch to the geodesic distance.

The geodesic distance $\gamma_{P_1P_2}$ between two neighbour points P_1 and P_2 is defined as follows:

$$\gamma_{P_1P_2} = \sqrt{\alpha^2_{P_1P_2} + \beta^2_{P_1P_2}} \tag{2}$$

where $\alpha_{P_1P_2}$ is the distance on the image plane, which is perpendicular to the viewing direction $\vec{n_v}$; $\beta_{P_1P_2}$ is the relative height between the points P_1 and P_2 in the viewing direction $\vec{n_v}$, which can be acquired from in the height map Φ_{HM} as mentioned before in this section.

The geodesic distance $\gamma_{P_1P_n}$ between two points P_1 and P_n connected by the path $path_{P_1P_n}$ is defined as follows:

$$\gamma_{P_1P_n} = \sum_{i=1}^{n-1} \gamma_{P_iP_{i+1}} \tag{3}$$

where $P_1, P_2, \ldots, P_n$ is the neighbour points along the path $path_{P_1P_n}$.

Assume there is N paths from the point O to the point P, the geodesic distance β_{OP} between O and P should be the shortest distance among these paths on the mesh-modeling.

$$\beta_{OP} = min\{path^1_{OP}, path^2_{OP}, \ldots, path^N_{OP}\}$$

Our solution to this shortest path problem is implemented by the following procedure (Dijkstra's algorithm) [5].

We now have all the ingredients to construct the histogram of β and γ for the surface patch centered at the point O.

In our experiment we construct a 10 by 10 bin histograms of β and γ for an image patch less than equal to 32 by 32 pixels. The exact pixel number depends on how many of them fall into the face region. The histogram is also normalised so as to be scale invariant.

As an additional step, we have performed PCA on spin image histograms to reduce data dimensionality. We normalise the bin contents of each spin image histogram to unity. The normalised bin contents of histograms are concatenated as a long-vector as follows:

$$\begin{bmatrix} \{O_1 : a_{11}\} & \{O_1 : a_{12}\} & \dots & \{O_1 : a_{1n}\} & \dots \\ \{O_1 : a_{n1}\} & \{O_1 : a_{n2}\} & \dots & \{O_1 : a_{nn}\} & \\ \{O_2 : a_{11}\} & \{O_2 : a_{12}\} & \dots & \{O_2 : a_{1n}\} & \dots \\ \{O_2 : a_{n1}\} & \{O_2 : a_{n2}\} & \dots & \{O_2 : a_{nn}\} & \\ \vdots & \vdots & \dots & \vdots & \\ \{O_M : a_{11}\} & \{O_M : a_{12}\} & \dots & \{O_M : a_{1n}\} & \dots \\ \{O_M : a_{n1}\} & \{O_M : a_{n2}\} & \dots & \{O_M : a_{nn}\} & \end{bmatrix} \tag{4}$$

where $\{O_1, \dots, O_M\}$ are spin image histograms, and $\{O_i : a_{jk}\}$ is the (j, k) bin of the ith histogram.

Dimensionality reduction is effected by projecting the long-vector onto the leading eigenvectors of the long-vector covariance matrix.

In the adaptation of spin image on surface normal, the computation cost is reduced to $O(n)$ instead of $O(n^2)$ in the original approach.

5 Recognition

In our preprocessing of the images to extract needle maps, we perform alignment. This means that we can apply a patch template to the extracted needle maps to decompose the face into regions. The patch template is constructed from the mean facial needle map, and consists of regions that are either wholly concave or wholly convex. The convexity/concavity test is made using the sign of the changes in surface normal direction. By performing the spin image analysis on these regions, we avoid problems associated with inflexion points when the approximations outlined in Sect. 2 are employed.

As an alternative to constructing the template from the mean needle map, we have explored constructing it from the needle map extracted from each facial image.

Our measure of facial similarity is based on the normalised correlation of the spin image histograms for corresponding template patches.

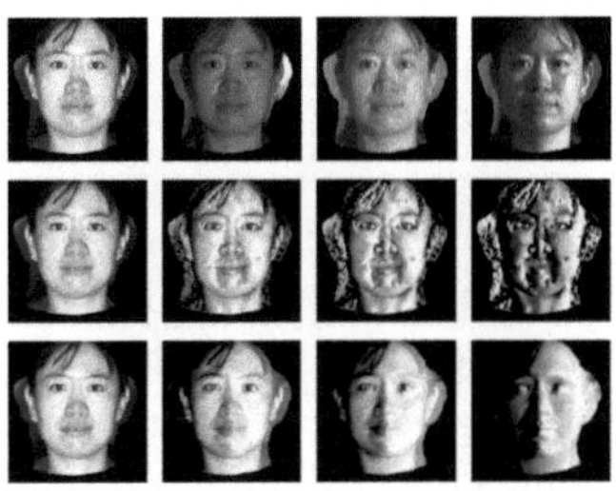

Fig. 3. The images in the first row are real images illuminated by the light sources from different directions. The images in the second/third row are the original needle maps[13]/the needle maps treated with Mean Needle Map Alignment (Sect. 2) respectively rendered by the frontal light source. The images in the third row are more photo-realistic and carry less noise than the ones in the second row.

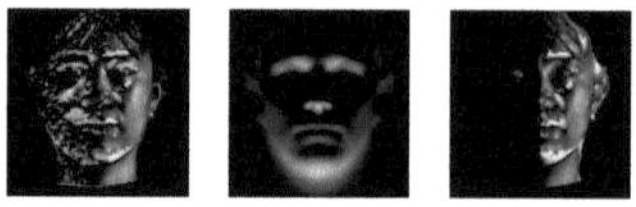

Fig. 4. The first image is the original needle map projected to the image plane. The second image is the mean needle map that we use as the template. The third image is the needle map projected to the image plane after the Mean Needle Map Alignment (Sect. 2). The third image compensate the concave/convex problem of the first one.

Johnson and Hebert use normalised correlation to evaluate spin image similarity[6]. The method assumes that spin images from proximal points on the surface for different views of an object will be linearly related. This is because the number of points that fall into corresponding bins will be similar (given that the distribution of points over the surface of the objects is the same). In our case, this assumption still holds. We hence use normalised correlation to compare the patch-based spin images. The correlation between two single patch spin image histograms x and y of different spin images X and Y is given by

$$r_{xy} = \frac{n\sum x_i y_i - \sum x_i \sum y_i}{\sqrt{(n\sum x_i^2 - (\sum x_i)^2)(n\sum y_i^2 - (\sum y_i)^2)}} \tag{5}$$

where r_{xy} is the correlation of two histograms x and y. n is the bin number of histograms, x_i and y_i are the ith bin contents of two histograms x and y respectively.

The correlation between two spin images X and Y is given by

$$r_{sum} = \sum_{i=1}^{min(M,N)} r_{m_i n_i} \tag{6}$$

where M and N are the number of histograms of two spin images X and Y, and m_i and n_i are the ith histograms x and y of two spin images X and Y respectively.

6 Experiments

We apply our method to the CMU PIE face database. We use only frontal-viewed face images in this paper. The sub-database contains $67 \times 7 = 469$ (67 subjects (1-67) and 7 lights (3,10,7,8,9,13,16)) images. We apply the two different patch segmentation strategies outlined above.

For the 7 images of the same subject illuminated by different lights, we use 3 of them for the training set and 4 of them for the test set. To perform recognition on the 67 subjects, we select a probe image from the training set and order the images in the test sets according to their similarities. The results of our experiments are summarised using the precision-recall curves shown in Fig. 5.

The plus-dotted curve shows the result of using a global histogram of curvature attributes extracted from the needle maps [8]. The circle-dotted curve shows the result of our rectangular patch spin image, and the star-dotted curve shows the corresponding spin image vector. The cross-dotted curve shows the result of our new irregular patch spin image, and the square-dotted curve shows the corresponding spin image vector. The irregular patch spin image vector gives the best performance.

In Table 1 we show the result of applying the various shape representations to the initial needle maps and the needle maps adjusted with Mean Needle Map Alignment (MNMA). In each case there is a significant improvement, and therefor in the following experiments we only use the needle maps adjusted with MNMA.

Table 1. Recognition rates using the initiate surface normal and the one adjusted with MNMA

	Original	MNMA
Global Histogram	37.50%	47.91%
Irregular Spin Image	75.23%	92.61%
Irregular Vector	78.10%	92.99%

Table 2. Recognition rates obtained by the rectangular/irregular patch spin image and the corresponding PCA spin image vector. The performances of PCA vector are slightly better.

	Rectangular Spin Image	Rectangular Vector	Irregular Spin Image	Irregular Vector
Face Components	27.81%	29.69%	30.42%	32.71%
Whole Face	80.67%	81.57%	92.61%	92.99%

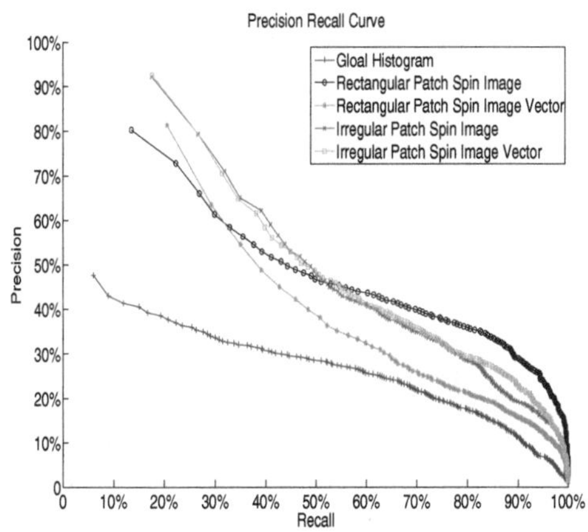

Fig. 5. There are five precision-recall curves of different approaches in this figure: the global histogram approach and the rectangular/irregular patch spin image/vector approaches. All these results are based on the surface normal processed by the Mean Needle Map Alignment because that approach has been proved improving the distinguishing ability in Table 1. Among them the irregular patch spin image vector approach gives the best performance.

In Table 2 we compare the recognition results obtained using the spin image and applying PCA to the spin image long-vectors. Performance is improved using PCA, and this can be ascribed to the fact that PCA effectively discards the histogram bins that are associated with insignificant variance.

Please notice the face component performance is obtained by only comparing the similarity of a single face component (cheek, nose, mouth, etc.) instead of the whole face, so the recognition rate will be reasonably low and can only be used to compare the performance of these four methods.

7 Conclusion and Future Work

In this paper we have proposed a more advanced approach comparing to our previous one [7] to extract spin images from 2D facial images using shape-from-shading. We made a few changes. First, we use geodesic distance to replace Euclidean distance in spin images to better describe the 3D shape. Second, we use the irregular patch to replace the rectangular patch to build spin images to exclude the background and unwanted face parts so that spin images can involve less surface normal errors. These make the irregular patch spin image obtain better result than the previous approach.

In the next step, we will apply this approach to faces with various poses to check its performance under more difficult circumstances, and also we will look for better way of similarity evaluation by adding weight for each histogram in spin image according to their locations or sizes.

References

1. Belhumeur, P., Kriegman, D.: What is the set of images of an object under all possible illumination conditions? International Journal of Computer Vision 28(3), 245–260 (1998)
2. Blanz,: Automatic face identification system using flexible appearance models. IVC 13(5), 393–401 (1995)
3. Blanz, V., Vetter, T.: Face recognition based on fitting a 3d morphable model. IEEE Transactions on Pattern Analysis and Machine Intelligence 25(9), 1063–1074 (2003)
4. Cootes, T.F., Edwards, G.J., Taylor, C.J.: Active appearance models. In: Burkhardt, H., Neumann, B. (eds.) ECCV 1998. LNCS, vol. 1407, pp. 484–498. Springer, Heidelberg (1998)
5. Dijkstra, E.W.: A note on two problems in connexion with graphs. In: Numerische Mathematik, vol. 1, pp. 269–271. Mathematisch Centrum, Amsterdam, The Netherlands (1959)
6. Johnson, A.E., Hebert, M.: Using spin images for efficient object recognition in cluttered 3d scenes. IEEE Transactions on Pattern Analysis and Machine Intelligence 21(5), 433–449 (1999)
7. Li, Y., Hancock, E.: Face recognition using shading-based curvature attributes. In: International Conference on Pattern Recognition ICPR, Cambridge, UK (August 2004)
8. Li, Y., Hancock, E.: Face recognition using shading-based curvature attributes. In: International Conference on Pattern Recognition ICPR, Cambridge, UK (August 2004)
9. Frankot, R.T., Chellappa, Z.: A method for enforcing integrability in shape from shading algorithms. IEEE Transactions in Pattern Recognition and Machine Intelligence 10, 439–451 (1988)
10. Sim, T., Kanade, T.: Combining models and exemplars for face recognition: An illuminating example. In: Proceedings of the CVPR 2001 Workshop on Models versus Exemplars in Computer Vision (December 2001)

11. Smith, W., Hancock, E.R.: Recovering facial shape and albedo using a statistical model of surface normal direction. In: Proc. ICCV, pp. 588–595 (2005)
12. Turk, M., Pentland, A.: Face recognition using eigenfaces. In: IEEE Conference on Computer Vision and Pattern Recognition (1991)
13. Worthington, P.L., Hancock, E.R.: New constraints on data-closeness and needle map consistency for shape-from-shaping. IEEE Transactions on Pattern Analysis and Machine Intelligence 21(12), 1250–1267 (1999)

Performance Evaluation of Adaptive Residual Interpolation, a Tool for Inter-layer Prediction in H.264/AVC Scalable Video Coding

Koen De Wolf, Davy De Schrijver, Jan De Cock, Wesley De Neve, and Rik Van de Walle

Ghent University – IBBT
Department of Electronics and Information Systems – Multimedia Lab
Gaston Crommenlaan 8 bus 201, B-9050 Ledeberg-Ghent, Belgium
koen.dewolf@ugent.be

Abstract. Inter-layer prediction is the most important technique for improving coding performance in spatial enhancement layers in Scalable Video Coding (SVC). In this paper we discuss Adaptive Residual Interpolation (ARI), a new approach to inter-layer prediction of residual data. This prediction method yields a higher coding performance. We integrated the ARI tool in the Joint Scalable Video Model software. Special attention was paid to the CABAC context model initialization. Further, the use, complexity, and coding performance of this technology is discussed. Three filters were tested for the interpolation of lower-layer residuals: a bi-linear filter, the H.264/AVC 6-tap filter, and a median filter. Tests have shown that ARI prediction results in an average bit rate reduction of 0.40 % for the tested configurations without a loss in visual quality. In a particular test case, a maximum bit rate reduction of 10.10 % was observed for the same objective quality.

1 Introduction

Scalable Video Coding (SVC) adds an extra dimension to traditional video compression schemes: adaptivity. SVC bit streams can be adapted by reducing the spatial resolution and/or temporal resolution and/or the quality level of the decoded video. This enables multimedia content providers to cope with the heterogeneity of networks and multimedia devices. Driven by standardization efforts of the Joint Video Team (JVT) – formed by the Moving Picture Experts Group (MPEG) and the ITU-T Video Coding Experts Group (VCEG), SVC is a hot topic in the domain of video coding standardization.

In this paper, we discuss and evaluate the novel concept of Adaptive Residual Interpolation (ARI), which is a new approach in scalable video coding for inter-layer prediction of residual data. The use of this tool results in bit rate savings without a loss in visual quality. A complexity and coding performance analysis is presented. Further, we elaborate on the signaling of this prediction method in the bit stream using Context-based Adaptive Binary Arithmetic Coding (CABAC).

B.K. Ersbøll and K.S. Pedersen (Eds.): SCIA 2007, LNCS 4522, pp. 740–749, 2007.

This paper is organized as follows. First, we introduce H.264/AVC Scalable Video Coding [1], focusing on the layered structure of this coding scheme and on the inter-layer residual prediction. The subsequent section describes the concept of ARI, its use, and its computational complexity. The conducted tests, the used interpolation filters, the initialization of the CABAC context model, and the coding performance are discussed in Sect. 4. Finally, this paper is concluded in Sect. 5.

2 H.264/AVC Scalable Video Coding

In this paper, we focus on the Joint Scalable Video Model (JSVM) as proposed by the JVT. In that context, scalability is formally defined in [2] as a functionality for the removal of parts of the coded video bit stream while achieving a Rate-Distortion (R-D) performance at any supported spatial, temporal, or Signal-to-Noise-Ratio (SNR) resolution that is comparable to the single-layer H.264/AVC coding scheme [3] at that particular resolution. In this requirements document, a coding efficiency penalty of 10 % in bit rate for the same perceptual quality is set as an upper limit. Based on this definition, we can identify three main types of scalability: spatial, temporal, and SNR scalability. Spatial scalability means that it should be possible to decode the input video at lower spatial resolutions. Temporal scalability means that frames can be dropped in the bit stream. This implies that not all encoded frames will be decoded, resulting in a lower frame rate of the decoded video sequence. Finally, SNR scalability means that the bit stream can be truncated in order to reduce the bit rate. Undeniably, this will also result in a decrease of the visual quality. These types of scalability can be embedded in the bit stream individually or as combinations, based on the requirements of the target application.

SVC can be classified as a layered video specification based on the single-layer H.264/AVC specification. Enhancement Layers (ELs) are added which contain information pertaining to the embedded spatial and SNR enhancements. Similar single-layer prediction techniques as in H.264/AVC are applied, in particular intra and motion-compensated prediction. However, additional inter-layer prediction mechanisms have been added for the minimization of redundant information between the different layers.

2.1 JSVM Structure

As mentioned, the JSVM is a layered extension of the H.264/AVC video coding specification. The structure of a possible JSVM encoder is shown in Fig. 1. In this figure, the original input video sequence is downscaled in order to obtain the pictures for all different spatial layers (resulting in spatial scalability). On each spatial layer, a motion-compensated pyramidal decomposition is performed taking into account the characteristics of each layer. This temporal decomposition results in a motion vector field on the one hand and residual texture data on the other hand. This information is coded by using similar techniques as in H.264/AVC extended with progressive SNR refinement features.

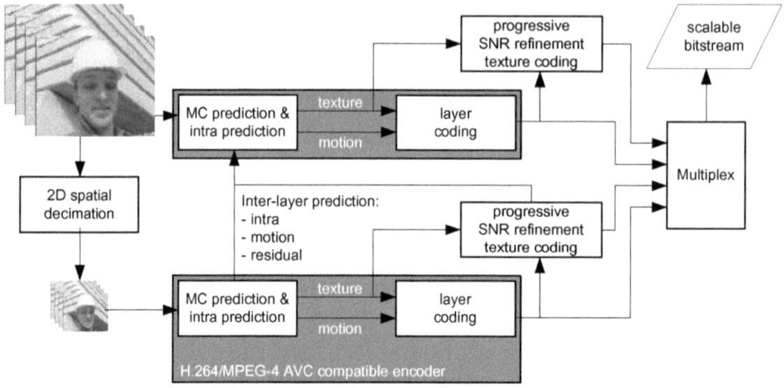

Fig. 1. SVC encoder structure supporting two spatial layers and progressive SNR refinement [4]

A typical JSVM bit stream contains hierarchical layers. A JSVM-aware adaptation engine is capable of extracting only these layers that are necessary for decoding the video at a targeted (possibly reduced) spatial resolution, frame rate, and/or bit rate.

Several methods that allow the reuse of coded information among different spatial resolution layers are under investigation by the JVT. In particular, the layered structure of the JSVM allows the reuse of motion vectors, residual data, and intra texture information of lower spatial and SNR layers for the prediction of higher-layer pictures in order to reduce inter-layer redundancy. In the next section, we elaborate on these inter-layer prediction methods. Also, note that other extended spatial scalability modes are considered by the JVT, such as cropping and non-dyadic scaling. In this paper, we will focus on inter-layer prediction using residual pictures from lower layers.

2.2 Inter-layer Residual Prediction

In SVC, residual information of inter-picture coded macroblocks (MBs) from a lower resolution layer can be used for the prediction of the residual of the current layer. A flag, indicating whether inter-layer residual prediction is used, is transmitted for each MB of an enhancement layer. When residual prediction is applied, the lower resolution layer residuals of the co-located sub-MBs are blockwise upsampled using a bi-linear filter with constant border extension. Doing this, only the difference between the residual of the current layer obtained after motion compensation (MC) and the upsampled residual of a lower resolution layer is coded. In Fig. 2, the picture reconstruction scheme using inter-layer residual prediction of the decoder is shown.

At the encoder side, the decoded and upsampled lower layer residuals are subtracted from the original image in the current layer preceding the Motion Estimation (ME) stage. This is illustrated in Fig. 3. By applying residual prediction in advance of ME, we are sure to obtain the most efficient motion vectors

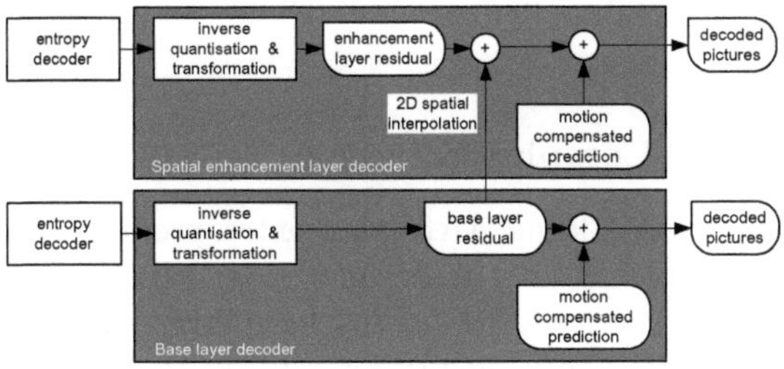

Fig. 2. Structure of the decoder using residual prediction

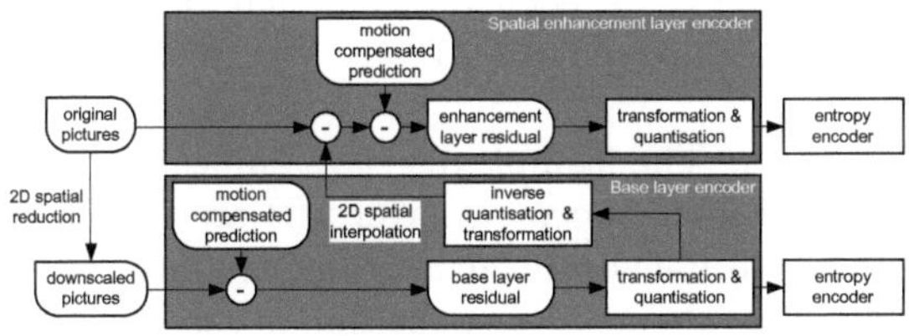

Fig. 3. Prediction of enhancement layer residual at the encoder

resulting in lower magnitudes in the enhancement layer residual which will result in a higher compression efficiency.

Inter-layer residual prediction has shown to be useful when the motion vectors of a block are aligned with the motion vectors of the corresponding block in a lower spatial resolution layer. This can be explained by the likeliness that after motion compensation, the residuals for those blocks will show a high resemblance.

3 Adaptive Residual Interpolation

3.1 Definition

ARI can be seen as an extension to the inter-layer residual prediction technique that is incorporated in the JSVM, (as explained in Sect. 2.2). In [5], we have presented ARI as a prediction method applied in the context of SVC which evaluates multiple interpolation filters used for the upsampling of lower layer residuals. This evaluation is performed on a MB basis in an R-D optimized sense. The latter means that a cost function is used to determine which filter should be used.

In [6], we have shown that the use of multiple pre-defined interpolation filters for the upsampling of base layer residuals, results in a better prediction of the enhancement layer residuals. As a result, the magnitudes of these remaining residuals are lower. The applied filter was chosen on a picture basis. Moreover, lower bit rates and Peak Signal-to-Noise Ratio (PSNR) gains can be achieved when the applied residual interpolation filter is changed on a MB basis, hereby selecting the best filter out of multiple pre-defined interpolation filters for that particular MB [5].

3.2 Computational Complexity

Decoder Complexity. Adding ARI as an extra prediction mode results in a limited change of complexity at the decoder side. When residual prediction is used for a particular MB, its corresponding base layer residual must be upsampled anyhow. So, the minor complexity introduced by ARI originates from the complexity of the filters being used. In particular, the computational complexity is reduced when the applied filters are less complex than the bi-linear filter (which is defined in the JSVM specification) or that the complexity is increased when using more complex interpolation filters.

Encoder Complexity. The complexity is significantly increased when all modes for the coding of a enhancement layer MB are evaluated at the encoder side in order to obtain an R-D optimized coding decision. This is caused by the fact that, when no fast mode-decision algorithm is used, a distinct ME needs to be performed for each interpolation filter and this for all possible coding modes. In general, implementations of contemporary video coding specifications will use fast mode-decision and fast ME algorithms to cope with the high computational complexity. These algorithms can be modified in order to support ARI. For example, by relying on hierarchical search algorithms and previously made coding decisions. For off-line scenarios, this extra complexity shouldn't be a burden.

4 Tests and Results

4.1 Test Setup

For the performance evaluation of ARI in terms of coding efficiency, we integrated the ARI tool in the JSVM software. We have implemented two additional filters next to the bi-linear filter which is already incorporated in the current JSVM specification (version 5). We have chosen the 6-tap filter already specified in both H.264/AVC and the JSVM (used there for sub-pixel ME) in combination with a median filter. A description of these filters is given in Sect. 4.2. Note that the interpolation is done on 4×4-blocks with constant border extension as specified in the JSVM.

In order to limit the encoder complexity, we limited the number of filters that can be chosen for residual interpolation, to two. In this test, we used two filter combinations: bi-linear & H.264/AVC 6-tap filter (combination A) and H.264/AVC 6-tap & median filter (combination B). Five CIF-resolution test sequences of 300 pictures at 30Hz with different motion characteristics were used: Crew, Foreman, Mobile & Calender, Mother & Daughter, and Stefan. A downsampled version was generated using the 11-tap FIR JSVM downsampling filter. This results in a base layer at QCIF resolution and a CIF spatial enhancement layer. We used 4 fixed Quantization Parameter (QPs) for the Base Layer

(QP_{BL}) and the Enhancement Layer (QP_{EL}): 12, 18, 24, 30. This leads to 16 tested QP-combinations. These QP-combinations are chosen to cover a broad range of possible applications. ARI is used for the prediction of P-pictures only. The achievable coding gains when ARI is used for B-pictures, are minimal. This is due to the fact that for most MBs in B-pictures no additional residual is coded. As such, enabling ARI in B-pictures will result in the coding of an extra flag without enhancing the coded residual.

4.2 Definition and Complexity of the Tested Interpolation Filters

AVC 6-tap Filter. For the derivation of pixel values at half-sample positions, we refer to the H.264/AVC specification and an overview paper by Wiegand *et al.* [7]. An in-depth complexity analysis with respect to this filter is performed in [8].

Median Filter. In Fig. 4, the positions labeled with upper-case letters represent the signal at full-sample locations inside a two-dimensional block of residual samples. The signals at half-sample locations (labeled with lower-case letters) are derived as follows.

1. The value of each sample labeled with double lower-case letters is determined by the median of the neighboring full-sample values; e.g., cc = Median(D, E, H, I);
2. The value of each sample labeled with a single lower-case letter is determined by the median of the two neighboring full-sample values and the two neighboring half-sample values as derived in Step 1; e.g., a = Median(D, E, aa, cc).

For samples at the border of a 4×4 block, the complexity of the sorting algorithm can be reduced when constant border extension is used for the construction of the samples outside the 4×4 block. In particular, in Fig. 4, the value for x is derived as

$$Med\left(\frac{(A+B)}{2}, A, B, aa\right) \quad . \tag{1}$$

This can be simplified as follows:

Let $A < B$,

$$x = \begin{cases} \frac{3A+B}{4} & \text{if } aa < A \ , \\ \frac{A+3B}{4} & \text{if } B < aa \ , \\ \frac{2aa+A+B}{4} & \text{if } aa \in [A..B] \ . \end{cases} \tag{2}$$

4.3 ARI Signaling

For each MB, the choice whether residual prediction is being used, has to be signaled. For ARI, an additional syntax element denoting which interpolation

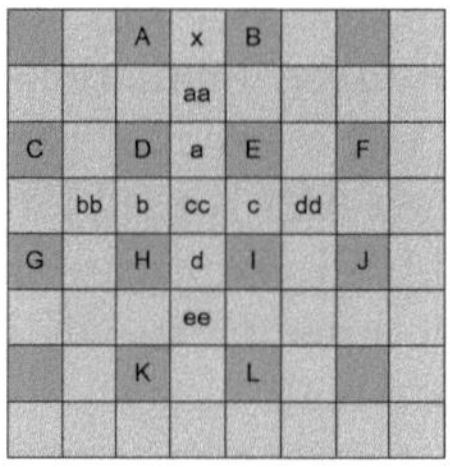

Fig. 4. Full-sample positions (dark-shaded blocks with upper-case letters) and half-sample positions (light-shaded blocks with lower-case letters)

filter is being applied, needs to be coded for each MB. In our test, we limited the number of filters to two. As such, this syntax element can take 0 or 1 as value. This syntax element (symbol) is coded using a CABAC entropy coder. Within this context-based coder, a model probability distribution is assigned to the symbols that need to be coded. Furthermore, this coding engine adaptively changes the state of the context models (i.e., probability model distribution) based on symbols already coded.

Such a context state of model γ is determined by two parameters: the probability of the Least Probable Symbol (LPS) σ_γ and the Most Probable Symbol (MPS) ϖ_γ. For the ARI context, MPS denotes which of both filters is most probably being used. The LPS probability indicates the probability of the other filter being used for the current context state.

The initial state for the context is determined by two parameters (μ_γ, ν_γ). These parameters are used to derive ϖ_γ and σ_γ, hereby taking into account the QP of the current slice For a detailed explanation on context state transition and context initialization, we refer to [9].

For this ARI context, μ_γ and ν_γ were deduced using linear regression on the results of a training set (115 tested configurations) that have been used to fit the different initial probability states for both filter combinations (i.e., combinations A and B). In Fig. 5, a QP-based probability of median filter usage for filter combination B (i.e., H.264/AVC 6-tap & median filter) is given. This probability is plotted against the QP of the enhancement layer. In this graph, we make a distinction between the different QPs of the base layer. Linear regression lines are drawn for all tested configurations with the same QP_{BL}. A linear regression line that takes all tested configurations into account is also drawn. It is clear from the graph that the QP of the base layer has an impact on the behavior of this regression line and thus on the initial probability σ.

Context model initialization in CABAC was designed for the single-layer H.264/AVC coding specification. As such, a model can not be initialized by using the QP of the base layer. Therefore the context model initialization for inter-layer prediction tools in the JSVM is sub-optimal. This shortcoming can

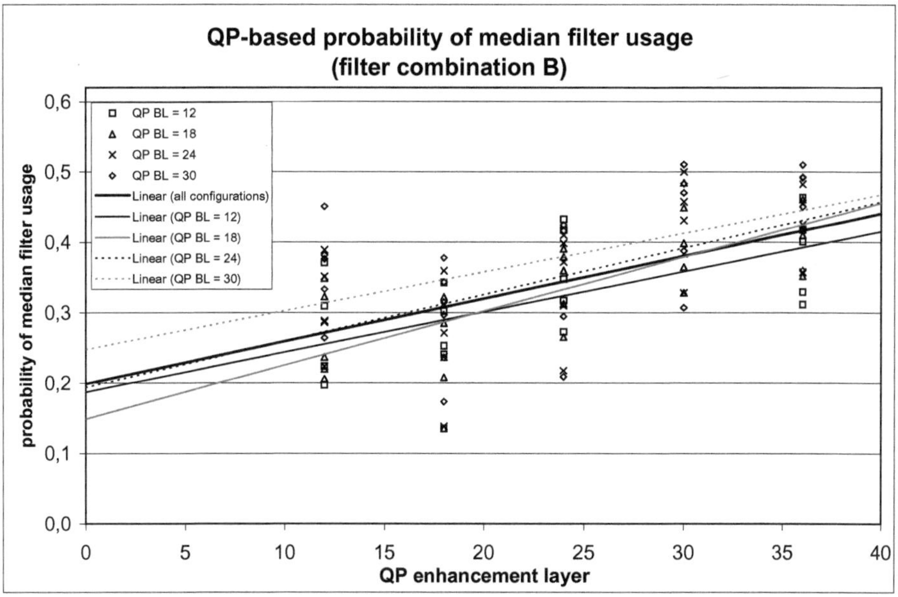

Fig. 5. QP-based probability of median filter usage for filter combination B

somehow be circumvented by using `cabac_init_idc`[1]. Doing so, three different initialization parameter pairs (μ_γ, ν_γ) can be defined per context model. The values of these pairs can be chosen, taking into account the base layer QP.

4.4 Results

The coding performance of the presented inter-layer residual prediction tool is evaluated using Bjøntegaard Delta Bit Rate (BD-BR) and Bjøntegaard Delta PSNR (BD-PSNR) [10], which are respectively the average bit rate difference and the PSNR difference between the original residual prediction algorithm (explained in Sect. 2.2) and the proposed ARI algorithm. BD-BR and BD-PSNR are derived from simulation results for each fixed QP_{BL} in combination with varying QP_{EL} = 28, 32, 36, 40.

From Table 1, we observe that the BD-PSNR of the tested configurations with ARI enabled, shows a slight gain (less than 0.1 dB). From Table 2, we see that the highest bit rate reductions are achieved for low QP_{BL} when filter combination B is used. This can be explained by the fact that when QP_{BL} is low, more information will be present in the coded residual, which can be used for inter-layer residual prediction on the one hand and the median filter preserves

[1] `cabac_init_idc` specifies the index for determining the initialization table used in the initialization process for context variables. The value of `cabac_init_idc` shall be in the range of 0 to 2, inclusive. [3]

Table 1. ARI performance: BD-PSNR (dB) for filter combinations A and B

QP_{BL}	Crew		Foreman		Mobile & Calender		Mother & Daughter		Stefan	
	A	B	A	B	A	B	A	B	A	B
12	0.05	0.07	0.02	0.03	0.02	0.03	0.06	0.11	0.03	0.04
18	0.03	0.04	0.02	0.02	0.02	0.02	0.04	0.05	0.03	0.02
24	-0.02	0.03	0.02	0.01	0.03	0.02	0.02	0.02	0.03	0.02
30	0.01	0.01	0.01	0.01	0.02	0.02	0.02	0.02	0.01	0.00

Table 2. ARI performance: BD-BR (%) for filter combinations A and B

QP_{BL}	Crew		Foreman		Mobile & Calender		Mother & Daughter		Stefan	
	A	B	A	B	A	B	A	B	A	B
12	-0.99	-1.15	-0.35	-0.55	-0.25	-0.28	-1.29	-2.01	-0.35	-0.38
18	-0.49	-0.57	-0.32	-0.32	-0.20	-0.15	-0.76	-0.92	-0.32	-0.23
24	-0.28	-0.37	-0.22	-0.20	-0.23	-0.16	-0.38	-0.41	-0.29	-0.22
30	-0.08	-0.08	-0.16	-0.08	-0.16	-0.15	-0.38	-0.35	-0.08	-0.04

borders in the residuals, whereas the bi-linear and H.264/AVC 6-tap filters rather smoothen these borders on the other hand.

Due to space limitations, we are unable to publish all results in detail. However, we have observed a clear relation between bit rate reduction and the values of QP_{BL} and QP_{EL}. When the BL is less quantized than the EL, the bit rate reduction increases when the difference between QP_{BL} and QP_{EL} increases.

The motion characteristics of the coded video sequences also have an important impact on the bit rate reduction. ARI proves to be very useful for low-motion sequences, such as the Mother & Daughter sequence. For the coding of this sequence, a maximum bit rate reduction of 10.10 % is achieved when filter combination B is used with $QP_{\mathrm{BL}} = 24$ and $QP_{\mathrm{EL}} = 36$ (not shown in this paper). However, complex-motion sequences, such as Mobile & Calender, still benefit from ARI.

The average bit rate reduction for the tested configurations is 0.4 % with an average PSNR increase of 0.03 dB.

5 Conclusion and Future Work

In this paper, we have presented ARI as a tool for inter-layer residual prediction in SVC. The use, computational complexity, CABAC context initialization, and coding performance of the tested filters were discussed. This tool proves to be especially useful for coding low-motion sequences when the combination of the 6-tap filter and the median filter is used for the interpolation of lower-layer residuals. Moreover, complex-motion sequences also benefit from ARI.

We have observed that impact of the newly introduced syntax element is not negligible. In future work, we will investigate a combined signaling of this syntax element in the slice header and MB header in order to further improve the coding performance of this inter-layer prediction tool.

Acknowledgements

The research activities that have been described in this paper were funded by Ghent University, the Interdisciplinary Institute for Broadband Technology (IBBT), the Institute for the Promotion of Innovation by Science and Technology in Flanders (IWT), the fund for Scientific Research-Flanders (FWO-Flanders), the Belgian Federal Science Policy Office (BFSPO), and the European Union.

References

1. Reichel, J., Schwarz, H., Wien, M.: Joint scalable video model JSVM 5, doc. JVT-R202 (2006)
2. ISO/IEC, JTC1/SC29/WG11: Applications and requirements for scalable video coding. ISO/IEC JTC1/SC29/WG11 N6880 (2005)
3. ITU-T, ISO/IEC JTC 1: Advanced video coding for generic audiovisual services, ITU-T Rec. H.264 and ISO/IEC 14496-10 AVC (2003)
4. Schwarz, H., Marpe, D., Wiegand, T.: Basics concepts for supporting spatial and SNR scalability in the scalable H.264/MPEG4-AVC extension. In: Proceedings of IEEE IWSSIP (2005)
5. De Wolf, K., De Schrijver, D., De Neve, W., Van de Walle, R.: Adaptive Residual Interpolation: a tool for efficient spatial scalability in digital video coding. In: Proceedings of International Conference on Image Processing, Computer Vision & Pattern Recognition (IPCV'06). vol. 1, pp. 131-137, Las Vegas (2006)
6. De Wolf, K., De Sutter, R., De Neve, W., Van de Walle, R.: Comparison of prediction schemes with motion information reuse for low complexity spatial scalability. In: Proceedings of SPIE/Visual Communications and Image Processing. vol. 5960, pp. 1911–1920, Beijing (2005)
7. Wiegand, T., Sullivan, G., Bjøntegaard, G., Luthra, A.: Overview of the H.264/AVC video coding standard. IEEE Trans. CS Video Technology 13, 560–576 (2003)
8. Horowitz, M., Joch, A., Kossentini, F., Hallapuro, A.: H.264/AVC baseline profile decoder complexity analysis. IEEE Trans. CS Video Technology 13, 704–716 (2003)
9. Marpe, D., Wiegand, T., Schwarz, H.: Context-based adaptive binary arithmetic coding in the H.264/AVC video compression standard. IEEE Trans. CS Video Technology 13, 620–636 (2003)
10. Bjøntegaard, G.: Calculation of average PSNR difference between RD-curves, VCEG-M33 (2001)

3D Deformable Registration for Monitoring Radiotherapy Treatment in Prostate Cancer

Borja Rodríguez-Vila[1], Johanna Pettersson[2], Magnus Borga[2], Feliciano García-Vicente[3], Enrique J. Gómez[1], and Hans Knutsson[2]

[1] Bioengineering and Telemedicine Group, Universidad Politécnica de Madrid, Spain
{brvila,egomez}@gbt.tfo.upm.es
[2] Department of Biomedical Engineering and Center for Medical Image Science and Visualization, Linköping University, Sweden
{johpe,knutte,magnus}@imt.liu.se
[3] Medical Physics, Radiotherapy Department, University Hospital La Princesa, Spain
fgarcia.hlpr@salud.madrid.org

Abstract. Two deformable registration methods, the Demons and the Morphon algorithms, have been used for registration of CT datasets to evaluate their usability in radiotherapy planning for prostate cancer. These methods were chosen because they can perform deformable registration in a fully automated way. The experiments show that for intrapatient registration both of the methods give useful results, although some differences exist in the way they deform the template. The Morphon method has, however, some advantageous compared to the Demons method. It is invariant to the image intensity and it does not distort the deformed data. The conclusion is therefore to recommend the Morphon method as a registration tool for this application. A more flexible regularization model is needed, though, in order to be able to catch the full range of deformations required to match the datasets.

1 Introduction

Prostate cancer is the third most common cause of death from cancer in men of all ages and it is the most common cause of death from cancer in men over the age of 75 [8]. External beam radiotherapy has shown to be an effective treatment for localized prostate cancer in early stages. Radiation dose to the prostate was earlier very limited because of concern about normal tissue toxicity. Now three-dimensional conformal radiotherapy has allowed safe dose escalation to treat prostate cancer, which makes it possible to apply higher radiation doses on cancerous tissues and, at the same time, reduce the dose in healthy tissues [7]. There are several side effects associated with radiation therapy, including for example rectal bleeding and hematuria (blood in urine), basically derived from the undesired but unavoidable radiation of rectum and bladder [2]. This implies that careful consideration must be taken to the dose-volume histograms of normal tissues to avoid excessive toxicity in these regions.

Before beginning radiotherapy treatment CT scans have to be obtained for computerized treatment planning, to determine the most appropriate way to deliver radiation therapy. The radiotherapy team selects the target volume and

B.K. Ersbøll and K.S. Pedersen (Eds.): SCIA 2007, LNCS 4522, pp. 750–759, 2007.

computes security margins using the CT scan and thereby obtains the volume to be radiated. External beam radiation therapy is usually performed 5 days a week for 6-8 weeks, and during this time there are some displacements and deformations of the prostate and high-risk organs that demand monitoring [1].

One of the problems to solve in the reduction of the radiated volume is the significant daily change in position of the prostate. The use of for example affine or projective registration algorithms, nowadays available in commercial radiotherapy software, allow quantification of the displacements of the prostate and reduction of the radiated volume considerably. However, the deformations of high-risk organs, such as rectum and bladder, where the dose evaluation is very important, remain an obstacle for the additional reduction of the margin. In addition, daily deformations pose a challenge in accurate tracking of the dose radiated to each point of the pelvis anatomy since it is not possible to quantify exactly the total radiation dose administered in several sessions. The use of deformable registration algorithms would allow setting an anatomical correspondence between the planning CT scan and the monitoring CT scan, and simulating the deformation of the anatomy more accurately.

The monitoring process is controlled with CT scans of the patient in different stages of the treatment, so the registration algorithm will be intrapatient and will match small deformations. The next step of our research is to perform the planning of a patient using a planning template and the CT scans, which involves interpatient registration with larger deformations. One example of this has been included in this evaluation.

2 Methods

There are numerous algorithms for medical image registration in the literature, normally oriented to a concrete clinical application and to the type of images implicated in the process, see for example [4]. In this work two selected deformable registration algorithms have been employed and evaluated for the radiotherapy planning application. The methods are fully automatic and work on 3D data. The first one is the Demons algorithm [5], previously tested on e.g. inter-patient MRI brain and SPECT cardiac images and already used in radiotherapy planning by e.g. Wang et al. [10]. The second one is the Morphon algorithm [6], which is a relatively new method that has shown to work well for automatic segmentation of for example hip fractures and the hippocampus [9,11].

2.1 The Demons Method

The Demons algorithm, proposed by J.P.Thirion [5], is a method to perform iterative deformable matching of two 3D medical images fully automatic. The Demons method presumes that the boundaries of an object in the reference image can be simulated as a semi-permeable membrane, while the template image is a deformable grid whose junctions are two kind of particles: inside or outside. Each demon acts locally, to push the deformable model particles perpendicularly to the contour, but the direction of the push depends on the nature (inside or outside) of the current estimate of the model at that point.

There are some different ways to perform the algorithm, depending on the voxels considered to be demons, on the type of deformations allowed and on the way to compute the magnitude and direction of the demons forces. The implementation of the Demons method used in this work is characterized by the use of demons in all the voxels of the reference image. Since for medical images, iso-intensity contours are closely related to the shapes of the objects, one demon can be associated to each voxel of the reference image where the gradient norm is not null and hence a 3D grid of demons acts to deform the image.

A totally free-form deformation is allowed, but a low pass filter of the deformation vector field is used to control the deformability of the model. With the assumption that the intensity of an object is constant along time over small displacements, we used a modification of the classic optical flow equation to compute the displacement field:

$$\boldsymbol{v}^T \left(\frac{\boldsymbol{\nabla} f + \boldsymbol{\nabla} m}{2} \right) = m - f \tag{1}$$

where f is the intensity of the reference image, m is the intensity of the template image, $\boldsymbol{\nabla} f$ and $\boldsymbol{\nabla} m$ are their corresponding gradients, and $\boldsymbol{v}$ is the motion which brings m closer to f.

From eq. 1, only the projection of $\boldsymbol{v}$ on $(\boldsymbol{\nabla} f + \boldsymbol{\nabla} m)$ is determined, so we have:

$$\boldsymbol{v} = \frac{2(m-f)(\boldsymbol{\nabla} f + \boldsymbol{\nabla} m)}{\|\boldsymbol{\nabla} f + \boldsymbol{\nabla} f\|^2} \tag{2}$$

In order to avoid very large deformation forces that can distort the template image, the forces of the demons are needed to be close to zero when one of the gradient norms is close to zero, but the expression in eq. 2 tends towards zero only when both gradients are small at the same time. Furthermore, eq. 2 is unstable when the gradient norms are small. In this case, a small variation of the intensity can push the end point of $\boldsymbol{v}$ to infinity in any direction. The solution proposed by Thirion is to multiply $\boldsymbol{v}$ with a similarity coefficient s, resulting in $\boldsymbol{d} = \boldsymbol{v} \cdot s$, where s is defined by:

$$s = \begin{cases} 0 & \text{if } \boldsymbol{\nabla} f^T \boldsymbol{\nabla} m \leq 0 \\ \frac{2 \cdot \boldsymbol{\nabla} f^T \boldsymbol{\nabla} m}{\|\boldsymbol{\nabla} f\|^2 + \|\boldsymbol{\nabla} m\|^2 + 2(m-f)^2} & \text{if } \boldsymbol{\nabla} f^T \boldsymbol{\nabla} m > 0 \end{cases} \tag{3}$$

By applying this regularization the displacement field, when $\boldsymbol{\nabla} f^T \boldsymbol{\nabla} m > 0$, is given by the iterative equation:

$$\boldsymbol{d}' = \boldsymbol{d} + \frac{4(T_n(m) - f)(\boldsymbol{\nabla} f + \boldsymbol{\nabla} T_n(m))(\boldsymbol{\nabla} f^T \boldsymbol{\nabla} T_n(m))}{(\|\boldsymbol{\nabla} f + \boldsymbol{\nabla} T_n(m)\|^2)(\|\boldsymbol{\nabla} f\|^2 + \|\boldsymbol{\nabla} T_n(m)\|^2 + 2(T_n(m) - f)^2)} \tag{4}$$

where $T_n(m)$ is the deformed template image after n iterations.

This expression tends to zero when one of the gradients norms tends to zero and also when the two gradients are very dissimilar. It is important to notice that this expression requires the computation of $\boldsymbol{\nabla} T_n(m)$ for each point and at each iteration.

Since this method uses the intensity difference between the images, in addition to the gradient of the intensity of the images, it is sensitive to intensity variations of the same anatomic point in different images. We have assumed that the intensity of an object is constant over time for small displacements, which is not always true for CT scans. Two images taken with the same device can have different values of intensity due to the noise in acquisition and reconstruction of CT scans. We have implemented a preprocessing that equalizes the histograms of both images in order to decrease the effect of this.

2.2 The Morphon Method

The Morphon algorithm has previously been presented in e.g. [6,11]. It is a non-rigid registration method in which a template is iteratively deformed to match a target in 2D or 3D using phase information. A dense deformation field is accumulated in each iteration under the influence of certainty measures. These certainty measures are associated with the displacement estimates found in each iteration and assure that the accumulated field is built from the most reliable estimates. The displacement estimates are derived using local image phase, which makes the method invariant to intensity variations between the template and target. Unlike the demons method, where the intensity values are considered to be consistent between the images, no preprocessing of the data is needed in the Morphon method.

The updated accumulated deformation field, $\boldsymbol{d}_{n+1}$, is found in each iteration by adding the displacement estimates from the current iteration, $\boldsymbol{d}$, to the accumulated deformation field from the previous iteration, $\boldsymbol{d}_n$. This sum is weighted with certainty measures associated with the accumulated field, c_n, and certainty measures associated with the displacement field form this iteration, c.

$$\boldsymbol{d}_{n+1} = \boldsymbol{d}_n + \frac{c}{c_n + c}\boldsymbol{d} \tag{5}$$

The displacement estimates are found from local phase difference. The local image phase can be derived using so called quadrature filters [3]. Quadrature filters are onedimensional and must be associated to a specific direction when used for multidimensional data. Therefore a set of quadrature filters, each one sensitive to structures in a certain direction, is applied to the target and template respectively. The output when convolving one quadrature filter f with the signal s is:

$$q = (f * s)(\boldsymbol{x}) = A(x)e^{i\phi(x)} \tag{6}$$

The phase difference between two signals can be found from the complex valued product between the filter output from the first signal, the target, and the complex conjugate (denoted by *) of the output from the second signal, the template:

$$q_1 q_2^* = A_1 A_2 e^{i\Delta\phi(x)} \tag{7}$$

The local displacement estimate d_i in a certain filter direction i is proportional to the local phase difference in that direction, which is found as the argument of this product, $\Delta\phi(\boldsymbol{x}) = \phi_1(\boldsymbol{x}) - \phi_2(\boldsymbol{x})$.

A displacement estimate is found for each pixel and for each filter in the filter set. Thus, a displacement field is obtained for each filter direction $\hat{\boldsymbol{n}}_i$. These fields are combined into one displacement field, covering all directions, by solving the following least square problem:

$$\min_{\mathbf{d}} \sum_i [c_i(\hat{\boldsymbol{n}}_i^T \boldsymbol{d} - d_i)]^2 \tag{8}$$

where $\boldsymbol{d}$ is the sought displacement field, $\hat{\boldsymbol{n}}$ is the direction of filter i, and c_i is the certainty measure (equal to the magnitude of the product in equation 7).

3 Results

Two CT scans of four different patients have been available for evaluating the deformable registration. For each patient intrapatient registration has been performed using the two deformable registration methods. Each one of the two patient scans has been used as template resulting in two registrations per patient and per method. In one scan of one of the patients contrast agent has been injected, which affects the registration process. This case is not comparable with the other datasets and is therefore handled separately in the discussion of the results.

To get an overview of the registration results we start by showing the correlation between the datasets before and after registration with the different methods. The graph in figure 1 shows the mean correlation per slice for the three datasets without contrast, before and after registration with the Demons and the Morphon respectively. The correlation has been computed for the edge information in the data instead of the original intensity information. This was chosen to obtain a measure that does not vary depending on the intensities in the homogeneous areas. Instead it reflects the deformation of the structures in the data better. From the graph we can see that there is not much difference between the two registration methods, they give similarly good results and have improved the alignment of the datasets considerably. While this gives an idea of how well the datasets have matched, one must also validate the deformed data visually to see how the algorithms have deformed the data. 2D slices from two of the registrations are shown in figure 2. These have been chosen to illustrate the result of the registration as well as the differences between the methods.

The Demons and the Morphon method deform the data in somewhat different ways. The implementation of Demons that has been used computes the displacement field based on the gradients of both reference and deformed template image. This makes it possible to avoid extremely large deformations that might destroy the structures and to e.g. fill holes that are not too big, as can be seen in figure 2 (a)-(e). This, however, often results in distortions of the data, which is demonstrated in the enlarged parts of the examples shown in figure 3. These images also show that the Morphon method gives a smoother deformation, keeping the structures of the template image. The Morphon method has, however, failed to deform the template sufficiently in parts of the soft tissue areas as can bee seen in figure 2 (f)-(j).

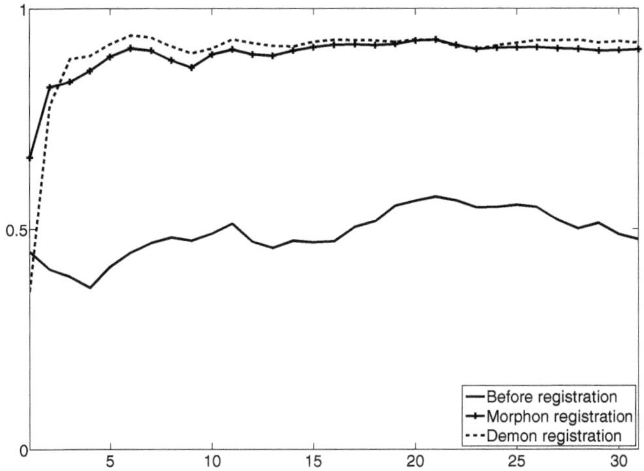

Fig. 1. Mean correlation per slice based on the edge information in each slice. The Demons and the Morphon method generate similar results.

A slice of the dataset with the contrast agent is shown in figure 4. The problem with this dataset is that the contrast agent that fills the urinary bladder makes the edge between this tissue and the surrounding tissue much stronger than in the dataset without contrast agent. For the Demons algorithm this causes problems in the whole process. The Demons method assumes the same anatomical point having the same intensity level in both images, and a histogram equalization is performed before the registration starts trying to obtain this. The contrast agent corresponds to a completely new intensity level in this datasets, not present in the other scan of the patient. The Demons preprocessing step tries to match the histograms of the two datasets anyway, which in the worst case may increase the intensity differences instead of decreasing it. This, in turn, deteriorates the registration result. The Morphon method is invariant to the intensity levels in the datasets since it uses the local image phase to estimate the displacements. The stronger edge in the contrast dataset is however recognized by the filters used for phase estimation, which makes the Morphon registration responsive to the contrast as well, although only in the neighborhood around this edge. In the Morphon result it is difficult to say how well the datasets have matched in the area covered by the contrast agent, although the rest of the image has matched well, figure 4(c). For the Demons result parts of the anatomy, such as the hip bones, have completely lost their shape. The soft tissue, such as the bladder, has not been registered correctly either, figure 4(d).

Finally an example of an interpatient registration is included as a first test of the next step of our research that will involve not only the monitoring but also the planning of the process. This registration process involves larger deformations than for intrapatient matching. The Morphon has shown to work well for this kind of data in previous work, for example when registering a template with a very simple description of the anatomy in the hip region to real data from a CT scan, containing a lot more structures [11]. A 2D slice of the

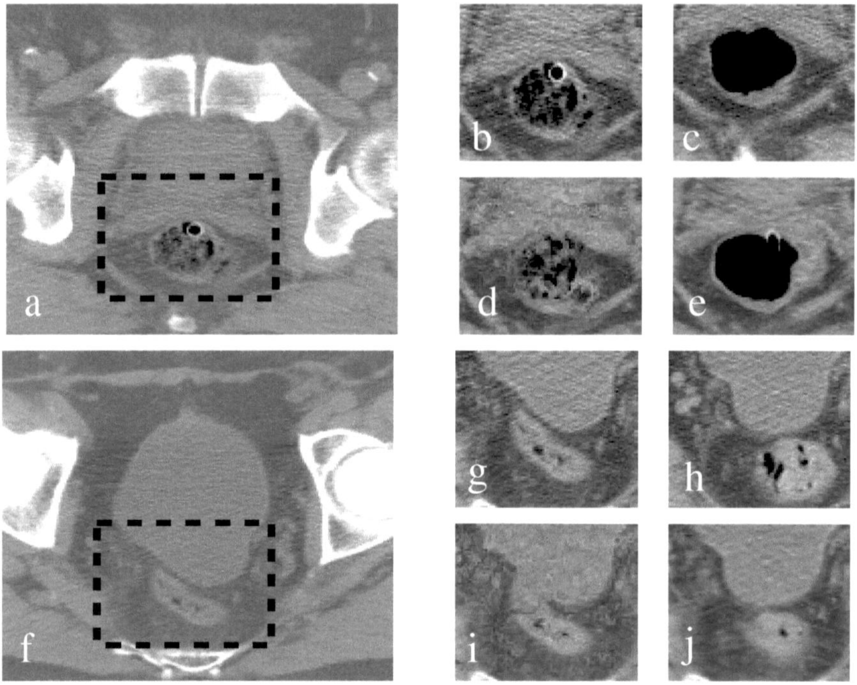

Fig. 2. Slice of dataset with a region of interest (roi) specified to highlight the difference in the Demons and the Morphon result. Images (a) and (f) show two different slices of the target with the roi marked with a dotted line. This region is enlarged in the right part of the figure, where (b) and(g) are the roi of the target for each slice with a bit more contrast to emphasize the structures, (c) and (h) are the roi of the template data. (d) and (i) are the deformed template resulting from Demons registration and (e) and (j), finally, are the result from Morphon registration. The Demons method has been able to better match the template compared to the Morphon method. The drawback is that it distorts the data.

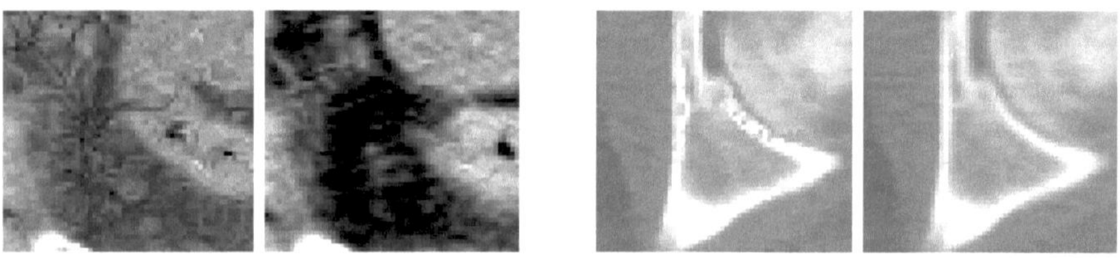

(a) Demons (left) and morphons (right) (b) Demons (left) and morphons (right)

Fig. 3. Enlargement of details of the result to show the distortion introduced by the Demons algorithm compared to the more smooth Morphon results

interpatient registration results for the Morphon method is shown in figure 5. The Demons algorithm, although being able to perform small interpatient deformations like in brain atlas-based registration, is not sufficient to solve this kind of

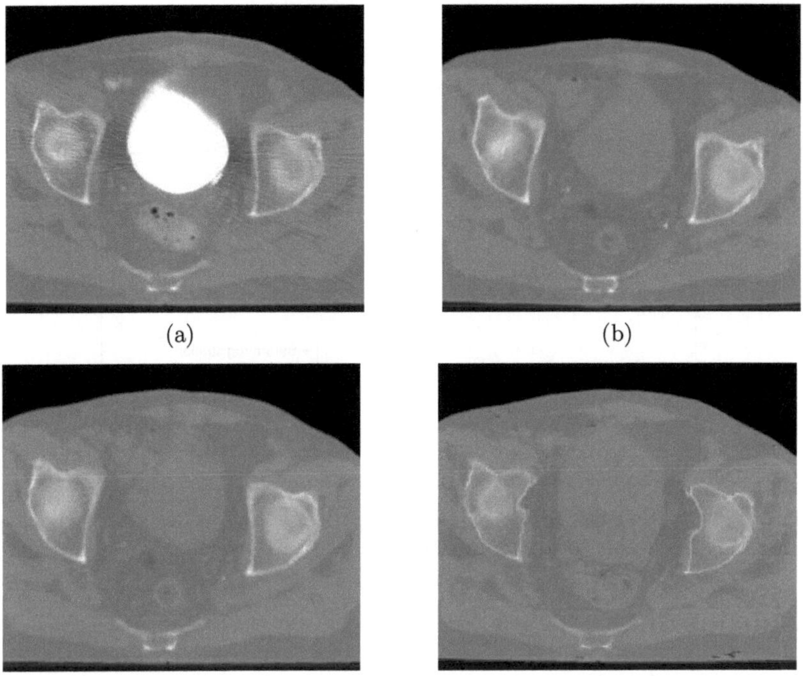

(a) (b) (c) (d)

Fig. 4. Result for registration of dataset with contrast agent. (a) Target, (b) template, (c) Morphon result and (d) Demons result.

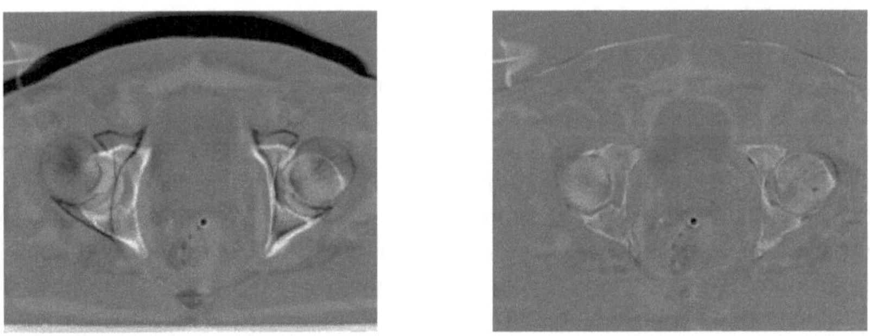

Fig. 5. Example of difference in one slice for the CT data for interpatient registration. The left image show the difference before registration and the right image show the difference after registration with the Morphon method. This image is ideally gray.

registration problems on pelvic images due to the large deformations and the intensity differences. Figure 6 shows the per slice correlation of datasets before and after registration. The Demons algorithm has not been able to match the datasets except for a few slices in the center of the stack of the data volume.

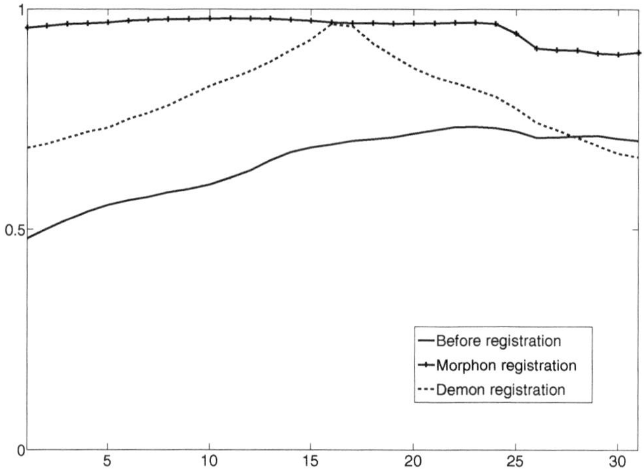

Fig. 6. The correlation (per slice) for the interpatient registration. The Demons algorithm has only been able to match a couple of slices in the center.

4 Conclusions

Two non-rigid registration methods have been used for fully automatic registration of CT volumes of the pelvic region, to evaluate their ability to catch the deformations of the anatomy between patient scans. The methods used were the Demons method and the Morphon method. They were chosen due to their capability of performing deformable registration in a fully automated way. We can conclude that the methods behave different in some aspects although the overall result is good for both of them when performing intrapatient registration. The Demons method allows a higher level of deformation of the template and may for example fill holes if it results in a better match to the target. One can discuss if this is a desirable feature in all situations, since this means that the anatomy present in the template has been modified considerably. The Demons algorithm also distorts the data to some extent, which may introduce strange artifacts in the deformed template. The Morphon method avoid the problem of distortion but has in these experiments been too stiff in some regions. Parameter settings can be changed to give the Morphon more degrees of freedom. This has been tested, and even though this results in a different deformation of the template, it did not result in a better registration of the datasets. Instead it would be usable to have different regularization models in different regions of the dataset. This feature is currently under development for the Morphon method and will be evaluated for this application in future work.

When either template or target dataset contain an intensity level not present in the other dataset, as for the dataset with contrast agent, the preprocessing step of the Demons algorithm will create an inaccurate output and the registration is misled. The Morphon method is not sensitive to the intensity levels in the same way, although the filters used for local phase estimation detects the strong edge, and the result is affected in the region of this edge.

For the experiment with interpatient registration, the Demons method was not able to give any usable result due to its sensitivity to image intensity differences between the images and the larger deformations involved in the process. The Morphon method gives adequate results, although the desire to have a more flexible regularization model exist also in this case.

Since we have used two completely different implementations of the algorithms (Matlab for the Morphon method and ITK for the Demons method) computational time is not a good measure of comparison at this stage, and is therefore left out in this discussion.

Although a more quantitative evaluation is required, the preliminary conclusion is that the Morphon method, with a more flexible regularization model, is the preferable method depending on two main causes. It is not sensitive to difference in image intensity between the datasets as is the Demons method, and it does not introduce the distortions in the deformed data seen in the Demons results.

References

1. Antolak, J.A., Rosen, I.I., Childress, C.H., Zagars, G.K., Pollack, A.: Prostate target volume variations during a course of radiotherapy. Int. Journal of Radiation Oncology Biol Phys 42(3), 661–672 (1998)
2. Cheung, R., Tucker, S.L., Ye, J.S., Dong, L., Liu, H., Huang, E., Mohan, R., Kuban, D.: Characterization of rectal normal tissue complication probability after high-dose external beam radiotherapy for prostate cancer. Int. Journal of Radiation Oncology Biol Phys 58(5), 1513–1519 (2004)
3. Granlund, G.H., Knutsson, H.: Signal Processing for Computer Vision. Kluwer Academic Publishers, Dordrecht (1995)
4. Hill, D., Batchelor, P., Holden, M., Hawkes, D.: Medical image registration. Physics in Medicine and Biology, 46(3), 1–45 (2001)
5. Thirion, J.-P.: Fast non-rigid matching of 3d medical images. Technical Report 2547, INRIA (May 1995)
6. Knutsson, H., Andersson, M.: Morphons: Paint on priors and elastic canvas for segmentation and registration. In: Kalviainen, H., Parkkinen, J., Kaarna, A. (eds.) SCIA 2005. LNCS, vol. 3540, Springer, Heidelberg (2005)
7. Martinez, A.A., Yan, D., Lockman, D., Brabbins, D., Kota, K., Sharpe, M., Jaffray, D.A., Vicini, F., Wong, J.: Improvement in dose escalation using the process of adaptive radiotherapy combined with three-dimensional conformal or intensity-modulated beams for prostate cancer. Int. Journal of Radiation Oncology Biol Phys. 50(5), 1226–1234 (2001)
8. National Library of Medicine. Medlineplus health information on prostate cancer. `http://www.nlm.nih.gov/medlineplus/ency/article/000380.htm` (November 2006)
9. Pettersson, J.: Automatic generation of patient specific models for hip surgery simulation. Lic. Thesis LiU-Tek-Lic-2006:24, Linköping University, Sweden, Thesis No. 1243 (April 2006)
10. Wang, H., Dong, L., Lii, M.F., Lee, A.L., de Crevoisier, R., Mohan, R., Cox, J.D., Kuban, D.A., Cheung, R.: Implementation and validation of a three-dimensional deformable registration algorithm for targeted prostate cancer radiotherapy. Int. Journal of Radiation Oncology Biol Phys 61(3), 725–735 (2005)
11. Wrangsjö, A., Pettersson, J., Knutsson, H.: Non-rigid registration using morphons. In: Kalviainen, H., Parkkinen, J., Kaarna, A. (eds.) SCIA 2005. LNCS, vol. 3540, Springer, Heidelberg (2005)

Reconstruction of 3D Curves for Quality Control

Hanna Martinsson[1], Francois Gaspard[2], Adrien Bartoli[2], and Jean-Marc Lavest[2]

[1] CEA, LIST, Boîte Courrier 94, F-91 191 Gif sur Yvette, France
hanna.martinsson@cea.fr
[2] LASMEA (CNRS/UBP), 24 avenue des Landais, F-63 177 Aubière, France
adrien.bartoli@gmail.fr

Abstract. In the area of quality control by vision, the reconstruction of 3D curves is a convenient tool to detect and quantify possible anomalies. Whereas other methods exist that allow us to describe surface elements, the contour approach will prove to be useful to reconstruct the object close to discontinuities, such as holes or edges.

We present an algorithm for the reconstruction of 3D parametric curves, based on a fixed complexity model, embedded in an iterative framework of control point insertion. The successive increase of degrees of freedom provides for a good precision while avoiding to over-parameterize the model. The curve is reconstructed by adapting the projections of a 3D NURBS snake to the observed curves in a multi-view setting.

1 Introduction

The use of optical sensors in metrology applications is a complicated task when dealing with complex or irregular structures. More precisely, projection of structured light allows for an accurate reconstruction of surface points but does not allow for a precise localization of the discontinuities of the object. This paper deals with the problem of reconstruction of 3D curves, given the CAD model, for the purpose of a control of conformity with respect to this model. We dispose of a set of images with given perspective projection matrices. The reconstruction will be accomplished by means of the observed contours and their matching, both across the images and to the model.

Algorithms based on active contours [9] allows for a local adjustment of the model and a precise reconstruction of primitives. More precisely, the method allows for an evolution of the reprojected model curves toward the image edges, thus to minimize the distance in the images between the predicted curves and the observed edges.

The parameterization of the curves as well as the optimization algorithms we use must yield an estimate that meets the requirements of accuracy and robustness necessary to perform a control of conformity. We have chosen to use NURBS curves [11], a powerful mathematical tool that is also widely used in industrial applications.

B.K. Ersbøll and K.S. Pedersen (Eds.): SCIA 2007, LNCS 4522, pp. 760–769, 2007.

In order to ensure stability, any method used ought to be robust to erroneous data, namely the primitives extracted from the images, since images of metallic objects incorporate numerous false edges due to reflections.

Although initially defined for ordered point clouds, active contours have been adapted to parametric curves. Cham and Cipolla propose a method based on affine epipolar geometry [3] that reconstructs a parametric curve in a canonical frame using stereo vision. The result is two coupled snakes, but without directly expressing the 3D points. In [15], Xiao and Li deal with the problem of reconstruction of 3D curves from two images. However, the NURBS curves are approximated by B-splines, which makes the problem linear, at the expense of loosing projective invariance. The reconstruction is based on a matching process using epipolar geometry followed by triangulation. The estimation of the curves is performed independently in the two images, that is, there is no interactivity between the 2D observations and the 3D curve in the optimization. Kahl and August introduce in [8] a coupling between matching and reconstruction, based on an a priori distribution of the curves and on an image formation model. The curves are expressed as B-splines and the optimization is done using gradient descent.

Other problems related to the estimation of parametric structures have come up in the area of surfaces. In [14], Siddiqui and Sclaroff present a method to reconstruct rational B-spline surfaces. Point correspondences are supposed given. In a first step, B-spline surface patches are estimated in each view, then the surface in 3D, together with the projection matrices, are computed using factorization. Finally, the surface and the projection matrices are refined iteratively by minimizing the 2D residual error. So as to avoid problems due to over-parameterization, the number of control points is limited initially, to be increased later on in a hierarchical process by control point insertion.

In the case of 2D curve estimation, other aspects of the problem are addressed. Cham and Cipolla adjust a spline curve to fit an image contour [4]. Control points are inserted iteratively using a new method called PERM (potential for energy-reduction maximization). An MDL (minimal description length [7]) strategy is used to define a stopping criterion. In order to update the curve, the actual curve is sampled and a line-search is performed in the image to localize the target shape. The optimization is performed by gradient descent. Brigger et al. present in [2] a B-spline snake method without internal energy, due to the intrinsic regularity of B-spline curves. The optimization is done on the knot points rather than on the control points, which allows the formulation of a system of equations that can be solved by digital filtering. So as to increase numerical stability, the method is embedded in a multi-resolution framework. Meegama and Rajapakse introduce in [10] an adaptive procedure for control point insertion and deletion, based on the euclidean distance between consecutive control points and on the curvature of the NURBS curve. Local control is ensured by adjustment of the weights. The control points evolve in each iteration in a small neighborhood (3×3 pixels). Yang et al. use a distance field computed a priori with the fast marching method in order to adjust a B-spline snake [16]. Control points are added in the segment presenting a large estimation error, due to a degree of freedom insufficient for a good fit of the curve. The procedure is repeated until the error is lower than a fixed threshold. Redundant control points are then removed, as long as the error remains lower than the threshold.

2 Problem Formulation

Given a set of images of an object, together with its CAD model, our goal is to reconstruct in 3D the curves observed in the images. In order to obtain a 3D curve that meets our requirements regarding regularity, rather than reconstructing a point cloud, we estimate a NURBS curve. The reconstruction is performed by minimizing the quadratic approximation error. The minimization problem is formulated for a set of M images and N sample points by

$$\mathbf{C}(\mathcal{P}) = \arg\min_{\mathcal{P}} \sum_{i=0}^{M-1} \sum_{j=0}^{N-1} (\mathbf{q}_{ij} - T_i(\mathbf{C}(\mathcal{P}, t_j)))^2, \tag{1}$$

where $\mathbf{q}_{ij}$ is a contour point associated with the curve point of parameter t_j, T_i is the projective operator for image i and $\mathcal{P}$ is the set of control points.

Our choice to use NURBS curves is justified by several reasons. First, NURBS curves have interesting geometrical properties, namely concerning regularity and continuity. An important geometrical property that will be of particular interest is the invariance under projective transformations.

3 Properties of NURBS Curves

Let $U = \{u_0, \cdots, u_m\}$ be an increasing vector, called the knot vector. A NURBS curve is a vector valued, piecewise rational polynomial over U, defined by

$$\mathbf{C}(t) = \sum_{i=0}^{n} \mathbf{P}_i R_{i,k}(t) \quad \text{with} \quad R_{i,k}(t) = \frac{w_i B_{i,k}(t)}{\sum_{j=0}^{n} w_j B_{j,k}(t)}, \tag{2}$$

where $\mathbf{P}_i$ are the control points, $B_{i,k}(t)$ the B-spline basis functions defined over U, w_i the associated weights and k the degree.

It is a common choice to take $k = 3$, which has proved to be a good compromise between required smoothness and the problem of oscillation, inherent to high degree polynomials. For our purposes, the parameterization of closed curves, we consider periodic knot vectors, that is, verifying $u_{j+m} = u_j$. Given all these parameters, the set of NURBS defined on U forms, together with the operations of point-wise addition and multiplication with a scalar, a vector space.

For details on NURBS curves and their properties, refer to [11].

3.1 Projective Invariance

According to the pinhole camera model, the perspective projection $T(\cdot)$ that transforms a world point into an image point is expressed in homogeneous coordinates by means of the transformation matrix $\mathbf{T}_{3\times4}$. Using weights associated with the control points, NURBS curves have the important property of being invariant under projective transformations. Indeed, the projection of (2) remains a NURBS, defined by its projected control points and their modified weights. The curve is written

$$\mathbf{c}(t) = T(\mathbf{C})(t) = \frac{\sum_{i=0}^{n} w_i' \, T(\mathbf{P}_i) \, B_{i,k}(t)}{\sum_{i=0}^{n} w_i' \, B_{i,k}(t)} = \sum_{i=0}^{n} T(\mathbf{P}_i) R_{i,k}'(t) \tag{3}$$

The new weights are given by

$$w_i' = (T_{3,1} X_i + T_{3,2} Y_i + T_{3,3} Z_i + T_{3,4}) \, w_i = \mathbf{n} \cdot (\mathbf{C}_{\mathrm{O}} - \mathbf{P}_i) \, w_i, \tag{4}$$

where $\mathbf{n}$ is a unit vector along the optical axis and $\mathbf{C}_{\mathrm{O}}$ the optical center of the camera.

3.2 Control Point Insertion

One of the fundamental geometric algorithms available for NURBS is the control point insertion. The key is the knot insertion, which is equivalent to adding one dimension to the vector space, consequently adapting the basis. Since the original vector space is included in the new one, there is a set of control points such that the curve remains unchanged.

Let $\bar{u} \in [u_j, u_{j+1})$. We insert $\bar{u}$ in U, forming the new knot vector $\bar{U} = \{\bar{u}_0 = u_0, \cdots, \bar{u}_j = u_j, \bar{u}_{j+1} = \bar{u}, \bar{u}_{j+2} = u_{j+1}, \cdots, \bar{u}_{m+1} = u_m\}$. The new control points $\bar{\mathbf{P}}_i$ are given by the linear system

$$\sum_{i=0}^{n} \mathbf{P}_i R_{i,k}(t) = \sum_{i=0}^{n+1} \bar{\mathbf{P}}_i \bar{R}_{i,k}(t). \tag{5}$$

We present the solution without proof. The new control points are written

$$\bar{\mathbf{P}}_i = \alpha_i \, \mathbf{P}_i + (1 - \alpha_i) \, \mathbf{P}_{i-1}, \tag{6}$$

with

$$\alpha_i = \begin{cases} 1 & i \leq j - k \\ \dfrac{\bar{u} - u_i}{u_{i+k} - u_i} & \text{if } \; j - k + 1 \leq i \leq j \; . \\ 0 & i \geq j + 1 \end{cases} \tag{7}$$

Note that only k new control points need to be computed, due to the local influence of splines.

4 Curve Estimation

Using the NURBS formulation, the minimization problem (1) is written

$$\min_{\{\hat{\mathbf{P}}_l\}} \sum_{i=0}^{M-1} \sum_{j=0}^{N-1} \left(\mathbf{q}_{ij} - \sum_{l=0}^{n} T_i(\hat{\mathbf{P}}_l) R_{l,k}^{(i)}(t_j) \right)^2, \tag{8}$$

where $R_{l,k}^{(i)}$ are the basis functions for the projected NURBS curve in image i.

The problem has two parts. First, the search for candidate edge points $\mathbf{q}_{ij}$ in the images, then the optimization of the 3D NURBS curve by optimization on the control points. The search for candidate points is carried out independently in the images using a method inspired by the one used by Drummond and Cipolla in [6]. For the optimization problem, we use the non-linear Levenberg-Marquardt minimization method. This optimization allows the control points to move in 3D, but does not change their number. In order to obtain an optimal reconstruction of the observed curve, we iteratively perform control point insertion.

4.1 Search for Image Contours

We sample the NURBS curve projected in the image, to use as starting points in the search for matching contour points. A line-search is performed in order to find the new position of the curve, ideally corresponding to an edge. Our approach is based solely on the contours. Due to the aperture problem, the component of motion of an edge, tangent to itself, is not detectable locally and we therefore restrict the search for the new edge position to the edge normal at each sample point. As we expect the motion to be small, we define a search range (typically in the order of 20 pixels) so as to limit computational cost. In order to find the new position of a sample point, for each point belonging to the normal within the range, we evaluate the gradient and compute a weight based on the intensity and the orientation of the gradient and the distance from the sample point. The weight function v_j for a sample point p_j and the candidate point p_ξ will be of the form

$$v_j(p_\xi) = \varphi_1(|\nabla I_\xi|) \cdot \varphi_2\left(\frac{\hat{n}_j \cdot \nabla I_\xi}{|\nabla I_\xi|}\right) \cdot \varphi_3(|p_j - p_\xi|),$$

where $\hat{n}_j$ is the normal of the projected curve at sample point j, ∇I_ξ is the gradient at the candidate point and the φ_k are functions to define. The weight function will be evaluated for each candidate p_ξ and the point p'_j with the highest weight, identified by its distance from the original point $d_j = |p_j - p'_j|$, will be retained as the candidate for the new position of the point.

The bounded search range and the weighting of the point based on their distance from the curve yield a robust behavior, close to that of an M-estimator.

4.2 Optimization on the Control Points

The first step of the optimization consists in projecting the curve in the images. Since the surface model is known, we can identify the visible parts of the curve in each image and retain only the parts corresponding to knot intervals that are completely visible. During the iterations, so as to keep the same cost function, the residual error must be evaluated in the same points in each iteration. Supposing small displacements, we can consider that visible pieces will remain visible throughout the optimization.

The optimization of (8) is done on the 3D control point coordinates, leaving the remaining parameters of the NURBS curve constant. The weights associated with the control points are modified by the projection giving 2D weights varying with the depth of each control point, according to the formula (4), but they are not subject to the optimization.

In order to avoid over-parameterization for stability reasons, the first optimization is carried out on a limited number of control points. Their number is then increased by iterative insertion, so that the estimated 3D curve fits correctly also high curvature regions. As mentioned earlier, the insertion of a control point is done without influence on the curve and a second optimization is thus necessary to estimate the curve. We finally need a criterion to decide when to stop the control point insertion procedure.

Control Point Insertion. Due to the use of NURBS, we have a method to insert control points. What remains is to decide where to place them. Several strategies have been used. Cham and Cipolla consider in [4] the dual problem of knot insertion. They define an error energy reduction potential and propose to place the knot point so as to maximize this potential. The control point is placed using the method described earlier. In our algorithm, since every insertion is followed by an optimization that locally adjusts the control points, we settle for choosing the interval where to place the point. Since the exact location within the interval is not critical, the point is placed at its midpoint. Dierckx suggests in [5] to place the new point at the interval that presents the highest error. This is consistent with an interpretation of the error as the result of a lack of degrees of freedom that inhibits a good description of the curve. If, however, the error derives from other sources, this solution is not always optimal. In our case, a significant error could also indicate the presence of parasite edges or that of a parallel structure close to the target curve. We have therefore chosen a heuristic approach, that consists in considering all the intervals of the NURBS curve, in order to retain the one that allows for the largest global error decrease.

Stopping Criterion. One of the motives for introducing parametric curves was to avoid treating all curve points, as only the control points are modified during the optimization. If the number of control points is close to the number of samples, the benefit is limited. Too many control points could also cause numerical instabilities, due to an over-parameterization of the curve on the one hand and the size of the non-linear minimization problem on the other hand. It is thus necessary to define a criterion that decides when to stop the control point insertion.

A strategy that aims to avoid the over-parameterization is the use of statistical methods inspired by the information theory. Based in a Maximum Likelihood environment, these methods combine a term equivalent, in the case of a normal distributed errors, to the sum of squares of the residual errors with a term penalizing the model complexity. Given two estimated models, in our case differentiated by their number of control points, the one with the lowest criterion will be retained. The first criterion of this type, called AIC (Akaike Information Criterion), was introduced by Akaike in [1] and is written

$$AIC = 2k + n \ln \frac{RSS}{n}, \tag{9}$$

where k is the number of control points, n is the number of observations and RSS is the sum of the squared residual errors. Another criterion, based on a bayesian formalism, is the BIC (Bayesian Information Criterion) presented by Schwarz [13]. It stresses the number of data points n, so as to ensure an asymptotic consistency and is written

$$BIC = 2k \ln n + n \ln \frac{RSS}{n}. \tag{10}$$

Another family of methods uses the MDL [12] formulation, which consists in associating a cost with the quantity of information necessary to describe the curve. Different criteria follow, depending on the formulation of the estimation problem. In the iterative control point insertion procedure of Cham et Cipolla [4], the stopping criterion is defined by means of MDL. The criterion depends, on the one hand on the number of control points and on the residual errors, on the other hand on the number of samples and on the covariance.

Yet another way of choosing an appropriate model complexity is the classical method of cross-validation. The models are evaluated based on their capacity to describe the data. A subset of the data is used to define a fixed complexity model, while the rest serve to validate it. The process is repeated and a model is retained if its performance is considered good enough.

We have chosen to use the BIC for this first version of our algorithm. A more thorough study of the influence of the stopping criterion in our setting will be performed at a later stage.

4.3 Algorithm

The algorithm we implemented has two parts. The optimization of a curve using a fixed complexity model is embedded in an iterative structure that aims to increase the number of control points. The non-linear optimization of the 3D curve is performed by the Levenberg-Marquardt algorithm, using a cost function based on a search for contour points in the images.

5 Experimental Evaluation

5.1 Virtual Images

In order to validate our algorithms for image data extraction and for curve reconstruction, we have performed a number of tests on virtual images. The virtual setting also allows us to simulate deformations of the target object.

We construct a simplified model of an object, based on a single target curve. We then apply our 3D reconstruction algorithm, starting at a modified "model curve", on a set of virtual views, see Fig. 1. The image size is 1284×1002 pixels. The starting curve has 10 control points, to which 11 new points are added. The sampling used for the computations is of 200 points. To fix the scale, note that at the mean distance from the object curve, one pixel corresponds roughly to 0.22 mm. The evaluation of the results is done by measuring the distance from

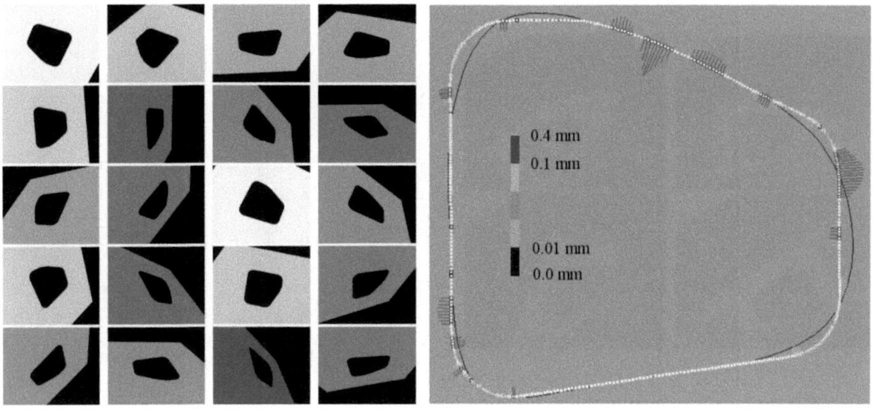

Fig. 1. *Left:* The virtual images used for the reconstruction of the central curve. *Right:* The distances from the sampled points from the reconstructed 3D curve to the model curve. The reconstruction is based on 20 virtual images. The cloud of sample points from the estimated curve is shown together with the target curve. The starting curve is shown in black. The differences are represented by lines with length proportional to the distance between the curve and the target, using a scale factor of 30.

a set of sampled points from the estimated curve to the target model curve. The distances from the target curve are shown in Fig. 1. We obtain the following results:

Mean error	0.0621 mm
Median error	0.0421 mm
Standard deviation	0.0528 mm

We note that the error corresponds to less than a pixel in the images, which indicates a sub-pixel image precision.

5.2 Real Images

We also consider a set of real images, see Fig. 2, with the same target curve, using the same starting "model curve" as in the virtual case. We now need to face the problem of noisy image data, multiple parallel structures and imprecision in the localization and the calibration of the views. The image size is 1392×1040 pixels. The starting curve has 10 control points, to which 11 new points are added. The sampling used for the computations is of 200 points. At the mean distance from the object curve, one pixel corresponds roughly to 0.28 mm. The distances from the target curve are shown in Fig. 2. We obtain the following results:

Mean error	0.157 mm
Median error	0.124 mm
Standard deviation	0.123 mm

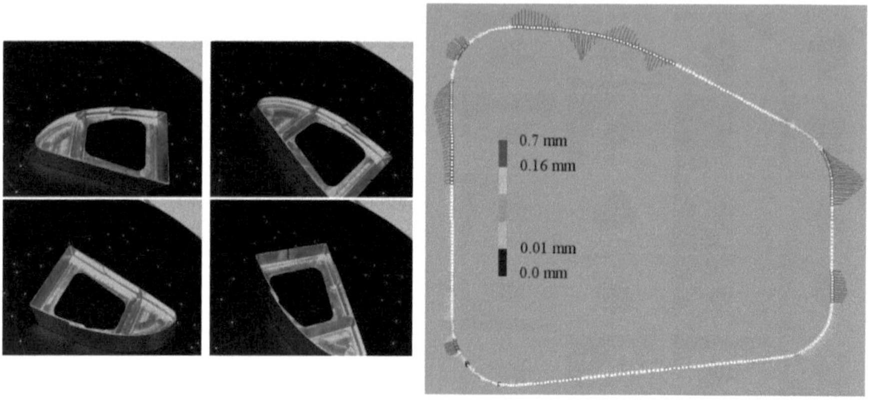

Fig. 2. *Left:* Four of the 18 real images used for the reconstruction of the curve describing the central hole. *Right:* The distances from the sampled points from the reconstructed 3D curve to the model curve. The reconstruction is based on 18 real images. The differences are represented by lines with length proportional to the distance between the curve and the target, using a scale factor of 20.

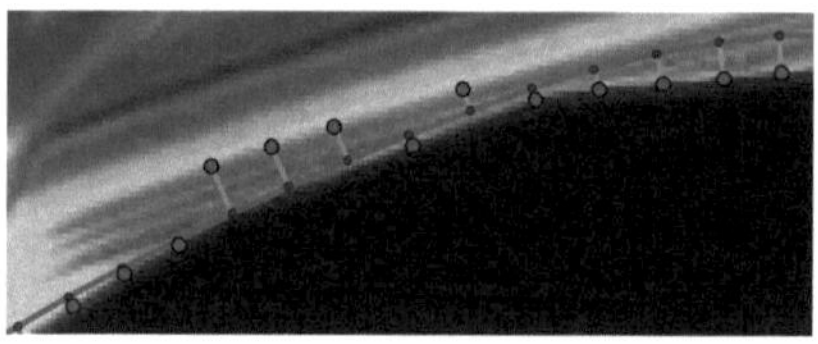

Fig. 3. Problems related to specularities and to the search for candidate points. Starting at the projection of the initial curve (in blue), some candidate points (in magenta) belong to a parasite edge.

Even if the errors are higher than in the case of virtual images, we note that they still correspond to less than a pixel in the images. The difference is partly explained by the noise and the parallel structures perturbing the edge tracking algorithm. An example of candidate points located on a parallel image contour, due to specularities, is given in Fig. 3.

6 Conclusions

We have presented an adaptive 3D reconstruction method using parametric curves, limiting the degrees of freedom of the problem. An algorithm for 3D reconstruction of curves using a fixed complexity model is embedded in an iterative framework, allowing an enhanced approximation by control point insertion. An experimental evaluation of the method, using virtual as well as real images, has let us validate its performance in some simple, nevertheless realistic, cases with specular objects subject to occlusions and noise.

Future work will be devoted to the integration of knowledge of the CAD model in the image based edge tracking. Considering the expected neighborhood of a sample point, the problem of parasite contours should be controlled and has limited impact on the obtained precision.

We also plan to do a deeper study around the stopping criterion used in the control point insertion process, using cross-validation.

References

1. Akaike, H.: A new look at the statistical model identification. IEEE Transactions on Automated Control 19(6), 716–723 (1974)
2. Brigger, P., Hoeg, J., Unser, M.: B-spline snakes: A flexible tool for parametric contour detection. IEEE Trans. on Image Processing 9(9), 1484–1496 (2000)
3. Cham, T.-J., Cipolla, R.: Stereo coupled active contours. In: Conference on Computer Vision and Pattern Recognition, pp. 1094–1099. IEEE Computer Society, Los Alamitos (1997)
4. Cham, T.-J., Cipolla, R.: Automated B-spline curve representation incorporating MDL and error-minimizing control point insertion strategies. IEEE Transactions on Pattern Analysis and Machine Intelligence 21(1), 49–53 (1999)
5. Dierckx, P.: Curve and Surface Fitting with Splines. Oxford University Press, Inc, New York, USA (1993)
6. Drummond, T., Cipolla, R.: Real-time visual tracking of complex structures. IEEE Transactions on Pattern Analysis and Machine Intelligence 7, 932–946 (2002)
7. Hansen, M.H., Yu, B.: Model selection and the principle of minimum description length. Journal of the American Statistical Association 96(454), 746–774 (2001)
8. Kahl, F., August, J.: Multiview reconstruction of space curves. In: 9th International Conference on Computer Vision, vol. 2, pp. 1017–1024 (2003)
9. Kass, M., Witkin, A., Terzopoulos, D.: Snakes: Active contour models. International Journal of Computer Vision 4(1), 321–331 (1987)
10. Meegama, R.G.N., Rajapakse, J.C.: NURBS snakes. Image and Vision Computing 21, 551–562 (2003)
11. Piegl, L., Tiller, W.: The NURBS book. In: Monographs in visual communication, 2nd edn. Springer, Heidelberg (1997)
12. Rissanen, J.: Modeling by shortest data description. Automatica 14, 465–471 (1978)
13. Schwarz, G.: Estimating the dimension of a model. Ann. of Stat. 6, 461–464 (1978)
14. Siddiqui, M., Sclaroff, S.: Surface reconstruction from multiple views using rational B-splines and knot insertion. In: First International Symposium on 3D Data Processing Visualization and Transmission, pp. 372–378 (2002)
15. Xiao, Y.J., Li, Y.F.: Stereo vision based on perspective invariance of NURBS curves. In: IEEE International Conference on Mechatronics and Machine Vision in Practice, vol. 2, pp. 51–56 (2001)
16. Yang, H., Wang, W., Sun, J.: Control point adjustment for B-spline curve approximation. Computer-Aided Design 36, 639–652 (2004)

Video Segmentation and Shot Boundary Detection Using Self-Organizing Maps

Hannes Muurinen and Jorma Laaksonen

Laboratory of Computer and Information Science*
Helsinki University of Technology
P.O. Box 5400, FIN-02015 TKK, Finland
{hannes.muurinen,jorma.laaksonen}@tkk.fi

Abstract. We present a video shot boundary detection (SBD) algorithm that spots discontinuities in visual stream by monitoring video frame trajectories on Self-Organizing Maps (SOMs). The SOM mapping compensates for the probability density differences in the feature space, and consequently distances between SOM coordinates are more informative than distances between plain feature vectors.

The proposed method compares two sliding best-matching unit windows instead of just measuring distances between two trajectory points, which increases the robustness of the detector. This can be seen as a variant of the adaptive threshold SBD methods. Furthermore, the robustness is increased by using a committee machine of multiple SOM-based detectors. Experimental evaluation made by NIST in the TRECVID evaluation confirms that the SOM-based SBD method works comparatively well in news video segmentation, especially in gradual transition detection.

Keywords: self-organizing map, video shot boundary detection.

1 Introduction

Consecutive frames within a shot, i.e. a continuous video segment that has been filmed in a single camera run, are usually visually similar. Shot boundaries on the other hand are characterised by discontinuities in the visual stream. Boundary detectors therefore often try to search for discontinuity peaks to spot boundaries. Various transition effects are available for video editors to bind shots together. The effects can be divided coarsely into *abrupt cuts* and *gradual transitions* based on their duration. In cut transitions there is an instantaneous change from one shot to another without any special effects, whereas in gradual transitions the shift has a nonzero duration, and there are transitional frames that do not belong exclusively to either one of the two shots.

Shot boundary detectors usually compute some distance measures between frames and the distances are compared to a threshold value to detect boundaries. The distance measures can be computed, for example, by comparing pixel

* Supported by the Academy of Finland in the projects *Neural methods in information retrieval based on automatic content analysis and relevance feedback* and *Finnish Centre of Excellence in Adaptive Informatics Research*.

B.K. Ersbøll and K.S. Pedersen (Eds.): SCIA 2007, LNCS 4522, pp. 770–779, 2007.

differences in consecutive motion-compensated frames, or by comparing some feature vectors or colour histograms calculated from the frames [1].

However, there can be differences of varying magnitude between consecutive frames within a shot, and the detector should be able to distinguish these from real shot boundaries. These differences are mainly caused by camera or object motion and lighting changes [2]. Especially camera flashlights are known to cause problems when handling news videos. A flashlight changes the illumination of the entire room for a short while, which is usually seen as two sharp discontinuity peaks in the visual flow. Also transmission errors might be seen as similar discontinuities. Ideally the detector should not react to this kind of differences, but only in the presence of a real boundary. In addition very slow gradual transitions induce further difficulties, since the detector should be sensitive enough to spot the slow transitions, but not too sensitive to react to fast visual changes within a shot. Adaptive threshold values have been used to resolve this challenge [3].

In this paper we will present a shot boundary detection method that compares two sliding frame windows on the Self-Organizing Map to spot the visual discontinuities. This can be seen as a variant of the adaptive threshold methods.

2 Shot Boundary Detection by Trajectory Monitoring

2.1 Trajectories

In our proposed technique the path that the data mapped on a Self-Organizing Map (SOM) [4] traverses as time advances is called a temporal trajectory. At each time step a feature vector is first calculated from the data, and then the vector is mapped to the best matching unit (BMU) on the map by finding the map node that has the most similar model vector. The process is illustrated in Figure 1. When visualising the trajectory, the consecutive BMUs are connected with line segments to form the full path as depicted in Figure 2.

Feature extraction methods usually produce similar feature vectors for visually similar frames, and similar feature vectors are mapped to nearby regions on a Self-Organizing Map. Thus similar frames are mapped close to each other on the SOM, and distances between consecutive trajectory points can be used to spot discontinuities in the visual stream. Abrupt cuts are characterised by large leaps while gradual transitions can be seen as more gradual drifts from one map region to another. The drift pattern obviously depends on the feature extraction methods and map properties.

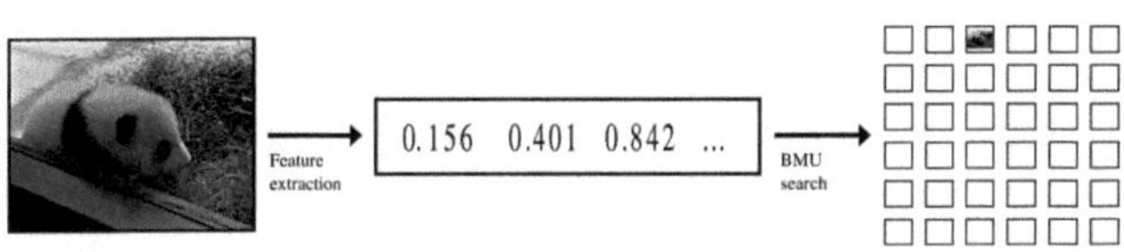

Fig. 1. A feature vector is calculated from the raw data. Then the vector is mapped on the SOM by finding the best-matching map unit.

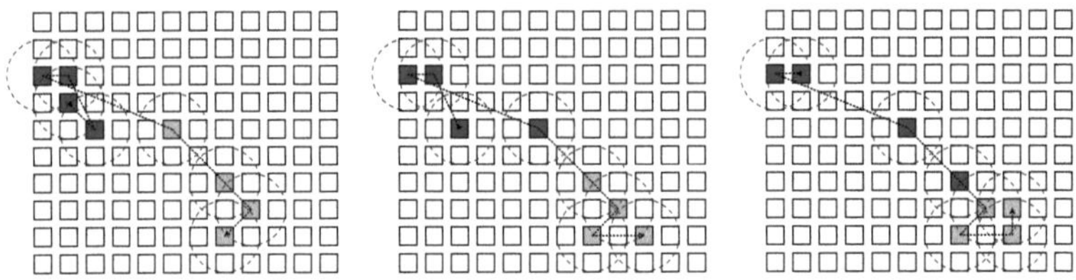

Fig. 2. Segments of a trajectory at three consecutive time steps. The SOM cells marked with dark gray colour represent trajectory points belonging to the set of preceding frames, and light gray cells represent the following frames. The trajectory is illustrated with a dotted black line, and the circles represent the area spanned by the preceding and following frame sets. The areas do not overlap in the two leftmost figures, but overlap at the third figure. This is interpreted as a one frame gradual transition.

2.2 Sliding Frame Windows

To increase the robustness of the proposed detector, distances between multiple frames before and after the current point of interest are compared instead of relying on map distances between two consecutive BMUs. Sliding frame windows on both sides of the point of interest are examined, and the corresponding map points on the trajectory are compared. The minimum distance between the map points of the two frame windows is identified and compared to a threshold value to determine the shot boundaries.

To express this formally, one has to define a distance measure that determines the distance between the two frame windows. If the lengths of the preceding and following frame windows are l_p and l_f, the sets of the preceding and the following frames, $S_p(t)$ and $S_f(t)$, at time instants $t \in \{l_p, l_p + 1, \ldots\}$ are

$$\begin{aligned} S_p(t) &= \{f_{t-l_p}, f_{t-l_p+1}, \ldots, f_{t-1}\} \\ S_f(t) &= \{f_t, f_{t+1}, \ldots, f_{t+l_f-1}\} , \end{aligned} \tag{1}$$

where f_n denotes frame n of the video. The distance measured at time t on SOM k can then be formulated:

$$d(k,t) = d(k; S_p(t), S_f(t)) = \min_{f_i, f_j} ||C_k(f_i) - C_k(f_j)||; f_i \in S_p(t), f_j \in S_f(t) \ . \tag{2}$$

Function $C_k(f_n)$ maps frame f_n on SOM k by calculating the corresponding map-specific feature vector value and finding the BMU. The function returns the discrete BMU coordinates (x, y).

The distances between the frame windows are then compared to a fixed SOM-specific threshold T_k, and if $d(k,t) > T_k$, we declare that there is a shot boundary between the frame windows $S_p(t)$ and $S_f(t)$, i.e. between frames f_{t-1} and f_t. Figure 2 illustrates how the preceding (dark gray) and following (light gray) frame windows are mapped on a SOM on one fictional trajectory at three consecutive time steps. The minimum distance of map coordinates between the two windows is then compared to the threshold value T_k to detect shot boundaries.

The method can also be described more intuitively by using the notion of areas. We may assume that separate shots occupy separate regions on the SOMs. A circular area that has radius of half the distance threshold value, $r_k = \frac{T_k}{2}$, is laid over each map point in the group of the preceding and following frames. When combined, the union of these areas form the *spanned area* of each frame window. If the two frame windows reside completely over separate shots, i.e. there is a boundary between the windows, the areas spanned by them do not overlap on the SOM. If the two windows contain frames that belong to the same shot, the areas probably do overlap. Therefore the separation of the areas can be used as an indicator of shot boundary locations. This is also illustrated in Figure 2. The BMU map points span an area around them, and the boundary classification can be done by checking if the circles around the light and dark gray map nodes overlap.

As mentioned, flashes and transmission error artifacts might cause sudden discontinuity peaks in the feature vector values. In a trajectory this can be seen as a jump to and back from some region on the SOM. If the lengths of the preceding and following frame windows are longer than the length of the distortion, the window method helps to prevent false positive detections during this kind of events.

Gradual transitions are characterised by multiple consecutive trajectory leaps. This kind of patterns can be detected using the same detector if the threshold values are suitably low. Consecutive boundary points are combined into a single data structure that specifies the boundary starting point and its length. Additionally we have defined a minimum length L for the shots. If the detector has observed two boundaries that are too close to each other, these are combined into a single gradual transition. This is because shots with length of a couple of frames are not observable by humans, and the two separate boundary observations are clearly mistakes.

2.3 Committee Machine of Detectors

Some features might be better at detecting some type of transitions, while others might be better with other types. Therefore a committee machine of several parallel Self-Organizing Maps trained with different features is utilised in the detector. Additionally, a committee machine helps to reduce misclassifications if each detector does somewhat different mistakes, but they mostly agree on the correct boundary decisions.

The separate detectors run in parallel and return the locations of the detected transitions. The results of the committee are then combined. For each point between two consecutive frames f_{t-1} and f_t we check how the detectors have classified them. The final verdict is made by letting the detectors vote if a given point belongs into a transition. Each detector k gives vote

$$v_k(t) = \begin{cases} 1 \text{ if the detector determines that the point belongs into a transition,} \\ 0 \text{ otherwise.} \end{cases} \tag{3}$$

The weighted vote result is then computed as

$$V(t) = \frac{\sum_{k=1}^{N} v_k(t) w_k}{\sum_{k=1}^{N} w_k} , \tag{4}$$

where w_k is the weight of the kth detector and N is the total number of detectors. $V(t)$ is compared to a threshold value $T_v \in [0, 1]$. If $V(t) > T_v$, we determine that the point between frames f_{t-1} and f_t is a boundary point. The transitions that are too close to each other are again combined, and the final starting locations and lengths of the transitions are obtained.

2.4 Parameter Selection

The shot boundary detection algorithm has quite a lot of parameters that can be tuned to change the behaviour of the detector. Some of the parameters are global, that is, they are shared by all the detectors in the committee, whereas some are detector-specific. Thus the number of free parameters increases as we increase the number of feature maps in the system. Table 1 summarises the global and map-specific parameters.

Some of the parameters were made detector-specific because optimal values for them are probably different for separate features. For example the distance threshold parameter, which determines the radius of the circles when comparing the areas spanned by the trajectory segments, should obviously be specific for each map because the lengths of significant jumps on trajectories might not be alike when using disparate features.

The parameter values were chosen using a variant of the discrete gradient descent method [5]. A set of training videos with human annotated shot boundary reference data were utilised in the parameter adjusting phase. The F_1 value [6] that combines precision P and recall R into a single scalar was used as the optimisation criterion. F_1 is a harmonic mean value that gives equal weight to both precision and recall, and it can be defined as

$$F_1 = \frac{2PR}{P + R} . \tag{5}$$

3 Features

Eleven feature extraction methods were applied to calculate feature vectors from all the frames in the videos. These were used to train feature-specific SOMs that

Table 1. The global and SOM-specific parameters that are freely selectable in the shot boundary detector. k is the SOM index number.

Global	Map-specific
Preceding frame window length l_p	Weight w_k
Following frame window length l_f	Distance threshold T_k
Minimum shot length L	
Vote result threshold T_v	

were used in the detectors. Five of the features were standard MPEG-7 [7] descriptors: *MPEG-7 Colour Structure*, *MPEG-7 Dominant Colour*, *MPEG-7 Scalable Colour*, *MPEG-7 Region Shape* and *MPEG-7 Edge Histogram*. The remaining six features were *Average Colour*, *Colour Moments*, *Texture Neighbourhood*, *Edge Histogram*, *Edge Co-occurrence* and *Edge Fourier*. [8]

The Average Colour vector contains the average RGB values of all the pixels in the frame. The Colour Moments feature extractor separates the three colour channels of the HSV colour representation of the image. The first three moments, mean, variance and skewness, are estimated for each channel to create a nine-dimensional feature vector.

The Texture Neighbourhood feature is calculated from the luminance component of the YIQ colour representation of an image. The 8-neighbourhood of each inner pixel is examined, and a probability estimate is calculated for the probabilities that the pixel in each of the surrounding relative position is brighter than the central pixel. The feature vector contains these eight estimates.

The Edge Histogram feature is not related to the MPEG-7 Edge Histogram descriptor. The feature vector consists of the histogram values of the four Sobel edge detectors. The Edge Co-occurrence vector contains values of the co-occurrence matrix calculated from the four Sobel edge detector outputs. The Edge Fourier feature vector contains 128 values computed from the Fast Fourier Transformation of the Sobel edge image of each frame. [9]

4 Experiments

4.1 TRECVID Evaluation by NIST

The performance of the shot boundary detector was evaluated by NIST in their annual TRECVID video retrieval evaluation, which consists of several video retrieval related tasks. In 2006 we participated in the shot boundary detection task for the first time [8]. The SBD task had 24 participating groups. Altogether there is slightly over 10 hours of news video in the SBD training and test data sets. The training data set also contains about 2.5 hours of NASA's educational programmes. The training files contain the total of 1341591 frames.

Ten separate detectors were trained using slightly differing portions of the training data and slightly differing combinations of feature extraction methods. The test data consisted of only news videos, and therefore we, for example, tested if omitting the NASA videos from the training set would improve the performance. Channel-specific detectors were also trained. We also tested if training detectors using only the cuts or only the gradual transitions would lead to better cut-specific and gradual-specific shot boundary detectors.

The best detector of our team turned out to be the one that had been trained with all the available data and with both cuts and gradual transitions. The cut-specific and gradual-specific detectors did not perform well. That is, when the detector had to make compromises between gradual and cut detection, its performance actually increased. This leads to suspicions of overlearning when using only a portion of the training data during the gradient descent optimisation.

Our detectors performed comparatively well in the evaluation. Table 2 shows the performance of our best detectors compared to the TRECVID median and average. Separate F_1 values were calculated for cut and gradual transition detection, and additionally a total F_1 score combining these two was computed. For gradual transitions also *frame recall*, *frame precision* and *frame* F_1 values were computed. These describe how accurately the true starting and ending locations of the gradual transitions were found, whereas it is possible to get good gradual recall and precision even when there is only one frame overlapping between the reference and submitted transition windows. Frame precision and recall are calculated by comparing the number of frames that overlap in the submitted and reference transition windows to the total number of frames in these windows. A more comprehensive interpretation of the TRECVID results is given in [8].

Table 2. The TRECVID shot boundary detection result comparison

Detector	Total F_1	Cut F_1	Gradual F_1	Frame F_1
Best single configuration	0.709	0.732	0.647	0.716
Best individual values out of ten	0.709	0.732	0.654	0.737
TRECVID median	0.747	0.792	0.625	0.734
TRECVID average	0.668	0.713	0.509	0.658
TRECVID standard deviation	0.199	0.214	0.250	0.212

4.2 Frame Window Experiment

In addition to the TRECVID experiments three detector configurations were compared to investigate how much the detector benefited from using frame windows instead of single frames in distance computations. The parameters of the first detector were optimised using all the training data, both cuts and gradual transitions, and the following six feature maps: Average Colour, Colour Moments, Texture Neighbourhood, Edge Histogram, MPEG-7 Dominant Colour and MPEG-7 Colour Structure. These features obtained the highest weights w_k in the TRECVID experiment. All the parameters including the window lengths were freely adjustable. The second detector was trained using the same features and training data, but we fixed the window lengths to $l_p = l_f = 1$. We also tested an unoptimised version of the one-frame detector by using the optimal parameters of the first configuration and just setting the window lengths to one.

Figure 3 depicts the precision and recall values of the three detectors. By comparing the performance values we can confirm that the frame windows contribute significantly to the overall performance of the detector. Rather good results could be obtained also using the optimised one-frame detector.

We can interpret the result of this experiment by referring to the idea of areas spanned by the frame windows. The optimised one-frame detector tries to model the area occupied by the given shot using the map location of only one frame. With a suitably large circle radius this can be harshly approximated, but the detector that models the area using multiple sequential map coordinates and the union of smaller circles can represent this area much more accurately.

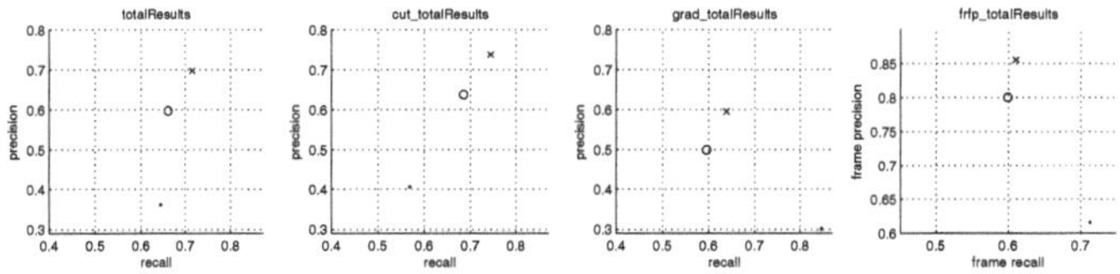

Fig. 3. The performance values of the detector with freely adjustable window lengths (cross), and the optimised (circle) and unoptimised (dot) detectors with window length of one

Comparing frame windows instead of distances between two frames can be also seen as a variant of the adaptive distance threshold methods. In fast-paced shots the area spanned by the frame window is much larger than during a slow-paced shot. This is comparable to the growth of an adaptive threshold value during content with high motion activity. If the frame window length is fixed to one, the distance threshold is constant. It can be argued that our method can be more accurate than using an adaptive one-dimensional threshold value. This is because in our method the adaptive threshold value is not a scalar that increases the threshold equally to all directions on the map, but there is an adaptively transforming two-dimensional threshold boundary that propagates selectively towards specific directions on the map.

4.3 Feature Number Experiment

In our last experiment twelve detectors using varying number of features were compared. First a detector with all the eleven features was trained, and then the number of features was decreased by discarding the feature that got the lowest weight w_k in the previous training run. Two one-feature detectors were also tested to compare the performance of the best edge feature to the performance of the best colour feature. The features were in the order of importance: Edge Histogram, Average Colour, Colour Moments, MPEG-7 Dominant Colour, Texture Neighbourhood, MPEG-7 Colour Structure, MPEG-7 Region Shape, MPEG-7 Edge Histogram, MPEG-7 Scalable Colour, Edge Co-occurrence and Edge Fourier.

The precision and recall values for the detectors with varying number of features are shown in Figure 4. It seems like the overall performance increases as the number of features increases. This was expected, although we anticipated overlearning when the number of features increased too much. This does not seem to happen with the maximum number of eleven features we have used. Using both cuts and gradual transitions in training may be one reason why the overlearning problem has not manifested itself.

The single colour feature seemed to be better in gradual transition detection than the single edge feature. This can be seen also in Figure 5 that shows a more detailed experiment in which the distance thresholds T_k were varied for the two one-feature detectors. Combining the two features clearly improves the results, which was expected because the amount of information available to the

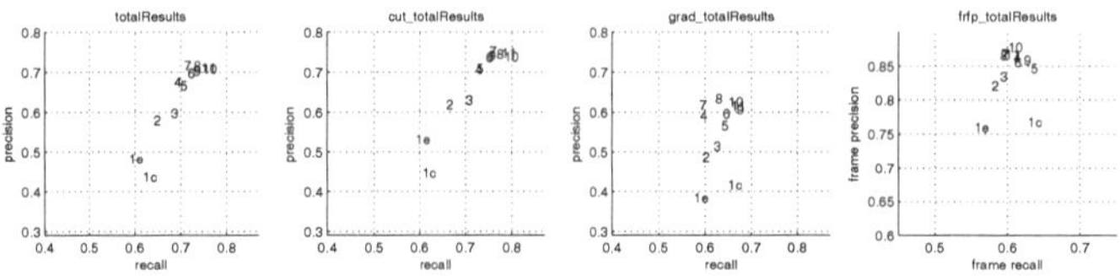

Fig. 4. Recall and precision for detectors using varying number of features. Detector 1e is a one feature detector using the Edge Histogram feature and detector 1c is a detector using the Average Colour feature.

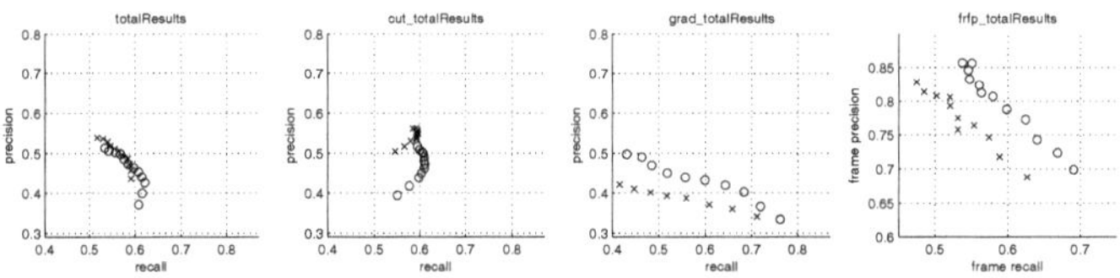

Fig. 5. Recall and precision for Average Colour (circle) and Edge Histogram (cross) based detectors with varying distance threshold parameters

detector increased. The cut detection performance values of the two detectors are similar. The edge feature configuration has better precision, while the colour feature configuration has slightly better recall.

5 Conclusions

A shot boundary detector using parallel Self-Organizing Maps was implemented during this work. We used it to participate in TRECVID shot boundary detection task, and our SOM-based method seemed to work quite well. Gradual transition detection performance was above the TRECVID median and cut detection performance above the TRECVID average.

The proposed SBD method tries to spot discontinuities in SOM trajectories to detect transitions. The novel idea is to use frame windows and areas spanned by the frame windows in detection. The areas occupied by preceding and following frame windows are approximated by using a union of circles centred on the map points corresponding to the frames. We assume that separate shots occupy separate regions on the map. If the areas occupied by the two frame windows do not overlap, they clearly belong to distinct shots, and a shot boundary between the frame windows can be declared. In the experiments this approach proved to be clearly better than the alternative method of just comparing the distances of two map points on the trajectory.

The feature vectors calculated from the frames could have been used on their own to detect shot boundaries, but in the implemented detector an additional step of mapping the vectors on a SOM was performed before measuring the distances between the frames. The transitions on a SOM are more informative than the distances between feature vectors because the nonlinear mapping of the SOM algorithm compensates the probability density differences in feature space, i.e the SOM takes into account that some regions in the feature space might be more probable than others.

Furthermore, in this algorithm the vector quantisation property of SOMs is helpful as it helps to suppress the little variations between the visually similar frames within a shot. Only significantly large changes in the input vectors are seen as movement on the maps. In general vector quantisation can cause problems since in the worst case scenario two vectors can be mapped to neighbouring nodes even when they reside right next to each other in the vector space. In practice this does not cause problems in our system because the distance threshold parameters always converged to such values that trajectory drifts to neighbouring SOM nodes are not interpreted as shot transitions.

The feature extraction methods utilised in our detector have been originally developed for image and video retrieval, and they might not be optimal in shot boundary detection task. Developing new SBD-optimised feature extraction methods might increase the performance of the detector. Especially lighting, translation and rotation invariance would be desired properties in shot boundary detection. Furthermore, new objects arriving from the edges of the screen should not change the feature vector values too much. Perhaps this could be achieved by using feature extraction methods that give gradually diminishing amount of weight to the pixels near the edges of the image.

References

1. Rui, Y., Huang, T.S., Mehrotra, S.: Exploring video structure beyond the shots. In: International Conference on Multimedia Computing and Systems. pp. 237–240 (1998)
2. Hanjalic, A.: Shot-boundary detection: unraveled and resolved? IEEE Trans. Circuits Syst. Video Techn. 12(2), 90–105 (2002)
3. Yeo, B.L., Liu, B.: Rapid scene analysis on compressed video. Circuits and Systems for Video Technology, IEEE Transactions 5(6), 533–544 (1995)
4. Kohonen, T.: Self-Organizing Maps, 3rd edn. Springer Series in Information Sciences. Springer, Heidelberg (2001)
5. Christensen, J., Marks, J., Shieber, S.: An empirical study of algorithms for point-feature label placement. ACM Trans. Graph. 14(3), 203–232 (1995)
6. Van Rijsbergen, C.J.: Information Retrieval, 2nd edn. Dept. of Computer Science, University of Glasgow (1979)
7. ISO/IEC: Information technology - Multimedia content description interface - Part 3: Visual, 15938-3:2002(E) (2002)
8. Sjöberg, M., Muurinen, H., Laaksonen, J., Koskela, M.: PicSOM experiments in TRECVID 2006. In: Proceedings of the TRECVID 2006 Workshop, Gaithersburg, MD, USA (November 2006)
9. Brandt, S., Laaksonen, J., Oja, E.: Statistical shape features for content-based image retrieval. Journal of Mathematical Imaging and Vision 17(2), 187–198 (2002)

Surface-to-Surface Registration Using Level Sets

Mads Fogtmann Hansen[1], Søren Erbou[1], Martin Vester-Christensen[1], Rasmus Larsen[1], Bjarne Ersbøll[1], and Lars Bager Christensen[2]

[1] Technical University of Denmark
[2] Danish Meat Research Institute

Abstract. This paper presents a general approach for surface-to-surface registration (S2SR) with the Euclidean metric using signed distance maps. In addition, the method is symmetric such that the registration of a shape A to a shape B is identical to the registration of the shape B to the shape A.

The S2SR problem can be approximated by the image registration (IR) problem of the signed distance maps (SDMs) of the surfaces confined to some narrow band. By shrinking the narrow bands around the zero level sets the solution to the IR problem converges towards the S2SR problem. It is our hypothesis that this approach is more robust and less prone to fall into local minima than ordinary surface-to-surface registration. The IR problem is solved using the inverse compositional algorithm.

In this paper, a set of 40 pelvic bones of Duroc pigs are registered to each other w.r.t. the Euclidean transformation with both the S2SR approach and iterative closest point approach, and the results are compared.

1 Introduction

This paper addresses the problem of shape registration or alignment which plays an essential role in shape analysis. Many registration procedures such as generalized Procrustes analysis [9,7] rely on a prior manual annotation of landmarks. The main drawback with these approaches is the reliance on manual annotation which becomes cumbersome and infeasible for larger 2Ddatasets and for 3D data.

Methods for explicitly deriving landmarks form training curves/surfaces based on information theoretic theory has been published [6]. Unfortunately these often suffer form exceeding use of computation time.

The iterative closest point (ICP) algorithm by Besl et al. [2] solves the problem of landmark dependence by iteratively updating the point correspondence after the closest point criterium. Since the introduction in 1992 many extensions and improvements of original ICP have been proposed in literature [8,11,10]. Most of these methods still require a good initial estimate in order not to converge to a local minimum. Furthermore, common for these methods are that they do not utilize the knowledge of the connectedness of the point cloud, which is available in many cases.

B.K. Ersbøll and K.S. Pedersen (Eds.): SCIA 2007, LNCS 4522, pp. 780–788, 2007.

The approach described in this paper is in many ways related to the approach presented by Darkner et al. [5], which aligns two point clouds by minimizing the sum of squared difference between the distance functions of the point clouds in some rectangular box domain. The problem with the scheme by Darkner et al. is that it is likely produce a suboptimal result when applied to concave shapes. That is, the concave parts of a shape will not propagate as far out in a distance map as the convex parts. As a consequence points placed on the convex parts of a shape are given more weight than points on concave parts. Our approach differs from the approach presented in [5], as it uses *signed* distance maps and minimizes the squared difference between the signed distance maps restricted to a shrinking narrow band. Thus, it does not suffer from the same defect as [5].

2 Theory

The registration of a surface S_x to a surface S_y w.r.t the Euclidean metric can be expressed as the minimization of the functional

$$F_1(\boldsymbol{p}) = \oint_{\mathcal{S}_x} d(W(\boldsymbol{x};\boldsymbol{p}), \mathcal{S}_y)^2 d\boldsymbol{x}, \tag{1}$$

where $W(\_;\boldsymbol{p})$ is the warp function.

A minor flaw with this approach is that the registration of S_x to S_y is not necessarily equivalent to the registration S_y to S_x. If $W(\_;\boldsymbol{p})$ is invertible (1) can be extended to

$$\begin{aligned} F_2(\boldsymbol{p}) &= \oint_{\mathcal{S}_x} d(W(\boldsymbol{x};\boldsymbol{p}), \mathcal{S}_y)^2 d\boldsymbol{x} + \oint_{\mathcal{S}_y} d(W(\boldsymbol{y};\boldsymbol{p})^{-1}, \mathcal{S}_x)^2 d\boldsymbol{y} \\ &= \oint_{\mathcal{S}_x} d_{,\mathcal{S}_y}(W(\boldsymbol{x};\boldsymbol{p}))^2 d\boldsymbol{x} + \oint_{\mathcal{S}_y} d_{\mathcal{S}_x}(W(\boldsymbol{y};\boldsymbol{p})^{-1})^2 d\boldsymbol{y}, \end{aligned} \tag{2}$$

where $d_{\mathcal{S}_x}$ and $d_{\mathcal{S}_y}$ are the distance maps of the surfaces $\mathcal{S}_x$ and $\mathcal{S}_y$, respectively. This energy functional ensures a symmetric registration.

A minimum of F_2 can be obtained by any gradient or Newton based optimization scheme. However, such schemes may very well get stuck in a local minimum instead of the global minimum. To overcome this problem we introduce a slightly different energy functional

$$F_3(\boldsymbol{p}) = \int_{U_x^r} (\Phi_y(W(\boldsymbol{x};\boldsymbol{p})) - \Phi_x(x))^2 d\boldsymbol{x} + \int_{U_y^r} (\Phi_x(W(\boldsymbol{y};\boldsymbol{p})^{-1}) - \Phi_y(y))^2 d\boldsymbol{y}, \tag{3}$$

where $\Phi_x(\boldsymbol{x})$ and $\Phi_y(\boldsymbol{y})$ are the signed distance maps (SDMs) of the surfaces $\mathcal{S}_x$ and $\mathcal{S}_y$, and $U^r = \{\boldsymbol{x} \mid \boldsymbol{x} \in \mathbb{R}^d, |\Phi(x)| < r\}$. And we note that

$$F_3 \to F_2 \text{ for } r \to 0. \tag{4}$$

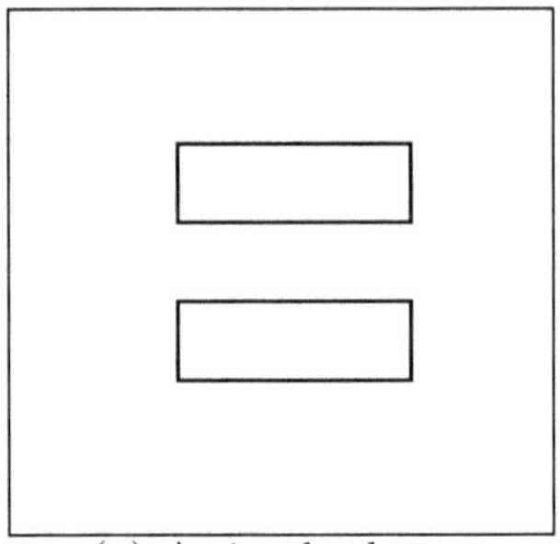

(a) A simple shape.

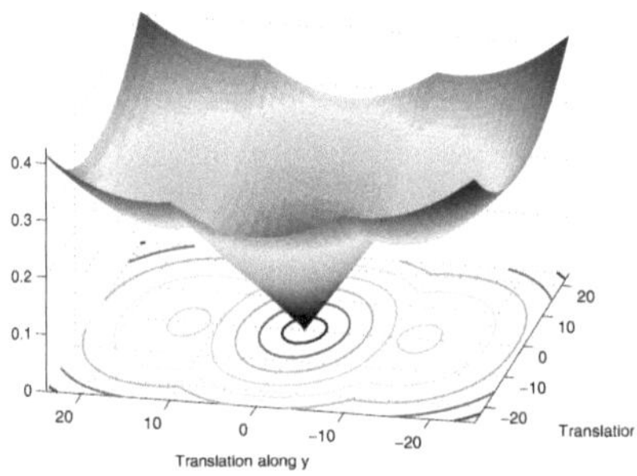

(b) F_2 cost functional.

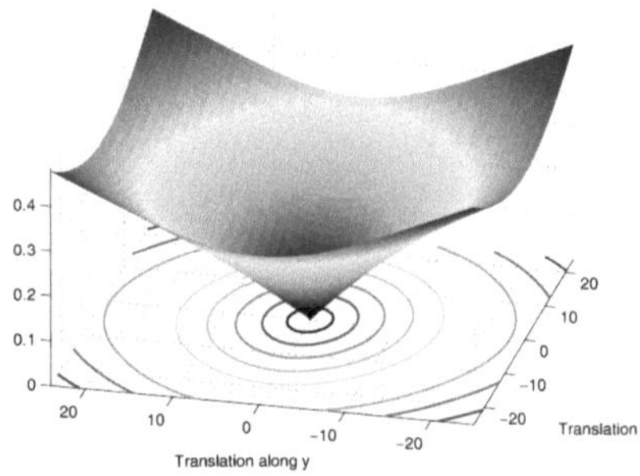

(c) F_3 cost functional.

Fig. 1. Cost as a function of translation in x and y direction

Now, consider the shape in Figure 1(a) consisting of two identical rectangles. If we translate the shape in both the x and the y direction between -25 and 25 pixels and calculate the energy in each position using F_2 and F_3 with $r = 25$ pixels, we get energy landscapes shown in Figure 1(b,c). In this case, the F_2 cost function produces three minima while the F_3 cost function produces only the global minimum.

3 Method

The energy functional defined in F_3 can be viewed as a image registration problem between two SDMs Φ_x and Φ_y, where the points to be warped are those inside the narrow bands U_x^r and U_y^r. The problem is solved using an extended version of the inverse compositional algorithm presented by Baker et al [1]. To preserve the same notation as in [1], we assume that Φ_x and Φ_y are discretized SDMs. Thus, F_3 becomes

$$F_4(\boldsymbol{p}) = \sum_{\boldsymbol{x} \in U_x^r} (\Phi_y(W(\boldsymbol{x};\boldsymbol{p})) - \Phi_x(\boldsymbol{x}))^2 + \sum_{\boldsymbol{y} \in U_y^r} (\Phi_x(W(\boldsymbol{y};\boldsymbol{p})^{-1}) - \Phi_y(\boldsymbol{y}))^2. \quad (5)$$

If the set of warps forms a *group* the minimization of F_4 is equivalent to the minimization of

$$F_5(\boldsymbol{p}) = \sum_{\boldsymbol{x} \in U_x^r} (\Phi_y(W(\boldsymbol{x};\boldsymbol{p})) - \Phi_x(W(\boldsymbol{x};\Delta\boldsymbol{p})))^2 + \sum_{\boldsymbol{y} \in U_y^r} (\Phi_x(W(\boldsymbol{y};\boldsymbol{p})^{-1}) - \Phi_y(W(\boldsymbol{y};\Delta\boldsymbol{p})^{-1}))^2. \qquad (6)$$

with the update rule $W(\boldsymbol{x};\boldsymbol{p}) \leftarrow W(\boldsymbol{x};\boldsymbol{p}) \circ W(\boldsymbol{x};\Delta\boldsymbol{p})^{-1}$. By applying the first order Taylor expansion to (6) we get

$$F_5(\boldsymbol{p}) \approx \sum_{\boldsymbol{x} \in U_x^r} \left(\Phi_y(W(\boldsymbol{x};\boldsymbol{p})) - \Phi_x(W(\boldsymbol{x};\boldsymbol{0})) - \nabla\Phi_x \frac{\partial W(\boldsymbol{x};\boldsymbol{0})}{\partial \boldsymbol{p}} \Delta\boldsymbol{p} \right)^2 + \sum_{\boldsymbol{y} \in U_y^r} \left(\Phi_x(W(\boldsymbol{y};\boldsymbol{p})^{-1}) - \Phi_y(W(\boldsymbol{y};\boldsymbol{0})^{-1}) - \nabla\Phi_y \frac{\partial W(\boldsymbol{y};\boldsymbol{0})^{-1}}{\partial \boldsymbol{p}} \Delta\boldsymbol{p} \right)^2 \qquad (7)$$

By taking the derivatives of F_5 w.r.t. $\Delta\boldsymbol{p}$ and setting them equal to zero we get the update equation

$$\Delta\boldsymbol{p} = -\boldsymbol{H}^{-1} \left(\sum_{\boldsymbol{x} \in U_x^r} \boldsymbol{S}_x^\top \boldsymbol{E}_{\boldsymbol{x}} + \sum_{\boldsymbol{y} \in U_y^r} \boldsymbol{S}_y^\top E_y \right), \qquad (8)$$

where $\boldsymbol{S}_x = \nabla\Phi_x \frac{\partial W(\boldsymbol{x};\boldsymbol{0})}{\partial \boldsymbol{p}}$, $\boldsymbol{S}_y = \nabla\Phi_y \frac{\partial W(\boldsymbol{y};\boldsymbol{0})^{-1}}{\partial \boldsymbol{p}}$, $E_x = \Phi_y(W(\boldsymbol{x};\boldsymbol{p})) - \Phi_x(\boldsymbol{x})$, $E_y = \Phi_x(W(\boldsymbol{y};\boldsymbol{p})^{-1}) - \Phi_y(\boldsymbol{y})$ and $\boldsymbol{H} = \sum_{\boldsymbol{x} \in U_x^r} \boldsymbol{S}_x^\top \boldsymbol{S}_x + \sum_{\boldsymbol{y} \in U_y^r} \boldsymbol{S}_y^\top \boldsymbol{S}_y$. Note that $\boldsymbol{H}^{-1}$, $\boldsymbol{S}_x$ and $\boldsymbol{S}_y$ only have to be computed once. A S2SR can be obtained with Algorithm 1.

The best sequence of r_i's is properly highly depended on the problem. We have applied the following scheme with success:

$$r_{i+1} \approx \frac{r_i}{2}. \qquad (9)$$

Selecting a suitable initial narrow band r_0 is however not entirely straight forward. If the choice of r_0 is too small the algorithm may get stuck in a local

Algorithm 1. S2SR

$\boldsymbol{r} = [r_1 \dots r_n]$; $\{ r_i > r_{i+1} \}$
for each $r_i \in \boldsymbol{r}$ **do**
  $k = 0$;
  repeat
    update $\boldsymbol{p}$ using (8) with $U_x^{r_i}$ and $U_y^{r_i}$;
    $k = k + 1$;
  until *convergence* or $k > k_{max}$
end for

minima, and if it is too large the algorithm will use an unnecessary amount of computational power. Note, that the computation time is much more depended on the radius of the initial narrow band than the number of narrow bands as the global minimum of F_3 for the narrow band r_i properly is relatively close to the global minimum of F_3 for r_{i+1}. In general, r_0 should be larger than the width of largest structure or feature in the image which might introduce a local minimum in F_3.

3.1 Extending Approach to Open Surfaces

SDMs are in principal only defined for closed surfaces as it is impossible to label the inside and outside of an open surface. Thus, our approach can only be applied to closed surfaces. To overcome this problem we introduce the notion of a pseudo SDM.

The pseudo SDM of triangle mesh of an open surface is computed using the following recipe:

1. Close the surface by triangulating all the holes in the triangle mesh.
2. Compute the SDM of the closed tiangle mesh. Bærentzen et al. [3,4] describe how to compute the SDM of a closed triangle mesh.
3. Set all voxels in the discretized SDMs with distances to the added faces to an undefined value.

Under the registration, voxels from one SDM may be warped to an undefined volume of the other SDM. In such cases, it is reasonable to assume that the distance in the undefined volume is 0, as we have no way of knowing whether the point is on the outside or inside of the shape. This hack allows for a bit of slack around the open areas of a surface. In many cases a surface is only open as it has been chosen to disregard a part of the shape - cutting away part of a shape in the exact *same place* is impossible. Furthermore, the gradient of a SDM at the borders between the defined and undefined volumes is likewise assumed to be equal to 0.

Sometimes, it is impossible to close an open surface with triangulation without introducing intersections between the new faces and the existing faces. Also, it might not be reasonable to close a surface if the hole is very large. In such cases, it might be more advisable simply to use unsigned distance maps instead of SDMs. The question of how large a hole in a surface can be, before the registration algorithm fails or produces suboptimal result with pseudo SDMs, needs to be investigated in the future.

4 Experiments

Two experiments were conducted to test the surface registration approach; (i) a toy example where the outline of the right and left hand of one of the authors were registered to each other, and (ii) a real example where 40 pelvic bones of Duroc pigs were registered with the ICP algorithm by Fitzgibbon [8] and with our S2SR algorithm.

4.1 Hand Example

To test the robustness of the S2SR algorithm a left and a right hand were traced on a piece of paper and scanned into a computer. The left hand was flipped horizontally, displaced 100 pixels in the x-direction and -25 pixels in the y-direction, and rotated 5 degrees counter clockwise. Figure 2 shows the initial position of the hands, the final position with regular S2SR[1] and the final position with our S2SR algorithm with the narrow bands $r = 30, 15, 7, 3, 1$. Evidently, the regular S2SR approach gets stuck in a local minimum or saddle point, while the shrinking narrow band S2SR approach registers the left and right hand perfectly.

(a) Initial position. (b) Final position for S2SR with narrow band r=1. (c) Final position for S2SR with narrow bands 30, 15, 7.5, 1.

Fig. 2. Rigid registration of left (green) and right hand (red)

4.2 Pelvic Bones

Half pig skeletons were automatic extracted from CT scans of half pig carcasses and fitted with implicit surfaces. From the implicit surfaces triangle meshes were created, and the pelvic bones were manually removed from the triangulated skeletons. An example of a pelvic bone can be found in Figure 3.

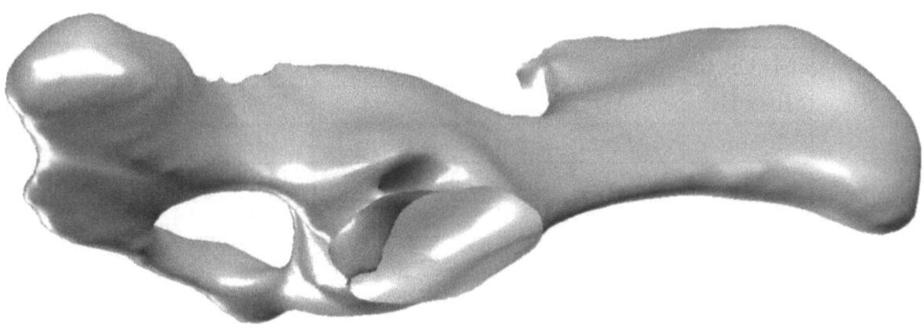

Fig. 3. Example of a pelvic bone from a Duroc pig

From the set of pelvic bone shapes a shape was selected to be the reference, and the remaining shapes were registered to the reference shape with ICP and our level set based S2SR algorithm with r = 20, 10, 5, 2.5, 1 mm. To compare

[1] Simulated with the small narrow band r=1.

the two registration approaches we use the mean squared error (MSE) and the maximum error (ME). As ICP minimizes the point-to-closest-point (CP) distance and our algorithm minimizes the surface-to-surface distance[2], we evaluate the performance of the registration algorithms using both distance concepts. Furthermore, as our registration algorithm does symmetric minimization of the squared distances and ICP does not, the MSE and the ME are calculated in the same direction as the ICP registration, in the other direction and in both directions combined. The registration results for ICP and S2SR can be found in Table 1 and 2, respectively. As no surprise, the ICP registration has a lower MSE and ME in the same direction as the registration, when we are using the CP distance. It is neither a surprise that our S2SR algorithm has lower MSEs and MEs in the opposite direction of the ICP registration and in both directions. It is however a bit of a surprise, that our S2SR algorithm has a smaller MSE than ICP when using tthe SDMs to extract distances. A possible explanation for this result is that our algorithm allows for a bit of slack around the open regions of the surface and is therefore better at fitting the remaining regions of the surface. Figure 4 illustrates this by color-coding the surfaces of two registered pelvic bones with the shortest distance.

Table 1. The MSE and ME averaged over the 39 registrations after ICP registration

Method	ICP ($A \rightarrow B$)					
Direction	$A \rightarrow B$		$A \leftarrow B$		$A \leftrightarrow B$	
Measure	$\sqrt{MSE}$	ME	$\sqrt{MSE}$	ME	$\sqrt{MSE}$	ME
SDM	11.92	20.13	12.70	27.04	12.34	27.24
CP	12.32	20.85	14.40	29.41	13.44	29.48

Table 2. The MSE and ME averaged over the 39 registrations after S2S registration

Method	S2SR ($A \leftrightarrow B$)					
Direction	$A \rightarrow B$		$A \leftarrow B$		$A \leftrightarrow B$	
Measure	$\sqrt{MSE}$	ME	$\sqrt{MSE}$	ME	$\sqrt{MSE}$	ME
SDM	11.69	24.11	12.18	24.03	11.90	26.28
CP	12.77	25.70	13.11	26.34	12.95	28.30

W.r.t. computation time, it can be mentioned that it took approximately 20 minutes to run the 39 registrations with ICP and approximately 50 minutes to run 39 registrations with S2SR on a standard Dell laptop with a 1.6Ghz Centrino CPU and 2Gb ram. It is difficult to compare the computation time of the two algorithm as the computation time for the S2SR algorithm is vastly depended

[2] The distances are found by interpolating the SDMs. To ensure fairness, when evaluating the MSE and ME, points, which are warped to an undefined area of a SDM, are ignored instead of receiving the distance 0.

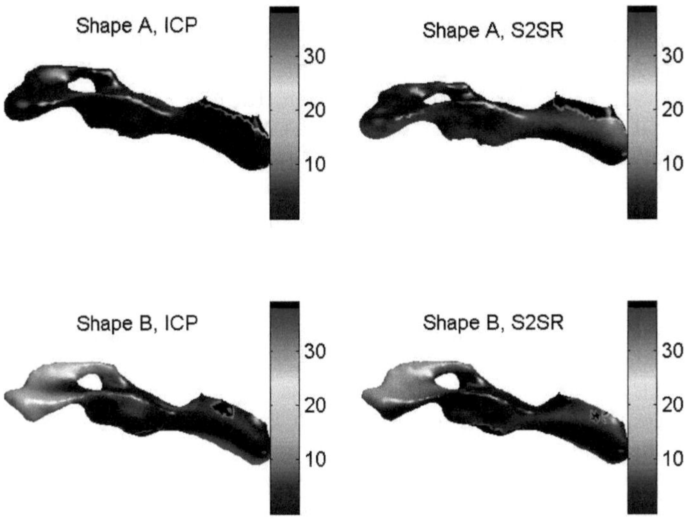

Fig. 4. Distance color-coded surfaces after registration. Black areas are areas where the distance could not be interpolated in the SDM because the point is situated in a undefined area.

on the chosen parameters, e.g. the chosen narrow bands and the resolution of the discretized SDMs.

5 Conclusion

This paper has presented a method for S2SR. The registration algorithm was tested on two examples, where its properties were highlighted; (i) it is less prone to fall into local minima than ordinary S2SR, (ii) and it does symmetric registration. As the method relies on SDMs it only works in theory on surfaces. Nevertheless, this paper has demonstrated that it can work on open surfaces by introducing a pseudo SDM, where distances are not defined in volumes *close* to the open regions of the surface.

In the future, we will use non-rigid transformations with the registration approach.

References

1. Baker, S., Matthews, I.: Lucas-Kanade 20 Years On: A unifying Framework". CMU. Part 1 (2002)
2. Besl, P.J., Mckay, N.D.: A Method for Registration of 3-D Shapes. IEEE Transaction on Pattern Analysis and Machine Intelligence 14, 239–256 (1992)
3. Bærentzen, A.: Robust Generation of Signed Distance Fields from Triangle Meshes. Fourth International Workshop on Volume Graphics (2005)

4. Bærentzen, A., Aanæs, H.: Signed Distance Computation using the Angle Weighted Pseudo-normal. IEEE Transactions on Visualization and Computer Graphics, 11, 243–253 (2005)
5. Darkner, S., Vester-Christensen, M., Larsen, R., Nielsen, C., Paulsen, R.R.: Automated 3D Rigid Registration of Open 2D Manifolds, MICCAI 2006 Workshop From Statistical Atlases to Personalized Models (2006)
6. Davies, R.H., Twining, C.J., Cootes, T.F., Waterton, J., Taylor, C.J.: A minimum description length approach to statistical shape modelling, IEEE Transactions on Medical Imaging (2002)
7. Dryden, I.L., Mardia, K.V.: Statistical Shape Analysis. John Wiley, New York (1998)
8. Fitzgibbon, A.: Robust Registration of 2D and 3D Point Sets. In: Proc. British Machine Vision Conference. vol. II (2001)
9. Gower, J.C.: Generalized procrustes analysis. Psychometrika (1975)
10. S. Granger, X. Pennec. "Multi-scale EM-ICP: A Fast and Robust Approach for Surface Registration." *ECCV 2002*, 2002.
11. Rusinkiewicz, S., Levoy, M.: Efficient Variants of the ICP Algorithm. 3DIM 2001 (2001)

Multiple Object Tracking Via Multi-layer Multi-modal Framework

Hang-Bong Kang and Kihong Chun

Dept. of Computer Eng., Catholic University of Korea
#43-1 Yokkok 2-dong Wonmi-Gu, Puchon, Kyonggi-Do Korea
hbkang@catholic.ac.kr

Abstract. In this paper, we propose a new multiple object tracking method via multi-layer multi-modal framework. To handle erroneous merge and labeling problem in multiple object tracking, we use a multi layer representation of dynamic Bayesian network and modified sampling method. For robust visual tracking, our dynamic Bayesian network based tracker fuses multi-modal features such as color and edge orientation histogram. The proposed method was evaluated under several real situations and promising results were obtained.

1 Introduction

Visual tracking in complex environment is an important task for surveillance, teleconferencing, and human computer interaction. It should be computationally efficient and robust to occlusion, changes in 3D pose and scale as well as distractions from background clutter. In particular, multiple object tracking of similar objects fails in the cases of adjacency or occlusion of tracked objects. For multiple identical or similar object tracking, it is necessary to solve merge and labeling problem correctly.

There have been some research works on multiple object tracking. In the multiple object tracking, it is very important to model the interaction among objects and solve the data association problem [1-6]. Usually, a joint state representation is used and joint data associations are inferred from the possible interactions between objects and observations. This approach requires a high computational cost due to the complexity of joint state representation. Yu *et* al. [4] propose collaboration approach among filters by modeling objects' joint prior using a Markov Random field. However, this approach has some limitations in correctly labeling objects. Qu *et* al. [5] suggest an interactively distributed multi-object tracking using a magnetic-inertia potential model. This approach is good for solving the labeling problem in identical multiple object tracking. However, the tracking performance suffers when the occlusion duration among objects gets a little longer.

For our multiple object tracking, we adopt a dynamic Bayesian network which has multi-layer representation and multi-modal features fusion. Dynamic Bayesian network (DBN) provides a unified probabilistic framework in integrating multi-modalities by using a graphical representation of the dynamic systems. The proposed tracker has the following characteristics. First, multiple modalities are integrated in the dynamic Bayesian network to evaluate the posterior of each feature such as color

B.K. Ersbøll and K.S. Pedersen (Eds.): SCIA 2007, LNCS 4522, pp. 789–797, 2007.

and edge orientation. Secondly, the erroneous merge and labeling problem can be solved in two phases of DBN framework.

The paper is organized as follows. Section 2 discusses a multi-layer representation of dynamic Bayesian network. Section 3 presents our proposed multi-modal multiple object tracking method in DBN framework. Section 4 shows experimental results of our proposed method.

2 Multi-layer Representation of DBN

The Dynamic Bayesian Network (DBN) provides a coherent and unified probabilistic framework to determine the target object state in each frame by integrating modalities such as the prior model of reference state and evidence in target object candidate [7]. To construct DBN for visual tracking, we must specify three kinds of information such as the prior distribution over state variables $p(x_0)$, the transition model $p(x_n | x_{n-1})$ and the observation model $p(y_n | x_n)$. The transition model $p(x_n | x_{n-1})$ describes how the state evolves over time. The observation model $p(y_n | x_n)$ describes how the evidence variables are affected by the actual state of the object tracking. The target object candidate is evaluated by the posterior probability through the integration of multiple cues in DBN.

$$p(x_n | y_n, x_{n-1}) \tag{1}$$

where x_n and x_{n-1} are the target object candidate and reference object state, respectively and y_n is the evidence of low-level features such as color and edge information from the target object candidate. In our visual tracking, we use two features such as color and edge orientation information. For evidence variables in our framework, we use color likelihood $p(c_n | x_n)$ and edge orientation likelihood $p(e_n | x_n)$ where c_n and e_n are the color and edge likelihood measurements at time n, respectively. The posterior probability like Eq. (1) is interpreted as

$$p(x_n | c_n, e_n, x_{n-1}) \propto p(c_n | x_n) p(e_n | x_n) p(x_n | x_{n-1}) \tag{2}$$

To deal with multiple object tracking, we denote the state of an individual object by x_n^i, where $i = 1.......m$ is the index of objects and n is the time index. To represent the interactive objects, we use a multi layer representation of DBN which is similar to hierarchical Hidden Markov Model. This is shown in Fig. 1. The top layer in DBN computes interactivity among hidden nodes by estimating the distance between a pair of objects. If the objects are adjacent or occluded, the interactivity among hidden nodes is computed. Using color and edge orientation likelihood information for the target object and the reference object, we classify the interactivities between two objects A and B into 5 cases such as "A and B are located near by"(A,B), "A is partially occluded by B"(A⊂B), "B is partially occluded by A"(B⊂A), "A is fully occluded by B "(B), and "B is fully occluded by A"(A). Based on objects' interactivities, the labeling node shown as a rectangle is activated or deactivated for each object state. After that, correct labels are assigned to each object state.

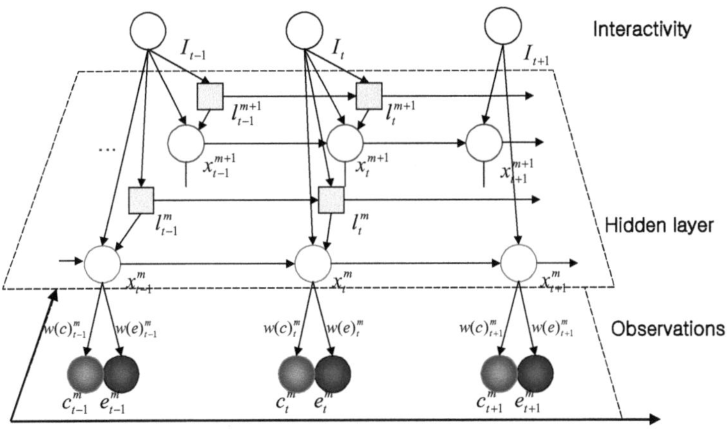

Fig. 1. Multi-layer multi-modal framework

3 Multiple Object Tracking

For multiple object tracking, it is necessary to solve erroneous merge problem where the tracker loses its target object and falsely coalesces with other trackers, and labeling problem where incorrect labels are assigned to the objects after occlusion.

3.1 Erroneous Merge Problem

To solve the erroneous merge problem, the accurate object state should be estimated when merge or split occurs. If two objects are merged (adjacent or partially occluded), it is very difficult to estimate exact position of the target object using conventional particle filtering because two objects are similar to each other. To reduce the effect of other object's presence, we use a Gaussian-weighted circular window in computing particle weight. For example, samples that are further away form the point having highest similarity value can be assigned smaller weights by employing a weighting function

$$\varphi(r) = \begin{cases} 1 - r^2 : r < 1 \\ 0 : otherwise \end{cases} \tag{3}$$

where r is the normalized distance from the highest similar point to the sample. Fig. 2(a) shows the original similarity distribution of samples for red object. By using weighting function like Eq. (3), the similarity distribution is transformed to that in Fig. 2(b). So, we can estimate exact target object position regardless of the green object in Fig. 2.

When the objects are merged, the tracker for the occluded object computes the similarity distribution of the reference model. If the similarity value is larger than the threshold value, weighting function like Eq. (3) is used to estimating the position of that object. Then, the tracker can solve the problem of false coalescence.

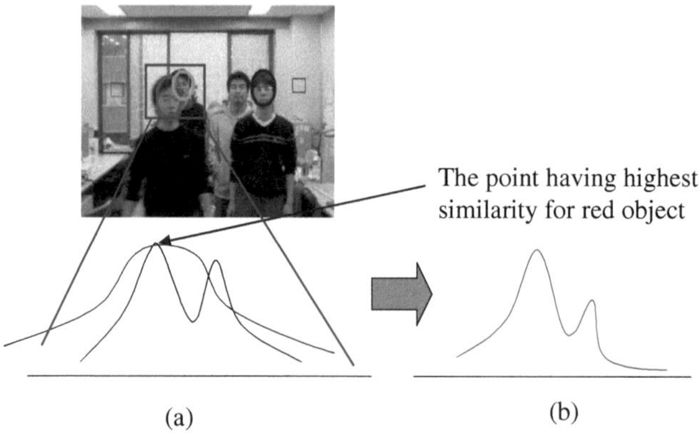

Fig. 2. (a) Original similarity distribution of samples for red object, (b) Transformed similarity distribution

3.2 Labeling Problem

To correctly assign a label to target object after occlusion, it is necessary to robustly discriminate target object candidates. We use multi-modal features such as color and edge orientation histogram.

The color likelihood $p(c_n \mid x_n)$ is defined as

$$p(c_n \mid x_n) = \left[\frac{1}{\sqrt{2\pi\sigma}} e^{-\frac{1-\sum_{u=1}^{m}\sqrt{p_i^{(u)}q^{(u)}}}{2\sigma^2}} \right] \tag{4}$$

where $p_i^{(u)}$ is the ith object candidate's color distribution and $q^{(u)}$ is color distribution of reference object.

The edge likelihood is computed from edge orientation histogram similar to SIFT [8]. To compute edge orientation, we detect edges using horizontal and vertical Sobel operators. After computing the strength and orientation of the edges, we apply threshold operation to remove outliers. The edge intensity and orientation of the target object is quantized into 8 bins (0 degree, 45 degree…) and displayed in each quarter plane. Peaks in the orientation histogram correspond to dominant directions of local gradients. After detecting the dominant orientation, edge descriptor for the reference model will be represented relative to this dominant direction and therefore achieve invariance to image rotation. We compute 16x 16 edge descriptor for the center point of the target object candidate. Fig. 3 shows the edge descriptor for a quarter plane of a target candidate. The edge likelihood between the target candidate and the reference model is computed as L2 distance between two edge orientation histograms. The edge likelihood $p(e_n \mid x_n)$ is defined as

$$p(e_n \mid x_n) = \left[\frac{1}{\sqrt{2\pi\sigma}} e^{-\frac{\rho^2(p_i^{(u)}, q^{(u)})}{2\sigma^2}} \right] \quad (5)$$

where $\rho(p,q)$ is Euclidean distance, $p_i^{(u)}$ is the ith object candidate's edge histogram distribution and $q^{(u)}$ is edge histogram distribution of reference object. Our edge likelihood is invariant to rotation and is discriminative against the background with confusing colors.

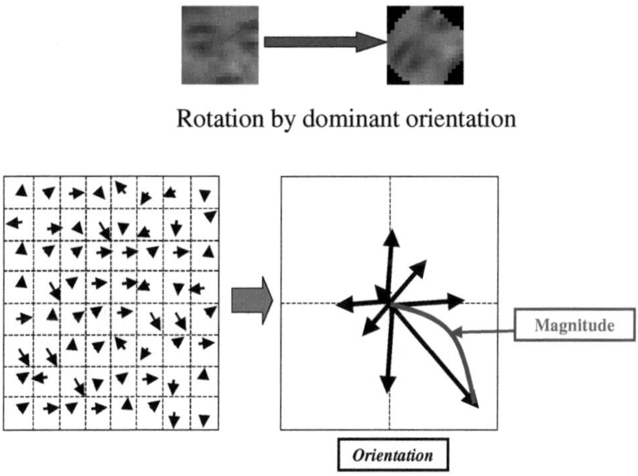

Fig. 3. Edge Orientation Histogram in quarter plane

The confidence weights for multi-cues such as color and edge likelihood are conditionally determined by previous weights. The observation likelihood is decided by each cue's weight. By adopting adaptive confidence weights, our tracker can discriminate target objects and then the correct label will be assigned to the target object.

For the approximate inference in DBN, we use the modified particle filtering since it seems to maintain a good approximation to the true posterior by using a constant number of samples [9]. In our proposed approach, the sampling-based method is executed as follows:

Step 1: N samples are created by sampling from the prior distribution at time 0.
Step 2: Each sample is propagated forward by sampling from the transition model like $p(x_n^i \mid x_{n-1}^i)$.
Step 3:Each sample is weighted by the log likelihood such as $k_1 p(c_n^i \mid x_n^i) + k_2 p(e_n^i \mid x_n^i)$ *where* k_1, k_2 *are the confidence weight of the likelihood of each sample. When the objects are adjacent or occluded, the log likelihood is* $k_1 p(c_n^i \mid x_n^i)\varphi() + k_2 p(e_n^i \mid x_n^i)\varphi()$, where $\varphi()$ is the Gaussian weighting function.
Step 4: The population is re-sampled to generate a new population of N samples with weighted-sample-with-replacement.

4 Experimental Results

Our proposed DBN-based visual tracking algorithm is implemented on a P4-3.0Ghz system with 320 x 240 image size. The number of particles is 400. The target object size is 16 x 16 pixels. For the implementation of the multi-layer DBN, we used Intel's Probabilistic Network Library (OpenPNL) [10] for building blocks. The input for DBN is the weighted sum of color and edge orientation likelihood.

(a)

(b)

Fig. 4. (a) Conventional Particle Filtering, (b) Proposed Method

(a)

(b)

Fig. 5. (a) Color-based tracking, (b) Color and edge-based tracking

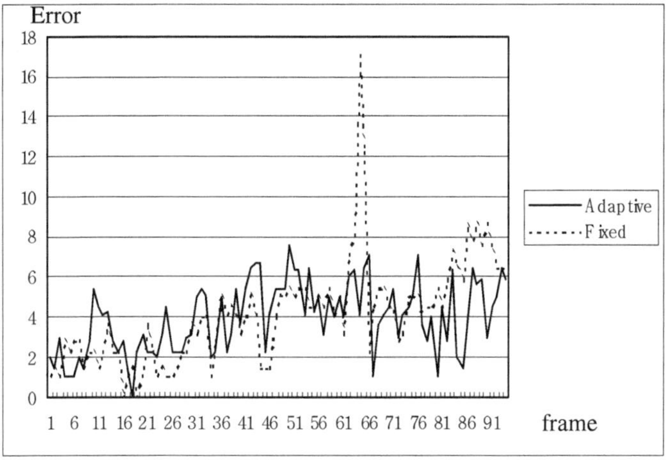

Fig 6. Tracking errors using fixed and adaptive confidence weights for color and edge likelihood

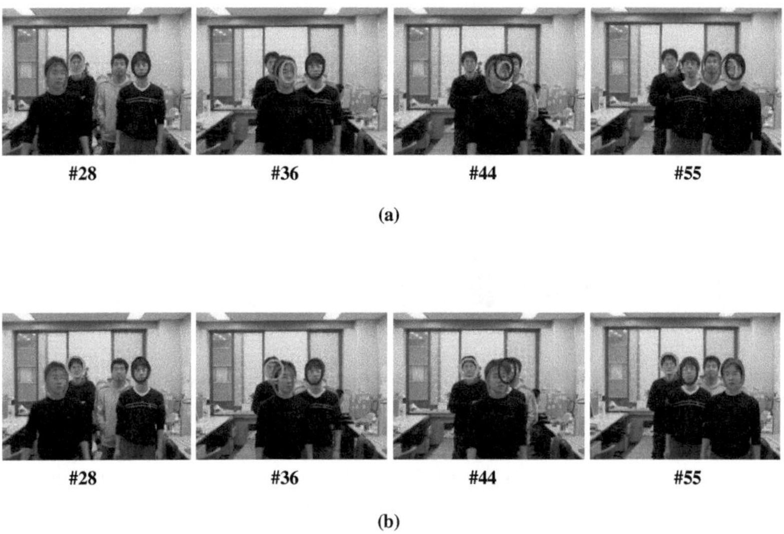

Fig. 7. Comparison of IDMOT[5] and our proposed tracker for office sequence (a) IDMOT result, (b) proposed tracker result

We made several experiments in a variety of environments to show the robustness of our proposed method. The first video sequence contains four moving persons in the office. This sequence is very difficult for multiple object tracking due to frequent occlusion. Fig. 4 shows the result of conventional particle filtering and our proposed modified particle filtering using Gaussian-weighted circular window. Our proposed method estimates accurate position of the target object.

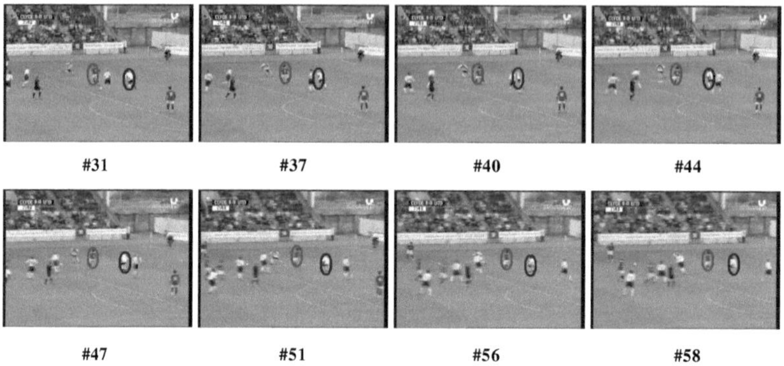

Fig. 8. Results of the proposed multi-modal multi-object tracker for the soccer game sequence

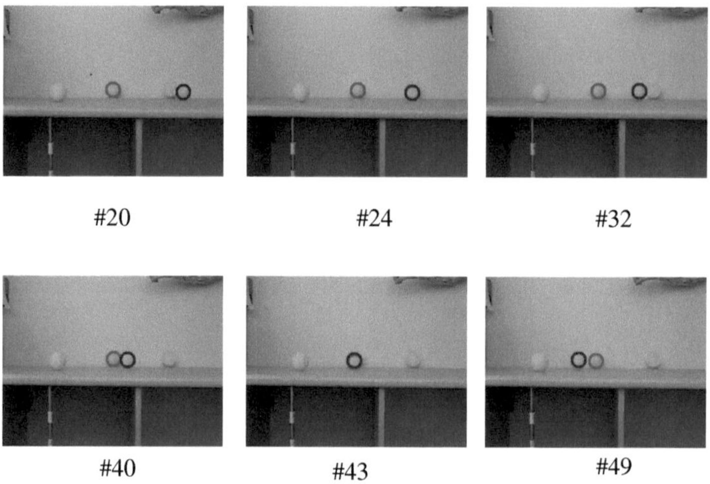

Fig. 9. Tracking result of tennis ball sequence

We experimented multi-cues approach in multiple object tracking. Fig. 5(a) shows the result of color only tracking and Fig. 5(b) shows the tracking result using our color and edge orientation likelihood. The confidence weights in color and edge likelihood are also important in robust tracking. Fig. 6 shows the errors occurred in tracking that used the fixed and adaptive confidence weights in our proposed method.

In Fig. 7, we compared our proposed method with other tracking method. IDMOT [5] suffers from labeling problems in long duration of occlusion (see Fig. 7(a)). However, our approach performs well solving both erroneous merge and labeling problem. This is shown in Fig. 7(b).

The second video sequence is the soccer game sequence. Each object moves independently. The image size is 320 x 240. Examples of the tracking results are shown in Fig. 8. Our proposed algorithm successfully tracked all objects throughout all frames. Fig. 9 shows the tracking result of tennis ball sequence. The labels in each ball are correct.

5 Conclusions

In this paper, multi layer representation of dynamic Bayesian network is proposed for multiple object tracking. For robust tracking, we implement a modified sampling and multi-modal tracking method that integrates color and edge orientation histogram. Our proposed tracker can handle erroneous merge and labeling problem in multiple object tracking. We have presented results from realistic scenarios to show the validity of the proposed approach. Compared to other tracking algorithms, our proposed system shows better and more robust tracking performance.

Acknowledgements

This work was supported by the Culture Research Center Project, the Ministry of Culture & Tourism and the KOCCA R&D program in Korea.

References

1. Mccormick, J., Blake, A.: A probabilistic exclusion principle for tracking multiple objects. Int. J. Comput. Vis (2000)
2. Isard, M., Mccormick, J.: Bramble: A Bayesian multiple blob tracker. In: Proc. ICCV 01 (2001)
3. Khan, Z., Balch, T., Dellaert, T.: An MCMC-based particle filter for tracking multiple interacting targets. In: Pajdla, T., Matas, J(G.) (eds.) ECCV 2004. LNCS, vol. 3021, Springer, Heidelberg (2004)
4. Yu, T., Wu, T.: Collaborative Tracking of Multiple Targets. In: Proc. CVPR'04 (2004)
5. Qu, W., Schonfeld, D., Mohamed, M.: Real-time Interactively Distributed Multi-Object Tracking Using a Magnetic-Inertia Potential Model. In: Proc. ICCV'05 (2005)
6. Yang, C., Duraiswami, R., Davis, R.: Fast Multiple Object Tracking via a Hierarchical Particle Filter. In: Proc. ICCV'05 (2005)
7. Kang, H-.B., Cho, S.: A Dynamic Bayesian Network-based Framework for Visual Tracking. In: Blanc-Talon, J., Philips, W., Popescu, D.C., Scheunders, P. (eds.) ACIVS 2005. LNCS, vol. 3708, pp. 603–610. Springer, Heidelberg (2005)
8. Lowe, D.: Distinctive image features from scale-invariant keypoints. Int'l J. Computer Vision, 91–110 (2004)
9. Nummiaro, K., Koller-Meier, E., Van Gool, L.: A Color-Based Particle Filter, First International Workshop on Generative-Model-Based Vision pp. 53–60 (2002)
10. Intel Open Source Probabilistic Network Library (OpnePNL) http://www.intel.com/research/mrl/pnl

Colorimetric and Multispectral Image Acquisition Using Model-Based and Empirical Device Characterization

Daniel Nyström

Dept. of Science and Technology (ITN), Linköping University
SE-60174 Norrköping, Sweden
danny@itn.liu.se

Abstract. The focus of the study is high quality image acquisition in colorimetric and multispectral formats. The aim is to combine the spatial resolution of digital images with the spectral resolution of color measurement instruments, to allow for accurate colorimetric and spectral measurements in each pixel of the acquired images. An experimental image acquisition system is used, which besides trichromatic RGB filters also provides the possibility of acquiring multi-channel images, using a set of narrowband filters. To derive mappings to colorimetric and multispectral representations, two conceptually different approaches are used. In the model-based characterization, the physical model describing the image acquisition process is inverted, to reconstruct spectral reflectance from the recorded device response. In the empirical characterization, the characteristics of the individual components are ignored, and the functions are derived by relating the device response for a set of test colors to the corresponding colorimetric and spectral measurements, using linear and polynomial least squares regression. The results indicate that for trichromatic imaging, accurate colorimetric mappings can be derived by the empirical approach, using polynomial regression to CIEXYZ and CIELAB. However, accurate spectral reconstructions requires for multi-channel imaging, with the best results obtained using the model-based approach.

Keywords: Multispectral imaging, Device characterization, Spectral reconstruction, Metamerism.

1 Introduction

The trichromatic principle of representing color has for a long time been dominating in color imaging. The reason is the trichromatic nature of human color vision, but as the characteristics of typical color imaging devices are different from those of human eyes, there is a need to go beyond the trichromatic approach. The interest for multi-channel imaging, i.e. increasing the number of color channels, has made it an active research topic with a substantial potential of application.

To achieve consistent color imaging, one needs to map the imaging-device data to the device-independent colorimetric representations CIEXYZ or CIELAB. As the color coordinates depend not only on the reflective spectrum of the object but also on

B.K. Ersbøll and K.S. Pedersen (Eds.): SCIA 2007, LNCS 4522, pp. 798–807, 2007.

the spectral properties of the illuminant, the colorimetric representation suffers from metamerism, i.e. objects of the same color under a specific illumination may appear different when they are illuminated by another light source. Furthermore, when the sensitivities of the imaging device differ from the CIE color matching functions, two spectra that appear different for human observers may result in identical device response. In multispectral imaging, color is represented by the object's spectral reflectance, which is illuminant independent. With multispectral imaging, different spectra are readily distinguishable, no matter they are metameric or not. The spectrum can then be transformed to any color space and be rendered under any illumination.

The focus of the paper is colorimetric and multispectral image acquisition, which requires methods for computing colorimetric and spectral data from the recorded device signals. Experiments are performed using trichromatic imaging as well as multi-channel imaging, using an experimental image acquisition system. Two conceptually different approaches for device characterization are evaluated: *model-based* and *empirical* characterization. In the model-based approach, the physical model describing the process by which the device captures color is inverted to reconstruct spectral reflectance. In the empirical approach, the device characteristics are ignored and the mappings are derived by correlating the device response for a set of reference colors to the corresponding colorimetric and spectral measurements, using least squares regression.

2 Model-Based Characterization

The linear model for the image acquisition process, describing the device response to a known input, is given in Eq.1. The device response, d_k, for the k:th channel is, for each pixel, given by:

$$d_k = \int_{\lambda \in V} I(\lambda) F_k(\lambda) R(\lambda) S(\lambda) d\lambda + \varepsilon_k \tag{1}$$

where $I(\lambda)$ is the spectral irradiance of the illumination, $F_k(\lambda)$ is the spectral transmittance of filter k, $R(\lambda)$ is the spectral reflectance of the object, $S(\lambda)$ is the spectral sensitivity function for the camera, ε_k is the measurement noise for channel k, and V is the spectral sensitivity region of the device.

The spectral characteristics of the illumination and the filters have been derived from direct measurements, using a spectroradiometer. The spectral sensitivity of the CCD camera has previously been estimated by relating the device response to the known spectral reflectance for a set of carefully selected color samples, using least-squares regression techniques [1]. The spectral properties of the components of the image acquisition system are given in Fig. 1.

Having obtained the forward characterization function of all the components in the image acquisition system, the known spectral characteristics of the system can be

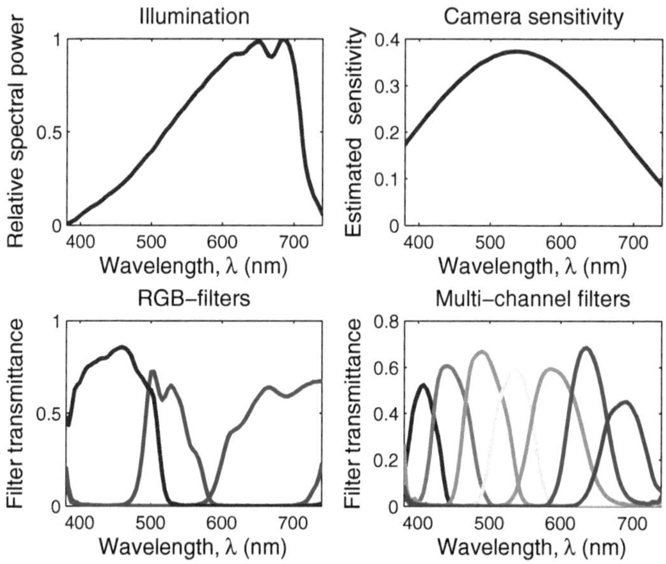

Fig. 1. Measured spectral power for the illuminant (*a*), estimated spectral sensitivity for the camera (*b*) and measured spectral transmittance for the RGB filters (*c*) and the 7 multi-channel filters (*d*)

represented by a spectral transfer function [2]. The spectral transfer function, $W_k(\lambda)$, describes the spectral characteristics for each channel k, as:

$$W_k(\lambda) = I(\lambda)F_k(\lambda)S(\lambda) . \quad (2)$$

Denote the spectral signal as a discrete N-component vector, sampled at wavelengths $\lambda_1,\ldots, \lambda_N$, and let $\mathbf{W}$ be the $N \times K$ matrix in which each column describes the spectral transfer function of channel k. Then the device response vector, $\mathbf{d}$, for a sample with spectral reflectance $\mathbf{r}$ is given by:

$$\mathbf{d} = \mathbf{W}^t\mathbf{r} . \quad (3)$$

When inverting the model, we seek the $N \times K$ reconstruction matrix $\mathbf{M}$ that reconstructs the spectral reflectance, $\breve{\mathbf{r}}$, from the camera response $\mathbf{d}$, as:

$$\breve{\mathbf{r}} = \mathbf{Md} . \quad (4)$$

The most straightforward approach to derive the reconstruction matrix is to simply invert Eq. 3, using the pseudo-inverse approach, giving the reconstruction operator:

$$\mathbf{M}_0 = (\mathbf{WW}^t)^{-1}\mathbf{W} = (\mathbf{W}^t)^{-} . \quad (5)$$

where $(\mathbf{W}^t)^-$ denotes for the More-Penrose pseudo-inverse of $\mathbf{W}^t$. Generally, the pseudo-inverse reconstruction is sensitive to noise, which makes the approach not always useful in practices. When $K < N$, i.e. the number of color channels K is less than the number of spectral sampling points N, the matrix $\mathbf{W}$ is of insufficient rank and the algebraic equations are underdetermined. Further more, this method minimizes the Euclidian distance in the camera response domain (i.e. between $\mathbf{d}$ and $\mathbf{W}^t\,\check{\mathbf{r}}$), which does not necessarily mean that the reconstructed spectrum will be close to the real spectrum [3].

Another approach is to instead seek another reconstruction matrix, $\mathbf{M}_1$, which minimizes the Euclidian distance between the reconstructed spectrum and the original spectrum [3]. By exploiting the *a priori* information that the vast majority of reflectance spectra for real and man-made surfaces are smooth functions of wavelength ([4], [5]), it can be assumed that the spectrum can be represented by a linear combination of a set of smooth basis functions, $\mathbf{B} = [\mathbf{b}_1,\mathbf{b}_2\ldots\mathbf{b}_p]$. This gives the reconstruction operator $\mathbf{M}_1$, which minimizes the RMS spectral difference of the reconstructed spectrum, as (refer to [3], for details):

$$\mathbf{M}_1 = \mathbf{B}\mathbf{B}^t\mathbf{W}\left(\mathbf{W}^t\mathbf{B}\mathbf{B}^t\mathbf{W}\right)^{-1}. \qquad (6)$$

The base functions, $\mathbf{B}$, can consist of a set of real, measured spectral reflectances, which then should be representative to the reflectance of samples that is likely to be encountered in the image acquisition system. An alternative to spectral basis is to simply let $\mathbf{B}$ consist of a set of Fourier basis functions.

3 Empirical Characterization

In empirical characterization, colorimetric and spectral data are derived using a "black box" approach, i.e. without explicitly modeling the device characteristics. By correlating the device response for a training set of color samples to the corresponding colorimetric or spectral values, the characterization functions are derived using least squares regression.

The characterization functions derived using empirical approaches will be optimized only for a specific set of conditions, including the illuminant, the media and the colorant. Once the conditions change, e.g. a different substrate or a different print mechanism, the characterization has to be re-derived in order to obtain good accuracy (see for example [6], [7]). The dependency on the illuminant is not an issue when the light source is fixed and can be considered as a property of the system. However, the fact that the characterization function is also media- and colorant dependent is a major drawback, preventing the characterization function from being applied to arbitrary combinations of media and colorants.

3.1 Spectral Regression

Even though empirical approaches are mainly used to derive mappings to colorimetric data, CIEXYZ or CIELAB, there have been attempts to reconstruct spectral reflectance [8]. The spectral reconstruction matrix, $\mathbf{M}$, is now derived entirely based on the recorded device response to a set of training samples, i.e. ignoring the spectral characteristics of the imaging system. If the spectral reflectance for a set of T training samples are collected into a $T \times N$ matrix $\mathbf{R} = [\mathbf{r}_1, \ldots, \mathbf{r}_T]$ and the corresponding device responses into a $T \times K$ matrix $\mathbf{D} = [\mathbf{d}_1, \ldots, \mathbf{d}_T]$, then the linear relationship is given by:

$$\mathbf{R} = \mathbf{DM}. \tag{7}$$

and the optimal $K \times N$ spectral reconstruction matrix $\mathbf{M}$ is then given by:

$$\mathbf{M} = (\mathbf{D}^t\mathbf{D})^{-1}\mathbf{D}^t\mathbf{R} = (\mathbf{D})^{-}\mathbf{R}. \tag{8}$$

In the same way as for the model-based approach, the reconstructed spectra can be represented as linear combinations of a set of basis functions.

3.2 Colorimetric Regression

A common approach to derive colorimetric data is to use polynomial regression from device values to CIEXYZ [9]. For example, the inverse characterization function of a 3-channel system, mapping RGB values to XYZ tristimulus values, is obtained by expressing XYZ as polynomial functions of R, G and B. As an example, a second order polynomial approximation is given by:

$$[X\,Y\,Z] = \left[1, R, G, B, R^2, RG, RB, G^2, GB, B^2\right] \begin{bmatrix} w_{X,1} & w_{Y,1} & w_{Z,1} \\ w_{X,2} & w_{Y,2} & w_{Z,2} \\ & \cdots & \\ w_{X,10} & w_{Y,10} & w_{Z,10} \end{bmatrix}. \tag{9}$$

or, generally:

$$\mathbf{c} = \mathbf{pA}. \tag{10}$$

where $\mathbf{c}$ is the colorimetric output vector, $\mathbf{p}$ is the Q-component vector of polynomial terms derived from the device data $\mathbf{d}$, and $\mathbf{A}$ is the $Q \times n$ matrix of polynomial weights. The optimal matrix of polynomial weights, $\mathbf{A}$, is then given by:

$$\mathbf{A} = (\mathbf{P}^t\mathbf{P})^{-1}\mathbf{P}^t\mathbf{C} = (\mathbf{P})^{-}\mathbf{C}. \tag{11}$$

The drawback with using regression to CIEXYZ is that the RMS error in XYZ color space, which is minimized in the regression, is not closely related to the perceived color difference. If the final aim is to derive data in CIELAB color space, it is therefore preferable to use regression directly in the CIELAB domain, i.e. to minimize the CIE 1976 color difference ΔE_{ab}, which provides a better correspondence to the visual color difference [3]. Since the relationship between device data and CIELAB is not linear, a non-linear pre-processing step of the device values using a cubic root function has been proposed, i.e. using $R^{1/3}$, $G^{1/3}$, $B^{1/3}$ in the regression [3]. The cubic root function originates from the CIELAB transformation, which involves a cubic root function of the XYZ tristimulus values.

4 The Image Acquisition System

The images are captured using a monochrome CCD camera with 12 bit dynamic range, specially designed for scientific imaging. The illumination is provided using a tungsten halogen lamp through optical fibers, which offers an adjustable angle of incidence, as well as the possibility of using a backlight setup. Color images are sequentially captured, using filters mounted in a filter wheel in front of the light source. By using this color sequential method, there is no need for any interpolation or de-mosaicing scheme, as is the case for conventional digital cameras. Besides the trichromatic RGB-filters, the filter wheel also contains a set of 7 interference filters, allowing for the acquisition of multi-channel images. The interference filters have been selected to cover the visible spectrum with equally spaced pass bands, see Fig. 1.

Since the accuracy of the characterization will always be limited by the stability and uniformity of a given device, the characterization procedure has been preceded by a thorough calibration of the system. All the components have been controlled with respect to linearity, temporal stability and spatial uniformity.

5 Experimental Setup

The evaluation of the spectral and colorimetric reconstructions requires for the acquisition of spectral and colorimetric data for a set of test colors, along with the corresponding device response. Spectral measurements of the color-patches are performed using a spectroradiometer, placed in the same optical axis as the CCD-camera, using the 45°/0° measurement geometry. For each color patch, the mean reflectance spectrum from 5 sequential measurements is computed. The colorimetric data have been computed using standard formulae under the D65 standard illuminant. Correspondingly, the camera response values have been acquired under identical conditions. Before the mean values are computed, the images are corrected for dark current and CCD gain.

For reference colors to evaluate the results of the model-based spectral reconstruction, 25 color patches from NCS are used. For the empirical characterization, a training set of 50 printed test colors are used to derive the characterization functions. For the evaluation, 50 independent colors are used, printed using the same substrate and conditions as the training set. Since characterization functions derived by least squares regression will always be optimized for the specific training set, it is important to use an independent set of evaluation colors to guard against a model that overfits the training set, giving unrealistically good results [10].

As basis functions we evaluate spectral basis, using a database of real spectra available from NCS, as well as Fourier basis. Five basis functions are used, corresponding to the first five Fourier basis functions and to the five singular vectors corresponding to the most significant singular values in the spectral autocorrelation function of the spectral database, using the principle eigenvector method [3].

6 Experimental Results

6.1 Spectral Reconstruction

Spectral data has been reconstructed from the recorded device response, using the pseudo-inverse (PI) method, as well as using spectral and Fourier basis, for the model-based and empirical approaches, respectively. The results are evaluated using the spectral RMS error, corresponding to the Euclidian distance in spectral reflectance space, between the original and the reconstructed spectra. The CIE 1976 color difference ΔE_{ab} is also computed, to provide a measure of the perceived color difference between the spectra.

Table 1 lists the mean and maximum reconstruction errors, for the different characterization methods. Examples of reconstructed spectra, compared to the corresponding measured spectra, are displayed in Fig 2.

The results show that for the model-based approach, trichromatic imaging is not sufficient to achieve spectral or colorimetric accuracy. For the multi-channel images, the results improve dramatically. Spectral basis and Fourier basis lead to equivalent results in terms of the RMS difference, while the colorimetric results are in favor of the spectral basis. The pseudo-inverse solution is somewhat noisy and suffers from larger RMS difference. However, the general shapes of the reconstructed spectra follow the real spectra well, resulting small colorimetric errors. Clearly, the PI-method produces spectral reconstructions that are close to metameric matches.

For the empirical characterization using trichromatic imaging, the pseudo-inverse method is superior to the corresponding model-based results. However, the improvement when applying the different basis functions is not as evident, and the best results for the model-based approach could not be achieved. The results using multi-channel imaging is comparable to the corresponding model-based approach in terms of spectral RMS difference, but produces larger colorimetric errors.

Table 1. Spectral reconstruction errors, in terms of spectral RMS error and ΔE_{ab}

			RMS		**ΔE_{ab}**	
	Data	*Method*	*Max*	*Mean*	*Max*	*Mean*
Model-based	RGB	PI	0.0706	0.0230	24.35	14.80
		Spectral	**0.0041**	**0.0014**	**15.87**	**4.170**
		Fourier	0.0155	0.0049	18.75	8.287
	Multi	PI	0.0092	0.0030	4.364	**1.529**
		Spectral	**0.0039**	0.0012	**4.218**	1.816
		Fourier	0.0040	**0.0011**	7.271	2.112
Empirical	RGB	PI	0.0082	**0.0023**	13.22	7.532
		Spectral	**0.0062**	0.0031	**12.49**	7.444
		Fourier	0.0072	0.0035	13.81	**6.897**
	Multi	PI	**0.0030**	**0.0004**	**6.899**	**3.908**
		Spectral	0.0040	0.0018	9.320	5.525
		Fourier	0.0052	0.0023	13.90	6.085

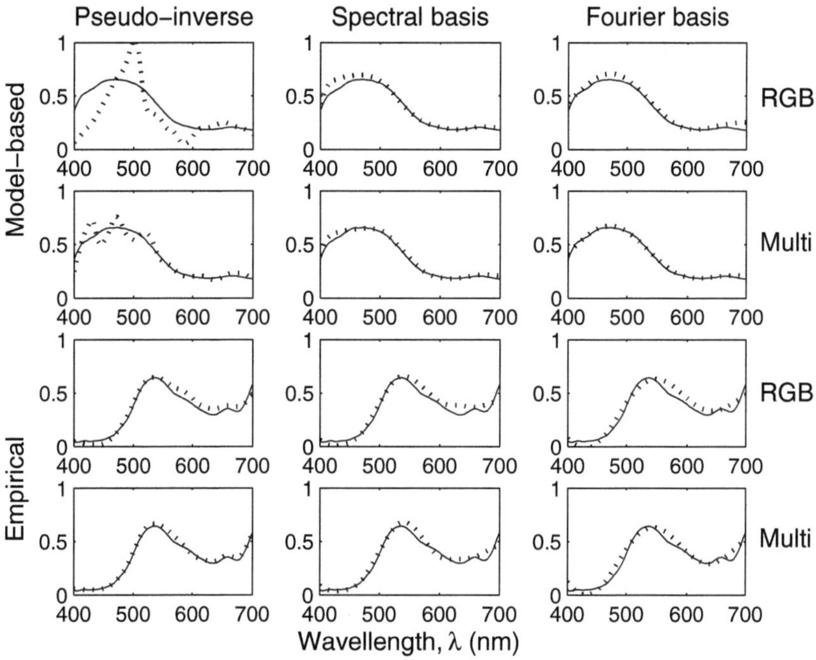

Fig. 2. Reconstructed spectral reflectance (*dashed lines*) compared to measured (*full lines*). Results for the model-based characterization using RGB (*first row*) and multi-channel imaging (*second row*), and the empirical characterization using RGB (*third row*) and multi-channel imaging (*last row*).

6.2 Colorimetric Regression

For the polynomial regression, there are numerous ways to build the approximation functions, **p**, and the number of terms, *Q*, increases rapidly for higher order polynomials. Among the different polynomials evaluated [11], the polynomials found to give the best results, for RGB and multi-channel imaging, respectively, were:

$$\mathbf{p}_{RGB} = [1, R, G, B, R^2, RG, RB, G^2, GB, B^2, R^3, R^2G, R^2B, RG^2, RGB, RB^2, G^3, G^2B, GB^2, B^3] \quad (12)$$

$$\mathbf{p}_{multi} = [1, M_1, M_2, M_3, M_4, M_5, M_6, M_7, M_1^2, M_2^2, M_3^2, M_4^2, M_5^2, M_6^2, M_7^2, M_1M_2M_3M_4M_5M_6M_7] \quad (13)$$

Table 2 lists the results for the colorimetric regression to CIEXYZ and CIELAB, using the polynomials according to Eqs. 12 & 13. For the regression directly to CIELAB, the non-linear pre-processing step has been used, which proved to be superior to regression to CIELAB using the unprocessed device response. To investigate the media-dependency, cross-media characterization was carried out, using the characterization functions derived for the printed training set to reconstruct colorimetric data for the NCS color patches.

The results show that colorimetric regression directly to CIEXYZ and CIELAB gives a good colorimetric accuracy. Noticeable is that the results from the trichromatic RGB images are comparable to the multi-channel results. However, for cross-media characterization, the reconstruction errors increase dramatically, illustrating the strong media dependency of the method.

Table 2. The results for the colorimetric regression to CIEXYZ and CIELAB

			ΔXYZ		ΔE_{ab}	
	Data	*Regression*	*Max*	*Mean*	*Max*	*Mean*
	RGB	CIEXYZ	3.453	0.904	4.558	2.086
		CIELAB			4.317	1.722
	Multi	CIEXYZ	3.240	0.945	3.765	1.942
		CIELAB			3.846	1.957
Cross-media	RGB	CIEXYZ	25.55	12.96	16.67	9.494
		CIELAB			18.29	8.417
	Multi	CIEXYZ	13.71	5.366	26.00	9.781
		CIELAB			18.76	8.877

7 Summary

The focus of this study has been colorimetric and multispectral image acquisition, using both trichromatic and multi-channel imaging. To reconstruct colorimetric and spectral data from the recorded device response, two conceptually different approaches have been investigated: model-based and empirical characterization. In the model-based approach, the spectral model of the image acquisition system is inverted. A priori knowledge on the smooth nature of spectral reflectance was utilized by representing the reconstructed spectra as linear combinations of basis functions, using Fourier basis and a set of real reflectance spectra. In the empirical approach, the spectral characteristics of the system are ignored and the mappings are derived by relating the recorded device response to colorimetric and spectral data for a set of training colors, using least squares regression techniques.

The results have showed that when only trichromatic imaging is available, the best method for colorimetric imaging is the empirical approach, using polynomial regression. However, because of the media-dependency, this requires for the characterization functions to be derived for each combination of media and colorants. For multispectral imaging, reconstructing the spectral reflectance of objects, multi-channel images are required to obtain the highest accuracy. The best results were obtained with the model-based approach, using multi-channel images combined with spectral basis. The model-based approach provides the additional advantage of being general, since it is derived based on the spectral characteristics of the image acquisition system, rather than on the characteristics of a set of color samples. However, the model-based approach requires for multi-channel imaging to obtain a satisfactory spectral or colorimetric accuracy.

References

1. Nyström, D., Kruse, B.: Colorimetric Device Characterization for Accurate Color Image Acquisition. In: Advances in Printing and Media Technology, vol. 33 (2006)
2. Farell, J.E., et al.: Estimating Spectral Reflectances of Digital Artwork. In: Proc. Chiba Conference of Multispectral Imaging (1999)
3. Hardeberg, J.Y.: Acquisition and Reproduction of Color Images: Colorimetric and Multispectral Approaches, Dissertation.com, ISBN 1-58112-135-0 (2001)
4. Maloney, L.T.: Evaluation of linear models of surface spectral reflectance with small numbers of parameters. In: J. Opt. Soc. Am. A, 3 (1986)
5. Connah, D. et al.: Recovering spectral information using digital camera systems. Coloration technology 117, 309–312 (2001)
6. Pan, Z., et al.: Color Scanner Characterization with Scan Targets of Different Media Types and Printing Mechanisms. In: Proc. Color Imaging: Device-Independent Color, Color Hardcopy, and Graphic Arts VI, pp. 58–63 (2001)
7. Andersson, M., et al.: The Substrate influence on color measurement, In: Proc. IS&T NIP 19, New Orleans (2003)
8. Solli, M., et al.: Color Measurements with a Consumer Digital Camera Using Spectral Estimation Techniques. In: Proc. Scandinavian Conference on Image Analysis, pp. 105–114 (2005)
9. Hong, G., Lou, M.R.: A Study of Digital Camera Colorimetric Characterization Based on Polynomial Modeling. In: COLOR research and application, vol. 26(1) (2001)
10. Cheung, V. et al.: Characterization of trichromatic color cameras by using a new multispectral imaging technique. J. Opt. Soc. Am. A, 22(7), 1231–1240 (2005)
11. Nyström, D.: Colorimetric and Multispectral Image Acquisition. Licentiate Thesis No. 1289, Linköping University, Sweden (2006)

Robust Pseudo-hierarchical Support Vector Clustering

Michael Sass Hansen[1], Karl Sjöstrand[1], Hildur Ólafsdóttir[1], Henrik B.W. Larsson[2], Mikkel B. Stegmann[3], and Rasmus Larsen[1]

[1] Informatics and Mathematical Modelling, Technical University of Denmark, Lyngby, Denmark
[2] Hospital of Glostrup, Glostrup, Denmark
[3] 3shape A/S, Copenhagen, Denmark

Abstract. Support vector clustering (SVC) has proven an efficient algorithm for clustering of noisy and high-dimensional data sets, with applications within many fields of research. An inherent problem, however, has been setting the parameters of the SVC algorithm. Using the recent emergence of a method for calculating the entire regularization path of the support vector domain description, we propose a fast method for robust pseudo-hierarchical support vector clustering (HSVC). The method is demonstrated to work well on generated data, as well as for detecting ischemic segments from multidimensional myocardial perfusion magnetic resonance imaging data, giving robust results while drastically reducing the need for parameter estimation.

1 Introduction

Support Vector Clustering (SVC) was introduced by Ben-Hur et al. [1]. SVC uses the one-class Support Vector Domain Description (SVDD) as the basis of the clustering algorithm. SVDD was introduced by Tax and Duin [2] in 1999, and it is often calculated with a Gaussian kernel replacing the Euclidian inner product. The SVDD description maps the points into a high dimensional feature space dividing inliers from outliers, where the decision boundary consists of contours enclosing clusters of the data points.

The clustering is done with no assumption on the number of clusters or the shape of the clusters. Ben-Hur et. al. proposed to vary the parameters of the SVDD in a manner that increases the number of clusters while keeping the number of outliers and bounded support vectors (BSV) low. Strictly hierarchical support vector clustering was presented by Ben-Hur in [3]. This algorithm applies SVC subsequently on subsets of the data contained in clusters, and thus achieves a hierarchy of clusters. The clustering, however, depends on the initial steps of the division process.

Yang et al. have proposed improvements to the cluster labelling using proximity graph modelling [4], similar to that of the presented method.

Recently Sjöstrand and Larsen showed that the entire regularization path of the SVDD can be calculated efficiently [5]. This result is the backbone of the

B.K. Ersbøll and K.S. Pedersen (Eds.): SCIA 2007, LNCS 4522, pp. 808–817, 2007.

presented method, and allows for a robust pseudo-hierarchical support vector clustering (HSVC). Given a scale parameter of the Gaussian kernel, a clustering can be estimated efficiently for all values of the regularization parameter. From this ensemble of clusterings a more robust clustering estimate is calculated. To validate the method, the clustering was tested on both artificially generated data, and a real work example of a high dimensional clustering problem.

2 Methods

As other SVC algorithms the basis of the current algorithm is the one-class support vector classification. The recently emerged method for an efficient calculation of the entire regularization path of the SSVD is described briefly for completeness. It is shown that between events the discrimination function varies monotonically, and it is concluded that the description is complete.

2.1 Support Vector Domain Description

The support vector domain description was presented by Tax and Duin[2], posing it as a quadratic optimization problem for a fixed value of the regularization parameter. The criterion to be maximized, given a point set x_i, can be formulated as

$$\min_{R^2,\ \mathbf{a},\ \xi_i} \sum_i \xi_i + \lambda R^2 \ , \qquad \text{Subject to}$$
$$(\mathbf{x}_i - \mathbf{a})(\mathbf{x}_i - \mathbf{a})^T \leq R^2 + \xi_i \quad \text{and} \quad \xi_i \geq 0 \ \forall\, i,$$

where the general idea is to find the minimal sphere that encapsulates the points, allowing some points to be outside the sphere. The regularization parameter λ penalizes the radius R^2 and for large values of λ the radius will tend to be smaller and vice versa. Some points, the outliers, are allowed to be outside the sphere, and the number of outliers is governed by the regularization parameter λ.

Using Lagrange multipliers this optimization problem can be restated as

$$\max_{\alpha_i} \sum_i \alpha_i \mathbf{x}_i \mathbf{x}_i^T - \frac{1}{\lambda} \sum_i \sum_j \alpha_i \alpha_j \mathbf{x}_i \mathbf{x}_j^T,$$
$$0 \leq \alpha_i \leq 1, \ \sum_i \alpha_i = \lambda, \tag{1}$$

where α_i are the Lagrange mulitpliers and as a consequence of the Karush-Kuhn-Tucker complimentary conditions is that for inliers $\alpha_i = 0$ and for outliers $\alpha_i = 1$. The dimensionality can be increased using a basis expansion and substituting the dot-product with an inner product, the inner products can be replaced by $K_{i,j} = K(\mathbf{x}_i, \mathbf{x}_j)$, where K is some suitable kernel function. In the presented work the Gaussian kernel $K(\mathbf{x}_i, \mathbf{x}_j) = e^{-\frac{\|\mathbf{x}_i - \mathbf{x}_j\|^2}{\gamma}}$ was used as a kernel function. The optimization problem is then given by

$$\max_{\alpha_i} \sum_i \alpha_i K_{i,i} - \frac{1}{\lambda} \sum_o \sum_j \alpha_i \alpha_j K_{i,j}$$
$$0 \leq \alpha_i \leq 1, \ \sum_i \alpha_i = \lambda. \tag{2}$$

For a given λ the squared distance from the center of the sphere to a point $\mathbf{x}$ is

$$f(\mathbf{x}; \lambda) = \|h(\mathbf{x}) - \mathbf{a}\|^2 = K(\mathbf{x}, \mathbf{x})$$
$$-\frac{2}{\lambda} \sum_i \alpha_i K(\mathbf{x}, \mathbf{x}_i) + \frac{1}{\lambda^2} \sum_i \sum_j \alpha_i \alpha_j K_{i,j} \tag{3}$$

The entire regularization path of the SVDD. Sjöstrand and Larsen have shown that the entire regularization path of the parameter λ can be calculated with approximately the same complexity as required for solving the initial optimization problem, posed by Tax and Duin [5]. This is because the regularization path of the parameters α_i is piecewise linear. This can be realized by examining the distance functions of two points on the boundary.

$$f(\mathbf{x}_h; \lambda) = f(\mathbf{x}_k; \lambda), \quad h, k \in B \tag{4}$$

where B is the set of points on the boundary. Formulating this equation for different points on the boundary and using the constraint of the sum of α_i gives a complete set of equations for estimating all the α_i. Let $\boldsymbol{\alpha}$ be a vector with the values α_i and let $\boldsymbol{p}$ and $\boldsymbol{q}$ be the slope and intersection respectively, then (refer to [5] for a detailed derivation)

$$\boldsymbol{\alpha} = \lambda \boldsymbol{p} + \boldsymbol{q}, \tag{5}$$

where $\boldsymbol{p}$ and $\boldsymbol{q}$ are constant on intervals $[\lambda_l; \lambda_{l+1}[$, which are defined as intervals between events where a point either leaves or joins the boundary. The division in inliers and outliers is illustrated in Figure 1.

2.2 Support Vector Clustering

The SVDD yields an explicit expression for the distance given by Eq. (3). Now R can be calculated by

$$R = f(\mathbf{x}_k; \lambda) = K_{k,k} - \frac{2}{\lambda} \sum_i \alpha_i K_{k,i} + \frac{1}{\lambda^2} \sum_i \sum_j \alpha_i \alpha_j K_{i,j}.$$

Consider an arbitrary point $\boldsymbol{x}$, and define the distance function $g(\boldsymbol{x}, \lambda)$, as the distance to the boundary.

$$g(\boldsymbol{x}, \lambda) = f(\boldsymbol{x}, \lambda) - R\ . \tag{6}$$

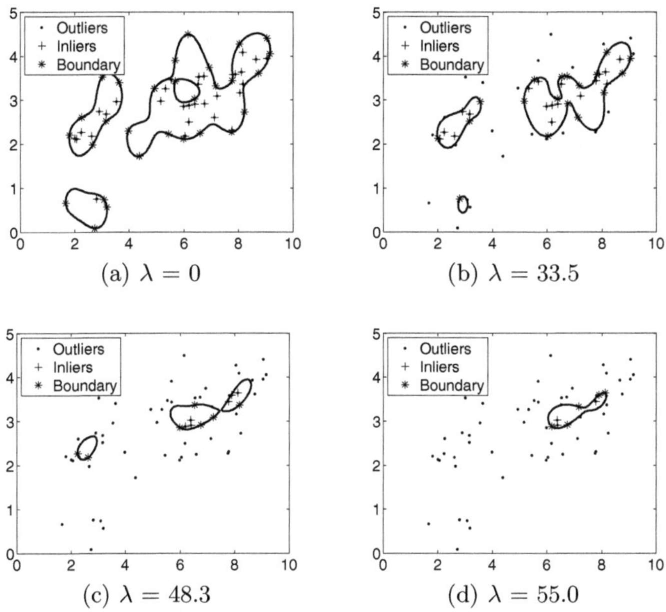

Fig. 1. SVDD calculated for the entire regularization path. The line marks the boundary between inliers and outliers, the generalized circle.

The function g is the decision criteria determining if a point is an inlier or an outlier. In Figure 1 the discriminating function g is calculated to create the contour dividing inliers from outliers. Though the optimization problem is to find a circle in the space of the expanded basis, the result appears very little like a circle in the input-space, which in this case has two dimensions. The different enclosed areas could be considered as clusters, denoted support vector clusters.

Assigning clusters. While evaluating $g(\boldsymbol{x}, \lambda)$ reveals if $\boldsymbol{x}$ is an inlier or outlier, it does not contain any specific information on the assignment of clusters. Inspired from Figure 1 it is observed that all paths connecting two points in two different clusters have some points outside the clusters. The current implementation uses an adjacency matrix to determine which points are connected, and which are not. The connection graph is sparsely built, similar to the approach chosen by Yang et. al. [4].

$$A_{ij} = \left\{ \begin{matrix} 1\,, & \text{if } g(\mathbf{x}_i + \mu(\mathbf{x}_j - \mathbf{x}_i)) < 0 \;\; \forall \;\; \mu \in [0;1] \\ 0\,, & \text{else} \end{matrix} \right\}, \tag{7}$$

Connected clusters are detected from the adjacency matrix by using standard graph theory concepts. Outliers are by definition not adjacent to any points, but are assigned to the closest detected cluster.

2.3 Regularized SVC Based on the Entire Regularization Path

Given a λ the clustering can be determined from the adjacency matrix (7), but λ on the interval [0;n] gives rise to changes in the distance function, and thus potentially the clustering. In Section 2.3 it is shown that the distance function (3) is monotonic in the interval $[\lambda_l; \lambda_{l+1}[$ between two events, which means that an almost complete description is obtained by detecting the clusters in the points of the events.

Completeness of the hierarchical description. To ensure that the complete description of the clustering path has been obtained, the distance function is analyzed as a function of the regularization parameter λ.

$$\begin{aligned} g(\boldsymbol{x}, \lambda) &= f(\boldsymbol{x}, \lambda) - R = f(\boldsymbol{x}, \lambda) - f(\boldsymbol{x}_k, \lambda) \ , \ k \in B, \\ &= K(\boldsymbol{x}, \boldsymbol{x}) - K_{k,k} - \frac{2}{\lambda} \sum_i \alpha_i (K(\boldsymbol{x}, \boldsymbol{x}_i) - K_{k,i}). \end{aligned} \tag{8}$$

Equation (5) states the linear relation between α and λ is given by $\boldsymbol{\alpha} = \lambda \boldsymbol{p} + \boldsymbol{q}$. Let each Lagrange multiplier be given by $\alpha_i = \lambda p_i + q_i$, and the derivative $\frac{\delta g}{\delta \lambda}$ can be calculated as

$$\begin{aligned} \frac{\delta g}{\delta \lambda} &= \frac{\delta}{\delta \lambda} \left[-2 \sum_i (p_i + \frac{q_i}{\lambda})(K(\boldsymbol{x}, \boldsymbol{x}_i) - K_{k,i}) \right] \\ &= \frac{2}{\lambda^2} \sum_i q_i (K(\boldsymbol{x}, \boldsymbol{x}_i) - K_{k,i}), \qquad \lambda \in]\lambda_l; \lambda_{l+1}[\ . \end{aligned} \tag{9}$$

The only dependence on λ in Eq. (9) is on a (inverse squared) multiplicative term. From this, it is concluded that $g(\boldsymbol{x}, \lambda)$ can only change sign once on the interval $[\lambda_l; \lambda_{l+1}]$, so all changes in the clustering are observed in the clustering calculated at every event.

2.4 Pseudo-hierarchical Support Vector Clustering

The calculated clusters are often only changing slowly with changes in the regularization parameter λ. When an event consists of a point leaving the boundary to become an outlier, this does not necessarily alter the boundary much elsewhere. Since the point is still close to the same cluster, and may be associated with this, still, many clusters are close to identical. The similarity can be observed in Figure 1. Moreover, the same clusters may appear again for a different value of the regularization parameter.

The idea presented in this paper, is to collect all the similar clusterings, and build a hierarchy of clusters, which can be thought of as being composed of other, and obviously bigger, clusters. The toy example illustrated in Figure 1 only has a few clusterings that are actually different, and there is a strong relation between the different clusterings of the data, which is illustrated in Figure 2.

There is, however, no guarantee that different clusters, calculated for different values of the regularization parameter, are nested in a strict hierarchical way. In

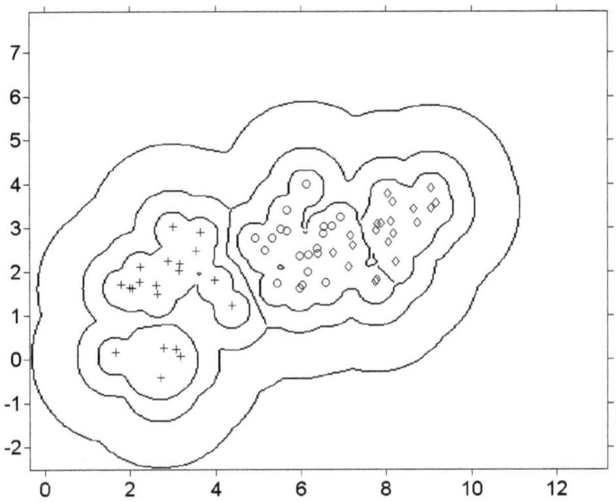

Fig. 2. Hierarchical clustering: From coarse to detailed description

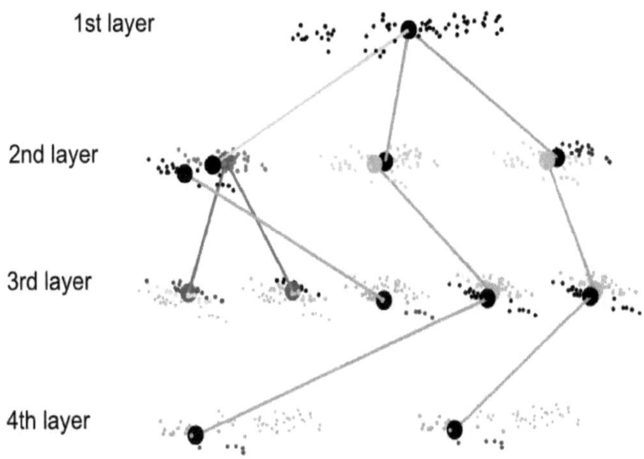

Fig. 3. Competing hierarchical clusterings, obtained from the entire regularization path of the SVDD. The lines show how a cluster is split into smaller clusters. The light gray pixels are the ones not included to describe the subclustering of the cluster. The gray and colored points form together the whole reference data set illustrated in Figure 1.

fact multiple different hierarchical clustering may be proposed. This is illustrated in Figure 3. Each branch of these different cluster representations demonstrate two or more ways, the cluster could be split in smaller clusters. For each cluster, it is known for which intervals of the regularization parameter, the cluster is present. Also it is possible to record if the points forming the cluster are inliers or outliers.

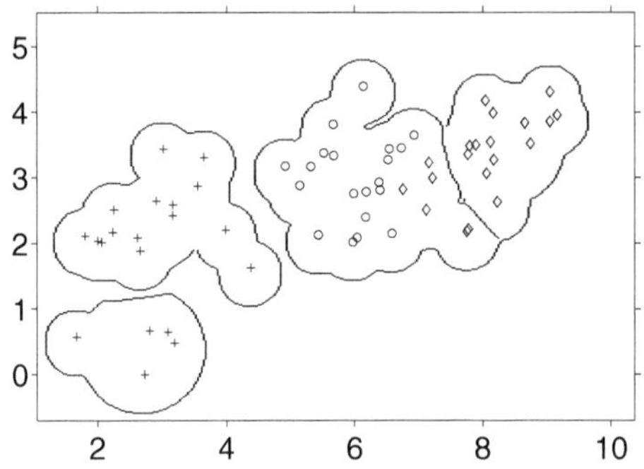

Fig. 4. Clustering of the reference data set by HSVC using the cluster discrimination feature, based on a generalized within covariance matrix

Quality measure of competing clusterings. The analysis described in the previous sections results in a number of competing cluster representations of the data. This analysis, however, does not directly indicate which clustering is the preferred one. We propose a scheme similar to using the 'within' and the 'between' covariance matrices, trace($S_W^{-1}S_B$). Instead of S_W we argue that a weighted within matrix S_W^* should be calculated, weighted by the length of the interval where a given point is an inlier, or an outlier associated with the cluster.

$$S_W^* = \sum_{j=1}^{n_{\text{clusters}}} \sum_{i \in \mathbf{C}_j} \frac{1}{\Pi_j} (\mathbf{x}_i - \mu_i)^T \Lambda_{j,i} (\mathbf{x}_i - \mu_i) , \tag{10}$$

where Λ_j defines the weighting of the point, which depends linearly on the length of the interval of λ where the point is an inlier and where it is an outlier. Π_j is a normalization constant. A potential clustering can now be assessed using the measure trace(${S_W^*}^{-1}S_B$), which evaluates the variance within clusters, compared to the introduced distance between clusters. In Figure 4 this is done for the same generated data that was used in Figures 1 and 2. The reference data is actually generated from three random independent distributions, generated as mixtures of Gaussian and uniform distributions. The three different sets are marked by the symbols '+', 'o' and '◊' respectively. It can be observed that the clusters 'o' and '◊' overlap to some extent, whereas '+' seems more separated from the other groups, and is split in two parts. In Figure 1 small values of λ, corresponding to a high confidence in the data, results in a separation of the two parts of the '+' cluster, whereas the other groups are merged into one cluster. This is opposite for high values of the regularization parameter, where the smaller clusters only appear to be outliers, but the two overlapping clusters are divided. The discrimination feature removes the need to select one value of λ, and appears to adapt to clusters of different variance. The criterion for accepting a subclustering is introduced as a threshold on the cluster separation, given by trace(${S_W^*}^{-1}S_B$). The lower the threshold, the more clusters are accepted.

2.5 Complexity

The complexity of the algorithm is vastly reduced by calculating the entire regularization path of the SVDD in an efficient sequential way, as described. The complexity for the referenced algorithm is $O(n^2)$ for each step between two events. For each event, the clusters are detected from the adjacency matrix, which can also be calculated with a complexity of the order of $O(n^2)$. Comparing with other clusters is done with complexity $O(n \cdot n_{clusters})$. Since the number of events is typically in the vicinity of 3-5 n the overall complexity is polynomial with a degree around 3. On the tested example, with about 500 points in 50 dimensions the algorithm took minutes.

3 Example Application: Detection of Ischemic Segments

To test the capability of the presented clustering algorithm, it has been applied to detect ischemic segments from perfusion MR images. In Figure 5 selected frames from a registered sequence of perfusion MR images of the myocardium are shown. The segmentation was performed previously, with satisfying results [6]. Intensity curves can be obtained pixel-wise from the intensity images, because of the pixel-wise correspondence. Previously ischemic segments have usually been detected using the measures *time-to-peak*, *maximum-upslope* and *peak value* [7]. In a previous study we showed that a generalized version of the distances obtained in the SVDD description corresponded well to the usual measures [8]. In Figure 6(a) the measures are illustrated. The developed HSVC method was applied on the data, which consisted of little less than 500 pixels, and 50 time steps were available for the intensity curve. HSVC divided data in a very few clusters, and in Figure6(b) the curves belonging to each cluster is colored in distinct colors.

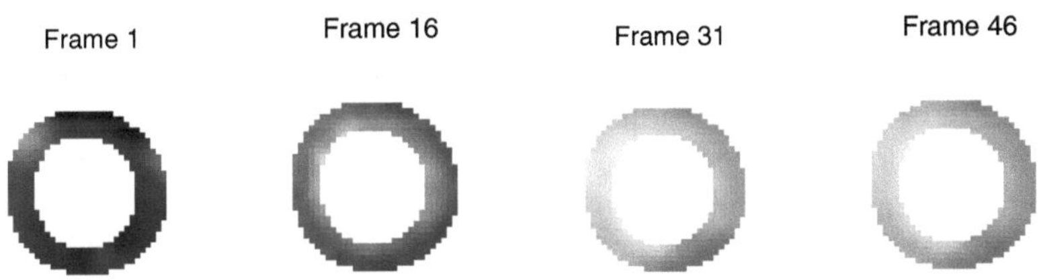

Fig. 5. Different registered frames of one of the slices of the perfusion MR images

The perfusion measures were calculated previously, and they are illustrated in Figure 7(a-c).

The correspondence between the areas is good, and the clustering is seen to provide a very good base for a simple cluster classification. All noise is suppressed by the HSVC, so the cluster covers a connected region in the image. It is worth noting that the only parameter which has been changed in this example instead of the previous example is the width of the Gaussian kernel. So using the statistical term trace$({S_W^*}^{-1} S_B)$ as cluster separation measure helps to reduce the dimensionality of the estimation problem.

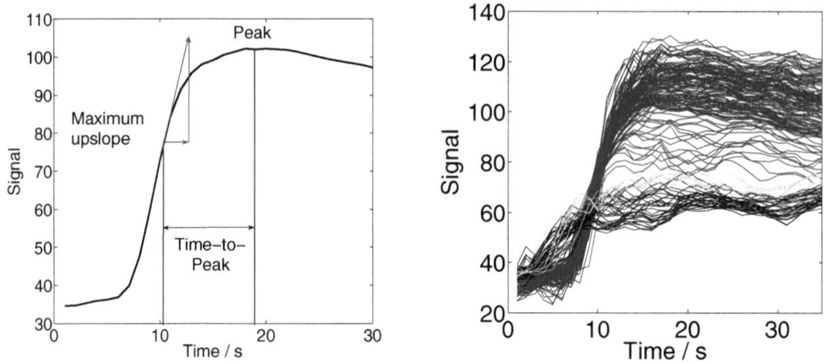

(a) Parameter description. (b) Intensity plots.

Fig. 6. Pixel-wise intensity plots. (a) Idealized plot, describing the perfusion parameters (b) Intensity curves for the 3 detected clusters, colors correspond to clusters.

(a) Max. upslope. (b) Peak values. (c) Time-to-peak. (d) Clustering.

Fig. 7. Ischemic segment detection with standard measures compared to clustering

4 Conclusion

The proposed robust pseudo-hierarchical support vector clustering (HSVC) is demonstrated to give good results on both a random data set and in real application, and this with the same parameters though the two data sets are very different in range, n and dimensionality.

The proposed clustering algorithm has only one parameter, which is the threshold for splitting clusters, and this parameter correlates strongly with the number of clusters (and their quality in terms of separation). We therefore believe that HSVC can be a very useful tool in many applications where it is possible to define a kernel.

References

1. Ben-Hur, A., Horn, D., Siegelmann, H.T., Vapnik, V.: Support vector clustering. Journal of Machine Learning 2, 125–137 (2001)
2. Tax, D.M.J., Duin, R.P.W.: Support vector domain description. Pattern Recognition Letters 20, 1191–1199 (1999)

3. Ben-Hur, A., Horn, D., Siegelmann, H.T., Vapnik, V.: A support vector clustering method. In: Proceedings of conference on Advances in Neural Information Processing Systems (2001)
4. Yang, J., Estivill-Castro, V., Chalup, S.K.: Support vector clustering through proximity graph modelling (2002)
5. Sjöstrand, K., Larsen, R.: The entire regularization path for the support vector domain description. In: Larsen, R., Nielsen, M., Sporring, J. (eds.) MICCAI 2006. LNCS, vol. 4190, Springer, Heidelberg (2006)
6. Ólafsdóttir, H., Stegmann, M.B., Larsson, H.B.: Automatic assessment of cardiac perfusion MRI. In: Barillot, C., Haynor, D.R., Hellier, P. (eds.) MICCAI 2004. LNCS, vol. 3217, pp. 1060–1061. Springer, Heidelberg (2004)
7. Stegmann, M.B., Ólafsdóttir, H., Larsson, H.B.W.: Unsupervised motion-compensation of multi-slice cardiac perfusion mri. Medical Image Analysis 9(4), 394–410 (2005)
8. Hansen, M.S., Ólafsdóttir, H., Sjöstrand, K., Erbou, S.G., Larsson, H.B., Stegmann, M.B., Larsen, R.: Ischemic segment detection using the support vector domain description. In for Optical Engineering (SPIE), T.I.S., ed.: International Symposium on Medical Imaging. (February 2007)

Using Importance Sampling for Bayesian Feature Space Filtering

Anders Brun[1,4], Björn Svensson[1,4], Carl-Fredrik Westin[2], Magnus Herberthson[3], Andreas Wrangsjö[1,4], and Hans Knutsson[1,4]

[1] Department of Biomedical Engineering, Linköping University, Sweden
andbr@imt.liu.se
[2] Department of Mathematics, Linköping University, Sweden
[3] Laboratory of Mathematics in Imaging, Harvard Medical School, Boston, USA
[4] Center for Medical Image Science and Visualization, Linköping University, Sweden

Abstract. We present a one-pass framework for filtering vector-valued images and unordered sets of data points in an N-dimensional feature space. It is based on a local Bayesian framework, previously developed for scalar images, where estimates are computed using expectation values and histograms. In this paper we extended this framework to handle N-dimensional data. To avoid the curse of dimensionality, it uses importance sampling instead of histograms to represent probability density functions. In this novel computational framework we are able to efficiently filter both vector-valued images and data, similar to e.g. the well-known bilateral, median and mean shift filters.

1 Introduction

In this paper we present a method for filtering of vector-valued images, $x(q) \in \mathbb{V} = \mathbb{R}^n$, where $\mathbb{V}$ is a feature vector space such as the RGB color space. For the purposes of this paper, q is a point in a spatial vector space, $q \in \mathbb{U} = \mathbb{R}^m$, e.g. $q \in \mathbb{R}^2$ for images. It is however easy to extend this filtering to a curved m-dimensional manifold, $q \in M$. We also show how a slight modification can generalize this method to be used for filtering unordered sets of data points in a feature space, $\{x_i\} \in \mathbb{V} = \mathbb{R}^n$.

The proposed method is inspired by previous work by Wrangsjö et al. [17], a local Bayesian framework for image denoising of scalar-valued images. That method was based on a computational framework involving histograms, which made it slow and nearly impossible to use for vector-valued images. In this paper we propose the novel use of a Monte Carlo method called *importance sampling* to overcome this difficulty. It makes this particular kind of Bayesian filtering feasible for vector-valued images and data.

2 Previous Work

In [17] the proposed filter is related to bilateral filters [6,10,15,16]. Other filters operating on local neighborhoods in images with similar characteristics include mean shift-filtering [4], median filters[2], total variation filters [14], diffusion based noise reduction [3,13] and steerable filters [5,9]. Several of these filters are compared in [12].

B.K. Ersbøll and K.S. Pedersen (Eds.): SCIA 2007, LNCS 4522, pp. 818–827, 2007.

3 The Bayesian Method

The method is founded on Bayesian theory and for this reason the *a posteriori* probability distribution function, $p_{S|X=x}(s)$, is important. If we let s be the true value and x be the measured value which is corrupted by noise then

$$p_{S|X=x}(s) = \frac{p_{X|S=s}(x)p_S(s)}{p_X(x)}.$$

In order to derive an estimate $\hat{s}$ of the true signal s from the above formula, the conditional expectation value of s may be calculated,

$$\hat{s} = \int_{s \in \mathbb{V}} s\, p_{S|X=x}(s)ds = E[S]|_{X=x} . \tag{1}$$

This is the Minimum Mean Squared Error estimate, which can be calculated if the different probability distributions are modelled appropriately.

3.1 Noise Models

The modelling of noise, how measurements are related to the true signal value, is important. For the general case, the conditional probability $p_{X|S=s}(x)$ need to be known and in many applications this is not a problem. For the special case of additive noise, $X = S + N$, where N can belong to e.g. a Gaussian or super-Gaussian distribution, some simplifications can be made,

$$\begin{aligned} p_{X|S=s}(x) &= \int_{t \in \mathbb{V}} \delta(x - t - s)p_N(t)dt \\ &= p_N(x - s). \end{aligned}$$

For some important special cases, in particular Rician noise which is present in Magnetic Resonance (MR) images, the additive model is however not valid unless the noise is approximated using a Gaussian distribution.

It should also be mentioned that the present method can only handle cases where the measurements can be considered to be independent and identically distributed (i.i.d.). This makes it difficult to handle e.g. speckle noise in ultrasound images efficiently.

3.2 Signal Models for Images

Most of the power of the method proposed in [17] is embedded in the *a priori* p.d.f., $p_S(s)$, which is derived from a local neighborhood around the pixel which is to be estimated. Without knowledge of the exact distribution, a kernel (Parsen window) estimate of $p_X(x)$ is used to model a suitable local prior:

$$p_S(s) = C_0 \left[\sum_i b_v(x_i - s)b_s(q_0 - q_i) \right]^\alpha \tag{2}$$

$$\approx C_0\, p_X(s)^\alpha \tag{3}$$

where $b_v(\cdot)$ is the kernel used to approximate density in $\mathbb{V}$, e.g. a Gaussian, and $b_s(\cdot)$ is a similar spatial weight which is used to favor samples which are close to q_0, the position of the pixel to be estimated. The normalizing constant C has no effect on the estimate, but the exponent $\alpha \geq 1$ make the histogram sharper and a higher value of α promote a harder bias towards the most probable mode in the distribution $p_X(x)$. This local modelling is ad hoc, but has proven to work surprisingly well in practice.

3.3 Signal Models for N-D Data Sets

For unordered data we need to slightly modify this approach. We propose a similar way to model the *a priori* distribution for unordered data, the difference being the lack of a spatial weight.

$$p_S(s) = C_1 \left[\sum_i b_v(x_i - s) \right]^\alpha \tag{4}$$

$$\approx C_2\, p_X(s)^\alpha \tag{5}$$

3.4 Estimation

In the original approach for scalar images, histograms were used to estimate the *a priori* density function. Since the continuous integrals could not be evaluated exactly, all integrations were performed numerically in this way. In this paper we instead propose a solution based on *importance sampling* to calculate Eq. 1 more efficiently.

4 Importance Sampling

In the original approach for scalar-valued images, discretized histograms were used to estimate the *a priori* density function in the numerical calculation of the estimate given by Eq. 1. This turned out to be infeasible for vector-valued images.

It is evident that the integral in Eq. 1 can be evaluated using Monte Carlo, by drawing samples s_i from $p_{S|X=x}(s)$ and calculate the expectation value numerically. This correspond to the upper left illustration in Fig. 1. Sampling from a distribution can however be tricky and we will now introduce the concepts *proper samples* and *importance sampling* which will give us some freedom.

4.1 Proper Samples

We define the following [1,7,8,11]. A set of weighted random samples $\{z_i, w_i\}$, $z_i \in p_Z$, is called proper with respect to a distribution p_X if for any square integrable function $h(\cdot)$,

$$E[w_i h(z_i)] = cE[h(x_i)] \Leftrightarrow$$

$$\int w(y)h(y)p_Z(y)dy = c \int h(y)p_X(y)dy,$$

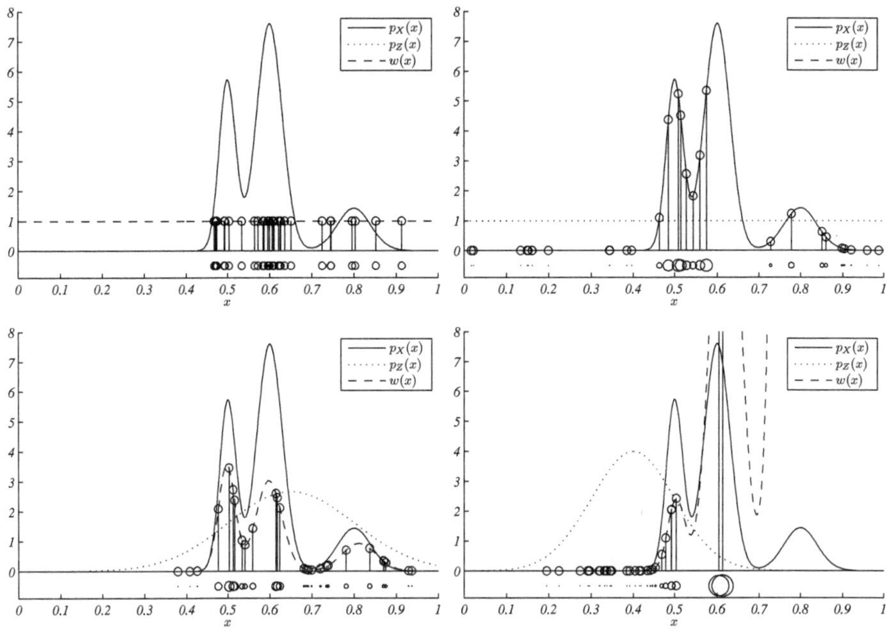

Fig. 1. Examples of sampling. T-L: Sampling from $p_X(x)$. T-R: Sampling from a uniform distribution, weighting with $w_i = p_X(x_i)$. B-L: Sampling using a Gaussian as a trial distribution. B-R: Sampling using a not so suitable Gaussian trial distribution.

for some constant c. Since this should be valid for any $h(\cdot)$, $w(y) = c p_X(y)/p_Z(y)$, and

$$\int c p_X(y) dy = \int w(y) p_Z(y) dy$$
$$c = E[w(z_i)].$$

4.2 Importance Sampling

The notion of proper samples now allow us to numerically calculate the expectation value of a distribution p_X using M samples from a trial distribution p_Z,

$$E[h(x_i)] = \frac{1}{c} E[w_i h(z_i)]$$
$$\approx \frac{1}{\sum_i^M w_i} \sum_i^M w_i h(z_i).$$

This is how expectation values are calculated in importance sampling. It can be used when sampling from p_X is difficult but sampling from p_Z is easy. This is the case if the trial distribution p_Z is e.g. a uniform distribution, a Gaussian or a mixture of Gaussians. For us it means that we can evaluate the integral in Eq. 1 by sampling from another distribution p_Z, if we choose the weights w_i appropriately. For the application at hand, we choose a trial distribution which is similar to the distribution of pixel-values found in the window defined by $b_s(\cdot)$.

In figure 1 some examples of proper sampling are shown. Note in particular that even though evaluation using importance sampling theoretically converge to the correct expectation value when $M \to \infty$, an unsuitable choice of trial distribution may give very slow convergence. Generically, the weight w_i for a sample z_i should be chosen so that $w_i = p_X(z_i)/p_Z(z_i)$. If these weights grow very large, it is an indication that convergence towards the true expectation value will be slow.

5 Implementation

The Bayesian feature space filtering method was implemented in Matlab and tested using various choices of trial functions. Two variants were derived, one for vector-valued images and one for unordered sets of data.

5.1 Vector-Valued Images

The filter was evaluated for each pixel in the image, x_i being the values of the pixels in a neighborhood large enough to fit the spatial weight function $b_s(q)$. In the following, x_0 is the measured value in the pixel to be estimated, located at position q_0. The function $b_v(x)$ is an isotropic Gaussian distribution with zero mean and standard deviation σ_v, corresponding to a kernel in the feature space used in the density estimation. In the spatial domain $b_s(q)$ is an isotropic Gaussian weight function with standard deviation σ_s. The noise of the pixel to be estimated, x_0, is modelled using $p_{X|S=z}(x_0)$, which is also an isotropic Gaussian distribution with standard deviation σ_n. The conditional expectation value of S can now be expressed using the stochastic variable Z which is distributed according to the trial distribution.

$$\begin{aligned}\bar{s} &= E[S]|_{X=x_0} \\ &= \int_{s \in U} s\, p_{S|X=x_0}(s) ds \\ &= E[Z\, w(Z)]/E[w(Z)]\end{aligned}$$

is approximated for a finite number of samples by

$$\hat{s} = \frac{1}{\sum_{i=1}^{M} w(z_i)} \sum_{i=1}^{M} z_i w(z_i).$$

The weight which should be used to guarantee proper samples is

$$\begin{aligned}w(z) &= \frac{p_{S|X=x_0}(z)}{p_Z(z)} \\ &= \frac{p_{X|S=z}(x_0) p_S(z)}{p_Z(z) p_X(x_0)},\end{aligned}$$

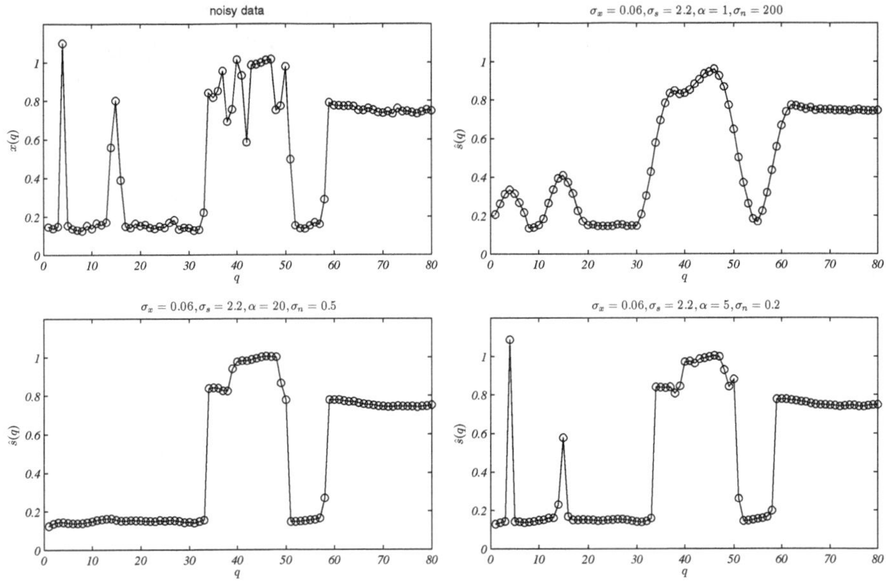

Fig. 2. Filtering a 1-D scalar signal. Parameters are shown in the figure.

where $p_X(x_0)$ is a consequence of Bayes rule in the derivation above, but in practice has no effect on the estimate. The prior $p_S(z)$ is modelled using Eq. 2 and the trial distribution used in the sampling is a mixture of Gaussians,

$$p_Z(z) = \frac{1}{C_3} \sum_i b_v(x_i - z), \tag{6}$$

which is fairly easy to sample from. In general the choice of trial distribution is very important when implementing importance sampling. In our experiments we found that this local estimate of p_X worked well in this particular application. Generically this distribution will contain the same modes and have the same support as the a posteriori distribution we are interested in. Ignoring all constants, the weights can be calculated,

$$w(z) = p_{X|S=z}(x_0) \left[\sum_i b_v(x_i - z) b_s(q_0 - q_i) \right]^{\alpha} / \sum_i b_v(x_i - z).$$

A non-stochastic alternative would have been to use the samples x_i themselves, in the neighborhood of x_0, as samples z_i and use the estimate of p_X in the neighborhood to approximate their probability density function. We implemented this variant and it worked well, but for the experiments on images reported in this paper we have actually used true importance sampling, with a neighborhood of 5×5 pixels and 125 samples z_i from the trial distribution Z in each neighborhood.

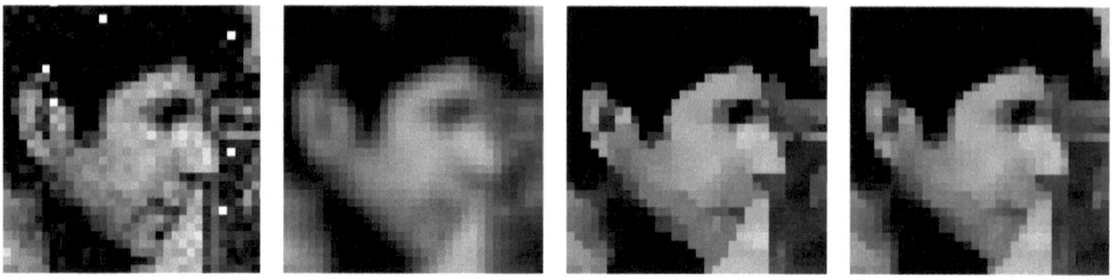

Fig. 3. Filtering a noisy 2-D scalar image with outliers. Left-Right: Noisy data. [$\sigma_v = 0.04$, $\sigma_n = 100$, $\sigma_s = 1.0$, $\alpha = 1$]. [$\sigma_v = 0.04$, $\sigma_n = 0.5$, $\sigma_s = 1.0$, $\alpha = 20$]. [$\sigma_v = 0.04$, $\sigma_n = 0.5$, $\sigma_s = 1.0$, $\alpha = 5$.]

Fig. 4. Filtering a noisy 2-D RGB image. Left-Right: Noisy data. [$\sigma_v = 0.04$, $\sigma_n = 100$, $\sigma_s = 0.8$, $\alpha = 2$]. [$\sigma_v = 0.14$, $\sigma_n = 0.6$, $\sigma_s = 2.0$, $\alpha = 20$]. [$\sigma_v = 0.04$, $\sigma_n = 0.2$, $\sigma_s = 0.8$, $\alpha = 6$].

5.2 Unordered N-D Data

For an unordered set of N-dimensional data, we use the prior defined in Eq. 4, i.e. we regard all elements in $\{x_i\}$ as "neighbors" to the point x_0 to be estimated, and repeat this procedure for each choice of $x_0 \in \{x_i\}$. The trial distribution from Eq. 6 is used and the lack of spatial weighting allow us to simplify the weight function,

$$w(z) = p_{X|S=z}(x_0) \left[\sum_i b_v(x_i - z) \right]^{\alpha - 1} .$$

Observing that the trial distribution used here is essentially the same as the distribution of points in $\{x_i\}$, we use approximated importance sampling in the implementation. This means that instead of sampling from the true trial distribution, we choose $z_i = x_i$.

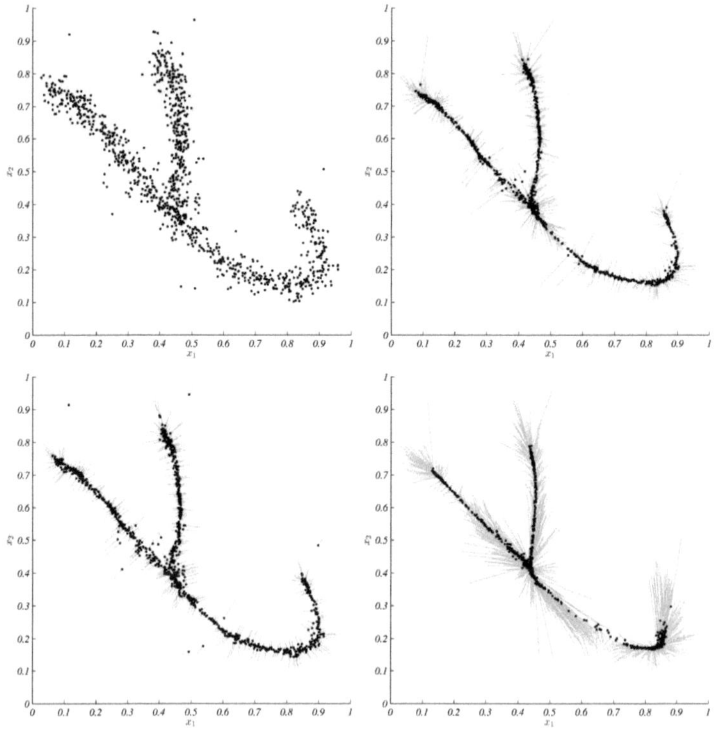

Fig. 5. Filtering unordered 2-D. The data is a 1-D "manifold" embedded in 2-D, corrupted by noise and outliers. The gray arrows show the how each point has moved in the resulting image. T-L: Noisy data. T-R: $\sigma_v = 0.05$, $\sigma_n = 0.05$, $\alpha = 6$. B-L: $\sigma_v = 0.15$, $\sigma_n = 0.06$, $\alpha = 1$. B-R: $\sigma_v = 0.1$, $\sigma_n = 0.08$, $\alpha = 20$.

This deterministic procedure turned out to give very similar results to true importance sampling when the number of data points was large enough.

6 Experiments

Some experiments are included to demonstrate the proposed method.

6.1 Scalar Signals

Experiments in Fig. 2 shows a simple example of filtering a 1-D signal. In Fig. 3 the method was tried out on a scalar image. These two experiments were included mainly to illustrate the behavior of the filter and show that it is similar to the previous filter proposed in [17].

6.2 Vector-Valued Signals

Next the filter was tested on 2D color images, encoded as pixels with RGB color vectors. The parameters of the filters were tuned manually and Fig. 4 show both good and bad results.

6.3 Unordered N-D Data

The filter was then tested on unordered 2-D and 3-D data, see Fig. 5 and Fig. 6. The data points in Fig. 6 were derived from the RGB-values of the boat image in Fig. 4.

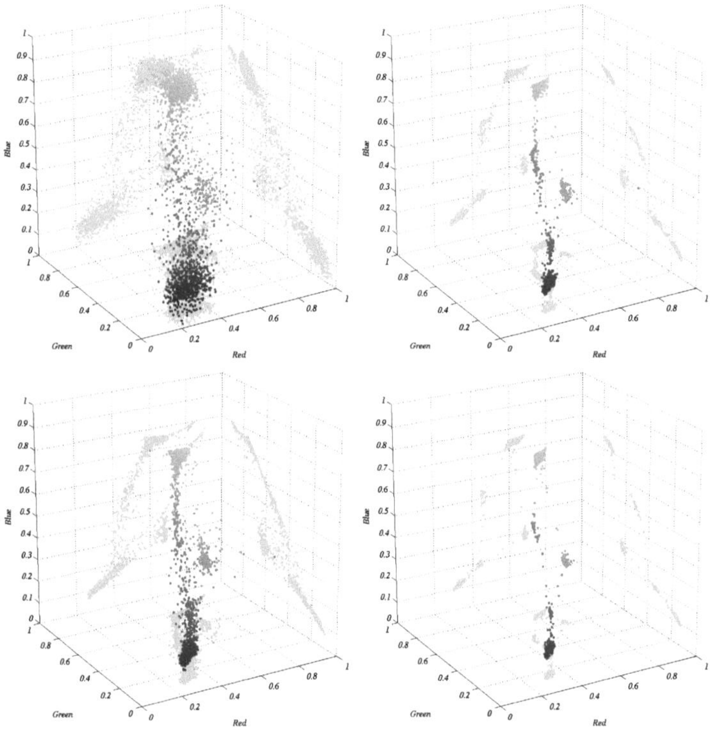

Fig. 6. Filtering unordered 3-D data. The data is the color values from Fig. 4. T-L: Noisy data. T-R: $\sigma_v = 0.05$, $\sigma_n = 0.05$, $\alpha = 10$. B-L: $\sigma_v = 0.05$, $\sigma_n = 0.1$, $\alpha = 0$. B-R: $\sigma_v = 0.05$, $\sigma_n = 0.05$, $\alpha = 20$.

7 Conclusion

We have presented a novel computational framework extending the previous method proposed in [17] from scalar to vector-valued images and data. The two implementations we have presented, for images and unordered data, are examples of stochastic and deterministic variants of the framework.

While the statistical modelling used here is quite simple, it should be noted that more sophisticated Bayesian modelling could be used within the same framework, for instance to model the noise more accurately for a specific application such as X-ray imaging or Diffusion Tensor MRI (DT-MRI).

It should also be noted that the proposed method based on importance sampling could also be useful for certain cases when images are scalar-valued and the dynamic range is so large that it is difficult to create histograms with the precision needed. This could be the case in computed tomography (CT).

A drawback with the method is the large number of parameters and future research will have to address this issue. Nevertheless we have found our method easy to tune and use in practice. The wide range of parameters can also be regarded as a feature since it allows the filter to change characteristics, spanning for instance both low-pass and median-like filter solutions.

References

1. Andrieu, C., Freitas, N., Doucet, A., Jordan, M.I.: An Introduction to MCMC for Machine Learning. Machine Learning (2002)
2. Borik, A.C., Huang, T.S., Munson, D.C.: A generalization of median filtering using combination of order statistics. IEEE Proceedings 71(31), 1342–1350 (1983)
3. Catte, F., Lions, P.L., Morel, J.M.: Image selective smoothing and edge detection by nonlinear diffusion. SIAM Journal on Numerical Analysis I(29), 182–193 (1992)
4. Comaniciu, D., Meer, P.: Mean shift: A robust approach toward feature space analysis. IEEE Trans. Pattern Anal. Mach. Intell. 24(5), 603–619 (2002)
5. Freeman, W.T., Adelson, E.H.: The design and use of steerable filters. IEEE Transactions on Pattern Analysis and Machine Intelligence 13(9), 891–906 (1991)
6. Godtliebsen, F., Spjøtvoll, E., Marron, J.S.: A nonlinear Gaussian filter applied to images with discontinuities. Nonparametric Statistics 8, 21–43 (1997)
7. Iba, Y.: Population Monte Carlo algorithms. Transactions of the Japanese Society for Artificial Intelligence 16(2), 279–286 (2001)
8. Isard, M., Blake, A.: Condensation – conditional density propagation for visual tracking. International Journal of Computer Vision 29(1), 5–28 (1998)
9. Knutsson, H., Wilson, R., Granlund, G.H.: Anisotropic non-stationary image estimation and its applications — Part I: Restoration of noisy images. IEEE Transactions on Communications 31(3), 388–397 (1983)
10. Lee, J.S.: Digital image smoothing and the sigma filter. Computer Vision, Graphics and Image Processing 24, 255–269 (1983)
11. Liu, J.S., Chen, R., Logvienko, T.: A Theoretical Framework for Sequential Importance Sampling and Resampling. In: Doucet, A., de Freitas, N., Gordon, N. (eds.) Sequential Monte Carlo Methods in Practice, Springer, Heidelberg (2001)
12. Mrázek, P., Weickert, J., Bruhn, A.: On robust estimation and smoothing with spatial and tonal kernels. In: Klette, R., Kozera, R., Noakes, L., Weickert, J. (eds.) Geometric properties for incomplete data, pp. 335–352. Springer, Heidelberg (2006)
13. Perona, P., Malik, J.: Scale space and edge diffusion using anisotropic diffusion. IEEE Transactions on Pattern Analysis and Machine Intelligence 12(7), 629–639 (1990)
14. Rudin, L.I., Osher, S., Fatemi, E.: Nonlinear total variation based noise removal algorithms. Physica D. 60, 259–268 (1992)
15. Smith, S., Brady, J.: SUSAN - a new approach to low level image processing. International Journal of Computer Vision 23(1), 45–78 (1997)
16. Tomasi, C., Manduchi, R.: Bilateral filtering for gray and color images. In: IEEE International Conference on Computer Vision 98, pp. 839–846, Bombay, India, January 1998. IEEE
17. Wrangsjö, A., Borga, M., Knutsson. H.: A bayesian approach to image restoration. In: IEEE International Symposium on Biomedical Imaging (ISBI'04), Arlington, USA (2004)

Robust Moving Region Boundary Extraction Using Second Order Statistics

Astrit Rexhepi and Farzin Mokhtarian

Centre for Vision, Speech, and Signal Processing
School of Electronics and Physical Sciences
University of Surrey
Guildford GU2 7XH
United Kingdom
a.rexhepi@surrey.ac.uk,
f.mokhtarian@surrey.ac.uk
http://www.surrey.ac.uk

Abstract. This paper describes a novel method of extracting moving region boundaries from the frames of an image sequence by filtering information of the first temporal cooccurrence matrix using the corresponding matrix from the next frame. The method described in this paper does not make use of any threshold and it is very robust and efficient with respect to noise.

1 Introduction

Cooccurrence matrices, originally called gray-tone spatial dependency matrices, were introduced by Haralick *et al.* [1], who used them to define textural properties of images.

Let I be an image whose pixel gray levels are in the range $[0, \ldots, 255]$. Let $\delta = (u, v)$ be an integer-valued displacement vector; δ specifies the relative position of the pixels at coordinates (x, y) and $(x + u, y + v)$. A *spatial cooccurrence matrix* M_δ of I is a 256×256 matrix whose (i, j) element is the number of pairs of pixels of I in relative position δ such that the first pixel has gray level i and the second one has gray level j. Any δ, or set of δ-s, can be used to define a spatial cooccurrence matrix. In what follows we will usually assume that δ is a set of unit horizontal or vertical displacement, so that M_δ involves counts of pairs of neighboring pixels.

In addition to their original use in defining textural properties, cooccurrence matrices have been used for image segmentation. Ahuja and Rosenfeld [2] observed that pairs of pixels in the interiors of smooth regions in I contribute to elements of M_δ near its main diagonal; thus in a histogram of the gray levels of the pixels that belong to such pairs, the peaks associated with the regions will be preserved, but the valleys associated with the boundaries between the regions will be suppressed, so that it becomes easier to select thresholds that separate the peaks and thus segment the image into the regions. In [3], Haddon and Boyce observed that homogeneous regions in I give rise to peaks (clusters of

B.K. Ersbøll and K.S. Pedersen (Eds.): SCIA 2007, LNCS 4522, pp. 828–837, 2007.

high-valued elements) near the main diagonal of M_δ, while boundaries between pairs of adjacent regions give rise to smaller peaks at off-diagonal locations; thus selecting the pixels that contribute to on-diagonal and off-diagonal peaks provides a segmentation of I into homogeneous regions and boundaries. The peaks that can be expected to occur in cooccurrence matrices will be further described in Section 2.3 (compare [3]).

Pairs of pixels in the same spatial position that have a given temporal separation in a sequence of images can be used to define *temporal cooccurrence matrices.* Let I and J be images acquired at times t and $t + dt$; thus dt is the temporal displacement between I and J. A temporal cooccurrence matrix M_{dt} is a 256×256 matrix whose (i, j) element is the number of pairs of pixels in corresponding positions in I and J such that the first pixel has gray level i and the second one has gray level j.

Boyce *et al.* [4] introduced temporal cooccurrence matrices and used them in conjunction with spatial cooccurrence matrices to make initial estimates of the optical flow in an image sequence. They demonstrated that an initial probability of a pixel being in the interior or on the boundary of a region that has smooth optical flow in a given direction in a pair of images could be derived from the positions of the peaks in a spatial cooccurrence matrix of one of the images for a displacement in the given direction, and in the temporal cooccurrence matrix of the pair of images. Borghys *et al.*[5] used temporal cooccurrence matrices to detect sensor motion in a moving target detection system by comparing the spatial cooccurrence matrix of one of the images with the temporal cooccurrence matrix of the pair of images.

The following is the organization of this paper: Section 2 describes the peaks (clusters of high values) that can be expected to occur in spatial and temporal cooccurrence matrices when the image contains smooth regions. Section 3 describes our new method of extracting moving region boundaries from an image sequence by suppressing information from first temporal cooccurrence matrix using the corresponding matrix from the next frame. In Sections 4 and 5 we propose and develop a filter design and a feed-back system for moving boundary enhancement and noise removal. Section 6 describes the extension of our method for static boundary detection. Section 7 summarizes our work and discusses possible extensions.

2 The Structure of Cooccurrence Matrices

In the next section we will describe methods of using temporal cooccurrence matrices to extract moving region boundaries from the images of a sequence. In this section we describe the peak structures that should be present in spatial and temporal cooccurrence matrices.

We assume that an image I is composed of regions in which (ignoring noise) the gray levels vary smoothly, and that if two regions are adjacent, they meet along a boundary at which the gray level changes significantly. It is well known (see [3]) that in a spatial cooccurrence matrix of I, each region (say having

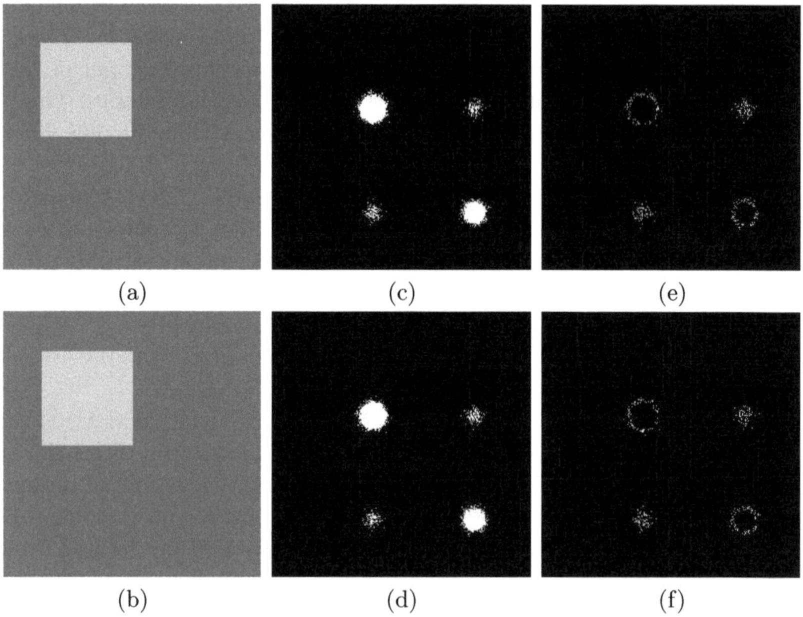

Fig. 1. (a and b) Two synthetic images(images I and J). (c and d) Their corresponding temporal cooccurrence matrices. (e and f) Their mutual suppression.

mean gray level g) should give rise to a peak centered on the main diagonal in approximate position (g, g); the sum of the element values in this cluster should be proportional to the area of the region. Similarly, each boundary between two adjacent regions (say having mean gray levels g and h) should give rise to a pair of off-diagonal peaks at approximate positions (g, h) and (h, g), and with value sum proportional to the length of the border. The same holds for temporal cooccurrence matrices when no motion occur, namely, we simply can assume that the whole frame is moving in the amount of one pixel. Thus, we can slide one of the frames in the amount of one pixel (top-down and left-right composed shift) and find the temporal cooccurrence matrices (in which ignoring border effects it is the same as spatial cooccurrence matrix),

Figure 1a is a test image I containing a solid square with gray level 200 on a background with gray level 100. This image is composed with noise having a normal distribution with mean value zero and variance equal four. Let J be another image(Figure 1b), the same as I but where the solid square has moved from its original position by one pixel, and having the same instance of noise generated by a random generator.

We can treat I and J as consecutive frames of an image sequence acquired by a stationary camera. In this case the frames show an object (solid square) moving against a stationary background at a rate of one pixel per frame. In the temporal cooccurrence matrix of I and J (Figure 1c), pairs of pixels that are in a moving region in both images will contribute to an on-diagonal peak. Similar, pairs of pixels that are in the background in both images will contribute also to another on-diagonal peak. Pairs of pixels that are covered up or uncovered by the motion will contribute to a pair of off-diagonal peaks.

3 Extracting Boundaries Using Cooccurrence Matrices

As discussed in Section 2, motion of an object against a contrasting background between two frames of an image sequence gives rise to off-diagonal peaks in a temporal cooccurrence matrix of the two frames. Thus it should be possible in principle to extract moving boundaries from a pair of successive frames of an image sequence by detecting off-diagonal peaks in the temporal cooccurrence matrix of the two frames and identifying the pixels in either of the frames that contributed to those peaks.

Unfortunately, off-diagonal peaks are not always easy to detect in cooccurrence matrices. Since the images are noisy, all the elements near the diagonal of a cooccurrence matrix tend to have high values, and the presence of these values makes it hard to detect off-diagonal peaks in the matrix that lie close to the diagonal since these peaks tend to have lower values. If we knew the standard deviation of the image noise, we could estimate how far the high values which are due to noise extend away from the diagonal of the cooccurrence matrix, and we could then look for peaks in the matrix that are farther than this from the diagonal; but information about the image noise level is usually not available.

In this section we describe a simple method of suppressing clusters of high-valued elements from a temporal cooccurrence matrix. As we will see, the suppressed matrix elements tend to lie near the diagonal of the matrix. Hence when the suppression process is applied to a temporal cooccurrence matrix the image pixels that contributed to the unsuppressed elements of the matrix tend to lie on the boundaries of moving regions.

Our method of suppressing clusters of high-valued elements from a cooccurrence matrix takes advantage of two observations:

(1) The matrix elements in the vicinity of a high-valued cluster almost certainly have nonzero values, so that the nonzero values in and near the cluster are "solid". On the other hand, it is more likely that there are zero-valued elements in and near a cluster of low-valued elements, so that the nonzero values in and near such a cluster are "sparse".
(2) As we saw in Section 2, the on-diagonal clusters in a cooccurrence matrix, which arise from regions in the image, can be expected to be symmetric around the main diagonal, and the off-diagonal clusters, which arise from motion, can be expected to occur in pairs whose means are symmetrically located around the main diagonal, since the noise in the image has zero mean. Hence if we have two cooccurrence matrices that are transposes of one another (see below), the clusters in these matrices should occur in the same approximate positions.

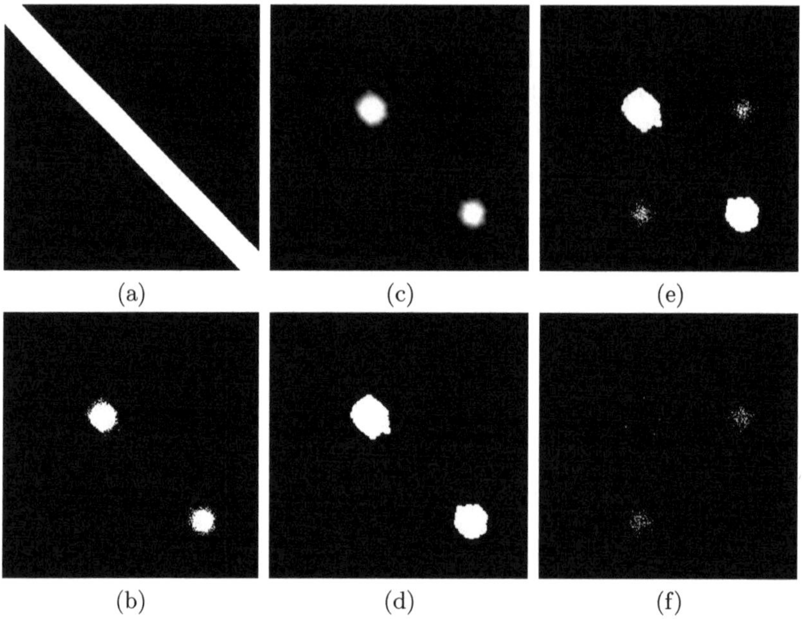

Fig. 2. (a) Filter F. (b) M+F. (c) Smoothed M+F. (d) Setting to one all the nonzero elements of the smoothed M+F yielding F'. (e) M+F'=M'. (f) N/M'.

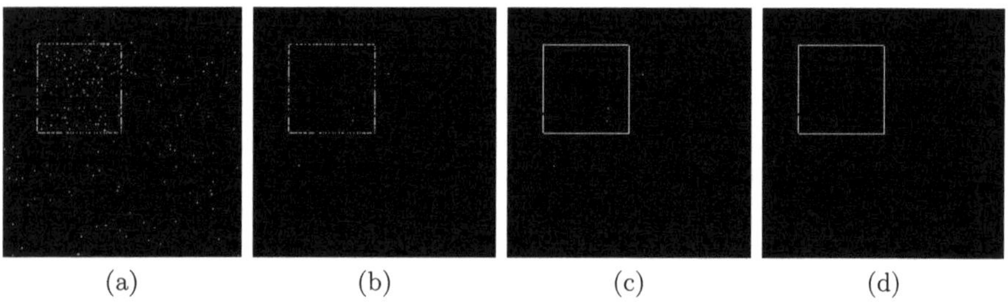

Fig. 3. (a) The result obtained by only Temporal Cooccurrence Matrices Suppression. (b) The result after filtering with F'. (c) The result after applying feedback-system. (d) The result after noise removal.

We can obtain temporal cooccurrence matrices that are transposes of one another by using reverse temporal displacements; i.e., if I and J are successive frames of an image sequence, we can use the temporal cooccurrence matrices of I and J (Figure 1c) and of J and I (Figure 1d). Evidently, Figure 1c and Figure 1d are transposes of each other.

Let M and N be two cooccurrence matrices that are transposes of one another. We suppress from M all elements that are nonzero in N (or vice versa). Elements of M that are in or near a "solid" cluster will almost certainly have nonzero values in N; hence these elements will almost certainly be suppressed from M. On the other hand, many of the elements of M that are in or near a "sparse" cluster will have zero values in N because the nonzero elements of these clusters in M and N are not in exactly symmetrical positions; hence many of these elements will not be eliminated by the suppression process.

Figure 1e shows the nonzero elements of Figure 1c that are zero in Figure 1d, and Figure 1f shows the nonzero elements of Figure 1d that are zero in Figure 1c. We see that the "solid" parts of the matrix have been suppressed and the "sparse" parts have survived. Figure 3a shows the pixels of Figure 1a that contributed to the nonzero elements in Figure 1e. Almost all of these pixels lie on region boundaries in Figure 3a.

4 The Problem of Residuals

As we saw in the previous section, using temporal cooccurrence matrices suppression we achieved some important results. Namely, we are now able to detect moving boundaries and we don't need to set any threshold to do this, which is in contrast with existing methods. It is obvious that at this stage the results we just obtained contain some spurious noise pixels (Figure 3a), but as we will show in this section, these spurious noise pixels we are able to eliminate successfully by developing some filters in a logical sense.

There are two reasons why these noise pixels appear in Figure 3a:

a) The on-diagonal clusters in temporal cooccurrence matrices are results of regions (static and inside moving objects regions). Thus, a homogenous region R having an area (in pixels) A will yield an on-diagonal cluster in the temporal cooccurrence matrix whose shape is circular centered at (k, k) where k is the mean value of R, and a radius r that depends on noise variance (which is unknown) and the area of R. Usually, the noise has normal (multinomial) distribution or it can be approximated by normal distribution, thus, this cluster will be denser near (k, k) and the density decreases as we go away from (k, k). Places where the density of the cluster is low will survive suppression and we will call them *residuals* that have a ring shape as seen in Figure 1e and Figure 1f. These residuals appear as noise pixels in Figure 3a.

b) Small regions having a unique gray level will yield (approximately) on-diagonal sparse clusters in the temporal cooccurrence. Hence, when suppression is applied most of the elements of these clusters will survive. These elements appear as noise in Figure 3a.

4.1 Designing the Filter to Suppress Residuals

One way of suppressing residuals (and this is the only one we show in this paper) is to develop a filter F as below:

Let M and N be the temporal cooccurrence matrices of I and J as shown in Figure 1c and Figure 1d, and let S be one of suppressed cooccurrence matrices, we put a filter-like-strip along the main diagonal in M whose width is calculated using the following logic:

Starting from the width equal to zero we increase it until the sum of nonzero elements of S that do not belong inside the filter F is equal(a place where the number of nonzero elements of M that do not belong inside F changes from being higher to smaller) the number of nonzero elements of S that belongs inside the filter F. After we find F (Figure 2a), the next task is to use this filter and place along the main diagonal of N. Nonzero elements of N belonging inside the filter F (Figure 2b) we smooth using an averaging filter-like-a-disk (or we can perform dilation instead) with radius 3 for all cases (Figure 2c), let us set all the nonzero elements of this matrix to *one* and denote it by F' as shown in Figure 2d. Now, if we set to *one* all nonzero elements of this processed N (Figure 2e) and suppress M from N (Figure 2f) the shape-like-a-ring residuals almost completely have disappeared, so let us denote it with S'. The corresponding pixels of I and J that contributed to nonzero elements of S' are shown in Figure 3b, so let us denote it by B'. It is obvious that Figure 3b is almost noiseless compared to Figure 3a.

5 Moving Boundaries Enhancement and Noise Removal

In the previous section we were able to reduce the noise pixels using filter F. The results obtained are impressive, but further processing still has to be done because moving boundaries appear to be broken and there are still some noise pixels to be cleaned. In this section we will be dealing with the above mentioned problems.

Broken boundaries in Figure 3b are result of suppression process. As discussed in Section 3, the off-diagonal clusters of M and N are results of moving boundaries. These clusters are sparse but it doesn't mean that when the suppression process is applied they will completely survive, something is possible to be suppressed. These suppressed elements in turn result in missing parts of boundaries in Figure 3b. To recover these missing parts of boundaries we will develop a *feedback system* that is based on the following assumption:

The density of nonzero elements in B' in and around boundaries is higher so that around nonzero elements of B' it is very likely to find elements (in the same spatial position) in I and J that contribute to nonzero elements of off-diagonal clusters of M/F' where / is a difference operator. On the other hand the density of nonzero elements of B' representing noise is very sparse, so that it is quite unlikely that around these points of B' we will find (in the same spatial position) elements in I and J that contribute to any of nonzero elements in M/F'. Thus, principally it would be possible to recover suppressed elements of off-diagonal clusters of S' by searching in the neighborhood of nonzero elements of B'. To do this, we developed a system as following:

- Smooth B' with an averaging filter of size 3x3.
- Develop B'' by setting to *one* the positive elements of smoothed B'.
- Create images $I' = I \diamond B''$ and $J' = J \diamond B''$.
- Find their temporal cooccurrence matrix M'.

- Find $T = (M'/F')$.
- Find elements of I and J that contributed to nonzero elements of T, yielding updated B'.
- Repeat above steps until there is no change between consecutive B' images.

where $\diamond$ stands for element-by-element multiplication. Usually the last condition is satisfied in the third iteration, and the corresponding results we obtain (Figure 3c) are very satisfactory. Finally, the remaining noise points we reduce by using an operator of size 5x5 and exclude positive elements (belonging to the center of the operator) of updated B' if the total sum of positive elements inside the operator is less than 2. The result is shown in Figure 3d, as we can see; the updated B' contains only the boundary of moving object. It might have been thought that the same results could have been achieved by using only M/F or by only using image difference and use the width of F as a threshold (a number that decides what is the boundary and what noise), but it is not possible in general because low contrast boundaries will be vanished and many noise clusters will be left which are not possible to be cleaned in image domain because of their high density.

6 Detecting Static Boundaries

Our method can be extended for *static boundary detection*. To do this, we simply can assume that the whole frame is moving in the amount of one pixel. Thus, we can slide one of the frames in the amount of one pixel (top-down and left-right composed shift) and find the temporal cooccurrence matrices (in which ignoring border effects it is the same as spatial cooccurrence matrix), after we do this we filter on-diagonal elements using filter F' (in a similar way we did for moving boundaries in the previous sections) and find the corresponding pixels in I and J that contributed to the nonzero elements of their filtered temporal cooccurrence matrix.

Examples are given for real images that include both moving and static boundaries (Figure 4, Figure 5, Figure 6, and a comparison to image subtraction in Figure 7) (please observe threshold values we needed to take in order to get clean moving boundaries for the case of image subtraction, and compare results taken without any threshold using our system in 4 and 5 respectively).

(a) (b) (c) (d)

Fig. 4. (a) and (b): Two successive frames of an image sequence showing a moving man. (c) and (d); Moving and static boundary detection using our system.

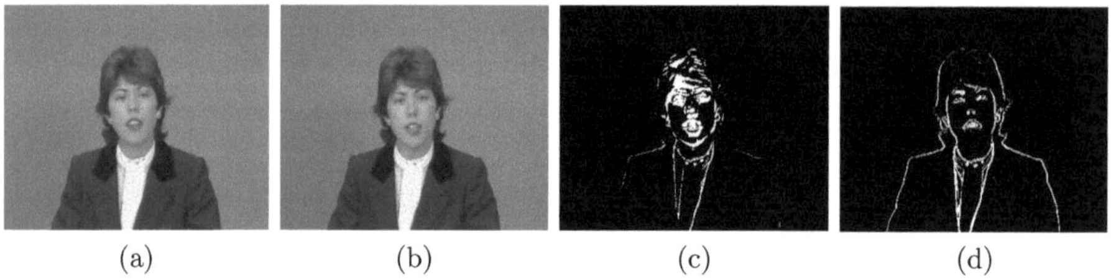

Fig. 5. (a) and (b): Two successive frames of an image sequence showing a moving speaker-woman. (c) and (d); Moving and static boundary detection using our system.

Fig. 6. (a) and (b): Two successive frames of an image sequence showing a moving woman and a child. (c) and (d); Moving and static boundary detection using our system.

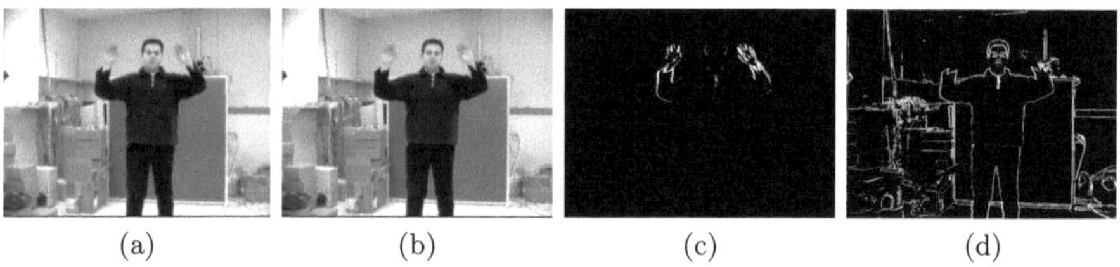

Fig. 7. (a) and (b): Two successive frames of an image sequence showing a man waiving his hands and arms. (c) and (d); Moving and static boundary detection using our system.

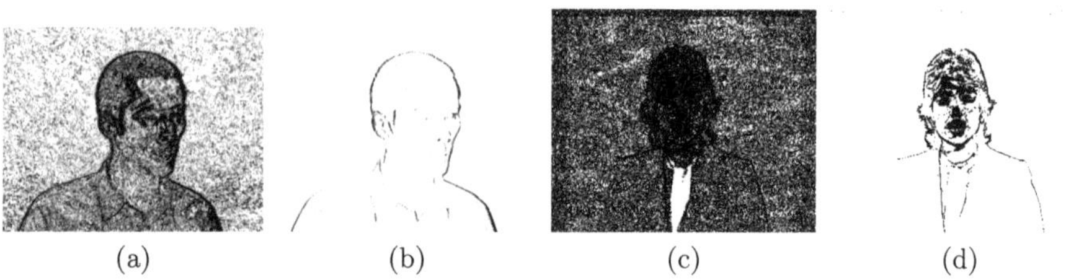

Fig. 8. (a): Image Subtraction of Figures 4(a and b). (b): Moving boundaries after applying a threshold $T = 40$. (c): Image Subtraction of Figures 5(a and b). (d) Moving boundaries after applying a threshold $T = 7$.

7 Conclusions

Moving Boundaries detection is a crucial step in computer vision that determines success or failure of the whole system. In this paper we presented an integrated system for moving boundary detection and its extension for static boundaries. This system is full-automatic (no human intervention is need) which has the following important properties:

– Does not make use of any threshold.
– Almost completely removes noise.
– In the same framework we can extract both moving and static boundaries.
– The complete system can be developed using only Boolean algebra.
– It is fast and very easy to realize even in hardware.

References

1. Haralick, R.M., Shanmugam, R., Dinstein, I.: Textural Features for Image Classification. IEEE Trans. on Systems, Man, and Cybernetics 3, 610–621 (1973)
2. Ahuja, N., Rosenfeld, A.: A Note on the Use of Second-order Gray-level Statistics for Threshold Selection. IEEE Trans. on Systems, Man, and Cybernetics 8, 895–898 (1978)
3. Haddon, J.F., Boyce, J.F.: Image Segmentation by Unifying Region and Boundary Information. IEEE Trans. on Pattern Analysis and Machine Intelligence 12, 929–948 (1990)
4. Boyce, J.F., Protheroe, S.R., Haddon, J.F.: A Relaxation Computation of Optic Flow from Spatial and Temporal Cooccurrence Matrices. International Conference on Pattern Recognition 3, 594–597 (1992)
5. Borghys, D., Verlinde, P., Perneel, C., Acheroy, M.: Long-range Target Detection in a Cluttered Environment Using Multi-sensor Image Sequences. Proc. SPIE 3068, 569–578 (1997)

A Linear Mapping for Stereo Triangulation

Klas Nordberg

Computer Vision Laboratory
Department of Electrical Engineering
Linköping University

Abstract. A novel and computationally simple method is presented for triangulation of 3D points corresponding to the image coordinates in a pair of stereo images. The image points are described in terms of homogeneous coordinates which are jointly represented as the outer products of these homogeneous coordinates. This paper derives a linear transformation which maps the joint representation directly to the homogeneous representation of the corresponding 3D point in the scene. Compared to the other triangulation methods this approach gives similar reconstruction error but is numerically faster, since it only requires linear operations. The proposed method is projective invariant in the same way as the optimal method of Hartley and Sturm. The methods has a "blind plane"; a plane through the camera focal points which cannot be reconstructed by this method. For "forward-looking" camera configurations, however, the blind plane can be placed outside the visible scene and does not constitute a problem.

1 Introduction

Reconstruction of 3D points from stereo images is a classical problem. The methods which address the problem fall coarsely into two classes; dense and sparse. The first class normally assumes a restricted camera configuration with camera viewing directions which are approximately parallel and a short camera baseline, resulting in approximately horizontal and relatively small displacements between corresponding points in the two images. As a consequence, a displacement or disparity field can be estimated between the two images for all points, although with low reliability for image points which, e.g. are located in a constant intensity region. Several methods for solving the disparity estimation problem have been proposed in the literature and an overview is presented, for example, in [9]. Given the estimated disparity field, the 3D reconstruction can be done using the inverse proportionality between depth in the scene and the disparity.

The second class allows general camera configurations, with the main restriction that they cameras depict a common scene. For a typical scene, however, this implies that only a smaller set of point may be visible in both images and can be robustly detected as corresponding points in both images. The proposed methods normally solve the reconstruction problem in two steps. First, two sets of points in each of the images are determined, and further investigated to produce point pairs where each pair corresponds to the same point in the 3D scene.

B.K. Ersbøll and K.S. Pedersen (Eds.): SCIA 2007, LNCS 4522, pp. 838–847, 2007.

A triangulation procedure can then be applied on each such pair to reconstruct the 3D point. In this paper we assume that the correspondence problem has been solved, and instead deal with the triangulation procedure.

In order to solve the triangulation problem, some related issues must first be addressed. Most solutions assume that the cameras involved in producing the stereo images can sufficiently accurately be modelled as pin-hole cameras. This implies that the mapping from 3D points to 2D coordinates in each of the two images can be described as a linear mapping on homogeneous representations of both 3D and 2D coordinates. Let $\mathbf{x}$ be a homogeneous representation of the coordinates of a 3D point (a 4-dimensional vector), let $\mathbf{y}_k$ be a homogeneous representation of the corresponding image coordinate in image k, (a 3-dimensional vector), and let $\mathbf{C}_k$ be the linear mapping which describes the mapping of camera k (a 3×4 matrix). The pin-hole camera model then implies that

$$\mathbf{y}_k \sim \mathbf{C}_k \, \mathbf{x}, \qquad k = 1, 2 \tag{1}$$

where $\sim$ denotes equality up to a scalar multiplication. Notice, that in this particular relation, the scalar in is only dependent on $\mathbf{x}$. The inverse mapping can be written as

$$\mathbf{x} \sim \mathbf{C}_k^{+} \mathbf{y}_k + \lambda_k \, \mathbf{n}_k, \qquad \lambda_k \in R \tag{2}$$

where $\mathbf{C}_k^{+}$ is the pseudo-inverse of $\mathbf{C}_k$ and $\mathbf{n}_k$ is the homogeneous representation of the focal point of camera k, i.e.,

$$\mathbf{C}_k \, \mathbf{n}_k = \mathbf{0} \tag{3}$$

From Equation (2) follows that the original 3D point must lie on a *projection line* through the image point and the focal point. In the following, we will assume that both camera matrices have been determined with a sufficient accuracy to be useful in the following computations.

Another issue is the so so-called epipolar constraint. It implies that two corresponding image points must satisfy the relation

$$\mathbf{y}_1^T \mathbf{F} \, \mathbf{y}_2 = 0 \tag{4}$$

where $\mathbf{F}$ is the so-called *fundamental matrix* which is determined from the two camera matrices [5]. Intuitively, we can think of this relation as a condition on $\mathbf{y}_1$ and $\mathbf{y}_2$ to assure that the corresponding projection lines intersect at the point $\mathbf{x}$. In practice, however, there is no guarantee that $\mathbf{y}_1$ and $\mathbf{y}_2$ satisfy Equation (4) exactly. For example, many point detection methods are based on finding local maxima or minima of some function, e.g., [3], typically producing integer valued image coordinates. As a consequence, the two projection lines do not always intersect.

Even if we assume that the cameras can be modeled as pin-hole cameras whose matrices are known with sufficient degree accuracy and that in some way or another all pairs of corresponding image points have been modified to satisfy Equation (4), we now face the ultimate problem of triangulation: finding the intersecting point of the two projection lines for each pair of corresponding image points. In principle, this is a trivial problem had it not been for the fact that Equation (4) is not always satisfied. The conceptually easiest approach is the *mid-point method* were we seek the mid-point of the shortest line segment which joins the projection lines of the two image points [1]. From an algebraic point of view, the problem can also be solved by using Equation (1) to obtain

$$\begin{aligned} \mathbf{y}_1 \times (\mathbf{C}_1\, \mathbf{x}) &= \mathbf{0} \\ \mathbf{y}_2 \times (\mathbf{C}_2\, \mathbf{x}) &= \mathbf{0} \end{aligned} \tag{5}$$

where "$\times$" denotes the vector cross product. These relations imply that we can establish six linear expressions in the elements of $\mathbf{x}$, an over-determined system of equations. On the other hand, if we know that $\mathbf{y}_1$ and $\mathbf{y}_2$ satisfy Equation (4) then there must be a unique solution $\mathbf{x}$ of Equation (5), disregarding a scalar multiplication. By rewriting Equation (5) as

$$\mathbf{M}\, \mathbf{x} = \mathbf{0} \tag{6}$$

we can either find $\mathbf{x}$ as the right singular vector of $\mathbf{M}$ corresponding to the smallest singular value, or by solving for the non-homogeneous coordinates of the 3D point from the corresponding 6×3 inhomogeneous equation using a least squares solution. These are the *homogeneous* and the *inhomogeneous* triangulation methods [5].

All three methods appear to work in most situations, but they have some practical differences, notably that the resulting 3D point is not the same for all three methods in the case that Equation (4) is not satisfied. The implementations of all three methods are relatively simple but they do not provide a simple closed form expression for $\mathbf{x}$ in terms of $\mathbf{y}_1, \mathbf{y}_2$. Also, the basic form of both the mid-point method and the inhomogeneous method cannot provide a robust estimate of the 3D point in the case that it is at a large of infinite distance from the camera.

There is also a difference in the accuracy of each method. This can be defined in terms of the 3D distance between the resulting 3D point and the correct 3D point, but from a statistical point of view it can also be argued that the accuracy should be defined in terms of the Euclidean 2D distances of the projection of these 3D points. This leads to the *optimal method* for triangulation which seeks two subsidiary image points $\hat{\mathbf{y}}_1, \hat{\mathbf{y}}_2$ which are at total smallest squared distance from $\mathbf{y}_1, \mathbf{y}_2$, measured in the 2D image planes, which in addition satisfy $\hat{\mathbf{y}}_1^T \mathbf{F}\, \hat{\mathbf{y}}_2 = 0$. Once $\hat{\mathbf{y}}_1, \hat{\mathbf{y}}_2$ are determined, any of the above mentioned methods can then produce the "optimal" estimate of $\mathbf{x}$. A computational approach for determining the subsidiary image points is presented in [7]. Although this method is not iterative, it is of relatively high complexity and does not give $\mathbf{x}$ as a closed form expression. On the other hand, this method can be shown to be projective invariant, meaning that the resulting point $\mathbf{x}$ is invariant to any projective transformation of the 3D space.

In summary, there exist a larger number of methods for solving the sparse triangulation problem, even besides the "standard" methods presented above. For an overview of sparse triangulation methods, [7,5] serve as good starting points. Most of the recent work in this area derives reconstruction methods based on various cost functions, for example see [4,8].

This paper takes a slightly different view on triangulation by leaving cost functions and optimization aside and instead focus on the basic problem of finding an algebraic inverse of the combined camera mapping from a 3D point to the two corresponding image points. It proves the existence of a *closed form expression* for the mapping from $\mathbf{y}_1$ and $\mathbf{y}_2$ to $\mathbf{x}$. The expression is in the form of a *second order polynomial* in the elements of $\mathbf{y}_1$ and $\mathbf{y}_2$, or more precisely, a linear mapping on the outer product $\mathbf{y}_1\mathbf{y}_2^T$. Beside its simple computation, another advantage is that it can be shown to be projective invariant, although this aspect is not discussed in detail here. An experimental evaluation of the proposed method shows that, on that specific data set, it has an reconstruction error which is comparable to the other standard methods, including optimal triangulation.

2 Derivation of a Reconstruction Operator

2.1 The Mapping to Y

Let $\mathbf{y}_k$ be the homogeneous representation of a point in image k, given by Equation (1). We can write this expression in element form as

$$y_{ki} = \alpha_k(\mathbf{x})\, \mathbf{c}_{ki} \cdot \mathbf{x} \tag{7}$$

where $\mathbf{c}_{ki}$ is the i-th row of camera matrix k and the product in the right-hand side is the inner product between the vectors $\mathbf{c}_{kl}$ and $\mathbf{x}$. Now, form the outer or tensor product between $\mathbf{y}_1$ and $\mathbf{y}_2$. The result can be seen as a 3×3 matrix $\mathbf{Y} = \mathbf{y}_1\mathbf{y}_2^T$. The elements of $\mathbf{Y}$ are given by

$$Y_{ij} = y_{1i}\, y_{2j} = \alpha_1(\mathbf{x})\, \alpha_2(\mathbf{x})\, (\mathbf{c}_{1i} \cdot \mathbf{x})(\mathbf{c}_{2i} \cdot \mathbf{x}) = \tag{8}$$

$$= \alpha_1(\mathbf{x})\, \alpha_2(\mathbf{x})\, (\mathbf{c}_{1i}\mathbf{c}_{2j}^T) \cdot (\mathbf{x}\mathbf{x}^T) = \alpha_1(\mathbf{x})\, \alpha_2(\mathbf{x})\, (\mathbf{c}_{1i}\mathbf{c}_{2j}^T) \cdot \mathbf{X} \tag{9}$$

We can interpret this as: element Y_{ij} is given by an inner product between $\mathbf{c}_{1i}\mathbf{c}_{2j}^T$ and $\mathbf{X} = \mathbf{x}\mathbf{x}^T$, defined by the previous relations. Notice that $\mathbf{X}$ is always a symmetric 4×4 matrix, which means that we can rewrite Y_{ij} as

$$Y_{ij} = \frac{\alpha_1(\mathbf{x})\, \alpha_2(\mathbf{x})}{2} (\mathbf{c}_{1i}\mathbf{c}_{2j}^T + \mathbf{c}_{2j}\mathbf{c}_{1i}^T) \cdot \mathbf{X} = \alpha_1(\mathbf{x})\, \alpha_2(\mathbf{x})\, \mathbf{B}_{ij} \cdot \mathbf{X} \tag{10}$$

where each $\mathbf{B}_{ij} = (\mathbf{c}_{1i}\mathbf{c}_{2j}^T + \mathbf{c}_{2j}\mathbf{c}_{1i}^T)/2$ is a symmetric 4×4 matrix.

2.2 The Set $\mathbf{B}_{ij}$

Let S denote the vector space of symmetric 4×4 matrices. Notice that $\mathbf{X} \in S$ and $\mathbf{B}_{ij} \in S$. The question now is: what kind of a set is $\mathbf{B}_{ij}$? Obviously, it cannot be a basis of S; the space is 10-dimensional and there are only 9 matrices. Recall that $\mathbf{n}_k$ is the homogeneous representation of the focal point for camera k, i.e.,

$\mathbf{C}_k\, \mathbf{n}_k = \mathbf{0}$. We assume that $\mathbf{n}_1 \neq \mathbf{n}_2$ (as projective elements). It then follows that

$$\mathbf{B}_{ij} \cdot (\mathbf{n}_k \mathbf{n}_k^T) = (\mathbf{c}_{1i} \cdot \mathbf{n}_k)(\mathbf{c}_{2j} \cdot \mathbf{n}_k) = 0 \tag{11}$$

i.e., $\mathbf{Q}_k = \mathbf{n}_k \mathbf{n}_k^T$ is perpendicular to all matrices $\mathbf{B}_{ij}$ for $k = 1, 2$. This implies that the 9 matrices at most span an 8-dimensional space, i.e., there exists at least one set of coefficients $\tilde{F}_{ij}$ such that

$$\sum_{ij} \tilde{F}_{ij}\, \mathbf{B}_{ij} = \mathbf{0} \tag{12}$$

Consider the expression

$$\sum_{ij} \tilde{F}_{ij} Y_{ij} = y_{1i}\, y_{2j}\, \tilde{F}_{ij} = \mathbf{y}_1^T \tilde{\mathbf{F}}\, \mathbf{y}_2 \tag{13}$$

where $\tilde{\mathbf{F}}$ is a 3×3 matrix with elements $\tilde{F}_{ij}$. We can now insert Equation (10) into the left-hand side of the last equation and get

$$\alpha_1(\mathbf{x})\, \alpha_2(\mathbf{x}) \sum_{ij} \tilde{F}_{ij} (\mathbf{B}_{ij} \cdot \mathbf{X}) = \alpha_1(\mathbf{x})\, \alpha_2(\mathbf{x})\, (\sum_{ij} \tilde{F}_{ij} \mathbf{B}_{ij}) \cdot \mathbf{X} = \mathbf{0} \cdot \mathbf{X} = 0 \tag{14}$$

Consequently, the right-hand side of Equation (13) vanishes, which is equivalent to the statement made in Equation (4), i.e., we can identify $\tilde{\mathbf{F}}$ with the fundamental matrix $\mathbf{F}$. Since this matrix is unique (disregarding scalar multiplications), it follows that the set of 9 matrices $\mathbf{B}_{ij}$ spans an (9 - 1 = 8)-dimensional subspace of S, denoted S_c. The matrices $\mathbf{B}_{ij}$ therefore form a *frame* [2] rather than a basis of S_c. Furthermore, $\mathbf{Q}_1$ and $\mathbf{Q}_2$ span the 2-dimensional subspace of S which is perpendicular to S_c.

2.3 The Dual Frame

Let us now focus on the subspace S_c. First of all, any matrix $\mathbf{S} \in S_c$ can be written as a linear combination of the 9 frame matrices $\mathbf{B}_{ij}$. Since these matrices are linearly dependent such a linear combination is not unique, but a particular linear combination can be found as

$$\mathbf{S} = \sum_{ij} (\mathbf{S} \cdot \tilde{\mathbf{B}}_{ij})\, \mathbf{B}_{ij} \tag{15}$$

where $\tilde{\mathbf{B}}_{ij}$ is the dual frame relative to $\mathbf{B}_{ij}$. In this case, we compute the dual frame in the following way:

1. Reshape each $\mathbf{B}_{ij}$ to a 16-dimensional vector $\mathbf{a}_I$ with label $I = i + 3j - 3$: , i.e., there are 9 such vectors.
2. Construct a 16×9 matrix $\mathbf{A}$ with each $\mathbf{a}_I$ in its columns.

3. Compute an SVD of $\mathbf{A}$, $\mathbf{A} = \mathbf{U}\mathbf{S}\mathbf{V}^T$, where $\mathbf{S}$ is a 9×9 diagonal matrix with the singular values. According to the discussion above, exactly one singular value must vanish since $\mathbf{f}\,\mathbf{A} = \mathbf{0}$ where $\mathbf{f}$ is the is the fundamental matrix reshaped to a 9-dimensional vector. Consequently, $\mathbf{A} = \tilde{\mathbf{U}}\,\tilde{\mathbf{S}}\,\tilde{\mathbf{V}}^T$, where $\tilde{\mathbf{S}}$ is an 8×8 diagonal matrix with only non-zero singular values.
4. Form the 16×9 matrix $\tilde{\mathbf{A}} = \tilde{\mathbf{U}}\,\tilde{\mathbf{S}}^{-1}\,\tilde{\mathbf{V}}^T$.
5. The columns of $\tilde{\mathbf{A}}$ now contains the dual frame. The corresponding matrices $\tilde{\mathbf{B}}_{ij}$ can be obtained by doing the inverse operations of steps 2 and 1.

By constructing $\tilde{\mathbf{A}}$ in this way it follows that $\mathbf{A}^T\tilde{\mathbf{A}}$ is an identity mapping on the subspace S_{c}, when the elements of this space are reshaped according to step 1 above.

2.4 Triangulation

We will now choose $\mathbf{S}$ in particular ways. Let $\mathbf{p}$ be the dual homogeneous representation of a plane which passes through the focal points of the two cameras. This implies that $\mathbf{p}\cdot\mathbf{n}_1 = \mathbf{p}\cdot\mathbf{n}_2 = 0$. Let $\mathbf{r}$ be an arbitrary 4-dimensional vector, and form the 4×4 matrix

$$\mathbf{S} = \mathbf{p}\mathbf{r}^T + \mathbf{r}\mathbf{p}^T \tag{16}$$

Clearly, $\mathbf{S} \in S$, but it is also the case that $\mathbf{S} \in S_{\mathrm{c}}$. This follows from

$$\mathbf{S}\cdot\mathbf{Q}_k = \mathbf{S}\cdot(\mathbf{n}_k\mathbf{n}_k^T) = 2\,(\mathbf{n}_k\cdot\mathbf{p})(\mathbf{n}_k\cdot\mathbf{r}) = 0 \tag{17}$$

which implies that $\mathbf{S}$ is perpendicular to $\mathbf{Q}_1$ and $\mathbf{Q}_2$. Consequently, this $\mathbf{S}$ can be written as in Equation (15). Consider the expression $\mathbf{S}\cdot\mathbf{X}$. By inserting this into Equation (15) and with the help of Equation (10) we get

$$\mathbf{S}\cdot\mathbf{X} = \sum_{ij}(\mathbf{S}\cdot\tilde{\mathbf{B}}_{ij})(\mathbf{B}_{ij}\cdot\mathbf{X}) = \frac{2}{\alpha_1(\mathbf{x})\,\alpha_2(\mathbf{x})}\sum_{ij}(\mathbf{S}\cdot\tilde{\mathbf{B}}_{ij})\,Y_{ij} \tag{18}$$

If we instead insert it into Equation (16), we get

$$\mathbf{S}\cdot\mathbf{X} = (\mathbf{p}\mathbf{r}^T + \mathbf{r}\mathbf{p}^T)\cdot(\mathbf{x}\mathbf{x}^T) = 2\,(\mathbf{x}\cdot\mathbf{p})(\mathbf{x}\cdot\mathbf{r}) \tag{19}$$

and by combining Equations (18) and (19)

$$\sum_{ij}(\mathbf{S}\cdot\tilde{\mathbf{B}}_{ij})\,Y_{ij} = \alpha_1(\mathbf{x})\,\alpha_2(\mathbf{x})\,(\mathbf{x}\cdot\mathbf{p})(\mathbf{x}\cdot\mathbf{r}) \tag{20}$$

Let $\mathbf{e}_l$ be the standard basis of R^4: $\mathbf{e}_l\cdot\mathbf{x} = x_l$, where x_l is the l-th element of $\mathbf{x}$. Define 4 matrices $\mathbf{S}_l$ according to

$$\mathbf{S}_l = \mathbf{p}\mathbf{e}_l^T + \mathbf{e}_l\mathbf{p}^T \tag{21}$$

where $\mathbf{r}$ now is replaced by $\mathbf{e}_l$ to produce $\mathbf{S}_l$ from $\mathbf{S}$ in Equation (16). As a consequence, Equation (20) becomes

$$\sum_{ij}(\mathbf{S}_l\cdot\tilde{\mathbf{B}}_{ij})\,Y_{ij} = (\mathbf{x}\cdot\mathbf{p})(\mathbf{x}\cdot\mathbf{e}_l) = \alpha_1(\mathbf{x})\,\alpha_2(\mathbf{x})\,(\mathbf{x}\cdot\mathbf{p})\,x_l \tag{22}$$

Notice that the factor $\alpha_1(\mathbf{x})\,\alpha_2(\mathbf{x})\,(\mathbf{x}\cdot\mathbf{p})$ is independent of l. Set $K_{lij} = \mathbf{S}_l \cdot \tilde{\mathbf{B}}_{ij}$. This is a $4 \times 3 \times 3$ array of scalars which can be seen as a linear transformation or an operator which maps $\mathbf{Y}$ to $\mathbf{x}$:

$$\mathbf{K}\,\mathbf{Y} = \alpha_1(\mathbf{x})\,\alpha_2(\mathbf{x})\;(\mathbf{x}\cdot\mathbf{p})\,\mathbf{x} \tag{23}$$

The factor $\alpha_1(\mathbf{x})\;\alpha_2(\mathbf{x})\;(\mathbf{x}\cdot\mathbf{p})$ vanishes when the 3D point $\mathbf{x}$ is in the plane $\mathbf{p}$. Assuming that this is not the case, we can disregard this factor and write $\mathbf{K}\,\mathbf{Y} \sim \mathbf{x}$. It should be noticed, however, that the existence of this factor in the derivations implies that the proposed method cannot reconstruct 3D points if they lie in or sufficiently close to the *blind plane* $\mathbf{p}$.

The blind plane can be a problem if the camera configuration is such that the cameras "see" each other, that is, the 3D line (base-line) which intersects both focal points is visible to both cameras. In this case, a plane $\mathbf{p}$ cannot be chosen which is not visible to both cameras. On the other hand, if both cameras are "forward-looking" in the sense that they are not seeing each other, it is possible to choose $\mathbf{p}$ so that it is not visible to the cameras. In this case, reconstruction of points in the blind plane will never occur in practice.

The reconstruction formula $\mathbf{x} \sim \mathbf{K}\,\mathbf{Y}$ is interesting since it means that $\mathbf{X}$ can be computed only by multiplying a 4×9 matrix on the 9-dimensional vector $\mathbf{Y}$ which, in turn is given by reshaping the outer product of $\mathbf{y}_1$ and $\mathbf{y}_2$. Alternatively, $\mathbf{x}$ can be computed by first reshaping $\mathbf{K}$ as a 12×3 matrix that is multiplied on $\mathbf{y}_1$, resulting in a 12-dimensional vector $\mathbf{x}'$. $\mathbf{x}$ is then obtained by reshaping $\mathbf{x}'$ as a 4×3 matrix and multiply it on $\mathbf{y}_2$. The computational complexity for computing $\mathbf{x}$ is therefore not more than 32 additions and 45 (or 48) multiplications, depending on which approach is used.

3 Experiments

A set of 72 3D points is defined by a calibration pattern placed on three planes at straight angels to each other. These points are viewed by two cameras which

Fig. 1. Images from the two cameras viewing the calibration pattern

Table 1. Reconstruction error for the triangulation operator **K**, the optimal method (O), the mid-point method (MP), the homogeneous method (H), and the inhomogeneous method (IH). All units are in mm. The last row shows a rough figure for the computation time relative to the proposed method.

	K	O	MP	H	IH
$\bar{e}$	1.08	1.07	1.07	1.12	1.12
e_{max}	2.74	2.79	2.79	3.44	3.45
σ_e	0.62	0.62	0.62	0.66	0.66
Time	1	34	15	5	3

are assumed to satisfy the pinhole camera model, Equation (1), see Figure 1. The calibration points are manually identified and measured in terms of their image coordinates in each of the two images. As a result, we have one set of 3D points $\{\mathbf{x}_k\}$ and two sets of *corresponding* 2D coordinates in the two images $\{\mathbf{y}_{1,k}\}$ and $\{\mathbf{y}_{2,k}\}$.

From these data sets, the two camera matrices $\mathbf{C}_1$ and $\mathbf{C}_2$ are estimated using the normalized DLT-method [5]. From $\mathbf{C}_1$ and $\mathbf{C}_2$, the corresponding focal points $\mathbf{n}_1$ and $\mathbf{n}_2$ can be determined using Equation (3), and a suitable blind plane $\mathbf{p}$ can be set up by choosing a third point so that this $\mathbf{p}$ is approximately perpendicular to the viewing directions of the two cameras. Given $\mathbf{C}_1$, $\mathbf{C}_2$ and $\mathbf{p}$, a triangulation operator $\mathbf{K}$ is computed according to the algorithm described in Section 2. This computation is made using coordinates which are normalized according to [6], followed by a proper re-normalization.

The triangulation operator $\mathbf{K}$ is evaluated on the data set together with the standard methods described in Section 1. The homogeneous and the inhomogeneous method is computed on normalized data and the result is transformed back to the original coordinates. For each of the 72 corresponding image points a 3D point in reconstructed, and the norm of the Euclidean reconstruction error is computed as $e = \|\mathbf{x}'_i - \mathbf{x}'_{i,rec}\|$ where $\mathbf{x}'_i$ is the i-th 3D coordinate and $\mathbf{x}'_{i,rec}$ is the corresponding reconstructed 3D coordinate. From the entire set of such errors, the mean $\bar{e}$, the maximum e_{max} and the standard deviation σ_e are estimated. The resulting values are presented in Table 1. All calculations are made in Matlab.

A few conclusions can be drawn from these figures. First, the triangulation operator $\mathbf{K}$ appears to be *comparable* in terms of accuracy with the best of the standard methods, which for this data set happens to be the optimal method. Given that the accuracy of the original 3D data is approximately ± 1 mm. and the 2D data has an accuracy of approximately ± 1 pixel, a more precise conclusion than this cannot be made based on this data. The last row of Table 1 shows very rough figures for the computation time of each of the methods as given by Matlab's profile function. The proposed method has a computational time which is significantly less than any of the standard methods, although it provides a comparable reconstruction error, at least for this data set.

4 Summary

This paper demonstrates that there exists a linear transformation $\mathbf{K}$ which maps the tensor or outer product $\mathbf{Y}$ of the homogeneous representations of two corresponding stereo image points to a homogeneous representation of the original 3D point. The main restriction of the proposed method is that the resulting linear transformation is dependent on an arbitrarily chosen 3D plane which includes the two focal points of the cameras, and only points which are not in the plane can be triangulated.

The proposed method is not derived from a perspective of optimality in either the 2D or 3D domains. The experiment presented above suggests, however, that at least the 3D reconstruction error is comparable even to the optimal method. In addition to this, the proposed method offers the following advantages

- It is can be implemented at a low computational cost; 32 additions and 45 multiplications for obtaining the homogeneous coordinates of the reconstructed 3D point. This is a critical factor in real-time applications or in RANSAC loops involving 3D matching.
- The existence of the reconstruction operator $\mathbf{K}$ implies that closed form expressions from homogeneous 2D coordinates to various homogeneous representations in 3D can be described. For example, given two pairs of corresponding points with their joint representations $\mathbf{Y}_1$ and $\mathbf{Y}_2$, two 3D points can be reconstructed as $\mathbf{x}_1 = \mathbf{K}\,\mathbf{Y}_1$ and $\mathbf{x}_2 = \mathbf{K}\,\mathbf{Y}_2$. The 3D line which passes through these two points can be represented by the anti-symmetric matrix $\mathbf{L} = \mathbf{x}_1 \otimes \mathbf{x}_2 - \mathbf{x}_2 \otimes \mathbf{x}_1$. A combination of these expressions then allows us to write $\mathbf{L} = (\mathbf{K} \otimes \mathbf{K})(\mathbf{Y}_1 \otimes \mathbf{Y}_2 - \mathbf{Y}_2 \otimes \mathbf{Y}_1)$, i.e., we can first combine the joint representations of corresponding points in the image domain and then transform this combination to the 3D space and directly get $\mathbf{L}$. This may only be of academic interest, but implies that $\mathbf{K}$ serves as some kind of stereo camera inverse (an inverse of a combination of both camera matrices).
- Although this property is not proven here, it follows from the construction of $\mathbf{K}$ that it is projective invariant in the sense described in [7]. This implies that the reconstructed point $\mathbf{x}$ is invariant, that is, it is the same point in space, independent of projective transformations of the coordinate system.

Acknowledgement

This work has been made within the VISCOS project funded by the Swedish Foundation for Strategic Research (SSF).

References

1. Beardsley, P.A., Zisserman, A., Murray, D.W.: Navigation using affine structure from motion. In: Eklundh, J.-O. (ed.) ECCV 1994. LNCS, vol. 801, pp. 85–96. Springer, Heidelberg (1994)

2. Christensen, O.: An Introduction to Frames and Riesz Bases. Birkhäuser (2003)
3. Harris, C., Stephens, M.: A combined corner and edge detector. In: Proc. 4th Alvey Vision Conference, Manchester, UK pp. 147–151 (1988)
4. Hartley, R., Schaffalitzky, F.: minimization in geometric reconstruction problems. In: Proceedings of the IEEE Conference on Computer Vision and Pattern Recognition, vol. I, pp. 769–775 (2004)
5. Hartley, R., Zisserman, A.: Multiple View Geometry in Computer Vision, 2nd edn. Cambridge University Press, Cambridge (2003)
6. Hartley, R.I.: In defence of the 8-point algorithm. IEEE Trans. on Pattern Recognition and Machine Intelligence 19(6), 580–593 (1997)
7. Hartley, R.I., Sturm, P.: Triangulation. Computer Vision and Image Understanding 68(2), 146–157 (1997)
8. Kahl, F., Henrion, D.: Globally optimal estimates for geometric reconstruction problems. In: Proceedings of International Conference on Computer Vision, vol. 2, pp. 978–985 (2005)
9. Scharstein, D., Szeliski, R.: A taxonomy and evaluation of dense two-frame stereo correspondence algorithms. International Journal in Computer Vision, 47(1–3):7–42 (2002)

Double Adaptive Filtering of Gaussian Noise Degraded Images

Tuan D. Pham[1,2]

[1] Bioinformatics Applications Research Centre
[2] School of Mathematics, Physics, and Information Technology
James Cook University
Townsville, QLD 4811, Australia
tuan.pham@jcu.edu.au

Abstract. Good estimate and simulation of the behavior of additive noise is central to the adaptive restoration of images corrupted with Gaussian noise. This paper presents a double adaptive filtering scheme in the sense that the filter is able to estimate the variance of additive noise in order to determine the filter gain for pixel updating, and also able to decide if the pixel should remain unfiltered. Experimental results obtained from the restoration of several images have shown the superiority of the proposed method to some benchmark image filters.

Keywords: Image restoration, Gaussian noise, adaptive filtering.

1 Introduction

The topic of restoring noisy images to its original verions is one of important research areas in image processing and computer vision. An optimal image restoration is usually based on some criteria for noise modeling and noise deduction. There are several types of noise that can degrade orginal images such as Gaussian (additive), Poisson, salt & pepper (on and off), and speckle (multiplicative). Thus, there are many methods proposed for dealing with different types of noise in images. Literature review in image restoration is huge but most methods are based on the methodologies of linear, non-linear, and adaptive filtering [5]. However, the problem of image restoration still remains challengingly open because image restoration is difficult since it is an ill-posed inverse problem – necessary information required about the degraded image to reconstruct the original is inherently imprecise. Therefore, methods for effective and better solutions continue to be developed.

Awate and Whitaker [1] have recently proposed an adaptive image filtering scheme for restoring degraded images by comparing pixel intensities having similar neighborhood in the image. Although this work makes no assumption about the properties of image signal and noise, it is empirically observed to perform best with additive noise. Other recently developed methods for image restoration are wavelet-based image denoising [10], mean-shift algorithms [2], and iterative function [16], pointspread function [3], and histogram-based fuzzy filter [15]. In

B.K. Ersbøll and K.S. Pedersen (Eds.): SCIA 2007, LNCS 4522, pp. 848–857, 2007.

particular, for additive noise, the principle of Wiener filtering is still one of the earliest and best known approaches for adaptive image filtering [5]. Based on this motivation, our contribution focuses on the reliable estimate of noise which is normally assumed in the Wiener filter process. We then extend the algorithm with an ability to detect noisy pixels to be adaptively updated. The rest of this paper is organized as follows. To facilitate the discussion of the new algorithms, basic formulation of adaptive Wiener filtering is outlined in Section 2. Section 3 presents a geostatistics-based method for noise estimation. In Section 4, the concept of spatial correlation in geostatistics is utilized to derive a decision rule for noise detection. Section 5 illustrates the performance of the proposed method and comparisons. In Section 6, we conclude our contribution and suggest further development of the proposed method.

2 Basic Adaptive Restoration Process

Let $f(x, y)$ be the ideal image, $g(x, y)$ the degraded image, and $v(x, y)$ the white noise which is additive, stationary, and has zero mean. The degraded image can be modeled as

$$g(x, y) = f(x, y) + v(x, y) \tag{1}$$

Given the observed image $g(x, y)$ and the noise statistics, the task of image restoration is to recover the signal as close as possible to the ideal image $f(x, y)$. Space-variant Wiener filters have been designed for such task, and a specific algorithm can be described as follows [9].

The power spectrum of the noise $v(x, y)$ with variance σ_v^2 is given by

$$P_v(\omega_x, \omega_y) = \sigma_v^2 \tag{2}$$

Assuming that $f(x, y)$ is stationary within a small local image $\mathcal{L}$, then $f(x, y)$ can be modeled by its local mean m_f, local standard deviation σ_f and the white noise $n(x, y)$ that has zero mean and unit variance:

$$f(x, y) = m_f + \sigma_f \, n(x, y) \tag{3}$$

The Wiener filter whose frequency reponse $H(\omega_x, \omega_y)$ within $\mathcal{L}$ is

$$\begin{aligned} H(\omega_x, \omega_y) &= \frac{P_f(\omega_x, \omega_y)}{P_f(\omega_x, \omega_y) + P_v(\omega_x, \omega_y)} \\ &= \frac{\sigma_f^2}{\sigma_f^2 + \sigma_v^2} \end{aligned} \tag{4}$$

From (5), a scaled impulse response $h(x, y)$ is given as

$$h(x, y) = \frac{\sigma_f^2}{\sigma_f^2 + \sigma_v^2} \delta(x, y) \tag{5}$$

From (5), the restored image $r(x,y) \in \mathcal{L}$ can be obtained by

$$r(x,y) = m_f + [g(x,y) - m_f] * \frac{\sigma_f^2}{\sigma_f^2 + \sigma_v^2}\, \delta(x,y) \tag{6}$$

$$= m_f + \frac{\sigma_f^2}{\sigma_f^2 + \sigma_v^2}[g(x,y) - m_f] \tag{7}$$

in which $*$ denotes the convolution operator.

If the local mean and standard deviation are updated at each pixel, we have [7]

$$r(x,y) = m_f(x,y) + k(x,y)\,[g(x,y) - m_f(x,y)] \tag{8}$$

where $k(x,y)$ is a filter gain and defined as

$$k(x,y) = \frac{\sigma_f^2(x,y)}{\sigma_f^2(x,y) + \sigma_v^2} \tag{9}$$

The local mean and local standard deviation described in (9) can be estimated using local statistics [7]

$$m_f(x,y) = \frac{1}{NM} \sum_{k,\, l \in w} g(k,l) \tag{10}$$

where w is a local $N \times M$ window of the image.

And

$$\sigma_f^2(x,y) = \frac{1}{NM} \sum_{k,\, l \in w} [g(k,l) - m_f(x,y)]^2 \tag{11}$$

If the noise variance is not known then it can be approximated as the mean of all the local variances in the image I, i.e.,

$$\sigma_v^2 \approx \frac{1}{NM} \sum_{k,\, l \in I} \sigma_f^2(k,l) \tag{12}$$

We are particularly interested in estimating the variance of white additive noise in images using the variogram function introduced in geostatistics, which models the spatial distribution of an image with a random function and estimates the noise variance as the expectation of the squared error of the random variables separated by a spatial distance. We then seek to formulate a rule to decide if a pixel value being corrupted with noise. This decision is based on the comparison of the long-range and local spatial probabilities of a pixel intensity.

3 Estimating Additive Noise with Geostatistics

The theory of geostatistics is based on the notion of regionalized variables and their random functions [6]. A variable $z(p)$ that is distributed in space Ω is said to be regionalized and represented as a realization of a random function $Z(p)$:

$$Z(p) = \{Z(p_i), \forall p_i \in \Omega\} \tag{13}$$

Let $z(p)$ and $z(p+h)$ be two real values at two locations p and $p+h$. The spatial variance of these two points is expressed by the variogram function $2\gamma(p, h)$, which is defined as the expectation of the squared difference of the two random variables

$$2\gamma(p, h) = E\{[Z(p) - Z(p+h)]^2\} \tag{14}$$

The estimation of (14) requires several realizations of the pair of random variables $[Z(p)-Z(p+h)]$, i.e., $[z_1(p), z_1(p+h)], [z_2(p), z_2(p+h)], \ldots, [z_k(p), z_k(p+h)]$. However, in many applications, only one realization $[z(p), z(p+h)]$ can be available, that is the actual measure of the values at point p and $p+h$. To overcome this problem, the *intrinsic hypothesis* [6] is introduced, which states that a random function $Z(p)$ is intrinsic when

- its mathematical expectation exists and does not depend on p:

$$E\{Z(p)]\} = m, \; \forall p$$

- the increment $[Z(p) - Z(p+h)]$ has a finite variance which does not depend on p:

$$Var\{Z(p) - Z(p+h)\} = E\{[Z(p) - Z(p+h)]^2\} = 2\gamma(h), \; \forall p$$

The variogram $2\gamma(h)$ is therefore constructed using the actual data as follows.

$$2\gamma(h) = \frac{1}{N(h)} \sum_{i=1}^{N(h)} [z(p_i) - z(p_i + h)]^2 \tag{15}$$

where $N(h)$ is the number of experimental pairs $[z(p_i) - z(p_i + h)]$ of the data separated by h.

As all the random variables of the random function have the same mean and variance, expanding (15) we have:

$$\begin{aligned}
\gamma(h) &= \frac{1}{2}E\{[Z(p+h) - Z(p)]^2\} \\
&= \frac{1}{2}E\{Z(p+h)^2\} + \frac{1}{2}E\{Z(p)^2\} - E\{Z(p+h)Z(p)\} && (16) \\
&= E\{Z^2\} - E\{Z(p+h)Z(p)\} && (17) \\
&= E\{Z^2\} - m^2 - [E\{Z(p+h)Z(p)\} - m^2] && (18) \\
&= \sigma^2 - Cov\{Z(p+h)Z(p)\} && (19)
\end{aligned}$$

Based on (13), (14), and (19), we can draw two conceptual aspects as follows.

1. The variogram can be viewed as an estimation variance in which an error occurs when the value at p is estimated by the value at $p+h$, i.e.,

$$\sigma_E^2 = E\{[Z(p) - Z(p+h)]^2\} \quad (20)$$

2. Given that the location $p+h$ is sufficiently close to p, then $z(p+h)$ can be considered to be another realization of $Z(p)$, i.e., $z(p+h) \approx z_k(p)$.

We therefore can approximate the variance of additive noise in an image as the variogram where $h = 1$ expressed in pixels:

$$\sigma_v^2 \approx \sigma_E^2 = 2\gamma(h = 1) \quad (21)$$

It is realized that we can deduce to estimate the noise variance in an image from the experimental variogram at h=0. Such estimation can be obtained by extrapolation [6], and has been applied for calculating the standard deviations of non-uniform images [12]. However, extrapolation is less accurate and the extrapolation proposed in [12] is not suitable for automatic processing.

4 Detection of Noisy Pixels

Let I be an image, and let f_i and f_j be the gray levels of pixels located at positions i and j in I respectively. The purpose is to determine the probability of the spatial non-randomness of the pixel considered for filtering within a neighborhood. This probability is called the local probability of spatial non-randomness. If this probability is higher than that of the long-range probability of non-randomness of the same pixel intensity, then its intensity value is considered as uncorrupted and therefore remains unchanged; otherwise the pixel will be filtered. We proceed the determination of these two types of probability as follows.

Let T be the gray-level threshold of I. The thresholded image of I at T is defined as

$$y_i(T) = \begin{cases} 1 & : \quad f_i = T \\ 0 & : \quad f_i \neq T \end{cases} \quad (22)$$

Utilizating the concepts of variograms and histograms, the long-range probability of spatial non-randomness of pixels having intensity value T separated by lag distance h over I, denoted as $P_G(T,h)$, can be defined by

$$P_G(T,h) = \frac{1}{N(h)} \sum_{(i,j)\in I|h_{ij}=h} \delta_{ij}(h) \quad (23)$$

where $N(h)$ is the total number of pairs of pixels in I separated by lag distance h, h_{ij} is the distance between f_i and f_j, and $\delta_{ij}(h)$ behaves like a *Kronecker delta* which is defined as

$$\delta_{ij}(h) = \begin{cases} 1 & : \quad y_i(T) = y_j(T),\ h_{ij} = h \\ 0 & : \quad y_i(T) \neq y_j(T),\ h_{ij} = h \end{cases} \tag{24}$$

Similarly, the local probability of the spatial non-randomness of pixels having intensity value T separated by lag distance h over some chosen window w_L, denoted as $P_L(T,h)$, can be defined by

$$P_L(T,h) = \frac{1}{M(h)} \sum_{(i,j)\in w_L | h_{ij}=h} \delta_{ij}(h) \tag{25}$$

where $M(h)$ is the total number of pairs of pixels in window w_L separated by lag distance h.

It is known that the variogram is very useful for exploring the features of spatial data, and can help to detect anomalies, inhomogeneities and randomness of sample values in spatial domain [14]. We use the conceptual framework of calculating the variogram and the histogram to compute the probability that indicates the chance of a pixel intensity being located closely to each other in space. We then make an assumption that if the local probability exceeds the global (long-range) probability, then the pixel is not corrupted with noise. Thus, the decision to either retain or update each pixel is according to the following rules:

$$\begin{array}{lcl} P_L(T,h) > P_G(T,h) & : & \text{no updating} \\ P_L(T,h) \leq P_G(T,h) & : & \text{updating} \end{array} \tag{26}$$

5 Experiments

We test the algorithm with several images as shown in Figure 1 which consists of 4 different original images. Those images are selected based on the good content of different details, texture, regions, and shading. To measure the performance of the restoration, the signal-to-noise ratio (SNR) improvement expressed in dB is used, which is defined by [9]

$$\text{SNR Improvement} = 10 \log_{10} \frac{\text{NMSE}(f,g)}{\text{NMSE}(f,r)} \tag{27}$$

where the normalized mean square error (NMSE) between the original image f and the restored image r is defined by

$$\text{NMSE}(f,r) = 100\, \frac{Var(f-r)}{Var(f)}\ \% \tag{28}$$

All images were processed with a 3×3 filter window (w) which was suggested by [7] as a fairly good choice. To compute the local probability, we set the size

Fig. 1. Original images: Top left: photographer, top right: Lenna, bottom left: moon, bottom right: saturn

Fig. 2. Top left: degraded image with Gaussian noise (0 mean, 0.05 variance), top right: adaptive Wiener filter, bottom left: mean filter, bottom right: double adaptive filter

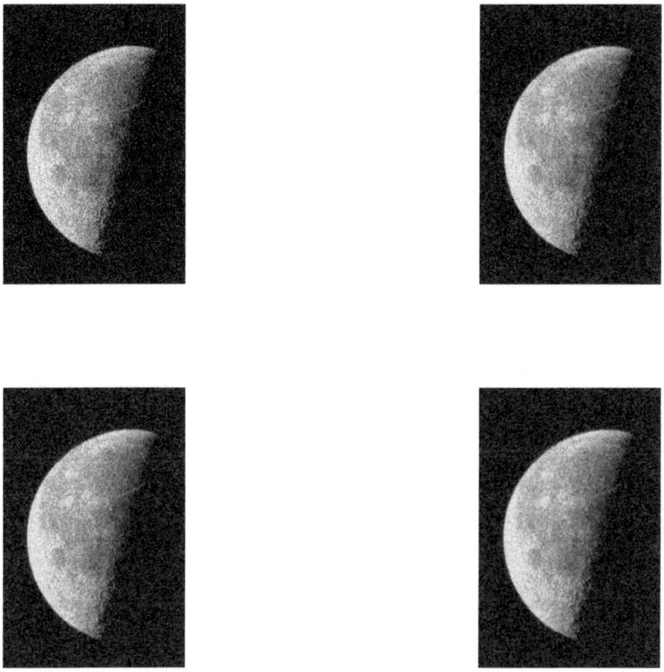

Fig. 3. Top left: degraded image with Gaussian noise (0 mean, 0.05 variance), top right: adaptive Wiener filter, bottom left: mean filter, bottom right: double adaptive filter

of the local window (w_L) to be 9×9 which is considered sufficient enough to calculate the spatial correlation (in geostatistics, the number of data points being greater than 25 can be considered sufficent to compute the semi-variogram). The original images were corrupted with zero-mean Gaussian white noise of variances of 0.02 and 0.05 respectively. Tables 1 and 2 show the SNR improvements of the processed images obtained by the adaptive Wiener filter method, the mean filter, and the proposed method where single adaptive does not employ the noise detection procedure but the double adaptive does. The proposed method gives the highest SNR improvements in all cases; whereas and the adaptive Wiener filter gives higher SNR improvements than the mean filter. Figure 2 shows the moon image degraded with zero-mean Gaussian noise of variances of 0.05 and its restored versions obtained from the three methods. In addition to its highest SNR, the restored image achieved from the proposed filter can be observed that noise is suppressed with smoother effect than the average filter, and the edges can still be better reserved than the adaptive Wiener filter.

To compare with other recently developed methods, we used the Lenna image whose size is larger than that used in UNITA method developed by Awate and

Table 1. SNR Improvements (dB) – Gaussian Noise (zero mean, 0.02 variance)

Image	Adaptive Wiener	Mean	Single Adaptive	Double Adaptive
Lena (222 × 208)	6.86	6.46	7.56	8.23
photographer (256 × 256)	6.16	5.89	6.31	7.45
moon (900 × 1200)	5.30	5.16	7.16	8.01
saturn (900 × 1201)	5.09	4.98	7.13	8.46
Logo database	5.13	4.64	5.29	6.77

Table 2. SNR Improvements (dB) – Gaussian Noise (zero mean, 0.05 variance)

Image	Adaptive Wiener	Mean	Single Adaptive	Double Adaptive
Lena (222 × 208)	6.96	6.76	8.17	9.82
photographer (256 × 256)	6.27	6.09	6.99	8.01
moon (900 × 1200)	5.24	5.18	7.11	9.14
saturn (900 × 1201)	5.16	5.12	7.17	9.10
Logo database	5.24	4.78	5.92	7.24

Table 3. RMSE ratios of different methods using Lenna image

UNITA	WF1	WF2	WF3	DA
0.46	0.36	0.38	0.41	0.38

Whitaker [1]. The image was simulated with the same level of the additive noise to carry out the restoration using the proposed double adaptive filter (denoted as DA). We then used the root-mean-square error (RMSE) ratio of the initial and restored RMSE errors to compare the results obtained from the UNITA, wavelet-based filter (denoted as WF1) developed by Portilla *et al.* [11], wavelet-based filter (denoted as WF2) developed by Sender and Selesnick [13], and wavelet-based filter (denoted as WF3) developed by Pizurica *et al.* [10]. The results from these three wavelet-based methods were taken from [1]. Table 3 show the error ratios obtained from the five models. The error ratio of the proposed method is equivalent to that of WF2, higher than WF1, and lower than UNITA and WF3. However, it was mentioned in [1] that although the wavelet-based approach gave lower error than that of the UNITA, it has ringing-like artifacts in smooth regions. Our proposed method yields lower error ratio than the UNITA filter and still does not suffer from that effect. We further tested the same image restoration methods using the logo database of the University of Maryland [4] that consists of 105 intensity logo images of scanned binary versions. The results obtained show similar analysis addressed earlier, where the proposed double filtering approach gives the best improvements.

6 Conclusions

The ideas of spatial relation and spatial randomness modeled by the theory of regionalized variables and geostatistics have been utilized to largely improve the well-known principle of adaptive Wiener filter for restoring images degraded with Guassian white noise. The experiments have shown that this algorithm provided the most favorable results by means of the SNR improvements and visual observation over some benchmark methods. The proposed method appears to be better or competitive with some other recently developed techniques. In addition, the computational procedure and speed of the proposed method is very easy for computer implementation, and can be extended for filtering other types of noise in images.

References

1. Awate, S.P., Whitaker, R.T.: Unsupervised, information-theoretic, adaptive image filtering for image restoration. IEEE Trans. PAMI 28, 364–376 (2006)
2. Barash, D., Comaniciu, D.: A Common Framework for Nonlinear Diffusion, Adaptive Smoothing, Bilateral Filtering and Mean Shift. Image Vision Computing 22, 73–81 (2004)
3. Ferreira, V.C., Mascarenhas, N.D.A.: Analysis of the Robustness of Iterative Restoration Methods with Respect to Variations of the Point Spread Function. In: Proc. ICIP'00 III, 789–792 (2000)
4. `ftp://ftp.cfar.umd.edu/pub/documents/contrib/database/UMDlogo_database.tar`
5. Gonzalez, R.C., Woods, R.E., Eddins, S.L.: Digital Image Processing using Matlab. Pearson Prentice Hall, Englewood Cliffs (2004)
6. Journel, A.G, Huijbregts, C.J.: Mining Geostatistics. Academic Press, Chicago (1978)
7. Lee, J.-S.: IEEE Trans. PAMI, vol. 2 pp.165–168 (1980)
8. Lee, J.-S., Hoppel, K.: Noise modeling and estimation of remotely-sensed images. In: Proc. Int. Conf. Geoscience and Remote Sensing 2, 1005–1008 (1989)
9. Lim, J.S.: Two-Dimensional Signal and Image Processing. Prentice-Hall, Englewood Cliffs (1990)
10. Pizurica, A., Philips, W., Lemahieu, I., Acheroy, M.: A joint inter and intrascale statistical model for Bayesian wavelet based image denoising. IEEE Trans. Image Processing 11, 545–557 (2002)
11. Portilla, J., Strela, V., Wainwright, M., Simoncelli, E.: Image denoising using scale mixtures of Gaussians in the wavelet domain. IEEE Trans. Image Processing 12, 1338–1351 (2003)
12. Sanchez-Brea, L.M., Bernabeu, E.: On the standard deviation in charge-coupled device cameras: A variogram-based technique for nonuniform images. J. Electronic Imaging 11, 121–126 (2002)
13. Sendur, L., Selesnick, I.: Bivariate shrinkage functions for wavelet-based denoising exploiting interscale dependency. IEEE Trans. SP 50, 2744–2756 (2002)
14. Wackernagel, H.: Multivariate Geostatistics. Springer, Heidelberg (2003)
15. Wang, J.H., Liu, W.J., Lin, L.D.: Histogram-based fuzzy filter for image restoration. IEEE Trans. SMC - Part B: Cybernetics 32, 230–238 (2002)
16. Xue, F., Liu, Q.S., Fan, W.H.: Iterative Image Restoration using a Non-Local Regularization Function and a Local Regularization Operator. In: Proc. ICPR'06 III, 766–769 (2006)

Automatic Extraction and Classification of Vegetation Areas from High Resolution Images in Urban Areas

Corina Iovan[1,2], Didier Boldo[1], Matthieu Cord[2], and Mats Erikson[3]

[1] Institut Géographique National, Laboratoire MATIS
2/4 avenue Pasteur, 94165 Saint-Mandé Cedex, France
{Corina.Iovan, Didier.Boldo}@ign.fr
[2] Université Pierre et Marie Curie, Laboratoire d'Informatique de Paris VI
8 rue du Capitaine Scott, 75015 Paris, France
Matthieu.Cord@lip6.fr
[3] ARIANA Research Group, CNRS/INRIA/UNSA
2004, route des Lucioles - BP 93, 06902 Sophia Antipolis Cedex, France
Mats.Eriksson@sophia.inria.fr

Abstract. This paper presents a complete high resolution aerial-images processing workflow to detect and characterize vegetation structures in high density urban areas. We present a hierarchical strategy to extract, analyze and delineate vegetation areas according to their height. To detect urban vegetation areas, we develop two methods, one using spectral indices and the second one based on a Support Vector Machines (SVM) classifier. Once vegetation areas detected, we differentiate lawns from treed areas by computing a texture operator on the Digital Surface Model (DSM). A robust region growing method based on the DSM is proposed for an accurate delineation of tree crowns. Delineation results are compared to results obtained by a Random Walk region growing technique for tree crown delineation. We evaluate the accuracy of the tree crown delineation results to a reference manual delineation. Results obtained are discussed and the influential factors are put forward.

1 Introduction

Automatic 3D reconstruction of urban areas is an active research topic in distinct application areas and an issue of primary importance in fields such as urban planning, disaster management or telecommunications planning. Significant progress has been made in recent years concerning the automatic reconstruction of man-made objects or environments from multiple aerial images [1]. Yet, a lot of challenge concerning the modelling of other objects present on the terrain surface, such as trees, shrubs, hedges or lawns still exists. An accurate automatic reconstruction of such types of vegetation areas is a real challenge due to their complex nature and to their intricate distribution between man-made objects in dense urban areas. Many researches deal with automatic tree crown delineation from aerial or satellite images. We can divide them into two classes: methods applied to forest stands and methods applied to urban environments.

B.K. Ersbøll and K.S. Pedersen (Eds.): SCIA 2007, LNCS 4522, pp. 858–867, 2007.

Several algorithms have been proposed for the segmentation of individual trees in forest stands. A first class uses local maxima information to estimate tree top position and the number of trunks [2][3]. A second class of methods exploits the shadows around the tree crowns to delineate their contour [4], such as valley-following algorithms [5] or region growing methods [6]. Other contour based methods use multi-scale analysis [7] or active contours [8] to delineate tree crowns. A third class of methods are object-based methods [9][10][11], modelling synthetic tree crown templates to find tree top positions.

Algorithms developed for the automatic extraction of tree crowns in urban environments firstly detect vegetation areas followed by a finer analysis of objects present therein. Depending on the input data, vegetation areas are detected either using vegetation responses in color infrared (CIR) images [12] or by computing surface roughness indicators on the DSM [13]. A finer analysis of treed areas is then performed and its goal ranges from simple tasks, such as estimating tree position [12][14], to more complex tasks such as tree species classification [6].

This study presents our approach for vegetation detection and segmentation in urban areas. A linear-kernel SVM classifier using a four dimensional radiometric feature vector is used to identify vegetation areas. Texture features computed on the DSM separate lawns from treed areas. A robust algorithm for tree crown delineation taking into account the trees height and shape characteristics is proposed to accurately delineate individual tree crowns.

The accuracy of the segmentation results is evaluated against a reference delineation and they are also compared to results obtained by a random walk tree crown delineation algorithm [6]. Results obtained using the proposed method are very promising and show their potential by improving delineation results obtained by the second method.

2 Study Area and Data

2.1 Study Area

The study area is located in the city of Marseille, situated in the south-east of France. Marseille's climate is Mediterranean, with a great variety of vegetation species. It's a dense urban area, with many greened and treed resting places, highly intermingled with buildings.

2.2 Data

In this study, tests were carried out on digital color infrared aerial images, taken in November 2004, with a ground resolution of 20cm per pixel. A DSM is derived from the stereoscopic aerial images using an algorithm based on a multiresolution implementation [15] of Cox and Roy's image matching algorithm [16] based on graph cuts. Figure 1 presents the input data for one of our study areas.

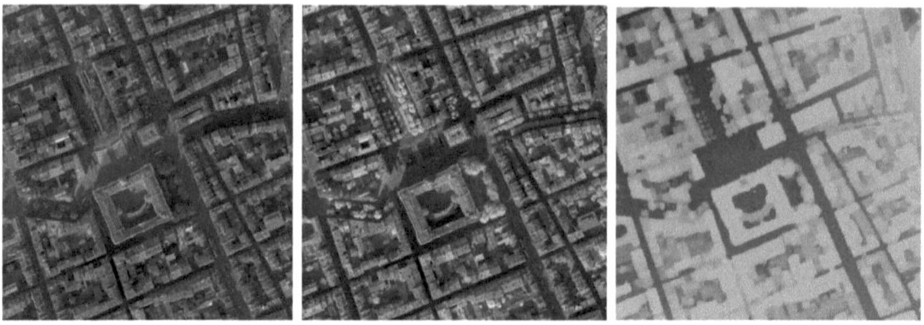

a) b) c)

Fig. 1. An aerial image of Marseille (France) representing a high density urban area, where 1 pixel corresponds to approximately 20cm a) RGB channels b) IR channel c) DSM for the same area

3 Methods

Our approach is based on a hierarchical strategy containing several steps: detection of vegetation areas, segmentation of vegetation types according to their height followed by individual tree crown delineation. In the following paragraphs we will describe the methods developed for each of these steps.

3.1 Radiometric Corrections

In a first step, the four channels of the raw images are radiometrically corrected. Radiometric corrections address variations in the pixel intensities that are not caused by the object or scene being scanned. Due to the sea vicinity, haze is often perturbing the signal penetration. Atmospheric haze reduction is a simple method that assumes that some image pixels should have a reflectance of zero. Actual values of zero pixels result from atmospheric scattering. Haze correction consists in subtracting the histogram offset from all pixels in a specific band. The result of the atmospheric correction is depicted in Fig. 2.

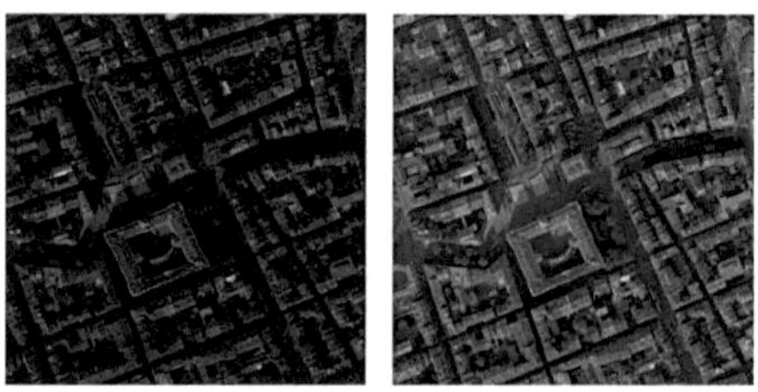

Fig. 2. RGB image representing downtown of Marseille before (left) and after (right) the atmospheric haze reduction

3.2 Vegetation Detection

Two methods were developed to identify vegetation areas. The first one is an unsupervised classification method based on different spectral indices. The second one is a supervised classification method using a linear-kernel Support Vector Machines (SVM) classifier.

Unsupervised Classification
The unsupervised classification method uses several spectral indices to identify vegetation areas. The first index computed for each pixel in our images is the NDVI (Normalised Difference Vegetation Index)[17]. It allows the creation of a gray-level image, the NDVI image (presented in Fig. 3 b)), by computing for each pixel the NDVI index, according to (1)

$$NDVI = \frac{\varphi_{IR} - \varphi_R}{\varphi_{IR} + \varphi_R} \tag{1}$$

where φ_{IR} and φ_R are the values of the pixels respectively in the infrared and the red band. This index highlights areas with a higher reflectance in the infrared band than in the red band (i.e. vegetation). Applying a threshold on the NDVI image gives a coarse segmentation of the urban scene in vegetation areas and non-vegetation areas. As there are also other materials present in an urban environment with a high reflectance in the infrared band, we refine vegetation classification results using a second spectral index computed for each pixel, according to (2)

$$SI = \frac{\varphi_R - \varphi_B}{\varphi_R + \varphi_B} \tag{2}$$

This is the saturation index (SI) [18] and the gray-level image obtained for this index for each pixel is presented in Fig. 3 c). The images obtained with these two spectral indices are binarized and used together to create the vegetation mask. The result presented in Fig. 3 d) emphasizes all vegetation areas.

Supervised Classification
Although the vegetation detection method based on spectral indices gives satisfying results, it is a method highly dependent on the spectral characteristics of the

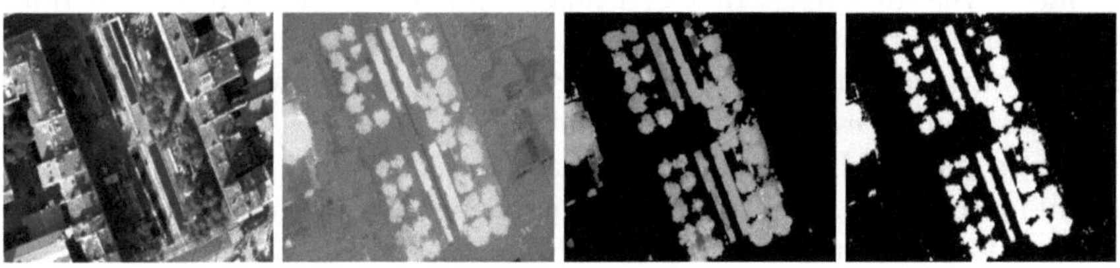

a) b) c) d)

Fig. 3. Spectral indices used for vegetation detection a) RGB input image b) NDVI image c) SI image d) Vegetation mask obtained with the NDVI and SI spectral indices data present in the study area. Our goal was to develop a method which performs well in any case. Therefore, we used a reliable supervised classification method based on SVM's. For all pixels in the training dataset, the feature vector contains four characteristics, namely, the reflectance values of each pixel in the infrared, red, green and blue bands. The choice of a linear-kernel for the classifier was motivated by the fact that the spectral indices we used in the first method are linear combinations of the image's channels and they perform well for distinguishing between vegetation and non vegetation areas. Therefore, instead of deciding where the separator between the two classes is, by combining different spectral indices, we decided to leave this task to the SVM and thus exploit its capacities in finding the optimal linear separator. Figure 4 b) presents the vegetation mask obtained by the SVM classifier for the test area presented in Fig. 4 a).

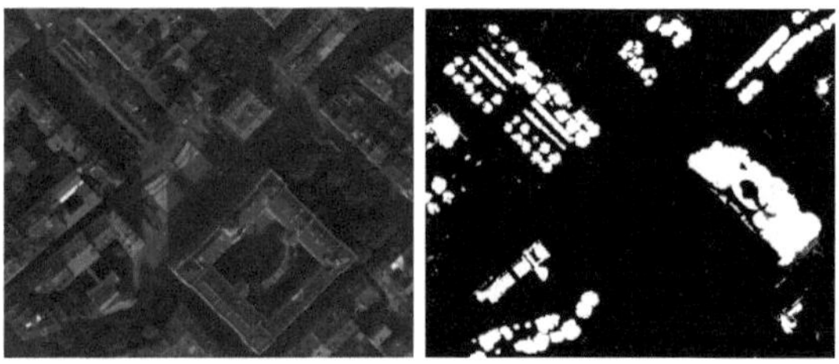

a) b)

Fig. 4. Vegetation detection based on linear-kernel SVM classification a)Input image b)Vegetation mask

Both methods presented in this section give pertinent results for vegetation detection but in the following of our study we use the vegetation mask obtained by the unsupervised classification.

3.3 Grass/Tree Segmentation

Once the regions of interest identified, i.e. vegetation areas, we proceed to a finer level of analysis of these vegetation structures, by performing texture analysis on the corresponding areas of the DSM. The goal of this second step is to differentiate lawns from trees. In a CIR image, grasses are characterized by ranges of coloration and texture. In the DSM, treed areas are characterized by a higher gray level variance compared to lawn areas. The method we developed to separate grass from trees takes into account this property by computing the local variance on the DSM. The resulting image is thresholded to obtain masks for grass and treed areas. Figure 5 shows the results obtained for grass/treed area separation for the test area depicted in Fig. 3 a).

3.4 Tree Crown Delineation

To separate tree crowns from each other, we developed a region growing method taking into account the treed areas previously detected. All region growing (*RG*)

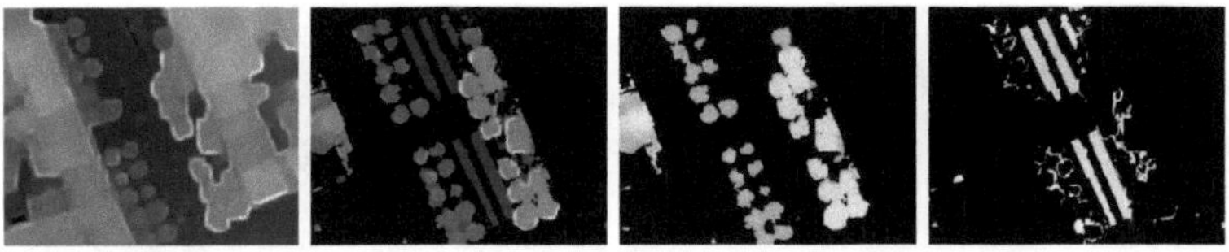

a) b) c) d)

Fig. 5. Differentiation between grass and treed areas a)Local variance computed on the DSM b)Vegetation areas on the DSM c)Treed areas d)Lawns

methods need seed points for each region before growing. The performance of these algorithms is highly dependent on the number of seed points initialising each region. The ideal case is to have one seed point per region.

Seed points. We aimed at achieving this goal: having one seed point for each tree. This information could be the information concerning the top of each tree. We use the DSM to estimate tree tops. To reduce the number of possible candidates for a tree top, we use a Gaussian filter as a smoothing filter for the DSM, with an empirical determined mask, approaching the average size of the trees in the image. To determine tree tops, we evaluate the maximum height of the trees present in the DSM and we consider all points having the same height as tree tops. In the first iteration we obtain points corresponding to the highest trees in the stand. Therefore, we iteratively decrease the analysis altitude, h. At each step, we analyse all points at higher heights than h and detect a new seed when a new region appears and it doesn't touch pixels previously labelized as seeds. A graphical ilustration of this algorithm is presented in Fig. 6.

Region growing. Starting from the previously determined tree tops, tree crown borders are obtained by a region growing approach based on geometric criteria of the trees. We used the DSM to evaluate the height of all points neighbour to a seed point. All pixels corresponding to a lower height point are aggregated to the region corresponding to the closest (in terms of height) tree top. The results

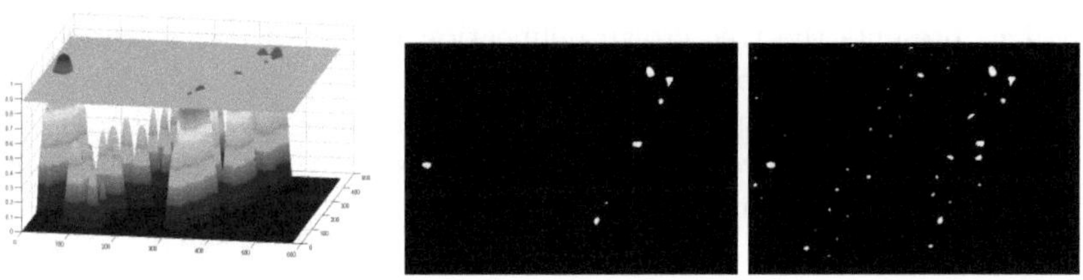

a) b) c)

Fig. 6. Detecting tree tops from the DSM a)3D view of the DSM: all points higher than the analysis altitude h are evaluated for tree top estimation b)2D view of the 30th iteration c)Seed points detected after the final iteration: we can notice that we obtain one seed region for each tree

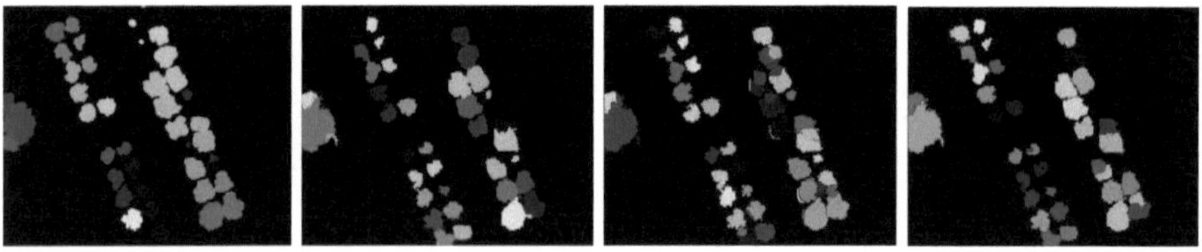

a) b) c) d)

Fig. 7. Tree crown delineation results a)Reference delineation of tree crowns b)Segmentation results for the Height-RG (HRG) method c)Segmentation results for the (RW) region growing method ($RWRG$) d) Tree crown delineation results for the ($RWRG$) method applied to height data ($H - RWRG$)

of this algorithm for tree crown delineation for the test area presented in Fig. 3 a) can be seen in Fig. 7 b).

We compared the method we developed with a random-walk (RW) region growing method, described in [6]. This method, briefly described below, is applied on an artificial image containing for each channel the DSM, instead of simply applying it on the colour image.

Seed points. To find seed points, the first band of the input image (DSM) is thresholded and a distance transform is performed on the resulting image. This distance image is smoothed with a Gaussian filter and local maxima on this image represent the seeds for the region growing algorithm.

Region growing. Each of the previously detected seeds is grown to become a region. A priority queue is established for the order the seeds are processed: seeds which are border pixels to a region are processed sooner than seeds which are not border pixels. The higher the value the border pixel is the sooner it will be connected to a region. This value is taken from a new image, a random walk image which is obtained from the original image by simulating random walks for each seed point. The value of each pixel represents the number of times the simulated particles have reached the pixel. A series of constraints decide on the rapidity of a pixel aggregation to a region. Further details can be found in [6]. Figure 7 c) presents the tree crown delineation results of this method for the same test area as the one presented in Fig. 3 a).

We also combined the two methods by using seed points found by the first method with the region growing method of the second method. The results of tree crown delineation for this case can be seen in Fig. 7 d).

4 Results

This section evaluates the results of the tree crown segmentation methods presented in the first part of this article and depicted in Fig. 7 b) - d). All segmentation results are compared to the reference manual delineation of the trees presented in Fig. 7 a). This manual delineation has been generated by an experienced photo interpreter by means of stereo restitution. It contains, for all trees visible in the CIR images, the exact delineation of tree crowns which will be considered as reference delineation in the following of the evaluation.

The accuracy assessment results are presented in table 1, where *Nt* denotes the number of trees and *Ratio*, the percentage of the total number of trees computed using the total number of trees in the stand.

Table 1. Comparison between the reference delineation of tree crowns and results obtained for tree crown delineation by the three methods

	Height Region Growing Method		RW Region Growing Method		Height-RW Region Growing Method	
	Nt	Ratio	Nt	Ratio	Nt	Ratio
Trees correctly segmented	32	78.0	23	56.1	30	73.1
Trees over-segmented	1	2.4	11	26.8	3	7.3
Trees under-segmented	4	9.7	4	9.7	4	9.7
Trees omitted	4	9.7	4	9.7	4	9.7
Total number of trees in the stand	41		41		41	
Total number of detected trees	37		51		37	

4.1 Evaluation Measures

The approach used for the evaluation is similar to the one presented in [14]. A statistical analysis is first performed taking into consideration the total number of trees in the ground truth and the omission (omitted trees) and commission errors (segments not associated with a tree). We take into consideration the following cases for the spatial analysis of the segmentation: pure segments, over-segmented trees, under-segmented trees. Pure segments correspond to correctly identified trees. We consider that a segment is 100% pure if it corresponds to one and only one segment in the ground truth and vice versa, with an overlap area greater than 80%. Over-segmented trees correspond to the case when more than one segment is associated with the ground truth delineation. Under segmented trees correspond to segments which include a significant part ($> 10\%$) of more than one tree.

4.2 Disscusion

The two methods for vegetation/non–vegetation classification give very good results. Surface classification rates are high for the two methods, from 87.5% for the spectral index based method to 98.5% for the SVM classification method.

Concerning the grass/lawn segmentation, the results were evaluated using a manual delineation and the results are very promising. More than 97% of the grass surface in the reference delineation was correctly classified as lawn.

Regarding the tree crown delineation, we notice that all the three methods previously described have detected most of the large trees. The H–RWRG and the HRG methods detect the same number of trees in the stand, and this is due to the fact that the same seeds are used for the region growing step. Omitted trees have in fact a low height, and due to the gaussian blurring of the DSM before finding seed points, these trees are not present in the DSM when the region growing part starts. The number of correctly segmented trees is higher for the HRG method, and this is due to the way the seeds are grown. The results of the RWRG method are improved by 17% when it uses one seed for region. These results show the good potential of the proposed method to find one seed for each tree.

5 Conclusion

Three region growing methods for tree crown delineation have been evaluated and show the capacity of having a realistic geometric description of tree crowns in urban areas. Our ongoing research deals with the improvement of the height–based tree crown delineation method by including information from the CIR images in the segmentation step. Extra information (tree crown diameter, height) can be extracted for each tree and it can be used for a 3D modelisation of trees. We will also consider evaluating the performances of the proposed method for tree crown delineation on laser DSM′s.

Acknowledgments. The fourth author thanks the Arc Mode De Vie Project at INRIA for financial support.

References

1. Taillandier, F., Vallet, B.: Fitting constrained 3d models in multiple aerial images. In: British Machine Vision Conference, Oxford, U.K. (August 2005)
2. Pinz, A.: Tree isolation and species classification. In: Hill, D., Leckie, D., eds.: Proc. of the International Forum on Automated Interpretation of High Spatial Resolution Digital Imagery for Forestry, Victoria, British Columbia, Canada pp. 127–139 (February 1998)
3. Wulder, M., White, J., Niemann, K., Nelson, T.: Comparison of airborne and satellite high spatial resolution data for the identifcation of individual trees with local maxima filtering. International Journal of Remote Sensing 25(11), 2225–2232 (2004)
4. Gougeon, F., Leckie, D.: Individual tree crown image analysis - a step towards precision forestry. In: Proc. of the First International Precision Forestry Symposium, Seattle, U.S (June 2001)
5. Gougeon, F.: A system for individual tree crown classification of conifer stands at high spatial resolution. In: Proc. of the 17th Cananadian Symposium on Remote Sensing, Saskatchewan, Canada pp. 635–642 (June 1995)
6. Erikson, M.: Segmentation and Classification of Individual Tree Crowns. PhD thesis, Swedish University of Agricultural Sciences, Uppsala, Sweden (2004)
7. Brandtberg, T., Walter, F.: Automatic delineation of individual tree crowns in high spatial resolution aerial images by multiple-scale analysis. Machine Vision and Applications 11(2), 64–73 (1998)

8. Horvath, P., Jermyn, I., Kato, Z., Zerubia, J.: A higher-order active contour model for tree detection. In: Proc. International Conference on Pattern Recognition (ICPR), Hong Kong (August (2006)
9. Pollock, R.: The Automatic Recognition of Individual Trees in Aerial Images of Forests based upon a Synthetic Tree Crown Image Model. PhD thesis, University of BC, Department of Computer Science, Vancouver, Canada (1996)
10. Larsen, M.: Crown modelling to find tree top positions in aerial photographs. In: Proc. of the Third Internationl Airborne Remote Sensing Conference and Exhibition. vol. 2, pp.428–435 (1997)
11. Perrin, G., Descombes, X., Zerubia, J.: 2d and 3d vegetation resource parameters assessment using marked point processes. In: Proc. International Conference on Pattern Recognition (ICPR). pp. 1–4 (2006)
12. Straub, B.M.: Automatic extraction of trees from aerial images and surface models. In: ISPRS Conference Photogrammetric Image Analysis (PIA). Volume XXXIV, Part 3/W834., Munich, Germany (September 2003)
13. Rottensteiner, F., Trinder, J., Clode, S., Kubik, K.: Using the dempster-shafer method for the fusion of lidar data and multi-spectral images for building detection. Information Fusion 6(4), 283–300 (2005)
14. Mei, C., Durrieu, S.: Tree crown delineation from digital elevation models and high resolution imagery. In: Proc. of the ISPRS Workshop 'Laser scanners for Forest and Landscape Assement'. vol. 36., Fribourg, Germany (October 2004)
15. Pierrot-Deseilligny, M., Paparoditis, N.: A multiresolution and optimization-based image matching approach: An application to surface reconstruction from spot5-hrs stereo imagery. In: Proc. of the ISPRS Conference Topographic Mapping From Space (With Special Emphasis on Small Satellites), Ankara, Turkey, ISPRS (February 2006)
16. Roy, S., Cox, I.: A maximum-flow formulation of the n-camera stereo correspondence problem. In: Proc. of the IEEE International Conference on Computer Vision, Bombay, India pp. 492–499 (January 1998)
17. Gong, P., Pu, R., Biging, G.S., Larrieu, M.R.: Estimation of forest leaf area index using vegetation indices derived from hyperion hyperspectral data. IEEE Transactions on Geoscience and Remote Sensing 41(6), 1355–1362 (2003)
18. Mathieu, R., Pouget, M., Cervelle, B., Escadafal, R.: Relationships between satellite based radiometric indices simulated using laboratory reflectance data and typic soil colour of an arid environment. Remote Sensing of Environment 66, 17–28 (1998)

An Intelligent Image Retrieval System Based on the Synergy of Color and Artificial Ant Colonies

Konstantinos Konstantinidis, Georgios Ch. Sirakoulis, and Ioannis Andreadis

Laboratory of Electronics, Dept. of Electrical and Computer Engineering, Democritus University of Thrace, 12 V. Sofias Str., 67100 Xanthi, Greece
{konkonst, gsirak, iandread}@ee.duth.gr

Abstract. In this paper a new image retrieval algorithm is proposed which aims to discard irrelevant images and increase the amount of relevant ones in a large database. This method utilizes a two-stage ant colony algorithm employing in parallel color, texture and spatial information. In the first stage, the synergy of the low-level descriptors is considered to be a group of ants seeking the optimal path to the "food" which is the most similar image to the query, whilst settling pheromone on each of the images that they confront in the high similarity zone. In the second stage additional queries are made by using the highest ranked images as new queries, resulting in an aggregate deposition of pheromone through which the final retrieval is performed. The results prove the system to be satisfactorily efficient as well as fast.

Keywords: Ant Colony Algorithm, Image Retrieval.

1 Introduction

Since multimedia technology is daily enhanced, vast image, video and audio databases are rapidly growing. These digital libraries are produced by various applications such as medicine, military, entertainment or education and belong to either private or public (World Wide Web) databases. The need for these databases to be indexed has led to an increasing interest on the research and development of automatic content-based image retrieval systems. Effective retrieval of image data is important for general multimedia information management. For an image to be retrievable, it has to be indexed by its content. Color can provide significant information about the content of an image. Among the methods that use color as a retrieval feature, the most popular one is probably that of color histograms [1,2]. The histogram is a global statistical feature which describes the intensity distribution for a given image [1]. Its main advantage is that low computational cost is required for its manipulation, storage and comparison. Moreover, it is insensitive to rotation and scale of the image scene and to any displacement of objects in the image. On the other hand it is also somewhat unreliable as it is sensitive even to small changes in the context of the image. Swain and Ballard [2] proposed a simple but nonetheless effective method of matching

B.K. Ersbøll and K.S. Pedersen (Eds.): SCIA 2007, LNCS 4522, pp. 868–877, 2007.

images through the intersection of their color histograms. Other low-level features widely used by researchers for indexing and retrieval of images, except color [1,2], are texture [1,3] and shape [1,4]. In order to exploit the strong aspects of each of these features while constructing an optimum and robust CBIR system, a plethora of methods, introduced over time, have been based on combinations of these features [4,5].

In this paper the synergy of such features, specifically color and texture, is performed by use of an artificial ant colony. This specific type of insect was selected since studies showed that when in groups, the ants show self-organization as well as adaptation which are desirable attributes in image retrieval. Artificial ant colonies have previously been used in text based site retrieval (i.e. search engines) [6] but also in texture classification [7]. To the best of our knowledge this is the first time that ant colony behavior is applied on image retrieval in very large databases which contain images of general interest, as is the case with the LabelMe database (80000 images and increasing) [8]. The main thrust of the proposed method is a two stage ant colony algorithm employing in parallel color, texture and spatial information which are extracted from the images themselves. The study of ant colony behavior and of their self-organizing abilities inspired the algorithm. In the first stage, the synergy of the low-level descriptors is considered to be a swarm of ants, seeking for the optimal path to the "food" which is actually the most similar image to the query, whilst settling pheromone on each of the images that it confronts in the high similarity zone represented by 200 images. The terrain on which the ants move is predefined through the low-level features of the query image. The features involved are a newly proposed spatially-biased color histogram, a color-texture histogram and a color histogram inspired by attributes of the Human Visual System (HVS). Although these particular descriptors are proposed for use in this paper, the ant colony algorithm can employ a variety of features depending on the implementer or user. In the second stage the terrain changes as additional queries are made by using the highest ranked images as additional query images. The results prove the system to be satisfactorily efficient as well as fast since the features have already been extracted.

Following the histogram creation procedures resembling the "food" in the proposed method which are illustrated in the next section, the Matusita distance metric is used in order to measure the similarity between the features of the query image and of the ones stored in the database. In this manner the pheromone is dispensed to the corresponding images which form the path to the image of highest similarity and through descending order to the less similar ones.

2 Feature Extraction

The three features which represent the ants seeking the closest "food" (images) are a spatially biased histogram, a color-texture histogram and a color histogram inspired by the attributes of the HVS.

2.1 Spatially-Biased Histogram

The use of global histograms for image retrieval has proven to be an efficient and robust retrieval method [1,2], as it describes the overall statistics of the color in the images and is insensitive to rotation and scaling of the images themselves. However, such techniques lack in cases in which the images have similar colors, but are spatially distributed differently. This led to the need of the adoption of global histograms with embedded local characteristics, such as the newly proposed spatially-biased histogram. The suggested histogram creation method has a two stage straightforward algorithm and only the hue component is enriched with spatial information so as to maintain the original histogram speed.

In the first stage a histogram is created with the hue component being divided into 16 regions, whereas saturation and value into 4 each. This unbalance is due to the fact that the hue component carries the majority of color information from the three in the HSV color space and is hence considered more important in this method. Finally, the three color components are interlinked, thus creating a (16x4x4) histogram of 256 bins. In the second and most important part of this method the spatial information is inserted into the final histogram via the use of the mask M illustrated below. This mask is used to collect the color information from a 5 pixel "radius" neighborhood for each separate pixel in the manner of a shattered cross so as to increase the speed of the system and to decrease the chance of taking into consideration random noise; although this is also covered by use of a threshold when checking the concentration.

$$M = \begin{matrix} 0&0&0&0&0&1&0&0&0&0&0 \\ 0&0&0&0&0&0&0&0&0&0&0 \\ 0&0&0&0&0&-1&0&0&0&0&0 \\ 0&0&0&0&0&0&0&0&0&0&0 \\ 0&0&0&0&0&1&0&0&0&0&0 \\ 1&0&-1&0&1&1&1&0&-1&0&1 \\ 0&0&0&0&0&1&0&0&0&0&0 \\ 0&0&0&0&0&0&0&0&0&0&0 \\ 0&0&0&0&0&-1&0&0&0&0&0 \\ 0&0&0&0&0&0&0&0&0&0&0 \\ 0&0&0&0&0&1&0&0&0&0&0 \end{matrix}$$

The negative values in the cross are used to counteract the fact that significantly different adjacent pixel values could have the same sum as identical pixel values. In this way the possibility of false positive is somewhat decreased, though maintaining the extraction speed. Thus the whole image is convolved with the M matrix, resulting in a new hue component which contains the color information for the neighborhood of each pixel. If the pixels which are included in the vicinity of the full length of the cross possess a color similar to the one of the central pixel, then an additional hue value is added to the extension of the final histogram resulting in 272 bins (256+16). The similarity of the surrounding pixels indicating the concentration is revealed through the absolute difference between the energy of the mask multiplied by the central pixel value, in respect with the convolved one.

2.2 Color-Texture Histogram

The second color-texture-based system is a newly proposed histogram creation method which uses the L*a*b* color space. The color-texture feature extracted from the a* and b* components is expressed through a single, 64-bin histogram.

In the first part of the method, all the images are transformed into the L*a*b* color space and only the a* and b* components are reserved. The color-texture feature is extracted from the image by means of convolution with Laws' energy masks [3] which were selected due to their speed as they possess low computational complexity. The a* and b* components are convolved with the 5x5 masks, hence producing the respective *ENa** and *ENb** components which represent the chromaticity texture energy of the image. The reason why having 25x2 color-texture components for each image does not pose a problem is that only the components with the greatest energy are reserved. The energy volume for each image is computed by Eq. 1.

$$E_{Vol} = \sum \left(\sum_{x=1,y=1}^{m,n} ENa^*(x,y), \sum_{x=1,y=1}^{m,n} ENb^*(x,y) \right) \qquad (1)$$

When the components with the greatest energy are found, the discretization process is activated, during which the two components (*ENa**, *ENb**) are divided into sections. The color-texture components are split up into 8 regions each, and by interlinking them a (8x8) 64 bin histogram is created.

2.3 Center-Surround Histogram

This method is based on the retinal signal processing of the Human Visual System. A center-surround operator C similar to the receptive fields of the ganglion cells of the retina is employed to create a Center-Surround Histogram (CSH) [9]. The contribution of the CSH to the proposed system is that it incorporates a degree of spatial information into the histogram in a dual manner that only the pixels around the edges are considered, and furthermore that these pixels are weighted in respect to whether they belong to a strong or weak edge. Moreover, it reduces the processed visual information by using only the colored area surrounding the zero-crossings of an image. The proposed histogram contains only the chromatic information of these areas and consists of 256 bins.

3 Ant Colonies

In the approach discussed in this paper we distribute the search activities over the so-called "ants," that is, agents with very simple basic capabilities which, to some extent, mimic the behavior of real ants, in order to advance research in image retrieval. The Ant Colony Optimization (ACO) heuristic has been used successfully to solve a wide variety of problems such as the traveling salesman problem [10]. The simple question arising from the usage of the ant colony optimization in the above indiscipline applications is how the ant algorithms work?

The ant algorithms are basically a colony of cooperative agents, designed to solve a particular problem. These algorithms are probabilistic in nature because they avoid the local minima entrapment and provide very good solutions close to the natural solution.

More specifically, one of the problems studied by ethnologists was to understand how almost blind animals like ants could manage to establish shortest route paths from their colony to feeding sources and back. It was found that the medium used to communicate information among individuals regarding paths, and used to decide where to go, consists of pheromone trails. A moving ant lays some pheromone (in varying quantities) on the ground, thus marking the path by a trail of this substance. While an isolated ant moves essentially at random, an ant encountering a previously laid trail can detect it and decide with high probability to follow it, thus reinforcing the trail with its own pheromone. The collective behavior that emerges is a form of autocatalytic behavior where the more the ants following a trail, the more attractive that trail becomes for being followed [10]. The process is thus characterized by a positive feedback loop, where the probability with which an ant chooses a path increases with the number of ants that previously chose the same path.

Consider for example the experimental setting shown in Fig. 1 [10]. There is a path along which ants are walking (for example from food source A to the nest E, and vice versa, see Fig. 1a). Suddenly an obstacle appears and the path is cut off. So at position B the ants walking from A to E (or at position D those walking in the opposite direction) have to decide whether to turn right or left (Fig. 1b). The choice is influenced by the intensity of the pheromone trails left by preceding ants. A higher level of pheromone on the right path gives an ant a stronger stimulus and thus a higher probability to turn right. The first ant reaching point B (or D) has the same probability to turn right or left (as there was no previous pheromone on the two alternative paths). Because path BCD is shorter than BHD, the first ant following it will reach D before the first ant following path BHD (Fig. 1c). The result is that an ant returning from E to D will find a stronger trail on path DCB, caused by the half of all the ants that by chance decided to approach the obstacle via DCBA and by the already arrived ones coming via BCD: they will therefore prefer (in probability) path DCB to path DHB. As a consequence, the number of ants following path BCD per unit of time will be higher than the number of ants following BHD. This causes the quantity of pheromone on the shorter path to grow faster than on the longer one, and therefore the probability with which any single ant chooses the path to follow is quickly biased towards the shorter one. The final result is that very quickly all ants will choose the shorter path.

Inspired by this probabilistic behavior of the real ants, ant algorithms are the software agents that coordinate by updating the information of a common memory similar to the pheromone trail of the real ants. When a number of these simple artificial agents coordinate based on the memory updating they are able to build good solutions to hard combinatorial optimization problems.

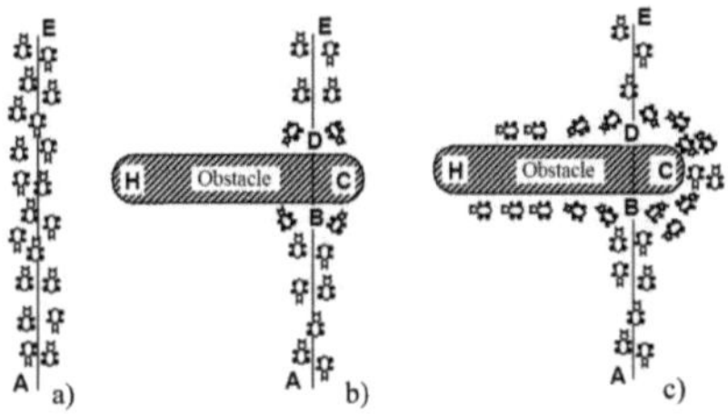

Fig. 1. An example with real ants [10]

As mentioned before, the artificial ant agents have many properties that differentiate them from the real ants and thus involve various ant algorithms based systems [10]. Along with these unique features that enhance the capabilities of the artificial agents there are other governing parameters such as the optimum number of ants, the pheromone decay rate, and the constants that make the solution to converge to the experimental results. As we are not interested in the simulation of ant colonies, but in the use of artificial ant colonies as an optimization tool in the field of image retrieval, the proposed system will have some major differences with a real (natural) one that will be discussed next. More specifically, Fig. 2 represents a generalized block diagram of the proposed ant algorithm. The problem is defined in the form of a network. All possible links between the components of the network and limiting criteria's are identified.

3.1 First Stage

Following the extraction of the three descriptors described beforehand from the images in the database and considering each histogram bin to be a virtual ant, a query is posed by mobilizing a sum of 592 (272+64+256) ants. The terrain of the "ground" where the ants "walk" depends strictly on the query in a way that, through comparison with the other images, it provides the information about the relative position and distance of the surrounding "food". On this terrain, the ants move in a straight line from image to image. In this first stage of the algorithm a comparison is performed using all three features, in other words the whole population of the ants, and an initial ranking of the images takes place.

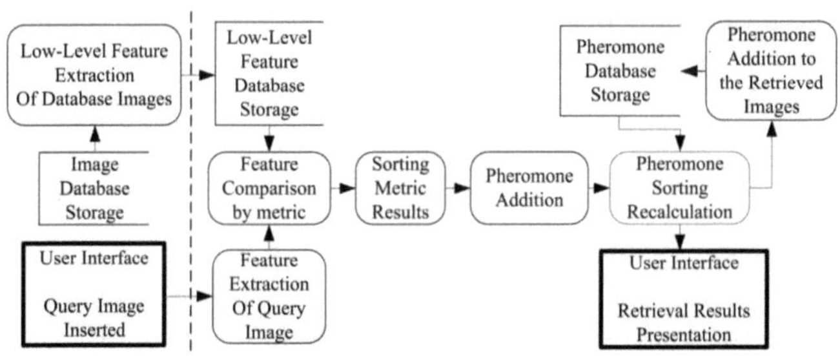

Fig. 2. Block diagram of the ant colony image retrieval system

Following numerous tests and simulations, it was concluded that the best similarity metric in order to compare the features was the Matusita distance expressed by the equation shown below:

$$M(H_Q, H_C) = \sqrt{\sum_i (\sqrt{H_Q(i)} - \sqrt{H_C(i)})^2} \tag{2}$$

where H_Q is the query histogram, H_C is the histogram to be compared and *(i)* is the number of bins. This distance is a separability measure which provides a reliable criterion presumably because as a function of class separability it behaves much more like probability of correct classification.

Following the initial ranking of all the images in the database, pheromone is attached to the 200 first most similar images in a descending manner according totheir ranking, meaning that the highest ranked image acquires the most pheromone. The aggregate pheromone deposition from all the ants results in the creation of a pool consisting of only 200 images, thus creating a new much smaller sub-terrain, in the vicinity of which the second stage takes place.

3.2 Second Stage

Having completed the pre-classification of the 80000 image database and concluded to a pool of just 200, the final retrieval takes place. Taking into consideration the possibility that the initial ant search can result in false positive results, the terrain is slightly altered by substituting the initial query image with the second highest ranked image from the pool and a new query takes place. In the new query, a new group of 592 ants is mobilized and the process from the first stage is repeated, although instead of having the whole of the database as a terrain, the ants are constrained strictly in the vicinity of the pool, thus changing the amount of pheromone attached to each of the previously ranked images. In order to restrain the overall time cost and to avoid false terrain alterations caused by false positives in the first stage, this new process is repeated for four iterations meaning that the sub-terrain of the pool is altered four times and that the first four images of the initial query are used as queries themselves, thus continuously altering the pheromone deposition in accordance to the new ranking of each repetition.

The final retrieval is not based upon the value produced by the metric stating the distance between the features of the images, but on the final amount of pheromone that has accumulated on each image at the conclusion of the two stages.

4 Performance Evaluation

We evaluate the total performance of our system in terms of retrieval accuracy and image recall versus precision. The database used to measure the system's effectiveness and efficiency was the LabelMe database [8], which consists of 80000 images (and increasing); is one of the largest databases available freely on the internet and due to its immense volume is adequate for CBIR testing (http://labelme.csail.mit.edu/).

The retrieval outcome is presented through a query session which produces images ranked in similarity according to the pheromone laid by the ants. The larger the amount of pheromone aggregation, the higher the similarity of that specific image. The measurement used in order to evaluate the system is the retrieval performance percentage, which is the percentage of actual similar images produced in the 20 to 50 first images retrieved by the system (the percentage displayed depends on the number of existing similar images, e.g. for the first image set there are only 20 relevant images in the database; hence the accuracy presentation reaches only the first 20 images). In Table 1 one can see a synopsis of the comparisons of the method's performance for eight different image sets and notice the high percentage in accuracy, which means that most of the existing similar images were retrieved in the first 20 to 30. These image sets are characteristic of the image database and also show the worst and best precision and recall performances of the proposed system. The nature of the images varies from natural scenes (e.g. sunsets, fields, people, animals, flowers), to indoor and outdoor images of buildings as well as city images (streets, cars, parks). It should be stated here that the whole of the database is used in each query and that no preclassification is performed to enhance the performance. The time cost requested to perform a single query, in the supposition that the descriptors have been extracted a priori, is 28 seconds, which is fairly reasonable considering that the database consists of 80000 images. In addition to the precision perspective, another aspect of retrieval performance is presented by the graph in Fig. 3: precision versus recall. Precision is the proportion of relevant images retrieved (similar to the query image) in respect to the total retrieved, whereas recall is the proportion of similar images retrieved in respect to the similar images that exist. Generally, precision and recall are used together in order to point out the change of the precision in respect to recall [11]. In most typical systems the precision drops as recall increases, hence, in order for an image retrieval system to be considered effective the precision values must be higher than the same recall ones, which is mostly the case in the current system. In order to provide a sense

Table 1. Retrieval efficiency of the proposed system

	Total images retrieved				
	1-10	11-20	21-30	31-40	41-50
Image set 1	100%	90%			
Image Set 2	70%	60%	63%	65%	62%
Image Set 3	90%	90%	77%	65%	58%
Image Set 4	100%	90%	63%	50%	
Image Set 5	90%	55%			
Image Set 6	90%	50%			
Image Set 7	100%	95%	90%	85%	80%
Image Set 8	100%	100%	93%	90%	88%

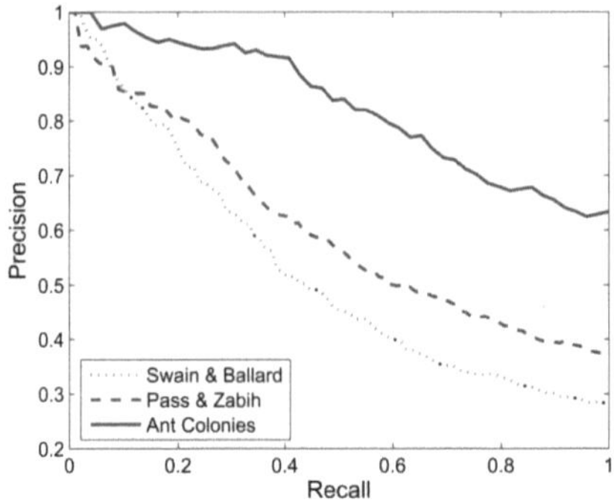

Fig. 3. Average precision versus recall graph for Swain and Ballard's Method, Pass and Zabih's Method and the proposed method

Table 2. Computational cost of the three features. n is the number of pixels, N is the number of nonzero elements of each respective mask (M for spatial, L for Law's masks and C for center-surround) and E is the number of pixels surrounding the edges.

Extraction Method	Operation	Additions	Multiplications
Spatially-Biased Histogram	1st Histogram	n	0
	2nd Histogram	n	0
	Convolution	nN_M	nN_M
Texture Histogram	Histogram	$2n$	0
	Convolution	$2nN_L$	nN_L
	Energy Volume	$2n$	0
Center-Surround Histogram	Histogram	E	0
	Convolution	nN_C	nN_C
Total		$(N_M + 2N_L + N_C + 6)n + E$	$(N_M + N_L + N_C)n$

of the proposed method's performance, the average precision versus recall factor of all the queries, is compared to the respective one of Swain and Ballard's method [2], as well as to Pass and Zabih's joint histogram method [5]. The last aspect considered is the computational cost of the three descriptors as shortly described in Table 2. We believe that the total cost of the operations performed during the extraction of the features is not a heavy price to pay, taking into account the significant increase in accuracy produced by the presented method.

5 Conclusions

In this paper we have presented a new two-stage content-based image retrieval system based on the behavior of ant colonies. The entities of these ants are represented through the basic elements of three image descriptors: a newly proposed spatially-biased histogram, a second histogram acquired through utilization of the chromaticity-texture-energy of the image and last a histogram which results through incorporation of certain attributes of the human visual system. Despite the bulk of the LabelMe database which consists of 80000 images, a precision versus recall graph for eight different image sets proves that the system exhibits an effective performance as the precision values are significantly higher compared to the respective recall ones.

Acknowledgements. This work is funded by the EU-EPEAEK Archimedes-II Project (2.2.3.z,subprogram 10).

References

1. del Bimbo, A.: Visual Information Retrieval. Academic Press, London (1999)
2. Swain, M.J., Ballard, D.H.: Color indexing. International Journal of Computer Vision 7(1), 11–32 (1991)
3. Laws, K.: Rapid texture identification. In: Image processing for missile guidance; Proceedings of the Seminar, San Diego, CA, July 29-August 1, 1980. (A81-39326 18-04) Bellingham, WA, Society of Photo-Optical Instrumentation Engineers, pp. 376-380 (1980)
4. Jain, A.K., Vailaya, A.: Image retrieval using color and shape. Pattern Recognition 29(8), 1233–1244 (1996)
5. Pass, G., Zabih, R.: Comparing images using joint histograms. Multimedia Systems 7(3), 234–240 (1999)
6. Kouzas, G., Kayafas, E., Loumos, V.: Ant seeker: An algorithm for enhanced web search. In: AIAI. pp. 649–656 (2006)
7. Ramos, V., Muge, F., Pina, P.: Self-organized data and image retrieval as a consequence of inter-dynamic synergistic relationships in artificial ant colonies. In: HIS. pp. 500–512 (2002)
8. Russell, B.C., Torralba, A., Murphy, K.P., Freeman, W.T.: Labelme: a database and web-based tool for image annotation. MIT AI Lab Memo AIM-2005-025 1, 1–10 (2005)
9. Panitsidis, G., Konstantinidis, K., Vonikakis, V., Andreadis, I., Gasteratos, A.: Fast image retrieval based on attributes of the human visual system. In: 7th Nordic Signal Processing Symposium (NORSIG 2006), Reykjavik, Iceland, Reykjavik, Iceland pp. 206–209 (2006)
10. Dorigo, M., Maniezzo, V., Colorni, A.: The ant system: Optimization by a colony of cooperating agents. IEEE Transactions on Systems, Man, and Cybernetics Part B: Cybernetics 26(1), 29–41 (1996)
11. Muller, H., Muller, W., Squire, D.M., Marchand-Maillet, S., Pun, T.: Performance evaluation in content-based image retrieval: overview and proposals. Pattern Recogn. Lett. 22(5), 593–601 (2001)

Filtering Video Volumes Using the Graphics Hardware

Andreas Langs and Matthias Biedermann

Universität Koblenz-Landau,
Universitätsstrasse 1, 56070 Koblenz, Germany
{alangs,mbmann}@uni-koblenz.de

Abstract. Denoising video is an important task, especially for videos captured in dim lighting environments. The filtering of video in a volumetric manner with time as the third dimension can improve the results significantly. In this work a 3D bilateral filter for edge preserving smoothing of video sequences exploiting commodity graphics hardware is presented. A hardware friendly streaming concept has been implemented to allow the processing of video sequences of arbitrary length. The clear advantage of time-based filtering compared to frame-by-frame filtering is presented as well as solutions to current limitations for volume filtering on graphics hardware. In addition, a significant speedup over a CPU based implementation is shown.

Keywords: Non-Linear Filtering, Graphics Hardware, Video Processing.

1 Introduction

Data acquired by any type of analog-digital converter, such as CCD and CMOS image sensors or sensors in CT/MRI scanners contains noise. In the case of commodity hardware, this can be observed especially in video sequences or images captured in dim lighting conditions. As a result, further processing, like segmentation, is difficult for these datasets.

To enhance the distorted data, filtering is a common first step in the workflow. There are three different types of operations that transfer a pixel of a source dataset or image into the corresponding pixel in the destination dataset: point operations where the destination pixel depends only on the source pixel; local operations where the destination pixel is dependent both on the source pixel and its distinct neighborhood; and finally global operations, where the value of the destination pixel is influenced by all pixels of the source image. These three operations can either be linear or nonlinear, where linear filters describe a convolution of the source image with a given filter kernel. A typical example is the Gaussian filter that resembles a linear local operation with a filter mask consisting of weights that have Gaussian distribution. Another example is the median filter as a nonlinear local operation where the destination pixel's value becomes the median of the source pixel and its neighbourhood. In order to enhance the

B.K. Ersbøll and K.S. Pedersen (Eds.): SCIA 2007, LNCS 4522, pp. 878–887, 2007.

distorted data we would like to smoothen the homogeneous regions while maintaining edges, thus performing an edge preserving smoothing. For reasons to be shown, we use the well-known bilateral filter [1]. Our target datasets are video sequences, i.e., a sequence of individual frames. Thus, we have the possibility of filtering each frame separately with a two-dimensional kernel. Trying to eliminate the noise as much as possible using this bilateral filter usually results in a comic style [2], however, which is caused by the evenly coloured areas with sharp edges in the resulting image. By choosing the parameters of the filter in a way that we get a more natural look, we cannot denoise the data sufficiently.

In our approach we regard the video sequence as a volume and use the bilateral filter in all three dimensions. Areas in the video sequence that are not changing or moving from one frame to the next are then homogeneous regions in time. Thus, we can choose the parameters of the bilateral filter in a way that we limit the influence of each frame and compensate for the lack of denoising with the now available temporal dimension. We are also able to eliminate the noise without introducing the comic look, but preserving a more natural appearence instead.

Representing video sequences as volumes and operations thereon is no new idea, as shown in [3], or [4], but advanced filtering is rarely used due to its high computational costs. To perform filtering on reasonably sized video sequences in an acceptable amount of time, we utilize recent graphic processing units (GPU) found on commodity hardware. Using the GPU for general purpose computations (GPGPU) is currently becoming more and more popular, motivated by the exponential growth of performance, exceeding the already fast progress of CPUs [5]. This way, we are able to perform high quality volumetric filtering of video sequences at PAL resolution in realtime.

The paper is organized as follows: In section 2.1 the underlying system used for data and shader handling and graphics hardware concepts are introduced, followed by implementation details of the GPU based bilateral filter. Subsequently, we explain our approach to filtering video sequences of arbitrary length (section 2.3), including a discussion of current limitations and workarounds. We conclude with quality and performance comparisons in section 3 and finally give an outlook on improvements and filtering in other domains in section 4.

2 Approach

2.1 Basic Concepts and Setup

Filtering images with custom filter kernels on the GPU has become possible quite recently with the introduction of programmable parts of the 3D rendering pipeline. It basically resembles a stream processor when performing general purpose computation in the graphics hardware. Input and output are streams (textures) on which kernels (shader programs) are performing various, user defineable operations. Important properties of these operations are data independency and data locality, which are both key factors for the high performance of graphics processors. This means that an output stream depends only on the input stream data, and no additional data besides the stream itself can be transferred between kernels. Imposing these programming restrictions, the GPU is able to compute several kernels in parallel. Although there are some limited exceptions from the mentioned stream processing concept, not all computation problems can be efficiently ported to the GPU without changing the algorithm completely.

In our implementation we used the Plexus framework that has been developed at the University of Koblenz-Landau during a collaborative project. With this graph-based tool we are able to easily model data flow between various nodes, called "devices", representing all kinds of functionality. The type of data flowing can be of an arbitrary nature, although the devices have to be capable of handling the appropriate data, of course. Plexus provides some basic functionality to transfer data between CPU and GPU, along with devices that are able to provide a stream of images from a video file, webcam, etc. By using this framework we have been able to adapt our approach quickly to various data sources and use it in different configurations.

The basic data type for representing 2D image data on the CPU in Plexus is "Image". These "Images" can be uploaded to the GPU where they become a "Texture2D" (for reading) or a "GPUStream" (for reading or writing), which are roughly equivalent. In addition, for the GPU there are data types like "Texture3D" and "FlatVolume", which will be described in more detail in section 2.3. All data types that have been mentioned before can have different numbers of channels and bit-depths per channel. For filtering video sequences, we used three channels (RGB) with eight bits each, which is sufficient for common video material and is also natively supported by a wide range of GPUs.

Concerning the architecture of GPUs, we only use one of currently three user programmable parts of the graphics pipeline: the fragment processing step. The vertex and geometry shader (the latter introduced with Shader Model 4.0 [6]) are not used in our approach, as is obvious for pure pixel data such as video frames. Therefore, we simply use the graphics API to draw a quadrilateral in size of the input data (i.e., video frame resolution) to initiate data processing. In our implementation the source video frame is the input data to be processed by the initial shader, i.e., the first render pass. The one-dimensional bilateral filter is then applied to this texture resulting in one filtered pixel value, which is written to the framebuffer (or some equivalent offscreen target). In our application, this result is used in subsequent processing steps to realize three-dimensional filtering.

2.2 Bilateral Filtering on the GPU

As image denoising is a common research topic, various filter types have been proposed. Among these approaches anisotropic diffusion and wavelet based filters are well studied and widely used, mainly due to their edge preserving properties. In the context of GPGPU, however, anisotropic diffusion as an iterative solution is not well suited for the GPU. Wavelet based filters on the other hand require several conversions to the wavelet domain and thus introduce a considerable overhead.

Barash has shown in [7] that bilateral filtering resembles anisotropic diffusion by using adaptive smoothing as link between the two approaches. Therefore, we are able to implement bilateral filtering as a more GPU-friendly algorithm because of its "local operation" nature, while maintaining their common basis. That is, it is possible to implement the bilateral filter, without significant changes, as a shader program using the OpenGL Shading Language.

The bilateral filter itself was first introduced in 1998 by C. Tomasi und R. Manduchi [1]. Its basic idea is to combine the domain and the range of an image. The domain describes the spatial location or closeness of two pixels, while the range describes their similarity, or the distance of the pixel values. In traditional filters only a pixel's location in the spatial domain is taken into account, resulting in less influence for more distant pixels, for example. The bilateral filter additionally takes the similarity into account.

The bilateral filter is defined as:

$$k(x) = \int_{-\infty}^{\infty} \int_{-\infty}^{\infty} c(\xi, x)\, s(f(\xi), f(x))\, d\xi$$

with the spatial closeness $c(\xi, x)$ between the neighborhood center x and a nearby point ξ, and the similarity $s(f(\xi), f(x))$ between x and ξ. The pixel value at x is therefore a combination of the domain c and the range s.

A simple and important case of the distance function in the domain is the gaussian weighted euclidean distance:

$$c(\xi, x) = e^{-\frac{1}{2}\left(\frac{d(\xi,x)}{\sigma_d}\right)^2}$$

with the euclidean distance function:

$$d(\xi, x) = d(\xi - x) = \|\xi - x\|$$

The distance function in the range is defined analog to c:

$$s(\xi, x) = e^{-\frac{1}{2}\left(\frac{\delta(f(\xi), f(x))}{\sigma_r}\right)^2}$$

with an appropriate distance measure for the intensity values ϕ and f:

$$\delta(\phi, f) = \delta(\phi - f) = \|\phi - f\|$$

In the simplest case, this can be the absolute pixel value. With parameters σ_d and σ_r, the filter result can be influenced. The bilateral filter is inherently non-linear because of its utilization of the image's range, and is therefore not directly separable.

In our implementation, however, we use the separated bilateral filter introduced by Pham et al. [8]. As described in their work, it is not a true separation of the original bilateral filter, thus leading to approximated rather than equal results. However, for the purpose of filtering video sequences the introduced error is negligible, especially considering the potential increase in performance. In our approach we convolve the source frame with the separated, two-dimensional bilateral filter, thus applying the one-dimensional kernel twice. The internal video volume is then built from these filtered slices and finally reused several times for the convolution in the third dimension. The whole procedure is described in more detail in the following section.

2.3 Streaming Approach

Filtering video data in a volumetric manner cannot be done directly for sequences of reasonable lengths as the entire data usually does not fit into the graphics memory. Therefore, we have implemented a streaming approach to allow for videos of arbitrary length. When choosing the parameters for the kernel so that n previous and n subsequent frames are used to filter the current frame, we have to keep $2n+1$ frames in the graphics cards memory at once. In every filtering step we upload the next video frame to the GPU, which then is processed separately in x and y direction using the separated bilateral filter. This two-dimensionally filtered slice is subsequently added to the video volume, thus replacing the oldest slice in the volume. The last step is to generate the processed video frame by filtering the video volume in z direction. The final video frame is then displayed and/or downloaded into the CPUs memory, to be written to a file or processed further. The whole process is depicted in the figure below:

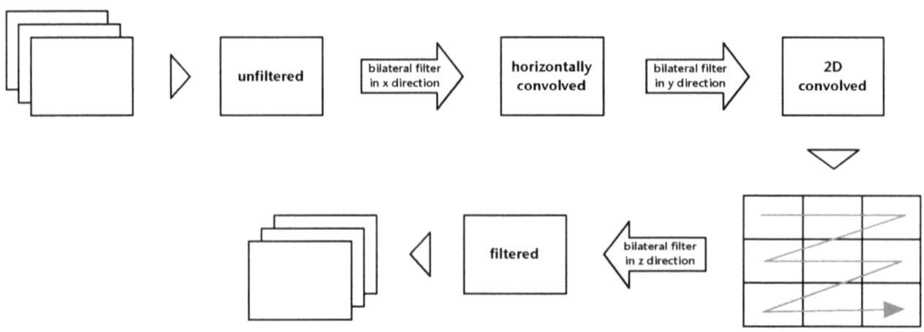

Fig. 1. Processing steps of our separable bilateral filtering approach

To minimize data transfers during copying to and from the GPU, and inside the GPU's memory, we use a ring buffer storage scheme for the video volume. Thus, it is possible to constantly add slices to the volume until the maximum number of slices is reached.

However, using this approach we encountered a severe limitation especially on NVIDIA graphics cards: the maximum size of 3D textures being 512^3 pixels, whereas only their most recent hardware (available since November 2006) is capable of larger 3D textures (2048^3). With this constraint we were not able to filter videos with the desired PAL resolution of 720×576 pixels using traditional 3D textures in our original implementation. To circumvent this limitation we used a 3D data representation known as "flat volume", introduced by Harris et al. [9].

The idea behind this technique is to place every slice of the volume side by side, resulting in a large 2D texture. This concept is also shown in figure 1, and exploits the graphics hardware's support for 2D texture sizes up to 4096^2 (latest hardware: 8192^2) pixels. We can then use a simple address translation to convert a pixel position in 3D (x, y, z) to the according pixel position in the 2D flat volume (x, y). As an additional benefit of using the "flat volume" we also measured a significant performance increase as described in section 3.2.

3 Results

3.1 Comparison 3D/2D Filtering

To assess the quality of our filtering – especially the difference between the 2D and 3D version – we did a *full reference* comparison. Therefore, we took a video sequence without noise (our reference sequence) and added synthetic noise to it by adding an evenly distributed random value between -50 and 50 to the R, G, and B channel of every pixel independently. This is motivated by the nature of noise of commodity sensors, especially when used in dim lighting conditions. We then compared the video sequences with two different measures by performing a frame-by-frame comparison and averaged the measured error values for the entire sequence. The first video sequence that has been compared is an outdoor video sequence captured in daylight without noticeable noise. The second video is a synthetic animation showing a typical pre-video countdown, combined with a testbar screen for broadcasting, thus being completely noise free.

The two measures we used are the SSIM (Structural Similarity Index Measure) introduced by Wang et al. [10] and the MSE (Mean squared error). The SSIM is a measure specially designed to compare two images in a perception based manner, whereas the MSE measure is purely mathematically based and straight forward, but is by its nature not capable of measuring the quality difference from an observer's point of view.

After computing an error value for the distorted video sequence in comparison to the original sequence, we are able to compare the distorted video sequence filtered with the 2D and 3D bilateral filter, respectively, with the reference sequence. The results can be seen in table 1. By the method's definition, a value of 1 means completely equal frames with the SSIM: the lower the value, the more

Table 1. Quality comparison

	outdoor		synthetic	
	SSIM	MSE	SSIM	MSE
unfiltered	0,393	669,6	0,311	626,3
filtered with "2D bilateral filter" (GPU-BF 2D)	0,737	135,0	0,720	125,7
filtered with "3D bilateral filter" (GPU-BF 3D/flat3D)	0,844	87,3	0,890	73,9
filtered with AfterEffects "Remove grain"	0,829	89,7	0,908	68,3
filtered with "Neat Video"	0,909	53,9	0,954	48,1

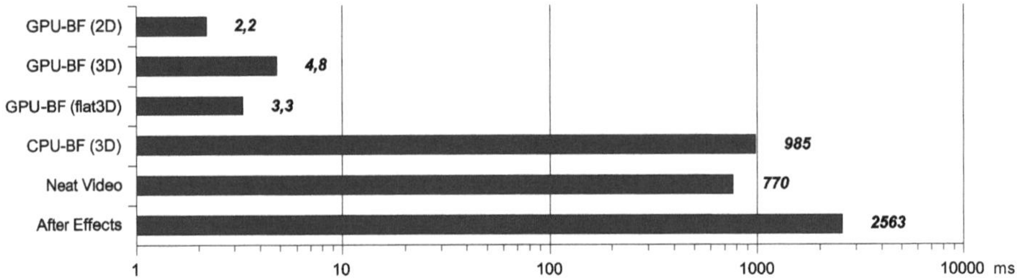

Fig. 2. Performance comparison (time per frame)

Fig. 3. Example "outdoor" for quality comparison

different the compared images are. In contrast, value 0 corresponds to no differences with the MSE. We also included the results of two commercial products: "Adobe AfterEffects 7.0" (i.e., "Remove grain" filter) and the relatively new "Neat Video Pro 1.5", which are both widely used and designed for this field of application. The filter methods used by these two programs are not disclosed. A visual comparison of the results is shown in figure 3.

As can be seen there is no significant difference of the filter quality between the two very different types of video sequences: both error metrics indicate the same relationship. The quality of 3D filtering surpasses the quality of the 2D filter, but is comparable to the quality of the "Adobe AfterEffects" filter. In contrast, the visual result of "Neat Video" is better than ours.

We also measured the time needed for filtering a single frame. Our application runs on a commodity Windows PC that features following components: Intel CoreDuo 6400 (2.13 GHz), 2 GB RAM, NVIDIA GeForce 8800 GTX 768 MB PCIe 16x. The video sequence "outdoor", used for comparison, has a resolution of 720×576 and is 1085 frames long. The denoted time per frame is an average over all frames of the filtering only, that is excluding the time needed for reading or writing the video file or uploading the data to the GPU. The results are given in figure 2. They obviously show that using the GPU for filtering yields a significant performance improvement, approximately three orders of magnitude faster than comparable functionality in commercial products.

3.2 Performance 3D/Flat Volume

The processing time for our implementation using the flat volume storage concept has a performance gain of approximately 25% on current hardware with respect to full 3D volume textures, as can be seen from figure 2. On previous graphics hardware the difference using the flat volume was even higher, up to a factor of two. This indicates some trend for improved and complete 3D texture processing in graphics hardware, but due to the very new hardware a detailed evaluation could not be taken into account in this work yet.

By using the flat volume textures, we were also able to filter PAL resolution video sequences even on older graphics hardware (esp. NVIDIA) in their original resolution. As the architecture of commodity graphics cards is mainly driven by the computer games industry, features like true 3D texture functionality have not yet been optimized to the same level as their 2D counterparts. This can also be experienced with API functions for 3D texture processing like OpenGL function calls which have not been hardware accelerated or are not available at all, until recently, as described before. Therefore, if no advanced texture access (e.g., trilinear filtering) is needed, flat volume textures are an appropriate alternative to native 3D textures. Despite the additional overhead for the address translation, they offer a considerable performance increase, especially with older graphics hardware.

4 Conclusion and Future Work

In this paper we have described that using a 3D bilateral filter for noise reduction on video sequences results in higher quality than with frame-by-frame 2D bilateral filtering. In figure 4 we give another real world example with comparable performance results, which depicts a video frame acquired in a very dark room, lit only indirectly by a projection screen. In addition, we have shown that using modern commodity graphics hardware clearly decreases processing time for this rather costly volumetric filter operation. In order to exploit their capabilities even better, we argued that using flat volumes for three-dimensional data

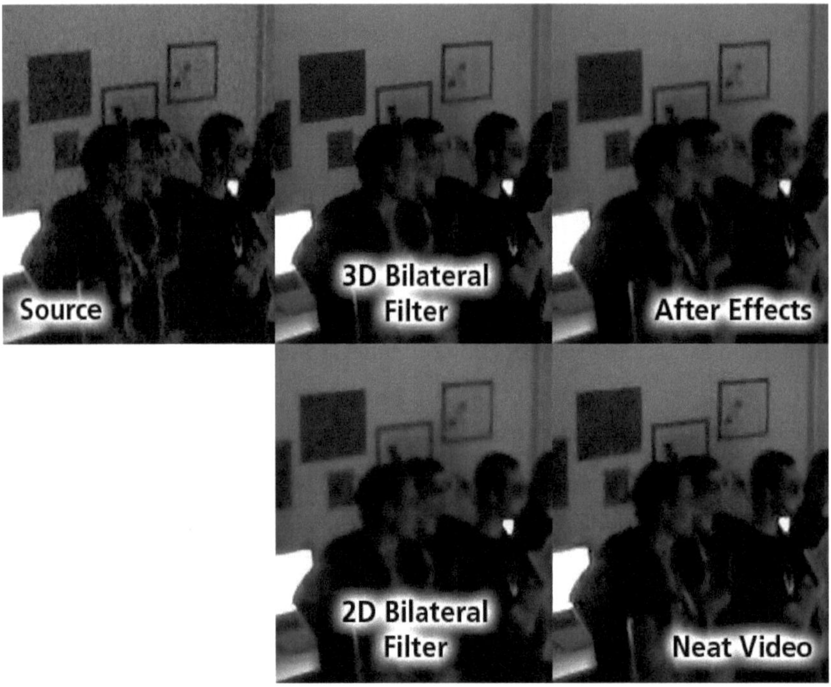

Fig. 4. Real world example in dim lighting conditions with natural noise

offers advantages in comparison to true 3D textures for a wide range of graphics hardware.

Based on this application, it is also possible to perform fast filtering of real 3D datasets, like medical volumes to enhance the results of subsequent processing steps like segmentation. However, the size of the volume is then directly limited by the available memory on the graphics card, so that other approaches like bricking have to be employed.

One natural disadvantage of smoothing is the loss of high frequencies which can be seen in some regions of the example material. To compensate for this, we would like to investigate other types of smoothing filters or sequences of filter operations. Therefore, we are planning to further generalize the concepts to other kinds of computations amenable to the stream processing nature of current graphics hardware.

The dramatic decrease in filtering time by using the GPU has been shown with video data of 8 bit per channel. It would be interesting to see how well our approach works on data with 16 bit per channel or even other representations, e.g., true high dynamic range data. As current graphics hardware is heavily optimized for floating point data, an equally, if not greater performance increase should be possible.

Acknowledgements

We would like to thank Tobias Ritschel, who has provided the Plexus framework developed at the University of Koblenz-Landau as a basis for our work, as well as Oliver Abert and Thorsten Grosch for helpful discussions and suggestions. Also, we would like to thank the reviewers for their helpful and motivating comments.

References

1. Tomasi, C., Manduchi, R.: Bilateral Filtering for Gray and Color Images. ICCV: 839-846 (1998)
2. Fischer, J., Bartz, D., Straßer, W.: Stylized Augmented Reality for Improved Immersion. In: Proceedings of IEEE Virtual Reality (VR 2005), pp. 195–202 (2005)
3. Hájek, J.: Timespace Reconstruction of Videosequences, 6th Central European Seminar on Computer Graphics (CESCG) (2002)
4. Daniel, G., Chen, M.: Visualising Video Sequences Using Direct Volume Rendering, Vision, Video, and Graphics, VVG 2003, University of Bath, UK, July 10-11th, 2003. In: Proceedings (2003)
5. Owens, J.D., Luebke, D., Govindaraju, N., Harris, M., Krüger, J., Lefohn, A.E., Purcell, T.J.: A Survey of General-Purpose Computation on Graphics Hardware. Computer Graphics Forum, vol. 26, (to appear) (2007)
6. Blythe, D.: The Direct3D 10 system. ACM Trans. Graph. 25(3), 724–734 (2006)
7. Barash, D.: Bilateral Filtering and Anisotropic Diffusion: Towards a Unified Viewpoint, Scale-Space '01. In: Proceedings of the Third International Conference on Scale-Space and Morphology in Computer Vision, pp. 273–280. Springer, Heidelberg (2001)
8. Pham, T.Q., van Vliet, L.J.: Separable bilateral filtering for fast video preprocessing. In: ICME 2005: IEEE International Conference on Multimedia & Expo, IEEE Computer Society Press, Los Alamitos (2005)
9. Harris, M.J., Baxter, W.V., Scheuermann, T., Lastra, A.: Simulation of cloud dynamics on graphics hardware. HWWS '03: In: Proceedings of the ACM SIGGRAPH/EUROGRAPHICS conference on Graphics hardware Eurographics Association
10. Wang, Z., Bovik, A.C., Sheikh, H.R., Simoncelli, E.P.: Image Quality Assessment: From Error Visibility to Structural Similarity. IEEE Trans. Image Processing, vol. 13 (2004)
11. Viola, I., Kanitsar, A., Gröller, M.E.: Hardware-Based Nonlinear Filtering and Segmentation using High-Level Shading Languages. In: Proceedings of IEEE Visualization, IEEE, New York (2003)

Performance Comparison of Techniques for Approximating Image-Based Lighting by Directional Light Sources

Claus B. Madsen and Rune E. Laursen

Laboratory of Computer Vision and Media Technology, Aalborg University, Denmark
{cbm,rul}@cvmt.aau.dk
http://www.cvmt.aau.dk

Abstract. Image-Based Lighting (IBL) has become a very popular approach in computer graphics. In essence IBL is based on capturing the illumination conditions in a scene in an omni-directional image, called a light probe image. Using the illumination information from such an image virtual objects can be rendered with consistent shading including global illumination effects such as color bleeding.

Rendering with light probe illumination is extremely time consuming. Therefore a range of techniques exist for approximating the incident radiance described in a light probe image by a finite number of directional light sources. We describe two such techniques from the literature and perform a comparative evaluation of them in terms of how well they each approximate the final irradiance. We demonstrate that there is significant difference in the performance of the two techniques.

Keywords: Augmented Reality, Image-Based Lighting, median cut, irradiance, real-time rendering, directional light sources.

1 Introduction

Image-based approaches have gained widespread popularity in computer graphics, [1]. Image-based techniques have been used for 3D modeling of real scenes (Image-Based Modeling), for rendering from a bank of images with no 3D model whatsoever (Image-Based Rendering), and for modeling the complex illumination conditions in real scenes (Image-Based Lighting).

Image-Based Lighting (IBL) has become a frequently used technique to render a virtual object with illumination conditions that are consistent with those of a real scene given that the scene is distant, [2]. The idea is simply to use a camera to measure the light arriving at some point in the scene, the point at which you want to insert a virtual object. In practice people most often use a polished steel ball, place it somewhere in a scene, and take an image of it with a tele-lens from some distance away. The image thus contains information about how much light arrives at the position of the ball from all possible directions. Figure 3 shows example light probe images re-mapped to the longitude-latitude format, where the full 360 degrees are represented along the horizontal (longitude) axis, and

B.K. Ersbøll and K.S. Pedersen (Eds.): SCIA 2007, LNCS 4522, pp. 888–897, 2007.

180 degrees are represented along the vertical (latitude) axis. For construction and remapping of light probe images we use HDRShop 2.09, [3]. Handling the dynamic range of the light in the scene is done by acquiring the same view at different exposures, gradually lowering the exposure time until no pixels in the image are saturated. These multiple exposure are then fused into a single High Dynamic Range (HDR) floating point image, [4].

The light probe is a map of the incident radiance at the acquisition point. Each pixel in the map corresponds to a certain direction and solid angle, and together all pixels cover the entire sphere around the acquisition point. A light probe can thus also be called a radiance map, or an environment map. With this information virtual objects can be rendered into the scene with scenario consistent illumination e.g., [2,5,6]. Light probes can also be used to estimate the reflectance distribution functions of surfaces from images, as demonstrated in [7,8]. For a review of illumination models in mixed reality see [9].

Actually using light probes for rendering is computationally very heavy. Using image-based lighting for a full global illumination rendering with path tracing is time consuming in order to reduce the noise level in the final rendering, simply because the light probe has to be treated as a spherical area light source enclosing the entire scene. To get a noise free estimate of the irradiance at a certain point requires thousands and thousands of samples of this area source, unless the light probe has very low frequency content.

To combat this problem several approaches have been proposed which take a light probe and attempt to approximate its illumination by a relatively low number of directional light sources. That is, the idea of these approaches is to find directions and the radiances of some number, say 64, directional light sources, such that the combined illumination from these sources approximate the combined illumination from the entire light probe. With such a directional light source approximation to a light probe, Image-Based Lighting using light probes can also be implemented in real-time applications taking into account that each source causes shadows to be cast.

The aim of the present paper is simply to test the performance of such approximation techniques in terms of how well they actually approximate the light probe for a given number of sources. The paper is organized as follows. Section 2 gives an overview of the approach and results in the paper. Section 3 introduces some concepts and terminology. Section 4 describes two completely different approaches to computing a directional light sources approximation. Section 5 then tests these two techniques in terms of their relationships between approximation error and number of sources used. Conclusions and directions for future research are given in section 6.

2 Overview of the Idea of This Work

Figure 1 shows a light probe together with the result from one of the approximation techniques studied in this paper. In this particular case the technique has been allocated 8 directional sources which it has then distributed across the

Fig. 1. Result from running the Median Cut approximation technique using 8 directional sources on Galileo's Tomb light probe. The light probe is obtained from [10]. Each rectangular region contains a red dot. This red dot marks the chosen direction for a particular directional source, and all the combined radiance from the region has been transferred to this particular source direction.

light probe longitude-latitude map in an attempt to capture the radiance distribution of the original light probe. Naturally, the accuracy of the approximation depends on the number of sources allocated. The original light probe is simply W times H directional sources, where W is the number of pixels in the longitude direction, and H is the number of pixels in the latitude direction.

We then run any given technique on some light probe image to produce approximations with 2, 4, 8, 16, etc. light sources. Given these sets of approximated sources we compute what the resulting error in irradiance is compared to ground truth, which in this case is the irradiance computed by using the radiance from all pixels in the light probe. Figure 2 shows an example of true and approximated irradiance. In section 5 we test and compare two different approximation techniques on three different light probes.

The irradiance is chosen as the error measure in order to have a compact, quantitative performance measure, which is independent of surface characteristics, and independent on view point. For highly glossy surfaces approximating the light probe by a relatively low number of light sources will obviously lead to visible errors, whereas reflected radiance from diffuse surfaces will be correct if the approximated irradiance is correct.

3 Terminology

Before we proceed with the actual techniques and their performances we need to establish a small theoretical basis. Above we have used the term "directional light source" a few times. There are correctness problems relating to using radiometric properties in conjunction with point and directional sources as they have no physical extent, [11,12]. The purpose of the remainder of this section is to establish a physically correct terminology regarding light probe images.

In formal terms the light probe image is a spatially discrete measurement of the continuous function describing the incident radiance (measured in $W/(m^2 \cdot$

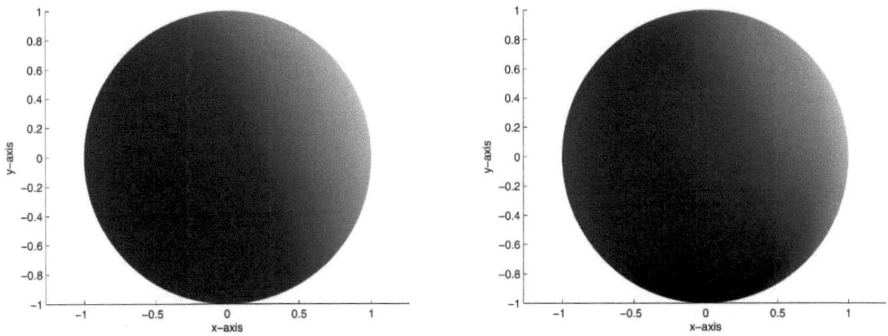

Fig. 2. Left: ground truth irradiance for around 20000 normals distributed evenly on a sphere computed for the Galileo's Tomb light probe, figure 1. Right: irradiances resulting from running the Median Cut source approximation technique to produce 8 directional sources. On print the difference may be visually subtle, but the average error is actually around 25 percent, and the maximum error is more than 75 percent.

Sr)), which in turn is a function of the incident direction. Let $\boldsymbol{n}$ be the normal of a differential area surface, and let $\Omega_{\boldsymbol{n}}$ be the hemi-sphere defined by this normal. By integrating the incident radiance, $L(\boldsymbol{\omega})$, from the direction $\boldsymbol{\omega}$ over the hemi-sphere the total irradiance, $E(\boldsymbol{\omega})$, can be computed:

$$E(\boldsymbol{n}) = \int_{\Omega_{\boldsymbol{n}}} L(\boldsymbol{\omega})(\boldsymbol{n} \cdot \boldsymbol{\omega}) d\boldsymbol{\omega} \tag{1}$$

which then is measured in W/m^2. The term $d\boldsymbol{\omega}$ signifies the differential solid angle $d\omega = |d\boldsymbol{\omega}|$ in the direction $\frac{d\boldsymbol{\omega}}{|d\boldsymbol{\omega}|}$.

For computational purposes it is beneficial to formulate these matters in terms of standard spherical coordinates. A direction in space is then written as $\boldsymbol{\omega}(\theta, \phi) = [\sin(\theta)\cos(\phi), \sin(\theta)\sin(\phi), \cos(\theta)]$, where θ is the angle the direction vector makes with the coordinate system z-axis (latitude), and ϕ is the angle the projection of the vector on the xy-plane makes with the x-axis. The irradiance from Eq. 1 then becomes:

$$E(\boldsymbol{n}) = \underset{(\theta,\phi) \in \Omega_{\boldsymbol{n}}}{\int\int} L(\theta, \phi)(\boldsymbol{n} \cdot \boldsymbol{\omega}(\theta, \phi)) \sin(\theta) d\theta d\phi \tag{2}$$

In this paper we will exclusively use the latitude-longitude mapping (LL mapping) of light probe images. Let the resolution of the LL light probe image be W by H pixels, and let u and v represent pixel coordinates in an image coordinate system with origin in the top left corner of the LL map, and v-axis oriented downwards. Thus the *middle row* in the image corresponds to the equator of the unit sphere, i.e, corresponds to $\theta = \pi/2$, the top row corresponds to $\theta = 0$ and the bottom row corresponds to $\theta = \pi$. Moreover $\phi = 0$ corresponds to the *leftmost column*. Each light probe pixel, $P(u, v)$, represents the radiance in

$W/(m^2 \cdot Sr)$ (if the light probe acquisition is radiometrically calibrated) from the direction given by $\boldsymbol{\omega}(u,v) = \boldsymbol{\omega}(\theta(v), \phi(u))$, where $\theta(v) = v\Delta_\theta$ and $\phi(u) = u\Delta_\phi$, where $\Delta_\theta = \pi/H$ and $\Delta_\phi = 2\pi/W$. The discrete version of Eq. 2 then becomes:

$$E(\boldsymbol{n}) \approx \sum_u \sum_v P(u,v)(\boldsymbol{n} \cdot \boldsymbol{\omega}(u,v)) \sin(\theta(v))\Delta_\theta\Delta_\phi \tag{3}$$

where the summations are subject to the constraint that $(\theta(v), \phi(u)) \in \Omega_{\boldsymbol{n}}$, i.e., that the combinations of u and v represent pixels inside the region corresponding to the hemi-sphere defined by the surface normal $\boldsymbol{n}$.

From Eq. 3 it is evident that if every pixel, $P(u,v)$, in the LL map is scaled with $\Delta_\theta \cdot \Delta_\phi = 2\pi^2/(W \cdot H)$ and weighted by $\sin(\theta(v))$, we get a very simple summation. We therefore produce a new LL map, where each pixel $Q(u,v) = 2\pi^2 P(u,v)\sin(\theta(v))/(W \cdot H)$. The irradiance for a given normal is then simply computed as:

$$E(\boldsymbol{n}) \approx \sum_u \sum_v Q(u,v)(\boldsymbol{n} \cdot \boldsymbol{\omega}(u,v)) \tag{4}$$

where the summations again are subject to the constraint that $(\theta(v), \phi(u)) \in \Omega_{\boldsymbol{n}}$.

To recapitulate in a different way: Each pixel in the LL map acts as a small area light source subtending a solid angle of $A_p = 2\pi^2/(W \cdot H)$ $[Sr/pixel]$. By weighting each pixel by $\sin(\theta(v))$ we achieve "permission" to treat all pixels equally in the sense that we cancel out the effect of the non-uniform sampling density of the LL mapping (poles are severely over-sampled). By subsequently scaling by A_p we convert the solid angle domain from steradians to pixels. I.e., each $Q(u,v) = 2\pi^2 P(u,v)\sin(\theta(v))/(W \cdot H)$ measures the radiance in $W/(m^2 \cdot pixel)$, such that by performing a simple cosine weighted sum of pixels we directly get the irradiance contributed by the pixels involved in the sum (Eq. 4). Another way of putting it is: each pixel $Q(u,v)$ is an area light source contributing $Q(u,v)(\boldsymbol{n} \cdot \boldsymbol{\omega}(v,u))$ irradiance to the differential area surface with normal $\boldsymbol{n}$.

In the remainder of the paper we take the meaning of a directional light source to be a very small area light source (there are normally a lot of pixels in a light probe image). The direction to such a source is taken to be the direction to its center, and it is assumed that for each such source we know its radiance and its area.

4 Light Probe Approximation Techniques

We have found three different approaches to finding a set of directional light sources which approximate a full radiance map in the form of a light probe. Two of the approaches are closely related and operate directly in the radiance space image domain of the light probe, in particular on the longitude-latitude mapping. The last technique is quite different in that it operates in *irradiance* space.

4.1 Median Cut

The LightGen, [13], and the Median Cut, [14], techniques operate directly in the image domain of the light probe. We have chosen to focus on the Median Cut technique since it is better documented than the LightGen technique, which only exists as a plugin to HDRShop. The median cut technique, [14], is conceptually wonderfully simple. The idea is to recursively split the LL map into regions of approximately equal summed radiance. Since the method splits all regions K times the techniques produces 2^K sources, i.e., 2, 4, 8, 16, 32, etc. Figure 1 illustrated the result of running the technique on a light probe. The algorithm is as follows:

1. Add the entire light probe image to the region list as a single region.
2. For each region in the list, subdivide along the longest dimension such that its light energy is divided evenly.
3. If the number of iterations is less than K, return to step 2.
4. Place a light source at the centroid of each region, and set the light source radiance to the sum of the pixel values within the region.

The strength of this approach is that it is so straight forward, computationally light and easy to implement. The problems with this approach lies in two issues. The first issue is that it subdivides all regions at each iteration and depending on K there can be a large jump in the number of sources, which may be disadvantageous for real-time rendering with the approximated sources, where one would like as many sources as possible, but at the same time there is a performance limit in the graphics hardware. The second issue relates to step 4, where, for small K, and thereby large regions, a lot of radiance is transferred quite large distances over the sphere without any cosine weighting. This transfer could be done physically correct, but only for a single known surface normal. For arbitrary normals there is no alternative to just transferring the radiance and hope the regions are small enough that it does not constitute a grave error. This is naturally an invalid assumption for very small K.

4.2 Irradiance Optimization

The Irradiance Optimization technique by Madsen et al., [15], is significantly different from the previous one. While the Median Cut technique operates entirely on a pixel level in the light probe image, i.e., operate in radiance space, the Irradiance Optimization method operates in *irradiance* space.

The method is based on first using the original light probe image to compute the ground truth irradiance for a large number (M) of normal directions uniformly distributed across the unit sphere using Eq. 4. These M irradiance values constitute the goal vector in an optimization to estimate the parameters of N directional sources. Each source is defined by five parameters (RGB radiances and two angles for direction).

Given an estimate of these 5 times N parameters it is possible to compute the approximated irradiances for the M normals. Let L_i be the radiance of the ith source, and let $\omega(\theta_i, \phi_i) = [\sin(\theta_i)\cos(\phi_i), \sin(\theta_i)\sin(\phi_i), \cos(\theta_i)]$ be the direction vector to the ith source. Furthermore, let A be a fixed, small area (in steradians) of each light source to accommodate physical correctness. A disc area light source of 1 degree has an area of $2.392 \cdot 10^{-4}$ steradian. Finally, let $\boldsymbol{n}_k$ be the kth normal. The irradiance on a differential area surface with normal $\boldsymbol{n}_k$ is then:

$$E(\boldsymbol{n}_k) = \sum_{i=1}^{N} L_i \cdot A \cdot (\boldsymbol{n}_k \cdot \boldsymbol{\omega}(\theta_i, \phi_i)) \tag{5}$$

By comparing the approximated irradiances to the ground truth irradiances an error vector is obtained, which can be converted to a parameter update vector. The source estimation process is thus an iterative, non-linear optimization process based on Newton's iterative method, since the Jacobian can be expressed analytically.

The strength of this method is that it inherently produces a configuration of directional source which result in irradiances that are similar to the irradiances achieved by full IBL using the entire light probe. Furthermore, this method can produce source configurations with any number of sources, not just a power of 2. The weakneses of the method is that it is significantly more complicated to implement than the Median Cut method, and it is computationally much heavier. The estimation part takes a few minutes in Matlab, but computing the ground truth irradiances can be time consuming. We use HDRShop, [3], to precompute the irradiance map, and it takes on the order of 24 hours for a 1024x512 resolution irradiance map. We tried using the fast diffuse convolution method implemented in HDRShop (executes in seconds), but this method in itself is approximative and has too many artifacts to be used as ground truth.

5 Comparative Evaluation

We have tested the two methods (Median Cut and Irradiance Optimization) on a number of qualitatively quite different light probe image shown in figure 3. We ran the methods on the test light probes, and produced directional source approximations with 2, 4, 8, 16, 32, 64, and 128 sources. The Irradiance Optimization technique can produce any number of sources, but was constrained to the source number cases which were also feasible for the Median Cut approach. We were unable to obtain a convergence on a 128 source solution with the Irradiance Optimization technique. This will be discussed later.

The evaluation is based on computing the irradiances resulting from the estimated set of sources for a large number (20480 resulting from a subdivision of an icosahedron) of surface normals evenly distributed on a unit sphere, and comparing them to the ground truth irradiances (computed using HDRShop). Figure 2 showed an example of true and approximated irradiance. For each color channel we then compute the mean and the maximum of the absolute differences between estimated and ground truth irradiances. Figure 4 shows curves representing average and maximum error for each of the two methods as a function of the number of light sources used. The errors are an average over RBG.

Fig. 3. The tested light probes. Top row: Galileo's tomb in Florence, Italy, and St. Peter's Cathedral, Rome, Italy. Both acquired from [10]. Bottom: Outdoor daylight scene. Everything below the horizon has been set to zero, so the light probe only accounts for the sun and the sky, not the ground. All probes are available via [10].

5.1 Discussion

The plots of mean and maximum irradiance errors in figure 4 clearly show that both approximation techniques, regardless of what light probe they are applied to, perform consistently better the more light sources the technique is allowed to employ. This simple fact means that in terms of accuracy it will always be better to choose a higher number of sources.

Secondly the experiments show that the Irradiance Optimization technique consistently performs much better than the Median Cut method. This is the case for both average and maximum error. Generally the Median Cut method requires 2 to 3 times as many sources to achieve the same error as the Irradiance Optimization technique. For rendering this is very important from a computational point of view, since it will always be an advantage to use as few sources as possible. Typically the Irradiance Optimization gets below 5 percent error for approximately 5 to 6 sources, whereas the Median Cut technique requires on the order of 20 sources or more.

The Irradiance Optimization technique could not converge when the number of sources comes above some threshold (64). This is due to the fact that when the number of sources grows too high there is too little energy for some sources to latch on to. Early in the iterations the dominant sources become stable, leaving ever smaller amounts of energy to distribute among the rest of the sources. At the same time the sources tend to repel each other when they distribute across the sphere, because each source has a semi-spherical "footprint" (each source is like shining a torch on a sphere) and sources will be reluctant to overlap footprints too much.

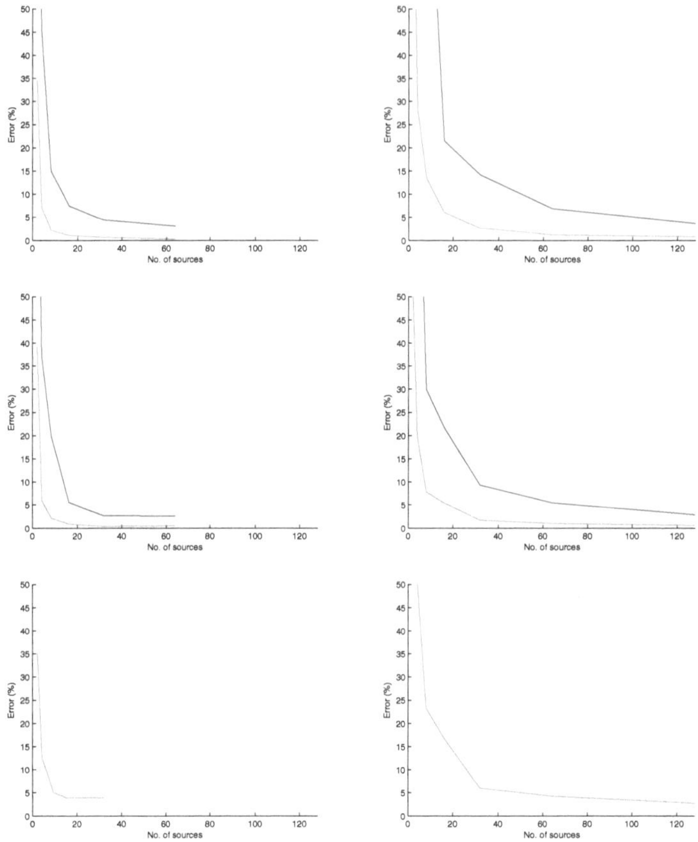

Fig. 4. Mean and max irradiance error in percent for the two tested techniques for three different light probes. Left column: Irradiance Optimization. Right column: Median Cut. Top to bottom row: Galileo, St. Peter's, and the Sky probe, respectively. See figure 3 for the three light probes.

6 Conclusions

We have demonstrated a significant difference in the performance of available techniques to approximate light probes with a set of directional light sources. Tests clearly demonstrate that the Irradiance Optimization technique requires substantially fewer sources to achieve the same error level as the other technique.

In terms of rendering speed it will always be an advantage to have as few light sources as possible and still achieve visibly acceptable performance. If a rendering is solely for visual purposes it may not be crucial whether the irradiance at a point is 1 percent or 5 percent wrong, but renderings can are also used in more radiometrically challenging contexts such as for inverse methods, aiming at estimating surface reflectance parameters from images, [7,8]. For inverse problems the accuracy in the approximation can be very important.

Acknowledgments

This research is funded by the CoSPE project (26-04-0171) under the Danish Research Agency. This support is gratefully acknowledged.

References

1. Oliveira, M.M.: Image-based modelling and rendering: A survey. RITA - Revista de Informatica Teorica a Aplicada, Brasillian journal, but paper is in English 9(2), 37–66 (2002)
2. Debevec, P.: Rendering synthetic objects into real scenes: Bridging traditional and image-based graphics with global illumination and high dynamic range photography. In: Proceedings: SIGGRAPH 1998, Orlando, Florida, USA (July 1998)
3. Debevec, P., et al.: Homepage of HDRShop, `www.hdrshop.com`
4. Debevec, P., Malik, J.: Recovering high dynamic range radiance maps from photographs. In: Proceedings: SIGGRAPH 1997, Los Angeles, CA, USA (August 1997)
5. Debevec, P.: Tutorial: Image-based lighting. IEEE Computer Graphics and Applications pp. 26 – 34 (March/April 2002)
6. Gibson, S., Cook, J., Howard, T., Hubbold, R.: Rapic shadow generation in real-world lighting environments. In: Proceedings: EuroGraphics Symposium on Rendering, Leuwen, Belgium (June 2003)
7. Yu, Y., Malik, J.: Recovering photometric properties of architectural scenes from photographs. In: Proceedings: SIGGRAPH 1998, Orlando, Florida, USA. pp. 207 – 217 (July 1998)
8. Yu, Y., Debevec, P., Malik, J., Hawkins, T.: Inverse global illumination: Recovering reflectance models of real scenes from photographs. In: Proceedings: SIGGRAPH 1999, Los Angeles, California, USA. pp. 215 – 224 (August 1999)
9. Jacobs, K., Loscos, C.: State of the art report on classification of illumination methods for mixed reality. In: EUROGRAPHICS, Grenoble, France (September 2004)
10. Debevec, P.: Homepage of Paul Debevec, `www.debevec.org/probes`
11. Phar, M., Humphreys, G.: Physicaly Based Rendering – From Theory to Implementation. Elsevier, Amsterdam (2004)
12. Jensen, H.W.: Realistic Image Synthesis Using Photon Mapping. A. K. Peters (2001)
13. Cohen, J.M., Debevec, P.: The LightGen HDRShop plugin.`www.hdrshop.com/main-pages/plugins.html` (2001)
14. Debevec, P.: A median cut algorithm for light probe sampling. In: Proceedings: SIGGRAPH 2005, Los Angeles, California, USA. Poster abstract (August 2005)
15. Madsen, C.B., Sørensen, M.K.D., Vittrup, M.: Estimating positions and radiances of a small number of light sources for real-time image-based lighting. In: Proceedings: Annual Conference of the European Association for Computer Graphics, EUROGRAPHICS 2003, Granada, Spain. pp. 37 – 44 (September 2003)

A Statistical Model of Head Asymmetry in Infants with Deformational Plagiocephaly

Stéphanie Lanche[1,2,3], Tron A. Darvann[1], Hildur Ólafsdóttir[2,1], Nuno V. Hermann[4,1], Andrea E. Van Pelt[5], Daniel Govier[5], Marissa J. Tenenbaum[5], Sybill Naidoo[5], Per Larsen[1], Sven Kreiborg[4,1], Rasmus Larsen[2], and Alex A. Kane[5]

[1] 3D-Laboratory, (School of Dentistry, University of Copenhagen; Copenhagen University Hospital; Informatics and Mathematical Modelling, Technical University of Denmark), Denmark
[2] Informatics and Mathematical Modelling, Technical University of Denmark, Denmark
[3] Ecole Supérieure de Chimie Physique Electronique de Lyon (ESCPE Lyon), France
[4] Department of Pediatric Dentistry and Clinical Genetics, School of Dentistry, University of Copenhagen, Denmark
[5] Division of Plastic & Reconstructive Surgery, Washington University School of Medicine, St. Louis, MO, USA

Abstract. Deformational plagiocephaly is a term describing cranial asymmetry and deformation commonly seen in infants. The purpose of this work was to develop a methodology for assessment and modelling of head asymmetry. The clinical population consisted of 38 infants for whom 3-dimensional surface scans of the head had been obtained both before and after their helmet orthotic treatment. Non-rigid registration of a symmetric template to each of the scans provided detailed point correspondence between scans. A new asymmetry measure was defined and was used in order to quantify and localize the asymmetry of each infant's head, and again employed to estimate the improvement of asymmetry after the helmet therapy. A statistical model of head asymmetry was developed (PCA). The main modes of variation were in good agreement with clinical observations, and the model provided an excellent and instructive quantitative description of the asymmetry present in the dataset.

1 Introduction

Deformational Plagiocephaly (DP) is a term describing cranial asymmetry and deformation commonly seen in infants. Its incidence has been estimated to be as high as 15% in the USA ([1]). The deformity is thought to result from protracted external intrauterine pressure to the skull, followed by continued postnatal molding due to infant positioning. The incidence has increased exponentially due to the "back to sleep" campaign to promote supine infant positioning to reduce sudden infant death syndrome. DP is manifested most commonly as either left-right asymmetry or brachycephaly (forshortening of the head). Both are treated

B.K. Ersbøll and K.S. Pedersen (Eds.): SCIA 2007, LNCS 4522, pp. 898–907, 2007.

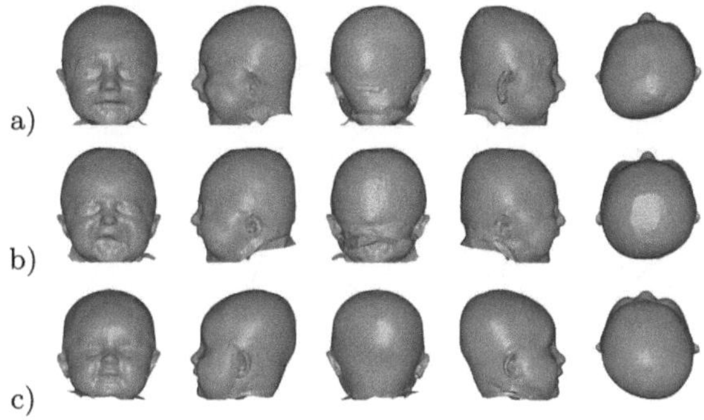

Fig. 1. Five different views of three of the captured 3D full head surfaces. a) Right-sided flattening posteriorly and left-sided flattening anteriorly. b) Brachycephaly. c) Left-sided flattening posteriorly and right-sided flattening anteriorly.

non-surgically. Treatments include parental education on how to prevent further deformation (e.g., alternating sleep positions [2]) and orthotic molding helmet therapy (e.g., [3] and [4]). It is widely held that correction is best accomplished in infancy due to the sequence of skull mineralization, however little is known concerning the outcomes from different treatment regimens. DP affects the occiput at the back of the head and, to a lesser extent, the forehead contour. Ear position is often skewed so that the ear is anteriorly positioned on the same side as the occipital flattening. When viewed from above, the head shape can be inscribed within a parallelogram. The purposes of this work were to develop a new methodology for head asymmetry assessment and to develop a statistical model of the asymmetry (using Principal Components Analysis) in order to quantify and localize the asymmetry of each infant's head before and after the helmet therapy and to determine the effect of helmet treatment.

2 Material

3D full-head surfaces of 38 patients with DP were captured both before and after treatment utilizing a 3dMD cranial system (www.3dMD.com) at the Division of Plastic & Reconstructive Surgery, Washington University School of Medicine, St. Louis, MO, USA. All infants commenced their helmet treatments before 6 months of age, and were treated for a maximum of 6 months. Figure 1 presents examples of these scans.

3 Methods

3.1 Template Matching

The method used for computation and modelling of asymmetry (described in the forthcoming sections) requires establishment of detailed point correspondence between surface points on the left and right sides of the head, respectively.

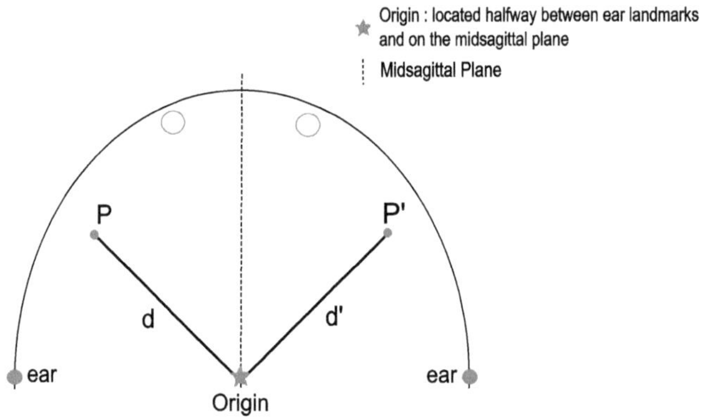

Fig. 2. Computation of the asymmetry: Illustration of the distances d and d' between the origin and the points P and P', respectively, in an axial view

This is achieved through a process of template matching, whereby a symmetric "ideal" head surface (template) is oriented and deformed to assume the shape of the patient's head surface [5]. The process consists of three steps:

1. Non-isotropic scaling of the template to the patient surface.
2. Rigid orientation of the patient surface to the scaled template surface, using ear-landmarks and nasion.
3. Non-rigid deformation of the scaled template surface to the oriented patient surface using a Thin Plate Spline (TPS) controlled by 22 manually placed facial and ear landmarks, and 40 constructed landmarks on the top of the head. The latter landmarks are determined by intersecting the surfaces with 40 radial lines (equidistant in terms of angle) originating from the midpoint between the ears. They are necessary in order to control the deformation at the top and back of the head where there are no visible anatomical landmarks.

3.2 Asymmetry Computation

The definition of the asymmetry A_P of a point P involves the computation of the ratio between two distances: 1) the distance d from the origin (midpoint between the ear landmarks) to the surface point P on one side of the midsagittal plane, and 2) the distance d' from the origin to the corresponding point P' on the other side of the midsagittal plane (Figure 2).

Since, intuitively, the amount of asymmetry at P and P' should be equal, except for a sign introduced in order to distinguish a point in a "bulged" area from a point in a "flattened" area, A_P and $A_{P'}$ are defined by:

$$\text{if } d > d' \text{ then } A_P = 1 - (\frac{d'}{d}) \text{ and } A_{P'} = -A_P \quad (1)$$

$$\text{if } d' > d \text{ then } A_{P'} = 1 - (\frac{d}{d'}) \text{ and } A_P = -A_{P'} \quad (2)$$

The change in head asymmetry is calculated as the difference between the asymmetry absolute values at the two stages:

$$\text{Change} = |A_{P,\mathbf{stage1}}| - |A_{P,\mathbf{stage2}}| \tag{3}$$

Hence, a positive change (improvement) implies that $A_{P,\mathbf{stage2}}$ is closer to 0 than $A_{P,\mathbf{stage1}}$ (i.e., closer to perfect symmetry).

3.3 Modelling Asymmetry Using Principal Components Analysis

PCA is a popular method for shape modelling (an excellent description is found in [6]). The PCA is performed as an eigenanalysis of the covariance matrix of the (aligned) asymmetry measures.

The asymmetry values for each scan are ordered according to the mesh points of the template scan (cf section 3.1.) and stored in a vector of size M:

$$\mathbf{a} = [|A_{P1}|, |A_{P2}|, \dots, |A_{PM/2}|, |A_{P'1}|, |A_{P'2}|, \dots, |A_{P'M/2}|] \tag{4}$$

Here the first and last $M/2$ elements are asymmetry values for the points on the right and left sides of the midsagittal plane, respectively. The maximum-likelihood estimate of the covariance matrix can be written as:

$$\mathbf{\Sigma}_a = \frac{1}{N}\sum_{i=1}^{N}(\mathbf{a}_i - \overline{\mathbf{a}}_i)(\mathbf{a}_i - \overline{\mathbf{a}}_i)^{\mathrm{T}} = \frac{1}{N}\mathbf{A}\,\mathbf{A}^{\mathrm{T}} \tag{5}$$

where $\overline{\mathbf{a}}$ is the maximum-likelihood estimate of the mean asymmetry of the N data-vectors. The principal axes of the M-dimensional point cloud of asymmetry are now given as eigenvectors, $\mathbf{\Phi}_a$, of the covariance matrix:

$$\mathbf{\Sigma}_a\,\mathbf{\Phi}_a = \mathbf{\Phi}_a\,\mathbf{\Lambda}_a \tag{6}$$

where $\mathbf{\Lambda}_a$ is a diagonal matrix containing the eigenvalues of the covariance matrix, and the columns of $\mathbf{\Phi}_a$ contain its eigenvectors. An asymmetry instance can be generated by modifying the mean asymmetry by adding a linear combination of eigenvectors:

$$\mathbf{a} = \overline{\mathbf{a}} + \mathbf{\Phi}_a\,\mathbf{b}_a \tag{7}$$

where $\mathbf{b}_a$ is a matrix containing the asymmetry model parameters.

As the number of observations ($N = 76$ scans) is much smaller than the number of surface points ($M = 190076$), the eigenanalysis is carried out using a reduced covariance matrix:

$$\mathbf{\Sigma}_{reduc} = \frac{1}{N}\mathbf{A}^{\mathrm{T}}\mathbf{A} \tag{8}$$

The eigenanalysis of this matrix gives the eigenvalues and eigenvectors of the covariance matrix. The eigenvalues and the eigenvectors may then be computed by:

$$\mathbf{\Lambda}_a = \mathbf{\Lambda}_{reduc} \tag{9}$$

$$\mathbf{\Phi}_a = \mathbf{A}\,\mathbf{\Phi}_{reduc} \tag{10}$$

In practice, the eigenanalysis may be carried out by Singular Value Decomposition (SVD).

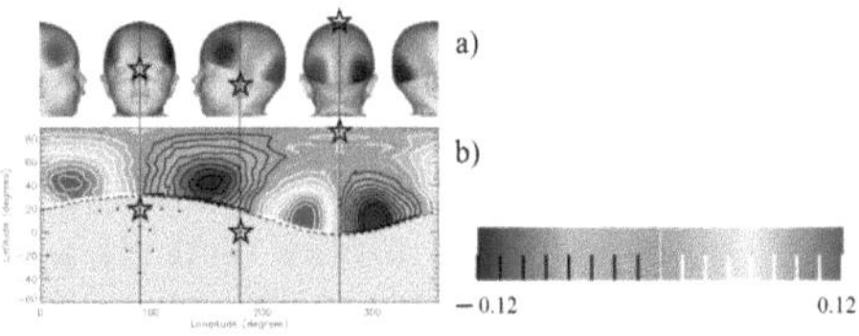

Fig. 3. Flat map construction. a) Asymmetry values in an example subject shown as color coding. b) Corresponding flat map with contours (black: negative, white: positive). Some landmarks are shown as star symbols. Lower limit of helmet region is shown as dashed curve.

3.4 Projection of 3D Surfaces into 2D Flat Maps

A more compact means of presentation is to construct a flat map (Figure 3b) by a simple transformation from rectangular to spherical coordinates. The flat map has right ear landmark at longitude = 0 degrees, midface at 90 degrees, left ear landmark at 180 degrees and center of the back of the head at 270 degrees. Regions below the helmet area are shown in light gray, below the dashed curve. Levels of asymmetry are indicated by contours in the flatmap. There are 16 contour intervals, spanning the range of asymmetry as indicated by the color bar. The contours are equidistant in terms of asymmetry and are drawn in black for negative values, and in white for positive values. Hence, black contours show "bulged" areas (negative), white contours "flattened" areas (positive), and areas exhibiting no asymmetry are displayed in light gray.

4 Results

4.1 Asymmetry

Figure 4 presents the results of the asymmetry computations in three example subjects. Top views of the head before (a) and after (b) treatment are shown together with corresponding asymmetry flat maps. In addition, a map of change (c) is shown.

Figure 4.1. shows an asymmetric DP patient with right-sided flattening posteriorly, as well as a left-sided flattening anteriorly (a). The typical parallelogram shape is also reflected in the asymmetry flat map. Note the improvement in asymmetry after treatment (b,c).

Figure 4.2. shows a typical brachycephalic patient (a). Brachycephalic patients are generally not very asymmetric, as their deformation mainly causes a foreshortning of the skull. Note improved shape after treatment (b,c).

The third patient, Figure 4.3, has left-sided flattening posteriorly as well as a right-sided flattening anteriorly (a). Note the improvement after treatment (b,c).

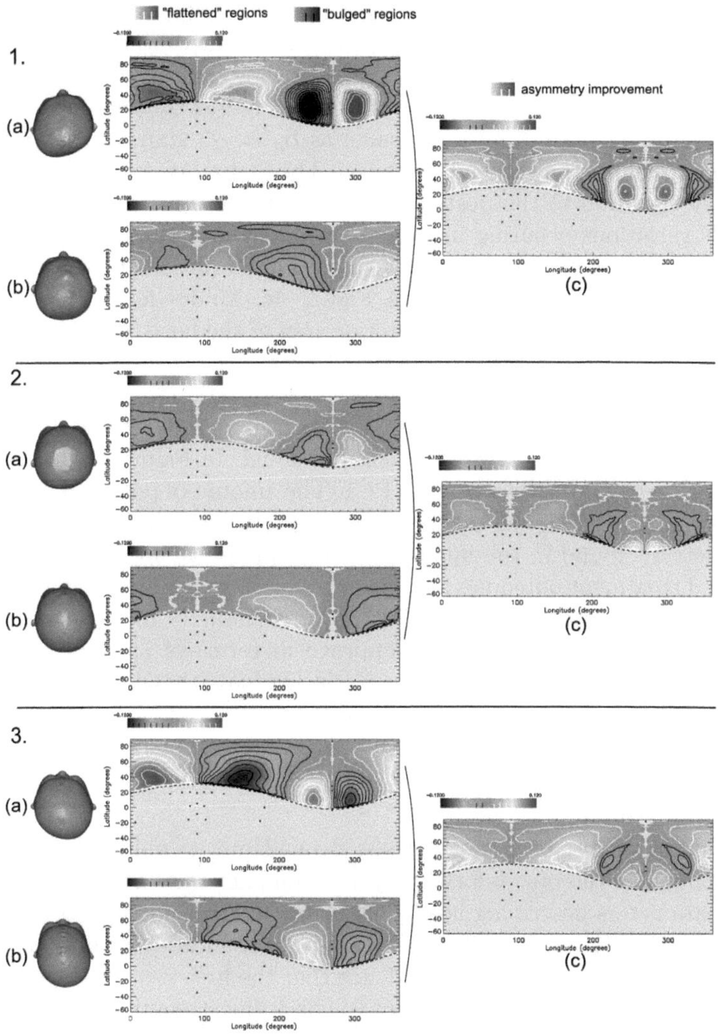

Fig. 4. Results of the asymmetry computation and changes for: 1. Right-sided flattening posteriorly and left-sided flattening anteriorly. 2. Brachycephaly. 3. Left-sided flattening posteriorly and right-sided flattening anteriorly. (a) Scans at stage 1. (b) Scans at stage 2. (c) Changes between the two stages. In the flat maps showing asymmetry (middle column), positive and negative values denote "flattening" and "bulging" respectively. In the flat maps of change, positive values denote improvement.

4.2 Statistical Model

A statistical model was created by performing PCA on the 76 scans. The input of the PCA was the vector of asymmetry measures at each point in the helmet region. The decay of eigenvalues (Figure 5a) indicates that 96 % of the asymmetry variation can be modelled using the first eight parameters. The mean asymmetry (Figure 5b) emphasizes posterior and anterior regions with high asymmetry, while the anterior parts exhibit smaller magnitude.

Figures 5(c-j) display the first eight modes showing only $\mathbf{\Phi b}_a$ (cf. Equation 7) with $\mathbf{b}_a = -3$ standard deviations. As the images corresponding to $\mathbf{b}_a = +3$ standard deviations are exactly the same as $\mathbf{b}_a = -3$ standard deviations but with opposite colors, they are not displayed. The first mode (c) localized the main asymmetry variation to the posterior region of the head. The second mode (d) represents variations occuring in the anterior region of the head, but spatially more spread out than the posterior region. The variations of the third mode occured above the ears, also seen in Figure 4c. Modes four (f) and five (g) revealed variability mainly in the posterior area of the head, probably the result of variation in the location of the affected area posteriorly. In general, higher modes represented higher spatial frequencies of variation.

The scores of the three first modes (Figure 6) demonstrate the direction and amount of asymmetry progress for each individual. In Figure 6a, the scores for PC2 are plotted against the scores for PC1. The amount of posterior and anterior asymmetry may be read off the x- and y-axes, respectively. The least amount of asymmetry is found in the upper-left corner of this figure. This is the region where good treatment outcomes are located, as well as the brachycephalic heads. Individuals that improve in terms of posterior asymmetry move leftwards in the diagram, whereras individuals that improve in terms of anterior asymmertry move upward. Analogously, in Figure 6b, individuals that improve in terms of asymmetry above the ear move downward.

4.3 Validation of the Asymmetry Model

The usefulness of the asymmetry model depends on its ability to capture and describe clinically relevant information in a compact way. Two of the most important parameters describing head asymmetry in DP could be stated as "magnitude of posterior asymmetry" and "magnitude of anterior asymmetry". In the previous section there was strong evidence that the first two modes were related to these particular clinical parameters. To check the strength of the relation between the model modes and the clinical parameters, a search for local extrema of asymmetry was conducted in the asymmetry flat maps. Figure 7 shows the correlation between scores and local minima.

5 Discussion

The computed asymmetry corresponded well (Figure 4) to observed asymmetry in the scans. Limitations of the method of establishing point correspondence between scans were the use of the ears (that are often affected in DP) for the registration, and the use of constructed landmarks instead of anatomical landmarks on top of the head. None of these limitations seem to have severely affected a valid asymmetry measurement. PCA is often used for summarizing data. The new variables created by PCA, however, are not guaranteed to be interpretable.

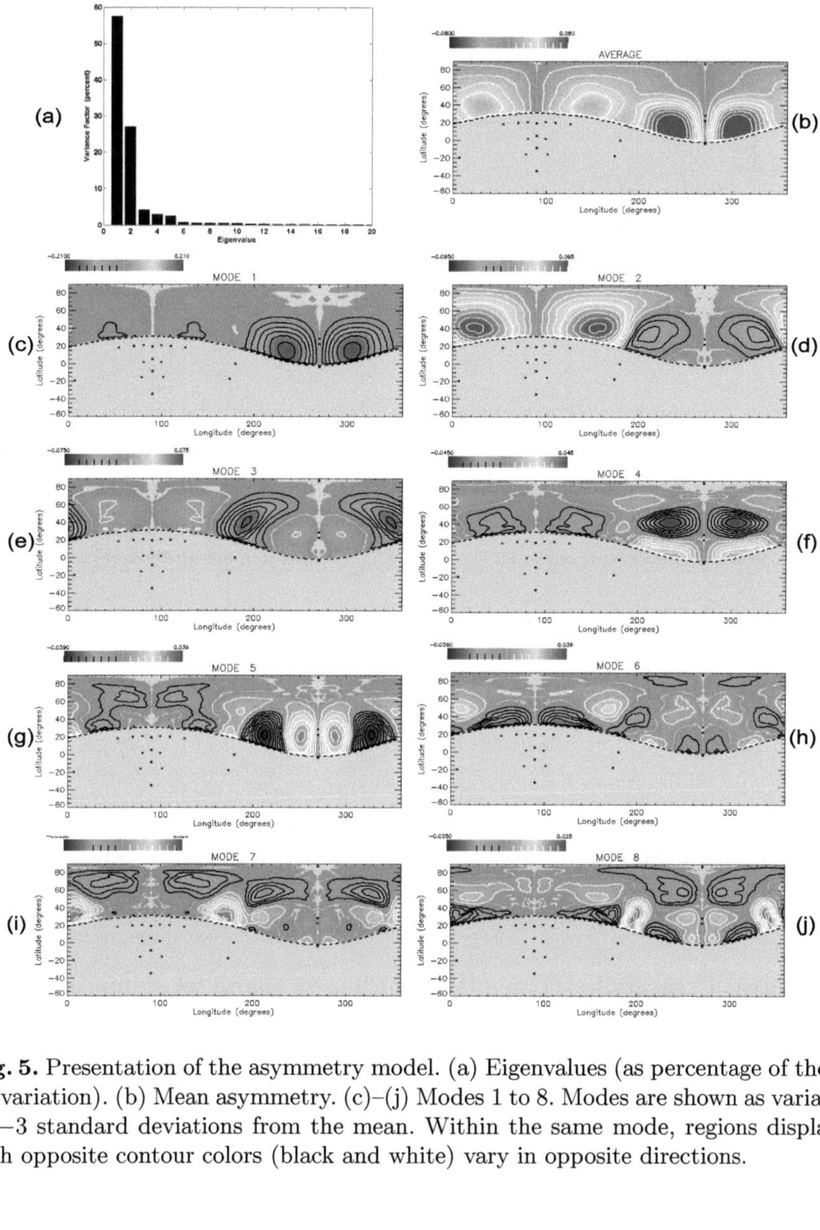

Fig. 5. Presentation of the asymmetry model. (a) Eigenvalues (as percentage of the total variation). (b) Mean asymmetry. (c)–(j) Modes 1 to 8. Modes are shown as variation at −3 standard deviations from the mean. Within the same mode, regions displayed with opposite contour colors (black and white) vary in opposite directions.

The success of the asymmetry model (Figure 7) could be due to the less complex, "global" types of asymmetry variation present in the DP dataset. The excellent properties of the model makes using the model attractive compared to other methods of asymmetry assessment. Other methods, as direct anthropometry of the head (e.g., [7]), measurement systems using a head ring or strip (e.g., [8], and [9]), or even measurements on 3D scans (e.g., [10]), produce a multitude of parameters, making the interpretation difficult in terms of asymmetry and

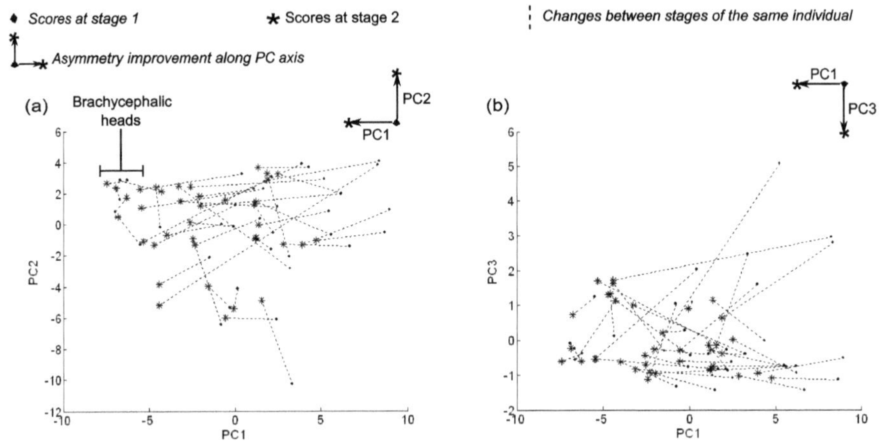

Fig. 6. Score plots of the asymmetry model: (a) PC1 vs. PC2. (b) PC1 vs. PC3.

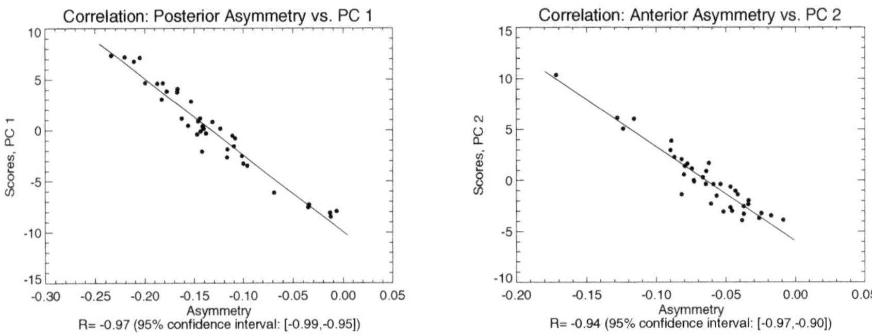

Fig. 7. Correlation between clinical parameters and model PC scores

less intuitive. Contrary to [11] and [12], which use a sparse set of inter-landmark distances, computing the asymmetry at every surface point provides the opportunity to create a high spatial resolution asymmetry model.

6 Conclusion

A new 3D asymmetry measure was developed, providing a detailed surface map of asymmetry covering the whole head. The asymmetry measure was seen to reflect observed asymmetry in DP very well. A statistical model was created by performing PCA on the asymmetry maps in 38 patients. PCA modes were seen to correspond very well to clinically relevant parameters. In particular, the first and second modes corresponded to variation at the back and front of the head, respectively. The method is suitable for monitoring asymmetry treatment in individuals, as well as for classifying asymmetry in population studies.

SL acknowledges financial support from the BIOP graduate school.

References

1. Littlefield, T.R.: Cranial remodeling devices: treatment of deformational plagiocephaly and postsurgical applications. Semin. Pediatr. Neurol. 11, 268–277 (2004)
2. Hummel, P., Fortado, D.: A parents' guide to improving head shape. Adv. Neonatal Care 5, 341–342 (2005)
3. American Academy of Pediatrics Task Force on Sudden Infant Death Syndrome. Pediatrics vol. 116, pp. 1245–1255 (2005)
4. Lee, W.T., Richards, K., Redhed, J., Papay, F.A.: A pneumatic orthotic cranial molding helmet for correcting positional plagiocephaly. J. Craniofac. Surg. 17, 139–144 (2006)
5. Darvann, T.A., Hermann, N.V., Tenenbaum, M.J., Govier, D., Naidoo, S., Larsen, P., Kreiborg, S., Kane, A.A.: Head shape development in positional plagiocephaly: Methods for registration of surface scans. In: proceedings: Darvann, T.A., Hermann, N.V., Larsen, P., Kreiborg, S. (eds.): "Craniofacial Image Analysis for Biology, Clinical Genetics, Diagnostics and Treatment", Workshop of the 9th MICCAI conference, Copenhagen, Denmark, pp. 59–66 (October 5) (2006)
6. Stegmann, M.B.: Active Appearance Models. IMM Technical Report, IMM-EKS 2000-25, Technical University of Denmark, Lyngby, Denmark (2000)
7. Kolar, J.C., Salter, E.M.: Craniofacial Anthropometry. Practical Measurement of the Head and Face for Clinical, Surgical and Research Use. Springfield, Illinois: Charles C. Thomas, Publisher (1997)
8. Chang, P.Y., Chang, N.C., Perng, D.B., Chien, Y.W., Huang, F.Y.: Computer-aided measurement of cranial asymmetry in children with and without torticollis. Clin. Orthod. Res. 4, 200–205 (2001)
9. van Vlimmeren, L.A., Takken, T., van Adrichen, L.N.A., van der Graaf, Y., Helders, P.J.M., Engelbert, R.H.H.: Plagiocephalometry: a non-invasive method to quantify asymmetry of the skull; a reliability study. Eur. J. Pediatr. 165, 149–157 (2006)
10. Plank, L.H., Giavedoni, B., Lombardo, J.R., Geil, M.D., Reisner, A.: Comparison of infant head shape changes in deformational plagiocephaly following treatment with a cranial remolding orthosis using a noninvasive laser shape digitizer. J. Craniofac. Surg. 17(6), 1084–1091 (2006)
11. Lele, R., Richtsmeier, T.: An Invariant Approach to Statistical Analysis of Shapes. Chapman & Hall/CRC (2001)
12. Bookstein, F.: Morphometric Tools for Landmark Data. Cambridge (1997)

Real-Time Visual Recognition of Objects and Scenes Using P-Channel Matching*

Michael Felsberg and Johan Hedborg

Computer Vision Laboratory, Linköping University, S-58183 Linköping, Sweden
mfe@isy.liu.se
http://www.cvl.isy.liu.se/~mfe

Abstract. In this paper we propose a new approach to real-time view-based object recognition and scene registration. Object recognition is an important sub-task in many applications, as e.g., robotics, retrieval, and surveillance. Scene registration is particularly useful for identifying camera views in databases or video sequences. All of these applications require a fast recognition process and the possibility to extend the database with new material, i.e., to update the recognition system online.

The method that we propose is based on P-channels, a special kind of information representation which combines advantages of histograms and local linear models. Our approach is motivated by its similarity to information representation in biological systems but its main advantage is its robustness against common distortions as clutter and occlusion. The recognition algorithm extracts a number of basic, intensity invariant image features, encodes them into P-channels, and compares the query P-channels to a set of prototype P-channels in a database. The algorithm is applied in a cross-validation experiment on the COIL database, resulting in nearly ideal ROC curves. Furthermore, results from scene registration with a fish-eye camera are presented.

Keywords: object recognition, scene registration, P-channels, real-time processing, view-based computer vision.

1 Introduction

Object and scene recognition is an important application area of methods from image processing, computer vision, and pattern recognition. Most recognition approaches are based on either of the following two paradigms: Model-based recognition or view-based recognition. As we believe that view-based recognition is better motivated from biological vision, we focus on the latter.

Hence, the problem that we consider in this paper is the following: Recognize a previously seen object or scene with a system which has seen many objects or

* This work has been supported by EC Grants IST-2003-004176 COSPAL and IST-2002-002013 MATRIS. This paper does not represent the opinion of the European Community, and the European Community is not responsible for any use which may be made of its contents.

B.K. Ersbøll and K.S. Pedersen (Eds.): SCIA 2007, LNCS 4522, pp. 908–917, 2007.

scenes from different poses. In case of scene recognition, different scenes rather means the same setting seen from different view angles and the task is to find a view with an approximately correct view angle. We consider both problems in this paper as we believe that there are many similarities in the problems and that our proposed approach solves both issues.

A further side condition of the problems is that they have to be solved in real-time. For object recognition, the real-time requirement is dependent on the task, but for scene recognition, video real-time is required, i.e., recognition in images with PAL resolution at 25 Hz. The real-time requirement rules out many powerful approaches for recognition. The situation becomes even worse if even the *learning has to be done on the fly* in terms of constantly adding new objects or scenes to the database. The latter requirement disqualifies all methods which rely on computationally expensive data analysis during a batch mode learning stage.

In the literature a vast amount of different recognition techniques are proposed, and we do not intend to give an exhaustive overview here. One prominent member of recognition systems is the one developed by Matas et. al., see e.g. [1] for a contribution to the indexing problem in the recognition scheme. The main purpose of the cited work is however to recognize objects as good as possible from a single view, whereas we propose a method which recognizes an object or a scene with an approximately correct pose.

This kind of problem is well-reflected by the COIL database [2], where 72 poses of each of the 100 objects are available. The main drawback of the COIL database is that the recognition task is fairly simple and perfect recognition has been reported for a subset COIL (30 images), with 36 views for learning and 36 views for evaluation [3]. This has been confirmed by later results, e.g., [4]. These methods are however not real-time capable and cannot perform on-the-fly learning. Furthermore, the recognition is very much intensity sensitive, as intensity respectively RGB channels are used for recognition. A more recent work [5] reaches real-time performance, but reported a significant decrease of the ROC (receiver operating characteristic) compared to the previously mentioned methods.

Our proposed method combines the following properties:

- Real-time recognition
- On-the-fly learning is possible
- Intensity invariance (to a certain degree)
- Few training views necessary (experiment: 12)
- State-of-the-art recognition performances (ROC)

This is achieved by a very efficient implementation of a sampled density estimator for the involved features hue, saturation, and orientation. The estimator is based on P-channels, a feature representation that is motivated from observations in biological systems. The density is then compared to reference densities by means of a modified Kullback-Leibler divergence, see e.g. [6]. The method for comparing P-channels is the main contribution of this paper. The resulting recognition method performs comparable to other, computationally much more demanding methods, which is shown in terms of ROC curves for experiments on the COIL database.

2 Methods for Density Estimation

Density estimation is a very wide field and similar to the field of recognitions methods, we do not intend to give an overview here. The interested reader is referred to standard text books as e.g. [7,8]. In this section we introduce a method for non-parametric density estimation, which is significantly faster than standard kernel density estimators and grid-based methods.

2.1 Channel Representations

The approach for density estimation that we will apply in what follows, is based on the biologically motivated channel representation [9,10]. The latter is based on the idea of placing local functions, the *channels*, pretty arbitrarily in space and to project the data onto the channels - i.e., we have some kind of (fuzzy) voting. The most trivial case are histograms, but their drawback of losing accuracy is compensated in the channel representation by knowledge about the algebraic relation between the channels.

The projections onto the channels result in tuples of numbers which - although often written as vectors (boldface letters) - do not form a vector space. In particular the value zero (in each component) has a special meaning, *no information*, and need not be stored in the memory. Note that channel representations are not just a way to re-represent data, but they allow advanced non-linear processing by means of linear operators, see Sect. 2.2.

Formally, the channel representation is obtained from a finite set of *channel projection operators* F_n. These are applied to the feature vectors $\mathbf{f}$ in a point-wise way to calculate the *channel values* p_n:

$$p_n = F_n(\mathbf{f}) \qquad n = 1, \ldots, N \; . \tag{1}$$

Each feature vector $\mathbf{f}$ is mapped to a vector $\mathbf{p} = (p_1, \ldots, p_N)$, the *channel vector*. If the considered feature is vector-valued, i.e., we would like to encode K feature vectors $\mathbf{f}_k$, we have to compute the outer product (tensor product) of the respective channel vectors $\mathbf{p}_k$:

$$\mathbf{P} = \bigotimes_{k=1}^{K} \mathbf{p}_k \; , \tag{2}$$

which is only feasible for small K, since the number of computations scales with a^K if a is the overlap between the channels.

2.2 Relation to Density Estimation

The projection operators can be of various form, e.g., $\cos^2$ functions, B-splines, or Gaussian functions [11]. The channel representation can be used in different contexts, but typically it is applied for associative learning [12] or robust smoothing [13].

In context of robust smoothing it has been shown that summing B-spline channel vectors of samples from a stochastic variable ξ results in a sampled kernel density estimate of the underlying distribution $p(\xi)$:

$$E\{\mathbf{p}\} = E\{[F_n(\xi)]\} = (B * p)(n) \ . \tag{3}$$

The global maximum of p is the most probable value for ξ and for locally symmetric distributions, it is equivalent to the maximum of $B * p$. The latter can be approximately extracted from the channel vector $\mathbf{p}$ using an implicit B-spline interpolation [13] resulting in an efficient semi-analytic method. The extraction of the maximum can therefore be considered as a functional inverse of the projection onto the channels.

In what follows, we name the projection operation also *channel encoding* and the maximum extraction *channel decoding*. In [14], advanced methods for channel decoding have been considered, which even allow the reconstruction of a complete density function. In this paper we concentrate on linear B-splines (B_1-kernels), such that no prefiltering according to [15] is necessary, and we just apply ordinary linear interpolation.

2.3 P-Channels

The idea of P-channels [16] is borrowed from projective geometry where homogeneous coordinates are used to represent translations as linear mappings and where vectors are invariant under global scalings. The P-channels are obtained by dropping the requirements for real-valued and smooth basis functions for channel representations. Instead, rectangular basis functions, i.e., ordinary histograms, are considered. Since rectangular basis functions do not allow exact reconstruction, a second component which stores the offset from the channel center is added. As a consequence, the channels become vector-valued (boldface letters) and the channel vector becomes a matrix (boldface capital).

A set of 1D values f_j is encoded into P-channels as follows. Without loss of generality the channels are located at integer positions. The values f_j are accounted respectively to the channels with the center $[f_j]$, where $[f_j]$ is the closest integer to f_j:

$$\mathbf{p}_i = \sum_j \delta(i - [f_j]) \begin{pmatrix} f_j - i \\ 1 \end{pmatrix} , \tag{4}$$

where δ denotes the Kronecker delta. Hence, the second component of the channel vector is an ordinary histogram counting the number of occurrences within the channel bounds. The first component of the channel vector contains the cumulated linear offset from the channel center.

Let $\mathbf{f}_j$ be a K-dimensional feature vector. The P-channel representation of a set of vectors $(\mathbf{f}_j)_j$ is defined as

$$\mathbf{p}_\mathbf{i} = \sum_j \delta(\mathbf{i} - [\mathbf{f}_j]) \begin{pmatrix} \mathbf{f}_j - \mathbf{i} \\ 1 \end{pmatrix} , \tag{5}$$

where $\mathbf{i}$ is a multi-index, i.e., a vector of indices $(i_1, i_2, \ldots)^T$, and $[\mathbf{f}]$ means $([f_1], [f_2], \ldots)^T$.

The main advantage of P-channels opposed to overlapping channels is the linear increase of complexity with growing number of dimensions. Whereas each input sample affects a^K channels if the channels overlap with a neighbors, it affects only $K+1$ P-channels. Hence, P-channel representations have a tendency to be extremely sparse and thus fast to compute.

3 Object Recognition Based on P-Channels Matching

The new contributions of this paper are: the special combination of image features, cf. Sect. 3.1, the conversion of P-channels to B_1-spline channels, cf. Sect. 3.2, and the matching scheme according to the Kullback-Leibler divergence, cf. Sect. 3.3.

3.1 Feature Space

The feature space is chosen in an ad-hoc manner based on the following requirements:

- Fast implementation
- Include color information
- Intensity invariance for a certain interval
- Inclusion of geometric information
- Stability
- Robustness

The first three requirements motivate the use of hue h and saturation s instead of RGB channels or some advanced color model. The value component v, which is not included in the feature vector, is used to derive the geometric information. Stability requirements suggest a linear operator and robustness (sensible behavior outside the model assumptions) induces a simple geometric descriptor. Therefore, we use an ordinary gradient estimate to determine the local orientation in double-angle representation, cf. [17]:

$$\theta = \arg((\partial_x v + i\partial_y v)^2) \ . \tag{6}$$

The feature vector is complemented with explicit spatial coordinates, such that the final vector to be encoded is five-dimensional: hue, saturation, orientation, x-, and y-coordinate. For each image point, such a vector is encoded into P-channels, where we used up to 8 channels per dimension. Note in this context that periodic entities (e.g. orientation) can be encoded in exactly the same way as linear ones (e.g. x-coordinate). Only for the conversion to overlapping channels (see below), the periodicity has to be taken into account.

3.2 Computing Overlapping Channels from P-Channels

In this section, we describe a way to convert P-channels into B_1-spline channels with linear complexity in the number of dimensions K. This appears to be surprising at first glance, as the volume of non-zero channels seems to increase by a factor of 2^K. This is surely true for a single channel, but in practice data clusters, and the overlapping channels result in making the clusters grow in diameter. As the cluster volume has a linear upper bound, cf. [16], and since isotropic clusters grow sub-linearly with the diameter, one can easily find an upper linear bound. This is however not true for very flat clusters, i.e., clusters which have nearly zero extent in several dimensions.

The efficient computational scheme to obtain linear B-spline channels is based on the separability of the K-D linear B-spline, i.e., the multi-linear interpolation. We start however with a short consideration of the 1D case: Two neighbored P-channels correspond to the local constant (h_1, h_2) respectively linear (o_1, o_2) kernels in Fig. 1. Combining them in a suitable way results in the linear B-spline:

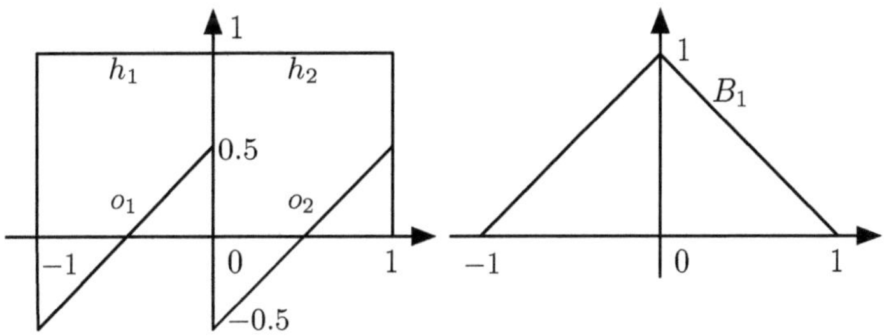

Fig. 1. Left: basis functions of two neighbored P-channels. Right: B_1-spline.

$$B_1 = \frac{h_1 + h_2}{2} + o_1 - o_2 \ . \tag{7}$$

The histogram components are averaged (2-boxfilter) and the offset components are differentiated (difference of neighbors), which implies also a shifting of the grid by 1/2. Consequently, non-periodic dimensions shrink by one channel. For periodic dimensions, one has to replicate the first channel after the last before computing the B_1-channels and the number of channels remains the same.

In order to convert multi-dimensional P-channels into multi-linear kernels, the histogram components have to be convolved with a multi-dimensional 2-boxfilter, which can be separated into 1D 2-boxfilters. Each offset component has to be convolved with the corresponding gradient operator and 2-boxfilters for all other dimensions than the one of the offset. This can be considered as a kind of div computation known from vector fields.

The resulting channel representation is identical to the one obtained from directly computing an overlapping channel representation with B_1-functions, if the underlying distribution of the data is locally independent. By the latter we mean that for each B_1-function support, the distribution is separable. In practice this means that the equality only holds approximately, but with the benefit of a much faster implementation. The final result is a sampled kernel density estimate with a multi-linear window function.

3.3 Matching of Densities

There are many possible ways to measure the (dis-) similarity of densities. In the standard linear framework, one could apply an SVD to the matrix of all densities in order to obtain the pseudoinverse. This method has the advantage of being the central step in any type of least-squares estimation method, e.g., if the aim is to extract not only the object type but also a pose interpolation. Unfortunately, we are not aware of any SVD algorithms which exploit and maintain the sparseness of a matrix, i.e., the computation becomes fairly demanding.

Until recently, we were not aware of the method for incremental SVD computation [18], and hence, we based our implementation below on the ordinary SVD. This means that our SVD-based recognition method cannot be used in a scenario with online learning, as e.g., in a cognitive system, but it is well suited for partly controlled environments where pose measurements are required, e.g., scene registration for augmented reality.

Same as the SVD, an ad hoc choice of (dis-) similarity measure as, e.g., the Euclidian distance, does not constrain the comparisons of prototype $\mathbf{p}$ and query $\mathbf{q}$ to be based on non-negative components. Due to the special structure of the problem, namely to compare estimates of densities, we believe that the Kullback-Leibler divergence

$$D(\mathbf{p},\mathbf{q}) = \sum_j p_j \log \frac{p_j}{q_j} \tag{8}$$

is most appropriate, as it combines maintaining sparseness, incremental updating, and non-negativity. In order to speed up the matching, one can precompute the terms that only depend on $\mathbf{p}$, i.e., the negative entropy of $\mathbf{p}$:

$$-H_{\mathbf{p}} = \sum_j p_j \log p_j \ , \tag{9}$$

such that the divergence can be computed by a scalar product:

$$D(\mathbf{p},\mathbf{q}) = -H_{\mathbf{p}} - \langle \mathbf{p} | \log \mathbf{q} \rangle \ . \tag{10}$$

All involved logarithms are typically regularized by adding an $\varepsilon > 0$ to $\mathbf{p}$ respective $\mathbf{q}$.

4 Recognition Experiments

In this section we present two experiments, one quantitative using the COIL database and reporting the ROC curves, and one qualitative experiment for scene registration.

4.1 Experiment on Object Recognition

The learning set for this experiment consists of 12 poses for each object from the COIL database. The test set consists of the remaining 60 views for each object, 6000 altogether. We evaluated the recognition based on three different feature sets: RGB colors (for comparison with [4]), orientation only (according to (6)), and the complete feature set described in Sect. 3.1. For each of these three recognition problems, we applied two different matching schemes: The one using SVD and the one using (10).

For each of the six combinations, we computed the ROC curves (Fig. 2) and their integrals, cf. Tab. 1.

Table 1. Results for object recognition (COIL database), using Kullback-Leibler divergence (KLB) and SVD. Three different features: orientation θ, RGB, and hue, saturation, orientation (hsθ).

Method	ROC integral
KLD, θ	0.9817
SVD, θ	0.9840
KLD, RGB	0.9983
SVD, RGB	0.9998
KLD, hsθ	0.9939
SVD, hsθ	1.0000

As it can be seen from the ROC curves and the integrals, the SVD performs marginally better than the KLD matching, both though close to the ideal result. However, we have not tried to improve the matching results by tuning the channel resolutions. We have fixed the resolutions by experience and practical constraints before the experiment was started. The number of P-channels is kept constant in all experiments.

4.2 Experiment on Scene Recognition

The task considered in this experiment was to find the view to a scene with the most similar view angle, using a fisheye lens. A first run was made on real data without known pose angles, i.e., the evaluation had to be done by hand. A second run was made on synthetic data with known pose angles. Example views for either experiment can be found in Fig. 3.

In either case, recognition rates were similar to those reported for object recognition above. A true positive was only reported if the recognized view had an angle close to the correct one, either by inspection (real data) or measured against the accompanying XML data (synthetic data). It virtually never happened that two false responses were generated in a row, which means with a one-frame delay, the scene is reliably registered.

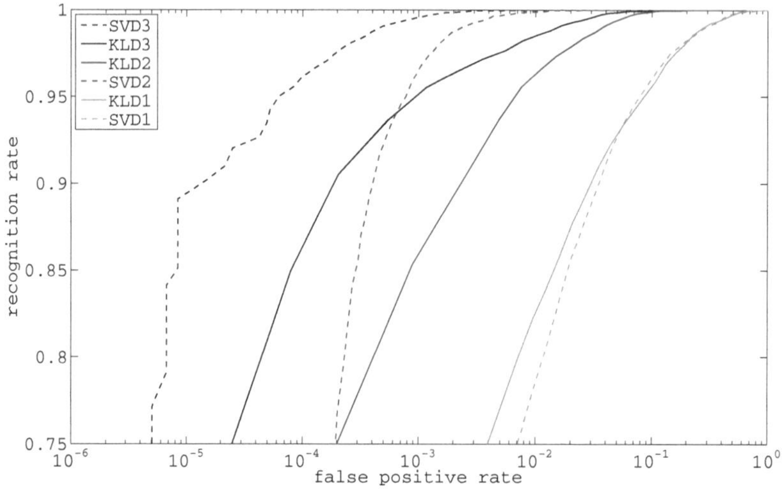

Fig. 2. ROC curves (semi logarithmic scale) for the six experiments described in the text. Index 1 refers to orientation only, index 2 to RGB, and index 3 to hsθ.

Fig. 3. Examples for fisheye images (scene registration). Left: real fisheye image. Right: synthetic image.

5 Conclusion

We have presented a novel approach to object recognition and scene registration suitable for real-time applications. The method is based on a sampled kernel density estimate, computed from the P-channel representation of the considered image features. The density estimates from the test set are classified according to the SVD of the training set and according to the Kullback-Leibler divergence. Both methods result in nearly perfect ROC curves for the COIL database, where the SVD approach performs slightly better. The divergence-based method is however suitable for online learning, i.e., adding new views and / or objects on the fly.

References

1. Obdržálek, Š., Matas, J.: Sub-linear indexing for large scale object recognition. In: Clocksin, W.F., Fitzgibbon, A.W., Torr, P.H.S., (eds.): BMVC 2005: Proceedings of the 16th British Machine Vision Conference. London, UK, BMVA Vol. 1, pp. 1–10 (September 2005)
2. Nene, S.A., Nayar, S.K., Murase, H.: Columbia object image library (coil-100). Technical Report CUCS-006-96 (1996)
3. Pontil, M., Verri, A.: Support vector machines for 3d object recognition. IEEE Transactions on Pattern Analysis and Machine Intelligence 20(6), 637–646 (1998)
4. Roobaert, D., Zillich, M., Eklundh, J.O.: A pure learning approach to background-invariant object recognition using pedagogical support vector learning. In: IEEE Computer Vision and Pattern Recognition. vol. 2, pp. 351–357 (2001)
5. Murphy-Chutorian, E., Aboutalib, S., Triesch, J.: Analysis of a biologically-inspired system for real-time object recognition. Cognitive Science Online 3, 1–14 (2005)
6. Cover, T.M., Thomas, J.A.: Elements of Information Theory. Wiley, New York (1991)
7. Bishop, C.M.: Neural Networks for Pattern Recognition. Oxford University Press, New York (1995)
8. Bishop, C.M.: Pattern Recognition and Machine Learning. Springer, Heidelberg (2006)
9. Granlund, G.H.: An Associative Perception-Action Structure Using a Localized Space Variant Information Representation. In: Proceedings of Algebraic Frames for the Perception-Action Cycle (AFPAC), Kiel, Germany (September 2000)
10. Snippe, H.P., Koenderink, J.J.: Discrimination thresholds for channel-coded systems. Biological Cybernetics 66, 543–551 (1992)
11. Forssén, P.E.: Low and Medium Level Vision using Channel Representations. PhD thesis, Linköping University, Sweden (2004)
12. Johansson, B., Elfving, T., Kozlov, V., Censor, Y., Forssén, P.E., Granlund, G.: The application of an oblique-projected landweber method to a model of supervised learning. Mathematical and Computer Modelling 43, 892–909 (2006)
13. Felsberg, M., Forssén, P.E., Scharr, H.: Channel smoothing: Efficient robust smoothing of low-level signal features. IEEE Transactions on Pattern Analysis and Machine Intelligence 28(2), 209–222 (2006)
14. Jonsson, E., Felsberg, M.: Reconstruction of probability density functions from channel representations. In: Proc. 14th Scandinavian Conference on Image Analysis (2005)
15. Unser, M.: Splines – a perfect fit for signal and image processing. IEEE Signal Processing Magazine 16, 22–38 (1999)
16. Felsberg, M., Granlund, G.: P-channels: Robust multivariate m-estimation of large datasets. In: International Conference on Pattern Recognition, Hong Kong (August 2006)
17. Granlund, G.H., Knutsson, H.: Signal Processing for Computer Vision. Kluwer Academic Publishers, Dordrecht (1995)
18. Brand, M.: Incremental singular value decomposition of uncertain data with missing values. Technical Report TR-2002-24, Mitsubishi Electric Research Laboratory (2002)

Graph Cut Based Segmentation of Soft Shadows for Seamless Removal and Augmentation

Michael Nielsen and Claus B. Madsen

Laboratory of Computer Vision and Media Technology, Aalborg University, Denmark
{mnielsen,cbm}@cvmt.dk
http://www.cvmt.dk

Abstract. This paper introduces a new concept within shadow segmentation for usage in shadow removal and augmentation through construction of a multiplicity alpha overlay shadow model. Previously, an image was considered to consist of shadow and non-shadow regions. This makes it difficult to seamlessly remove shadows and insert augmented shadows that overlap real shadows. We construct a model that accounts for sunlit, umbra and penumbra regions by estimating the degree of shadow. The model is based on theories about color constancy, daylight, and the geometry that causes penumbra. A graph cut energy minimization is applied to estimate the alpha parameter. Overlapping shadow augmentation and removal is also demonstrated. The approach is demonstrated on natural complex image situations. The results are convincing, and the quality of augmented shadows overlapping real shadows and removed shadows depends on the quality of the estimated alpha gradient in penumbra.

Keywords: shadow segmentation, graph cuts, augmented reality.

1 Introduction

The methods that are investigated in this paper are part of an idea to augment real images with virtual objects. If these have to look believable their shadows must look like the real shadows. For this purpose the light sources must be known and they can be estimated by detecting the real shadows. It is also necessary to know the degree of shadow for any given pixel, otherwise augmented shadows will stack upon real shadows.

There are many applications that benefit from shadow detection. For example segmentation of foreground objects without obstruction from shadows, classification of e.g. faces with shadows that could make it difficult to find the best match, and extraction of illumination such as light source direction and color. We want to find a model that can be used for shadow segmentation as well as shadow synthesis.

Our aim is to make a purely pixel driven method which works on single images in un-augmented scenes with no geometric knowledge about the scene. However, we will assume outdoor illumination.

B.K. Ersbøll and K.S. Pedersen (Eds.): SCIA 2007, LNCS 4522, pp. 918–927, 2007.

1.1 State of the Art

Salvador [1] distinguished between cast shadows (onto the ground plane) and self shadow. The detection relied on the edge image of a linearized chromaticity image. They considered dark pixels a-priori to be shadows and corrected this belief using heuristics concerning the edges of the real image and edges of the chromaticity image. This worked well in images with controlled simple geometry. It was tested with still images (of fruit) and video (moving people and cars).

Madsen [2] described shadows as an RGB alpha overlay. It is not just a black layer with an alpha channel, because the shadows are not only darker versions of the illuminated areas, but there is a change of hue, caused by the difference in hue between direct and ambient light. There is a fixed alpha for any given region. α can be described as the degree of shadow and the overlay color relates to the the tonal and intensity change of the shadow. Furthermore, shadows are characterized as full shadow, *umbra*, and half shadow *penumbra*, assuming only one light source. Multiple light sources would generate more complex grades of shadow regions.

Finlayson [3] takes advantage of planckian light and retinex theory. Assuming a single direct light source and another ambient light (different hue) computes a 1-d invariant image from the known path (illuminant direction) the shadow imposes on a 2-d log-ratio chromaticity plot. Note that ambient occlusion and surface normal direction is not taken into account in this model. The known path/offset is to be pre-calibrated. The edge maps of this invariant image can be used just like in [1] or to set a threshold in a retinex path. The results were images that looked flat and unsaturated with attenuated shadows and blur around the boundaries. The detected boundaries were high quality. Later an automatic initialization was developed using entropy minimization [4].

Cheng Lu [5] continued Finlayson's work using graph cuts for optimizing the shadow mask. Their method finds a binary shadow mask and use the illumination invariant chromaticity transform [3] as a static clue for computation of the capacities of the capacities in the graph model. They do not use any data term but considered the brightness changes in the means of windows around the supposed shadow edges as well as the chromaticity "shift" caused by the illumination color. It is not tested for difficult scenes and it does require knowledge of log illumination direction.

Previous work segmented shadows as a binary mask and used the edges as clues. They tested their algorithms in simplistic setups and had strict requirements to their cameras. We will use an α-overlay shadow model based on Finlayson's color theory, which is the most versatile approach. Our model must be invertible so that it can be used to generate shadows and to remove shadows. The degree of shadow should be adjustable through the α-parameter. The color theory does not explain what happens in the penumbra region. In the following the main theory is presented (for more detail refer to [3]) and the penumbra region is investigated. Followed by a graph cut algorithm for estimation of the α-channel for a series of natural images.

2 Methods

The color of a surface in full shadow is assumed to be a product of its color in sunlight and a fixed shading factor [6]:

$$\begin{bmatrix} R_{shad} \\ G_{shad} \\ B_{shad} \end{bmatrix} = \begin{bmatrix} \alpha R_{sun} \\ \beta G_{sun} \\ \gamma B_{sun} \end{bmatrix} \quad (1)$$

This relation holds for all pixels in the image. It follows from this generalization that in log chromaticity space there is an illumination direction vector added to the sunlit pixel, because $log(a * b) = log(a) + log(b)$. See figure 1. However, it will be necessary to be able to weight the shading effect by an *alpha* in the penumbra areas. The left side shows the illumination direction in log chromaticity space shows the direction that a surface color changes in the 2-d plot when the color temperature changes from sun to shadow. It follows a straight line. Surface 1 (S^1) is plotted in sun and shadow (umbra). We extend the model to account for varying degrees of shadow. Surface 2 (S^2) is plotted in sun, half shadow (penumbra), and shadow (umbra). However, tonal changes from ambient occlusions and inter-reflections do not follow the straight line. Ambient occlusions adjusts the color towards $[0, 0]$, while inter-reflections ajdusts the color toward the chromaticities of the reflecting surfaces.

The right side (in fig. figure 1) shows the geometry that causes umbra and penumbra regions and how our α overlay model should respond to those regions (shown at the bottom of the figure). In the sun $\alpha = 0$ and in the shadow $\alpha = 1$, and the in penumbra region α is proportional to the visible area of the sun, hence an S-shaped curve[1]. In the corner at the first box is a situation where the shadow becomes darker because the hemisphere is less accessible from those locations. This is not accounted for in the model.

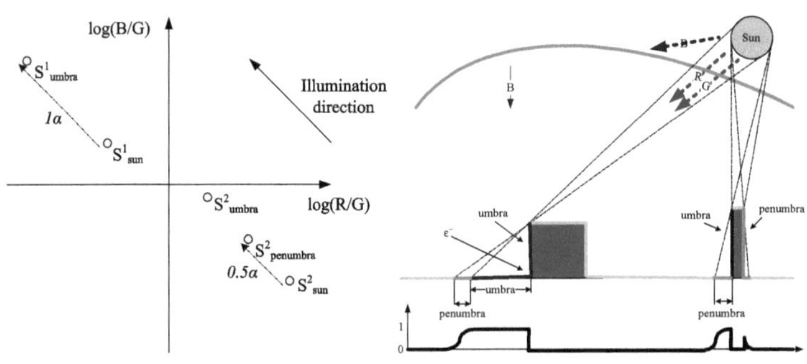

Fig. 1. [Left] 2-d log chromaticity space and illumination direction. [Right] Geometry causing penumbra and umbra. ϵ^- denotes ambient occlusion, which is not handled by the theory.

[1] For more detailed explanation [7]

The shadow model has two main tasks: 1. Shadow Augmentation for which the shadow region as a function of the sunlit region is needed. 2. Shadow Removal for which the sunlit region as a function of the shadow region is needed. The following model takes advantage of the relationship described in equation 1. It is adapted to control the degree of shadow with α in equation 2 (for each pixel p).

$$\rho_k^{shad} = (1 - S(\alpha)O_k)\rho_k^{sun}, \quad k = R, G, B \tag{2}$$

The model is easily inverted to represent ρ^{sun} (also called "shadow free image") as a function of α and ρ^{shad}. The S-shaped profile is modeled by a sigmoid function (S) in eq. 3.

$$S(\alpha) = (\frac{1}{(1 + e^{-(\alpha - 0.5)*6.\bar{6}})} - b) * a \tag{3}$$

Scaling factors $a = 1.074$ and $b = -0.0344$ were chosen such that $S(\alpha) = 0$ when $\alpha = 0$, and $S(\alpha) = 1$ when $\alpha = 1$.

The only initialization consists of finding the overlay color. The notation of the following investigation of optimal overlay color ($O = \{o_r, o_g, o_b\}$) is simplified: The surface color (albedo) will be denoted A. The irradiance from the sky will be E_{sky} and the irradiance from the sun will be E_{sun}. We consider a sunlit pixel to be $A(E_{sky} + E_{sun})$ and an umbra pixel to be AE_{sky}. Furthermore, ambient occlusion and sunlight direction is assumed to be fixed at zero.

$$\begin{aligned} AE_{sky} &= (1 - \alpha_{max}O)A(E_{sky} + E_{sun}) \\ O &= \frac{(1 - \frac{AE_{sky}}{A(E_{sky}+E_{sun})})}{\alpha_{max}} = \frac{(1 - \frac{E_{sky}}{E_{sky}+E_{sun}})}{\alpha_{max}} \end{aligned} \tag{4}$$

where $\alpha_{max} = 1$. We conclude that O is an umbra pixel divided by its corresponding sunlit pixel.

2.1 Estimation of α Via Graph Cuts

We treated the α estimation as a piecewise smooth labeling problem. An energy function would be minimized using graph cuts. If an energy function can be described as binary variables with regular energy terms with robust metrics, it is fast to find a strong local minimum [8] if the energy terms are regular.

The α channel of the overlay was estimated through α-expansions (thus reducing the problem to binary labels). For natural conditions where the illumination and camera settings are not fixed, it is not desirable to rely on pre-calibration of the camera that requires fixed settings for all images. We wish to find an initialization for the algorithm that is tailored for the image at hand without knowledge of the camera parameters. This process should be automatic, but at this stage the color of the overlay was given by manual initialization by handpicking a sunlit surface and its shadow counterpart. The mean red, green, and blue for each region was computed and the overlay color was given by $1 - \mu_{shadow}/\mu_{sun}$ as in equation 4.

The graph construction is given in [8]. The energy terms for the total energy (equation 5) remains to be defined. There were a number of ideas how to use the data term and the simplest method is to use a constant, because there is no way to tell if a given pixel is shadow. A sunlit black matte surface is darker than most bright surfaces in shadow. We construct the energy implementing three assumptions: 1. Piecewise smoothness as the standard smoothness term (V), 2. Attenuation. Reward α cuts such that shadow edge attenuation occurs (A), and 3. Legality. Punish illegal α cuts that creates edges that were not present in the original photo (L).

$$E(f) = w^- V^{p,q}(f) + w^+ A^{p,q}(f) + w^- L^{p,q}(f) \tag{5}$$

where $E(f)$ is the total energy of the configuration f of all variables (pixels). The terms are weighted by $w^+ = |p - q|$ or $w^- = N_{max} + 1 - |p - q|$. N_{max} is the maximum $|p - q|$.

In addition some heuristics about chromaticity edges was used. The overlay color ($\boldsymbol{O}$) corresponds to the illumination direction vector in [3]($\boldsymbol{d}$ eq. 6). All edges between pixels (p and their neighbors (q_b) the direction from the log chromaticity plot (i^p, eq. 7) of the lightest pixel (i^+) to the log chromaticity plot of the darkest pixel(i^+) are found as chromaticity edges ($\boldsymbol{d^{pq}} = i^- - i^+$). Then the difference between the illumination direction vector ($\boldsymbol{d}$) and the chromaticity edge was found (eq. 8). If an intensity edge ($\delta i = |i(p) - i(q)|$) between p and q was over a certain threshold (T_i) and the difference from the illuminant direction was under a certain threshold (T_a) then it was considered a shadow edge. This information was stored in a multi-layered shadow edge image. Each layer (bit) represented a neighborhood. Bit b for pixel p was set if pixel p in neighbor pixel q_b was labeled as a shadow edge.

$$\boldsymbol{d}_k = log\left(\frac{1 - O_k}{1 - O_r}\right) \quad k \neq r \tag{6}$$

$$\boldsymbol{i^p}_k = log\left(\frac{\rho_{shad}(p)_k}{\rho_{shad}(p)_r}\right) \quad k \neq r \tag{7}$$

$$I_{SE}(p_b) = (\delta i > T_i)\&\left(\frac{\boldsymbol{d} \times \boldsymbol{d^{pq}}}{|\boldsymbol{d}||\boldsymbol{d^{pq}}|}\right) < T_a\&(\boldsymbol{d} \bullet \boldsymbol{d^{pq}} > 0) \tag{8}$$

The resulting shadow edge image was morphologically dilated (bitwise) as to smoothen the penumbra regions. This allowed for a lower angle threshold (T_a) and made the window based sampling used in [5] unnecessary.

Pott's energy term (eq. 9) was used as a global smoothness term (V) and is applied for all p, q in the neighborhood of N.

$$V_{p,q \in N}(f(p), f(q)) = \begin{cases} 0 \text{ , if } & f(q) = f(p) \\ \lambda \text{ , if } & f(p) \neq f(q) \end{cases} \tag{9}$$

The attenuation term was applied where shadow edges (in I_{SE}) were detected. The energy was given by the absolute difference between the neighboring pixels in the estimated shadow free image (inverted equation 2) without the use of window sampling unlike [5].

$$A_{p,q \in N \cap I_{SE}}(f(p), f(q)) = min(0, |\rho_{sun}(p) - \rho_{sun}(q)|_{max} - K) \quad (10)$$

where $\rho_{sun}(p)$ and $\rho_{sun}(q)$ are the reconstructed shadow free image at pixels p and q given a labeling f(p) and f(q).

This smoothness term is not a regular metric at the shadow edges, so it is made regular through truncation. When building the graph, regularity is tested $(A(1,0) + A(0,1) \geq A(0,0) + A(1,1))$. When it is not regular, the energies are manipulated in such a way that the information given by the shadow edge is still maintained while $A(1,0) + A(0,1) = A(0,0) + A(1,1)$.

Improbable alpha cuts are punished by the mean gradient of the color bands $|\rho_{sun}(p) - \rho_{sun}(q)|_\mu$ in eq. 11.

$$L_{p,q \in N}(f(p), f(q)) = \begin{cases} 0 & \text{, if } \; f(p) = f(q) \\ |\rho_{sun}(p) - \rho_{sun}(q)|_\mu & \text{, if } \; f(p) \neq f(q) \end{cases} \quad (11)$$

To avoid problems with soft edges a neighborhood system that resembles pyramid scale space jumps was used. It was based on a 4-connected neighborhood (having two neighborhood relations; $x - 1$ and $y - 1$) and was extended to 4-layers by adding neighborhood relations at -2, -4, and -8. This should cover soft shadow edges 8 pixels wide. If the penumbras are wider than that, more layers can be added (-16, etc.).

V and L are weighted inverse to the distance because the farther away a neighbor is, the more likely it is that it is not supposed to be smooth. A is weighted more if the neighbor is far away in order to maximize the inner shadow degree of the shadow region with a broad penumbra.

Neighboring surface colors that are mapped into the same constellation in the chromaticity plot will be a source for error in the method.

2.2 Augmentation

It is easy to augment shadows once the initial shadow degree (α_0) and its corresponding shadow free image is known. A new soft shadow region ($\alpha_j, j > 0$) is constructed by choosing the geometric shape of the shadow. There should be a linear slope from 0 to 1 through penumbra. The new region should be added to the initial shadow region.

$$\alpha_{aug} = min(1, \alpha_0 + \sum_{j \in A} \alpha_j) \quad (12)$$

where α_{aug} is the augmented *alpha* map and A is the set of regions ($A = \{a_1, a_2, .., a_N\}$) that should be included in the augmented *alpha* map.

To remove a region of shadow it can simply be subtracted.

$$\alpha_{aug} = max(0, \alpha_0 - \sum_{l \in D} \alpha_l) \quad (13)$$

where D is the set of regions ($D = \{d_1, d_2, .., d_N\}$) that should be deleted in the augmented *alpha* map.

In the experiment the augmenting shadow regions were constructed by painting shadows using the mouse to add or subtract a pyramid shape where the mouse pointer was. It is easy to find connected shadow pixels using traditional connected pixels labeling algorithm on the alpha map and delete entire shadow regions by pointing out individual regions.

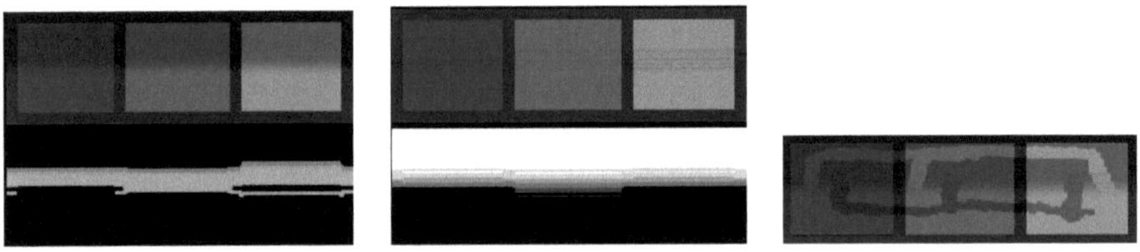

Fig. 2. Results trying to remove the shadow from a synthetic shadow image. [Left] Image (top) and detected shadow edges (bottom). [Middle] Estimated shadow free image (top) Estimated α (bottom) Black is $\alpha = 0$ and white is $\alpha = 1$. [Right] Augmented shadow image.

3 Results

The test was performed using $T_i = 10$, $T_a = 0.5$, dilation 5x5 structural element. $K = 255$ and the α resolution is 10 degrees of shadow. Shadow estimation figures consists of four sections: (left-top) The Image. (Left bottom) detected shadow edges. (right top) Estimated shadow free image. (right bottom) Estimated shadow degree. Black is no shadow and white is full shadow (umbra). Anything in between is penumbra. The linear professional camera was Canon EOS-1D Mark II. Images were stored in RAW format and converted into 24 bit bitmaps.

Figure 2 shows a synthetic image that resembles three squares of a Macbeth color checker pattern. The shadow (on top) is generated with the sigmoid shadow model. The overlay color was chosen from the mean of the shadow area and the mean of the sun area. The algorithm is applied to estimate the shadow degree and reconstruct the shadow free image. The algorithm is not cheated by the near-black albedo. When the shadow was estimated, some of the shadow was erased and new shadow was augmented on top of the old shadow.

Figures 3 through 4 shows natural examples captured by a linear professional camera in RAW format converted to 8bit per channel bitmap images. They are typical images with concrete and grass surfaces. The result is not very sensitive to lambda (fig. 3). The grass is not really diffuse surfaces, but the results are good anyway. Windows are very difficult. Especially when they reflect something in shadow, but is it really a false positive when this is estimated as shadow? The same question stands for the dark bottom of a cloud in fig. 4. The windows also cause gross inter-reflections, causing an increase of light in both shadow and sunlit regions.

Fig. 3. Professional camera. RAW converted into bitmap. Manual settings. [Left] Low lambda ($\lambda = 1$). [Middle] Higher lambda ($\lambda = 3$). The photographers leg is segmented as full shadow, but the shadow from the small thin pole in front of the blue container is deleted by the over-smoothed non shadow region. [Right] The shadow of the photographer has been removed.

Fig. 4. Professional camera. RAW converted into bitmap. Manual settings.

Fig. 5. Holiday photos. JPG compression. Automatic Settings.

The overlay color was optimally chosen from a neutral surface with little hue such as road or pavement when applicable. In fig. 4(right) the grass was used for initialization.

Figure 5 shows results from holiday photos captured by a consumer camera (Minolta Dimage 414). JPG compressed and automatic settings regarding white balance, focus, exposure, and shutter. Figure 5(right) was initialized differently

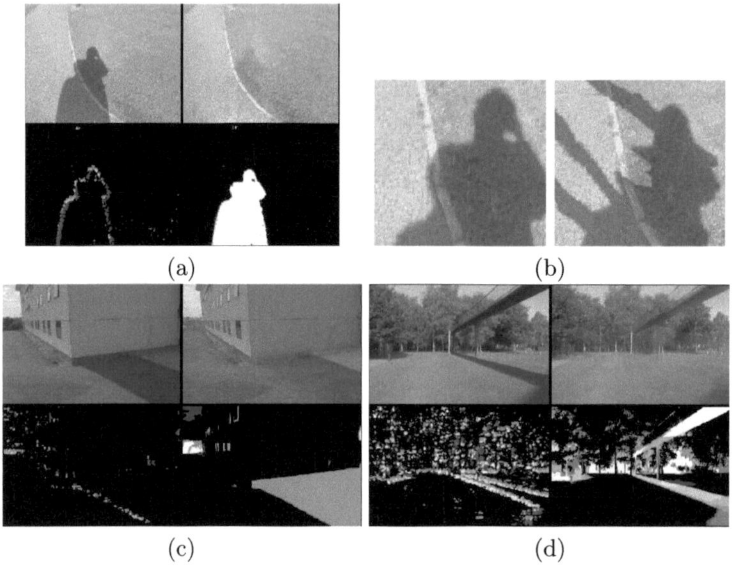

Fig. 6. (a,c,d) Our algorithm used on images from [3][6]. Screendumps from PDF papers, JPG compressed. (b) Seamless erased and augmented shadows.

from the others to see how it worked; the underexposed jacket was used. The result was good despite this bad initialization.

Figure 6 shows results on images from [3][6] for comparison. It shows clearly that our method yields more natural looking results; near perfection when the shadow degree is found correctly. $T_i = 30$ in fig. 6(a) otherwise the high frequency texture was obstructive. The lower right corner of the bridge image shows that the shadow degree is underestimated when penumbra is too wide for the neighborhood system to cover.

4 Conclusions

Our contribution estimates a mask of arbitrary levels of shadow for a variety of natural conditions. The model was usable for augmentation and removal of shadows in natural images. The theoretical limitation of the model is that it is constrained to diffuse surfaces and takes no account for ambient occlusion and surface normals. However, qualitatively the method estimates the degree of shadow such that it efficiently minimizes the error from these limitations. Better approximization would require the overlay color not to be fixed. The results in this paper was based on manual initialization of overlay color. However, ongoing research using gaussian mixtures fitted to the log chromaticity plot in combination with the entropy minimization method in [4] shows promising results for future research. High frequency texture is a potential problem, which calls for chromaticity- and edge-preserving smoothing.

Acknowledgments

This research is funded by the CoSPE project (26-04-0171) under the Danish Research Agency. This support is gratefully acknowledged.

References

1. Salvador, E., Cavallaro, A., Ebrahimi, T.: Cast shadow segmentation using invariant color features. Comput. Vis. Image Underst. 95(2), 238–259 (2004)
2. Madsen, C.B.: Using real shadows to create virtual ones. In: Bigun, J., Gustavsson, T. (eds.) SCIA 2003. LNCS, vol. 2749, pp. 820–827. Springer, Heidelberg (2003)
3. Finlayson, G.D., Hordley, S.D., Drew, M.S.: Removing shadows from images. In: Heyden, A., Sparr, G., Nielsen, M., Johansen, P. (eds.) ECCV(4) 2002. LNCS, vol. 2353, pp. 823–836. Springer, Heidelberg (2002)
4. Finlayson, G.D., Drew, M.S., Lu, C.: Intrinsic images by entropy minimization. In: Pajdla, T., Matas, J., eds.: In: Proc. 8th European Conf. on Computer Vision, Praque. pp. 582–595 (2004)
5. Lu, C., Drew, M.S.: Shadow segmentation and shadow-free chromaticity via markov random fields. In: IS&T/SID 13th Color Imaging Conference (2005)
6. Finlayson, G.D., Hordley, S.D., Drew, M.S.: Removing shadows from images using retinex. In: Color Imaging Conference, IS&T - The Society for Imaging Science and Technology pp. 73–79 (2002)
7. Nielsen, M., Madsen, C.B.: Segmentation of soft shadows based on a daylight- and penumbra model. In: Gagalowicz, A., Philips, W. (eds.) Proceedings of Mirage 2007, Springer, Heidelberg (2007)
8. Kolmogorov, V., Zabih, R.: What energy functions can be minimized via graph cuts? IEEE Trans. Pattern Anal. Mach. Intell. 26(2), 147–159 (2004)

Shadow Resistant Direct Image Registration

Daniel Pizarro[1] and Adrien Bartoli[2]

[1] Department of Electronics, University of Alcala,
Esc. Politecnica, 28871 Alcala de Henares, Spain
pizarro@depeca.uah.es
[2] LASMEA, Blaise Pascal University,
24 avenue des Landais, 63411 Aubiere, France
adrien.bartoli@univ-bpclermont.fr

Abstract. Direct image registration methods usually treat shadows as outliers. We propose a method which registers images in a 1D shadow invariant space. Shadow invariant image formation is possible by projecting color images, expressed in a log-chromaticity space, onto an 'intrinsic line'. The slope of the line is a camera dependent parameter, usually obtained in a prior calibration step. In this paper, calibration is avoided by jointly determining the 'invariant slope' with the registration parameters. The method deals with images taken by different cameras by using a different slope for each image and compensating for photometric variations. Prior information about the camera is, thus, not required. The method is assessed on synthetic and real data.

Keywords: Direct Registration, Shadow Invariant, Photometric Camera Calibration.

1 Introduction

The registration of image pairs consists in finding the transformation that best fits two images. That has been a key issue in computer vision, robotics, augmented reality and medical imagery. Although it was thoroughly studied in the past decades, there remain several open problems. Roughly speaking, there are two kinds of approaches: direct and feature based methods. The formers rely on fiducial points described by local properties, which allows matching despite geometric and photometric transformations. The geometric registration is thus formed by minimizing an error between the fiducials position expressed in pixels. As opposed to the local approach, direct methods use pixel discrepancy as a registration error measure. The brightness constancy assumption states that pixel values are equivalent under the sought after transformation. The warp relating two images consists of some geometrical transformation (*e.g.* an homography or an affine transformation) and some photometric model (*e.g.* channel intensity bias and gain or full affine channel mixing).

One of the main problems that arises in direct methods is the existence of partial illumination or shadow changes in the scene to register. In such cases,

B.K. Ersbøll and K.S. Pedersen (Eds.): SCIA 2007, LNCS 4522, pp. 928–937, 2007.

the brightness constancy assumption is violated. This paper addresses the problem of directly registering such kind of images. Our proposal is based on expressing the error in a transformed space different from the usual one based on image intensities. In this space which is onedimensional the change of illumination or shadows are removed. This invariant space is governed by a single parameter which is camera-dependent and defines a transformation between the log-chromaticity values of the original RGB image and the invariant image. We propose a method for jointly computing the sought after geometric registration and the parameters defining the shadow invariant space for each image.

Paper Organization

We review previous work and give some background in §2. In §3 we state our error function and give an algorithm for effectively registering images in §4. Results on synthetic and real images are presented in §5. Finally, conclusions are presented in §6.

2 Previous Work

The content of this section is divided into two major parts. First, some previous work about direct image registration is briefly described. The general approach and the most common problems are described. Secondly, some background on color image formation is presented, necessary for describing the process of shadow invariant image formation, which is finally stated.

2.1 Direct Image Registration

The registration of two images is a function $\mathcal{P}$, which models the transformation between a source image, $\mathcal{S}$ and a target image $\mathcal{T}$ over a region of interest $\mathcal{R}$. Function $\mathcal{P}(\mathcal{T}(q), q; \phi)$ is parametrized by a vector ϕ composed of geometric and photometric parameters in the general case.

The error function to be minimized make is the sum of square differences of intensity values, over the parameter vector ϕ.

The problem is formally stated as:

$$\min_{\phi} \sum_{q \in \mathcal{R}} \|\mathcal{S}(q) - \mathcal{P}(\mathcal{T}(q), q; \phi)\|^2. \tag{1}$$

A linearization of each residual, which allows to solve it in an iterative Linear Least Square fashion, was popularized by the Lucas-Kanade algorithm [1]. There exist remarkably fast approaches for warps functions forming groups. It is known as the Inverse Compositional algorithm [2] and it has been successfully applied with geometric transformations and in [3] an affine photometric model is also included.

The presence of shadows or illumination changes affect the applicability of equation (1), producing registration errors or divergence in the algorithm. There exist plenty of proposals in the literature to extend the direct registration with a certain grade of immunity against perturbations. The most common approach is to mark shadow areas as outliers. The use of robust kernels inside the minimization process [4] allows the algorithm to reach a solution. Other approaches try to model the shadows and changes of illumination. In [5] a learning approach is used to tackle illumination changes by using a linear appearance basis.

2.2 Background on Color Image Formation

We present the physical model used to describe the image formation process. The theory of invariant images is described later in terms and under the assumptions stated below.

We consider that all the surfaces are lambertian, that the lights follow a planckian model and that the camera sensor is narrow-band. The RGB color obtained at a pixel is modeled by the following physical model:

$$\rho_k = \sigma S(\lambda_k) E(\lambda_k, T) Q_k \delta(\lambda - \lambda_k) \quad k = 1, 2, 3, \tag{2}$$

where $\sigma S(\lambda_k)$ represents the surface spectral reflectance functions times the lambertian factor. The term $Q_k \delta(\lambda - \lambda_k)$ represents the sensor spectral response function for each color channel k centered at wavelength λ_k. $E(\lambda_k, T)$ is the spectral power distribution of the light in the planckian model. This is modeled by the following expression:

$$E(\lambda, T) = I c_1 \lambda^{(-5)} e^{\left(\frac{-c_2}{T\lambda}\right)} \tag{3}$$

This model holds for a high rank of color temperatures $T = [2500^o, 10000^o]$. The term I is a global light intensity and the constants c_1 and c_2 are fixed.

According to this model, the value obtained by the camera at any pixel ρ_k is directly obtained by:

$$\rho_k = \sigma I c_1 (\lambda_k)^{-5} e^{\left(\frac{-c_2}{T\lambda_k}\right)} S(\lambda_k) Q_k. \tag{4}$$

2.3 Shadow Invariant Image Theory

The transformation which allows invariant image formation is based on the original work of [6] in which a method for obtaining an illumination invariant, intrinsic image from an input color image is developed . The method relies on the above presented image formation model, based on the assumption of lambertian surfaces, narrow-band sensors and planckian illuminants.

Given the three channel color components $\rho = (\rho_1, \rho_2, \rho_3)$ described in (4), the logarithm of chromaticity ratios are formed.

$$\begin{aligned} \mathcal{X}_1 &= \log\left(\frac{\rho_1}{\rho_3}\right) = \log(s_1/s_3) + (e_1 - e_3)/T \\ \mathcal{X}_2 &= \log\left(\frac{\rho_2}{\rho_3}\right) = \log(s_2/s_3) + (e_2 - e_3)/T, \end{aligned} \tag{5}$$

where $e_k = -c_2/\lambda_k$ only depends on camera spectral response and not on the surface and $s_k = c_1\lambda_k^{(-5)}S(\lambda_k)Q_k$ does not depend on color temperature T.

The pair of values $\mathcal{X}_1$ and $\mathcal{X}_2$ lie on a line with direction vector $\bar{e} = (e_1 - e_3, e_2 - e_3)$. Across different illumination temperature T, vector $\mathcal{X} = (\mathcal{X}_1, \mathcal{X}_2)$ moves along the line.

An illumination invariant quantity can be formed by projecting any vector $\mathcal{X}$ onto the orthogonal line defined by $\bar{e}^{\perp} = (\cos(\theta), \sin(\theta))$. Therefore, two pixels from the same surface viewed under different illuminations get projected at the same place.

To reduce the arbitrary election of the chromaticity ratios, in [6], is proposed a method to use the geometrical mean $(\rho_1\rho_2\rho_3)^{(1/3)}$ of the three channel values as denominator. A vector of three linearly dependent coordinates is obtained. By choosing a proper decomposition, a twodimensional equivalent vector $\mathcal{X}$ is obtained that preserve the essential properties of equation (5).

The transformation $\mathcal{L}$ is simply obtained by projecting vector $\mathcal{X}$ onto the invariant line parametrized by its slope angle θ:

$$\mathcal{L}(\rho, \theta) = \mathcal{X}_1(\rho)\cos(\theta) + \mathcal{X}_2(\rho)\sin(\theta) \tag{6}$$

This transformation, as it has been previously stated, represents the mapping between a color image and its corresponding shadow invariant representation. By explicitly describing the whole color image $\mathcal{S}$ as an input in (6), the result of $\mathcal{L}(\mathcal{S}, \theta)$ is a 1D shadow invariant image. The transformation is therefore global so it does not depend on pixel position $q \in \mathbb{R}^2$, but only on its color value.

The slope angle θ of the invariant line only depends on camera spectral properties, so it varies across different cameras. In [6] it is presented a method to obtain the slope by a calibration step using a color pattern or by a set of preregistered images from the same camera under illumination changes. In [7] an autocalibration approach is presented by finding the slope for which the entropy of the invariant image is minimum. The later method is proved to be capable to find the correct slope with only one image.

The entropy based method unless simple and powerful requires images in which remarkable shadow areas are present. In the case of images in which the change of illumination is global and no shadow is present, the method is not able to produce the correct slope.

3 Joint Image Registration and Photometric Camera Calibration

Image registration is proposed under invariant transformation $\mathcal{L}(\mathcal{S}(q), \theta)$, applied to both the target and the source image.

The new cost function, is expressed as follows:

$$\min_{\phi,\theta_1,\theta_2} \sum_{q\in\mathcal{R}} \|\mathcal{L}(\mathcal{S}(q), \theta_1) - \mathcal{L}(\mathcal{P}(\mathcal{T}, q; \phi), \theta_2)\|^2 \tag{7}$$

As previously stated, the slope parameter θ is, in general, different in both images (θ_1 and θ_2). According to the camera model, any change of illumination intensity and temperature will be discarded in the invariant image once parameter θ is obtained. Therefore, in theory, $\mathcal{P}$ can only be geometrical.

The validity of (7) is based on the assumption that in different cameras, intrinsic images are comparable. However as is stated in this section, in general such an hypothesis does not hold due to differences in camera response functions. A photometric model is proposed for compensating such differences.

3.1 Camera Response Dependent Parameters

In this section it will be shown that besides the slope, between two cameras it is of importance the inclusion of photometric parameters over RGB space so that the invariant space of two images is directly comparable. Such parameters will not try to compensate for global illumination as in previous attempts [3], but instead they will represent a compensation between different camera responses.

Multiple Gain Compensation. Assuming that each camera has similar spectral response, so that the values of λ_k are similar, the slope and surface reflectance will produce similar values. However for different channel gains Q_k the log-chromaticity values are affected.

It is thus reasonable to include multiple gains compensation a_k per channel for the target image before computing its log-chromaticity values:

$$\mathcal{X}_k = \log\left(\frac{a_k \rho_k}{a_3 \rho_3}\right) = \log(a_k/a_3) + \log(\rho_k/\rho_3) \tag{8}$$

According to (6), the projection reduces the photometric compensation into a one dimensional offset $d_{\mathcal{L}}$.

$$\mathcal{L}(\rho, \theta) = \log\left(\frac{\rho_1}{\rho_3}\right)\cos(\theta) + \log\left(\frac{\rho_2}{\rho_3}\right)\sin(\theta) + d_{\mathcal{L}}, \tag{9}$$

where $d_{\mathcal{L}} = \log(a_1/a_3)\cos(\theta) + \log(a_2/a_3)\sin(\theta)$.

In the case where both cameras where different by only constant gains, it is still enough as a way to compensate camera responses, to compute a single offset.

Multiple gain and bias for each channel compensation. As stated in [8], the real response for most digital cameras is not linear. Under certain range of values we can consider that the camera response can be approximated by a gain and bias function. The presence of bias over RGB represents a problem since the assumption of the invariant line is no longer valid.

Adding bias and gain over RGB results in the following invariant representation:

$$\mathcal{L}(\rho, \theta) = \log\left(\frac{\rho_1 + b_1}{\rho_3 + b_3}\right)\cos(\theta) + \log\left(\frac{\rho_2 + b_2}{\rho_3 + b_3}\right)\sin(\theta) + d_{\mathcal{L}}. \tag{10}$$

Where $d_{\mathcal{L}}$ is the same commented in the multiple gain model, and b_k are biases added to color values.

The total number of required photometric parameters is four under the assumption that only one of the cameras suffer from the bias problem. In the case that both images are suitable to be affected, an extra bias model is introduced for the source image. For such critical case the number of parameters is increased to seven.

The new cost function, which includes photometric parameters (ϕ_p^1, ϕ_p^2) in source and target images, is presented:

$$\min_{\phi,\theta_1,\theta_2,\phi_p^1,\phi_p^2} \sum_{q\in\mathcal{R}} \|\mathcal{L}(\mathcal{S}(q),\theta_1,\phi_p^1) - \mathcal{L}(\mathcal{P}(\mathcal{T},q;\phi),\theta_2,\phi_p^2)\|^2 \tag{11}$$

Besides the commented models, specially amateur cameras suffer from many artificial perturbations which includes saturation boosting, channel mixing and digital filters applied to the raw image sensed.

4 Minimizing the Error Function

In this section, the optimization process involved in obtaining image registration and invariant space parameters is presented in details.

Given the more general expression (11), which includes a photometric model, a Gauss-Newton approach is derived as an optimization method.

A first order approximation of the warped image around current estimation of parameters $\Phi = (\phi, \theta_1, \theta_2, \phi_S, \phi_T)$ is obtained. The residue in the error function is also renamed by using two different functions $\mathcal{W}_1$ and $\mathcal{W}_2$ depending on vector Φ.

The renamed cost function becomes:

$$\min_{\Phi} \sum_{q\in\mathcal{R}} \|\mathcal{W}_1(\mathcal{S}(q),\Phi) - \mathcal{W}_2(\mathcal{T}(q),q,\Phi)\|^2. \tag{12}$$

The Gauss-Newton approximation is:

$$\epsilon^2 \approx \sum_{q\in\mathcal{R}} \|\mathcal{W}_1(\mathcal{S}(q),\Phi) - \mathcal{W}_2(\mathcal{T}(q),q,\Phi) + (L_{\mathcal{W}_1}(q) + L_{\mathcal{W}_2}(q))\Delta\Phi\|^2, \tag{13}$$

where $L_{\mathcal{W}_1}(q)$ and $L_{\mathcal{W}_2}(q)$ represents respectively the first derivatives of functions $\mathcal{W}_1$ and $\mathcal{W}_2$. ϵ^2 is the residual error from the cost function to minimize.

The parameter increment $\Delta\Phi$ is given by solving the following linear system:

$$E_\Phi \Delta_\Phi = b_\Phi, \tag{14}$$

where E_Φ represents the approximated Hessian of the error function:

$$E_\Phi = \sum_{q\in\mathcal{R}} (L_{\mathcal{W}_1}(q) + L_{\mathcal{W}_2}(q))(L_{\mathcal{W}_1}(q) + L_{\mathcal{W}_2}(q))^T. \tag{15}$$

The right hand side of the linear system b_Φ includes the error image:

$$b_\Phi = \sum_{q\in\mathcal{R}} (L_{\mathcal{W}_1}(q) + L_{\mathcal{W}_2}(q))(\mathcal{W}_1(\mathcal{S}(q),\Phi) - \mathcal{W}_2(\mathcal{T}(q),q,\Phi)). \tag{16}$$

Once the increment $\Delta\Phi$ is obtained Φ is updated accordingly with each model.

4.1 Using an Homography for the Geometric Model

The used geometric model consists of an homography transformation. Homographies are fully representative as global geometrical models. They are suitable for registering planar scenes or under camera rotation. It is a groupwise homogeneous transformation represented by a full rank 3×3 matrix H with eight degrees of freedom. The homography is applied to the homogeneous coordinates q in the target image for composing the warp. It is assumed that the first eight coordinates of vector Φ represent the values of ΔH at each iteration.

5 Experimental Results

In this section some of the results are presented in order to validate the proposal. The experiments are designed to compare the convergence properties of our algorithm and to test the photometric models we proposed.

5.1 Synthetic Image Registration

A set of synthetic images is generated according to the model presented in §2. Each image consists of a set of quadrangular color patches under different illuminations, covering a range of color temperatures from 2500° to 10000°. By modifying camera response parameters we are able to simulate images taken by different cameras. Two models are considered for the experiment.

- *Multiple gains*: described by only one bias parameter $d_{\mathcal{L}}$ in the invariant space.
- *Multiple gains and biases*: Considering the complete case, the model has five parameters for the target image $\phi_{p\mathcal{T}}$ and three parameters $\phi_{p\mathcal{S}}$ for the source image.

compared Algorithms. In all the experiments the following algorithms are compared:

- DRSI-NP: Direct Registration in Shadow Invariant space with No Photometric model to compensate between cameras.
- DRSI-MG: Direct Registration in Shadow Invariant space with Multiple Gains as a photometric model to compensate between cameras.
- DRSI-MGB: Direct Registration in Shadow Invariant with Multiple Gains and Biases in target image and only bias in source image.
- DR: Direct Registration over greylevel values

Simulation Setup. Given two differently illuminated sequences of patches, we simulate a 2D homography by displacing the corners in the target image in random directions by some value γ with default value of 5 pixels. The target image is contaminated by gaussian noise with variance σ and a default value of 25.5. The value of photometric parameters for target and source image has been chosen fixed for the simulations: $\phi_{\mathcal{T}} = (b_1 = 3.2, b_2 = 2.1, b_3 = 1)$ and $\phi_{\mathcal{S}} = (b_1 = 0.8, b_2 = 4, b_3 = 3.5)$. Both target and source image has a slope parameter of $\theta_1 = \theta_2 = 169.23^o$. Interest area $\mathcal{R}$ is obtained by using strong edges in greylevel image and dilating them by a factor of 8.

Fig. 1. Pair of synthetic images

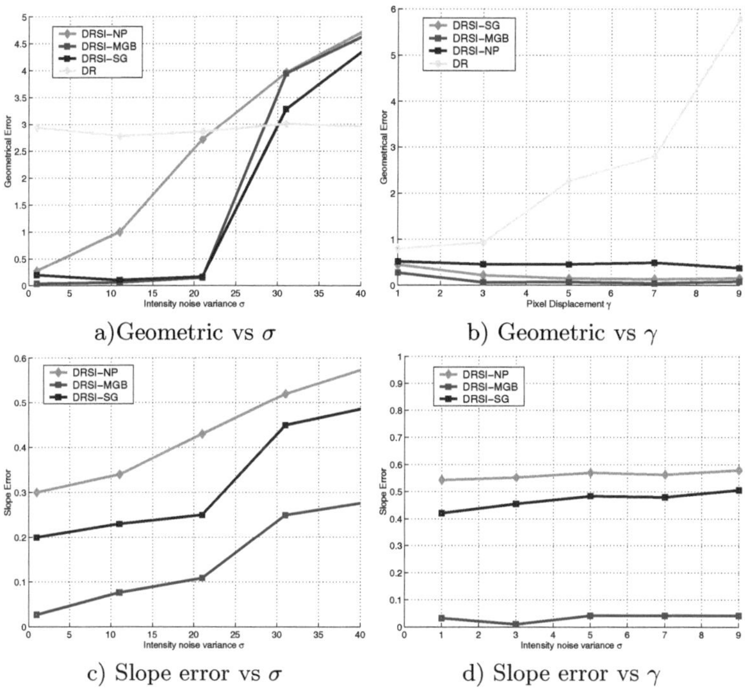

a)Geometric vs σ

b) Geometric vs γ

c) Slope error vs σ

d) Slope error vs γ

Fig. 2. Residual Error vs σ and Geometrical Error vs γ

Results. In Figure 2.a and 2.b the geometric error is presented against noise variance σ and initial pixel displacement γ. In Figure 2.c and 2.d the slope angle error is presented against noise variance σ and initial pixel displacement γ.

5.2 Real Image Registration

For testing the presented proposal with real images, the same planar surface is acquired with two different low-cost commercial cameras. By manually clicking in the four corners of the planar shape, an interest area $\mathcal{R}$ is obtained. If the two regions are far from 10 pixels of displacement a pre-registration is used using manual clicked points. The results presented show the error measured between a pair of images transformed into its respective invariant spaces across the iterations.

a) Source Image b) Target Image c) DRSI-MGB

a) DRSI-SGB b) DRSI-NP c) DR2

Fig. 3. Real images and its resulting registration

6 Conclusions

A new method to achieve direct image registration in the presence of shadows is proposed. The approach is based on minimizing the registration error directly in a transformed space from RGB space. The new space is parametrized by a single camera dependent parameter, the invariant line slope. Such parameter is in general different from each camera, so it is included in the optimization stage. Solving registration parameters in the invariant space from images taken by different cameras offers difficulties due to the response function of each camera. In this paper, two models are proposed to compensate such differences: Multiple Gain compensation and Multiple Gain and Bias. Results on synthetic data show that the last one obtains better registration performance against pixel displacement and noise. In real images, under some conditions the use of the multiple gains and biases model is crucial to achieve registration. If both cameras are of similar response, the simple algorithm which avoid photometric model calculations is the best choice. The use of invariant space in direct methods allows to avoid shadows directly without the need of using complex methods which use illumination modeling or robust kernel optimization.

References

[1] Lucas, B., Kanade, T.: An iterative image registration technique with an application to stereo vision. In: Proceedings of the International Joint Conference on Artificial Intelligence (1981)
[2] Baker, S., Matthews, I.: Lucas-Kanade 20 years on: A unifying framework. International Journal of Computer Vision 56(3), 221–255 (2004)
[3] Bartoli, A.: Groupwise Geometric and Photometric Direct Image Registration. In: BMVC'06 Proceedings of the Seventeenth British Machine Vision Conference, Edinburgh, UK, pp.11–20 (September 2006)
[4] Baker, S., Gross, R., Matthews, I., Ishikawa, T.: Lucas-Kanade 20 Years ON: A Unifying Framework: Part 2, tech. report CMU-RI-TR-03-01, Robotics Institute, Carnegie Mellon University (February 2003)
[5] Baker, S., Gross, R., Matthews, I., Ishikawa, T.: Face Recognition Across Pose and Illumination. In: Li, S.Z., Jain, A.K. (eds.) Handbook of Face Recognition, Springer, Heidelberg (2004)
[6] Finlayson, G.D., Hordley, S.D., Drew, M.S.: Removing shadows from images. In: ECCV 2002: European Conference on Computer Vision, vol. 4, pp. 823–836 (2002)
[7] Finlayson, G., Drew, M., Lu, C.: Intrinsic Images by Entropy Minimization. In: Proceedings of 8th European Conference on Computer Vision, Praque, pp. 582–595 (2004)
[8] Barnard, K., Funt, B.: Camera characterization for color research. Color Research and Application 27(3), 153–164 (2002)

Classification of Biological Objects Using Active Appearance Modelling and Color Cooccurrence Matrices

Anders Bjorholm Dahl[1,2], Henrik Aanæs[1], Rasmus Larsen[1], and Bjarne K. Ersbøll[1]

[1] Informatics and Mathematical Modelling, Technical University of Denmark
[2] Dralle A/S - Cognitive Systems, Copenhagen, Denmark
abd@imm.dtu.dk
www.dralle.dk

Abstract. We use the popular active appearance models (AAM) for extracting discriminative features from images of biological objects. The relevant discriminative features are combined principal component (PCA) vectors from the AAM and texture features from cooccurrence matrices. Texture features are extracted by extending the AAM's with a textural warp guided by the AAM shape. Based on this, texture cooccurrence features are calculated. We use the different features for classifying the biological objects to species using standard classifiers, and we show that even though the objects are highly variant, the AAM's are well suited for extracting relevant features, thus obtaining good classification results. Classification is conducted on two real data sets, one containing various vegetables and one containing different species of wood logs.

1 Introduction

Object recognition is one of the fundamental problems in computer visions, and plays a vital role in constructing 'intelligent' machines. Our initial motivation for this work is the construction of an automated forestry system, which needs to keep track of wood logs. Many of the objects in our daily environment in general, and in our motivating problem in particular, are biological, and pose special challenges to a computer vision system. The origin of these challenges are the high degree of intraclass variation, which we as humans are very good at dealing with, e.g. consider the multitudes of ways a face or a potato can look. To enable biological variation to be handled in a classification system, we have to find methods for extracting discriminative features, from the depicted objects. AAM's have proven very well suited for addressing the challenge of handling biological variation in the case of image registration, cf. [4]. It is thus highly interesting if this property of the AAM's also proves well for classification of objects, and how this should be implemented. Therefore, we have investigated AAM's for extracting discriminative features by conducting the following experiments:

1. Classification based on **Multiple AAM's**, i.e. building an AAM for each class and assigning images of unknown objects to the most probable AAM.

B.K. Ersbøll and K.S. Pedersen (Eds.): SCIA 2007, LNCS 4522, pp. 938–947, 2007.

2. Classification based on a **global AAM**, i.e. building one single AAM and using model parameters for assigning images of unknown objects to the most probable class.
3. Identify relevant discriminative patches from the use of an AAM. The object is identified by shape alignment from the **AAM and texture** is extracted and used for second order texture analysis.

Two data sets have been investigated in this paper, one containing vegetables and one containing wood logs. Experiment 1 and 2 have been conducted for both data sets and experiment 3 has been conducted only for the wood log data set.

1.1 Related Work

The environment plays a vital role in solving object recognition problems. In a natural environment objects may be seen from many different angles, they may be occluded, light may change etc. Efforts on solving this type of problem have been put in identifying local object features invariant to the changing conditions, cf. e.g. [16,14,13], and the way to match these features to a model, cf. e.g. [5,6].

Controlling the environment in some way, gives the opportunity of easing the flexibility constraints of the object recognition system. In some situation object recognition on whole objects is a reasonable approach, giving the option of e.g. extracting global PCA features. This is done for face recognition by e.g. Turk & Pentland [20] with the eigenface, and Belhumeuer *et al.* [1] for their fisherface based on Fishers Linear Discriminant Analysis. No shape information is included with these methods. Edwards *et al.* [7] introduces the use of an AAM for face recognition based on both shape and texture information. Fagertun *et al.* [9] improves this method by the use of canonical discriminant analysis. AAM's have been used for related recognition problems, e.g. eye tracking by Stegmann [18] and Hansen *et al.* [10].

Pure texture has also been used for object recognition. The second order texture statistics based on cooccurrence matrices, was originally developed by Haralick *et al.* [11] for pixel classification. This method has been extended to object recognition by e.g. Chang & Krumm [3] using color cooccurrence histograms. Palm [15] does classification of different textures, including wood bark textures, using color cooccurrence matrices. He extends from gray level to color images and improves the classification.

In this paper we focus on object recognition in an environment with some degree of controlled conditions. We use a black background, controlled lighting, and we make sure that the whole object is visible.

2 AAM and Texture Features

In the following we describe the methods for extracting the discriminative features used in the three experiments.

2.1 AAM

The AAM model - in 2D as it will be used here - is a description of an object in an image via it's contour or shape and it's texture. Each of these entities can be represented as a vector, i.e. $\mathbf{s}_i$ and $\mathbf{t}_i$ respectively, where the subscript, i, denotes an instance (image). The parameters of the AAM is, however, a lower dimensional vector, $\mathbf{c}_i$, and a specific AAM consists of an linear mapping for $\mathbf{c}_i$ to $\mathbf{s}_i$ and $\mathbf{t}_i$, i.e.

$$\mathbf{m}_i = \begin{bmatrix} \mathbf{Ws} \\ \mathbf{t} \end{bmatrix}_i = \mathbf{\Phi c}_i \ , \tag{1}$$

where $\mathbf{\Phi}$ is a matrix representing the linear map. The AAM or $\mathbf{\Phi}$ is estimated from an annotated training set. By optimizing the AAM to a new depicted object, an image close to the original is synthesized, and the model parameters $\mathbf{c}_i$ is a vector of principal components describing the unknown object with regards to the shape and texture of the object. The interested reader is referred to Cootes and Taylor [4] for a more detailed description and Stegmann *et al.* [19] for a detailed description of the model implementation.

Features from multiple AAM's. In this case an AAM, $\mathbf{\Phi}_j$, is fitted to each class $\mathcal{C}_j$, i.e. the training set is divided into its component classes, and one AAM is fitted to each.

Here there is a feature vector $\mathbf{c}_i$ i specific for each model, and these features are not comparable, because they belong to a specific model and can not be used directly for classification.

Given an AAM for each class $\mathcal{C}_j$, you would expect the optimization of an image i to perform best for the class that the object belongs to. Therefore, a goodness of fit would be a reasonable measure for classifying the object. For a given unknown object image textural difference between the object texture $g_{i_{obj}}$ and the model instance $\bar{g}_{mod}$ is calculated:

$$E = \sum_{i=1}^{n} (g_{i_{obj}} - \bar{g}_{mod})^2 = ||g_{i_{obj}} - \bar{g}_{mod}||_2^2, \tag{2}$$

where E is the difference between the model image and the measured image by the squared 2-norm.

Features from a global AAM. In this case a single global AAM, $\mathbf{\Phi}$, is fitted to instances of all classes. Following this, the $\mathbf{c}_i$ are calculated for each instance in the training set. The elements of $\mathbf{c}_i$, containing both shape and texture information, are used in a linear classifier, see section 2.3.

Textural warp. The basis for making a textural warp is knowledge of the log localization in the image. This comes from the AAM shape alignment. The warp is done by sampling pixels along elliptic curves on the surface of the logs using bilinear interpolation, see Figure 1. The elliptic curves are calculated from the shape of the end face of the wood log, and guided by the shape of the sides

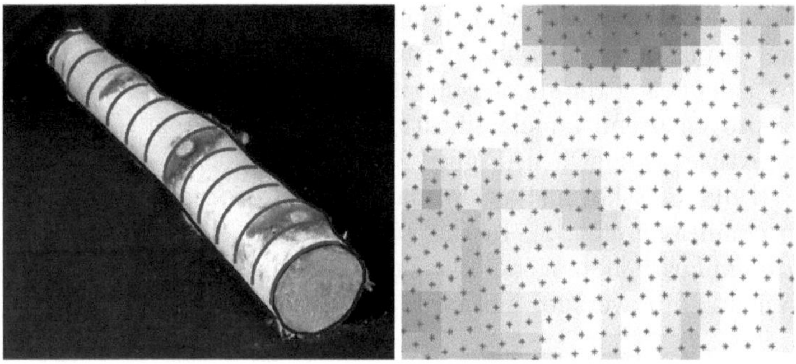

Fig. 1. Illustration of bark texture warp. Left is an image of a Birch log shown with a few of the elliptic sampling curves shown in red. Blue lines show the AAM shape alignment. The right image is a close up of sampling points.

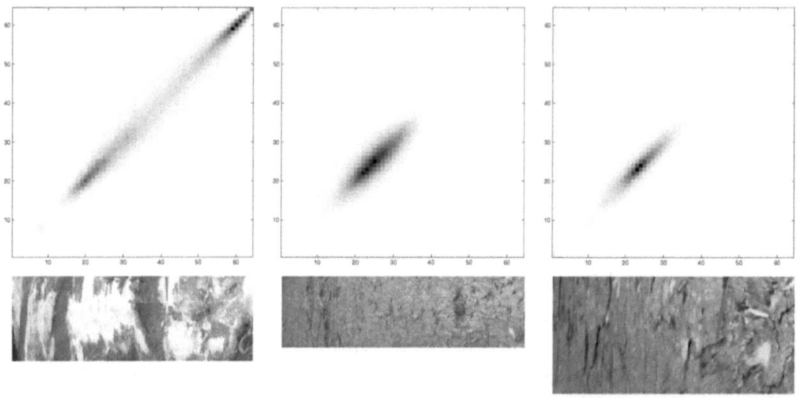

Fig. 2. Illustration of cooccurrence matrix (*top*) of bark texture (*bottom*). Higher intensity illustrates larger cooccurrence. From left to right: Birch, Spruce, and Pine. The displacement is $(1, 1)$ in a 64 level image.

of the log. A square image of the bark texture is the result of the warp, which is illustrated in Figure 2. One end of the log is usually smaller than the other, resulting in a difference in the sampling intensity in the warped bark image. Other shape variations may result in the same kind of sampling variation. These small variations have not been considered as a problem for the texture analysis.

2.2 Color Cooccurrence Matrices

As mentioned above, the AAM clasification, is extended by texture clasification, where the texture is obtained via texture warp as described in Section 2.1. This classification is done via second order textural statistics in the form of cooccurrence matrices (CM) cf. [2,11,15]. The fundamental element of calculating

CM's is the probability of a certain change in pixel intensity classes (k, l) given a certain pixel displacement $\mathbf{h}$ equivalent to $Pr(k, l|\mathbf{h})$. The CM's can be extended to color, by calculating the displacements in each band and across bands. The CM's have proven useful for classification cf. [15]. Sample CM's for the relevant bark textures is shown in Figure 2. In this paper the textures have been preprocessed by Gaussian histogram matching, in order to increase robustness to lighting conditions, cf. [2]. In this paper we use the following CM classes: contrast, dissimilarity, homogeneity, energy, entropy, maximum, correlation, diagonal moment, sum average, sum entropy, and sum variance.

2.3 Classifier

The classifiers used in experiments 1 to 3 are as follows:

1. **Multiple AAM's.** The model minimizing (2) is chosen.
2. **Global AAM.** Here three different classifications schemes based on the AAM feature vector are evaluated. These are: Bayes classifier, Canonical discriminant analysis, and LARS-EN classifier cf. [8,12,21].
3. **AAM and Texture.** Here LARS-EN is applied to the texture, obtained via the AAM based warp described in Section 2.1.

3 Data

Experiments were conducted for two groups of biological objects: vegetables, cf. Figure 3 and wood logs cf. Figure 1. The vegetables are are apples, carrots and potatoes and consist of 189 images totally where 27 are used for training the models, i.e. 9 from each group. The wood log data consists of the three species Scotch Pine, Norway Spruce and Birch. There was a total of 164 wood log images, 18 from each group were used for training. Also a reduced wood log data set, consisting of the 30 most characteristic logs from each group (90 in all) was used.

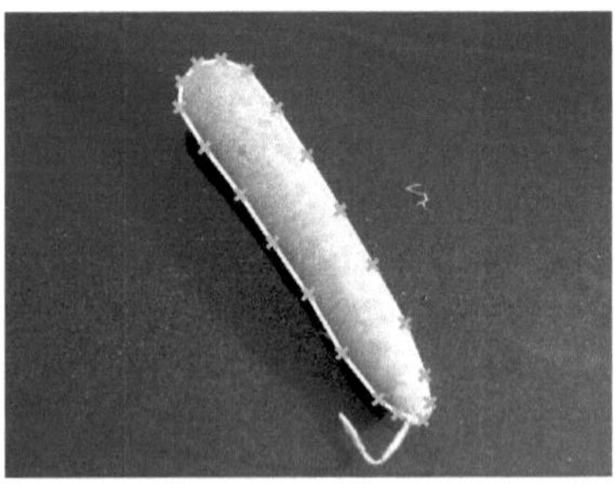

Fig. 3. Illustration of AAM alignment. The light blue line illustrates the shape and the blue crosses mark the annotated points.

4 Experiments

All three experiments are illustrated in Figure 4. The procedure is as follows.

Experiment 1. For each class there is built an AAM based on the training images. All models are matched to each of the test images giving model textures for all classes. The model texture is then compared to the original image by calculating the texture difference, see section 2.1. Classification is done by assigning the test image in question to the model giving the least texture difference.

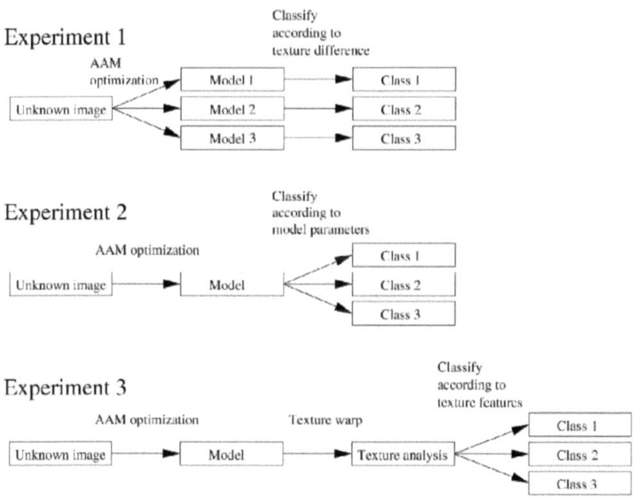

Fig. 4. Schematic representation of the three experiments

Experiment 2. In this experiment one AAM is built based on training images from all classes. The parameters from matching the model to a test image are used for classification. Based on these parameters the image is assigned to the most probable class as described in section 2.3.

Experiment 3. As in experiment 2, one AAM is matched to a test image, but here the alignment is used for extracting texture features. To enable a calculation of cooccurrence features from the bark texture, a warp of the bark area of the image is conducted.

4.1 Results and Discussion

Results of our three experiments are presented in table 1 and 2.

Experiment 1. This experiment gave rather stable and good results. In the vegetable data only one image of a potato was misclassified as an apple, giving an average classification rate of 99.3%. The wood log data set gave stable classifications around 83% except for the whole log model, where many Spruce logs were misclassified as Pine logs and visa versa.

Table 1. Classification rates of experiment 1 and 2 using the different classifiers. In the Vegetable and Whole log experiments, the shape covers the entire object, whereas in the rest, only the end part of the object is covered. Large images refer to the use of higher resolution images. Reduced refers to the reduced data set.

Experiment	1	2		
Model	Texture difference	Bayes	Canonical	LARS-EN
Vegetable	99.3%	100.0%	93.7%	100.0%
Log end	82.8%	70.2%	71.0%	75.5%
Large images	83.5%	64.1%	48.3%	82.1%
Whole log	67.8%	72.8%	71.1%	72.8%
Log end, reduced	82.5%	71.4%	38.1%	85.7%

Table 2. Classification rates of experiment 3 using LARS-EN for classification based on texture features

Model	LARS-EN	
Data set	*whole*	*reduced*
Gray level	78.4%	93.7%
Gray level directional	89.9%	96.8%
Color	89.5%	90.5%

Experiment 2. For this experiment three different classifiers have been tried. The vegetable experiment obtains 100% correct classification with the Bayes classifier and LARS-EN but the canonical discriminant analysis gives only a classification rate of 93.7%. The canonical discriminant analysis is also very unstable for the wood log data set. The Bayes classifier gives around 70% correct classification and LARS-EN around 80%.

AAM results in a relatively large number of features, and therefore, it is necessary to have many observations for training a classifier. A limited number of observations could be one reason for the relatively poor performance of the classifiers.

LARS-EN gives a good indication of the importance of the features in a linear model. For both data sets, the first two principal components are the most discriminative features. But there is large difference in the discriminative capabilities of the parameters from the two data sets, which also would be expected when the features are plotted, see Figure 5. The rest of the principal components are selected somewhat randomly, showing that feature reduction using PCA is not necessarily in accordance with classification criterions.

A problem encountered using canonical discriminant analysis, is a good separation of the training data, but a poor performance of assigning the unknown objects to the right classes. For the vegetable data, where we have a very good separation, this becomes very clear. We would expect to retain the separation, but that is not the case because of variation in training data. This sensitivity towards variation is a clear limitation to the canonical discriminant analysis.

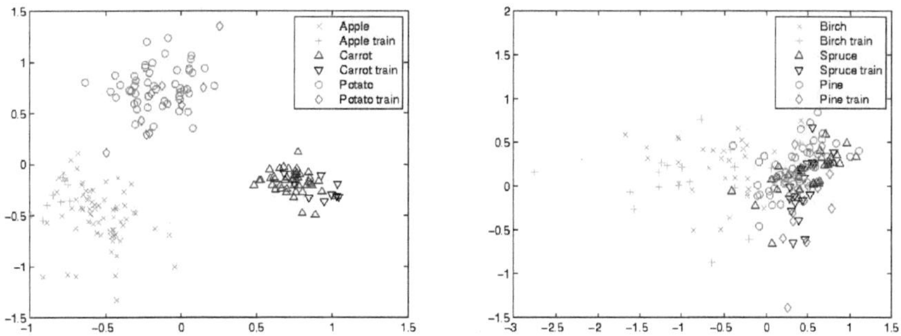

Fig. 5. Plot of the first two principal components from the AAM in the vegetable experiment (*left*) and wood log experiment (*right*)

The AAM's in the reduced data set is based on only 9 images of each class. This is probably too few to get good estimates of $\mathbf{\Phi}$, and could be a cause of the poor AAM classification performance.

Experiment 3. The best performance for the wood logs is achieved by the texture analysis in experiment 3, reaching close to 90% correct classification rates for the whole data set, and 96.8% correct classification in the best experiment of the reduced data set. This shows an accordance between what we see as humans and the predictions of the model.

A varying number of features are calculated, because of differences in distances, directions, and across color bands in the three models. In the first model we have 33 features (*only different distances*), in the second 132 features (*different distance and direction*), and 528 features in the third (*varying all three*).

Looking at which features are selected by the LARS-EN classifier, we can see a pattern in the features selected for classification. The sum average and the diagonal moment, are the most frequently used features, even though, there is not a clear pattern in which features works best for classifying the wood logs. In contrast to Palms [15] investigations, the performance in our experiments is not improved by extending the analysis to include color images.

A hard task using the LARS-EN algorithm is to find the right stopping criterion [21]. The results presented here is the number of features giving the best classification rates. Therefore, the LARS-EN algorithm will be problematic to implement in a real world application.

In these experiments we have used logs of young trees where some important biological characteristics have not yet developed, e.g. the colored core of Scotch Pine, which could improve the hard distinguishing of Pine and Spruce.

5 Conclusion

We have investigated the use of active appearance models (AAM's, cf. [4]) for classification of biological objects, and shown that this approach is well suited for different objects. Two data set, one of vegetables and one of wood logs have been investigated.

In experiment 1 an AAM is built for each class, and we obtain results close to 100% correct classification for the vegetable data, and around 80% classification rates for wood logs.

In experiment 2 one AAM for all classes is built, and model parameters for test images are used for classification. Most models gave 100% correct classification for the vegetables. On average the classification for the wood logs was not as good as experiment 1, and especially canonical discriminant analysis gave very poor results.

In experiment 3 LARS-EN has been used for classifying texture features, where only the wood log data is investigated. This experiment gave the best results classifying about 90% of the test set correct in the best cases of the whole data set, and up to 96.7% correct classification using a reduced data set.

It is hard to find a good stopping criterion for the LARS-EN model, which is problematic for classification. Therefore, we conclude that the most promising classification model is the texture difference used in experiment 1.

Acknowledgements

Thanks to Mikkel B. Stegmann for the AAM API and Karl Skoglund for the LARS-EN classifier, cf. [17,19]. Also thanks to Dralle A/S for partial financial support.

References

1. Belhumeur, P.N., Hespanha, J.P., Kriegman, D.J.: Eigenfaces vs. fisherfaces: recognition using class specific linear projection. Pattern Analysis and Machine Intelligence, IEEE Transactions on 19(7), 711–720 (1997)
2. Carstensen, J.M.: Description and Simulation of Visual Texture. PhD thesis, Institure of Mahematical Statistics and Operations Research, Technical University of Denmark, Kgs. Lyngby (1992)
3. Chang, P., Krumm, J.: Object Recognition with Color Cooccurrence Histogram (1999)
4. Cootes, T.F., Taylor, C.J.: Statistical models of appearance for medical image analysis and computer vision, In: Proc. SPIE Medical Imaging (2004)
5. Demirci, M., Shokoufandeh, A., Dickinson, S., Keselman, Y., Bretzner, L.: Manyto -many feature matching using spherical coding of directed graphs (2004)
6. Dickinson, S., Bretzner, L., Keselman, Y., Shokoufandeh, A., Demirci, M.F.: Object recognition as many-to-many feature matching. International Journal of Computer Vision 69(2), 203–222 (2006)
7. Edwards, G.J., Cootes, T.F., Taylor, C.J.: Face recognition using active appearance models. In: Burkhardt, H., Neumann, B. (eds.) ECCV 1998. LNCS, vol. 1407, pp. 581–595. Springer, Heidelberg (1998)
8. Efron, B., Hastie, T., Johnstone, I., Tibshirani, R.: Least angle regression. Annals of Statistics 32(2), 407–451 (2004)

9. Fagertun, J., Gomez, D.D., Ersbøll, B.K., Larsen, R.: A face recognition algorithm based on multiple individual discriminative models. In: Olsen, S.I., (ed.), Dansk Selskab for Genkendelse af Mønstre (Danish Pattern Recognition Society) DSAGM 2005, DIKU Technical report 2005/06, pp. 69–75, Universitetsparken 1, 2100 København Ø, aug DIKU, University of Copenhagen (2005)
10. Hansen, D.W., Nielsen, M., Hansen, J.P., Johansen, A.S., Stegmann, M.B.: Tracking eyes using shape and appearance. In: IAPR Workshop on Machine Vision Applications - MVA, pp. 201–204 (December 2002)
11. Haralick, R.M., Shanmugam, K., Dinstein, I.: Textural features for image classification. IEEE Trans Syst Man. Cybern SMC-3(6), 610–621 (1973)
12. Hastie, T., Tibshirani, J., Friedman, J.: The Elements of Statistical Learning, Data Mining, Inference and Prediction. Springer, Heidelberg (2001)
13. Lowe, D.G.: Distinctive image features from scale-invariant keypoints. International Journal of Computer Vision 60(2), 91–110 (2004)
14. Lowe, D.G.: Object recognition from local scale-invariant features. Computer Vision, 1999. In: The Proceedings of the Seventh IEEE International Conference on 2, 1150–1157 (1999)
15. Palm, C.: Color texture classification by integrative co-occurrence matrices. Pattern Recognition 37(5), 965–976 (2004)
16. Schmid, C., Mohr, R.: Local grayvalue invariants for image retrieval. Pattern Analysis and Machine Intelligence, IEEE Transactions on 19(5), 530–535 (1997)
17. Skoglund, K.: The lars-en algorithm for elastic net regression - matlab implementation, `kas@imm.dtu.dk` (2006)
18. Stegmann, M.B.: Object tracking using active appearance models. In: Olsen, S.I., (ed.). Proc. 10th Danish Conference on Pattern Recognition and Image Analysis, Copenhagen, Denmark, DIKU vol. 1, pp. 54–60, (July 2001)
19. Stegmann, M.B., Ersbøll, B.K., Larsen, R.: FAME - a flexible appearance modelling environment. IEEE Transactions on Medical Imaging 22(10), 1319–1331 (2003)
20. Turk, M.A., Pentland, A.P.: Face recognition using eigenfaces. In: Computer Vision and Pattern Recognition. Proceedings CVPR '91, pp. 586–591. IEEE Computer Society, Washington, DC (1991)
21. Zou, H., Hastie, T.: Regularization and variable selection via the elastic net. Journal of the Royal Statistical Society. Series B, Statistical Methodology 67(2), 301 (2005)

Estimation of Non-Cartesian Local Structure Tensor Fields

Björn Svensson, Anders Brun, Mats Andersson, and Hans Knutsson

Department of Biomedical Engineering, Medical Informatics
Center for Medical Image Science and Visualization
Linköping University, Sweden
{bjosv, andbr, matsa, knutte}@.imt.liu.se

Abstract. In medical imaging, signals acquired in non-Cartesian coordinate systems are common. For instance, CT and MRI often produce significantly higher resolution within scan planes, compared to the distance between two adjacent planes. Even oblique sampling occurs, by the use of gantry tilt. In ultrasound imaging, samples are acquired in a polar coordinate system, which implies a spatially varying metric.

In order to produce a geometrically correct image, signals are generally resampled to a Cartesian coordinate system. This paper concerns estimation of local structure tensors directly from the non-Cartesian coordinate system, thus avoiding deteriorated signal and noise characteristics caused by resampling. In many cases processing directly in the warped coordinate system is also less time-consuming.

A geometrically correct tensor must obey certain transformation rules originating from fundamental differential geometry. Subsequently, this fact also affects the tensor estimation. As the local structure tensor is estimated using filters, a change of coordinate system also change the shape of the spatial support of these filters. Implications and limitations brought on by sampling require the filter design criteria to be adapted to the coordinate system.

1 Introduction

Tensor fields occur frequently in image processing, either as physical measurements for instance diffusion tensors and strain tensors, or as features representing local signal characteristics such as local structure, orientation, junctions, corners and curvature. A common view, is to look upon the tensor as a matrix. For warped coordinate this view is however in conflict with the viewpoint of this paper, i.e. a tensor is a geometric object. Most algorithms for processing or estimating tensor fields from sampled signals in non-Cartesian coordinate systems face the same problem, which can be summarized in three points and is also illustrated by Fig. 1.

- A tensor definition is required obey the transformation rules brought on by differential geometry.
- Any filters used needs to be consistent with this definition, at least if the continuous signal is available. As a consequence the resulting tensor will

B.K. Ersbøll and K.S. Pedersen (Eds.): SCIA 2007, LNCS 4522, pp. 948–957, 2007.

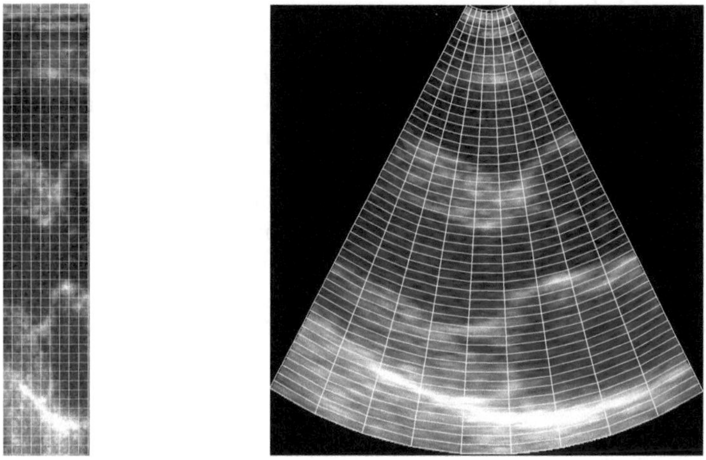

Fig. 1. Ultrasound images is one obvious example of when acquisition is performed in an non-Cartesian coordinate system. Image statistics are often described in the correct geometry, i.e. a Cartesian coordinate system (**right**) while noise statistics in general are easier to describe and thus also to suppress in the original (polar) coordinate system (**left**). A filter response is obtained by integrating the signal in a small neighborhood, usually defined in the correct geometry. Carrying out the same calculation in the polar grid requires reshaping the area to integrate.

be geometrically correct even if it is processed or estimated in a warped coordinate system.
- Discrete sampling implies that issues such as aliasing must be considered. Implementation of efficient filters which approximate the ideal behavior is required.

To obtain a true geometric description it is essential that the imaging device provides a signal sampled in a Cartesian coordinate system. In medical imaging this condition is in general not fulfilled. Resampling before computation or processing of the tensor field is the standard approach to compensate for such geometric distortion. However, resampling deteriorates signal and noise characteristics, which significantly complicates estimation or processing of the tensor field.

The concept of representing local image structure with tensors introduced in [1], has proved to be useful for estimating e.g. orientation, velocity, curvature and diffusion. Adaptive anisotropic filtering [2] and motion compensation [3] are examples of algorithms that rely on local structure tensor fields. Estimating local structure for curved manifolds or samples acquired from non-Cartesian coordinate system is a relatively unexplored area of research. Steps in this direction has been taken in [4,5]. In the former, affine filters are used in the framework of adaptive filtering for simultaneous resampling and compensation for geometric distortion. The latter concerns transforming tensor fields in space-time geometry used for motion estimation.

2 Tensor Calculus

To introduce the notation and to emphasize important differences between Cartesian and non-Cartesian tensors, this section gives a brief introduction to tensor calculus. Since this paper concerns local structure tensors, second order tensors are frequently used in the examples.

A contravariant tensor $T \in V \otimes V$ with components T^{ij} is said to be of order $(2,0)$ with two upper indices. A covariant tensor $T \in V^* \otimes V^*$ of order $(0,2)$ has two lower indices, where V^* is the dual space of the vector space V. Moreover, a contravariant vector v is a tensor of order $(1,0)$, i.e. an element $v \in V$. Consequently, a covariant vector $w \in V^*$ is a tensor of order $(0,1)$. The d-dimensional vectors v and w with components v^i, w_i are expressed in the basis $\frac{\partial}{\partial x^i}$ and dx^i in Eq. 1.

$$v = v^1 \frac{\partial}{\partial x^1} + v^2 \frac{\partial}{\partial x^2} + \cdots + v^d \frac{\partial}{\partial x^d} \qquad w = w_1 dx^1 + w_2 dx^2 + \cdots + w_d x^d \tag{1}$$

In tensor calculus the Einstein summation convention is frequently used to streamline algebraic expressions. It means that indices occurring more than once in an expression are implicitly summed over, which allows a convenient notation, e.g. second order contravariant and covariant tensors S, T can be compactly written as in Eq. 2.

$$S = S^{ij} \frac{\partial}{\partial x^i} \frac{\partial}{\partial x^j} = \sum_{i,j} S^{ij} \frac{\partial}{\partial x^i} \frac{\partial}{\partial x^j} \qquad T = T_{ij} dx^i dx^j = \sum_{i,j} T_{ij} dx^i dx^j \tag{2}$$

The metric tensor g defines an inner product in V, which makes it possible to identify elements in V^* with elements in V. The metric tensor is a covariant second order tensor, with elements g_{ij}. Consequently the metric allows us to move between covariant and contravariant tensors illustrated in Eq. 3.

$$g(u,v) = \langle u, v \rangle = g_{ij} u^i v^j \qquad w = g(\cdot, v) = w_i dx^i \quad \Longrightarrow \quad w_i = g_{ij} v^j \tag{3}$$

The metric tensor in the Euclidean space is the Kronecker delta, i.e. $g_{ij} = \delta_{ij}$. Consequently, covariant and contravariant tensors are equivalent in a Cartesian coordinate system. Such tensors are called Cartesian tensors. A change of coordinate system from x to $\tilde{x}$ yields new components expressed in a new basis. Hence, the transformation behavior for a tensor second order contravariant tensor T with components T^{ij} and basis $\frac{\partial}{\partial x^i} \frac{\partial}{\partial x^j}$ is easily derived. The tensor components T^{ij} transforms to $\tilde{T}^{ij}$ by the chain-rule according to Eq. 4.

$$T = \tilde{T}^{ij} \frac{\partial}{\partial \tilde{x}^i} \frac{\partial}{\partial \tilde{x}^j} = T^{ij} \frac{\partial}{\partial x^i} \frac{\partial}{\partial x^j} \quad \Longrightarrow \quad \tilde{T}^{ij} = \frac{\partial \tilde{x}^i}{\partial x^k} \frac{\partial \tilde{x}^j}{\partial x^l} T^{kl} \tag{4}$$

In the same way a covariant second order tensor T with components T_{ij} transforms to $\tilde{T}_{ij}$ by Eq. 5.

$$T = \tilde{T}_{ij} d\tilde{x}^i d\tilde{x}^j = T_{ij} dx^i dx^j \quad \Longrightarrow \quad \tilde{T}_{ij} = \frac{\partial x^k}{\partial \tilde{x}^i} \frac{\partial x^l}{\partial \tilde{x}^j} T_{kl} \tag{5}$$

A uniform stretch of the coordinate system, i.e. the transformation $\tilde{x} = ax$ with $a > 1$ causes covariant tensor components T_{ij} to shrink, while the contravariant components T^{ij} grow. Thus, it is of outmost importance to know whether the tensor is contravariant or covariant.

The contraction of a tensor of order (p, q) yields a tensor of order $(p-1, q-1)$, e.g. the trace of $T^i j$ in Eq. 6 gives a scalar (a tensor of order 0). Note also that a contraction is only defined for mixed tensors, i.e. tensors with both contravariant and covariant indices.

$$\text{trace}\,(T^i{}_j) = T^i{}_i = \sum_i T^i{}_i \tag{6}$$

Another example of a contraction is the eigenvalue equation tensors in Eq. 7. Note again that the eigenvalue equation is only defined for mixed tensors, i.e. if the tensor is not mixed, eigenvalue decomposition can not be performed without a metric.

$$T^i{}_j v^j = g^{ik} T_{kj} v^j = \sum_{j,k} g^{ik} T_{kj} v^j = \lambda v^i \tag{7}$$

With a metric g^{ij} equal to the identity operator, i.e. a Cartesian coordinate system, the standard eigenvalue equation is obtained. Now consider solving the eigenvalue equation for non-Cartesian tensors. The eigenvalues will be invariant to a change of coordinate system, while the eigenvectors are not.

3 Local Structure Tensor Fields

An orientation tensor, a covariant second order tensor $T \in V^* \otimes V^*$, is defined for simple signal neighborhoods $s = f(u_i x^i + \theta)$. A simple signal s is a scalar function intrinsically 1-dimensional, i.e. constant in the direction orthogonal to u. In most practical applications V is the Euclidean space spanned by its natural basis $\frac{\partial}{\partial x^i}$. Given a simple signal, T is defined by Eq. 8, i.e. the outer product of the signal direction u.

$$T_{ij} = \lambda u_i u_j, \quad \lambda \geq 0 \tag{8}$$

As opposed to a vector representation, signal orientation can be continuously represented by a tensor. Using vectors, there is no distinction between a signal with direction $-u_k$ and a signal in the direction of u_k. The outer product however yields a representation which maps both u_k and $-u_k$ to the same geometric object, the orientation tensor [1].

To represent orientation, the tensor must meet two fundamental requirements. Firstly, the tensor norm must be invariant to rotation, i.e. $\|T\|$ is independent of the direction of u. Secondly, the tensor must be phase-invariant, i.e. T is independent of any phase-shift θ of the signal s. These partly contradictory requirements are hard to accomplish for sampled signals. Hence, several different approaches how to estimate the tensor has been suggested (e.g. [6,7,8,9,10]). Note that all of them does not claim to meet these requirements. The behavior for non-simple signals are entirely defined by the estimator used. An entire class of

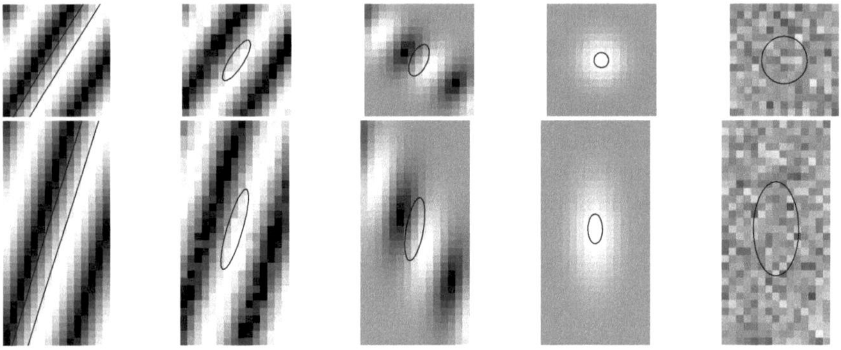

Fig. 2. A few samples of two-dimensional signal neighborhoods. They represent a simple signal (**a**), a simple signal corrupted with white noise (**b**), a non-simple signal (**c**), a signal with no orientation (**d**), and white noise (**e**). The patches in the **top row** are sampled in a Cartesian coordinate system, while the **bottom row** shows the same signals, sampled more densely in the vertical direction. The tensor glyph $T_{ij}v^iv^j = 1$ is displayed. Note how the glyph is stretched in the same way as the coordinate system, which is a correct transformation behavior.

functions that meets the fundamental requirements of the orientation tensor are derived in [11].

For non-simple signals, the orientation tensor does however carry more information about the local signal neighborhood, like for instance degree of anisotropy and how planar/linear a 3-dimensional neighborhood is. The term local structure tensor in Eq. 9 is a wider interpretation of the orientation tensor, representing local signal neighborhoods that are concentrated to a k-dimensional subspace in the Fourier domain, spanned by the orthonormal basis $u_1, u_2, \cdots, u_k$.

$$T_{ij} = \lambda^k u_{ik} u_{jk}, \quad \lambda^1, \lambda^2, \cdots, \lambda^k \geq 0 \tag{9}$$

For displaying purposes, glyphs are a common way to visualize a tensor field. The isometric tensor glyph in Eq. 10, can be used to represent a second order covariant tensor, which is positive semidefinite. Moreover this glyph is invariant to change of coordinate system. The size of the glyph is inversely proportional to $\sqrt{\lambda}$.

$$T_{ij}v^iv^j = 1 \tag{10}$$

In Fig. 2 a few sample patches are shown in two different coordinate systems. The bottom row is the same continuous signal sampled more densely in the vertical direction, i.e. the top row shows Cartesian tensors while the bottom row shows non-Cartesian tensors, sampled in the coordinate system $\tilde{x}^2 = ax^2$, with $a > 1$. Drawing the local structure tensor in this way implies that the local structure tensor can be interpreted as a local image metric, describing how the signal energy measure is locally distributed. The tensor glyph is then the isocurve where this measure is constant. This interpretation presumes a positive semidefinite tensor.

4 Tensor Estimation

In this paper quadrature filters are used to estimate the local structure tensor. Without going through the details, a quadrature filter responses is a phase-invariant signal energy measure projected on a discrete number of directions n_k. The projections, i.e. the filter responses $\{q_1, q_2, \cdots, q_k\}$ (scalar values) for simple signals of single frequencies are given by Eq. 11.

$$|q_k| = \lambda(u_i n_k{}^i)^2 \tag{11}$$

A tensor estimate in arbitrary direction is then obtained Eq. 12 by linearly combine the filter responses by the use of a tensors $\{M^1{}_{ij}, M^2{}_{ij}, \cdots M^k{}_{ij}\}$, where each M_{ij} is dual to the outer product of its corresponding filter direction $n^i n^j$.

$$T_{ij} = |q_k| M^k{}_{ij} \tag{12}$$

A quadrature-based estimate of the local structure tensor is approximately phase-invariant for signals of low bandwidth, i.e. the tensor is insensitive to small shifts of the signal s. Since the tensors T, M are all elements in $V^* \otimes V^*$, a local structure estimate $\tilde{T}_{ij}$ in a non-Cartesian coordinate system is obtained by applying quadrature filters adjusted to the new coordinate system and compute the estimate in Eq. 12 with $\tilde{M}_{ij}$ obtained by transforming M_{ij}.

Constructing the tensor in this way is consistent with the tensor definition, if the filter responses q_k are invariant to a change of coordinate system. The convolution integral in Eq. 13 reveals that $q(\tilde{x}^k) = q(x^k)$ if the filter is transformed in the same way as the signal, i.e. $h(\tilde{x}^k) = h(a^i{}_k x^k)$. This is not at all surprising, since a filter produce a scalar value and can be interpreted as a projection. It is however important to realize that the spatial support Ω changes, e.g. an isotropic spatial support becomes anisotropic in the warped coordinate system.

$$q(\tilde{x}^k) = \int_{\tilde{\Omega}} s(\tilde{y}^k) h(\tilde{y}^k - \tilde{x}^k) \frac{1}{|\det(a^i{}_k)|} d\tilde{y}^k = \int_{\Omega} s(y^k) h(y^k - x^k) dy^k = q(x^k) \tag{13}$$

Since estimators are usually defined in the frequency domain, it is desirable to translate this result to the frequency domain. If $a^i{}_k$ is an affine transform the frequency response of the filter $h(\tilde{x}^k) = h(a^i{}_k x)$ relates to the frequency response of $h(x^k)$ by Eq. 14.

$$H(\tilde{u}_i) = \mathcal{F}\{h(\tilde{x})\} = \int_{\mathbb{R}^d} h(\tilde{x}^k) \exp(-i\tilde{u}_k \tilde{x}^k) dx^k = |\det(a_i{}^k)| H(a^{-1}{}_i{}^k u_k) \tag{14}$$

Thus, an estimate consistent with the definition in Eq. 8 is obtained if $\mathcal{F}\{h(\tilde{x})\} = |\det(a_i{}^k)| H(a^{-1}{}_i{}^k u_k)$. But so far ignored the effect of sampling has been ignored. The limitations brought on by sampling makes it difficult to estimate a tensor, which is strictly consistent with the definition due to aliasing effects and limited resolution.

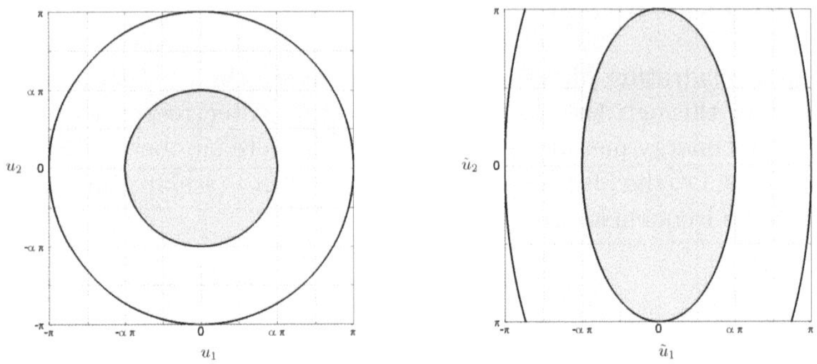

Fig. 3. In a Cartesian coordinate system (**left**) the filters must suppress the corners of the frequency domain, to ensure that $\|T\|$ does not vary with signal orientation. Aliasing occurs for frequencies which is above the sampling frequency ($\|u\| > \pi$). Taking fewer samples in the vertical direction implies that aliasing occurs for $\tilde{u}_2 > \pi$ which correspond to $u_2 > \alpha\pi$, where $\alpha < 1$. To simultaneously avoid aliasing and still get an unbiased tensor estimate in the non-Cartesian coordinate system (**right**) the filters must suppress frequencies outside $\|u\| < \alpha\pi$ (dark-gray), a region which is isotropic to the left and consequently anisotropic to the right.

5 Filter Design

For a sampled signal it is enough to study the frequencies contained in $\mathbb{U} = \{u : |u_i| \leq \pi\}$, since frequencies outside $\mathbb{U}$ can not be represented and will suffer from aliasing. A change of basis then implies that the sampling distance varies with direction and $\mathbb{U}$ is reshaped. In fact this phenomenon actually occurs already in the Cartesian coordinate system. To get an unbiased estimate of local structure the filters then must suppress the corners of the Fourier domain , i.e. directions in which frequencies above the Nyquist frequency ($\|u\| > \pi$).

To avoid aliasing, the Shannon sampling theorem states that a signal can be perfectly reconstructed if the signal is band-limited, in this case if the signal is contained in $\mathbb{U}$. Reshaping $\mathbb{U}$ then might imply a limited ability to represent signals as illustrated in Fig. 3. Restricting the transformations to be affine as in the previous section ensures that the Shannon sampling theorem is applicable for the warped coordinate system.

Another important observation is that $h(x)$ and $h(\tilde{x})$ no longer can be the same functions, since samples are acquired at integer positions $z^k \in \mathbb{Z}^d$ pointing out different locations $z^i \frac{\partial}{\partial x^i}$ and $z^i \frac{\partial}{\partial \tilde{x}^i}$ if the coordinate system is changed. It is however possible to choose the values of the filter coefficients c, $\tilde{c}$ at these positions such that $h(x^k)$ and $\tilde{h}(\tilde{x}^k)$ defined by Eq. 15 approximate the same frequency response $H(u_i)$ and $H(\tilde{u}_i)$ derived in the previous section.

$$h(x) = c_k \delta(x^k - z^k) \qquad \tilde{h}(\tilde{x}^k) = \tilde{c}_k \delta(\tilde{x}^k - z^k) \tag{15}$$

Solving the optimization problem in Eq. 16 yields the spatial kernel $\tilde{h}(\tilde{x}^k)$ with the closest fit to the desired frequency response $H(\tilde{u}_i)$.

$$\min_{c} \varepsilon = \|H(\tilde{u}_i) - \mathcal{F}\{\tilde{h}(\tilde{x}^k)\}\|^2_W + \lambda\|\tilde{h}(\tilde{x}^k)\|^2_w \tag{16}$$

The weighted norm is used to favor a close fit for the most common frequencies and the regularization term is used to favor a small spatial size of the filter kernel [12].

6 Experiments

The following experiment was setup to compare estimation of local structure before resampling to the more common approach where the signal first is resampled using bilinear interpolation to a geometrically correct grid from which the local structure is estimated.

The continuous simple signal $s = \cos(u_i x^i + \theta)$ shown in Fig. 2 corrupted with additive white Gaussian noise was sampled in a coordinate system where the vertical axis was shortened a factor 2, i.e. a patch of size 17×9 pixels. 60000 signal neighborhoods with 2000 different orientations uniformly distributed and random phase-shift θ was used in the experiments. Two different local structure tensors were estimated for the neighborhood origin for all patches. One estimated directly in the non-Cartesian coordinate system, one estimated after resampling the patch to a spatial size of 17×17 pixels, i.e. a Cartesian coordinate system. As a reference a tensor was also estimated from a signal with all samples in the 17×17 patch available, i.e. a signal with twice the amount of information compared to the two former estimates.

Since s is a simple signal the largest eigenvector u of the tensor should correspond to the signal direction v. Measuring the root-mean-square angular error $\Delta\varphi_{rms}$ reveals the squared deviation between v and u. Fig. 4 shows the error-measures defined by Eq. 17 for the three different tensor estimations. It is also interesting to study how the estimate u varies with the true signal direction v calculating the angular bias $\Delta\varphi$ for all sampled signal orientations.

$$\Delta\varphi_{rms} = \cos^{-1}\left(\sqrt{\frac{1}{n}\sum_u \langle u, v\rangle^2}\right) \quad \Delta\varphi = \sin^{-1}\left(\frac{1}{n}\sum_u (u \times v)\mathrm{sign}(\langle u, v\rangle)\right) \tag{17}$$

Estimating the dominant orientation directly in the non-Cartesian system yields a better accuracy compared to the approach where the signal is resampled prior to tensor estimation. A small bias is obtained for noise-free signals due to filter approximation errors. This bias is however very small in comparison to the bias introduced by resampled noise. It should also be mentioned that estimating the non-Cartesian tensor requires no resampling and is roughly 2 times faster due to smaller filters.

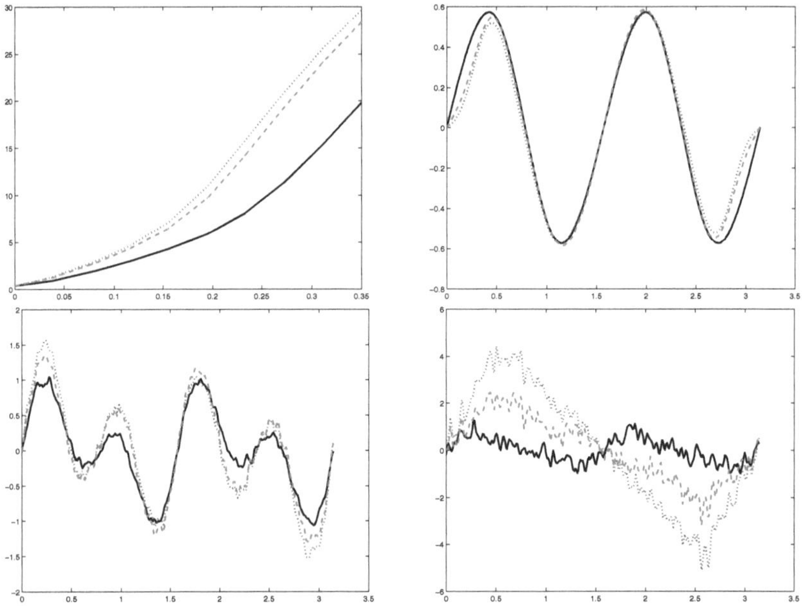

Fig. 4. The angular root mean square error $\Delta\varphi_{rms}$ in degrees is shown as a function of noise level (**upper left**), going from a signal-to-noise ratio of $SNR = \infty$dB to a signal-to-noise ratio of $SNR = 0$dB. Estimating the dominant orientation directly in the non-Cartesian coordinate system (dashed) yields a smaller angular error compared to estimation after resampling (dotted) as the noise level increases. The angular deviation $\Delta\varphi$ as a function of signal orientation φ reveals a biased estimate, the mean angular deviation from the true signal varies with signal orientation. For noise-free signal patches (**upper right**), the deviation is very small and the difference between the non-Cartesian (dashed), resampled(dotted) and the reference is not significant. However for moderate (**lower left**) and high (**lower right**) noise levels ($SNR = 9.3$dB, $SNR = 0$dB) the non-Cartesian estimation shows better performance compared to resampling.

7 Discussion

A framework for estimation of local structure directly from sampled signals in warped coordinate systems was presented. This approach can be advantageous in situations where resampling deteriorates signal and noise characteristics. Experiments show that for the test pattern used more accurate estimates can be obtained with less amount of processing. The presented work is also applicable to curved manifolds, provided that they are locally flat, i.e. a relatively low curvature.

Designing filters for estimating non-Cartesian local structure tensors might seem straightforward. However, it becomes more difficult to meet the contradictory demands on locality in both the spatial and the frequency domain, as the signal of interest moves closer to the Nyquist frequency it becomes harder to avoid aliasing and at the same time achieve unbiased tensor estimates.

The presented work relies on that the transformation between the coordinate systems can be approximated by an affine transformation, at least locally. This is required for the Shannon sampling theorem to be applicable. However, there are sampling theorems that allow more general transformations as long as the signal band-limitedness is preserved.

References

1. Knutsson, H.: Representing local structure using tensors. In: The 6th Scandinavian Conference on Image Analysis, Oulu, Finland, June 1989. Report LiTH–ISY–I–1019, Computer Vision Laboratory, Linköping University, Sweden pp. 244–251 (1989)
2. Westin, C-F., Knutsson, H., Kikinis, R.: Adaptive image filtering. In: Bankman (ed.) Handbook of Medical Imaging - Processing and Analysis, Academic press, San Diego (2000)
3. Hemmendorff, M.: Motion Estimation and Compensation in Medical Imaging. PhD thesis, Linköping University, Sweden, SE-581 85 Linköping, Sweden, Dissertation No 703 (2001)
4. Westin, C.-F., Richolt, J., Moharir, V., Kikinis, R.: Affine adaptive filtering of CT data. Medical Image Analysis 4(2), 161–172 (2000)
5. Andersson, M., Knutsson, H.: Transformation of local spatio-temporal structure tensor fields. In: IEEE International Conference on Acoustics Speech and Signal Processing (ICASSP), Presented at ICIP 2003 in Barcelona, Spain, (September 20 2003)
6. Knutsson, H.: Filtering and Reconstruction in Image Processing. PhD thesis, Linköping University, Sweden, Diss. No. 88 (1982)
7. Bigün, J., Granlund, G.H., Wiklund, J.: Multidimensional orientation: texture analysis and optical flow. IEEE Transactions on Pattern Analysis and Machine Intelligence, PAMI–13(8) (August 1991)
8. Farnebäck, G.: Orientation estimation based on weighted projection onto quadratic polynomials. In: Girod, B., Greiner, G., Niemann, H., Seidel, H.-P. (eds.), Vision, Modeling, and Visualization 2000: proceedings, pp. 89–96, Saarbrücken (November 2000)
9. Knutsson, H., Andersson, M.: Implications of invariance and uncertainty for local structure analysis filter sets. Signal Processing: Image Communications 20(6), 569–581 (2005)
10. Felsberg, M., Köthe, U.: Get: The connection between monogenic scale-space and gaussian derivatives. In: Kimmel, R., Sochen, N., Weickert, J. (eds.) Scale-Space 2005. LNCS, vol. 3459, pp. 192–203. Springer, Heidelberg (2005)
11. Nordberg, K., Farnebäck, G.: Estimation of orientation tensors for simple signals by means of second-order filters. Signal Processing: Image Communication 20(6), 582–594 (2005)
12. Knutsson, H., Andersson, M., Wiklund, J.: Advanced filter design. In: Proceedings of the 11th Scandinavian Conference on Image Analysis, Greenland, SCIA (June 1999)

Similar Pattern Discrimination by Filter Mask Learning with Probabilistic Descent

Yoshiaki Kurosawa

Toshiba Solutions Co., Advanced Technology Development, Platform Solutions Div.
1-15 Musashidai, 1, Fuchu-shi, Tokyo 183-8532, Japan

Abstract. The purpose of this research was to examine the learning system for a feature extraction unit in OCR. Average Risk Criterion and Probabilistic Descent (basic model of MCE/GPD) are employed in the character recognition system which consists of feature extraction with filters and Euclidian distance. The learning process was applied to the similar character discrimination problem and the effects were shown as the accuracy improvement.

Key words: OCR, Feature, Filter, Learning, GPD, MCE.

1 Introduction

Similar character discrimination has become a primal factor to improve the OCR performance in recent years. One of the solutions for this problem is to develop an individual discrimination process for each similar pattern pair. But difficulties with this approach emerge when the number of similar pairs becomes too large. A new learning technique for fixing feature extraction parameters will be effective in such a case. This report describes a technique to fix the filter mask parameters in the feature extraction of OCR. For this purpose, Averaged Risk Criterion and Probabilistic Descent (AR/PD) were adopted as a learning method.

AR/PD was proposed by Amari [1], and improved by Katagiri and others [2] as Minimum Classification Error Criterion (MCE) with Generalized Probabilistic Descent (GPD). As the method of MCE/GPD, a simpler version which might be called AR/PD was used in this research. The learning target is the feature extraction in the system and this approach was proposed as Discriminative Feature Extraction (DFE) [3].

There are three types of target in DFE: Filter Parameter Learning, Filter Selector Learning and Filter Mask Learning. In the Filter Parameter Learning [4] [5] [6], the filters are described by some parameters like the sigma and mean value of a Gaussian filter. In the Filter Mask Learning, the learning target is the mask value of the filters. In the Filter Selector Learning [7], the filter is not used for feature extraction itself in comparison with above two approach, the filters are used for reducing the dimension of the feature vector which has been extracted by the other feature extractor.

The filter mask is the most popular technique for the feature extraction in the OCR field, and this is the reason why the Filter Mask Learning is examined in this research.

B.K. Ersbøll and K.S. Pedersen (Eds.): SCIA 2007, LNCS 4522, pp. 958–967, 2007.

The learning system adopted here resembles 3 layer neural nets [8]. The intermediate layer of those NN corresponds to the filters in the system adopted here. Also, Cognitron [9] is one of the related systems. There have been other feature learning systems [10] [11] and they have shown good performance respectively against complex and difficult problems.

Through these previous works, the filters generated in the learning system have not been considered as filters in some cases, and have not been observed visually in many cases. It has been too complex to analyze the inner workings in some cases. The purpose of this research is to investigate the learning process of feature extraction filters, the basic effect of this learning, the generated filter's performance and the human visual observation of generated filters. This report describes the successive research of the preliminary investigation [12] on this subject.

2 Overview of a Learning System

The recognition system used in this research is shown in Fig. 1. The local regions are assigned in the input image region and the local filter masks are defined and correspond to each local region. The inner products calculated by the local region vector consists of pixel values and local filter masks are treated as the elements of feature vector. The nonlinear transformation is executed after the inner products calculation. This feature vector is provided to the Euclidian distance unit and the distance is calculated with reference vectors.

If the local filter denoted F is a linear transformation, z is an input vector and $\hat{x}$ is a transformed vector, the transformation is described as $\hat{x} = Fz$.

Let the element value of x and $\hat{x}$ be x_i and $\hat{x}_i$ respectively, the nonlinear conversion is defined as $x_i = \rho(\hat{x}_i)$ and let it be denoted as $x = \rho(\hat{x})$. By this notation, the feature extraction from z is described as follows.

$$x = \rho(Fz). \tag{1}$$

If the reference vector is denoted as φ, the Euclidian distance S is described as

$$S = ||x - \varphi||^2. \tag{2}$$

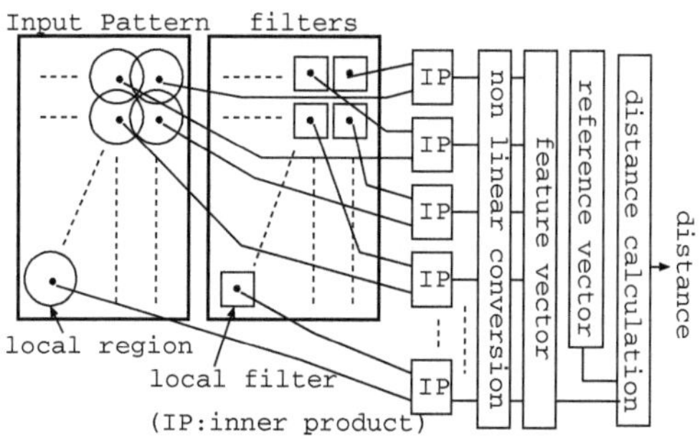

Fig. 1. Outline of recognition system used for filter mask learning

The recognition result is determined by the category whose reference have the minimum distance.

The update terms $\Delta F, \Delta\varphi$ are calculated from F and φ by AR/PD theory and F and φ are changed so that the system has better performance through this updating. The changed filter masks F are arranged and displayed to be observed visually as shown in Fig. 1.

The update equation is obtained as described below by AR/PD. Let $l(d)$ be the loss function and $l'(d)$ represent the differential function. Here, d is defined by $d = S_{ok} - S_{err}$ where S_{ok} is the distance between the correct category's reference and the input, and S_{err} is the one between the other best category's reference and input (see (2)). The update equations are

$$\Delta F = \mp 2\epsilon_f l'(d)\rho'(\hat{x})(x-\varphi)z^T, \tag{3}$$

$$\Delta\varphi = \pm 2\epsilon_r l'(d)(x-\varphi), \tag{4}$$

where ϵ_f, ϵ_r are the values which define the strength of the learning. The upper part of $\pm$ or $\mp$ is for updating when an input category is same as reference's one and the lower is for other category's input.

3 Experiments

System of Experiments. An input character image's size is normalized to be a 40x40 pattern and is provided to the feature extractor. Fig. 2 shows arrangements of local filter masks at the upper left portion of the input image area and the small square represents one pixel. The positions $a, \ldots, p$ means the center of each filter. The filter size is 7x7 and it is shown as a square in broken line. The filter F is the filter whose center is f, the filter K is the filter whose center is k. The filters are located at intervals of 3 pixels for horizontal and vertical directions. The output feature vector size is 14x14.

The initial values $f_{x,y}$ of filters are given by

$$f_{x,y} = A \cdot exp(-1/2(x^2/\sigma_x + y^2/\sigma_y)), \tag{5}$$

where (x,y) is the pixel position and A is a constant for normalization.

The concrete values of the initial filter used in this research are shown in Fig. 3. The upper left 4x4 pixels of the 7x7 mask which corresponds to (5) are shown. In this filter, σ_x is slightly larger than σ_y because it shows better performance when comparing the case of $\sigma_x = \sigma_y$ in the preliminary test. The initial reference vectors are calculated by these initial filters.

The value $\hat{x}_i$ is obtained as an inner product of the filter mask and the input pattern's pixel values in the local region. Then it is converted by a nonlinear function. The function adopted here is

$$\rho(\hat{x}_i) = 1/(1 + e^{(b - a\hat{x}_i)}). \tag{6}$$

The values a, b are set to $a = 9.4, b = 2$ so that the output is from 0.12 to 0.88 when the input is from 0 to 0.42.

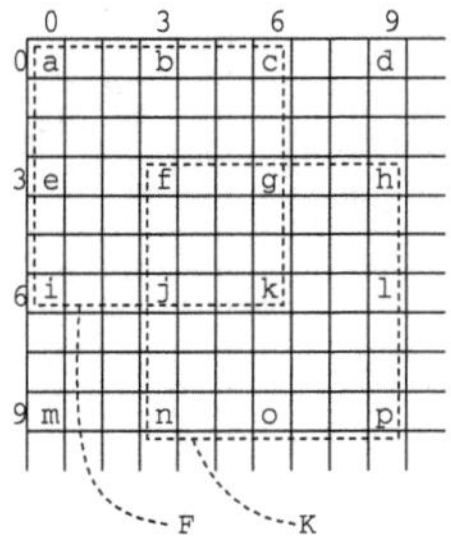

Fig. 2. Filter size and arrangement

.00000	.00005	.00027	.00049
.00007	.00132	.00769	.01385
.00051	.00973	.05684	.10235
.00100	.01896	.11070	.19935

Fig. 3. Upper left 4x4 values of initial filter

The sigmoid function is adopted as a loss function and it's differential form are

$$l(d) = 1/(1 + e^{-\delta d}), \quad l'(d) = \delta l(d)(1 - l(d)). \tag{7}$$

Environment of Experiments. There are two similar character pairs used in the experiments, one is "bo" and "po", and the other is "ma" and "mon". These are all Japanese handwritten characters. The example of "bo" and "po" are shown in Fig. 5, 6 and the difference between those characters is only the upper right portion. The example of "ma" and "mon" are shown in Fig. 8, 9 and the difference between those characters is only the lower center portion.
There are 2000 "bo-po" patterns for learning and 2000 for testing.
There are 2000 "ma-mon" patterns for learning and 1300 for testing.

The number of iteration was 3540 times for "bopo", and 2020 times for "ma-mon" where one cycle learning needed approximately 1.5 seconds with 2.8G PenIV PC.

Results of Experiment for "bo-po". Fig. 4 (A) shows the initial filters used in this research. The filters are arranged with 14 pieces horizontally and 14 pieces vertically (just 4 columns are shown). Each of the filters has 7x7 pixels. The pixel values are represented by a black circle. A large black circle means a large plus value and a small black circle means a large minus value. The value zero corresponds to the circles observed at the rim of the filters. Here, even if the filters in the figure seem to be slightly different, all filters are the same and this might be the effect of image capture and printing.

Fig. 4 (B) shows the generated filters after learning. Almost all filters except the filters located in the upper right corner seem to be the same as the initial filter. The generated filters in the upper right corner are considered to be the vertical line extraction filter. They resemble a Gabor filter or a Mexican hat filter. These results show that the learning has made the filter emphasize the difference between the two category images.

Fig. 5 and 6 show the input patterns in the upper array of the figure, the feature patterns obtained by initial filters in the middle array, and the feature patterns obtained by generated filters in the bottom array. The feature patterns

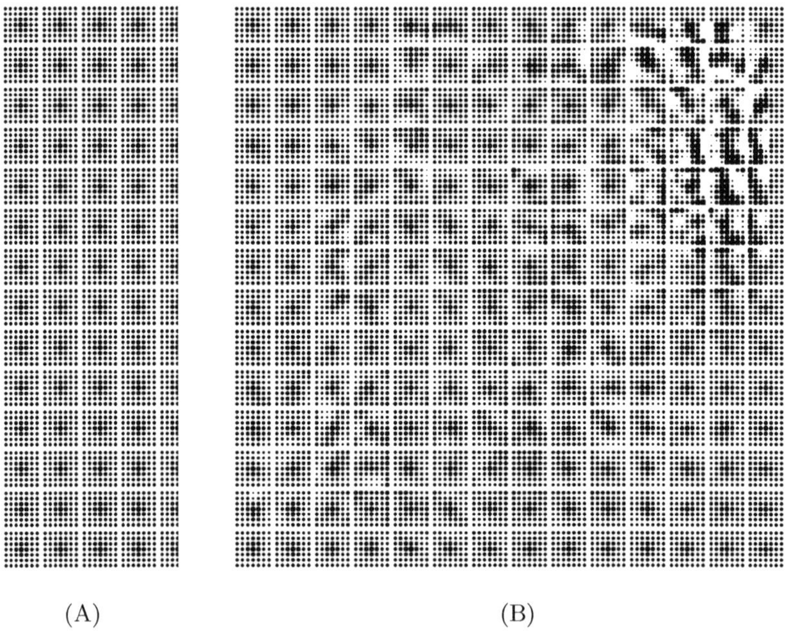

Fig. 4. (A) Initial filters before learning (left side part: 4 columns), (B) Filters generated by "bo","po" character learning

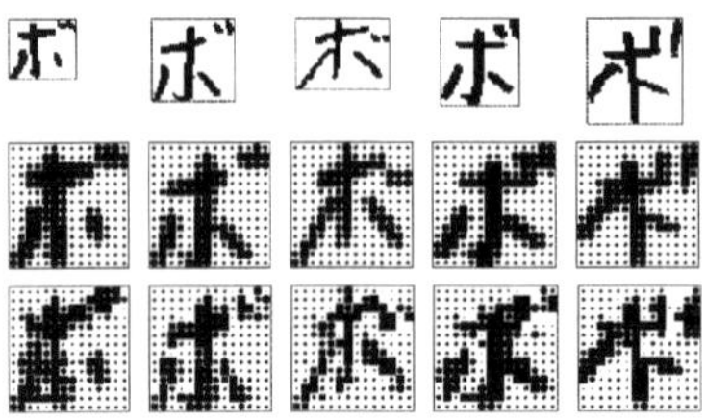

Fig. 5. "bo": Original (upper), Gauss (middle), Generated (bottom)

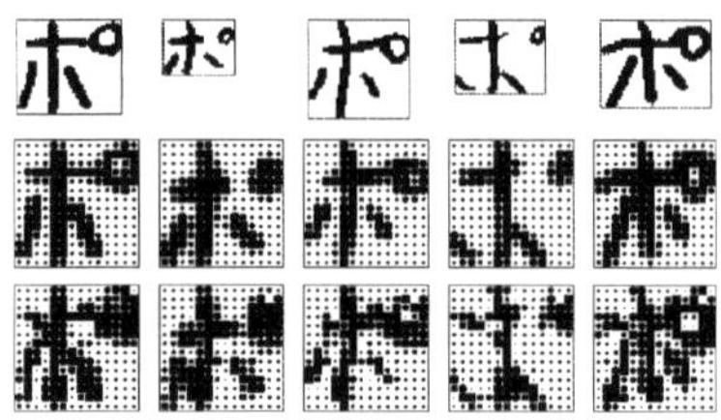

Fig. 6. "po": same order as the patterns on the left ("bo")

become noisy and the right upper part is emphasized more after learning. Some of the "bo" pattern's two dots in the right upper part are separated while they are touching in the initial one. In the "po" case, the upper part seems slightly larger than the initial one.

Results of Experiment for "ma-mon". Fig. 7 (B) shows the generated filters after learning. Fig. 8 and 9 shows the input patterns and feature patterns in the same manner as the "bo-po" results. In these figures, the lower center part of the generated patterns seem to be emphasized and they are noisy compared with the initial patterns.

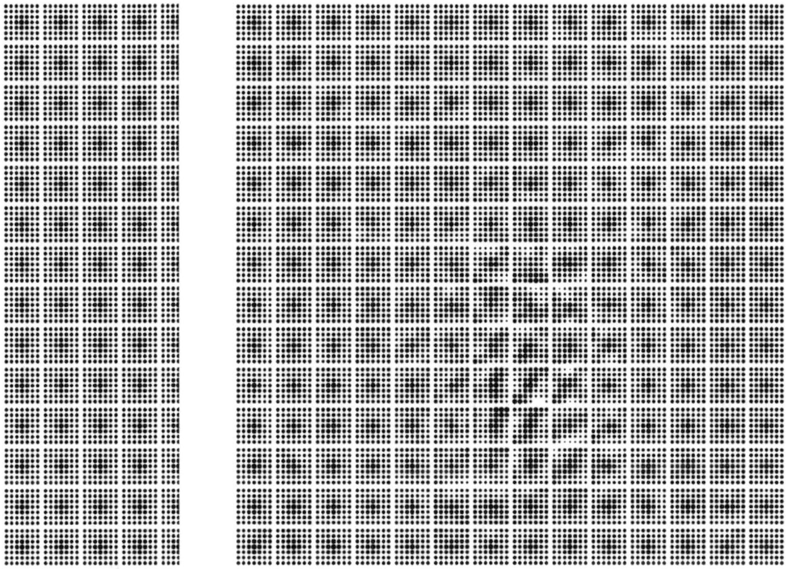

(A) (B)

Fig. 7. (A) Initial filters before learning (left side part: 4 columns), (B) Filters generated by "ma","mon" character learning

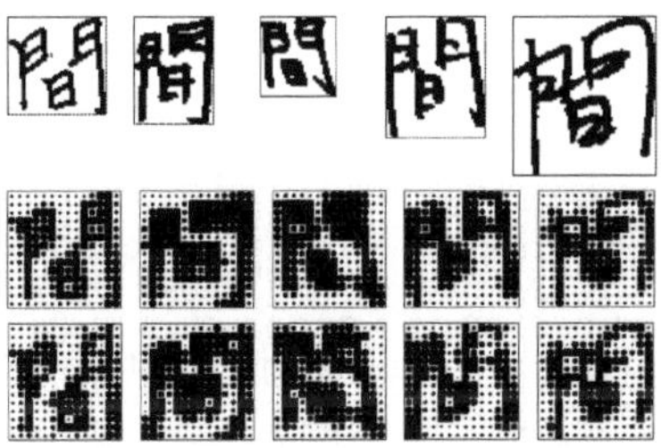

Fig. 8. "ma": Original (upper), Gauss (middle), Generated (bottom)

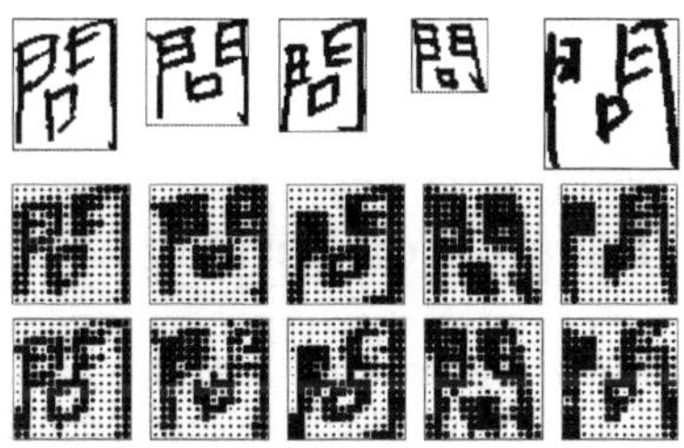

Fig. 9. "mon":same order as the patterns on the left ("ma")

Evaluation of Performance of Generated Filters. In the case of "bo-po" discrimination, the learning performance in terms of recognition accuracy was tested by comparing it to the Gabor feature extraction [13]. The Gabor feature extraction adopted here has a simple structure that has 8x8 local regions and 4 directions in each local region, totaling 256 output elements. This kind of Gabor filter is often used for character recognition and considered to have reasonable processing speed and good performance. As for the recognition method, LVQ, SVM, MQDF (modified quadratic discriminant functions) were tested. LVQ [14] is the version of a differentiated sigmoid window function, the system "svmlight" [15] is used as SVM and MQDF is constructed based on the reference [16].

Table 1. Error rate for "bo-po"

Method	Gauss	Learning	Gabor
FML		2.7%	
LVQ	6.2%	2.8%	3.8%
SVM	5.5%	2.7%	3.9%
MQDF	6.6%	3.2%	4.7%
Average	6.1%	2.9%	4.1%

Table 2. Error rate for "ma-mon"

Method	Gauss	Learning	Gabor
FML		8.3%	
LVQ	12.8%	8.2%	16.1%
SVM	8.7%	7.3%	11.5%
MQDF	12.9%	9.8%	14.3%
Average	11.5%	8.4%	14.0%

The results of "bo-po" are shown in Table. 1. The values in the table are the error rate. The column labeled "Gauss" means the filter is Gaussian, that is the initial filter. "Learning" means the filter is the filter obtained by filter mask learning. "Gabor" means the filter is Gabor. The array of "FML" shows the results of the filter mask learning, consequently it shows the system's performance of itself after learning.

The array of "LVQ" shows the results of LVQ with 3 types of filters, and "SVM", "MQDF" are the same. The final "Average" shows the average results of LVQ, SVM and MQDF.

From this comparison, "FML" had better performance compared to the conventional recognition system with a Gabor filter. And also, as for the filter itself, the "FML" filter had better performance compared with the Gabor filter in 3 different recognition systems.

These results are considered reasonable because the portion where the similar patterns are different is strongly focused in the generated filters. The results for "ma-mon" are also shown in Table. 2.

4 Parameter Setting

Relation Between ϵ_f and ϵ_r. The parameter setting is important to make a learning process stable and effective, especially the parameters ϵ_f and ϵ_r are essential. Both parameters have a relationship to each other and need to have different values in some cases for effective learning.

Fig. 10 shows a simple evaluation function as an example. This figure represents a contour map of some function on the x-y plane. The lower value is better and position B has the lowest value. Let the start position be A, and let the optimization process be a descent algorithm. In this case, the first direction to change position is shown by arrow C which is the normal line to the contour line. In this figure, let's imagine that the ellipse's long axis is 1000 times longer than the short axis, like the Grand Canyon. It might be difficult to come close to B from A because there is too much distance between A and B in comparison with the adequate length of C for stable learning. The direction of C is completely different from the desired direction. This problem could be related to the research [17], but a simple approach is adopted here.

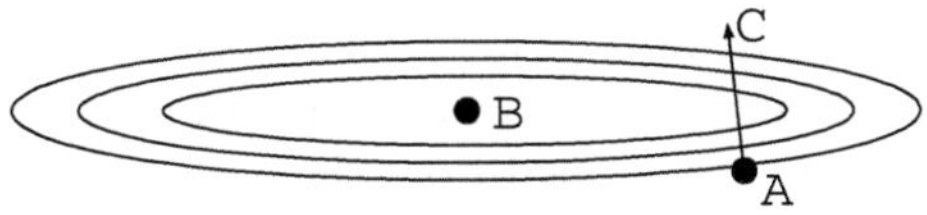

Fig. 10. Grand Canyon problem in descent algorithm

To avoid this problem, we need to convert such a super thin ellipse to the adequate size ellipse to avoid this difficulty. This conversion is virtually realized by the technique in which the parameters ϵ_f and ϵ_r have different values.

For this purpose, the proper value of ϵ was estimated as follows. The elements of the update filter (3) and vector (4) are described as

$$\Delta F_{ij} = \mp 2\epsilon_f l'(d)\rho'(\hat{x_i})(x_i - \varphi_i)z_j, \quad \Delta\varphi_i = \pm 2\epsilon_r l'(d)(x_i - \varphi_i). \tag{8}$$

Here, $\overline{|a|}$ denotes the estimated average of absolute value of a, and $\langle\langle F_{ij}\rangle\rangle$, $\langle\langle \Delta F_{ij}\rangle\rangle$, $\langle\langle \varphi_i\rangle\rangle$ and $\langle\langle \Delta\varphi_i\rangle\rangle$ are defined by follows. $\langle\langle F_{ij}\rangle\rangle = \overline{|F_{ij}|}$, $\langle\langle \varphi_i\rangle\rangle = \overline{|\varphi_i|}$,

$$\langle\langle \Delta F_{ij}\rangle\rangle = 2\epsilon_f\overline{|l'(d)|}\,\overline{|\rho'(\hat{x_i})|}\,\overline{|(x_i - \varphi_i)|}\,\overline{|z_j|}, \ \langle\langle \Delta\varphi_i\rangle\rangle = 2\epsilon_r\overline{|l'(d)|}\,\overline{|(x_i - \varphi_i)|}. \tag{9}$$

These estimated average values are determined by human observation of experimental data. The following results are obtained by those values.

$$\langle\langle F_{ij}\rangle\rangle = 0.2, \quad \langle\langle \Delta F_{ij}\rangle\rangle = 1200\epsilon_f, \quad \langle\langle \varphi_i\rangle\rangle = 32, \quad \langle\langle \Delta\varphi_i\rangle\rangle = 8\epsilon_r. \tag{10}$$

The final result ϵ_f/ϵ_r is calculated by (10) and the assumption (11).

$$\langle\langle \Delta F_{ij}\rangle\rangle / \langle\langle F_{ij}\rangle\rangle = \langle\langle \Delta\varphi_i\rangle\rangle / \langle\langle \varphi_i\rangle\rangle. \tag{11}$$

$$\epsilon_f/\epsilon_r = 40 \times 10^{-6}. \tag{12}$$

This estimated result is compared with the experimental results in Fig. 11. Several values of ϵ_r are tested with a certain value of ϵ_f, and the best value of ϵ_r is determined for each value of ϵ_f. This best value is calculated as the average of the good values with which the system has showed good performance. These best values are plotted in Fig. 11 as "result". The line "estimated" refers to equation (12). This estimation fits well to the obtained values as the best values up to the value $\epsilon_r = 0.03$.

As a result, the above mentioned approach was proved to be useful to estimate the ratio of ϵ.

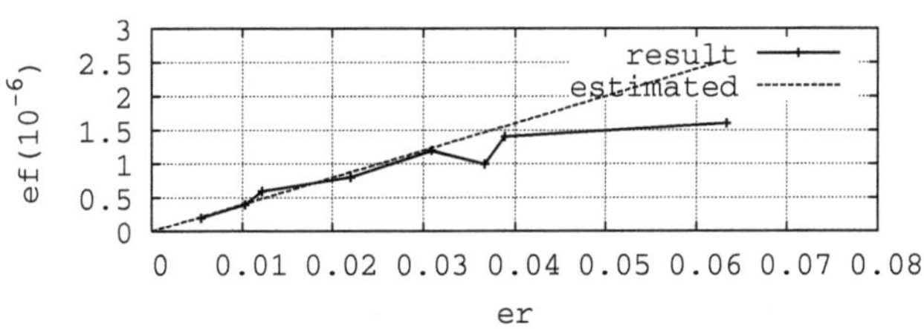

Fig. 11. Estimated ϵ_f/ϵ_r and the experimental results

Fixing δ in Sigmoid Function. The recognition accuracy after learning depends on the parameter δ in (7). This value is determined by human observation on the distribution of value d obtained from the actual recognition system. This value was determined so that about 1/5 or 1/10 of the patterns contribute to the learning. After that, several values around the above determined value are tested in preliminary attempts, and the best value is finally determined as δ .

5 Conclusion

Through these experiments, the AR/PD approach for Filter Mask Learning was proved to be effective in terms of the system's recognition accuracy and the generated filters had good performance even if it was embedded in the other recognition systems. The filter images and the feature patterns prepared for human observation showed that the learning was focused on the important part for discrimination of similar patterns.

The multiple filters at the local regions and the multi layer filter structure will be considered in the further research and they are expected to become useful technology in pattern recognition and provide new knowledge about intelligence and learning.

References

1. Amari, S.: A Theory of Adaptive Pattern Classifiers. IEEE Trans. EC 16(3), 299–307 (1967)
2. Katagiri, S., Lee, C.-H., Juang, B.-H.: A Generalized Probabilistic Descent Method, ASJ, Fall Conf. 2-p-6, Nagoya, Japan pp. 141–142 (1990)
3. Paliwal, K.K., Bacchiani, M., Sagisaka, Y.: Minimum Classification Error Training Algorithm for Feature Extractor and Pattern Classifier in Speech Recognition. EUROSPEECH '95 1, 541–544 (1995)
4. Biem, A., Katagiri, S.: Feature Extraction Based on Minimum Classification Error / Generalized Probabilistic Descent Method. In: Proc. IEEE Int. Conf. Acoust. Speech, Signal Processing 2, 275–278 (1993)
5. Biem, A., Katagiri, S.: Filter Bank Design Based on Discriminative Feature Extraction. In: Proc. IEEE Int. Conf. Acoust. Speech, Signal Processing 1, 485–488 (1994)
6. Bacchiani, M., Aikawa, K.: Optimization of Time-Frequency Masking Filters Using the Minimum Classification Error Criterion. In: Proc. IEEE Int. Conf. Acoust. Speech, Signal Processing 2, 197–200 (1994)
7. Kawamura, A. and Nitta, T.: Feature-Extraction-Based Character Recognition Using Minimum Classification Error Training, IEICE Trans. Information and Systems, Vol.J81-D-II, No.12 2749-2756, (in Japanese) (1998)
8. LeCun, Y., Bottou, L., Haffner, P.: Gradient-Based Learning Applied to Document Recognition. In: Proc. of the IEEE 86(11), 2278–2324 (1998)
9. Fukushima, K.: Neocognitron: a Self-Organizing Neural Network Model for a Mechanism of Pattern Recognition Unaffected by Shift in Position. Biol. Cybernet. 36, 193–202 (1980)

10. Lampinen, J., Oja, E.: Distortion Tolerant Pattern Recognition Based on Self-Organizing Feature Extraction. IEEE Trans. Neural Networks 6(3), 539–547 (1995)
11. Koichi, I., Tanaka, H., Tanaka, K., Kyuma, K.: Learning Algorithm by Reinforcement Signals for the Automatic Recognition System, SMC2004 pp. 4844–4848 (2004)
12. Kurosawa, Y.: Filter Learning Method for Feature Extraction in Character Recognition System, MIRU2006, (in Japanese) (2006)
13. Hamamoto, Y., Uchimura, S., Watanabe, M., Yasuda, T., Mitani, Y., Tomita, S.: A Gabor Filter-Based Method for Recognizing Handwritten Numerals, Pattern Recognition 31(4), 395–400 (1998)
14. Kohonen, T.: Learning Vector Quantization, Helsinki University of Technology, Laboratory of Computer and Information Science, Report TKK-F-A601 (1986)
15. Joachims, T.: Making Large-Scale SVM Learning Practical, Advances in Kernel Methods - Support Vector Learning. MIT-Press, Cambridge (1999)
16. Kimura, F., Takashina, K., Tsuruoka, S., Miyake, Y.: Modified Quadratic Discriminant Functions and the Application to Chinese Character Recognition. IEEE Trans. PAMI 9(1), 149–153 (1987)
17. Amari, S.: Natural Gradient Works Efficiently in Learning, Neural Computation, Vol. Neural Computation 10(2), 251–276 (1998)

Robust Pose Estimation Using the SwissRanger SR-3000 Camera

Sigurjón Árni Guðmundsson, Rasmus Larsen, and Bjarne K. Ersbøll

Technical University of Denmark, Informatics and Mathematical Modelling
Building 321, Richard Petersens Plads, DTU DK-2800 Kgs. Lyngby
sag@imm.dtu.dk
http://www2.imm.dtu.dk/~sag

Abstract. In this paper a robust method is presented to classify and estimate an objects pose from a real time range image and a low dimensional model. The model is made from a range image training set which is reduced dimensionally by a nonlinear manifold learning method named Local Linear Embedding (LLE). New range images are then projected to this model giving the low dimensional coordinates of the object pose in an efficient manner. The range images are acquired by a state of the art SwissRanger SR-3000 camera making the projection process work in real-time.

1 Introduction

The pose says how the object is oriented in space. Detecting an objects pose is an active research field in computer vision and closely related to other very active topics such as object recognition, classification and face recognition.

In this work a statistical machine learning method of dimensionality reduction of data is proposed to detect the pose of an object; on which side it lies and the planar angle from the cameras view. A low dimensional model is trained using images of an object in various positions. The model then gives the low dimensional pose coordinates of the data points. Finally a new image can be projected onto the model resulting in the objects pose coordinates. Analyzing intensity image data with manifold methods have shown good results before [1,2,3]. In the present work a model is constructed from range images from a SwissRanger camera that produces range images in real time.

1.1 Related Research

The pose estimation problem has been approached in numerous ways. Many industrial methods are based on matching objects to a CAD model [9]. Other methods such as extended gaussian images [8] recognize the pose by analyzing the surface normal behavior of the object. Machine learning has been a "hot" subject in academic research in this field, especially dimensionality reduction techniques. Eigen shapes and optimal component projections have shown good results on a wide range of data [10]. In recent years manifold learning methods

B.K. Ersbøll and K.S. Pedersen (Eds.): SCIA 2007, LNCS 4522, pp. 968–975, 2007.

such as isomap, Laplacian eigenmaps and locally linear embedding (LLE) have shown how various data lie on nonlinear manifolds that can be uncovered [1,3,4]. These methods strive at the same goal but have different different properties with regard to classification and projecting new data etc. In this work the LLE will be investigated and shown how its properties suit the pose problem.

1.2 Contributions

The potential of this nonlinear pose estimator based on range data is considerable. LLE interprets the change between the training images in the simplest fashion while at the same time giving a possibility to project a new data point onto the model without needing to go through the heavy eigen - calculations.

The effect of using range data from the SwissRanger camera also gives the model added robustness; eliminating the effect of variable lighting and giving a possibility of dimensionally scaling new images before projecting onto the model.

1.3 Outline of Paper

This paper proposes a pose estimation technique based on the locally linear embedding (LLE). It is composed of four main sections. In section 2 the the theoretical background of LLE is given and the SwissRanger is introduced. In section 3 data, the experiments and their results are discussed. Finally section 4 shows the direction for further research.

2 Technical Approach

The approach presented is based on the following:

- Images of an object in different poses lie on a non-linear manifold in the high dimensional pixel space.
- LLE finds the intrinsic dimension of the manifold.
- New data points are projected onto the embedded dimensions using Locally Linear Projecting (LLP)
- Range data adds robustness to the model and simplifies preprocessing of both training data and the new data.

2.1 Reduction of Dimensionality

One of the hottest topics in statistical machine learning is dealing with high dimensional data. A sequence of 100×100 pixel images can be seen as points in a 10000 dimensional pixel space. If these images are in some way correlated it is likely that a much lower dimensional feature space can be found; where the features in some way describe the sequence.

The classical approach to dimensionality reduction is Principal Component Analysis (PCA). PCA linearly maps the dataset to the maximum variance subspace by eigen-analyses of the covariance matrix, where the eigenvectors are the principal axes of the subspace, the eigenvalues give the projected variance of the data and the number of the significant eigenvalues gives the "true" dimensionality. PCA is an optimal linear approach and through its usage nonlinear structures in the data can be lost. It has been shown ([4,1,3]) that many very interesting data types lie on nonlinear low dimensional manifolds in the high dimensional space and several so called nonlinear manifold learning methods have been developed to describe the data in terms of these manifolds. One example of such methods is Locally Linear Embedding (LLE) which is the subject of the next section.

2.2 Local Linear Embedding

LLE assumes that locally each data-points' close neighborhood is linear and optimally preserves the geometry in these neighborhoods. It is an elegant unsupervised, non-iterative method that avoids the local minima problems that plague many other methods. For a dataset $\mathbf{X}$ with N points in D dimensions $(D \times N)$, an output $\mathbf{Y}$ of N, d dimensional points is found $(d \times N)$ where $d \ll D$.

The algorithm has three basic steps. In the first step each points K-nearest neighbors are found, second the points are approximated by a linear weighted combination of its neighbors. Finally a linear mapping is found through an eigenvalue problem to reduce the dimensionality of the approximated points to the embedded dimensions d. A full description and proofs of the algorithm is given in [2].

LLE Properties: LLE is a "natural" classifier. The separate classes' samples are mapped to separate dimensions as they are seen as lying on separate manifolds in the data. With C classes all samples of a certain class are mapped onto a single point in $C-1$ dimensions. The choice of dimensions for the third step in LLE is then twofold; choice of the local intrinsic dimensionality d_L which is the dimension of the manifold each class lies on, and the choice of global intrinsic dimensionality which is then used in step 3:

$$d_G = C \cdot d_L + (C - 1) \tag{1}$$

Finding d_L can be a complicated matter as the residual variances of each component cannot be measured such as is done in PCA and Isomap. On the other hand the intrinsic dimension can be found by using Isomap or by experiment in LLE to find what d_L describes the manifold. It has been shown in research ([1,3]) that the dimensionality of a dataset made of images of a 3D object rotated 360°, then 2D are needed to describe the change between the points, i.e. the rotation is described by a circle.

LLE is not easily extended to out-of-sample data and new images x_n cannot simply be mapped to the low dimensional feature space. Calculating new LLE coordinates for each new image with a large training set is way too heavy computationally and therefore out of the question for most fast or real time applications.

A simple method to map new data is sometimes called Local Linear Projecting (LLP, [6,7,2]). It utilizes the first two steps in LLE and omits the expensive third step. First it finds K neighbors' of x_n in the training set $\mathbf{X}$, then the weights are calculated as in *step 2* and these are used to make a weighted combination of the neighbors embeddings in $\mathbf{Y}$. The method thus exploits the local geometrical preservation quality of LLE. On the other hand the method is sensitive to translated input but this effect can be eliminated by preprocessing. This has been proved as a quick and effective method with good results.

2.3 The SwissRanger SR-3000 Camera and Data

The SwissRanger [11] camera is designed on the criteria to be a cost-efficient and eye-safe range finder solution.

Fig. 1. The CSEM SwissRanger SR-3000 Camera

Basically it has an amplitude modulated light source and a two dimensional sensor built in a miniaturized package (see fig. 1). The light source is an array of 55 near-infrared diodes (wavelength 850nm) that are modulated by a sinusoidal at 20MHz. This light is invisible to the naked eye.

The sensor is a 176 × 144 custom designed $0.8\mu m$ CMOS/CCD chip where each pixel in the sensor demodulates the reflected light by a lock-in pixel method, taking four measurement samples 90° apart for every period. From these samples the returning signal can be reconstructed. From the reconstructed signal two images are generated: An intensity (gray scale) image and a range image derived from the amplitude and the phase offset of the signal in each pixel.

The quality of the range-images is not as high as in many other range imaging devices. The spatial resolution is low and due to low emitted power the accuracy is dependent on the reflection properties of the subject and in some situations the results can be distorted by high noise and effects such as multiple-paths. This makes these images difficult to use in some conventional 3D pose recognition methods where higher accuracy is needed. On the other hand the low latency and high frame rate giving near real-time images with near field accuracy within 2 mm makes the camera excellently suited in many scenarios. For the purposes of this experiment this camera was an excellent choice.

3 Testing Methodology

The experiment data was made of images of a cardboard box with wooden knobs attached to it (figure 2). This object can lie on 3 sides and is without symmetrical features. The images were acquired of the box while it was rotated 360° in each lying pose with 2° intervals (540 images in each dataset). The dataset was purposely made to make a model that could detect on which side a new image of the object was lying and detecting its orientation at the same time. Two models were constructed; one using the range data the other using intensity data for comparison purposes. The depth data was aligned so that the a smoothed version of the data's closest point was centralized.

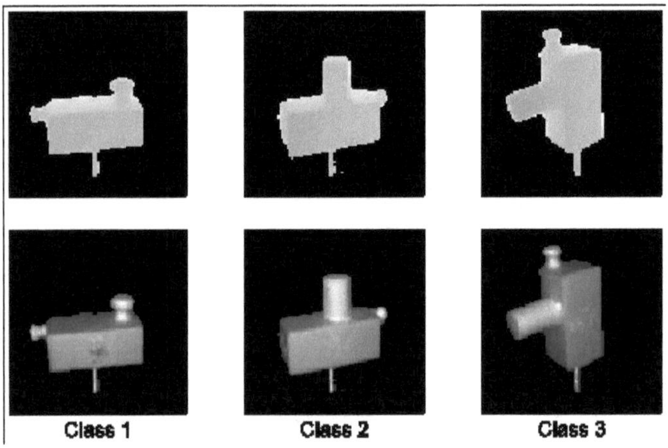

Fig. 2. The box dataset. The upper row is from the range dataset and the lower is from the intensity dataset. One image from each class is shown.

The LLP projection of new data points was tested by choosing a subset of the images and using them as a test set on the model made from the remaining images. As the test points positions are known relative to the training points it can be measured if the points are mapped between the correct points. The test points where aligned according to the depth in the same manor as the training points. In this way the depth is used to minimize the input translation problem of LLP.

3.1 Results

The two first principal components of both datasets in figure 3A shows that the classes can be separated but the method completely fails to capture the regularity of the objects change in orientation. Still these two components contain 95% of the variance of the points.

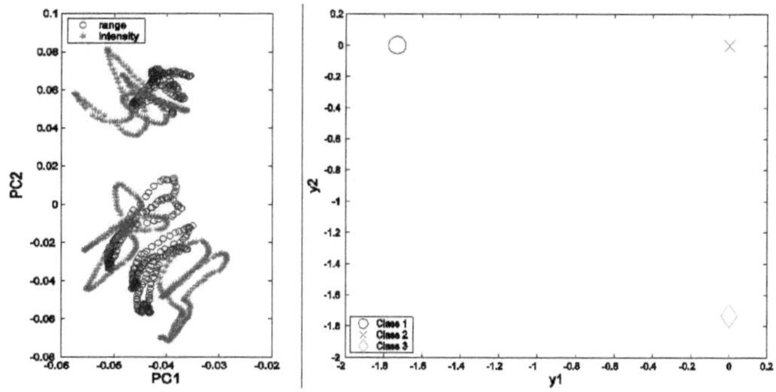

Fig. 3. A:) Two first principal components of the box data. B:)The the "Class" dimensions of the LLE mapping of the data.

The plot in figure 3B shows that the LLE embedding in 2D gives a perfect separation by mapping all the 180 data points of each class to three points. The three 2D local dimensional embeddings are shown in figure 4.

Figure 4 shows the local dimensions of each class using the range data on one hand and intensity data on the other. Both models model the 360° rotation of the object with a circle and in the case of range data the circles are almost perfectly smooth while the intensity model's curves are less smooth.

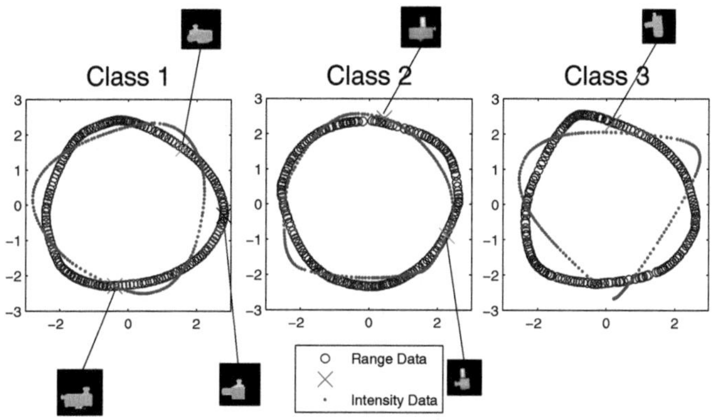

Fig. 4. The local intrinsic dimensions for each class. Six examples of image to intrinsic dimension mappings are shown.

Experiments on LLP by projecting small subsets of the image data gave 100% correct pose estimations. Figure 5 shows the results when only half of the data is used to find the embedding dimensions and the other half is projected one by one into these dimensions. This sparser model results in less smooth curves than before but the test points are still projected to correct positions in all cases. The

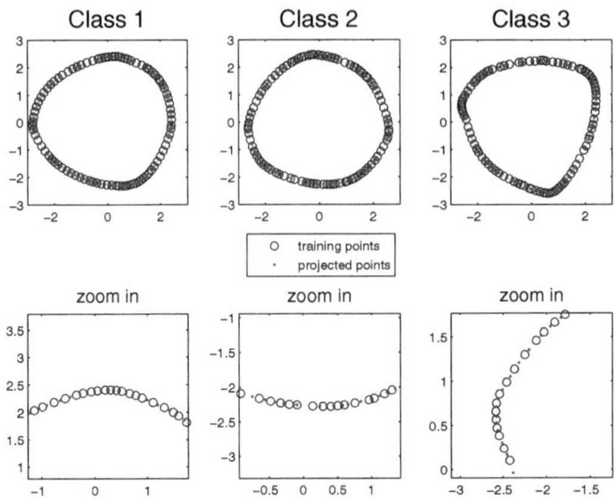

Fig. 5. *The Range Model.* An embedding made from 270 range images, the remaining 270 are then projected onto the embedding one by one.

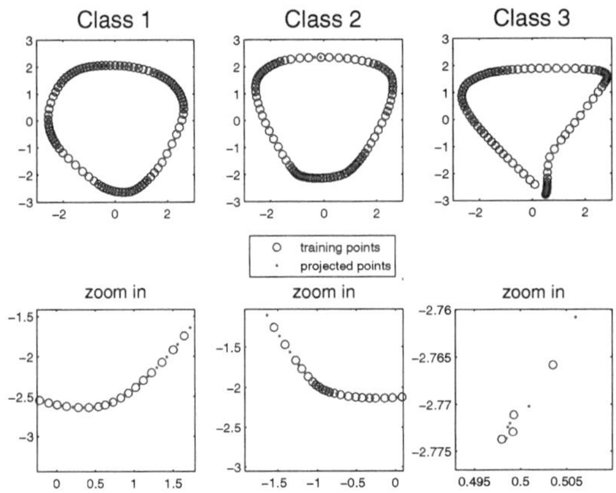

Fig. 6. *The Intensity Model.* An embedding made from 270 intensity images, the remaining 270 are then projected onto the embedding one by one.

intensity model in figure 6 shows poorer performance with not as smooth curves and in the case of class 3 it has problems with connecting the starting and final points of the rotation, resulting in almost erratic projections i.e. points have a third close neighbor on the embedding.

4 Summary

LLE proved to capture the nature of the change between the points in a very efficient manner; both finding on which side the object is lying and the planar angle. LLP is also a very efficient way to quickly project an image to a pre-calculated model. Creating a model from range images also showed advantages over intensity images; i.e. smoother curves better capturing the nature of the points.

References

1. Roweis, S., Saul, L.: Nonlinear dimensionality reduction by locally linear embedding. Science 290, 2323–2326 (2000)
2. Roweis, S., Saul, L.: Think globally, fit locally: Unsupervised learning of low dimensional manifolds. Journal of Machine Learning Research 4, 119–155 (2003)
3. Tenenbaum, J.B., de Silva, V., Langford, J.C.: A global geometric framework for nonlinear dimension reduction. Science 290, 2319–2323 (2000)
4. Belkin, M., Niyogi, P.: Laplacian eigenmaps for dimensionality reduction and data representation. MIT Press, Neural Computation 15, 1373–1396 (2003)
5. Bengio, Y., Paiement, J., Vincent, P.: Out-of-sample extensions for lle, isomap, mds, eigenmaps, and spectral clustering. Technical Report 1238, Université de Montréal (2004)
6. Ham, J., Lin, Y., Lee, D.: Learning nonlinear appearance manifolds for robot localization. IEEE Pacific Rim Conference on Communications, Computers and signal Processing, pp. 2971–2976 (2005)
7. de Ridder, D., Duin, R.: Locally linear embedding for classification. TUDelft, Pattern Recognition Group Technical Report Series, PH-2002-01 (2002)
8. Horn, B.K.P.: Robot Vision (MIT Electrical Engineering and Computer Science). The MIT Press, Cambridge (1986)
9. Balslev, I., Larsen, R.: Scape vision, a vision system for flexible binpicking in industry. IMM Industrial Visiondays, DTU (2006)
10. Turk, M., Pentland, A.: Eigenfaces for recognition. Journal for Cognitive Neuroscience 3, 71–86 (1991)
11. Blanc, N., Lustenberger, F., Oggier, T.: Miniature 3D TOF Camera for Real-Time Imaging. In: André, E., Dybkjær, L., Minker, W., Neumann, H., Weber, M. (eds.) PIT 2006. LNCS (LNAI), vol. 4021, pp. 212–216. Springer, Heidelberg (2006)

Pseudo-real Image Sequence Generator for Optical Flow Computations

Vladimír Ulman and Jan Hubený

Centre for Biomedical Image Analysis
Masaryk University, Brno 621 00, Czech Republic
xulman@fi.muni.cz

Abstract. The availability of ground-truth flow field is crucial for quantitative evaluation of any optical flow computation method. The fidelity of test data is also important when artificially generated. Therefore, we generated an artificial flow field together with an artificial image sequence based on real-world sample image. The presented framework benefits of a two-layered approach in which user-selected foreground was locally moved and inserted into an artificially generated background. The background is visually similar to input sample image while the foreground is extracted from original and so is the same. The framework is capable of generating 2D and 3D image sequences of arbitrary length. Several examples of the version tuned to simulate real fluorescent microscope images are presented. We also provide a brief discussion.

1 Introduction

There is hardly any new method being used in image analysis that hadn't been tested thoroughly beforehand. The situation is not different in the field of optical flow computing methods. The optical flow is, according to Horn and Schunck [1], the distribution of apparent velocities of movement of brightness patterns in an image. In other words, the outcome of an optical flow method is a flow field in which a velocity vector is assigned to every voxel in the image. The vector represents movement of a given voxel. Thus, this is often used for representation of a movement in the sequence of images [2]. In particular, optical flow methods are used for analysis of such time-lapse image sequences acquired from fluorescence optical microscope [3, 4].

The most common approach to the validation of a method for computing optical flow is to compare its result to some certain flow field [5], which is commonly termed as ground-truth flow field. Unfortunately, we don't have the ground-truth information at hand when testing some method on real data. Therefore, the automatic generation of pseudo-real high-fidelity data together with correct flow field is very useful. Not only it can produce vast amount of unbiased data, it also may speed up the tuning of an existing or newly developed optical flow computation method by allowing for its immediate evaluation over close-to-real data. This may lead to a real improvement of the investigated method's precision.

B.K. Ersbøll and K.S. Pedersen (Eds.): SCIA 2007, LNCS 4522, pp. 976–985, 2007.

We propose a framework in this paper that allows for the evaluation of optical flow computing methods. The primary aim is to create a pair of new grayscale images similar to the given real-world image (e.g. Fig. 3A) together with appropriate flow field. The framework should be flexible: it should handle global cell motion together with independent local motions of selected intracellular structures which is a phenomenon often observed in the field of fluorescence microscopy. It should be accurate: the created images should perfectly resemble the real-world images as well as created flow field should describe the movements that are displayed in the image data. Since we considered 3D image as a stack of 2D images, we didn't have to utilize any 3D-to-2D projection – the flow field remained three-dimensional in this case. Hence, we referred to such a 2D or 3D flow field as to a ground-truth flow field. Last but not least, the framework should be fast and simple to use too.

The next section describes our proposed framework. It also gives the motivation to the adopted solution by means of very decent overview of some possible approaches. The third section will describe its behaviour and present few sample images coming out of the system tuned to fluorescence optical microscopy. The paper is concluded in the last section.

2 The Framework

Basically, there are just two possible approaches to obtain image sequences with ground-truth flow fields. One may inspect the real data and manually determine the flow field. Despite the bias [6] and possible errors, this usually leads to a tedious work, especially, when inspecting 3D image sequences. The other way is to generate sequences of artificial images from scratch by exploiting some prior knowledge of a generated scene. This is most often accomplished by taking 2D snapshots of a changing 3D scene [7, 5, 8]. The prior knowledge is encoded in models which control everything from the shape of objects, movements, generation of textures, noise simulation, etc. [9, 10]. This may involve a determination of many parameters as well as proper understanding of the modeled system. Once the two consecutive images are created, the information about movement between these two can be extracted from the underlying model and represented in the flow field.

We have adopted, in fact, the latter approach. Every created artificial image consisted of two layers. The bottom background layer contained an artificial background generated by the algorithm that will be described later. The background area was given by a mask image $\mathbf{M}$. The parameters of the algorithm were determined online from a real-world image which we will refer to as the sample input image $\mathbf{I}$. The foreground layer contained exact copies of regions of the sample input image. The foreground regions were defined by a mask image $\mathbf{m}$. All images had to be of the same size. We denote the value of voxel intensity in image $\mathbf{I}$ at position $\mathbf{x}$ as $\mathbf{I}(\mathbf{x})$. Similarly, the flow field $\mathbf{FF}$ holds a vector $\mathbf{FF}(\mathbf{x})$ at each position $\mathbf{x}$.

The background was subject to a global movement while the foreground was subject to global and independent local movements. The ground-truth flow field was a composition of these. Since both background and foreground were given only by mask images, we accomplished the movements using the backward transformation technique [11]. Let us write $\mathbf{O} = \text{BackT}(\mathbf{I}, \mathbf{FF})$ and say that the image $\mathbf{O}$ is the image $\mathbf{I}$ transformed according to the flow field $\mathbf{FF}$. Basically, the backward technique translates $\mathbf{I}(\mathbf{x}+\mathbf{FF}(\mathbf{x}))$ to $\mathbf{O}(\mathbf{x})$ for every $\mathbf{x}$ – voxel values move "against" the flow field. There exists a forward transformation technique too, refer to [11] for more detailed explanation. Figures 1 and 2 show the major pitfalls of both techniques. The artifacts occur when the flow field is not smooth enough. Unfortunately, that was the case owing to the local independent movements. Hence, we developed the following framework in order to avoid that. We use the notation $\mathbf{O} = \text{Copy}(\mathbf{I}, \mathbf{m})$ to state that only a regions given by mask $\mathbf{m}$ are copied from $\mathbf{I}$, the rest of $\mathbf{O}$ remains untouched.

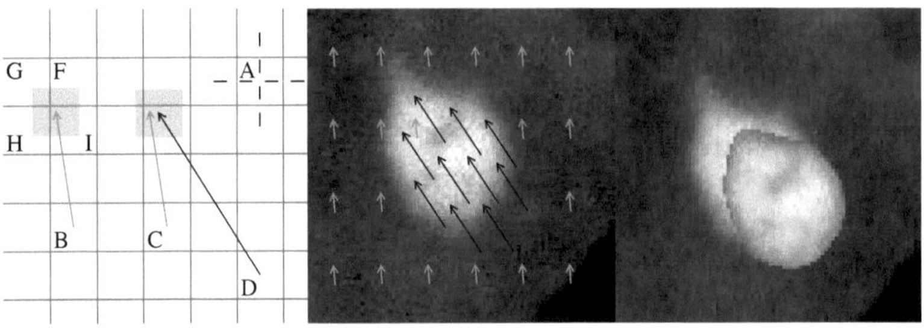

Fig. 1. The backward transformation. Left section: The grid represents voxels and their boundaries. The position of coordinate of voxel A is illustrated by dashed lines. The transformation moves value from vector's end to its beginning. In the case of real-valued vector (as is the one originating from voxel B), the moved value is the weighted sum of the nearest voxels' values with weights given by the portion of the gray area (values of voxels F, G, H and I). More vectors from distant places (as demonstrated with vectors originating from C and D) may fetch almost the same value when the flow is not smooth enough. This drawback results in the "copy" effect. Middle section: An example of input image with non-smooth flow field. Right section: A result of the backward transformation with the "copy" effect.

The framework's input was a sample input image $\mathbf{I}$, the background mask $\mathbf{M}$ and the foreground mask $\mathbf{m}$. The output would consist of images $\mathbf{I}_{1st}$, $\mathbf{I}_{2nd}$ and ground-truth flow field $\mathbf{gtFF}$ between $\mathbf{I}_{1st}$, $\mathbf{I}_{2nd}$ which denoted the first and the second image in the created sequence, respectively. It would hold: if $\forall \mathbf{x}\colon \mathbf{gtFF}(\mathbf{x})$ is an integer valued vector then $\forall \mathbf{x}\colon \mathbf{I}_{1st}(\mathbf{x}) = \mathbf{I}_{2nd}(\mathbf{x} + \mathbf{gtFF}(\mathbf{x}))$.

The preliminary step of the algorithm was to prepare a pool $\mathbf{R}$ of voxel intensities. Only voxels $\mathbf{x}$ satisfying $\mathbf{M}(\mathbf{x}) > 0$ and $\mathbf{m}(\mathbf{x}) = 0$ (exactly the background voxels) were copied into the pool. The mean value μ was computed within $\mathbf{R}$ since we observed that histogram of the background voxels resembled Gaussian-shaped distribution (Fig. 3B). Because of that, voxels with intensities i for which $i \notin (\mu - \sigma, \mu + k\sigma)$ were removed from the pool. We chose $\sigma = 11$ and $k = 3/2$ to fit the histogram better. This interval is shown as the white strip in Fig. 3B.

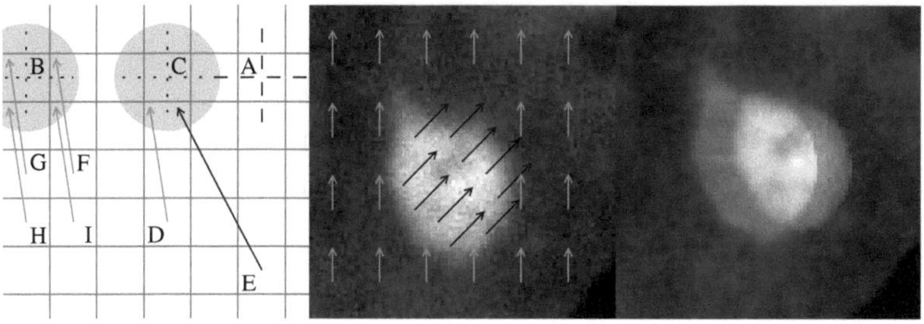

Fig. 2. The forward transformation. Left section: The grid represents voxels and their boundaries. The position of coordinate of voxel A is illustrated by dashed lines. The transformation moves value from vector's beginning to its end. In the case of real-valued vectors (as are those originating from voxels F, G, H and I), the value for particular voxel must be searched for in the close vicinity (illustrated by gray region around voxel B). The value is then weighted with weights being the distance of the nearest vectors' ends in each quadrant of the marked area. More vectors from distant places (as demonstrated with vectors originating from D and E) may end up in almost the same location when the flow is not smooth enough. This drawback results in the "averaging" effect. Middle section: An example of input image with non-smooth flow field. Right section: A result of the forward transformation with the "averaging" effect.

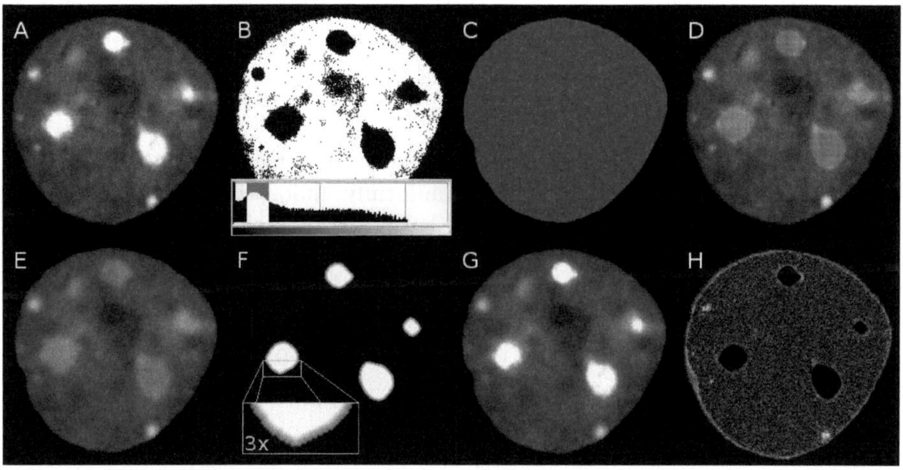

Fig. 3. Example of an image formation. A) The input image **I** of fluorescently marked HP1 proteins in the HL60 cell. B) The intensity histogram of the input image at the bottom, binary image above displays voxels with intensities in the white strip of the histogram. C) Outcome of the rand(**R**) generator. D) **fI** from (4). E) $\mathbf{I}_{2nd}$ from (5). F) The weights of the extended foreground mask,brighter intensity shows higher weights, see section 3. G) $\mathbf{I}_{2nd}$ from (6). H) The map of intensity differences between **I** and $\mathbf{I}_{2nd}$, the maximal brightness shows the value of 30. All images were enhanced for the purpose of displaying.

An artificial background was generated in two steps. First, the foreground was replaced by interpolated values. For each $\mathbf{x} = (x, y, z)$, $\mathbf{m}(\mathbf{x}) > 0$ find x_1 and x_2 such that

$$x_1 = \max\left(x' < x \text{ and } \mathbf{x}' = (x', y, z) \text{ and } \mathbf{m}(\mathbf{x}') = 0\right), \tag{1}$$
$$x_2 = \min\left(x' > x \text{ and } \mathbf{x}' = (x', y, z) \text{ and } \mathbf{m}(\mathbf{x}') = 0\right) \tag{2}$$

and set $V_{x_1} = \mathbf{I}(x_1, y, z)$ and $V_{x_2} = \mathbf{I}(x_2, y, z)$. If there were no such x_1, which happens exceptionally only when a mask touches the image border, then set $V_{x_1} = \mu$ and x_1 to the leftmost coordinate in $\mathbf{I}$. The x_2 was treated in the similar fashion. The value for $\mathbf{I}(\mathbf{x})$, proportionally along x-axis, was

$$V_x = (V_{x_2} - V_{x_1}) \cdot \frac{x - x_1}{l_x + 1} + V_{x_1} \tag{3}$$

with $l_x + 1 = x_2 - x_1$. The V_y and V_z values were obtained in the similar fashion. The replacing of foreground was finished by assigning (Fig. 3D):

$$\forall \mathbf{x}\colon \mathbf{fI}(\mathbf{x}) = \begin{cases} \frac{l_y l_z V_x + l_x l_z V_y + l_x l_y V_z}{l_y l_z + l_x l_z + l_x l_y} & \text{if } \mathbf{m}(\mathbf{x}) > 0 \\ \mathbf{I}(\mathbf{x}) & \text{otherwise} \ . \end{cases} \tag{4}$$

Second, the new artificial background was generated. The $\mathbf{fI}$ image was convolved with separable averaging kernel. We used the filter $\frac{1}{9}(1, 1, 1, 1, 1, 1, 1, 1, 1)$ for each axis. The new background image was then computed as

$$\forall \mathbf{x}\colon \mathbf{I}_{\text{2nd}}(\mathbf{x}) = \text{rand}(\mathbf{R}) + (\mathbf{fI}(\mathbf{x}) - \mu) \tag{5}$$

where rand() is a generator of random numbers obeying uniform distribution. The effect of this term in (5) was to uniformly choose intensity values from the pool $\mathbf{R}$. This ensured the generated background to share similar statistics, including intensity fluctuations and noise. The last term in (5) enabled to display intracellular structures in the background, e.g. nucleolus as in Fig. 3E. Finally, the image was Gaussian blurred with sigma set to 0.7px.

To finish the output image $\mathbf{I}_{\text{2nd}}$, the foreground was overlaid over the artificial background:

$$\mathbf{I}_{\text{2nd}} = \text{Copy}(\mathbf{I}, \mathbf{m}) \ . \tag{6}$$

The ground-truth flow field for global movement of the whole image was created into $\mathbf{gtFF}$. We utilized an arbitrary rotation around arbitrary centre together with arbitrary translation. The flow field was created regardless of masks $\mathbf{m}$ and $\mathbf{M}$. In fact, any flow field could have been used provided it is reasonably smooth. The images were transformed:

$$\mathbf{I}' = \text{BackT}(\mathbf{I}, \mathbf{gtFF}), \tag{7}$$
$$\mathbf{M} = \text{BackT}(\mathbf{M}, \mathbf{gtFF}), \tag{8}$$
$$\mathbf{m} = \text{BackT}(\mathbf{m}, \mathbf{gtFF}) \ . \tag{9}$$

We repeated the process of artificial background generation in the new position with $\mathbf{I}'$ instead of $\mathbf{I}$. The result was stored into $\mathbf{I}_{1st}$. Note that we used the same intensity pool $\mathbf{R}$.

A random translational vector, say v_i, was assigned to each component i of the mask $\mathbf{m}$. For each i, we created flow field $\mathbf{FF}_i$ and mask image $\mathbf{m}_i$

$$\forall \mathbf{x}: \quad \mathbf{FF}_i(\mathbf{x}) = v_i, \tag{10}$$

$$\forall \mathbf{x}: \quad \mathbf{m}_i(\mathbf{x}) = \begin{cases} 1 & \text{if } \mathbf{x} \text{ belongs to component } i \\ 0 & \text{otherwise} \end{cases} . \tag{11}$$

Note that $\mathbf{FF}_i$ is uniformly filled what guarantees a smooth flow field. Independent local movements are embedded into $\mathbf{gtFF}$ by computing the following equations:

$$\forall \mathbf{x}: \quad \mathbf{gFF}(\mathbf{x}) = \mathbf{gtFF}(\mathbf{x}), \tag{12}$$

$$\forall i: \quad \mathbf{m}_i' = \text{BackT}(\mathbf{m}_i, \mathbf{FF}_i), \tag{13}$$

$$\forall i: \quad \mathbf{I}_i' = \text{BackT}(\mathbf{I}', \mathbf{FF}_i), \tag{14}$$

$$\forall i: \quad \mathbf{gtFF} = \text{Copy}(\text{BackT}(\mathbf{gFF}, \mathbf{FF}_i), \mathbf{m}_i'), \tag{15}$$

$$\forall i: \quad \mathbf{FF}_i = \text{Copy}(\mathbf{0}, 1 - \mathbf{m}_i'), \tag{16}$$

$$\forall \mathbf{x}: \forall i: \quad \mathbf{gtFF}(\mathbf{x}) = \mathbf{gtFF}(\mathbf{x}) + \mathbf{FF}_i(\mathbf{x}) \tag{17}$$

with the following interpretations: backup $\mathbf{gtFF}$ (12), translate the patch corresponding to each component in $\mathbf{gtFF}$ (15) together with its mask (13) according to its flow field, zero the component's flow field outside of its moved mask (16) and add the result to the $\mathbf{gtFF}$ (17). Equations (15) to (17), in fact, concatenate global and local flow fields since the movement of the foreground consisted of global (9) and then local (13) movement. The image of each component was separately moved according to its flow field (14). Moving the entire image $\mathbf{I}'$ according to the final $\mathbf{gtFF}$ would produce image corrupted by the "copy" effect of non-smooth flow field.

Finally, the output image $\mathbf{I}_{1st}$ was computed:

$$\forall i: \mathbf{I}_{1st} = \text{Copy}(\mathbf{I}_i', \mathbf{m}_i') . \tag{18}$$

The Copy() operation just overlaid the moved foreground regions over the artificially generated background. Optionally, the ground-truth flow field could be trimmed:

$$\mathbf{gtFF} = \text{Copy}(\mathbf{0}, 1 - \mathbf{M}) . \tag{19}$$

The presented framework also allows for generation of an arbitrary long time-lapse image sequence. Due to the property of the backward transformation technique, the generation proceeds from the last image $\mathbf{I}_{nth}$ of the sequence, given some $n \geq 2$, towards the first image $\mathbf{I}_{1st}$. Clearly, the last image is the artificial substitute for the sample input image and so $\mathbf{I}$ can be used as a sample without any modification. For the other images in the generated sequence, $\mathbf{I}$ must be transformed to the actual position. Instead of iteratively moving the image, we suggest to hold the flow fields that prescribe the transformations required to get **I** to the demanded position. Two flow fields should be enough. Let $\mathbf{gtFF}_{i,j}$, $i < j$ denote the flow between images $\mathbf{I}_{i\text{th}}$ and $\mathbf{I}_{j\text{th}}$. For the purpose of consecutive generation of $\mathbf{I}_{k\text{th}}$, $k = n-2\ldots 1$, compute

$$\forall \mathbf{x}\colon \mathbf{gtFF}_{k,n}(\mathbf{x}) = \mathbf{gtFF}_{k,k+1}(\mathbf{x}) + \mathrm{BackT}(\mathbf{gtFF}_{k+1,n}, \mathbf{gtFF}_{k,k+1})(\mathbf{x}), \quad (20)$$

$$\forall \mathbf{x}\colon \mathbf{glFF}_{k,n}(\mathbf{x}) = \mathbf{glFF}_{k,k+1}(\mathbf{x}) + \mathrm{BackT}(\mathbf{glFF}_{k+1,n}, \mathbf{glFF}_{k,k+1})(\mathbf{x}) \quad (21)$$

where $\mathbf{glFF}_{k,k+1}$ is the flow field corresponding to just the global component of $\mathbf{gtFF}_{k,k+1}$. Then, repeat the second part of the proposed framework from (7), as if $\mathbf{I}_{1\text{st}}$ should be created, with the following exceptions: for the background generation use $\mathbf{glFF}_{k,n}$ in (7) while for the foreground movements prepare $\mathbf{I}'$ with $\mathbf{gtFF}_{k,n}$ in (7) and set $\mathbf{gtFF} = \mathbf{gtFF}_{k,k+1}$. Also start the background generation from its second step with the convolution of **fI**.

3 Results and Discussion

We implemented and tested presented framework in variants utilizing both backward and forward transformations. The framework was designed to transform images only according to the smooth flow fields. This and the definition of both transformations justify the ground-truth property of the created flow field.

Table 1. Comparisons of images **I** and $\mathbf{I}_{2\text{nd}}$. The column heading "Ext." shows the number of dilations performed on the foreground mask **m**. The mask controlled the foreground extraction as well as its plain overlaying[1] or weighted merging[2] (explained in section 3). A) and B) Comparisons over two 2D images. C) Comparison over a 3D image. D) Comparison over the same 3D image, separate pools of voxel intensities were used for each 2D slice during the formation of the artificial background.

	Ext.	Corr.[1]	Avg. diff.[1]	RMS[1]	Corr.[2]	Avg. diff.[2]	RMS[2]
A	0	0.989	3.87	5.13	0.989	3.87	5.12
	1	0.989	3.80	5.03	0.989	3.85	5.05
	2	0.989	3.73	4.94	0.989	3.82	5.00
	3	0.989	3.68	4.90	0.989	3.83	4.98
B	0	0.992	2.76	3.83	0.992	2.77	3.85
	1	0.992	2.62	3.69	0.992	2.74	3.75
	2	0.993	2.41	3.46	0.992	2.62	3.58
	3	0.993	2.33	3.40	0.992	2.64	3.57
C	0	0.980	3.67	4.79	0.980	3.67	4.79
	1	0.980	3.73	4.89	0.980	3.81	4.92
	2	0.981	3.53	4.69	0.981	3.70	4.77
	3	0.981	3.42	4.59	0.981	3.66	4.72
D	0	0.982	3.15	4.16	0.982	3.16	4.17
	1	0.983	3.07	4.08	0.982	3.13	4.11
	2	0.983	3.00	4.03	0.983	3.11	4.08
	3	0.984	2.92	3.96	0.983	3.10	4.05

The masks were generated by thresholding of sample input image with manually chosen threshold. The thresholded output was considered as a starting point for advanced segmentation method [12] which produced final cell and foreground masks. The generator was tested on several different 2D real-world images and one such 3D image.

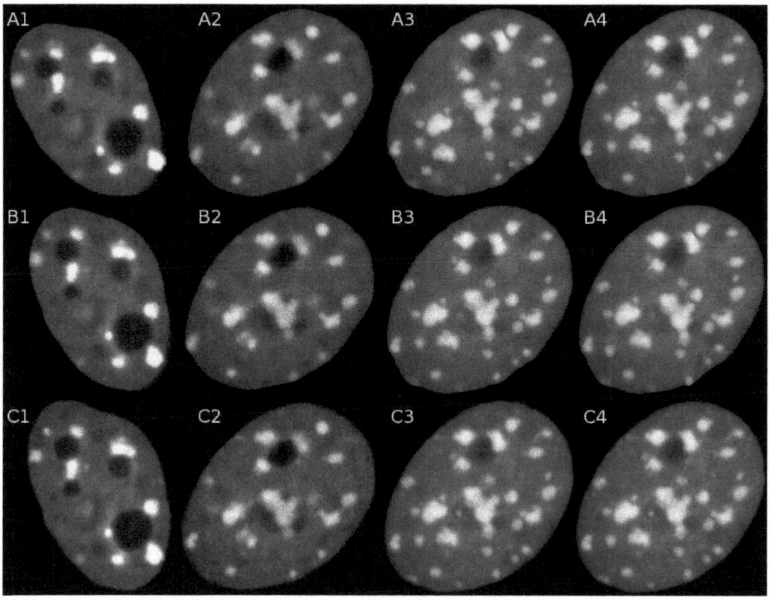

Fig. 4. Examples of generated pseudo-real 2D images. The $\mathbf{I}_{1st}$, the $\mathbf{I}_{2nd}$ and the sample input image $\mathbf{I}$ are shown in rows A), B) and C), respectively. Notice the similarity between rows B) and C) in columns 1), 2) and 3). Images A4) and C4) should be similar too. Foreground objects (the white spots) in each cell were subject to additional local movements. 1) An example of the cell rotated 9 degrees clock-wise around its edge. 2) An example of another cell rotated 9 degrees clock-wise around its centre. 3) An example of a similar cell with more foreground objects and with no global motion. 4) The same as 3) but the generator based on the forward transformation was used. All images were enhanced for the purpose of displaying.

All generated images were inspected. Since every generated image arose from some supplied sample image $\mathbf{I}$, we could compare $\mathbf{I}$ and $\mathbf{I}_{2nd}$. For each pair, we computed the correlation coefficient (Corr.), average absolute difference (Avg. diff.) and root mean squared difference (RMS). The results are summarized in Table 1. The generator achieved minimal value of 0.98 for correlation, see Fig. 3H. This quantitatively supports our observations that generated images were very similar to their originals. A few 2D examples are shown in Fig. 4. Decent improvement was observed when artificial background of 3D images was formed in a slice-per-slice manner what is also acknowledged in Table 1.

The image $\mathbf{I}_{1\text{st}}$ could not be evaluated quantitatively for the obvious reason. Nevertheless, the ratio would be definitely worse since all moved images are blurred a little. This is a feature of both backward and forward transformations when processing flow fields containing vectors with non-integer elements. In order to make both output images appear the same, we suggest to let $\mathbf{I}_{2\text{nd}}$ image perform the translation along vector $(0.5, 0.5, 0.5)$ and modify the **gtFF** correspondingly.

Inappropriately created foreground mask may emphasize the borders of extracted foreground when inserted into artificial background. We replaced the Copy() operation in eqs. (6) and (18) by the following sequence of operations: extend the foreground mask by several dilations (the "Ext." column in Table 1), compute the distance transform (we used [13]) on the mask and threshold it (see Fig. 3F), insert the foreground according to the weights (for details refer to [14]). We generally observed visually better results with this modification. According to Table 1, just 2 dilations achieved qualitatively better results in comparison to overlaying of foreground driven by unmodified input mask $\mathbf{m}$.

We also tried the local movements mask which permitted the foreground to translate only inside this mask. This should prevent the structures from moving into the regions where there were not supposed to be, i.e. outside the cell. The masks are simple to create, for example by extending the foreground mask into demanded directions. The generated images became even more real.

We argue against further iterations of the framework to get $\mathbf{I}_{k\text{th}}$ from $\mathbf{I}_{(k+1)\text{th}}$. When proceeding towards smaller k, transforming images iteratively leads to worse quality images because of the smoothing effect (Fig. 4A) of both transformations. Our suggested solution guarantees not more than two transformations of sample input image when creating $\mathbf{I}_{k\text{th}}$ for arbitrary $k \in \langle 1, n-1 \rangle$.

We implemented the algorithm in C++. We confirm that forward variant is up to two orders of magnitude slower than backward variant for 2D images. This is mainly because of greater complexity of forward transformation in contrast to backward transformation.

4 Conclusion

We have proposed a framework for generating time-lapse pseudo-real image data. It allows for automatic synthesis of unbiased sequences of 2D and 3D images. By suppling real-world sample image we could force images in the sequence to look more realistic. The background mask of the cell and the foreground mask of selected intracellular structures were supplied too. This gave us a layered control over the regions where global and local movements should occur. The aim was to automatically generate a vast amount of data together with corresponding flow field, that we called ground-truth, in order to evaluate methods for foreground tracking as the next step. The methodology was targeted at fluorescence optical microscopy.

We have tested the framework mainly in 2D. From Table 1 we may conclude that it generated images very similar to the sample image. The foreground was a copy from the sample image which implicitly assured its quality. The background voxels posed the same statistics since they were generated to do so. Theoretically, the presented framework has ambitions to work reliably on arbitrary data comprising of unimodal background distribution. The framework is also less sensitive to errors in the foreground segmentation. This is due to the seamless overlaying of the foreground. We also made use of local movements mask which gave us ultimate control over the foreground movements.

Acknowledgements. This work has been supported by the Ministry of Education of the Czech Republic (Grant No. MSM0021622419, LC535 and 2B06052).

References

[1] Horn, B.K.P., Schunck, B.G.: Determining optical flow. Artificial Intelligence 17, 185–203 (1981)
[2] Cédras, C., Shah, M.A.: Motion based recognition: A survey. Image and Vision Computing 13(2), 129–155 (1995)
[3] Gerlich, D., Mattes, J., Eils, R.: Quantitative motion analysis and visualization of cellular structures. Methods 29(1), 3–13 (2003)
[4] Eils, R., Athale, C.: Computational imaging in cell biology. The Journal of Cell Biology 161, 447–481 (2003)
[5] Barron, J.L., Fleet, D.J., Beauchemin, S.S.: Performance of optical flow techniques. Int. J. Comput. Vision 12(1), 43–77 (1994)
[6] Webb, D., Hamilton, M.A., Harkin, G.J., Lawrence, S., Camper, A.K., Lewandowski, Z.: Assessing technician effects when extracting quantities from microscope images. Journal of Microbiological Methods 53(1), 97–106 (2003)
[7] Galvin, B., McCane, B., Novins, K., Mason, D., Mills, S.: Recovering motion fields: An evaluation of eight optical flow algorithms. In: Proc. of the 9th British Mach. Vis. Conf (BMVC '98) 1, 195–204 (1998)
[8] Beauchemin, S.S., Barron, J.L.: The computation of optical flow. ACM Comput. Surv. 27(3), 433–466 (1995)
[9] Lehmussola, A., Selinummi, J., Ruusuvuori, P., Niemisto, A., Yli-Harja, O.: Simulating fluorescent microscope images of cell populations. In: IEEE Engineering in Medicine and Biology 27th Annual Conference pp. 3153–3156 (2005)
[10] Young, I.: Quantitative microscopy. IEEE Engineering in Medicine and Biology Magazine 15(1), 59–66 (1996)
[11] Lin, T., Barron, J.: Image reconstruction error for optical flow. In: Vision Interface. pp. 73–80 (1994)
[12] Hubený, J., Matula, P.: Fast and robust segmentation of low contrast biomedical images. In: Proceedings of the Sixth IASTED International Conference VIIP. vol. 8 (2006)
[13] Saito, T., Toriwaki, J.I.: New algorithms for Euclidean distance transformations of an n-dimensional digitized picture with applications. Pattern Recognition 27, 1551–1565 (1994)
[14] Ulman, V.: Mosaicking of high-resolution biomedical images acquired from wide-field optical microscope. In: EMBEC'05: Proceedings of the 3rd European Medical & Biological Engineering Conference. Vol. 11 (2005)

Author Index

Lecture Notes in Computer Science

For information about Vols. 1–4427

please contact your bookseller or Springer

Vol. 4542: P. Sawyer, B. Paech, P. Heymans (Eds.), Requirements Engineering: Foundation for Software Quality. IX, 384 pages. 2007.

Vol. 4539: N.H. Bshouty, C. Gentile (Eds.), Learning Theory. XII, 634 pages. 2007. (Sublibrary LNAI).

Vol. 4538: F. Escolano, M. Vento (Eds.), Graph-Based Representations in Pattern Recognition. XII, 416 pages. 2007.

Vol. 4534: I. Tomkos, F. Neri, J. Solé Pareta, X. Masip Bruin, S. Sánchez Lopez (Eds.), Optical Network Design and Modeling. XI, 460 pages. 2007.

Vol. 4531: J. Indulska, K. Raymond (Eds.), Distributed Applications and Interoperable Systems. XI, 337 pages. 2007.

Vol. 4530: D.H. Akehurst, R. Vogel, R.F. Paige (Eds.), Model Driven Architecture- Foundations and Applications. X, 219 pages. 2007.

Vol. 4529: P. Melin, O. Castillo, L.T. Aguilar, J. Kacprzyk, W. Pedrycz (Eds.), Foundations of Fuzzy Logic and Soft Computing. XIX, 830 pages. 2007. (Sublibrary LNAI).

Vol. 4526: M. Malek, M. Reitenspieß, A. van Moorsel (Eds.), Service Availability. X, 155 pages. 2007.

Vol. 4525: C. Demetrescu (Ed.), Experimental Algorithms. XIII, 448 pages. 2007.

Vol. 4524: M. Marchiori, J.Z. Pan, C.d.S. Marie (Eds.), Web Reasoning and Rule Systems. XI, 382 pages. 2007.

Vol. 4523: Y.-H. Lee, H.-N. Kim, J. Kim, Y. Park, L.T. Yang, S.W. Kim (Eds.), Embedded Software and Systems. XIX, 829 pages. 2007.

Vol. 4522: B.K. Ersbøll, K.S. Pedersen (Eds.), Image Analysis. XVIII, 989 pages. 2007.

Vol. 4521: J. Katz, M. Yung (Eds.), Applied Cryptography and Network Security. XIII, 498 pages. 2007.

Vol. 4519: E. Franconi, M. Kifer, W. May (Eds.), The Semantic Web: Research and Applications. XVIII, 830 pages. 2007.

Vol. 4517: F. Boavida, E. Monteiro, S. Mascolo, Y. Koucheryavy (Eds.), Wired/Wireless Internet Communications. XIV, 382 pages. 2007.

Vol. 4515: M. Naor (Ed.), Advances in Cryptology - EUROCRYPT 2007. XIII, 591 pages. 2007.

Vol. 4514: S.N. Artemov, A. Nerode (Eds.), Logical Foundations of Computer Science. XI, 513 pages. 2007.

Vol. 4510: P. Van Hentenryck, L. Wolsey (Eds.), Integration of AI and OR Techniques in Constraint Programming for Combinatorial Optimization Problems. X, 391 pages. 2007.

Vol. 4509: Z. Kobti, D. Wu (Eds.), Advances in Artificial Intelligence. XII, 552 pages. 2007. (Sublibrary LNAI).

Vol. 4508: M.-Y. Kao, X.-Y. Li (Eds.), Algorithmic Aspects in Information and Management. VIII, 428 pages. 2007.

Vol. 4506: D. Zeng, I. Gotham, K. Komatsu, C. Lynch, M. Thurmond, D. Madigan, B. Lober, J. Kvach, H. Chen (Eds.), Intelligence and Security Informatics: Biosurveillance. XI, 234 pages. 2007.

Vol. 4505: G. Dong, X. Lin, W. Wang, Y. Yang, J.X. Yu (Eds.), Advances in Data and Web Management. XXII, 896 pages. 2007.

Vol. 4504: J. Huang, R. Kowalczyk, Z. Maamar, D. Martin, I. Müller, S. Stoutenburg, K.P. Sycara (Eds.), Service-Oriented Computing: Agents, Semantics, and Engineering. X, 175 pages. 2007.

Vol. 4501: J. Marques-Silva, K.A. Sakallah (Eds.), Theory and Applications of Satisfiability Testing – SAT 2007. XI, 384 pages. 2007.

Vol. 4500: N. Streitz, A. Kameas, I. Mavrommati (Eds.), The Disappearing Computer. XVIII, 304 pages. 2007.

Vol. 4496: N.T. Nguyen, A. Grzech, R.J. Howlett, L.C. Jain (Eds.), Agent and Multi-Agent Systems: Technologies and Applications. XXI, 1046 pages. 2007. (Sublibrary LNAI).

Vol. 4495: J. Krogstie, A. Opdahl, G. Sindre (Eds.), Advanced Information Systems Engineering. XVI, 606 pages. 2007.

Vol. 4493: D. Liu, S. Fei, Z. Hou, H. Zhang, C. Sun (Eds.), Advances in Neural Networks – ISNN 2007, Part III. XXVI, 1215 pages. 2007.

Vol. 4492: D. Liu, S. Fei, Z. Hou, H. Zhang, C. Sun (Eds.), Advances in Neural Networks – ISNN 2007, Part II. XXVII, 1321 pages. 2007.

Vol. 4491: D. Liu, S. Fei, Z.-G. Hou, H. Zhang, C. Sun (Eds.), Advances in Neural Networks – ISNN 2007, Part I. LIV, 1365 pages. 2007.

Vol. 4490: Y. Shi, G.D. van Albada, J. Dongarra, P.M.A. Sloot (Eds.), Computational Science – ICCS 2007, Part IV. XXXVII, 1211 pages. 2007.

Vol. 4489: Y. Shi, G.D. van Albada, J. Dongarra, P.M.A. Sloot (Eds.), Computational Science – ICCS 2007, Part III. XXXVII, 1257 pages. 2007.

Vol. 4488: Y. Shi, G.D. van Albada, J. Dongarra, P.M.A. Sloot (Eds.), Computational Science – ICCS 2007, Part II. XXXV, 1251 pages. 2007.

Vol. 4487: Y. Shi, G.D. van Albada, J. Dongarra, P.M.A. Sloot (Eds.), Computational Science – ICCS 2007, Part I. LXXXI, 1275 pages. 2007.

Vol. 4486: M. Bernardo, J. Hillston (Eds.), Formal Methods for Performance Evaluation. VII, 469 pages. 2007.

Vol. 4485: F. Sgallari, A. Murli, N. Paragios (Eds.), Scale Space and Variational Methods in Computer Vision. XV, 931 pages. 2007.

Vol. 4484: J.-Y. Cai, S.B. Cooper, H. Zhu (Eds.), Theory and Applications of Models of Computation. XIII, 772 pages. 2007.

Vol. 4483: C. Baral, G. Brewka, J. Schlipf (Eds.), Logic Programming and Nonmonotonic Reasoning. IX, 327 pages. 2007. (Sublibrary LNAI).

Vol. 4482: A. An, J. Stefanowski, S. Ramanna, C.J. Butz, W. Pedrycz, G. Wang (Eds.), Rough Sets, Fuzzy Sets, Data Mining and Granular Computing. XIV, 585 pages. 2007. (Sublibrary LNAI).

Vol. 4481: J. Yao, P. Lingras, W.-Z. Wu, M. Szczuka, N.J. Cercone, D. Ślęzak (Eds.), Rough Sets and Knowledge Technology. XIV, 576 pages. 2007. (Sublibrary LNAI).

Vol. 4480: A. LaMarca, M. Langheinrich, K.N. Truong (Eds.), Pervasive Computing. XIII, 369 pages. 2007.

Vol. 4479: I.F. Akyildiz, R. Sivakumar, E. Ekici, J.C.d. Oliveira, J. McNair (Eds.), NETWORKING 2007. Ad Hoc and Sensor Networks, Wireless Networks, Next Generation Internet. XXVII, 1252 pages. 2007.

Vol. 4478: J. Martí, J.M. Benedí, A.M. Mendonça, J. Serrat (Eds.), Pattern Recognition and Image Analysis, Part II. XXVII, 657 pages. 2007.

Vol. 4477: J. Martí, J.M. Benedí, A.M. Mendonça, J. Serrat (Eds.), Pattern Recognition and Image Analysis, Part I. XXVII, 625 pages. 2007.

Vol. 4476: V. Gorodetsky, C. Zhang, V.A. Skormin, L. Cao (Eds.), Autonomous Intelligent Systems: Multi-Agents and Data Mining. XIII, 323 pages. 2007. (Sublibrary LNAI).

Vol. 4475: P. Crescenzi, G. Prencipe, G. Pucci (Eds.), Fun with Algorithms. X, 273 pages. 2007.

Vol. 4474: G. Prencipe, S. Zaks (Eds.), Structural Information and Communication Complexity. XI, 342 pages. 2007.

Vol. 4472: M. Haindl, J. Kittler, F. Roli (Eds.), Multiple Classifier Systems. XI, 524 pages. 2007.

Vol. 4471: P. Cesar, K. Chorianopoulos, J.F. Jensen (Eds.), Interactive TV: a Shared Experience. XIII, 236 pages. 2007.

Vol. 4470: Q. Wang, D. Pfahl, D.M. Raffo (Eds.), Software Process Dynamics and Agility. XI, 346 pages. 2007.

Vol. 4468: M.M. Bonsangue, E.B. Johnsen (Eds.), Formal Methods for Open Object-Based Distributed Systems. X, 317 pages. 2007.

Vol. 4467: A.L. Murphy, J. Vitek (Eds.), Coordination Models and Languages. X, 235 pages. 2007.

Vol. 4466: F.B. Sachse, G. Seemann (Eds.), Functional Imaging and Modeling of the Heart. XV, 486 pages. 2007.

Vol. 4465: T. Chahed, B. Tuffin (Eds.), Network Control and Optimization. XIII, 305 pages. 2007.

Vol. 4464: E. Dawson, D.S. Wong (Eds.), Information Security Practice and Experience. XIII, 361 pages. 2007.

Vol. 4463: I. Măndoiu, A. Zelikovsky (Eds.), Bioinformatics Research and Applications. XV, 653 pages. 2007. (Sublibrary LNBI).

Vol. 4462: D. Sauveron, K. Markantonakis, A. Bilas, J.-J. Quisquater (Eds.), Information Security Theory and Practices. XII, 255 pages. 2007.

Vol. 4459: C. Cérin, K.-C. Li (Eds.), Advances in Grid and Pervasive Computing. XVI, 759 pages. 2007.

Vol. 4453: T. Speed, H. Huang (Eds.), Research in Computational Molecular Biology. XVI, 550 pages. 2007. (Sublibrary LNBI).

Vol. 4452: M. Fasli, O. Shehory (Eds.), Agent-Mediated Electronic Commerce. VIII, 249 pages. 2007. (Sublibrary LNAI).

Vol. 4451: T.S. Huang, A. Nijholt, M. Pantic, A. Pentland (Eds.), Artifical Intelligence for Human Computing. XVI, 359 pages. 2007. (Sublibrary LNAI).

Vol. 4450: T. Okamoto, X. Wang (Eds.), Public Key Cryptography – PKC 2007. XIII, 491 pages. 2007.

Vol. 4448: M. Giacobini et al. (Ed.), Applications of Evolutionary Computing. XXIII, 755 pages. 2007.

Vol. 4447: E. Marchiori, J.H. Moore, J.C. Rajapakse (Eds.), Evolutionary Computation,Machine Learning and Data Mining in Bioinformatics. XI, 302 pages. 2007.

Vol. 4446: C. Cotta, J. van Hemert (Eds.), Evolutionary Computation in Combinatorial Optimization. XII, 241 pages. 2007.

Vol. 4445: M. Ebner, M. O'Neill, A. Ekárt, L. Vanneschi, A.I. Esparcia-Alcázar (Eds.), Genetic Programming. XI, 382 pages. 2007.

Vol. 4444: T. Reps, M. Sagiv, J. Bauer (Eds.), Program Analysis and Compilation, Theory and Practice. X, 361 pages. 2007.

Vol. 4443: R. Kotagiri, P.R. Krishna, M. Mohania, E. Nantajeewarawat (Eds.), Advances in Databases: Concepts, Systems and Applications. XXI, 1126 pages. 2007.

Vol. 4440: B. Liblit, Cooperative Bug Isolation. XV, 101 pages. 2007.

Vol. 4439: W. Abramowicz (Ed.), Business Information Systems. XV, 654 pages. 2007.

Vol. 4438: L. Maicher, A. Sigel, L.M. Garshol (Eds.), Leveraging the Semantics of Topic Maps. X, 257 pages. 2007. (Sublibrary LNAI).

Vol. 4433: E. Şahin, W.M. Spears, A.F.T. Winfield (Eds.), Swarm Robotics. XII, 221 pages. 2007.

Vol. 4432: B. Beliczynski, A. Dzielinski, M. Iwanowski, B. Ribeiro (Eds.), Adaptive and Natural Computing Algorithms, Part II. XXVI, 761 pages. 2007.

Vol. 4431: B. Beliczynski, A. Dzielinski, M. Iwanowski, B. Ribeiro (Eds.), Adaptive and Natural Computing Algorithms, Part I. XXV, 851 pages. 2007.

Vol. 4430: C.C. Yang, D. Zeng, M. Chau, K. Chang, Q. Yang, X. Cheng, J. Wang, F.-Y. Wang, H. Chen (Eds.), Intelligence and Security Informatics. XII, 330 pages. 2007.

Vol. 4429: R. Lu, J.H. Siekmann, C. Ullrich (Eds.), Cognitive Systems. X, 161 pages. 2007. (Sublibrary LNAI).

MIX
Papier aus verantwortungsvollen Quellen
Paper from responsible sources
FSC® C105338

If you have any concerns about our products,
you can contact us on
ProductSafety@springernature.com

In case Publisher is established outside the EU,
the EU authorized representative is:
Springer Nature Customer Service Center GmbH
Europaplatz 3, 69115 Heidelberg, Germany

Printed by Libri Plureos GmbH
in Hamburg, Germany